Statistical Inference via Convex Optimization

Princeton Series in Applied Mathematics

The Princeton Series in Applied Mathematics features high-quality advanced texts and monographs in all areas of applied mathematics. The series includes books of a theoretical and general nature as well as those that deal with the mathematics of specific applications and real-world scenarios. For a full list of titles in the series, go to https://press.princeton.edu/series/princeton-series-in-applied-mathematics

Statistical Inference via Convex Optimization, Anatoli Juditsky and Arkadi Nemirovski

A Dynamical Systems Theory of Thermodynamics, Wassim M. Haddad

Formal Verification of Control System Software, Pierre-Loic Garoche

Rays, Waves, and Scattering: Topics in Classical Mathematical Physics, John A. Adam

Mathematical Methods in Elasticity Imaging, Habib Ammari, Elie Bretin, Josselin Garnier, Hyeonbae Kang, Hyundae Lee, and Abdul Wahab

Hidden Markov Processes: Theory and Applications to Biology, M. Vidyasagar

Topics in Quaternion Linear Algebra, Leiba Rodman

Mathematical Analysis of Deterministic and Stochastic Problems in Complex Media Electromagnetics, G. F. Roach, I. G. Stratis, and A. N. Yannacopoulos

Stability and Control of Large-Scale Dynamical Systems: A Vector Dissipative Systems Approach, Wassim M. Haddad and Sergey G. Nersesov

Matrix Completions, Moments, and Sums of Hermitian Squares, Mihály Bakonyi and Hugo J. Woerdeman

Modern Anti-windup Synthesis: Control Augmentation for Actuator Saturation, Luca Zaccarian and Andrew R. Teel

Totally Nonnegative Matrices, Shaun M. Fallat and Charles R. Johnson

Graph Theoretic Methods in Multiagent Networks, Mehran Mesbahi and Magnus Egerstedt

Matrices, Moments and Quadrature with Applications, Gene H. Golub and Gérard Meurant

Control Theoretic Splines: Optimal Control, Statistics, and Path Planning, Magnus Egerstedt and Clyde Martin

Robust Optimization, Aharon Ben-Tal, Laurent El Ghaoui, and Arkadi Nemirovski

Distributed Control of Robotic Networks: A Mathematical Approach to Motion Coordination Algorithms, Francesco Bullo, Jorge Cortés, and Sonia Martinez

Algebraic Curves over a Finite Field, J.W.P. Hirschfeld, G. Korchmáros, and F. Torres

Wave Scattering by Time-Dependent Perturbations: An Introduction, G. F. Roach

Genomic Signal Processing, Ilya Shmulevich and Edward R. Dougherty

The Traveling Salesman Problem: A Computational Study, David L. Applegate, Robert E. Bixby, Vašek Chvátal, and William J. Cook

Positive Definite Matrices, Rajendra Bhatia

Impulsive and Hybrid Dynamical Systems: Stability, Dissipativity, and Control, Wassim M. Haddad, VijaySekhar Chellaboina, and Sergey G. Nersesov

Statistical Inference via Convex Optimization

Anatoli Juditsky
Arkadi Nemirovski

Princeton University Press
Princeton and Oxford

Published by Princeton University Press
41 William Street, Princeton, New Jersey 08540
6 Oxford Street, Woodstock, Oxfordshire OX20 1TR

press.princeton.edu

ISBN 978-0-691-19729-6
ISBN (e-book) 978-0-691-20031-6

British Library Cataloging-in-Publication Data is available

Editorial: Susannah Shoemaker and Lauren Bucca
Production Editorial: Nathan Carr
Production: Jacquie Poirier
Publicity: Matthew Taylor and Katie Lewis
Jacket/Cover Credit: Adapted from François de Kresz, "Excusez-moi, excusez-moi...," 1974
Copyeditor: Bhisham Bherwani

The publisher would like to acknowledge the authors of this volume for providing the camera-ready copy from which this book was printed.

This book has been composed in LaTeX

Printed on acid-free paper.

Printed in the United States of America

10 9 8 7 6 5 4 3 2 1

Contents

List of Figures

PREFACE

When speaking about links between Statistics and Optimization, what comes to mind first is the indispensable role played by optimization algorithms in the "computational toolbox" of Statistics (think about the numerical implementation of the fundamental Maximum Likelihood method). However, on a second thought, we should conclude that no matter how significant this role could be, the fact that it comes to our mind first primarily reflects the weaknesses of Optimization rather than its strengths; were optimization algorithms which are used in Statistics as efficient and as reliable as, say, Linear Algebra techniques, nobody would think about special links between Statistics and Optimization, just as nobody usually thinks about special links between Statistics and Linear Algebra. When computational, rather than methodological, issues are concerned, we start to think about links of Statistics with Optimization, Linear Algebra, Numerical Analysis, etc. only when computational tools offered to us by these disciplines do not work well and need the attention of experts in these disciplines.

The goal of this book is to present other types of links between Optimization and Statistics, those which have little in common with algorithms and number-crunching. What we are speaking about, are the situations where Optimization theory (theory, not algorithms!) seems to be of methodological value in Statistics, acting as the source of statistical inferences with provably optimal, or nearly so, performance. In this context, we focus on utilizing Convex Programming theory, mainly due to its power, but also due to the desire to end up with inference routines reducing to solving convex optimization problems and thus implementable in a computationally efficient fashion. Therefore, while we do not mention computational issues explicitly, we do remember that at the end of the day we need a number, and in this respect, intrinsically computationally friendly convex optimization models are the first choice.

The three topics we intend to consider are:

A. Sparsity-oriented Compressive Sensing. Here the role of Convex Optimization theory as a creative tool motivating the construction of inference procedures is relatively less important than in the two other topics. This being said, its role is by far non-negligible in the analysis of Compressive Sensing routines (it allows, e.g., to derive from "first principles" the necessary and sufficient conditions for the validity of ℓ_1 recovery). On account of this, and also due to its popularity and the fact that now it is one of the major "customers" of advanced convex optimization algorithms, we believe that Compressive Sensing is worthy of being considered.
B. Pairwise and Multiple Hypothesis Testing, including sequential tests, estimation of linear functionals, and some rudimentary design of experiments.
C. Recovery of signals from noisy observations of their linear images.

B and C are the topics where, as of now, the approaches we present in this book appear to be the most successful.

The exposition does *not* require prior knowledge of Statistics and Optimization; as far as these disciplines are concerned, all necessary facts and concepts are incorporated into the text. The actual prerequisites are basic Calculus, Probability, and Linear Algebra.

Selection and treatment of our topics are inspired by a kind of "philosophy" which can be explained to an expert as follows. Compare two well-known results of nonparametric statistics ("$\langle ... \rangle$" marks fragments irrelevant to the discussion to follow):

Theorem A [I. Ibragimov & R. Khasminskii [124], 1979] *Given α, L, k, let $\mathcal{X}$ be the set of all functions $f : [0,1] \to \mathbf{R}$ with (α, L)-Hölder continuous k-th derivative. For a given t, the minimax risk of estimating $f(t)$, $f \in \mathcal{X}$, from noisy observations $y = f\big|_{\Gamma_n} + \xi$, $\xi \sim \mathcal{N}(0; I_n)$ taken along n-point equidistant grid Γ_n, up to a factor $C(\beta) = \langle ... \rangle$, $\beta := k + \alpha$, is $(Ln^{-\beta})^{1/(2\beta+1)}$, and the upper risk bound is attained at the affine in y estimate explicitly given by $\langle ... \rangle$.*

Theorem B [D. Donoho [64], 1994] *Let $\mathcal{X} \subset \mathbf{R}^N$ be a convex compact set, A be an $n \times N$ matrix, and $g(\cdot)$ be a linear form on $\mathcal{X}$. The minimax, over $f \in \mathcal{X}$, risk of recovering $g(f)$ from the noisy observations $y = Af + \xi$, $\xi \sim \mathcal{N}(0, I_n)$, within factor 1.2 is attained at an affine in y estimate which, along with its risk, can be built efficiently by solving convex optimization problem $\langle ... \rangle$.*

In many respects, **A** and **B** are similar: both are theorems on minimax optimal estimation of a given linear form of an unknown "signal" f known to belong to a given convex set $\mathcal{X}$ from observations, corrupted by Gaussian noise, of the image of f under linear mapping,[1] and both are associated with efficiently computable near-optimal—in a minimax sense—estimators which happen to be affine in observations. There is, however, a significant structural difference: **A** gives an explicit "closed form" analytic description of the minimax risk as a function of n and smoothness parameters of f, along with explicit description of the near-optimal estimator. Numerous results of this type—let us call them *descriptive*—form the backbone of the deep and rich theory of Nonparametric Statistics. This being said, strong "explanation power" of descriptive results has its price: we need to impose assumptions, sometimes quite restrictive, on the entities involved. For example, **A** says nothing about what happens with the minimax risk/estimate when in addition to smoothness other a priori information on f, like monotonicity or convexity, is available, and/or when "direct" observations of $f|_{\Gamma_n}$ are replaced with observations of a linear image of f (say, convolution of f with a given kernel; more often than not, this is what happens in applications), and descriptive answers to the questions just posed require a dedicated (and sometimes quite problematic) investigation more or less "from scratch." In contrast, the explanation power of **B** is basically nonexistent: the statement presents "closed form" expressions neither for the near-optimal estimate, nor for its worst-case risk. As a compensation, **B** makes only (relatively) mild general structural assumptions about the model (convexity and compactness of $\mathcal{X}$, linear dependence of y on f), and all the rest—the near-optimal estimate and its risk—can be found by *efficient* computation. Moreover, we know in advance that the risk, whatever it happens to be, is within 20% of the actual minimax risk achievable under the circumstances. In this respect, **B** is an *operational*, rather than a descriptive, result: it explains *how to act* to achieve the (nearly) best possible performance, with no a priori prediction of what this performance will be. This hardly is a "big issue" in applications—with huge computational power readily available, efficient computability is, basically, as good as a "simple explicit formula." We

[1] Infinite dimensionality of $\mathcal{X}$ in **A** is of no importance—nothing changes when replacing the original $\mathcal{X}$ with its n-dimensional image under the mapping $f \mapsto f|_{\Gamma_n}$.

strongly believe that as far as applications of high-dimensional statistics are concerned, operational results, possessing much broader scope than their descriptive counterparts, are of significant importance and potential. Our main motivation when writing this book was to contribute to the body of operational results in Statistics, and this is what Chapters 2–5 to follow are about.

Anatoli Juditsky & Arkadi Nemirovski
March 6, 2019

ACKNOWLEDGEMENTS

We are greatly indebted to H. Edwin Romeijn who initiated creating the Ph.D. course "Topics in Data Science." The Lecture Notes for this course form the seed of the book to follow. We gratefully acknowledge support from SF Grant CC-1523768 *Statistical Inference via Convex Optimization*; this research project is the source of basically all novel results presented in Chapters 2–5. Our deepest gratitude goes to Lucien Birge, who encouraged us to write this monograph, and to Stephen Boyd, who many years ago taught one of the authors "operational philosophy," motivating the research we are presenting.

Our separate thanks to those who decades ago guided our first steps along the road which led to this book—Rafail Khasminskii, Yakov Tsypkin, and Boris Polyak. We are deeply indebted to our colleagues Alekh Agarwal, Aharon Ben-Tal, Fabienne Comte, Arnak Dalalyan, David Donoho, Céline Duval, Valentine Genon-Catalot, Alexander Goldenshluger, Yuri Golubev, Zaid Harchaoui, Gérard Kerkyacharian, Vladimir Koltchinskii, Oleg Lepski, Pascal Massart, Eric Moulines, Axel Munk, Aleksander Nazin, Yuri Nesterov, Dominique Picard, Alexander Rakhlin, Philippe Rigollet, Alex Shapiro, Vladimir Spokoiny, Alexandre Tsybakov, and Frank Werner for their advice and remarks.

We would like to thank Elitsa Marielle, Andrey Kulunchakov and Hlib Tsyntseus for their assistance when preparing the manuscript. It was our pleasure to collaborate with Princeton University Press on this project. We highly appreciate valuable comments of the anonymous referees, which helped to improve the initial text. We are greatly impressed by the professionalism of Princeton University Press editors, and in particular, Lauren Bucca, Nathan Carr, and Susannah Shoemaker, and also by their care and patience.

Needless to say, responsibility for all drawbacks of the book is ours.

A. J. & A. N.

ACKNOWLEDGMENTS

NOTATIONAL CONVENTIONS

Vectors and matrices. By default, all vectors are column ones; to write them down, we use "Matlab notation": $\begin{bmatrix} 1 \\ 2 \\ 3 \end{bmatrix}$ is written as $[1; 2; 3]$. More generally, for vectors/matrices $A, B, ..., Z$ of the same "width" (or vectors/matrices $A, B, C, ..., Z$ of the same "height"), $[A; B; C; ...; D]$ is the matrix obtained by vertical (or horizontal) concatenation of A, B, C, etc. Examples: For what in the "normal" notation is written down as $A = \begin{bmatrix} 1 & 2 \\ 3 & 4 \end{bmatrix}$, $B = \begin{bmatrix} 5 & 6 \end{bmatrix}$, $C = \begin{bmatrix} 7 \\ 8 \end{bmatrix}$, we have

$$[A; B] = \begin{bmatrix} 1 & 2 \\ 3 & 4 \\ 5 & 6 \end{bmatrix} = [1, 2; 3, 4; 5, 6], \ [A, C] = \begin{bmatrix} 1 & 2 & 7 \\ 3 & 4 & 8 \end{bmatrix} = [1, 2, 7; 3, 4, 8].$$

Blanks in matrices replace (blocks of) zero entries. For example,

$$\begin{bmatrix} 1 & & \\ 2 & & \\ 3 & 4 & 5 \end{bmatrix} = \begin{bmatrix} 1 & 0 & 0 \\ 2 & 0 & 0 \\ 3 & 4 & 5 \end{bmatrix}.$$

$\text{Diag}\{A_1, A_2, ..., A_k\}$ stands for a block-diagonal matrix with diagonal blocks A_1, A_2, ..., A_k. For example,

$$\text{Diag}\{1, 2, 3\} = \begin{bmatrix} 1 & & \\ & 2 & \\ & & 3 \end{bmatrix}, \ \text{Diag}\{[1, 2]; [3; 4]\} = \begin{bmatrix} 1 & 2 & \\ & & 3 \\ & & 4 \end{bmatrix}.$$

For an $m \times n$ matrix A, $\text{dg}(A)$ is the diagonal of A—a vector of dimension $\min[m, n]$ with entries A_{ii}, $1 \leq i \leq \min[m, n]$.

Standard linear spaces in our book are $\mathbf{R}^n$ (the space of n-dimensional column vectors), $\mathbf{R}^{m \times n}$ (the space of $m \times n$ real matrices), and $\mathbf{S}^n$ (the space of $n \times n$ real symmetric matrices). All these linear spaces are equipped with the standard inner product:

$$\langle A, B \rangle = \sum_{i,j} A_{ij} B_{ij} = \text{Tr}(AB^T) = \text{Tr}(BA^T) = \text{Tr}(A^T B) = \text{Tr}(B^T A);$$

in the case when $A = a$ and $B = b$ are column vectors, this simplifies to $\langle a, b \rangle = a^T b = b^T a$, and when A, B are symmetric, there is no need to write B^T in $\text{Tr}(AB^T)$.

Usually, we denote vectors by lowercase, and matrices by uppercase letters; sometimes, however, lowercase letters are used also for matrices.

Given a linear mapping $\mathcal{A}(x) : E_x \to E_y$, where E_x, E_y are standard linear spaces, one can define the *conjugate* mapping $\mathcal{A}^*(y) : E_y \to E_x$ via the identity

$$\langle \mathcal{A}(x), y \rangle = \langle x, \mathcal{A}^*(y) \rangle \ \forall (x \in E_x, y \in E_y).$$

One always has $(\mathcal{A}^*)^* = \mathcal{A}$. When $E_x = \mathbf{R}^n$, $E_y = \mathbf{R}^m$ and $\mathcal{A}(x) = Ax$, one has $\mathcal{A}^*(y) = A^T y$; when $E_x = \mathbf{R}^n$, $E_y = \mathbf{S}^m$, so that $\mathcal{A}(x) = \sum_{i=1}^n x_i A_i$, $A_i \in \mathbf{S}^m$, we have

$$\mathcal{A}^*(Y) = [\mathrm{Tr}(A_1 Y); ...; \mathrm{Tr}(A_n Y)].$$

$\mathbf{Z}^n$ is the set of n-dimensional integer vectors.

Norms. For $1 \leq p \leq \infty$ and for a vector $x = [x_1; ...; x_n] \in \mathbf{R}^n$, $\|x\|_p$ is the standard p-norm of x:

$$\|x\|_p = \begin{cases} \left(\sum_{i=1}^n |x_i|^p\right)^{1/p}, & 1 \leq p < \infty, \\ \max_i |x_i| = \lim_{p' \to \infty} \|x\|_{p'}, & p = \infty. \end{cases}$$

The spectral norm (the largest singular value) of a matrix A is denoted by $\|A\|_{2,2}$; notation for other norms of matrices is specified when used.

Standard cones. $\mathbf{R}_+$ is the nonnegative ray on the real axis; $\mathbf{R}^n_+$ stands for the *n-dimensional nonnegative orthant*, the cone comprised of all entrywise nonnegative vectors from $\mathbf{R}^n$; $\mathbf{S}^n_+$ stands for the *positive semidefinite cone in* $\mathbf{S}^n$, the cone comprised of all positive semidefinite matrices from $\mathbf{S}^n$.

Miscellaneous.
• For matrices A, B, relation $A \preceq B$, or, equivalently, $B \succeq A$, means that A, B are symmetric matrices of the same size such that $B - A$ is positive semidefinite; we write $A \succeq 0$ to express the fact that A is a symmetric positive semidefinite matrix. Strict version $A \succ B$ ($\Leftrightarrow B \prec A$) of $A \succeq B$ means that $A - B$ is positive definite (and, as above, A and B are symmetric matrices of the same size).
• Linear Matrix Inequality (LMI, a.k.a. *semidefinite constraint*) in variables x is the constraint on x stating that a symmetric matrix affinely depending on x is positive semidefinite. When $x \in \mathbf{R}^n$, LMI reads

$$A_0 + \sum_i x_i A_i \succeq 0 \qquad [A_i \in \mathbf{S}^m, 0 \leq i \leq n].$$

• $\mathcal{N}(\mu, \Theta)$ stands for the Gaussian distribution with mean μ and covariance matrix Θ. Poisson(μ) denotes Poisson distribution with parameter $\mu \in \mathbf{R}_+$, i.e., the distribution of a random variable taking values $i = 0, 1, 2, ...$ with probabilities $\frac{\mu^i}{i!}\mathrm{e}^{-\mu}$. Uniform($[a, b]$) is the uniform distribution on segment $[a, b]$.
• For a probability distribution P,

- $\xi \sim P$ means that ξ is a random variable with distribution P. Sometimes we express the same fact by writing $\xi \sim p(\cdot)$, where p is the density of P taken w.r.t. some reference measure (the latter always is fixed by the context);
- $\mathbf{E}_{\xi \sim P}\{f(\xi)\}$ is the expectation of $f(\xi)$, $\xi \sim P$; when P is clear from the context, this notation can be shortened to $\mathbf{E}_\xi\{f(\xi)\}$, or $\mathbf{E}_P\{f(\xi)\}$, or even $\mathbf{E}\{f(\xi)\}$. Similarly, $\mathrm{Prob}_{\xi \sim P}\{...\}$, $\mathrm{Prob}_\xi\{...\}$, $\mathrm{Prob}_P\{...\}$, and $\mathrm{Prob}\{...\}$ denote the P-probability of the event specified inside the braces.

• $O(1)$'s stand for positive *absolute* constants—positive reals with numerical values (completely independent of the parameters of the situation at hand) which we do not want or are too lazy to write down explicitly, as in $\sin(x) \leq O(1)|x|$.
• $\int_\Omega f(\xi)\Pi(d\xi)$ stands for the integral, taken w.r.t. measure Π over domain Ω, of function f.

ABOUT PROOFS

The book is basically self-contained in terms of proofs of the statements to follow. Simple proofs usually are placed immediately after the corresponding statements; more technical proofs are transferred to dedicated sections titled "Proof of ..." at the end of each chapter, and this is where a reader should look for "missing" proofs.

ON COMPUTATIONAL TRACTABILITY

In the main body of the book, one can frequently meet sentences like "$\Phi(\cdot)$ is an efficiently computable convex function," or "X is a computationally tractable convex set," or "(P) is an explicit, and therefore efficiently solvable, convex optimization problem." For an "executive summary" on what these words actually mean, we refer the reader to the Appendix.

Statistical Inference via Convex Optimization

Chapter One

Sparse Recovery via ℓ_1 Minimization

In this chapter, we overview basic results of *Compressed Sensing*, a relatively new and rapidly developing area in Statistics and Signal Processing dealing with recovering signals (vectors x from some $\mathbf{R}^n$) from their noisy observations $Ax + \eta$ (A is a given $m \times n$ *sensing matrix*, η is observation noise) in the case when the number of observations m is much smaller than the signal's dimension n, but is essentially larger than the "true" dimension—the number of nonzero entries—in the signal. This setup leads to a deep, elegant and highly innovative theory and possesses quite significant application potential. It should be added that along with the plain sparsity (small number of nonzero entries), Compressed Sensing deals with other types of "low-dimensional structure" hidden in high-dimensional signals, most notably, with the case of *low rank matrix recovery*—when the signal is a matrix, and sparse signals are matrices with low ranks—and the case of *block sparsity*, where the signal is a block vector, and sparsity means that only a small number of blocks are nonzero. In our presentation, we do *not* consider these extensions, and restrict ourselves to the simplest sparsity paradigm.

1.1 COMPRESSED SENSING: WHAT IS IT ABOUT?

1.1.1 Signal Recovery Problem

One of the basic problems in Signal Processing is the problem of recovering a *signal* $x \in \mathbf{R}^n$ from noisy observations

$$y = Ax + \eta \tag{1.1}$$

of a linear image of the signal under a given *sensing mapping* $x \mapsto Ax : \mathbf{R}^n \to \mathbf{R}^m$; in (1.1), η is the *observation error.* Matrix A in (1.1) is called *sensing matrix.*

Recovery problems of the outlined types arise in many applications, including, but *by far* not reducing to,

- *communications*, where x is the signal sent by the transmitter, y is the signal recorded by the receiver, and A represents the communication channel (reflecting, e.g., dependencies of decays in the signals' amplitude on the transmitter-receiver distances); η here typically is modeled as the standard (zero mean, unit covariance matrix) m-dimensional Gaussian noise;[1]

[1]While the "physical" noise indeed is often Gaussian with zero mean, its covariance matrix is not necessarily the unit matrix. Note, however, that a zero mean Gaussian noise η always can be represented as $Q\xi$ with standard Gaussian ξ. Assuming that Q is known and is nonsingular (which indeed is so when the covariance matrix of η is positive definite), we can rewrite (1.1) equivalently as

$$Q^{-1}y = [Q^{-1}A]x + \xi$$

and treat $Q^{-1}y$ and $Q^{-1}A$ as our new observation and new sensing matrix; the new observation

- *image reconstruction*, where the signal x is an image—a 2D array in the usual photography, or a 3D array in tomography—and y is data acquired by the imaging device. Here η in many cases (although not always) can again be modeled as the standard Gaussian noise;
- *linear regression*, arising in a wide range of applications. In linear regression, one is given m pairs "input $a^i \in \mathbf{R}^n$" to a "black box," with output $y_i \in \mathbf{R}$. Sometimes we have reason to believe that the output is a corrupted by noise version of the "existing in nature," but unobservable, "ideal output" $y_i^* = x^T a^i$ which is just a linear function of the input (this is called "linear regression model," with inputs a^i called "regressors"). Our goal is to convert actual observations (a^i, y_i), $1 \leq i \leq m$, into estimates of the *unknown* "true" vector of parameters x. Denoting by A the matrix with the rows $[a^i]^T$ and assembling individual observations y_i into a single observation $y = [y_1; ...; y_m] \in \mathbf{R}^m$, we arrive at the problem of recovering vector x from noisy observations of Ax. Here again the most popular model for η is the standard Gaussian noise.

1.1.2 Signal Recovery: Parametric and nonparametric cases

Recovering signal x from observation y would be easy if there were no observation noise ($\eta = 0$) and the rank of matrix A were equal to the dimension n of the signals. In this case, which arises only when $m \geq n$ ("more observations than unknown parameters"), and is typical in this range of m and n, the desired x would be the unique solution to the system of linear equations, and to find x would be a simple problem of Linear Algebra. Aside from this trivial "enough observations, no noise" case, people over the years have looked at the following two versions of the recovery problem:

Parametric case: $m \gg n$, η is nontrivial noise with zero mean, say, standard Gaussian. This is the classical statistical setup with the emphasis on how to use numerous available observations in order to suppress in the recovery, to the extent possible, the influence of observation noise.

Nonparametric case: $m \ll n$.[2] If addressed literally, this case seems to be senseless: when the number of observations is less that the number of unknown parameters, even in the noiseless case we arrive at the necessity to solve an undetermined (fewer equations than unknowns) system of linear equations. Linear Algebra says that if solvable, the system has infinitely many solutions. Moreover, the solution set (an affine subspace of positive dimension) is unbounded, meaning that the solutions are in no sense close to each other. A typical way to make the case of $m \ll n$ meaningful is to add to the observations (1.1) some a priori information about the signal. In traditional Nonparametric Statistics, this additional information is summarized in a *bounded convex set* $X \subset \mathbf{R}^n$, given to us in advance, known to contain the true signal x. This set usually is such that *every signal* $x \in X$ *can be approximated by a linear combination of* $s = 1, 2, ..., n$ *vectors*

noise ξ is indeed standard. Thus, in the case of Gaussian zero mean observation noise, to assume the noise standard Gaussian is the same as to assume that its covariance matrix is known.

[2]Of course, this is a blatant simplification—the nonparametric case covers also a variety of important and by far nontrivial situations in which m is comparable to n or larger than n (or even $\gg n$). However, this simplification is very convenient, and we will use it in this introduction.

from a properly selected basis known to us in advance ("dictionary" in the slang of signal processing) *within accuracy* $\delta(s)$, where $\delta(s)$ is a function, known in advance, approaching 0 as $s \to \infty$. In this situation, with appropriate A (e.g., just the unit matrix, as in the denoising problem), we can select some $s \ll m$ and try to recover x *as if* it were a vector from the linear span E_s of the *first* s *vectors* of the outlined basis [54, 86, 124, 112, 208]. In the "ideal case," $x \in E_s$, recovering x in fact reduces to the case where the dimension of the signal is $s \ll m$ rather than $n \gg m$, and we arrive at the well-studied situation of recovering a signal of low (compared to the number of observations) dimension. In the "realistic case" of x $\delta(s)$-close to E_s, deviation of x from E_s results in an additional component in the recovery error ("bias"); a typical result of traditional Nonparametric Statistics quantifies the resulting error and minimizes it in s [86, 124, 178, 222, 223, 230, 239]. Of course, this outline of the traditional approach to "nonparametric" (with $n \gg m$) recovery problems is extremely sketchy, but it captures the most important fact in our context: with the traditional approach to nonparametric signal recovery, one assumes that after representing the signals by vectors of their coefficients in properly selected base, the n-dimensional signal to be recovered can be well approximated by an s-sparse (at most s nonzero entries) signal, with $s \ll n$, *and this sparse approximation can be obtained by zeroing out all but the first* s *entries in the signal vector.* The assumption just formulated indeed takes place for signals obtained by discretization of *smooth* uni- and multivariate functions, and this class of signals for several decades was the main, if not the only, focus of Nonparametric Statistics.

Compressed Sensing. The situation changed dramatically around the year 2000 as a consequence of important theoretical breakthroughs due to D. Donoho, T. Tao, J. Romberg, E. Candes, and J.-J. Fuchs, among many other researchers [49, 44, 45, 46, 48, 67, 68, 69, 70, 93, 94]; as a result of these breakthroughs, a novel and rich area of research, called *Compressed Sensing*, emerged.

In the Compressed Sensing (CS) setup of the Signal Recovery problem, as in the traditional Nonparametric Statistics approach to the $m \ll n$ case, it is assumed that after passing to an appropriate basis, the signal to be recovered is s-sparse (has $\leq s$ nonzero entries, with $s \ll m$), or is well approximated by an s-sparse signal. The difference with the traditional approach is that now we assume *nothing* about the location of the nonzero entries. Thus, the a priori information about the signal x both in the traditional and in the CS settings is summarized in a set X known to contain the signal x we want to recover. The difference is that in the traditional setting, X is a bounded convex and "nice" (well approximated by its low-dimensional cross-sections) set, while in CS this set is, computationally speaking, a "monster": already in the simplest case of recovering *exactly* s*-sparse* signals, X is the union of all s-dimensional coordinate planes, which is a heavily combinatorial entity.

Note that, in many applications we indeed can assume that the true vector of parameters x is sparse. Consider, e.g., the following story about signal detection. *There are* n *locations where signal transmitters could be placed, and* m *locations with the receivers. The contribution of a signal of unit magnitude originating in location* j *to the signal measured by receiver* i *is a known quantity* A_{ij}, *and signals originating in different locations merely sum up in the receivers. Thus, if* x *is the* n*-dimensional vector with entries* x_j *representing the magnitudes of signals transmitted in locations* $j = 1, 2, ..., n$, *then the* m*-dimensional vector* y *of measurements of the* m *receivers is* $y =$

$Ax + \eta$, where η is the observation noise. Given y, we intend to recover x.

Now, if the receivers are, say, hydrophones registering noises emitted by submarines in a certain part of the Atlantic, tentative positions of "submarines" being discretized with resolution 500 m, the dimension of the vector x (the number of points in the discretization grid) may be in the range of tens of thousands, if not tens of millions. At the same time, presumably, there is only a handful of "submarines" (i.e., nonzero entries in x) in the area.

To "see" sparsity in everyday life, look at the 256×256 image at the top of Figure 1.1. The image can be thought of as a $256^2 = 65,536$-dimensional vector comprised of the pixels' intensities in gray scale, and there is not much sparsity in this vector. However, when representing the image in the *wavelet basis*, whatever it means, we get a "nearly sparse" vector of wavelet coefficients (this is true for typical "non-pathological" images). At the bottom of Figure 1.1 we see what happens when we zero out all but a small percentage of the wavelet coefficients largest in magnitude and replace the true image by its sparse—in the wavelet basis—approximations.

This simple visual illustration along with numerous similar examples shows the "everyday presence" of sparsity and the possibility to utilize it when compressing signals. The difficulty, however, is that simple compression—compute the coefficients of the signal in an appropriate basis and then keep, say, 10% of the largest in magnitude coefficients—requires us to start with digitalizing the signal—representing it as an array of all its coefficients in some orthonormal basis. These coefficients are inner products of the signal with vectors of the basis; for a "physical" signal, like speech or image, these inner products are computed by analogous devices, with subsequent discretization of the results. After the measurements are discretized, processing the signal (denoising, compression, storing, etc.) can be fully computerized. The major (to some extent, already actualized) advantage of Compressed Sensing is in the possibility to reduce the "analogous effort" in the outlined process: instead of computing analogously n linear forms of n-dimensional signal x (its coefficients in a basis), we use an analog device to compute $m \ll n$ other linear forms of the signal and then use the signal's sparsity in a basis known to us in order to recover the signal reasonably well from these m observations.

In our "picture illustration" this technology would work (in fact, works—it is called "single pixel camera" [83]; see Figure 1.2) as follows: in reality, the digital 256×256 image on the top of Figure 1.1 was obtained by an analog device—a digital camera which gets on input an analog signal (light of varying intensity along the field of view caught by camera's lens) and discretizes the light's intensity in every pixel to get the digitalized image. We then can compute the wavelet coefficients of the digitalized image, compress its representation by keeping, say, just 10% of leading coefficients, etc., but "the damage is already done"—we have already spent our analog resources to get the entire digitalized image. The technology utilizing Compressed Sensing would work as follows: instead of measuring and discretizing light intensity in each of the 65,536 pixels, we compute (using an analog device) the integral, taken over the field of view, of the product of light intensity and an analog-generated "mask." We repeat it for, say, 20,000 different masks, thus obtaining measurements of 20,000 linear forms of our 65,536-dimensional signal. Next we utilize, via the Compressed Sensing machinery, the signal's sparsity in the wavelet basis in order to recover the signal from these 20,000 measurements. With this approach, we reduce the "analog component" of signal processing effort,

1% of leading wavelet coefficients (97.83 % of energy) kept

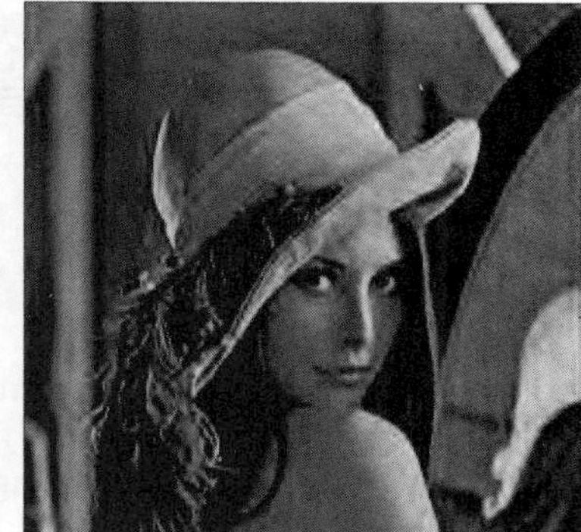

5% of leading wavelet coefficients (99.51 % of energy) kept

10% of leading wavelet coefficients (99.82% of energy) kept

25% of leading wavelet coefficients (99.97% of energy) kept

Figure 1.1: Top: true 256×256 image; bottom: sparse in the wavelet basis approximations of the image. Wavelet basis is orthonormal, and a natural way to quantify near-sparsity of a signal is to look at the fraction of total energy (sum of squares of wavelet coefficients) stored in the leading coefficients; these are the "energy data" presented in the figure.

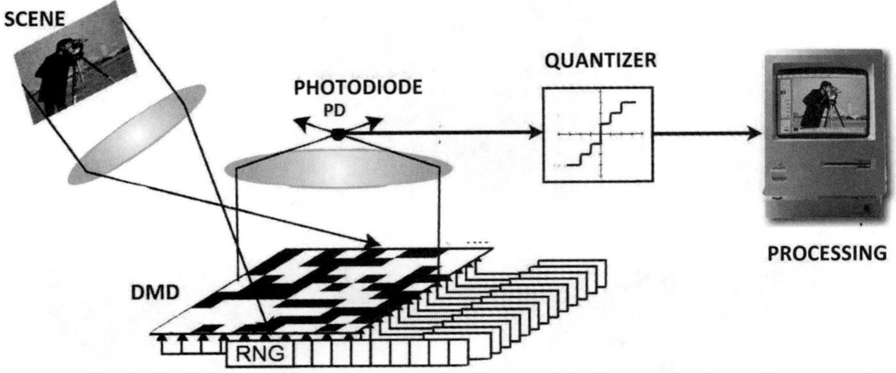

Figure 1.2: Singe-pixel camera.

at the price of increasing the "computerized component" of the effort (instead of ready-to-use digitalized image directly given by 65,536 analog measurements, we need to recover the image by applying computationally nontrivial decoding algorithms to our 20,000 "indirect" measurements). When taking pictures with your camera or iPad, the game is not worth the candle—the analog component of taking usual pictures is cheap enough, and decreasing it at the cost of nontrivial decoding of the digitalized measurements would be counterproductive. There are, however, important applications where the advantages stemming from reduced "analog effort" outweigh significantly the drawbacks caused by the necessity to use nontrivial computerized decoding [96, 176].

1.1.3 Compressed Sensing via ℓ_1 minimization: Motivation

1.1.3.1 Preliminaries

In principle there is nothing surprising in the fact that under reasonable assumption on the $m \times n$ sensing matrix A we may hope to recover from noisy observations of Ax an s-sparse signal x, with $s \ll m$. Indeed, assume for the sake of simplicity that there are no observation errors, and let $\mathrm{Col}_j[A]$ be j-th column in A. If we knew the locations $j_1 < j_2 < ... < j_s$ of the nonzero entries in x, identifying x could be reduced to solving the system of linear equations $\sum_{\ell=1}^{s} x_{i_\ell} \mathrm{Col}_{j_\ell}[A] = y$ with m equations and $s \ll m$ unknowns; assuming every s columns in A to be linearly independent (a quite unrestrictive assumption on a matrix with $m \geq s$ rows), the solution to the above system is unique, and is exactly the signal we are looking for. Of course, the assumption that we know the locations of nonzeros in x makes the recovery problem completely trivial. However, it suggests the following course of action: given noiseless observation $y = Ax$ of an s-sparse signal x, let us solve the combinatorial optimization problem

$$\min_z \{\|z\|_0 : Az = y\}, \tag{1.2}$$

where $\|z\|_0$ is the number of nonzero entries in z. Clearly, the problem has a solution with the value of the objective at most s. Moreover, it is immediately seen that if every $2s$ columns in A are linearly independent (which again is a very unrestrictive assumption on the matrix A provided that $m \geq 2s$), then the true signal x is the unique optimal solution to (1.2).

What was said so far can be extended to the case of noisy observations and "nearly s-sparse" signals x. For example, assuming that the observation error is "uncertain-but-bounded," specifically some known norm $\|\cdot\|$ of this error does not exceed a given $\epsilon > 0$, and that the true signal is s-sparse, we could solve the combinatorial optimization problem

$$\min_z \{\|z\|_0 : \|Az - y\| \leq \epsilon\}. \tag{1.3}$$

Assuming that every $m \times 2s$ submatrix $\bar{A}$ of A is not just with linearly independent columns (i.e., with trivial kernel), but is reasonably well conditioned,

$$\|\bar{A}w\| \geq C^{-1}\|w\|_2$$

for all $(2s)$-dimensional vectors w, with some constant C, it is immediately seen that the true signal x underlying the observation and the optimal solution $\widehat{x}$ of (1.3) are close to each other within accuracy of order of ϵ: $\|x - \widehat{x}\|_2 \leq 2C\epsilon$. It is easily seen that the resulting error bound is basically as good as it could be.

We see that the difficulties with recovering sparse signals stem not from the lack of information; they are of purely computational nature: (1.2) is a difficult combinatorial problem. As far as known theoretical complexity guarantees are concerned, they are not better than "brute force" search through all guesses on where the nonzeros in x are located—by inspecting first the only option that there are no nonzeros in x at all, then by inspecting n options that there is only one nonzero, for every one of n locations of this nonzero, then $n(n-1)/2$ options that there are exactly two nonzeros, etc., until the current option results in a solvable system of linear equations $Az = y$ in variables z with entries restricted to vanish outside the locations prescribed by the current option. The running time of this "brute force" search, beyond the range of small values of s and n (by far too small to be of any applied interest), is by many orders of magnitude larger than what we can afford in reality.[3]

A partial remedy is as follows. Well, if we do not know how to minimize the "bad" objective $\|z\|_0$ under linear constraints, as in (1.2), let us "approximate" this objective with one which we do know how to minimize. The true objective is separable: $\|z\| = \sum_{i=1}^n \xi(z_j)$, where $\xi(s)$ is the function on the axis equal to 0 at the origin and equal to 1 otherwise. As a matter of fact, the separable functions which

[3]When $s = 5$ and $n = 100$, a sharp upper bound on the number of linear systems we should process before termination in the "brute force" algorithm is $\approx 7.53\text{e}7$—a lot, but perhaps doable. When $n = 200$ and $s = 20$, the number of systems to be processed jumps to $\approx 1.61\text{e}27$, which is by many orders of magnitude beyond our "computational grasp"; we would be unable to carry out that many computations even if the fate of the mankind were at stake. And from the perspective of Compressed Sensing, $n = 200$ still is a completely toy size, 3–4 orders of magnitude less than we would like to handle.

we do know how to minimize under linear constraints are sums of *convex* functions of $z_1, \ldots, z_n$. The most natural candidate to the role of *convex* approximation of $\xi(s)$ is $|s|$; with this approximation, (1.2) converts into the ℓ_1 *minimization problem*

$$\min_z \left\{ \|z\|_1 := \sum\nolimits_{i=1}^n |z_j| : Az = y \right\}, \tag{1.4}$$

and (1.3) becomes the convex optimization problem

$$\min_z \{ \|z\|_1 : \|Az - y\| \leq \epsilon \}. \tag{1.5}$$

Both problems are efficiently solvable, which is nice; the question, however, is how relevant these problems are in our context—whether it is true that they do recover the "true" s-sparse signals in the noiseless case, or "nearly recover" these signals when the observation error is small. Since we want to be able to handle *any* s-sparse signal, the validity of ℓ_1 recovery—its ability to recover well *every* s-sparse signal—depends solely on the sensing matrix A. Our current goal is to understand which sensing matrices are "good" in this respect.

1.2 VALIDITY OF SPARSE SIGNAL RECOVERY VIA ℓ_1 MINIMIZATION

What follows is based on the standard basic results of Compressed Sensing theory originating from [19, 49, 45, 44, 46, 47, 48, 67, 69, 70, 93, 94, 232] and augmented by the results of [129, 130, 132, 133].[4]

1.2.1 Validity of ℓ_1 minimization in the noiseless case

The minimal requirement on sensing matrix A which makes ℓ_1 minimization valid is to guarantee the correct recovery of *exactly* s-sparse signals in the *noiseless* case, and we start with investigating this property.

1.2.1.1 Notational convention

From now on, for a vector $x \in \mathbf{R}^n$

- $I_x = \{j : x_j \neq 0\}$ stands for the *support* of x; we also set

$$I_x^+ = \{j : x_j > 0\}, \; I_x^- = \{j : x_j < 0\} \qquad [\Rightarrow I_x = I_x^+ \cup I_x^-];$$

- for a subset I of the index set $\{1, \ldots, n\}$, x_I stands for the vector obtained from x by zeroing out entries with indices *not* in I, and I^o for the complement of I:

$$I^o = \{i \in \{1, \ldots, n\} : i \notin I\};$$

- for $s \leq n$, x^s stands for the vector obtained from x by zeroing out all but the s

[4]In fact, in the latter source, an extension of the sparsity, the so-called block sparsity, is considered; in what follows, we restrict the results of [130] to the case of plain sparsity.

entries largest in magnitude.[5] Note that x^s is the best s-sparse approximation of x in all ℓ_p norms, $1 \le p \le \infty$;

- for $s \le n$ and $p \in [1, \infty]$, we set

$$\|x\|_{s,p} = \|x^s\|_p;$$

note that $\|\cdot\|_{s,p}$ is a norm.

1.2.1.2 *s-Goodness*

Definition of s-goodness. Let us say that an $m \times n$ sensing matrix A is *s-good* if whenever the true signal x underlying *noiseless* observations is s-sparse, this signal will be recovered *exactly* by ℓ_1 minimization. In other words, A is s-good if whenever y in (1.4) is of the form $y = Ax$ with s-sparse x, x is the unique optimal solution to (1.4).

Nullspace property. There is a simply-looking *necessary and sufficient* condition for a sensing matrix A to be s-good—the *nullspace property* originating from [70]. After this property is guessed, it is easy to see that it indeed is necessary and sufficient for s-goodness; we, however, prefer to *derive* this condition from the "first principles," which can be easily done via Convex Optimization. Thus, in the case in question, as in many other cases, there is no necessity to be smart to arrive at the truth via a "lucky guess"; it suffices to be knowledgeable and use the standard tools.

Let us start with necessary condition for A to be such that whenever x is s-sparse, x is an optimal solution (perhaps not the unique one) of the optimization problem

$$\min_z \{\|z\|_1 : Az = Ax\}; \qquad (P[x])$$

we refer to the latter property of A as *weak s-goodness.* Our first observation is as follows:

Proposition 1.1. *If A is weakly s-good, then the following condition holds true: whenever I is a subset of $\{1, ..., n\}$ of cardinality $\le s$, we have*

$$\forall w \in \mathrm{Ker} A \;\; \|w_I\|_1 \le \|w_{I^o}\|_1. \qquad (1.6)$$

Proof is immediate. Assume A is weakly s-good, and let us verify (1.6). Let I be an s-element subset of $\{1, ..., n\}$, and x be an s-sparse vector with support I. Since A is weakly s-good, x is an optimal solution to $(P[x])$. Rewriting the latter problem in the form of LP, that is, as

$$\min_{z,t}\{\sum_j t_j : t_j + z_j \ge 0, t_j - z_j \ge 0, Az = Ax\},$$

and invoking LP optimality conditions, the necessary and sufficient condition for

[5] Note that in general x^s is not uniquely defined by x and s, since the s-th largest among the magnitudes of entries in x can be achieved at several entries. In our context, it does not matter how ties of this type are resolved; for the sake of definiteness, we can assume that when ordering the entries in x according to their magnitudes, from the largest to the smallest, entries of equal magnitude are ordered in the order of their indices.

$z = x$ to be the z-component of an optimal solution is the existence of λ_j^+, λ_j^-, $\mu \in \mathbf{R}^m$ (Lagrange multipliers for the constraints $t_j - z_j \geq 0$, $t_j + z_j \geq 0$, and $Az = Ax$, respectively) such that

$$\begin{array}{lrcl} (a) & \lambda_j^+ + \lambda_j^- & = & 1\,\forall j, \\ (b) & \lambda^+ - \lambda^- + A^T\mu & = & 0, \\ (c) & \lambda_j^+(|x_j| - x_j) & = & 0\,\forall j, \\ (d) & \lambda_j^-(|x_j| + x_j) & = & 0\,\forall j, \\ (e) & \lambda_j^+ & \geq & 0\,\forall j, \\ (f) & \lambda_j^- & \geq & 0\,\forall j. \end{array} \tag{1.7}$$

From (c, d), we have $\lambda_j^+ = 1, \lambda_j^- = 0$ for $j \in I_x^+$ and $\lambda_j^+ = 0, \lambda_j^- = 1$ for $j \in I_x^-$. From (a) and nonnegativity of $\lambda_j^\pm$ it follows that for $j \notin I_x$ we should have $-1 \leq \lambda_j^+ - \lambda_j^- \leq 1$. With this in mind, the above optimality conditions admit eliminating λ's and reduce to the following conclusion:

(!) *x is an optimal solution to $(P[x])$ if and only if there exists vector $\mu \in \mathbf{R}^m$ such that the j-th entry of $A^T\mu$ is -1 if $x_j > 0$, $+1$ if $x_j < 0$, and a real from $[-1, 1]$ if $x_j = 0$.*

Now let $w \in \operatorname{Ker} A$ be a vector with the same signs of entries w_i, $i \in I$, as these of the entries in x. Then

$$\begin{array}{l} 0 = \mu^T A w = [A^T\mu]^T w = \sum_j [A^T\mu]_j w_j \\ \Rightarrow \sum_{j \in I_x} |w_j| = \sum_{j \in I_x} [A^T\mu]_j w_j = -\sum_{j \notin I_x} [A^T\mu]_j w_j \leq \sum_{j \notin I_x} |w_j| \end{array}$$

(we have used the fact that $[A^T\mu]_j = \operatorname{sign} x_j = \operatorname{sign} w_j$ for $j \in I_x$ and $|[A^T\mu]_j| \leq 1$ for all j). Since I can be an arbitrary s-element subset of $\{1, ..., n\}$ and the pattern of signs of an s-sparse vector x supported on I can be arbitrary, (1.6) holds true. □

1.2.1.3 Nullspace property

In fact, it can be shown that (1.6) is not only a necessary, but also sufficient condition for weak s-goodness of A; we, however, skip this verification, since our goal so far was to *guess* the condition for s-goodness, and this goal has already been achieved—from what we already know it immediately follows that a *necessary* condition for s-goodness is for the inequality in (1.6) to be strict whenever $w \in \operatorname{Ker} A$ is nonzero. Indeed, we already know that if A is s-good, then for every I of cardinality s and every nonzero $w \in \operatorname{Ker} A$ it holds

$$\|w_I\|_1 \leq \|w_{I^o}\|_1.$$

If the latter inequality for some I and w in question holds true as equality, then A clearly is *not* s-good, since the s-sparse signal $x = w_I$ is *not* the unique optimal solution to $(P[x])$—the vector $-w_{I^o}$ is a different feasible solution to the same problem and with the same value of the objective. We conclude that for A to be s-good, a necessary condition is

$$\forall (0 \neq w \in \operatorname{Ker} A, I, \operatorname{Card}(I) \leq s) : \|w_I\|_1 < \|w_{I^o}\|_1.$$

By the standard compactness argument, this is the same as the existence of $\gamma \in (0,1)$ such that

$$\forall (w \in \operatorname{Ker} A, I, \operatorname{Card}(I) \le s) : \|w_I\|_1 \le \gamma \|w_{I^o}\|_1,$$

or—which is the same—existence of $\kappa \in (0, 1/2)$ such that

$$\forall (w \in \operatorname{Ker} A, I, \operatorname{Card}(I) \le s) : \|w_I\|_1 \le \kappa \|w\|_1.$$

Finally, the supremum of $\|w_I\|_1$ over I of cardinality s is the norm $\|w\|_{s,1}$ (the sum of s largest magnitudes of entries) of w, so that the condition we are processing finally can be formulated as

$$\exists \kappa \in (0, 1/2) : \|w\|_{s,1} \le \kappa \|w\|_1 \ \forall w \in \operatorname{Ker} A. \tag{1.8}$$

The resulting *nullspace condition* in fact is necessary *and sufficient* for A to be s-good:

Proposition 1.2. *Condition* (1.8) *is necessary and sufficient for A to be s-good.*

Proof. We have already seen that the nullspace condition is necessary for s-goodness. To verify sufficiency, let A satisfy the nullspace condition, and let us prove that A is s-good. Indeed, let x be an s-sparse vector, and y be an optimal solution to $(P[x])$; all we need is to prove that $y = x$. Let I be the support of x, and $w = y - x$, so that $w \in \operatorname{Ker} A$. By the nullspace property we have

$$\begin{array}{ll} & \|w_I\|_1 \le \kappa \|w\|_1 = \kappa[\|w_I\|_1 + \|w_{I^o}\|_1] = \kappa[\|w_I\|_1 + \|y_{I^o}\|_1 \\ \Rightarrow & \|w_I\|_1 \le \frac{\kappa}{1-\kappa} \|y_{I^o}\|_1 \\ \Rightarrow & \|x\|_1 = \|x_I\|_1 = \|y_I - w_I\|_1 \le \|y_I\|_1 + \frac{\kappa}{1-\kappa} \|y_{I^o}\|_1 \le \|y_I\|_1 + \|y_{I^o}\|_1 = \|y\|_1 \end{array}$$

where the concluding $\le$ is due to $\kappa \in [0, 1/2)$. Since x is a feasible, and y is an optimal solution to $(P[x])$, the resulting inequality $\|x\|_1 \le \|y\|_1$ must be equality, which, again due to $\kappa \in [0, 1/2)$, is possible only when $y_{I^o} = 0$. Thus, y has the same support I as x, and $w = x - y \in \operatorname{Ker} A$ is supported on s-element set I; by nullspace property, we should have $\|w_I\|_1 \le \kappa \|w\|_1 = \kappa \|w_I\|_1$, which is possible only when $w = 0$. □

1.2.2 Imperfect ℓ_1 minimization

We have found a necessary and sufficient condition for ℓ_1 minimization to recover *exactly s-sparse signals* in the *noiseless* case. More often than not, both these assumptions are violated: instead of s-sparse signals, we should speak about "nearly s-sparse" ones, quantifying the deviation from sparsity by the distance from the signal x underlying the observations to its best s-sparse approximation x^s. Similarly, we should allow for nonzero observation noise. With noisy observations and/or imperfect sparsity, we cannot hope to recover the signal exactly. All we may hope for, is to recover it with some error depending on the level of observation noise and "deviation from s-sparsity," and tending to zero as the level and deviation tend to 0. We are about to quantify the nullspace property to allow for instructive "error analysis."

1.2.2.1 *Contrast matrices and quantifications of Nullspace property*

By itself, the nullspace property says something about the signals from the kernel of the sensing matrix. We can reformulate it equivalently to say something important about *all* signals. Namely, observe that given sparsity s and $\kappa \in (0, 1/2)$, the nullspace property

$$\|w\|_{s,1} \leq \kappa \|w\|_1 \;\forall w \in \operatorname{Ker} A \tag{1.9}$$

is satisfied if and only if for a properly selected constant C one has[6]

$$\|w\|_{s,1} \leq C\|Aw\|_2 + \kappa\|w\|_1 \;\forall w. \tag{1.10}$$

Indeed, (1.10) clearly implies (1.9); to get the inverse implication, note that for every h orthogonal to $\operatorname{Ker} A$ it holds

$$\|Ah\|_2 \geq \sigma \|h\|_2,$$

where $\sigma > 0$ is the minimal positive singular value of A. Now, given $w \in \mathbf{R}^n$, we can decompose w into the sum of $\bar{w} \in \operatorname{Ker} A$ and $h \in (\operatorname{Ker} A)^\perp$, so that

$$\begin{array}{l} \|w\|_{s,1} \leq \|\bar{w}\|_{s,1} + \|h\|_{s,1} \leq \kappa\|\bar{w}\|_1 + \sqrt{s}\|h\|_{s,2} \leq \kappa[\|w\|_1 + \|h\|_1] + \sqrt{s}\|h\|_2 \\ \leq \kappa\|w\|_1 + [\kappa\sqrt{n} + \sqrt{s}]\|h\|_2 \leq \underbrace{\sigma^{-1}[\kappa\sqrt{n} + \sqrt{s}]}_{C} \underbrace{\|Ah\|_2}_{=\|Aw\|_2} + \kappa\|w\|_1, \end{array}$$

as required in (1.10).

Condition $\mathbf{Q}_1(s, \kappa)$. For our purposes, it is convenient to present the condition (1.10) in the following flexible form:

$$\|w\|_{s,1} \leq s\|H^T Aw\| + \kappa\|w\|_1, \tag{1.11}$$

where H is an $m \times N$ *contrast* matrix and $\|\cdot\|$ is some norm on $\mathbf{R}^N$. Whenever a pair $(H, \|\cdot\|)$, called *contrast pair*, satisfies (1.11), we say that $(H, \|\cdot\|)$ *satisfies condition* $\mathbf{Q}_1(s, \kappa)$. From what we have seen, *If* A possesses nullspace property with some sparsity level s and some $\kappa \in (0, 1/2)$, *then* there are many ways to select pairs $(H, \|\cdot\|)$ satisfying $\mathbf{Q}_1(s, \kappa)$, e.g., to take $H = CI_m$ with appropriately large C and $\|\cdot\| = \|\cdot\|_2$.

Conditions $\mathbf{Q}_q(s, \kappa)$. As we will see in a while, it makes sense to embed the condition $\mathbf{Q}_1(s, \kappa)$ into a parametric family of conditions $\mathbf{Q}_q(s, \kappa)$, where the parameter q runs through $[1, \infty]$. Specifically,

> *Given an $m \times n$ sensing matrix A, sparsity level $s \leq n$, and $\kappa \in (0, 1/2)$, we say that $m \times N$ matrix H and a norm $\|\cdot\|$ on $\mathbf{R}^N$ satisfy condition $\mathbf{Q}_q(s, \kappa)$ if*
>
> $$\|w\|_{s,q} \leq s^{\frac{1}{q}}\|H^T Aw\| + \kappa s^{\frac{1}{q}-1}\|w\|_1 \;\forall w \in \mathbf{R}^n. \tag{1.12}$$

Let us make two immediate observations on relations between the conditions:

A. When a pair $(H, \|\cdot\|)$ satisfies condition $\mathbf{Q}_q(s, \kappa)$, the pair satisfies also all conditions $\mathbf{Q}_{q'}(s, \kappa)$ with $1 \leq q' \leq q$.

---

[6] Note that (1.9) is exactly the $\phi^2(s, \kappa)$-Compatibility condition of [231] with $\phi(s, \kappa) = C/\sqrt{s}$; see also [232] for the analysis of relationships of this condition with other assumptions (e.g., a similar Restricted Eigenvalue assumption of [20]) used to analyse ℓ_1-minimization procedures.

Indeed in the situation in question for $1 \le q' \le q$ it holds

$$\begin{array}{rcl}\|w\|_{s,q'} & \le & s^{\frac{1}{q'}-\frac{1}{q}}\|w\|_{q,s} \le s^{\frac{1}{q'}-\frac{1}{q}}\left[s^{\frac{1}{q}}\|H^TAw\| + \kappa s^{\frac{1}{q}-1}\|w\|_1\right] \\ & = & s^{\frac{1}{q'}}\|H^TAw\| + \kappa s^{\frac{1}{q'}-1}\|w\|_1,\end{array}$$

where the first inequality is the standard inequality between ℓ_p-norms of the s-dimensional vector w^s.

B. When a pair $(H, \|\cdot\|)$ satisfies condition $\mathbf{Q}_q(s,\kappa)$ and $1 \le s' \le s$, the pair $((s/s')^{\frac{1}{q}}H, \|\cdot\|)$ satisfies the condition $\mathbf{Q}_q(s',\kappa)$.

Indeed, in the situation in question we clearly have for $1 \le s' \le s$:

$$\|w\|_{s',q} \le \|w\|_{s,q} \le (s')^{\frac{1}{q}}\|\left[(s/s')^{\frac{1}{q}}H\right]Aw\| + \kappa \underbrace{s^{\frac{1}{q}-1}}_{\le (s')^{\frac{1}{q}-1}} \|w\|_1.$$

1.2.3 Regular ℓ_1 recovery

Given the observation scheme (1.1) with an $m \times n$ sensing matrix A, we define the *regular ℓ_1 recovery* of x via observation y as

$$\widehat{x}_{\rm reg}(y) \in \operatorname*{Argmin}_u \left\{\|u\|_1 : \|H^T(Au-y)\| \le \rho\right\}, \tag{1.13}$$

where the *contrast matrix* $H \in \mathbf{R}^{m\times N}$, the norm $\|\cdot\|$ on $\mathbf{R}^N$ and $\rho > 0$ are parameters of the construction.

The role of $\mathbf{Q}$-conditions we have introduced is clear from the following

Theorem 1.3. *Let s be a positive integer, $q \in [1,\infty]$ and $\kappa \in (0,1/2)$. Assume that a pair $(H,\|\cdot\|)$ satisfies the condition $\mathbf{Q}_q(s,\kappa)$ associated with A, and let*

$$\Xi_\rho = \{\eta : \|H^T\eta\| \le \rho\}. \tag{1.14}$$

Then for all $x \in \mathbf{R}^n$ and $\eta \in \Xi_\rho$ one has

$$\|\widehat{x}_{\rm reg}(Ax+\eta) - x\|_p \le \frac{4(2s)^{\frac{1}{p}}}{1-2\kappa}\left[\rho + \frac{\|x-x^s\|_1}{2s}\right], \ 1 \le p \le q. \tag{1.15}$$

The above result can be slightly strengthened by replacing the assumption that $(H,\|\cdot\|)$ satisfies $\mathbf{Q}_q(s,\kappa)$ with some $\kappa < 1/2$, with a weaker—by observation **A** from Section 1.2.2.1—assumption that $(H,\|\cdot\|)$ satisfies $\mathbf{Q}_1(s,\varkappa)$ with $\varkappa < 1/2$ and satisfies $\mathbf{Q}_q(s,\kappa)$ with some (perhaps large) κ:

Theorem 1.4. *Given A, integer $s > 0$, and $q \in [1,\infty]$, assume that $(H,\|\cdot\|)$ satisfies the condition $\mathbf{Q}_1(s,\varkappa)$ with $\varkappa < 1/2$ and the condition $\mathbf{Q}_q(s,\kappa)$ with some $\kappa \ge \varkappa$, and let Ξ_ρ be given by (1.14). Then for all $x \in \mathbf{R}^n$ and $\eta \in \Xi_\rho$ it holds:*

$$\|\widehat{x}_{\rm reg}(Ax+\eta)-x\|_p \le \frac{4(2s)^{\frac{1}{p}}[1+\kappa-\varkappa]^{\frac{q(p-1)}{p(q-1)}}}{1-2\varkappa}\left[\rho + \frac{\|x-x^s\|_1}{2s}\right], \ 1 \le p \le q. \tag{1.16}$$

For proofs of Theorems 1.3 and 1.4, see Section 1.5.1.

Before commenting on the above results, let us present their alternative versions.

1.2.4 Penalized ℓ_1 recovery

Penalized ℓ_1 recovery of signal x from its observation (1.1) is

$$\widehat{x}_{\rm pen}(y) \in \mathop{\rm Argmin}_u \left\{ \|u\|_1 + \lambda \|H^T(Au - y)\| \right\}, \tag{1.17}$$

where $H \in \mathbf{R}^{m\times N}$, a norm $\|\cdot\|$ on $\mathbf{R}^N$, and a positive real λ are parameters of the construction.

Theorem 1.5. *Given A, positive integer s, and $q \in [1,\infty]$, assume that $(H, \|\cdot\|)$ satisfies the conditions $\mathbf{Q}_q(s,\kappa)$ and $\mathbf{Q}_1(s,\varkappa)$ with $\varkappa < 1/2$ and $\kappa \geq \varkappa$. Then*

(i) *Let $\lambda \geq 2s$. Then for all $x \in \mathbf{R}^n$, $y \in \mathbf{R}^m$ it holds:*

$$\|\widehat{x}_{\rm pen}(y) - x\|_p \leq \frac{4\lambda^{\frac{1}{p}}}{1-2\varkappa}\left[1 + \frac{\kappa\lambda}{2s} - \varkappa\right]^{\frac{q(p-1)}{p(q-1)}} \left[\|H^T(Ax - y)\| + \frac{\|x - x^s\|_1}{2s}\right], \ 1 \leq p \leq q. \tag{1.18}$$

In particular, with $\lambda = 2s$ we have:

$$\|\widehat{x}_{\rm pen}(y) - x\|_p \leq \frac{4(2s)^{\frac{1}{p}}}{1-2\varkappa}\left[1 + \kappa - \varkappa\right]^{\frac{q(p-1)}{p(q-1)}} \left[\|H^T(Ax - y)\| + \frac{\|x - x^s\|_1}{2s}\right], \ 1 \leq p \leq q. \tag{1.19}$$

(ii) *Let $\rho \geq 0$, and let Ξ_ρ be given by (1.14). Then for all $x \in \mathbf{R}^n$ and all $\eta \in \Xi_\rho$ one has:*

$$\begin{array}{ll} \lambda \geq 2s & \Rightarrow \\ & \|\widehat{x}_{\rm pen}(Ax+\eta) - x\|_p \leq \frac{4\lambda^{\frac{1}{p}}}{1-2\varkappa}\left[1 + \frac{\kappa\lambda}{2s} - \varkappa\right]^{\frac{q(p-1)}{p(q-1)}} \left[\rho + \frac{\|x - x^s\|_1}{2s}\right], \ 1 \leq p \leq q; \\ \lambda = 2s & \Rightarrow \\ & \|\widehat{x}_{\rm pen}(Ax+\eta) - x\|_p \leq \frac{4(2s)^{\frac{1}{p}}}{1-2\varkappa}\left[1 + \kappa - \varkappa\right]^{\frac{q(p-1)}{p(q-1)}} \left[\rho + \frac{\|x - x^s\|_1}{2s}\right], \ 1 \leq p \leq q. \end{array} \tag{1.20}$$

For proof, see Section 1.5.2.

1.2.5 Discussion

Some remarks are in order.

A. Qualitatively speaking, Theorems 1.3, 1.4, and 1.5 say the same thing: when $\mathbf{Q}$-conditions are satisfied, the regular or penalized recoveries reproduce the true signal *exactly* when there is no observation noise and the signal is s-sparse. In the presence of observation error η and imperfect sparsity, the signal is recovered within the error which can be upper-bounded by the sum of two terms, one proportional to the magnitude of observation noise and one proportional to the deviation $\|x - x^s\|_1$ of the signal from s-sparse ones. In the penalized recovery, the observation error is measured in the scale given by the contrast matrix and the norm $\|\cdot\|$—as $\|H^T\eta\|$—and in the regular recovery by an a priori upper bound ρ on $\|H^T\eta\|$; when $\rho \geq \|H^T\eta\|$, η belongs to Ξ_ρ and thus the bounds (1.15) and (1.16) are applicable to the actual observation error η. Clearly, in qualitative terms, an error bound of this type is the best we may hope for. Now let us look at the quantitative aspect. Assume that in the regular recovery we use $\rho \approx \|H^T\eta\|$, and in the penalized one $\lambda = 2s$. In this case, error bounds (1.15), (1.16), and (1.20), up to factors C depending solely on $\varkappa$ and κ, are the same, specifically,

$$\|\widehat{x} - x\|_p \leq C s^{1/p}[\|H^T\eta\| + \|x - x^s\|_1/s], \ 1 \leq p \leq q. \tag{!}$$

Is this error bound bad or good? The answer depends on many factors, including on how well we select H and $\|\cdot\|$. To get a kind of orientation, consider the trivial case of *direct* observations, where matrix A is square and, moreover, is proportional to the unit matrix: $A = \alpha I$. Let us assume in addition that x is exactly s-sparse. In this case, the simplest way to ensure condition $\mathbf{Q}_q(s,\kappa)$, even with $\kappa = 0$, is to take $\|\cdot\| = \|\cdot\|_{s,q}$ and $H = s^{-1/q}\alpha^{-1}I$, so that (!) becomes

$$\|\widehat{x} - x\|_p \leq C\alpha^{-1}s^{1/p-1/q}\|\eta\|_{s,q},\ 1 \leq p \leq q. \tag{!!}$$

As far as the dependence of the bound on the magnitude $\|\eta\|_{s,q}$ of the observation noise is concerned, this dependence is as good as it can be—even if we knew in advance the positions of the s entries of x of largest magnitudes, we would be unable to recover x in q-norm with error $\leq \alpha^{-1}\|\eta\|_{s,q}$. In addition, with the s largest magnitudes of entries in η equal to each other, the $\|\cdot\|_p$-norm of the recovery error clearly cannot be guaranteed to be less than $\alpha^{-1}\|\eta\|_{s,p} = \alpha^{-1}s^{1/p-1/q}\|\eta\|_{s,q}$. Thus, at least for s-sparse signals x, our error bound is, basically, the best one can get already in the "ideal" case of direct observations.

B. Given that $(H, \|\cdot\|)$ obeys $\mathbf{Q}_1(s,\varkappa)$ with some $\varkappa < 1/2$, the larger the q such that the pair $(H, \|\cdot\|)$ obeys the condition $\mathbf{Q}_q(s,\kappa)$ with a given $\kappa \geq \varkappa$ (recall that κ can be $\geq 1/2$) and s, the larger the range $p \leq q$ of values of p where the error bounds (1.16) and (1.20) are applicable. This is in full accordance with the fact that if a pair $(H, \|\cdot\|)$ obeys condition $\mathbf{Q}_q(s,\kappa)$, it obeys also all conditions $\mathbf{Q}_{q'}(s,\kappa)$ with $1 \leq q' \leq q$ (item **A** in Section 1.2.2.1).

C. The flexibility offered by contrast matrix H and norm $\|\cdot\|$ allows us to adjust, to some extent, the recovery to the "geometry of observation errors." For example, when η is "uncertain but bounded," say, when all we know is that $\|\eta\|_2 \leq \delta$ with some given δ, all that matters (on the top of the requirement for $(H, \|\cdot\|)$ to obey $\mathbf{Q}$-conditions) is how large $\|H^T\eta\|$ could be when $\|\eta\|_2 \leq \delta$. In particular, when $\|\cdot\| = \|\cdot\|_2$, the error bound "is governed" by the spectral norm of H. Consequently, if we have a technique allowing us *to design* H such that $(H, \|\cdot\|_2)$ obeys $\mathbf{Q}$-condition(s) with given parameters, it makes sense to look for a design with as small a spectral norm of H as possible. In contrast to this, in the case of Gaussian noise the most interesting for applications,

$$y = Ax + \eta,\ \eta \sim \mathcal{N}(0, \sigma^2 I_m), \tag{1.21}$$

looking at the spectral norm of H, with $\|\cdot\|_2$ in the role of $\|\cdot\|$, is counterproductive, since a typical realization of η is of Euclidean norm of order of $\sqrt{m}\sigma$ and thus is quite large when m is large. In this case to quantify "the magnitude" of $H^T\eta$ by the product of the spectral norm of H and the Euclidean norm of η is *completely misleading*—in typical cases, this product will grow rapidly with the number of observations m, completely ignoring the fact that η is random with zero mean.[7] What is much better suited for the case of Gaussian noise, is the $\|\cdot\|_\infty$ norm in the role of $\|\cdot\|$ and the norm of H which is "the maximum of $\|\cdot\|_2$-norms of the columns

[7]The simplest way to see the difference is to look at a particular entry $h^T\eta$ in $H^T\eta$. Operating with spectral norms, we upper-bound this entry by $\|h\|_2\|\eta\|_2$, and the second factor for $\eta \sim \mathcal{N}(0, \sigma^2 I_m)$ is typically as large as $\sigma\sqrt{m}$. This is in sharp contrast to the fact that typical values of $h^T\eta$ are of order of $\sigma\|h\|_2$, independently of what m is!

in H," denoted by $\|H\|_{1,2}$. Indeed, with $\eta \sim \mathcal{N}(0, \sigma^2 I_m)$, the entries in $H^T\eta$ are Gaussian with zero mean and variance bounded by $\sigma^2\|H\|_{1,2}^2$, so that $\|H^T\eta\|_\infty$ is the maximum of magnitudes of N zero mean Gaussian random variables with standard deviations bounded by $\sigma\|H\|_{1,2}$. As a result,

$$\mathrm{Prob}\{\|H^T\eta\|_\infty \geq \rho\} \leq 2N\mathrm{Erfc}\left(\frac{\rho}{\sigma\|H\|_{1,2}}\right) \leq N\mathrm{e}^{-\frac{\rho^2}{2\sigma^2\|H\|_{1,2}^2}}, \tag{1.22}$$

where

$$\mathrm{Erfc}(s) = \mathrm{Prob}_{\xi\sim\mathcal{N}(0,1)}\{\xi \geq s\} = \frac{1}{\sqrt{2\pi}}\int_s^\infty \mathrm{e}^{-t^2/2}dt$$

is the (slightly rescaled) complementary error function.

It follows that the typical values of $\|H^T\eta\|_\infty$, $\eta \sim \mathcal{N}(0, \sigma^2 I_m)$ are of order of at most $\sigma\sqrt{\ln(N)}\|H\|_{1,2}$. In applications we consider in this chapter, we have $N = O(m)$, so that with σ and $\|H\|_{1,2}$ given, typical values $\|H^T\eta\|_\infty$ are nearly independent of m. The bottom line is that ℓ_1 minimization is capable of handling large-scale Gaussian observation noise incomparably better than "uncertain-but-bounded" observation noise of similar magnitude (measured in Euclidean norm).

D. As far as comparison of regular and penalized ℓ_1 recoveries with the same pair $(H, \|\cdot\|)$ is concerned, the situation is as follows. Assume for the sake of simplicity that $(H, \|\cdot\|)$ satisfies $\mathbf{Q}_q(s,\kappa)$ with some s and $\kappa < 1/2$, and let the observation error be random. Given $\epsilon \in (0,1)$, let

$$\rho_\epsilon[H, \|\cdot\|] = \min\left\{\rho : \ \mathrm{Prob}\left\{\eta : \|H^T\eta\| \leq \rho\right\} \geq 1-\epsilon\right\}; \tag{1.23}$$

this is nothing but the smallest ρ such that

$$\mathrm{Prob}\{\eta \in \Xi_\rho\} \geq 1-\epsilon \tag{1.24}$$

(see (1.14)), and thus the smallest ρ for which the error bound (1.15) for the regular ℓ_1 recovery holds true with probability $1-\epsilon$ (or at least the smallest ρ for which the latter claim is supported by Theorem 1.3). With $\rho = \rho_\epsilon[H, \|\cdot\|]$, the regular ℓ_1 recovery guarantees (and that is the best guarantee one can extract from Theorem 1.3) that

(#) *For some set* Ξ, $\mathrm{Prob}\{\eta \in \Xi\} \geq 1-\epsilon$, *of "good" realizations of* $\eta \sim \mathcal{N}(0, \sigma^2 I_m)$, *one has*

$$\|\widehat{x}(Ax+\eta) - x\|_p \leq \frac{4(2s)^{\frac{1}{p}}}{1-2\kappa}\left[\rho_\epsilon[H, \|\cdot\|] + \frac{\|x - x^s\|_1}{2s}\right], \ 1 \leq p \leq q, \tag{1.25}$$

whenever $x \in \mathbf{R}^n$ *and* $\eta \in \Xi_\rho$.

The error bound (1.19) (where we set $\varkappa = \kappa$) says that *(#) holds true for the penalized* ℓ_1 *recovery with* $\lambda = 2s$. The latter observation suggests that the penalized ℓ_1 recovery associated with $(H, \|\cdot\|)$ and $\lambda = 2s$ is better than its regular counterpart, the reason being twofold. First, in order to ensure (#) with the regular recovery, the "built in" parameter ρ of this recovery should be set to $\rho_\epsilon[H, \|\cdot\|]$, and the latter quantity is not always easy to identify. In contrast to this, the construction of penalized ℓ_1 recovery is completely independent of a priori assumptions on the structure of observation errors, while automatically ensuring (#) for the error model we use. Second, and more importantly, for the penalized recovery the bound (1.25) is no more than the "worst, with confidence $1-\epsilon$, case," while the typical values of the quantity $\|H^T\eta\|$ which indeed participates in the error bound (1.18) may be essentially smaller than $\rho_\epsilon[H, \|\cdot\|]$. Numerical experience fully supports the above claim: the difference in observed performance of the two routines in question, although not dramatic, is definitely in favor of the penalized recovery. The only potential disadvantage of the latter routine is that the penalty parameter λ should be tuned to the level s of sparsity we aim at, while the regular recovery is free of any guess of this type. Of course, the "tuning" is rather loose—all we need (and experiments show that we indeed need this) is the relation $\lambda \geq 2s$, so that a rough upper bound on s will do. However, that bound (1.18) deteriorates as λ grows.

Finally, we remark that when H is $m \times N$ and $\eta \sim \mathcal{N}(0, \sigma^2 I_m)$, we have

$$\rho_\epsilon[H, \|\cdot\|_\infty] \leq \sigma \mathrm{ErfcInv}(\frac{\epsilon}{2N})\|H\|_{1,2} \leq \sigma\sqrt{2\ln(N/\epsilon)}\|H\|_{1,2}$$

(see (1.22)); here ErfcInv(δ) is the inverse complementary error function:

$$\mathrm{Erfc}(\mathrm{ErfcInv}(\delta)) = \delta,\ 0 < \delta < 1. \tag{1.26}$$

How it works. Here we present a small numerical illustration. We observe in Gaussian noise $m = n/2$ randomly selected terms in n-element "time series" $z = (z_1, ..., z_n)$ and want to recover this series under the assumption that the series is "nearly s-sparse in frequency domain," that is, that

$$z = Fx \text{ with } \|x - x^s\|_1 \leq \delta,$$

where F is the matrix of $n \times n$ the Inverse Discrete Cosine Transform, x^s is the vector obtained from x by zeroing out all but the s entries of largest magnitudes and δ upper-bounds the distance from x to s-sparse signals. Denoting by A the $m \times n$ submatrix of F corresponding to the time instants t where z_t is observed, our observation becomes

$$y = Ax + \sigma\xi,$$

where ξ is the standard Gaussian noise. After the signal in frequency domain, that is, x, is recovered by ℓ_1 minimization, let the recovery be $\widehat{x}$, we recover the signal in the time domain as $\widehat{z} = F\widehat{x}$. In Figure 1.3, we present four test signals, of different (near-)sparsity, along with their regular and penalized ℓ_1 recoveries. The data in Figure 1.3 clearly show how the quality of ℓ_1 recovery deteriorates as the number s of "essential nonzeros" of the signal in the frequency domain grows. It is seen also that the penalized recovery meaningfully outperforms the regular one in the range of sparsities up to 64.

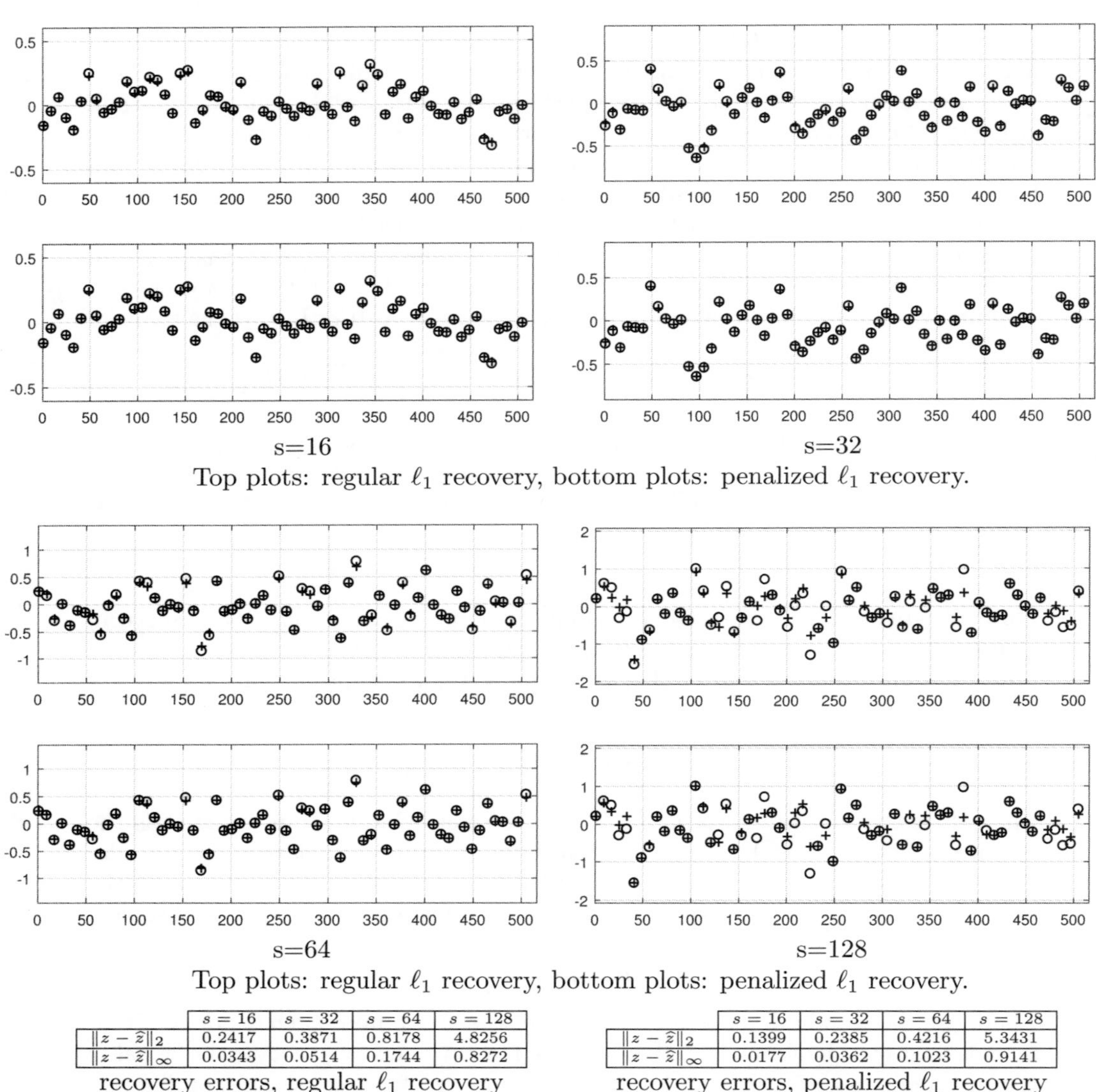

s=16 s=32

Top plots: regular ℓ_1 recovery, bottom plots: penalized ℓ_1 recovery.

s=64 s=128

Top plots: regular ℓ_1 recovery, bottom plots: penalized ℓ_1 recovery.

	$s = 16$	$s = 32$	$s = 64$	$s = 128$
$\|z - \widehat{z}\|_2$	0.2417	0.3871	0.8178	4.8256
$\|z - \widehat{z}\|_\infty$	0.0343	0.0514	0.1744	0.8272

recovery errors, regular ℓ_1 recovery

	$s = 16$	$s = 32$	$s = 64$	$s = 128$
$\|z - \widehat{z}\|_2$	0.1399	0.2385	0.4216	5.3431
$\|z - \widehat{z}\|_\infty$	0.0177	0.0362	0.1023	0.9141

recovery errors, penalized ℓ_1 recovery

Figure 1.3: Regular and penalized ℓ_1 recovery of nearly s-sparse signals. o: true signals, +: recoveries (to make the plots readable, one per eight consecutive vector's entries is shown). Problem sizes are $m = 256$ and $n = 2m = 512$, noise level is $\sigma = 0.01$, deviation from s-sparsity is $\|x - x^s\|_1 = 1$, contrast pair is $(H = \sqrt{n/m}A, \|\cdot\|_\infty)$. In penalized recovery, $\lambda = 2s$, parameter ρ of regular recovery is set to $\sigma \cdot \mathrm{ErfcInv}(0.005/n)$.

1.3 VERIFIABILITY AND TRACTABILITY ISSUES

The good news about ℓ_1 recovery stated in Theorems 1.3, 1.4, and 1.5 is "conditional"—we assume that we are smart enough to point out a pair $(H, \|\cdot\|)$ satisfying condition $\mathbf{Q}_1(s, \varkappa)$ with $\varkappa < 1/2$ (and condition $\mathbf{Q}_q(s, \kappa)$ with a "moderate" $\varkappa$ [8]). The related issues are twofold:

1. First, we do not know in which range of s, m, and n these conditions, or even the weaker than $\mathbf{Q}_1(s, \varkappa)$, $\varkappa < 1/2$, nullspace property can be satisfied; and without the nullspace property, ℓ_1 minimization becomes useless, at least when we want to guarantee its validity whatever be the s-sparse signal we want to recover;
2. Second, it is unclear how to verify whether a given sensing matrix A satisfies the nullspace property for a given s, or a given pair $(H, \|\cdot\|)$ satisfies the condition $\mathbf{Q}_q(s, \kappa)$ with given parameters.

What is known about these crucial issues can be outlined as follows.

1. It is known that for given m, n with $m \ll n$ (say, $m/n \leq 1/2$), there exist $m \times n$ sensing matrices which are s-good for the values of s "nearly as large as m," specifically, for $s \leq O(1)\frac{m}{\ln(n/m)}$.[9] Moreover, there are natural families of matrices where this level of goodness "is a rule." E.g., when drawing an $m \times n$ matrix at random from Gaussian or Rademacher distributions (i.e., when filling the matrix with independent realizations of a random variable which is either a standard (zero mean, unit variance) Gaussian one, or takes values ± 1 with probabilities 0.5), the result will be s-good, for the outlined value of s, with probability approaching 1 as m and n grow. All this remains true when instead of speaking about matrices A satisfying "plain" nullspace properties, we are speaking about matrices A for which it is easy to point out a pair $(H, \|\cdot\|)$ satisfying the condition $\mathbf{Q}_2(s, \varkappa)$ with, say, $\varkappa = 1/4$.
   The above results can be considered as a good news. A bad news is that we do *not* know how to check efficiently, given an s and a sensing matrix A, that the matrix is s-good, just as we do not know how to check that A admits good (i.e., satisfying $\mathbf{Q}_1(s, \varkappa)$ with $\varkappa < 1/2$) pairs $(H, \|\cdot\|)$. Even worse: we do not know an efficient recipe allowing us to build, given m, an $m \times 2m$ matrix A^m which is provably s-good for s larger than $O(1)\sqrt{m}$, which is a much smaller "level of goodness" than the one promised by theory for randomly generated matrices.[10] The "common life" analogy of this situation would be as follows: you know that 90% of bricks in your wall are made of gold, and at the same time, you do not know how to tell a golden brick from a usual one.
2. There exist *verifiable sufficient conditions* for s-goodness of a sensing matrix, similarly to verifiable sufficient conditions for a pair $(H, \|\cdot\|)$ to satisfy condition

---

[8] $\mathbf{Q}_q(s, \kappa)$ is always satisfied with "large enough" κ, e.g., $\kappa = s$, but such values of κ are of no interest: the associated bounds on p-norms of the recovery error are straightforward consequences of the bounds on the $\|\cdot\|_1$-norm of this error yielded by the condition $\mathbf{Q}_1(s, \varkappa)$.

[9] Recall that $O(1)$'s denote positive *absolute constants*—appropriately chosen numbers like 0.5, or 1, or perhaps 100,000. We could, in principle, replace all $O(1)$'s with specific numbers; following the standard mathematical practice, we do not do it, partly out of laziness, partly because particular values of these numbers in our context are irrelevant.

[10] Note that the naive algorithm "generate $m \times 2m$ matrices at random until an s-good, with s promised by the theory, matrix is generated" is *not* an efficient recipe, since we still do not know how to check s-goodness efficiently.

$\mathbf{Q}_q(s,\kappa)$. The bad news is that when $m \ll n$, these verifiable sufficient conditions can be satisfied only when $s \leq O(1)\sqrt{m}$—once again, in a much more narrow range of values of s than when typical randomly selected sensing matrices are s-good. In fact, $s = O(\sqrt{m})$ is so far *the best* known sparsity level for which we know individual s-good $m \times n$ sensing matrices with $m \leq n/2$.

1.3.1 Restricted Isometry Property and s-goodness of random matrices

There are several sufficient conditions for s-goodness, equally difficult to verify, but provably satisfied for typical random sensing matrices. The best known of them is the *Restricted Isometry Property* (RIP) defined as follows:

Definition 1.6. *Let k be an integer and $\delta \in (0,1)$. We say that an $m \times n$ sensing matrix A possesses the Restricted Isometry Property with parameters δ and k,* RIP(δ,k), *if for every k-sparse $x \in \mathbf{R}^n$ one has*

$$(1-\delta)\|x\|_2^2 \leq \|Ax\|_2^2 \leq (1+\delta)\|x\|_2^2. \tag{1.27}$$

It turns out that for natural ensembles of random $m \times n$ matrices, a typical matrix from the ensemble satisfies RIP(δ,k) with small δ and k "nearly as large as m," and that RIP$(\frac{1}{6},2s)$ implies the nullspace condition, and more. The simplest versions of the corresponding results are as follows.

Proposition 1.7. *Given $\delta \in (0,\frac{1}{5}]$, with properly selected positive $c = c(\delta)$, $d = d(\delta)$, $f = f(\delta)$ for all $m \leq n$ and all positive integers k such that*

$$k \leq \frac{m}{c\ln(n/m)+d} \tag{1.28}$$

the probability for a random $m \times n$ matrix A with independent $\mathcal{N}(0,\frac{1}{m})$ entries to satisfy RIP(δ,k) *is at least $1-\exp\{-fm\}$.*

For proof, see Section 1.5.3.

Proposition 1.8. *Let $A \in \mathbf{R}^{m\times n}$ satisfy* RIP$(\delta,2s)$ *for some $\delta < 1/3$ and positive integer s. Then*

(i) *The pair $\left(H = \frac{s^{-1/2}}{\sqrt{1-\delta}}I_m, \|\cdot\|_2\right)$ satisfies the condition $\mathbf{Q}_2\left(s,\frac{\delta}{1-\delta}\right)$ associated with A;*

(ii) *The pair $(H = \frac{1}{1-\delta}A, \|\cdot\|_\infty)$ satisfies the condition $\mathbf{Q}_2\left(s,\frac{\delta}{1-\delta}\right)$ associated with A.*

For proof, see Section 1.5.4.

1.3.2 Verifiable sufficient conditions for $\mathbf{Q}_q(s,\kappa)$

When speaking about verifiable sufficient conditions for a pair $(H,\|\cdot\|)$ to satisfy $\mathbf{Q}_q(s,\kappa)$, it is convenient to restrict ourselves to the case where H, like A, is an $m \times n$ matrix, and $\|\cdot\| = \|\cdot\|_\infty$.

Proposition 1.9. *Let A be an $m \times n$ sensing matrix, and $s \leq n$ be a sparsity level.*

Given an $m \times n$ matrix H and $q \in [1, \infty]$, let us set

$$\nu_{s,q}[H] = \max_{j \leq n} \|\mathrm{Col}_j[I - H^T A]\|_{s,q}, \tag{1.29}$$

where $\mathrm{Col}_j[C]$ is j-th column of matrix C. Then

$$\|w\|_{s,q} \leq s^{1/q}\|H^T Aw\|_\infty + \nu_{s,q}[H]\|w\|_1 \;\forall w \in \mathbf{R}^n, \tag{1.30}$$

implying that the pair $(H, \|\cdot\|_\infty)$ satisfies the condition $\mathbf{Q}_q(s, s^{1-\frac{1}{q}}\nu_{s,q}[H])$.

Proof is immediate. Setting $V = I - H^T A$, we have

$$\begin{array}{l}\|w\|_{s,q} = \|[H^T A + V]w\|_{s,q} \leq \|H^T Aw\|_{s,q} + \|Vw\|_{s,q} \\ \leq s^{1/q}\|H^T Aw\|_\infty + \sum_j |w_j|\|\mathrm{Col}_j[V]\|_{s,q} \leq s^{1/q}\|H^T A\|_\infty + \nu_{s,q}[H]\|w\|_1.\end{array} \qquad \square$$

Observe that the function $\nu_{s,q}[H]$ is an efficiently computable convex function of H, so that the set

$$\mathcal{H}^\kappa_{s,q} = \{H \in \mathbf{R}^{m\times n} : \nu_{s,q}[H] \leq s^{\frac{1}{q}-1}\kappa\} \tag{1.31}$$

is a computationally tractable convex set. When this set is nonempty for some $\kappa < 1/2$, every point H in this set is a contrast matrix such that $(H, \|\cdot\|_\infty)$ satisfies the condition $\mathbf{Q}_q(s, \kappa)$, that is, we can find contrast matrices making ℓ_1 minimization valid. Moreover, we can *design* contrast matrix, e.g., by minimizing over $\mathcal{H}^\kappa_{s,q}$ the function $\|H\|_{1,2}$, thus optimizing the sensitivity of the corresponding ℓ_1 recoveries to Gaussian observation noise; see items **C**, **D** in Section 1.2.5.

Explanation. The sufficient condition for s-goodness of A stated in Proposition 1.9 looks as if coming out of thin air; in fact it is a particular case of a simple and general construction as follows. Let $f(x)$ be a real-valued convex function on $\mathbf{R}^n$, and $X \subset \mathbf{R}^n$ be a nonempty bounded polytope represented as

$$X = \{x \in \mathrm{Conv}\{g_1, ..., g_N\} : Ax = 0\},$$

where $\mathrm{Conv}\{g_1, ..., g_N\} = \{\sum_i \lambda_i g_i : \lambda \geq 0, \sum_i \lambda_i = 1\}$ is the convex hull of vectors $g_1, ..., g_N$. Our goal is to upper-bound the maximum $\mathrm{Opt} = \max_{x\in X} f(x)$; this is a meaningful problem, since precisely maximizing a convex function over a polyhedron typically is a computationally intractable task. Let us act as follows: clearly, for any matrix H of the same size as A we have $\max_{x\in X} f(x) = \max_{x\in X} f([I - H^T A]x)$, since on X we have $[I - H^T A]x = x$. As a result,

$$\begin{array}{rcl}\mathrm{Opt} & := & \max\limits_{x\in X} f(x) = \max\limits_{x\in X} f([I - H^T A]x) \\ & \leq & \max\limits_{x\in \mathrm{Conv}\{g_1,...,g_N\}} f([I - H^T A]x) \\ & = & \max\limits_{j\leq N} f([I - H^T A]g_j).\end{array}$$

We get a parametric—the parameter being H—upper bound on Opt, namely, the bound $\max_{j\leq N} f([I - H^T A]g_j)$. This parametric bound is convex in H, and thus is well suited for minimization over this parameter.

The result of Proposition 1.9 is inspired by this construction as applied to the nullspace property: given an $m \times n$ sensing matrix A and setting

$$X = \{x \in \mathbf{R}^n : \|x\|_1 \leq 1, Ax = 0\} = \{x \in \text{Conv}\{\pm e_1, ..., \pm e_n\} : Ax = 0\}$$

(e_i are the basic orths in $\mathbf{R}^n$), A is s-good if and only if

$$\text{Opt}_s := \max_{x \in X} \{f(x) := \|x\|_{s,1}\} < 1/2.$$

A verifiable sufficient condition for this, as yielded by the above construction, is the existence of an $m \times n$ matrix H such that

$$\max_{j \leq n} \max[f([I_n - H^T A]e_j), f(-[I_n - H^T A]e_j)] < 1/2,$$

or, which is the same,

$$\max_j \|\text{Col}_j[I_n - H^T A]\|_{s,1} < 1/2.$$

This observation brings to our attention the matrix $I - H^T A$ with varying H and the idea of expressing sufficient conditions for s-goodness and related properties in terms of this matrix.

1.3.3 Tractability of $\mathbf{Q}_\infty(s, \kappa)$

As we have already mentioned, the conditions $\mathbf{Q}_q(s, \kappa)$ are intractable, in the sense that we do not know how to verify whether a given pair $(H, \|\cdot\|)$ satisfies the condition. Surprisingly, this is *not* the case with the strongest of these conditions, the one with $q = \infty$. Namely,

Proposition 1.10. *Let A be an $m \times n$ sensing matrix, s be a sparsity level, and $\kappa \geq 0$. Then whenever a pair $(\bar{H}, \|\cdot\|)$ satisfies the condition $\mathbf{Q}_\infty(s, \kappa)$, there exists an $m \times n$ matrix H such that*

$$\|\text{Col}_j[I_n - H^T A]\|_{s,\infty} = \|\text{Col}_j[I_n - H^T A]\|_\infty \leq s^{-1}\kappa, \ 1 \leq j \leq n$$

(so that $(H, \|\cdot\|_\infty)$ satisfies $\mathbf{Q}_\infty(s, \kappa)$ by Proposition 1.9), and also

$$\|H^T \eta\|_\infty \leq \|\bar{H}^T \eta\| \ \forall \eta \in \mathbf{R}^m. \tag{1.32}$$

In addition, the $m \times n$ contrast matrix H such that the pair $(H, \|\cdot\|_\infty)$ satisfies the condition $\mathbf{Q}_\infty(s, \kappa)$ with as small κ as possible can be found as follows. Consider n LP programs

$$\text{Opt}_i = \min_{\nu, h} \left\{\nu : \|A^T h - e^i\|_\infty \leq \nu\right\}, \tag{$\#_i$}$$

where e^i is i-th basic orth of $\mathbf{R}^n$. Let Opt_i, h_i, $i = 1, ..., n$ be optimal solutions to these problems; we set $H = [h_1, ..., h_n]$; the corresponding value of κ is

$$\kappa_* = s \max_i \text{Opt}_i.$$

Besides this, there exists a transparent alternative description of the quantities Opt_i

(and thus of κ_*); specifically,

$$\mathrm{Opt}_i = \max_x \{x_i : \|x\|_1 \leq 1, Ax = 0\}. \tag{1.33}$$

For proof, see Section 1.5.5.

Taken along with (1.32) and error bounds of Theorems 1.3, 1.4, and 1.5, Proposition 1.10 says that

> *As far as the condition $\mathbf{Q}_\infty(s,\kappa)$ is concerned, we lose nothing when restricting ourselves with pairs $(H \in \mathbf{R}^{m\times n}, \|\cdot\|_\infty)$ and contrast matrices H satisfying the condition*
>
> $$|[I_n - H^T A]_{ij}| \leq s^{-1}\kappa, \tag{1.34}$$
>
> *implying that $(H, \|\cdot\|_\infty)$ satisfies $\mathbf{Q}_\infty(s,\kappa)$.*

The good news is that (1.34) is an explicit convex constraint on H (in fact, even on H and κ), so that we can solve the *design problems*, where we want to optimize a convex function of H under the requirement that $(H, \|\cdot\|_\infty)$ satisfies the condition $\mathbf{Q}_\infty(s,\kappa)$ (and, perhaps, additional convex constraints on H and κ).

1.3.3.1 Mutual Incoherence

The simplest (and up to some point in time, the only) verifiable sufficient condition for s-goodness of a sensing matrix A is expressed in terms of *mutual incoherence* of A, defined as

$$\mu(A) = \max_{i\neq j} \frac{|\mathrm{Col}_i^T[A]\mathrm{Col}_j[A]|}{\|\mathrm{Col}_i[A]\|_2^2}. \tag{1.35}$$

This quantity is well defined whenever A has no zero columns (otherwise A is not even 1-good). Note that when A is normalized to have all columns of equal $\|\cdot\|_2$-lengths,[11] $\mu(A)$ is small when the columns of A are nearly mutually orthogonal. The standard related result is that

> *Whenever A and a positive integer s are such that $\frac{2\mu(A)}{1+\mu(A)} < \frac{1}{s}$, A is s-good.*

It is immediately seen that the latter condition is weaker than what we can get with the aid of (1.34):

Proposition 1.11. *Let A be an $m \times n$ matrix, and let the columns of $m \times n$ matrix H be given by*

$$\mathrm{Col}_j(H) = \frac{1}{(1+\mu(A))\|\mathrm{Col}_j(A)\|_2^2}\mathrm{Col}_j(A),\ 1 \leq j \leq n.$$

Then

$$|[I_m - H^T A]_{ij}| \leq \frac{\mu(A)}{1+\mu(A)}\ \forall i,j. \tag{1.36}$$

---

[11]As far as ℓ_1 minimization is concerned, this normalization is non-restrictive: we always can enforce it by diagonal scaling of the signal underlying observations (1.1), and ℓ_1 minimization in scaled variables is the same as weighted ℓ_1 minimization in original variables.

In particular, when $\frac{2\mu(A)}{1+\mu(A)} < \frac{1}{s}$, *A is s-good.*

Proof. With H as above, the diagonal entries in $I - H^T A$ are equal to $1 - \frac{1}{1+\mu(A)} = \frac{\mu(A)}{1+\mu(A)}$, while by definition of mutual incoherence the magnitudes of the off-diagonal entries in $I - H^T A$ are $\leq \frac{\mu(A)}{1+\mu(A)}$ as well, implying (1.36). The "in particular" claim is given by (1.36) combined with Proposition 1.9. □

1.3.3.2 From RIP to conditions $\mathbf{Q}_q(\cdot, \kappa)$

It turns out that when A is RIP(δ, k) and $q \geq 2$, it is easy to point out pairs $(H, \|\cdot\|)$ satisfying $\mathbf{Q}_q(t, \kappa)$ with a desired $\kappa > 0$ and properly selected t:

Proposition 1.12. *Let A be an* $m \times n$ *sensing matrix satisfying* RIP$(\delta, 2s)$ *with some s and some* $\delta \in (0,1)$, *and let* $q \in [2, \infty]$ *and* $\kappa > 0$ *be given. Then*

(i) *Whenever a positive integer t satisfies*

$$t \leq \min\left[\left[\frac{\kappa(1-\delta)}{\delta}\right]^{\frac{q}{q-1}}, s^{\frac{q-2}{q-1}}\right] s^{\frac{q}{2q-2}}, \tag{1.37}$$

the pair $(H = \frac{t^{-\frac{1}{q}}}{\sqrt{1-\delta}} I_m, \|\cdot\|_2)$ *satisfies* $\mathbf{Q}_q(t, \kappa)$*;*

(ii) *Whenever a positive integer t satisfies* (1.37), the pair $(H = \frac{s^{\frac{1}{2}} t^{-\frac{1}{q}}}{1-\delta} A, \|\cdot\|_\infty)$ satisfies $\mathbf{Q}_q(t, \kappa)$.

For proof, see Section 1.5.4.

The most important consequence of Proposition 1.12 deals with the case of $q = \infty$ and states that *when s-goodness of a sensing matrix A can be ensured by the difficult to verify condition* RIP$(\delta, 2s)$ *with, say,* $\delta = 0.2$, *the somehow worse level of sparsity,* $t = O(1)\sqrt{s}$ *with properly selected absolute constant* $O(1)$, *can be certified via condition* $\mathbf{Q}_\infty(t, \frac{1}{3})$*—there exists a pair* $(H, \|\cdot\|_\infty)$ *satisfying this condition.* The point is that by Proposition 1.10, if the condition $\mathbf{Q}_\infty(t, \frac{1}{3})$ can at all be satisfied, a pair $(H, \|\cdot\|_\infty)$ satisfying this condition can be found efficiently.

Unfortunately, the significant "dropdown" in the level of sparsity when passing from unverifiable RIP to verifiable $\mathbf{Q}_\infty$ is inevitable; this bad news is what is on our agenda now.

1.3.3.3 Limits of performance of verifiable sufficient conditions for goodness

Proposition 1.13. *Let A be an* $m \times n$ *sensing matrix which is "essentially non-square," specifically, such that* $2m \leq n$, *and let* $q \in [1, \infty]$. *Whenever a positive integer s and an* $m \times n$ *matrix H are linked by the relation*

$$\|\mathrm{Col}_j[I_n - H^T A]\|_{s,q} < \tfrac{1}{2} s^{\frac{1}{q}-1}, \; 1 \leq j \leq n, \tag{1.38}$$

one has

$$s \leq \sqrt{m}. \tag{1.39}$$

As a result, the sufficient condition for the validity of $\mathbf{Q}_q(s, \kappa)$ *with* $\kappa < 1/2$ *from Proposition 1.9 can never be satisfied when* $s > \sqrt{m}$. *Similarly, the verifiable sufficient condition* $\mathbf{Q}_\infty(s, \kappa)$, $\kappa < 1/2$, *for s-goodness of A cannot be satisfied*

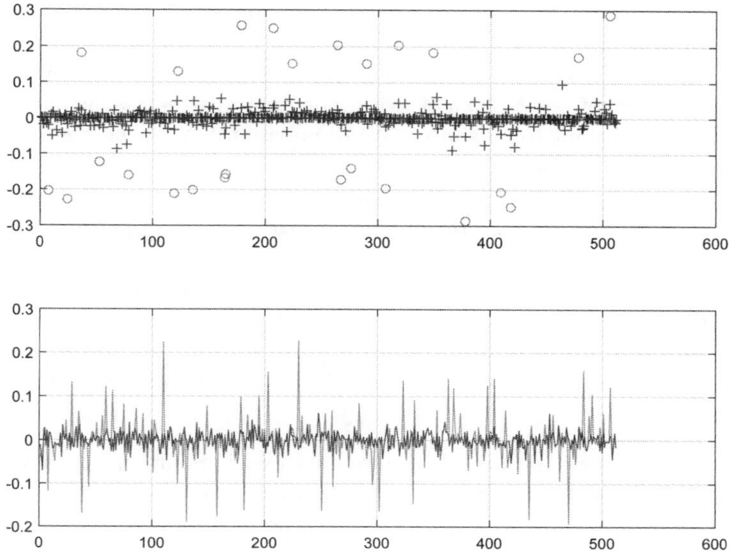

Figure 1.4: Erroneous ℓ_1 recovery of a 25-sparse signal, no observation noise. Top: frequency domain, o – true signal, + – recovery. Bottom: time domain.

when $s > \sqrt{m}$.

For proof, see Section 1.5.6.

We see that unless A is "nearly square," our (same as all others known to us) verifiable sufficient conditions for s-goodness are unable to justify this property for "large" s. This unpleasant fact is in full accordance with the already mentioned fact that no individual provably s-good "essentially nonsquare" $m \times n$ matrices with $s \geq O(1)\sqrt{m}$ are known.

Matrices for which our verifiable sufficient conditions do establish s-goodness with $s \leq O(1)\sqrt{m}$ do exist.

How it works: Numerical illustration. Let us apply our machinery to the 256×512 randomly selected submatrix A of the matrix of 512×512 Inverse Discrete Cosine Transform which we used in experiments reported in Figure 1.3. These experiments exhibit nice performance of ℓ_1 minimization when recovering sparse (even nearly sparse) signals with as many as 64 nonzeros. *In fact, the level of goodness of A is at most 24*, as is witnessed in Figure 1.4.

In order to upper-bound the level of goodness of a matrix A, one can try to maximize the convex function $\|w\|_{s,1}$ over the set $W = \{w : Aw = 0, \|w\|_1 \leq 1\}$: if, for a given s, the maximum of $\|\cdot\|_{s,1}$ over W is $\geq 1/2$, the matrix is not s-good—it does not possess the nullspace property. Now, while global maximization of the convex function $\|w\|_{s,1}$ over W is difficult, we can try to find suboptimal solutions as follows. Let us start with a vector $w_1 \in W$ of $\|\cdot\|_1$-norm 1, and let u^1 be obtained from w_1 by replacing the s entries in w_1 of largest magnitudes by the signs of these entries and zeroing out all other entries, so that $w_1^T u^1 = \|w_1\|_{s,1}$. After u^1 is found, let us solve the LO program $\max_w \{[u^1]^T w : w \in W\}$. w_1 is a feasible solution to this problem, so that for the optimal solution w_2 we have $[u^1]^T w_2 \geq [u^1]^T w_1 =$

$\|w_1\|_{s,1}$; this inequality, by virtue of what u^1 is, implies that $\|w_2\|_{s,1} \geq \|w_1\|_{s,1}$, and, by construction, $w_2 \in W$. We now can iterate the construction, with w_2 in the role of w_1, to get $w_3 \in W$ with $\|w_3\|_{s,1} \geq \|w_2\|_{s,1}$, etc. Proceeding in this way, we generate a sequence of points from W with monotonically increasing value of the objective $\|\cdot\|_{s,1}$ we want to maximize. We terminate this recurrence either when the achieved value of the objective becomes $\geq 1/2$ (then we know for sure that A is not s-good, and can proceed to investigating s-goodness for a smaller value of s) or when the recurrence gets stuck—the observed progress in the objective falls below a given threshold, say, 10^{-6}. When it happens, we can restart the process from a new starting point randomly selected in W, after getting stuck, restart again, etc., until we exhaust our time budget. The output of the process is the best of the points we have generated—that of the largest $\|\cdot\|_{s,1}$. Applying this approach to the matrix A in question, in a couple of minutes it turns out that the matrix is at most 24-good.

One can ask how it may happen that previous experiments with recovering 64-sparse signals went fine, when in fact some 25-sparse signals cannot be recovered by ℓ_1 minimization even in the ideal noiseless case. The answer is simple: in our experiments, we dealt with *randomly selected* signals, and typical randomly selected data are much nicer, whatever be the purpose of a numerical experiment, than the worst-case data.

It is interesting to understand also which goodness we can certify using our verifiable sufficient conditions. Computations show that the fully verifiable (and strongest in our scale of sufficient conditions for s-goodness) condition $\mathbf{Q}_\infty(s,\varkappa)$ can be satisfied with $\varkappa < 1/2$ when s is as large as 7 and $\varkappa = 0.4887$, and can*not* be satisfied with $\varkappa < 1/2$ when $s = 8$. As for Mutual Incoherence, it can only justify 3-goodness, no more. We can hardly be happy with the resulting bounds—goodness at least 7 and at most 24; however, it could be worse.

1.4 EXERCISES FOR CHAPTER 1

Exercise 1.1.

The k-th Hadamard matrix, $\mathcal{H}_k$ (here k is a nonnegative integer), is the $n_k \times n_k$ matrix, $n_k = 2^k$, given by the recurrence

$$\mathcal{H}_0 = [1]; \mathcal{H}_{k+1} = \left[\begin{array}{c|c} \mathcal{H}_k & \mathcal{H}_k \\ \hline \mathcal{H}_k & -\mathcal{H}_k \end{array}\right].$$

In the sequel, we assume that $k > 0$. Now comes the exercise:

1. Check that $\mathcal{H}_k$ is a symmetric matrix with entries ± 1, and columns of the matrix are mutually orthogonal, so that $\mathcal{H}_k/\sqrt{n_k}$ is an orthogonal matrix.
2. Check that when $k > 0$, $\mathcal{H}_k$ has just two distinct eigenvalues, $\sqrt{n_k}$ and $-\sqrt{n_k}$, each of multiplicity $m_k := 2^{k-1} = n_k/2$.
3. Prove that whenever f is an eigenvector of $\mathcal{H}_k$, one has

$$\|f\|_\infty \leq \|f\|_1/\sqrt{n_k}.$$

Derive from this observation the conclusion as follows:

Let $a_1, ..., a_{m_k} \in \mathbf{R}^{n_k}$ be unit vectors orthogonal to each other which are eigenvectors of $\mathcal{H}_k$ with eigenvalues $\sqrt{n_k}$ (by the above, the dimension of the eigenspace of $\mathcal{H}_k$ associated with the eigenvalue $\sqrt{n_k}$ is m_k, so that the required $a_1, ..., a_{m_k}$ do exist), and let A be the $m_k \times n_k$ matrix with the rows $a_1^T, ..., a_{m_k}^T$. For every $x \in \operatorname{Ker} A$ it holds

$$\|x\|_\infty \leq \frac{1}{\sqrt{n_k}}\|x\|_1,$$

whence A satisfies the nullspace property whenever the sparsity s satisfies $2s < \sqrt{n_k} = \sqrt{2m_k}$. Moreover, there exists (and can be found efficiently) an $m_k \times n_k$ contrast matrix $H = H_k$ such that for every $s < \frac{1}{2}\sqrt{n_k}$, the pair $(H_k, \|\cdot\|_\infty)$ satisfies the condition $\mathbf{Q}_\infty(s, \kappa_s = \underbrace{s/\sqrt{n_k}}_{<1/2})$ associated with A, and the $\|\cdot\|_2$-norms of columns of H_k do not exceed $\sqrt{2\frac{\sqrt{n_k}+1}{\sqrt{n_k}}}$.

Note that the above conclusion yields a sequence of individual $(m_k = 2^{k-1}) \times (n_k = 2^k)$ sensing matrices, $k = 1, 2, ...$, with "size ratio" $n_k/m_k = 2$, which make an efficiently verifiable condition for s-goodness, say, $\mathbf{Q}_\infty(s, \frac{1}{3})$, satisfiable in basically the entire range of values of s allowed by Proposition 1.13. It would be interesting to get similar "fully constructive" results for other size ratios, like $m : n = 1 : 4$, $m : n = 1 : 8$, etc.

Exercise 1.2.

[Follow-up to Exercise 1.1] Exercise 1.1 provides us with an explicitly given $(m = 512) \times (n = 1024)$ sensing matrix $\bar{A}$ such that the efficiently verifiable condition $\mathbf{Q}_\infty(15, \frac{15}{32})$ is satisfiable; in particular, $\bar{A}$ is 15-good. With all we know about limits of performance of verifiable sufficient conditions for goodness, how should we evaluate this specific sensing matrix? Could we point out a sensing matrix of the same size which is provably s-good for a value of s larger (or "much larger") than 15?

We do not know the answer, and you are requested to explore some possibilities, including (but not reducing to—you are welcome to investigate more options!) the following ones.

1. Generate at random a sample of $m \times n$ sensing matrices A, compute their mutual incoherences, and look at how large are the goodness levels certified by these incoherences. What happens when the matrices are Gaussian (independent $\mathcal{N}(0, 1)$ entries) and Rademacher (independent entries taking values ± 1 with probabilities 1/2)?
2. Generate at random a sample of $m \times n$ matrices with independent $\mathcal{N}(0, 1/m)$ entries. Proposition 1.7 suggests that a sample matrix A has good chances to satisfy $\mathrm{RIP}(\delta, k)$ with some $\delta < 1/3$ and some k, and thus to be s-good (and even more than this, see Proposition 1.8) for every $s \leq k/2$. Of course, given A we cannot check whether the matrix indeed satisfies $\mathrm{RIP}(\delta, k)$ with given δ, k; what we can try to do is to certify that $\mathrm{RIP}(\delta, k)$ does *not* take place. To this end, it suffices to select at random, say, 200 $m \times k$ submatrices $\tilde{A}$ of A and compute the eigenvalues of $\tilde{A}^T\tilde{A}$; if A possesses $\mathrm{RIP}(\delta, k)$, all these eigenvalues should belong to the segment $[1 - \delta, 1 + \delta]$, and if in reality this does not happen, A definitely is not $\mathrm{RIP}(\delta, k)$.

Exercise 1.3.

Let us start with a preamble. Consider a finite Abelian group; the only thing which matters for us is that such a group G is specified by a collection of $k \geq 1$ of positive integers $\nu_1, ..., \nu_k$ and is comprised of all collections $\omega = (\omega_1, ..., \omega_k)$ where every ω_i is an integer from the range $\{0, 1, ..., \nu_k - 1\}$; the group operation, denoted by $\oplus$, is

$$(\omega_1, ..., \omega_k) \oplus (\omega_1', ..., \omega_k') = ((\omega_1 + \omega_1') \bmod \nu_1, ..., (\omega_k + \omega_k') \bmod \nu_k),$$

where $a \bmod b$ is the remainder, taking values in $\{0, 1, ..., b-1\}$, in the division of an integer a by a positive integer b, e.g., $5 \bmod 3 = 2$ and $6 \bmod 3 = 0$. Clearly, the cardinality of the above group G is $n_k = \nu_1 \nu_2 ... \nu_k$. A *character* of group G is a homomorphism acting from G into the multiplicative group of complex numbers of modulus 1, or, in simple words, a complex-valued function $\chi(\omega)$ on G such that $|\chi(\omega)| = 1$ for all $\omega \in G$ and $\chi(\omega \oplus \omega') = \chi(\omega)\chi(\omega')$ for all $\omega, \omega' \in G$. Note that characters themselves form a group w.r.t. pointwise multiplication; clearly, all characters of our G are functions of the form

$$\chi((\omega_1, ..., \omega_k)) = \mu_1^{\omega_1} ... \mu_k^{\omega_k},$$

where μ_i are restricted to be roots of degree ν_i from 1: $\mu_i^{\nu_i} = 1$. It is immediately seen that the group G_* of characters of G is of the same cardinality $n_k = \nu_1 ... \nu_k$ as G. We can associate with G the matrix $\mathcal{F}$ of size $n_k \times n_k$; the columns in the matrix are indexed by the elements ω of G, the rows by the characters $\chi \in G_*$ of G, and the element in cell (χ, ω) is $\chi(\omega)$. The standard example here corresponds to $k = 1$, in which case $\mathcal{F}$ clearly is the $\nu_1 \times \nu_1$ matrix of the Discrete Fourier Transform.

Now comes the exercise:

1. Verify that the above $\mathcal{F}$ is, up to factor $\sqrt{n_k}$, a unitary matrix: denoting by $\bar{a}$ the complex conjugate of a complex number a, $\sum_{\omega \in G} \chi(\omega)\overline{\chi'}(\omega)$ is n_k or 0 depending on whether $\chi = \chi'$ or $\chi \neq \chi'$.
2. Let $\bar{\omega}, \bar{\omega}'$ be two elements of G. Prove that there exists a permutation Π of elements of G which maps $\bar{\omega}$ into $\bar{\omega}'$ and is such that

$$\mathrm{Col}_{\Pi(\omega)}[\mathcal{F}] = D\mathrm{Col}_{\omega}[\mathcal{F}] \;\forall \omega \in G,$$

   where D is diagonal matrix with diagonal entries $\chi(\bar{\omega}')/\chi(\bar{\omega})$, $\chi \in G_*$.
3. Consider the special case of the above construction where $\nu_1 = \nu_2 = ... = \nu_k = 2$. Verify that in this case $\mathcal{F}$, up to permutation of rows and permutation of columns (these permutations depend on how we assign the elements of G and G_* their serial numbers), is exactly the Hadamard matrix $\mathcal{H}_k$.
4. Extract from the above the following fact: let m, k be positive integers such that $m \leq n_k := 2^k$, and let sensing matrix A be obtained from $\mathcal{H}_k$ by selecting m distinct rows. Assume we want to find an $m \times n_k$ contrast matrix H such that the pair $(H, \|\cdot\|_\infty)$ satisfies the condition $\mathbf{Q}_\infty(s, \kappa)$ with as small a κ as possible; by Proposition 1.10, to this end we should solve n LP programs

$$\mathrm{Opt}_i = \min_h \|e^i - A^T h\|_\infty,$$

   where e^i is i-th basic orth in $\mathbf{R}^n$. Prove that with A coming from $\mathcal{H}_k$, all these problems have the same optimal value, and optimal solutions to all of the problems are readily given by the optimal solution to just one of them.

Exercise 1.4.

Proposition 1.13 states that the verifiable condition $\mathbf{Q}_\infty(s,\kappa)$ can certify s-goodness of an "essentially nonsquare" (with $m \leq n/2$) $m \times n$ sensing matrix A only when s is small as compared to m, namely, $s \leq \sqrt{2m}$. The exercise to follow is aimed at investigating what happens when $m \times n$ "low" (with $m < n$) sensing matrix A is "nearly square", meaning that $m^o = n - m$ is small as compared to n. Specifically, you should prove that for properly selected individual $(n-m^o)\times n$ matrices A the condition $\mathbf{Q}_\infty(s,\kappa)$ with $\kappa < 1/2$ is satisfiable when s is as large as $O(1)n/\sqrt{m^o}$.

1. Let $n = 2^k p$ with positive integer p and integer $k \geq 1$, and let $m^o = 2^{k-1}$. Given a $2m^o$-dimensional vector u, let u^+ be an n-dimensional vector built as follows: we split indexes from $\{1,...,n = 2^k p\}$ into 2^k consecutive groups $I_1,...,I_{2^k}$, p elements per group, and all entries of u^+ with indexes from I_i are equal to the i-th entry, u_i, of vector u. Now let U be the linear subspace in $\mathbf{R}^{2^k}$ comprised of all eigenvectors, with eigenvalue $\sqrt{2^k}$, of the Hadamard matrix $\mathcal{H}_k$—see Exercise 1.1—so that the dimension of U is $2^{k-1} = m^o$, and let L be given by
$$L = \{u^+ : u \in U\} \subset \mathbf{R}^n.$$
Clearly, L is a linear subspace in $\mathbf{R}^n$ of dimension m^o. Prove that
$$\forall x \in L : \|x\|_\infty \leq \frac{\sqrt{2m^o}}{n}\|x\|_1.$$
Conclude that if A is an $(n-m^o)\times n$ sensing matrix with $\operatorname{Ker} A = L$, then the verifiable sufficient condition $\mathbf{Q}_\infty(s,\kappa)$ does certify s-goodness of A whenever
$$1 \leq s < \frac{n}{2\sqrt{2m^o}}.$$
2. Let L be an m^o-dimensional subspace in $\mathbf{R}^n$. Prove that L contains a nonzero vector x with
$$\|x\|_\infty \geq \frac{\sqrt{m^o}}{n}\|x\|_1,$$
so that the condition $\mathbf{Q}_\infty(s,\kappa)$ cannot certify s-goodness of an $(n-m^o)\times n$ sensing matrix A whenever $s > O(1)n/\sqrt{m^o}$, for properly selected absolute constant $O(1)$.

Exercise 1.5.

Utilize the results of Exercise 1.3 in a numerical experiment as follows.

- select n as an integer power 2^k of 2, say, $n = 2^{10} = 1024$;
- select a "representative" sequence M of values of m, $1 \leq m < n$, including values of m close to n and "much smaller" than n, say,
$$M = \{2,5,8,16,32,64,128,256,512,7,896,960,992,1008,1016,1020,1022,1023\};$$
- for every $m \in M$,

- generate at random an $m \times n$ submatrix A of the $n \times n$ Hadamard matrix $\mathcal{H}_k$ and utilize the result of item 4 of Exercise 1.3 in order to find the largest s such that the s-goodness of A can be certified via the condition $\mathbf{Q}_\infty(\cdot,\cdot)$; call $s(m)$ the resulting value of s;
- generate a moderate sample of Gaussian $m \times n$ sensing matrices A_i with independent $\mathcal{N}(0,1/m)$ entries and use the construction from Exercise 1.2 to upper-bound the largest s for which a matrix from the sample satisfies RIP$(1/3, 2s)$; call $\widehat{s}(m)$ the largest—over your A_i's—of the resulting upper bounds.

The goal of the exercise is to compare the computed values of $s(m)$ and $\widehat{s}(m)$; in other words, we again want to understand how "theoretically perfect" RIP compares to "conservative restricted scope" condition $\mathbf{Q}_\infty$.

1.5 PROOFS

1.5.1 Proofs of Theorem 1.3, 1.4

All we need is to prove Theorem 1.4, since Theorem 1.3 is the particular case $\varkappa = \kappa < 1/2$ of Theorem 1.4.

Let us fix $x \in \mathbf{R}^n$ and $\eta \in \Xi_\rho$, and let us set $\widehat{x} = \widehat{x}_{\text{reg}}(Ax+\eta)$. Let also $I \subset \{1,...,n\}$ be the set of indexes of the s entries in x of largest magnitudes, I^o be the complement of I in $\{1,...,n\}$, and, for $w \in \mathbf{R}^n$, w_I and w_{I^o} be the vectors obtained from w by zeroing entries with indexes $j \notin I$ and $j \notin I^o$, respectively, and keeping the remaining entries intact. Finally, let $z = \widehat{x} - x$.

1°. By the definition of Ξ_ρ and due to $\eta \in \Xi_\rho$, we have

$$\|H^T([Ax+\eta] - Ax)\| \leq \rho, \tag{1.40}$$

so that x is a feasible solution to the optimization problem specifying $\widehat{x}$, whence $\|\widehat{x}\|_1 \leq \|x\|_1$. We therefore have

$$\begin{array}{rcl}\|\widehat{x}_{I^o}\|_1 & = & \|\widehat{x}\|_1 - \|\widehat{x}_I\|_1 \leq \|x\|_1 - \|\widehat{x}_I\|_1 = \|x_I\|_1 + \|x_{I^o}\|_1 - \|\widehat{x}_I\|_1 \\ & \leq & \|z_I\|_1 + \|x_{I^o}\|_1,\end{array} \tag{1.41}$$

and therefore

$$\|z_{I^o}\|_1 \leq \|\widehat{x}_{I^o}\|_1 + \|x_{I^o}\|_1 \leq \|z_I\|_1 + 2\|x_{I^o}\|_1.$$

It follows that

$$\|z\|_1 = \|z_I\|_1 + \|z_{I^o}\|_1 \leq 2\|z_I\|_1 + 2\|x_{I^o}\|_1. \tag{1.42}$$

Further, by definition of $\widehat{x}$ we have $\|H^T([Ax+\eta] - A\widehat{x})\| \leq \rho$, which combines with (1.40) to imply that

$$\|H^T A(\widehat{x} - x)\| \leq 2\rho. \tag{1.43}$$

2°. Since $(H, \|\cdot\|)$ satisfies $\mathbf{Q}_1(s,\varkappa)$, we have

$$\|z\|_{s,1} \leq s\|H^T Az\| + \varkappa\|z\|_1.$$

By (1.43), it follows that $\|z\|_{s,1} \leq 2s\rho + \varkappa\|z\|_1$, which combines with the evident inequality $\|z_I\| \leq \|z\|_{s,1}$ (recall that $\mathrm{Card}(I) = s$) and with (1.42) to imply that

$$\|z_I\|_1 \leq 2s\rho + \varkappa\|z\|_1 \leq 2s\rho + 2\varkappa\|z_I\|_1 + 2\varkappa\|x_{I^o}\|_1,$$

whence

$$\|z_I\|_1 \leq \frac{2s\rho + 2\varkappa\|x_{I^o}\|_1}{1-2\varkappa}.$$

Invoking (1.42), we conclude that

$$\|z\|_1 \leq \frac{4s}{1-2\varkappa}\left[\rho + \frac{\|x_{I^o}\|_1}{2s}\right]. \tag{1.44}$$

3°. Since $(H, \|\cdot\|)$ satisfies $\mathbf{Q}_q(s,\kappa)$, we have

$$\|z\|_{s,q} \leq s^{\frac{1}{q}}\|H^TAz\| + \kappa s^{\frac{1}{q}-1}\|z\|_1,$$

which combines with (1.44) and (1.43) to imply that

$$\|z\|_{s,q} \leq s^{\frac{1}{q}}2\rho + \kappa s^{\frac{1}{q}}\frac{4\rho+2s^{-1}\|x_{I^o}\|_1}{1-2\varkappa} \leq \frac{4s^{\frac{1}{q}}[1+\kappa-\varkappa]}{1-2\varkappa}\left[\rho + \frac{\|x_o\|_1}{2s}\right] \tag{1.45}$$

(we have taken into account that $\varkappa < 1/2$ and $\kappa \geq \varkappa$). Let θ be the $(s+1)$-st largest magnitude of entries in z, and let $w = z - z^s$. Now (1.45) implies that

$$\theta \leq \|z\|_{s,q}s^{-\frac{1}{q}} \leq \frac{4[1+\kappa-\varkappa]}{1-2\varkappa}\left[\rho + \frac{\|x_{I^o}\|_1}{2s}\right].$$

Hence invoking (1.44) we have

$$\begin{array}{rcl}\|w\|_q & \leq & \|w\|_\infty^{\frac{q-1}{q}}\|w\|_1^{\frac{1}{q}} \leq \theta^{\frac{q-1}{q}}\|z\|_1^{\frac{1}{q}} \\ & \leq & \theta^{\frac{q-1}{q}}\frac{(4s)^{\frac{1}{q}}}{[1-2\varkappa]^{\frac{1}{q}}}\left[\rho + \frac{\|x_{I^o}\|_1}{2s}\right]^{\frac{1}{q}} \\ & \leq & \frac{4s^{\frac{1}{q}}[1+\kappa-\varkappa]^{\frac{q-1}{q}}}{1-2\varkappa}\left[\rho + \frac{\|x_{I^o}\|_1}{2s}\right].\end{array}$$

Taking into account (1.45) and the fact that the supports of z^s and w do not intersect, we get

$$\begin{array}{rcl}\|z\|_q & \leq & 2^{\frac{1}{q}}\max[\|z^s\|_q, \|w\|_q] = 2^{\frac{1}{q}}\max[\|z\|_{s,q}, \|w\|_q] \\ & \leq & \frac{4(2s)^{\frac{1}{q}}[1+\kappa-\varkappa]}{1-2\varkappa}\left[\rho + \frac{\|x_{I^o}\|_1}{2s}\right].\end{array}$$

This bound combines with (1.44), the Moment inequality,[12] and with the relation $\|x_{I^o}\|_1 = \|x - x^s\|_1$ to imply (1.16). □

---

[12]The Moment inequality states that if (Ω, μ) is a space with measure and f is a μ-measurable real-valued function on Ω, then $\phi(\rho) = \ln\left(\int_\Omega |f(\omega)|^{\frac{1}{\rho}}\mu(d\omega)\right)^\rho$ is a convex function of ρ on every segment $\Delta \subset [0,1]$ such that $\phi(\cdot)$ is well defined at the endpoints of Δ. As a corollary, when $x \in \mathbf{R}^n$ and $1 \leq p \leq q \leq \infty$, one has $\|x\|_p \leq \|x\|_1^{\frac{q-p}{p(q-1)}}\|x\|_q^{\frac{q(p-1)}{p(q-1)}}$.

1.5.2 Proof of Theorem 1.5

Let us prove (i). Let us fix $x \in \mathbf{R}^n$ and η, and let us set $\widehat{x} = \widehat{x}_{\text{pen}}(Ax+\eta)$. Let also $I \subset \{1,...,K\}$ be the set of indexes of the s entries in x of largest magnitudes, I^o be the complement of I in $\{1,...,n\}$, and, for $w \in \mathbf{R}^n$, w_I and w_{I^o} be the vectors obtained from w by zeroing out all entries with indexes not in I and not in I^o, respectively. Finally, let $z = \widehat{x} - x$ and $\nu = \|H^T\eta\|$.

1°. We have

$$\|\widehat{x}\|_1 + \lambda\|H^T(A\widehat{x} - Ax - \eta)\| \leq \|x\|_1 + \lambda\|H^T\eta\|$$

and

$$\|H^T(A\widehat{x} - Ax - \eta)\| = \|H^T(Az - \eta)\| \geq \|H^TAz\| - \|H^T\eta\|,$$

whence

$$\|\widehat{x}\|_1 + \lambda\|H^TAz\| \leq \|x\|_1 + 2\lambda\|H^T\eta\| = \|x\|_1 + 2\lambda\nu. \tag{1.46}$$

We have

$$\begin{array}{rcl} \|\widehat{x}\|_1 & = & \|x+z\|_1 = \|x_I + z_I\|_1 + \|x_{I^o} + z_{I^o}\|_1 \\ & \geq & \|x_I\|_1 - \|z_I\|_1 + \|z_{I^o}\|_1 - \|x_{I^o}\|_1, \end{array}$$

which combines with (1.46) to imply that

$$\|x_I\|_1 - \|z_I\|_1 + \|z_{I^o}\|_1 - \|x_{I^o}\|_1 + \lambda\|H^TAz\| \leq \|x\|_1 + 2\lambda\nu,$$

or, which is the same,

$$\|z_{I^o}\|_1 - \|z_I\|_1 + \lambda\|H^TAz\| \leq 2\|x_{I^o}\|_1 + 2\lambda\nu. \tag{1.47}$$

Since $(H, \|\cdot\|)$ satisfies $\mathbf{Q}_1(s,\varkappa)$, we have

$$\|z_I\|_1 \leq \|z\|_{s,1} \leq s\|H^TAz\| + \varkappa\|z\|_1,$$

so that

$$(1-\varkappa)\|z_I\|_1 - \varkappa\|z_{I^o}\|_1 - s\|H^TAz\| \leq 0. \tag{1.48}$$

Taking a weighted sum of (1.47) and (1.48), the weights being 1 and 2, respectively, we get

$$(1-2\varkappa)\left[\|z_I\|_1 + \|z_{I^o}\|_1\right] + (\lambda - 2s)\|H^TAz\| \leq 2\|x_{I^o}\|_1 + 2\lambda\nu,$$

whence, due to $\lambda \geq 2s$,

$$\|z\|_1 \leq \frac{2\lambda\nu + 2\|x_{I^o}\|_1}{1-2\varkappa} \leq \frac{2\lambda}{1-2\varkappa}\left[\nu + \frac{\|x_{I^o}\|_1}{2s}\right]. \tag{1.49}$$

Further, by (1.46) we have

$$\lambda\|H^TAz\| \leq \|x\|_1 - \|\widehat{x}\|_1 + 2\lambda\nu \leq \|z\|_1 + 2\lambda\nu,$$

which combines with (1.49) to imply that

$$\lambda\|HA^Tz\| \leq \frac{2\lambda\nu + 2\|x_{I^o}\|_1}{1-2\varkappa} + 2\lambda\nu = \frac{2\lambda\nu(2-2\varkappa) + 2\|x_{I^o}\|_1}{1-2\varkappa}. \tag{1.50}$$

From $\mathbf{Q}_q(s,\kappa)$ it follows that

$$\|z\|_{s,q} \leq s^{\frac{1}{q}}\|H^TAz\| + \kappa s^{\frac{1}{q}-1}\|z\|_1,$$

which combines with (1.50) and (1.49) to imply that

$$\begin{array}{rl}\|z\|_{s,q} & \leq s^{\frac{1}{q}-1}\left[s\|H^TAz\| + \kappa\|z\|_1\right] \leq s^{\frac{1}{q}-1}\left[\frac{4s\nu(1-\varkappa)+\frac{2s}{\lambda}\|x_{I^o}\|_1}{1-2\varkappa} + \frac{\kappa[2\lambda\nu+\frac{\lambda}{s}\|x_{I^o}\|_1]}{1-2\varkappa}\right]\\ & = s^{\frac{1}{q}}\frac{[4(1-\varkappa)+2s^{-1}\lambda\kappa]\nu+[2\lambda^{-1}+\kappa s^{-2}\lambda]\|x_{I^o}\|_1}{1-2\varkappa} \leq 4\frac{s^{\frac{1}{q}}}{1-2\varkappa}\left[1+\frac{\kappa\lambda}{2s}-\varkappa\right]\left[\nu+\frac{\|x_{I^o}\|_1}{2s}\right]\end{array} \tag{1.51}$$

(recall that $\lambda \geq 2s$, $\kappa \geq \varkappa$, and $\varkappa < 1/2$). It remains to repeat the reasoning following (1.45) in item 3° of the proof of Theorem 1.4. Specifically, denoting by θ the $(s+1)$-st largest magnitude of entries in z, (1.51) implies that

$$\theta \leq s^{-1/q}\|z\|_{s,q} \leq \frac{4}{1-2\varkappa}[1+\kappa\frac{\lambda}{2s}-\varkappa]\left[\nu+\frac{\|x_{I^o}\|_1}{2s}\right], \tag{1.52}$$

so that for the vector $w = z - z^s$ one has

$$\|w\|_q \;\leq\; \theta^{1-\frac{1}{q}}\|w\|_1^{\frac{1}{q}} \leq \frac{4(\lambda/2)^{\frac{1}{q}}}{1-2\varkappa}\left[1+\kappa\frac{\lambda}{2s}-\varkappa\right]^{\frac{q-1}{q}}\left[\nu+\frac{\|x_{I^o}\|_1}{2s}\right]$$

(we have used (1.52) and (1.49)). Hence, taking into account that z^s and w have nonintersecting supports,

$$\begin{array}{rcl}\|z\|_q & \leq & 2^{\frac{1}{q}}\max[\|z^s\|_q,\|w\|_q] = 2^{\frac{1}{q}}\max[\|z\|_{s,q},\|w\|_q]\\ & \leq & \frac{4\lambda^{\frac{1}{q}}}{1-2\varkappa}\left[1+\kappa\frac{\lambda}{2s}-\varkappa\right]\left[\nu+\frac{\|x_{I^o}\|_1}{2s}\right]\end{array}$$

(we have used (1.51) along with $\lambda \geq 2s$ and $\kappa \geq \varkappa$). This combines with (1.49) and the Moment inequality to imply (1.18). All remaining claims of Theorem 1.5 are immediate corollaries of (1.18). □

1.5.3 Proof of Proposition 1.7

1°. Assuming $k \leq m$ and selecting a set I of k indices from $\{1,...,n\}$ distinct from each other, consider an $m \times k$ submatrix A_I of A comprised of columns with indexes from I, and let u be a unit vector in $\mathbf{R}^k$. The entries in the vector $m^{1/2}A_Iu$ are independent $\mathcal{N}(0,1)$ random variables, so that for the random variable $\zeta_u = \sum_{i=1}^m (m^{1/2}A_Iu)_i^2$ and $\gamma \in (-1/2,1/2)$ it holds (in what follows, expectations and probabilities are taken w.r.t. our ensemble of random A's)

$$\ln\left(\mathbf{E}\{\exp\{\gamma\zeta\}\}\right) = m\ln\left(\frac{1}{\sqrt{2\pi}}\int e^{\gamma t^2-\frac{1}{2}t^2}ds\right) = -\frac{m}{2}\ln(1-2\gamma).$$

Given $\alpha \in (0,0.1]$ and selecting γ in such a way that $1-2\gamma = \frac{1}{1+\alpha}$, we get $0<\gamma<1/2$ and therefore

$$\begin{array}{l}\mathrm{Prob}\{\zeta_u > m(1+\alpha)\} \leq \mathbf{E}\{\exp\{\gamma\zeta_u\}\}\exp\{-m\gamma(1+\alpha)\}\\ = \exp\{-\frac{m}{2}\ln(1-2\gamma) - m\gamma(1+\alpha)\}\\ = \exp\{\frac{m}{2}[\ln(1+\alpha)-\alpha]\} \leq \exp\{-\frac{m}{5}\alpha^2\},\end{array}$$

and similarly, selecting γ in such a way that $1-2\gamma = \frac{1}{1-\alpha}$, we get $-1/2 < \gamma < 0$ and therefore

$$\begin{array}{l}\mathrm{Prob}\{\zeta_u < m(1-\alpha)\} \leq \mathbf{E}\{\exp\{\gamma\zeta_u\}\}\exp\{-m\gamma(1-\alpha)\}\\ = \exp\{-\frac{m}{2}\ln(1-2\gamma) - m\gamma(1-\alpha)\}\\ = \exp\{\frac{m}{2}[\ln(1-\alpha)+\alpha]\} \leq \exp\{-\frac{m}{5}\alpha^2\},\end{array}$$

and we end up with

$$u \in \mathbf{R}^k, \|u\|_2 = 1 \Rightarrow \left\{\begin{array}{l}\mathrm{Prob}\{A : \|A_I u\|_2^2 > 1+\alpha\} \leq \exp\{-\frac{m}{5}\alpha^2\}\\ \mathrm{Prob}\{A : \|A_I u\|_2^2 < 1-\alpha\} \leq \exp\{-\frac{m}{5}\alpha^2\}\end{array}\right. . \tag{1.53}$$

2°. As above, let $\alpha \in (0, 0.1]$, let

$$M = 1+2\alpha, \epsilon = \frac{\alpha}{2(1+2\alpha)},$$

and let us build an ϵ-net on the unit sphere S in $\mathbf{R}^k$ as follows. We start with a point $u_1 \in S$; after $\{u_1, ..., u_t\} \subset S$ is already built, we check whether there is a point in S at the $\|\cdot\|_2$-distance from all points of the set $> \epsilon$. If it is the case, we add such a point to the net built so far and proceed with building the net; otherwise we terminate with the net $\{u_1, ..., u_t\}$. By compactness of S and due to $\epsilon > 0$, this process eventually terminates; upon termination, we have at our disposal the collection $\{u_1, ..., u_N\}$ of unit vectors such that every two of them are at $\|\cdot\|_2$-distance $> \epsilon$ from each other, and every point from S is at distance at most ϵ from some point of the collection. We claim that the cardinality N of the resulting set can be bounded as

$$N \leq \left[\frac{2+\epsilon}{\epsilon}\right]^k = \left[\frac{4+9\alpha}{\alpha}\right]^k \leq \left(\frac{5}{\alpha}\right)^k. \tag{1.54}$$

Indeed, the interiors of the $\|\cdot\|_2$-balls of radius $\epsilon/2$ centered at the points $u_1, ..., u_N$ are mutually disjoint, and their union is contained in the $\|\cdot\|_2$-ball of radius $1+\epsilon/2$ centered at the origin; comparing the volume of the union and that of the ball, we arrive at (1.54).

3°. Consider event E comprised of all realizations of A such that for all k-element subsets I of $\{1, ..., n\}$ and all $t \leq n$ it holds

$$1-\alpha \leq \|A_I u_t\|_2^2 \leq 1+\alpha. \tag{1.55}$$

By (1.53) and the union bound,

$$\mathrm{Prob}\{A \notin E\} \leq 2N\binom{n}{k}\exp\{-\frac{m}{5}\alpha^2\}. \tag{1.56}$$

We claim that

$$A \in E \Rightarrow (1-2\alpha) \leq \|A_I u\|_2^2 \leq 1+2\alpha \;\forall \left(\begin{array}{l}I \subset \{1, ..., n\} : \mathrm{Card}(I) = k\\ u \in \mathbf{R}^k : \|u\|_2 = 1\end{array}\right). \tag{1.57}$$

Indeed, let $A \in E$, let us fix $I \in \{1,...,n\}$, $\mathrm{Card}(I) = k$, and let M be the maximal value of the quadratic form $f(u) = u^T A_I^T A_I u$ on the unit $\|\cdot\|_2$-ball B, centered at the origin, in $\mathbf{R}^k$. In this ball, f is Lipschitz continuous with constant $2M$ w.r.t. $\|\cdot\|_2$; denoting by $\bar{u}$ a maximizer of the form on B, we lose nothing when assuming that $\bar{u}$ is a unit vector. Now let u_s be the point of our net which is at $\|\cdot\|_2$-distance at most ϵ from $\bar{u}$. We have

$$M = f(\bar{u}) \leq f(u_s) + 2M\epsilon \leq 1 + \alpha + 2M\epsilon,$$

whence

$$M \leq \frac{1+\alpha}{1-2\epsilon} = 1 + 2\alpha,$$

implying the right inequality in (1.57). Now let u be unit vector in $\mathbf{R}^k$, and u_s be a point in the net at $\|\cdot\|$-distance $\leq \epsilon$ from u. We have

$$f(u) \geq f(u_s) - 2M\epsilon \geq 1 - \alpha - 2\frac{1+\alpha}{1-2\epsilon}\epsilon = 1 - 2\alpha,$$

justifying the first inequality in (1.57).

The bottom line is:

$$\begin{array}{l} \delta \in (0, 0.2], 1 \leq k \leq n \\ \Rightarrow \mathrm{Prob}\{A : A \text{ does not satisfy } \mathrm{RIP}(\delta, k)\} \leq \underbrace{2\left(\frac{10}{\delta}\right)^k}_{\leq (\frac{20}{\delta})^k} \binom{n}{k} \exp\{-\frac{m\delta^2}{20}\}. \end{array} \tag{1.58}$$

Indeed, setting $\alpha = \delta/2$, we have seen that whenever $A \notin E$, we have $(1-\delta) \leq \|Au\|_2^2 \leq (1+\delta)$ for all unit k-sparse u, which is nothing but $\mathrm{RIP}(\delta, k)$; with this in mind, (1.58) follows from (1.56) and (1.54).

4°. It remains to verify that with properly selected—depending solely on δ—positive quantities c, d, f, for every $k \geq 1$ satisfying (1.28) the right-hand side in (1.58) is at most $\exp\{-fm\}$. Passing to logarithms, our goal is to ensure the relation

$$\begin{array}{c} G := a(\delta)m - b(\delta)k - \ln\binom{n}{k} \geq mf(\delta) > 0 \\ \left[a(\delta) = \frac{\delta^2}{20}, b(\delta) = \ln\left(\frac{20}{\delta}\right)\right] \end{array} \tag{1.59}$$

provided that $k \geq 1$ satisfies (1.28).

Let k satisfy (1.28) with some c, d to be specified later, and let $y = k/m$. Assuming $d \geq 3$, we have $0 \leq y \leq 1/3$. Now, it is well known that

$$C := \ln\binom{n}{k} \leq n\left[\frac{k}{n}\ln(\frac{n}{k}) + \frac{n-k}{n}\ln(\frac{n}{n-k})\right],$$

whence

$$\begin{array}{rl} C & \leq n\left[\frac{m}{n}y\ln(\frac{n}{my}) + \frac{n-k}{n}\underbrace{\ln(1+\frac{k}{n-k})}_{\leq \frac{k}{n-k}}\right] \\ & \leq n\left[\frac{m}{n}y\ln(\frac{n}{my}) + \frac{k}{n}\right] = m\left[y\ln(\frac{n}{my}) + y\right] \leq 2my\ln(\frac{n}{my}) \end{array}$$

(recall that $n \geq m$ and $y \leq 1/3$). It follows that

$$\begin{array}{rcl} G & = & a(\delta)m - b(\delta)k - C \geq a(\delta)m - b(\delta)ym - 2my\ln(\frac{n}{my}) \\ & = & m\underbrace{\left[a(\delta) - b(\delta)y - 2y\ln(\frac{n}{m}) - 2y\ln(\frac{1}{y})\right]}_{H}, \end{array}$$

and all we need is to select c, d in such a way that (1.28) would imply that $H \geq f$ with some positive $f = f(\delta)$. This is immediate: we can find $u(\delta) > 0$ such that when $0 \leq y \leq u(\delta)$, we have $2y\ln(1/y) + b(\delta)y \leq \frac{1}{3}a(\delta)$; selecting $d(\delta) \geq 3$ large enough, (1.28) would imply $y \leq u(\delta)$, and thus would imply

$$H \geq \frac{2}{3}a(\delta) - 2y\ln(\frac{n}{m}).$$

Now we can select $c(\delta)$ large enough for (1.28) to ensure that $2y\ln(\frac{n}{m}) \leq \frac{1}{3}a(\delta)$. With the c, d just specified, (1.28) implies that $H \geq \frac{1}{3}a(\delta)$, and we can take the latter quantity as $f(\delta)$. □

1.5.4 Proof of Propositions 1.8 and 1.12

Let $x \in \mathbf{R}^n$, and let $x^1, ..., x^q$ be obtained from x by the following construction: x^1 is obtained from x by zeroing all but the s entries of largest magnitudes; x^2 is obtained by the same procedure applied to $x - x^1$; x^3—by the same procedure applied to $x - x^1 - x^2$; and so on; the process is terminated at the first step q when it happens that $x = x^1 + ... + x^q$. Note that for $j \geq 2$ we have $\|x^j\|_\infty \leq s^{-1}\|x^{j-1}\|_1$ and $\|x^j\|_1 \leq \|x^{j-1}\|_1$, whence also $\|x^j\|_2 \leq \sqrt{\|x^j\|_\infty \|x^j\|_1} \leq s^{-1/2}\|x^{j-1}\|_1$. It is easily seen that if A is RIP$(\delta, 2s)$, then for every two s-sparse vectors u, v with nonoverlapping supports we have

$$|v^T A^T A u| \leq \delta \|u\|_2 \|v\|_2. \qquad (*)$$

Indeed, for s-sparse u, v, let I be the index set of cardinality $\leq 2s$ containing the supports of u and v, so that, denoting by A_I the submatrix of A comprised of columns with indexes from I, we have $v^T A^T A u = v_I^T [A_I^T A_I] u_I$. By RIP, the eigenvalues $\lambda_i = 1 + \mu_i$ of the symmetric matrix $Q = A_I^T A_I$ are in-between $1 - \delta$ and $1 + \delta$; representing u_I and v_I by vectors w and z of their coordinates in the orthonormal eigenbasis of Q, we get $|v^T A^T A u| = |\sum_i \lambda_i w_i z_i| = |\sum_i w_i z_i + \sum_i \mu_i w_i z_i| \leq |w^T z| + \delta \|w\|_2 \|z\|_2$. It remains to note that $w^T z = u_I^T v_I = 0$ and $\|w\|_2 = \|u\|_2$, $\|z\|_2 = \|v\|_2$.

We have

$$\begin{array}{rl} & \|Ax^1\|_2 \|Ax\|_2 \geq [x^1]^T A^T A x = \|Ax^1\|_2^2 + \sum_{j=2}^q [x^1]^T A^T A x^j \\ & \geq \|Ax^1\|_2^2 - \delta \sum_{j=2}^q \|x^1\|_2 \|x^j\|_2 \text{ [by } (*)] \\ & \geq \|Ax^1\|_2^2 - \delta s^{-1/2} \|x^1\|_2 \sum_{j=2}^q \|x^{j-1}\|_1 \geq \|Ax^1\|_2^2 - \delta s^{-1/2} \|x^1\|_2 \|x\|_1 \\ \Rightarrow & \|Ax^1\|_2^2 \leq \|Ax^1\|_2 \|Ax\|_2 + \delta s^{-1/2} \|x^1\|_2 \|x\|_1 \\ \Rightarrow & \|x^1\|_2 = \frac{\|x^1\|_2}{\|Ax^1\|_2^2} \|Ax^1\|_2^2 \leq \frac{\|x^1\|_2}{\|Ax^1\|_2} \|Ax\|_2 + \delta s^{-1/2} \left(\frac{\|x^1\|_2}{\|Ax^1\|_2}\right)^2 \|x\|_1 \\ \Rightarrow & \|x\|_{s,2} = \|x^1\|_2 \leq \frac{1}{\sqrt{1-\delta}} \|Ax\|_2 + \frac{\delta s^{-1/2}}{1-\delta} \|x\|_1 \qquad (!) \\ & \text{[by RIP}(\delta, 2s)] \end{array}$$

and we see that the pair $\left(H = \frac{s^{-1/2}}{\sqrt{1-\delta}} I_m, \|\cdot\|_2\right)$ satisfies $\mathbf{Q}_2(s, \frac{\delta}{1-\delta})$, as claimed in Proposition 1.8.i. Moreover, when $q \geq 2$, $\kappa > 0$, and integer $t \geq 1$ satisfy $t \leq s$ and $\kappa t^{1/q-1} \geq \frac{\delta s^{-1/2}}{1-\delta}$, by (!) we have

$$\|x\|_{t,q} \leq \|x\|_{s,q} \leq \|x\|_{s,2} \leq \frac{1}{\sqrt{1-\delta}} \|Ax\|_2 + \kappa t^{1/q-1} \|x\|_1,$$

or, equivalently,

$$\begin{array}{rl} & 1 \leq t \leq \min\left[\left[\frac{\kappa(1-\delta)}{\delta}\right]^{\frac{q}{q-1}}, s^{\frac{q-2}{2q-2}}\right] s^{\frac{q}{2q-2}} \\ \Rightarrow & (H = \frac{t^{-\frac{1}{q}}}{\sqrt{1-\delta}} I_m, \|\cdot\|_2) \text{ satisfies } \mathbf{Q}_q(t, \kappa), \end{array}$$

as required in Proposition 1.12.i.

Next, we have

$$\begin{array}{rl} & \|x^1\|_1 \|A^T Ax\|_\infty \geq [x^1]^T A^T Ax = \|Ax^1\|_2^2 + \sum_{j=2}^q [x^1]^T A^T Ax^j \\ & \geq \|Ax^1\|_2^2 - \delta s^{-1/2} \|x^1\|_2 \|x\|_1 \text{ [exactly as above]} \\ \Rightarrow & \|Ax^1\|_2^2 \leq \|x^1\|_1 \|A^T Ax\|_\infty + \delta s^{-1/2} \|x^1\|_2 \|x\|_1 \\ \Rightarrow & (1-\delta)\|x^1\|_2^2 \leq \|x^1\|_1 \|A^T Ax\|_\infty + \delta s^{-1/2} \|x^1\|_2 \|x\|_1 \text{ [by RIP}(\delta, 2s)] \\ & \leq s^{1/2} \|x^1\|_2 \|A^T Ax\|_\infty + \delta s^{-1/2} \|x^1\|_2 \|x\|_1 \\ \Rightarrow & \|x\|_{s,2} = \|x^1\|_2 \leq \frac{s^{1/2}}{1-\delta} \|A^T Ax\|_\infty + \frac{\delta}{1-\delta} s^{-1/2} \|x\|_1 \qquad (!!) \end{array}$$

and we see that the pair $\left(H = \frac{1}{1-\delta} A, \|\cdot\|_\infty\right)$ satisfies the condition $\mathbf{Q}_2\left(s, \frac{\delta}{1-\delta}\right)$, as required in Proposition 1.8.ii. Moreover, when $q \geq 2$, $\kappa > 0$, and integer $t \geq 1$ satisfy $t \leq s$ and $\kappa t^{1/q-1} \geq \frac{\delta}{1-\delta} s^{-1/2}$, we have by (!!)

$$\|x\|_{t,q} \leq \|x\|_{s,q} \leq \|x\|_{s,2} \leq \frac{1}{1-\delta} s^{1/2} \|A^T Ax\|_\infty + \kappa t^{1/q-1} \|x\|_1,$$

or, equivalently,

$$\begin{array}{l} 1 \leq t \leq \min\left[\left[\frac{\kappa(1-\delta)}{\delta}\right]^{\frac{q}{q-1}}, s^{\frac{q-2}{2q-2}}\right] s^{\frac{q}{2q-2}} \\ \Rightarrow (H = \frac{s^{\frac{1}{2}} t^{-\frac{1}{q}}}{1-\delta} A, \|\cdot\|_\infty) \text{ satisfies } \mathbf{Q}_q(t, \kappa), \end{array}$$

as required in Proposition 1.12.ii. □

1.5.5 Proof of Proposition 1.10

(i): Let $\bar{H} \in \mathbf{R}^{m \times N}$ and $\|\cdot\|$ satisfy $\mathbf{Q}_\infty(s, \kappa)$. Then for every $k \leq n$ we have

$$|x_k| \leq \|\bar{H}^T Ax\| + s^{-1} \kappa \|x\|_1,$$

or, which is the same by homogeneity,

$$\min_x \left\{ \|\bar{H}^T Ax\| - x_k : \|x\|_1 \leq 1 \right\} \geq -s^{-1} \kappa.$$

In other words, the optimal value Opt_k of the conic optimization problem[13]

$$\mathrm{Opt}_k = \min_{x,t} \left\{ t - [e^k]^T x : \|\bar{H}^T A x\| \le t, \|x\|_1 \le 1 \right\},$$

where $e^k \in \mathbf{R}^n$ is k-th basic orth, is $\geq -s^{-1}\kappa$. Since the problem clearly is strictly feasible, this is the same as saying that the dual problem

$$\max_{\mu \in \mathbf{R}, g \in \mathbf{R}^n, \eta \in \mathbf{R}^N} \left\{ -\mu : A^T \bar{H} \eta + g = e^k, \|g\|_\infty \le \mu, \|\eta\|_* \le 1 \right\},$$

where $\|\cdot\|_*$ is the norm conjugate to $\|\cdot\|$,

$$\|u\|_* = \max_{\|h\| \le 1} h^T u,$$

has a feasible solution with the value of the objective $\geq -s^{-1}\kappa$. It follows that there exist $\eta = \eta^k$ and $g = g^k$ such that

$$\begin{array}{l} (a) : e^k = A^T h^k + g^k, \\ (b) : h^k := \bar{H} \eta^k, \|\eta^k\|_* \le 1, \\ (c) : \|g^k\|_\infty \le s^{-1}\kappa. \end{array} \tag{1.60}$$

Denoting $H = [h^1, ..., h^n]$, $V = I - H^T A$, we get

$$\mathrm{Col}_k[V^T] = e^k - A^T h^k = g^k,$$

implying that $\|\mathrm{Col}_k[V^T]\|_\infty \le s^{-1}\kappa$. Since the latter inequality is true for all $k \le n$, we conclude that

$$\|\mathrm{Col}_k[V]\|_{s,\infty} = \|\mathrm{Col}_k[V]\|_\infty \le s^{-1}\kappa, \ 1 \le k \le n,$$

whence, by Proposition 1.9, $(H, \|\cdot\|_\infty)$ satisfies $\mathbf{Q}_\infty(s, \kappa)$. Moreover, for every $\eta \in \mathbf{R}^m$ and every $k \le n$ we have, in view of (b) and (c),

$$|[h^k]^T \eta| = |[\eta^k]^T \bar{H}^T \eta| \le \|\eta^k\|_* \|\bar{H}^T \eta\|,$$

whence $\|H^T \eta\|_\infty \le \|\bar{H}^T \eta\|$.

Now let us prove the "In addition" part of the proposition. Let $H = [h_1, ..., h_n]$ be the contrast matrix specified in this part. We have

$$|[I_m - H^T A]_{ij}| = |[[e^i]^T - h_i^T A]_j| \le \|[e^i]^T - h_i^T A\|_\infty = \|e^i - A^T h_i\|_\infty \le \mathrm{Opt}_i,$$

implying by Proposition 1.9 that $(H, \|\cdot\|_\infty)$ does satisfy the condition $\mathbf{Q}_\infty(s, \kappa_*)$ with $\kappa_* = s \max_i \mathrm{Opt}_i$. Now assume that there exists a matrix H' which, taken along with some norm $\|\cdot\|$, satisfies the condition $\mathbf{Q}_\infty(s, \kappa)$ with $\kappa < \kappa_*$, and let us lead this assumption to a contradiction. By the already proved first part of Proposition 1.10, our assumption implies that there exists an $m \times n$ matrix $\bar{H} = [\bar{h}_1, ..., \bar{h}_n]$ such that $\|\mathrm{Col}_j[I_n - \bar{H}^T A]\|_\infty \le s^{-1}\kappa$ for all $j \le n$, implying that $|[[e^i]^T - \bar{h}_i^T A]_j| \le s^{-1}\kappa$ for all i and j, or, which is the same, $\|e^i - A^T \bar{h}_i\|_\infty \le s^{-1}\kappa$ for all i. Due to the origin of Opt_i, we have $\mathrm{Opt}_i \le \|e^i - A^T \bar{h}_i\|_\infty$ for all i,

[13] For a summary on conic programming, see Section 4.1.

and we arrive at $s^{-1}\kappa_* = \max_i \mathrm{Opt}_i \leq s^{-1}\kappa$, that is, $\kappa_* \leq \kappa$, which is a desired contradiction.

It remains to prove (1.33), which is just an exercise on LP duality: denoting by $\mathbf{e}$ an n-dimensional all-ones vector, we have

$$\begin{array}{rcl}\mathrm{Opt}_i & := & \min_h \|e^i - A^T h\|_\infty = \min_{h,t}\left\{t : e^i - A^T h \leq t\mathbf{e}, A^T h - e^i \leq t\mathbf{e}\right\}\\ & = & \max_{\lambda,\mu}\left\{\lambda_i - \mu_i : \lambda, \mu \geq 0, A[\lambda - \mu] = 0, \sum_i \lambda_i + \sum_i \mu_i = 1\right\}\\ & & \text{[LP duality]}\\ & = & \max_{x:=\lambda-\mu}\left\{x_i : Ax = 0, \|x\|_1 \leq 1\right\}\end{array}$$

where the concluding equality follows from the fact that vectors x representable as $\lambda - \mu$ with $\lambda, \mu \geq 0$ satisfying $\|\lambda\|_1 + \|\mu\|_1 = 1$ are exactly vectors x with $\|x\|_1 \leq 1$. □

1.5.6 Proof of Proposition 1.13

Let H satisfy (1.38). Since $\|v\|_{s,1} \leq s^{1-1/q}\|v\|_{s,q}$, it follows that H satisfies for some $\alpha < 1/2$ the condition

$$\|\mathrm{Col}_j[I_n - H^T A]\|_{s,1} \leq \alpha,\ 1 \leq j \leq n, \tag{1.61}$$

whence, as we know from Proposition 1.9,

$$\|x\|_{s,1} \leq s\|H^T Ax\|_\infty + \alpha\|x\|_1\ \forall x \in \mathbf{R}^n.$$

It follows that $s \leq m$, since otherwise there exists a nonzero s-sparse vector x with $Ax = 0$; for this x, the inequality above cannot hold true.

Let us set $\bar{n} = 2m$, so that $\bar{n} \leq n$, and let $\bar{H}$ and $\bar{A}$ be the $m \times \bar{n}$ matrices comprised of the first $2m$ columns of H and A, respectively. Relation (1.61) implies that the matrix $V = I_{\bar{n}} - \bar{H}^T\bar{A}$ satisfies

$$\|\mathrm{Col}_j[V]\|_{s,1} \leq \alpha < 1/2, 1 \leq j \leq \bar{n}. \tag{1.62}$$

Now, since the rank of $\bar{H}^T\bar{A}$ is $\leq m$, at least $\bar{n} - m$ singular values of V are ≥ 1, and therefore the squared Frobenius norm $\|V\|_F^2$ of V is at least $\bar{n} - m$. On the other hand, we can upper-bound this squared norm as follows. Observe that for every $\bar{n}$-dimensional vector f one has

$$\|f\|_2^2 \leq \max\left[\frac{\bar{n}}{s^2}, 1\right]\|f\|_{s,1}^2. \tag{1.63}$$

Indeed, by homogeneity it suffices to verify the inequality when $\|f\|_{s,1} = 1$; besides, we can assume w.l.o.g. that the entries in f are nonnegative, and that $f_1 \geq f_2 \geq \ldots \geq f_{\bar{n}}$. We have $f_s \leq \|f\|_{s,1}/s = \frac{1}{s}$; in addition, $\sum_{j=s+1}^{\bar{n}} f_j^2 \leq (\bar{n} - s)f_s^2$. Now, due to $\|f\|_{s,1} = 1$, for fixed $f_s \in [0, 1/s]$ we have

$$\sum_{j=1}^{s} f_j^2 \leq f_s^2 + \max_t\left\{\sum_{j=1}^{s-1} t_j^2 : t_j \geq f_s, j \leq s-1, \sum_{j=1}^{s-1} t_j = 1 - f_s\right\}.$$

The maximum on the right-hand side is the maximum of a convex function over a bounded polytope; it is achieved at an extreme point, that is, at a point where one of the t_j is equal to $1-(s-1)f_s$, and all remaining t_j are equal to f_s. As a result,

$$\sum_j f_j^2 \le [(1-(s-1)f_s)^2 + (s-1)f_s^2] + (\bar{n}-s)f_s^2 \le (1-(s-1)f_s)^2 + (\bar{n}-1)f_s^2.$$

The right-hand side in the latter inequality is convex in f_s and thus achieves its maximum over the range $[0,1/s]$ of allowed values of f_s at an endpoint, yielding $\sum_j f_j^2 \le \max[1,\bar{n}/s^2]$, as claimed.

Applying (1.63) to the columns of V and recalling that $\bar{n}=2m$, we get

$$\|V\|_F^2 = \sum_{j=1}^{2m} \|\mathrm{Col}_j[V]\|_2^2 \le \max\left[1,\frac{2m}{s^2}\right] \sum_{j=1}^{2m} \|\mathrm{Col}_j[V]\|_{s,1}^2 \le 2\alpha^2 m \max\left[1,\frac{2m}{s^2}\right].$$

The left hand side in this inequality, as we remember, is $\ge \bar{n}-m=m$, and we arrive at

$$m \le 2\alpha^2 m \max[1,2m/s^2].$$

Since $\alpha<1/2$, this inequality implies $2m/s^2 \ge 2$, whence $s \le \sqrt{m}$.

It remains to prove that when $m \le n/2$, the condition $\mathbf{Q}_\infty(s,\kappa)$ with $\kappa<1/2$ can be satisfied only when $s \le \sqrt{m}$. This is immediate: by Proposition 1.10, assuming $\mathbf{Q}_\infty(s,\kappa)$ satisfiable, there exists an $m\times n$ contrast matrix H such that $|[I_n - H^T A]_{ij}| \le \kappa/s$ for all i,j, which, by the already proved part of Proposition 1.13, is impossible when $s > \sqrt{m}$. □

Chapter Two

Hypothesis Testing

Disclaimer for experts. In what follows, we allow for "general" probability and observation spaces, general probability distributions, etc., which, formally, would make it necessary to address the related measurability issues. In order to streamline our exposition, and taking into account that we do not expect our target audience to be experts in formal nuances of the measure theory, we decided to omit in the text comments (always self-evident for an expert) on measurability and replace them with a "disclaimer" as follows:
Below, unless the opposite is explicitly stated,

- all probability and observation spaces are Polish (complete separable metric) spaces equipped with σ-algebras of Borel sets;
- all random variables (i.e., functions from a probability space to some other space) take values in Polish spaces; these variables, like other functions we deal with, are Borel;
- all probability distributions we are dealing with are σ-additive Borel measures on the respective probability spaces; the same is true for all reference measures and probability densities taken w.r.t. these measures.

When an entity (a random variable, or a probability density, or a function, say, a test) is part of the data, the Borel property is a default assumption; e.g., the sentence "Let random variable η be a deterministic transformation of random variable ξ" should be read as "let $\eta = f(\xi)$ for some Borel function f," and the sentence "Consider a test $\mathcal{T}$ deciding on hypotheses $H_1, ..., H_L$ via observation $\omega \in \Omega$" should be read as "Consider a Borel function $\mathcal{T}$ on Polish space Ω, the values of the function being subsets of the set $\{1, ..., L\}$." When an entity is built by us rather than being part of the data, the Borel property is (an always straightforwardly verifiable) property of the construction. For example, the statement "The test $\mathcal{T}$ given by ... is such that ..." should be read as "The test $\mathcal{T}$ given by ... is a Borel function of observations and is such that ...."

On several occasions, we still use the word "Borel"; those not acquainted with the notion are welcome to just ignore this word.

2.1 PRELIMINARIES FROM STATISTICS: HYPOTHESES, TESTS, RISKS

2.1.1 Hypothesis Testing Problem

Hypothesis Testing is one of fundamental problems of Statistics. Informally, this is the problem where one is given an *observation*—a realization of a random variable with unknown (at least partly) probability distribution—and wants to decide, based on this observation, on two or more hypotheses on the actual distribution of the observed variable. A formal setting convenient for us is as follows:

Given are:

- *Observation space* Ω, where the observed random variable (r.v.) takes its values;
- *L families* $\mathcal{P}_\ell$ of probability distributions on Ω. We associate with these families L hypotheses $H_1, ..., H_L$, with H_ℓ stating that the probability distribution P of the observed r.v. belongs to the family $\mathcal{P}_\ell$ (shorthand: $H_\ell : P \in \mathcal{P}_\ell$). We shall say that the distributions from $\mathcal{P}_\ell$ *obey* hypothesis H_ℓ.
  Hypothesis H_ℓ is called *simple* if $\mathcal{P}_\ell$ is a singleton, and is called *composite* otherwise.

Our goal is, given an observation—a realization ω of the r.v. in question—to decide which of the hypotheses is true.

2.1.2 Tests

Informally, a *test* is an inference procedure one can use in the above testing problem. Formally, a test for this testing problem is a function $\mathcal{T}(\omega)$ of $\omega \in \Omega$; the value $\mathcal{T}(\omega)$ of this function at a point ω is some subset of the set $\{1, ..., L\}$:

$$\mathcal{T}(\omega) \subset \{1, ..., L\}.$$

Given observation ω, the test accepts all hypotheses H_ℓ with $\ell \in \mathcal{T}(\omega)$ and rejects all hypotheses H_ℓ with $\ell \notin \mathcal{T}(\omega)$. We call a test *simple* if $\mathcal{T}(\omega)$ is a singleton for every ω, that is, whatever be the observation, the test accepts exactly one of the hypotheses $H_1, ..., H_L$ and rejects all other hypotheses.

Note: What we have defined is a *deterministic* test. Sometimes we shall consider also *randomized* tests, where the set of accepted hypotheses is a (deterministic) function of an observation ω *and* a realization θ of a random parameter (which w.l.o.g. can be assumed to be uniformly distributed on $[0, 1]$) independent of ω. Thus, in a randomized test, the inference depends both on the observation ω and the outcome θ of "flipping a coin," while in a deterministic test the inference depends on observation only. In fact, randomized testing can be reduced to deterministic testing. To this end it suffices to pass from our "actual" observation ω to the new observation $\omega_+ = (\omega, \theta)$, where $\theta \sim \text{Uniform}[0, 1]$ is independent of ω; the ω-component of our new observation ω_+ is, as before, generated "by nature," and the θ-component is generated by us. Now, given families $\mathcal{P}_\ell$, $1 \leq \ell \leq L$, of probability distributions on the original observation space Ω, we can associate with them families $\mathcal{P}_{\ell,+} = \{P \times \text{Uniform}[0, 1] : P \in \mathcal{P}_\ell\}$ of probability distributions on our new observation space $\Omega_+ = \Omega \times [0, 1]$. Clearly, to decide on the hypotheses associated with the families $\mathcal{P}_\ell$ via observation ω is the same as to decide on the hypotheses associated with the families $\mathcal{P}_{\ell,+}$ of our new observation ω_+, and deterministic tests for the latter testing problem are exactly the randomized tests for the former one.

2.1.3 Testing from repeated observations

There are situations where an inference can be based on several observations $\omega_1, ..., \omega_K$ rather than on a single one. Our related setup is as follows:

We are given L families $\mathcal{P}_\ell$, $\ell = 1, ..., L$, of probability distributions on the observation space Ω and a collection

$$\omega^K = (\omega_1, ..., \omega_K) \in \Omega^K = \underbrace{\Omega \times ... \times \Omega}_{K}$$

and want to make conclusions on how the distribution of ω^K "is positioned" w.r.t. the families $\mathcal{P}_\ell$, $1 \leq \ell \leq L$.

We will be interested in three situations of this type, specifically, as follows.

2.1.3.1 *Stationary K-repeated observations*

In the case of stationary K-repeated observations, $\omega_1, ..., \omega_K$ are *independently of each other* drawn from a distribution P. Our goal is to decide, given ω^K, on the hypotheses $P \in \mathcal{P}_\ell$, $\ell = 1, ..., L$.

Equivalently: Families $\mathcal{P}_\ell$ of probability distributions of $\omega \in \Omega$, $1 \leq \ell \leq L$, give rise to the families

$$\mathcal{P}_\ell^{\odot,K} = \{P^K = \underbrace{P \times ... \times P}_{K} : P \in \mathcal{P}_\ell\}$$

of probability distributions on Ω^K; we refer to the family $\mathcal{P}_\ell^{\odot,K}$ as the *K-th diagonal power* of the family $\mathcal{P}_\ell$. Given observation $\omega^K \in \Omega^K$, we want to decide on the hypotheses

$$H_\ell^{\odot,K} : \omega^K \sim P^K \in \mathcal{P}_\ell^{\odot,K},\ 1 \leq \ell \leq L.$$

2.1.3.2 *Semi-stationary K-repeated observations*

In the case of semi-stationary K-repeated observations, "nature" selects somehow a sequence $P_1, ..., P_K$ of distributions on Ω, and then draws, *independently across* k, observations ω_k, $k = 1, ..., K$, from these distributions:

$$\omega_k \sim P_k,\ \omega_k \text{ are independent across } k \leq K.$$

Our goal is to decide, given $\omega^K = (\omega_1, ..., \omega_K)$, on the hypotheses $\{P_k \in \mathcal{P}_\ell, 1 \leq k \leq K\}$, $\ell = 1, ..., L$.

Equivalently: Families $\mathcal{P}_\ell$ of probability distributions of $\omega \in \Omega$, $1 \leq \ell \leq L$, give rise to the families

$$\mathcal{P}_\ell^{\oplus,K} = \{P^K = P_1 \times ... \times P_K : P_k \in \mathcal{P}_\ell,\ 1 \leq k \leq K\}$$

of probability distributions on Ω^K. Given observation $\omega^K \in \Omega^K$, we want to decide on the hypotheses

$$H_\ell^{\oplus,K} : \omega^K \sim P^K \in \mathcal{P}_\ell^{\oplus,K},\ 1 \leq \ell \leq L.$$

In the sequel, we refer to families $\mathcal{P}_\ell^{\oplus,K}$ as the *K-th direct powers* of the families

$\mathcal{P}_\ell$. A closely related notion is that of the *direct product*

$$\mathcal{P}_\ell^{\oplus,K} = \bigoplus_{k=1}^{K} \mathcal{P}_{\ell,k}$$

of K families $\mathcal{P}_{\ell,k}$, of probability distributions on Ω_k, over $k = 1, ..., K$. By definition,

$$\mathcal{P}_\ell^{\oplus,K} = \{P^K = P_1 \times ... \times P_K : P_k \in \mathcal{P}_{\ell,k}, 1 \le k \le K\}.$$

2.1.3.3 Quasi-stationary K-repeated observations

Quasi-stationary K-repeated observations $\omega_1 \in \Omega, ..., \omega_K \in \Omega$ stemming from a family $\mathcal{P}$ of probability distributions on an observation space Ω are generated as follows:

> "In nature" there exists random sequence $\zeta^K = (\zeta_1, ..., \zeta_K)$ of "driving factors" (or states) such that for every k, ω_k is a deterministic function of $\zeta_1, ..., \zeta_k$,
>
> $$\omega_k = \theta_k(\zeta_1, ..., \zeta_k),$$
>
> and the conditional distribution $P_{\omega_k|\zeta_1,...,\zeta_{k-1}}$ of ω_k given $\zeta_1, ..., \zeta_{k-1}$ always (i.e., for all $\zeta_1, ..., \zeta_{k-1}$) belongs to $\mathcal{P}$.

With the above mechanism, the collection $\omega^K = (\omega_1, ..., \omega_K)$ has some distribution P^K which depends on the distribution of driving factors and functions $\theta_k(\cdot)$. We denote by $\mathcal{P}^{\otimes,K}$ the family of all distributions P^K which can be obtained in this fashion, and we refer to random observations ω^K with distribution P^K of the type just defined as the *quasi-stationary K-repeated observations stemming from $\mathcal{P}$*. The quasi-stationary version of our hypothesis testing problem reads: Given L families $\mathcal{P}_\ell$ of probability distributions $\mathcal{P}_\ell$, $\ell = 1, ..., L$, on Ω and an observation $\omega^K \in \Omega^K$, decide on the hypotheses

$$H_\ell^{\otimes,K} = \{P^K \in \mathcal{P}_\ell^{\otimes,K}\}, \ 1 \le \ell \le K$$

on the distribution P^K of the observation ω^K.

A related notion is that of the *quasi-direct product*

$$\mathcal{P}_\ell^{\otimes,K} = \bigotimes_{k=1}^{K} \mathcal{P}_{\ell,k}$$

of K families $\mathcal{P}_{\ell,k}$, of probability distributions on Ω_k, over $k = 1, ..., K$. By definition, $\mathcal{P}_\ell^{\otimes,K}$ is comprised of all probability distributions of random sequences $\omega^K = (\omega_1, ..., \omega_K)$, $\omega_k \in \Omega_k$, which can be generated as follows: "in nature" there exists a random sequence $\zeta^K = (\zeta_1, ..., \zeta_K)$ of "driving factors" such that for every $k \le K$, ω_k is a deterministic function of $\zeta^k = (\zeta_1, ..., \zeta_k)$, and the conditional distribution of ω_k given ζ^{k-1} always belongs to $\mathcal{P}_{\ell,k}$.

The description of quasi-stationary K-repeated observations seems to be too complicated. However, this is exactly what happens in some important applications, e.g., in *hidden Markov chains.* Suppose that $\Omega = \{1, ..., d\}$ is a finite set, and that $\omega_k \in \Omega$, $k = 1, 2, ...$, are generated as follows: "in nature" there exists a Markov chain with D-element state space $\mathcal{S}$ split into d nonoverlapping bins, and ω_k is the index $\beta(\eta)$ of the bin to which the state η_k of the chain belongs. Every column Q^j of the transition matrix Q of the chain (this column is a probability distribution on $\{1,...,D\}$) generates a probability distribution P_j on Ω, specifically, the distribution of $\beta(\eta)$, $\eta \sim Q^j$. Now, a family $\mathcal{P}$ of distributions on Ω induces a family $\mathcal{Q}[\mathcal{P}]$ of all $D \times D$ stochastic matrices Q for which all D distributions P^j, $j = 1,...,D$, belong to $\mathcal{P}$. When $Q \in \mathcal{Q}[\mathcal{P}]$, observations ω_k, $k = 1,2,...$, clearly are given by the above "quasi-stationary mechanism" with η_k in the role of driving factors and $\mathcal{P}$ in the role of $\mathcal{P}_\ell$. Thus, in the situation in question, given L families $\mathcal{P}_\ell$, $\ell = 1,...,L$, of probability distributions on $\mathcal{S}$, deciding on hypotheses $Q \in \mathcal{Q}[\mathcal{P}_\ell]$, $\ell = 1,...,L$, on the transition matrix Q of the Markov chain underlying our observations reduces to hypothesis testing via quasi-stationary K-repeated observations.

2.1.4 Risk of a simple test

Let $\mathcal{P}_\ell$, $\ell = 1,...,L$, be families of probability distributions on observation space Ω; these families give rise to hypotheses

$$H_\ell : P \in \mathcal{P}_\ell,\ \ell = 1,...,L$$

on the distribution P of a random observation $\omega \sim P$. We are about to define the *risks* of a *simple* test $\mathcal{T}$ deciding on the hypotheses H_ℓ, $\ell = 1,...,L$, via observation ω. Recall that simplicity means that as applied to an observation, our test accepts exactly one hypothesis and rejects all other hypotheses.

Partial risks $\mathrm{Risk}_\ell(\mathcal{T}|H_1,...,H_L)$ are the worst-case, over $P \in \mathcal{P}_\ell$, P-probabilities of $\mathcal{T}$ rejecting the ℓ-th hypothesis when it is true, that is, when $\omega \sim P$:

$$\mathrm{Risk}_\ell(\mathcal{T}|H_1,...,H_L) = \sup_{P \in \mathcal{P}_\ell} \mathrm{Prob}_{\omega \sim P}\left\{\omega : \mathcal{T}(\omega) \neq \{\ell\}\right\},\ \ell = 1,...,L.$$

Obviously, for ℓ fixed, the ℓ-th partial risk depends on how we order the hypotheses; when reordering them, we should reorder risks as well. In particular, for a test $\mathcal{T}$ deciding on two hypotheses H, H' we have

$$\mathrm{Risk}_1(\mathcal{T}|H,H') = \mathrm{Risk}_2(\mathcal{T}|H',H).$$

Total risk $\mathrm{Risk}_{\mathrm{tot}}(\mathcal{T}|H_1,...,H_L)$ is the sum of all L partial risks:

$$\mathrm{Risk}_{\mathrm{tot}}(\mathcal{T}|H_1,...,H_L) = \sum_{\ell=1}^{L} \mathrm{Risk}_\ell(\mathcal{T}|H_1,...,H_L).$$

Risk $\mathrm{Risk}(\mathcal{T}|H_1,...,H_L)$ is the maximum of all L partial risks:

$$\mathrm{Risk}(\mathcal{T}|H_1,...,H_L) = \max_{1 \le \ell \le L} \mathrm{Risk}_\ell(\mathcal{T}|H_1,...,H_L).$$

Note that *at first glance*, we have defined risks for single-observation tests only; in fact, we have defined them for tests based on stationary, semi-stationary, and quasi-stationary K-repeated observations as well, since, as we remember from Section

2.1.3, the corresponding testing problems, after redefining observations and families of probability distributions (ω^K in the role of ω and, say, $\mathcal{P}_\ell^{\oplus,K} = \bigoplus_{k=1}^K \mathcal{P}_\ell$ in the role of $\mathcal{P}_\ell$), become single-observation testing problems.
Pay attention to the following two important observations:

- Partial risks of a simple test are defined in the worst-case fashion: as the maximal, over the true distributions P of observations compatible with the hypothesis in question, probability to reject this hypothesis.
- Risks of a simple test say what happens, statistically speaking, when the true distribution P of observation obeys one of the hypotheses in question, and *say nothing about what happens when P does not obey any of the L hypotheses.*

Remark 2.1. "The smaller are the hypotheses, the less are the risks." Specifically, given families of probability distributions $\mathcal{P}_\ell \subset \mathcal{P}'_\ell$, $\ell = 1, ..., L$, on observation space Ω, along with hypotheses $H_\ell : P \in \mathcal{P}_\ell$, $H'_\ell : P \in \mathcal{P}'_\ell$ on the distribution P of an observation $\omega \in \Omega$, every test $\mathcal{T}$ deciding on the "larger" hypotheses $H'_1, ..., H'_L$ can be considered as a test deciding on the smaller hypotheses $H_1, ..., H_L$ as well, and the risks of the test when passing from larger hypotheses to smaller ones can only drop down:

$$\mathcal{P}_\ell \subset \mathcal{P}'_\ell,\ 1 \leq \ell \leq L \Rightarrow \text{Risk}(\mathcal{T}|H_1, ..., H_L) \leq \text{Risk}(\mathcal{T}|H'_1, ..., H'_L).$$

For example, families of probability distributions $\mathcal{P}_\ell$, $1 \leq \ell \leq L$, on Ω and a positive integer K induce three families of hypotheses on a distribution P^K of K-repeated observations:

$$H_\ell^{\odot,K} K : P^K \in \mathcal{P}_\ell^{\odot,K}, \quad H_\ell^{\oplus,K} : P^K \in \mathcal{P}_\ell^{\oplus,K} = \bigoplus_{k=1}^K \mathcal{P}_\ell,$$
$$H_\ell^{\otimes,K} : P^K \in \mathcal{P}_\ell^{\otimes,K} = \bigotimes_{k=1}^K \mathcal{P}_\ell,\ 1 \leq \ell \leq L$$

(see Section 2.1.3), and clearly

$$\mathcal{P}_\ell^K \subset \mathcal{P}_\ell^{\oplus,K} \subset \mathcal{P}_\ell^{\otimes,K}.$$

It follows that when passing from quasi-stationary K-repeated observations to semi-stationary K-repeated observations, and then to stationary K-repeated observations, the risks of a test can only go down.

2.1.5 Two-point lower risk bound

The following well-known [162, 164] observation is nearly evident:

Proposition 2.2. *Consider two simple hypotheses $H_1 : P = P_1$ and $H_2 : P = P_2$ on the distribution P of observation $\omega \in \Omega$, and assume that P_1, P_2 have densities p_1, p_2 w.r.t. some reference measure Π on Ω.[1] Then for any simple test $\mathcal{T}$ deciding*

[1] *This assumption is w.l.o.g.—we can take, as Π, the sum of the measures P_1 and P_2.*

on H_1, H_2 it holds

$$\mathrm{Risk}_{\mathrm{tot}}(\mathcal{T}|H_1,H_2) \geq \int_\Omega \min[p_1(\omega),p_2(\omega)]\Pi(d\omega). \tag{2.1}$$

Note that the right-hand side in this relation is independent of how Π is selected.

Proof. Consider a simple test $\mathcal{T}$, perhaps a randomized one, and let $\pi(\omega)$ be the probability for this test to accept H_1 and reject H_2 when the observation is ω. Since the test is simple, the probability for $\mathcal{T}$ to accept H_2 and to reject H_1, the observation being ω, is $1-\pi(\omega)$. Consequently,

$$\begin{array}{rcl}\mathrm{Risk}_1(\mathcal{T}|H_1,H_2) & = & \int_\Omega (1-\pi(\omega))p_1(\omega)\Pi(d\omega),\\ \mathrm{Risk}_2(\mathcal{T}|H_1,H_2) & = & \int_\Omega \pi(\omega)p_2(\omega)\Pi(d\omega),\end{array}$$

whence

$$\begin{array}{rcl}\mathrm{Risk}_{\mathrm{tot}}(\mathcal{T}|H_1,H_2) & = & \int_\Omega [(1-\pi(\omega))p_1(\omega)+\pi(\omega)p_2(\omega)]\Pi(d\omega)\\ & \geq & \int_\Omega \min[p_1(\omega),p_2(\omega)]\Pi(d\omega). \qquad \square\end{array}$$

Remark 2.3. Note that the lower risk bound *(2.1)* is achievable; given an observation ω, the corresponding test $\mathcal{T}$ accepts H_1 with probability 1 (i.e., $\pi(\omega)=1$) when $p_1(\omega)>p_2(\omega)$, accepts H_2 when $p_1(\omega)<p_2(\omega)$ (i.e., $\pi(\omega)=0$ when $p_1(\omega)<p_2(\omega)$) and accepts H_1 and H_2 with probabilities 1/2 in the case of a tie (i.e., $\pi(\omega)=1/2$ when $p_1(\omega)=p_2(\omega)$). This is nothing but the likelihood ratio test naturally adjusted to account for ties.

Example 2.1. Let $\Omega=\mathbf{R}^d$, let the reference measure Π be the Lebesgue measure on $\mathbf{R}^d$, and let $p_\chi(\cdot)=\mathcal{N}(\mu_\chi,I_d)$, be the Gaussian densities on $\mathbf{R}^d$ with unit covariance and means μ_χ, $\chi=1,2$. In this case, assuming $\mu_1\neq\mu_2$, the recipe from Remark 2.3 reduces to the following:

Let

$$\phi_{1,2}(\omega)=\tfrac{1}{2}[\mu_1-\mu_2]^T[\omega-w],\ w=\tfrac{1}{2}[\mu_1+\mu_2]. \tag{2.2}$$

Consider the simple test $\mathcal{T}$ which, given an observation ω, accepts $H_1: p=p_1$ (and rejects $H_2: p=p_2$) when $\phi_{1,2}(\omega)\geq 0$, and accepts H_2 (and rejects H_1) otherwise. For this test,

$$\begin{array}{l}\mathrm{Risk}_1(\mathcal{T}|H_1,H_2)=\mathrm{Risk}_2(\mathcal{T}|H_1,H_2)=\mathrm{Risk}(\mathcal{T}|H_1,H_2)\\ \quad =\tfrac{1}{2}\mathrm{Risk}_{\mathrm{tot}}(\mathcal{T}|H_1,H_2)=\mathrm{Erfc}(\tfrac{1}{2}\|\mu_1-\mu_2\|_2)\end{array} \tag{2.3}$$

(see *(1.22)* for the definition of Erfc), and the test is optimal in terms of its risk and its total risk.

Note that optimality of $\mathcal{T}$ in terms of total risk is given by Proposition 2.2 and Remark 2.3; optimality in terms of risk is ensured by optimality in terms of total risk combined with the first equality in *(2.3)*.

Example 2.1 admits an immediate and useful extension [36, 37, 84, 128]:

Example 2.2. Let $\Omega=\mathbf{R}^d$, let the reference measure Π be the Lebesgue measure on $\mathbf{R}^d$, and let M_1 and M_2 be two nonempty closed convex sets in $\mathbf{R}^d$ with empty

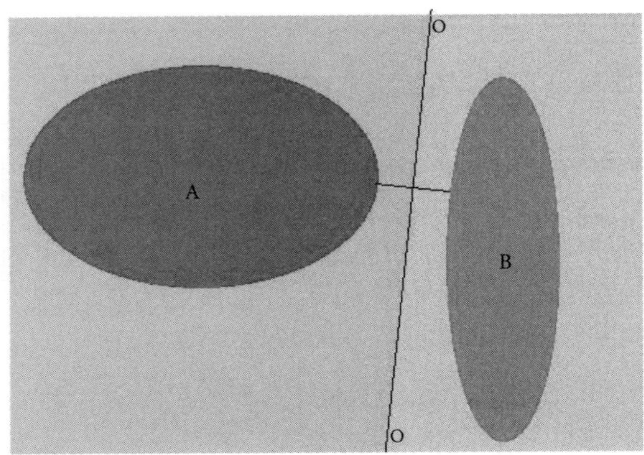

Figure 2.1: "Gaussian Separation" (Example 2.5): Optimal test deciding on whether the mean of Gaussian r.v. belongs to the domain A (H_1) or to the domain B (H_2). Hyperplane o-o separates the acceptance domains for H_1 ("left" half-space) and for H_2 ("right" half-space).

intersection and such that the convex optimization program

$$\min_{\mu_1,\mu_2} \{\|\mu_1 - \mu_2\|_2 : \mu_\chi \in M_\chi,\ \chi = 1, 2\} \tag{$*$}$$

has an optimal solution μ_1^*, μ_2^* (this definitely is the case when at least one of the sets M_1, M_2 is bounded). Let

$$\phi_{1,2}(\omega) = \tfrac{1}{2}[\mu_1^* - \mu_2^*]^T[\omega - w],\ w = \tfrac{1}{2}[\mu_1^* + \mu_2^*], \tag{2.4}$$

and let the simple test $\mathcal{T}$ deciding on the hypotheses

$$H_1 : p = \mathcal{N}(\mu, I_d) \text{ with } \mu \in M_1, \quad H_2 : p = \mathcal{N}(\mu, I_d) \text{ with } \mu \in M_2$$

be as follows (see Figure 2.1): given an observation ω, $\mathcal{T}$ accepts H_1 (and rejects H_2) when $\phi_{1,2}(\omega) \geq 0$, and accepts H_2 (and rejects H_1) otherwise. Then

$$\begin{array}{l} \text{Risk}_1(\mathcal{T}|H_1, H_2) = \text{Risk}_2(\mathcal{T}|H_1, H_2) = \text{Risk}(\mathcal{T}|H_1, H_2) \\ \quad = \frac{1}{2}\text{Risk}_{\text{tot}}(\mathcal{T}|H_1, H_2) = \text{Erfc}(\frac{1}{2}\|\mu_1^* - \mu_2^*\|_2), \end{array} \tag{2.5}$$

and the test is optimal in terms of its risk and its total risk.

Justification of Example 2.2 is immediate. Let e be the $\|\cdot\|_2$-unit vector in the direction of $\mu_1^* - \mu_2^*$, and let $\xi[\omega] = e^T(\omega - w)$. From optimality conditions for ($*$) it follows that

$$e^T\mu \geq e^T\mu_1^*\ \forall \mu \in M_1\ \&\ e^T\mu \leq e^T\mu_2^*\ \forall \mu \in M_2.$$

As a result, if $\mu \in M_1$ and the density of ω is $p_\mu = \mathcal{N}(\mu, I_d)$, the random variable $\xi[\omega]$ is a scalar Gaussian random variable with unit variance and expectation $\geq \delta := \frac{1}{2}\|\mu_1^* - \mu_2^*\|_2$, implying that the p_μ-probability for $\xi[\omega]$ to be negative (which is exactly the same as the p_μ-probability for $\mathcal{T}$ to reject H_1 and accept H_2) is at most

$\mathrm{Erfc}(\delta)$. Similarly, when $\mu \in M_2$ and the density of ω is $p_\mu = \mathcal{N}(\mu, I_d)$, $\xi[\omega]$ is a scalar Gaussian random variable with unit variance and expectation $\leq -\delta$, implying that the p_μ-probability for $\xi[\omega]$ to be nonnegative (which is exactly the same as the probability for $\mathcal{T}$ to reject H_2 and accept H_1) is at most $\mathrm{Erfc}(\delta)$. These observations imply the validity of (2.5). The test optimality in terms of risks follows from the fact that the risks of a simple test deciding on our now—composite—hypotheses H_1, H_1 on the density p of observation ω can be only larger than the risks of a simple test deciding on two simple hypotheses $p = p_{\mu_1^*}$ and $p = p_{\mu_2^*}$. In other words, the quantity $\mathrm{Erfc}(\frac{1}{2}\|\mu_1^* - \mu_2^*\|_2)$—see Example 2.1—is a lower bound on the risk and half of the total risk of a test deciding on H_1, H_2. With this in mind, the announced optimalities of $\mathcal{T}$ in terms of risks are immediate consequences of (2.5).

We remark that the (nearly self-evident) result stated in Example 2.2 seems to have first been noticed in [36].

Example 2.2 allows for substantial extensions in two directions: first, it turns out that the "Euclidean separation" underlying the test built in this example can be used to decide on hypotheses on the location of a "center" of d-dimensional distribution far beyond the Gaussian observation model considered in this example. This extension will be our goal in the next section, based on the recent paper [110]. Less straightforward and, we believe, more instructive extensions, originating from [102], will be considered in Section 2.3.

2.2 HYPOTHESIS TESTING VIA EUCLIDEAN SEPARATION

2.2.1 Situation

In this section, we will be interested in testing hypotheses

$$H_\ell : P \in \mathcal{P}_\ell, \ell = 1, ..., L \tag{2.6}$$

on the probability distribution of a random observation ω in the situation where the families of distributions $\mathcal{P}_\ell$ are obtained from a given family $\mathcal{P}$ of probability distributions by shifts. Specifically, we are given

- a family $\mathcal{P}$ of probability distributions on $\Omega = \mathbf{R}^d$ such that all distributions from $\mathcal{P}$ possess densities with respect to the Lebesgue measure on $\mathbf{R}^n$, and these densities are even functions on $\mathbf{R}^d$;[2]
- a collection $X_1, ..., X_L$ of nonempty closed and convex subsets of $\mathbf{R}^d$, with at most one of the sets unbounded.

These data specify L families $\mathcal{P}_\ell$ of distributions on $\mathbf{R}^d$; $\mathcal{P}_\ell$ is comprised of distributions of random vectors of the form $x + \xi$, where $x \in X_\ell$ is deterministic, and ξ is random with distribution from $\mathcal{P}$. Note that with this setup, deciding upon hypotheses (2.6) via observation $\omega \sim P$ is exactly the same as deciding, given observation

$$\omega = x + \xi, \tag{2.7}$$

[2] Allowing for a slight abuse of notation, we write $P \in \mathcal{P}$, where P is a probability distribution, to express the fact that P belongs to $\mathcal{P}$ (no abuse of notation so far), and write $p(\cdot) \in \mathcal{P}$ (this is an abuse of notation), where $p(\cdot)$ is the density of the probability distribution P, to express the fact that $P \in \mathcal{P}$.

where x is a deterministic "signal" and ξ is random noise with distribution P known to belong to $\mathcal{P}$, on the "position" of x w.r.t. $X_1, ..., X_L$, the ℓ-th hypothesis H_ℓ saying that $x \in X_\ell$. The latter allows us to write down the ℓ-th hypothesis as $H_\ell : x \in X_\ell$ (of course, this shorthand makes sense only within the scope of our current "signal plus noise" setup).

2.2.2 Pairwise Hypothesis Testing via Euclidean Separation

2.2.2.1 The simplest case

Consider nearly the simplest case of the situation from Section 2.2.1 where $L = 2$, $X_1 = \{x^1\}$ and $X_2 = \{x^2\}$, $x^1 \neq x^2$, are singletons, and $\mathcal{P}$ also is a singleton. Let the probability density of the only distribution from $\mathcal{P}$ be of the form

$$p(u) = f(\|u\|_2),\ f(\cdot)\text{ is a strictly monotonically decreasing function on the nonnegative ray.} \tag{2.8}$$

This situation is a generalization of the one considered in Example 2.1, where we dealt with the special case of f, namely, with

$$p(u) = (2\pi)^{-d/2} e^{-u^T u/2}.$$

In the case in question our goal is to decide on two simple hypotheses $H_\chi : p(u) = f(\|u - x^\chi\|_2)$, $\chi = 1, 2$, on the density of observation (2.7). Let us set

$$\delta = \tfrac{1}{2}\|x^1 - x^2\|_2,\ e = \frac{x^1 - x^2}{\|x^1 - x^2\|_2},\ \phi(\omega) = e^T\omega - \underbrace{\tfrac{1}{2}e^T[x^1 + x^2]}_{c}, \tag{2.9}$$

and consider the test $\mathcal{T}$ which, given observation $\omega = x + \xi$, accepts the hypothesis $H_1 : x = x^1$ when $\phi(\omega) \geq 0$, and accepts the hypothesis $H_2 : x = x^2$ otherwise.

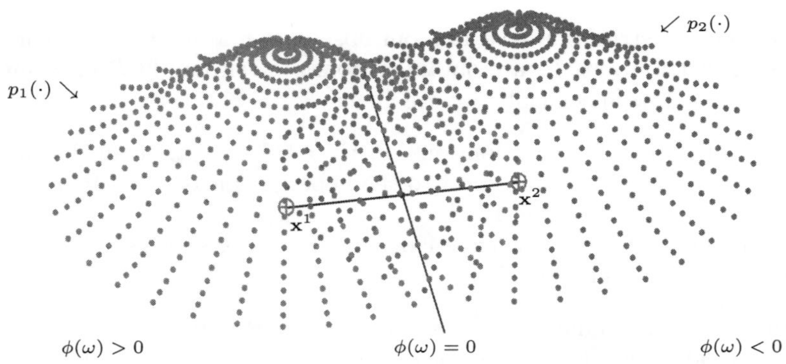

We have (cf. Example 2.1)

$$\begin{aligned}\mathrm{Risk}_1(\mathcal{T}|H_1, H_2) &= \int\limits_{\omega:\phi(\omega)<0} p_1(\omega)d\omega = \int\limits_{u:e^T u<-\delta} f(\|u\|_2)du \\ &= \int\limits_{u:e^T u\geq\delta} f(\|u\|_2)du = \int\limits_{\omega:\phi(\omega)\geq 0} p_2(\omega)d\omega = \mathrm{Risk}_2(\mathcal{T}|H_1, H_2).\end{aligned}$$

Since $p(u)$ is strictly decreasing function of $\|u\|_2$, we have also

$$\min[p_1(u),p_2(u)] = \left\{\begin{array}{ll} p_1(u), & \phi(u)<0 \\ p_2(u), & \phi(u)\geq 0 \end{array}\right.,$$

whence

$$\begin{array}{rcl}\mathrm{Risk}_1(\mathcal{T}|H_1,H_2)+\mathrm{Risk}_2(\mathcal{T}|H_1,H_2) &=& \int\limits_{\omega:\phi(\omega)<0} p_1(\omega)d\omega + \int\limits_{\omega:\phi(\omega)\geq 0} p_2(\omega)d\omega \\ &=& \int\limits_{\mathbf{R}^d} \min[p_1(u),p_2(u)]du \end{array}$$

Invoking Proposition 2.2, we conclude that *the test $\mathcal{T}$ is the minimum risk simple test deciding on H_1, H_2, and the risk of this test is*

$$\mathrm{Risk}(\mathcal{T}|H_1,H_2) = \int\limits_{u:e^Tu\geq\delta} f(\|u\|_2)du. \tag{2.10}$$

2.2.2.2 *Extension*

Now consider a slightly more complicated case of the situation from Section 2.2.1 with $L = 2$ so that X_1 and X_2 are nonempty and nonintersecting closed convex sets, one of the sets being bounded. As for $\mathcal{P}$, we still assume that it is a singleton, and the density of the only distribution from $\mathcal{P}$ is of the form (2.8). Our current situation is an extension of that in Example 2.2. For exactly the same reasons as in the latter example, with X_1, X_2 as above, the convex minimization problem

$$\mathrm{Opt} = \min_{x^1\in X_1, x^2\in X_2} \tfrac{1}{2}\|x^1-x^2\|_2 \tag{2.11}$$

is solvable, and denoting by (x^1_*, x^2_*) an optimal solution and setting

$$\phi(\omega) = e^T\omega - c,\ e = \frac{x^1_*-x^2_*}{\|x^1_*-x^2_*\|_2},\ c = \tfrac{1}{2}e^T[x^1_*+x^2_*], \tag{2.12}$$

the stripe $\{\omega : -\mathrm{Opt} \leq \phi(x) \leq \mathrm{Opt}\}$ separates X_1 and X_2:

$$\phi(x^1) \geq \phi(x^1_*) = \mathrm{Opt}\,\forall x^1 \in X_1 \ \&\ \phi(x^2) \leq \phi(x^2_*) = -\mathrm{Opt}\,\forall x^2 \in X_2 \tag{2.13}$$

Proposition 2.4. *Let X_1, X_2 be nonempty and nonintersecting closed convex sets in $\mathbf{R}^d$, one of the sets being bounded. With* Opt *and $\phi(\cdot)$ given by* (2.11) and (2.12), *let us split the width* 2Opt *of the stripe $\{\omega : -\mathrm{Opt} \leq \phi(\omega) \leq \mathrm{Opt}\}$ separating X_1 and X_2 into two nonnegative parts:*

$$\delta_1 \geq 0, \delta_2 \geq 0,\ \delta_1+\delta_2 = 2\mathrm{Opt} \tag{2.14}$$

and consider the simple test $\mathcal{T}$ which decides on the hypotheses $H_1 : x \in X_1$ and

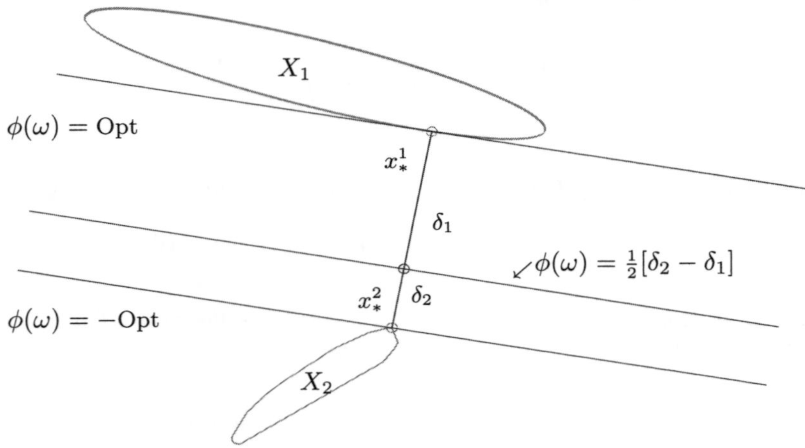

Figure 2.2: Drawing for Proposition 2.4.

$H_2 : x \in X_2$ via observation (2.7) *accepting H_1 when*

$$\phi(\omega) \geq \tfrac{1}{2}[\delta_2 - \delta_1]$$

and accepting H_2 otherwise. Then

$$\text{Risk}_\chi(\mathcal{T}|H_1, H_2) \leq \int_{\delta_\chi}^{\infty} \gamma(s)ds, \;\; \chi = 1, 2, \tag{2.15}$$

where $\gamma(\cdot)$ is the univariate marginal density of ξ, that is, the probability density of the scalar random variable $h^T\xi$, where $\|h\|_2 = 1$ (note that due to (2.8)*, $\gamma(\cdot)$ is independent of how we select h with $\|h\|_2 = 1$).*

In addition, when $\delta_1 = \delta_2 = \text{Opt}$, $\mathcal{T}$ is the minimum risk test deciding on H_1, H_2. The risk of this test is

$$\text{Risk}(\mathcal{T}|H_1, H_2) = \int_{\text{Opt}}^{\infty} \gamma(s)ds. \tag{2.16}$$

Proof. By (2.8) and (2.13), for $x \in X_1$ we have (see Figure 2.2):

$$\text{Prob}_{\xi \sim p(\cdot)}\left\{\phi(x + \xi) < \tfrac{1}{2}[\delta_2 - \delta_1]\right\} \leq \text{Prob}_{\xi \sim p(\cdot)}\left\{[-e]^T\xi \geq \delta_1\right\} = \int_{\delta_1}^{\infty} \gamma(s)ds.$$

By the "symmetric" reasoning, for $x \in X_2$ we have

$$\text{Prob}_{\xi \sim p(\cdot)}\left\{\phi(x + \xi) \geq \tfrac{1}{2}[\delta_2 - \delta_1]\}\right] \leq \text{Prob}_{\xi \sim p(\cdot)}\left\{e^T\xi \geq \delta_2\right\} = \int_{\delta_2}^{\infty} \gamma(s)ds,$$

and we arrive at (2.15). The fact that in the case of $\delta_1 = \delta_2 = \mathrm{Opt}$ our test $\mathcal{T}$ becomes the minimum risk test deciding on composite hypotheses H_1, H_2 is readily given by the analysis in Section 2.2.2.1: the minimal over all possible tests risk of deciding on two simple hypotheses $H_1' : x = x_*^1$, $H_2' : x = x_*^2$ is given by (2.10), i.e., it is equal to $\int_{\mathrm{Opt}}^{\infty} \gamma(s)ds$. In the case of $\delta_1 = \delta_2 = \mathrm{Opt}$ this is exactly the upper bound (2.16) on the risk of the test $\mathcal{T}$ deciding on the composite hypotheses H_χ, $\chi = 1, 2$, larger than H_χ'. □

2.2.2.3 Further extensions: spherical families of distributions

As in Section 2.2.2.2, we continue to assume that we are in the situation of Section 2.2.1 with $L = 2$ and nonempty closed, convex, and nonintersecting X_1, X_2, one of the sets being bounded. Our next goal is to relax the restrictions on the family $\mathcal{P}$ of noise distributions, which in Section 2.2.2.2 was just a singleton with density which is a strictly decreasing function of the $\|\cdot\|_2$-norm. Observe that as far as the density $p(\cdot)$ of noise is concerned, justification of the upper risk bound (2.15) in Proposition 2.4 used only the fact that whenever $h \in \mathbf{R}^d$ is a $\|\cdot\|_2$-unit vector and $\delta \geq 0$, we have $\int_{h^T u \geq \delta} p(u)du \leq \int_\delta^\infty \gamma(s)ds$, with the even univariate probability density $\gamma(\cdot)$ specified in the proposition. We use this observation to extend our construction to *spherical families of probability densities.*

2.2.2.3.A. Spherical families of probability densities. Let $\gamma(\cdot)$ be an even probability density on the axis such that there is no neighborhood of the origin where $\gamma = 0$ almost surely. We associate with γ a *spherical family of densities* $\mathcal{P}_\gamma^d$ comprised of all probability densities $p(\cdot)$ on $\mathbf{R}^d$ such that

A. $p(\cdot)$ is even

B. Whenever $e \in \mathbf{R}^d$, $\|e\|_2 = 1$, and $\delta \geq 0$, we have

$$\mathrm{Prob}_{\xi \sim P}\{\xi : e^T\xi \geq \delta\} \leq P_\gamma(\delta) := \int_\delta^\infty \gamma(s)ds. \tag{2.17}$$

Geometrically: the $p(\cdot)$-probability for $\xi \sim p(\cdot)$ to belong to a half-space not containing the origin does not exceed $P_\gamma(\delta)$, where δ is the $\|\cdot\|_2$-distance from the origin to the half-space.

Note that density (2.8) belongs to the family $\mathcal{P}_\gamma^d$ with $\gamma(\cdot)$ defined in Proposition 2.4; the resulting γ, in addition to being an even density, is strictly monotonically decreasing on the nonnegative ray. When speaking about general-type spherical families $\mathcal{P}_\gamma^d$, we do *not* impose monotonicity requirements on $\gamma(\cdot)$. If a spherical family $\mathcal{P}_\gamma^d$ includes a density $p(\cdot)$ of the form (2.8) *such that* $\gamma(\cdot)$ *is the univariate marginal density induced by* $p(\cdot)$, as in Proposition 2.4, we say that $\mathcal{P}_\gamma^d$ *has a cap,* and this cap is $p(\cdot)$.

2.2.2.3.B. Example: Gaussian mixtures. Let $\eta \sim \mathcal{N}(0, \Theta)$, where the $d \times d$ covariance matrix Θ satisfies $\Theta \preceq I_d$, and let Z be a positive scalar random variable independent of η. The *Gaussian mixture* of Z and η (or, better said, of the distribution P_Z of Z and the distribution $\mathcal{N}(0, \Theta)$) is the probability distribution of the random vector $\xi = \sqrt{Z}\eta$. Examples of Gaussian mixtures [89, 153] include

- Gaussian distribution $\mathcal{N}(0, \Theta)$ (take Z identically equal to 1),

- multidimensional Student's t-distribution with $\nu \in \{1,2,...\}$ degrees of freedom and "covariance structure" Θ; here Z is given by the requirement that ν/Z has χ^2-distribution with ν degrees of freedom.

An immediate observation (see Exercise 2.2) is that with γ given by the distribution P_Z of Z according to

$$\gamma_Z(s) = \int_{z>0} \frac{1}{\sqrt{2\pi z}} \mathrm{e}^{-\frac{s^2}{2z}} P_Z(dz), \tag{2.18}$$

the distribution of random variable $\sqrt{Z}\eta$, with $\eta \sim \mathcal{N}(0,\Theta)$, $\Theta \preceq I_d$, independent of Z, belongs to the family $\mathcal{P}^d_{\gamma_Z}$. The family $\mathcal{P}^d_{\gamma_Z}$ has a cap, specifically, the Gaussian mixture of P_Z and $\mathcal{N}(0, I_d)$.

Another example of this type: the Gaussian mixture of a distribution P_Z of random variable Z taking values in $(0,1]$ and a distribution $\mathcal{N}(0,\Theta)$ with $\Theta \preceq I_d$ belongs to the spherical family $\mathcal{P}^d_{\gamma_\mathcal{G}}$ associated with the standard univariate Gaussian density

$$\gamma_\mathcal{G}(s) = \frac{1}{\sqrt{2\pi}} \mathrm{e}^{-s^2/2}.$$

This family has a cap, specifically, the standard Gaussian d-dimensional distribution $\mathcal{N}(0, I_d)$.

2.2.2.3.C. Main result. Observing the proof of Proposition 2.4, we arrive at the following.

Proposition 2.5. *Let X_1 and X_2 be nonempty and nonintersecting closed convex sets in $\mathbf{R}^d$, one of the sets being bounded, and let $\mathcal{P}^d_\gamma$ be a spherical family of probability distributions. With* Opt *and $\phi(\cdot)$ given by (2.11)—(2.12), let us split the width* 2Opt *of the stripe $\{\omega : -\mathrm{Opt} \leq \phi(\omega) \leq \mathrm{Opt}\}$ separating X_1 and X_2 into two nonnegative parts:*

$$\delta_1 \geq 0,\ \delta_2 \geq 0,\ \delta_1 + \delta_2 = 2\mathrm{Opt}. \tag{2.19}$$

Let us consider a simple test $\mathcal{T}$ deciding on the hypotheses $H_1 : x \in X_1$, $H_2 : x \in X_2$ via observation (2.7) accepting H_1 when

$$\phi(\omega) \geq \tfrac{1}{2}[\delta_2 - \delta_1]$$

and accepting H_2 otherwise. Then

$$\mathrm{Risk}_\chi(\mathcal{T}|H_1, H_2) \leq \int_{\delta_\chi}^{\infty} \gamma(s)ds,\ \ \chi = 1, 2. \tag{2.20}$$

In addition, when $\delta_1 = \delta_2 =$ Opt and $\mathcal{P}^d_\gamma$ has a cap, $\mathcal{T}$ is the minimum risk test deciding on H_1, H_2. The risk of this test is given by

$$\mathrm{Risk}(\mathcal{T}|H_1, H_2) = P_\gamma(\mathrm{Opt}) := \int_{\mathrm{Opt}}^{\infty} \gamma(s)ds. \tag{2.21}$$

To illustrate the power of Proposition 2.5, consider the case when γ is the function (2.18) stemming from Student's t-distribution on $\mathbf{R}^d$ with ν degrees of freedom. It is known that in this case γ is the density of univariate Student's t-distribution with ν degrees of freedom [153]:

$$\gamma(s) = \frac{\Gamma\left(\frac{\nu+1}{2}\right)}{\Gamma\left(\frac{\nu}{2}\right)\sqrt{\pi\nu}}(1+s^2/\nu)^{-\frac{\nu+1}{2}},$$

where $\Gamma(\cdot)$ is Euler's Gamma function. When $\nu = 1$, $\gamma(\cdot)$ is just the heavy tailed (no expectation!) standard Cauchy density $\frac{1}{\pi}(1+s^2)^{-1}$. As in this "extreme case," multidimensional Student's distributions have relatively heavy tails (the heavier, the less is ν) and as such are of interest in statistical application in Finance.

2.2.3 Euclidean Separation, Repeated Observations, and Majority Tests

Assume that X_1, X_2 and $\mathcal{P}^d_\gamma$ are as in the premise of Proposition 2.5 and K-repeated observations are allowed, $K > 1$. An immediate attempt to reduce the situation to the single-observation case by calling the K-repeated observation $\omega^K = (\omega_1, ..., \omega_K)$ our new observation and thus reducing testing via repeated observations to the single-observation case seemingly fails: already in the simplest case of stationary K-repeated observations this reduction would require replacing the family $\mathcal{P}^d_\gamma$ with the family of product distributions $\underbrace{P \times ... \times P}_{K}$ stemming from $P \in \mathcal{P}^d_\gamma$, and it is unclear how to apply to the resulting single-observation testing problem our machinery based on Euclidean separation. Instead, we will use the *K-step majority test*.

2.2.3.1 Preliminaries: Repeated observations in "signal plus noise" observation model

We are in the situation where our inference should be based on observations

$$\omega^K = (\omega_1, \omega_2, ..., \omega_K), \tag{2.22}$$

and decide on hypotheses $\mathcal{H}_1$, $\mathcal{H}_2$ on the distribution Q^K of ω^K, and we are interested in the following three cases:

S [*stationary K-repeated observations*, cf. Section 2.1.3.1]: $\omega_1, ..., \omega_K$ are drawn independently of each other from the same distribution Q, that is, Q^K is the product distribution $Q \times ... \times Q$. Further, under hypothesis $\mathcal{H}_\chi$, $\chi = 1, 2$, Q is the distribution of random variable $\omega = x + \xi$, where $x \in X_\chi$ is deterministic, and the distribution P of ξ belongs to the family $\mathcal{P}^d_\gamma$;

SS [*semi-stationary K-repeated observations*, cf. Section 2.1.3.2]: there are two deterministic sequences, one of signals $\{x_k\}_{k=1}^K$, another of distributions $\{P_k \in \mathcal{P}^d_\gamma\}_{k=1}^K$, and $\omega_k = x_k + \xi_k$, $1 \le k \le K$, with $\xi_k \sim P_k$ independent across k. Under hypothesis $\mathcal{H}_\chi$, *all* signals x_k, $k \le K$, belong to X_χ.

QS [*quasi-stationary K-repeated observations*, cf. Section 2.1.3.3]: "in nature" there exists a random sequence of driving factors $\zeta^K = (\zeta_1, ..., \zeta_K)$ such that observation ω_k, for every k, is a deterministic function of $\zeta^k = (\zeta_1, ..., \zeta_k)$: $\omega_k = \theta_k(\zeta^k)$. On top of that, under ℓ-th hypothesis $\mathcal{H}_\ell$, for all $k \le K$ and all ζ^{k-1}, the conditional distribution of ω_k given ζ^{k-1} belongs to the family $\mathcal{P}_\ell$ of distributions of all random vectors of the form $x+\xi$, where $x \in X_\ell$ is deterministic, and ξ is random noise with distribution from $\mathcal{P}^d_\gamma$.

2.2.3.2 *Majority Test*

2.2.3.2.A. The construction of the K-observation majority test is very natural. We use Euclidean separation to build a simple single-observation test $\mathcal{T}$ deciding on hypotheses $H_\chi : x \in X_\chi$, $\chi = 1,2$, via observation $\omega = x+\xi$, where x is deterministic, and the distribution of noise ξ belongs to $\mathcal{P}^d_\gamma$. $\mathcal{T}$ is given by the construction from Proposition 2.5 applied with $\delta_1 = \delta_2 = \mathrm{Opt}$. The summary of our actions is as follows:

$$
\begin{array}{rl}
X_1, X_2 & \Rightarrow \left\{ \begin{array}{c} \mathrm{Opt} = \min_{x^1\in X_1, x^2\in X_2} \frac{1}{2}\|x^1-x^2\|_2 \\ (x^1_*, x^2_*) \in \mathrm{Argmin}_{x^1\in X_1, x^2\in X_2} \frac{1}{2}\|x^1-x^2\|_2 \end{array} \right. \\
& \Rightarrow e = \dfrac{x^1_* - x^2_*}{\|x^1_*-x^2_*\|_2},\ c = \frac{1}{2}e^T[x^1_*+x^2_*] \\
& \Rightarrow \phi(\omega) = e^T\omega - c.
\end{array}
$$

The Majority test $\mathcal{T}^{\mathrm{maj}}_K$, as applied to the K-repeated observation $\omega^K = (\omega_1, ..., \omega_K)$, computes the K reals $v_k = \phi(\omega_k)$; if at least $K/2$ of these reals are nonnegative, the test accepts $\mathcal{H}_1$ and rejects $\mathcal{H}_2$; otherwise the test accepts $\mathcal{H}_2$ and rejects $\mathcal{H}_1$.

2.2.3.2.B. Risk analysis. We are to carry out the risk analysis for the case **QS** of quasi-stationary K-repeated observations; this analysis automatically applies to the cases **S** of stationary and **SS** of semi-stationary K-repeated observations, which are special cases of **QS**.

Proposition 2.6. *With $X_1, X_2, \mathcal{P}^d_\gamma$ obeying the premise of Proposition 2.5, in the case* **QS** *of quasi-stationary observations the risk of the K-observation Majority test $\mathcal{T}^{\mathrm{maj}}_K$ can be bounded as*

$$
\mathrm{Risk}(\mathcal{T}^{\mathrm{maj}}_K|\mathcal{H}_1,\mathcal{H}_2) \le \epsilon_K \equiv \sum_{K/2\le k\le K} \binom{K}{k} \epsilon_\star^k (1-\epsilon_\star)^{K-k},\ \epsilon_\star = \int_{\mathrm{Opt}}^{\infty} \gamma(s)ds. \tag{2.23}
$$

Proof. *For the sake of clarity, here we restrict ourselves to the case* **SS** *of semi-stationary K-repeated observations.* In "full generality," that is, in the case **QS** of quasi-stationary K-repeated observations, the proposition is proved in Section 2.11.2.

Assume that $\mathcal{H}_1$ takes place, so that (recall that we are in the **SS** case!) $\omega_k = x_k + \xi_k$ with some deterministic $x_k \in X_1$ and noises $\xi_k \sim P_k$ independent across k, for some deterministic sequence $P_k \in \mathcal{P}^d_\gamma$. Let us fix $\{x_k \in X_1\}_{k=1}^K$ and $\{P_k \in \mathcal{P}^d_\gamma\}_{k=1}^K$. Then the random reals $v_k = \phi(\omega_k = x_k+\xi_k)$ are independent across k, and so are the Boolean random variables

$$
\chi_k = \left\{ \begin{array}{ll} 1, & v_i < 0, \\ 0, & v_i \ge 0; \end{array} \right.
$$

$\chi_k = 1$ if and only if test $\mathcal{T}$, as applied to observation ω_k, rejects hypothesis $H_1 : x_k \in X_1$. By Proposition 2.5, P_k-probability p_k of the event $\chi_k = 1$ is at most $\epsilon_\star$. Further, by construction of the Majority test, if $\mathcal{T}_K^{\rm maj}$ rejects the true hypothesis $\mathcal{H}_1$, then the number of k's with $\chi_k = 1$ is $\geq K/2$. Thus, with $x_k \in X_1$ and $P_k \in \mathcal{P}_\gamma^d$, $1 \leq k \leq K$, the probability of rejecting $\mathcal{H}_1$ is not greater than the probability of the event

> In K independent coin tosses, with probability $p_k \leq \epsilon_*$ of getting heads in k-th toss, the total number of heads is $\geq K/2$.

The probability of this event clearly does not exceed the right-hand side in (2.23), implying that $\mathrm{Risk}_1(\mathcal{T}_K^{\rm maj}|\mathcal{H}_1, \mathcal{H}_2) \leq \epsilon_K$. A "symmetric" reasoning yields

$$\mathrm{Risk}_2(\mathcal{T}_K^{\rm maj}|\mathcal{H}_1, \mathcal{H}_2) \leq \epsilon_K,$$

completing the proof of (2.23). □

Corollary 2.7. *Under the premise of Proposition 2.6, the upper bounds ϵ_K on the risk of the K-observation Majority test goes to 0 exponentially fast as $K \to \infty$.*

Indeed, we are in the situation of $\mathrm{Opt} > 0$, so that $\epsilon_\star < \frac{1}{2}$.[3]

Remark 2.8. When proving (the ***SS*** version of) Proposition 2.6, we have used an "evident" observation as follows:

> (#) Let χ_1,..., χ_K be independent random variables taking values 0 and 1, and let the probabilities p_k for χ_k to take value 1 be upper-bounded by some $\epsilon \in [0, 1]$ for all k. Then for every fixed M the probability of the event *"at least M of $\chi_1, ..., \chi_K$ are equal to 1"* is upper-bounded by the probability $\sum_{M \leq k \leq K} \binom{K}{k} \epsilon^k (1-\epsilon)^{K-k}$ of the same event in the case when $p_k = \epsilon$ for all k.

If there are evident facts in Math, (#) definitely is one of them. Nevertheless, it requires a proof; this proof (finally, not completely evident) can be found in Section 2.11.2.

2.2.3.2.C. Near-optimality. We are about to show that under appropriate assumptions, the majority test $\mathcal{T}_K^{\rm maj}$ is near-optimal. The precise statement is as follows:

Proposition 2.9. *Let $X_1, X_2, \mathcal{P}_\gamma^d$ obey the premise of Proposition 2.5. Assume that the spherical family $\mathcal{P}_\gamma$ and positive reals D, α, β are such that*

$$\beta D \leq \tfrac{1}{4}, \tag{2.24}$$

$$\int_0^\delta \gamma(s) ds \geq \beta\delta, \;\; 0 \leq \delta \leq D, \tag{2.25}$$

and $\mathcal{P}_\gamma$ contains a density $q(\cdot)$ such that

$$\int_{\mathbf{R}^n} \sqrt{q(\xi - e) q(\xi + e)} d\xi \geq \exp\{-\alpha e^T e\} \;\; \forall (e : \|e\|_2 \leq D). \tag{2.26}$$

---

[3] Recall that we have assumed from the very beginning that γ is an even probability density on the axis, and there is no neighbourhood of the origin where $\gamma = 0$ a.s.

Let also the sets X_1 *and* X_2 *be such that* Opt *as given by* (2.11) *satisfies the relation*

$$\text{Opt} \le D. \tag{2.27}$$

Given tolerance $\epsilon \in (0, 1/5)$, *the risk of* K*-observation majority test* $\mathcal{T}_K^{\text{maj}}$ *utilizing* **QS** *observations ensures the relation*

$$K \ge K^* := \left\rfloor \frac{\ln(1/\epsilon)}{2\beta^2 \text{Opt}^2} \right\lfloor \Rightarrow \text{Risk}(\mathcal{T}_K^{\text{maj}} | \mathcal{H}_1, \mathcal{H}_2) \le \epsilon \tag{2.28}$$

(here $\rfloor x \lfloor$ *stands for the smallest integer* $\ge x \in \mathbf{R}$*). In addition, for every* K*-observation test* $\mathcal{T}_K$ *utilizing stationary repeated observations and satisfying*

$$\text{Risk}(\mathcal{T}_K | \mathcal{H}_1, \mathcal{H}_2) \le \epsilon$$

it holds

$$K \ge K_* := \frac{\ln\left(\frac{1}{4\epsilon}\right)}{2\alpha \text{Opt}^2}. \tag{2.29}$$

As a result, the majority test $\mathcal{T}_{K^*}^{\text{maj}}$ *(which by* (2.28) *has risk* $\le \epsilon$*) is near-optimal in terms of the required number of observations among all tests with risk* $\le \epsilon$*: the number* K *of observations in such a test satisfies the relation*

$$K^*/K \le \theta := K^*/K_* = O(1)\frac{\alpha}{\beta^2}.$$

Proof of the proposition is the subject of Exercise 2.3.

Illustration. Given $\nu \ge 1$, consider the case when $\mathcal{P} = \mathcal{P}_\gamma$ is the spherical family with n-variate (spherical) Student's distribution in the role of the cap, so that

$$\gamma(s) = \frac{\Gamma\left(\frac{\nu+1}{2}\right)}{\Gamma\left(\frac{\nu}{2}\right)(\pi\nu)^{1/2}} \left[1 + s^2/\nu\right]^{-(\nu+1)/2}. \tag{2.30}$$

It is easily seen (cf. Exercise 2.3) that $\mathcal{P}$ contains the $\mathcal{N}(0, \frac{1}{2}I_n)$ density $q(\cdot)$, implying that setting

$$D = 1,\ \alpha = 1,\ \beta = \tfrac{1}{7},$$

one ensures relations (2.24), (2.25) and (2.27). As a result, when Opt as yielded by (2.11) is ≤ 1, the nonoptimality factor θ of the majority test $\mathcal{T}_{K^*}^{\text{maj}}$ as defined in Proposition 2.9 is $O(1)$.

2.2.4 From Pairwise to Multiple Hypotheses Testing

2.2.4.1 Situation

Assume we are given L families of probability distributions $\mathcal{P}_\ell$, $1 \le \ell \le L$, on observation space Ω, and observe a realization of random variable $\omega \sim P$ taking values in Ω. Given ω, we want to decide on the L hypotheses

$$H_\ell : P \in \mathcal{P}_\ell,\ 1 \le \ell \le L. \tag{2.31}$$

Our *ideal goal* would be to find a low-risk simple test deciding on the hypotheses. However, it may happen that this " ideal goal" cannot be achieved, for example, when some pairs of families $\mathcal{P}_\ell$ have nonempty intersections. When $\mathcal{P}_\ell \cap \mathcal{P}_{\ell'} \neq \emptyset$ for some $\ell \neq \ell'$, there is no way to decide on the hypotheses with risk $< 1/2$.

But: *Impossibility to decide reliably on all L hypotheses "individually" does* not *mean that no meaningful inferences can be made.* For example, consider the three rectangles on the plane

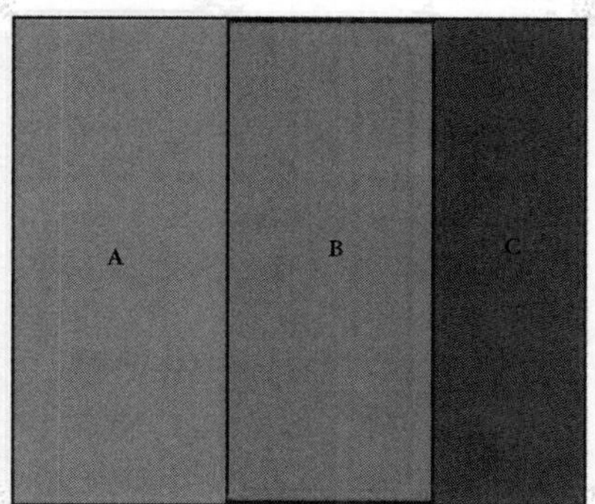

and three hypotheses, with H_ℓ, $\ell \in \{A, B, C\}$, stating that our observation is $\omega = x+\xi$ with deterministic "signal" x belonging to rectangle ℓ and $\xi \sim \mathcal{N}(0, \sigma^2 I_2)$. However small σ is, no test can decide on the three hypotheses with risk $< 1/2$; e.g., there is no way to decide reliably on H_A vs. H_B. However, we may hope that when σ is small (or when repeated observations are allowed), observations allow us to discard reliably at least some of the hypotheses. For instance, when the signal belongs to rectangle A (i.e., H_A holds true), we hardly can discard reliably the hypothesis H_B stating that the signal belongs to rectangle B, but hopefully can reliably discard H_C (that is, infer that the signal is *not* in rectangle C).

When handling multiple hypotheses which cannot be reliably decided upon "as they are," it makes sense to speak about *testing the hypotheses "up to closeness."*

2.2.4.2 Closeness relation and "up to closeness" risks

Closeness relation, or simply *closeness* $\mathcal{C}$ on a collection of L hypotheses $H_1, ..., H_L$ is defined as some *set of pairs* (ℓ, ℓ') with $1 \leq \ell, \ell' \leq L$. We interpret the relation $(\ell, \ell') \in \mathcal{C}$ as the fact that the hypotheses H_ℓ and H'_ℓ are close to each other. Sometimes we shall use the words "ℓ and ℓ' are/are not $\mathcal{C}$-close to each other" as an equivalent form of "hypotheses H_ℓ, $H_{\ell'}$ are/are not $\mathcal{C}$-close to each other."
We always assume that

- $\mathcal{C}$ contains all "diagonal pairs" (ℓ, ℓ), $1 \leq \ell \leq L$ ("every hypothesis is close to itself");
- $(\ell, \ell') \in \mathcal{C}$ if and only if $(\ell', \ell) \in \mathcal{C}$ ("closeness is a symmetric relation").

Note that by symmetry of $\mathcal{C}$, the relation $(\ell, \ell') \in \mathcal{T}$ is in fact a property of *unordered* pair $\{\ell, \ell'\}$.

"Up to closeness" risks. Let $\mathcal{T}$ be a test deciding on L hypotheses $H_1, ..., H_L$; see (2.31). Given observation ω, $\mathcal{T}$ accepts all hypotheses H_ℓ with indexes $\ell \in \mathcal{T}(\omega)$

and rejects all other hypotheses. We say that the *ℓ-th partial $\mathcal{C}$-risk of test $\mathcal{T}$ is* $\leq \epsilon$ if *whenever H_ℓ is true: $\omega \sim P \in \mathcal{P}_\ell$, the P-probability of the event*

$\mathcal{T}$ *accepts* H_ℓ*:* $\ell \in \mathcal{T}(\omega)$
and
all hypotheses $H_{\ell'}$ *accepted by* $\mathcal{T}$ *are* $\mathcal{C}$*-close to* H_ℓ*:* $(\ell, \ell') \in \mathcal{C}, \forall \ell' \in \mathcal{T}(\omega)$

is *at least* $1 - \epsilon$.

The ℓ-th partial $\mathcal{C}$-risk $\mathrm{Risk}_\ell^{\mathcal{C}}(\mathcal{T}|H_1, ..., H_L)$ *of $\mathcal{T}$* is the smallest ϵ with the outlined property, or, equivalently,

$$\mathrm{Risk}_\ell^{\mathcal{C}}(\mathcal{T}|H_1, ..., H_L) = \sup_{P \in \mathcal{P}_\ell} \mathrm{Prob}_{\omega \sim P} \left\{ [\ell \notin \mathcal{T}(\omega)] \text{ or } [\exists \ell' \in \mathcal{T}(\omega) : (\ell, \ell') \notin \mathcal{C}] \right\}.$$

$\mathcal{C}$-risk $\mathrm{Risk}^{\mathcal{C}}(\mathcal{T}|H_1, ..., H_L)$ *of $\mathcal{T}$* is the largest of the partial $\mathcal{C}$-risks of the test:

$$\mathrm{Risk}^{\mathcal{C}}(\mathcal{T}|H_1, ..., H_L) = \max_{1 \leq \ell \leq L} \mathrm{Risk}_\ell^{\mathcal{C}}(\mathcal{T}|H_1, ..., H_L).$$

Observe that when $\mathcal{C}$ is the "strictest possible" closeness, that is, $(\ell, \ell') \in \mathcal{C}$ if and only if $\ell = \ell'$, then a test $\mathcal{T}$ deciding on $H_1, ..., H_L$ up to closeness $\mathcal{C}$ with risk ϵ is, *basically*, the same as a simple test deciding on $H_1, ..., H_L$ with risk $\leq \epsilon$. Indeed, a test with the latter property clearly decides on $H_1, ..., H_L$ with $\mathcal{C}$-risk $\leq \epsilon$. The inverse statement, *taken literally*, is not true, since even with our "as strict as possible" closeness, a test $\mathcal{T}$ with $\mathcal{C}$-risk $\leq \epsilon$ is not necessarily simple. However, we can enforce $\mathcal{T}$ to be simple, specifically, to accept a once and forever fixed hypothesis, say, H_1, and only it, when the set of hypotheses accepted by $\mathcal{T}$ "as is" is not a singleton, and otherwise accept exactly the same hypothesis as $\mathcal{T}$. The modified test already is simple, and clearly its $\mathcal{C}$-risk does not exceed that of $\mathcal{T}$.

2.2.4.3 Multiple Hypothesis Testing via pairwise tests

Assume that for every *unordered* pair $\{\ell, \ell'\}$ with $(\ell, \ell') \notin \mathcal{C}$ we are given a *simple* test $\mathcal{T}_{\{\ell,\ell'\}}$ deciding on H_ℓ vs. $H_{\ell'}$ via observation ω.
Our goal is to "assemble" the tests $\mathcal{T}_{\{\ell,\ell'\}}$, $(\ell, \ell') \notin \mathcal{C}$, into a test $\mathcal{T}$ deciding on $H_1 ..., H_L$ up to closeness $\mathcal{C}$.

The construction we intend to use is as follows:

- For $1 \leq \ell, \ell' \leq L$, we define functions $T_{\ell\ell'}(\omega)$ as follows:
  - when $(\ell, \ell') \in \mathcal{C}$, we set $T_{\ell\ell'}(\cdot) \equiv 0$;
  - when $(\ell, \ell') \notin \mathcal{C}$, so that $\ell \neq \ell'$, we set

$$T_{\ell\ell'}(\omega) = \begin{cases} 1, & \mathcal{T}_{\{\ell,\ell'\}}(\omega) = \{\ell\} \\ -1, & \mathcal{T}_{\{\ell,\ell'\}}(\omega) = \{\ell'\} \end{cases}. \tag{2.32}$$

  Note that $\mathcal{T}_{\{\ell,\ell'\}}$ is a simple test, so that $T_{\ell\ell'}(\cdot)$ is well defined and takes values ± 1 when $(\ell, \ell') \notin \mathcal{C}$ and 0 when $(\ell, \ell') \in \mathcal{C}$.

Note that by construction and since $\mathcal{C}$ is symmetric, we have

$$T_{\ell\ell'}(\omega) \equiv -T_{\ell'\ell}(\omega),\ 1 \le \ell, \ell' \le L. \tag{2.33}$$

- The test $\mathcal{T}$ is as follows: *Given observation ω, we build the $L \times L$ matrix $T(\omega) = [T_{\ell\ell'}(\omega)]$ and accept exactly those of the hypotheses H_ℓ for which the ℓ-th row in $T(\omega)$ is nonnegative.*

Observation 2.10. When $\mathcal{T}$ accepts a hypothesis H_ℓ, all hypotheses accepted by $\mathcal{T}$ are $\mathcal{C}$-close to H_ℓ.

Indeed, if ω is such that $\ell \in \mathcal{T}(\omega)$, then the ℓ-th row in $T(\omega)$ is nonnegative. If now ℓ' is *not* $\mathcal{C}$-close to ℓ, we have $T_{\ell\ell'}(\omega) \ge 0$ and $T_{\ell\ell'}(\omega) \in \{-1, 1\}$, whence $T_{\ell\ell'}(\omega) = 1$. Consequently, by (2.33) it holds $T_{\ell'\ell}(\omega) = -1$, implying that the ℓ'-th row in $T(\omega)$ is *not* nonnegative, and thus $\ell' \notin \mathcal{T}(\omega)$. □

Risk analysis. For $(\ell, \ell') \notin \mathcal{C}$, let

$$\begin{array}{rcl} \epsilon_{\ell\ell'} & = & \mathrm{Risk}_1(\mathcal{T}_{\{\ell,\ell'\}}|H_\ell, H_{\ell'}) = \sup\limits_{P \in \mathcal{P}_\ell} \mathrm{Prob}_{\omega\sim P}\{\ell \notin \mathcal{T}_{\{\ell,\ell'\}}(\omega)\} \\ & = & \sup\limits_{P \in \mathcal{P}_\ell} \mathrm{Prob}_{\omega\sim P}\{T_{\ell\ell'}(\omega) = -1\} = \sup\limits_{P \in \mathcal{P}_\ell} \mathrm{Prob}_{\omega\sim P}\{T_{\ell'\ell}(\omega) = 1\} \\ & = & \sup\limits_{P \in \mathcal{P}_\ell} \mathrm{Prob}_{\omega\sim P}\{\ell' \in \mathcal{T}_{\{\ell,\ell'\}}(\omega)\} = \mathrm{Risk}_2(\mathcal{T}_{\{\ell,\ell'\}}|H_{\ell'}, H_\ell). \end{array} \tag{2.34}$$

Proposition 2.11. *For the test $\mathcal{T}$ just defined it holds*

$$\forall \ell \le L: \mathrm{Risk}_\ell^{\mathcal{C}}(\mathcal{T}|H_1, ..., H_L) \le \epsilon_\ell := \sum_{\ell':(\ell,\ell')\notin\mathcal{C}} \epsilon_{\ell\ell'}. \tag{2.35}$$

Proof. Let us fix ℓ, let H_ℓ be true, and let $P \in \mathcal{P}_\ell$ be the distribution of observation ω. Set $I = \{\ell' \le L : (\ell, \ell') \notin \mathcal{C}\}$. For $\ell' \in I$, let $E_{\ell'}$ be the event

$$\{\omega : T_{\ell\ell'}(\omega) = -1\}.$$

We have $\mathrm{Prob}_{\omega\sim P}(E_{\ell'}) \le \epsilon_{\ell\ell'}$ (by definition of $\epsilon_{\ell\ell'}$), whence

$$\mathrm{Prob}_{\omega\sim P}\big(\underbrace{\cup_{\ell'\in I} E_{\ell'}}_{E}\big) \le \epsilon_\ell.$$

When the event E does *not* take place, we have $T_{\ell\ell'}(\omega) = 1$ for all $\ell' \in I$, so that $T_{\ell\ell'}(\omega) \ge 0$ for all ℓ', $1 \le \ell' \le L$, whence $\ell \in \mathcal{T}(\omega)$. By Observation 2.10, the latter inclusion implies that

$$\{\ell \in \mathcal{T}(\omega)\}\ \&\ \{(\ell, \ell') \in \mathcal{C}\ \forall \ell' \in \mathcal{T}(\omega)\}.$$

Invoking the definition of partial $\mathcal{C}$-risk, we get

$$\mathrm{Risk}_\ell^{\mathcal{C}}(\mathcal{T}|H_1, ..., H_L) \le \mathrm{Prob}_{\omega\sim P}(E) \le \epsilon_\ell.$$

□

2.2.4.4 *Testing Multiple Hypotheses via Euclidean separation*

Situation. We are given L nonempty and closed convex sets $X_\ell \subset \Omega = \mathbf{R}^d$, $1 \leq \ell \leq L$, with at least $L-1$ of the sets being bounded, and a spherical family of probability distributions $\mathcal{P}^d_\gamma$. These data define L families $\mathcal{P}_\ell$ of probability distributions on $\mathbf{R}^d$, the family $\mathcal{P}_\ell$, $1 \leq \ell \leq L$, comprised of probability distributions of all random vectors of the form $x + \xi$, where deterministic x ("signal") belongs to X_ℓ, and ξ is random noise with distribution from $\mathcal{P}^d_\gamma$. Given positive integer K, we can speak about L hypotheses on the distribution P^K of K-repeated observation $\omega^K = (\omega_1, ..., \omega_K)$, with $\mathcal{H}_\ell$ stating that ω^K is a quasi-stationary K-repeated observation associated with $\mathcal{P}_\ell$. In other words $\mathcal{H}_\ell = H_\ell^{\otimes,K}$; see Section 2.1.3.3. Finally, we are given a closeness $\mathcal{C}$.

Our goal is to decide on the hypotheses $\mathcal{H}_1, ..., \mathcal{H}_L$ up to closeness $\mathcal{C}$ via K-repeated observation ω^K. Note that this is a natural extension of the case **QS** of pairwise testing from repeated observations considered in Section 2.2.3 (there $L = 2$ and $\mathcal{C}$ is the only meaningful closeness on a two-hypotheses set: $(\ell, \ell') \in \mathcal{C}$ if and only if $\ell = \ell'$).

The Standing Assumption which we assume to hold by default everywhere in this section is:

Whenever ℓ, ℓ' are not $\mathcal{C}$-close: $(\ell, \ell') \notin \mathcal{C}$, the sets X_ℓ, $X_{\ell'}$ do not intersect.

Strategy: We intend to attack the above testing problem by assembling pairwise Euclidean separation Majority tests via the construction from Section 2.2.4.3.

Building blocks to be assembled are Euclidean separation K-observation pairwise Majority tests constructed for the pairs $\mathcal{H}_\ell$, $\mathcal{H}_{\ell'}$ of hypotheses with ℓ and ℓ' *not* close to each other, that is, with $(\ell, \ell') \notin \mathcal{C}$. These tests are built as explained in Section 2.2.3.2; for the reader's convenience, here is the construction. For a pair $(\ell, \ell') \notin \mathcal{C}$, we

1. Find the optimal value $\mathrm{Opt}_{\ell\ell'}$ and an optimal solution $(u_{\ell\ell'}, v_{\ell\ell'})$ to the convex optimization problem
$$\mathrm{Opt}_{\ell\ell'} = \min_{u \in X_\ell, v \in X_{\ell'}} \tfrac{1}{2}\|u - v\|_2. \tag{2.36}$$
The latter problem is solvable, since we have assumed from the very beginning that X_ℓ, X'_ℓ are nonempty, closed, and convex, and that at least one of these sets is bounded;
2. Set
$$e_{\ell\ell'} = \frac{u_{\ell\ell'} - v_{\ell\ell'}}{\|u_{\ell\ell'} - v_{\ell\ell'}\|_2}, \; c_{\ell\ell'} = \tfrac{1}{2} e^T_{\ell\ell'}[u_{\ell\ell'} + v_{\ell\ell'}], \; \phi_{\ell\ell'}(\omega) = e^T_{\ell\ell'}\omega - c_{\ell\ell'}.$$
Note that the construction makes sense, since by our Standing Assumption for the ℓ, ℓ' in question X_ℓ and $X_{\ell'}$ do not intersect. Further, $e_{\ell\ell'}$ and $c_{\ell\ell'}$ clearly depend solely on (ℓ, ℓ'), but not on how we select an optimal solution $(u_{\ell\ell'}, v_{\ell\ell'})$ to (2.36). Finally, we have
$$e_{\ell\ell'} = -e_{\ell'\ell}, \; c_{\ell\ell'} = -c_{\ell'\ell}, \; \phi_{\ell\ell'}(\cdot) \equiv -\phi_{\ell'\ell}(\cdot).$$

3. We consider separately the case of $K = 1$ and the case of $K > 1$. Specifically,

a) when $K = 1$, we select nonnegative reals $\delta_{\ell\ell'}$, $\delta_{\ell'\ell}$ such that

$$\delta_{\ell\ell'} + \delta_{\ell'\ell} = 2\mathrm{Opt}_{\ell\ell'} \tag{2.37}$$

and specify the single-observation simple test $\mathcal{T}_{\ell\ell'}$ deciding on the hypotheses $\mathcal{H}_\ell$, $\mathcal{H}_{\ell'}$ according to

$$\mathcal{T}_{\ell\ell'}(\omega) = \begin{cases} \{\ell\}, & \phi_{\ell\ell'}(\omega) \geq \frac{1}{2}[\delta_{\ell'\ell} - \delta_{\ell\ell'}], \\ \{\ell'\}, & \text{otherwise.} \end{cases}$$

Note that by Proposition 2.5, setting

$$P_\gamma(\delta) = \int\limits_\delta^\infty \gamma(s)ds, \tag{2.38}$$

we have

$$\begin{array}{rcl} \mathrm{Risk}_1(\mathcal{T}_{\ell\ell'}|\mathcal{H}_\ell, \mathcal{H}_{\ell'}) & \leq & P_\gamma(\delta_{\ell\ell'}), \\ \mathrm{Risk}_2(\mathcal{T}_{\ell\ell'}|\mathcal{H}_\ell, \mathcal{H}_{\ell'}) & \leq & P_\gamma(\delta_{\ell'\ell}), \\ \mathrm{Risk}_1(\mathcal{T}_{\ell'\ell}|\mathcal{H}_{\ell'}, \mathcal{H}_\ell) & \leq & P_\gamma(\delta_{\ell'\ell}), \\ \mathrm{Risk}_2(\mathcal{T}_{\ell'\ell}|\mathcal{H}_{\ell'}, \mathcal{H}_\ell) & \leq & P_\gamma(\delta_{\ell\ell'}). \end{array} \tag{2.39}$$

b) when $K > 1$, we specify the K-observation simple test $\mathcal{T}_{\ell\ell' K}$ deciding on $\mathcal{H}_\ell$, $\mathcal{H}_{\ell'}$ according to

$$\mathcal{T}_{\ell\ell'}(\omega^k = (\omega_1, ..., \omega_k)) = \begin{cases} \{\ell\}, & \mathrm{Card}\{k \leq K : \phi_{\ell\ell'} \geq 0\} \geq K/2, \\ \{\ell'\}, & \text{otherwise} \end{cases}.$$

Note that by Proposition 2.6 we have

$$\begin{array}{rcl} \mathrm{Risk}(\mathcal{T}_{\ell\ell' K}|\mathcal{H}_\ell, \mathcal{H}'_\ell) & \leq & \epsilon_{\ell\ell' K} := \sum\limits_{K/2 \leq k \leq K} \binom{K}{k} \epsilon^k_{\star\ell\ell'}(1 - \epsilon_{\star\ell\ell'})^{K-k}, \\ \epsilon_{\star\ell\ell'} & = & P_\gamma(\mathrm{Opt}_{\ell\ell'}) = \epsilon_{\star\ell'\ell}. \end{array}$$

Assembling the building blocks, case of $K = 1$. In the case of $K = 1$, we specify the simple pairwise tests $\mathcal{T}_{\{\ell,\ell'\}}$, $(\ell, \ell') \notin \mathcal{C}$, participating in the construction of the multi-hypothesis test presented in Section 2.2.4.3, as follows. Given unordered pair $\{\ell, \ell'\}$ with $(\ell, \ell') \notin \mathcal{C}$ (which is exactly the same as $(\ell', \ell) \notin \mathcal{C}$), we arrange ℓ, ℓ' in an ascending order, thus arriving at ordered pair $(\bar{\ell}, \bar{\ell}')$, and set

$$\mathcal{T}_{\{\ell,\ell'\}}(\cdot) = \mathcal{T}_{\bar{\ell}\bar{\ell}'}(\cdot),$$

with the right-hand side tests defined as explained above. We then assemble, as explained in Section 2.2.4.3, the tests $\mathcal{T}_{\{\ell,\ell'\}}$ into a single-observation test $\mathcal{T}_1$ deciding on hypotheses $\mathcal{H}_1, ..., \mathcal{H}_L$. From (2.34) and (2.39) we conclude that for the tests $\mathcal{T}_{\{\ell,\ell'\}}$ just defined and the quantities $\epsilon_{\ell\ell'}$ associated with the tests $\mathcal{T}_{\{\ell,\ell'\}}$, via (2.34), it holds

$$(\ell, \ell') \notin \mathcal{C} \Rightarrow \epsilon_{\ell\ell'} \leq P_\gamma(\delta_{\ell\ell'}). \tag{2.40}$$

Invoking Proposition 2.11, we get

Proposition 2.12. *In the situation described at the beginning of Section 2.2.4.4 and under Standing Assumption, the $\mathcal{C}$-risks of the test $\mathcal{T}_1$ just defined—whatever the choice of nonnegative $\delta_{\ell\ell'}$, $(\ell,\ell')\not\in\mathcal{C}$, satisfying* (2.37)*—can be upper-bounded as*

$$\mathrm{Risk}_\ell^{\mathcal{C}}(\mathcal{T}_1|\mathcal{H}_1,...,\mathcal{H}_L)\leq\sum_{\ell':(\ell,\ell')\not\in\mathcal{C}}P_\gamma(\delta_{\ell\ell'}),\tag{2.41}$$

with $P_\gamma(\cdot)$ given by (2.38).

Case of $K=1$ (continued): Optimizing the construction. We can try to optimize the risk bounds (2.41) over the parameters $\delta_{\ell\ell'}$ of the construction. The first question to be addressed here is what to minimize—we have defined several risks. A natural model here may be as follows. Let us fix a nonnegative $M\times L$ *weight matrix* W and an M-dimensional positive *profile vector* w, and solve the optimization problem

$$\min_{t,\{\delta_{\ell\ell'}:(\ell,\ell')\not\in\mathcal{C}}\left\{t:\begin{array}{l}W\cdot\left[\sum_{\ell':(\ell,\ell')\not\in\mathcal{C}}P_\gamma(\delta_{\ell\ell'})\right]_{\ell=1}^L\leq tw\\ \delta_{\ell\ell'}\geq0,\delta_{\ell\ell'}+\delta_{\ell'\ell}=2\mathrm{Opt}_{\ell\ell'},\ (\ell,\ell')\not\in\mathcal{C}\end{array}\right\}.\tag{2.42}$$

For instance, when $M=1$ and $w=1$, we minimize a weighted sum of (upper bounds on) partial $\mathcal{C}$-risks of our test; when W is a diagonal matrix with positive diagonal entries and w is the all-ones vector, we minimize the largest of scaled partial risks. Note that when $P_\gamma(\cdot)$ is convex on $\mathbf{R}_+$, or, which is the same, $\gamma(\cdot)$ is nonincreasing in $\mathbf{R}_+$, (2.42) is a convex, and thus efficiently solvable, problem.

Assembling building blocks, case of $K>1$. We again pass from our building blocks—K-observation simple pairwise tests $\mathcal{T}_{\ell\ell'K}$, $(\ell,\ell')\not\in\mathcal{C}$, we have already specified—to tests $\mathcal{T}_{\{\ell,\ell'\}}=\mathcal{T}_{\bar\ell\bar\ell'K}$, with $\bar\ell=\min[\ell,\ell']$ and $\bar\ell'=\max[\ell,\ell']$, and then apply to the resulting tests the construction from Section 2.2.4.3, arriving at the K-observation multi-hypothesis test $\mathcal{T}_K$. By Proposition 2.6, the quantities $\epsilon_{\ell\ell'}$ associated with the tests $\mathcal{T}_{\{\ell,\ell'\}}$ via (2.34) satisfy the relation

$$(\ell,\ell')\not\in\mathcal{C}\Rightarrow\epsilon_{\ell\ell'}\leq\sum_{K/2\leq k\leq K}\binom{K}{k}[P_\gamma(\mathrm{Opt}_{\ell\ell'})]^k[1-P_\gamma(\mathrm{Opt}_{\ell\ell'})]^{K-k},\tag{2.43}$$

which combines with Proposition 2.11 to imply

Proposition 2.13. *Consider the situation described at the beginning of Section 2.2.4.4, and let $K>1$. Under Standing Assumption, the $\mathcal{C}$-risks of the test $\mathcal{T}_K$ just defined can be upper-bounded as*

$$\mathrm{Risk}_\ell^{\mathcal{C}}(\mathcal{T}_K|\mathcal{H}_1,...,\mathcal{H}_L)\leq\sum_{\ell':(\ell,\ell')\not\in\mathcal{C}}\ \sum_{K/2\leq k\leq K}\binom{K}{k}[P_\gamma(\mathrm{Opt}_{\ell\ell'})]^k[1-P_\gamma(\mathrm{Opt}_{\ell\ell'})]^{K-k},\tag{2.44}$$

with $P_\gamma(\cdot)$ given by (2.38) *and $\mathrm{Opt}_{\ell\ell'}$ given by* (2.36).

Note that by Standing Assumption the quantities $P_\gamma(\mathrm{Opt}_{\ell\ell'})$ for $(\ell,\ell')\not\in\mathcal{C}$ are $<1/2$, so that the risks $\mathrm{Risk}_\ell^{\mathcal{C}}(\mathcal{T}_K|H_1,...,H_L)$ go to 0 exponentially fast as $K\to\infty$.

2.3 DETECTORS AND DETECTOR-BASED TESTS

2.3.1 Detectors and their risks

Let Ω be an observation space, and $\mathcal{P}_\chi$, $\chi = 1, 2$, be two families of probability distributions on Ω. By definition, a *detector* associated with Ω is a real-valued function $\phi(\omega)$ of Ω. We associate with a detector ϕ and families $\mathcal{P}_\chi$, $\chi = 1, 2$, *risks* defined as follows:

$$\begin{array}{llcl}(a) & \mathrm{Risk}_-[\phi|\mathcal{P}_1] & = & \sup_{P\in\mathcal{P}_1}\int_\Omega \exp\{-\phi(\omega)\}P(d\omega)\\ (b) & \mathrm{Risk}_+[\phi|\mathcal{P}_2] & = & \sup_{P\in\mathcal{P}_2}\int_\Omega \exp\{\phi(\omega)\}P(d\omega)\\ (c) & \mathrm{Risk}[\phi|\mathcal{P}_1,\mathcal{P}_2] & = & \max[\mathrm{Risk}_-[\phi|\mathcal{P}_1], \mathrm{Risk}_+[\phi|\mathcal{P}_2]]\end{array} \tag{2.45}$$

Given a detector ϕ, we can associate with it a simple test $\mathcal{T}_\phi$ deciding via observation $\omega \sim P$ on the hypotheses

$$H_1 : P \in \mathcal{P}_1,\ H_2 : P \in \mathcal{P}_2. \tag{2.46}$$

Namely, given observation $\omega \in \Omega$, the test $\mathcal{T}_\phi$ accepts H_1 (and rejects H_2) whenever $\phi(\omega) \geq 0$, and accepts H_2 and rejects H_1 otherwise.

Let us make the following immediate observation:

Proposition 2.14. *Let Ω be an observation space, $\mathcal{P}_\chi$, $\chi = 1, 2$, be two families of probability distributions on Ω, and ϕ be a detector. The risks of the test $\mathcal{T}_\phi$ associated with this detector satisfy*

$$\begin{array}{lcl}\mathrm{Risk}_1(\mathcal{T}_\phi|H_1, H_2) & \leq & \mathrm{Risk}_-[\phi|\mathcal{P}_1],\\ \mathrm{Risk}_2(\mathcal{T}_\phi|H_1, H_2) & \leq & \mathrm{Risk}_+[\phi|\mathcal{P}_2].\end{array} \tag{2.47}$$

Proof. Let $\omega \sim P \in \mathcal{P}_1$. Then the P-probability of the event $\{\omega : \phi(\omega) < 0\}$ does not exceed $\mathrm{Risk}_-[\phi|\mathcal{P}_1]$, since on the set $\{\omega : \phi(\omega) < 0\}$ the integrand in (2.45.a) is > 1, and this integrand is nonnegative everywhere, so that the integral in (2.45.a) is $\geq P\{\omega : \phi(\omega) < 0\}$. Recalling what $\mathcal{T}_\phi$ is, we see that the P-probability to reject H_1 is at most $\mathrm{Risk}_-[\phi|\mathcal{P}_1]$, implying the first relation in (2.47). By a similar argument, with (2.45.b) in the role of (2.45.a), when $\omega \sim P \in \mathcal{P}_2$, the P-probability of the event $\{\omega : \phi(\omega) \geq 0\}$ is upper-bounded by $\mathrm{Risk}_+[\phi|\mathcal{P}_2]$, implying the second relation in (2.47). □

2.3.2 Detector-based tests

Our current goal is to establish some basic properties of detector-based tests.

2.3.2.1 Structural properties of risks

Observe that the fact that ϵ_1 and ϵ_2 are upper bounds on the risks of a detector are expressed by a system of *convex* constraints

$$\begin{array}{lcl}\sup_{P\in\mathcal{P}_1}\int_\Omega \exp\{-\phi(\omega)\}P(d\omega) & \leq & \epsilon_1\\ \sup_{P\in\mathcal{P}_2}\int_\Omega \exp\{\phi(\omega)\}P(d\omega) & \leq & \epsilon_2\end{array}$$

on ϵ_1, ϵ_2 and $\phi(\cdot)$. This observation is interesting, but not very useful, since the convex constraints in question usually are infinite-dimensional when $\phi(\cdot)$ is so, and are semi-infinite (suprema—over parameters ranging in an infinite set—of parametric families of convex constraints) provided $\mathcal{P}_1$ or $\mathcal{P}_2$ are of infinite cardinalities; constraints of this type can be intractable computationally.

Another important observation is that the distributions P enter the constraints linearly; as a result, *when passing from families of probability distributions $\mathcal{P}_1$, $\mathcal{P}_2$ to their convex hulls, the risks of a detector remain intact.*

2.3.2.2 Renormalization

Let Ω, $\mathcal{P}_1$, and $\mathcal{P}_2$ be the same as in Section 2.3.1, and let ϕ be a detector. When shifting this detector by a real a—passing from ϕ to the detector

$$\phi_a(\omega) = \phi(\omega) - a$$

— the risks clearly update according to:

$$\begin{array}{rcl} \mathrm{Risk}_-[\phi_a|\mathcal{P}_1] & = & \mathrm{e}^a \mathrm{Risk}_-[\phi|\mathcal{P}_1], \\ \mathrm{Risk}_+[\phi_a|\mathcal{P}_2] & = & \mathrm{e}^{-a} \mathrm{Risk}_+[\phi|\mathcal{P}_2]. \end{array} \tag{2.48}$$

We see that

When speaking about risks of a detector, what matters is the product

$$\mathrm{Risk}_-[\phi|\mathcal{P}_1]\,\mathrm{Risk}_+[\phi|\mathcal{P}_2]$$

of the risks, not these risks individually: by shifting the detector, we can redistribute this product between the factors in any way we want. In particular, we can always shift a detector to make it balanced, i.e., satisfying

$$\mathrm{Risk}_-[\phi|\mathcal{P}_1] = \mathrm{Risk}_+[\phi|\mathcal{P}_2] = \mathrm{Risk}[\phi|\mathcal{P}_1, \mathcal{P}_2].$$

When deciding on the hypotheses

$$H_1 : P \in \mathcal{P}_1, \;\; H_2 : P \in \mathcal{P}_2$$

on the distribution P of observation, the risk of the test $\mathcal{T}_\phi$ associated with a balanced detector ϕ is bounded by the risk $\mathrm{Risk}[\phi|\mathcal{P}_1, \mathcal{P}_2]$ *of the detector:*

$$\begin{array}{rcl} \mathrm{Risk}(\mathcal{T}_\phi|H_1, H_2) & := & \max\left[\mathrm{Risk}_1(\mathcal{T}_\phi|H_1, H_2), \mathrm{Risk}_2(\mathcal{T}_\phi|H_1, H_2)\right] \\ & \leq & \mathrm{Risk}[\phi|\mathcal{P}_1, \mathcal{P}_2]. \end{array}$$

2.3.2.3 Detector-based testing from repeated observations

We are about to show that detector-based tests are perfectly well suited for passing from inferences based on a *single* observation to those based on *repeated* observations.

Given K observation spaces Ω_k, $1 \leq k \leq K$, each equipped with a pair $\mathcal{P}_{k,1}$, $\mathcal{P}_{k,2}$ of families of probability distributions, we can build a new observation space

$$\Omega^K = \Omega_1 \times \ldots \times \Omega_K = \{\omega^K = (\omega_1, \ldots, \omega_K) : \omega_k \in \Omega_k, k \leq K\}$$

and equip it with two families $\mathcal{P}_\chi^K$, $\chi = 1, 2$, of probability distributions; distributions from $\mathcal{P}_\chi^K$ are exactly the product-type distributions $P = P_1 \times ... \times P_K$ with all factors P_k taken from $\mathcal{P}_{k,\chi}$. Observations $\omega^K = (\omega_1, ..., \omega_K)$ from Ω^K drawn from a distribution $P = P_1 \times .... \times P_K \in \mathcal{P}_\chi^K$ are nothing but collections of observations ω_k, $k = 1, ..., K$, drawn, independently of each other, from distributions P_k. Now, given detectors $\phi_k(\cdot)$ on observation spaces Ω_k and setting

$$\phi^{(K)}(\omega^K) = \sum\nolimits_{k=1}^K \phi_k(\omega_k) : \Omega^K \to \mathbf{R},$$

we clearly have

$$\begin{array}{rcl} \text{Risk}_-[\phi^{(K)}|\mathcal{P}_1^K] & = & \prod\limits_{k=1}^K \text{Risk}_-[\phi_k|\mathcal{P}_{k,1}], \\ \text{Risk}_+[\phi^{(K)}|\mathcal{P}_2^K] & = & \prod\limits_{k=1}^K \text{Risk}_+[\phi_k|\mathcal{P}_{k,2}]. \end{array} \tag{2.49}$$

Let us look at some useful consequences of (2.49).

Stationary K-repeated observations. Consider the case of Section 2.1.3.1: we are given an observation space Ω and a positive integer K, and what we observe is a sample $\omega^K = (\omega_1, ..., \omega_K)$ with $\omega_1, ..., \omega_K$ drawn, independently of each other, from some distribution P on Ω. Let now $\mathcal{P}_1$, $\mathcal{P}_2$, be two families of probability distributions on Ω; we can associate with these families two hypotheses, $H_1^{\odot,K}$, $H_2^{\odot,K}$, on the distribution of K-repeated observation $\omega^K = (\omega_1, ..., \omega_K)$, with $H_\chi^{\odot,K}$ stating that $\omega_1, ..., \omega_K$ are drawn, independently of each other, from a distribution $P \in \mathcal{P}_\chi$. Given a detector ϕ on Ω, we can associate with it the detector

$$\phi^{(K)}(\omega^K) = \sum_{k=1}^K \phi(\omega_k)$$

on

$$\Omega^K : \underbrace{\Omega \times ... \times \Omega}_{K}.$$

Combining (2.49) and Proposition 2.14, we arrive at the following nice result:

Proposition 2.15. *Consider the simple test $\mathcal{T}_{\phi^{(K)}}$ deciding, given K-repeated observation $\omega^K = (\omega_1, ..., \omega_K)$, on the hypotheses*

$$\begin{array}{l} H_1^{\odot,K} : \omega_k,\ k \leq K, \textit{ are drawn from } P \in \mathcal{P}_1 \textit{ independently of each other} \\ H_2^{\odot,K} : \omega_k,\ k \leq K, \textit{ are drawn from } P \in \mathcal{P}_2 \textit{ independently of each other} \end{array}$$

according to the rule

$$\phi^{(K)}(\omega^K) := \sum_{k=1}^K \phi(\omega_k) \left\{ \begin{array}{lcl} \geq 0 & \Rightarrow & \textit{accept } H_1^{\odot,K}, \\ < 0 & \Rightarrow & \textit{accept } H_2^{\odot,K}. \end{array} \right.$$

The risks of $\mathcal{T}_{\phi^{(K)}}$ admit the upper bounds

$$\begin{array}{rcl} \text{Risk}_1(\mathcal{T}_{\phi^{(K)}}|H_1^{\odot,K}, H_2^{\odot,K}) & \leq & (\text{Risk}_-[\phi|\mathcal{P}_1])^K, \\ \text{Risk}_2(\mathcal{T}_{\phi^{(K)}}|H_1^{\odot,K}, H_2^{\odot,K}) & \leq & (\text{Risk}_+[\phi|\mathcal{P}_2])^K. \end{array}$$

Semi- and Quasi-Stationary K-repeated observations. Recall that Semi-Stationary and Quasi-Stationary K-repeated observations associated with a family $\mathcal{P}$ of distributions on observation space Ω were defined in Sections 2.1.3.2 and 2.1.3.3, respectively. It turns out that Proposition 2.15 extends to quasi-stationary K-repeated observations:

Proposition 2.16. *Let Ω be an observation space, $\mathcal{P}_\chi$, $\chi = 1, 2$, be families of probability distributions on Ω, $\phi : \Omega \to \mathbf{R}$ be a detector, and K be a positive integer.*

Families $\mathcal{P}_\chi$, $\chi = 1, 2$, give rise to two hypotheses on the distribution P^K of quasi-stationary K-repeated observation ω^K,

$$H_\chi^{\otimes,K} : P^K \in \mathcal{P}_\chi^{\otimes,K} = \bigotimes_{k=1}^{K} \mathcal{P}_\chi,\ \chi = 1, 2$$

(see Section 2.1.3.3), and ϕ gives rise to the detector

$$\phi^{(K)}(\omega^K) := \sum_{k=1}^{K} \phi(\omega_k).$$

The risks of the detector $\phi^{(K)}$ on the families $\mathcal{P}_\chi^{\otimes,K}$, $\chi = 1, 2$, can be upper-bounded as follows:

$$\begin{array}{rcl} \mathrm{Risk}_-[\phi^{(K)}|\mathcal{P}_1^{\otimes,K}] & \leq & (\mathrm{Risk}_-[\phi|\mathcal{P}_1])^K, \\ \mathrm{Risk}_+[\phi^{(K)}|\mathcal{P}_2^{\otimes,K}] & \leq & (\mathrm{Risk}_+[\phi|\mathcal{P}_2])^K. \end{array} \tag{2.50}$$

Furthermore, the detector $\phi^{(K)}$ induces simple test $\mathcal{T}_{\phi^{(K)}}$ deciding on $H_\chi^{\otimes,K}$, $\chi = 1, 2$, as follows: given ω^K, the test accepts $H_1^{\otimes,K}$ when $\phi^{(K)}(\omega^K) \geq 0$, and accepts $H_2^{\otimes,K}$ otherwise. The risks of this test can be upper-bounded as

$$\begin{array}{rcl} \mathrm{Risk}_1(\mathcal{T}_{\phi^{(K)}}|H_1^{\otimes,K}, H_2^{\otimes,K}) & \leq & (\mathrm{Risk}_-[\phi|\mathcal{P}_1])^K, \\ \mathrm{Risk}_2(\mathcal{T}_{\phi^{(K)}}|H_1^{\otimes,K}, H_2^{\otimes,K}) & \leq & (\mathrm{Risk}_+[\phi|\mathcal{P}_2])^K. \end{array}$$

Finally, the above results remain intact when passing from quasi-stationary to semi-stationary K-repeated observations (that is, when replacing $\mathcal{P}_\chi^{\otimes,K}$ with $\mathcal{P}_\chi^{\oplus,K} = \bigoplus_{k=1}^{K} \mathcal{P}_\chi$ and $H_\chi^{\otimes,K}$ with the hypotheses $H_\chi^{\oplus,K}$ stating that the distribution of ω^K belongs to $\mathcal{P}_\chi^{\oplus,K}$, $\chi = 1, 2$).

Proof. All we need to verify is (2.50)—in view of Proposition 2.14, all other claims in Proposition 2.16 are immediate consequences of (2.50) and the inclusions $\mathcal{P}_\chi^{\oplus,K} \subset \mathcal{P}_\chi^{\otimes,K}$, $\chi = 1, 2$. Verification of (2.50) is as follows. Let $P^K \in \mathcal{P}_1^{\otimes,K}$, so that by definition of $\mathcal{P}_1^{\otimes,K}$ P^K is the distribution of random sequence $\omega^K = (\omega_1, ..., \omega_K)$ such that there exists a random sequence of driving factors $\zeta_1, ..., \zeta_K$ such that ω_k is a deterministic function of $\zeta^k = (\zeta_1, ..., \zeta_k)$,

$$\omega_k = \theta_k(\zeta_1, ..., \zeta_k),$$

and the conditional distribution $P_{\omega_k|\zeta^{k-1}}$ given $\zeta_1, ..., \zeta_{k-1}$ belongs to $\mathcal{P}_1$. Let P_{ζ^k} be the distribution of the first k driving factors, and $P_{\zeta_k|\zeta^{k-1}}$ be the conditional distribution of ζ_k given $\zeta_1, ...\zeta_{k-1}$. Let us put

$$\psi^{(k)}(\zeta_1, ..., \zeta_k) = \sum_{t=1}^{k} \phi(\theta_t(\zeta_1, ..., \zeta_t)),$$

so that

$$\int_{\Omega^K} \exp\{-\phi^{(K)}(\omega^k)\} P^K(d\omega^K) = \int \exp\{-\psi^{(K)}(\zeta^K)\} P_{\zeta^K}(d\zeta^K). \tag{2.51}$$

On the other hand, denoting $C_0 = 1$, we have

$$\begin{array}{rcl} C_k & := & \int e^{-\psi^{(k)}(\zeta^k)} P_{\zeta^k}(d\zeta^k) = \int \exp\{-\psi^{(k-1)}(\zeta^{k-1}) - \phi(\theta_k(\zeta^k))\} P_{\zeta^k}(d\zeta^k) \\ & = & \int e^{-\psi^{(k-1)}(\zeta^{k-1})} \underbrace{\left[\int e^{-\phi(\theta_k(\zeta^k))} P_{\zeta_k|\zeta^{k-1}}(d\zeta_k) \right]}_{= \int_\Omega e^{-\phi(\omega_k)} P_{\omega_k|\zeta^{k-1}}(d\omega_k)} P_{\zeta^{k-1}}(d\zeta^{k-1}) \\ & \underbrace{\leq}_{(*)} & \text{Risk}_-[\phi|\mathcal{P}_1] \int e^{-\psi^{(k-1)}(\zeta^{k-1})} P_{\zeta^{k-1}}(d\zeta^{k-1}) = \text{Risk}_-[\phi|\mathcal{P}_1] C_{k-1} \end{array}$$

where $(*)$ is due to the fact that the distribution $P_{\omega_k|\zeta^{k-1}}$ belongs to $\mathcal{P}_1$. From the resulting recurrence we get

$$C_K \leq (\text{Risk}_-[\phi|\mathcal{P}_1])^K,$$

which combines with (2.51) to imply that

$$\int_{\Omega^K} e^{-\phi^{(K)}(\omega^k)} P^K(d\omega^K) \leq (\text{Risk}_-[\phi|\mathcal{P}_1])^K.$$

The latter inequality holds true for every distribution $P^K \in \mathcal{P}_\chi^{\otimes,K}$, and the first inequality in (2.50) follows. The second inequality in (2.50) is given by a completely similar reasoning, with $\mathcal{P}_2$ in the role of $\mathcal{P}_1$, and $-\phi$, $-\phi^{(K)}$ in the roles of ϕ, $\phi^{(K)}$, respectively. □

The fact that observations ω_k under hypotheses $H_\ell^{\otimes,K}$, $\ell = 1, 2$, are related to "constant in time" families $\mathcal{P}_\ell$ has no importance here, and in fact the proof of Proposition 2.16 after absolutely evident modifications of wording allows us to justify the following "non-stationary" version of the proposition:

Proposition 2.17. *For $k = 1, ..., K$, let Ω_k be observation spaces, $\mathcal{P}_{\chi,k}$, $\chi = 1, 2$, be families of probability distributions on Ω_k, and $\phi_k : \Omega_k \to \mathbf{R}$ be detectors.*

Families $\mathcal{P}_{\chi,k}$, $\chi = 1, 2$, give rise to quasi-direct products (see Section 2.1.3.3) $\mathcal{P}_\chi^{\otimes,K} = \bigotimes_{k=1}^{K} \mathcal{P}_{\chi,k}$ of the families $\mathcal{P}_{\chi,k}$ over $1 \leq k \leq K$, and thus to two hypotheses on the distribution P^K of observation $\omega^K = (\omega_1, ..., \omega_K) \in \Omega^K = \Omega_1 \times ... \times \Omega_K$:

$$H_\chi^{\otimes,K} : P^K \in \mathcal{P}_\chi^{\otimes,K}, \ \chi = 1, 2.$$

Detectors ϕ_k, $1 \le k \le K$, induce the detector

$$\phi^K(\omega^K) := \sum_{k=1}^K \phi_k(\omega_k).$$

The risks of the detector ϕ^K on the families $\mathcal{P}_\chi^{\otimes,K}$, $\chi = 1, 2$, can be upper-bounded as follows:

$$\begin{array}{rcl} \text{Risk}_-[\phi^K|\mathcal{P}_1^{\otimes,K}] & \le & \prod_{k=1}^K \text{Risk}_-[\phi_k|\mathcal{P}_{1,k}], \\ \text{Risk}_+[\phi^K|\mathcal{P}_2^{\otimes,K}] & \le & \prod_{k=1}^K \text{Risk}_+[\phi_k|\mathcal{P}_{2,k}]. \end{array}$$

Further, the detector ϕ^K induces simple test $\mathcal{T}_{\phi^{(K)}}$ deciding on $H_\chi^{\otimes,K}$, $\chi = 1, 2$, as follows: given ω^K, the test accepts $H_1^{\otimes,K}$ when $\phi^K(\omega^K) \ge 0$, and accepts $H_2^{\otimes,K}$ otherwise. The risks of this test can be upper-bounded as

$$\begin{array}{rcl} \text{Risk}_1(\mathcal{T}_{\phi^K}|H_1^{\otimes,K}, H_2^{\otimes,K}) & \le & \prod_{k=1}^K \text{Risk}_-[\phi_k|\mathcal{P}_{1,k}], \\ \text{Risk}_2(\mathcal{T}_{\phi^{(K)}}|H_1^{\otimes,K}, H_2^{\otimes,K}) & \le & \prod_{k=1}^K \text{Risk}_+[\phi_k|\mathcal{P}_{2,k}]. \end{array}$$

Finally, the above results remain intact when passing from quasi-direct products to direct products of the families of distributions in question (that is, when replacing $\mathcal{P}_\chi^{\otimes,K}$ with $\mathcal{P}_\chi^{\oplus,K} = \bigoplus_{k=1}^K \mathcal{P}_{\chi,k}$ and $H_\chi^{\otimes,K}$ with the hypotheses $H_\chi^{\oplus,K}$ stating that the distribution of ω^K belongs to $\mathcal{P}_\chi^{\oplus,K}$, $\chi = 1, 2$).

2.3.2.4 Limits of performance of detector-based tests

We are about to demonstrate that as far as limits of performance of pairwise simple detector-based tests are concerned, these tests are nearly as good as simple tests can be.

Proposition 2.18. *Let Ω be an observation space, and $\mathcal{P}_\chi$, $\chi = 1, 2$, be families of probability distributions on Ω. Assume that for some $\epsilon \in (0, 1/2)$ "in nature" there exists a simple test (deterministic or randomized) deciding on the hypotheses*

$$H_1 : P \in \mathcal{P}_1,\ H_2 : P \in \mathcal{P}_2$$

on the distribution P of observation ω with risks $\le \epsilon$:

$$\text{Risk}_1(\mathcal{T}|H_1, H_2) \le \epsilon \ \&\ \text{Risk}_2(\mathcal{T}|H_1, H_2) \le \epsilon.$$

Then there exists a detector-based test $\mathcal{T}_\phi$ deciding on the same pair of hypotheses with the risk "comparable" with ϵ:

$$\text{Risk}_1(\mathcal{T}_\phi|H_1, H_2) \le \epsilon^+ \ \&\ \text{Risk}_2(\mathcal{T}_\phi|H_1, H_2) \le \epsilon^+,\ \ \epsilon^+ = 2\sqrt{\epsilon(1-\epsilon)}. \quad (2.52)$$

Proof. Let us prove the claim in the case when the test $\mathcal{T}$ is deterministic; the case when this test is randomized is the subject of Exercise 2.11.

For $\chi = 1, 2$, let Ω_χ be the set of $\omega \in \Omega$ such that $\mathcal{T}$ if "fed" with observation ω accepts H_χ. Since $\mathcal{T}$ is simple, Ω_1, Ω_2 split Ω into two nonoverlapping parts, and since the risks of $\mathcal{T}$ are $\leq \epsilon$, we have

$$\begin{array}{l}\epsilon_2(P) := P\{\Omega_2\} \leq \epsilon\, \forall P \in \mathcal{P}_1,\\ \epsilon_1(P) := P\{\Omega_1\} \leq \epsilon\, \forall P \in \mathcal{P}_2.\end{array}$$

Let $\delta = \sqrt{(1-\epsilon)/\epsilon}$, so that $\delta \geq 1$ due to $0 < \epsilon \leq 1/2$, and let

$$\psi(\omega) = \left\{\begin{array}{ll}\delta, & \omega \in \Omega_1\\ 1/\delta, & \omega \in \Omega_2\end{array}\right., \ \phi(\omega) = \ln(\psi(\omega)).$$

When $P \in \mathcal{P}_1$ we have

$$\int_\Omega \exp\{-\phi(\omega)\}P(d\omega) = \frac{1}{\delta}P\{\Omega_1\} + \delta P\{\Omega_2\} = \frac{1}{\delta} + \underbrace{\left[\delta - \frac{1}{\delta}\right]}_{\geq 0}\epsilon_2(P) \leq \frac{1}{\delta} + \left[\delta - \frac{1}{\delta}\right]\epsilon = \epsilon^+,$$

whence $\mathrm{Risk}_-[\phi|\mathcal{P}_1] \leq \epsilon^+$. Similarly, when $P \in \mathcal{P}_2$ we have

$$\int_\Omega \exp\{\phi(\omega)\}P(d\omega) = \delta P\{\Omega_1\} + \frac{1}{\delta}P\{\Omega_2\} = \underbrace{\left[\delta - \frac{1}{\delta}\right]}_{\geq 0}\epsilon_1(P) + \frac{1}{\delta} \leq \left[\delta - \frac{1}{\delta}\right]\epsilon + \frac{1}{\delta} = \epsilon^+,$$

whence $\mathrm{Risk}_+[\phi|\mathcal{P}_2] \leq \epsilon^+$. □

Discussion. Proposition 2.18 states that we can restrict ourselves to detector-based tests at the cost of passing from risk ϵ exhibited by "the best test existing in nature" to "comparable" risk $\epsilon^+ = 2\sqrt{\epsilon(1-\epsilon)}$. What we buy when sticking to detector-based tests are the nice properties listed in Sections 2.3.2.1–2.3.2.3 and the possibility to compute *under favorable circumstances*—see below—the best detector-based tests in terms of their risk. Optimizing risk of a detector-based test turns out to be an essentially more realistic task than optimizing risk of a general-type test. This being said, one can argue that treating ϵ and ϵ^+ as "comparable" is too optimistic. For example, risk level $\epsilon = 0.01$ seems to be much more attractive than $[0.01]^+ \approx 0.2$. While passing from a test $\mathcal{T}$ with risk 0.01 to a detector-based test $\mathcal{T}_\phi$ with risk 0.2 could indeed be a "heavy toll"; there is some comfort in the fact that passing from a single observation to three of them (i.e., to a 3-repeated version of the original observation scheme, stationary or non-stationary alike), we can straightforwardly convert $\mathcal{T}_\phi$ into a test with risk $(0.2)^3 = 0.008 < 0.01$, and passing to six observations, to risk less than 0.0001. On the other hand, seemingly the only way to convert a general-type single-observation test $\mathcal{T}$ with risk 0.01 into a multi-observation test with essentially smaller risk is to pass to a Majority version of $\mathcal{T}$; see Section 2.2.3.2.[4] Computation shows that with $\epsilon_\star = 0.01$, to make the risk of the majority test ≤ 0.0001 takes five observations, which is only marginally better than the six observations needed in the detector-based construction.

[4]In Section 2.2.3.2, we dealt with "signal plus noise" observations and with a specific test $\mathcal{T}$ given by Euclidean separation. However, a straightforward inspection of the construction and the proof of Proposition 2.6 make it clear that the construction is applicable to any simple test $\mathcal{T}$, and that the risk of the resulting multi-observation test obeys the upper bound in (2.23), with the risk of $\mathcal{T}$ in the role of $\epsilon_\star$.

2.4 SIMPLE OBSERVATION SCHEMES

2.4.1 Simple observation schemes—Motivation

A natural conclusion one can extract from the previous section is that it makes sense, to say the least, to learn how to build detector-based tests with minimal risk. Thus, we arrive at the following design problem:

> Given an observation space Ω and two families, $\mathcal{P}_1$ and $\mathcal{P}_2$, of probability distributions on Ω, solve the optimization problem
>
> $$\mathrm{Opt} = \min_{\phi:\Omega\to\mathbf{R}} \max\Big[\underbrace{\sup_{P\in\mathcal{P}_1}\int_\Omega \mathrm{e}^{-\phi(\omega)}P(d\omega)}_{F[\phi]}, \underbrace{\sup_{P\in\mathcal{P}_2}\int_\Omega \mathrm{e}^{\phi(\omega)}P(d\omega)}_{G[\phi]}\Big]. \tag{2.53}$$

While being convex, problem (2.53) is typically computationally intractable. First, it is infinite-dimensional—candidate solutions are multivariate functions; how do we represent them on a computer, not to mention, how do we optimize over them? Besides, the objective to be optimized is expressed in terms of suprema of infinitely many (provided $\mathcal{P}_1$ and/or $\mathcal{P}_2$ are infinite) expectations, and computing just a single expectation can be a difficult task .... We are about to consider "favorable" cases—*simple observation schemes*—where (2.53) is efficiently solvable.

To arrive at the notion of a simple observation scheme, consider the case when all distributions from $\mathcal{P}_1$, $\mathcal{P}_2$ admit densities taken w.r.t. some reference measure Π on Ω, and these densities are parameterized by a "parameter" μ running through some parameter space $\mathcal{M}$. In other words, $\mathcal{P}_1$ is comprised of all distributions with densities $p_\mu(\cdot)$ and μ belonging to some subset M_1 of $\mathcal{M}$, while $\mathcal{P}_2$ is comprised of distributions with densities $p_\mu(\cdot)$ and μ belonging to another subset, M_2, of $\mathcal{M}$. To save words, we shall identify distributions with their densities taken w.r.t. Π, so that

$$\mathcal{P}_\chi = \{p_\mu : \mu \in M_\chi\},\ \chi = 1, 2,$$

where $\{p_\mu(\cdot) : \mu \in \mathcal{M}\}$ is a given "parametric" family of probability densities. The quotation marks in "parametric" reflect the fact that at this point in time, the "parameter" μ can be infinite-dimensional (e.g, we can parameterize a density by itself), so that assuming "parametric" representation of the distributions from $\mathcal{P}_1$, $\mathcal{P}_2$ in fact does not restrict the generality.

Our first observation is that in our "parametric" setup, we can rewrite problem (2.53) equivalently as

$$\ln(\mathrm{Opt}) = \min_{\phi:\Omega\to\mathbf{R}} \sup_{\mu\in M_1,\nu\in M_2} \underbrace{\frac{1}{2}\left[\ln\left(\int_\Omega \mathrm{e}^{-\phi(\omega)}p_\mu(\omega)\Pi(d\omega)\right) + \ln\left(\int_\Omega \mathrm{e}^{\phi(\omega)}p_\nu(\omega)\Pi(d\omega)\right)\right]}_{\Phi(\phi;\mu,\nu)}. \tag{2.54}$$

> Indeed, when shifting ϕ by a constant, $\phi(\cdot) \mapsto \phi(\cdot) - a$, the positive quantities $F[\phi]$ and $G[\phi]$ participating in (2.53) are multiplied by e^a and e^{-a}, respectively, and their product remains intact. It follows that to minimize over ϕ the maximum of $F[\phi]$ and $G[\phi]$ (this is what (2.53) wants of us) is exactly the same as to minimize over ϕ the quantity $H[\phi] := \sqrt{F[\phi]G[\phi]}$. Indeed, a candidate solution ϕ to the problem $\min_\phi H[\phi]$ can be *balanced*—shifted by a constant to ensure $F[\phi] = G[\phi]$, and this balancing does not change

$H[\cdot]$. As a result, minimizing H over all ϕ is the same as minimizing H over balanced ϕ, and the latter problem clearly is equivalent to (2.53). It remains to note that (2.54) is nothing but the problem of minimizing $\ln(H[\phi])$.

Now, (2.54) is a min-max problem—a problem of the generic form

$$\min_{u\in U}\max_{v\in V}\Psi(u,v).$$

Problems of this type (at least, finite-dimensional ones) are computationally tractable when the domain of the minimization argument is convex and the cost function Ψ is convex in the minimization argument (this indeed is the case for (2.54)), and, moreover, the domain of the maximization argument is convex, and the cost function is concave in this argument (this not necessarily is the case for (2.54)). *Simple observation schemes* we are about to define are, essentially, the schemes where the requirements of finite dimensionality and convexity-concavity just outlined indeed are met.

2.4.2 Simple observation schemes—The definition

Consider the situation in which we are given

1. A Polish (complete separable metric) *observation space* Ω equipped with a σ-finite σ-additive Borel reference measure Π such that the support of Π is the entire Ω.
   Those not fully comfortable with some of the notions from the previous sentence can be assured that the only observation spaces we indeed shall deal with are pretty simple:
   - $\Omega = \mathbf{R}^d$ equipped with the Lebesgue measure Π, and
   - a finite or countable set Ω which is discrete (distances between distinct points are equal to 1) and is equipped with the counting measure Π.
2. A parametric family $\{p_\mu(\cdot) : \mu \in \mathcal{M}\}$ of probability densities, taken w.r.t. Π, such that
   - the space $\mathcal{M}$ of parameters is a convex set in some $\mathbf{R}^n$ which coincides with its relative interior,
   - the function $p_\mu(\omega) : \mathcal{M}\times\Omega \to \mathbf{R}$ is continuous in (μ,ω) and positive everywhere.
3. A finite-dimensional linear subspace $\mathcal{F}$ of the space of continuous functions on Ω such that
   - $\mathcal{F}$ contains constants,
   - all functions of the form $\ln(p_\mu(\omega)/p_\nu(\omega))$ with $\mu,\nu\in\mathcal{M}$ are contained in $\mathcal{F}$,
   - for every $\phi(\cdot)\in\mathcal{F}$, the function

   $$\ln\left(\int_\Omega e^{\phi(\omega)} p_\mu(\omega)\Pi(d\omega)\right)$$

   is real-valued and *concave* on $\mathcal{M}$.

In this situation we call the collection

$$(\Omega, \Pi; \{p_\mu : \mu \in \mathcal{M}\}; \mathcal{F})$$

a *simple observation scheme* (s.o.s. for short).

Nondegenerate simple o.s. We call a simple observation scheme *nondegenerate*, if the mapping $\mu \mapsto p_\mu$ is an embedding: whenever $\mu, \mu' \in \mathcal{M}$ and $\mu \neq \mu'$, we have $p_\mu \neq p_{\mu'}$.

2.4.3 Simple observation schemes—Examples

We are about to list the basic examples of s.o.s.'s.

2.4.3.1 Gaussian observation scheme

In Gaussian o.s.,

- the observation space (Ω, Π) is the space $\mathbf{R}^d$ with Lebesgue measure;
- the family $\{p_\mu(\cdot) : \mu \in \mathcal{M}\}$ is the family of Gaussian densities $\mathcal{N}(\mu, \Theta)$, with fixed positive definite covariance matrix Θ; distributions from the family are parameterized by their expectations μ. Thus,

$$\mathcal{M} = \mathbf{R}^d,\ p_\mu(\omega) = \frac{1}{(2\pi)^{d/2}\sqrt{\mathrm{Det}(\Theta)}} \exp\left\{-\tfrac{1}{2}(\omega - \mu)^T \Theta^{-1}(\omega - \mu)\right\};$$

- the family $\mathcal{F}$ is the family of all affine functions on $\mathbf{R}^d$.

It is immediately seen that Gaussian o.s. meets all requirements imposed on a simple o.s. For example,

$$\ln(p_\mu(\omega)/p_\nu(\omega)) = (\nu - \mu)^T \Theta^{-1}\omega + \tfrac{1}{2}\left[\nu^T \Theta^{-1}\nu - \mu^T \Theta^{-1}\mu\right]$$

is an affine function of ω and thus belongs to $\mathcal{F}$. Besides this, a function $\phi(\cdot) \in \mathcal{F}$ is affine: $\phi(\omega) = a^T\omega + b$, implying that

$$\begin{array}{rcl} f(\mu) & := & \ln\left(\int_{\mathbf{R}^d} e^{\phi(\omega)} p_\mu(\omega)d\omega\right) = \ln\left(\mathbf{E}_{\xi\sim\mathcal{N}(0,I_d)}\left\{\exp\{a^T(\Theta^{1/2}\xi + \mu) + b\}\right\}\right) \\ & = & a^T\mu + b + \text{const}, \\ \text{const} & = & \ln\left(\mathbf{E}_{\xi\sim\mathcal{N}(0,I_d)}\left\{\exp\{a^T\Theta^{1/2}\xi\}\right\}\right) = \tfrac{1}{2}a^T\Theta a \end{array}$$

is an affine (and thus a concave) function of μ.

As we remember from Chapter 1, Gaussian o.s. is responsible for the standard *signal processing* model where one is given a noisy observation

$$\omega = Ax + \xi \qquad [\xi \sim \mathcal{N}(0, \Theta)]$$

of the image Ax of unknown signal $x \in \mathbf{R}^n$ under linear transformation with known $d \times n$ *sensing matrix*, and the goal is to infer from this observation some knowledge about x. In this situation, a hypothesis that x belongs to some set X translates into the hypothesis that the observation ω is drawn from Gaussian distribution with known covariance matrix Θ and expectation known to belong to the set $M = \{\mu = Ax : x \in X\}$. Therefore, deciding upon various hypotheses on where x is located reduces to deciding on hypotheses on the distribution of observations in Gaussian o.s.

2.4.3.2 Poisson observation scheme

In Poisson o.s.,

- the observation space Ω is the set $\mathbf{Z}^d_+$ of d-dimensional vectors with nonnegative integer entries, and this set is equipped with the counting measure;
- the family $\{p_\mu(\cdot) : \mu \in \mathcal{M}\}$ is the family of product-type Poisson distributions with positive parameters, i.e.,

$$\mathcal{M} = \{\mu \in \mathbf{R}^d : \mu > 0\}, p_\mu(\omega) = \frac{\mu_1^{\omega_1}\mu_2^{\omega_2}...\mu_d^{\omega_d}}{\omega_1!\omega_2!...\omega_d!}\mathrm{e}^{-\mu_1-\mu_2-...-\mu_d},\ \omega \in \mathbf{Z}^d_+.$$

  In other words, random variable $\omega \sim p_\mu$, $\mu \in \mathcal{M}$, is a d-dimensional vector with independent random entries, with the i-th entry $\omega_i \sim$ Poisson(μ_i);
- the space $\mathcal{F}$ is comprised of affine functions on $\mathbf{Z}^+_d$.

It is immediately seen that Poisson o.s. is simple. For example,

$$\ln(p_\mu(\omega)/p_\nu(\omega)) = \sum_{i=1}^d \ln(\mu_i/\nu_i)\omega_i - \sum_{i=1}^d [\mu_i - \nu_i]$$

is an affine function of ω and thus belongs to $\mathcal{F}$. Besides this, a function $\phi \in \mathcal{F}$ is affine: $\phi(\omega) = a^T\omega + b$, implying that the function

$$\begin{aligned} f(\mu) \quad &:= \quad \ln\left(\int_\Omega \mathrm{e}^{\phi(\omega)}p_\mu(\omega)\Pi(d\omega)\right) = \ln\left(\sum_{\omega\in\mathbf{Z}^d_+} \mathrm{e}^{a^T\omega+b}\prod_{i=1}^d \frac{\mu_i^{\omega_i}\mathrm{e}^{-\mu_i}}{\omega_i!}\right) \\ &= \quad b + \ln\left(\prod_{i=1}^d\left[\mathrm{e}^{-\mu_i}\sum_{s=0}^\infty \frac{[e^{a_i}\mu_i]^s}{s!}\right]\right) = b + \sum_{i=1}^d \ln(\exp\{\mathrm{e}^{a_i}\mu_i - \mu_i\}) \\ &= \quad \sum_i [\mathrm{e}^{a_i} - 1]\mu_i + b \end{aligned}$$

is an affine (and thus a concave) function of μ.

The Poisson observation scheme is responsible for *Poisson Imaging*. This is the situation where there are n "sources of customers;" arrivals of customers at source i are independent of what happens at other sources, and inter-arrival times at source j are independent random variables with exponential distribution, with parameter λ_j, so that the number of customers arriving at source j in a unit time interval is a Poisson random variable with parameter λ_j. Now, there are d "servers," and a customer arriving at source j is dispatched to server i with some given probability A_{ij}, $\sum_i A_{ij} \leq 1$; with probability $1 - \sum_i A_{ij}$, such a customer leaves the system. The dispatches are independent of each other and of the arrival processes. What we observe is the vector $\omega = (\omega_1, ..., \omega_d)$, where ω_i is the number of customers dispatched to server i on the time horizon $[0, 1]$. It is easy to verify that in the situation just described, the entries ω_i in ω indeed are independent of the other

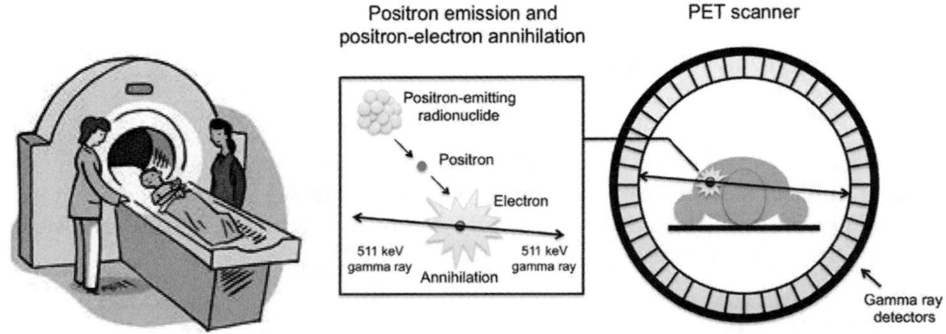

Figure 2.3: Positron Emission Tomography (PET)

Poisson random variables with Poisson parameters

$$\mu_i = \sum_{j=1}^{n} A_{ij}\lambda_j.$$

In what is called *Poisson Imaging*, one is given a random observation ω of the above type along with *sensing matrix* $A = [A_{ij}]$, and the goal is to use the observation to infer conclusions on the parameter $\mu = A\lambda$ and the "signal" λ underlying this parameter.

Poisson imaging has several important applications,[5] for example, in Positron Emission Tomography (PET). In PET (see Figure 2.3), a patient is injected with a radioactive tracer and is placed in a PET tomograph, which can be thought of as a cylinder with the surface split into small detector cells. The tracer disintegrates, and every disintegration act produces a positron which immediately annihilates with a nearby electron, giving rise to two γ-quants flying at the speed of light in two opposite directions along a line ("line of response" – LOR) with completely random orientation. Eventually, each of the γ-quants hits its own detector cell. When two detector cells are "simultaneously" hit (in fact, hit within a short time interval, like 10^{-8} sec), this event—*coincidence*—and the serial number of the *bin* (pair of detectors) where the hits were observed are registered. Observing a coincidence in some bin, we know that somewhere on the line linking the detector cells from the bin a disintegration act took place. The data collected in a PET study are the numbers of coincidences registered in each bin. When discretizing the field of view (patient's body) into small 3D cubes (voxels) we arrive at an accurate enough model of the data which is a realization ω of a random vector with independent Poisson entries $\omega_i \sim \text{Poisson}(\mu_i)$, with μ_i given by

$$\mu_i = \sum_{j=1}^{n} p_{ij}\lambda_j$$

[5]In all these applications, the signal λ we ultimately are interested in is an image; this is where "Imaging" comes from.

where λ_j is proportional to the amount of the tracer in voxel j, and p_{ij} is the probability for LOR emanating from voxel j to be registered in bin i (these probabilities can be computed given the geometry of the PET device). The tracer is selected to concentrate in the areas of interest (say, the areas of high metabolic activity when a tumor is sought), and the goal of the study is to infer from the observation ω the density of the tracer. The characteristic feature of PET as compared to other types of tomography is that with a properly selected tracer this technique allows us to visualize metabolic activity, and not only the anatomy of tissues in the body. Now, PET fits perfectly well the "dispatching customers" story above, with disintegration acts taking place in voxel j in the role of customers arriving at location j and bins in the role of servers. The arrival intensities are (proportional to) the amounts λ_j of the tracer in voxels, and the random dispatch of customers to servers corresponds to the random orientation of the LORs (in reality, nature draws their directions from the uniform distribution on the unit sphere in 3D).

It is worth noting that there are two other real-life applications of Poisson Imaging: Large Binocular Telescope and Nanoscale Fluorescent Microscopy.[6]

2.4.3.3 Discrete observation scheme

In Discrete o.s.,

- the observation space is a finite set $\Omega = \{1, ..., d\}$ equipped with a counting measure;
- the family $\{p_\mu(\cdot) : \mu \in \mathcal{M}\}$ is comprised of all nonvanishing distributions on Ω, that is,

$$\mathcal{M} = \left\{\mu \in \mathbf{R}^d : \mu > 0, \sum_{\omega\in\Omega} \mu_\omega = 1\right\}, \; p_\mu(\omega) = \mu_\omega, \; \omega \in \Omega;$$

- $\mathcal{F}$ is the space of all real-valued functions on the finite set Ω.

Clearly, Discrete o.s. is simple; the function

$$f(\mu) := \ln\left(\int_\Omega e^{\phi(\omega)} p_\mu(\omega)\Pi(d\omega)\right) = \ln\left(\sum_{\omega\in\Omega} e^{\phi(\omega)}\mu_\omega\right)$$

indeed is concave in $\mu \in \mathcal{M}$.

2.4.3.4 Direct products of simple observation schemes

Given K simple observation schemes

$$\mathcal{O}_k = \left(\Omega_k, \Pi_k; \{p_{\mu,k}(\cdot) : \mu \in \mathcal{M}_k\}; \mathcal{F}_k\right), \; 1 \le k \le K,$$

[6]Large Binocular Telescope [16, 17] is a cutting-edge instrument for high-resolution optical/infrared astronomical imaging; it is the subject of a huge ongoing international project; see http://www.lbto.org. Nanoscale Fluorescent Microscopy (a.k.a. Poisson Biophotonics) is a revolutionary tool for cell imaging trigged by the advent of techniques [18, 113, 117, 211] (2014 Nobel Prize in Chemistry) allowing us to break the diffraction barrier and to view biological molecules "at work" at a resolution of 10–20 nm, yielding entirely new insights into the signalling and transport processes within cells.

we can define their *direct product*

$$\mathcal{O}^K = \prod_{k=1}^{K} \mathcal{O}_k = (\Omega^K, \Pi^K; \{p_\mu : \mu \in \mathcal{M}^K\}; \mathcal{F}^K)$$

by modeling the situation where our observation is a tuple $\omega^K = (\omega_1, ..., \omega_K)$ with components ω_k yielded, independently of each other, by observation schemes $\mathcal{O}_k$, namely, as follows:

- The observation space Ω^K is the direct product of observations spaces $\Omega_1, ..., \Omega_K$, and the reference measure Π^K is the product of the measures $\Pi_1, ..., \Pi_K$;
- The parameter space $\mathcal{M}^K$ is the direct product of partial parameter spaces $\mathcal{M}_1, ..., \mathcal{M}_K$, and the distribution $p_\mu(\omega^K)$ associated with parameter

  $$\mu = (\mu_1, \mu_2, ..., \mu_K) \in \mathcal{M}^K = \mathcal{M}_1 \times ... \times \mathcal{M}_K$$

  is the probability distribution on Ω^K with the density

  $$p_\mu(\omega^K) = \prod_{k=1}^{K} p_{\mu,k}(\omega_k)$$

  w.r.t. Π^K. In other words, random observation $\omega^K \sim p_\mu$ is a sample of observations $\omega_1, ..., \omega_K$, drawn, independently of each other, from the distributions $p_{\mu_1,1}, p_{\mu_2,2}, ..., p_{\mu_K,K}$;
- The space $\mathcal{F}^K$ is comprised of all *separable* functions

  $$\phi(\omega^K) = \sum_{k=1}^{K} \phi_k(\omega_k)$$

  with $\phi_k(\cdot) \in \mathcal{F}_k$, $1 \le k \le K$.

It is immediately seen that the direct product of simple observation o.s.'s is simple.

When all factors $\mathcal{O}_k$, $1 \le k \le K$, are the identical simple o.s.

$$\mathcal{O} = (\Omega, \Pi; \{p_\mu : \mu \in \mathcal{M}\}; \mathcal{F}),$$

the direct product of the factors can be "truncated" to yield the *K-th power* (called also the *stationary K-repeated version*) of $\mathcal{O}$, denoted by

$$[\mathcal{O}]^K = (\Omega^K, \Pi^K; \{p_\mu^{(K)} : \mu \in \mathcal{M}\}; \mathcal{F}^{(K)})$$

and defined as follows.

- Ω^K and Π^K are exactly the same as in the direct product:

  $$\Omega^K = \underbrace{\Omega \times ... \times \Omega}_{K}, \; \Pi^K = \underbrace{\Pi \times ... \times \Pi}_{K};$$

- the parameter space is $\mathcal{M}$ rather than the direct product of K copies of $\mathcal{M}$, and the densities are

$$p_\mu^{(K)}(\omega^K = (\omega_1, ..., \omega_K)) = \prod_{k=1}^K p_\mu(\omega_k).$$

In other words, random observations $\omega^K \sim p_\mu^{(K)}$ are K-element samples with components drawn, independently of each other, from p_μ;

- the space $\mathcal{F}^{(K)}$ is comprised of separable functions

$$\phi^{(K)}(\omega^K) = \sum_{k=1}^K \phi(\omega_k)$$

with identical components belonging to $\mathcal{F}$ (i.e., $\phi \in \mathcal{F}$).

It is immediately seen that a power of a simple o.s. is simple.

Remark 2.19. Gaussian, Poisson, and Discrete o.s.'s clearly are nondegenerate. It is also clear that the direct product of nondegenerate o.s.'s is nondegenerate.

2.4.4 Simple observation schemes—Main result

We are about to demonstrate that when deciding on *convex*, in some precise sense to be specified below, hypotheses in *simple* observation schemes, optimal detectors can be found efficiently by solving *convex-concave saddle point problems.*

We start with an "executive summary" of convex-concave saddle point problems.

2.4.4.1 Executive summary of convex-concave saddle point problems

The results to follow are absolutely standard, and their proofs can be found in all textbooks on the subject, see, e.g., [221] or [15, Section D.4].

Let U and V be nonempty sets, and let $\Phi : U \times V \to \mathbf{R}$ be a function. These data define an *antagonistic game* of two players, I and II, where player I selects a point $u \in U$, and player II selects a point $v \in V$; as an outcome of these selections, player I pays to player II the sum $\Phi(u, v)$. Clearly, player I is interested in minimizing this payment, and player II in maximizing it. The data U, V, Φ are known to the players in advance, and the question is, what should be their selections?

When player I makes his selection u first, and player II makes his selection v with u already known, player I should be ready to pay for a selection $u \in U$ a toll as large as

$$\overline{\Phi}(u) = \sup_{v \in V} \Phi(u, v).$$

In this situation, a risk-averse player I would select u by minimizing the above worst-case payment, by solving the *primal* problem

$$\mathrm{Opt}(P) = \inf_{u \in U} \overline{\Phi}(u) = \inf_{u \in U} \sup_{v \in V} \Phi(u, v) \tag{P}$$

associated with the data U, V, Φ.

Similarly, if player II makes his selection v first, and player I selects u after v becomes known, player II should be ready to get, as a result of selecting $v \in V$, the amount as small as

$$\underline{\Phi}(v) = \inf_{u \in U} \Phi(u, v).$$

In this situation, a risk-averse player II would select v by maximizing the above worst-case payment, by solving the *dual* problem

$$\mathrm{Opt}(D) = \sup_{v \in V} \underline{\Phi}(v) = \sup_{v \in V} \inf_{u \in U} \Phi(u, v). \qquad (D)$$

Intuitively, the first situation is less preferable for player I than the second one, so that his guaranteed payment in the first situation, that is, Opt(P), should be $\geq$ his guaranteed payment, Opt(D), in the second situation:

$$\mathrm{Opt}(P) := \inf_{u \in U} \sup_{v \in V} \Phi(u, v) \geq \sup_{v \in V} \inf_{u \in U} \Phi(u, v) =: \mathrm{Opt}(D).$$

This fact, called *Weak Duality*, indeed is true.

The central question related to the game is what should the players do when making their selections simultaneously, with no knowledge of what is selected by the adversary. There is a case when this question has a completely satisfactory answer—this is the case where Φ has a *saddle point* on $U \times V$.

Definition 2.20. *A point* $(u_*, v_*) \in U \times V$ *is called a* saddle point[7] *of function* $\Phi(u, v) : U \times V \to \mathbf{R}$ *if* Φ *as a function of* $u \in U$ *attains at this point its minimum, and as a function of* $v \in V$ *its maximum, that is, if*

$$\Phi(u, v_*) \geq \Phi(u_*, v_*) \geq \Phi(u_*, v) \;\; \forall (u \in U, v \in V).$$

From the viewpoint of our game, a saddle point (u_*, v_*) is an equilibrium: when one of the players sticks to the selection stemming from this point, the other one has no incentive to deviate from his selection stemming from the point. Indeed, if player II selects v_*, there is no reason for player I to deviate from selecting u_*, since with another selection, his loss (the payment) can only increase; similarly, when player I selects u_*, there is no reason for player II to deviate from v_*, since with any other selection, his gain (the payment) can only decrease. As a result, if the cost function Φ has a saddle point on $U \times V$, this saddle point (u_*, v_*) can be considered as a solution to the game, as the pair of preferred selections of rational players. It can be easily seen that while Φ can have many saddle points, the values of Φ at all these points are equal to each other; we denote this common value by SadVal. If (u_*, v_*) is a saddle point and player I selects $u = u_*$, his worst loss, over selections $v \in V$ of player II, is SadVal, and if player I selects any $u \in U$, his worst-case loss, over the selections of player II can be only $\geq$ SadVal. Similarly, when player II selects $v = v_*$, his worst-case gain, over the selections of player I, is SadVal, and if player II selects any $v \in V$, his worst-case gain, over the selections of player I, can be only $\leq$ SadVal.

Existence of saddle points of Φ (min in $u \in U$, max in $v \in V$) can be expressed in terms of the primal problem (P) and the dual problem (P):

[7] *More precisely, "saddle point (*min *in* $u \in U$, max *in* $v \in V$*)"; we will usually skip the clarification in parentheses, since it always will be clear from the context what are the minimization variables and what are the maximization ones.*

Proposition 2.21. *Φ has a saddle point (*min *in $u \in U$,* max *in $v \in V$) if and only if problems (P) and (D) are solvable with equal optimal values:*

$$\mathrm{Opt}(P) := \inf_{u\in U}\sup_{v\in V}\Phi(u,v) = \sup_{v\in V}\inf_{u\in U}\Phi(u,v) =: \mathrm{Opt}(D). \tag{2.55}$$

Whenever this is the case, the saddle points of Φ are exactly the pairs (u_*, v_*) comprised of optimal solutions to problems (P) and (D), and the value of Φ at every one of these points is the common value SadVal *of* Opt(P) *and* Opt(D).

Existence of a saddle point of a function is a "rare commodity," and the standard sufficient condition for it is convexity-concavity of Φ coupled with convexity of U and V. The precise statement is as follows:

Theorem 2.22. *[Sion-Kakutani; see, e.g., [221] or [15, Theorems D.4.3, D.4.4]] Let $U \subset \mathbf{R}^m, V \subset \mathbf{R}^n$ be nonempty closed convex sets, with V bounded, and let $\Phi : U \times V \to \mathbf{R}$ be a continuous function which is convex in $u \in U$ for every fixed $v \in V$, and is concave in $v \in V$ for every fixed $u \in U$. Then the equality* (2.55) *holds true (although it may happen that* Opt$(P) =$ Opt$(D) = -\infty$*).*

If, in addition, Φ is coercive in u, meaning that the level sets

$$\{u \in U : \Phi(u,v) \le a\}$$

*are bounded for every $a \in \mathbf{R}$ and $v \in V$ (equivalently: for every $v \in V$, $\Phi(u_i, v) \to +\infty$ along every sequence $u_i \in U$ going to ∞: $\|u_i\| \to \infty$ as $i \to \infty$), then Φ admits saddle points (*min *in $u \in U$,* max *in $v \in V$).*

Note that the "true" Sion-Kakutani Theorem is a bit stronger than Theorem 2.22; the latter, however, covers all our related needs.

2.4.4.2 Main result

Theorem 2.23. *Let*

$$\mathcal{O} = (\Omega, \Pi; \{p_\mu : \mu \in \mathcal{M}\}; \mathcal{F})$$

be a simple observation scheme, and let M_1, M_2 be nonempty compact convex subsets of $\mathcal{M}$. Then

(i) *The function*

$$\Phi(\phi,[\mu;\nu]) = \tfrac{1}{2}\left[\ln\left(\int_\Omega e^{-\phi(\omega)}p_\mu(\omega)\Pi(d\omega)\right) + \ln\left(\int_\Omega e^{\phi(\omega)}p_\nu(\omega)\Pi(d\omega)\right)\right] : \mathcal{F}\times(M_1\times M_2)\to\mathbf{R} \tag{2.56}$$

*is continuous on its domain, convex in $\phi(\cdot) \in \mathcal{F}$, concave in $[\mu;\nu] \in M_1 \times M_2$, and possesses a saddle point (*min *in $\phi \in \mathcal{F}$,* max *in $[\mu;\nu] \in M_1 \times M_2$) $(\phi_*(\cdot), [\mu_*;\nu_*])$ on $\mathcal{F} \times (M_1 \times M_2)$. W.l.o.g. ϕ_* can be assumed to satisfy the relation*[8]

$$\int_\Omega \exp\{-\phi_*(\omega)\}p_{\mu_*}(\omega)\Pi(d\omega) = \int_\Omega \exp\{\phi_*(\omega)\}p_{\nu_*}(\omega)\Pi(d\omega). \tag{2.57}$$

[8]Note that $\mathcal{F}$ contains constants, and shifting by a constant the ϕ-component of a saddle point of Φ and keeping its $[\mu;\nu]$-component intact, we clearly get another saddle point of Φ.

Denoting the common value of the two quantities in (2.57) *by* $\varepsilon_\star$, *the saddle point value*

$$\min_{\phi\in\mathcal{F}} \max_{[\mu;\nu]\in M_1\times M_2} \Phi(\phi,[\mu;\nu])$$

is $\ln(\varepsilon_\star)$. *Besides this, setting* $\phi_*^a(\cdot)=\phi_*(\cdot)-a$, *one has*

$$\begin{array}{lrcl}(a) & \int_\Omega \exp\{-\phi_*^a(\omega)\}p_\mu(\omega)\Pi(d\omega) & \leq & \exp\{a\}\varepsilon_\star\ \forall \mu\in M_1,\\ (b) & \int_\Omega \exp\{\phi_*^a(\omega)\}p_\nu(\omega)\Pi(d\omega) & \leq & \exp\{-a\}\varepsilon_\star\ \forall \nu\in M_2.\end{array} \tag{2.58}$$

In view of Proposition 2.14 this implies that when deciding via an observation $\omega\in\Omega$ *on the hypotheses*

$$H_\chi:\omega\sim p_\mu \text{ with } \mu\in M_\chi,\quad \chi=1,2,$$

the risks of the simple test $\mathcal{T}_{\phi_*^a}$ *based on the detector* ϕ_*^a *can be upper-bounded as follows:*

$$\mathrm{Risk}_1(\mathcal{T}_{\phi_*^a}|H_1,H_2)\leq \exp\{a\}\varepsilon_\star,\ \mathrm{Risk}_2(\mathcal{T}_{\phi_*^a}|H_1,H_2)\leq \exp\{-a\}\varepsilon_\star.$$

Moreover, $\phi_*,\varepsilon_\star$ *form an optimal solution to the optimization problem*

$$\min_{\phi,\epsilon}\left\{\epsilon: \begin{array}{l}\int_\Omega e^{-\phi(\omega)}p_\mu(\omega)\Pi(d\omega)\leq \epsilon\,\forall \mu\in M_1\\ \int_\Omega e^{\phi(\omega)}p_\mu(\omega)\Pi(d\omega)\leq \epsilon\,\forall \mu\in M_2\end{array}\right\} \tag{2.59}$$

(the minimum in (2.59) *is taken over all* $\epsilon>0$ *and all* Π*-measurable functions* $\phi(\cdot)$*, not just over* $\phi\in\mathcal{F}$*).*

(ii) *The dual problem associated with the saddle point data* Φ, $\mathcal{F}$, $M_1\times M_2$ *is*

$$\max_{\mu\in M_1,\nu\in M_2}\left\{\underline{\Phi}(\mu,\nu):=\inf_{\phi\in\mathcal{F}}\Phi(\phi;[\mu;\nu])\right\}. \tag{D}$$

The objective in this problem is in fact the logarithm of the Hellinger affinity of p_μ *and* p_ν,

$$\underline{\Phi}(\mu,\nu)=\ln\left(\int_\Omega\sqrt{p_\mu(\omega)p_\nu(\omega)}\Pi(d\omega)\right), \tag{2.60}$$

and this objective is concave and continuous on $M_1\times M_2$.

The (μ,ν)*-components of saddle points of* Φ *are exactly the maximizers* (μ_*,ν_*) *of the concave function* $\underline{\Phi}$ *on* $M_1\times M_2$. *Given such a maximizer* $[\mu_*;\nu_*]$ *and setting*

$$\phi_*(\omega)=\tfrac{1}{2}\ln(p_{\mu_*}(\omega)/p_{\nu_*}(\omega)) \tag{2.61}$$

we get a saddle point $(\phi_*,[\mu_*;\nu_*])$ *of* Φ *satisfying* (2.57).

(iii) *Let* $[\mu_*;\nu_*]$ *be a maximizer of* $\underline{\Phi}$ *over* $M_1\times M_2$. *Let, further,* $\epsilon\in[0,1/2]$ *be such that there exists any (perhaps randomized) test for deciding via observation* $\omega\in\Omega$ *on two simple hypotheses*

$$(A):\omega\sim p(\cdot):=p_{\mu_*}(\cdot),\quad (B):\omega\sim q(\cdot):=p_{\nu_*}(\cdot) \tag{2.62}$$

with total risk $\leq 2\epsilon$. *Then*

$$\varepsilon_\star\leq 2\sqrt{\epsilon(1-\epsilon)}.$$

In other words, if the simple hypotheses (A), (B) *can be decided, by any test, with total risk* 2ϵ, *the risks of the simple test with detector* ϕ_* *given by* (2.61) *on the* *composite hypotheses* H_1*,* H_2 *do not exceed* $2\sqrt{\epsilon(1-\epsilon)}$.

For proof, see Section 2.11.3.

Remark 2.24. Assume that we are under the premise of Theorem 2.23 and that the simple o.s. in question is nondegenerate (see Section 2.4.2). Then $\varepsilon_\star < 1$ if and only if the sets M_1 and M_2 do not intersect.

Indeed, by Theorem 2.23.i, $\ln(\varepsilon_\star)$ is the saddle point value of $\Phi(\phi, [\mu; \nu])$ on $\mathcal{F} \times (M_1 \times M_2)$, or, which is the same by Theorem 2.23.ii, the maximum of the function (2.60) on $M_1 \times M_2$; since saddle points exist, this maximum is achieved at some pair $[\mu; \nu] \in M_1 \times M_2$. Since (2.60) clearly is ≤ 0, we conclude that $\varepsilon_\star \leq 1$ and the equality takes place if and only if $\int_\Omega \sqrt{p_\mu(\omega) p_\nu(\omega)} \Pi(d\omega) = 1$ for some $\mu \in M_1$ and $\nu \in M_2$, or, which is the same, $\int_\Omega (\sqrt{p_\mu(\omega)} - \sqrt{p_\nu(\omega)})^2 \Pi(d\omega) = 0$ for these μ and ν. Since $p_\mu(\cdot)$ and $p_\nu(\cdot)$ are continuous and the support of Π is the entire Ω, the latter can happen if and only if $p_\mu = p_\nu$ for our μ, ν, or, by nondegeneracy of $\mathcal{O}$, if and only if $M_1 \cap M_2 \neq \emptyset$. □

2.4.5 Simple observation schemes—Examples of optimal detectors

Theorem 2.23.i states that when the observation scheme

$$\mathcal{O} = (\Omega, \Pi; \{p_\mu : \mu \in \mathcal{M}\}; \mathcal{F})$$

is simple and we are interested in deciding on a pair of hypotheses on the distribution of observation $\omega \in \Omega$,

$$H_\chi : \omega \sim p_\mu \text{ with } \mu \in M_\chi, \ \chi = 1, 2$$

and *the hypotheses are convex*, meaning that the underlying parameter sets M_χ are convex and compact, building an optimal, in terms of its risk, detector ϕ_*—that is, solving (in general, a semi-infinite and infinite-dimensional) optimization problem (2.59)—reduces to solving a finite-dimensional convex problem. Specifically, an optimal solution $(\phi_*, \varepsilon_\star)$ can be built as follows:

1. We solve optimization problem

$$\text{Opt} = \max_{\mu \in M_1, \nu \in M_2} \left[\underline{\Phi}(\mu, \nu) := \ln \left(\int_\Omega \sqrt{p_\mu(\omega) p_\nu(\omega)} \Pi(d\omega) \right) \right] \tag{2.63}$$

   of maximizing Hellinger affinity (the quantity under the logarithm) of a pair of distributions obeying H_1 and H_2, respectively; for a simple o.s., the objective in this problem is concave and continuous, and optimal solutions do exist;
2. (Any) optimal solution $[\mu_*; \nu_*]$ to (2.63) gives rise to an optimal detector ϕ_* and its risk $\varepsilon_\star$, according to

$$\phi_*(\omega) = \frac{1}{2} \ln \left(\frac{p_{\mu_*}(\omega)}{p_{\nu_*}(\omega)} \right), \ \varepsilon_\star = \exp\{\text{Opt}\}.$$

   The risks of the simple test $\mathcal{T}_{\phi_*}$ associated with the above detector and deciding on H_1, H_2, satisfy the bounds

$$\max \left[\text{Risk}_1(\mathcal{T}_{\phi_*} | H_1, H_2), \text{Risk}_2(\mathcal{T}_{\phi_*} | H_1, H_2) \right] \leq \varepsilon_\star,$$

and the test is *near-optimal*, meaning that whenever the hypotheses H_1, H_2 (and in fact even two simple hypotheses stating that $\omega \sim p_{\mu_*}$ and $\omega \sim p_{\nu_*}$, respectively) can be decided upon by a test with total risk $\leq 2\epsilon \leq 1$, $\mathcal{T}_{\phi_*}$ exhibits a "comparable" risk:

$$\varepsilon_\star \leq 2\sqrt{\epsilon(1-\epsilon)}. \tag{2.64}$$

The test $\mathcal{T}_{\phi_*}$ is just the maximum likelihood test induced by the probability densities p_{μ_*} and p_{ν_*}.

Note that after we know that $(\phi_*, \varepsilon_\star)$ form an optimal solution to (2.59), some kind of near-optimality of the test $\mathcal{T}_{\phi_*}$ is guaranteed already by Proposition 2.18. By this proposition, whenever in nature there exists a simple test $\mathcal{T}$ which decides on H_1, H_2 with risks $\mathrm{Risk}_1, \mathrm{Risk}_2$ bounded by some $\epsilon \leq 1/2$, the upper bound $\varepsilon_\star$ on the risks of $\mathcal{T}_{\phi_*}$ can be bounded according to (2.64). Our now near-optimality statement is slightly stronger: first, we allow $\mathcal{T}$ to have the total risk $\leq 2\epsilon$, which is weaker than to have both risks $\leq \epsilon$; second, and more important, now 2ϵ should upper-bound the total risk of $\mathcal{T}$ on a pair of *simple* hypotheses "embedded" into the hypotheses H_1, H_2; both these modifications extend the family of tests $\mathcal{T}$ to which we compare the test $\mathcal{T}_{\phi_*}$, and thus enrich the comparison.

Let us look how the above recipe works for our basic simple o.s.'s.

2.4.5.1 Gaussian o.s.

When $\mathcal{O}$ is a Gaussian o.s., that is, $\{p_\mu : \mu \in \mathcal{M}\}$ are Gaussian densities with expectations $\mu \in \mathcal{M} = \mathbf{R}^d$ and common positive definite covariance matrix Θ, and $\mathcal{F}$ is the family of affine functions on $\Omega = \mathbf{R}^d$,

- M_1, M_2 can be arbitrary nonempty convex compact subsets of $\mathbf{R}^d$,
- problem (2.63) becomes the convex optimization problem

$$\mathrm{Opt} = -\min_{\mu\in M_1, \nu\in M_2} \tfrac{1}{8}(\mu-\nu)^T\Theta^{-1}(\mu-\nu), \tag{2.65}$$

- the optimal detector ϕ_* and the upper bound $\varepsilon_\star$ on its risks given by an optimal solution (μ_*, ν_*) to (2.65) are

$$\begin{array}{rcl} \phi_*(\omega) &=& \frac{1}{2}[\mu_*-\nu_*]^T\Theta^{-1}[\omega - w],\ w = \frac{1}{2}[\mu_*+\nu_*] \\ \varepsilon_\star &=& \exp\{-\frac{1}{8}[\mu_*-\nu_*]\Theta^{-1}[\mu_*-\nu_*]\}. \end{array} \tag{2.66}$$

Note that when $\Theta = I_d$, the test $\mathcal{T}_{\phi_*}$ becomes exactly the optimal test from Example 2.1. The upper bound on the risks of this test established in Example 2.1 (in our present notation, this bound is $\mathrm{Erfc}(\frac{1}{2}\|\mu_*-\nu_*\|_2)$) is slightly better than the bound $\varepsilon_\star = \exp\{-\|\mu_*-\nu_*\|_2^2/8\}$ given by (2.66) when $\Theta = I_d$. Note, however, that when speaking about the distance $\delta = \|\mu_*-\nu_*\|_2$ between M_1 and M_2 allowing for a test with risks $\leq \epsilon \ll 1$, the results of Example 2.1 and (2.66) say nearly the same thing: Example 2.1 says that δ should be $\geq 2\mathrm{ErfcInv}(\epsilon)$, with ErfcInv defined in (1.26), and (2.66) says that δ should be $\geq 2\sqrt{2\ln(1/\epsilon)}$. When $\epsilon \to +0$, the ratio of these two lower bounds on δ tends to 1.

It should be noted that our general construction of optimal detectors as applied to Gaussian o.s. and a pair of convex hypotheses results in *exactly* an optimal test and can be analyzed directly, without any "science" (see Example 2.1).

2.4.5.2 Poisson o.s.

When $\mathcal{O}$ is a Poisson o.s., that is, $\mathcal{M} = \mathbf{R}^d_{++}$ is the interior of the nonnegative orthant in $\mathbf{R}^d$, and p_μ, $\mu \in \mathcal{M}$, is the density

$$p_\mu(\omega) = \prod_i \left(\frac{\mu_i^{\omega_i}}{\omega_i!} \mathrm{e}^{-\mu_i} \right), \ \omega = (\omega_1, ..., \omega_d) \in \mathbf{Z}^d_+$$

taken w.r.t. the counting measure Π on $\Omega = \mathbf{Z}^d_+$, and $\mathcal{F}$ is the family of affine functions on Ω, the recipe from the beginning of Section 2.4.5 reads as follows:

- M_1, M_2 can be arbitrary nonempty convex compact subsets of $\mathbf{R}^d_{++} = \{x \in \mathbf{R}^d : x > 0\}$;
- problem (2.63) becomes the convex optimization problem

$$\mathrm{Opt} = - \min_{\mu \in M_1, \nu \in M_2} \frac{1}{2} \sum_{i=1}^d (\sqrt{\mu_i} - \sqrt{\nu_i})^2; \tag{2.67}$$

- the optimal detector ϕ_* and the upper bound $\varepsilon_\star$ on its risks given by an optimal solution (μ^*, ν^*) to (2.67) are

$$\phi_*(\omega) = \frac{1}{2} \sum_{i=1}^d \ln \left(\frac{\mu_i^*}{\nu_i^*} \right) \omega_i + \frac{1}{2} \sum_{i=1}^d [\nu_i^* - \mu_i^*], \quad \varepsilon_\star = \mathrm{e}^{\mathrm{Opt}}.$$

2.4.5.3 Discrete o.s.

When $\mathcal{O}$ is a Discrete o.s., that is, $\Omega = \{1, ..., d\}$, Π is a counting measure on Ω, $\mathcal{M} = \{\mu \in \mathbf{R}^d : \mu > 0, \sum_i \mu_i = 1\}$, and

$$p_\mu(\omega) = \mu_\omega, \ \omega = 1, ..., d, \ \mu \in \mathcal{M},$$

the recipe from the beginning of Section 2.4.5 reads as follows:[9]

- M_1, M_2 can be arbitrary nonempty convex compact subsets of the relative interior $\mathcal{M}$ of the probabilistic simplex,
- problem (2.63) *is equivalent* to the convex program

$$\varepsilon_\star = \max_{\mu \in M_1, \nu \in M_2} \sum_{i=1}^d \sqrt{\mu_i \nu_i}; \tag{2.68}$$

- the optimal detector ϕ_* given by an optimal solution (μ^*, ν^*) to (2.67) is

$$\phi_*(\omega) \quad = \quad \tfrac{1}{2} \ln \left(\tfrac{\mu_\omega^*}{\nu_\omega^*} \right), \tag{2.69}$$

  and the upper bound $\varepsilon_\star$ on the risks of this detector is given by (2.68).

[9]It should be mentioned that the results of this section as applied to the Discrete observation scheme are a simple particular case—that of finite Ω—of the results of [21, 22, 25] on distinguishing convex sets of probability distributions.

2.4.5.4 K-th power of a simple o.s.

Recall that K-th power of a simple o.s. $\mathcal{O} = (\Omega, \Pi; \{p_\mu : \mu \in \mathcal{M}\}; \mathcal{F})$ (see Section 2.4.3.4) is the o.s.

$$[\mathcal{O}]^K = (\Omega^K, \Pi^K; \{p_\mu^{(K)} : \mu \in \mathcal{M}\}; \mathcal{F}^{(K)})$$

where Ω^K is the direct product of K copies of Ω, Π^K is the product of K copies of Π, the densities $p_\mu^{(K)}$ are product densities induced by K copies of the density p_μ, $\mu \in \mathcal{M}$,

$$p_\mu^{(K)}(\omega^K = (\omega_1, ..., \omega_K)) = \prod_{k=1}^K p_\mu(\omega_k),$$

and $\mathcal{F}^{(K)}$ is comprised of functions

$$\phi^{(K)}(\omega^K = (\omega_1, ..., \omega_K)) = \sum_{k=1}^K \phi(\omega_k)$$

stemming from functions $\phi \in \mathcal{F}$. Clearly, $[\mathcal{O}]^K$ is the observation scheme describing the stationary K-repeated observations $\omega^K = (\omega_1, ..., \omega_K)$ with ω_k stemming from the o.s. $\mathcal{O}$; see Section 2.3.2.3. As we remember, $[\mathcal{O}]^K$ is simple provided that $\mathcal{O}$ is so.

Assuming $\mathcal{O}$ simple, it is immediately seen that as applied to the o.s. $[\mathcal{O}]^K$, the recipe from the beginning of Section 2.4.5 reads as follows:

- M_1, M_2 can be arbitrary nonempty convex compact subsets of $\mathcal{M}$, and the corresponding hypotheses, H_χ^K, $\chi = 1, 2$, state that the components ω_k of observation $\omega^K = (\omega_1, ..., \omega_K)$ are independently of each other drawn from distribution p_μ with $\mu \in M_1$ (hypothesis H_1^K) or $\mu \in M_2$ (hypothesis H_2^K);
- problem (2.63) is the convex program

$$\mathrm{Opt}(K) = \max_{\mu \in M_1, \nu \in M_2} \underbrace{\ln \left(\int_{\Omega^K} \sqrt{p_\mu^{(K)}(\omega^K) p_\nu^{(K)}(\omega^K)} \Pi^K(d\Omega) \right)}_{\equiv K \ln \left(\int_\Omega \sqrt{p_\mu(\omega) p_\nu(\omega)} \Pi(d\omega) \right)} \qquad (D_K)$$

  implying that any optimal solution to the "single-observation" problem (D_1) associated with M_1, M_2 is optimal for the "K-observation" problem (D_K) associated with M_1, M_2, and $\mathrm{Opt}(K) = K\mathrm{Opt}(1)$;
- the optimal detector $\phi_*^{(K)}$ given by an optimal solution (μ_*, ν_*) to (D_1) (this solution is optimal for (D_K) as well) is

$$\phi_*^{(K)}(\omega^K) = \sum_{k=1}^K \phi_*(\omega_k), \quad \phi_*(\omega) = \frac{1}{2} \ln \left(\frac{p_{\mu_*}(\omega)}{p_{\nu_*}(\omega)} \right), \qquad (2.70)$$

  and the upper bound $\varepsilon_\star(K)$ on the risks of the detector $\phi_*^{(K)}$ on the pair of families of distributions obeying hypotheses H_1^K or H_2^K is

$$\varepsilon_\star(K) = \mathrm{e}^{\mathrm{Opt}(K)} = \mathrm{e}^{K\mathrm{Opt}(1)} = [\epsilon_\star(1)]^K. \qquad (2.71)$$

The just outlined results on powers of simple observation schemes allow us to express near-optimality of detector-based tests in simple o.s.'s in a nicer form.

Proposition 2.25. *Let $\mathcal{O} = (\Omega, \Pi; \{p_\mu : \mu \in \mathcal{M}\}; \mathcal{F})$ be a simple observation scheme, M_1, M_2 be two nonempty convex compact subsets of $\mathcal{M}$, and (μ_*, ν_*) be an optimal solution to the convex optimization problem (cf. Theorem 2.23)*

$$\mathrm{Opt} = \max_{\mu \in M_1, \nu \in M_2} \ln\left(\int_\Omega \sqrt{p_\mu(\omega) p_\nu(\omega)} \Pi(d\omega)\right).$$

Let ϕ_* and ϕ_*^K be single- and K-observation detectors induced by (μ_*, ν_*) via (2.70).

Let $\epsilon \in (0, 1/2)$, and assume that for some positive integer K in nature there exists a simple test $\mathcal{T}^K$ deciding via K i.i.d. observations $\omega^K = (\omega_1, ..., \omega_K)$ with $\omega_k \sim p_\mu$, for some unknown $\mu \in \mathcal{M}$, on the hypotheses

$$H_\chi^{(K)} : \mu \in M_\chi, \ \chi = 1, 2,$$

with risks Risk_1, Risk_2 *not exceeding ϵ. Then setting*

$$K_+ = \left\rfloor \frac{2}{1 - \ln(4(1-\epsilon))/\ln(1/\epsilon)} K \right\lfloor,$$

the simple test $\mathcal{T}_{\phi_*^{(K_+)}}$ utilizing K_+ i.i.d. observations decides on $H_1^{(K_+)}$, $H_2^{(K_+)}$ with risks $\leq \epsilon$. Note that K_+ "is of the order of K": $K_+/K \to 2$ as $\epsilon \to +0$.

Proof. Applying item (iii) of Theorem 2.23 to the simple o.s. $[\mathcal{O}]^K$, we see that what above was called $\varepsilon_\star(K)$ satisfies

$$\varepsilon_\star(K) \leq 2\sqrt{\epsilon(1-\epsilon)}.$$

By (2.71), we conclude that $\varepsilon_\star(1) \leq \left(2\sqrt{\epsilon(1-\epsilon)}\right)^{1/K}$, whence, by the same (2.71), $\varepsilon_\star(T) \leq \left(2\sqrt{\epsilon(1-\epsilon)}\right)^{T/K}$, $T = 1, 2, ...$ When plugging in this bound $T = K_+$, we get the inequality $\varepsilon_\star(K_+) \leq \epsilon$. It remains to recall that $\varepsilon_\star(K_+)$ upper-bounds the risks of the test $\mathcal{T}_{\phi_*^{(K_+)}}$ when deciding on $H_1^{(K_+)}$ vs. $H_2^{(K_+)}$. □

2.5 TESTING MULTIPLE HYPOTHESES

So far, we have focused on detector-based tests deciding on pairs of hypotheses, and our "constructive" results were restricted to pairs of *convex* hypotheses dealing with a simple o.s.

$$\mathcal{O} = (\Omega, \Pi; \{p_\mu : \mu \in \mathcal{M}\}; \mathcal{F}), \tag{2.72}$$

convexity of a hypothesis meaning that the family of probability distributions obeying the hypothesis is $\{p_\mu : \mu \in X\}$, associated with a convex (in fact, convex compact) set $X \subset \mathcal{M}$.

In this section, we will be interested in pairwise testing *unions* of convex hypotheses and testing *multiple* (more than two) hypotheses.

2.5.1 Testing unions

2.5.1.1 Situation and goal

Let Ω be an observation space, and assume we are given two finite collections of families of probability distributions on Ω: families of *red* distributions $\mathcal{R}_i$, $1 \le i \le r$, and families of *blue* distributions $\mathcal{B}_j$, $1 \le j \le b$. These families give rise to r red and b blue hypotheses on the distribution P of an observation $\omega \in \Omega$, specifically,

$$R_i : P \in \mathcal{R}_i \text{ (red hypotheses) and } B_j : P \in \mathcal{B}_j \text{ (blue hypotheses)}.$$

Assume that for every $i \le r$, $j \le b$ we have at our disposal a simple detector-based test $\mathcal{T}_{ij}$ capable of deciding on R_i vs. B_j. What we want is to assemble these tests into a test $\mathcal{T}$ deciding on the union R of red hypotheses vs. the union B of blue ones:

$$R : P \in \mathcal{R} := \bigcup_{i=1}^{r} \mathcal{R}_i, \quad B : P \in \mathcal{B} := \bigcup_{j=1}^{b} \mathcal{B}_j.$$

Here P, as always, stands for the probability distribution of observation $\omega \in \Omega$.

Our motivation primarily stems from the case where R_i and B_j are convex hypotheses in a simple o.s. (2.72):

$$\mathcal{R}_i = \{p_\mu : \mu \in M_i\}, \, \mathcal{B}_j = \{p_\mu : \mu \in N_j\},$$

where M_i and N_j are convex compact subsets of $\mathcal{M}$. In this case we indeed know how to build near-optimal tests deciding on R_i vs. B_j, and the question we have posed becomes, how do we assemble these tests into a test deciding on R vs. B, with

$$\begin{array}{ll} R : P \in \mathcal{R} = \{p_\mu : \mu \in X\}, & X = \bigcup_i M_i, \\ B : P \in \mathcal{B} = \{p_\mu : \mu \in Y\}, & Y = \bigcup_j N_j? \end{array}$$

While the structure of R, B is similar to that of R_i, B_j, there is a significant difference: the sets X, Y are, in general, nonconvex, and therefore the techniques we have developed fail to address testing R vs. B directly.

2.5.1.2 The construction

In the situation just described, let ϕ_{ij} be the detectors underlying the tests $\mathcal{T}_{ij}$; w.l.o.g., we can assume these detectors balanced (see Section 2.3.2.2) with some risks ϵ_{ij}:

$$\left.\begin{array}{ll} \int_\Omega e^{-\phi_{ij}(\omega)} P(d\omega) \le \epsilon_{ij} & \forall P \in \mathcal{R}_i \\ \int_\Omega e^{\phi_{ij}(\omega)} P(d\omega) \le \epsilon_{ij} & \forall P \in \mathcal{B}_j \end{array}\right\}, 1 \le i \le r, 1 \le j \le b. \tag{2.73}$$

Let us assemble the detectors ϕ_{ij} into a detector for R, B as follows:

$$\phi(\omega) = \max_{1 \le i \le r} \min_{1 \le j \le b} [\phi_{ij}(\omega) - \alpha_{ij}], \tag{2.74}$$

where the *shifts* α_{ij} are parameters of the construction.

Proposition 2.26. *The risks of ϕ on R, B can be bounded as*

$$\begin{array}{ll} \forall P \in \mathcal{R}: & \int_\Omega e^{-\phi(\omega)} P(d\omega) \le \max_{i \le r} \left[\sum_{j=1}^b \epsilon_{ij} e^{\alpha_{ij}}\right], \\ \forall P \in \mathcal{B}: & \int_\Omega e^{\phi(\omega)} P(d\omega) \le \max_{j \le b} \left[\sum_{i=1}^r \epsilon_{ij} e^{-\alpha_{ij}}\right]. \end{array} \tag{2.75}$$

Therefore, the risks of ϕ on R, B are upper-bounded by the quantity

$$\varepsilon_\star = \max\left[\max_{i\le r}\left[\sum\nolimits_{j=1}^{b}\epsilon_{ij}e^{\alpha_{ij}}\right], \max_{j\le b}\left[\sum\nolimits_{i=1}^{r}\epsilon_{ij}e^{-\alpha_{ij}}\right]\right], \tag{2.76}$$

whence the risks of the simple test $\mathcal{T}_\phi$, based on the detector ϕ, deciding on R, B are upper-bounded by $\varepsilon_\star$.

Proof. Let $P \in \mathcal{R}$, so that $P \in \mathcal{R}_{i_*}$ for some $i_* \le r$. Then

$$\begin{array}{l}
\int_\Omega e^{-\phi(\omega)}P(d\omega) = \int_\Omega e^{\min_{i\le r}\max_{j\le b}[-\phi_{ij}(\omega)+\alpha_{ij}]}P(d\omega) \\
\quad\le \int_\Omega e^{\max_{j\le b}[-\phi_{i_*j}(\omega)+\alpha_{i_*j}]}P(d\omega) \le \sum_{j=1}^{b}\int_\Omega e^{-\phi_{i_*j}(\omega)+\alpha_{i_*j}}P(d\omega) \\
\quad= \sum_{j=1}^{b} e^{\alpha_{i_*j}}\int_\Omega e^{-\phi_{i_*j}(\omega)}P(d\omega) \\
\quad\le \sum_{j=1}^{b}\epsilon_{i_*j}e^{\alpha_{i_*j}} \text{ [by (2.73) due to } P \in \mathcal{R}_{i_*}] \\
\quad\le \max_{i\le r}\left[\sum_{j=1}^{b}\epsilon_{ij}e^{\alpha_{ij}}\right].
\end{array}$$

Now let $P \in \mathcal{B}$, so that $P \in \mathcal{B}_{j_*}$ for some j_*. We have

$$\begin{array}{l}
\int_\Omega e^{\phi(\omega)}P(d\omega) = \int_\Omega e^{\max_{i\le r}\min_{j\le b}[\phi_{ij}(\omega)-\alpha_{ij}]}P(d\omega) \\
\quad\le \int_\Omega e^{\max_{i\le r}[\phi_{ij_*}(\omega)-\alpha_{ij_*}]}P(d\omega) \le \sum_{i=1}^{r}\int_\Omega e^{\phi_{ij_*}(\omega)-\alpha_{ij_*}}P(d\omega) \\
\quad= \sum_{i=1}^{r} e^{-\alpha_{ij_*}}\int_\Omega e^{\phi_{ij_*}(\omega)}P(d\omega) \\
\quad\le \sum_{i=1}^{r}\epsilon_{ij_*}e^{-\alpha_{ij_*}} \text{ [by (2.73) due to } P \in \mathcal{B}_{j_*}] \\
\quad\le \max_{j\le b}\left[\sum_{i=1}^{r}\epsilon_{ij}e^{-\alpha_{ij}}\right].
\end{array}$$

(2.75) is proved. The remaining claims of the proposition are readily given by (2.75) combined with Proposition 2.14. □

Optimal choice of shift parameters. The detector and the test considered in Proposition 2.26, like the resulting risk bound $\varepsilon_\star$, depend on the shifts α_{ij}. Let us optimize the risk bound w.r.t. these shifts. To this end, consider the $r \times b$ matrix

$$E = [\epsilon_{ij}]_{\substack{i\le r\\ j\le b}}$$

and the symmetric $(r+b)\times(r+b)$ matrix

$$\mathcal{E} = \left[\begin{array}{c|c} & E \\ \hline E^T & \end{array}\right].$$

As is well known, the eigenvalues of the symmetric matrix $\mathcal{E}$ are comprised of the pairs $(\sigma_s, -\sigma_s)$, where σ_s are the singular values of E, and several zeros; in particular, the leading eigenvalue of $\mathcal{E}$ is the spectral norm $\|E\|_{2,2}$ (the largest singular value) of matrix E. Further, E is a matrix with positive entries, so that $\mathcal{E}$ is a symmetric entrywise nonnegative matrix. By the Perron-Frobenius Theorem, the leading eigenvector of this matrix can be selected to be nonnegative. Denoting this nonnegative eigenvector $[g; h]$ with r-dimensional g and b-dimensional h, and setting $\rho = \|E\|_{2,2}$, we have

$$\begin{array}{rcl} \rho g &=& Eh \\ \rho h &=& E^T g. \end{array} \tag{2.77}$$

Observe that $\rho > 0$ (evident), whence both g and h are nonzero (since otherwise (2.77) would imply $g = h = 0$, which is impossible—the eigenvector $[g; h]$ is nonzero). Since h and g are nonzero nonnegative vectors, $\rho > 0$ and E is entrywise positive, (2.77) says that g and h are strictly positive vectors. The latter allows us to define shifts α_{ij} according to

$$\alpha_{ij} = \ln(h_j/g_i). \tag{2.78}$$

With these shifts, we get

$$\max_{i\leq r}\left[\sum_{j=1}^{b}\epsilon_{ij}e^{\alpha_{ij}}\right] = \max_{i\leq r}\sum_{j=1}^{b}\epsilon_{ij}h_j/g_i = \max_{i\leq r}(Eh)_i/g_i = \max_{i\leq r}\rho = \rho$$

(we have used the first relation in (2.77)), and

$$\max_{j\leq b}\left[\sum_{i=1}^{r}\epsilon_{ij}e^{-\alpha_{ij}}\right] = \max_{j\leq b}\sum_{i=1}^{r}\epsilon_{ij}g_i/h_j = \max_{j\leq b}[E^Tg]_j/h_j = \max_{j\leq b}\rho = \rho$$

(we have used the second relation in (2.77)). The bottom line is as follows:

Proposition 2.27. *In the situation and the notation of Section 2.5.1.1, the risks of the detector* (2.74) *with shifts* (2.77), (2.78) *on the families $\mathcal{R}$, $\mathcal{B}$ do not exceed the quantity*

$$\|E := [\epsilon_{ij}]_{i\leq r, j\leq b}\|_{2,2}.$$

As a result, the risks of the simple test $\mathcal{T}_\phi$ deciding on the hypotheses R, B, does not exceed $\|E\|_{2,2}$ as well.

In fact, the shifts in the above proposition are the best possible; this is an immediate consequence of the following simple fact:

Proposition 2.28. *Let $\mathcal{E} = [e_{ij}]$ be a nonzero entrywise nonnegative $n \times n$ symmetric matrix. Then the optimal value in the optimization problem*

$$\mathrm{Opt} = \min_{\alpha_{ij}}\left\{\max_{i\leq n}\sum_{j=1}^{n}e_{ij}e^{\alpha_{ij}} : \alpha_{ij} = -\alpha_{ji}\right\} \tag{$*$}$$

is equal to $\|\mathcal{E}\|_{2,2}$. When the Perron-Frobenius eigenvector f of $\mathcal{E}$ can be selected positive, the problem is solvable, and an optimal solution is given by

$$\alpha_{ij} = \ln(f_j/f_i),\ 1 \leq i,j \leq n. \tag{2.79}$$

Proof. Let us prove that $\mathrm{Opt} \leq \rho := \|\mathcal{E}\|_{2,2}$. Given $\epsilon > 0$, we clearly can find an entrywise nonnegative symmetric matrix $\mathcal{E}'$ with entries e'_{ij} inbetween e_{ij} and $e_{ij} + \epsilon$ such that the Perron-Frobenius eigenvector f of $\mathcal{E}'$ can be selected positive (it suffices, e.g., to set $e'_{ij} = e_{ij} + \epsilon$). Selecting α_{ij} according to (2.79), we get a feasible solution to $(*)$ such that

$$\forall i : \sum_j e_{ij}e^{\alpha_{ij}} \leq \sum_j e'_{ij}f_j/f_i = \|\mathcal{E}'\|_{2,2},$$

implying that $\mathrm{Opt} \leq \|\mathcal{E}'\|_{2,2}$. Passing to limit as $\epsilon \to +0$, we get $\mathrm{Opt} \leq \|\mathcal{E}\|_{2,2}$. As a byproduct of our reasoning, if $\mathcal{E}$ admits a positive Perron-Frobenius eigenvector f, then (2.79) yields a feasible solution to $(*)$ with the value of the objective equal to $\|\mathcal{E}\|_{2,2}$.

It remain to prove that $\text{Opt} \geq \|\mathcal{E}\|_{2,2}$. Assume that this is not the case, so that $(*)$ admits a feasible solution $\widehat{\alpha}_{ij}$ such that

$$\widehat{\rho} := \max_i \sum_j e_{ij} e^{\widehat{\alpha}_{ij}} < \rho := \|\mathcal{E}\|_{2,2}.$$

By an arbitrarily small perturbation of $\mathcal{E}$, we can make this matrix symmetric and entrywise positive, and still satisfying the above strict inequality; to save notation, assume that already the original $\mathcal{E}$ is entrywise positive. Let f be a positive Perron-Frobenius eigenvector of $\mathcal{E}$, and let, as above, $\alpha_{ij} = \ln(f_j/f_i)$, so that

$$\sum_j e_{ij} e^{\alpha_{ij}} = \sum_j e_{ij} f_j/f_i = \rho \ \forall i.$$

Setting $\delta_{ij} = \widehat{\alpha}_{ij} - \alpha_{ij}$, we conclude that the convex functions

$$\theta_i(t) = \sum_j e_{ij} e^{\alpha_{ij} + t\delta_{ij}}$$

all are equal to ρ as $t = 0$, and all are $\leq \widehat{\rho} < \rho$ as $t = 1$, implying that $\theta_i(1) < \theta_i(0)$ for every i. The latter, in view of convexity of $\theta_i(\cdot)$, implies that

$$\theta_i'(0) = \sum_j e_{ij} e^{\alpha_{ij}} \delta_{ij} = \sum_j e_{ij} (f_j/f_i) \delta_{ij} < 0 \ \forall i.$$

Multiplying the resulting inequalities by f_i^2 and summing up over i, we get

$$\sum_{i,j} e_{ij} f_i f_j \delta_{ij} < 0,$$

which is impossible: we have $e_{ij} = e_{ji}$ and $\delta_{ij} = -\delta_{ji}$, implying that the left-hand side in the latter inequality is 0. □

2.5.2 Testing multiple hypotheses "up to closeness"

So far, we have considered detector-based simple tests deciding on pairs of hypotheses, specifically, convex hypotheses in simple o.s.'s (Section 2.4.4) and unions of convex hypotheses (Section 2.5.1).[10] Now we intend to consider testing of multiple (perhaps more than 2) hypotheses "up to closeness"; the latter notion was introduced in Section 2.2.4.2.

[10]Strictly speaking, in Section 2.5.1 it was not explicitly stated that the unions under consideration involve convex hypotheses in simple o.s.'s; our emphasis was on how to decide on a pair of union-type hypotheses *given pairwise detectors for "red" and "blue" components of the unions from the pair.* However, for now, the only situation where we indeed have at our disposal good pairwise detectors for red and blue components is that in which these components are convex hypotheses in a good o.s.

2.5.2.1 Situation and goal

Let Ω be an observation space, and let a collection $\mathcal{P}_1, ..., \mathcal{P}_L$ of families of probability distributions on Ω be given. As always, families $\mathcal{P}_\ell$ give rise to hypotheses

$$H_\ell : P \in \mathcal{P}_\ell$$

on the distribution P of observation $\omega \in \Omega$. Assume also that we are given a *closeness relation* $\mathcal{C}$ on $\{1, ..., L\}$. Recall that, formally, a closeness relation is some set of pairs of indices $(\ell, \ell') \in \{1, ..., L\}$; we interpret the inclusion $(\ell, \ell') \in \mathcal{C}$ as the fact that hypothesis H_ℓ "is close" to hypothesis $H_{\ell'}$. When $(\ell, \ell') \in \mathcal{C}$, we say that ℓ' is close (or $\mathcal{C}$-close) to ℓ. We always assume that

- $\mathcal{C}$ contains the diagonal: $(\ell, \ell) \in \mathcal{C}$ for every $\ell \leq L$ ("each hypothesis is close to itself"), and
- $\mathcal{C}$ is symmetric: whenever $(\ell, \ell') \in \mathcal{C}$, we have also $(\ell', \ell) \in \mathcal{C}$ ("if the ℓ-th hypothesis is close to the ℓ'-th one, then the ℓ'-th hypothesis is close to the ℓ-th one").

Recall that a test $\mathcal{T}$ deciding on the hypotheses $H_1, ..., H_L$ via observation $\omega \in \Omega$ is a procedure which, given on input $\omega \in \Omega$, builds some set $\mathcal{T}(\omega) \subset \{1, ..., L\}$, accepts all hypotheses H_ℓ with $\ell \in \mathcal{T}(\omega)$, and rejects all other hypotheses.

Risks of an "up to closeness" test. The notion of $\mathcal{C}$-risk of a test was introduced in Section 2.2.4.2, we reproduce it here for the reader's convenience. Given closeness $\mathcal{C}$ and a test $\mathcal{T}$, we define the *$\mathcal{C}$-risk*

$$\mathrm{Risk}^{\mathcal{C}}(\mathcal{T}|H_1, ..., H_L)$$

of $\mathcal{T}$ as the smallest $\epsilon \geq 0$ such that

> *Whenever an observation ω is drawn from a distribution $P \in \bigcup_\ell \mathcal{P}_\ell$, and ℓ_* is such that $P \in \mathcal{P}_{\ell_*}$ (i.e., hypothesis H_{ℓ_*} is true), the P-probability of the event $\ell_* \notin \mathcal{T}(\omega)$ ("true hypothesis H_{ℓ_*} is not accepted") <u>or</u> there exists ℓ' <u>not close to ℓ</u> such that $H_{\ell'}$ is accepted" is <u>at most</u> ϵ.*

Equivalently:

> $\mathrm{Risk}^{\mathcal{C}}(\mathcal{T}|H_1, ..., H_L) \leq \epsilon$ if and only if the following takes place:
>
> *Whenever an observation ω is drawn from a distribution $P \in \bigcup_\ell \mathcal{P}_\ell$, and ℓ_* is such that $P \in \mathcal{P}_{\ell_*}$ (i.e., hypothesis H_{ℓ_*} is true), the P-probability of the event*
>
> > *$\ell_* \in \mathcal{T}(\omega)$ ("the true hypothesis H_{ℓ_*} is accepted") <u>and</u> $\ell' \in \mathcal{T}(\omega)$ implies that $(\ell, \ell') \in \mathcal{C}$ ("all accepted hypotheses are $\mathcal{C}$-close to the true hypothesis H_{ℓ_*}") is <u>at least</u> $1 - \epsilon$.*

For example, consider nine polygons presented on Figure 2.4 and associate with them nine hypotheses on a 2D "signal plus noise" observation $\omega = x + \xi$, $\xi \sim \mathcal{N}(0, I_2)$, with the ℓ-th hypothesis stating that x belongs to the ℓ-th polygon. We define closeness $\mathcal{C}$ on the collection of hypotheses presented on Figure 2.4 as

"two hypotheses are close if and only if the corresponding polygons intersect," like **A** and **B**, or **A** and **E**. Now the fact that a test $\mathcal{T}$ has $\mathcal{C}$-risk ≤ 0.01 would imply, in particular, that if the probability distribution P underlying the observed

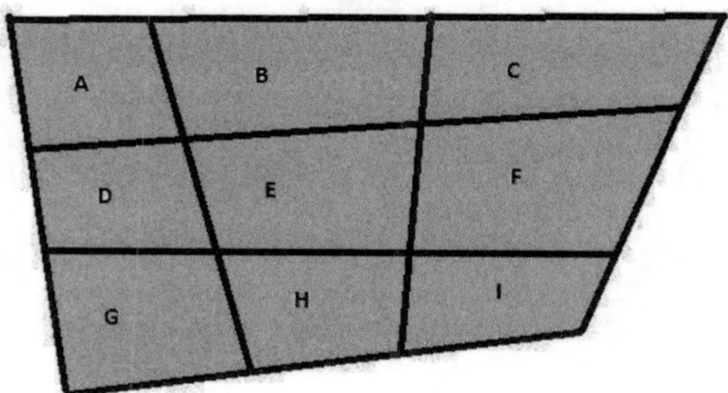

Figure 2.4: Nine hypotheses on the location of the mean μ of observation $\omega \sim \mathcal{N}(\mu, I_2)$, each stating that μ belongs to a specific polygon.

ω obeys hypothesis **A** (i.e., the mean of P belongs to the polygon **A**), then with P-probability at least 0.99 the list of accepted hypotheses includes hypothesis **A**, and the only other hypotheses in this list are among hypotheses **B**, **D**, and **E**.

2.5.2.2 "Building blocks" and construction

The construction we are about to present is, essentially, that used in Section 2.2.4.3 as applied to detector-generated tests. This being said, the presentation to follow is self-contained.

The building blocks of our construction are pairwise detectors $\phi_{\ell\ell'}(\omega)$, $1 \leq \ell < \ell' \leq L$, for pairs $\mathcal{P}_\ell$, $\mathcal{P}_{\ell'}$ along with (upper bounds on) the risks $\epsilon_{\ell\ell'}$ of these detectors:

$$\left.\begin{array}{ll}\forall (P \in \mathcal{P}_\ell): & \int_\Omega \mathrm{e}^{-\phi_{\ell\ell'}(\omega)} P(d\omega) \leq \epsilon_{\ell\ell'} \\ \forall (P \in \mathcal{P}_{\ell'}): & \int_\Omega \mathrm{e}^{\phi_{\ell\ell'}(\omega)} P(d\omega) \leq \epsilon_{\ell\ell'}\end{array}\right\}, \ 1 \leq \ell < \ell' \leq L.$$

Setting

$$\phi_{\ell'\ell}(\omega) = -\phi_{\ell\ell'}(\omega),\ \epsilon_{\ell'\ell} = \epsilon_{\ell\ell'},\ 1 \leq \ell < \ell' \leq L,\ \phi_{\ell\ell}(\omega) \equiv 0, \epsilon_{\ell\ell} = 1,\ 1 \leq \ell \leq L,$$

we get what we refer to as a *balanced system of detectors $\phi_{\ell\ell'}$ and risks $\epsilon_{\ell\ell'}$, $1 \leq \ell, \ell' \leq L$, for the collection $\mathcal{P}_1, ..., \mathcal{P}_L$*, meaning that

$$\begin{array}{ll}\phi_{\ell\ell'}(\omega) + \phi_{\ell'\ell}(\omega) \equiv 0,\ \epsilon_{\ell\ell'} = \epsilon_{\ell'\ell}, & 1 \leq \ell, \ell' \leq L, \\ \forall P \in \mathcal{P}_\ell : \int_\Omega \mathrm{e}^{-\phi_{\ell\ell'}(\omega)} P(d\omega) \leq \epsilon_{\ell\ell'}, & 1 \leq \ell, \ell' \leq L.\end{array} \tag{2.80}$$

Given closeness $\mathcal{C}$, we associate with it the symmetric $L \times L$ matrix $\mathbf{C}$ given by

$$\mathbf{C}_{\ell\ell'} = \left\{\begin{array}{ll} 0, & (\ell, \ell') \in \mathcal{C}, \\ 1, & (\ell, \ell') \notin \mathcal{C}.\end{array}\right. \tag{2.81}$$

Test $\mathcal{T}_\mathcal{C}$. Let a collection of shifts $\alpha_{\ell\ell'} \in \mathbf{R}$ satisfying the relation

$$\alpha_{\ell\ell'} = -\alpha_{\ell'\ell},\ 1 \leq \ell, \ell' \leq L \tag{2.82}$$

be given. The detectors $\phi_{\ell\ell'}$ and the shifts $\alpha_{\ell\ell'}$ specify a test $\mathcal{T}_\mathcal{C}$ deciding on hypotheses $H_1, ..., H_L$. Precisely, given an observation ω, the test $\mathcal{T}_\mathcal{C}$ accepts exactly those hypotheses H_ℓ for which $\phi_{\ell\ell'}(\omega) - \alpha_{\ell\ell'} > 0$ whenever ℓ' is *not* $\mathcal{C}$-close to ℓ:

$$\mathcal{T}_\mathcal{C}(\omega) = \{\ell : \phi_{\ell\ell'}(\omega) - \alpha_{\ell\ell'} > 0 \ \forall(\ell' : (\ell, \ell') \notin \mathcal{C})\}.$$

Proposition 2.29. (i) *The $\mathcal{C}$-risk of the test $\mathcal{T}_\mathcal{C}$ just defined is upper-bounded by the quantity*

$$\varepsilon[\alpha] = \max_{\ell \leq L} \sum_{\ell'=1}^{L} \epsilon_{\ell\ell'} \mathbf{C}_{\ell\ell'} \mathrm{e}^{\alpha_{\ell\ell'}}$$

with $\mathbf{C}$ *given by* (2.81).

(ii) *The infimum, over shifts α satisfying (2.82), of the risk bound $\varepsilon[\alpha]$ is the quantity*

$$\varepsilon_\star = \|\mathcal{E}\|_{2,2},$$

where the $L \times L$ symmetric entrywise nonnegative matrix $\mathcal{E}$ is given by

$$\mathcal{E} = [e_{\ell\ell'} := \epsilon_{\ell\ell'} \mathbf{C}_{\ell\ell'}]_{\ell,\ell' \leq L}.$$

Assuming $\mathcal{E}$ admits a strictly positive Perron-Frobenius vector f, an optimal choice of the shifts is

$$\alpha_{\ell\ell'} = \ln(f_{\ell'}/f_\ell), 1 \leq \ell, \ell' \leq L,$$

resulting in $\varepsilon[\alpha] = \varepsilon_\star = \|\mathcal{E}\|_{2,2}$.

Proof. (i): Setting

$$\bar{\phi}_{\ell\ell'}(\omega) = \phi_{\ell\ell'}(\omega) - \alpha_{\ell\ell'},\ \bar{\epsilon}_{\ell\ell'} = \epsilon_{\ell\ell'} \mathrm{e}^{\alpha_{\ell\ell'}},$$

(2.80) and (2.82) imply that

$$\begin{array}{lll} (a) & \bar{\phi}_{\ell\ell'}(\omega) + \bar{\phi}_{\ell'\ell}(\omega) \equiv 0, & 1 \leq \ell, \ell' \leq L \\ (b) & \forall P \in \mathcal{P}_\ell : \int_\Omega \mathrm{e}^{-\bar{\phi}_{\ell\ell'}(\omega)} P(d\omega) \leq \bar{\epsilon}_{\ell\ell'}, & 1 \leq \ell, \ell' \leq L. \end{array} \tag{2.83}$$

Now let ℓ_* be such that the distribution P of observation ω belongs to $\mathcal{P}_{\ell_*}$. Then for every ℓ' the P-probability of the event $\bar{\phi}_{\ell_*\ell'}(\omega) \leq 0$ is $\leq \bar{\epsilon}_{\ell_*\ell'}$ by (2.83.b), whence the P-probability of the event

$$E_* = \left\{\omega : \exists \ell' : (\ell_*, \ell') \notin \mathcal{C} \ \& \ \bar{\phi}_{\ell_*\ell'}(\omega) \leq 0\right\}$$

is upper-bounded by

$$\sum_{\ell' : (\ell_*, \ell') \notin \mathcal{C}} \bar{\epsilon}_{\ell_*\ell'} = \sum_{\ell'=1}^{L} \mathbf{C}_{\ell_*\ell'} \epsilon_{\ell_*\ell'} \mathrm{e}^{\alpha_{\ell_*\ell'}} \leq \varepsilon[\alpha].$$

Assume that E_* does not take place (as we have seen, this indeed is so with P-probability $\geq 1 - \varepsilon[\alpha]$). Then $\bar{\phi}_{\ell_*\ell'}(\omega) > 0$ for all ℓ' such that $(\ell_*, \ell') \notin \mathcal{C}$, implying, first, that H_{ℓ_*} is accepted by our test. Second, $\bar{\phi}_{\ell'\ell_*}(\omega) = -\bar{\phi}_{\ell_*\ell'}(\omega) < 0$ whenever $(\ell_*, \ell') \notin \mathcal{C}$, or, due to the symmetry of closeness, whenever $(\ell', \ell_*) \notin \mathcal{C}$, implying that the test $\mathcal{T}_\mathcal{C}$ rejects the hypothesis $H_{\ell'}$ when ℓ' is not $\mathcal{C}$-close to ℓ_*. Thus, the P-probability of the event "H_{ℓ_*} is accepted, and all accepted hypotheses are $\mathcal{C}$-close to H_{ℓ_*}" is at least $1-\varepsilon[\alpha]$. We conclude that the $\mathcal{C}$-risk $\mathrm{Risk}^{\mathcal{C}}(\mathcal{T}_{\mathcal{C}}|H_1,...,H_L)$ of the test $\mathcal{T}_{\mathcal{C}}$ is at most $\varepsilon[\alpha]$. (i) is proved. (ii) is readily given by Proposition 2.28. □

2.5.2.3 Testing multiple hypotheses via repeated observations

In the situation of Section 2.5.2.1, given a balanced system of detectors $\phi_{\ell\ell'}$ and risks $\epsilon_{\ell\ell'}$, $1 \leq \ell,\ell' \leq L$, for the collection $\mathcal{P}_1,...,\mathcal{P}_L$ (see (2.80)) *and* a positive integer K, we can

- pass from detectors $\phi_{\ell\ell'}$ and risks $\epsilon_{\ell\ell'}$ to the entities

$$\phi^{(K)}_{\ell\ell'}(\omega^K=(\omega_1,...,\omega_K))=\sum_{k=1}^K \phi_{\ell\ell'}(\omega_k),\ \epsilon^{(K)}_{\ell\ell'}=\epsilon^K_{\ell\ell'},\ 1\leq \ell,\ell'\leq L;$$

- associate with the families $\mathcal{P}_\ell$ families $\mathcal{P}^{(K)}_\ell$ of probability distributions underlying quasi-stationary K-repeated versions of observations $\omega\sim P\in\mathcal{P}_\ell$—see Section 2.3.2.3—and thus arrive at hypotheses $H^K_\ell=\mathcal{H}^{\otimes,K}_\ell$ stating that the distribution P^K of K-repeated observation $\omega^K=(\omega_1,...,\omega_K)$, $\omega_k\in\Omega$, belongs to the family $\mathcal{P}^{\otimes,K}_\ell=\bigotimes\limits_{k=1}^K \mathcal{P}_\ell$, associated with $\mathcal{P}_\ell$; see Section 2.1.3.3.

By Proposition 2.16 and (2.80), we arrive at the following analog of (2.80):

$$\begin{array}{ll}\phi^{(K)}_{\ell\ell'}(\omega^K)+\phi^{(K)}_{\ell'\ell}(\omega^K)\equiv 0,\ \epsilon^{(K)}_{\ell\ell'}=\epsilon^{(K)}_{\ell'\ell}=\epsilon^K_{\ell\ell'}, & 1\leq \ell,\ell'\leq L\\ \forall P^K\in\mathcal{P}^{(K)}_\ell:\int_{\Omega^K}\mathrm{e}^{-\phi^{(K)}_{\ell\ell'}(\omega^K)}P^K(d\omega^K)\leq\epsilon^{(K)}_{\ell\ell'}, & 1\leq \ell,\ell'\leq L.\end{array}$$

Given shifts $\alpha_{\ell\ell'}$ satisfying (2.82) and applying the construction from Section 2.5.2.2 to these shifts and our newly constructed detectors and risks, we arrive at the test $\mathcal{T}^K_{\mathcal{C}}$ deciding on hypotheses $H^K_1,...,H^K_L$ via K-repeated observation ω^K. Specifically, given an observation ω^K, the test $\mathcal{T}^K_{\mathcal{C}}$ accepts exactly those hypotheses H^K_ℓ for which $\phi^{(K)}_{\ell\ell'}(\omega^K)-\alpha_{\ell\ell'}>0$ whenever ℓ' is *not* $\mathcal{C}$-close to ℓ:

$$\mathcal{T}^K_{\mathcal{C}}(\omega^K)=\{\ell:\phi^{(K)}_{\ell\ell'}(\omega^K)-\alpha_{\ell\ell'}>0\ \forall(\ell':(\ell,\ell')\notin\mathcal{C})\}.$$

Invoking Proposition 2.29, we arrive at

Proposition 2.30. (i) *The $\mathcal{C}$-risk of the test $\mathcal{T}^K_{\mathcal{C}}$ just defined is upper-bounded by the quantity*

$$\varepsilon[\alpha,K]=\max_{\ell\leq L}\sum_{\ell'=1}^L \epsilon^K_{\ell\ell'}\mathbf{C}_{\ell\ell'}\mathrm{e}^{\alpha_{\ell\ell'}}.$$

(ii) *The infimum, over shifts α satisfying (2.82), of the risk bound $\varepsilon[\alpha,K]$ is the quantity*

$$\varepsilon_\star(K)=\|\mathcal{E}^{(K)}\|_{2,2},$$

where the $L\times L$ symmetric entrywise nonnegative matrix $\mathcal{E}^{(K)}$ is given by

$$\mathcal{E}^{(K)}=\left[e^{(K)}_{\ell\ell'}:=\epsilon^K_{\ell\ell'}\mathbf{C}_{\ell\ell'}\right]_{\ell,\ell'\leq L}.$$

Assuming $\mathcal{E}^{(K)}$ admits a strictly positive Perron-Frobenius vector f, an optimal *choice of the shifts is*

$$\alpha_{\ell\ell'} = \ln(f_\ell/f_{\ell'}), 1 \le \ell, \ell' \le L,$$

resulting in $\varepsilon[\alpha, K] = \varepsilon_\star(K) = \|\mathcal{E}^{(K)}\|_{2,2}$.

2.5.2.4 Consistency and near-optimality

Observe that when closeness $\mathcal{C}$ is such that $\epsilon_{\ell\ell'} < 1$ whenever ℓ, ℓ' are *not* $\mathcal{C}$-close to each other, the entries on the matrix $\mathcal{E}^{(K)}$ go to 0 as $K \to \infty$ exponentially fast, whence the $\mathcal{C}$-risk of test $\mathcal{T}_{\mathcal{C}}^K$ also goes to 0 as $K \to \infty$, meaning that test $\mathcal{T}_{\mathcal{C}}^K$ is *consistent.* When, in addition, $\mathcal{P}_\ell$ correspond to convex hypotheses in a simple o.s., the test $\mathcal{T}_{\mathcal{C}}^K$ possesses the property of near-optimality similar to that stated in Proposition 2.25:

Proposition 2.31. *Consider the special case of the situation from Section 2.5.2.1 where, given a simple o.s.* $\mathcal{O} = (\Omega, \Pi; \{p_\mu : \mu \in \mathcal{M}\}; \mathcal{F})$, *the families* $\mathcal{P}_\ell$ *of probability distributions are of the form* $\mathcal{P}_\ell = \{p_\mu : \mu \in N_\ell\}$, *where* N_ℓ, $1 \le \ell \le L$, *are nonempty convex compact subsets of* $\mathcal{M}$. *Let also the pairwise detectors* $\phi_{\ell\ell'}$ *and their risks* $\epsilon_{\ell\ell'}$ *underlying the construction from Section 2.5.2.2 be obtained by applying Theorem 2.23 to the pairs* $N_\ell, N_{\ell'}$, *so that for* $1 \le \ell < \ell' \le L$ *one has*

$$\phi_{\ell\ell'}(\omega) = \tfrac{1}{2}\ln(p_{\mu_{\ell,\ell'}}(\omega)/p_{\nu_{\ell,\ell'}}(\omega)),\ \epsilon_{\ell\ell'} = \exp\{\mathrm{Opt}_{\ell\ell'}\}$$

where

$$\mathrm{Opt}_{\ell\ell'} = \min_{\mu\in N_\ell, \nu \in N_{\ell'}} \ln\left(\int_\Omega \sqrt{p_\mu(\omega)p_\nu(\omega)}\Pi(d\omega)\right),$$

and $(\mu_{\ell\ell'}, \nu_{\ell\ell'})$ *form an optimal solution to the optimization problem on the right-hand side.*

Assume that for some positive integer K_* *in nature there exists a test* $\mathcal{T}^{K_*}$ *which decides with* $\mathcal{C}$*-risk* $\epsilon \in (0, 1/2)$, *via stationary* K_**-repeated observation* ω^{K_*}, *on the hypotheses* $H_\ell^{(K_*)}$, *stating that the components in* ω^{K_*} *are drawn, independently of each other, from a distribution* $P \in \mathcal{P}_\ell$, $\ell = 1, ..., L$, *and let*

$$K = \left\rfloor 2\frac{1+\ln(L-1)/\ln(1/\epsilon)}{1-\ln(4(1-\epsilon))/\ln(1/\epsilon)} K_* \right\lfloor. \tag{2.84}$$

Then the test $\mathcal{T}_{\mathcal{C}}^K$ *yielded by the construction from Section 2.5.2.2 as applied to the above* $\phi_{\ell\ell'}$, $\epsilon_{\ell\ell'}$, *and trivial shifts* $\alpha_{\ell\ell'} \equiv 0$, *decides on the hypotheses* H_ℓ^K*—see Section 2.5.2.3—via quasi-stationary* K*-repeated observations* ω^K, *with* $\mathcal{C}$*-risk* $\le \epsilon$.

Note that $K/K_* \to 2$ *as* $\epsilon \to +0$.

Proof. Let

$$\bar{\epsilon} = \max_{\ell,\ell'} \{\epsilon_{\ell\ell'} : \ell < \ell', \text{ and } \ell, \ell' \text{ are not } \mathcal{C}\text{-close to each other}\}.$$

Denoting by (ℓ_*, ℓ'_*) the corresponding maximizer, note that $\mathcal{T}^{K_*}$ induces a simple test $\mathcal{T}$ able to decide via stationary K_*-repeated observations ω^K on the pair of hypotheses $H_{\ell_*}^{(K_*)}$, $H_{\ell'_*}^{(K_*)}$ with risks $\le \epsilon$ (it suffices to make $\mathcal{T}$ to accept the first of the hypotheses in the pair and reject the second one whenever $\mathcal{T}^{K_*}$ on the same observation accepts $H_{\ell_*}^{(K_*)}$; otherwise $\mathcal{T}$ rejects the first hypothesis in the pair and accepts the second one). This observation, by the same argument as in the proof of Proposition 2.25, implies that $\bar{\epsilon}^{K_*} \leq 2\sqrt{\epsilon(1-\epsilon)} < 1$, whence all entries in the matrix $\mathcal{E}^{(K)}$ do not exceed $\bar{\epsilon}^{(K/K_*)}$, implying by Proposition 2.29 that the $\mathcal{C}$-risk of the test $\mathcal{T}_\mathcal{C}^K$ does not exceed

$$\epsilon(K) := (L-1)[2\sqrt{\epsilon(1-\epsilon)}]^{K/K_*}.$$

It remains to note that for K given by (2.84) one has $\epsilon(K) \leq \epsilon$. □

Remark 2.32. Note that tests $\mathcal{T}_\mathcal{C}$ and $\mathcal{T}_\mathcal{C}^K$ we have built may, depending on observations, accept no hypotheses at all, which sometimes is undesirable. Clearly, every test deciding on multiple hypotheses up to $\mathcal{C}$-closeness always can be modified to ensure that a hypothesis always is accepted. To this end, it suffices, for instance, that the modified test accepts exactly those hypotheses, if any, which are accepted by our original test, and accepts, say, hypothesis # 1 when the original test accepts no hypotheses. It is immediate to see that the $\mathcal{C}$-risk of the modified test cannot be larger than the risk of the original test.

2.5.3 Illustration: Selecting the best among a family of estimates

Let us illustrate our machinery for multiple hypothesis testing by applying it to the situation as follows:

We are given:

- a simple nondegenerate observation scheme $\mathcal{O} = (\Omega, \Pi; \{p_\mu(\cdot) : \mu \in \mathcal{M}\}; \mathcal{F})$,
- a seminorm $\|\cdot\|$ on $\mathbf{R}^n$,[11]
- a convex compact set $X \subset \mathbf{R}^n$ along with a collection of M points $x_i \in \mathbf{R}^n$, $1 \leq i \leq M$, and a positive D such that the $\|\cdot\|$-diameter of the set $X^+ = X \cup \{x_i : 1 \leq i \leq M\}$ is at most D:

$$\|x - x'\| \leq D \ \forall (x, x' \in X^+),$$

- an affine mapping $x \mapsto A(x)$ from $\mathbf{R}^n$ into the embedding space of $\mathcal{M}$ such that $A(x) \in \mathcal{M}$ for all $x \in X$,
- a tolerance $\epsilon \in (0,1)$.

We observe a K-element sample $\omega^K = (\omega_1, ..., \omega_K)$ of observations

$$\omega_k \sim p_{A(x_*)},\ 1 \leq k \leq K, \tag{2.85}$$

independent across k, where $x_* \in \mathbf{R}^n$ is an unknown signal known to belong to X. Our "ideal goal" is to use ω^K in order to identify, with probability $\geq 1-\epsilon$, the $\|\cdot\|$-closest to x_* point among the points $x_1, ..., x_M$.

The goal just outlined may be too ambitious, and in the sequel we focus on the relaxed goal as follows:

---

[11] A seminorm on $\mathbf{R}^n$ is defined by exactly the same requirements as a norm, except that now we allow zero seminorms for some nonzero vectors. Thus, a seminorm on $\mathbf{R}^n$ is a nonnegative function $\|\cdot\|$ which is even and homogeneous: $\|\lambda x\| = |\lambda|\|x\|$ and satisfies the triangle inequality $\|x+y\| \leq \|x\| + \|y\|$. A universal example is $\|x\| = \|Bx\|_o$, where $\|\cdot\|_o$ is a norm on some $\mathbf{R}^m$ and B is an $m \times n$ matrix; whenever this matrix has a nontrivial kernel, $\|\cdot\|$ is a seminorm rather than a norm.

Given a positive integer N and a "resolution" $\theta > 1$, consider the grid

$$\Gamma = \{r_j = D\theta^{-j}, 0 \leq j \leq N\}$$

and let

$$\rho(x) = \min\left\{\rho_j \in \Gamma : \rho_j \geq \min_{1\leq i\leq M} \|x - x_i\|\right\}.$$

Given the design parameters $\alpha \geq 1$ and $\beta \geq 0$, we want to specify a volume of observations K and an inference routine $\omega^K \mapsto i_{\alpha,\beta}(\omega^K) \in \{1, ..., M\}$ such that

$$\forall(x_* \in X) : \text{Prob}\{\|x_* - x_{i_{\alpha,\beta}(\omega^K)}\| > \alpha\rho(x_*) + \beta\} \geq 1 - \epsilon. \tag{2.86}$$

Note that when passing from the "ideal" to the relaxed goal, the simplification is twofold: first, instead of the precise distance $\min_i \|x_* - x_i\|$ from x_* to $\{x_1, ..., x_M\}$ we look at the best upper bound $\rho(x_*)$ on this distance from the grid Γ; second, we allow factor α and additive term β in mimicking the (discretized) distance $\rho(x_*)$ by $\|x_* - x_{i_{\alpha,\beta}(\omega^K)}\|$.

The problem we have posed is quite popular in Statistics and originates from the estimate aggregation problem [185, 229, 101] as follows: let x_i be candidate estimates of x_* yielded by a number of a priori "models" of x_* and perhaps some preliminary noisy observations of x_*. Given x_i and a matrix B, we want to select among the vectors Bx_i the (nearly) best approximation of Bx_* w.r.t. a given norm $\|\cdot\|_o$, utilizing additional observations ω^K of the signal. To bring this problem into our framework, it suffices to specify the seminorm as $\|x\| = \|Bx\|_o$. We shall see in the meantime that in the context of this problem, the "discretization of distances" is, for all practical purposes, irrelevant: the dependence of the volume of observations on N is just logarithmic, so that we can easily handle a fine grid, like the one with $\theta = 1.001$ and $\theta^{-N} = 10^{-10}$. As for factor α and additive term β, they indeed could be "expensive in terms of applications," but the "nearly ideal" goal of making α close to 1 and β close to 0 may be unattainable.

2.5.3.1 The construction

Let us associate with $i \leq M$ and j, $0 \leq j \leq N$, the hypothesis H_{ij} stating that the observations ω_k independent across k—see (2.85)—stem from

$$x_* \in X_{ij} := \{x \in X : \|x - x_i\| \leq r_j\}.$$

Note that the sets X_{ij} are convex and compact. We denote by $\mathcal{J}$ the set of all pairs (i, j), for which $i \in \{1, ..., M\}$, $j \in \{0, 1, ..., N\}$, and $X_{ij} \neq \emptyset$. Further, we define closeness $\mathcal{C}_{\alpha,\beta}$ on the set of hypotheses H_{ij}, $(i, j) \in \mathcal{J}$, as follows:

$(ij, i'j') \in \mathcal{C}_{\alpha\beta}$ *if and only if*

$$\|x_i - x_{i'}\| \leq \bar{\alpha}(r_j + r_{j'}) + \beta,\ \bar{\alpha} = \frac{\alpha - 1}{2} \tag{2.87}$$

(here and in what follows, $k\ell$ denotes the ordered pair (k, ℓ)).

Applying Theorem 2.23, we can build, in a computation-friendly fashion, the system $\phi_{ij,i'j'}(\omega)$, $ij, i'j' \in \mathcal{J}$, of optimal balanced detectors for the hypotheses H_{ij} along with the risks of these detectors, so that

$$\begin{array}{ll} \phi_{ij,i'j'}(\omega) \equiv -\phi_{i'j',ij}(\omega) & \forall (ij, i'j' \in \mathcal{J}), \\ \int_\Omega e^{-\phi_{ij,i'j'}(\omega)} p_{A(x)}(\omega)\Pi(d\omega) \leq \epsilon_{ij,i'j'} & \forall (ij \in \mathcal{J}, i'j' \in \mathcal{J}, x \in X_{ij}). \end{array}$$

Let us say that a pair (α, β) is *admissible* if $\alpha \geq 1$, $\beta \geq 0$, and

$$\forall ((i,j) \in \mathcal{J}, (i',j') \in \mathcal{J}, (ij, i'j') \notin \mathcal{C}_{\alpha,\beta}) : A(X_{ij}) \cap A(X_{i'j'}) = \emptyset.$$

Note that checking admissibility of a given pair (α, β) is a computationally tractable task.

Given an admissible pair (α, β), we associate with it a positive integer $K = K(\alpha, \beta)$ and inference $\omega^K \mapsto i_{\alpha,\beta}(\omega^K)$ as follows:

1. $K = K(\alpha, \beta)$ is the smallest integer such that the detector-based test $\mathcal{T}^K_{\mathcal{C}_{\alpha,\beta}}$ yielded by the machinery of Section 2.5.2.3 decides on the hypotheses H_{ij}, $ij \in \mathcal{J}$, with $\mathcal{C}_{\alpha,\beta}$-risk not exceeding ϵ. Note that by admissibility, $\epsilon_{ij,i'j'} < 1$ whenever $(ij, i'j') \notin \mathcal{C}_{\alpha,\beta}$, so that $K(\alpha, \beta)$ is well defined.
2. Given observation ω^K, $K = K(\alpha, \beta)$, we define $i_{\alpha,\beta}(\omega^K)$ as follows:
   a) We apply to ω^K the test $\mathcal{T}^K_{\mathcal{C}_{\alpha,\beta}}$. If the test accepts no hypothesis (case A), $i_{\alpha\beta}(\omega^K)$ is undefined. The observations ω^K resulting in case A comprise some set, which we denote by $\mathcal{B}$; given ω^K, we can recognize whether or not $\omega^K \in \mathcal{B}$.
   b) When $\omega^K \notin \mathcal{B}$, the test $\mathcal{T}^K_{\mathcal{C}_{\alpha,\beta}}$ accepts some of the hypotheses H_{ij}, let the set of their indices ij be $\mathcal{J}(\omega^K)$; we select from the pairs $ij \in \mathcal{J}(\omega^K)$ the one with the largest j, and set $i_{\alpha,\beta}(\omega^K)$ to be equal to the first component, and $j_{\alpha,\beta}(\omega^K)$ to be equal to the second component of the selected pair.

We have the following:

Proposition 2.33. *Assuming (α, β) admissible, for the inference $\omega^K \mapsto i_{\alpha,\beta}(\omega^K)$ just defined and for every $x_* \in X$, denoting by $P^K_{x_*}$ the distribution of stationary K-repeated observation ω^K stemming from x_* one has*

$$\|x_* - x_{i_{\alpha,\beta}(\omega^K)}\| \leq \alpha\rho(x_*) + \beta \tag{2.88}$$

with $P^K_{x_*}$-probability at least $1 - \epsilon$.

Proof. Let us fix $x_* \in X$, and let $j_* = j_*(x_*)$ be the largest $j \leq N$ such that

$$r_j \geq \min_{i \leq M} \|x_* - x_i\|;$$

note that j_* is well defined due to $r_0 = D \geq \|x_* - x_1\|$, and that

$$r_{j_*} = \rho(x_*).$$

We specify $i_* = i_*(x_*) \leq M$ in such a way that

$$\|x_* - x_{i_*}\| \leq r_{j_*}. \tag{2.89}$$

Note that i_* is well defined and that observations (2.85) stemming from x_* obey the hypothesis $H_{i_*j_*}$.

Let $\mathcal{E}$ be the set of those ω^K for which the predicate

$\mathcal{P}$: As applied to observation ω^K, the test $\mathcal{T}^K_{\mathcal{C}_{\alpha,\beta}}$ accepts $H_{i_*j_*}$, and all hypotheses accepted by the test are $\mathcal{C}_{\alpha,\beta}$-close to $H_{i_*j_*}$

holds true. Taking into account that the $\mathcal{C}_{\alpha,\beta}$-risk of $\mathcal{T}^K_{\mathcal{C}_{\alpha,\beta}}$ does not exceed ϵ and that the hypothesis $H_{i_*j_*}$ is true, the $P^K_{x_*}$-probability of the event $\mathcal{E}$ is at least $1-\epsilon$.

Let observation ω^K satisfy

$$\omega^K \in \mathcal{E}. \tag{2.90}$$

Then

1. The test $\mathcal{T}^K_{\mathcal{C}_{\alpha,\beta}}$ accepts the hypothesis $H_{i_*j_*}$, that is, $\omega^K \notin \mathcal{B}$. By construction of $i_{\alpha,\beta}(\omega^K)j_{\alpha,\beta}(\omega^K)$ (see the rule 2b above) and due to the fact that $\mathcal{T}^K_{\mathcal{C}_{\alpha,\beta}}$ accepts $H_{i_*j_*}$, we have $j_{\alpha,\beta}(\omega^K) \geq j_*$.
2. The hypothesis $H_{i_{\alpha,\beta}(\omega^K)j_{\alpha,\beta}(\omega^K)}$ is $\mathcal{C}_{\alpha,\beta}$-close to $H_{i_*j_*}$, so that

   $$\|x_{i_*} - x_{i_{\alpha,\beta}(\omega^K)}\| \leq \bar{\alpha}(r_{j_*} + r_{j_{\alpha,\beta}(\omega^K)}) + \beta \leq 2\bar{\alpha}r_{j_*} + \beta = 2\bar{\alpha}\rho(x_*) + \beta,$$

   where the concluding inequality is due to the fact that, as we have already seen, $j_{\alpha,\beta}(\omega^K) \geq j_*$ when (2.90) takes place.

Invoking (2.89), we conclude that with $P^K_{x_*}$-probability at least $1-\epsilon$ it holds

$$\|x_* - x_{i_{\alpha,\beta}(\omega^K)}\| \leq (2\bar{\alpha}+1)\rho(x_*) + \beta = \alpha\rho(x_*) + \beta. \qquad \square$$

2.5.3.2 A modification

From the computational viewpoint, an obvious shortcoming of the construction presented in the previous section is the necessity to operate with $M(N+1)$ hypotheses, which might require computing as many as $O(M^2N^2)$ detectors. We are about to present a modified construction, where we deal at most $N+1$ times with just M hypotheses at a time (i.e., with the total of at most $O(M^2N)$ detectors). The idea is to replace simultaneously processing all hypotheses H_{ij}, $ij \in \mathcal{J}$, with processing them in *stages* $j = 0, 1, \ldots$, with stage j operating only with the hypotheses H_{ij}, $i = 1, \ldots, M$.

The implementation of this idea is as follows. In the situation of Section 2.5.3, given the same entities Γ, (α, β), H_{ij}, X_{ij}, $ij \in \mathcal{J}$, as at the beginning of Section 2.5.3.1 and specifying closeness $\mathcal{C}_{\alpha,\beta}$ according to (2.87), we now act as follows.

Preprocessing. For $j = 0, 1, \ldots, N$

1. we identify the set $\mathcal{I}_j = \{i \leq M : X_{ij} \neq \emptyset\}$ and stop if this set is empty. If this set is nonempty,
2. we specify the closeness $\mathcal{C}^j_{\alpha\beta}$ on the set of hypotheses H_{ij}, $i \in \mathcal{I}_j$, as a "slice" of the closeness $\mathcal{C}_{\alpha,\beta}$:

   H_{ij} and $H_{i'j}$ (equivalently, i and i') are $\mathcal{C}^j_{\alpha,\beta}$-close to each other if $(ij, i'j)$ are $\mathcal{C}_{\alpha,\beta}$-close, that is,

   $$\|x_i - x_{i'}\| \leq 2\bar{\alpha}r_j + \beta, \ \bar{\alpha} = \frac{\alpha - 1}{2}.$$

3. We build the optimal detectors $\phi_{ij,i'j}$, along with their risks $\epsilon_{ij,i'j}$, for all $i, i' \in \mathcal{I}_j$ such that $(i, i') \notin \mathcal{C}^j_{\alpha,\beta}$. If $\epsilon_{ij,i'j} = 1$ for a pair i, i' of the latter type, that is,

$A(X_{ij}) \cap A(X_{i'j}) \neq \emptyset$, we claim that (α, β) is inadmissible and stop. Otherwise we find the smallest $K = K_j$ such that the spectral norm of the symmetric $M \times M$ matrix E^{jK} with the entries

$$E^{jK}_{ii'} = \begin{cases} \epsilon^K_{ij,i'j}, & i \in \mathcal{I}_j, i' \in \mathcal{I}_j, (i,i') \notin \mathcal{C}^j_{\alpha,\beta} \\ 0, & \text{otherwise} \end{cases}$$

does not exceed $\bar{\epsilon} = \epsilon/(N+1)$. We then use the machinery of Section 2.5.2.3 to build detector-based test $\mathcal{T}^{K_j}_{\mathcal{C}^j_{\alpha,\beta}}$, which decides on the hypotheses H_{ij}, $i \in \mathcal{I}_j$, with $\mathcal{C}^j_{\alpha,\beta}$-risk not exceeding $\bar{\epsilon}$.

It may happen that the outlined process stops when processing some value $\bar{j}$ of j; if this does not happen, we set $\bar{j} = N + 1$. Now, if the process does stop, and stops with the claim that (α, β) is inadmissible, we call (α, β) inadmissible and terminate—in this case we fail to produce a desired inference; note that if this is the case, (α, β) is inadmissible in the sense of Section 2.5.3.1 as well. When we do not stop with the inadmissibility claim, we call (α, β) admissible, and in this case we do produce an inference, specifically, as follows.

Processing observations:

1. We set $\bar{\mathcal{J}} = \{0, 1, ..., \widehat{j} = \bar{j} - 1\}$, $K = K(\alpha, \beta) = \max\limits_{0 \leq j \leq \widehat{j}} K^j$. Note that $\bar{\mathcal{J}}$ is nonempty due to $\bar{j} > 0$.[12]

2. Let $\omega^K = (\omega_1, ..., \omega_K)$ with independent across k components stemming from unknown signal $x_* \in X$ according to (2.85). We put $\widehat{\mathcal{I}}_{-1}(\omega^K) = \{1, ..., M\} = \mathcal{I}_0$.

   a) For $j = 0, 1, ..., \widehat{j}$, we act as follows. When processing j, we have at our disposal subsets $\widehat{\mathcal{I}}_k(\omega^K) \subset \{1, ..., M\}$, $-1 \leq k < j$. To build the set $\widehat{\mathcal{I}}_j(\omega^K)$

      i. we apply the test $\mathcal{T}^{K_j}_{\mathcal{C}^j_{\alpha,\beta}}$ to the initial K_j components of the observation ω^K. Let $\mathcal{I}^+_j(\omega^K)$ be the set of hypotheses H_{ij}, $i \in \mathcal{I}_j$, accepted by the test;

      ii. it may happen that $\mathcal{I}^+_j(\omega^K) = \emptyset$; if it is so, we terminate;

      iii. if $\mathcal{I}^+_j(\omega^K)$ is nonempty, we look, one by one, at indices $i \in \mathcal{I}^+_j(\omega^K)$ and call the index i good if for every $\ell \in \{-1, 0, ..., j-1\}$, $i \in \widehat{\mathcal{I}}_\ell(\omega^K)$;

      iv. we define $\widehat{\mathcal{I}}_j(\omega^K)$ as the set of good indices of $\mathcal{I}^+_j(\omega^K)$ if this set is not empty and proceed to the next value of j (if $j < \widehat{j}$), or terminate (if $j = \widehat{j}$). We terminate if there are no good indices in $\mathcal{I}^+_j(\omega^K)$.

   b) Upon termination, we have at our disposal a collection $\widehat{\mathcal{I}}_j(\omega^K)$, $0 \leq j \leq \widetilde{j}(\omega^K)$, of all sets $\widehat{\mathcal{I}}_j(\omega^K)$ we have built (this collection can be empty, which we encode by setting $\widetilde{j}(\omega^K) = -1$). When $\widetilde{j}(\omega^K) = -1$, our inference remains undefined. Otherwise we select from the set $\widehat{\mathcal{I}}_{\widetilde{j}(\omega^K)}(\omega^K)$ an index $i_{\alpha,\beta}(\omega^K)$, say, the smallest one, and claim that the point $x_{i_{\alpha,\beta}(\omega^K)}$ is the point among

[12] All the sets X_{i0} contain X and thus are nonempty, so that $\mathcal{I}_0 = \{1, ..., M\} \neq \emptyset$, and thus we cannot stop at step $j = 0$ due to $\mathcal{I}_0 = \emptyset$; the other possibility to stop at step $j = 0$ is ruled out by the fact that we are in the case where (α, β) is admissible.

$x_1, ..., x_M$ "nearly closest" to x_*.

We have the following analog of Proposition 2.33:

Proposition 2.34. *Assuming (α, β) admissible, for the inference $\omega^K \mapsto i_{\alpha,\beta}(\omega^K)$ just defined and for every $x_* \in X$, denoting by $P^K_{x_*}$ the distribution of stationary K-repeated observation ω^K stemming from x_* one has*

$$P^K_{x_*}\left\{\omega^K : i_{\alpha,\beta}(\omega^K) \text{ is well defined and } \|x_* - x_{i_{\alpha,\beta}(\omega^K)}\| \leq \alpha\rho(x_*) + \beta\right\} \geq 1 - \epsilon.$$

Proof. Let us fix the signal $x_* \in X$ underlying observations ω^K. As in the proof of Proposition 2.33, let j_* be such that $\rho(x_*) = r_{j_*}$, and let $i_* \leq M$ be such that $x_* \in X_{i_*j_*}$. Clearly, i_* and j_* are well defined, and the hypotheses H_{i_*j}, $0 \leq j \leq j_*$, are true. In particular, $X_{i_*j} \neq \emptyset$ when $j \leq j_*$, implying that $i_* \in \mathcal{I}_j$, $0 \leq j \leq j_*$, whence also $\widehat{j} \geq j_*$.

For $0 \leq j \leq j_*$, let $\mathcal{E}_j$ be the set of all realizations of ω^K such that

$$i_* \in \mathcal{I}_j^+(\omega^K) \;\&\; \{(i_*, i) \in \mathcal{C}^j_{\alpha,\beta}\, \forall i \in \mathcal{I}_j^+(\omega^K)\}.$$

Since the $\mathcal{C}^j_{\alpha,\beta}$-risk of the test $\mathcal{T}^{K_j}_{\mathcal{C}^j_{\alpha,\beta}}$ is $\leq \bar{\epsilon}$, we conclude that the $P^K_{x_*}$-probability of $\mathcal{E}_j$ is at least $1 - \bar{\epsilon}$, whence the $P^K_{x_*}$-probability of the event

$$\mathcal{E} = \bigcap_{j=0}^{j_*} \mathcal{E}_j$$

is at least $1 - (N+1)\vec{\epsilon} = 1 - \epsilon$.

Now let

$$\omega^K \in \mathcal{E}.$$

Then,

- By the definition of $\mathcal{E}_j$, when $j \leq j_*$, we have $i_* \in \mathcal{I}_j^+(\omega^K)$, whence, by evident induction in j, $i_* \in \widehat{\mathcal{I}}_j(\omega^K)$ for all $j \leq j_*$.
- We conclude from the above that $\widetilde{j}(\omega^K) \geq j_*$. In particular, $i := i_{\alpha,\beta}(\omega^K)$ is well defined and turned out to be good at step $\widetilde{j} \geq j_*$, implying that $i \in \widehat{\mathcal{I}}_{j_*}(\omega^K) \subset \mathcal{I}^+_{j_*}(\omega^K)$.

Thus, $i \in \mathcal{I}^+_{j_*}(\omega^K)$, which combines with the definition of $\mathcal{E}_{j_*}$ to imply that i and i_* are $\mathcal{C}^{j_*}_{\alpha,\beta}$-close to each other, whence

$$\|x_{i(\alpha,\beta)(\omega^K)} - x_{i_*}\| \leq 2\bar{\alpha}r_{j_*} + \beta = 2\bar{\alpha}\rho(x_*) + \beta,$$

resulting in the desired relation

$$\|x_{i(\alpha,\beta)(\omega^K)} - x_*\| \leq 2\bar{\alpha}\rho(x_*) + \beta + \|x_{i_*} - x_*\| \leq [2\bar{\alpha}+1]\rho(x_*) + \beta = \alpha\rho(x_*) + \beta. \;\square$$

2.5.3.3 "Near-optimality"

We augment the above constructions with the following

Proposition 2.35. *Let for some positive integer $\bar{K}$, $\epsilon \in (0, 1/2)$, and a pair $(a, b) \geq$*

0 *there exist an inference* $\omega^{\bar K} \mapsto i(\omega^{\bar K}) \in \{1,...,M\}$ *such that whenever* $x_* \in X$, *we have*

$$\mathrm{Prob}_{\omega^{\bar K}\sim P_{x_*}^{\bar K}}\{\|x_* - x_{i(\omega^{\bar K})}\| \leq a\rho(x_*) + b\} \geq 1-\epsilon.$$

Then the pair ($\alpha = 2a + 3, \beta = 2b$) *is admissible in the sense of Section 2.5.3.1 (and thus—in the sense of Section 2.5.3.2), and for the constructions in Sections 2.5.3.1 and 2.5.3.2 one has*

$$K(\alpha,\beta) \leq \mathrm{Ceil}\left(2\frac{1+\ln(M(N+1))/\ln(1/\epsilon)}{1-\frac{\ln(4(1-\epsilon))}{\ln(1/\epsilon)}}\bar K\right); \tag{2.91}$$

Proof. Consider the situation of Section 2.5.3.1 (the situation of Section 2.5.3.2 can be processed in a completely similar way). Observe that with α, β as above, there exists a simple test deciding on a pair of hypotheses H_{ij}, $H_{i'j'}$ which are *not* $\mathcal{C}_{\alpha,\beta}$-close to each other via stationary $\bar K$-repeated observation $\omega^{\bar K}$ with risk $\leq \epsilon$. Indeed, the desired test $\mathcal{T}$ is as follows: given $ij \in \mathcal{J}$, $i'j' \in \mathcal{J}$, and observation $\omega^{\bar K}$, we compute $i(\omega^{\bar K})$ and accept H_{ij} if and only if $\|x_{i(\omega^{\bar K})} - x_i\| \leq (a+1)r_j + b$, and accept $H_{i'j'}$ otherwise. Let us check that the risk of this test indeed is at most ϵ. Assume, first, that H_{ij} takes place. The $P_{x_*}^{\bar K}$-probability of the event

$$\mathcal{E}: \|x_{i(\omega^{\bar K})} - x_*\| \leq a\rho(x_*) + b$$

is at least $1-\epsilon$ due to the origin of $i(\cdot)$, and $\|x_i - x_*\| \leq r_j$ since H_{ij} takes place, implying that $\rho(x_*) \leq r_j$ by the definition of $\rho(\cdot)$. Thus, in the case of $\mathcal{E}$ it holds

$$\|x_{i(\omega^{\bar K})} - x_i\| \leq \|x_{i(\omega^{\bar K})} - x_*\| + \|x_i - x_*\| \leq a\rho(x_*) + b + r_j \leq (a+1)r_j + b.$$

We conclude that if H_{ij} is true and $\omega^{\bar K} \in \mathcal{E}$, then the test $\mathcal{T}$ accepts H_{ij}, and thus the $P_{x_*}^{\bar K}$-probability for the simple test $\mathcal{T}$ not to accept H_{ij} when the hypothesis takes place is $\leq \epsilon$.

Now let $H_{i'j'}$ take place, and let $\mathcal{E}$ be the same event as above. When $\omega^{\bar K} \in \mathcal{E}$, which happens with the $P_{x_*}^{\bar K}$-probability at least $1-\epsilon$, for the same reasons as above, we have $\|x_{i(\omega^{\bar K})} - x_{i'}\| \leq (a+1)r_{j'} + b$. It follows that when $H_{i'j'}$ takes place and $\omega^{\bar K} \in \mathcal{E}$, we have $\|x_{i(\omega^{\bar K})} - x_i\| > (a+1)r_j + b$, since otherwise we would have

$$\begin{array}{rcl}\|x_i - x_{i'}\| & \leq & \|x_{i(\omega^{\bar K})} - x_i\| + \|x_{i(\omega^{\bar K})} - x_{i'}\| \leq (a+1)r_j + b + (a+1)r_{j'} + b \\ & = & (a+1)(r_j + r_{j'}) + 2b = \frac{\alpha-1}{2}(r_j + r_{j'}) + \beta,\end{array}$$

which contradicts the fact that ij and $i'j'$ are not $\mathcal{C}_{\alpha,\beta}$-close. Thus, whenever $H_{i'j'}$ holds true and $\mathcal{E}$ takes place, we have $\|x_{i(\omega^{\bar K})} - x_i\| > (a+1)r_j + b$, implying by the definition of $\mathcal{T}$ that $\mathcal{T}$ accepts $H_{i'j'}$. Thus, the $P_{x_*}^{\bar K}$-probability not to accept $H_{i'j'}$ when the hypotheses is true is at most ϵ. From the fact that whenever $(ij, i'j') \notin \mathcal{C}_{\alpha,\beta}$, the hypotheses H_{ij}, $H_{i'j'}$ can be decided upon, via $\bar K$ observations, with risk $\leq \epsilon < 0.5$ it follows that for the $ij, i'j'$ in question, the sets $A(X_{ij})$ and $A(X_{i'j'})$ do not intersect, so that (α, β) is an admissible pair.

As in the proof of Proposition 2.31, by basic properties of simple observation schemes, the fact that the hypotheses H_{ij}, $H_{i'j'}$ with $(ij, i'j') \notin \mathcal{C}_{\alpha,\beta}$ can be decided upon via $\bar K$-repeated observations (2.85) with risk $\leq \epsilon < 1/2$ implies that $\epsilon_{ij,i'j'} \leq [2\sqrt{\epsilon(1-\epsilon)}]^{1/\bar K}$, whence, again by basic results on simple observation schemes (look once again at the proof of Proposition 2.31), the $\mathcal{C}_{\alpha,\beta}$-risk of K-observation detector-based test $\mathcal{T}_K$ deciding on the hypotheses H_{ij}, $ij \in \mathcal{J}$, up to closeness $\mathcal{C}_{\alpha,\beta}$ does not exceed $\mathrm{Card}(\mathcal{J})[2\sqrt{\epsilon(1-\epsilon)}]^{K/\bar{K}} \leq M(N+1)[2\sqrt{\epsilon(1-\epsilon)}]^{K/\bar{K}}$, and (2.91) follows. □

Comment. Proposition 2.35 says that in our problem, the "statistical toll" for quite large values of N and M is quite moderate: with $\epsilon = 0.01$, resolution $\theta = 1.001$ (which for all practical purposes is the same as no discretization of distances at all), D/r_N as large as 10^{10}, and M as large as 10,000, (2.91) reads $K = \mathrm{Ceil}(10.7\bar{K})$—not a disaster! The actual statistical toll of our construction is in replacing the "existing in nature" a and b with a $\alpha = 2\alpha + 3$ and $\beta = 2b$. And of course there is a huge computational toll for large M and N: we need to operate with large (albeit polynomial in M, N) number of hypotheses and detectors.

2.5.3.4 Numerical illustration

As an illustration of the approach presented in this section consider the following (toy) problem:

> A signal $x_* \in \mathbf{R}^n$ (one may think of x_* as of the restriction on the equidistant n-point grid in $[0,1]$ of a function of continuous argument $t \in [0,1]$) is observed according to
>
> $$\omega = Ax_* + \xi,\ \xi \sim \mathcal{N}(0, \sigma^2 I_n), \tag{2.92}$$
>
> where A is a "discretized integration":
>
> $$(Ax)_s = \frac{1}{n}\sum_{j=1}^{s} x_s,\ s = 1, ..., n.$$
>
> We want to approximate x in the discrete version of L_1-norm
>
> $$\|y\| = \frac{1}{n}\sum_{s=1}^{n} |y_s|,\ y \in \mathbf{R}^n$$
>
> by a low-order polynomial.

In order to build the approximation, we use a single observation ω as in (2.92), to build five candidate estimates x_i, $i = 1, ..., 5$, of x_*. Specifically, x_i is the Least Squares polynomial—of degree $\leq i-1$—approximation of x:

$$x_i = \underset{y \in \mathcal{P}_{i-1}}{\mathrm{argmin}}\, \|Ay - \omega\|_2^2,$$

where $\mathcal{P}_\kappa$ is the linear space of algebraic polynomials, of degree $\leq \kappa$, of discrete argument s varying in $\{1, 2, ..., n\}$. After the candidate estimates are built, we use additional K observations (2.92) "to select the model"—to select among our estimates the $\|\cdot\|$-closest to x_*.

In the experiment reported below we use $n = 128$ and $\sigma = 0.01$. The true signal x_* is a discretization of a piecewise linear function of continuous argument $t \in [0,1]$, with slope 2 to the left of $t = 0.5$, and with slope -2 to the right of $t = 0.5$; at $t = 0.5$, the function has a jump. The a priori information on the true signal is that

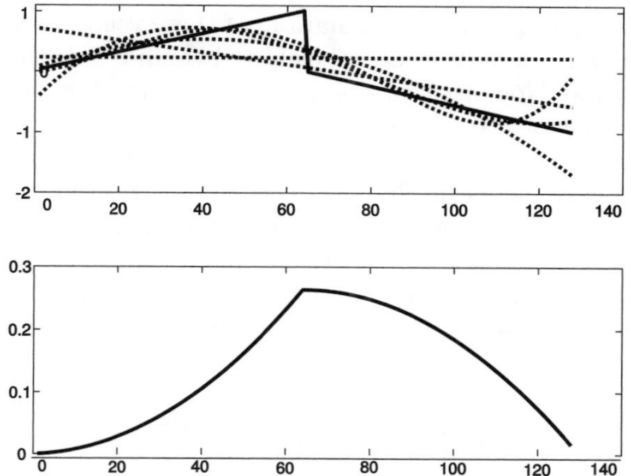

Figure 2.5: Signal (top, solid) and candidate estimates (top, dotted). Bottom: the primitive of the signal.

it belongs to the box $\{x \in \mathbf{R}^n : \|x\|_\infty \leq 1\}$. The signal and sample polynomial approximations x_i of x_*, $1 \leq i \leq 5$, are presented on the top plot in Figure 2.5; their actual $\|\cdot\|$-distances to x_* are as follows:

i	1	2	3	4	5
$\|x_i - x_*\|$	0.534	0.354	0.233	0.161	0.172

Setting $\epsilon = 0.01$, $N = 22$, and $\theta = 2^{1/4}$, $\alpha = 3$ and $\beta = 0.05$ resulted in $K = 3$. In a series of 1,000 simulations of the resulting inference, *all* 1,000 results correctly identified the candidate estimate x_4 $\|\cdot\|$-closest to x_*, in spite of the factor $\alpha = 3$ in (2.88). Surprisingly, the same holds true when we use the resulting inference with the reduced values of K, namely, $K = 1$ and $K = 2$, although the theoretical guarantees deteriorate: with $K = 1$ and $K = 2$, the theory guarantees the validity of (2.88) with probabilities 0.77 and 0.97, respectively.

2.6 SEQUENTIAL HYPOTHESIS TESTING

2.6.1 Motivation: Election polls

Let us consider the following "practical" question.

> One of L candidates for an office is about to be selected by a population-wide majority vote. Every member of the population votes for exactly one candidate. How do we predict the winner via an opinion poll?

A (naive) model of the situation could be as follows. Let us represent the preference of a particular voter by his *preference vector*—a basic orth e in $\mathbf{R}^L$ with unit entry in a position ℓ meaning that the voter is about to vote for the ℓ-th candidate. The entries μ_ℓ in the average μ, over the population, of these vectors are the fractions of votes in favor of the ℓ-th candidate, and the elected candidate is the one "indexing" the largest of the μ_ℓ's. Now assume that we select at random, from the uniform distribution, a member of the population and observe his preference vector. Our observation ω is a realization of a discrete random variable taking values in the set $\Omega = \{e_1, ..., e_L\}$ of basic orths in $\mathbf{R}^L$, and μ is the distribution of ω (technically, the density of this distribution w.r.t. the counting measure Π on Ω). Selecting a small threshold δ and assuming that the true—unknown to us—μ is such that the largest entry in μ is at least by δ larger than every other entry and that $\mu_\ell \geq \frac{1}{N}$ for all ℓ, N being the population size,[13] we can model the population preference for the ℓ-th candidate with

$$\begin{array}{rcl}\mu \in M_\ell & = & \{\mu \in \mathbf{R}^d : \mu_i \geq \frac{1}{N}, \sum_i \mu_i = 1, \mu_\ell \geq \mu_i + \delta\,\forall (i \neq \ell)\}\\ & \subset & \mathcal{M} = \{\mu \in \mathbf{R}^d : \mu > 0, \sum_i \mu_i = 1\}.\end{array}$$

In an (idealized) poll, we select at random a number K of voters and observe their preferences, thus arriving at a sample $\omega^K = (\omega_1, ..., \omega_K)$ of observations drawn, independently of each other, from an unknown distribution μ on Ω, with μ known to belong to $\bigcup_{\ell=1}^L M_\ell$. Therefore, to predict the winner is the same as to decide on L convex hypotheses, $H_1, ..., H_L$, in the Discrete o.s., with H_ℓ stating that $\omega_1, ..., \omega_K$ are drawn, independently of each other, from a distribution $\mu \in M_\ell$. What we end up with, is the problem of deciding on L convex hypotheses in the Discrete o.s. with L-element Ω via stationary K-repeated observations.

Illustration. Consider two-candidate elections; now the goal of a poll is, given K independent of each other realizations $\omega_1, ..., \omega_K$ of random variable ω taking value $\chi = 1, 2$ with probability μ_χ, $\mu_1 + \mu_2 = 1$, to decide what is larger, μ_1 or μ_2. As explained above, we select somehow a threshold δ and impose on the unknown μ an a priori assumption that the gap between the largest and the next largest (in our case, just the smallest) entry of μ is at least δ, thus arriving at two hypotheses,

$$H_1 : \mu_1 \geq \mu_2 + \delta, \quad H_2 : \mu_2 \geq \mu_1 + \delta,$$

which is the same as

$$\begin{array}{l}H_1 : \mu \in M_1 = \{\mu : \mu_1 \geq \frac{1+\delta}{2}, \mu_2 \geq 0, \mu_1 + \mu_2 = 1\},\\ H_2 : \mu \in M_2 = \{\mu : \mu_2 \geq \frac{1+\delta}{2}, \mu_1 \geq 0, \mu_1 + \mu_2 = 1\}.\end{array}$$

We now want to decide on these two hypotheses via a stationary K-repeated observation. We are in the case of a simple (specifically, Discrete) o.s.; the optimal detector as given by Theorem 2.23 stems from the optimal solution (μ^*, ν^*) to the convex optimization problem

$$\varepsilon_\star = \max_{\mu \in M_1, \nu \in M_2} [\sqrt{\mu_1 \nu_1} + \sqrt{\mu_2 \nu_2}]; \tag{2.93}$$

the optimal balanced single-observation detector is

$$\phi_*(\omega) = f_*^T \omega, \; f_* = \tfrac{1}{2}[\ln(\mu_1^*/\nu_1^*); \ln(\mu_2^*/\nu_2^*)]$$

13 With the size N of population in the range of tens of thousands and δ as $1/N$, both these assumptions seem to be quite realistic.

(recall that we encoded observations ω_k by basic orths from $\mathbf{R}^2$), the risk of this detector being $\varepsilon_\star$. In other words,

$$\mu^* = [\tfrac{1+\delta}{2}; \tfrac{1-\delta}{2}],\ \nu^* = [\tfrac{1-\delta}{2}; \tfrac{1+\delta}{2}],\ \varepsilon_\star = \sqrt{1-\delta^2},$$
$$f_* = \tfrac{1}{2}\left[\ln((1+\delta)/(1-\delta)); \ln((1-\delta)/(1+\delta))\right].$$

The optimal balanced K-observation detector and its risk are

$$\phi_*^{(K)}(\underbrace{\omega_1, ..., \omega_K}_{\omega^K}) = f_*^T(\omega_1 + ... + \omega_K),\ \varepsilon_\star^{(K)} = (1-\delta^2)^{K/2}.$$

The near-optimal K-observation test $\mathcal{T}_{\phi_*}^K$ accepts H_1 and rejects H_2 if $\phi_*^{(K)}(\omega^K) \geq 0$; otherwise it accepts H_2 and rejects H_1. Both risks of this test do not exceed $\varepsilon_\star^{(K)}$.

Given risk level ϵ, we can identify the minimal "poll size" K for which the risks Risk_1, Risk_2 of the test $\mathcal{T}_{\phi*}^K$ do not exceed ϵ. This poll size depends on ϵ and on our a priori "hypotheses separation" parameter $\delta : K = K_\epsilon(\delta)$. Some impression on this size can be obtained from Table 2.1, where, as in all subsequent "election illustrations," ϵ is set to 0.01.

We see that while poll sizes for "landslide" elections are surprisingly low, reliable prediction of the results of "close run" elections requires surprisingly high sizes of the polls. Note that this phenomenon reflects reality (to the extent to which the reality is captured by our model).[14] Indeed, from Proposition 2.25 we know that our poll size is within an explicit factor, depending solely on ϵ, from the "ideal" poll sizes—the smallest ones which allow to decide upon H_1, H_2 with risk $\leq \epsilon$. For $\epsilon = 0.01$, this factor is about 2.85, meaning that when $\delta = 0.01$, the ideal poll size is larger than 32,000. In fact, we can easily construct more accurate "numerical" lower bounds on the sizes of ideal polls, specifically, as follows. When computing the optimal detector ϕ_*, we get, as a byproduct, two distributions, μ^*, ν^* obeying H_1, H_2, respectively. Denoting by μ_K^* and ν_K^* the distributions of K-element i.i.d. samples drawn from μ^* and ν^*, the risk of deciding on two simple hypotheses on the distribution of ω^K—stating that this distribution is μ_K^* and ν_K^*, respectively—can be only smaller than the risk of deciding on H_1, H_2 via K-repeated stationary observations. On the other hand, the former risk can be lower-bounded by one half of the total risk of deciding on our two simple hypotheses, and the latter risk admits a sharp lower bound given by Proposition 2.2, namely,

$$\sum_{i_1,\ldots,i_K \in \{1,2\}} \min\left[\prod_\ell \mu_{i_\ell}^*, \prod_\ell \nu_{i_\ell}^*\right] = \mathbf{E}_{(i_1,\ldots,i_K)}\left\{\min\left[\prod_\ell (2\mu_{i_\ell}^*), \prod_\ell (2\nu_{i_\ell}^*)\right]\right\},$$

with the expectation taken w.r.t independent tuples of K integers taking values

[14] In actual opinion polls, additional information is used. For instance, in reality voters can be split into groups according to their age, sex, education, income, etc., with variability of preferences within a group essentially lower than across the entire population. When planning a poll, respondents are selected at random within these groups, with a prearranged number of selections in every group, and their preferences are properly weighted, yielding more accurate predictions as compared to the case when the respondents are selected from the uniform distribution. In other words, in actual polls a nontrivial a priori information on the "true" distribution of preferences is used—something we do not have in our naive model.

δ	0.5623	0.3162	0.1778	0.1000	0.0562	0.0316	0.0177	0.0100
$K_{0.01}(\delta), L = 2$	25	88	287	917	2908	9206	29118	92098
$K_{0.01}(\delta), L = 5$	32	114	373	1193	3784	11977	37885	119745

Table 2.1: Sample of values of poll size $K_{0.01}(\delta)$ as a function of δ for 2-candidate ($L = 2$) and 5-candidate ($L = 5$) elections. Values of δ form a geometric progression with ratio $10^{-1/4}$.

1 and 2 with probabilities 1/2. Of course, when K is in the range of a few tens and more, we cannot compute the 2^K-term sum above exactly; however, we can use Monte Carlo simulation in order to estimate the sum reliably with moderate accuracy, like 0.005, and use this estimate to lower-bound the value of K for which an "ideal" K-observation test decides on H_1, H_2 with risks ≤ 0.01. Here are the resulting lower bounds (along with upper bounds from Table 2.1):

δ	0.5623	0.3162	0.1778	0.1000	0.0562	0.0316	0.0177	0.0100
$\underline{K}/\overline{K}$	$\frac{14}{25}$	$\frac{51}{88}$	$\frac{166}{287}$	$\frac{534}{917}$	$\frac{1699}{2908}$	$\frac{5379}{9206}$	$\frac{17023}{29122}$	$\frac{53820}{92064}$

Lower ($\underline{K}$) and upper ($\overline{K}$) bounds on the "ideal" poll sizes

We see that the poll sizes as yielded by our machinery are within factor 2 of the "ideal" poll sizes. Clearly, the outlined approach can be extended to L-candidate elections with $L \geq 2$. In our model of the corresponding problem we decide, via stationary K-repeated observations drawn from unknown probability distribution μ on L-element set, on L hypotheses

$$H_\ell : \mu \in M_\ell = \left\{ \mu \in \mathbf{R}^d : \mu_i \geq \frac{1}{N}, i \leq L, \sum_i \mu_i = 1, \mu_\ell \geq \mu_{\ell'} + \delta \,\forall (\ell' \neq \ell) \right\}, \ \ell \leq L. \tag{2.94}$$

Here $\delta > 0$ is a threshold selected in advance small enough to believe that the actual preferences of the voters correspond to $\mu \in \bigcup_\ell M_\ell$. Defining closeness $\mathcal{C}$ in the strongest possible way—H_ℓ is close to $H_{\ell'}$ if and only if $\ell = \ell'$—predicting the outcome of elections with risk ϵ becomes the problem of deciding upon our multiple hypotheses with $\mathcal{C}$-risk $\leq \epsilon$. Thus, we can use pairwise detectors yielded by Theorem 2.23 to identify the smallest possible $K = K_\epsilon$ such that the test $\mathcal{T}_\mathcal{C}^K$ from Section 2.5.2.3 is capable of deciding upon our L hypotheses with $\mathcal{C}$-risk $\leq \epsilon$. A numerical illustration of the performance of this approach in 5-candidate elections is presented in Table 2.1 (where ϵ is set to 0.01).

2.6.2 Sequential hypothesis testing

In view of the above analysis, when predicting outcomes of "close run" elections, huge poll sizes are necessary. It, however, does not mean that nothing can be done in order to build more reasonable opinion polls. The classical related statistical idea, going back to Wald [236], is to pass to *sequential tests* where the observations are processed one by one, and at every instant we either accept some of our hypotheses and terminate, or conclude that the observations obtained so far are insufficient to make a reliable inference and pass to the next observation. The idea is that a properly built sequential test, while still ensuring a desired risk, will be able to make "early decisions" in the case when the distribution underlying observations is "well inside" the true hypothesis and thus is far from the alternatives. Let us show

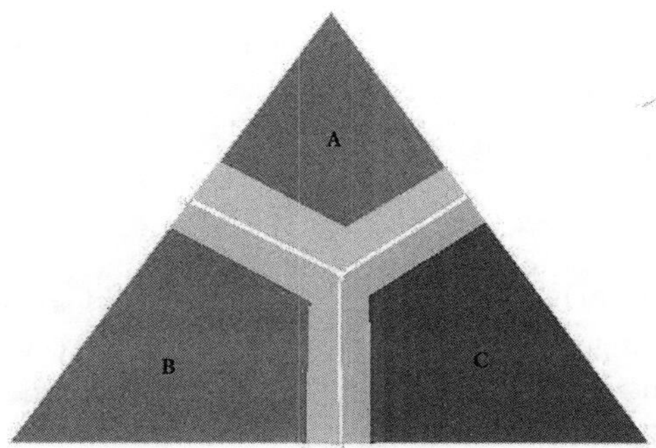

Figure 2.6: 3-candidate hypotheses in probabilistic simplex $\mathbf{\Delta}_3$

[area A]	M_1	dark tetragon + light border strip: candidate A wins with margin $\geq \delta_S$
[area A]	M_1^s	dark tetragon: candidate A wins with margin $\geq \delta_s > \delta_S$
[area B]	M_2	dark tetragon + light border strip: candidate B wins with margin $\geq \delta_S$
[area B]	M_2^s	dark tetragon: candidate B wins with margin $\geq \delta_s > \delta_S$
[area C]	M_3	dark tetragon + light border strip: candidate C wins with margin $\geq \delta_S$
[area C]	M_3^s	dark tetragon: candidate C wins with margin $\geq \delta_s > \delta_S$

$\mathcal{C}_s$ closeness: hypotheses in the tuple $\{G_{2\ell-1}^s : \mu \in M_\ell, G_{2\ell}^s : \mu \in M_\ell^s, 1 \leq \ell \leq 3\}$ are *not* $\mathcal{C}_s$-close to each other if the corresponding M-sets belong to different areas and at least one of the sets is painted dark, like M_1^s and M_2, but not M_1 and M_2.

how our machinery can be utilized to conceive a sequential test for the problem of predicting the outcome of L-candidate elections. Thus, our goal is, given a small threshold δ, to decide upon L hypotheses (2.94). Let us act as follows.

1. We select a factor $\theta \in (0,1)$, say, $\theta = 10^{-1/4}$, and consider thresholds $\delta_1 = \theta$, $\delta_2 = \theta\delta_1$, $\delta_3 = \theta\delta_2$, and so on, until for the first time we get a threshold $\leq \delta$; to save notation, we assume that this threshold is exactly δ, and let the number of the thresholds be S.
2. We split somehow (e.g., equally) the risk ϵ which we want to guarantee into S portions ϵ_s, $1 \leq s \leq S$, so that ϵ_s are positive and

$$\sum_{s=1}^{S} \epsilon_s = \epsilon.$$

3. For $s \in \{1, 2, ..., S\}$, we define, along with the hypotheses H_ℓ, the hypotheses

$$H_\ell^s : \mu \in M_\ell^s = \{\mu \in M_\ell : \mu_\ell \geq \mu_{\ell'} + \delta_s, \forall(\ell' \neq \ell)\}, \ell = 1, ..., L,$$

(see Figure 2.6), and introduce $2L$ hypotheses $G_{2\ell-1}^s = H_\ell$, and $G_{2\ell}^s = H_\ell^s$, $1 \leq \ell \leq L$. It is convenient to color these hypotheses in L colors, with $G_{2\ell-1}^s = H_\ell$ and $G_{2\ell}^s = H_\ell^s$ assigned color ℓ. We define also *s-th closeness* $\mathcal{C}_s$ as follows:

> *When* $s < S$, hypotheses G_i^s and G_j^s are $\mathcal{C}_s$-close to each other if either they are of the same color, or they are of different colors and both of them have odd indices (that is, one of them is H_ℓ, and another one is $H_{\ell'}$ with $\ell \neq \ell'$).

When $s = S$ (in this case $G^S_{2\ell-1} = H_\ell = G^S_{2\ell}$), hypotheses G^S_ℓ and $G^S_{\ell'}$ are $\mathcal{C}_S$-close to each other if and only if they are of the same color, i.e., both coincide with the same hypothesis H_ℓ.

Observe that G^s_i is a convex hypothesis:

$$G^s_i : \mu \in Y^s_i \qquad [Y^s_{2\ell-1} = M_\ell, Y^s_{2\ell} = M^s_\ell]$$

The key observation is that when G^s_i and G^s_j are *not* $\mathcal{C}_s$-close, sets Y^s_i and Y^s_j are "separated" by at least δ_s, meaning that for some vector $e \in \mathbf{R}^L$ with just two nonvanishing entries, equal to 1 and -1, we have

$$\min_{\mu \in Y^s_i} e^T\mu \geq \delta_s + \max_{\mu \in Y^s_j} e^T\mu. \tag{2.95}$$

Indeed, let G^s_i and G^s_j not be $\mathcal{C}_s$-close to each other. That means that the hypotheses are of different colors, say, ℓ and $\ell' \neq \ell$, and at least one of them has even index. W.l.o.g. we can assume that the even-indexed hypothesis is G^s_i, so that

$$Y^s_i \subset \{\mu : \mu_\ell - \mu_{\ell'} \geq \delta_s\},$$

while Y^s_j is contained in the set $\{\mu : \mu_{\ell'} \geq \mu_\ell\}$. Specifying e as the vector with just two nonzero entries, ℓ-th equal to 1 and ℓ'-th equal to -1, we ensure (2.95).

4. For $1 \leq s \leq S$, we apply the construction from Section 2.5.2.3 to identify the smallest $K = K(s)$ for which the test $\mathcal{T}_s$ yielded by this construction as applied to a stationary K-repeated observation allows us to decide on the hypotheses $G^s_1, ..., G^s_{2L}$ with $\mathcal{C}_s$-risk $\leq \epsilon_s$. The required K exists due to the already mentioned separation of members in a pair of not $\mathcal{C}_s$-close hypotheses G^s_i, G^s_j. It is easily seen that $K(1) \leq K(2) \leq ... \leq K(S-1)$. However, it may happen that $K(S-1) > K(S)$, the reason being that $\mathcal{C}_S$ is defined differently than $\mathcal{C}_s$ with $s < S$. We set

$$\mathcal{S} = \{s \leq S : K(s) \leq K(S)\}.$$

For example, here is what we get in L-candidate Opinion Poll problem when $S = 8$, $\delta = \delta_S = 0.01$, and for properly selected ϵ_s with $\sum_{s=1}^8 \epsilon_s = 0.01$:

L	$K(1)$	$K(2)$	$K(3)$	$K(4)$	$K(5)$	$K(6)$	$K(7)$	$K(8)$
2	177	617	1829	5099	15704	49699	153299	160118
5	208	723	2175	6204	19205	60781	188203	187718

$S = 8$, $\delta_s = 10^{-s/4}$.

$\mathcal{S} = \{1, 2, ..., 8\}$ when $L = 2$ and $\mathcal{S} = \{1, 2, ..., 6\} \cup \{8\}$ when $L = 5$.

5. Our sequential test $\mathcal{T}_{\text{seq}}$ works in *attempts* (stages) $s \in \mathcal{S}$—it tries to make conclusions after observing $K(s)$, $s \in \mathcal{S}$, realizations ω_k of ω. At the s-th attempt, we apply the test $\mathcal{T}_s$ to the collection $\omega^{K(s)}$ of observations obtained so far to decide on hypotheses $G^s_1, ..., G^s_{2L}$. If $\mathcal{T}_s$ accepts some of these hypotheses *and all accepted hypotheses are of the same color*—let it be ℓ—the sequential test accepts the hypothesis H_ℓ and terminates; otherwise we continue to observe the realizations of ω (if $s < S$) or terminate with no hypotheses accepted/rejected (if $s = S$).

It is easily seen that the risk of the outlined sequential test $\mathcal{T}_{\text{seq}}$ does not exceed ϵ, meaning that whatever be the distribution $\mu \in \bigcup_{\ell=1}^L M_\ell$ underlying observations

$\omega_1, \omega_2, \ldots\omega_{K(S)}$ and ℓ_* such that $\mu \in M_{\ell_*}$, the μ-probability of the event

$$\mathcal{T}_{\rm seq} \textit{ accepts exactly one hypothesis, namely, } H_{\ell_*}$$

is at least $1 - \epsilon$.

Indeed, observe, first, that the sequential test always accepts at most one of the hypotheses $H_1, ..., H_L$. Second, let $\omega_k \sim \mu$ with μ obeying H_{ℓ_*}. Consider events E_s, $s \in \mathcal{S}$, defined as follows:

- when $s < S$, E_s is the event "the test $\mathcal{T}_s$ as applied to observation $\omega^{K(s)}$ does not accept the true hypothesis $G^s_{2\ell_*-1} = H_{\ell_*}$";
- E_S is the event "as applied to observation $\omega^{K(S)}$, the test $\mathcal{T}_S$ does not accept the true hypothesis $G^S_{2\ell_*-1} = H_{\ell_*}$ or accepts a hypothesis not $\mathcal{C}_S$-close to $G^S_{2\ell_*-1}$."

Note that by our selection of $K(s)$'s, the μ-probability of E_s does not exceed ϵ_s, so that the μ-probability of *none* of the events E_s, $s \in \mathcal{S}$, taking place is at least $1 - \epsilon$. To justify the above claim on the risk of the sequential test, all we need to verify is that *when none of the events E_s, $s \in \mathcal{S}$, takes place, the sequential test accepts the true hypothesis H_{ℓ_*}*. Verification is immediate: let the observations be such that none of the E_s's takes place. We claim that in this case

(a) The sequential test does accept a hypothesis—if this does not happen at the s-th attempt with some $s < S$, it definitely happens at the S-th attempt.

Indeed, since E_S does not take place, $\mathcal{T}_S$ accepts $G^S_{2\ell_*-1}$ and all other hypotheses, if any, accepted by $\mathcal{T}_S$ are $\mathcal{C}_S$-close to $G^S_{2\ell_*-1}$, implying by construction of $\mathcal{C}_S$ that $\mathcal{T}_S$ does accept hypotheses, and all these hypotheses are of the same color. That is, the sequential test at the S-th attempt does accept a hypothesis.

(b) The sequential test does *not* accept a wrong hypothesis.

Indeed, assume that the sequential test accepts a wrong hypothesis, $H_{\ell'}$, $\ell' \neq \ell_*$, and it happens at the s-th attempt, and let us lead this assumption to a contradiction. Observe that under our assumption the test $\mathcal{T}_s$ as applied to observation $\omega^{K(s)}$ does accept some hypothesis G^s_i, but does *not* accept the true hypothesis $G^s_{2\ell_*-1} = H_{\ell_*}$. Indeed, assuming $G^s_{2\ell_*-1}$ to be accepted, its color, which is ℓ_*, should be the same as the color ℓ' of G^s_i—we are in the case where the sequential test accepts $H_{\ell'}$ at the s-th attempt! Since in fact $\ell' \neq \ell_*$, the above assumption leads to a contradiction. On the other hand, we are in the case where E_s does not take place, that is, $\mathcal{T}_s$ does accept the true hypothesis $G^s_{2\ell_*-1}$, and we arrive at the desired contradiction.

(a) and (b) provide us with the verification we were looking for.

Discussion and illustration. It can be easily seen that when $\epsilon_s = \epsilon/S$ for all s, the worst-case duration $K(S)$ of our sequential test is within a logarithmic in the SL factor of the duration of any other test capable of deciding on our L hypotheses with risk ϵ. At the same time it is easily seen that when the distribution μ of our observation is "deeply inside" some set M_ℓ, specifically, $\mu \in M^s_\ell$ for some $s \in \mathcal{S}$, $s < S$, then the μ-probability to terminate not later than just after $K(s)$ realizations ω_k of $\omega \sim \mu$ are observed and to infer correctly what is the true hypothesis is at least $1 - \epsilon$. Informally speaking, in the case of "landslide" elections, a reliable prediction of the elections' outcome will be made after a relatively small number of respondents are interviewed.

Indeed, let $s \in \mathcal{S}$ and $\omega_k \sim \mu \in M^s_\ell$, so that μ obeys the hypothesis $G^s_{2\ell}$. Consider the s events E_t, $1 \leq t \leq s$, defined as follows:

- For $t < s$, E_t occurs when the sequential test terminates at attempt t by accepting, instead of H_ℓ, the wrong hypothesis $H_{\ell'}$, $\ell' \neq \ell$. Note that E_t can take place only when $\mathcal{T}_t$ does not accept the true hypothesis $G^t_{2\ell-1} = H_\ell$, and the

μ-probability of this outcome is $\leq \epsilon_t$.

- E_s occurs when $\mathcal{T}_s$ does not accept the true hypothesis $G^s_{2\ell}$ or accepts it along with some hypothesis G^s_j, $1 \leq j \leq 2L$, of color different from ℓ. Note that we are in the situation where the hypothesis $G^s_{2\ell}$ is true, and, by construction of $\mathcal{C}_s$, all hypotheses $\mathcal{C}_s$-close to $G^s_{2\ell}$ are of the same color ℓ as $G^s_{2\ell}$. Recalling what $\mathcal{C}_s$-risk is and that the $\mathcal{C}_s$-risk of $\mathcal{T}_s$ is $\leq \epsilon_s$, we conclude that the μ-probability of E_s is at most ϵ_s.

The bottom line is that the μ-probability of the event $\bigcup_{t\leq s} E_t$ is at most $\sum_{t=1}^s \epsilon_t \leq \epsilon$; by construction of the sequential test, if the event $\bigcup_{t\leq s} E_t$ does *not* take place, the test terminates in the course of the first s attempts by accepting the correct hypothesis H_ℓ. Our claim is justified.

Numerical illustration. To get an impression of the "power" of sequential hypothesis testing, here are the data on the durations of non-sequential and sequential tests with risk $\epsilon = 0.01$ for various values of δ; in the sequential tests, $\theta = 10^{-1/4}$ is used. The worst-case data for 2-candidate and 5-candidate elections are as follows (below, "volume" stands for the number of observations used by the test)

δ	0.5623	0.3162	0.1778	0.1000	0.0562	0.0316	0.0177	0.0100
$K, L=2$	25	88	287	917	2908	9206	29118	92098
$S\,/\,K(S)$, $L=2$	$\frac{1}{25}$	$\frac{2}{152}$	$\frac{3}{499}$	$\frac{4}{1594}$	$\frac{5}{5056}$	$\frac{6}{16005}$	$\frac{7}{50624}$	$\frac{8}{160118}$
$K, L=5$	32	114	373	1193	3784	11977	37885	119745
$S\,/\,K(S)$, $L=5$	$\frac{1}{32}$	$\frac{2}{179}$	$\frac{3}{585}$	$\frac{4}{1870}$	$\frac{5}{5931}$	$\frac{6}{18776}$	$\frac{7}{59391}$	$\frac{8}{187720}$

Volume K of non-sequential test, number S of stages, and worst-case volume $K(S)$ of sequential test as functions of threshold $\delta = \delta_S$. Risk ϵ is set to 0.01.

As it should be, the worst-case volume of the sequential test is significantly larger than the volume of the non-sequential test.[15] This being said, look at what happens in the "average," rather than the worst, case; specifically, let us look at the empirical distribution of the volume when the distribution μ of observations is selected in the L-dimensional probabilistic simplex $\mathbf{\Delta}_L = \{\mu \in \mathbf{R}^L : \mu \geq 0, \sum_\ell \mu_\ell = 1\}$ at random. Here are the empirical statistics of test volume obtained when drawing μ from the uniform distribution on $\bigcup_{\ell\leq L} M_\ell$ and running the sequential test[16] on observations drawn from the selected μ:

L	risk	median	mean	60%	65%	70%
2	0.0010	177	9182	177	397	617
5	0.0040	1449	18564	2175	4189	6204
L	75%	80%	85%	90%	95%	100%
2	617	1223	1829	8766	87911	160118
5	12704	19205	39993	60781	124249	187718

Parameters (columns "median, mean") and quantiles (columns "60%,..., 100%") of the sample distribution of the observation volume of the Sequential test for a given empirical risk (column "risk") .

The data in the table are obtained from 1,000 experiments. We see that with the Sequential test, "typical" numbers of observations before termination are much

[15]The reason is twofold: first, for $s < S$ we pass from deciding on L hypotheses to deciding on $2L$ of them; second, the desired risk ϵ is now distributed among several tests, so that each of them should be more reliable than the non-sequential test with risk ϵ.

[16]Corresponding to $\delta = 0.01$, $\theta = 10^{-1/4}$ and $\epsilon = 0.01$.

less than the worst-case values of these numbers. For example, in as much as 80% of experiments these numbers were below quite reasonable levels, at least in the case $L = 2$. Of course, what is "typical," and what is not, depends on how we generate μ's (this is called "prior Bayesian distribution"). Were our generation more likely to produce "close run" distributions, the advantages of sequential decision making would be reduced. This ambiguity is, however, unavoidable when attempting to go beyond worst-case-oriented analysis.

2.6.3 Concluding remarks

Application of our machinery to sequential hypothesis testing is in no sense restricted to the simple election model considered so far. A natural general setup we can handle is as follows:

> We are given a simple observation scheme $\mathcal{O}$ and a number L of related convex hypotheses, colored in d colors, on the distribution of an observation, with distributions obeying hypotheses of different colors being distinct from each other. Given the risk level ϵ, we want to decide $(1-\epsilon)$-reliably on the color of the distribution underlying observations (i.e., the color of the hypothesis obeyed by this distribution) from stationary K-repeated observations, utilizing as small a number of observations as possible.

For detailed description of related constructions and results, an interested reader is referred to [134].

2.7 MEASUREMENT DESIGN IN SIMPLE OBSERVATION SCHEMES

2.7.1 Motivation: Opinion polls revisited

Consider the same situation as in Section 2.6.1—we want to use an opinion poll to predict the winner in a population-wide election with L candidates. When addressing this situation earlier, no essential a priori information on the distribution of voters' preferences was available. Now consider the case when the population is split into I groups (according to age, sex, income, etc., etc.), with the i-th group forming the fraction θ_i of the entire population, and we have at our disposal, at least for some i, nontrivial a priori information about the distribution p^i of the preferences across group # i (the ℓ-th entry p^i_ℓ in p^i is the fraction of voters of group i voting for candidate ℓ). For instance, we could know in advance that at least 90% of members of group #1 vote for candidate #1, and at least 85% of members of group #2 vote for candidate #2; no information of this type for group #3 is available. In this situation it would be wise to select respondents in the poll via a two-stage procedure, first selecting at random, with probabilities $q_1, ..., q_I$, the group from which the next respondent will be picked, and second selecting the respondent from this group at random according to the uniform distribution on the group. When the q_i are proportional to the sizes of the groups (i.e., $q_i = \theta_i$ for all i), we come back to selecting respondents at random from the uniform distribution over the entire population. The point, however, is that in the presence of a priori information, it makes sense to use q_i different from θ_i, specifically, to make the ratios q_i/θ_i "large" or "small" depending on whether a priori information on group $\#i$ is poor or rich.

The story we have just told is an example of a situation in which we can "design measurements"—draw observations from a distribution which partly is under our control. Indeed, what in fact happens in the story is the following. "In nature" there exist I probabilistic vectors $p^1, ..., p^I$ of dimension L representing distributions of voting preferences within the corresponding groups; the distribution of preferences across the entire population is $p = \sum_i \theta_i p^i$. With the two-stage selection of respondents, the outcome of a particular interview becomes a pair (i, ℓ), with i identifying the group to which the respondent belongs, and ℓ identifying the candidate preferred by this respondent. In subsequent interviews, the pairs (i, ℓ)—these are our observations—are drawn, independently of each other, from the probability distribution on the pairs (i, ℓ), $i \leq I$, $\ell \leq L$, with the probability of an outcome (i, ℓ) equal to

$$p(i, \ell) = q_i p^i_\ell.$$

Thus, we find ourselves in the situation of stationary repeated observations stemming from the Discrete o.s. with observation space Ω of cardinality IL; the distribution from which the observations are drawn is a probabilistic vector μ of the form

$$\mu = Ax,$$

where

- $x = [p^1; ...; p^I]$ is the "signal" underlying our observations and representing the preferences of the population; this signal is selected by nature in the set $\mathcal{X}$ known to us defined in terms of our a priori information on $p^1, ..., p^I$:

$$\mathcal{X} = \{x = [x^1; ...; x^I] : x^i \in \Pi_i, 1 \leq i \leq I\}, \tag{2.96}$$

  where the Π_i are the sets, given by our a priori information, of possible values of the preference vectors p^i of the voters from i-th group. In the sequel, we assume that the Π_i are convex compact subsets of the positive part $\mathbf{\Delta}^o_L = \{p \in \mathbf{R}^L : p > 0, \sum_\ell p_\ell = 1\}$ of the L-dimensional probabilistic simplex;
- A is a "sensing matrix" which, to some extent, is under our control; specifically,

$$A[x^1; ...; x^I] = [q_1 x^1; q_2 x^2; ...; q_I x^I], \tag{2.97}$$

  with $q = [q_1; ...; q_I]$ fully controlled by us (up to the fact that q must be a probabilistic vector).

Note that in the situation under consideration the hypotheses we want to decide upon can be represented by convex sets *in the space of signals,* with a particular hypothesis stating that the observations stem from a distribution μ on Ω, with μ belonging to the image of some convex compact set $X_\ell \subset \mathcal{X}$ under the mapping $x \mapsto \mu = Ax$. For example, when $\nu = \sum_i \theta_i x^i$, the hypotheses

$$H_\ell : \nu \in M_\ell = \left\{ \nu \in \mathbf{R}^L : \sum_j \nu_j = 1, \nu_j \geq \tfrac{1}{N}, \nu_\ell \geq \nu_{\ell'} + \delta, \ell' \neq \ell \right\}, 1 \leq \ell \leq L,$$

considered in Section 2.6.1 can be expressed in terms of the signal $x = [x^1; ...; x^I]$:

$$H_\ell : \mu = Ax,\ x \in X_\ell = \left\{ x = [x^1; ...; x^I] : \begin{array}{l} x^i \geq 0, \sum_\ell x^i_\ell = 1 \forall i \leq I \\ \sum_i \theta_i x^i_\ell \geq \sum_i \theta_i x^i_{\ell'} + \delta \forall (\ell' \neq \ell) \\ \sum_i \theta_i x^i_j \geq \frac{1}{N}, \forall j \end{array} \right\}. \tag{2.98}$$

The challenge we intend to address is as follows: so far, we were interested in inferences from observations drawn from distributions selected "by nature." Now our goal is to make inferences from observations drawn from a distribution selected partly by nature and partly by us: nature selects the signal x, we select from some set matrix A, and the observations are drawn from the distribution Ax. As a result, we arrive at a question completely new for us: how do we utilize the freedom in selecting A in order to improve our inferences (this is somewhat similar to what is called "design of experiments" in Statistics)?

2.7.2 Measurement Design: Setup

In what follows we address measurement design in simple observation schemes, and our setup is as follows (to make our intensions transparent, we illustrate our general setup by explaining how it should be specified to cover the outlined two-stage Opinion Poll Design (OPD) problem).

Given are

- simple observation scheme $\mathcal{O} = (\Omega, \Pi; \{p_\mu : \mu \in \mathcal{M}\}; \mathcal{F})$, specifically, Gaussian, Poisson, or Discrete, with $\mathcal{M} \subset \mathbf{R}^d$.
  In OPD, $\mathcal{O}$ is the Discrete o.s. with $\Omega = \{(i, \ell) : 1 \leq i \leq I, 1 \leq \ell \leq L\}$, that is, points of Ω are the potential outcomes "reference group, preferred candidate" of individual interviews.
- a nonempty closed convex *signal space* $\mathcal{X} \subset \mathbf{R}^n$, along with L nonempty convex compact subsets X_ℓ of $\mathcal{X}$, $\ell = 1, ..., L$.
  In OPD, $\mathcal{X}$ is the set (2.96) comprised of tuples of allowed distributions of voters' preferences from various groups, and X_ℓ are the sets (2.98) of signals associated with the hypotheses H_ℓ we intend to decide upon.
- a nonempty convex compact set $\mathcal{Q}$ in some $\mathbf{R}^N$ along with a continuous mapping $q \mapsto A_q$ acting from $\mathcal{Q}$ into the space of $d \times n$ matrices such that

$$\forall (x \in \mathcal{X}, q \in \mathcal{Q}) : A_q x \in \mathcal{M}. \tag{2.99}$$

  In OPD, $\mathcal{Q}$ is the set of probabilistic vectors $q = [q_1; ...; q_I]$ specifying our measurement design, and A_q is the matrix of the mapping (2.97).
- a closeness $\mathcal{C}$ on the set $\{1, ..., L\}$ (that is, a set $\mathcal{C}$ of pairs (i, j) with $1 \leq i, j \leq L$ such that $(i, i) \in \mathcal{C}$ for all $i \leq L$ and $(j, i) \in \mathcal{C}$ whenever $(i, j) \in \mathcal{C}$), and a positive integer K.
  In OPD, the closeness $\mathcal{C}$ is as strict as it could be—i is close to j if and only if $i = j$,[17] and K is the total number of interviews in the poll.

[17]This closeness makes sense when the goal of the poll is to predict the winner; a less ambitious goal, e.g., to decide whether the winner will or will not belong to a particular set of candidates, would require weaker closeness.

We associate with $q \in \mathcal{Q}$ and X_ℓ, $\ell \leq L$, the nonempty convex compact sets M_ℓ^q in the space $\mathcal{M}$,

$$M_\ell^q = \{A_q x : x \in X_\ell\},$$

and hypotheses H_ℓ^q on K-repeated stationary observations $\omega^K = (\omega_1, ..., \omega_K)$, H_ℓ^q stating that the ω_k, $k = 1, ..., K$, are drawn, independently of each other, from a distribution $\mu \in M_\ell^q$, $\ell = 1, ..., L$. Closeness $\mathcal{C}$ can be thought of as closeness on the collection of hypotheses $H_1^q, H_2^q, ..., H_L^q$. Given $q \in \mathcal{Q}$, we can use the construction from Section 2.5.2 in order to build the test $\mathcal{T}_{\phi_*}^K$ deciding on the hypotheses H_ℓ^q up to closeness $\mathcal{C}$, the $\mathcal{C}$-risk of the test being the smallest allowed by the construction. Note that this $\mathcal{C}$-risk depends on q; the "Measurement Design" (MD for short) problem we are about to consider is to select $q \in \mathcal{Q}$ which minimizes the $\mathcal{C}$-risk of the associated test $\mathcal{T}_{\phi_*}^K$.

2.7.3 Formulating the MD problem

By Proposition 2.30, the $\mathcal{C}$-risk of the test $\mathcal{T}_{\phi_*}^K$ is upper-bounded by the spectral norm of the symmetric entrywise nonnegative $L \times L$ matrix

$$E^{(K)}(q) = [\epsilon_{\ell\ell'}(q)]_{\ell,\ell'},$$

and this is what we intend to minimize in our MD problem. In the above formula, $\epsilon_{\ell\ell'}(q) = \epsilon_{\ell'\ell}(q)$ are zeros if $(\ell, \ell') \in \mathcal{C}$. For $(\ell, \ell') \notin \mathcal{C}$ and $1 \leq \ell < \ell' \leq L$, the quantities $\epsilon_{\ell\ell'}(q) = \epsilon_{\ell'\ell}(q)$ are defined depending on what the simple o.s. is $\mathcal{O}$. Specifically,

- In the case of the *Gaussian* observation scheme (see Section 2.4.5.1), restriction (2.99) does not restrain the dependence A_q on q at all (modulo the default constraint that A_q is a $d \times n$ matrix continuous in $q \in \mathcal{Q}$), and

  $$\epsilon_{\ell\ell'}(q) = \exp\{K\text{Opt}_{\ell\ell'}(q)\}$$

  where

  $$\text{Opt}_{\ell\ell'}(q) = \max_{x \in X_\ell, y \in X_{\ell'}} -\tfrac{1}{8}[A_q(x-y)]^T \Theta^{-1}[A_q(x-y)] \qquad (G_q)$$

  and Θ is the common covariance matrix of the Gaussian densities forming the family $\{p_\mu : \mu \in \mathcal{M}\}$;
- In the case of Poisson o.s. (see Section 2.4.5.2), restriction (2.99) requires of $A_q x$ to be a positive vector whenever $q \in \mathcal{Q}$ and $x \in \mathcal{X}$, and

  $$\epsilon_{\ell\ell'}(q) = \exp\{K\text{Opt}_{\ell\ell'}(q)\},$$

  where

  $$\text{Opt}_{\ell\ell'}(q) = \max_{x \in X_\ell, y \in X_{\ell'}} \sum_i \left[\sqrt{[A_q x]_i [A_q y]_i} - \tfrac{1}{2}[A_q x]_i - \tfrac{1}{2}[A_q y]_i\right]; \qquad (P_q)$$

- In the case of Discrete o.s. (see Section 2.4.5.3), restriction (2.99) requires of $A_q x$ to be a positive probabilistic vector whenever $q \in \mathcal{Q}$ and $x \in \mathcal{X}$, and

  $$\epsilon_{\ell\ell'}(q) = [\text{Opt}_{\ell\ell'}(q)]^K,$$

where

$$\mathrm{Opt}_{\ell\ell'}(q) = \max_{x\in X_\ell, y\in X_{\ell'}} \sum_i \sqrt{[A_q x]_i [A_q y]_i}. \tag{D_q}$$

The summary of the above observations is as follows. The norm $\|E^{(K)}\|_{2,2}$—the quantity we are interested in minimizing in $q \in \mathcal{Q}$—as a function of $q \in \mathcal{Q}$ is of the form

$$\Psi(q) = \psi(\underbrace{\{\mathrm{Opt}_{\ell\ell'}(q) : (\ell,\ell') \notin \mathcal{C}\}}_{\overline{\mathrm{Opt}}(q)}) \tag{2.100}$$

where the outer function ψ is an explicitly given real-valued function on $\mathbf{R}^N$ (N is the cardinality of the set of pairs (ℓ,ℓ'), $1 \leq \ell, \ell' \leq L$, with $(\ell,\ell') \notin \mathcal{C}$) which is convex and nondecreasing in each argument. Indeed, denoting by $\Gamma(S)$ the spectral norm of the $d \times d$ matrix S, note that Γ is a convex function of S, and this function is nondecreasing in every one of the entries of S, provided that S is restricted to be entrywise nonnegative.[18] $\psi(\cdot)$ is obtained from $\Gamma(S)$ by substituting for the entries $S_{\ell\ell'}$ of S, certain—explicit everywhere—convex, nonnegative and nondecreasing functions of variables $z = \{z_{\ell\ell'} : (\ell,\ell') \notin \mathcal{C}, 1 \leq \ell,\ell' \leq L\}$. Namely,

- when $(\ell,\ell') \in \mathcal{C}$, we set $S_{\ell\ell'}$ to zero;
- when $(\ell,\ell') \notin \mathcal{C}$, we set $S_{\ell\ell'} = \exp\{K z_{\ell\ell'}\}$ in the case of Gaussian and Poisson o.s.'s, and set $S_{\ell\ell'} = \max[0, z_{\ell\ell'}]^K$, in the case of Discrete o.s.

As a result, we indeed get a convex and nondecreasing, in every argument, function ψ of $z \in \mathbf{R}^N$.

Now, the Measurement Design problem we want to solve reads

$$\mathrm{Opt} = \min_{q\in\mathcal{Q}} \psi(\overline{\mathrm{Opt}}(q)). \tag{2.101}$$

As we remember, the entries in the inner function $\overline{\mathrm{Opt}}(q)$ are optimal values of solvable *convex* optimization problems and as such are efficiently computable. When these entries are also *convex* functions of $q \in \mathcal{Q}$, the objective in (2.101), due to the already established convexity and monotonicity properties of ψ, is a convex function of q, meaning that (2.101) is a convex and thus efficiently solvable problem. On the other hand, when some of the entries in $\overline{\mathrm{Opt}}(q)$ are nonconvex in q, we can hardly expect (2.101) to be easy to solve. Unfortunately, convexity of the entries in $\overline{\mathrm{Opt}}(q)$ in q turns out to be a "rare commodity." For example, we can verify by inspection that the objectives in (G_q), (P_q), and (D_q) *as functions of* A_q (not of q!) are *concave* rather than convex. Thus, the optimal values in the problems, as functions of q, are maxima, over the parameters, of parametric families of concave functions of A_q (the parameters in these parametric families are the optimization variables in $(G_q) - (D_q)$) and as such can hardly be convex *as functions of* A_q. And indeed, as a matter of fact, the MD problem usually is nonconvex and difficult to solve. We intend to consider a "Simple case" where this difficulty does not arise, i.e., the case where the objectives of the optimization problems specifying $\mathrm{Opt}_{\ell\ell'}(q)$ are *affine in* q. In this case, $\mathrm{Opt}_{\ell\ell'}(q)$ as a function of q is the maximum, over the

[18]The monotonicity follows from the fact that for an entrywise nonnegative S, we have

$$\|S\|_{2,2} = \max_{x,y}\{x^T S y : \|x\|_2 \leq 1, \|y\|_2 \leq 1\} = \max_{x,y}\{x^T S y : \|x\|_2 \leq 1, \|y\|_2 \leq 1, x \geq 0, y \geq 0\}.$$

parameters (optimization variables in the corresponding problems), of parametric families of affine functions of q and as such is convex.

Our current goal is to understand what our sufficient condition for tractability of the MD problem—affinity in q of the objectives in the respective problems (G_q), (P_q), and (D_q)—actually means, and to show that this, by itself quite restrictive, assumption indeed takes place in some important applications.

2.7.3.1 Simple case, Discrete o.s.

Looking at the optimization problem (D_q), we see that the simplest way to ensure that its objective is affine in q is to assume that

$$A_q = \mathrm{Diag}\{Bq\}A, \tag{2.102}$$

where A is some fixed $d \times n$ matrix, and B is some fixed $d \times (\dim q)$ matrix such that Bq is positive whenever $q \in \mathcal{Q}$. On the top of this, we should ensure that when $q \in \mathcal{Q}$ and $x \in \mathcal{X}$, $A_q x$ is a positive probabilistic vector; this amounts to some restrictions linking $\mathcal{Q}$, $\mathcal{X}$, A, and B.

Illustration. The Opinion Poll Design problem of Section 2.7.1 provides an instructive example of the Simple case of Measurement Design in Discrete o.s.: recall that in this problem the voting population is split into I groups, with the i-th group constituting fraction θ_i of the entire population. The distribution of voters' preferences in the i-th group is represented by an unknown L-dimensional probabilistic vector $x^i = [x^i_1; ...; x^i_L]$ (L is the number of candidates, x^i_ℓ is the fraction of voters in the i-th group intending to vote for the ℓ-th candidate), known to belong to a given convex compact subset Π_i of the "positive part" $\boldsymbol{\Delta}^o_L = \{x \in \mathbf{R}^L : x > 0, \sum_\ell x_\ell = 1\}$ of the L-dimensional probabilistic simplex. We are given threshold $\delta > 0$ and want to decide on L hypotheses H_1,..., H_L, with H_ℓ stating that the population-wide vector $y = \sum_{i=1}^I \theta_i x^i$ of voters' preferences belongs to the closed convex set

$$Y_\ell = \left\{ y = \sum_{i=1}^I \theta_i x^i : x^i \in \Pi_i,\ 1 \le i \le I, y_\ell \ge y_{\ell'} + \delta,\ \forall (\ell' \neq \ell) \right\}.$$

Note that Y_ℓ is the image, under the linear mapping

$$[x^1; ...; x^I] \mapsto y(x) = \sum_i \theta_i x^i,$$

of the compact convex set

$$X_\ell = \left\{ x = [x^1; ...; x^I] : x^i \in \Pi_i,\ 1 \le i \le I, y_\ell(x) \ge y_{\ell'}(x) + \delta,\ \forall (\ell' \neq \ell) \right\},$$

which is a subset of the convex compact set

$$\mathcal{X} = \{x = [x^1; ...; x^I] : x^i \in \Pi_i,\ 1 \le i \le I\}.$$

The k-th poll interview is organized as follows:

> We draw at random a group among the I groups of voters, with probability q_i to draw i-th group, and then draw at random, from the uniform distribution on the group, the respondent to be interviewed. The outcome of

the interview—our observation ω_k—is the pair (i, ℓ), where i is the group to which the respondent belongs, and ℓ is the candidate preferred by the respondent.

This results in a sensing matrix A_q—see (2.97)—which is in the form of (2.102), namely,

$$A_q = \mathrm{Diag}\{q_1 I_L, q_2 I_L, ..., q_I I_L\}, \qquad [q \in \mathbf{\Delta}_I]$$

and the outcome of k-th interview is drawn at random from the discrete probability distribution $A_q x$, where $x \in \mathcal{X}$ is the "signal" summarizing voters' preferences in the groups.

Given the total number of observations K, our goal is to decide with a given risk ϵ on our L hypotheses. Whether this goal is or is not achievable depends on K and on A_q. What we want is to find q for which the above goal can be attained with as small a K as possible; in the case in question, this reduces to solving, for various trial values of K, problem (2.101), which under the circumstances is an explicit *convex* optimization problem.

To get an impression of the potential of Measurement Design, we present a sample of numerical results. In all reported experiments, we use $\delta = 0.05$, $\epsilon = 0.01$ and equal fractions $\theta_i = I^{-1}$ for all groups. The sets Π_i, $1 \leq i \leq I$, are generated as follows: we pick at random a probabilistic vector $\bar{p}^i$ of dimension L, and define Π_i as the intersection of the box $\{p : \bar{p}_\ell - u_i \leq p_\ell \leq \bar{p}_\ell + u_i\}$ centered at $\bar{p}$ with the probabilistic simplex $\mathbf{\Delta}_L$, where the u_i, $i = 1, ..., I$, are prescribed "uncertainty levels." Note that uncertainty level $u_i \geq 1$ is the same as absence of any a priori information on the preferences of voters from the i-th group.

The results of our numerical experiments are as follows:

L	I	Uncertainty levels u	$K_{\rm ini}$	$q_{\rm opt}$	$K_{\rm opt}$
2	2	[0.03;1.00]	1212	[0.437;0.563]	1194
2	2	[0.02;1.00]	2699	[0.000;1.000]	1948
3	3	[0.02;0.03;1.00]	3177	[0.000;0.455;0.545]	2726
5	4	[0.02;0.02;0.03;1.00]	2556	[0.000;0.131;0.322;0.547]	2086
5	4	[1.00;1.00;1.00;1.00]	4788	[0.250;0.250;0.250;0.250]	4788

Effect of measurement design: poll sizes required for 0.99-reliable winner prediction when $q = \theta$ (column $K_{\rm ini}$) and $q = q_{\rm opt}$ (column $K_{\rm opt}$).

We see that measurement design allows us to reduce (for some data, quite significantly) the volume of observations as compared to the straightforward selecting of the respondents uniformly across the entire population. To compare our current model and results with those from Section 2.6.1, note that now we have more a priori information on the true distribution of voting preferences due to some a priori knowledge of preferences within groups, which allows us to reduce the poll sizes with both straightforward and optimal measurement designs.[19] On the other hand, the difference between $K_{\rm ini}$ and $K_{\rm opt}$ is fully due to the measurement design.

Comparative drug study. A Simple case of the Measurement Design in Discrete o.s. related to OPD and perhaps more interesting is as follows. Suppose that now,

[19]To illustrate this point, look at the last two lines in the table: utilizing a priori information allows us to reduce the poll size from 4,7,88 to 2,556 even with the straightforward measurement design.

instead of L competing candidates running for an office we have L competing drugs, and the population of patients the drugs are aimed at rather than the population of voters. For the sake of simplicity, assume that when a particular drug is administered to a particular patient, the outcome is binary: (positive) "effect" or "no effect" (what follows can be easily extended to the case of non-binary categorial outcomes, like "strong positive effect," "weak positive effect," "negative effect," and alike). Our goal is to organize a clinical study in order to decide on comparative drug efficiency, measured by the percentage of patients on which a particular drug has effect. The difference with organizing an opinion poll is that now we cannot just ask a respondent what his or her preferences are; we may only administer to a participant of the study a single drug of our choice and look at the result.

As in the OPD problem, we assume that the population of patients is split into I groups, with the i-th group comprising a fraction θ_i of the entire population.

We model the situation as follows. We associate with a patient a Boolean vector of dimension $2L$, with the ℓ-th entry in the vector equal to 1 or 0 depending on whether drug # ℓ has effect on the patient, and the $(L+\ell)$-th entry complementing the ℓ-th one to 1 (that is, if the ℓ-th entry is χ, then the $(L+\ell)$-th entry is $1-\chi$). Let x^i be the average of these vectors over patients from group i. We define "signal" x underlying our measurements as the vector $[x^1; ...; x^I]$ and assume that our a priori information allows us to localize x in a closed convex subset $\mathcal{X}$ of the set

$$\mathcal{Y} = \{x = [x^1; ...; x^I] : x^i \geq 0, x^i_\ell + x^i_{L+\ell} = 1,\ 1 \leq i \leq I, 1 \leq \ell \leq L\}$$

to which all our signals belong by construction. Note that the vector

$$y = Bx = \sum_i \theta_i x^i$$

can be treated as a "population-wise distribution of drug effects:" y_ℓ, $\ell \leq L$, is the fraction, in the entire population of patients, of those patients on whom drug ℓ has effect, and $y_{L+\ell} = 1 - y_\ell$. As a result, typical hypotheses related to comparison of the drugs, like "drug ℓ has effect on a larger fraction, at least by margin δ, of patients than drug ℓ'," become convex hypotheses on the signal x. In order to test hypotheses of this type, we can use a two-stage procedure for observing drug effects, namely, as follows.

To get a particular observation, we select at random, with probability $q_{i\ell}$, a pair (i, ℓ) from the set $\{(i, \ell) : 1 \leq i \leq I, 1 \leq \ell \leq L\}$, select a patient from group i according to the uniform distribution on the group, administer to the patient the drug ℓ, and check whether the drug has effect. Thus, a single observation is a triple (i, ℓ, χ), where $\chi = 0$ if the administered drug has no effect on the patient, and $\chi = 1$ otherwise. The probability of getting observation $(i, \ell, 1)$ is $q_{i\ell}x^i_\ell$, and the probability of getting observation $(i, \ell, 0)$ is $q_{i\ell}x^i_{L+\ell}$. Thus, we arrive at the Discrete o.s. where the distribution μ of observations is of the form $\mu = A_q x$, with the rows in A_q indexed by triples $\omega = (i, \ell, \chi) \in \Omega := \{1, 2, ..., I\} \times \{1, 2, ..., L\} \times \{0, 1\}$ and given by

$$(A_q[x^1; ...; x^I])_{i,\ell,\chi} = \begin{cases} q_{i\ell}x^i_\ell & \chi = 1, \\ q_{i\ell}x^i_{L+\ell} & \chi = 0. \end{cases}$$

Specifying the set $\mathcal{Q}$ of admissible measurement designs as a closed convex subset of the set of all nonvanishing discrete probability distributions on the set $\{1, 2, ..., I\} \times \{1, 2, ..., L\}$, we find ourselves in the Simple case of Discrete o.s., as defined by

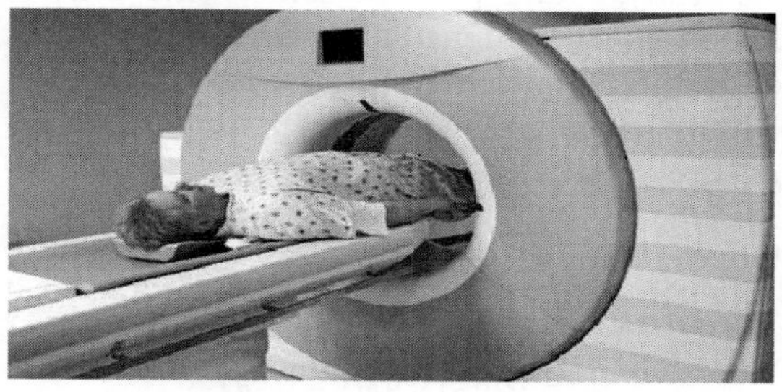

Figure 2.7: PET scanner

(2.102), and $A_q x$ is a probabilistic vector whenever $q \in \mathcal{Q}$ and $x \in \mathcal{Y}$.

2.7.3.2 Simple case, Poisson o.s.

Looking at the optimization problem (P_q), we see that the simplest way to ensure its objective is, as in the case of Discrete o.s., to assume that

$$A_q = \mathrm{Diag}\{Bq\}A,$$

where A is some fixed $d \times n$ matrix, and B is some fixed $d \times (\dim q)$ matrix such that Bq is positive whenever $q \in \mathcal{Q}$. On the top of this, we should ensure that when $q \in \mathcal{Q}$ and $x \in \mathcal{X}$, $A_q x$ is a positive vector; this amounts to some restrictions linking $\mathcal{Q}$, $\mathcal{X}$, A, and B.

Application Example: PET with time control. Positron Emission Tomography was already mentioned, as an example of Poisson o.s., in Section 2.4.3.2. As explained in the section, in PET we observe a random vector $\omega \in \mathbf{R}^d$ with independent entries $[\omega]_i \sim \mathrm{Poisson}(\mu_i)$, $1 \leq i \leq d$, where the vector of parameters $\mu = [\mu_1; ...\mu_d]$ of the Poisson distributions is the linear image $\mu = A\lambda$ of an unknown "signal" λ (the tracer's density in patient's body) belonging to some known subset Λ of $\mathbf{R}^D_+$, with entrywise nonnegative matrix A. Our goal is to make inferences about λ. Now, in an actual PET scan, the patient's position w.r.t. the scanner is not the same during the entire study; the position is kept fixed for an i-th time period, $1 \leq i \leq I$, and changes from period to period in order to expose to the scanner the entire "area of interest." For example, with the scanner shown on Figure 2.7, during the PET study the imaging table with the patient will be shifted several times along the axis of the scanning ring. As a result, the observed vector ω can be split into blocks ω^i, $i = 1, ..., I$, of data acquired during the i-th period, $1 \leq i \leq I$. On closer inspection, the corresponding block μ^i in μ is

$$\mu^i = q_i A_i \lambda,$$

where A_i is an entrywise nonnegative matrix known in advance, and q_i is the duration of the i-th period. In principle, the q_i could be treated as nonnegative design variables subject to the "budget constraint" $\sum_{i=1}^I q_i = T$, where T is the total duration of the study,[20] and perhaps some other convex constraints, say, positive lower bounds on q_i. It is immediately seen that the outlined situation is exactly as is required in the Simple case of Poisson o.s.

2.7.3.3 Simple case, Gaussian o.s.

Looking at the optimization problem (G_q), we see that the simplest way to ensure that its objective is affine in q is to assume that the covariance matrix Θ is diagonal, and

$$A_q = \mathrm{Diag}\{\sqrt{q_1}, ..., \sqrt{q_d}\}A \tag{2.103}$$

where A is a fixed $d \times n$ matrix, and q runs through a convex compact subset of $\mathbf{R}^d_+$.

It turns out that there are situations where assumption (2.103) makes perfect sense. Let us start with a preamble. In Gaussian o.s.

$$\begin{array}{c}\omega = Ax + \xi \\ \left[A \in \mathbf{R}^{d\times n}, \xi \sim \mathcal{N}(0, \Sigma), \, \Sigma = \mathrm{Diag}\{\sigma_1^2, ..., \sigma_d^2\}\right]\end{array} \tag{2.104}$$

the "physics" behind the observations in many cases is as follows. There are d sensors (receivers), the i-th registering the continuous time analogous input depending linearly on the underlying observations signal x. On the time horizon on which the measurements are taken, this input is constant in time and is registered by the i-th sensor on time interval Δ_i. The deterministic component of the measurement registered by sensor i is the integral of the corresponding input taken over Δ_i, and the stochastic component of the measurement is obtained by integrating white Gaussian noise over the same interval. As far as this noise is concerned, what matters is that when the white noise affecting the i-th sensor is integrated over a time interval Δ_i, the result is a Gaussian random variable with zero mean and variance $\sigma_i^2|\Delta_i|$ (here $|\Delta_i|$ is the length of Δ_i), and the random variables obtained by integrating white noise over nonoverlapping segments are independent. Besides this, we assume that the noisy components of measurements are independent across the sensors.

Now, there could be two basic versions of the situation just outlined, both leading to the same observation model (2.104). In the first, "parallel," version, all d sensors work in parallel on the same time horizon of duration 1. In the second, "sequential," version, the sensors are activated and scanned one by one, each working unit time; thus, here the full time horizon is d, and the sensors are registering their respective inputs on consecutive time intervals of duration 1 each. In this second "physical" version of Gaussian o.s., we can, in principle, allow for sensors to register their inputs on consecutive time segments of varying durations $q_1 \geq 0$, $q_2 \geq 0$, ..., $q_d \geq 0$, with the additional to nonnegativity restriction that our total time budget is respected: $\sum_i q_i = d$ (perhaps with some other convex constraints on q_i). Let us look what the observation scheme we end up with is. Assuming that (2.104) represents correctly our observations in the reference case where all the $|\Delta_i|$ are equal to 1, the deterministic component of the measurement registered by sensor i in time interval of duration q_i will be $q_i \sum_j a_{ij}x_j$, and the standard deviation of the noisy component will be $\sigma_i\sqrt{q_i}$, so that the measurements

[20] T cannot be too large; aside from other considerations, the tracer disintegrates, and its density can be considered as nearly constant only on a properly restricted time horizon.

become

$$z_i = \sigma_i \sqrt{q_i} \zeta_i + q_i \sum_j a_{ij} x_j, \; i = 1, ..., d,$$

with standard (zero mean, unit variance) Gaussian noises ζ_i independent of each other. Now, since we know q_i, we can scale the latter observations by making the standard deviation of the noisy component the same σ_i as in the reference case. Specifically, we lose nothing when assuming that our observations are

$$\omega_i = z_i / \sqrt{q_i} = \underbrace{\sigma_i \zeta_i}_{\xi_i} + \sqrt{q_i} \sum_j a_{ij} x_j,$$

or, equivalently,

$$\omega = \xi + \underbrace{\mathrm{Diag}\{\sqrt{q_1}, ..., \sqrt{q_d}\} A}_{A_q} x, \; \xi \sim \mathcal{N}(0, \mathrm{Diag}\{\sigma_1^2, ..., \sigma_d^2\}) \qquad [A = [a_{ij}]]$$

where q runs through a convex compact subset $\mathcal{Q}$ of the simplex $\{q \in \mathbf{R}^d_+ : \sum_i q_i = d\}$. Thus, if the "physical nature" of a Gaussian o.s. is sequential, then, making the activity times of the sensors our design variables, as is natural under the circumstances, we arrive at (2.103), and, as a result, end up with an easy-to-solve Measurements Design problem.

2.8 AFFINE DETECTORS BEYOND SIMPLE OBSERVATION SCHEMES

On a closer inspection, the "common denominator" of our basic simple o.s.'s—Gaussian, Poisson and Discrete ones—is that in all these cases the minimal risk detector for a pair of convex hypotheses is *affine*. At first glance, this indeed is so for Gaussian and Poisson o.s.'s, where $\mathcal{F}$ is comprised of affine functions on the corresponding observation space Ω ($\mathbf{R}^d$ for Gaussian o.s., and $\mathbf{Z}^d_+ \subset \mathbf{R}^d$ for Poisson o.s.), but is *not* so for Discrete o.s.—in that case, $\Omega = \{1, ..., d\}$, and $\mathcal{F}$ is comprised of all functions on Ω, while "affine functions on $\Omega = \{1, ..., d\}$" make no sense. Note, however, that we can encode (and from now on this is what we do) the points $i = 1, ..., d$ of a d-element set by basic orths $e_i = [0; ...; 0; 1; 0; ...; 0] \in \mathbf{R}^d$ in $\mathbf{R}^d$, thus making our observation space Ω a subset of $\mathbf{R}^d$. With this encoding, *every* real valued function on $\{1, ..., d\}$ becomes a restriction on Ω of an affine function. Note that when passing from our basic simple o.s.'s to their direct products, the minimum risk detectors for pairs of convex hypotheses remain affine.

Now, in our context the following two properties of simple o.s.'s are essential:

A) the best—with the smallest possible risk—*affine* detector, like its risk, can be efficiently computed;
B) the smallest risk *affine* detector from A) is the best detector, in terms of risk, available under the circumstances, so that the associated test is near-optimal.

Note that as far as practical applications of the detector-based hypothesis testing are concerned, one "can survive" without B) (near-optimality of the constructed detectors), while A) *is a requisite.*

In this section we focus on families of probability distributions obeying A). This class turns out to be incomparably larger than what was defined as simple o.s.'s in Section 2.4; in particular, it includes nonparametric families of distributions. Staying within this much broader class, we still are able to construct in a computationally efficient way the best affine detectors, in certain precise sense, for a pair of "convex" hypotheses, along with valid upper bounds on the risks of the detectors. What we, in general, can*not* claim anymore, is that the tests associated with such detectors are near-optimal. This being said, we believe that investigating possibilities for building tests and quantifying their performance in a computationally friendly manner is of value even when we cannot provably guarantee near-optimality of these tests. The results to follow originate from [135, 136].

2.8.1 Situation

In what follows, we fix an *observation space* $\Omega = \mathbf{R}^d$, and let $\mathcal{P}_j$, $1 \leq j \leq J$, be given families of probability distributions on Ω. Put broadly, our goal still is, given a random observation $\omega \sim P$, where $P \in \bigcup\limits_{j\leq J} \mathcal{P}_j$, to decide upon the hypotheses $H_j : P \in \mathcal{P}_j$, $j = 1, ..., J$. We intend to address this goal in the case when the families $\mathcal{P}_j$ are *simple*—they are comprised of distributions for which moment-generating functions admit an explicit upper bound.

2.8.1.1 Preliminaries: Regular data and associated families of distributions

Definition 2.36. *A.* ***Regular data*** *is as a triple $\mathcal{H}, \mathcal{M}, \Phi(\cdot,\cdot)$, where*

- *$\mathcal{H}$ is a nonempty closed convex set in $\Omega = \mathbf{R}^d$ symmetric w.r.t. the origin,*
- *$\mathcal{M}$ is a closed convex set in some $\mathbf{R}^n$,*
- *$\Phi(h;\mu) : \mathcal{H}\times\mathcal{M} \to \mathbf{R}$ is a continuous function convex in $h \in \mathcal{H}$ and concave in $\mu \in \mathcal{M}$.*

B. Regular data $\mathcal{H}, \mathcal{M}, \Phi(\cdot,\cdot)$ define two families of probability distributions on Ω:

- *the family of* ***regular*** *distributions*

$$\mathcal{R} = \mathcal{R}[\mathcal{H}, \mathcal{M}, \Phi]$$

comprised of all probability distributions P on Ω such that

$$\forall h \in \mathcal{H}\ \exists \mu \in \mathcal{M} : \ln\left(\int_\Omega \exp\{h^T\omega\}P(d\omega)\right) \leq \Phi(h;\mu).$$

- *the family of* ***simple*** *distributions*

$$\mathcal{S} = \mathcal{S}[\mathcal{H}, \mathcal{M}, \Phi]$$

comprised of probability distributions P on Ω such that

$$\exists \mu \in \mathcal{M} : \forall h \in \mathcal{H} : \ln\left(\int_\Omega \exp\{h^T\omega\}P(d\omega)\right) \leq \Phi(h;\mu). \tag{2.105}$$

For a probability distribution $P \in \mathcal{S}[\mathcal{H}, \mathcal{M}, \Phi]$, every $\mu \in \mathcal{M}$ satisfying (2.105) is referred to as a parameter *of P w.r.t. $\mathcal{S}$. Note that a distribution may have many parameters different from each other.*

Recall that beginning with Section 2.3, the starting point in all our constructions is a "plausibly good" detector-based test which, given two families $\mathcal{P}_1$ and $\mathcal{P}_2$ of distributions with common observation space, and repeated observations $\omega_1, ..., \omega_t$ drawn from a distribution $P \in \mathcal{P}_1 \cup \mathcal{P}_2$, decides whether $P \in \mathcal{P}_1$ or $P \in \mathcal{P}_2$. Our interest in the families of regular/simple distributions stems from the fact that when the families $\mathcal{P}_1$ and $\mathcal{P}_2$ are of this type, building such a test reduces to solving a convex-concave saddle point problem and thus can be carried out in a computationally efficient manner. We postpone the related construction and analysis to Section 2.8.2, and continue with presenting some basic examples of families of simple and regular distributions along with a simple "calculus" of these families.

2.8.1.2 Basic examples of simple families of probability distributions

2.8.1.2.A. Sub-Gaussian distributions: Let $\mathcal{H} = \Omega = \mathbf{R}^d$, let $\mathcal{M}$ be a closed convex subset of the set $\mathcal{G}_d = \{\mu = (\theta, \Theta) : \theta \in \mathbf{R}^d, \Theta \in \mathbf{S}^d_+\}$, where $\mathbf{S}^d_+$ is a cone of positive semidefinite matrices in the space $\mathbf{S}^d$ of symmetric $d \times d$ matrices, and let

$$\Phi(h; \theta, \Theta) = \theta^T h + \tfrac{1}{2} h^T \Theta h.$$

Recall that a distribution P on $\Omega = \mathbf{R}^d$ is called sub-Gaussian with sub-Gaussianity parameters $\theta \in \mathbf{R}^d$ and $\Theta \in \mathbf{S}^d_+$ if

$$\mathbf{E}_{\omega \sim P}\{\exp\{h^T \omega\}\} \leq \exp\{\theta^T h + \tfrac{1}{2} h^T \Theta h\} \;\; \forall h \in \mathbf{R}^d. \tag{2.106}$$

Whenever this is the case, θ is the expected value of P. We shall use the notation $\xi \sim \mathcal{SG}(\theta, \Theta)$ as a shortcut for the sentence "random vector ξ is sub-Gaussian with parameters θ, Θ." It is immediately seen that when $\xi \sim \mathcal{N}(\theta, \Theta)$, we also have $\xi \sim \mathcal{SG}(\theta, \Theta)$, and (2.106) in this case is an identity rather than an inequality.

With Φ as above, $\mathcal{S}[\mathcal{H}, \mathcal{M}, \Phi]$ clearly contains every sub-Gaussian distribution P on $\mathbf{R}^d$ with sub-Gaussianity parameters (forming a parameter of P w.r.t. $\mathcal{S}$) from $\mathcal{M}$. In particular, $\mathcal{S}[\mathcal{H}, \mathcal{M}, \Phi]$ contains all Gaussian distributions $\mathcal{N}(\theta, \Theta)$ with $(\theta, \Theta) \in \mathcal{M}$.

2.8.1.2.B. Poisson distributions: Let $\mathcal{H} = \Omega = \mathbf{R}^d$, let $\mathcal{M}$ be a closed convex subset of d-dimensional nonnegative orthant $\mathbf{R}^d_+$, and let

$$\Phi(h = [h_1; ...; h_d]; \mu = [\mu_1; ...; \mu_d]) = \sum_{i=1}^{d} \mu_i [\exp\{h_i\} - 1] : \mathcal{H} \times \mathcal{M} \to \mathbf{R}.$$

The family $\mathcal{S} = \mathcal{S}[\mathcal{H}, \mathcal{M}, \Phi]$ contains all Poisson distributions Poisson$[\mu]$ with vectors μ of parameters belonging to $\mathcal{M}$; here Poisson$[\mu]$ is the distribution of a random d-dimensional vector with entries independent of each other, the i-th entry being a Poisson random variable with parameter μ_i. μ is a parameter of Poisson$[\mu]$ w.r.t. $\mathcal{S}$.

2.8.1.2.C. Discrete distributions. Consider a discrete random variable taking values in d-element set $\{1, 2, ..., d\}$, and let us think of such a variable as of random variable taking values $e_i \in \mathbf{R}^d$, $i = 1, ..., d$, where $e_i = [0; ...; 0; 1; 0; ...; 0]$ (1 in position i) are standard basic orths in $\mathbf{R}^d$. The probability distribution of such a variable can be identified with a point $\mu = [\mu_1; ...; \mu_d]$ from the d-dimensional probabilistic simplex

$$\boldsymbol{\Delta}_d = \left\{ \nu \in \mathbf{R}^d_+ : \sum_{i=1}^d \nu_i = 1 \right\},$$

where μ_i is the probability for the variable to take value e_i. With these identifications, setting $\mathcal{H} = \mathbf{R}^d$, specifying $\mathcal{M}$ as a closed convex subset of $\boldsymbol{\Delta}_d$, and setting

$$\Phi(h = [h_1; ...; h_d]; \mu = [\mu_1; ...; \mu_d]) = \ln \left(\sum_{i=1}^d \mu_i \exp\{h_i\} \right),$$

the family $\mathcal{S} = \mathcal{S}[\mathcal{H}, \mathcal{M}, \Phi]$ contains distributions of all discrete random variables taking values in $\{1, ..., d\}$ with probabilities $\mu_1, ..., \mu_d$ comprising a vector from $\mathcal{M}$. This vector is a parameter of the corresponding distribution w.r.t. $\mathcal{S}$.

2.8.1.2.D. Distributions with bounded support. Consider the family $\mathcal{P}[X]$ of probability distributions supported on a closed and bounded convex set $X \subset \Omega = \mathbf{R}^d$, and let

$$\phi_X(h) = \max_{x \in X} h^T x$$

be the support function of X. We have the following result (to be refined in Section 2.8.1.3):

Proposition 2.37. *For every $P \in \mathcal{P}[X]$ it holds*

$$\forall h \in \mathbf{R}^d: \ \ln \left(\int_{\mathbf{R}^d} \exp\{h^T \omega\} P(d\omega) \right) \leq h^T e[P] + \tfrac{1}{8} \left[\phi_X(h) + \phi_X(-h) \right]^2, \quad (2.107)$$

where $e[P] = \int_{\mathbf{R}^d} \omega P(d\omega)$ is the expectation of P, and the function in the right-hand side of (2.107) is convex. As a result, setting

$$\mathcal{H} = \mathbf{R}^d, \ \mathcal{M} = X, \ \Phi(h; \mu) = h^T \mu + \tfrac{1}{8} \left[\phi_X(h) + \phi_X(-h) \right]^2$$

we obtain regular data such that $\mathcal{P}[X] \subset \mathcal{S}[\mathcal{H}, \mathcal{M}, \Phi]$, $e[P]$ being a parameter of a distribution $P \in \mathcal{P}[X]$ w.r.t. $\mathcal{S}$.

For proof, see Section 2.11.4

2.8.1.3 Calculus of regular and simple families of probability distributions

Families of regular and simple distributions admit "fully algorithmic" calculus, with the main calculus rules as follows.

2.8.1.3.A. Direct summation. For $1 \leq \ell \leq L$, let regular data $\mathcal{H}_\ell \subset \Omega_\ell = \mathbf{R}^{d_\ell}$,

$\mathcal{M}_\ell \subset \mathbf{R}^{n_\ell}$, $\Phi_\ell(h_\ell;\mu_\ell) : \mathcal{H}_\ell \times \mathcal{M}_\ell \to \mathbf{R}$ be given. Let us set

$$\begin{array}{rcl}\Omega & = & \Omega_1 \times ... \times \Omega_L = \mathbf{R}^d,\ d = d_1 + ... + d_L,\\ \mathcal{H} & = & \mathcal{H}_1 \times ... \times \mathcal{H}_L = \{h = [h^1; ...; h^L] : h^\ell \in \mathcal{H}_\ell, \ell \leq L\},\\ \mathcal{M} & = & \mathcal{M}_1 \times ... \times \mathcal{M}_L = \{\mu = [\mu^1; ...; \mu^L] : \mu^\ell \in \mathcal{M}^\ell, \ell \leq L\} \subset \mathbf{R}^n,\\ & & \qquad n = n_1 + ... + n_L,\\ \multicolumn{3}{l}{\Phi(h = [h^1; ...; h^L]; \mu = [\mu^1; ...; \mu^L]) = \sum_{\ell=1}^L \Phi_\ell(h^\ell; \mu^\ell) : \mathcal{H} \times \mathcal{M} \to \mathbf{R}.}\end{array}$$

Then $\mathcal{H}$ is a closed convex set in $\Omega = \mathbf{R}^d$, symmetric w.r.t. the origin, $\mathcal{M}$ is a nonempty closed convex set in $\mathbf{R}^n$, $\Phi : \mathcal{H} \times \mathcal{M} \to \mathbf{R}$ is a continuous convex-concave function, and clearly

- the family $\mathcal{R}[\mathcal{H}, \mathcal{M}, \Phi]$ contains all product-type distributions $P = P_1 \times ... \times P_L$ on $\Omega = \Omega_1 \times ... \times \Omega_L$ with $P_\ell \in \mathcal{R}[\mathcal{H}_\ell, \mathcal{M}_\ell, \Phi_\ell]$, $1 \leq \ell \leq L$;
- the family $\mathcal{S} = \mathcal{S}[\mathcal{H}, \mathcal{M}, \Phi]$ contains all product-type distributions $P = P_1 \times ... \times P_L$ on $\Omega = \Omega_1 \times ... \times \Omega_L$ with $P_\ell \in \mathcal{S}_\ell = \mathcal{S}[\mathcal{H}_\ell, \mathcal{M}_\ell, \Phi_\ell]$, $1 \leq \ell \leq L$, a parameter of P w.r.t. $\mathcal{S}$ being the vector of parameters of P_ℓ w.r.t. $\mathcal{S}_\ell$.

2.8.1.3.B. Mixing. For $1 \leq \ell \leq L$, let regular data $\mathcal{H}_\ell \subset \Omega = \mathbf{R}^d$, $\mathcal{M}_\ell \subset \mathbf{R}^{n_\ell}$, $\Phi_\ell(h_\ell; \mu_\ell) : \mathcal{H}_\ell \times \mathcal{M}_\ell \to \mathbf{R}$ be given, with compact $\mathcal{M}_\ell$. Let also $\nu = [\nu_1; ...; \nu_L]$ be a probabilistic vector. For a tuple $P^L = \{P_\ell \in \mathcal{R}[\mathcal{H}_\ell, \mathcal{M}_\ell, \Phi_\ell]\}_{\ell=1}^L$, let $\Pi[P^L, \nu]$ be the ν-mixture of distributions $P_1, ..., P_L$ defined as the distribution of random vector $\omega \sim \Omega$ generated as follows: we draw at random, from probability distribution ν on $\{1, ..., L\}$, index ℓ, and then draw ω at random from the distribution P_ℓ. Finally, let $\mathcal{P}$ be the set of all probability distributions on Ω which can be obtained as $\Pi[P^L, \nu]$ from the outlined tuples P^L and vectors ν running through the probabilistic simplex $\mathbf{\Delta}_L = \{\mu \in \mathbf{R}^L : \nu \geq 0, \sum_\ell \nu_\ell = 1\}$.

Let us set

$$\begin{array}{rcl}\mathcal{H} & = & \bigcap\limits_{\ell=1}^{L} \mathcal{H}_\ell,\\ \Psi_\ell(h) & = & \max\limits_{\mu_\ell \in \mathcal{M}_\ell} \Phi_\ell(h; \mu_\ell) : \mathcal{H}_\ell \to \mathbf{R},\\ \Phi(h; \nu) & = & \ln\left(\sum_{\ell=1}^L \nu_\ell \exp\{\Psi_\ell(h)\}\right) : \mathcal{H} \times \mathbf{\Delta}_L \to \mathbf{R}.\end{array}$$

Then $\mathcal{H}, \mathbf{\Delta}_L, \Phi$ clearly is regular data (recall that all $\mathcal{M}_\ell$ are compact sets), and for every $\nu \in \mathbf{\Delta}_L$ and tuple P^L of the above type one has

$$P = \Pi[P^L, \nu] \Rightarrow \ln\left(\int_\Omega \mathrm{e}^{h^T\omega} P(d\omega)\right) \leq \Phi(h; \nu)\ \forall h \in \mathcal{H}, \tag{2.108}$$

implying that $\mathcal{P} \subset \mathcal{S}[\mathcal{H}, \mathbf{\Delta}_L, \Phi]$, ν being a parameter of $P = \Pi[P^L, \nu] \in \mathcal{P}$.

Indeed,(2.108) is readily given by the fact that for $P = \Pi[P^L, \nu] \in \mathcal{P}$ and $h \in \mathcal{H}$ it holds

$$\ln\left(\mathbf{E}_{\omega\sim P}\left\{\mathrm{e}^{h^T\omega}\right\}\right) = \ln\left(\sum_{\ell=1}^L \nu_\ell \mathbf{E}_{\omega\sim P_\ell}\{\mathrm{e}^{h^T\omega}\}\right) \leq \ln\left(\sum_{\ell=1}^L \nu_\ell \exp\{\Psi_\ell(h)\}\right) = \Phi(h; \nu),$$

with the concluding inequality given by $h \in \mathcal{H} \subset \mathcal{H}_\ell$ and $P_\ell \in \mathcal{R}[\mathcal{H}_\ell, \mathcal{M}_\ell, \Phi_\ell]$, $1 \leq \ell \leq L$.

We have built a simple family of distributions $\mathcal{S} := \mathcal{S}[\mathcal{H}, \mathbf{\Delta}_L, \Phi]$ which contains all mixtures of distributions from given regular families $\mathcal{R}_\ell := \mathcal{R}[\mathcal{H}_\ell, \mathcal{M}_\ell, \Phi_\ell]$, $1 \leq \ell \leq L$, which makes $\mathcal{S}$ a simple outer approximation of mixtures of distributions from the simple families $\mathcal{S}_\ell := \mathcal{S}[\mathcal{H}_\ell, \mathcal{M}_\ell, \Phi_\ell]$, $1 \leq \ell \leq L$. In this latter capacity, $\mathcal{S}$ has a drawback—the only parameter of the mixture $P = \Pi[P^L, \nu]$ of distributions $P_\ell \in \mathcal{S}_\ell$ is ν, while the parameters of P_ℓ's disappear. In some situations, this makes the outer approximation $\mathcal{S}$ of $\mathcal{P}$ too conservative. We are about to get rid, to some extent, of this drawback.

A modification. In the situation described at the beginning of 2.8.1.3.B, let a vector $\bar{\nu} \in \mathbf{\Delta}_L$ be given, and let

$$\bar{\Phi}(h; \mu_1, ..., \mu_L) = \sum_{\ell=1}^{L} \bar{\nu}_\ell \Phi_\ell(h; \mu_\ell) : \mathcal{H} \times (\mathcal{M}_1 \times ... \times \mathcal{M}_L) \to \mathbf{R}.$$

Let $d \times d$ matrix $Q \succeq 0$ satisfy

$$\left(\Phi_\ell(h; \mu_\ell) - \bar{\Phi}(h; \mu_1, ..., \mu_L)\right)^2 \leq h^T Q h \;\forall (h \in \mathcal{H}, \ell \leq L, \mu \in \mathcal{M}_1 \times ... \times \mathcal{M}_L), \tag{2.109}$$

and let

$$\Phi(h; \mu_1, ..., \mu_L) = \tfrac{3}{5} h^T Q h + \bar{\Phi}(h; \mu_1, ..., \mu_L) : \mathcal{H} \times (\mathcal{M}_1 \times ... \times \mathcal{M}_L) \to \mathbf{R}. \tag{2.110}$$

Φ clearly is convex-concave and continuous on its domain, whence $\mathcal{H} = \bigcap_\ell \mathcal{H}_\ell, \mathcal{M}_1 \times ... \times \mathcal{M}_L, \Phi$ is regular data.

Proposition 2.38. *In the situation just defined, denoting by $\mathcal{P}_{\bar{\nu}}$ the family of all probability distributions $P = \Pi[P^L, \bar{\nu}]$, stemming from tuples*

$$P^L = \{P_\ell \in \mathcal{S}[\mathcal{H}_\ell, \mathcal{M}_\ell, \Phi_\ell]\}_{\ell=1}^L, \tag{2.111}$$

one has

$$\mathcal{P}_{\bar{\nu}} \subset \mathcal{S}[\mathcal{H}, \mathcal{M}_1 \times ... \times \mathcal{M}_L, \Phi].$$

As a parameter of distribution $P = \Pi[P^L, \bar{\nu}] \in \mathcal{P}_{\bar{\nu}}$ with P^L as in (2.111), one can take $\mu^L = [\mu_1; ....; \mu_L]$.

Proof. It is easily seen that

$$\mathrm{e}^a \leq a + \mathrm{e}^{\frac{3}{5}a^2}, \;\forall a.$$

As a result, when a_ℓ, $\ell = 1, ..., L$, satisfy $\sum_\ell \bar{\nu}_\ell a_\ell = 0$, we have

$$\sum_\ell \bar{\nu}_\ell \mathrm{e}^{a_\ell} \leq \sum_\ell \bar{\nu}_\ell a_\ell + \sum_\ell \bar{\nu}_\ell \mathrm{e}^{\frac{3}{5}a_\ell^2} \leq \mathrm{e}^{\frac{3}{5}\max_\ell a_\ell^2}. \tag{2.112}$$

Now let P^L be as in (2.111), and let $h \in \mathcal{H} = \bigcap_L \mathcal{H}_\ell$. Setting $P = \Pi[P^L, \bar{\nu}]$, we have

$$\begin{array}{l}\ln\left(\int_\Omega \mathrm{e}^{h^T\omega} P(d\omega)\right) = \ln\left(\sum_\ell \bar{\nu}_\ell \int_\Omega \mathrm{e}^{h^T\omega} P_\ell(d\omega)\right) = \ln\left(\sum_\ell \bar{\nu}_\ell \exp\{\Phi_\ell(h, \mu_\ell)\}\right)\\ = \bar{\Phi}(h; \mu_1, ...\mu_L) + \ln\left(\sum_\ell \bar{\nu}_\ell \exp\{\Phi_\ell(h, \mu_\ell) - \bar{\Phi}(h; \mu_1, ...\mu_L)\}\right)\\ \underbrace{\leq}_{a} \bar{\Phi}(h; \mu_1, ...\mu_L) + \tfrac{3}{5}\max_\ell[\Phi_\ell(h, \mu_\ell) - \bar{\Phi}(h; \mu_1, ...\mu_L)]^2 \underbrace{\leq}_{b} \Phi(h; \mu_1, ..., \mu_L),\end{array}$$

where a is given by (2.112) as applied to $a_\ell = \Phi_\ell(h, \mu_\ell) - \bar{\Phi}(h; \mu_1, ...\mu_L)$, and b is due to (2.109) and (2.110). The resulting inequality, which holds true for all $h \in \mathcal{H}$, is all we need. □

2.8.1.3.C. I.i.d. summation. Let $\Omega = \mathbf{R}^d$ be an observation space, $(\mathcal{H}, \mathcal{M}, \Phi)$ be regular data on this space, and $\lambda = \{\lambda_\ell\}_{\ell=1}^K$ be a collection of reals. We can associate with the outlined entities new data $(\mathcal{H}_\lambda, \mathcal{M}, \Phi_\lambda)$ on Ω by setting

$$\mathcal{H}_\lambda = \{h \in \Omega : \|\lambda\|_\infty h \in \mathcal{H}\}, \quad \Phi_\lambda(h; \mu) = \sum_{\ell=1}^L \Phi(\lambda_\ell h; \mu) : \mathcal{H}_\lambda \times \mathcal{M} \to \mathbf{R}.$$

Now, given a probability distribution P on Ω, we can associate with it and with the above λ a new probability distribution P^λ on Ω as follows: P^λ is the distribution of $\sum_\ell \lambda_\ell \omega_\ell$, where $\omega_1, \omega_2, ..., \omega_L$ are drawn, independently of each other, from P. An immediate observation is that the data $(\mathcal{H}_\lambda, \mathcal{M}, \Phi_\lambda)$ is regular, and

whenever a probability distribution P belongs to $\mathcal{S}[\mathcal{H}, \mathcal{M}, \Phi]$, the distribution P^λ belongs to $\mathcal{S}[\mathcal{H}_\lambda, \mathcal{M}, \Phi_\lambda]$, and every parameter of P is a parameter of P^λ. In particular, when $\omega \sim P \in \mathcal{S}[\mathcal{H}, \mathcal{M}, \Phi]$ the distribution P^L of the sum of L independent copies of ω belongs to $\mathcal{S}[\mathcal{H}, \mathcal{M}, L\Phi]$.

2.8.1.3.D. Semi-direct summation. For $1 \leq \ell \leq L$, let regular data $\mathcal{H}_\ell \subset \Omega_\ell = \mathbf{R}^{d_\ell}$, $\mathcal{M}_\ell$, Φ_ℓ be given. To avoid complications, we assume that for every ℓ,

- $\mathcal{H}_\ell = \Omega_\ell$,
- $\mathcal{M}_\ell$ is bounded.

Let also an $\epsilon > 0$ be given. We assume that ϵ is small, namely, $L\epsilon < 1$.

Let us aggregate the given regular data into a new one by setting

$$\mathcal{H} = \Omega := \Omega_1 \times ... \times \Omega_L = \mathbf{R}^d, \; d = d_1 + ... + d_L, \;\; \mathcal{M} = \mathcal{M}_1 \times ... \times \mathcal{M}_L,$$

and let us define function $\Phi(h; \mu) : \Omega^d \times \mathcal{M} \to \mathbf{R}$ as follows:

$$\begin{array}{l} \Phi(h = [h^1; ...; h^L]; \mu = [\mu^1; ...; \mu^L]) = \inf_{\lambda \in \mathbf{\Delta}^\epsilon} \sum_{\ell=1}^d \lambda_\ell \Phi_\ell(h^\ell / \lambda_\ell; \mu^\ell), \\ \mathbf{\Delta}^\epsilon = \{\lambda \in \mathbf{R}^d : \lambda_\ell \geq \epsilon \, \forall \ell \; \& \; \sum_{\ell=1}^L \lambda_\ell = 1\}. \end{array} \tag{2.113}$$

For evident reasons, the infimum in the description of Φ is achieved, and Φ is continuous. In addition, Φ is convex in $h \in \mathbf{R}^d$ and concave in $\mu \in \mathcal{M}$. Postponing for a moment verification, the consequences are that $\mathcal{H} = \Omega = \mathbf{R}^d$, $\mathcal{M}$, and Φ form regular data. We claim that

Whenever $\omega = [\omega^1; ...; \omega^L]$ is a random variable taking values in $\Omega = \mathbf{R}^{d_1} \times ... \times \mathbf{R}^{d_L}$, and the marginal distributions P_ℓ, $1 \leq \ell \leq L$, of ω belong to the families $\mathcal{S}_\ell = \mathcal{S}[\mathbf{R}^{d_\ell}, \mathcal{M}_\ell, \Phi_\ell]$ for all $1 \leq \ell \leq L$, the distribution P of ω belongs to $\mathcal{S} = \mathcal{S}[\mathbf{R}^d, \mathcal{M}, \Phi]$, a parameter of P w.r.t. $\mathcal{S}$ being the vector comprised of parameters of P_ℓ w.r.t. $\mathcal{S}_\ell$.

Indeed, since $P_\ell \in \mathcal{S}[\mathbf{R}^{d_\ell}, \mathcal{M}_\ell, \Phi_\ell]$, there exists $\widehat{\mu}^\ell \in \mathcal{M}_\ell$ such that

$$\ln(\mathbf{E}_{\omega^\ell \sim P_\ell}\{\exp\{g^T \omega^\ell\}\}) \leq \Phi_\ell(g; \widehat{\mu}^\ell) \; \forall g \in \mathbf{R}^{d_\ell}.$$

Let us set $\widehat{\mu} = [\widehat{\mu}^1; ...; \widehat{\mu}^L]$, and let $h = [h^1; ...; h^L] \in \Omega$ be given. We can find $\lambda \in \mathbf{\Delta}^\epsilon$ such that

$$\Phi(h; \widehat{\mu}) = \sum_{\ell=1}^{L} \lambda_\ell \Phi_\ell(h^\ell/\lambda_\ell; \widehat{\mu}^\ell).$$

Applying the Hölder inequality, we get

$$\mathbf{E}_{[\omega^1;...;\omega^L]\sim P}\left\{\exp\{\sum_\ell [h^\ell]^T\omega^\ell\}\right\} \leq \prod_{\ell=1}^{L}\left(\mathbf{E}_{\omega^\ell\sim P_\ell}\left\{[h^\ell]^T\omega^\ell/\lambda_\ell\}\right)^{\lambda_\ell},$$

whence

$$\ln\left(\mathbf{E}_{[\omega^1;...;\omega^L]\sim P}\left\{\exp\{\sum_\ell [h^\ell]^T\omega^\ell\}\right\}\right) \leq \sum_{\ell=1}^{L} \lambda_\ell \Phi_\ell(h^\ell/\lambda_\ell; \widehat{\mu}^\ell) = \Phi(h; \widehat{\mu}).$$

We see that

$$\ln\left(\mathbf{E}_{[\omega^1;...;\omega^L]\sim P}\left\{\exp\{\sum_\ell [h^\ell]^T\omega^\ell\}\right\}\right) \leq \Phi(h; \widehat{\mu})\ \forall h \in \mathcal{H} = \mathbf{R}^d,$$

and thus $P \in \mathcal{S}[\mathbf{R}^d, \mathcal{M}, \Phi]$, as claimed.

It remains to verify that the function Φ defined by (2.113) indeed is convex in $h \in \mathbf{R}^d$ and concave in $\mu \in \mathcal{M}$. Concavity in μ is evident. Further, functions $\lambda_\ell \Phi_\ell(h^\ell/\lambda_\ell; \mu)$ (as perspective transformations of convex functions $\Phi_\ell(\cdot; \mu)$) are jointly convex in λ and h^ℓ, and so is $\Psi(\lambda, h; \mu) = \sum_{\ell=1}^{L} \lambda_\ell \Phi_\ell(h^\ell/\lambda_\ell, \mu)$. Thus $\Phi(\cdot; \mu)$, obtained by partial minimization of Ψ in λ, indeed is convex.

2.8.1.3.E. Affine image. Let $\mathcal{H}$, $\mathcal{M}$, Φ be regular data, Ω be the embedding space of $\mathcal{H}$, and $x \mapsto Ax + a$ be an affine mapping from Ω to $\bar{\Omega} = \mathbf{R}^{\bar{d}}$, and let us set

$$\bar{\mathcal{H}} = \{\bar{h} \in \mathbf{R}^{\bar{d}} : A^T\bar{h} \in \mathcal{H}\},\ \bar{\mathcal{M}} = \mathcal{M},\ \bar{\Phi}(\bar{h}; \mu) = \Phi(A^T\bar{h}; \mu) + a^T\bar{h} : \bar{\mathcal{H}} \times \bar{\mathcal{M}} \to \mathbf{R}.$$

Note that $\bar{\mathcal{H}}$, $\bar{\mathcal{M}}$, $\bar{\Phi}$ is regular data. It is immediately seen that

> *Whenever the probability distribution P of a random variable ω belongs to $\mathcal{R}[\mathcal{H}, \mathcal{M}, \Phi]$ (or belongs to $\mathcal{S}[\mathcal{H}, \mathcal{M}, \Phi]$), the distribution $\bar{P}[P]$ of the random variable $\bar{\omega} = A\omega + a$ belongs to $\mathcal{R}[\bar{\mathcal{H}}, \bar{\mathcal{M}}, \bar{\Phi}]$ (respectively, belongs to $\mathcal{S}[\bar{\mathcal{H}}, \bar{\mathcal{M}}, \bar{\Phi}]$, and every parameter of P is a parameter of $\bar{P}[P]$).*

2.8.1.3.F. Incorporating support information. Consider the situation as follows. We are given regular data $\mathcal{H} \subset \Omega = \mathbf{R}^d, \mathcal{M}, \Phi$ and are interested in a family $\mathcal{P}$ of distributions known to belong to $\mathcal{S}[\mathcal{H}, \mathcal{M}, \Phi]$. In addition, we know that all distributions P from $\mathcal{P}$ are supported on a given closed convex set $X \subset \mathbf{R}^d$. How could we incorporate this domain information to pass from the family $\mathcal{S}[\mathcal{H}, \mathcal{M}, \Phi]$ containing $\mathcal{P}$ to a smaller family of the same type still containing $\mathcal{P}$? We are about to give an answer in the simplest case of $\mathcal{H} = \Omega$. When denoting by $\phi_X(\cdot)$ the support function of X and selecting somehow a closed convex set $G \subset \mathbf{R}^d$ containing the origin, let us set

$$\widehat{\Phi}(h;\mu) = \inf_{g\in G}\left[\Phi^+(h,g;\mu) := \Phi(h-g;\mu) + \phi_X(g)\right],$$

where $\Phi(h;\mu):\mathbf{R}^d\times\mathcal{M}\to\mathbf{R}$ is the continuous convex-concave function participating in the original regular data. Assuming that $\widehat{\Phi}$ is real-valued and continuous on the domain $\mathbf{R}^d\times\mathcal{M}$ (which definitely is the case when G is a compact set such that ϕ_X is finite and continuous on G), note that $\widehat{\Phi}$ is convex-concave on this domain, so that $\mathbf{R}^d,\mathcal{M},\widehat{\Phi}$ is regular data. We claim that

The family $\mathcal{S}[\mathbf{R}^d,\mathcal{M},\widehat{\Phi}]$ contains $\mathcal{P}$, provided the family $\mathcal{S}[\mathbf{R}^d,\mathcal{M},\Phi]$ does so, and the first of these two families is smaller than the second one.

Verification of the claim is immediate. Let $P\in\mathcal{P}$, so that for properly selected $\mu=\mu_P\in\mathcal{M}$ and for all $e\in\mathbf{R}^d$ it holds

$$\ln\left(\int_{\mathbf{R}^d}\exp\{e^T\omega\}P(d\omega)\right)\leq\Phi(e;\mu_P).$$

On the other hand, for every $g\in G$ we have $\phi_X(g)-g^T\omega\geq 0$ on the support of P, whence for every $h\in\mathbf{R}^d$ one has

$$\begin{array}{l}\ln\left(\int_{\mathbf{R}^d}\exp\{h^T\omega\}P(d\omega)\right)\leq\ln\left(\int_{\mathbf{R}^d}\exp\{h^T\omega+\phi_X(g)-g^T\omega\}P(d\omega)\right)\\ \leq\phi_X(g)+\Phi(h-g;\mu_P).\end{array}$$

Since the resulting inequality holds true for all $g\in G$, we get

$$\ln\left(\int_{\mathbf{R}^d}\exp\{h^T\omega\}P(d\omega)\right)\leq\widehat{\Phi}(h;\mu_P)\ \forall h\in\mathbf{R}^d,$$

implying that $P\in\mathcal{S}[\mathbf{R}^d,\mathcal{M},\widehat{\Phi}]$; because $P\in\mathcal{P}$ is arbitrary, the first part of the claim is justified. The inclusion $\mathcal{S}[\mathbf{R}^d,\mathcal{M},\widehat{\Phi}]\subset\mathcal{S}[\mathbf{R}^d,\mathcal{M},\Phi]$ is readily given by the inequality $\widehat{\Phi}\leq\Phi$, and the latter is due to $\widehat{\Phi}(h,\mu)\leq\Phi(h-0,\mu)+\phi_X(0)$.

Illustration: Distributions with bounded support revisited. In Section 2.8.1.2, given a convex compact set $X\subset\mathbf{R}^d$ with support function ϕ_X, we checked that the data $\mathcal{H}=\mathbf{R}^d$, $\mathcal{M}=X$, $\Phi(h;\mu)=h^T\mu+\frac{1}{8}[\phi_X(h)+\phi_X(-h)]^2$ is regular and the family $\mathcal{S}[\mathbf{R}^d,\mathcal{M},\Phi]$ contains the family $\mathcal{P}[X]$ of all probability distributions supported on X. Moreover, for every $\mu\in\mathcal{M}=X$, the family $\mathcal{S}[\mathbf{R}^d,\{\mu\},\Phi|_{\mathbf{R}^d\times\{\mu\}}]$ contains all distributions supported on X with the expectations $e[P]=\mu$. Note that $\Phi(h;e[P])$ describes well the behavior of the logarithm $F_P(h)=\ln\left(\int_{\mathbf{R}^d}e^{h^T\omega}P(d\omega)\right)$ of the moment-generating function of $P\in\mathcal{P}[X]$ when h is small (indeed, $F_P(h)=h^Te[P]+O(\|h\|^2)$ as $h\to 0$), and by far overestimates $F_P(h)$ when h is large. Utilizing the above construction, we replace Φ with the real-valued, convex-concave, and continuous on $\mathbf{R}^d\times\mathcal{M}=\mathbf{R}^d\times X$ (see Exercise 2.22) function

$$\begin{array}{rcl}\widehat{\Phi}(h;\mu) & = & \inf_g\left[\widehat{\Psi}(h,g;\mu) := (h-g)^T\mu+\frac{1}{8}[\phi_X(h-g)+\phi_X(-h+g)]^2+\phi_X(g)\right]\\ & \leq & \Phi(h;\mu).\end{array}\qquad(2.114)$$

It is easy to see that $\widehat{\Phi}(\cdot;\cdot)$ still ensures the inclusion $P\in\mathcal{S}[\mathbf{R}^d,\{e[P]\},\widehat{\Phi}|_{\mathbf{R}^d\times\{e[P]\}}]$ for every distribution $P\in\mathcal{P}[X]$ and "reproduces $F_P(h)$ reasonably well" for both small and large h. Indeed, since $F_P(h)\leq\widehat{\Phi}(h;e[P])\leq\Phi(h;e[P])$, for small h $\widehat{\Phi}(h;e[P])$ reproduces $F_P(h)$ even better than $\Phi(h;e[P])$, and we clearly have

$$\widehat{\Phi}(h;\mu)\leq\left[(h-h)^T\mu+\tfrac{1}{8}[\phi_X(h-h)+\phi_X(-h+h)]^2+\phi_X(h)\right]=\phi_X(h)\ \forall\mu,$$

and $\phi_X(h)$ is a correct description of $F_P(h)$ for large h.

2.8.2 Main result

2.8.2.1 Situation & Construction

Assume we are given two collections of regular data with common $\Omega=\mathbf{R}^d$ and common $\mathcal{H}$, specifically, the collections $(\mathcal{H},\mathcal{M}_\chi,\Phi_\chi)$, $\chi=1,2$. We start with constructing a specific detector for the associated families of regular probability distributions

$$\mathcal{P}_\chi=\mathcal{R}[\mathcal{H},\mathcal{M}_\chi,\Phi_\chi],\ \chi=1,2.$$

When building the detector, we impose on the regular data in question the following

Assumption I: *The regular data* $(\mathcal{H},\mathcal{M}_\chi,\Phi_\chi)$, $\chi=1,2$, *are such that the convex-concave function*

$$\Psi(h;\mu_1,\mu_2)=\tfrac{1}{2}\left[\Phi_1(-h;\mu_1)+\Phi_2(h;\mu_2)\right]:\mathcal{H}\times(\mathcal{M}_1\times\mathcal{M}_2)\to\mathbf{R}\tag{2.115}$$

*has a saddle point (*min *in* $h\in\mathcal{H}$, max *in* $(\mu_1,\mu_2)\in\mathcal{M}_1\times\mathcal{M}_2$*).*

A simple sufficient condition for existence of a saddle point of (2.115) is

Condition A: *The sets* $\mathcal{M}_1$ *and* $\mathcal{M}_2$ *are compact, and the function*

$$\overline{\Phi}(h)=\max_{\mu_1\in\mathcal{M}_1,\mu_2\in\mathcal{M}_2}\Phi(h;\mu_1,\mu_2)$$

is coercive on $\mathcal{H}$*, meaning that* $\overline{\Phi}(h_i)\to\infty$ *along every sequence* $h_i\in\mathcal{H}$ *with* $\|h_i\|_2\to\infty$ *as* $i\to\infty$*.*

Indeed, under Condition A by the Sion-Kakutani Theorem (Theorem 2.22) it holds

$$\mathrm{SadVal}[\Phi]:=\inf_{h\in\mathcal{H}}\underbrace{\max_{\mu_1\in M_1,\mu_2\in\mathcal{M}_2}\Phi(h;\mu_1,\mu_2)}_{\overline{\Phi}(h)}=\sup_{\mu_1\in M_1,\mu_2\in\mathcal{M}_2}\underbrace{\inf_{h\in\mathcal{H}}\Phi(h;\mu_1,\mu_2)}_{\underline{\Phi}(\mu_1,\mu_2)},$$

so that the optimization problems

$$\begin{array}{ll}(P) & \mathrm{Opt}(P)=\min\limits_{h\in\mathcal{H}}\overline{\Phi}(h)\\ (D) & \mathrm{Opt}(D)=\max\limits_{\mu_1\in\mathcal{M}_1,\mu_2\in\mathcal{M}_2}\underline{\Phi}(\mu_1,\mu_2)\end{array}$$

have equal optimal values. Under Condition A, problem (P) clearly is a problem of minimizing a continuous coercive function over a closed set and as such is solvable; thus, $\mathrm{Opt}(P)=\mathrm{Opt}(D)$ is a real. Problem (D) clearly is the problem of maximizing over a compact set an upper semi-continuous (since Φ is continuous) function taking real values and, perhaps, value $-\infty$, and not identically equal to $-\infty$ (since $\mathrm{Opt}(D)$ is a real), and thus (D) is solvable. As a result, (P) and (D) are solvable with common optimal values, and therefore Φ has a saddle point.

2.8.2.2 Main Result

An immediate (and essential) observation is as follows:

Proposition 2.39. *In the situation of Section 2.8.2.1, let $h \in \mathcal{H}$ be such that the quantities*

$$\Psi_1(h) = \sup_{\mu_1 \in \mathcal{M}_1} \Phi_1(-h;\mu_1), \;\; \Psi_2(h) = \sup_{\mu_2 \in \mathcal{M}_2} \Phi_2(h;\mu_2)$$

are finite. Consider the affine detector

$$\phi_h(\omega) = h^T\omega + \underbrace{\tfrac{1}{2}[\Psi_1(h) - \Psi_2(h)]}_{\varkappa}.$$

Then

$$\mathrm{Risk}[\phi_h|\mathcal{R}[\mathcal{H},\mathcal{M}_1,\Phi_1],\mathcal{R}[\mathcal{H},\mathcal{M}_2,\Phi_2]] \leq \exp\{\tfrac{1}{2}[\Psi_1(h)+\Psi_2(h)]\}.$$

Proof. Let h satisfy the premise of the proposition. For every $\mu_1 \in \mathcal{M}_1$, we have $\Phi_1(-h;\mu_1) \leq \Psi_1(h)$, and for every $P \in \mathcal{R}[\mathcal{H},\mathcal{M}_1,\Phi_1]$ we have

$$\int_\Omega \exp\{-h^T\omega\}P(d\omega) \leq \exp\{\Phi_1(-h;\mu_1)\}$$

for properly selected $\mu_1 \in \mathcal{M}_1$. Thus,

$$\int_\Omega \exp\{-h^T\omega\}P(d\omega) \leq \exp\{\Psi_1(h)\} \;\forall P \in \mathcal{R}[\mathcal{H},\mathcal{M}_1,\Phi_1],$$

whence also

$$\int_\Omega \exp\{-h^T\omega - \varkappa\}P(d\omega) \leq \exp\{\Psi_1(h) - \varkappa\} = \exp\{\tfrac{1}{2}[\Psi_1(h)+\Psi_2(h)]\} \;\forall P \in \mathcal{R}[\mathcal{H},\mathcal{M}_1,\Phi_1].$$

Similarly, for every $\mu_2 \in \mathcal{M}_2$, we have $\Phi_2(h;\mu_2) \leq \Psi_2(h)$, and for every $P \in \mathcal{R}[\mathcal{H},\mathcal{M}_2,\Phi_2]$, we have

$$\int_\Omega \exp\{h^T\omega\}P(d\omega) \leq \exp\{\Phi_2(h;\mu_2)\}$$

for properly selected $\mu_2 \in \mathcal{M}_2$. Thus,

$$\int_\Omega \exp\{h^T\omega\}P(d\omega) \leq \exp\{\Psi_2(h)\} \;\forall P \in \mathcal{R}[\mathcal{H},\mathcal{M}_2,\Phi_2],$$

and

$$\int_\Omega \exp\{h^T\omega + \varkappa\}P(d\omega) \leq \exp\{\Psi_2(h) + \varkappa\} = \exp\{\tfrac{1}{2}[\Psi_1(h)+\Psi_2(h)]\} \;\forall P \in \mathcal{R}[\mathcal{H},\mathcal{M}_2,\Phi_2].$$

□

An immediate corollary is as follows:

Proposition 2.40. *In the situation of Section 2.8.2.1 and under Assumption I, let us associate with a saddle point* $(h_*; \mu_1^*, \mu_2^*)$ *of the convex-concave function* (2.115) *the following entities:*

- *the* risk
$$\epsilon_\star := \exp\{\Psi(h_*; \mu_1^*, \mu_2^*)\}; \tag{2.116}$$
this quantity is uniquely defined by the saddle point value of Ψ *and thus is independent of how we select a saddle point;*
- *the* detector $\phi_*(\omega)$*—the affine function of* $\omega \in \mathbf{R}^d$ *given by*
$$\phi_*(\omega) = h_*^T\omega + a, \; a = \tfrac{1}{2}\left[\Phi_1(-h_*; \mu_1^*) - \Phi_2(h_*; \mu_2^*)\right]. \tag{2.117}$$

Then
$$\mathrm{Risk}[\phi_*|\mathcal{R}[\mathcal{H}, \mathcal{M}_1, \Phi_1], \mathcal{R}[\mathcal{H}, \mathcal{M}_2, \Phi_2]] \leq \epsilon_\star.$$

Consequences. Assume we are given L collections $(\mathcal{H}, \mathcal{M}_\ell, \Phi_\ell)$ of regular data on a common observation space $\Omega = \mathbf{R}^d$ and with common $\mathcal{H}$, and let

$$\mathcal{P}_\ell = \mathcal{R}[\mathcal{H}, \mathcal{M}_\ell, \Phi_\ell]$$

be the corresponding families of regular distributions. Assume also that for every pair (ℓ, ℓ'), $1 \leq \ell < \ell' \leq L$, the pair $(\mathcal{H}, \mathcal{M}_\ell, \Phi_\ell)$, $(\mathcal{H}, \mathcal{M}_{\ell'}, \Phi_{\ell'})$ of regular data satisfies Assumption I, so that the convex-concave functions

$$\Psi_{\ell\ell'}(h; \mu_\ell, \mu_{\ell'}) = \tfrac{1}{2}\left[\Phi_\ell(-h; \mu_\ell) + \Phi_{\ell'}(h; \mu_{\ell'})\right] : \mathcal{H} \times (\mathcal{M}_\ell \times \mathcal{M}_{\ell'}) \to \mathbf{R} \qquad [1 \leq \ell < \ell' \leq L]$$

have saddle points $(h_{\ell\ell'}^*; (\mu_\ell^*, \mu_{\ell'}^*))$ (min in $h \in \mathcal{H}$, max in $(\mu_\ell, \mu_{\ell'}) \in \mathcal{M}_\ell \times \mathcal{M}_{\ell'}$). These saddle points give rise to the affine detectors

$$\phi_{\ell\ell'}(\omega) = [h_{\ell\ell'}^*]^T\omega + \tfrac{1}{2}\left[\Phi_\ell(-h_{\ell\ell'}^*; \mu_\ell^*) - \Phi_{\ell'}(h_*; \mu_{\ell'}^*)\right] \qquad [1 \leq \ell < \ell' \leq L]$$

and the quantities

$$\epsilon_{\ell\ell'} = \exp\left\{\tfrac{1}{2}\left[\Phi_\ell(-h_{\ell\ell'}^*; \mu_\ell^*) + \Phi_{\ell'}(h_*; \mu_{\ell'}^*)\right]\right\}; \qquad [1 \leq \ell < \ell' \leq L]$$

by Proposition 2.40, $\epsilon_{\ell\ell'}$ are upper bounds on the risks, taken w.r.t. $\mathcal{P}_\ell, \mathcal{P}_{\ell'}$, of the detectors $\phi_{\ell\ell'}$:

$$\int_\Omega \mathrm{e}^{-\phi_{\ell\ell'}(\omega)} P(d\omega) \leq \epsilon_{\ell\ell'} \;\forall P \in \mathcal{P}_\ell \;\&\; \int_\Omega \mathrm{e}^{\phi_{\ell\ell'}(\omega)} P(d\omega) \leq \epsilon_{\ell\ell'} \;\forall P \in \mathcal{P}_{\ell'}. \qquad [1 \leq \ell < \ell' \leq L]$$

Setting $\phi_{\ell\ell'}(\cdot) = -\phi_{\ell'\ell}(\cdot)$ and $\epsilon_{\ell\ell'} = \epsilon_{\ell'\ell}$ when $L \geq \ell > \ell' \geq 1$ and $\phi_{\ell\ell}(\cdot) \equiv 0$, $\epsilon_{\ell\ell} = 1$, $1 \leq \ell \leq L$, we get a system of detectors and risks satisfying (2.80) and, consequently, can use these "building blocks" in the machinery developed so far for pairwise and multiple hypothesis testing from single and repeated observations (stationary, semi-stationary, and quasi-stationary).

Numerical example. To get some impression of how Proposition 2.40 extends the grasp of our computation-friendly machinery of test design consider a toy problem as follows:

We are given an observation

$$\omega = Ax + \sigma A\mathrm{Diag}\left\{\sqrt{x_1}, ..., \sqrt{x_n}\right\}\xi, \tag{2.118}$$

where

- unknown signal x is known to belong to a given convex compact subset M of the *interior* of $\mathbf{R}^n_+$;
- A is a given $n \times n$ matrix of rank n, $\sigma > 0$ is a given noise intensity, and $\xi \sim \mathcal{N}(0, I_n)$.

Our goal is to decide via a K-repeated version of observations (2.118) on the pair of hypotheses $x \in X_\chi$, $\chi = 1, 2$, where X_1, X_2 are given nonempty convex compact subsets of M.

Note that an essential novelty, as compared to the standard Gaussian o.s., is that now we deal with zero mean Gaussian noise with covariance matrix

$$\Theta(x) = \sigma^2 A\mathrm{Diag}\{x\}A^T$$

depending on the true signal—the larger the signal, the greater the noise.

We can easily process the situation in question utilizing the machinery developed in this section. Namely, let us set

$$\begin{array}{c} \mathcal{H}_\chi = \mathbf{R}^n,\ \mathcal{M}_\chi = \{(x, \mathrm{Diag}\{x\}) : x \in X_\chi\} \subset \mathbf{R}^n_+ \times \mathbf{S}^n_+, \\ \Phi_\chi(h; x, \Xi) = h^TA^Tx + \frac{\sigma^2}{2}h^T[A\Xi A^T]h : \mathcal{M}_\chi \to \mathbf{R}. \end{array} \qquad [\chi = 1, 2]$$

It is immediately seen that for $\chi = 1, 2$, $\mathcal{H}, \mathcal{M}_\chi, \Phi_\chi$ is regular data, and that the distribution P of observation (2.118) stemming from a signal $x \in X_\chi$ belongs to $\mathcal{S}[\mathcal{H}, \mathcal{M}_\chi, \Phi_\chi]$, so that we can use Proposition 2.40 to build an affine detector for the families $\mathcal{P}_\chi$, $\chi = 1, 2$, of distributions of observations (2.118) stemming from signals $x \in X_\chi$. The corresponding recipe boils down to the necessity to find a saddle point $(h_*; x_*, y_*)$ of the simple convex-concave function

$$\Psi(h; x, y) = \frac{1}{2}\left[h^TA(y - x) + \frac{\sigma^2}{2}h^TA\mathrm{Diag}\{x + y\}A^Th\right]$$

(min in $h \in \mathbf{R}^n$, max in $(x, y) \in X_1 \times X_2$). Such a point clearly exists and is easily found, and gives rise to affine detector

$$\phi_*(\omega) = h_*^T\omega + \underbrace{\tfrac{1}{4}\sigma^2h_*^TA\mathrm{Diag}\{x_* - y_*\}A^Th_* - \tfrac{1}{2}h_*^TA[x_* + y_*]}_{a}$$

such that

$$\mathrm{Risk}[\phi_*|\mathcal{P}_1, \mathcal{P}_2] \leq \exp\left\{\frac{1}{2}\left[h_*^TA[y_* - x_*] + \frac{\sigma^2}{2}h_*^TA\mathrm{Diag}\{x_* + y_*\}A^Th_*\right]\right\}. \tag{2.119}$$

Note that we could also process the situation when defining the regular data as $\mathcal{H}, \mathcal{M}^+_\chi = X_\chi, \Phi^+_\chi$, $\chi = 1, 2$, where

$$\Phi^+_\chi(h; x) = h^TAx + \frac{\sigma^2\theta}{2}h^TAA^Th \qquad [\theta = \max_{x \in X_1 \cup X_2} \|x\|_\infty],$$

which, basically, means passing from our actual observations (2.118) to the "more noisy" observations given by Gaussian o.s.

$$\omega = Ax + \eta, \ \eta \sim \mathcal{N}(0, \sigma^2 \theta A A^T). \tag{2.120}$$

It is easily seen that, for this Gaussian o.s., the risk $\mathrm{Risk}[\phi_\#|\mathcal{P}_1, \mathcal{P}_2]$ of the optimal, detector $\phi_\#$ can be upper-bounded by the risk $\mathrm{Risk}[\phi_\#|\mathcal{P}_1^+, \mathcal{P}_2^+]$ known to us, where $\mathcal{P}_\chi^+$ is the family of distributions of observations (2.120) induced by signals $x \in X_\chi$. Note that $\mathrm{Risk}[\phi_\#|\mathcal{P}_1^+, \mathcal{P}_2^+]$ is seemingly the best risk bound available for us "within the realm of detector-based tests in simple o.s.'s." The goal of the small numerical experiment we are about to report on is to understand how our new risk bound (2.119) compares to the "old" bound $\mathrm{Risk}[\phi_\#|\mathcal{P}_1^+, \mathcal{P}_2^+]$. We use

$$\begin{aligned} n = 16, \ X_1 &= \left\{ x \in \mathbf{R}^{16} : \begin{array}{l} 0.001 \le x_1 \le \delta \\ 0.001 \le x_i \le 1, \ 2 \le i \le 16 \end{array} \right\}, \\ X_2 &= \left\{ x \in \mathbf{R}^{16} : \begin{array}{l} 2\delta \le x_1 \le 1 \\ 0.001 \le x_i \le 1, \ 2 \le i \le 16 \end{array} \right\} \end{aligned}$$

and $\sigma = 0.1$. The "separation parameter" δ is set to 0.1. Finally, the 16×16 matrix A has condition number 100 (singular values $0.01^{(i-1)/15}$, $1 \le i \le 16$) and randomly oriented systems of left- and right singular vectors. With this setup, a typical numerical result is as follows:

- the right-hand side in (2.119) is 0.4346, implying that with detector ϕ_*, a 6-repeated observation is sufficient to decide on our two hypotheses with risk $\le$ 0.01;
- the quantity $\mathrm{Risk}[\phi_\#|\mathcal{P}_1^+, \mathcal{P}_2^+]$ is 0.8825, meaning that with detector $\phi_\#$, we need at least a 37-repeated observation to guarantee risk ≤ 0.01.

When the separation parameter δ participating in the descriptions of X_1, X_2 is reduced to 0.01, the risks in question grow to 0.9201 and 0.9988, respectively (a 56-repeated observation to decide on the hypotheses with risk 0.01 when ϕ_* is used vs. a 3685-repeated observation needed when $\phi_\#$ is used). The bottom line is that the new developments can indeed improve significantly the performance of our inferences.

2.8.2.3 Sub-Gaussian and Gaussian cases

For $\chi = 1, 2$, let U_χ be a nonempty closed convex set in $\mathbf{R}^d$, and $\mathcal{V}_\chi$ be a compact convex subset of the interior of the positive semidefinite cone $\mathbf{S}_+^d$. We assume that U_1 is compact. Setting

$$\begin{aligned} \mathcal{H}_\chi = \Omega = \mathbf{R}^d, \ \mathcal{M}_\chi = U_\chi \times \mathcal{V}_\chi, \\ \Phi_\chi(h; \theta, \Theta) = \theta^T h + \tfrac{1}{2} h^T \Theta h : \mathcal{H}_\chi \times \mathcal{M}_\chi \to \mathbf{R}, \chi = 1, 2, \end{aligned} \tag{2.121}$$

we get two collections $(\mathcal{H}, \mathcal{M}_\chi, \Phi_\chi)$, $\chi = 1, 2$, of regular data. As we know from Section 2.8.1.2, for $\chi = 1, 2$, the families of distributions $\mathcal{S}[\mathbf{R}^d, \mathcal{M}_\chi, \Phi_\chi]$ contain the families $\mathcal{SG}[U_\chi, \mathcal{V}_\chi]$ of sub-Gaussian distributions on $\mathbf{R}^d$ with sub-Gaussianity parameters $(\theta, \Theta) \in U_\chi \times \mathcal{V}_\chi$ (see (2.106)), as well as families $\mathcal{G}[U_\chi, \mathcal{V}_\chi]$ of Gaussian distributions on $\mathbf{R}^d$ with parameters (θ, Θ) (expectation and covariance matrix) running through $U_\chi \times \mathcal{V}_\chi$. Besides this, the pair of regular data in question clearly satisfies Condition A. Consequently, the test $\mathcal{T}_*^K$ given by the above construction as applied to the collections of regular data (2.121) is well defined and allows to decide on hypotheses

$$H_\chi : P \in \mathcal{R}[\mathbf{R}^d, U_\chi, \mathcal{V}_\chi],\ \chi = 1, 2,$$

on the distribution P underlying K-repeated observation ω^K. The same test can be also used to decide on stricter hypotheses H^G_χ, $\chi = 1, 2$, stating that the observations $\omega_1, ..., \omega_K$ are i.i.d. and drawn from a Gaussian distribution P belonging to $\mathcal{G}[U_\chi, \mathcal{V}_\chi]$. Our goal now is to process in detail the situation in question and to refine our conclusions on the risk of the test $\mathcal{T}^1_*$ when the *Gaussian* hypotheses H^G_χ are considered and the situation is *symmetric*, that is, when $\mathcal{V}_1 = \mathcal{V}_2$.

Observe, first, that the convex-concave function Ψ from (2.115) in the current setting becomes

$$\Psi(h; \theta_1, \Theta_1, \theta_2, \Theta_2) = \tfrac{1}{2} h^T[\theta_2 - \theta_1] + \tfrac{1}{4} h^T \Theta_1 h + \tfrac{1}{4} h^T \Theta_2 h. \tag{2.122}$$

We are interested in solutions to the saddle point problem

$$\min_{h\in\mathbf{R}^d} \max_{\substack{\theta_1\in U_1, \theta_2\in U_2\\ \Theta_1\in\mathcal{V}_1, \Theta_2\in\mathcal{V}_2}} \Psi(h; \theta_1, \Theta_1, \theta_2, \Theta_2) \tag{2.123}$$

associated with the function (2.122). From the structure of Ψ and compactness of U_1, $\mathcal{V}_1$, $\mathcal{V}_2$, combined with the fact that $\mathcal{V}_\chi$, $\chi = 1, 2$, are comprised of positive definite matrices, it immediately follows that saddle points do exist, and a saddle point $(h_*; \theta_1^*, \Theta_1^*, \theta_2^*, \Theta_2^*)$ satisfies the relations

$$\begin{array}{ll}(a) & h_* = [\Theta_1^* + \Theta_2^*]^{-1}[\theta_1^* - \theta_2^*],\\ (b) & h_*^T(\theta_1 - \theta_1^*) \geq 0\ \forall \theta_1 \in U_1,\ \ h_*^T(\theta_2^* - \theta_2) \geq 0\ \forall \theta_2 \in U_2,\\ (c) & h_*^T \Theta_1 h_* \leq h_*^T \Theta_1^* h_*\ \forall \Theta_1 \in \mathcal{V}_1,\ \ h_*^T \Theta_2 h_* \leq h_* \Theta_2^* h_*\ \forall \Theta_2 \in \mathcal{V}_2.\end{array} \tag{2.124}$$

From (2.124.a) it immediately follows that the affine detector $\phi_*(\cdot)$ and risk $\epsilon_\star$, as given by (2.116) and (2.117), are

$$\begin{array}{rcl}\phi_*(\omega) & = & h_*^T[\omega - w_*] + \frac{1}{2} h_*^T[\Theta_1^* - \Theta_2^*]h_*,\ w_* = \frac{1}{2}[\theta_1^* + \theta_2^*];\\ \epsilon_\star & = & \exp\{-\frac{1}{4}[\theta_1^* - \theta_2^*]^T[\Theta_1^* + \Theta_2^*]^{-1}[\theta_1^* - \theta_2^*]\}\\ & = & \exp\{-\frac{1}{4} h_*^T[\Theta_1^* + \Theta_2^*]h_*\}.\end{array} \tag{2.125}$$

Note that in the *symmetric case* (where $\mathcal{V}_1 = \mathcal{V}_2$), there always exists a saddle point of Ψ with $\Theta_1^* = \Theta_2^*$,[21] and the test $\mathcal{T}^1_*$ associated with such saddle point is quite transparent: it is the maximum likelihood test for two Gaussian distributions, $\mathcal{N}(\theta_1^*, \Theta_*)$, $\mathcal{N}(\theta_2^*, \Theta_*)$, where Θ_* is the common value of Θ_1^* and Θ_2^*. The bound $\epsilon_\star$ on the risk of the test is nothing but the Hellinger affinity of these two Gaussian distributions, or, equivalently,

$$\epsilon_\star = \exp\left\{-\tfrac{1}{8}[\theta_1^* - \theta_2^*]^T \Theta_*^{-1}[\theta_1^* - \theta_2^*]\right\}.$$

[21]Indeed, from (2.122) it follows that when $\mathcal{V}_1 = \mathcal{V}_2$, the function $\Psi(h; \theta_1, \Theta_1, \theta_2, \Theta_2)$ is symmetric w.r.t. Θ_1, Θ_2, implying similar symmetry of the function $\underline{\Psi}(\theta_1, \Theta_1, \theta_2, \Theta_2) = \min_{h\in\mathcal{H}} \Psi(h; \theta_1, \Theta_1, \theta_2, \Theta_2)$. Since $\underline{\Psi}$ is concave, the set M of its maximizers over $\mathcal{M}_1 \times \mathcal{M}_2$ (which, as we know, is nonempty) is symmetric w.r.t. the swap of Θ_1 and Θ_2 and is convex, implying that if $(\theta_1, \Theta_1, \theta_2, \Theta_2) \in M$, then $(\theta_1, \frac{1}{2}[\Theta_1 + \Theta_2], \theta_2, \frac{1}{2}[\Theta_1 + \Theta_2]) \in M$ as well, and the latter point is the desired component of the saddle point of Ψ with $\Theta_1 = \Theta_2$.

We arrive at the following result:

Proposition 2.41. *In the symmetric sub-Gaussian case (i.e., in the case of* (2.121) *with $\mathcal{V}_1 = \mathcal{V}_2$), saddle point problem* (2.122), (2.123) *admits a saddle point of the form $(h_*; \theta_1^*, \Theta_*, \theta_2^*, \Theta_*)$, and the associated affine detector and its risk are given by*

$$\begin{array}{rcl} \phi_*(\omega) & = & h_*^T[\omega - w_*], \; w_* = \frac{1}{2}[\theta_1^* + \theta_2^*]; \\ \epsilon_\star & = & \exp\{-\frac{1}{8}[\theta_1^* - \theta_2^*]^T \Theta_*^{-1}[\theta_1^* - \theta_2^*]\}. \end{array}$$

As a result, when deciding, via ω^K, on "sub-Gaussian hypotheses" H_χ, $\chi = 1, 2$, the risk of the test $\mathcal{T}_*^K$ associated with $\phi_*^{(K)}(\omega^K) := \sum_{t=1}^K \phi_*(\omega_t)$ is at most $\epsilon_\star^K$.

In the symmetric single-observation Gaussian case, that is, when $\mathcal{V}_1 = \mathcal{V}_2$ and we apply the test $\mathcal{T}_* = \mathcal{T}_*^1$ to observation $\omega \equiv \omega_1$ in order to decide on the hypotheses H_χ^G, $\chi = 1, 2$, the above risk bound can be improved:

Proposition 2.42. *Consider the symmetric case $\mathcal{V}_1 = \mathcal{V}_2 = \mathcal{V}$, let $(h_*; \theta_1^*; \Theta_1^*, \theta_2^*, \Theta_2^*)$ be the "symmetric"—with $\Theta_1^* = \Theta_2^* = \Theta_*$—saddle point of function Ψ given by* (2.122), *and let ϕ_* be the affine detector given by* (2.124) *and* (2.125)*:*

$$\phi_*(\omega) = h_*^T[\omega - w_*], \;\; h_* = \tfrac{1}{2}\Theta_*^{-1}[\theta_1^* - \theta_2^*], \;\; w_* = \tfrac{1}{2}[\theta_1^* + \theta_2^*].$$

Let also

$$\delta = \sqrt{h_*^T \Theta_* h_*} = \tfrac{1}{2}\sqrt{[\theta_1^* - \theta_2^*]^T \Theta_*^{-1}[\theta_1^* - \theta_2^*]},$$

so that

$$\delta^2 = h_*^T[\theta_1^* - w_*] = h_*^T[w_* - \theta_2^*] \;\; \textit{and} \; \epsilon_\star = \exp\left\{-\tfrac{1}{2}\delta^2\right\}. \tag{2.126}$$

Let, further, $\alpha \le \delta^2$, $\beta \le \delta^2$. Then

$$\begin{array}{ll} (a) & \forall (\theta \in U_1, \Theta \in \mathcal{V}) : \mathrm{Prob}_{\omega \sim \mathcal{N}(\theta, \Theta)}\{\phi_*(\omega) \le \alpha\} \le \mathrm{Erfc}(\delta - \alpha/\delta), \\ (b) & \forall (\theta \in U_2, \Theta \in \mathcal{V}) : \mathrm{Prob}_{\omega \sim \mathcal{N}(\theta, \Theta)}\{\phi_*(\omega) \ge -\beta\} \le \mathrm{Erfc}(\delta - \beta/\delta). \end{array} \tag{2.127}$$

In particular, when deciding, via a single observation ω, on Gaussian hypotheses H_χ^G, $\chi = 1, 2$, with H_χ^G stating that $\omega \sim \mathcal{N}(\theta, \Theta)$ with $(\theta, \Theta) \in U_\chi \times \mathcal{V}$, the risk of the test $\mathcal{T}_*^1$ associated with ϕ_* is at most $\mathrm{Erfc}(\delta)$.

Proof. Let us prove (a) (the proof of (b) is completely similar). For $\theta \in U_1$, $\Theta \in \mathcal{V}$ we have

$$\begin{array}{rl} \mathrm{Prob}_{\omega \sim \mathcal{N}(\theta,\Theta)}\{\phi_*(\omega) \le \alpha\} & = \mathrm{Prob}_{\omega \sim \mathcal{N}(\theta,\Theta)}\{h_*^T[\omega - w_*] \le \alpha\} \\ & = \mathrm{Prob}_{\xi \sim \mathcal{N}(0,I)}\{h_*^T[\theta + \Theta^{1/2}\xi - w_*] \le \alpha\} \\ & = \mathrm{Prob}_{\xi \sim \mathcal{N}(0,I)}\{[\Theta^{1/2}h_*]^T\xi \le \alpha - \underbrace{h_*^T[\theta - w_*]}_{\substack{\ge h_*^T[\theta_1^* - w_*] = \delta^2 \\ \text{by (2.124.}b\text{),(2.126)}}}\} \\ & \le \mathrm{Prob}_{\xi \sim \mathcal{N}(0,I)}\{[\Theta^{1/2}h_*]^T\xi \le \alpha - \delta^2\} \\ & = \mathrm{Erfc}([\delta^2 - \alpha]/\|\Theta^{1/2}h_*\|_2) \\ & \le \mathrm{Erfc}([\delta^2 - \alpha]/\|\Theta_*^{1/2}h_*\|_2) \\ & \quad [\text{due to } \delta^2 - \alpha \ge 0 \text{ and } h_*^T\Theta h_* \le h_*^T\Theta_* h_* \text{ by (2.124.}c\text{)}] \\ & = \mathrm{Erfc}([\delta^2 - \alpha]/\delta). \end{array}$$

The "in particular" part of Proposition is readily given by (2.127) as applied with $\alpha = \beta = 0$. □

Note that the progress, as compared to our results on the minimum risk detectors for convex hypotheses in Gaussian o.s., is that we do *not* assume anymore that the covariance matrix is once and forever fixed. Now neither the mean *nor* the covariance matrix of the observed Gaussian random variable are known in advance. In this setting, the mean is running through a closed convex set (depending on the hypothesis), and the covariance is running, independently of the mean, through a given convex compact subset of the interior of the positive definite cone, and this subset should be common for both hypotheses we are deciding upon.

2.9 BEYOND THE SCOPE OF AFFINE DETECTORS: LIFTING THE OBSERVATIONS

2.9.1 Motivation

The detectors considered in Section 2.8 were affine functions of observations. Note, however, that what an observation is, to some extent depends on us. To give an instructive example, consider the Gaussian observation

$$\zeta = A[u;1] + \xi \in \mathbf{R}^n,$$

where u is an unknown signal known to belong to a given set $U \subset \mathbf{R}^n$, $u \mapsto A[u;1]$ is a given affine mapping from $\mathbf{R}^n$ into the observation space $\mathbf{R}^d$, and ξ is zero mean Gaussian observation noise with covariance matrix Θ known to belong to a given convex compact subset $\mathcal{V}$ of the interior of the positive semidefinite cone $\mathbf{S}^d_+$. Treating the observation "as is," affine in the observation detector is affine in $[u;\xi]$. On the other hand, we can treat as our observation the image of the actual observation ζ under any deterministic mapping, e.g., the "quadratic lifting" $\zeta \mapsto (\zeta, \zeta\zeta^T)$. A detector affine in the new observation is quadratic in u and ξ—we get access to a wider set of detectors as compared to those affine in ζ! At first glance, applying our "affine detectors" machinery to appropriate "nonlinear liftings" of actual observations we can handle quite complicated detectors, e.g., polynomial, of arbitrary degree, in ζ. The bottleneck here stems from the fact that in general it is difficult to "cover" the distribution of a "nonlinearly lifted" observation ζ (even as simple as the Gaussian observation above) by an explicitly defined family of regular distributions, and such a "covering" is what we need in order to apply to the lifted observation our affine detector machinery. It turns out, however, that in some important cases the desired covering is achievable. We are about to demonstrate that this takes place in the case of the quadratic lifting $\zeta \mapsto (\zeta, \zeta\zeta^T)$ of (sub)Gaussian observation ζ, and the resulting quadratic detectors allow us to handle some important inference problems which are far beyond the grasp of "genuinely affine" detectors.

2.9.2 Quadratic lifting: Gaussian case

Given positive integer d, we define $\mathcal{E}^d$ as the linear space $\mathbf{R}^d \times \mathbf{S}^d$ equipped with the inner product

$$\langle (z,S),(z',S')\rangle = s^T z' + \tfrac{1}{2}\mathrm{Tr}(SS').$$

Note that the quadratic lifting $z \mapsto (z, zz^T)$ maps the space $\mathbf{R}^d$ into $\mathcal{E}^d$.

In the sequel, an instrumental role is played by the following result.

Proposition 2.43.

(i) *Assume we are given*

- *a nonempty and bounded subset U of $\mathbf{R}^n$;*
- *a convex compact set $\mathcal{V}$ contained in the interior of the cone $\mathbf{S}^d_+$ of positive semidefinite $d \times d$ matrices;*
- *a $d \times (n+1)$ matrix A.*

These data specify the family $\mathcal{G}_A[U,\mathcal{V}]$ of distributions of quadratic liftings $(\zeta, \zeta\zeta^T)$ of Gaussian random vectors $\zeta \sim \mathcal{N}(A[u;1],\Theta)$ stemming from $u \in U$ and $\Theta \in \mathcal{V}$.

Let us select some

1. *$\gamma \in (0,1)$,*
2. *convex compact subset $\mathcal{Z}$ of the set $\mathcal{Z}^n = \{Z \in \mathbf{S}^{n+1} : Z \succeq 0, Z_{n+1,n+1} = 1\}$ such that*
$$Z(u) := [u;1][u;1]^T \in \mathcal{Z}\ \forall u \in U, \tag{2.128}$$
3. *positive definite $d \times d$ matrix $\Theta_* \in \mathbf{S}^d_+$ and $\delta \in [0,2]$ such that*
$$\Theta_* \succeq \Theta\ \forall \Theta \in \mathcal{V}\ \&\ \|\Theta^{1/2}\Theta_*^{-1/2} - I_d\| \leq \delta\ \ \forall \Theta \in \mathcal{V}, \tag{2.129}$$
where $\|\cdot\|$ is the spectral norm,[22]

and set

$$\begin{array}{rcl}\mathcal{H} = \mathcal{H}^\gamma &:=& \{(h,H) \in \mathbf{R}^d \times \mathbf{S}^d : -\gamma\Theta_*^{-1} \preceq H \preceq \gamma\Theta_*^{-1}\},\\ \Phi_{A,\mathcal{Z}}(h,H;\Theta) &=& -\frac{1}{2}\ln\mathrm{Det}(I - \Theta_*^{1/2}H\Theta_*^{1/2}) + \frac{1}{2}\mathrm{Tr}([\Theta - \Theta_*]H)\\ && + \frac{\delta(2+\delta)}{2(1-\|\Theta_*^{1/2}H\Theta_*^{1/2}\|)}\|\Theta_*^{1/2}H\Theta_*^{1/2}\|_F^2\\ && + \frac{1}{2}\phi_{\mathcal{Z}}\left(B^T\left[\left[\begin{array}{c|c} H & h \\ \hline h^T & \end{array}\right] + [H,h]^T[\Theta_*^{-1} - H]^{-1}[H,h]\right]B\right) :\\ && \mathcal{H} \times \mathcal{V} \to \mathbf{R},\end{array} \tag{2.130}$$

where B is given by

$$B = \left[\begin{array}{c} A \\ [0,...,0,1] \end{array}\right], \tag{2.131}$$

the function

$$\phi_{\mathcal{Z}}(Y) := \max_{Z \in \mathcal{Z}} \mathrm{Tr}(ZY) \tag{2.132}$$

is the support function of $\mathcal{Z}$, and $\|\cdot\|_F$ is the Frobenius norm.

Function $\Phi_{A,\mathcal{Z}}$ is continuous on its domain, convex in $(h,H) \in \mathcal{H}$ and concave

---

[22] It is easily seen that with $\delta = 2$, the second relation in (2.129) is satisfied for all Θ such that $0 \preceq \Theta \preceq \Theta_*$, so that the restriction $\delta \leq 2$ is w.l.o.g..

in $\Theta \in \mathcal{V}$, *so that* $(\mathcal{H}, \mathcal{V}, \Phi_{A,\mathcal{Z}})$ *is regular data. Besides this,*

(#) Whenever $u \in \mathbf{R}^n$ *is such that* $[u;1][u;1]^T \in \mathcal{Z}$ *and* $\Theta \in \mathcal{V}$, *the Gaussian random vector* $\zeta \sim \mathcal{N}(A[u;1], \Theta)$ *satisfies the relation*

$$\forall (h,H) \in \mathcal{H}: \ \ln\left(\mathbf{E}_{\zeta\sim\mathcal{N}(A[u;1],\Theta)}\left\{e^{\frac{1}{2}\zeta^T H\zeta + h^T\zeta}\right\}\right) \leq \Phi_{A,\mathcal{Z}}(h,H;\Theta). \quad (2.133)$$

The latter relation combines with (2.128) *to imply that*

$$\mathcal{G}_A[U,\mathcal{V}] \subset \mathcal{S}[\mathcal{H},\mathcal{V},\Phi_{A,\mathcal{Z}}].$$

In addition, $\Phi_{A,\mathcal{Z}}$ *is coercive in* (h,H): $\Phi_{A,\mathcal{Z}}(h_i,H_i;\Theta) \to +\infty$ *as* $i \to \infty$ *whenever* $\Theta \in \mathcal{V}$, $(h_i,H_i) \in \mathcal{H}$, *and* $\|(h_i,H_i)\| \to \infty$, $i \to \infty$.

(ii) *Let two collections of entities from* (i), $(\mathcal{V}_\chi, \Theta_*^{(\chi)}, \delta_\chi, \gamma_\chi, A_\chi, \mathcal{Z}_\chi)$, $\chi = 1,2$, *with common* d *be given, giving rise to the sets* $\mathcal{H}_\chi$, *matrices* B_χ, *and functions* $\Phi_{A_\chi,\mathcal{Z}_\chi}(h,H;\Theta)$, $\chi = 1,2$. *These collections specify the families of normal distributions*

$$\mathcal{G}_\chi = \{\mathcal{N}(v,\Theta) : \Theta \in \mathcal{V}_\chi \ \& \ \exists u \in U : v = A_\chi[u;1]\}, \ \chi = 1,2.$$

Consider the convex-concave saddle point problem

$$\mathcal{SV} = \min_{(h,H)\in\mathcal{H}_1\cap\mathcal{H}_2} \max_{\Theta_1\in\mathcal{V}_1,\Theta_2\in\mathcal{V}_2} \underbrace{\tfrac{1}{2}\left[\Phi_{A_1,\mathcal{Z}_1}(-h,-H;\Theta_1) + \Phi_{A_2,\mathcal{Z}_2}(h,H;\Theta_2)\right]}_{\Phi(h,H;\Theta_1,\Theta_2)}. \quad (2.134)$$

A saddle point $(H_*, h_*; \Theta_1^*, \Theta_2^*)$ *in this problem does exist, and the induced quadratic detector*

$$\phi_*(\omega) = \tfrac{1}{2}\omega^T H_*\omega + h_*^T\omega + \underbrace{\tfrac{1}{2}\left[\Phi_{A_1,\mathcal{Z}_1}(-h_*,-H_*;\Theta_1^*) - \Phi_{A_2,\mathcal{Z}_2}(h_*,H_*;\Theta_2^*)\right]}_{a}, \quad (2.135)$$

when applied to the families of Gaussian distributions $\mathcal{G}_\chi$, $\chi = 1,2$, *has the risk*

$$\mathrm{Risk}[\phi_*|\mathcal{G}_1,\mathcal{G}_2] \leq \epsilon_\star := e^{\mathcal{SV}},$$

that is,

$$\begin{array}{lll}(a) & \int_{\mathbf{R}^d} e^{-\phi_*(\omega)}P(d\omega) \leq \epsilon_\star & \forall P \in \mathcal{G}_1, \\ (b) & \int_{\mathbf{R}^d} e^{\phi_*(\omega)}P(d\omega) \leq \epsilon_\star & \forall P \in \mathcal{G}_2.\end{array} \quad (2.136)$$

For proof, see Section 2.11.5.

Remark 2.44. Note that the computational effort to solve *(2.134)* reduces dramatically in the "*easy case*" of the situation described in item (ii) of Proposition 2.43 where

- the observations are *direct*, meaning that $A_\chi[u;1] \equiv u$, $u \in \mathbf{R}^d$, $\chi = 1,2$;
- the sets $\mathcal{V}_\chi$ are comprised of positive definite diagonal matrices, and matrices $\Theta_*^{(\chi)}$ are diagonal as well, $\chi = 1,2$;
- the sets $\mathcal{Z}_\chi$, $\chi = 1,2$, are convex compact sets of the form

$$\mathcal{Z}_\chi = \{Z \in \mathbf{S}_+^{d+1} : Z \succeq 0, \ \mathrm{Tr}(ZQ_j^\chi) \leq q_j^\chi, \ 1 \leq j \leq J_\chi\}$$

with *diagonal* matrices Q_j^χ,[23] and these sets intersect the interior of the positive semidefinite cone $\mathbf{S}_+^{d+1}$.

In this case, the convex-concave saddle point problem *(2.134)* admits a saddle point $(h_*, H_*; \Theta_1^*, \Theta_2^*)$ where $h_* = 0$ and H_* is diagonal.

Justifying the remark. In the easy case, we have $B_\chi = I_{d+1}$ and therefore

$$\begin{array}{rcl} M_\chi(h,H) & := & B_\chi^T\left[\left[\begin{array}{c|c} H & h \\ \hline h^T & \end{array}\right] + [H,h]^T\left[[\Theta_*^{(\chi)}]^{-1} - H\right]^{-1}[H,h]\right]B_\chi \\ & = & \left[\begin{array}{c|c} H + H\left[[\Theta_*^{(\chi)}]^{-1} - H\right]^{-1}H & h + H[[\Theta_*^{(\chi)}]^{-1} - H]^{-1}h \\ \hline h^T + h^T\left[[\Theta_*^{(\chi)}]^{-1} - H\right]^{-1}H & h^T\left[[\Theta_*^{(\chi)}]^{-1} - H\right]^{-1}h \end{array}\right] \end{array}$$

and

$$\begin{array}{rcl} \phi_{\mathcal{Z}_\chi}(Z) & = & \max\limits_W \left\{\mathrm{Tr}(ZW) : W \succeq 0,\, \mathrm{Tr}(WQ_j^\chi) \leq q_j^\chi,\, 1 \leq j \leq J_\chi\right\} \\ & = & \min_\lambda \left\{\sum_j q_j^\chi \lambda_j : \lambda \geq 0,\, Z \preceq \sum_j \lambda_j Q_j^\chi\right\}, \end{array}$$

where the last equality is due to semidefinite duality.[24] From the second representation of $\phi_{\mathcal{Z}_\chi}(\cdot)$ and the fact that all Q_j^χ are diagonal it follows that $\phi_{\mathcal{Z}_\chi}(M_\chi(-h,H)) = \phi_{\mathcal{Z}_\chi}(M_\chi(h,H))$ (indeed, with diagonal Q_j^χ, if λ is feasible for the minimization problem participating in the representation when $Z = M_\chi(h,H)$, it clearly remains feasible when Z is replaced with $M_\chi(-h,H)$). This, in turn, combines straightforwardly with (2.130) to imply that when replacing h_* with 0 in a saddle point $(h_*, H_*; \Theta_1^*, \Theta_2^*)$ of (2.134), we end up with another saddle point of (2.134). In other words, when solving (2.134), we can from the very beginning set h to 0, thus converting (2.134) into the convex-concave saddle point problem

$$\mathcal{SV} = \min_{H:(0,H)\in\mathcal{H}_1\cap\mathcal{H}_2}\ \max_{\Theta_1\in\mathcal{V}_1,\Theta_2\in\mathcal{V}_2} \Phi(0,H;\Theta_1,\Theta_2). \tag{2.137}$$

Taking into account that we are in the case where all matrices from the sets $\mathcal{V}_\chi$, like the matrices $\Theta_*^{(\chi)}$ and all the matrices Q_j^χ, $\chi = 1,2$, are diagonal, it is immediate to verify that $\Phi(0,H;\Theta_1,\Theta_2) = \Phi(0,EHE;\Theta_1,\Theta_2)$ for any $d \times d$ diagonal matrix E with diagonal entries ± 1. Due to convexity-concavity of Φ this implies that (2.137) admits a saddle point $(0, H_*; \Theta_1^*, \Theta_2^*)$ with H_* invariant w.r.t. transformations $H_* \mapsto EH_*E$ with the above E, that is, with diagonal H_*, as claimed. □

2.9.3 Quadratic lifting—Does it help?

Assume that for $\chi = 1,2$, we are given

- affine mappings $u \mapsto \mathcal{A}_\chi(u) = A_\chi[u;1] : \mathbf{R}^{n_\chi} \to \mathbf{R}^d$,
- nonempty convex compact sets $U_\chi \subset \mathbf{R}^{n_\chi}$,
- nonempty convex compact sets $\mathcal{V}_\chi \subset \mathrm{int}\,\mathbf{S}_+^d$.

These data define families $\mathcal{G}_\chi$ of Gaussian distributions on $\mathbf{R}^d$: $\mathcal{G}_\chi$ is comprised of all

---

[23] In terms of the sets U_χ, this assumption means that the latter sets are given by linear inequalities on the *squares* of entries in u.

[24] See Section 4.1 (or [187, Section 7.1] for more details).

distributions $\mathcal{N}(\mathcal{A}_\chi(u),\Theta)$ with $u\in U_\chi$ and $\Theta\in\mathcal{V}_\chi$. The data define also families $\mathcal{SG}_\chi$ of sub-Gaussian distributions on $\mathbf{R}^d$: $\mathcal{SG}_\chi$ is comprised of all sub-Gaussian distributions with parameters $(\mathcal{A}_\chi(u),\Theta)$ with $(u,\Theta)\in U_\chi\times\mathcal{V}_\chi$.

Assume we observe random variable $\zeta\in\mathbf{R}^d$ drawn from a distribution P known to belong to $\mathcal{G}_1\cup\mathcal{G}_2$, and our goal is to decide from a stationary K-repeated version of our observation on the pair of hypotheses $H_\chi: P\in\mathcal{G}_\chi$, $\chi=1,2$; we refer to this situation as the *Gaussian case*, and we assume from now on that we are in this case.[25]

At present, we have developed two approaches to building detector-based tests for H_1, H_2:

A. Utilizing the *affine* in ζ detector $\phi_{\rm aff}$ given by solution to the saddle point problem (see (2.122), (2.123) and set $\theta_\chi=\mathcal{A}_\chi(u_\chi)$ with u_χ running through U_χ)

$$\mathrm{SadVal}_{\rm aff}=\min_{h\in\mathbf{R}^d}\max_{\substack{u_1\in U_1,u_2\in U_2\\ \Theta_1\in\mathcal{V}_1,\Theta_2\in\mathcal{V}_2}}\tfrac{1}{2}\left[h^T[\mathcal{A}_2(u_2)-\mathcal{A}_1(u_1)]+\tfrac{1}{2}h^T\left[\Theta_1+\Theta_2\right]h\right];$$

this detector satisfies the risk bound

$$\mathrm{Risk}[\phi_{\rm aff}|\mathcal{G}_1,\mathcal{G}_2]\leq\exp\{\mathrm{SadVal}_{\rm aff}\}.$$

Q. Utilizing the quadratic in ζ detector $\phi_{\rm lift}$ given by Proposition 2.43.ii, with the risk bound

$$\mathrm{Risk}[\phi_{\rm lift}|\mathcal{G}_1,\mathcal{G}_2]\leq\exp\{\mathrm{SadVal}_{\rm lift}\},$$

with $\mathrm{SadVal}_{\rm lift}$ given by (2.134).

A natural question is, which of these options results in a better risk bound. Note that we cannot just say "clearly, the second option is better, since there are more quadratic detectors than affine ones"—the difficulty is that the key relation (2.133), in the context of Proposition 2.43, is inequality rather than equality.[26] We are about to show that under reasonable assumptions, the second option indeed is better:

Proposition 2.45. *In the situation in question, assume that the sets $\mathcal{V}_\chi$, $\chi=1,2$, contain the $\succeq$-largest elements, and that these elements are taken as the matrices $\Theta_*^{(\chi)}$ participating in Proposition 2.43.ii. Let, further, the convex compact sets $\mathcal{Z}_\chi$ participating in Proposition 2.43.ii satisfy*

$$\mathcal{Z}_\chi\subset\bar{\mathcal{Z}}_\chi:=\left\{Z=\left[\begin{array}{c|c}W&u\\\hline u^T&1\end{array}\right]\succeq0,u\in U_\chi\right\}\tag{2.138}$$

(this assumption does not restrict generality, since $\bar{\mathcal{Z}}_\chi$ is, along with U_χ, a closed convex set which clearly contains all matrices $[u;1][u;1]^T$ with $u\in U_\chi$). Then

$$\mathrm{SadVal}_{\rm lift}\leq\mathrm{SadVal}_{\rm aff},\tag{2.139}$$

*that is, option **Q** is at least as efficient as option **A**.*

---

[25]It is easily seen that what follows can be straightforwardly extended to the *sub-Gaussian* case, where the hypotheses we would decide upon state that $P\in\mathcal{SG}_\chi$.

[26]One cannot make (2.133) an equality by redefining the right-hand side function—it will lose the convexity-concavity properties required in our context.

ρ	σ_1	σ_2	unrestricted H and h	$H = 0$	$h = 0$
0.5	2	2	0.31	0.31	1.00
0.5	1	4	0.24	0.39	0.62
0.01	1	4	0.41	1.00	0.41

Table 2.2: Risk of quadratic detector $\phi(\zeta) = h^T\zeta + \frac{1}{2}\zeta^T H\zeta + \varkappa$.

Proof. Let $A_\chi = [\bar{A}_\chi, a_\chi]$. Looking at (2.122) (where one should substitute $\theta_\chi = \mathcal{A}_\chi(u_\chi)$ with u_χ running through U_χ) and taking into account that $\Theta_\chi \preceq \Theta_*^{(\chi)} \in \mathcal{V}_\chi$ when $\Theta_\chi \in \mathcal{V}_\chi$, we conclude that

$$\mathrm{SadVal_{aff}} = \min_h \max_{u_1\in U_1, u_2\in U_2} \tfrac{1}{2}\left[h^T[\bar{A}_2u_2 - \bar{A}_1u_1 + a_2 - a_1] + \tfrac{1}{2}h^T\left[\Theta_*^{(1)} + \Theta_*^{(2)}\right]h\right]. \tag{2.140}$$

At the same time, we have by Proposition 2.43.ii:

$$\begin{array}{rl}
\mathrm{SadVal_{lift}} & = \min\limits_{(h,H)\in\mathcal{H}_1\cap\mathcal{H}_2} \max\limits_{\Theta_1\in\mathcal{V}_1,\Theta_2\in\mathcal{V}_2} \frac{1}{2}\left[\Phi_{A_1,\mathcal{Z}_1}(-h,-H;\Theta_1) + \Phi_{A_2,\mathcal{Z}_2}(h,H;\Theta_2)\right] \\
& \leq \min\limits_{h\in\mathbf{R}^d} \max\limits_{\Theta_1\in\mathcal{V}_1,\Theta_2\in\mathcal{V}_2} \frac{1}{2}\left[\Phi_{A_1,\mathcal{Z}_1}(-h,0;\Theta_1) + \Phi_{A_2,\mathcal{Z}_2}(h,0;\Theta_2)\right] \\
& = \min\limits_{h\in\mathbf{R}^d} \max\limits_{\Theta_1\in\mathcal{V}_1,\Theta_2\in\mathcal{V}_2} \frac{1}{2}\left[\frac{1}{2}\max\limits_{Z_1\in\mathcal{Z}_1} \mathrm{Tr}\left(Z_1\left[\begin{array}{c|c} & -\bar{A}_1^Th \\ \hline -h^T\bar{A}_1 & -2h^Ta_1 + h^T\Theta_*^{(1)}h\end{array}\right]\right)\right. \\
& \qquad \left. + \frac{1}{2}\max\limits_{Z_2\in\mathcal{Z}_2} \mathrm{Tr}\left(Z_2\left[\begin{array}{c|c} & \bar{A}_2^Th \\ \hline h^T\bar{A}_2 & 2h^Ta_2 + h^T\Theta_*^{(2)}h\end{array}\right]\right)\right] \\
& \qquad\qquad [\text{by direct computation utilizing (2.130)}] \\
& \leq \min\limits_{h\in\mathbf{R}^d} \frac{1}{2}\left[\frac{1}{2}\max\limits_{u_1\in U_1}\left[-2u_1^T\bar{A}_1^Th - 2a_1^Th + h^T\Theta_*^{(1)}h\right] + \right. \\
& \qquad \left. \frac{1}{2}\max\limits_{u_2\in U_2}\left[2u_2^T\bar{A}_2^Th + 2a_2^Th + h^T\Theta_*^{(2)}h\right]\right] \\
& \qquad\qquad [\text{due to (2.138)}] \\
& = \mathrm{SadVal_{aff}},
\end{array}$$

where the concluding equality is due to (2.140). □

Numerical illustration. To get an impression of the performance of quadratic detectors as compared to affine ones under the premise of Proposition 2.45, we present here the results of an experiment where $U_1 = U_1^\rho = \{u \in \mathbf{R}^{12} : u_i \geq \rho, 1 \leq i \leq 12\}$, $U_2 = U_2^\rho = -U_1^\rho$, $A_1 = A_2 \in \mathbf{R}^{8\times 13}$, and $\mathcal{V}_\chi = \{\Theta_*^{(\chi)} = \sigma_\chi^2 I_8\}$ are singletons. The risks of affine, quadratic and "purely quadratic" (with h set to 0) detectors on the associated families $\mathcal{G}_1, \mathcal{G}_2$ are given in Table 2.2.

We see that

- when deciding on families of Gaussian distributions with a common covariance matrix and expectations varying in the convex sets associated with the families, passing from affine detectors described by Proposition 2.41 to quadratic detectors does not affect the risk (first row in the table). This should be expected: we are in the scope of Gaussian o.s., where minimum risk affine detectors are optimal among all possible detectors.
- When deciding on families of Gaussian distributions in the case where distributions from different families can have close expectations (third row in the table), affine detectors are useless, while the quadratic ones are not, provided that $\Theta_*^{(1)}$

differs from $\Theta_*^{(2)}$. This is how it should be—we are in the case where the first moments of the distribution of observation bear no definitive information on the family to which this distribution belongs, making affine detectors useless. In contrast, quadratic detectors are able to utilize information (valuable when $\Theta_*^{(1)} \neq \Theta_*^{(2)}$) "stored" in the second moments of the observation.

- "In general" (second row in the table), both affine and purely quadratic components in a quadratic detector are useful; suppressing one of them may increase significantly the attainable risk.

2.9.4 Quadratic lifting: Sub-Gaussian case

The sub-Gaussian version of Proposition 2.43 is as follows:

Proposition 2.46.

(i) *Assume we are given*

- *a nonempty and bounded subset U of $\mathbf{R}^n$;*
- *a convex compact set $\mathcal{V}$ contained in the interior of the cone $\mathbf{S}^d_+$ of positive semidefinite $d \times d$ matrices;*
- *a $d \times (n+1)$ matrix A.*

These data specify the family $\mathcal{SG}_A[U,\mathcal{V}]$ of distributions of quadratic liftings $(\zeta, \zeta\zeta^T)$ of sub-Gaussian random vectors ζ with sub-Gaussianity parameters $A[u;1], \Theta$ stemming from $u \in U$ and $\Theta \in \mathcal{V}$.

Let us select some

1. *reals γ, γ^+ such that $0 < \gamma < \gamma^+ < 1$,*
2. *convex compact subset $\mathcal{Z}$ of the set $\mathcal{Z}^n = \{Z \in \mathbf{S}^{n+1} : Z \succeq 0, Z_{n+1,n+1} = 1\}$ such that relation (2.128) takes place,*
3. *positive definite $d \times d$ matrix $\Theta_* \in \mathbf{S}^d_+$ and $\delta \in [0,2]$ such that (2.129) takes place.*

These data specify the closed convex sets

$$\begin{array}{rcl}
\mathcal{H} & = & \mathcal{H}^\gamma := \{(h,H) \in \mathbf{R}^d \times \mathbf{S}^d : -\gamma\Theta_*^{-1} \preceq H \preceq \gamma\Theta_*^{-1}\}, \\
\widehat{\mathcal{H}} & = & \widehat{\mathcal{H}}^{\gamma,\gamma^+} = \left\{(h,H,G) \in \mathbf{R}^d \times \mathbf{S}^d \times \mathbf{S}^d : \left\{\begin{array}{c} -\gamma\Theta_*^{-1} \preceq H \preceq \gamma\Theta_*^{-1} \\ 0 \preceq G \preceq \gamma^+\Theta_*^{-1},\ H \preceq G \end{array}\right.\right\}
\end{array}$$

and the functions

$$\begin{array}{rcl}
\Psi_{A,\mathcal{Z}}(h,H,G) & = & -\frac{1}{2}\ln \mathrm{Det}(I - \Theta_*^{1/2} G \Theta_*^{1/2}) \\
& & +\frac{1}{2}\phi_{\mathcal{Z}}\left(B^T\left[\left[\begin{array}{c|c} H & h \\ \hline h^T & \end{array}\right] + [H,h]^T[\Theta_*^{-1} - G]^{-1}[H,h]\right]B\right) : \\
& & \qquad \widehat{\mathcal{H}} \times \mathcal{Z} \to \mathbf{R}, \\
\Psi^\delta_{A,\mathcal{Z}}(h,H,G;\Theta) & = & -\frac{1}{2}\ln \mathrm{Det}(I - \Theta_*^{1/2} G \Theta_*^{1/2}) \\
& & +\frac{1}{2}\mathrm{Tr}([\Theta - \Theta_*]G) + \frac{\delta(2+\delta)}{2(1 - \|\Theta_*^{1/2} G \Theta_*^{1/2}\|)}\|\Theta_*^{1/2} G \Theta_*^{1/2}\|_F^2 \\
& & +\frac{1}{2}\phi_{\mathcal{Z}}\left(B^T\left[\left[\begin{array}{c|c} H & h \\ \hline h^T & \end{array}\right] + [H,h]^T[\Theta_*^{-1} - G]^{-1}[H,h]\right]B\right) : \\
& & \qquad \widehat{\mathcal{H}} \times \{0 \preceq \Theta \preceq \Theta_*\} \to \mathbf{R}
\end{array} \tag{2.141}$$

where B is given by (2.131) and $\phi_{\mathcal{Z}}(\cdot)$ is the support function of $\mathcal{Z}$ given by (2.132), *along with*

$$\begin{array}{rcl}\Phi_{A,\mathcal{Z}}(h,H) &=& \min\limits_{G}\left\{\Psi_{A,\mathcal{Z}}(h,H,G) : (h,H,G)\in\widehat{\mathcal{H}}\right\} : \mathcal{H}\to\mathbf{R},\\ \Phi^{\delta}_{A,\mathcal{Z}}(h,H;\Theta) &=& \min\limits_{G}\left\{\Psi^{\delta}_{A,\mathcal{Z}}(h,H,G;\Theta) : (h,H,G)\in\widehat{\mathcal{H}}\right\} : \mathcal{H}\times\{0\preceq\Theta\preceq\Theta_*\}\to\mathbf{R},\end{array}$$

$\Phi_{A,\mathcal{Z}}(h,H)$ *is convex and continuous on its domain, and* $\Phi^{\delta}_{A,\mathcal{Z}}(h,H;\Theta)$ *is continuous on its domain, convex in* $(h,H)\in\mathcal{H}$ *and concave in* $\Theta\in\{0\preceq\Theta\preceq\Theta_*\}$. *Besides this,*

(##) *Whenever* $u\in\mathbf{R}^n$ *is such that* $[u;1][u;1]^T\in\mathcal{Z}$ *and* $\Theta\in\mathcal{V}$, *the sub-Gaussian random vector* ζ, *with parameters* $(A[u;1],\Theta)$, *satisfies the relation*

$$\begin{array}{ll}\forall(h,H)\in\mathcal{H}: & \\ (a) & \ln\left(\mathbf{E}_{\zeta}\left\{e^{\frac{1}{2}\zeta^TH\zeta+h^T\zeta}\right\}\right)\leq\Phi_{A,\mathcal{Z}}(h,H),\\ (b) & \ln\left(\mathbf{E}_{\zeta}\left\{e^{\frac{1}{2}\zeta^TH\zeta+h^T\zeta}\right\}\right)\leq\Phi^{\delta}_{A,\mathcal{Z}}(h,H;\Theta),\end{array}\tag{2.142}$$

which combines with (2.128) *to imply that*

$$\mathcal{SG}_A[U,\mathcal{V}]\subset\mathcal{S}[\mathcal{H},\mathcal{V},\Phi_{A,\mathcal{Z}}]\ \&\ \mathcal{SG}_A[U,\mathcal{V}]\subset\mathcal{S}[\mathcal{H},\mathcal{V},\Phi^{\delta}_{A,\mathcal{Z}}].\tag{2.143}$$

In addition, $\Phi_{A,\mathcal{Z}}$ *and* $\Phi^{\delta}_{A,\mathcal{Z}}$ *are coercive in* (h,H): $\Phi_{A,\mathcal{Z}}(h_i,H_i)\to+\infty$ *and* $\Phi^{\delta}_{A,\mathcal{Z}}(h_i,H_i;\Theta)\to+\infty$ *as* $i\to\infty$ *whenever* $\Theta\in\mathcal{V}$, $(h_i,H_i)\in\mathcal{H}$, *and* $\|(h_i,H_i)\|\to\infty$, $i\to\infty$.

(ii) *Let two collections of data from (i):* $(\mathcal{V}_\chi,\Theta_*^{(\chi)},\delta_\chi,\gamma_\chi,\gamma_\chi^+,A_\chi,\mathcal{Z}_\chi)$, $\chi=1,2$, *with common d be given, giving rise to the sets* $\mathcal{H}_\chi$, *matrices* B_χ, *and functions* $\Phi_{A_\chi,\mathcal{Z}_\chi}(h,H)$, $\Phi^{\delta_\chi}_{A_\chi,\mathcal{Z}_\chi}(h,H;\Theta)$, $\chi=1,2$. *These collections specify the families* $\mathcal{SG}_\chi=\mathcal{SG}_{A_\chi}[U_\chi,\mathcal{V}_\chi]$ *of sub-Gaussian distributions.*

Consider the convex-concave saddle point problem

$$\mathcal{SV}=\min_{(h,H)\in\mathcal{H}_1\cap\mathcal{H}_2}\ \max_{\Theta_1\in\mathcal{V}_1,\Theta_2\in\mathcal{V}_2}\underbrace{\tfrac{1}{2}\left[\Phi^{\delta_1}_{A_1,\mathcal{Z}_1}(-h,-H;\Theta_1)+\Phi^{\delta_2}_{A_2,\mathcal{Z}_2}(h,H;\Theta_2)\right]}_{\Phi^{\delta_1,\delta_2}(h,H;\Theta_1,\Theta_2)}.\tag{2.144}$$

A saddle point $(H_*,h_*;\Theta_1^*,\Theta_2^*)$ *in this problem does exist, and the induced quadratic detector*

$$\phi_*(\omega)=\tfrac{1}{2}\omega^TH_*\omega+h_*^T\omega+\underbrace{\tfrac{1}{2}\left[\Phi^{\delta_1}_{A_1,\mathcal{Z}_1}(-h_*,-H_*;\Theta_1^*)-\Phi^{\delta_2}_{A_2,\mathcal{Z}_2}(h_*,H_*;\Theta_2^*)\right]}_{a},$$

when applied to the families of sub-Gaussian distributions $\mathcal{SG}_\chi$, $\chi=1,2$, *has the risk*

$$\mathrm{Risk}[\phi_*|\mathcal{SG}_1,\mathcal{SG}_2]\leq\epsilon_\star:=e^{\mathcal{SV}}.$$

As a result,

$$\begin{array}{lrl}(a) & \int_{\mathbf{R}^d}e^{-\phi_*(\omega)}P(d\omega)\leq\epsilon_\star & \forall P\in\mathcal{SG}_1,\\ (b) & \int_{\mathbf{R}^d}e^{\phi_*(\omega)}P(d\omega)\leq\epsilon_\star & \forall P\in\mathcal{SG}_2.\end{array}$$

Similarly, the convex minimization problem

$$\mathrm{Opt} = \min_{(h,H)\in\mathcal{H}_1\cap\mathcal{H}_2} \underbrace{\tfrac{1}{2}\left[\Phi_{A_1,\mathcal{Z}_1}(-h,-H) + \Phi_{A_2,\mathcal{Z}_2}(h,H)\right]}_{\Phi(h,H)}. \tag{2.145}$$

is solvable, and the quadratic detector induced by its optimal solution (h_*, H_*)

$$\phi_*(\omega) = \tfrac{1}{2}\omega^T H_*\omega + h_*^T\omega + \underbrace{\tfrac{1}{2}\left[\Phi_{A_1,\mathcal{Z}_1}(-h_*,-H_*) - \Phi_{A_2,\mathcal{Z}_2}(h_*,H_*)\right]}_{a}, \tag{2.146}$$

when applied to the families of sub-Gaussian distributions $\mathcal{SG}_\chi$, $\chi = 1,2$, *has the risk*

$$\mathrm{Risk}[\phi_*|\mathcal{SG}_1,\mathcal{SG}_2] \le \epsilon_\star := \mathrm{e}^{\mathrm{Opt}},$$

so that relation (2.145) *takes place for the* ϕ_* *and* $\epsilon_\star$ *just defined.*

For proof, see Section 2.11.6.

Remark 2.47. Proposition 2.46 offers two options for building quadratic detectors for the families $\mathcal{SG}_1$, $\mathcal{SG}_2$, *those based on the saddle point of (2.144) and on the optimal solution to (2.145). Inspecting the proof, the number of options can be increased to 4: we can replace any of the functions* $\Phi^{\delta_\chi}_{A_\chi,\mathcal{Z}_\chi}$, $\chi = 1,2$ *(or both these functions simultaneously), with* $\Phi_{A_\chi,\mathcal{Z}_\chi}$. *The second of the original two options is exactly what we get when replacing both* $\Phi^{\delta_\chi}_{A_\chi,\mathcal{Z}_\chi}$, $\chi = 1,2$, *with* $\Phi_{A_\chi,\mathcal{Z}_\chi}$. *It is easily seen that depending on the data, each of these four options can be the best—result in the smallest risk bound. Thus, it makes sense to keep all these options in mind and to use the one which, under the circumstances, results in the best risk bound. Note that the risk bounds are efficiently computable, so that identifying the best option is easy.*

2.9.5 Generic application: Quadratically constrained hypotheses

Propositions 2.43 and 2.46 operate with Gaussian/sub-Gaussian observations ζ with matrix parameters Θ running through convex compact subsets $\mathcal{V}$ of $\mathrm{int}\,\mathbf{S}^d_+$, and means of the form $A[u;1]$, with "signals" u running through given sets $U \subset \mathbf{R}^n$. The constructions, however, involved additional entities—convex compact sets $\mathcal{Z} \subset \mathcal{Z}^n := \{Z \in \mathbf{S}^{n+1}_+ : Z_{n+1,n+1} = 1\}$ containing quadratic liftings $[u;1][u;1]^T$ of all signals $u \in U$. Other things being equal, the smaller the $\mathcal{Z}$, the smaller the associated function $\Phi_{A,\mathcal{Z}}$ (or $\Phi^\delta_{A,\mathcal{Z}}$), and consequently, the smaller the (upper bounds on the) risks of the quadratic in ζ detectors we end up with. In order to implement these constructions, we need to understand how to build the required sets $\mathcal{Z}$ in an "economical" way. There is a relatively simple case when it is easy to get reasonable candidates for the role of $\mathcal{Z}$—the case of *quadratically constrained* signal set U:

$$U = \{u \in \mathbf{R}^n : f_k(u) := u^T Q_k u + 2q_k^T u \le b_k,\ 1 \le k \le K\}. \tag{2.147}$$

Indeed, the constraints $f_k(u) \le b_k$ are just linear constraints on the quadratic lifting $[u;1][u;1]^T$ of u:

$$u^T Q_k u + 2q_k^T u \le b_k \Leftrightarrow \mathrm{Tr}(F_k[u;1][u;1]^T) \le b_k,\ F_k = \left[\begin{array}{c|c} Q_k & q_k \\ \hline q_k^T & \end{array}\right] \in \mathbf{S}^{n+1}.$$

Consequently, in the case of (2.147), the simplest candidate on the role of $\mathcal{Z}$ is the set

$$\mathcal{Z} = \{Z \in \mathbf{S}^n : Z \succeq 0, Z_{n+1,n+1} = 1, \mathrm{Tr}(F_k Z) \leq b_k, \ 1 \leq k \leq K\}. \tag{2.148}$$

This set clearly is closed and convex (the latter even when U itself is not convex), and indeed contains the quadratic liftings $[u;1][u;1]^T$ of all points $u \in U$. We need also the compactness of $\mathcal{Z}$; the latter definitely takes place when the quadratic constraints describing U contain the constraint of the form $u^T u \leq R^2$, which, in turn, can be ensured, basically "for free," when U is bounded. It should be stressed that the "ideal" choice of $\mathcal{Z}$ would be the convex hull $\mathcal{Z}[U]$ of all rank 1 matrices $[u;1][u;1]^T$ with $u \in U$—this definitely is the smallest convex set which contains the quadratic liftings of all points from U. Moreover, $\mathcal{Z}[U]$ is closed and bounded, provided U is so. The difficulty is that $\mathcal{Z}[U]$ can be computationally intractable (and thus useless in our context) already for pretty simple sets U of the form (2.147). The set (2.148) is a simple outer approximation of $\mathcal{Z}[U]$, and this approximation can be very loose: for instance, when $U = \{u : -1 \leq u_k \leq 1, 1 \leq k \leq n\}$ is just the unit box in $\mathbf{R}^n$, the set (2.148) is

$$\{Z \in \mathbf{S}^{n+1} : Z \succeq 0, Z_{n+1,n+1} = 1, |Z_{k,n+1}| \leq 1, \ 1 \leq k \leq n\};$$

this set even is not bounded, while $\mathcal{Z}[U]$ clearly is bounded. There is, essentially, just one generic case when the set (2.148) is *exactly equal* to $\mathcal{Z}[U]$—the case where

$$U = \{u : u^T Q u \leq c\}, Q \succ 0$$

is an ellipsoid centered at the origin; the fact that in this case the set given by (2.148) is *exactly* $\mathcal{Z}[U]$ is a consequence of what is called $\mathcal{S}$-Lemma.

Though, in general, the set $\mathcal{Z}$ can be a very loose outer approximation of $\mathcal{Z}[U]$, this does not mean that this construction cannot be improved. As an instructive example, let $U = \{u \in \mathbf{R}^n : \|u\|_\infty \leq 1\}$. We get an approximation of $\mathcal{Z}[U]$ much better than the one above when applying (2.148) to an equivalent description of the box by *quadratic* constraints:

$$U := \{u \in \mathbf{R}^n : \|u\|_\infty \leq 1\} = \{u \in \mathbf{R}^n : u_k^2 \leq 1, 1 \leq k \leq n\}.$$

Applying the recipe of (2.148) to the latter description of U, we arrive at a significantly less conservative outer approximation of $\mathcal{Z}[U]$, specifically,

$$\mathcal{Z} = \{Z \in \mathbf{S}^{n+1} : Z \succeq 0, Z_{n+1,n+1} = 1, Z_{kk} \leq 1, \ 1 \leq k \leq n\}.$$

Not only the resulting set $\mathcal{Z}$ is bounded; we can get a reasonable "upper bound" on the discrepancy between $\mathcal{Z}$ and $\mathcal{Z}[U]$. Namely, denoting by Z^o the matrix obtained from a symmetric $n \times n$ matrix Z by zeroing out the entry $Z_{n+1,n+1}$ and keeping the remaining entries intact, we have

$$\mathcal{Z}^o[U] := \{Z^o : Z \in \mathcal{Z}[U]\} \subset \mathcal{Z}^o := \{Z^o : Z \in \mathcal{Z}\} \subset O(1)\ln(n+1)\mathcal{Z}^o.$$

This is a particular case of a general result (which goes back to [191]; we shall get this result as a byproduct of our forthcoming considerations, specifically, Proposition 4.6) as follows:

Let U be a bounded set given by a system of convex quadratic constraints without linear terms:

$$U = \{u \in \mathbf{R}^n : u^T Q_k u \le c_k,\ 1 \le k \le K\},\ Q_k \succeq 0, 1 \le k \le K,$$

and let $\mathcal{Z}$ be the associated set (2.148):

$$\mathcal{Z} = \{Z \in \mathbf{S}^{n+1} : Z \succeq 0, Z_{n+1,n+1} = 1,\ \mathrm{Tr}(Z\mathrm{Diag}\{Q_k, 1\}) \le c_k,\ 1 \le k \le K\}$$

Then

$$\mathcal{Z}^o[U] := \{Z^o : Z \in \mathcal{Z}[U]\} \subset \mathcal{Z}^o := \{Z^o : Z \in \mathcal{Z}\} \subset 3\ln(\sqrt{3}(K+1))\mathcal{Z}^o[U].$$

Note that when $K = 1$ (i.e., U is an ellipsoid centered at the origin), the factor $4\ln(5(K+1))$, as it was already mentioned, can be replaced by 1.

One can think that the factor $3\ln(\sqrt{3}(K+1))$ is too large to be of interest; well, this is nearly the best factor one can get under the circumstances, and a nice fact is that the factor is "nearly independent" of K.

Finally, we remark that, as in the case of a box, we can try to reduce the conservatism of the outer approximation (2.148) of $\mathcal{Z}[U]$ by passing from the initial description of U to an equivalent one. The standard recipe here is to replace linear constraints in the description of U by their quadratic consequences; for example, we can augment a pair of linear constraints $q_i^T u \le c_i$, $q_j^T u \le c_j$, assuming there is such a pair, with the quadratic constraint $(c_i - q_i^T u)(c_j - q_j^T u) \ge 0$. While this constraint is redundant, as far as the description of U itself is concerned, adding this constraint reduces, and sometimes significantly, the set given by (2.148). Informally speaking, transition from (2.147) to (2.148) is by itself "too stupid" to utilize the fact (known to every kid) that the product of two nonnegative quantities is nonnegative; when augmenting linear constraints in the description of U by their pairwise products, we somehow compensate for this stupidity. Unfortunately, while "computationally tractable" assistance of this type allows us to reduce the conservatism of (2.148), it usually does not allow us to eliminate it completely: a grave "fact of life" is that even in the case of the unit box U, the set $\mathcal{Z}[U]$ is computationally intractable. Scientifically speaking: maximizing quadratic forms over the unit box U is provably an NP-hard problem; were we able to get a computationally tractable description of $\mathcal{Z}[U]$, we would be able to solve this NP-hard problem efficiently, implying that P=NP. While we do not know for sure that the latter is not the case, "informal odds" are strongly against this possibility.

The bottom line is that while the approach we are discussing in *some* situations could result in quite conservative tests, "some" is by far not the same as "always"; on the positive side, this approach allows us to process some important problems. We are about to present a simple and instructive illustration.

2.9.5.1 Simple change detection

In Figure 2.8, you see a sample of frames from a "movie" in which a noisy picture of a dog gradually transforms into a noisy picture of a lady; several initial frames differ just by realizations of noise, and starting from some instant, the "signal" (the deterministic component of the image) starts to drift from the dog towards the lady. What, in your opinion, is the change point—the first time instant where

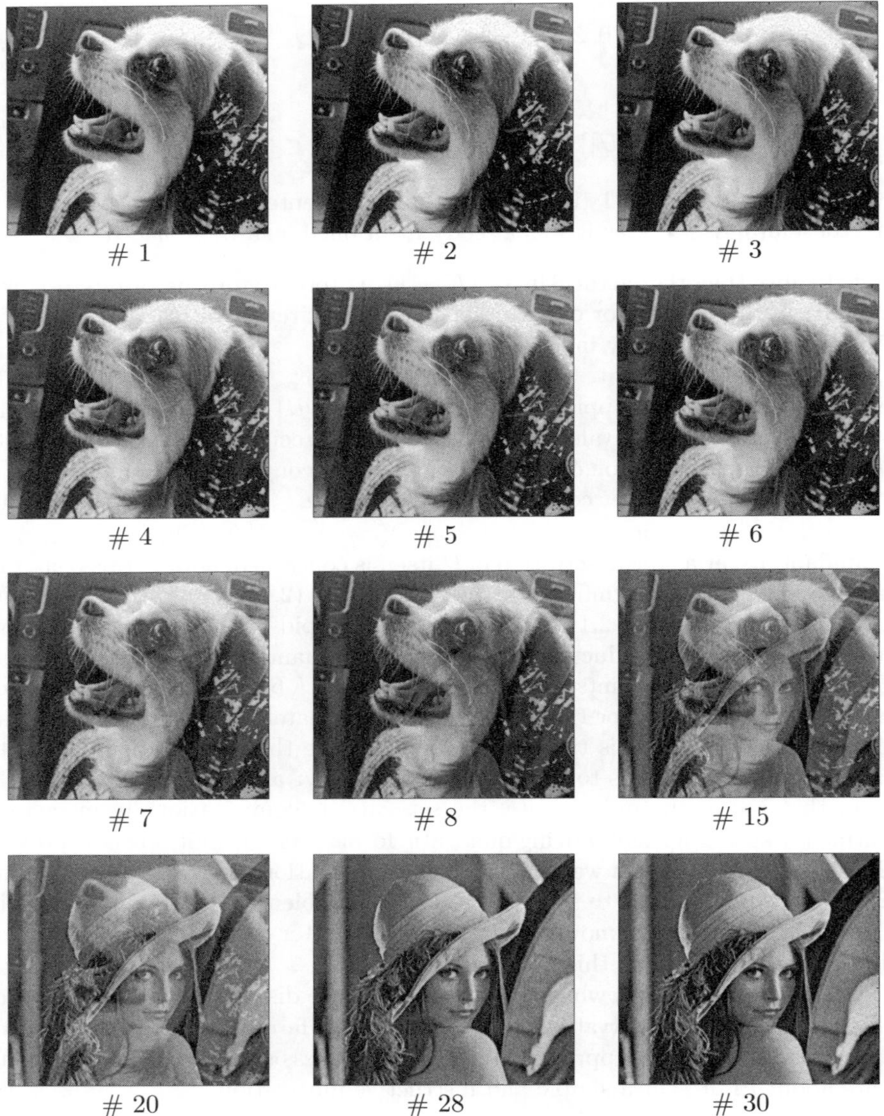

Figure 2.8: Frames from a "movie"

the signal component of the image differs from the signal component of the initial image?

A simple model of the situation is as follows: we observe, one by one, vectors (in fact, 2D arrays, but we can "vectorize" them)

$$\omega_t = x_t + \xi_t,\ t = 1, 2, ..., K, \tag{2.149}$$

where the x_t are deterministic components of the observations and the ξ_t are random noises. It may happen that for some $\tau \in \{2, 3, ..., K\}$, the vectors x_t are independent of t when $t < \tau$, and x_τ differs from $x_{\tau-1}$ ("τ is a change point"); if it is the case, τ is uniquely defined by $x^K = (x_1, ..., x_K)$. An alternative is that x_t is independent of t, for all $1 \leq t \leq K$ ("no change"). The goal is to decide, based on observation $\omega^K = (\omega_1, .., \omega_K)$, whether there was a change point, and if yes, then, perhaps, to localize it.

The model we have just described is the simplest case of "change detection," where, given noisy observations on some time horizon, one is interested in detecting a "change" in some time series underlying the observations. In our simple model, this time series is comprised of deterministic components x_t of observations, and "change at time τ" is understood in the most straightforward way—as the fact that x_τ differs from preceding x_t's equal to each other. In more complicated situations, our observations are obtained from the underlying time series $\{x_t\}$ by a non-anticipative transformation, like

$$\omega_t = \sum_{s=1}^{t} A_{ts} x_s + \xi_t,\ t = 1, ..., K,$$

and we still want to detect the change, if any, in the time series $\{x_t\}$. As an instructive example, consider observations, taken along an equidistant time grid, of the positions of an aircraft which "normally" flies with constant velocity, but at some time instant can start to maneuver. In this situation, the underlying time series is comprised of the velocities of the aircraft at consecutive time instants, observations are obtained from this time series by integration, and to detect a maneuver means to detect that on the observation horizon, there was a change in the series of velocities.

Change detection is the subject of a huge literature dealing with a wide range of models differing from each other in

- whether we deal with direct observations of the time series of interest, as in (2.149), or with indirect ones (in the latter case, there is a wide spectrum of options related to how the observations depend on the underlying time series),
- what are the assumptions on the noise,
- what happens with the x_t's after the change—do they jump from their common value prior to time τ to a new common value starting with this time, or start to depend on time (and if yes, then how), etc.

A significant role in change detection is played by hypothesis testing; as far as affine/quadratic-detector-based techniques developed in this section are concerned, their applications in the context of change detection are discussed in [50]. In what follows, we focus on the simplest of these applications.

Situation and goal. We consider the situation as follows:

1. Our observations are given by (2.149) with noises $\xi_t \sim \mathcal{N}(0, \sigma^2 I_d)$ independent across $t = 1, ..., K$. We do not known σ a priori; what we know is that σ is independent of t and belongs to a given segment $[\underline{\sigma}, \overline{\sigma}]$, with $0 < \underline{\sigma} \leq \overline{\sigma}$;
2. Observations (2.149) arrive one by one, so that at time t, $2 \leq t \leq K$, we have at our disposal observation $\omega^t = (\omega_1, ..., \omega_t)$. Our goal is to build a system of inferences $\mathcal{T}_t$, $2 \leq t \leq K$, such that $\mathcal{T}_t$ as applied to ω^t either infers that there was a change at time t or earlier, in which case we terminate, or infers that so far there has been no change, in which case we either proceed to time $t + 1$ (if $t < K$), or terminate (if $t = K$) with a "no change" conclusion.
   We are given $\epsilon \in (0, 1)$ and want our collection of inferences to satisfy the bound ϵ on the probability of *false alarm* (i.e., on the probability of terminating somewhere on time horizon $2, 3, ..., K$ with a "there was a change" conclusion in the situation where there was no change: $x_1 = ... = x_K$). Under this restriction, we want to make as small as possible the probability of a *miss* (of not detecting the change at all in the situation where there was a change).

The "small probability of a miss" desire should be clarified. When the noise is nontrivial, we have no chances to detect very small changes *and* respect the bound on the probability of false alarm. A realistic goal is to make as small as possible the probability of missing a *not too small* change, which can be formalized as follows. Given $\rho > 0$, and tolerances $\epsilon, \varepsilon \in (0, 1)$, let us look for a system of inferences $\{\mathcal{T}_t : 2 \leq t \leq K\}$ such that

- the probability of false alarm is at most ϵ, and
- the probability of "ρ-miss"—the probability of detecting no change when there was a change of energy $\geq \rho^2$ (i.e., when there was a change a time τ, and, moreover, it holds $\|x_\tau - x_1\|_2^2 \geq \rho^2$) is at most ε.

What we are interested in, is to achieve the goal just formulated with as small a ρ as possible.

Construction. Let us select a large "safety parameter" R, like $R = 10^8$ or even $R = 10^{80}$, so that we can assume that for all time series we are interested in it holds $\|x_t - x_\tau\|_2^2 \leq R^2$.[27] Let us associate with $\rho > 0$ "signal hypotheses" H_t^ρ, $t = 2, 3, ..., K$, on the distribution of observation ω^K given by (2.149), with H_t^ρ stating that at time t there is a change, of energy at least ρ^2, in the time series $\{x_t\}_{t=1}^K$ underlying the observation ω^K:

$$x_1 = x_2 = ... = x_{t-1} \ \& \ \|x_t - x_{t-1}\|_2^2 = \|x_t - x_1\|_2^2 \geq \rho^2$$

(and on top of that, $\|x_t - x_\tau\|_2^2 \leq R^2$ for all t, τ). Let us augment these hypotheses by the null hypothesis H_0 stating that there is no change at all—the observation ω^K stems from a stationary time series $x_1 = x_2 = ... = x_K$. We are about to use our machinery of detector-based tests in order to build a system of tests deciding, with partial risks ϵ and ε, on the null hypothesis vs. the "signal alternative" $\bigcup_t H_t^\rho$ for as small a ρ as possible.

The implementation is as follows. Given $\rho > 0$ such that $\rho^2 < R^2$, consider two

[27] R is needed only to make the domains we are working with bounded, thus allowing us to apply the theory we have developed so far. The actual value of R does *not* enter our constructions and conclusions.

hypotheses, G_1 and G_2^ρ, on the distribution of observation

$$\zeta = x + \xi \in \mathbf{R}^d. \tag{2.150}$$

Both hypotheses state that $\xi \sim \mathcal{N}(0, \sigma^2 I_d)$ with unknown σ known to belong to a given segment $\Delta := [\sqrt{2}\underline{\sigma}, \sqrt{2}\overline{\sigma}]$. In addition, G_1 states that $x = 0$, and G_2^ρ that $\rho^2 \leq \|x\|_2^2 \leq R^2$. We can use the result of Proposition 2.43.ii to build a detector quadratic in ζ for the families of distributions $\mathcal{P}_1$, $\mathcal{P}_2^\rho$ obeying the hypotheses G_1, G_2^ρ, respectively. To this end it suffices to apply the proposition to the collections

$$\mathcal{V}_\chi = \{\sigma^2 I_d : \sigma \in \Delta\}, \Theta_*^{(\chi)} = 2\overline{\sigma}^2 I_d, \delta_\chi = 1 - \underline{\sigma}/\overline{\sigma}, \gamma_\chi = 0.999, A_\chi = I_d, \mathcal{Z}_\chi, \quad [\chi = 1, 2]$$

where

$$\begin{array}{rcl}\mathcal{Z}_1 & = & \{[0;...;0;1][0;...;0;1]^T\} \subset \mathbf{S}_+^{d+1}, \\ \mathcal{Z}_2 & = & \mathcal{Z}_2^\rho = \{Z \in \mathbf{S}_+^{d+1} : Z_{d+1,d+1} = 1, 1 + R^2 \geq \mathrm{Tr}(Z) \geq 1 + \rho^2\}.\end{array}$$

The (upper bound on the) risk of the quadratic in ζ detector yielded by a saddle point of function (2.134), as given by Proposition 2.43.ii, is immediate: by the same argument as used when justifying Remark 2.44, in the situation in question one can look for a saddle point with $h = 0$, $H = \eta I_d$, and identifying the required η reduces to solving the univariate convex problem

$$\begin{array}{c}\mathrm{Opt}(\rho) = \min\limits_\eta \frac{1}{2}\Big\{ -\frac{d}{2}\ln(1 - \widehat{\sigma}^4\eta^2) - \frac{d}{2}\widehat{\sigma}^2(1 - \underline{\sigma}^2/\overline{\sigma}^2)\eta + \frac{d\delta(2+\delta)\widehat{\sigma}^4\eta^2}{1+\widehat{\sigma}^2\eta} \\ \qquad + \frac{\rho^2\eta}{2(1-\widehat{\sigma}^2\eta)} : -\gamma \leq \widehat{\sigma}^2\eta \leq 0\Big\} \\ \left[\widehat{\sigma} = \sqrt{2}\overline{\sigma},\ \delta = 1 - \underline{\sigma}/\overline{\sigma}\right]\end{array}$$

which can be done in no time by Bisection. The resulting detector and the upper bound on its risk are given by the optimal solution $\eta(\rho)$ to the latter problem according to

$$\phi_\rho^*(\zeta) = \tfrac{1}{2}\eta(\rho)\zeta^T\zeta + \underbrace{\frac{d}{4}\left[\ln\left(\frac{1 - \widehat{\sigma}^2\eta(\rho)}{1 + \widehat{\sigma}^2\eta(\rho)}\right) - \widehat{\sigma}^2(1 - \underline{\sigma}^2/\overline{\sigma}^2)\eta(\rho) - \frac{\rho^2\eta(\rho)}{d(1 - \widehat{\sigma}^2\eta(\rho))}\right]}_{a(\rho)}$$

with

$$\mathrm{Risk}[\phi_\rho^*|\mathcal{P}_1, \mathcal{P}_2] \leq \mathrm{Risk}(\rho) := \mathrm{e}^{\mathrm{Opt}(\rho)}$$

(observe that R appears neither in the definition of the optimal detector nor in the risk bound). It is immediately seen that $\mathrm{Opt}(\rho) \to 0$ as $\rho \to +0$ and $\mathrm{Opt}(\rho) \to -\infty$ as $\rho \to +\infty$, implying that given $\kappa \in (0, 1)$, we can easily find by bisection $\rho = \rho(\kappa)$ such that $\mathrm{Risk}(\rho) = \kappa$; in what follows, we assume w.l.o.g. that $R > \rho(\kappa)$ for the value of κ we end with; see below. Next, let us pass from the detector $\phi_{\rho(\kappa)}^*(\cdot)$ to its shift

$$\phi^{*,\kappa}(\zeta) = \phi_{\rho(\kappa)}^*(\zeta) + \ln(\varepsilon/\kappa),$$

so that for the simple test $\mathcal{T}^\kappa$ which, given observation ζ, accepts G_1 and rejects

$G_2^{\rho(\kappa)}$ whenever $\phi^{*,\kappa}(\zeta) \geq 0$, and accepts $G_2^{\rho(\kappa)}$ and rejects G_1 otherwise, it holds

$$\mathrm{Risk}_1(\mathcal{T}^\kappa|G_1, G_2^{\rho(\kappa)}) \leq \frac{\kappa^2}{\varepsilon}, \quad \mathrm{Risk}_2(\mathcal{T}^\kappa|G_1, G_2^{\rho(\kappa)}) \leq \varepsilon; \tag{2.151}$$

see Proposition 2.14 and (2.48).

We are nearly done. Given $\kappa \in (0,1)$, consider the system of tests $\mathcal{T}_t^\kappa$, $t = 2, 3, ..., K$, as follows. At time $t \in \{2, 3, ..., K\}$, given observations $\omega_1, ..., \omega_t$ stemming from (2.149), let us form the vector

$$\zeta_t = \omega_t - \omega_1$$

and compute the quantity $\phi^{*,\kappa}(\zeta_t)$. If this quantity is negative, we claim that the change has already taken place and terminate; otherwise we claim that so far, there was no change, and proceed to time $t+1$ (if $t < K$) or terminate (if $t = K$).

The risk analysis for the resulting system of inferences is immediate. Observe that

(!) *For every* $t = 2, 3, ..., K$:

- *if there is no change on time horizon* $1, ..., t$: $x_1 = x_2 = ... = x_t$ *(case A) the probability for* $\mathcal{T}_t^\kappa$ *to conclude that there was a change is at most* κ^2/ε*;*
- *if, on the other hand,* $\|x_t - x_1\|_2^2 \geq \rho^2(\kappa)$ *(case B), then the probability for* $\mathcal{T}_t^\kappa$ *to conclude that so far there was no change is at most* ε*.*

Indeed, we clearly have

$$\zeta_t = [x_t - x_1] + \xi^t,$$

where $\xi^t = \xi_t - \xi_1 \sim \mathcal{N}(0, \sigma^2 I_d)$ with $\sigma \in [\sqrt{2}\underline{\sigma}, \sqrt{2}\overline{\sigma}]$. Our action at time t is nothing but application of the test $\mathcal{T}^\kappa$ to the observation ζ_t. In case A the distribution of this observation obeys the hypothesis G_1, and the probability for $\mathcal{T}_t^\kappa$ to claim that there was a change is at most κ^2/ε by the first inequality in (2.151). In case B, the distribution of ζ_t obeys the hypothesis $G_2^{\rho(\kappa)}$, and thus the probability for $\mathcal{T}_t^\kappa$ to claim that there was no change on time horizon $1, ..., t$ is $\leq \varepsilon$ by the second inequality in (2.151).

In view of (!), the probability of false alarm for the system of inferences $\{\mathcal{T}_t^\kappa\}_{t=2}^K$ is at most $(K-1)\kappa^2/\varepsilon$, and specifying κ as

$$\kappa = \sqrt{\epsilon\varepsilon/(K-1)},$$

we make this probability $\leq \epsilon$. The resulting procedure, by the same (!), detects a change at time $t \in \{2, 3, ..., K\}$ with probability at least $1 - \varepsilon$, provided that the energy of this change is at least ρ_*^2, with

$$\rho_* = \rho\left(\sqrt{\epsilon\varepsilon/(K-1)}\right). \tag{2.152}$$

In fact we can say a bit more:

Proposition 2.48. *Let the deterministic sequence* $x_1, ..., x_K$ *underlying observations* (2.149) *be such that for some* t *it holds* $\|x_t - x_1\|_2^2 \geq \rho_*^2$*, with* ρ_* *given by*

(2.152). *Then the probability for the system of inferences we have built to detect a change at time t or earlier is at least* $1-\varepsilon$.

Indeed, under the premise of the proposition, the probability for $\mathcal{T}_t^\kappa$ to claim that a change already took place is at least $1-\varepsilon$, and this probability can be only smaller than the probability to detect change on time horizon $2, 3, ..., t$.

How it works. As applied to the "movie" story we started with, the outlined procedure works as follows. The images in question are of the size 256×256, so that we are in the case of $d = 256^2 = 65536$. The images are represented by 2D arrays in gray scale, that is, as 256×256 matrices with entries in the range $[0, 255]$. In the experiment to be reported (just as in the movie) we assumed the maximal noise intensity $\overline{\sigma}$ to be 10, and used $\underline{\sigma} = \overline{\sigma}/\sqrt{2}$. The reliability tolerances ϵ, ε were set to 0.01, and K was set to 9, resulting in

$$\rho_*^2 = 7.38 \cdot 10^6,$$

which corresponds to the per pixel energy $\rho_*^2/65536 = 112.68$—just 12% above the allowed expected per pixel energy of noise (the latter is $\overline{\sigma}^2 = 100$). The resulting detector is

$$\phi_*(\zeta) = -2.7138\frac{\zeta^T\zeta}{10^5} + 366.9548.$$

In other words, test $\mathcal{T}_t^\kappa$ claims that the change took place when the average, over pixels, per pixel energy in the difference $\omega_t - \omega_1$ was at least 206.33, which is pretty close to the expected per pixel energy (200.0) in the noise $\xi_t - \xi_1$ affecting the difference $\omega_t - \omega_1$.

Finally, this is how the system of inferences just described worked in simulations. The underlying sequence of images is obtained from the "basic sequence"

$$\bar{x}_t = D + 0.0357(t-1)(L-D), t = 1, 2, ...^{28} \tag{2.153}$$

where D is the image of the dog and L is the image of the lady (up to noise, these are the first and the last frames on Figure 2.8). To get the observations in a particular simulation, we augment this sequence from the left by a random number of images D in such a way that with probability 1/2 there was no change of image on the time horizon 1,2, ..., 9, and with probability 1/2 there was a change at time instant τ chosen at random from the uniform distribution on $\{2, 3, ..., 9\}$. The observation is obtained by taking the first nine images in the resulting sequence, and adding to them observation noises independent across the images drawn at random from $\mathcal{N}(0, 100I_{65536})$.

In the series of 3,000 simulations of this type we have not observed a *single* false alarm, while the empirical probability of a miss was 0.0553. Besides, the change at time t, *if detected*, was *never* detected with a delay more than 1.

Finally, in the particular "movie" in Figure 2.8 the change takes place at time $t = 3$, and the system of inferences we have just developed discovered the change at time 4. How does this compare to the time when you managed to detect the change?

"Numerical near-optimality." Recall that beyond the realm of simple o.s.'s we

[28]The coefficient 0.0357 corresponds to a 28-frame linear transition from D to L.

have no theoretical guarantees of near-optimality for the inferences we are developing. This does not mean, however, that we cannot quantify the conservatism of our techniques numerically. To give an example, let us forget, for the sake of simplicity, about change detection per se and focus on the auxiliary problem we have introduced above, that of deciding upon hypotheses G_1 and G_2^ρ via observation (2.150), and suppose that we want to decide on these two hypotheses from a single observation with risk $\leq \epsilon$, for a given $\epsilon \in (0,1)$. Whether this is possible or not depends on ρ; let us denote by ρ^+ the smallest ρ for which we can meet the risk specification with our detector-based approach (ρ^+ is nothing but what was above called $\rho(\epsilon)$), and by $\underline{\rho}$ the smallest ρ for which there exists "in nature" a simple test deciding on G_1 vs. G_2^ρ with risk $\leq \epsilon$. We can consider the ratio $\rho^+/\underline{\rho}$ as the "index of conservatism" of our approach. Now, ρ^+ is given by an efficient computation; what about $\underline{\rho}$? Well, there is a simple way to get a *lower bound* on $\underline{\rho}$, namely, as follows. Observe that if the composite hypotheses G_1, G_2^ρ can be decided upon with risk $\leq \epsilon$, the same holds true for two simple hypotheses stating that the distribution of observation (2.150) is P_1 or P_2 respectively, where P_1, P_2 correspond to the cases where

- (P_1): ζ is drawn from $\mathcal{N}(0, 2\overline{\sigma}^2 I_d)$
- (P_2): ζ is obtained by adding $\mathcal{N}(0, 2\underline{\sigma}^2 I_d)$-noise to a random signal u, independent of the noise, uniformly distributed on the sphere $\{\|u\|_2 = \rho\}$.

Indeed, P_1 obeys hypothesis G_1, and P_2 is a mixture of distributions obeying G_2^ρ; as a result, a simple test $\mathcal{T}$ deciding $(1-\epsilon)$-reliably on G_1 vs. G_2^ρ would induce a test deciding equally reliably on P_1 vs. P_2, specifically, the test which, given observation ζ, accepts P_1 if $\mathcal{T}$ on the same observation accepts G_1, and accepts P_2 otherwise.

We can now use a two-point lower bound (Proposition 2.2) to lower-bound the risk of deciding on P_1 vs. P_2. Because both distributions are spherically symmetric, computing this bound reduces to computing a similar bound for the univariate distributions of $\zeta^T\zeta$ induced by P_1 and P_2, and these univariate distributions are easy to compute. The resulting lower risk bound depends on ρ, and we can find the smallest ρ for which the bound is ≥ 0.01, and use this ρ in the role of $\underline{\rho}$; the associated indexes of conservatism can be only larger than the true ones. Let us look at what these indexes are for the data used in our change detection experiment, that is, $\epsilon = 0.01$, $d = 256^2 = 65536$, $\overline{\sigma} = 10$, $\underline{\sigma} = \overline{\sigma}/\sqrt{2}$. Computation shows that in this case we have

$$\rho^+ = 2702.4, \quad \rho^+/\underline{\rho} \leq 1.04$$

—nearly no conservatism at all! When eliminating the uncertainty in the intensity of noise by increasing $\underline{\sigma}$ from $\overline{\sigma}/\sqrt{2}$ to $\overline{\sigma}$, we get

$$\rho^+ = 668.46, \quad \rho^+/\underline{\rho} \leq 1.15$$

—still not that much of conservatism!

2.10 EXERCISES FOR CHAPTER 2

2.10.1 Two-point lower risk bound

Exercise 2.1.

Let p and q be two probability distributions distinct from each other on d-element observation space $\Omega = \{1, ..., d\}$, and consider two simple hypotheses on the distribution of observation $\omega \in \Omega$, $H_1 : \omega \sim p$, and $H_2 : \omega \sim q$.

1. Is it true that there always exists a simple deterministic test deciding on H_1, H_2 with risk $< 1/2$?
2. Is it true that there always exists a simple randomized test deciding on H_1, H_2 with risk $< 1/2$?
3. Is it true that when quasi-stationary K-repeated observations are allowed, one can decide on H_1, H_2 with any small risk, provided K is large enough?

2.10.2 Around Euclidean Separation

Exercise 2.2.

Justify the "immediate observation" in Section 2.2.2.3.B.

Exercise 2.3.

1) Prove Proposition 2.9.
Hint: You can find useful the following simple observation (prove it, provided you indeed use it):

Let $f(\omega)$, $g(\omega)$ be probability densities taken w.r.t. a reference measure P on an observation space Ω, and let $\epsilon \in (0, 1/2]$ be such that

$$2\bar{\epsilon} := \int_\Omega \min[f(\omega), g(\omega)] P(d\omega) \leq 2\epsilon.$$

Then

$$\int_\Omega \sqrt{f(\omega)g(\omega)} P(d\omega) \leq 2\sqrt{\epsilon(1-\epsilon)}.$$

2) Justify the illustration in Section 2.2.3.2.C.

2.10.3 Hypothesis testing via ℓ_1-separation

Let d be a positive integer, and the observation space Ω be the finite set $\{1, ..., d\}$ equipped with the counting reference measure.[29] Probability distributions on Ω can be identified with points p of d-dimensional *probabilistic simplex*

$$\mathbf{\Delta}_d = \{p \in \mathbf{R}^d : p \geq 0, \sum_i p_i = 1\};$$

[29]Counting measure is the measure on a discrete (finite or countable) set Ω which assigns every point of Ω with mass 1, so that the measure of a subset of Ω is the cardinality of the subset when it is finite and is $+\infty$ otherwise.

the i-th entry p_i in $p \in \mathbf{\Delta}_d$ is the probability for the random variable distributed according to p to take value $i \in \{1, ..., d\}$. With this interpretation, p is the probability density taken w.r.t. the counting measure on Ω.

Assume B and W are two nonintersecting nonempty closed convex subsets of $\mathbf{\Delta}_d$; we interpret B and W as the sets of black and white probability distributions on Ω, and our goal is to find the optimal, in terms of its total risk, test deciding on the hypotheses

$$H_1 : p \in B,\ H_2 : p \in W$$

via a single observation $\omega \sim p$.

Warning: Everywhere in this section, "test" means "simple test."

Exercise 2.4.

Our first goal is to find the optimal test, in terms of its total risk, deciding on the hypotheses H_1, H_2 via a *single* observation $\omega \sim p \in B \cup W$. To this end we consider the convex optimization problem

$$\mathrm{Opt} = \min_{p \in B, q \in W} \left[f(p,q) := \sum_{i=1}^{d} |p_i - q_i| \right] \tag{2.154}$$

and let (p^*, q^*) be an optimal solution to this problem (it clearly exists).

1. Extract from optimality conditions that there exist reals $\rho_i \in [-1, 1]$, $1 \leq i \leq n$, such that
$$\rho_i = \begin{cases} 1, & p_i^* > q_i^* \\ -1, & p_i^* < q_i^* \end{cases} \tag{2.155}$$
and
$$\rho^T(p - p^*) \geq 0 \,\forall p \in B \ \&\ \rho^T(q - q^*) \leq 0 \ \forall q \in W. \tag{2.156}$$

2. Extract from the previous item that the test $\mathcal{T}$ which, given an observation $\omega \in \{1, ..., d\}$, accepts H_1 with probability $\pi_\omega = (1 + \rho_\omega)/2$ and accepts H_2 with complementary probability, has its total risk equal to
$$\sum_{\omega \in \Omega} \min[p_\omega^*, q_\omega^*], \tag{2.157}$$
and thus is minimax optimal in terms of the total risk.

Comments. Exercise 2.4 describes an efficiently computable and, *in terms of worst-case total risk, optimal* simple test deciding on a pair of "convex" composite hypotheses on the distribution of a discrete random variable. While it seems an attractive result, we believe *by itself* this result is useless, since typically in the testing problem in question a *single* observation by far is not enough for a reasonable inference; such an inference requires observing *several* independent realizations $\omega_1, ..., \omega_K$ of the random variable in question. And the construction presented in Exercise 2.4 says nothing on how to adjust the test to the case of repeated observation. Of course, when $\omega^K = (\omega_1, ..., \omega_K)$ is a K-element i.i.d. sample drawn from a probability distribution p on $\Omega = \{1, ..., d\}$, ω^K can be thought of as a single observation of a discrete random variable taking value in the set $\Omega^K = \underbrace{\Omega \times ... \times \Omega}_{K}$, the probability distribution p^K of ω^K being readily given by p. So, why not to apply the construction from Exercise 2.4 to ω^K in the role of ω? On a close inspection, this idea fails. One of the reasons for this failure is that the cardinality of Ω^K (which, among other factors, is responsible for the computational complexity of implementing the test in Exercise 2.4) blows up exponentially as K grows. Another, even more serious, complication is that p^K depends on p nonlinearly, so that the family of distributions p^K of ω^K induced by a convex family of distributions p of ω—convexity meaning that the p's in question fill a *convex* subset of the probabilistic simplex—is not convex; and convexity of the sets B, W in the context of Exercise 2.4 is crucial. Thus, passing from a single realization of discrete random variable to the sample of $K > 1$ independent realizations of the variable results in severe structural and quantitative complications "killing," at least at first glance, the approach undertaken in Exercise 2.4.[30]

In spite of the above pessimistic conclusions, the single-observation test from Exercise 2.4 admits a meaningful multi-observation modification, which is the subject of our next exercise.

Exercise 2.5.

There is a straightforward way to use the optimal–in terms of its total risk—single-observation test built in Exercise 2.4 in the "multi-observation" environment. Specifically, following the notation from the exercise 2.4, let $\rho \in \mathbf{R}^d, p^*, q^*$ be the entities built in this Exercise, so that $p^* \in B$, $q^* \in W$, all entries in ρ belong to $[-1, 1]$, and

$$\begin{array}{c}\{\rho^T p \geq \alpha := \rho^T p^* \;\; \forall p \in B\} \;\&\; \{\rho^T q \leq \beta := \rho^T q^* \, \forall q \in W\} \\ \&\; \alpha - \beta = \rho^T[p^* - q^*] = \|p^* - q^*\|_1.\end{array}$$

Given an i.i.d. sample $\omega^K = (\omega_1, ..., \omega_K)$ with $\omega_t \sim p$, where $p \in B \cup W$, we could try to decide on the hypotheses $H_1 : p \in B$, $H_2 : p \in W$ as follows. Let us set $\zeta_t = \rho_{\omega_t}$. For large K, given ω^K, the observable quantity $\zeta^K := \frac{1}{K}\sum_{t=1}^K \zeta_t$, by the Law of Large Numbers, will be with overwhelming probability close to $\mathbf{E}_{\omega\sim p}\{\rho_\omega\} = \rho^T p$, and the latter quantity is $\geq \alpha$ when $p \in B$ and is $\leq \beta < \alpha$ when $p \in W$. Consequently, selecting a "comparison level" $\ell \in (\beta, \alpha)$, we can decide on the hypotheses $p \in B$ vs. $p \in W$ by computing ζ^K, comparing the result to ℓ, accepting the hypothesis $p \in B$ when $\zeta^K \geq \ell$, and accepting the alternative $p \in W$ otherwise. The goal of this exercise is to quantify the above qualitative considerations. To this end let us fix $\ell \in (\beta, \alpha)$ and K and ask ourselves the following questions:

A. For $p \in B$, how do we upper-bound the probability $\mathrm{Prob}_{p_K}\{\zeta^K \leq \ell\}$?
B. For $p \in W$, how do we upper-bound the probability $\mathrm{Prob}_{p_K}\{\zeta^K \geq \ell\}$?

Here p_K is the probability distribution of the i.i.d. sample $\omega^K = (\omega_1, ..., \omega_K)$ with $\omega_t \sim p$.

The simplest way to answer these questions is to use Bernstein's bounding scheme. Specifically, to answer question A, let us select $\gamma \geq 0$ and observe that for

[30]Though directly extending the optimal single-observation test to the case of repeated observations encounters significant technical difficulties, it was carried on in some specific situations. For instance, in [122, 123] such an extension has been proposed for the case of sets B and W of distributions which are dominated by bi-alternating capacities (see, e.g., [8, 12, 35], and references therein); explicit constructions of the test were proposed for some special sets of distributions [121, 196, 209].

every probability distribution p on $\{1, 2, ..., d\}$ it holds

$$\underbrace{\mathrm{Prob}_{p^K}\left\{\zeta^K \le \ell\right\}}_{\pi_{K,-}[p]} \exp\{-\gamma\ell\} \le \mathbf{E}_{p^K}\left\{\exp\{-\gamma\zeta^K\}\right\} = \left[\sum_{i=1}^d p_i \exp\left\{-\frac{1}{K}\gamma\rho_i\right\}\right]^K,$$

whence

$$\ln(\pi_{K,-}[p]) \le K \ln\left(\sum_{i=1}^d p_i \exp\left\{-\frac{1}{K}\gamma\rho_i\right\}\right) + \gamma\ell,$$

implying, via substitution $\gamma = \mu K$, that

$$\forall \mu \ge 0 : \ln(\pi_{K,-}[p]) \le K\psi_-(\mu, p),\ \ \psi_-(\mu, p) = \ln\left(\sum_{i=1}^d p_i \exp\{-\mu\rho_i\}\right) + \mu\ell.$$

Similarly, setting $\pi_{K,+}[p] = \mathrm{Prob}_{p^K}\left\{\zeta^K \ge \ell\right\}$, we get

$$\forall \nu \ge 0 : \ln(\pi_{K,+}[p]) \le K\psi_+(\nu, p),\ \ \psi_+(\nu, p) = \ln\left(\sum_{i=1}^d p_i \exp\{\nu\rho_i\}\right) - \nu\ell.$$

Now comes the exercise:

1. Extract from the above observations that

$$\mathrm{Risk}(\mathcal{T}^{K,\ell}|H_1, H_2) \le \exp\{K\varkappa\},\ \ \varkappa = \max\left[\max_{p\in B}\inf_{\mu\ge 0}\psi_-(\mu, p), \max_{q\in W}\inf_{\nu\ge 0}\psi_+(\nu, q)\right],$$

   where $\mathcal{T}^{K,\ell}$ is the K-observation test which accepts the hypothesis $H_1 : p \in B$ when $\zeta^K \ge \ell$ and accepts the hypothesis $H_2 : p \in W$ otherwise.
2. Verify that $\psi_-(\mu, p)$ is convex in μ and concave in p, and similarly for $\psi_+(\nu, q)$, so that

$$\max_{p\in B}\inf_{\mu\ge 0}\psi_-(\mu, p) = \inf_{\mu\ge 0}\max_{p\in B}\psi_-(\mu, p),\ \ \max_{q\in W}\inf_{\nu\ge 0}\psi_+(\nu, q) = \inf_{\nu\ge 0}\max_{q\in W}\psi_+(\nu, q).$$

   Thus, computing $\varkappa$ reduces to minimizing on the nonnegative ray the convex functions $\phi_-(\mu) = \max_{p\in B}\psi_+(\mu, p)$ and $\phi_+(\nu) = \max_{q\in W}\psi_+(\nu, q)$.
3. Prove that when $\ell = \frac{1}{2}[\alpha + \beta]$, one has

$$\varkappa \le -\frac{1}{12}\Delta^2,\ \ \Delta = \alpha - \beta = \|p^* - q^*\|_1.$$

Note that the above test and the quantity $\varkappa$ responsible for the upper bound on its risk depend, as on a parameter, on the "acceptance level" $\ell \in (\beta, \alpha)$. The simplest way to select a reasonable value of ℓ is to minimize $\varkappa$ over an equidistant grid $\Gamma \subset (\beta, \alpha)$, of small cardinality, of values of ℓ.

Now, let us consider an alternative way to pass from a "good" single-observation test to its multi-observation version. Our "building block" now is the minimum risk randomized single-observation test[31] and its multi-observation modification is just

[31] This test can differ from the test built in Exercise 2.4—the latter test is optimal in terms of the sum, rather than the maximum, of its partial risks.

the majority version of this building block. Our first observation is that building the minimum risk single-observation test reduces to solving a *convex* optimization problem.

Exercise 2.6.

Let, as above, B and W be nonempty nonintersecting closed convex subsets of probabilistic simplex $\mathbf{\Delta}_d$. Show that the problem of finding the best—in terms of its risk—randomized single-observation test deciding on $H_1 : p \in B$ vs. $H_2 : p \in W$ via observation $\omega \sim p$ reduces to solving a convex optimization problem. Write down this problem as an explicit LO program when B and W are polyhedral sets given by polyhedral representations:

$$\begin{array}{rcl} B &=& \{p : \exists u : P_B p + Q_B u \leq a_B\}, \\ W &=& \{p : \exists u : P_W p + Q_W u \leq a_W\}. \end{array}$$

We see that the "ideal building block"—the minimum-risk single-observation test—can be built efficiently. What is at this point unclear is whether this block is of any use for majority modifications, that is, whether the risk of this test $< 1/2$—this is what we need for the majority version of the minimum-risk single-observation test to be consistent.

Exercise 2.7.

Extract from Exercise 2.4 that in the situation of this section, denoting by Δ the optimal value in the optimization problem (2.154), one has

1. The risk of any single-observation test, deterministic or randomized, is $\geq \frac{1}{2} - \frac{\Delta}{4}$
2. There exists a single-observation randomized test with risk $\leq \frac{1}{2} - \frac{\Delta}{8}$, and thus the risk of the minimum risk single-observation test given by Exercise 2.6 does not exceed $\frac{1}{2} - \frac{\Delta}{8} < 1/2$ as well.

Pay attention to the fact that $\Delta > 0$ (since, by assumption, B and W do not intersect).

The bottom line is that in the situation of this section, given a target value ϵ of risk and assuming stationary repeated observations are allowed, we have (at least) three options to meet the risk specifications:

1. To start with the optimal—in terms of its total risk—single-observation detector as explained in Exercise 2.4, and then to pass to its multi-observation version built in Exercise 2.5;
2. To use the majority version of the minimum-risk randomized single-observation test built in Exercise 2.6;
3. To use the test based on the minimum risk detector for B, W, as explained in the main body of Chapter 2.

In all cases, we have to specify the number K of observations which guarantees that the risk of the resulting multi-observation test is at most a given target ϵ. A bound on K can be easily obtained by utilizing the results on the risk of a detector-based test in a Discrete o.s. from the main body of Chapter 2 along with risk-related results of Exercises 2.5, 2.6, and 2.7.

Exercise 2.8.

Run numerical experiments to see if one of the three options above always dominates the others (that is, requires a smaller sample of observations to ensure the same risk).

Let us now focus on a theoretical comparison of the detector-based test and the majority version of the minimum-risk single-observation test (options 1 and 2 above) in the general situation described at the beginning of Section 2.10.3. Given $\epsilon \in (0,1)$, the corresponding sample sizes K_d and K_m are completely determined by the relevant "measure of closeness" between B and W. Specifically,

- For K_d, the closeness measure is

$$\rho_d(B,W) = 1 - \max_{p\in B, q\in W} \sum_\omega \sqrt{p_\omega q_\omega}; \tag{2.158}$$

  $1-\rho_d(B,W)$ is the minimal risk of a detector for B, W, and for $\rho_d(B,W)$ and ϵ small, we have $K_d \approx \ln(1/\epsilon)/\rho_d(B,W)$ (why?).
- Given ϵ, K_m is fully specified by the minimal risk ρ of simple randomized single-observation test $\mathcal{T}$ deciding on the hypotheses associated with B, W. By Exercise 2.7, we have $\rho = \frac{1}{2} - \delta$, where δ is within absolute constant factor of the optimal value $\Delta = \min_{p\in B, q\in W}\|p-q\|_1$ of (2.154). The risk bound for the K-observation majority version of $\mathcal{T}$ is the probability to get at least $K/2$ heads in K independent tosses of coin with probability to get heads in a single toss equal to $\rho = 1/2 - \delta$. When ρ is not close to 0 and ϵ is small, the $(1-\epsilon)$-quantile of the number of heads in our K coin tosses is $K\rho + O(1)\sqrt{K\ln(1/\epsilon)} = K/2 - \delta K + O(1)\sqrt{K\ln(1/\epsilon)}$ (why?). K_m is the smallest K for which this quantile is $< K/2$, so that K_m is of the order of $\ln(1/\epsilon)/\delta^2$, or, which is the same, of the order of $\ln(1/\epsilon)/\Delta^2$. We see that the closeness between B and W "responsible for K_m" is

$$\rho_m(B,W) = \Delta^2 = \left[\min_{p\in B, q\in W}\|p-q\|_1\right]^2,$$

  and K_m is of the order of $\ln(1/\epsilon)/\rho_m(B,W)$.

The goal of the next exercise is to compare ρ_b and ρ_m.

Exercise 2.9.

Prove that in the situation of this section one has

$$\tfrac{1}{8}\rho_m(B,W) \le \rho_d(B,W) \le \tfrac{1}{2}\sqrt{\rho_m(B,W)}. \tag{2.159}$$

Relation (2.159) suggests that while K_d never is "much larger" than K_m (this we know in advance: in repeated versions of Discrete o.s., a properly built detector-based test provably is nearly optimal), K_m might be much larger than K_d. This indeed is the case:

Exercise 2.10.

Given $\delta \in (0,1/2)$, let $B = \{[\delta; 0; 1-\delta]\}$ and $W = \{[0;\delta;1-\delta]\}$. Verify that in this case the numbers of observations K_d and K_m, resulting in a given risk $\epsilon \ll 1$ of multi-observation tests, as functions of δ are proportional to $1/\delta$ and $1/\delta^2$, respectively. Compare the numbers when $\epsilon = 0.01$ and $\delta \in \{0.01; 0.05; 0.1\}$.

2.10.4 Miscellaneous exercises

Exercise 2.11.

Prove that the conclusion in Proposition 2.18 remains true when the test $\mathcal{T}$ in the premise of the proposition is randomized.

Exercise 2.12.

Let $p_1(\omega), p_2(\omega)$ be two positive probability densities, taken w.r.t. a reference measure Π on an observation space Ω, and let $\mathcal{P}_\chi = \{p_\chi\}$, $\chi = 1, 2$. Find the optimal—in terms of its risk—balanced detector for $\mathcal{P}_\chi$, $\chi = 1, 2$.

Exercise 2.13.

Recall that the exponential distribution on $\Omega = \mathbf{R}_+$, with parameter $\mu > 0$, is the distribution with the density $p_\mu(\omega) = \mu e^{-\mu\omega}$, $\omega \geq 0$. Given positive reals $\alpha < \beta$, consider two families of exponential distributions, $\mathcal{P}_1 = \{p_\mu : 0 < \mu \leq \alpha\}$, and $\mathcal{P}_2 = \{p_\mu : \mu \geq \beta\}$. Build the optimal—in terms of its risk—balanced detector for $\mathcal{P}_1$, $\mathcal{P}_2$. What happens with the risk of the detector you have built when the families $\mathcal{P}_\chi$, $\chi = 1, 2$, are replaced with their convex hulls?

Exercise 2.14.

[Follow-up to Exercise 2.13] Assume that the "lifetime" ζ of a lightbulb is a realization of random variable with exponential distribution (i.e., the density $p_\mu(\zeta) = \mu e^{-\mu\zeta}$, $\zeta \geq 0$; in particular, the expected lifespan of a lightbulb in this model is $1/\mu$).[32] Given a lot of lightbulbs, you should decide whether they were produced under normal conditions (resulting in $\mu \leq \alpha = 1$) or under abnormal ones (resulting in $\mu \geq \beta = 1.5$). To this end, you can select at random K lightbulbs and test them. How many lightbulbs should you test in order to make a 0.99-reliable conclusion? Answer this question in the situations when the observation ω in a test is

1. the lifespan of a lightbulb (i.e., $\omega \sim p_\mu(\cdot)$);
2. the minimum $\omega = \min[\zeta, \delta]$ of the lifespan $\zeta \sim p_\mu(\cdot)$ and the allowed duration $\delta > 0$ of your test (i.e., if the lightbulb you are testing does not "die" on time horizon δ, you terminate the test);
3. $\omega = \chi_{\zeta<\delta}$, that is, $\omega = 1$ when $\zeta < \delta$, and $\omega = 0$ otherwise; here, as above, $\zeta \sim p_\mu(\cdot)$ is the random lifespan of a lightbulb, and $\delta > 0$ is the allowed test duration (i.e., you observe whether or not a lightbulb "dies" on time horizon δ, but do not register the lifespan when it is $< \delta$).

Consider the values $0.25, 0.5, 1, 2, 4$ of δ.

Exercise 2.15.

[32] In Reliability, probability distribution of the lifespan ζ of an organism or a technical device is characterized by the *failure rate* $\lambda(t) = \lim_{\delta\to+0} \frac{\mathrm{Prob}\{t\leq\zeta\leq t+\delta\}}{\delta\cdot\mathrm{Prob}\{\zeta\geq t\}}$ (so that for small δ, $\lambda(t)\delta$ is the conditional probability to "die" in the time interval $[t, t+\delta]$ provided the organism or device is still "alive" at time t). The exponential distribution corresponds to the case of failure rate independent of t; in applications, this indeed is often the case except for "very small" and "very large" values of t.

[Follow-up to Exercise 2.14] In the situation of Exercise 2.14, build a sequential test for deciding on null hypothesis "the lifespan of a lightbulb from a given lot is $\zeta \sim p_\mu(\cdot)$ with $\mu \leq 1$" (recall that $p_\mu(z)$ is the exponential density $\mu e^{-\mu z}$ on the ray $\{z \geq 0\}$) vs. the alternative "the lifespan is $\zeta \sim p_\mu(\cdot)$ with $\mu > 1$." In this test, you can select a number K of lightbulbs from the lot, switch them on at time 0 and record the actual lifetimes of the lightbulbs you are testing. As a result at the end of (any) observation interval $\Delta = [0, \delta]$, you observe K independent realizations of r.v. $\min[\zeta, \delta]$, where $\zeta \sim p_\mu(\cdot)$ with some unknown μ. In your sequential test, you are welcome to make conclusions at the endpoints $\delta_1 < \delta_2 < ... < \delta_S$ of several observation intervals.

Note: We deliberately skip details of the problem's setting; how you decide on these missing details is part of your solution to the exercise.

Exercise 2.16.

In Section 2.6, we consider a model of elections where every member of the population was supposed to cast a vote. Enrich the model by incorporating the option for a voter not to participate in the elections at all. Implement Sequential test for the resulting model and run simulations.

Exercise 2.17.

Work out the following extension of the Opinion Poll Design problem. You are given two finite sets, $\Omega_1 = \{1, ..., I\}$ and $\Omega_2 = \{1, ..., M\}$, along with L nonempty closed convex subsets Y_ℓ of the set

$$\mathbf{\Delta}_{IM} = \left\{ [y_{im} > 0]_{i,m} : \sum_{i=1}^{I} \sum_{m=1}^{M} y_{im} = 1 \right\}$$

of all nonvanishing probability distributions on $\Omega = \Omega_1 \times \Omega_2 = \{(i, m) : 1 \leq i \leq I, 1 \leq m \leq M\}$. Sets Y_ℓ are such that all distributions from Y_ℓ have a common marginal distribution $\theta^\ell > 0$ of i:

$$\sum_{m=1}^{M} y_{im} = \theta^\ell_i,\ 1 \leq i \leq I,\ \forall y \in Y_\ell,\ 1 \leq \ell \leq L.$$

Your observations $\omega_1, \omega_2, ...$ are sampled, independently of each other, from a distribution partly selected "by nature," and partly by you. Specifically, nature selects $\ell \leq L$ and a distribution $y \in Y_\ell$, and you select a positive an I-dimensional probabilistic vector q from a given convex compact subset $\mathcal{Q}$ of the positive part of I-dimensional probabilistic simplex. Let $y_{|i}$ be the conditional distribution of $m \in \Omega_2$ given i induced by y, so that $y_{|i}$ is the M-dimensional probabilistic vector with entries

$$[y_{|i}]_m = \frac{y_{im}}{\sum_{\mu \leq M} y_{i\mu}} = \frac{y_{im}}{\theta^\ell_i}.$$

In order to generate $\omega_t = (i_t, m_t) \in \Omega$, you draw i_t at random from the distribution q, and then nature draws m_t at random from the distribution $y_{|i_t}$.

Given closeness relation $\mathcal{C}$, your goal is to decide, up to closeness $\mathcal{C}$, on the hypotheses $H_1, ..., H_L$, with H_ℓ stating that the distribution y selected by nature belongs to Y_ℓ. Given an "observation budget" (a number K of observations ω_k you can use), you want to find a probabilistic vector q which results in the test with as small a $\mathcal{C}$-risk as possible. Pose this Measurement Design problem as an efficiently solvable convex optimization problem.

Exercise 2.18.

[Probabilities of deviations from the mean] The goal of what follows is to present the most straightforward application of simple families of distributions—bounds on probabilities of deviations of random vectors from their means. Let $\mathcal{H} \subset \Omega = \mathbf{R}^d, \mathcal{M}, \Phi$ be regular data such that $0 \in \operatorname{int} \mathcal{H}$, $\mathcal{M}$ is compact, $\Phi(0;\mu) = 0 \,\forall \mu \in \mathcal{M}$, and $\Phi(h;\mu)$ is differentiable at $h = 0$ for every $\mu \in \mathcal{M}$. Let, further, $\bar{P} \in \mathcal{S}[\mathcal{H}, \mathcal{M}, \Phi]$ and let $\bar{\mu} \in \mathcal{M}$ be a parameter of $\bar{P}$. Prove that

1. $\bar{P}$ possesses expectation $e[\bar{P}]$, and
$$e[\bar{P}] = \nabla_h \Phi(0;\bar{\mu})$$
2. For every linear form $e^T\omega$ on Ω it holds
$$\begin{array}{rcl} \pi & := & \bar{P}\{\omega : e^T(\omega - e[\bar{P}]) \geq 1\} \\ & \leq & \exp\left\{ \inf\limits_{t \geq 0 : te \in \mathcal{H}} \left[\Phi(te;\bar{\mu}) - te^T \nabla_h \Phi(0;\bar{\mu}) - t \right] \right\}. \end{array} \tag{2.160}$$

What are the consequences of (2.160) for sub-Gaussian distributions?

Exercise 2.19.

[testing convex hypotheses on mixtures] Consider the situation as follows. For given positive integers K and L and for $\chi = 1, 2$, given are

- nonempty convex compact *signal sets* $U_\chi \subset \mathbf{R}^{n_\chi}$,
- regular data $\mathcal{H}^\chi_{k\ell} \subset \mathbf{R}^{d_k}, \mathcal{M}^\chi_{k\ell}, \Phi^\chi_{k\ell}$, and affine mappings
$$u_\chi \mapsto A^\chi_{k\ell}[u_\chi; 1] : \mathbf{R}^{n_\chi} \to \mathbf{R}^{d_k}$$
such that
$$u_\chi \in U_\chi \Rightarrow A^\chi_{k\ell}[u_\chi; 1] \in \mathcal{M}^\chi_{k\ell},$$
$1 \leq k \leq K$, $1 \leq \ell \leq L$,
- probabilistic vectors $\mu^k = [\mu^k_1; ...; \mu^k_L]$, $1 \leq k \leq K$.

We can associate with the outlined data families of probability distributions $\mathcal{P}_\chi$ on the observation space $\Omega = \mathbf{R}^{d_1} \times ... \times \mathbf{R}^{d_K}$ as follows. For $\chi = 1, 2$, $\mathcal{P}_\chi$ is comprised of all probability distributions P of random vectors $\omega^K = [\omega_1; ...; \omega_K] \in \Omega$ generated as follows:
We select

- a signal $u_\chi \in U_\chi$,
- a collection of probability distributions $P_{k\ell} \in \mathcal{S}[\mathcal{H}^\chi_{k\ell}, \mathcal{M}^\chi_{k\ell}, \Phi^\chi_{k\ell}]$, $1 \leq k \leq K$, $1 \leq \ell \leq L$, in such a way that $A^\chi_{k\ell}[u_\chi; 1]$ is a parameter of $P_{k\ell}$:
$$\forall h \in \mathcal{H}^\chi_{k\ell} : \ln\left(\mathbf{E}_{\omega_k \sim P_{k\ell}} \{ e^{h^T \omega_k} \} \right) \leq \Phi^\chi_{k\ell}(h_k; A^\chi_{k\ell}[u_\chi; 1]);$$
- we generate the components ω_k, $k = 1, ..., K$, *independently across* k, from μ^k-mixture $\Pi[\{P_{k\ell}\}_{\ell=1}^L, \mu]$ of distributions $P_{k\ell}$, $\ell = 1, ..., L$, that is, draw at random,

from distribution μ^k on $\{1,...,L\}$, index ℓ, and then draw ω_k from the distribution $P_{k\ell}$.

Prove that when setting

$$\begin{array}{rcl}\mathcal{H}_\chi & = & \{h=[h_1;...;h_K]\in\mathbf{R}^{d=d_1+...+d_K}: h_k\in\bigcap\limits_{\ell=1}^L \mathcal{H}^\chi_{k\ell}, 1\le k\le K\},\\ \mathcal{M}_\chi & = & \{0\}\subset\mathbf{R},\\ \Phi_\chi(h;\mu) & = & \sum_{k=1}^K\ln\left(\sum_{\ell=1}^L\mu_\ell^k\exp\left\{\max\limits_{u_\chi\in U_\chi}\Phi^\chi_{k\ell}(h_k;A^\chi_{k\ell}[u_\chi;1])\right\}\right): \mathcal{H}_\chi\times\mathcal{M}_\chi\to\mathbf{R},\end{array}$$

we obtain the regular data such that

$$\mathcal{P}_\chi\subset\mathcal{S}[\mathcal{H}_\chi,\mathcal{M}_\chi,\Phi_\chi].$$

Explain how to use this observation to compute via Convex Programming an affine detector and its risk for the families of distributions $\mathcal{P}_1$ and $\mathcal{P}_2$.

Exercise 2.20.

[Mixture of sub-Gaussian distributions] Let P_ℓ be sub-Gaussian distributions on $\mathbf{R}^d$ with sub-Gaussianity parameters θ_ℓ,Θ, $1\le\ell\le L$, with a common Θ-parameter, and let $\nu=[\nu_1;...;\nu_L]$ be a probabilistic vector. Consider the ν-mixture $P=\Pi[P^L,\nu]$ of distributions P_ℓ, so that $\omega\sim P$ is generated as follows: we draw at random from distribution ν index ℓ and then draw ω at random from distribution P_ℓ. Prove that P is sub-Gaussian with sub-Gaussianity parameters $\bar\theta=\sum_\ell\nu_\ell\theta_\ell$ and $\bar\Theta$, with (any) $\bar\Theta$ chosen to satisfy

$$\bar\Theta\succeq\Theta+\frac{6}{5}[\theta_\ell-\bar\theta][\theta_\ell-\bar\theta]^T\,\forall\ell,$$

in particular, according to any one of the following rules:

1. $\bar\Theta=\Theta+\left(\frac{6}{5}\max_\ell\|\theta_\ell-\bar\theta\|_2^2\right)I_d$,
2. $\bar\Theta=\Theta+\frac{6}{5}\sum_\ell(\theta_\ell-\bar\theta)(\theta_\ell-\bar\theta)^T$,
3. $\bar\Theta=\Theta+\frac{6}{5}\sum_\ell\theta_\ell\theta_\ell^T$, provided that $\nu_1=...=\nu_L=1/L$.

Exercise 2.21.

The goal of this exercise is to give a simple sufficient condition for quadratic lifting "to work" in the Gaussian case. Namely, let $\mathcal{A}_\chi$, U_χ, $\mathcal{V}_\chi$, $\mathcal{G}_\chi$, $\chi=1,2$, be as in Section 2.9.3, with the only difference that now we do *not* assume the compact sets $\mathcal{U}_\chi$ to be convex, and let $\mathcal{Z}_\chi$ be convex compact subsets of the sets $\mathcal{Z}^{n_\chi}$—see item i.2. in Proposition 2.43—such that

$$[u_\chi;1][u_\chi;1]^T\in\mathcal{Z}_\chi\;\forall u_\chi\in U_\chi,\;\chi=1,2.$$

Augmenting the above data with $\Theta_\chi^{(*)}$, δ_χ such that $\mathcal{V}=\mathcal{V}_\chi$, $\Theta_*=\Theta_*^{(\chi)}$, $\delta=\delta_\chi$ satisfy (2.129), $\chi=1,2$, and invoking Proposition 2.43.ii, we get at our disposal a quadratic detector $\phi_{\rm lift}$ such that

$$\mathrm{Risk}[\phi_{\rm lift}|\mathcal{G}_1,\mathcal{G}_2]\le\exp\{\mathrm{SadVal}_{\rm lift}\},$$

with $\mathrm{SadVal}_{\rm lift}$ given by (2.134). A natural question is, when $\mathrm{SadVal}_{\rm lift}$ is negative, meaning that our quadratic detector indeed "is working"—its risk is <1, implying that when repeated observations are allowed, tests based upon this detector are consistent—able to decide on the hypotheses $H_\chi : P \in \mathcal{G}_\chi$, $\chi = 1,2$, on the distribution of observation $\zeta \sim P$ with any small desired risk $\epsilon \in (0,1)$. With our computation-oriented ideology, this is not too important a question, since we can answer it via efficient computation. This being said, there is no harm in a "theoretical" answer which could provide us with an additional insight. The goal of the exercise is to justify a simple result on the subject. Here is the exercise:

In the situation in question, assume that $\mathcal{V}_1 = \mathcal{V}_2 = \{\Theta_*\}$, which allows us to set $\Theta_*^{(\chi)} = \Theta_*$, $\delta_\chi = 0$, $\chi = 1,2$. Prove that in this case a necessary and sufficient condition for $\mathrm{SadVal}_{\mathrm{lift}}$ to be negative is that the convex compact sets

$$\mathcal{U}_\chi = \{B_\chi Z B_\chi^T : Z \in \mathcal{Z}_\chi\} \subset \mathbf{S}_+^{d+1}, \ \chi = 1,2$$

do not intersect with each other.

Exercise 2.22.

Prove that if X is a nonempty convex compact set in $\mathbf{R}^d$, then the function $\widehat{\Phi}(h;\mu)$ given by (2.114) is real-valued and continuous on $\mathbf{R}^d \times X$ and is convex in h and concave in μ.

Exercise 2.23.

The goal of what follows is to refine the change detection procedure (let us refer to it as the"basic" one) developed in Section 2.9.5.1. The idea is pretty simple. With the notation from Section 2.9.5.1, in the basic procedure, when testing the null hypothesis H_0 vs. signal hypothesis H_t^ρ, we look at the difference $\zeta_t = \omega_t - \omega_1$ and try to decide whether the energy of the deterministic component $x_t - x_1$ of ζ_t is 0, as is the case under H_0, or is $\geq \rho^2$, as is the case under H_t^ρ. Note that if $\sigma \in [\underline{\sigma}, \overline{\sigma}]$ is the actual intensity of the observation noise, then the noise component of ζ_t is $\mathcal{N}(0, 2\sigma^2 I_d)$; other things being equal, the larger is the noise in ζ_t, the larger should be ρ to allow for a reliable—with a given reliability level—decision. Now note that under the hypothesis H_t^ρ, we have $x_1 = ... = x_{t-1}$, so that the deterministic component of the difference $\zeta_t = \omega_t - \omega_1$ is exactly the same as for the difference $\widetilde{\zeta}_t = \omega_t - \frac{1}{t-1}\sum_{s=1}^{t-1}\omega_s$, while the noise component in $\widetilde{\zeta}_t$ is $\mathcal{N}(0, \sigma_t^2 I_d)$ with $\sigma_t^2 = \sigma^2 + \frac{1}{t-1}\sigma^2 = \frac{t}{t-1}\sigma^2$. Thus, the intensity of noise in $\widetilde{\zeta}_t$ is at most the same as in ζ_t, and this intensity, in contrast to that for ζ_t, decreases as t grows. Here comes the exercise:

Let reliability tolerances $\epsilon, \varepsilon \in (0,1)$ be given, and let our goal be to design a system of inferences $\mathcal{T}_t$, $t = 2,3,...,K$, which, when used in the same fashion as tests $\mathcal{T}_t^\kappa$ were used in the basic procedure, results in false alarm probability at most ϵ and in probability to miss a change of energy $\geq \rho^2$ at most ε. Needless to say, we want to achieve this goal with as small a ρ as possible. Think how to utilize the above observation to refine the basic procedure eventually reducing (and provably not increasing) the required value of ρ. Implement the basic and the refined change detection procedures and compare their quality (the resulting values of ρ), e.g., on the data used in the experiment reported in Section 2.9.5.1.

2.11 PROOFS

2.11.1 Proof of the observation in Remark 2.8

We have to prove that if $p = [p_1; ...; p_K] \in B = [0, 1]^K$ then the probability $P_M(p)$ of the event

> The total number of heads in K independent coin tosses, with probability p_k to get heads in k-th toss, is at least M

is a nondecreasing function of p: if $p' \le p''$, $p', p'' \in B$, then $P_M(p') \le P_M(p'')$. To see it, let us associate with $p \in B$ a subset of B, specifically, $B_p = \{x \in B : 0 \le x_k \le p_k, 1 \le k \le K\}$, and a function $\chi_p(x) : B \to \{0, 1\}$ which is equal to 0 at every point $x \in B$ where the number of entries x_k satisfying $x_k \le p_k$ is less than M, and is equal to 1 otherwise. It is immediately seen that

$$P_M(p) \equiv \int_B \chi_p(x)dx \tag{2.161}$$

(since with respect to the uniform distribution on B, the events $E_k = \{x \in B : x_k \le p_k\}$ are independent across k and have probabilities p_k, and the right-hand side in (2.161) is exactly the probability, taken w.r.t. the uniform distribution on B, of the event "at least M of the events E_1,..., E_K take place"). But the right-hand side in (2.31) clearly is nondecreasing in $p \in B$, since χ_p, by construction, is the characteristic function of the set

$$B[p] = \{x : \text{ at least } M \text{ of the entries } x_k \text{ in } x \text{ satisfy } x_k \le p_k\},$$

and these sets clearly grow when p increases entrywise. □

2.11.2 Proof of Proposition 2.6 in the case of quasi-stationary K-repeated observations

2.11.2.A Situation and goal. We are in the case **QS**—see Section 2.2.3.1—of the setting described at the beginning of Section 2.2.3. It suffices to verify that if $\mathcal{H}_\ell$, $\ell \in \{1, 2\}$, is true then the probability for $\mathcal{T}_K^{\text{maj}}$ to reject $\mathcal{H}_\ell$ is at most the quantity ϵ_K defined in (2.23). Let us verify this statement in the case of $\ell = 1$; the reasoning for $\ell = 2$ "mirrors" the one to follow.

It is clear that our situation and goal can be formulated as follows:

- "In nature" there exists a random sequence $\zeta^K = (\zeta_1, ..., \zeta_K)$ of driving factors and a collection of deterministic functions $\theta_k(\zeta^k = (\zeta_1, ..., \zeta_k))$[33] taking values in $\Omega = \mathbf{R}^d$ such that our k-th observation is $\omega_k = \theta_k(\zeta^k)$. Additionally, the conditional distribution $P_{\omega_k|\zeta^{k-1}}$ of ω_k given ζ^{k-1} always belongs to the family $\mathcal{P}_1$ comprised of distributions of random vectors of the form $x + \xi$, where deterministic x belongs to X_1 and the distribution of ξ belongs to $\mathcal{P}_\gamma^d$.
- There exist deterministic functions $\chi_k : \Omega \to \{0, 1\}$ and integer M, $1 \le M \le K$, such that the test $\mathcal{T}_K^{\text{maj}}$, as applied to observation $\omega^K = (\omega_1, ..., \omega_K)$, rejects $\mathcal{H}_1$

[33] As always, given a K-element sequence, say, $\zeta_1, ..., \zeta_K$, we write ζ^t, $t \le K$, as a shorthand for the fragment $\zeta_1, ..., \zeta_t$ of this sequence.

if and only if the number of 1's among the quantities $\chi_k(\omega_k)$, $1 \le k \le K$, is at least M.
In the situation of Proposition 2.6, $M = \rfloor K/2 \lfloor$ and $\chi_k(\cdot)$ are in fact independent of k: $\chi_k(\omega) = 1$ if and only if $\phi(\omega) \le 0$.[34]

- What we know is that the conditional probability of the event $\chi_k(\omega_k = \theta_k(\zeta^k)) = 1$, ζ^{k-1} being given, is at most $\epsilon_\star$:

$$P_{\omega_k|\zeta^{k-1}}\{\omega_k : \chi_k(\omega_k) = 1\} \le \epsilon_\star \, \forall \zeta^{k-1}.$$

Indeed, $P_{\omega_k|\zeta^{k-1}} \in \mathcal{P}_1$. As a result,

$$\begin{array}{rcl} P_{\omega_k|\zeta^{k-1}}\{\omega_k : \phi_k(\omega_k) = 1\} &=& P_{\omega_k|\zeta^{k-1}}\{\omega_k : \phi(\omega_k) \le 0\} \\ &=& P_{\omega_k|\zeta^{k-1}}\{\omega_k : \phi(\omega_k) < 0\} \le \epsilon_\star, \end{array}$$

where the second equality is due to the fact that $\phi(\omega)$ is a nonconstant affine function and $P_{\omega_k|\zeta^{k-1}}$, along with all distributions from $\mathcal{P}_1$, has density, and the inequality is given by the origin of $\epsilon_\star$ which upper-bounds the risk of the single-observation test underlying $\mathcal{T}_K^{\rm maj}$.

What we want to prove is that under the circumstances we have just summarized, we have

$$\begin{array}{l} P_{\omega^K}\{\omega^K = (\omega_1, ..., \omega_K) : \mathrm{Card}\{k \le K : \chi_k(\omega_k) = 1\} \ge M\} \\ \qquad \le \epsilon_M = \sum_{M \le k \le K} \binom{K}{k} \epsilon_\star^k (1 - \epsilon_\star)^{K-k}, \end{array} \tag{2.162}$$

where P_{ω^K} is the distribution of $\omega^K = \{\omega_k = \theta_k(\zeta^{k-1})\}_{k=1}^K$ induced by the distribution of hidden factors. There is nothing to prove when $\epsilon_\star = 1$, since in this case $\epsilon_M = 1$. Thus, we assume from now on that $\epsilon_\star < 1$.

2.11.2.B Achieving the goal, step 1. Our reasoning, inspired by that used to justify Remark 2.8, is as follows. Consider a sequence of random variables η_k, $1 \le k \le K$, uniformly distributed on $[0, 1]$ and independent of each other and of ζ^K, and consider new driving factors $\lambda_k = [\zeta_k; \eta_k]$ and new observations[35]

$$\mu_k = [\omega_k = \theta_k(\zeta^k); \eta_k] = \Theta_k(\lambda^k = (\lambda_1, ..., \lambda_k)) \tag{2.163}$$

driven by these new driving factors, and let

$$\psi_k(\mu_k = [\omega_k; \eta_k]) = \chi_k(\omega_k).$$

It is immediately seen that

- $\mu_k = [\omega_k = \theta_k(\zeta^k); \eta_k]$ is a deterministic function, $\Theta_k(\lambda^k)$, of λ^k, and the con-

[34]In fact, we need to write $\phi(\omega) < 0$ instead of $\phi(\omega) \le 0$; we replace the strict inequality with its nonstrict version in order to make our reasoning applicable to the case of $\ell = 2$, where nonstrict inequalities do arise. Clearly, replacing in the definition of χ_k strict inequality with the nonstrict one, we only increase the "rejection domain" of $\mathcal{H}_1$, so that the upper bound on the probability of this domain we are about to get automatically is valid for the true rejection domain.

[35]In this display, as in what follows, whenever some of the variables $\lambda, \omega, \zeta, \eta, \mu$ appear in the same context, it should always be understood that ζ_t and η_t are components of $\lambda_t = [\zeta_t; \eta_t]$, $\mu_t = [\omega_t; \eta_t] = \Theta_t(\lambda^t)$, and $\omega_t = \theta_t(\zeta^t)$. To remind us about these "hidden relations," we sometimes write something like $\phi(\omega_k = \theta_k(\zeta^k))$ to stress that we are speaking about the value of function ϕ at the point $\omega_k = \theta_k(\zeta^k)$.

ditional distribution $P_{\mu_k|\lambda^{k-1}}$ of μ_k given $\lambda^{k-1} = [\zeta^{k-1}; \eta^{k-1}]$ is the product distribution $P_{\omega_k|\zeta^{k-1}} \times U$ on $\Omega \times [0,1]$, where U is the uniform distribution on $[0,1]$. In particular,

$$\begin{array}{rcl}\pi_k(\lambda^{k-1}) & := & P_{\mu_k|\lambda^{k-1}}\{\mu_k = [\omega_k; \eta_k] : \chi_k(\omega_k) = 1\} \\ & = & P_{\omega_k|\zeta^{k-1}}\{\omega_k : \chi_k(\omega_k) = 1\} \leq \epsilon_\star.\end{array} \tag{2.164}$$

- We have

$$\begin{array}{l}P_{\lambda^K}\{\lambda^K : \mathrm{Card}\{k \leq K : \psi_k(\mu_k = \Theta_k(\lambda^k)) = 1\} \geq M\} \\ \quad = P_{\omega^K}\{\omega^K = (\omega_1, ..., \omega_K) : \mathrm{Card}\{k \leq K : \chi_k(\omega_k) = 1\} \geq M\}\end{array} \tag{2.165}$$

  where P_{ω^K} is as in (2.162), and $\Theta_k(\cdot)$ is defined in (2.163).

Now let us define $\psi_k^+(\lambda^k)$ as follows:

- when $\psi_k(\Theta_k(\lambda^k)) = 1$, or, which is the same, $\chi_k(\omega_k = \theta_k(\zeta^k)) = 1$, we set $\psi_k^+(\lambda^k) = 1$ as well;
- when $\psi_k(\Theta_k(\lambda^k)) = 0$, or, which is the same, $\chi_k(\omega_k = \theta_k(\zeta^k)) = 0$, we set $\psi_k^+(\lambda^k) = 1$ whenever

$$\eta_k \leq \gamma_k(\lambda^{k-1}) := \frac{\epsilon_\star - \pi_k(\lambda^{k-1})}{1 - \pi_k(\lambda^{k-1})}$$

  and $\psi_k^+(\lambda^k) = 0$ otherwise.

Let us make the following immediate observations:

(A) Whenever λ^k is such that $\psi_k(\mu_k = \Theta_k(\lambda^k)) = 1$, we also have $\psi_k^+(\lambda^k) = 1$;

(B) The conditional probability of the event

$$\psi_k^+(\lambda^k) = 1,$$

given $\lambda^{k-1} = [\zeta^{k-1}; \eta^{k-1}]$ is exactly $\epsilon_\star$.
Indeed, let $P_{\lambda_k|\lambda^{k-1}}$ be the conditional distribution of λ_k given λ^{k-1}. Let us fix λ^{k-1}. The event $E = \{\lambda_k : \psi_k^+(\lambda^k) = 1\}$, by construction, is the union of two nonoverlapping events:

$$\begin{array}{rcl}E_1 & = & \{\lambda_k = [\zeta_k; \eta_k] : \chi_k(\theta_k(\zeta^k)) = 1\}, \\ E_2 & = & \{\lambda_k = [\zeta_k; \eta_k] : \chi_k(\theta_k(\zeta^k)) = 0, \eta_k \leq \gamma_k(\lambda^{k-1})\}.\end{array}$$

Taking into account that the conditional distribution of $\mu_k = [\omega_k = \theta_k(\zeta^k); \eta_k]$, λ^{k-1} being fixed, is the product distribution $P_{\omega_k|\zeta^{k-1}} \times U$, we conclude in view of (2.164) that

$$\begin{array}{rcl}P_{\lambda_k|\lambda^{k-1}}\{E_1\} & = & P_{\omega_k|\zeta^{k-1}}\{\omega_k : \chi_k(\omega_k) = 1\} = \pi_k(\lambda^{k-1}), \\ P_{\lambda_k|\lambda^{k-1}}\{E_2\} & = & P_{\omega_k|\zeta^{k-1}}\{\omega_k : \chi_k(\omega_k) = 0\}U\{\eta \leq \gamma_k(\lambda^{k-1})\} \\ & = & (1 - \pi_k(\lambda^{k-1}))\gamma_k(\lambda^{k-1}),\end{array}$$

which combines with the definition of $\gamma_k(\cdot)$ to imply (B).

2.11.2.C Achieving the goal, step 2. By (A) combined with (2.165) we have

$$\begin{array}{l} P_{\omega^K}\{\omega^K : \mathrm{Card}\{k \le K : \chi_k(\omega_k) = 1\} \ge M\} \\ \quad = P_{\lambda^K}\{\lambda^K : \mathrm{Card}\{k \le K : \psi_k(\mu_k = \Theta_k(\lambda^k)) = 1\} \ge M\} \\ \quad \le P_{\lambda^K}\{\lambda^K : \mathrm{Card}\{k \le K : \psi_k^+(\lambda^k) = 1\} \ge M\}, \end{array}$$

and all we need to verify is that the first quantity in this chain is upper-bounded by the quantity ϵ_M given by (2.162). Invoking (B), it is enough to prove the following claim:

> (!) Let $\lambda^K = (\lambda_1, ..., \lambda_K)$ be a random sequence with probability distribution P, let $\psi_k(\lambda^k)$ take values 0 and 1 only, and let for every $k \le K$ the conditional probability for $\psi_k^+(\lambda^k)$ to take value 1, λ^{k-1} being fixed, be equal to $\epsilon_\star$, for all λ^{k-1}. Then the P-probability of the event
>
> $$\{\lambda^K : \mathrm{Card}\{k \le K : \psi_k^+(\lambda_k) = 1\} \ge M\}$$
>
> is equal to ϵ_M given by (2.162).

This is immediate. For integers k, m, $1 \le k \le K$, $m \ge 0$, let $\chi_m^k(\lambda^k)$ be the characteristic function of the event

$$\{\lambda^k : \mathrm{Card}\{t \le k : \psi_t^+(\lambda^t) = 1\} = m\},$$

and let

$$\pi_m^k = P\{\lambda^K : \chi_m^k(\lambda^k) = 1\}.$$

We have the following evident recurrence:

$$\chi_m^k(\lambda^k) = \chi_m^{k-1}(\lambda^{k-1})(1 - \psi_k^+(\lambda^k)) + \chi_{m-1}^{k-1}(\lambda^{k-1})\psi_k^+(\lambda^k),\ k = 1, 2, ...$$

augmented by the "boundary conditions" $\chi_m^0 = 0$, $m > 0$, $\chi_0^0 = 1$, $\chi_{-1}^{k-1} = 0$ for all $k \ge 1$. Taking expectation w.r.t. P and utilizing the fact that conditional expectation of $\psi_k^+(\lambda^k)$ given λ^{k-1} is, identically in λ^{k-1}, equal to $\epsilon_\star$, we get

$$\begin{array}{rcl} \pi_m^k &=& \pi_m^{k-1}(1 - \epsilon_\star) + \pi_{m-1}^{k-1}\epsilon_\star,\ k = 1, ..., K, \\ \pi_m^0 &=& \left\{\begin{array}{ll} 1, & m = 0, \\ 0, & m > 0, \end{array}\right. \quad \pi_{-1}^{k-1} = 0,\ k = 1, 2, ... \end{array}$$

whence

$$\pi_m^k = \left\{\begin{array}{ll} \binom{k}{m}\epsilon_\star^m(1 - \epsilon_\star)^{k-m}, & m \le k, \\ 0, & m > k. \end{array}\right.$$

Therefore,

$$P\{\lambda^K : \mathrm{Card}\{k \le K : \psi_k^+(\lambda^k) = 1\} \ge M\} = \sum_{M \le k \le K} \pi_k^K = \epsilon_M,$$

as required. □

2.11.3 Proof of Theorem 2.23

1°. Since $\mathcal{O}$ is a simple o.s., the function $\Phi(\phi, [\mu;\nu])$ given by (2.56) is a well-defined real-valued function on $\mathcal{F}\times(\mathcal{M}\times\mathcal{M})$ which is concave in $[\mu;\nu]$; convexity of the function in $\phi\in\mathcal{F}$ is evident. Since both $\mathcal{F}$ and $\mathcal{M}$ are convex sets coinciding with their relative interiors, convexity-concavity and real valuedness of Φ on $\mathcal{F}\times(\mathcal{M}\times\mathcal{M})$ imply the continuity of Φ on the indicated domain. As a consequence, Φ is a convex-concave continuous real-valued function on $\mathcal{F}\times(M_1\times M_2)$.

Now let

$$\underline{\Phi}(\mu,\nu) = \inf_{\phi\in\mathcal{F}} \Phi(\phi,[\mu;\nu]). \tag{2.166}$$

Note that $\underline{\Phi}$, being the infimum of a family of concave functions of $[\mu;\nu]\in\mathcal{M}\times\mathcal{M}$, is concave on $\mathcal{M}\times\mathcal{M}$. We claim that for $\mu,\nu\in\mathcal{M}$ the function

$$\phi_{\mu,\nu}(\omega) = \tfrac{1}{2}\ln(p_\mu(\omega)/p_\nu(\omega))$$

(which, by definition of a simple o.s., belongs to $\mathcal{F}$) is an optimal solution to the right-hand side minimization problem in (2.166), so that

$$\begin{array}{l}\forall(\mu\in M_1,\nu\in M_2):\\ \underline{\Phi}([\mu;\nu]) := \inf_{\phi\in\mathcal{F}}\Phi(\phi,[\mu;\nu]) = \Phi(\phi_{\mu,\nu},[\mu;\nu]) = \ln\left(\int_\Omega \sqrt{p_\mu(\omega)p_\nu(\omega)}\Pi(d\omega)\right).\end{array} \tag{2.167}$$

Indeed, we have

$$\exp\{-\phi_{\mu,\nu}(\omega)\}p_\mu(\omega) = \exp\{\phi_{\mu,\nu}(\omega)\}p_\nu(\omega) = g(\omega) := \sqrt{p_\mu(\omega)p_\nu(\omega)}, \tag{2.168}$$

whence $\Phi(\phi_{\mu,\nu},[\mu;\nu]) = \ln\left(\int_\Omega g(\omega)\Pi(d\omega)\right)$. On the other hand, for $\phi(\cdot) = \phi_{\mu,\nu}(\cdot) + \delta(\cdot)\in\mathcal{F}$ we have

$$\begin{array}{ll} & \int_\Omega g(\omega)\Pi(d\omega) = \int_\Omega\left[\sqrt{g(\omega)}\exp\{-\delta(\omega)/2\}\right]\left[\sqrt{g(\omega)}\exp\{\delta(\omega)/2\}\right]\Pi(d\omega)\\ (a) & \le \left(\int_\Omega g(\omega)\exp\{-\delta(\omega)\}\Pi(d\omega)\right)^{1/2}\left(\int_\Omega g(\omega)\exp\{\delta(\omega)\}\Pi(d\omega)\right)^{1/2}\\ & = \left(\int_\Omega \exp\{-\phi(\omega)\}p_\mu(\omega)\Pi(d\omega)\right)^{1/2}\left(\int_\Omega \exp\{\phi(\omega)\}p_\nu(\omega)\Pi(d\omega)\right)^{1/2} \text{ [by (2.168)]}\\ (b) & \Rightarrow \ln\left(\int_\Omega g(\omega)\Pi(d\omega)\right) \le \Phi(\phi,[\mu;\nu]),\end{array}$$

and thus $\Phi(\phi_{\mu,\nu},[\mu;\nu]) \le \Phi(\phi,[\mu;\nu])$ for every $\phi\in\mathcal{F}$.

Remark 2.49. Note that the above reasoning did not use the fact that the minimization on the right-hand side of *(2.166)* is over $\phi\in\mathcal{F}$; in fact, this reasoning shows that $\phi_{\mu,\nu}(\cdot)$ minimizes $\Phi(\phi,[\mu;\nu])$ over all functions ϕ for which the integrals $\int_\Omega \exp\{-\phi(\omega)\}p_\mu(\omega)\Pi(d\omega)$ and $\int_\Omega\exp\{\phi(\omega)\}p_\nu(\omega)\Pi(d\omega)$ exist.

Remark 2.50. Note that the inequality in (b) can be equality only when the inequality in (a) is so. In other words, if $\bar\phi$ is a minimizer of $\Phi(\phi,[\mu;\nu])$ over $\phi\in\mathcal{F}$, setting $\delta(\cdot) = \bar\phi(\cdot) - \phi_{\mu,\nu}(\cdot)$, the functions $\sqrt{g(\omega)}\exp\{-\delta(\omega)/2\}$ and $\sqrt{g(\omega)}\exp\{\delta(\omega)/2\}$, considered as elements of $L_2[\Omega,\Pi]$, are proportional to each other. Since g is positive and g,δ are continuous, while the support of Π is the entire Ω, this "L_2-proportionality" means that the functions in question differ by a constant factor, or, which is the same, that $\delta(\cdot)$ is constant. Thus, *the minimizers of $\Phi(\phi,[\mu;\nu])$ over $\phi\in\mathcal{F}$ are exactly the functions of the form $\phi(\omega) = \phi_{\mu,\nu}(\omega) + \mathrm{const}$.*

2°. Let us verify that $\Phi(\phi,[\mu;\nu])$ has a saddle point (min in $\phi\in\mathcal{F}$, max in $[\mu;\nu]\in M_1\times M_2$). First, observe that on the domain of Φ it holds

$$\Phi(\phi(\cdot)+a,[\mu;\nu])=\Phi(\phi(\cdot),[\mu;\nu])\ \forall(a\in\mathbf{R},\phi\in\mathcal{F}). \tag{2.169}$$

Let us select some $\bar\mu\in\mathcal{M}$, and let $\overline{P}$ be the measure on Ω with density $p_{\bar\mu}$ w.r.t. Π. For $\phi\in\mathcal{F}$, the integrals $\int_\Omega \mathrm{e}^{\pm\phi(\omega)}\overline{P}(d\omega)$ are finite (since $\mathcal{O}$ is simple), implying that $\phi\in L_1[\Omega,\overline{P}]$; note also that $\overline{P}$ is a probabilistic measure. Let now $\mathcal{F}_0=\{\phi\in\mathcal{F}:\int_\Omega\phi(\omega)\overline{P}(d\omega)=0\}$, so that $\mathcal{F}_0$ is a linear subspace in $\mathcal{F}$, and all functions $\phi\in\mathcal{F}$ can be obtained by shifts of functions from $\mathcal{F}_0$ by constants. Now, by (2.169), to prove the existence of a saddle point of Φ on $\mathcal{F}\times(M_1\times M_2)$ is exactly the same as to prove the existence of a saddle point of Φ on $\mathcal{F}_0\times(M_1\times M_2)$. Let us verify that $\Phi(\phi,[\mu;\nu])$ indeed has a saddle point on $\mathcal{F}_0\times(M_1\times M_2)$. Because $M_1\times M_2$ is a convex compact set, and Φ is continuous on $\mathcal{F}_0\times(M_1\times M_2)$ and convex-concave, invoking the Sion-Kakutani Theorem we see that all we need in order to prove the existence of a saddle point is to verify that Φ is coercive in the first argument. In other words, we have to show that for every fixed $[\mu;\nu]\in M_1\times M_2$ one has $\Phi(\phi,[\mu;\nu])\to+\infty$ as $\phi\in\mathcal{F}_0$ and $\|\phi\|\to\infty$ (whatever be the norm $\|\cdot\|$ on $\mathcal{F}_0$; recall that $\mathcal{F}_0$ is a finite-dimensional linear space). Setting

$$\Theta(\phi)=\Phi(\phi,[\mu;\nu])=\frac{1}{2}\left[\ln\left(\int_\omega \mathrm{e}^{-\phi(\omega)}p_\mu(\omega)\Pi(d\omega)\right)+\ln\left(\int_\omega \mathrm{e}^{\phi(\omega)}p_\nu(\omega)\Pi(d\omega)\right)\right]$$

and taking into account that Θ is convex and finite on $\mathcal{F}_0$, in order to prove that Θ is coercive, it suffices to verify that $\Theta(t\phi)\to\infty$, $t\to\infty$, for every nonzero $\phi\in\mathcal{F}_0$, which is evident: since $\int_\Omega\phi(\omega)\overline{P}(d\omega)=0$ and ϕ is nonzero, we have $\int_\Omega\max[\phi(\omega),0]\overline{P}(d\omega)=\int_\Omega\max[-\phi(\omega),0]\overline{P}(d\omega)>0$, whence $\phi>0$ and $\phi<0$ on sets of Π-positive measure, so that $\Theta(t\phi)\to\infty$ as $t\to\infty$ due to the fact that both $p_\mu(\cdot)$ and $p_\nu(\cdot)$ are continuous and everywhere positive.

3°. Now let $(\phi_*(\cdot);[\mu_*;\nu_*])$ be a saddle point of Φ on $\mathcal{F}\times(M_1\times M_2)$. Shifting, if necessary, $\phi_*(\cdot)$ by a constant (by (2.169), this does not affect the fact that $(\phi_*,[\mu_*;\nu_*])$ is a saddle point of Φ), we can assume that

$$\varepsilon_\star:=\int_\Omega\exp\{-\phi_*(\omega)\}p_{\mu_*}(\omega)\Pi(d\omega)=\int_\Omega\exp\{\phi_*(\omega)\}p_{\nu_*}(\omega)\Pi(d\omega),$$

so that the saddle point value of Φ is

$$\Phi_*:=\max_{[\mu;\nu]\in M_1\times M_2}\min_{\phi\in\mathcal{F}}\Phi(\phi,[\mu;\nu])=\Phi(\phi_*,[\mu_*;\nu_*])=\ln(\varepsilon_\star), \tag{2.170}$$

as claimed in item (i) of the theorem.

Now let us prove (2.58). For $\mu\in M_1$, we have

$$\begin{array}{rcl}\ln(\varepsilon_\star)&=&\Phi_*\geq\Phi(\phi_*,[\mu;\nu_*])\\&=&\frac{1}{2}\ln\left(\int_\Omega\exp\{-\phi_*(\omega)\}p_\mu(\omega)\Pi(d\omega)\right)+\frac{1}{2}\ln\left(\int_\Omega\exp\{\phi_*(\omega)\}p_{\nu_*}(\omega)\Pi(d\omega)\right)\\&=&\frac{1}{2}\ln\left(\int_\Omega\exp\{-\phi_*(\omega)\}p_\mu(\omega)P(d\omega)\right)+\frac{1}{2}\ln(\varepsilon_\star).\end{array}$$

Hence,

$$\begin{array}{rcl}\ln\left(\int_\Omega\exp\{-\phi_*^a(\omega)\}p_\mu(\omega)\Pi(d\omega)\right)&=&\ln\left(\int_\Omega\exp\{-\phi_*(\omega)\}p_\mu(\omega)P(d\omega)\right)+a\\&\leq&\ln(\varepsilon_\star)+a,\end{array}$$

and $(2.58.a)$ follows. Similarly, when $\nu \in M_2$, we have

$$\begin{array}{rcl}\ln(\varepsilon_\star) & = & \Phi_* \geq \Phi(\phi_*, [\mu_*; \nu]) \\ & = & \frac{1}{2}\ln\left(\int_\Omega \exp\{-\phi_*(\omega)\}p_{\mu_*}(\omega)\Pi(d\omega)\right) + \frac{1}{2}\ln\left(\int_\Omega \exp\{\phi_*(\omega)\}p_\nu(\omega)\Pi(d\omega)\right) \\ & = & \frac{1}{2}\ln(\varepsilon_\star) + \frac{1}{2}\ln\left(\int_\Omega \exp\{\phi_*(\omega)\}p_\nu(\omega)\Pi(d\omega)\right),\end{array}$$

so that

$$\begin{array}{rcl}\ln\left(\int_\Omega \exp\{\phi_*^a(\omega)\}p_\nu(\omega)\Pi(d\omega)\right) & = & \ln\left(\int_\Omega \exp\{\phi_*(\omega)\}p_\nu(\omega)\Pi(d\omega)\right) - a \\ & \leq & \ln(\varepsilon_\star) - a,\end{array}$$

and $(2.58.b)$ follows.

We have proved all statements of item (i), except for the claim that the $\phi_*, \varepsilon_\star$ just defined form an optimal solution to (2.59). Note that by (2.58) as applied with $a = 0$, the pair in question is feasible for (2.59). Assuming that the problem admits a feasible solution $(\bar\phi, \epsilon)$ with $\epsilon < \varepsilon_\star$, let us lead this assumption to a contradiction. Note that $\bar\phi$ should be such that

$$\int_\Omega \mathrm{e}^{-\bar\phi(\omega)}p_{\mu_*}(\omega)\Pi(d\omega) < \varepsilon_\star \;\&\; \int_\Omega \mathrm{e}^{\bar\phi(\omega)}p_{\nu_*}(\omega)\Pi(d\omega) < \varepsilon_\star,$$

and consequently $\Phi(\bar\phi, [\mu_*; \nu_*]) < \ln(\varepsilon_\star)$. On the other hand, Remark 2.49 says that $\Phi(\bar\phi, [\mu_*; \nu_*])$ cannot be less than $\min\limits_{\phi\in\mathcal{F}} \Phi(\phi, [\mu_*; \nu_*])$, and the latter quantity is $\Phi(\phi_*, [\mu_*; \nu_*])$ because $(\phi_*, [\mu_*; \nu_*])$ is a saddle point of Φ on $\mathcal{F} \times (M_1 \times M_2)$. Thus, assuming that the optimal value in (2.59) is $< \varepsilon_\star$, we conclude that $\Phi(\phi_*, [\mu_*; \nu_*]) \leq \Phi(\bar\phi, [\mu_*; \nu_*]) < \ln(\varepsilon_\star)$, contradicting (2.170). Item (i) of Theorem 2.23 is proved.

4°. Let us prove item (ii) of Theorem 2.23. Relation (2.60) and concavity of the right-hand side of this relation in $[\mu; \nu]$ were already proved; moreover, these relations were proved in the range $\mathcal{M} \times \mathcal{M}$ of $[\mu; \nu]$. Since this range coincides with its relative interior, the real-valued concave function $\underline{\Phi}$ is continuous on $\mathcal{M} \times \mathcal{M}$ and thus is continuous on $M_1 \times M_2$. Next, let ϕ_* be the ϕ-component of a saddle point of Φ on $\mathcal{F} \times (M_1 \times M_2)$ (we already know that such a saddle point exists). By Proposition 2.21, the $[\mu; \nu]$-components of saddle points of Φ on $\mathcal{F} \times (M_1 \times M_2)$ are exactly the maximizers of $\underline{\Phi}$ on $M_1 \times M_2$; let $[\mu_*; \nu_*]$ be such a maximizer. By the same proposition, $(\phi_*, [\mu_*; \nu_*])$ is a saddle point of Φ, whence $\Phi(\phi, [\mu_*; \nu_*])$ attains its minimum over $\phi \in \mathcal{F}$ at $\phi = \phi_*$. We have also seen that $\Phi(\phi, [\mu_*; \nu_*])$ attains its minimum over $\phi \in \mathcal{F}$ at $\phi = \phi_{\mu_*,\nu_*}$. These observations combine with Remark 2.50 to imply that ϕ_* and ϕ_{μ_*,ν_*} differ by a constant, which, in view of (2.169), means that $(\phi_{\mu_*,\nu_*}, [\mu_*; \nu_*])$ is a saddle point of Φ along with $(\phi_*, [\mu_*; \nu_*])$. (ii) is proved.

5°. It remains to prove item (iii) of Theorem 2.23. In the notation from (iii), simple hypotheses (A) and (B) can be decided with the total risk $\leq 2\epsilon$, and therefore, by Proposition 2.2,

$$2\bar\epsilon := \int_\Omega \min[p(\omega), q(\omega)]\Pi(d\omega) \leq 2\epsilon.$$

On the other hand, we have seen that the saddle point value of Φ is $\ln(\varepsilon_\star)$; since $[\mu_*; \nu_*]$ is a component of a saddle point of Φ, it follows that $\min_{\phi\in\mathcal{F}} \Phi(\phi, [\mu_*; \nu_*]) = \ln(\varepsilon_\star)$. The left-hand side in this equality, by item 1°, is $\Phi(\phi_{\mu_*,\nu_*}, [\mu_*; \nu_*])$, and we arrive at

$$\ln(\varepsilon_\star) = \Phi(\tfrac{1}{2}\ln(p_{\mu_*}(\cdot)/p_{\nu_*}(\cdot)), [\mu_*; \nu_*]) = \ln\left(\int_\Omega \sqrt{p_{\mu_*}(\omega)p_{\nu_*}(\omega)}\Pi(d\omega)\right),$$

so that

$$\varepsilon_\star = \int_\Omega \sqrt{p_{\mu_*}(\omega)p_{\nu_*}(\omega)}\Pi(d\omega) = \int_\Omega \sqrt{p(\omega)q(\omega)}\Pi(d\omega).$$

We now have

$$\begin{array}{rl}\varepsilon_\star & = \int_\Omega \sqrt{p(\omega)q(\omega)}\Pi(d\omega) = \int_\Omega \sqrt{\min[p(\omega),q(\omega)]}\sqrt{\max[p(\omega),q(\omega)]}\Pi(d\omega)\\ & \leq \left(\int_\Omega \min[p(\omega),q(\omega)]\Pi(d\omega)\right)^{1/2}\left(\int_\Omega \max[p(\omega),q(\omega)]\Pi(d\omega)\right)^{1/2}\\ & = \left(\int_\Omega \min[p(\omega),q(\omega)]\Pi(d\omega)\right)^{1/2}\left(\int_\Omega (p(\omega)+q(\omega)-\min[p(\omega),q(\omega)])\Pi(d\omega)\right)^{1/2}\\ & = \sqrt{2\bar\epsilon(2-2\bar\epsilon)} \leq 2\sqrt{(1-\epsilon)\epsilon},\end{array}$$

where the concluding inequality is due to $\bar\epsilon \leq \epsilon \leq 1/2$. (iii) is proved, and the proof of Theorem 2.23 is complete. □

2.11.4 Proof of Proposition 2.37

All we need is to verify (2.107) and to check that the right-hand side function in this relation is convex. The latter is evident, since $\phi_X(h) + \phi_X(-h) \geq 2\phi_X(0) = 0$ and $\phi_X(h) + \phi_X(-h)$ is convex. To verify (2.107), let us fix $P \in \mathcal{P}[X]$ and $h \in \mathbf{R}^d$ and set

$$\nu = h^T e[P],$$

so that ν is the expectation of $h^T\omega$ with $\omega \sim P$. Note that for $\omega \sim P$ we have $h^T\omega \in [-\phi_X(-h), \phi_X(h)]$ with P-probability 1, whence $-\phi_X(-h) \leq \nu \leq \phi_X(h)$. In particular, when $\phi_X(h) + \phi_X(-h) = 0$, $h^T\omega = \nu$ with P-probability 1, so that (2.107) definitely holds true. Now let

$$\eta := \tfrac{1}{2}\left[\phi_X(h) + \phi_X(-h)\right] > 0,$$

and let

$$a = \tfrac{1}{2}\left[\phi_X(h) - \phi_X(-h)\right], \ \ \beta = (\nu - a)/\eta.$$

Denoting by P_h the distribution of $h^T\omega$ induced by the distribution P of ω and noting that this distribution is supported on $[-\phi_X(-h), \phi_X(h)] = [a-\eta, a+\eta]$ and has expectation ν, we get

$$\beta \in [-1, 1]$$

and

$$\gamma := \int \exp\{h^T\omega\}P(d\omega) = \int_{a-\eta}^{a+\eta}[\mathrm{e}^s - \lambda(s-\nu)]P_h(ds)$$

for all $\lambda \in \mathbf{R}$. Hence,

$$\begin{array}{rcl}\ln(\gamma) & \leq & \inf\limits_\lambda \ln\left(\max\limits_{a-\eta\leq s\leq a+\eta}[\mathrm{e}^s - \lambda(s-\nu)]\right)\\ & = & a + \inf\limits_\rho \ln\left(\max\limits_{-\eta\leq t\leq \eta}[\mathrm{e}^t - \rho(t-[\nu-a])]\right) \ \text{[substituting } \lambda = \mathrm{e}^a\rho, s = a+t]\\ & = & a + \inf\limits_\rho \ln\left(\max\limits_{-\eta\leq t\leq \eta}[\mathrm{e}^t - \rho(t-\eta\beta)]\right) \leq a + \ln\left(\max\limits_{-\eta\leq t\leq \eta}[\mathrm{e}^t - \bar\rho(t-\eta\beta)\right)\end{array}$$

with $\bar{\rho} = (2\eta)^{-1}(\mathrm{e}^{\eta} - \mathrm{e}^{-\eta})$. The function $g(t) = \mathrm{e}^t - \bar{\rho}(t - \eta\beta)$ is convex on $[-\eta, \eta]$, and

$$g(-\eta) = g(\eta) = \cosh(\eta) + \beta \sinh(\eta),$$

which combines with the above computation to yield the relation

$$\ln(\gamma) \leq a + \ln(\cosh(\eta) + \beta \sinh(\eta)). \tag{2.171}$$

Thus, all we need to verify is that

$$\forall(\eta > 0, \beta \in [-1, 1]) : \ \beta\eta + \tfrac{1}{2}\eta^2 - \ln(\cosh(\eta) + \beta \sinh(\eta)) \geq 0. \tag{2.172}$$

Indeed, if (2.172) holds true (2.171) implies that

$$\ln(\gamma) \leq a + \beta\eta + \tfrac{1}{2}\eta^2 = \nu + \tfrac{1}{2}\eta^2,$$

which, recalling what γ, ν, and η are, is exactly what we want to prove.

Verification of (2.172) is as follows. The left-hand side in (2.172) is convex in β for $\beta > -\frac{\cosh(\eta)}{\sinh(\eta)}$ containing, due to $\eta > 0$, the range of β in (2.172). Furthermore, the minimum of the left-hand side of (2.172) over $\beta > -\coth(\eta)$ is attained at $\beta = \frac{\sinh(\eta) - \eta\cosh(\eta)}{\eta \sinh(\eta)}$ and is equal to

$$r(\eta) = \tfrac{1}{2}\eta^2 + 1 - \eta \coth(\eta) - \ln(\sinh(\eta)/\eta).$$

All we need to prove is that the latter quantity is nonnegative whenever $\eta > 0$. We have

$$r'(\eta) = \eta - \coth(\eta) - \eta(1 - \coth^2(\eta)) - \coth(\eta) + \eta^{-1} = (\eta \coth(\eta) - 1)^2 \eta^{-1} \geq 0,$$

and since $r(+0) = 0$, we get $r(\eta) \geq 0$ when $\eta > 0$. □

2.11.5 Proof of Proposition 2.43

2.11.5.A Proof of Proposition 2.43.i

1°. Let $b = [0; ...; 0; 1] \in \mathbf{R}^{n+1}$, so that $B = \begin{bmatrix} A \\ b^T \end{bmatrix}$, and let $\mathcal{A}(u) = A[u; 1]$. For any $u \in \mathbf{R}^n$, $h \in \mathbf{R}^d$, $\Theta \in \mathbf{S}^d_+$, and $H \in \mathbf{S}^d$ such that $-I \prec \Theta^{1/2} H \Theta^{1/2} \prec I$ we have

$$\begin{array}{l}
\Psi(h, H; u, \Theta) := \ln\left(\mathbf{E}_{\zeta \sim \mathcal{N}(\mathcal{A}(u), \Theta)}\left\{\exp\{h^T\zeta + \tfrac{1}{2}\zeta^T H \zeta\}\right\}\right) \\
= \ln\left(\mathbf{E}_{\xi \sim \mathcal{N}(0,I)}\left\{\exp\{h^T[\mathcal{A}(u) + \Theta^{1/2}\xi] + \tfrac{1}{2}[\mathcal{A}(u) + \Theta^{1/2}\xi]^T H [\mathcal{A}(u) + \Theta^{1/2}\xi]\}\right\}\right) \\
= -\tfrac{1}{2}\ln \mathrm{Det}(I - \Theta^{1/2} H \Theta^{1/2}) + h^T \mathcal{A}(u) + \tfrac{1}{2}\mathcal{A}(u)^T H \mathcal{A}(u) \\
\qquad + \tfrac{1}{2}[H\mathcal{A}(u) + h]^T \Theta^{1/2}[I - \Theta^{1/2} H \Theta^{1/2}]^{-1}\Theta^{1/2}[H\mathcal{A}(u) + h] \\
= -\tfrac{1}{2}\ln \mathrm{Det}(I - \Theta^{1/2} H \Theta^{1/2}) + \tfrac{1}{2}[u;1]^T \left[bh^T A + A^T h b^T + A^T H A\right][u;1] \\
\qquad + \tfrac{1}{2}[u;1]^T \left[B^T [H, h]^T \Theta^{1/2}[I - \Theta^{1/2} H \Theta^{1/2}]^{-1}\Theta^{1/2}[H, h] B\right][u;1]
\end{array} \tag{2.173}$$

due to

$$h^T \mathcal{A}(u) = [u;1]^T b h^T A[u;1] = [u;1]^T A^T h b^T [u;1]$$

and $H\mathcal{A}(u) + h = [H, h]B[u; 1]$.

Observe that when $(h,H)\in\mathcal{H}^\gamma$, we have

$$\Theta^{1/2}[I-\Theta^{1/2}H\Theta^{1/2}]^{-1}\Theta^{1/2}=[\Theta^{-1}-H]^{-1}\preceq[\Theta_*^{-1}-H]^{-1},$$

so that (2.173) implies that for all $u\in\mathbf{R}^n$, $\Theta\in\mathcal{V}$, and $(h,H)\in\mathcal{H}^\gamma$,

$$\begin{array}{l}\Psi(h,H;u,\Theta)\leq-\frac{1}{2}\ln\mathrm{Det}(I-\Theta^{1/2}H\Theta^{1/2})\\ +\frac{1}{2}[u;1]^T\underbrace{\left[bh^TA+A^Thb^T+A^THA+B^T[H,h]^T[\Theta_*^{-1}-H]^{-1}[H,h]B\right]}_{Q[H,h]}[u;1]\\ \quad=-\frac{1}{2}\ln\mathrm{Det}(I-\Theta^{1/2}H\Theta^{1/2})+\frac{1}{2}\mathrm{Tr}(Q[H,h]Z(u))\\ \quad\leq-\frac{1}{2}\ln\mathrm{Det}(I-\Theta^{1/2}H\Theta^{1/2})+\Gamma_{\mathcal{Z}}(h,H),\\ \Gamma_{\mathcal{Z}}(h,H)=\frac{1}{2}\phi_{\mathcal{Z}}(Q[H,h])\end{array}\tag{2.174}$$

(we have taken into account that $Z(u)\in\mathcal{Z}$ when $u\in U$, the premise of the proposition, and therefore $\mathrm{Tr}(Q[H,h]Z(u))\leq\phi_{\mathcal{Z}}(Q[H,h])$). Note that the above function $Q[H,h]$ is nothing but

$$Q[H,h]=B^T\left(\left[\begin{array}{c|c}H&h\\\hline h^T&\end{array}\right]+[H,h]^T[\Theta_*^{-1}-H]^{-1}[H,h]\right)B.\tag{2.175}$$

2°. We need the following:

Lemma 2.51. *Let Θ_* be a $d\times d$ symmetric positive definite matrix, let $\delta\in[0,2]$, and let $\mathcal{V}$ be a closed convex subset of $\mathbf{S}^d_+$ such that*

$$\Theta\in\mathcal{V}\Rightarrow\{\Theta\preceq\Theta_*\}\ \&\ \{\|\Theta^{1/2}\Theta_*^{-1/2}-I_d\|\leq\delta\}\tag{2.176}$$

(cf. (2.129)). Let also $\mathcal{H}^o:=\{H\in\mathbf{S}^d:-\Theta_*^{-1}\prec H\prec\Theta_*^{-1}\}$. Then

$$\begin{array}{l}\forall(H,\Theta)\in\mathcal{H}^o\times\mathcal{V}:\\ G(H;\Theta):=-\frac{1}{2}\ln\mathrm{Det}(I-\Theta^{1/2}H\Theta^{1/2})\\ \quad\leq G^+(H;\Theta):=-\frac{1}{2}\ln\mathrm{Det}(I-\Theta_*^{1/2}H\Theta_*^{1/2})+\frac{1}{2}\mathrm{Tr}([\Theta-\Theta_*]H)\\ \qquad\qquad+\frac{\delta(2+\delta)}{2(1-\|\Theta_*^{1/2}H\Theta_*^{1/2}\|)}\|\Theta_*^{1/2}H\Theta_*^{1/2}\|_F^2,\end{array}\tag{2.177}$$

where $\|\cdot\|$ is the spectral, and $\|\cdot\|_F$ the Frobenius norm of a matrix. In addition, $G^+(H,\Theta)$ is a continuous function on $\mathcal{H}^o\times\mathcal{V}$ which is convex in $H\in H^o$ and concave (in fact, affine) in $\Theta\in\mathcal{V}$

Proof. Let us set

$$d(H)=\|\Theta_*^{1/2}H\Theta_*^{1/2}\|,$$

so that $d(H)<1$ for $H\in\mathcal{H}^o$. For $H\in\mathcal{H}^o$ and $\Theta\in\mathcal{V}$ fixed we have

$$\begin{array}{rcl}\|\Theta^{1/2}H\Theta^{1/2}\|&=&\|[\Theta^{1/2}\Theta_*^{-1/2}][\Theta_*^{1/2}H\Theta_*^{1/2}][\Theta^{1/2}\Theta_*^{-1/2}]^T\|\\ &\leq&\|\Theta^{1/2}\Theta_*^{-1/2}\|^2\|\Theta_*^{1/2}H\Theta_*^{1/2}\|\leq\|\Theta_*^{1/2}H\Theta_*^{1/2}\|=d(H)\end{array}\tag{2.178}$$

(we have used the fact that $0\preceq\Theta\preceq\Theta_*$ implies $\|\Theta^{1/2}\Theta_*^{-1/2}\|\leq1$). Noting that

$\|AB\|_F \le \|A\|\|B\|_F$, a computation completely similar to the one in (2.178) yields

$$\|\Theta^{1/2}H\Theta^{1/2}\|_F \le \|\Theta_*^{1/2}H\Theta_*^{1/2}\|_F =: D(H). \tag{2.179}$$

Besides this, setting $F(X) = -\ln \mathrm{Det}(X) : \mathrm{int}\,\mathbf{S}^d_+ \to \mathbf{R}$ and equipping $\mathbf{S}^d$ with the Frobenius inner product, we have $\nabla F(X) = -X^{-1}$, so that with $R_0 = \Theta_*^{1/2}H\Theta_*^{1/2}$, $R_1 = \Theta^{1/2}H\Theta^{1/2}$, and $\Delta = R_1 - R_0$, we have for properly selected $\lambda \in (0,1)$ and $R_\lambda = \lambda R_0 + (1-\lambda)R_1$:

$$\begin{array}{rcl} F(I-R_1) & = & F(I-R_0-\Delta) = F(I-R_0) + \langle \nabla F(I-R_\lambda), -\Delta\rangle \\ & = & F(I-R_0) + \langle (I-R_\lambda)^{-1}, \Delta\rangle \\ & = & F(I-R_0) + \langle I, \Delta\rangle + \langle (I-R_\lambda)^{-1} - I, \Delta\rangle. \end{array}$$

We conclude that

$$F(I-R_1) \le F(I-R_0) + \mathrm{Tr}(\Delta) + \|I-(I-R_\lambda)^{-1}\|_F\|\Delta\|_F. \tag{2.180}$$

Denoting by μ_i the eigenvalues of R_λ and noting that $\|R_\lambda\| \le \max[\|R_0\|, \|R_1\|] = d(H)$ (see (2.178)), we have $|\mu_i| \le d(H)$, and therefore eigenvalues $\nu_i = 1 - \frac{1}{1-\mu_i} = -\frac{\mu_i}{1-\mu_i}$ of $I-(I-R_\lambda)^{-1}$ satisfy $|\nu_i| \le |\mu_i|/(1-\mu_i) \le |\mu_i|/(1-d(H))$, whence

$$\|I-(I-R_\lambda)^{-1}\|_F \le \|R_\lambda\|_F/(1-d(H)).$$

Noting that $\|R_\lambda\|_F \le \max[\|R_0\|_F, \|R_1\|_F] \le D(H)$—see (2.179)—we conclude that $\|I-(I-R_\lambda)^{-1}\|_F \le D(H)/(1-d(H))$, so that (2.180) yields

$$F(I-R_1) \le F(I-R_0) + \mathrm{Tr}(\Delta) + D(H)\|\Delta\|_F/(1-d(H)). \tag{2.181}$$

Further, by (2.129) the matrix $D = \Theta^{1/2}\Theta_*^{-1/2} - I$ satisfies $\|D\| \le \delta$, whence

$$\Delta = \underbrace{\Theta^{1/2}H\Theta^{1/2}}_{R_1} - \underbrace{\Theta_*^{1/2}H\Theta_*^{1/2}}_{R_0} = (I+D)R_0(I+D^T) - R_0 = DR_0 + R_0D^T + DR_0D^T.$$

Consequently,

$$\begin{array}{rcl} \|\Delta\|_F & \le & \|DR_0\|_F + \|R_0D^T\|_F + \|DR_0D^T\|_F \le [2\|D\| + \|D\|^2]\|R_0\|_F \\ & \le & \delta(2+\delta)\|R_0\|_F = \delta(2+\delta)D(H). \end{array}$$

This combines with (2.181) and the relation

$$\mathrm{Tr}(\Delta) = \mathrm{Tr}(\Theta^{1/2}H\Theta^{1/2} - \Theta_*^{1/2}H\Theta_*^{1/2}) = \mathrm{Tr}([\Theta-\Theta_*]H)$$

to yield

$$\begin{array}{rcl} F(I-R_1) & \le & F(I-R_0) + \mathrm{Tr}([\Theta-\Theta_*]H) + \frac{\delta(2+\delta)}{1-d(H)}D(H) \\ & = & F(I-R_0) + \mathrm{Tr}([\Theta-\Theta_*]H) + \frac{\delta(2+\delta)}{1-\|\Theta_*^{1/2}H\Theta_*^{1/2}\}}\|\Theta_*^{1/2}H\Theta_*^{1/2}\|_F^2, \end{array}$$

and we arrive at (2.177). It remains to prove that $G^+(H;\Theta)$ is convex-concave and continuous on $\mathcal{H}^o \times \mathcal{V}$. The only component of this claim which is not completely evident is convexity of the function in $H \in \mathcal{H}^o$. To see that it is the case, note that $\ln \mathrm{Det}(S)$ is concave on the interior of the semidefinite cone, the function

$f(u,v) = \frac{u^2}{1-v}$ is convex and nondecreasing in u, v in the convex domain $\Pi = \{(u,v) : u \geq 0, v < 1\}$, and the function $\frac{\|\Theta_*^{1/2} H \Theta_*^{1/2}\|_F^2}{1-\|\Theta_*^{1/2} H \Theta_*^{1/2}\|}$ is obtained from f by convex substitution of variables $H \mapsto (\|\Theta_*^{1/2} H \Theta_*^{1/2}\|_F, \|\Theta_*^{1/2} H \Theta_*^{1/2}\|)$ mapping $\mathcal{H}^o$ into Π. $\square$

3°. Combining (2.177), (2.174), and (2.130) and the origin of Ψ—see (2.173)—we arrive at

$$\begin{array}{l}\forall ((u,\Theta) \in U \times \mathcal{V}, (h,H) \in \mathcal{H}^\gamma = \mathcal{H}) : \\ \ln\left(\mathbf{E}_{\zeta \sim \mathcal{N}(A[u;1],\Theta)} \left\{\exp\{h^T\zeta + \frac{1}{2}\zeta^T H \zeta\}\right\}\right) \leq \Phi_{A,\mathcal{Z}}(h,H;\Theta),\end{array}$$

as claimed in (2.133).

4°. Now let us check that $\Phi_{A,\mathcal{Z}}(h,H;\Theta) : \mathcal{H} \times \mathcal{V} \to \mathbf{R}$ is continuous and convex-concave. Recalling that the function $G^+(H;\Theta)$ from (2.177) is convex-concave and continuous on $\mathcal{H}^o \times \mathcal{V}$, all we need to verify is that $\Gamma_{\mathcal{Z}}(h,H)$ is convex and continuous on $\mathcal{H}$. Recalling that $\mathcal{Z}$ is a nonempty compact set, the function $\phi_{\mathcal{Z}}(\cdot) : \mathbf{S}^{d+1} \to \mathbf{R}$ is continuous, implying the continuity of $\Gamma_{\mathcal{Z}}(h,H) = \frac{1}{2}\phi_{\mathcal{Z}}(Q[H,h])$ on $\mathcal{H} = \mathcal{H}^\gamma$ ($Q[H,h]$ is defined in (2.175)). To prove convexity of $\Gamma_{\mathcal{Z}}$, note that $\mathcal{Z}$ is contained in $\mathbf{S}_+^{n+1}$, implying that $\phi_{\mathcal{Z}}(\cdot)$ is convex and $\succeq$-monotone. On the other hand, by the Schur Complement Lemma, we have

$$\begin{array}{rcl} S & := & \{(h,H,G) : G \succeq Q[H,h], (h,H) \in \mathcal{H}^\gamma\} \\ & = & \left\{(h,H,G) : \left[\begin{array}{c|c} G - [bh^TA + A^Thb^T + A^THA] & B^T[H,h]^T \\ \hline [H,h]B & \Theta_*^{-1} - H \end{array}\right] \succeq 0, \right. \\ & & \left. \hfill (h,H) \in \mathcal{H}^\gamma \right\}, \end{array}$$

implying that S is convex. Since $\phi_{\mathcal{Z}}(\cdot)$ is $\succeq$-monotone, we have

$$\begin{array}{l}\{(h,H,\tau) : (h,H) \in \mathcal{H}^\gamma,\ \tau \geq \Gamma_{\mathcal{Z}}(h,H)\} \\ = \{(h,H,\tau) :\ \exists G : G \succeq Q[H,h],\ 2\tau \geq \phi_{\mathcal{Z}}(G),\ (h,H) \in \mathcal{H}^\gamma\},\end{array}$$

and we see that the epigraph of $\Gamma_{\mathcal{Z}}$ is convex (since the set S and the epigraph of $\phi_{\mathcal{Z}}$ are so), as claimed.

5°. It remains to prove that $\Phi_{A,\mathcal{Z}}$ is coercive in H, h. Let $\Theta \in \mathcal{V}$ and $(h_i, H_i) \in \mathcal{H}^\gamma$ with $\|(h_i,H_i)\| \to \infty$ as $i \to \infty$, and let us prove that $\Phi_{A,\mathcal{Z}}(h_i,H_i;\Theta) \to \infty$. Looking at the expression for $\Phi_{A,\mathcal{Z}}(h_i,H_i;\Theta)$, it is immediately seen that all terms in this expression, except for the terms coming from $\phi_{\mathcal{Z}}(\cdot)$, remain bounded as i grows, so that all we need to verify is that the $\phi_{\mathcal{Z}}(\cdot)$-term goes to ∞ as $i \to \infty$. Observe that H_i are uniformly bounded due to $(h_i,H_i) \in \mathcal{H}^\gamma$, implying that $\|h_i\|_2 \to \infty$ as $i \to \infty$. Denoting $e = [0;...;0;1] \in \mathbf{R}^{d+1}$ and, as before, $b = [0;...;0;1] \in \mathbf{R}^{n+1}$, note that, by construction, $B^Te = b$. Now let $W \in \mathcal{Z}$, so that $W_{n+1,n+1} = 1$. Taking into account that the matrices $[\Theta_*^{-1} - H_i]^{-1}$ satisfy $\alpha I_d \preceq [\Theta_*^{-1} - H_i]^{-1} \preceq \beta I_d$ for some positive α, β due to $H_i \in \mathcal{H}^\gamma$, observe that

$$\underbrace{\left[\left[\begin{array}{c|c} H_i & h_i \\ \hline h_i^T & \end{array}\right] + [H_i,h_i]^T\left[\Theta_*^{-1} - H_i\right]^{-1}[H_i,h_i]\right]}_{Q_i = Q[H_i,h_i]} = \underbrace{[h_i^T[\Theta_*^{-1} - H_i]^{-1}h_i]}_{\alpha_i\|h_i\|_2^2} ee^T + R_i,$$

where $\alpha_i \geq \alpha > 0$ and $\|R_i\|_F \leq C(1+\|h_i\|_2)$. As a result,

$$\begin{array}{rcl}\phi_{\mathcal{Z}}(B^TQ_iB) & \geq & \mathrm{Tr}(WB^TQ_iB) = \mathrm{Tr}(WB^T[\alpha_i\|h_i\|_2^2ee^T + R_i]B) \\ & = & \alpha_i\|h_i\|_2^2 \underbrace{\mathrm{Tr}(Wbb^T)}_{=W_{n+1,n+1}=1} - \|BWB^T\|_F\|R_i\|_F \\ & \geq & \alpha\|h_i\|_2^2 - C(1+\|h_i\|_2)\|BWB^T\|_F,\end{array}$$

and the concluding quantity tends to ∞ as $i \to \infty$ due to $\|h_i\|_2 \to \infty$, $i \to \infty$. Part (i) is proved.

2.11.5.B Proof of Proposition 2.43.ii

By (i) the function $\Phi(h, H; \Theta_1, \Theta_2)$, as defined in (2.134), is continuous and convex-concave on the domain $\underbrace{(\mathcal{H}_1 \cap \mathcal{H}_2)}_{\mathcal{H}} \times \underbrace{(\mathcal{V}_1 \times \mathcal{V}_2)}_{\mathcal{V}}$ and is coercive in (h, H), $\mathcal{H}$ and $\mathcal{V}$ are closed and convex, and $\mathcal{V}$ in addition is compact, so that saddle point problem (2.134) is solvable (Sion-Kakutani Theorem, a.k.a. Theorem 2.22). Now let $(h_*, H_*; \Theta_1^*, \Theta_2^*)$ be a saddle point. To prove (2.136), let $P \in \mathcal{G}_1$, that is, $P = \mathcal{N}(A_1[u;1], \Theta_1)$ for some $\Theta_1 \in \mathcal{V}_1$ and some u with $[u;1][u;1]^T \in \mathcal{Z}_1$. Applying (2.133) to the first collection of data, with a given by (2.135), we get the first $\leq$ in the following chain:

$$\begin{array}{l}\ln\left(\int \mathrm{e}^{-\frac{1}{2}\omega^TH_*\omega - \omega^Th_* - a}P(d\omega)\right) \leq \Phi_{A_1,\mathcal{Z}_1}(-h_*, -H_*; \Theta_1) - a \\ \underbrace{\leq}_{(a)} \Phi_{A_1,\mathcal{Z}_1}(-h_*, -H_*; \Theta_1^*) - a \underbrace{=}_{(b)} \mathcal{SV},\end{array}$$

where (a) is due to the fact that $\Phi_{A_1,\mathcal{Z}_1}(-h_*, -H_*; \Theta_1) + \Phi_{A_2,\mathcal{Z}_2}(h_*, H_*; \Theta_2)$ attains its maximum over $(\Theta_1, \Theta_2) \in \mathcal{V}_1 \times \mathcal{V}_2$ at the point (Θ_1^*, Θ_2^*), and (b) is due to the origin of a and the relation $\mathcal{SV} = \frac{1}{2}[\Phi_{A_1,\mathcal{Z}_1}(-h_*, -H_*; \Theta_1^*) + \Phi_{A_2,\mathcal{Z}_2}(h_*, H_*; \Theta_2^*)]$. The bound in (2.136.a) is proved. Similarly, let $P \in \mathcal{G}_2$, that is, $P = \mathcal{N}(A_2[u;1], \Theta_2)$ for some $\Theta_2 \in \mathcal{V}_2$ and some u with $[u;1][u;1]^T \in \mathcal{Z}_2$. Applying (2.133) to the second collection of data, with the same a as above, we get the first $\leq$ in the following chain:

$$\begin{array}{l}\ln\left(\int \mathrm{e}^{\frac{1}{2}\omega^TH_*\omega + \omega^Th_* + a}P(d\omega)\right) \leq \Phi_{A_2,\mathcal{Z}_2}(h_*, H_*; \Theta_2) + a \\ \underbrace{\leq}_{(a)} \Phi_{A_2,\mathcal{Z}_2}(h_*, H_*; \Theta_2^*) + a \underbrace{=}_{(b)} \mathcal{SV},\end{array}$$

with exactly the same justification of (a) and (b) as above. The bound in (2.136.b) is proved. □

2.11.6 Proof of Proposition 2.46

2.11.6.A Preliminaries

We start with the following result:

Lemma 2.52. *Let $\bar{\Theta}$ be a positive definite $d \times d$ matrix,* $B = \left[\begin{array}{c} A \\ 0,\ldots,0,1\end{array}\right]$, *and let*

$$u \mapsto \mathcal{C}(u) = A[u;1]$$

be an affine mapping from $\mathbf{R}^n$ *into* $\mathbf{R}^d$. *Finally, let* $h \in \mathbf{R}^d$, $H \in \mathbf{S}^d$ *and* $P \in \mathbf{S}^d$ *satisfy the relations*

$$0 \preceq P \prec I_d \ \&\ P \succeq \bar\Theta^{1/2} H \bar\Theta^{1/2}. \tag{2.182}$$

Then, $\zeta \sim \mathcal{SG}(\mathcal{C}(u), \bar\Theta)$ *and for every* $u \in \mathbf{R}^n$ *it holds*

$$\begin{array}{l}\ln\left(\mathbf{E}_\zeta\left\{\mathrm{e}^{h^T\zeta + \frac{1}{2}\zeta^T H \zeta}\right\}\right) \leq -\frac{1}{2}\ln \mathrm{Det}(I - P) \\ +\frac{1}{2}[u;1]^T B^T \left[\left[\begin{array}{c|c} H & h \\ \hline h^T & \end{array}\right] + [H,h]^T \bar\Theta^{1/2}[I-P]^{-1}\bar\Theta^{1/2}[H,h]\right] B[u;1]\end{array} \tag{2.183}$$

Equivalently (set $G = \bar\Theta^{-1/2} P \bar\Theta^{-1/2}$*): whenever* $h \in \mathbf{R}^d$, $H \in \mathbf{S}^d$ *and* $G \in \mathbf{S}^d$ *satisfy the relations*

$$0 \preceq G \prec \bar\Theta^{-1} \ \&\ G \succeq H, \tag{2.184}$$

one has for $\zeta \sim \mathcal{SG}(\mathcal{C}(u), \bar\Theta)$ *and every for every* $u \in \mathbf{R}^n$*:*

$$\begin{array}{l}\ln\left(\mathbf{E}_\zeta\left\{\mathrm{e}^{h^T\zeta + \frac{1}{2}\zeta^T H \zeta}\right\}\right) \leq -\frac{1}{2}\ln \mathrm{Det}(I - \bar\Theta^{1/2} G \bar\Theta^{1/2}) \\ +\frac{1}{2}[u;1]^T B^T \left[\left[\begin{array}{c|c} H & h \\ \hline h^T & \end{array}\right] + [H,h]^T [\bar\Theta^{-1} - G]^{-1} [H,h]\right] B[u;1].\end{array} \tag{2.185}$$

Proof. 1°. Let us start with the following observation:

Lemma 2.53. *Let* $\Theta \in \mathbf{S}^d_+$ *and* $S \in \mathbf{R}^{d\times d}$ *be such that* $S\Theta S^T \prec I_d$. *Then for every* $\nu \in \mathbf{R}^d$ *one has*

$$\begin{array}{l}\ln\left(\mathbf{E}_{\xi\sim\mathcal{SG}(0,\Theta)}\left\{\mathrm{e}^{\nu^T S\xi + \frac{1}{2}\xi^T S^T S \xi}\right\}\right) \leq \ln\left(\mathbf{E}_{\eta\sim\mathcal{N}(\nu,I_d)}\left\{\mathrm{e}^{\frac{1}{2}\eta^T S\Theta S^T \eta}\right\}\right) \\ = -\frac{1}{2}\ln\mathrm{Det}(I_d - S\Theta S^T) + \frac{1}{2}\nu^T\left[S\Theta S^T(I_d - S\Theta S^T)^{-1}\right]\nu.\end{array} \tag{2.186}$$

Indeed, let $\xi \sim \mathcal{SG}(0,\Theta)$ and $\eta \sim \mathcal{N}(\nu, I_d)$ be independent. We have

$$\begin{array}{l}\mathbf{E}_\xi\left\{\mathrm{e}^{\nu^T S\xi + \frac{1}{2}\xi^T S^T S\xi}\right\} \underbrace{=}_{a} \mathbf{E}_\xi\left\{\mathbf{E}_\eta\left\{\mathrm{e}^{[S\xi]^T\eta}\right\}\right\} = \mathbf{E}_\eta\left\{\mathbf{E}_\xi\left\{\mathrm{e}^{[S^T\eta]^T\xi}\right\}\right\} \\ \underbrace{\leq}_{b} \mathbf{E}_\eta\left\{\mathrm{e}^{\frac{1}{2}\eta^T S\Theta S^T\eta}\right\},\end{array}$$

where a is due to $\eta \sim \mathcal{N}(\nu, I_d)$ and b is due to $\xi \sim \mathcal{SG}(0,\Theta)$. We have verified the inequality in (2.186); the equality in (2.186) is given by direct computation. □

2°. Now, in the situation described in Lemma 2.52, by continuity it suffices to prove (2.183) in the case when $P \succeq 0$ in (2.182) is replaced with $P \succ 0$. Under the premise of the lemma, given $u \in \mathbf{R}^n$ and assuming $P \succ 0$, let us set $\mu = \mathcal{C}(u) = A[u;1]$, $\nu = P^{-1/2}\bar\Theta^{1/2}[H\mu + h]$, and $S = P^{1/2}\bar\Theta^{-1/2}$, so that $S\bar\Theta S^T = P \prec I_d$, and let $G = \bar\Theta^{-1/2}P\bar\Theta^{-1/2}$, so that $G \succeq H$. Let $\zeta \sim \mathcal{SG}(\mu, \bar\Theta)$. Representing ζ as $\zeta = \mu + \xi$

with $\xi \sim \mathcal{SG}(0,\bar{\Theta})$, we have

$$\begin{array}{l}
\ln\left(\mathbf{E}_\zeta\left\{e^{h^T\zeta+\frac{1}{2}\zeta^TH\zeta}\right\}\right)=h^T\mu+\frac{1}{2}\mu^TH\mu+\ln\left(\mathbf{E}_\xi\left\{e^{[h+H\mu]^T\xi+\frac{1}{2}\xi^TH\xi}\right\}\right)\\
\leq h^T\mu+\frac{1}{2}\mu^TH\mu+\ln\left(\mathbf{E}_\xi\left\{e^{[h+H\mu]^T\xi+\frac{1}{2}\xi^TG\xi}\right\}\right)\\
\hfill[\text{since } G\succeq H]\\
=h^T\mu+\frac{1}{2}\mu^TH\mu+\ln\left(\mathbf{E}_\xi\left\{e^{\nu^TS\xi+\frac{1}{2}\xi^TS^TS\xi}\right\}\right)\\
\hfill[\text{since } S^T\nu=h+H\mu \text{ and } G=S^TS]\\
\leq h^T\mu+\frac{1}{2}\mu^TH\mu-\frac{1}{2}\ln\mathrm{Det}(I_d-S\bar{\Theta}S^T)+\frac{1}{2}\nu^T\left[S\bar{\Theta}S^T(I_d-S\bar{\Theta}S^T)^{-1}\right]\nu\\
\hfill[\text{by Lemma 2.53 with } \Theta=\bar{\Theta}]\\
=h^T\mu+\frac{1}{2}\mu^TH\mu-\frac{1}{2}\ln\mathrm{Det}(I_d-P)+\frac{1}{2}[H\mu+h]^T\bar{\Theta}^{1/2}(I_d-P)^{-1}\bar{\Theta}^{1/2}[H\mu+h]\\
\hfill[\text{plugging in } S \text{ and } \nu].
\end{array}$$

It is immediately seen that the concluding quantity in this chain is nothing but the right-hand side quantity in (2.183). □

2.11.6.B Completing the proof of Proposition 2.46.

1°. Let us prove (2.142.a). By Lemma 2.52 (see (2.185)) applied with $\bar{\Theta}=\Theta_*$, setting $\mathcal{C}(u)=A[u;1]$, we have

$$\begin{array}{l}
\forall\left((h,H)\in\mathcal{H},G:0\preceq G\preceq\gamma^+\Theta_*^{-1},G\succeq H,u\in\mathbf{R}^n:[u;1][u;1]^T\in\mathcal{Z}\right):\\
\ln\left(\mathbf{E}_{\zeta\sim\mathcal{SG}(\mathcal{C}(u),\Theta_*)}\left\{e^{h^T\zeta+\frac{1}{2}\zeta^TH\zeta}\right\}\right)\leq-\frac{1}{2}\ln\mathrm{Det}(I-\Theta_*^{1/2}G\Theta_*^{1/2})\\
\qquad+\frac{1}{2}[u;1]^TB^T\left[\left[\begin{array}{c|c}H&h\\\hline h^T&\end{array}\right]+[H,h]^T\left[\Theta_*^{-1}-G\right]^{-1}[H,h]\right]B[u;1]\\
\leq-\frac{1}{2}\ln\mathrm{Det}(I-\Theta_*^{1/2}G\Theta_*^{1/2})\\
+\frac{1}{2}\phi_{\mathcal{Z}}\left(B^T\left[\left[\begin{array}{c|c}H&h\\\hline h^T&\end{array}\right]+[H,h]^T\left[\Theta_*^{-1}-G\right]^{-1}[H,h]\right]B\right)=\Psi_{A,\mathcal{Z}}(h,H,G),
\end{array}\tag{2.187}$$

implying, due to the origin of $\Phi_{A,\mathcal{Z}}$, that under the premise of (2.187) we have

$$\ln\left(\mathbf{E}_{\zeta\sim\mathcal{SG}(\mathcal{C}(u),\Theta_*)}\left\{e^{h^T\zeta+\frac{1}{2}\zeta^TH\zeta}\right\}\right)\leq\Phi_{A,\mathcal{Z}}(h,H),\ \forall(h,H)\in\mathcal{H}.$$

Taking into account that when $\zeta\sim\mathcal{SG}(\mathcal{C}(u),\Theta)$ with $\Theta\in\mathcal{V}$, we have also $\zeta\sim\mathcal{SG}(\mathcal{C}(u),\Theta_*)$; (2.142.$a$) follows.

2°. Now let us prove (2.142.b). All we need is to verify the relation

$$\begin{array}{l}
\forall\left((h,H)\in\mathcal{H},G:0\preceq G\preceq\gamma^+\Theta_*^{-1},G\succeq H,u\in\mathbf{R}^n:[u;1][u;1]^T\in\mathcal{Z},\Theta\in\mathcal{V}\right):\\
\qquad\ln\left(\mathbf{E}_{\zeta\sim\mathcal{SG}(\mathcal{C}(u),\Theta)}\left\{e^{h^T\zeta+\frac{1}{2}\zeta^TH\zeta}\right\}\right)\leq\Psi^\delta_{A,\mathcal{Z}}(h,H,G;\Theta);
\end{array}\tag{2.188}$$

with this relation at our disposal (2.142.b) can be obtained by the same argument as the one we used in item 1° to derive (2.142.a).

To establish (2.188), let us fix h,H,G,u,Θ satisfying the premise of (2.188); recall that under the premise of Proposition 2.46.i, we have $0\preceq\Theta\preceq\Theta_*$. Now let $\lambda\in(0,1)$, and let $\Theta_\lambda=\Theta+\lambda(\Theta_*-\Theta)$, so that $0\prec\Theta_\lambda\preceq\Theta_*$, and let $\delta_\lambda=\|\Theta_\lambda^{1/2}\Theta_*^{-1/2}-I_d\|$, implying that $\delta_\lambda\in[0,2]$. We have $0\preceq G\preceq\gamma^+\Theta_*^{-1}\preceq\gamma^+\Theta_\lambda^{-1}$, that is, H,G satisfy (2.184) w.r.t. $\bar{\Theta}=\Theta_\lambda$. As a result, for our h,G,H,u, the $\bar{\Theta}$

just defined and the $\zeta \sim \mathcal{SG}(\mathcal{C}(u), \Theta_\lambda)$ relation (2.185) hold true:

$$\begin{array}{l}\ln\left(\mathbf{E}_\zeta\left\{e^{h^T\zeta+\frac{1}{2}\zeta^T H\zeta}\right\}\right) \leq -\frac{1}{2}\ln\mathrm{Det}(I-\Theta_\lambda^{1/2}G\Theta_\lambda^{1/2})\\ \quad +\frac{1}{2}[u;1]^T B^T\left[\left[\begin{array}{c|c}H & h\\ \hline h^T & \end{array}\right]+[H,h]^T\left[\Theta_\lambda^{-1}-G\right]^{-1}[H,h]\right]B[u;1]\\ \leq -\frac{1}{2}\ln\mathrm{Det}(I-\Theta_\lambda^{1/2}G\Theta_\lambda^{1/2})\\ \quad +\frac{1}{2}\phi_{\mathcal{Z}}\left(B^T\left[\left[\begin{array}{c|c}H & h\\ \hline h^T & \end{array}\right]+[H,h]^T\left[\Theta_\lambda^{-1}-G\right]^{-1}[H,h]\right]B\right)\end{array} \tag{2.189}$$

(recall that $[u;1][u;1]^T \in \mathcal{Z}$). As a result,

$$\begin{array}{l}\ln\left(\mathbf{E}_{\zeta\sim\mathcal{SG}(\mathcal{C}(u),\Theta)}\left\{e^{h^T\zeta+\frac{1}{2}\zeta^T H\zeta}\right\}\right) \leq -\frac{1}{2}\ln\mathrm{Det}(I-\Theta_\lambda^{1/2}G\Theta_\lambda^{1/2})\\ \quad +\frac{1}{2}\phi_{\mathcal{Z}}\left(B^T\left[\left[\begin{array}{c|c}H & h\\ \hline h^T & \end{array}\right]+[H,h]^T\left[\Theta_*^{-1}-G\right]^{-1}[H,h]\right]B\right).\end{array} \tag{2.190}$$

When deriving (2.190) from (2.189), we have used that
— $\Theta \preceq \Theta_\lambda$, so that when $\zeta \sim \mathcal{SG}(\mathcal{C}(u), \Theta)$, we have also $\zeta \sim \mathcal{SG}(\mathcal{C}(u), \Theta_\lambda)$,
— $0 \preceq \Theta_\lambda \preceq \Theta_*$ and $G \prec \Theta_*^{-1}$, whence $[\Theta_\lambda^{-1} - G]^{-1} \preceq [\Theta_*^{-1} - G]^{-1}$,
— $\mathcal{Z} \subset \mathbf{S}_+^{n+1}$, whence $\phi_{\mathcal{Z}}$ is $\succeq$-monotone: $\phi_{\mathcal{Z}}(M) \leq \phi_{\mathcal{Z}}(N)$ whenever $M \preceq N$.

By Lemma 2.51 applied with Θ_λ in the role of Θ and δ_λ in the role of δ, we have

$$\begin{array}{l}-\frac{1}{2}\ln\mathrm{Det}(I-\Theta_\lambda^{1/2}G\Theta_\lambda^{1/2}) \leq -\frac{1}{2}\ln\mathrm{Det}(I-\Theta_*^{1/2}G\Theta_*^{1/2})+\frac{1}{2}\mathrm{Tr}([\Theta_\lambda-\Theta_*]G)\\ \quad +\frac{\delta_\lambda(2+\delta_\lambda)}{2(1-\|\Theta_*^{1/2}G\Theta_*^{1/2}\|)}\|\Theta_*^{1/2}G\Theta_*^{1/2}\|_F^2.\end{array}$$

Consequently, (2.190) implies that

$$\begin{array}{l}\ln\left(\mathbf{E}_{\zeta\sim\mathcal{SG}(\mathcal{C}(u),\Theta)}\left\{e^{h^T\zeta+\frac{1}{2}\zeta^T H\zeta}\right\}\right) \leq -\frac{1}{2}\ln\mathrm{Det}(I-\Theta_*^{1/2}G\Theta_*^{1/2})+\frac{1}{2}\mathrm{Tr}([\Theta_\lambda-\Theta_*]G)\\ \quad +\frac{\delta_\lambda(2+\delta_\lambda)}{2(1-\|\Theta_*^{1/2}G\Theta_*^{1/2}\|)}\|\Theta_*^{1/2}G\Theta_*^{1/2}\|_F^2\\ \quad +\frac{1}{2}\phi_{\mathcal{Z}}\left(B^T\left[\left[\begin{array}{c|c}H & h\\ \hline h^T & \end{array}\right]+[H,h]^T\left[\Theta_*^{-1}-G\right]^{-1}[H,h]\right]B\right).\end{array}$$

The resulting inequality holds true for all small positive λ; taking lim inf of the right-hand side as $\lambda \to +0$, and recalling that $\Theta_0 = \Theta$, we get

$$\begin{array}{l}\ln\left(\mathbf{E}_{\zeta\sim\mathcal{SG}(\mathcal{C}(u),\Theta)}\left\{e^{h^T\zeta+\frac{1}{2}\zeta^T H\zeta}\right\}\right) \leq -\frac{1}{2}\ln\mathrm{Det}(I-\Theta_*^{1/2}G\Theta_*^{1/2})+\frac{1}{2}\mathrm{Tr}([\Theta-\Theta_*]G)\\ \quad +\frac{\delta(2+\delta)}{2(1-\|\Theta_*^{1/2}G\Theta_*^{1/2}\|)}\|\Theta_*^{1/2}G\Theta_*^{1/2}\|_F^2\\ \quad +\frac{1}{2}\phi_{\mathcal{Z}}\left(B^T\left[\left[\begin{array}{c|c}H & h\\ \hline h^T & \end{array}\right]+[H,h]^T\left[\Theta_*^{-1}-G\right]^{-1}[H,h]\right]B\right)\end{array}$$

(note that under the premise of Proposition 2.46.i we clearly have $\liminf_{\lambda\to+0}\delta_\lambda \leq \delta$). The right-hand side of the resulting inequality is nothing but $\Psi^\delta_{A,\mathcal{Z}}(h, H, G; \Theta)$—see (2.141)—and we arrive at the inequality required in the conclusion of (2.188).

3°. To complete the proof of Proposition 2.46.i, it remains to show that functions $\Phi_{A,\mathcal{Z}}$, $\Phi^\delta_{A,\mathcal{Z}}$, as announced in the proposition, possess continuity, convexity-concavity, and coerciveness properties. Let us verify that this indeed is so for $\Phi^\delta_{A,\mathcal{Z}}$; the reasoning which follows, with obvious simplifications, is applicable to $\Phi_{A,\mathcal{Z}}$ as well.

Observe, first, that for exactly the same reasons as in item $\mathbf{4}^o$ of the proof of Proposition 2.43, the function $\Psi^\delta_{A,\mathcal{Z}}(h,H,G;\Theta)$ is real-valued, continuous and convex-concave on the domain

$$\widehat{\mathcal{H}}\times\mathcal{V}=\{(h,H,G):-\gamma^+\Theta_*^{-1}\preceq H\preceq\gamma^+\Theta_*^{-1},0\preceq G\preceq\gamma^+\Theta_*^{-1},H\preceq G\}\times\mathcal{V}.$$

The function $\Phi^\delta_{A,\mathcal{Z}}(h,H;\Theta):\mathcal{H}\times\mathcal{V}\to\mathbf{R}$ is obtained from $\Psi^\delta_{A,\mathcal{Z}}(h,H,G;\Theta)$ by the following two operations: we first minimize $\Psi^\delta_{A,\mathcal{Z}}(h,H,G;\Theta)$ over G linked to (h,H) by the convex constraints $0\preceq G\preceq\gamma^+\Theta_*^{-1}$ and $G\succeq H$, thus obtaining a function

$$\bar{\Phi}(h,H;\Theta):\underbrace{\{(h,H):-\gamma^+\Theta_*^{-1}\preceq H\preceq\gamma^+\Theta_*^{-1}\}}_{\bar{\mathcal{H}}}\times\mathcal{V}\to\mathbf{R}\cup\{+\infty\}\cup\{-\infty\}.$$

Second, we restrict the function $\bar{\Phi}(h,H;\Theta)$ from $\bar{\mathcal{H}}\times\mathcal{V}$ onto $\mathcal{H}\times\mathcal{V}$. For $(h,H)\in\bar{\mathcal{H}}$, the set of G's linked to (h,H) by the above convex constraints clearly is a nonempty compact set; as a result, $\bar{\Phi}$ is a real-valued convex-concave function on $\bar{\mathcal{H}}\times\mathcal{V}$. From continuity of $\Psi^\delta_{A,\mathcal{Z}}$ on its domain it immediately follows that $\Psi^\delta_{A,\mathcal{Z}}$ is bounded and uniformly continuous on every bounded subset of this domain. This implies that $\bar{\Phi}(h,H;\Theta)$ is bounded in every domain of the form $\bar{B}\times\mathcal{V}$, where $\bar{B}$ is a bounded subset of $\bar{\mathcal{H}}$, and is continuous on $\bar{B}\times\mathcal{V}$ in $\Theta\in\mathcal{V}$ with properly selected modulus of continuity independent of $(h,H)\in\bar{B}$. Furthermore, by construction, $\mathcal{H}\subset\operatorname{int}\bar{\mathcal{H}}$, implying that if B is a convex compact subset of $\mathcal{H}$, it belongs to the interior of a properly selected convex compact subset $\bar{B}$ of $\bar{\mathcal{H}}$. Since $\bar{\Phi}$ is bounded on $\bar{B}\times\mathcal{V}$ and is convex in (h,H), the function $\bar{\Phi}$ is a Lipschitz continuous in $(h,H)\in B$ with Lipschitz constant which can be selected to be independent of $\Theta\in\mathcal{V}$. Taking into account that $\mathcal{H}$ is convex and closed, the bottom line is that $\Phi^\delta_{A,\mathcal{Z}}$ is not just real-valued convex-concave function on the domain $\mathcal{H}\times\mathcal{V}$, but is also continuous on this domain.

Coerciveness of $\Phi^\delta_{A,\mathcal{Z}}(h,H;\Theta)$ in (h,H) is proved in exactly the same way as the similar property of function (2.130); see item 5° in the proof of Proposition 2.43. The proof of item (i) of Proposition 2.46 is complete.

$\mathbf{4}^o$. Item (ii) of Proposition 2.46 can be derived from item (i) of the proposition following the steps of the proof of (ii) of Proposition 2.43. □

Chapter Three

From Hypothesis Testing to Estimating Functionals

In this chapter we extend the techniques developed in Chapter 2 beyond the hypothesis testing problem and apply them to estimating properly structured scalar functionals of the unknown signal, specifically:

- In simple observation schemes—linear (and more generally, N-*convex*; see Section 3.2) functionals on unions of convex sets (Sections 3.1 and 3.2);
- Beyond simple observation schemes—linear and quadratic functionals on convex sets (Sections 3.3 and 3.4).

3.1 ESTIMATING LINEAR FORMS ON UNIONS OF CONVEX SETS

The key to the subsequent developments in this section and in Sections 3.3 and 3.4 is the following simple observation. Let $\mathcal{P} = \{P_x : x \in \mathcal{X}\}$ be a parametric family of distributions on $\mathbf{R}^d$, $\mathcal{X}$ being a convex subset of some $\mathbf{R}^m$. Suppose that given a linear form $g^T x$ on $\mathbf{R}^m$ and an observation $\omega \sim P_x$ stemming from unknown signal $x \in \mathcal{X}$, we want to recover $g^T x$, and intend to use for this purpose an affine function $h^T\omega + \kappa$ of the observation. How do we ensure that the recovery, with a given probability $1 - \epsilon$, deviates from $g^T x$ by at most a given margin ρ, for all $x \in \mathcal{X}$?

Let us focus on one "half" of the answer: how to ensure that the probability of the event $h^T\omega + \kappa > g^T x + \rho$ does not exceed $\epsilon/2$, for every $x \in \mathcal{X}$. The answer becomes easy when assuming that we have at our disposal an upper bound on the exponential moments of the distributions from the family—a function $\Phi(h; x)$ such that

$$\ln\left(\int e^{h^T\omega} P_x(d\omega)\right) \leq \Phi(h; x)\ \forall(h \in \mathbf{R}^n, x \in \mathcal{X}).$$

Indeed, for obvious reasons, in this case the P_x-probability of the event $h^T\omega + \kappa - g^T x > \rho$ is at most

$$\exp\{\Phi(h; x) - [g^T x + \rho - \kappa]\}.$$

To add some flexibility, note that when $\alpha > 0$, the event in question is the same as the event $(h/\alpha)^T\omega + \kappa/\alpha > [g^T x + \rho]/\alpha$; thus we arrive at a parametric family of upper bounds

$$\exp\{\Phi(h/\alpha; x) - [g^T x + \rho - \kappa]/\alpha\},\ \alpha > 0,$$

on the P_x-probability of our "bad" event. It follows that a sufficient condition for this probability to be $\leq \epsilon/2$, *for a given* $x \in \mathcal{X}$, is the existence of $\alpha > 0$ such that

$$\exp\{\Phi(h/\alpha; x) - [g^T x + \rho - \kappa]/\alpha\} \leq \epsilon/2,$$

or

$$\Phi(h/\alpha; x) - [g^T x + \rho - \kappa]/\alpha \le \ln(\epsilon/2),$$

or, which again is the same, the existence of $\alpha > 0$ such that

$$\alpha\Phi(h/\alpha; x) + \alpha\ln(2/\epsilon) - g^T x \le \rho - \kappa.$$

In other words, a sufficient condition for the relation

$$\text{Prob}_{\omega\sim P_x}\{h^T\omega + \kappa > g^T x + \rho\} \le \epsilon/2$$

is

$$\inf_{\alpha>0}[\alpha\Phi(h/\alpha; x) + \alpha\ln(2/\epsilon) - g^T x] \le \rho - \kappa.$$

If we want the bad event in question to take place with P_x-probability $\le \epsilon/2$ *whatever be* $x \in \mathcal{X}$, the sufficient condition for this is

$$\sup_{x\in\mathcal{X}}\inf_{\alpha>0}[\alpha\Phi(h/\alpha; x) + \alpha\ln(2/\epsilon) - g^T x] \le \rho - \kappa. \tag{3.1}$$

Now assume that $\mathcal{X}$ is convex and compact, and $\Phi(h; x)$ is continuous, convex in h, and concave in x. In this case the function $\alpha\Phi(h/\alpha; x)$ is convex in (h, α) in the domain $\alpha > 0$ [1] and is concave in x, so that we can switch sup and inf, thus arriving at the sufficient condition

$$\exists\alpha > 0 : \max_{x\in\mathcal{X}}\left[\alpha\Phi(h/\alpha; x) + \alpha\ln(2/\epsilon) - g^T x\right] \le \rho - \kappa, \tag{3.2}$$

for the validity of the relation

$$\forall x \in \mathcal{X} : \text{Prob}_{\omega\sim P_x}\left\{h^T\omega + \kappa - g^T x \le \rho\right\} \ge 1 - \epsilon/2.$$

Note that our sufficient condition is expressed in terms of a *convex* constraint on h, κ, ρ, α. Consider also the dramatic simplification allowed by the convexity-concavity of Φ: in (3.1), every $x \in \mathcal{X}$ should be "served" by its own α, so that (3.1) is an infinite system of constraints on h, ρ, κ. In contrast, in (3.2) all $x \in \mathcal{X}$ are "served" by a *single* α.

The developments in this section and Sections 3.3 and 3.4 are no more than implementations, under various circumstances, of the simple idea we have just outlined.

3.1.1 The problem

Let $\mathcal{O} = (\Omega, \Pi, \{p_\mu(\cdot) : \mu \in \mathcal{M}\}, \mathcal{F})$ be a simple observation scheme (see Section 2.4.2). The problem we consider in this section is as follows:

> We are given a positive integer K and I nonempty convex compact sets $X_j \subset \mathbf{R}^n$, along with affine mappings $A_j(\cdot) : \mathbf{R}^n \to \mathbf{R}^M$ such that $A_j(x) \in \mathcal{M}$ whenever $x \in X_j$, $1 \le j \le I$. In addition, we are given a linear function

[1] This is due to the following standard fact: if $f(h)$ is a convex function, then the *projective transformation* $\alpha f(h/\alpha)$ of f is convex in (h, α) in the domain $\alpha > 0$.

$g^T x$ on $\mathbf{R}^n$. Given random observation

$$\omega^K = (\omega_1, ..., \omega_K)$$

with ω_k drawn, independently across k, from $p_{A_j(x)}$ with $j \leq I$ and $x \in X_j$, we want to recover $g^T x$.

It should be stressed that *we do not know j and x underlying our observation.*

Given reliability tolerance $\epsilon \in (0,1)$, we quantify the performance of a candidate estimate—a Borel function $\widehat{g}(\cdot) : \Omega \to \mathbf{R}$—by the worst-case, over j and x, width of a $(1-\epsilon)$-confidence interval. Precisely, we say that $\widehat{g}(\cdot)$ is (ρ,ϵ)-reliable if

$$\forall (j \leq I, x \in X_j) : \mathrm{Prob}_{\omega \sim p_{A_j(x)}}\{|\widehat{g}(\omega) - g^T x| > \rho\} \leq \epsilon. \tag{3.3}$$

We define the ϵ-risk of the estimate as

$$\mathrm{Risk}_\epsilon[\widehat{g}] = \inf\{\rho : \widehat{g} \text{ is } (\rho,\epsilon)\text{-reliable}\},$$

i.e., $\mathrm{Risk}_\epsilon[\widehat{g}]$ is the smallest ρ such that $\widehat{g}$ is (ρ,ϵ)-reliable.

The technique we are about to develop originates from [131] where estimating a linear form on a convex compact set in a simple o.s. (i.e., the case $I = 1$ of the problem at hand) was considered, and where it was proved that in this situation the estimate

$$\widehat{g}(\omega^K) = \sum_k \phi(\omega_k) + \varkappa$$

with properly selected $\phi \in \mathcal{F}$ and $\kappa \in \mathbf{R}$ is near-optimal. The problem of estimating linear functionals of a signal in Gaussian o.s. has a long history; see, e.g., [38, 40, 124, 125, 125, 127, 126, 170, 179] and references therein. In particular, in the case of $I = 1$, using different techniques, a similar fact was proved by D. Donoho [64] in 1991; related results in the case of $I > 1$ are available in [41, 42].

3.1.2 The estimate

In the sequel, we associate with the simple o.s. $\mathcal{O} = (\Omega, \Pi, \{p_\mu(\cdot) : \mu \in \mathcal{M}\}, \mathcal{F})$ in question the function

$$\Phi_{\mathcal{O}}(\phi;\mu) = \ln\left(\int e^{\phi(\omega)} p_\mu(\omega)\Pi(d\omega)\right), \quad (\phi,\mu) \in \mathcal{F} \times \mathcal{M}.$$

Recall that by definition of a simple o.s., this function is real-valued on $\mathcal{F} \times \mathcal{M}$, concave in $\mu \in \mathcal{M}$, convex in $\phi \in \mathcal{F}$, and continuous on $\mathcal{F} \times \mathcal{M}$ (the latter follows from convexity-concavity and relative openness of $\mathcal{M}$ and $\mathcal{F}$).

Let us associate with a pair (i,j), $1 \leq i,j \leq I$, the functions

$$\begin{array}{rcl}
\Phi_{ij}(\alpha,\phi;x,y) & = & \frac{1}{2}\big[K\alpha\Phi_{\mathcal{O}}(\phi/\alpha; A_i(x)) + K\alpha\Phi_{\mathcal{O}}(-\phi/\alpha; A_j(y)) \\
& & +g^T(y-x) + 2\alpha\ln(2I/\epsilon)\big] : \ \{\alpha > 0, \phi \in \mathcal{F}\} \times [X_i \times X_j] \to \mathbf{R}, \\
\Psi_{ij}(\alpha,\phi) & = & \max\limits_{x \in X_i, y \in X_j} \Phi_{ij}(\alpha,\phi;x,y) \\
& = & \frac{1}{2}\left[\Psi_{i,+}(\alpha,\phi) + \Psi_{j,-}(\alpha,\phi)\right] : \ \{\alpha > 0\} \times \mathcal{F} \to \mathbf{R}
\end{array}$$

where

$$\begin{array}{rcl}\Psi_{\ell,+}(\beta,\psi) & = & \max\limits_{x\in X_\ell}\left[K\beta\Phi_{\mathcal{O}}(\psi/\beta;A_\ell(x)) - g^Tx + \beta\ln(2I/\epsilon)\right]: \\ & & \qquad\qquad\{\beta>0,\psi\in\mathcal{F}\}\to\mathbf{R}, \\ \Psi_{\ell,-}(\beta,\psi) & = & \max\limits_{x\in X_\ell}\left[K\beta\Phi_{\mathcal{O}}(-\psi/\beta;A_\ell(x)) + g^Tx + \beta\ln(2I/\epsilon)\right]: \\ & & \qquad\qquad\{\beta>0,\psi\in\mathcal{F}\}\to\mathbf{R}.\end{array}$$

Note that the function $\alpha\Phi_{\mathcal{O}}(\phi/\alpha;A_i(x))$ is obtained from the continuous convex-concave function $\Phi_{\mathcal{O}}(\cdot,\cdot)$ by projective transformation in the convex argument, and affine substitution in the concave argument, so that the former function is convex-concave and continuous on the domain $\{\alpha>0,\phi\in\mathcal{X}\}\times X_i$. By similar argument, the function $\alpha\Phi_{\mathcal{O}}(-\phi/\alpha;A_j(y))$ is convex-concave and continuous on the domain $\{\alpha>0,\phi\in\mathcal{F}\}\times X_j$. These observations combine with compactness of X_i and X_j to imply that $\Psi_{ij}(\alpha,\phi)$ is a real-valued continuous convex function on the domain

$$\mathcal{F}^+=\{\alpha>0\}\times\mathcal{F}.$$

Observe that functions $\Psi_{ii}(\alpha,\phi)$ are nonnegative on $\mathcal{F}^+$. Indeed, selecting some $\bar{x}\in X_i$, and setting $\mu=A_i(\bar{x})$, we have

$$\begin{array}{rcl}\Psi_{ii}(\alpha,\phi) & \geq & \Phi_{ii}(\alpha,\phi;\bar{x},\bar{x}) = \alpha\left[\frac{1}{2}[\Phi_{\mathcal{O}}(\phi/\alpha;\mu)+\Phi_{\mathcal{O}}(-\phi/\alpha;\mu)]K+\ln(2I/\epsilon)\right] \\ & \geq & \alpha\Big[\underbrace{\Phi_{\mathcal{O}}(0;\mu)}_{=0}K+\ln(2I/\epsilon)\Big] = \alpha\ln(2I/\epsilon)\geq 0\end{array}$$

(we have used convexity of $\Phi_{\mathcal{O}}$ in the first argument).

Functions Ψ_{ij} give rise to convex and feasible optimization problems

$$\mathrm{Opt}_{ij}=\mathrm{Opt}_{ij}(K)=\min_{(\alpha,\phi)\in\mathcal{F}^+}\Psi_{ij}(\alpha,\phi). \tag{3.4}$$

By its origin, Opt_{ij} is either a real, or $-\infty$; by the observation above, Opt_{ii} are nonnegative. Our estimate is as follows.

1. For $1\leq i,j\leq I$, we select some feasible solutions α_{ij},ϕ_{ij} to problems (3.4) (the less the values of the corresponding objectives, the better) and set

$$\begin{array}{rcl}\rho_{ij} & = & \Psi_{ij}(\alpha_{ij},\phi_{ij})=\frac{1}{2}\left[\Psi_{i,+}(\alpha_{ij},\phi_{ij})+\Psi_{j,-}(\alpha_{ij},\phi_{ij})\right] \\ \varkappa_{ij} & = & \frac{1}{2}\left[\Psi_{j,-}(\alpha_{ij},\phi_{ij})-\Psi_{i,+}(\alpha_{ij},\phi_{ij})\right] \\ g_{ij}(\omega^K) & = & \sum_{k=1}^K\phi_{ij}(\omega_k)+\varkappa_{ij} \\ \rho & = & \max_{1\leq i,j\leq I}\rho_{ij}.\end{array} \tag{3.5}$$

2. Given observation ω^K, we specify the estimate $\widehat{g}(\omega^K)$ as follows:

$$\begin{array}{rcl}r_i & = & \max_{j\leq I}g_{ij}(\omega^K) \\ c_j & = & \min_{i\leq I}g_{ij}(\omega^K) \\ \widehat{g}(\omega^K) & = & \frac{1}{2}\left[\min_{i\leq I}r_i+\max_{j\leq I}c_j\right].\end{array} \tag{3.6}$$

3.1.3 Main result

Proposition 3.1. *The ϵ-risk of the estimate $\widehat{g}(\omega^K)$ can be upper-bounded as follows:*

$$\mathrm{Risk}_\epsilon[\widehat{g}] \leq \rho. \tag{3.7}$$

Proof. Let the common distribution p of components ω_k independent across k in observation ω^K be $p_{A_\ell(u)}$ for some $\ell \leq I$ and $u \in X_\ell$. Let us fix these ℓ and u; we denote $\mu = A_\ell(u)$ and let p^K stand for the distribution of ω^K.

1°. We have

$$\begin{array}{l}
\Psi_{\ell,+}(\alpha_{\ell j}, \phi_{\ell j}) = \max_{x\in X_\ell}\left[K\alpha_{\ell j}\Phi_{\mathcal{O}}(\phi_{\ell j}/\alpha_{\ell j}, A_\ell(x)) - g^Tx\right] + \alpha_{\ell j}\ln(2I/\epsilon)\\
\geq K\alpha_{\ell j}\Phi_{\mathcal{O}}(\phi_{\ell j}/\alpha_{\ell j}, \mu) - g^Tu + \alpha_{\ell j}\ln(2I/\epsilon) \text{ [since } u\in X_\ell \text{ and } \mu = A_\ell(u)]\\
= K\alpha_{\ell j}\ln\left(\int \exp\{\phi_{\ell j}(\omega)/\alpha_{\ell j}\}p_\mu(\omega)\Pi(d\omega)\right) - g^Tu + \alpha_{\ell j}\ln(2I/\epsilon)\\
\hfill \text{[by definition of } \Phi_{\mathcal{O}}]\\
= \alpha_{\ell j}\ln\left(\mathbf{E}_{\omega^K\sim p^K}\left\{\exp\{\alpha_{\ell j}^{-1}\sum_k\phi_{\ell j}(\omega_k)\}\right\}\right) - g^Tu + \alpha_{\ell j}\ln(2I/\epsilon)\\
= \alpha_{\ell j}\ln\left(\mathbf{E}_{\omega^K\sim p^K}\left\{\exp\{\alpha_{\ell j}^{-1}[g_{\ell j}(\omega^K) - \varkappa_{\ell j}]\}\right\}\right) - g^Tu + \alpha_{\ell j}\ln(2I/\epsilon)\\
= \alpha_{\ell j}\ln\left(\mathbf{E}_{\omega^K\sim p^K}\left\{\exp\{\alpha_{\ell j}^{-1}[g_{\ell j}(\omega^K) - g^Tu - \rho_{\ell j}]\}\right\}\right) + \rho_{\ell j} - \varkappa_{\ell j} + \alpha_{\ell j}\ln(2I/\epsilon)\\
\geq \alpha_{\ell j}\ln\left(\mathrm{Prob}_{\omega^K\sim p^K}\left\{g_{\ell j}(\omega^K) > g^Tu + \rho_{\ell j}\right\}\right) + \rho_{\ell j} - \varkappa_{\ell j} + \alpha_{\ell j}\ln(2I/\epsilon)\\
\Rightarrow\\
\alpha_{\ell j}\ln\left(\mathrm{Prob}_{\omega^K\sim p^K}\left\{g_{\ell j}(\omega^K) > g^Tu + \rho_{\ell j}\right\}\right) \leq \Psi_{\ell,+}(\alpha_{\ell j}, \phi_{\ell j}) + \varkappa_{\ell j} - \rho_{\ell j} + \alpha_{\ell j}\ln(\frac{\epsilon}{2I})\\
= \alpha_{\ell j}\ln(\frac{\epsilon}{2I}) \text{ [by (3.5)]},
\end{array}$$

and we arrive at

$$\mathrm{Prob}_{\omega^K\sim p^K}\left\{g_{\ell j}(\omega^K) > g^Tu + \rho_{\ell j}\right\} \leq \frac{\epsilon}{2I}. \tag{3.8}$$

Similarly,

$$\begin{array}{l}
\Psi_{\ell,-}(\alpha_{i\ell}, \phi_{i\ell}) = \max_{y\in X_\ell}\left[K\alpha_{i\ell}\Phi_{\mathcal{O}}(-\phi_{i\ell}/\alpha_{i\ell}, A_\ell(y)) + g^Ty\right] + \alpha_{i\ell}\ln(2I/\epsilon)\\
\geq K\alpha_{i\ell}\Phi_{\mathcal{O}}(-\phi_{i\ell}/\alpha_{i\ell}, \mu) + g^Tu + \alpha_{i\ell}\ln(2I/\epsilon) \text{ [since } u\in X_\ell \text{ and } \mu = A_\ell(u)]\\
= K\alpha_{i\ell}\ln\left(\int \exp\{-\phi_{i\ell}(\omega)/\alpha_{i\ell}\}p_\mu(\omega)\Pi(d\omega)\right) + g^Tu + \alpha_{i\ell}\ln(2I/\epsilon)\\
\hfill \text{[by definition of } \Phi_{\mathcal{O}}]\\
= \alpha_{i\ell}\ln\left(\mathbf{E}_{\omega^K\sim p^K}\left\{\exp\{-\alpha_{i\ell}^{-1}\sum_k\phi_{i\ell}(\omega_k)\}\right\}\right) + g^Tu + \alpha_{i\ell}\ln(2I/\epsilon)\\
= \alpha_{i\ell}\ln\left(\mathbf{E}_{\omega^K\sim p^K}\left\{\exp\{\alpha_{i\ell}^{-1}[-g_{i\ell}(\omega^K) + \varkappa_{i\ell}]\}\right\}\right) + g^Tu + \alpha_{i\ell}\ln(2I/\epsilon)\\
= \alpha_{i\ell}\ln\left(\mathbf{E}_{\omega^K\sim p^K}\left\{\exp\{\alpha_{i\ell}^{-1}[-g_{i\ell}(\omega^K) + g^Tu - \rho_{i\ell}]\}\right\}\right) + \rho_{i\ell} + \varkappa_{i\ell} + \alpha_{i\ell}\ln(2I/\epsilon)\\
\geq \alpha_{i\ell}\ln\left(\mathrm{Prob}_{\omega^K\sim p^K}\left\{g_{i\ell}(\omega^K) < g^Tu - \rho_{i\ell}\right\}\right) + \rho_{i\ell} + \varkappa_{i\ell} + \alpha_{i\ell}\ln(2I/\epsilon)\\
\Rightarrow\\
\alpha_{i\ell}\ln\left(\mathrm{Prob}_{\omega^K\sim p^K}\left\{g_{i\ell}(\omega^K) < g^Tu - \rho_{i\ell}\right\}\right) \leq \Psi_{\ell,-}(\alpha_{i\ell}, \phi_{i\ell}) - \varkappa_{i\ell} - \rho_{i\ell} + \alpha_{i\ell}\ln(\frac{\epsilon}{2I})\\
= \alpha_{i\ell}\ln(\frac{\epsilon}{2I}) \text{ [by (3.5)]},
\end{array}$$

and we arrive at

$$\mathrm{Prob}_{\omega^K\sim p^K}\left\{g_{i\ell}(\omega^K) < g^Tu - \rho_{i\ell}\right\} \leq \frac{\epsilon}{2I}. \tag{3.9}$$

2°. Let

$$\mathcal{E} = \{\omega^K : g_{\ell j}(\omega^K) \leq g^Tu + \rho_{\ell j},\ g_{i\ell}(\omega^K) \geq g^Tu - \rho_{i\ell}, 1 \leq i, j \leq I\}.$$

From (3.8) and (3.9) and the union bound it follows that p^K-probability of the event $\mathcal{E}$ is $\geq 1 - \epsilon$. As a result, all we need to complete the proof of the proposition is to verify that

$$\omega^K \in \mathcal{E} \Rightarrow |\widehat{g}(\omega^K) - g^T u| \leq \rho_\ell := \max[\max_i \rho_{i\ell}, \max_j \rho_{\ell j}], \tag{3.10}$$

since clearly $\rho_\ell \leq \rho := \max_{i,j} \rho_{ij}$. To this end, let us fix $\omega^K \in \mathcal{E}$, and let E be the $I \times I$ matrix with entries $E_{ij} = g_{ij}(\omega^K)$, $1 \leq i, j \leq I$. The quantity r_i—see (3.6)—is the maximum of the entries in the i-th row of E, while the quantity c_j is the minimum of the entries in the j-th column of E. In particular, $r_i \geq E_{ij} \geq c_j$ for all i, j, implying that $r_i \geq c_j$ for all i, j. Now, let

$$\Delta = [g^T u - \rho_\ell, g^T u + \rho_\ell].$$

Since $\omega^K \in \mathcal{E}$, we have $E_{\ell\ell} = g_{\ell\ell}(\omega^K) \geq g^T u - \rho_{\ell\ell} \geq g^T u - \rho_\ell$ and $E_{\ell j} = g_{\ell j}(\omega^K) \leq g^T u + \rho_{\ell j} \leq g^T u + \rho_\ell$ for all j, implying that $r_\ell = \max_j E_{\ell j} \in \Delta$. Similarly, $\omega^K \in \mathcal{E}$ implies that $E_{\ell\ell} = g_{\ell\ell}(\omega^K) \leq g^T u + \rho_\ell$ and $E_{i\ell} = g_{i\ell}(\omega^K) \geq g^T u - \rho_{i\ell} \geq g^T u - \rho_\ell$ for all i, implying that $c_\ell = \min_i E_{i\ell} \in \Delta$. We see that both r_ℓ and c_ℓ belong to Δ; since $r_* := \min_i r_i \leq r_\ell$ and, as have already seen, $r_i \geq c_\ell$ for all i, we conclude that $r_* \in \Delta$. By a similar argument, $c_* := \max_j c_j \in \Delta$ as well. By construction, $\widehat{g}(\omega^K) = \frac{1}{2}[r_* + c_*]$, that is, $\widehat{g}(\omega^K) \in \Delta$, and the conclusion in (3.10) indeed takes place. □

Remark 3.2. Let us consider a special case of $I = 1$. In this case, given a K-repeated observation of the signal in a simple o.s., our construction yields an estimate of a linear form $g^T x$ of unknown signal x, known to belong to a given convex compact set X_1. This estimate is

$$\widehat{g}(\omega^K) = \sum_{k=1}^{K} \phi(\omega_k) + \kappa, \tag{3.11}$$

and is associated with the optimization problem

$$\begin{aligned}
&\min_{\alpha>0, \phi \in \mathcal{F}} \left\{\Psi(\alpha, \phi) := \tfrac{1}{2}\left[\Psi_+(\alpha, \phi) + \Psi_-(\alpha, \phi)\right]\right\}, \\
&\Psi_+(\alpha, \phi) = \max_{x \in X_1} \left[K\alpha\Phi_{\mathcal{O}}(\phi/\alpha, A_1(x)) - g^T x + \alpha \ln(2/\epsilon)\right], \\
&\Psi_-(\alpha, \phi) = \max_{x \in X_1} \left[K\alpha\Phi_{\mathcal{O}}(-\phi/\alpha, A_1(x)) + g^T x + \alpha \ln(2/\epsilon)\right].
\end{aligned}$$

By Proposition 3.1, when α, ϕ is a feasible solution to the problem and

$$\kappa = \tfrac{1}{2}[\Psi_-(\alpha, \phi) - \Psi_+(\alpha, \phi)],$$

the ϵ-risk of estimate (3.11) does not exceed $\Psi(\alpha, \phi)$.

3.1.4 Near-optimality

Observe that by properly selecting ϕ_{ij} and α_{ij} we can make, in a computationally efficient manner, the upper bound ρ on the ϵ-risk of the above estimate arbitrarily close to

$$\mathrm{Opt}(K) = \max_{1 \leq i,j \leq I} \mathrm{Opt}_{ij}(K).$$

We are about to demonstrate that the quantity $\mathrm{Opt}(K)$ "nearly lower-bounds" the minimax optimal ϵ-risk

$$\mathrm{Risk}^*_\epsilon(K) = \inf_{\widehat{g}(\cdot)} \mathrm{Risk}_\epsilon[\widehat{g}],$$

the infimum being taken over all estimates (all Borel functions of ω^K). The precise statement is as follows:

Proposition 3.3. *In the situation of this section, let $\epsilon \in (0,1/2)$ and $\overline{K}$ be a positive integer. Then for every integer K satisfying*

$$K/\overline{K} > \frac{2\ln(2I/\epsilon)}{\ln\left(\frac{1}{4\epsilon(1-\epsilon)}\right)} \tag{3.12}$$

one has

$$\mathrm{Opt}(K) \leq \mathrm{Risk}^*_\epsilon(\overline{K}). \tag{3.13}$$

In addition, in the special case where for every i,j there exists $x_{ij} \in X_i \cap X_j$ such that $A_i(x_{ij}) = A_j(x_{ij})$ one has

$$K \geq \overline{K} \Rightarrow \mathrm{Opt}(K) \leq \frac{2\ln(2I/\epsilon)}{\ln\left(\frac{1}{4\epsilon(1-\epsilon)}\right)}\mathrm{Risk}^*_\epsilon(\overline{K}). \tag{3.14}$$

For proof, see Section 3.6.1.

3.1.5 Illustration

We illustrate our construction with the simplest possible example in which $X_i = \{x_i\}$ are singletons in $\mathbf{R}^n$, $i = 1, ..., I$, and the observation scheme is Gaussian. Thus, setting $y_i = A_i(x_i) \in \mathbf{R}^m$, the observation's components ω_k, $1 \leq k \leq K$, stemming from the signal x_i, are drawn, independently of each other, from the normal distribution $\mathcal{N}(y_i, I_m)$. The family $\mathcal{F}$ of functions ϕ associated with Gaussian o.s. is the family of all affine functions $\phi(\omega) = \phi_0 + \varphi^T\omega$ on the observation space (which at present is $\mathbf{R}^m$); we identify $\phi \in \mathcal{F}$ with the pair (ϕ_0, φ). The function $\Psi_{\mathcal{O}}$ associated with the Gaussian observation scheme with m-dimensional observations is

$$\Phi_{\mathcal{O}}(\phi; \mu) = \phi_0 + \varphi^T\mu + \tfrac{1}{2}\varphi^T\varphi : (\mathbf{R} \times \mathbf{R}^m) \times \mathbf{R}^m \to \mathbf{R};$$

a straightforward computation shows that in the case in question, setting

$$\theta = \ln(2I/\epsilon),$$

we have

$$\begin{array}{rcl}\Psi_{i,+}(\alpha,\phi) & = & K\alpha\left[\phi_0+\varphi^Ty_i/\alpha+\frac{1}{2}\varphi^T\varphi/\alpha^2\right]+\alpha\theta-g^Tx_i\\ & = & K\alpha\phi_0+K\varphi^Ty_i-g^Tx_i+\frac{K}{2\alpha}\varphi^T\varphi+\alpha\theta,\\ \Psi_{j,-}(\alpha,\phi) & = & -K\alpha\phi_0-K\varphi^Ty_j+g^Tx_j+\frac{K}{2\alpha}\varphi^T\varphi+\alpha\theta,\\ \mathrm{Opt}_{ij} & = & \inf\limits_{\alpha>0,\phi}\frac{1}{2}\left[\Psi_{i,+}(\alpha,\phi)+\Psi_{j,-}(\alpha,\phi)\right]\\ & = & \frac{1}{2}g^T[x_j-x_i]+\inf\limits_{\varphi}\left[\frac{K}{2}\varphi^T[y_i-y_j]+\inf\limits_{\alpha>0}\left[\frac{K}{2\alpha}\varphi^T\varphi+\alpha\theta\right]\right]\\ & = & \frac{1}{2}g^T[x_j-x_i]+\inf\limits_{\varphi}\left[\frac{K}{2}\varphi^T[y_i-y_j]+\sqrt{2K\theta}\|\varphi\|_2\right]\\ & = & \left\{\begin{array}{ll}\frac{1}{2}g^T[x_j-x_i], & \|y_i-y_j\|_2\leq 2\sqrt{2\theta/K}\\ -\infty, & \|y_i-y_j\|_2>2\sqrt{2\theta/K}.\end{array}\right.\end{array}$$

We see that we can put $\phi_0=0$, and that setting

$$\mathcal{I}=\{(i,j):\|y_i-y_j\|_2\leq 2\sqrt{2\theta/K}\},$$

$\mathrm{Opt}_{ij}(K)$ is finite if and only if $(i,j)\in\mathcal{I}$ and is $-\infty$ otherwise. In both cases, the optimization problem specifying Opt_{ij} has no optimal solution.[2] Indeed, this clearly is the case when $(i,j)\notin\mathcal{I}$; when $(i,j)\in\mathcal{I}$, a minimizing sequence is, e.g., $\phi_0\equiv 0$, $\varphi\equiv 0$, $\alpha_i\to 0$, but its limit is not in the minimization domain (on this domain, α should be positive). In this particular case, the simplest way to overcome the difficulty is to restrict the optimization domain $\mathcal{F}^+$ in (3.4) with its compact subset $\{\alpha\geq 1/R,\phi_0=0,\|\varphi\|_2\leq R\}$ with large R, like $R=10^{10}$ or 10^{20}. Then we specify the entities participating in (3.5) as

$$\begin{array}{rcl}\phi_{ij}(\omega) & = & \varphi_{ij}^T\omega,\ \varphi_{ij}=\left\{\begin{array}{ll}0, & (i,j)\in\mathcal{I}\\ -R[y_i-y_j]/\|y_i-y_j\|_2, & (i,j)\notin\mathcal{I}\end{array}\right.\\ \alpha_{ij} & = & \left\{\begin{array}{ll}1/R, & (i,j)\in\mathcal{I}\\ \sqrt{\frac{K}{2\theta}}R, & (i,j)\notin\mathcal{I}\end{array}\right.\end{array}$$

resulting in

$$\begin{array}{rl}\varkappa_{ij} & =\frac{1}{2}\left[\Psi_{j,-}(\alpha_{ij},\phi_{ij})-\Psi_{i,+}(\alpha_{ij},\phi_{ij})\right]\\ & =\frac{1}{2}\left[-K\varphi_{ij}^Ty_j+g^Tx_j+\frac{K}{2\alpha_{ij}}\varphi_{ij}^T\varphi_{ij}+\alpha_{ij}\theta-K\varphi_{ij}^Ty_i+g^Tx_i-\frac{K}{2\alpha_{ij}}\varphi_{ij}^T\varphi_{ij}-\alpha_{ij}\theta\right]\\ & =\frac{1}{2}g^T[x_i+x_j]-\frac{K}{2}\varphi_{ij}^T[y_i+y_j]\end{array}$$

[2]Handling this case was exactly the reason why in our construction we required ϕ_{ij},α_{ij} to be feasible, and not necessary optimal, solutions to the optimization problems (3.4).

and

$$\begin{array}{rcl}
\rho_{ij} & = & \frac{1}{2}\left[\Psi_{i,+}(\alpha_{ij},\phi_{ij})+\Psi_{j,-}(\alpha_{ij},\phi_{ij})\right] \\
& = & \frac{1}{2}\left[K\varphi_{ij}^T y_i - g^T x_i + \frac{K}{2\alpha_{ij}}\varphi_{ij}^T\varphi_{ij} + \alpha_{ij}\theta - K\varphi_{ij}^T y_j + g^T x_j + \frac{K}{2\alpha_{ij}}\varphi_{ij}^T\varphi_{ij} + \alpha_{ij}\theta\right] \\
& = & \frac{K}{2\alpha_{ij}}\varphi_{ij}^T\phi_{ij} + \alpha_{ij}\theta + \frac{1}{2}g^T[x_j - x_i] + \frac{K}{2}\varphi_{ij}^T[y_i - y_j] \\
& = & \left\{\begin{array}{ll} \frac{1}{2}g^T[x_j-x_i] + R^{-1}\theta, & (i,j)\in\mathcal{I}, \\ \frac{1}{2}g^T[x_j-x_i] + [\sqrt{2K\theta} - \frac{K}{2}\|y_i-y_j\|_2]R, & (i,j)\notin\mathcal{I}. \end{array}\right.
\end{array} \tag{3.15}$$

In the numerical experiment we report on we use $n = 20$, $m = 10$, and $I = 100$, with x_i, $i \leq I$, drawn independently of each other from $\mathcal{N}(0, I_n)$, and $y_i = Ax_i$ with randomly generated matrix A (specifically, matrix with independent $\mathcal{N}(0,1)$ entries normalized to have unit spectral norm). The linear form to be recovered is the first coordinate of x, the confidence parameter is set to $\epsilon = 0.01$, and $R = 10^{20}$. Results of a typical experiment are presented in Figure 3.1.

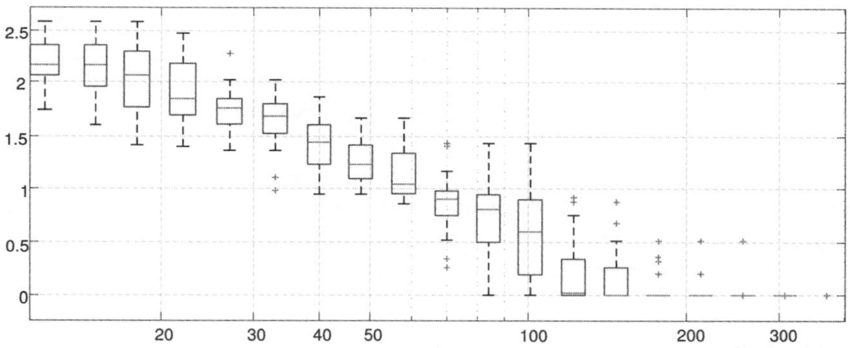

Figure 3.1: Boxplot of empirical distributions, over 20 random estimation problems, of the upper 0.01-risk bounds $\max_{1\leq i,j\leq 100}\rho_{ij}$ (as in (3.15)) for different observation sample sizes K.

3.2 ESTIMATING N-CONVEX FUNCTIONS ON UNIONS OF CONVEX SETS

In this section, we apply our testing machinery to the estimation problem as follows.

Given are:

- a simple o.s. $\mathcal{O} = (\Omega, \Pi; \{p_\mu : \mu \in \mathcal{M}\}; \mathcal{F})$,
- a *signal space* $X \subset \mathbf{R}^n$ along with the affine mapping $x \mapsto A(x) : X \to \mathcal{M}$,
- a real-valued function f on X.

Given observation $\omega \sim p_{A(x_*)}$ stemming from unknown signal x_* known to belong to X, we want to recover $f(x_*)$.

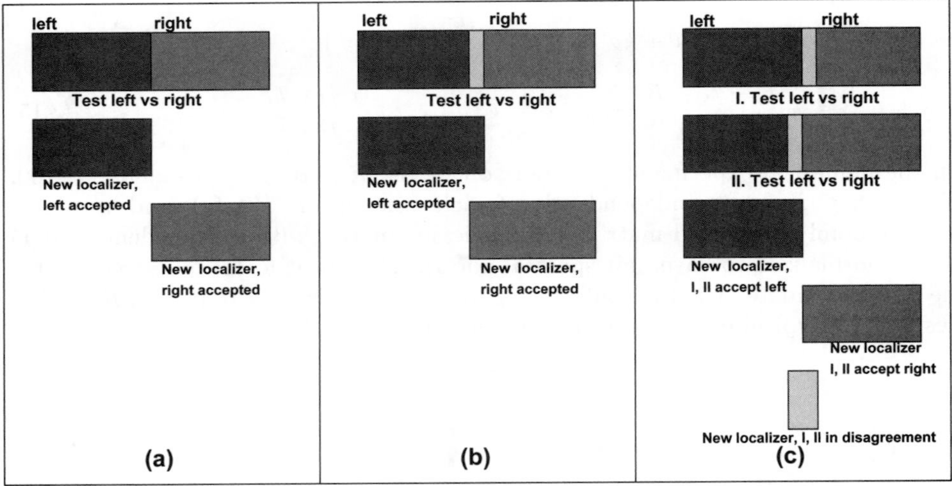

Figure 3.2: Bisection via Hypothesis Testing.

Our approach imposes severe restrictions on f (satisfied, e.g., when f is linear, or linear-fractional, or is the maximum of several linear functions); as a compensation, we allow for rather "complex" X—finite unions of convex sets.

3.2.1 Outline

Though the estimator we develop is, in a nutshell, quite simple, its formal description turns out to be rather involved.[3] For this reason we start its presentation with an informal outline, which exposes some simple ideas underlying its construction.

Consider the situation where the signal space X is the 2D rectangle as presented on the top of Figure 3.2.(a), and let the function to be recovered be $f(x) = x_1$. Thus, "nature" has somehow selected $x = [x_1, x_2]$ in the rectangle, and we observe a Gaussian random vector with the mean $A(x)$ and known covariance matrix, where $A(\cdot)$ is a given affine mapping. Note that hypotheses $f(x) \geq b$ and $f(x) \leq a$ translate into convex hypotheses on the expectation of the observed Gaussian r.v., so that we can use our hypothesis testing machinery to decide on hypotheses of this type and to localize $f(x)$ in a (hopefully, small) segment by a Bisection-type process. Before describing the process, let us make a terminological agreement. In the sequel we shall use pairwise hypothesis testing in the situation where it may happen that *neither* of the hypotheses we are deciding upon is true. In this case, we will say that the outcome of a test is correct if the rejected hypothesis indeed is wrong (the accepted hypothesis can be wrong as well, but the latter can happen

[3] It should be mentioned that the proposed estimation procedure is a "close relative" of the binary search algorithm of [77].

only in the case when both our hypotheses are wrong).

This is what the Bisection might look like.

1. Were we able to decide reliably on the left and the right hypotheses in Figure 3.2.(a), that is, to understand via observations whether x belongs to the left or to the right half of the original rectangle, our course of actions would be clear: depending on this decision, we would replace our original rectangle with a smaller rectangle localizing x, as shown in Figure 3.2.(a), and then iterate this process. The difficulty, of course, is that our left and right hypotheses intersect, so that is impossible to decide on them reliably.

2. In order to make the left and right hypotheses distinguishable from each other, we could act as shown in Figure 3.2.(b), by shrinking the left and the right rectangles and inserting a rectangle in the middle ("no man's land"). Assuming that the width of the middle rectangle allows to decide reliably on our new left and right hypotheses and utilizing the available observation, we can localize x either in the left, or in the right rectangle as shown in Figure 3.2.(b). Specifically, assume that our "left vs. right" test rejected correctly the right hypothesis. Then x can be located either in the left, or in the middle rectangle shown on the top, and thus x is in the new left localizer which is the union of the left and the middle original rectangles. Similarly, if our test rejects correctly the left hypothesis, then we can take, as the new localizer of x, the union of the original right and middle rectangles. Note that our localization is as reliable as our test is, and that it reduces the width of the localizer by a factor close to 2, provided the width of the middle rectangle is small compared to the width of the original localizer of x. We can iterate this process, until we arrive at a localizer so narrow that the corresponding separator—"no man's land" (this part cannot be too narrow, since it should allow for a reliable decision on the current left and right hypotheses)—becomes too large to allow reducing significantly the localizer's width.

Note that in this implementation of the binary search (same as in the implementation proposed in [77]), starting from the second step of the Bisection, the hypotheses to decide upon depend on the observations (e.g., when x belongs to the middle part of the three-rectangle localizer in Figure 3.2, deciding on "left vs. right" can, depending on observation, result in accepting either the left or the right hypothesis, leading to different updated localizers). Analysing this situation usually brings about complications we would like to avoid.

3. A simple modification of the Bisection allows us to circumvent the difficulties related to testing random hypotheses. Indeed, let us consider the following construction: given the current localizer for x (at the first step the initial rectangle), we consider two "three-rectangle" partitions of it as presented in Figure 3.2.(c). In the first partition, the left rectangle is the left half of the original rectangle, in the second partition the right rectangle is the right half of the original rectangle. We then run *two* "left vs. right" tests, the first on the pair of left and right hypotheses stemming from the first partition, and the second on the pair of left and right hypotheses stemming from the second partition. Assuming that in both tests the rejected hypotheses indeed were wrong, the results of these tests allow us to make the following conclusions:

- when both tests reject the right hypotheses from the corresponding pairs, x is located in the left half of the initial rectangle (since otherwise in the second test the rejected hypothesis were in fact true, contradicting to the assumption that both tests make no wrong rejections);

- when both tests reject the left hypotheses from the corresponding pairs, x is located in the right half of the original rectangle (for the exactly same reasons as in the previous case);
- when the tests "disagree," rejecting hypotheses of different types (like left in the firsts, and right in the second test), x is located in the union of the two middle rectangles we deal with. Indeed, otherwise x should be either in the left rectangles of both our three-rectangle partitions, or in the right rectangles of both of them. Since we have assumed that in both tests no wrong rejections took place, in the first case both tests must reject the right hypotheses, and both should reject the left hypotheses in the second, while none of these events took place.

Now, in the first two cases we can safely say to which of the "halves"—left or right—of the initial rectangle x belongs, and take this half as the new localizer. In the third case, we take as a new localizer for x the middle rectangle shown at the bottom of Figure 3.2 *and terminate our estimation process*—the new localizer already is narrow! In the proposed algorithm, unless we terminate at the very first step, we carry out the second step exactly in the same way as the first one, with the localizer of x yielded by the first step in the role of the initial localizer, then carry out, in the same way, the third step, etc., until termination either due to running into a disagreement, or due to reaching a prescribed number of steps. Upon termination, we return the last localizer for x which we have built, and claim that $f(x) = x_1$ belongs to the projection of this localizer onto the x_1-axis. *In all tests from the above process, we use the same observation.* Note that in the present situation, in contrast to that discussed earlier, reutilizing a single observation creates no difficulties, since *with no wrong rejections in the pairwise tests we use, the pairs of hypotheses participating in the tests are not random at all—they are uniquely defined by $f(x) = x_1$.* Indeed, with no wrong rejections, prior to termination everything is *as if* we were running deterministic Bisection, that is, were updating subsequent rectangles Δ_t containing x according to the rules

- Δ_1 is a rectangle containing x given in advance,
- Δ_{t+1} is precisely the half of Δ_t containing x (say, the left half in the case of a tie).

Thus, given x and assuming that there are no wrong rejections, the situation is as if a single observation were used in L tests running in "parallel" rather than sequentially. The only elaboration caused by the sequential nature of our process is the "risk accumulation"—we want the probability of error *in one or more of our L tests* to be less than the desired risk ϵ of wrong "bracketing" of $f(x)$, implying, in the absence of something better, that the risks of the individual tests should be at most ϵ/L. These risks, in turn, define the allowed width of separators and thus – the accuracy to which $f(x)$ can be estimated. It should be noted that the number L of steps of Bisection always is a moderate integer (since otherwise the width of "no man's land," which at the concluding Bisection steps is of order of 2^{-L}, would be too small to allow for deciding on the concluding pairs of our hypotheses with risk ϵ/L, at least when our observations possess non-negligible volatility). As a result, "the cost" of Bisection turns out to be significantly lower than in the case where every test uses its own observation.

From the above sketch of our construction it is clear that all that matters is our ability to decide on the pairs of hypotheses $\{x \in X : f(x) \leq a\}$ and $\{x \in X : f(x) \geq b\}$, with a and b given, via observation drawn from $p_{A(x)}$. In our outline, these were convex hypotheses in Gaussian o.s., and in this case we can use detector-based pairwise tests yielded by Theorem 2.23. Applying the machinery developed in Section 2.5.1, we could also handle the case when the sets $\{x \in X : f(x) \leq a\}$ and $\{x \in X : f(X) \geq b\}$ are unions of a moderate number of convex sets (e.g., f is affine, and X is the union of a number of convex sets), the o.s. in question still being simple, and this is the situation we intend to consider.

3.2.2 Estimating N-convex functions: Problem setting

In the rest of this section, we consider the situation as follows. We are given

1. simple o.s. $\mathcal{O} = (\Omega, P, \{p_\mu(\cdot) : \mu \in \mathcal{M}\}, \mathcal{F})$,
2. convex compact set $\mathcal{X} \subset \mathbf{R}^n$ along with a collection of I convex compact sets $X_i \subset \mathcal{X}$,
3. affine mapping $x \mapsto A(x) : \mathcal{X} \to \mathcal{M}$,
4. a continuous function $f(x) : \mathcal{X} \to \mathbf{R}$ which is N-*convex*, meaning that for every $a \in \mathbf{R}$ the sets $\mathcal{X}^{a,\geq} = \{x \in \mathcal{X} : f(x) \geq a\}$ and $\mathcal{X}^{a,\leq} = \{x \in \mathcal{X} : f(x) \leq a\}$ can be represented as the unions of at most N closed convex sets $\mathcal{X}^{a,\geq}_\nu$, $\mathcal{X}^{a,\leq}_\nu$:

$$\mathcal{X}^{a,\geq} = \bigcup_{\nu=1}^{N} \mathcal{X}^{a,\geq}_\nu, \ \ \mathcal{X}^{a,\leq} = \bigcup_{\nu=1}^{N} \mathcal{X}^{a,\leq}_\nu.$$

For some *unknown* x known to belong to $X = \bigcup\limits_{i=1}^{I} X_i$, we have at our disposal observation $\omega^K = (\omega_1, ..., \omega_K)$ with i.i.d. $\omega_t \sim p_{A(x)}(\cdot)$, and our goal is to estimate from this observation the quantity $f(x)$.

Given tolerances $\rho > 0$, $\epsilon \in (0,1)$, let us call a candidate estimate $\widehat{f}(\omega^K)$ (ρ,ϵ)-*reliable* (cf. (3.3)) if for every $x \in X$, with the $p_{A(x)}$-probability at least $1-\epsilon$, it holds $|\widehat{f}(\omega^K) - f(x)| \leq \rho$ or, which is the same,

$$\forall (x \in X) : \mathrm{Prob}_{\omega^K \sim p_{A(x)} \times \ldots \times p_{A(x)}} \left\{ |\widehat{f}(\omega^K) - f(x)| > \rho \right\} \leq \epsilon.$$

3.2.2.1 Examples of N-convex functions

Example 3.1. [Minima and maxima of linear-fractional functions] Every function which can be obtained from linear-fractional functions $\frac{g_\nu(x)}{h_\nu(x)}$ (g_ν, h_ν are affine functions on $\mathcal{X}$ and h_ν are positive on $\mathcal{X}$) by taking maxima and minima is N-convex for appropriately selected N due to the following immediate observations:

- linear-fractional function $\frac{g(x)}{h(x)}$ with denominator positive on $\mathcal{X}$ is 1-convex on $\mathcal{X}$;
- if $f(x)$ is N-convex, so is $-f(x)$;

- if $f_i(x)$ is N_i-convex, $i = 1, 2, ..., I$, then $f(x) = \max\limits_i f_i(x)$ is N-convex with

$$N = \max\left[\prod_i N_i, \sum_i N_i\right],$$

due to

$$\begin{array}{rcl}\{x \in \mathcal{X} : f(x) \leq a\} &=& \bigcap\limits_{i=1}^{I} \{x : f_i(x) \leq a\},\\ \{x \in \mathcal{X} : f(x) \geq a\} &=& \bigcup\limits_{i=1}^{I} \{x : f_i(x) \geq a\}.\end{array}$$

Note that the first set is the intersection of I unions of convex sets with N_i components in i-th union, and thus is the union of $\prod_i N_i$ convex sets. The second set is the union of I unions of convex sets with N_i elements in the i-th union, and thus is the union of $\sum_i N_i$ convex sets.

Example 3.2. [Conditional quantile] Let $S = \{s_1 < s_2 < ... < s_M\} \subset \mathbf{R}$. For a nonvanishing probability distribution q on S and $\alpha \in [0, 1]$, let $\chi_\alpha[q]$ be the *regularized* α-quantile of q defined as follows: we pass from q to the distribution on $[s_1, s_M]$ by spreading uniformly the mass q_ν, $1 < \nu \leq M$, over $[s_{\nu-1}, s_\nu]$, and assigning mass q_1 to the point s_1; $\chi_\alpha[q]$ is the usual α-quantile of the resulting distribution $\bar{q}$:

$$\chi_\alpha[q] = \min\{s \in [s_1, s_M] : \bar{q}\{[s_1, s]\} \geq \alpha\}.$$

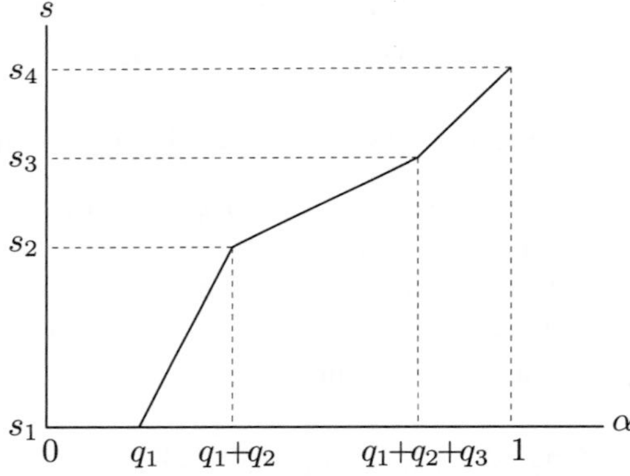

Regularized quantile as function of α, $M = 4$

Given, along with S, a finite set T, let $\mathcal{X}$ be a convex compact set in the space of nonvanishing probability distributions on $S \times T$. For $\tau \in T$, consider the conditional to $t = \tau$, distribution $p_\tau(\cdot)$ of $s \in S$ induced by a distribution $p(\cdot, \cdot) \in \mathcal{X}$:

$$p_\tau(\mu) = \frac{p(\mu, \tau)}{\sum_{\nu=1}^{M} p(\nu, \tau)}, \; 1 \leq \mu \leq M,$$

where $p(\mu, \tau)$ is the p-probability for (s, t) to take value (s_μ, τ), and $p_\tau(\mu)$ is the p_τ-probability for s to take value s_μ, $1 \leq \mu \leq M$.

The function $\chi_\alpha[p_\tau] : \mathcal{X} \to \mathbf{R}$ turns out to be 1-convex; for verification see Section 3.6.2.

3.2.3 Bisection estimate: Construction

While the construction to be presented admits numerous refinements, we focus here on its simplest version.

3.2.3.1 Preliminaries

Upper and lower feasibility/infeasibility, sets $Z_i^{a,\geq}$ and $Z_i^{a,\leq}$. Let a be a real. We associate with a a collection of *upper a-sets* defined as follows: we look at the sets $X_i \cap \mathcal{X}_\nu^{a,\geq}$, $1 \leq i \leq I$, $1 \leq \nu \leq N$, and arrange the nonempty sets from this family into a sequence $Z_i^{a,\geq}$, $1 \leq i \leq I_{a,\geq}$. Here $I_{a,\geq} = 0$ if all sets in the family are empty; in the latter case, we call a *upper-infeasible*, and call it *upper-feasible* otherwise. Similarly, we associate with a the collection of *lower a-sets* $Z_i^{a,\leq}$, $1 \leq i \leq I_{a,\leq}$, by arranging into a sequence all nonempty sets from the family $X_i \cap \mathcal{X}_\nu^{a,\leq}$, and call a lower-feasible or lower-infeasible depending on whether $I_{a,\leq}$ is positive or zero. Note that upper and lower a-sets are nonempty convex compact sets, and

$$
\begin{array}{rcl}
X^{a,\geq} & := & \{x \in X : f(x) \geq a\} = \bigcup\limits_{1 \leq i \leq I_{a,\geq}} Z_i^{a,\geq}, \\
X^{a,\leq} & := & \{x \in X : f(x) \leq a\} = \bigcup\limits_{1 \leq i \leq I_{a,\leq}} Z_i^{a,\leq}.
\end{array}
$$

Right tests. Given a segment $\Delta = [a, b]$ of positive length with lower-feasible a, we associate with this segment a *right test*—a function $\mathcal{T}_{\Delta,\mathrm{r}}^K(\omega^K)$ taking values `right` and `left`, and risk $\sigma_{\Delta,\mathrm{r}} \geq 0$—as follows:

1. if b is upper-infeasible, $\mathcal{T}_{\Delta,\mathrm{r}}^K(\cdot) \equiv$ `left` and $\sigma_{\Delta,\mathrm{r}} = 0$;
2. if b is upper-feasible, the collections of "right sets" $\{A(Z_i^{b,\geq})\}_{i \leq I_{b,\geq}}$ and of "left sets" $\{A(Z_j^{a,\leq})\}_{j \leq I_{a,\leq}}$ are nonempty, and the test is given by the construction from Section 2.5.1 *as applied to these sets and the stationary K-repeated version of* $\mathcal{O}$, specifically,
   - for $1 \leq i \leq I_{b,\geq}$, $1 \leq j \leq I_{a,\leq}$, we build the detectors

     $$\phi_{ij\Delta}^K(\omega^K) = \sum_{t=1}^K \phi_{ij\Delta}(\omega_t),$$

     with $\phi_{ij\Delta}(\omega)$ given by

     $$
     \begin{array}{rcl}
     (r_{ij\Delta}, s_{ij\Delta}) & \in & \mathop{\mathrm{Argmin}}\limits_{r \in Z_i^{b,\geq}, s \in Z_j^{a,\leq}} \ln\left(\int_\Omega \sqrt{p_{A(r)}(\omega) p_{A(s)}(\omega)} \Pi(d\omega)\right), \\
     \phi_{ij\Delta}(\omega) & = & \frac{1}{2} \ln\left(p_{A(r_{ij\Delta})}(\omega) / p_{A(s_{ij\Delta})}(\omega)\right).
     \end{array}
     $$

     We set

     $$\epsilon_{ij\Delta} = \int_\Omega \sqrt{p_{A(r_{ij\Delta})}(\omega) p_{A(s_{ij\Delta})}(\omega) \Pi(d\omega)}$$

     and build the $I_{b,\geq} \times I_{a,\leq}$ matrix $E_{\Delta,\mathrm{r}} = [\epsilon_{ij\Delta}^K]_{\substack{1 \leq i \leq I_{b,\geq} \\ 1 \leq j \leq I_{a,\leq}}}$;
   - we define $\sigma_{\Delta,\mathrm{r}}$ as the spectral norm of $E_{\Delta,\mathrm{r}}$. We compute the Perron-Frobenius eigenvector $[g^{\Delta,\mathrm{r}}; h^{\Delta,\mathrm{r}}]$ of the matrix $\left[\begin{array}{c|c} & E_{\Delta,\mathrm{r}} \\ \hline E_{\Delta,\mathrm{r}}^T & \end{array}\right]$, so that (see Section

2.5.1.2)

$$g^{\Delta,\mathrm{r}} > 0,\ h^{\Delta,\mathrm{r}} > 0,\ \sigma_{\Delta,\mathrm{r}} g^{\Delta,\mathrm{r}} = E_{\Delta,\mathrm{r}} h^{\Delta,\mathrm{r}},\ \sigma_{\Delta,\mathrm{r}} h^{\Delta,\mathrm{r}} = E^T_{\Delta,\mathrm{r}} g^{\Delta,\mathrm{r}}.$$

Finally, we define the matrix-valued function

$$D_{\Delta,\mathrm{r}}(\omega^K) = [\phi^K_{ij\Delta}(\omega^K) - \ln(h_j^{\Delta,\mathrm{r}}) + \ln(g_i^{\Delta,\mathrm{r}})]_{\substack{1\leq i\leq I_{b,\geq}\\ 1\leq j\leq I_{a,\leq}}}.$$

Test $\mathcal{T}^K_{\Delta,\mathrm{r}}(\omega^K)$ takes value `right` iff the matrix $D_{\Delta,\mathrm{r}}(\omega^K)$ has a nonnegative row, and takes value `left` otherwise.

Given $\delta > 0$, $\varkappa > 0$, we call segment $\Delta = [a, b]$ *δ-good (right)* if a is lower-feasible, $b > a$, and $\sigma_{\Delta,\mathrm{r}} \leq \delta$. We call a δ-good (right) segment $\Delta = [a, b]$ *$\varkappa$-maximal* if the segment $[a, b - \varkappa]$ is not δ-good (right).

Left tests. The "mirror" version of the above is as follows. Given a segment $\Delta = [a, b]$ of positive length with upper-feasible b, we associate with this segment a *left test*—a function $\mathcal{T}^K_{\Delta,\mathrm{l}}(\omega^K)$ taking values `right` and `left`, and risk $\sigma_{\Delta,\mathrm{l}} \geq 0$—as follows:

1. if a is lower-infeasible, $\mathcal{T}^K_{\Delta,\mathrm{l}}(\cdot) \equiv$ `right` and $\sigma_{\Delta,\mathrm{l}} = 0$;
2. if a is lower-feasible, we set $\mathcal{T}^K_{\Delta,\mathrm{l}} \equiv \mathcal{T}^K_{\Delta,\mathrm{r}}$, $\sigma_{\Delta,\mathrm{l}} = \sigma_{\Delta,\mathrm{r}}$.

Given $\delta > 0$, $\varkappa > 0$, we call segment $\Delta = [a, b]$ *δ-good (left)* if b is upper-feasible, $b > a$, and $\sigma_{\Delta,\mathrm{l}} \leq \delta$. We call a δ-good (left) segment $\Delta = [a, b]$ *$\varkappa$-maximal* if the segment $[a + \varkappa, b]$ is not δ-good (left).

Explanation: When $a < b$ and a is lower-feasible, b is upper-feasible, so that the sets

$$X^{a,\leq} = \{x \in X : f(x) \leq a\},\ X^{b,\geq} = \{x \in X : f(x) \geq b\}$$

are nonempty, the right and the left tests $\mathcal{T}^K_{\Delta,\mathrm{l}}$, $\mathcal{T}^K_{\Delta,\mathrm{r}}$ are identical to each other and coincide with the minimal risk test, built as explained in Section 2.5.1, deciding, via stationary K-repeated observations, on the "location" of the distribution $p_{A(x)}$ underlying the observations—whether this location is `left` (*left* hypothesis stating that $x \in X$ and $f(x) \leq a$, whence $A(x) \in \bigcup\limits_{1\leq i\leq I_{a,\leq}} A(Z_i^{a,\leq})$), or `right` (*right* hypothesis stating that $x \in X$ and $f(x) \geq b$, whence $A(x) \in \bigcup\limits_{1\leq i\leq I_{b,\geq}} A(Z_i^{b,\geq})$). When a is lower-feasible and b is *not* upper-feasible, the right hypothesis is empty, and the left test associated with $[a, b]$, naturally, always accepts the left hypothesis; similarly, when a is lower-infeasible and b is upper-feasible, the right test associated with $[a, b]$ always accepts the right hypothesis.

A segment $[a, b]$ with $a < b$ is δ-good (left) if the right hypothesis corresponding to the segment is nonempty, and the left test $\mathcal{T}^K_{\Delta\ell}$ associated with $[a, b]$ decides on the right and the left hypotheses with risk $\leq \delta$, and similarly for the δ-good (right) segment $[a, b]$.

3.2.4 Building Bisection estimate

3.2.4.1 Control parameters

The control parameters of the Bisection estimate are

1. positive integer L—the maximum allowed number of bisection steps,
2. tolerances $\delta \in (0,1)$ and $\varkappa > 0$.

3.2.4.2 Bisection estimate: Construction

The estimate of $f(x)$ (x is the signal underlying our observations: $\omega_t \sim p_{A(x)}$) is given by the following recurrence run on the observation $\bar{\omega}^K = (\bar{\omega}_1, ..., \bar{\omega}_K)$ at our disposal:

1. **Initialization.** We find a valid upper bound b_0 on $\max_{u\in X} f(u)$ and valid lower bound a_0 on $\min_{u\in X} f(u)$ and set $\Delta_0 = [a_0, b_0]$. We assume w.l.o.g. that $a_0 < b_0$; otherwise the estimation is trivial.
   Note: $f(x) \in \Delta_0$.
2. **Bisection Step ℓ, $1 \le \ell \le L$.** Given the *localizer* $\Delta_{\ell-1} = [a_{\ell-1}, b_{\ell-1}]$ with $a_{\ell-1} < b_{\ell-1}$, we act as follows:
   a) We set $c_\ell = \frac{1}{2}[a_{\ell-1} + b_{\ell-1}]$. If c_ℓ is not upper-feasible, we set $\Delta_\ell = [a_{\ell-1}, c_\ell]$ and pass to 2e, and if c_ℓ is not lower-feasible, we set $\Delta_\ell = [c_\ell, b_{\ell-1}]$ and pass to 2e.
   Note: When the rule requires us to pass to 2e, the set $\Delta_{\ell-1}\backslash\Delta_\ell$ does not intersect with $f(X)$; in particular, in such a case $f(x) \in \Delta_\ell$ provided that $f(x) \in \Delta_{\ell-1}$.
   b) When c_ℓ is both upper- and lower-feasible, we check whether the segment $[c_\ell, b_{\ell-1}]$ is δ-good (right). If it is not the case, we terminate and claim that $f(x) \in \bar{\Delta} := \Delta_{\ell-1}$; otherwise find v_ℓ, $c_\ell < v_\ell \le b_{\ell-1}$, such that the segment $\Delta_{\ell,\mathrm{rg}} = [c_\ell, v_\ell]$ is δ-good (right) $\varkappa$-maximal.
   Note: In terms of the outline of our strategy presented in Section 3.2.1, termination when the segment $[c_\ell, b_{\ell-1}]$ is not δ-good (right) corresponds to the case when the current localizer is too small to allow for the "no-man's land" wide enough to ensure low-risk decision on the left and the right hypotheses.
   To find v_ℓ, we check the candidates with $v_\ell^k = b_{\ell-1} - k\varkappa$, $k = 0, 1, ...$ until arriving for the first time at segment $[c_\ell, v_\ell^k]$, which is not δ-good (right), and take as v_ℓ the quantity v^{k-1} (because $k \ge 1$ the resulting value of v_ℓ is well-defined and clearly meets the above requirements).
   c) Similarly, we check whether the segment $[a_{\ell-1}, c_\ell]$ is δ-good (left). If it is not the case, we terminate and claim that $f(x) \in \bar{\Delta} := \Delta_{\ell-1}$; otherwise find u_ℓ, $a_{\ell-1} \le u_\ell < c_\ell$, such that the segment $\Delta_{\ell,\mathrm{lf}} = [u_\ell, c_\ell]$ is δ-good (left) $\varkappa$-maximal.
   Note: The rules for building u_ℓ are completely similar to those for v_ℓ.
   d) We compute $\mathcal{T}^K_{\Delta_{\ell,\mathrm{rg}},\mathrm{r}}(\bar{\omega}^K)$ and $\mathcal{T}^K_{\Delta_{\ell,\mathrm{lf}},\mathrm{l}}(\bar{\omega}^K)$. If $\mathcal{T}^K_{\Delta_{\ell,\mathrm{rg}},\mathrm{r}}(\bar{\omega}^K) = \mathcal{T}^K_{\Delta_{\ell,\mathrm{lf}},\mathrm{l}}(\bar{\omega}^K)$ ("consensus"), we set

$$\Delta_\ell = [a_\ell, b_\ell] = \begin{cases} [c_\ell, b_{\ell-1}], & \mathcal{T}^K_{\Delta_{\ell,\mathrm{rg}},\mathrm{r}}(\bar{\omega}^K) = \texttt{right}, \\ [a_{\ell-1}, c_\ell], & \mathcal{T}^K_{\Delta_{\ell,\mathrm{rg}},\mathrm{r}}(\bar{\omega}^K) = \texttt{left} \end{cases} \tag{3.16}$$

   and pass to 2e. Otherwise ("disagreement") we terminate and claim that

$f(x) \in \bar{\Delta} = [u_\ell, v_\ell]$.

e) We pass to step $\ell+1$ when $\ell < L$; otherwise we terminate with the claim that $f(x) \in \bar{\Delta} := \Delta_L$.

3. **Output of the estimation procedure** is the segment $\bar{\Delta}$ built upon termination and claimed to contain $f(x)$ (see rules 2b–2e) the midpoint of this segment is the estimate of $f(x)$ yielded by our procedure.

3.2.5 Bisection estimate: Main result

Our main result on Bisection is as follows:

Proposition 3.4. *Consider the situation described at the beginning of Section 3.2.2, and let* $\epsilon \in (0, 1/2)$ *be given. Then*

(i) [reliability of Bisection] For every positive integer L *and every* $\kappa > 0$, *Bisection with control parameters*

$$L,\ \delta = \frac{\epsilon}{2L},\ \kappa \tag{3.17}$$

is $(1-\epsilon)$*-reliable: for every* $x \in X$, *the* $p_{A(x)}$*-probability of the event*

$$f(x) \in \bar{\Delta}$$

($\bar{\Delta}$ *is the Bisection output as defined above) is at least* $1 - \epsilon$.

(ii) [near-optimality] Let $\rho > 0$ *and positive integer* $\overline{K}$ *be such that "in nature" there exists a* (ρ, ϵ)*-reliable estimate* $\widehat{f}(\cdot)$ *of* $f(x)$, $x \in X := \bigcup_{i \le I} X_i$, *via stationary* $\overline{K}$*-repeated observation* $\omega^{\overline{K}}$ *with* $\omega_k \sim p_{A(x)}$, $1 \le k \le \overline{K}$. *Given* $\widehat{\rho} > 2\rho$, *the Bisection estimate utilizing stationary* K*-repeated observations, with*

$$K \ge \frac{2\ln(2LNI/\epsilon)}{\ln\left(\frac{1}{4\epsilon(1-\epsilon)}\right)}\overline{K}, \tag{3.18}$$

the control parameters of the estimate being

$$L = \left\rfloor \log_2\left(\frac{b_0 - a_0}{2\widehat{\rho}}\right)\right\lfloor,\ \delta = \frac{\epsilon}{2L},\ \varkappa = \widehat{\rho} - 2\rho, \tag{3.19}$$

is $(\widehat{\rho}, \epsilon)$*-reliable. Note that* K *is only "slightly larger" than* $\overline{K}$.

For proof, see Section 3.6.3.

Note that the running time K of the Bisection estimate as given by (3.18) is just by (at most) logarithmic in N, I, L, and $1/\epsilon$ factor larger than $\overline{K}$; note also that L is just logarithmic in $1/\widehat{\rho}$. Assume, e.g., that for some $\gamma > 0$ "in nature" there exist $(\epsilon^\gamma, \epsilon)$-reliable estimates, parameterized by $\epsilon \in (0, 1/2)$, utilizing $\overline{K} = \overline{K}(\epsilon)$ observations. Then Bisection with the volume of observation and control parameters given by (3.18) and (3.19), where $\widehat{\rho} = 3\rho = 3\epsilon^\gamma$ and $\overline{K} = \overline{K}(\epsilon)$, is $(3\epsilon^\gamma, \epsilon)$-reliable and requires $K = K(\epsilon)$-repeated observations with $\overline{\lim}_{\epsilon \to +0} K(\epsilon)/\overline{K}(\epsilon) \le 2$.

3.2.6 Illustration

To illustrate bisection-based estimation of an N-convex function, consider the following situation.[4] There are M devices ("receivers") recording a signal u known to belong to a given convex compact and nonempty set $U \subset \mathbf{R}^n$; the output of the i-th receiver is the vector

$$y_i = A_i u + \sigma\xi \in \mathbf{R}^m \qquad [\xi \sim \mathcal{N}(0, I_m)]$$

where A_i are given $m \times n$ matrices (you may think of M allowed positions for a single receiver, and of y_i as the output of the receiver when the latter is in position i). Our observation ω is one of the vectors y_i, $1 \leq i \leq M$, *with index i unknown to us* ("we observe a noisy record of a signal, but do not know the position in which this record was taken"). Given ω, we want to recover a given linear function $g(x) = e^T u$ of the signal.

The problem can be modeled as follows. Consider M sets

$$X_i = \{x = [x^1; ...; x^M] \in \mathbf{R}^{Mn} = \underbrace{\mathbf{R}^n \times ... \times \mathbf{R}^n}_{M} : x^j = 0, j \neq i; x^i \in U\}$$

along with the linear mapping

$$A[x^1; ...; x^M] = \sum_{i=1}^{M} A_i x^i : \mathbf{R}^{Mn} \to \mathbf{R}^m$$

and linear function

$$f([x^1; ...; x^M]) = e^T \sum_i x^i : \mathbf{R}^{Mn} \to \mathbf{R}.$$

Let $\mathcal{X}$ be a convex compact set in $\mathbf{R}^{Mn}$ containing all the sets X_i, $1 \leq i \leq m$. Observe that the problem we are interested in is nothing but the problem of recovering $f(x)$ via observation

$$\omega = Ax + \sigma\xi, \;\; \xi \sim \mathcal{N}(0, I_m), \tag{3.20}$$

where the unknown signal x is known to belong to the union $\bigcup_{i=1}^{M} X_i$ of known convex compact sets X_i. As a result, our problem can be solved via the machinery developed in this section.

Numerical illustration. In the numerical experiments to be reported, we use $n = 128$, $m = 64$ and $M = 2$. The data is generated as follows:

- The set $U \subset \mathbf{R}^{128}$ of candidate signals is comprised of restrictions onto the equidistant ($n = 128$)-point grid in $[0, 1]$ of twice differentiable functions $h(t)$ of continuous argument $t \in [0, 1]$ satisfying the relations $|h(0)| \leq 1$, $|h'(0)| \leq 1$, $|h''(t)| \leq 1$, $0 \leq t \leq 1$. For the discretized signal $u = [h(0); h(1/n); ...; h(1 - 1/n)]$ this translates into the system of convex constraints

$$|u_1| \leq 1, n|u_2 - u_1| \leq 1, \, n^2|u_{i+1} - 2u_i + u_{i-1}| \leq 1, \, 2 \leq i \leq n - 1.$$

[4]Our goal is to illustrate a mathematical construction rather than to work out a particular application; the reader is welcome to invent a plausible "covering story."

Characteristic	min	median	mean	max
error bound	0.008	0.015	0.014	0.015
actual error	0.001	0.002	0.002	0.005
# of Bisection steps	5	7.00	6.60	8

Table 3.1: Data of 10 Bisection experiments, $\sigma = 0.01$. In the table, "error bound" is the half-length of the final localizer, which is an 0.99-reliable upper bound on the estimation error; the "actual error" is the actual estimation error.

- We look to estimate the discretized counterpart of the integral $\int_0^1 h(t)dt$, specifically, the quantity $e^T u = \alpha \sum_{i=1}^n u_i$. The normalizing constant α is selected to ensure $\max_{u\in U} e^T u = 1$, $\min_{u\in U} e^T u = -1$, allowing us to run Bisection over $\Delta_0 = [-1; 1]$.
- We generate A_1 as an $(m = 64)\times(n = 128)$ matrix with singular values $\sigma_i = \theta^{i-1}$, $1 \le i \le m$, with θ selected from the requirement $\sigma_m = 0.1$. The system of left singular vectors of A_1 is obtained from the system of basic orths in $\mathbf{R}^n$ by random rotation.
  Matrix A_2 was selected as $A_2 = A_1 S$, where S is a symmetry w.r.t. the axis e, that is,
  $$Se = e \;\&\; Sh = -h \text{ whenever } h \text{ is orthogonal to } e. \tag{3.21}$$
  Signals u underlying the observations are selected at random in U.
- The reliability $1 - \epsilon$ of the estimate is set to 0.99, while the maximal allowed number L of Bisection steps is set to 8. We use single observation (3.20) (i.e., use $K = 1$ in our general scheme) with $\sigma = 0.01$.

The results of our experiments are presented in Table 3.1. Observe that in the considered problem there exists an intrinsic obstacle for high accuracy estimation even in the case of noiseless observations and invertible matrices A_i, $i = 1, 2$ (recall that we are in the case of $M = 2$). Indeed, assume that there exist $u \in U$, $u' \in U$ such that $A_1 u = A_2 u'$ and $e^T u \neq e^T u'$. Since we do not know which of the matrices, A_1 or A_2, underlies the observation and $A_1 u = A_2 u'$, there is no way to distinguish between the two cases we have described, implying that the quantity

$$\rho = \max_{u,u'\in U} \left\{ \tfrac{1}{2}|e^T(u - u')| : A_1 u = A_2 u' \right\} \tag{3.22}$$

is a lower bound on the worst-case, over signals from U, error of a reliable recovery of $e^T u$, independently of how small the noise is. In the reported experiments, we used $A_2 = A_1 S$ with S linked to e (see (3.21)); with this selection of S, e, and A_2, *and invertible* A_1, the lower bound ρ would be trivial—just zero. Note that the selected A_1 is not invertible, resulting in a positive ρ. However, computation shows that with our data, this positive ρ is negligibly small (about $2.0e - 5$).

When we destroy the link between e and S, the estimation problem can become intrinsically more difficult, and the performance of our estimation procedure can deteriorate. Let us look at what happens when we keep A_1 and $A_2 = A_1 S$ exactly as they are, but replace the linear form to be estimated with $e^T u$, e being randomly selected.[5] The corresponding results are presented in Table 3.2. The data in the

[5]In the experiments to be reported, e is selected as follows: we start with a random unit vector drawn from the uniform distribution on the unit sphere in $\mathbf{R}^n$ and then normalize it to

Characteristic	min	median	mean	max
error bound	0.057	0.457	0.441	1.000
actual error	0.001	0.297	0.350	1.000
# of Bisection steps	1	1.00	2.20	5

"Difficult" signals, data over 10 experiments

ρ	0.022	0.028	0.154	0.170	0.213	0.248	0.250	0.500	0.605	0.924
error bound	0.057	0.063	0.219	0.239	0.406	0.508	0.516	0.625	0.773	1.000

Error bound vs. ρ, experiments sorted according to the values of ρ

Characteristic	min	median	mean	max
error bound	0.016	0.274	0.348	1.000
actual error	0.005	0.066	0.127	0.556
# of Bisection steps	1	2.00	2.80	7

Random signals, data over 10 experiments

ρ	0.010	0.085	0.177	0.243	0.294	0.334	0.337	0.554	0.630	0.762
error bound	0.016	0.182	0.376	0.438	0.602	0.029	0.031	0.688	0.125	1.000

Error bound vs. ρ, experiments sorted according to the values of ρ

Table 3.2: Results of experiments with randomly selected linear form, $\sigma = 0.01$.

top part of the table match "difficult" signals u—those participating in forming the lower bound (3.22) on the recovery error, while the data in the bottom part of the table correspond to randomly selected signals.[6] Observe that when estimating a randomly selected linear form, the error bounds indeed deteriorate, as compared to those in Table 3.1. We see also that the resulting error bounds are in a reasonably good agreement with the lower bound ρ, illustrating the basic property of nearly optimal estimates: the guaranteed performance of an estimate can be bad or good, but it is always nearly as good as is possible under the circumstances. As for actual estimation errors, they in some experiments are significantly less than the error bounds, especially when random signals are used.

3.2.7 Estimating N-convex functions: An alternative

Observe that the problem of estimating an N-convex function on the union of convex sets posed in Section 3.2.2 can be processed not only by Bisection. An alternative is as follows. In the notation of Section 3.2.2, we start with computing the range Δ of function f on the set $X = \bigcup\limits_{i\leq I} X_i$, that is, we compute the quantities

$$\underline{f} = \min_{x\in X} f(x), \;\; \overline{f} = \max_{x\in X} f(x)$$

have $\max_{u\in U} e^T u - \min_{u\in U} e^T u = 2$.

[6]Precisely, to generate a signal u, we draw a point $\bar{u}$ at random, from the uniform distribution on the sphere of radius $10\sqrt{n}$, and take as u the point of U $\|\cdot\|_2$-closest to $\bar{u}$.

and set $\Delta = [\underline{f}, \overline{f}]$. We assume that this segment is not a singleton; otherwise estimating f is trivial. Let $L \in \mathbf{Z}_+$ and let $\delta_L = (\overline{f} - \underline{f})/L$ be the desired estimation accuracy. We split Δ into L segments Δ_ℓ of equal length δ_L and consider the sets

$$X_{i\ell} = \{x \in X_i : f(x) \in \Delta_\ell\},\ 1 \leq i \leq I, 1 \leq \ell \leq L.$$

Since f is N-convex, each set $X_{i\ell}$ is a union of $M_{i\ell} \leq N^2$ convex compact sets $X_{i\ell j}$, $1 \leq j \leq M_{i\ell}$. Thus, we have at our disposal a collection of at most ILN^2 convex compact sets; let us eliminate from this collection empty sets and arrange the nonempty ones into a sequence $Y_1, ..., Y_M$, $M \leq ILN^2$. Note that $\bigcup_{s \leq M} Y_s = X$, so that the goal set in Section 3.2.2 can be reformulated as follows:

> For some *unknown* x known to belong to $X = \bigcup\limits_{s=1}^{M} Y_s$, we have at our disposal observation $\omega^K = (\omega_1, ..., \omega_K)$ with i.i.d. $\omega_t \sim p_{A(x)}(\cdot)$; we aim at estimating the quantity $f(x)$ from this observation.

The sets Y_s give rise to M hypotheses $H_1, ..., H_M$ on the distribution of the observations ω_t, $1 \leq t \leq K$; according to H_s, $\omega_t \sim p_{A(x)}(\cdot)$ with some $x \in Y_s$.

Let us define a closeness $\mathcal{C}$ on the set of our M hypotheses as follows. Given $s \leq M$, the set Y_s is some $X_{i(s)\ell(s)j(s)}$; we say that two hypotheses, H_s and $H_{s'}$, are $\mathcal{C}$-close if the segments $\Delta_{\ell(s)}$ and $\Delta_{\ell(s')}$ intersect. Observe that when H_s and $H_{s'}$ are *not* $\mathcal{C}$-close, the convex compact sets Y_s and Y_s' do not intersect, since the values of f on Y_s belong to $\Delta_{\ell(s)}$, the values of f on $Y_{s'}$ belong to $\Delta_{\ell(s')}$, and the segments $\Delta_{\ell(s)}$ and $\Delta_{\ell(s')}$ do not intersect.

Now let us apply to the hypotheses $H_1, ..., H_M$ our machinery for testing up to closeness $\mathcal{C}$; see Section 2.5.2. Assuming that whenever H_s and $H_{s'}$ are not $\mathcal{C}$-close, the risks $\epsilon_{ss'}$ defined in Section 2.5.2.2 are < 1,[7] we, given tolerance $\epsilon \in (0, 1)$, can find $K = K(\epsilon)$ such that stationary K-repeated observation ω^K allows us to decide $(1-\epsilon)$-reliably on $H_1, ..., H_M$ up to closeness $\mathcal{C}$. As applied to ω^K, the corresponding test $\mathcal{T}^K$ will accept some (perhaps, none) of the hypotheses, let the indexes of the accepted hypotheses form set $S = S(\omega^K)$. We convert S into an estimate $\widehat{f}(\omega^K)$ of $f(x)$, $x \in X = \bigcup_{s \leq M} Y_s$ being the signal underlying our observation, as follows:

- when $S = \emptyset$ the estimate is, say $(\overline{f} + \underline{f})/2$;
- when S is nonempty we take the union $\Delta(S)$ of the segments $\Delta_{\ell(s)}$, $s \in S$, and our estimate is the average of the largest and the smallest elements of $\Delta(S)$.

It is immediately seen that if the signal x underlying our stationary K-repeated observation ω^K belongs to some Y_{s_*}, so that the hypothesis H_{s_*} is true, and the outcome S of $\mathcal{T}^K$ contains s_* and is such that for all $s \in S$ H_s and H_{s_*} are $\mathcal{C}$-close to each other, we have $|f(x) - \widehat{f}(\omega^K)| \leq \delta_L$. Note that since the $\mathcal{C}$-risk of $\mathcal{T}^K$ is $\leq \epsilon$, the $p_{A(x)}$-probability to get such a "good" outcome, and thus to get $|f(x) - \widehat{f}(\omega^K)| \leq \delta_L$, is at least $1 - \epsilon$.

---

[7] In standard simple o.s.'s, this is the case whenever for s, s' in question the images of Y_s and $Y_{s'}$ under the mapping $x \mapsto A(x)$ do not intersect. Because for s, s', Y_s and $Y_{s'}$ do not intersect, this definitely is the case when $A(\cdot)$ is an embedding.

3.2.7.1 Numerical illustration

Our illustration deals with the situation when $I = 1$, $X = X_1$ is a convex compact set, and $f(x)$ is fractional-linear: $f(x) = a^T x / c^T x$ with positive on X denominator. Specifically, assume we are given noisy measurements of voltages V_i at *some* nodes i and currents I_{ij} in *some* arcs (i,j) of an electric circuit, and want to recover the resistance of a particular arc (i_*, j_*):

$$r_{i_* j_*} = \frac{V_{j_*} - V_{i_*}}{I_{i_* j_*}}.$$

The observation noises are assumed to be $\mathcal{N}(0, \sigma^2)$ and independent across the measurements.

In our experiment, we work with the data as follows:

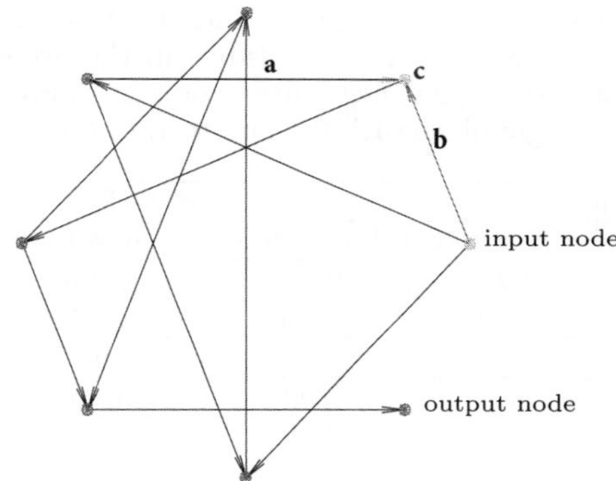

$x = $ [voltages at nodes; currents in arcs]
$Ax = $ [observable voltages; observable currents]

- Currents are measured in all arcs except for a, b
- Voltages are measured at all nodes except for c
- We want to recover resistance of arc b
- X : { *conservation of current, except for input/output nodes*; *zero voltage at input node, nonnegative currents*; *current in arc b at least 1, total of currents at most 33*; *Ohm's Law, resistances of arcs between 1 and 10* }

We are in the situation of $N = 1$ and $I = 1$, implying $M = L$. When using $L = 8$, the projections of the sets Y_s, $1 \leq s \leq L = 8$, onto the 2D plane of variables

$(V_{j_*} - V_{i_*}, I_{i_*j_*})$ are the "stripes" shown below:

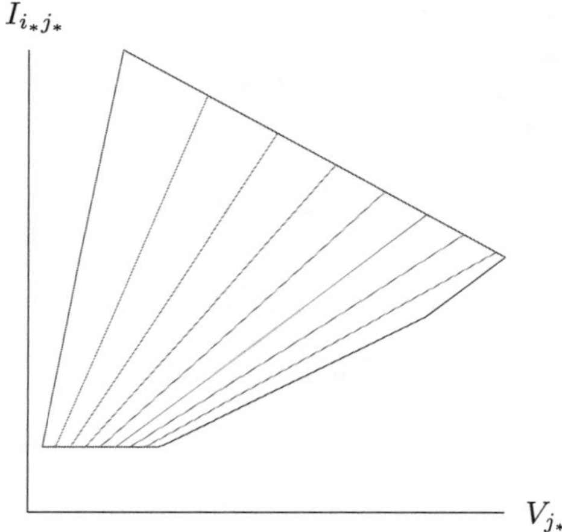

The range of the unknown resistance turns out to be $\Delta = [1, 10]$.

We set $\epsilon = 0.01$, and instead of looking for K such that the K-repeated observation allows us to recover 0.99-reliably the resistance in the arc of interest within accuracy $|\Delta|/L$, we look for the largest observation noise σ allowing us to achieve the desired recovery with a single observation. The results for $L = 8, 16, 32$ are as follows:

L	8	16	32
δ_L	$9/8 \approx 1.13$	$9/16 \approx 0.56$	$9/32 \approx 0.28$
σ	$\underline{0.024}$	$\underline{0.010}$	$\underline{0.005}$
$\sigma_{\rm opt}/\sigma \leq$	$\underline{1.31}$	$\underline{1.31}$	$\underline{1.33}$
σ	*0.031*	*0.013*	*0.006*
$\sigma_{\rm opt}/\sigma \leq$	*1.01*	*1.06*	*1.08*

In the above table:

- $\sigma_{\rm opt}$ is the largest σ for which "in nature" there exists a test deciding on $H_1, ..., H_L$ with $\mathcal{C}$-risk ≤ 0.01;
- Underlined data: Risks $\epsilon_{ss'}$ of pairwise tests are bounded via risks of optimal detectors; $\mathcal{C}$-risk of $\mathcal{T}$ is bounded by

$$\left\|[\epsilon_{ss'}\chi_{ss'}]_{s,s'=1}^{L}\right\|_{2,2}, \ \chi_{ss'} = \left\{\begin{array}{ll} 1, & (s,s') \notin \mathcal{C}, \\ 0, & (s,s') \in \mathcal{C}; \end{array}\right.$$

  see Proposition 2.29;
- "Slanted" data: Risks $\epsilon_{ss'}$ of pairwise tests are bounded via the error function; $\mathcal{C}$-risk of $\mathcal{T}$ is bounded by

$$\max_s \sum_{s':(s,s')\notin\mathcal{C}} \epsilon_{ss'}$$

  (it is immediately seen that in the case of Gaussian o.s., this indeed is a legitimate risk bound).

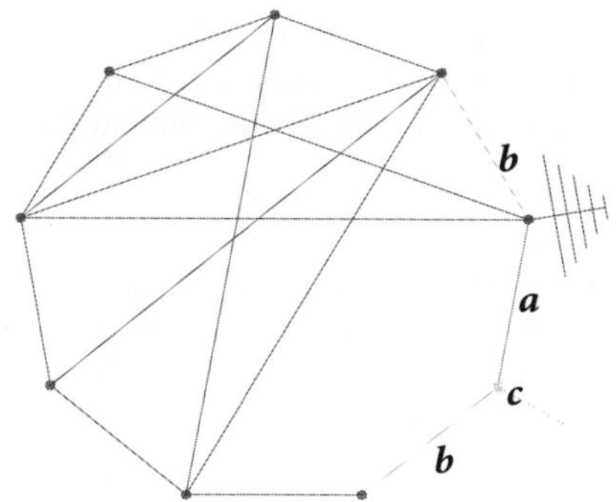

Figure 3.3: A circuit (nine nodes and 16 arcs). a: arc of interest; b: arcs with measured currents; c: input node where external current and voltage are measured.

3.2.7.2 Estimating dissipated power

The alternative approach to estimating N-convex functions proposed in Section 3.2.7 can be combined with the quadratic lifting described in Section 2.9 to yield, under favorable circumstances, estimates of quadratic and quadratic fractional functions. We are about to consider an instructive example of this type. Figure 3.3 represents a DC circuit. We have access to repeated noisy measurements of currents in some arcs and voltages at some nodes, with the voltage of the ground node equal to 0. The arcs are oriented; this orientation, however, is of no relevance in our context and therefore is not displayed. Our goal is to use these observations to estimate the power dissipated in a given "arc of interest." The a priori information is as follows:

- the (unknown) arc resistances are known to belong to a given range $[r, R]$, with $0 < r < R < \infty$;
- the currents and the voltages are linked by Kirchhoff's laws:
  - at every node, the sum of currents in the outgoing arcs is equal to the sum of currents in the incoming arcs plus the external current at the node.
    In our circuit, there are just two external currents, one at the ground node and one at the input node c.
  - the voltages and the currents are linked by Ohm's law: for every (inner) arc γ, we have
    $$I_\gamma r_\gamma = V_{j(\gamma)} - V_{i(\gamma)}$$
    where I_γ is the current in the arc, r_γ is the arc's resistance, V_s is the voltage at node s, and $i(\gamma)$, $j(\gamma)$ are the initial and the terminal nodes linked by arc γ;
- magnitudes of all currents and voltages are bounded by 1.

We assume that the measurements of observable currents and voltages are affected by zero mean Gaussian noise with scalar covariance matrix $\theta^2 I$, with unknown θ from a given range $[\underline{\sigma}, \overline{\sigma}]$.

Processing the problem. We specify the "signal" underlying our observation as a collection u of the voltages at nine nodes and currents I_γ in 16 (inner) arcs γ of the circuit, augmented by the external current I_o at the input node (so that $-I_o$ is the external current at the ground node). Thus, our single-time observation is

$$\zeta = Au + \theta\xi, \tag{3.23}$$

where A extracts from u four entries (currents in two arcs b and external current and voltage at the input node c), $\xi \sim \mathcal{N}(0, I_4)$, and $\theta \in [\underline{\sigma}, \overline{\sigma}]$. Our a priori information on u states that u belongs to the compact set U given by the quadratic constraints, namely, as follows:

$$U = \left\{ u = \{I_\gamma, I_o, V_i\} : \begin{array}{r} I_\gamma^2 \leq 1, V_i^2 \leq 1 \;\forall \gamma, i; \;\; u^T J^T J u \leq 0 \\ \left.\begin{array}{r} [V_{j(\gamma)} - V_{i(\gamma)}]^2/R - I_\gamma[V_{j(\gamma)} - V_{i(\gamma)}] \leq 0 \\ I_\gamma[V_{j(\gamma)} - V_{i(\gamma)}] - [V_{j(\gamma)} - V_{i(\gamma)}]^2/r \leq 0 \end{array}\right\} \forall\gamma \quad (a) \\ \left.\begin{array}{r} rI_\gamma^2 - I_\gamma[V_{j(\gamma)} - V_{i(\gamma)}] \leq 0 \\ I_\gamma[V_{j(\gamma)} - V_{i(\gamma)}] - RI_\gamma^2 \leq 0 \end{array}\right\} \forall\gamma \quad (b) \end{array} \right\} \tag{3.24}$$

where $Ju = 0$ expresses the first Kirchhoff's law, and quadratic constraints (a) and (b) account for Ohm's law in the situation when we do not know the exact resistances but only their range $[r, R]$. Note that groups (a) and (b) of constraints in (3.24) are "logical consequences" of each other, and thus one of groups seems to be redundant. However, on closer inspection, quadratic inequalities valid on U do not tighten the outer approximation $\mathcal{Z}$ of $\mathcal{Z}[U]$ and thus are redundant in our context only when these inequalities can be obtained from the inequalities we do include into the description of $\mathcal{Z}$ "in a linear fashion"—by taking weighted sums with nonnegative coefficients. This is *not* how (b) is obtained from (a). As a result, to get a smaller $\mathcal{Z}$, it makes sense to keep both (a) and (b).

The dissipated power we are interested in estimating is the quadratic function

$$f(u) = I_{\gamma_*}[V_{j_*} - V_{i_*}] = [u; 1]^T G[u; 1]$$

where $\gamma_* = (i_*, j_*)$ is the arc of interest, and $G \in \mathbf{S}^{n+1}$, $n = \dim u$, is a properly built matrix.

In order to build an estimate, we "lift quadratically" the observations

$$\zeta \mapsto \omega = (\zeta, \zeta\zeta^T)$$

and pass from the domain U of actual signals to the outer approximation $\mathcal{Z}$ of the quadratic lifting of U:

$$\begin{array}{rcl} \mathcal{Z} & := & \{Z \in \mathbf{S}^{n+1} : Z \succeq 0, Z_{n+1,n+1} = 1, \mathrm{Tr}(Q_s Z) \leq c_s, \; 1 \leq s \leq S\} \\ & \supset & \left\{[u; 1][u; 1]^T : u \in \mathcal{V}\right\}. \end{array}$$

Here the matrix $Q_s \in \mathbf{S}^{n+1}$ represents the left-hand side $F_s(u)$ of the s-th quadratic constraint in the description (3.24) of U: $F_s(u) \equiv [u; 1]^T Q_s[u; 1]$, and c_s is the right-hand side of the s-th constraint.

We process the problem similarly to what was done in Section 3.2.7.1, where our goal was to estimate a fractional-linear function. Specifically,

1. We compute the range of f on U; the smallest value $\underline{f}$ of f on U clearly is zero, and an upper bound on the maximum of $f(u)$ over $u \in U$ is the optimal value

in the convex optimization problem

$$\overline{f} = \max_{Z \in \mathcal{Z}} \mathrm{Tr}(GZ).$$

2. Given a positive integer L, we split the range $[\underline{f}, \overline{f}]$ into L segments $\Delta_\ell = [a_{\ell-1}, a_\ell]$ of equal length $\delta_L = (\overline{f} - \underline{f})/L$ and define convex compact sets

$$\mathcal{Z}_\ell = \{Z \in \mathcal{Z} : a_{\ell-1} \le \mathrm{Tr}(GZ) \le a_\ell\},\ 1 \le \ell \le L,$$

so that

$$u \in U, f(u) \in \Delta_\ell \Rightarrow [u; 1][u; 1]^T \in \mathcal{Z}_\ell,\ 1 \le \ell \le L.$$

3. We specify L quadratically constrained hypotheses $H_1, ..., H_L$ on the distribution of observation (3.23), with H_ℓ stating that $\zeta \sim \mathcal{N}(Au, \theta^2 I_4)$ with some $u \in U$ satisfying $f(u) \in \Delta_\ell$ (so that $[u; 1][u; 1]^T \in \mathcal{Z}_\ell$), and θ belongs to the above segment $[\underline{\sigma}, \overline{\sigma}]]$.
   We equip our hypotheses with a closeness relation $\mathcal{C}$; specifically, we consider H_ℓ and $H_{\ell'}$ $\mathcal{C}$-close if and only if the segments Δ_ℓ and $\Delta_{\ell'}$ intersect.
4. We use Propositions 2.43.ii and 2.40 to build detectors $\phi_{\ell\ell'}$ quadratic in ζ for the families of distributions obeying H_ℓ and $H_{\ell'}$, respectively, along with upper bounds $\epsilon_{\ell\ell'}$ on the risks of these detectors. Finally, we use the machinery from Section 2.5.2 to find the smallest K and a test $\mathcal{T}_\mathcal{C}^K$, based on a stationary K-repeated version of observation (3.23), able to decide on $H_1, ..., H_L$ with $\mathcal{C}$-risk $\le \epsilon$, where $\epsilon \in (0, 1)$ is a given tolerance.

Finally, given stationary K-repeated observation (3.23), we apply to it test $\mathcal{T}_\mathcal{C}^K$, look at the hypotheses, if any, accepted by the test, and build the union Δ of the corresponding segments Δ_ℓ. If $\Delta = \emptyset$, we estimate $f(u)$ as the midpoint of the power range $[\underline{f}, \overline{f}]$; otherwise the estimate is the mean of the largest and the smallest points in Δ. It is easily seen that for this estimate, the probability for the estimation error to be $> \delta_\ell$ is $\le \epsilon$.

The numerical results we present here correspond to the circuit presented in Figure 3.3. We set $\overline{\sigma} = 0.01$, $\underline{\sigma} = \overline{\sigma}/\sqrt{2}$, $[r, R] = [1, 2]$, $\epsilon = 0.01$, and $L = 8$. The simulation setting is as follows: the computed range $[\underline{f}, \overline{f}]$ of the dissipated power is $[0, 0.821]$, so that the estimate built recovers the dissipated power within accuracy 0.103 and reliability 0.99. The resulting value of K is $K = 95$.

In *all* 500 simulation runs, the actual recovery error was less than the bound 0.103, and the average error was as small as 0.041.

3.3 ESTIMATING LINEAR FORMS BEYOND SIMPLE OBSERVATION SCHEMES

We are about to show that the techniques developed in Section 2.8 can be applied to building estimates of linear and quadratic forms of the parameters of observed distributions. As compared to the machinery of Section 3.2, our new approach has somewhat restricted scope: we do not estimate general N-convex functions nor handle domains which are unions of convex sets; now we need the function to be linear (perhaps, after quadratic lifting of observations) and the domain to be convex.[8] As a compensation, we are not limited to simple observation schemes anymore—our approach is in fact a natural extension of the approach developed in Section 3.1 beyond simple o.s.'s.

In this section, we focus on estimating linear forms; estimating quadratic forms will be our subject in Section 3.4.

3.3.1 Situation and goal

Consider the situation as follows: given are Euclidean spaces $\Omega = \mathcal{E}_H$, $\mathcal{E}_M$, $\mathcal{E}_X$ along with

- regular data (see Section 2.8.1.1) $\mathcal{H} \subset \mathcal{E}_H, \mathcal{M} \subset \mathcal{E}_M, \Phi(\cdot;\cdot) : \mathcal{H} \times \mathcal{M} \to \mathbf{R}$, with $0 \in \operatorname{int} \mathcal{H}$,
- a nonempty convex compact set $\mathcal{X} \subset \mathcal{E}_X$,
- an affine mapping $x \mapsto \mathcal{A}(x) : \mathcal{E}_X \to \mathcal{E}_M$ such that $\mathcal{A}(\mathcal{X}) \subset \mathcal{M}$,
- a continuous convex *calibrating function* $\upsilon(x) : \mathcal{X} \to \mathbf{R}$,
- a vector $g \in \mathcal{E}_X$ and a constant c specifying the linear form $G(x) = \langle g, x \rangle + c : \mathcal{E}_X \to \mathbf{R}$,[9]
- a tolerance $\epsilon \in (0, 1)$.

These data specify, in particular, the family

$$\mathcal{P} = \mathcal{S}[\mathcal{H}, \mathcal{M}, \Phi]$$

of probability distributions on $\Omega = \mathcal{E}_H$; see Section 2.8.1.1. Given random observation

$$\omega \sim P(\cdot) \tag{3.25}$$

where $P \in \mathcal{P}$ is such that

$$\forall h \in \mathcal{H} : \ln \left(\int_{\mathcal{E}_H} e^{\langle h, \omega \rangle} P(d\omega) \right) \leq \Phi(h; \mathcal{A}(x)) \tag{3.26}$$

for some $x \in \mathcal{X}$ (that is, $\mathcal{A}(x)$ is a parameter, as defined in Section 2.8.1.1, of distribution P), we want to recover the quantity $G(x)$.

ϵ-risk. Given $\rho > 0$, we call an estimate $\widehat{g}(\cdot) : \mathcal{E}_H \to \mathbf{R}$ $(\rho, \epsilon, \upsilon(\cdot))$-*accurate* if for all pairs $x \in \mathcal{X}$, $P \in \mathcal{P}$ satisfying (3.26) it holds

$$\text{Prob}_{\omega \sim P} \{ |\widehat{g}(\omega) - G(x)| > \rho + \upsilon(x) \} \leq \epsilon.$$

If ρ_* is the infimum of those ρ for which estimate $\widehat{g}$ is $(\rho, \epsilon, \upsilon(\cdot))$-accurate, then clearly $\widehat{g}$ is $(\rho_*, \epsilon, \upsilon(\cdot))$-accurate; we shall call ρ_* the ϵ-*risk* of the estimate $\widehat{g}$ taken

[8] The latter is just for the sake of simplicity, to not overload the presentation to follow. An interested reader will certainly be able to reproduce the corresponding construction of Section 3.1 in the situation of this section.

[9] From now on, $\langle u, v \rangle$ denotes the inner product of vectors u, v belonging to a Euclidean space; what this space is will always be clear from the context.

w.r.t. the data $G(\cdot)$, $\mathcal{X}$, $\upsilon(\cdot)$, and $(\mathcal{A},\mathcal{H},\mathcal{M},\Phi)$:

$$\mathrm{Risk}_\epsilon(\widehat{g}(\cdot)|G,\mathcal{X},\upsilon,\mathcal{A},\mathcal{H},\mathcal{M},\Phi)=\min\left\{\rho:\mathrm{Prob}_{\omega\sim P}\{\omega:|\widehat{g}(\omega)-G(x)|>\rho+\upsilon(x)\}\leq\epsilon\right.$$
$$\left.\forall(x,P):\left\{\begin{array}{l}P\in\mathcal{P},x\in\mathcal{X}\\ \ln\left(\int e^{h^T\omega}P(d\omega)\right)\leq\Phi(h;\mathcal{A}(x))\;\forall h\in\mathcal{H}\end{array}\right.\right\}.\tag{3.27}$$

When $G,\mathcal{X},\upsilon,\mathcal{A},\mathcal{H},\mathcal{M}$, and Φ are clear from the context, we shorten

$$\mathrm{Risk}_\epsilon(\widehat{g}(\cdot)|G,\mathcal{X},\upsilon,\mathcal{A},\mathcal{H},\mathcal{M},\Phi)$$

to $\mathrm{Risk}_\epsilon(\widehat{g}(\cdot))$.

Given the data listed at the beginning of this section, we are about to build, in a computationally efficient fashion, an affine estimate $\widehat{g}(\omega)=\langle h_*,\omega\rangle+\varkappa$ along with ρ_* such that the estimate is $(\rho_*,\epsilon,\upsilon(\cdot))$-accurate.

3.3.2 Construction and main results

Let us set

$$\mathcal{H}^+=\{(h,\alpha):h\in\mathcal{E}_H,\alpha>0,h/\alpha\in\mathcal{H}\}$$

so that $\mathcal{H}^+$ is a nonempty convex set in $\mathcal{E}_H\times\mathbf{R}_+$, and let

$$\begin{array}{lrcl}(a)&\Psi_+(h,\alpha)&=&\sup\limits_{x\in\mathcal{X}}\left[\alpha\Phi(h/\alpha,\mathcal{A}(x))-G(x)-\upsilon(x)\right]:\mathcal{H}^+\to\mathbf{R},\\ (b)&\Psi_-(h,\beta)&=&\sup\limits_{x\in\mathcal{X}}\left[\beta\Phi(-h/\beta,\mathcal{A}(x))+G(x)-\upsilon(x)\right]:\mathcal{H}^+\to\mathbf{R},\end{array}\tag{3.28}$$

so that $\Psi_\pm$ are convex real-valued functions on $\mathcal{H}^+$ (recall that Φ is convex-concave and continuous on $\mathcal{H}\times\mathcal{M}$, while $\mathcal{A}(\mathcal{X})$ is a compact subset of $\mathcal{M}$).

Our starting point is quite simple:

Proposition 3.5. *Given $\epsilon\in(0,1)$, let $\bar{h}$, $\bar{\alpha}$, $\bar{\beta}$, $\bar{\varkappa}$, $\bar{\rho}$ be a feasible solution to the system of convex constraints*

$$\begin{array}{lrcl}(a_1)&(h,\alpha)&\in&\mathcal{H}^+\\ (a_2)&(h,\beta)&\in&\mathcal{H}^+\\ (b_1)&\alpha\ln(\epsilon/2)&\geq&\Psi_+(h,\alpha)-\rho+\varkappa\\ (b_2)&\beta\ln(\epsilon/2)&\geq&\Psi_-(h,\beta)-\rho-\varkappa\end{array}\tag{3.29}$$

in variables h, α, β, ρ, $\varkappa$. Setting

$$\widehat{g}(\omega)=\langle\bar{h},\omega\rangle+\bar{\varkappa},$$

we obtain an estimate with ϵ-risk at most $\bar{\rho}$.

Proof. Let $\epsilon\in(0,1)$, $\bar{h}$, $\bar{\alpha}$, $\bar{\beta}$, $\bar{\varkappa}$, $\bar{\rho}$ satisfy the premise of the proposition, and let $x\in\mathcal{X},P$ satisfy (3.26). We have

$$\begin{array}{rrcl}&\mathrm{Prob}_{\omega\sim P}\{\widehat{g}(\omega)>G(x)+\bar{\rho}+\upsilon(x)\}&=&\mathrm{Prob}_{\omega\sim P}\left\{\frac{\langle\bar{h},\omega\rangle}{\bar{\alpha}}>\frac{G(x)+\bar{\rho}-\bar{\varkappa}+\upsilon(x)}{\bar{\alpha}}\right\}\\ \Rightarrow&\mathrm{Prob}_{\omega\sim P}\{\widehat{g}(\omega)>G(x)+\bar{\rho}+\upsilon(x)\}&\leq&\left[\int e^{\langle\bar{h},\omega\rangle/\bar{\alpha}}P(d\omega)\right]e^{-\frac{G(x)+\bar{\rho}-\bar{\varkappa}+\upsilon(x)}{\bar{\alpha}}}\\ &&\leq&e^{\Phi(\bar{h}/\bar{\alpha},\mathcal{A}(x))}e^{-\frac{G(x)+\bar{\rho}-\bar{\varkappa}+\upsilon(x)}{\bar{\alpha}}}.\end{array}$$

As a result,

$$\begin{array}{l}\bar{\alpha}\ln\left(\mathrm{Prob}_{\omega\sim P}\{\widehat{g}(\omega)>G(x)+\bar{\rho}+\upsilon(x)\}\right)\\ \quad\leq\bar{\alpha}\Phi(\bar{h}/\bar{\alpha},\mathcal{A}(x))-G(x)-\bar{\rho}+\bar{\varkappa}-\upsilon(x)\\ \quad\leq\Psi_+(\bar{h},\bar{\alpha})-\bar{\rho}+\bar{\varkappa}\ [\text{by definition of }\Psi_+\text{ and due to }x\in\mathcal{X}]\\ \quad\leq\bar{\alpha}\ln(\epsilon/2)\ [\text{by }(3.29.b_1)]\end{array}$$

so that

$$\mathrm{Prob}_{\omega\sim P}\{\widehat{g}(\omega)>G(x)+\bar{\rho}+\upsilon(x)\}\leq\epsilon/2.$$

Similarly

$$\begin{array}{rl}\mathrm{Prob}_{\omega\sim P}\{\widehat{g}(\omega)<G(x)-\bar{\rho}-\upsilon(x)\} & =\mathrm{Prob}_{\omega\sim P}\left\{\frac{-\langle\bar{h},\omega\rangle}{\bar{\beta}}>\frac{-G(x)+\bar{\rho}+\bar{\varkappa}+\upsilon(x)}{\bar{\beta}}\right\}\\ \Rightarrow\mathrm{Prob}_{\omega\sim P}\{\widehat{g}(\omega)<G(x)-\bar{\rho}-\upsilon(x)\} & \leq\left[\int\mathrm{e}^{-\langle\bar{h},\omega\rangle/\bar{\beta}}P(d\omega)\right]\mathrm{e}^{-\frac{-G(x)+\bar{\rho}+\bar{\varkappa}+\upsilon(x)}{\bar{\beta}}}\\ & \leq\mathrm{e}^{\Phi(-\bar{h}/\bar{\beta},\mathcal{A}(x))}\mathrm{e}^{\frac{G(x)-\bar{\rho}-\bar{\varkappa}-\upsilon(x)}{\bar{\beta}}}.\end{array}$$

Thus

$$\begin{array}{l}\bar{\beta}\ln\left(\mathrm{Prob}_{\omega\sim P}\{\widehat{g}(\omega)<G(x)-\bar{\rho}-\upsilon(x)\}\right)\\ \quad\leq\bar{\beta}\Phi(-\bar{h}/\bar{\beta},\mathcal{A}(x))+G(x)-\bar{\rho}-\bar{\varkappa}-\upsilon(x)\\ \quad\leq\Psi_-(\bar{h},\bar{\beta})-\bar{\rho}-\bar{\varkappa}\ [\text{by definition of }\Psi_-\text{ and due to }x\in\mathcal{X}]\\ \quad\leq\bar{\beta}\ln(\epsilon/2)\ [\text{by }(3.29.b_2)]\end{array}$$

and

$$\mathrm{Prob}_{\omega\sim P}\{\widehat{g}(\omega)<G(x)-\bar{\rho}-\upsilon(x)\}\leq\epsilon/2.$$ □

Corollary 3.6. *In the situation described in Section 3.3.1, let Φ satisfy the relation*

$$\Phi(0;\mu)\geq 0\ \forall\mu\in\mathcal{M}. \tag{3.30}$$

Then

$$\begin{array}{ll}(a) & \widehat{\Psi}_+(h):=\inf_\alpha\left\{\Psi_+(h,\alpha)+\alpha\ln(2/\epsilon):\alpha>0,(h,\alpha)\in\mathcal{H}^+\right\}\\ & \quad=\sup_{x\in\mathcal{X}}\inf_{\alpha>0,(h,\alpha)\in\mathcal{H}^+}\left[\alpha\Phi(h/\alpha,\mathcal{A}(x))-G(x)-\upsilon(x)+\alpha\ln(2/\epsilon)\right],\\ (b) & \widehat{\Psi}_-(h):=\inf_\alpha\left\{\Psi_-(h,\alpha)+\alpha\ln(2/\epsilon):\alpha>0,(h,\alpha)\in\mathcal{H}^+\right\}\\ & \quad=\sup_{x\in\mathcal{X}}\inf_{\alpha>0,(h,\alpha)\in\mathcal{H}^+}\left[\alpha\Phi(-h/\alpha,\mathcal{A}(x))+G(x)-\upsilon(x)+\alpha\ln(2/\epsilon)\right],\end{array} \tag{3.31}$$

and functions $\widehat{\Psi}_\pm:\mathcal{E}_H\to\mathbf{R}$ are convex. Furthermore, let $\bar{h}$, $\bar{\varkappa}$, $\widetilde{\rho}$ be a feasible solution to the system of convex constraints

$$\widehat{\Psi}_+(h)\leq\rho-\varkappa,\ \ \widehat{\Psi}_-(h)\leq\rho+\varkappa \tag{3.32}$$

in variables h, ρ, $\varkappa$. Then the estimate

$$\widehat{g}(\omega)=\langle\bar{h},\omega\rangle+\bar{\varkappa}$$

of $G(x)$, $x\in X$, has the ϵ-risk at most $\widetilde{\rho}$:

$$\mathrm{Risk}_\epsilon(\widehat{g}(\cdot)|G,X,\upsilon,\mathcal{A},\mathcal{H},\mathcal{M},\Phi)\leq\widetilde{\rho}. \tag{3.33}$$

Relation (3.32) *(and thus the risk bound* (3.33)*) clearly holds true when $\bar{h}$ is a* *candidate solution to the convex optimization problem*

$$\mathrm{Opt} = \min_h \left\{ \widehat{\Psi}(h) := \tfrac{1}{2}\left[\widehat{\Psi}_+(h) + \widehat{\Psi}_-(h)\right]\right\}, \tag{3.34}$$

$\widetilde{\rho} = \widehat{\Psi}(\bar{h})$, *and*

$$\bar{\varkappa} = \tfrac{1}{2}\left[\widehat{\Psi}_-(\bar{h}) - \widehat{\Psi}_+(\bar{h})\right].$$

As a result, by properly selecting $\bar{h}$*, we can make (an upper bound on) the* ϵ*-risk of estimate* $\widehat{g}(\cdot)$ *arbitrarily close to* Opt*, and equal to* Opt *when optimization problem* (3.34) *is solvable.*

Proof. Let us first verify the identities in (3.31). The function

$$\Theta_+(h,\alpha;x) = \alpha\Phi(h/\alpha, \mathcal{A}(x)) - G(x) - \upsilon(x) + \alpha\ln(2/\epsilon) : \mathcal{H}^+ \times \mathcal{X} \to \mathbf{R}$$

is convex-concave and continuous, and $\mathcal{X}$ is compact, whence by the Sion-Kakutani Theorem

$$\begin{array}{rcl}\widehat{\Psi}_+(h) & := & \inf_\alpha \left\{\Psi_+(h,\alpha) + \alpha\ln(2/\epsilon) : \alpha > 0, (h,\alpha)\in\mathcal{H}^+\right\}\\ & = & \inf_{\alpha>0,(h,\alpha)\in\mathcal{H}^+}\max_{x\in\mathcal{X}}\Theta_+(h,\alpha;x)\\ & = & \sup_{x\in\mathcal{X}}\inf_{\alpha>0,(h,\alpha)\in\mathcal{H}^+}\Theta_+(h,\alpha;x)\\ & = & \sup_{x\in\mathcal{X}}\inf_{\alpha>0,(h,\alpha)\in\mathcal{H}^+}\left[\alpha\Phi(h/\alpha,\mathcal{A}(x)) - G(x) - \upsilon(x) + \alpha\ln(2/\epsilon)\right],\end{array}$$

as required in (3.31.a). As we know, $\Psi_+(h,\alpha)$ is real-valued continuous function on $\mathcal{H}^+$, so that $\widehat{\Psi}_+$ is convex on $\mathcal{E}_H$, provided that the function is real-valued. Now, let $\bar{x}\in\mathcal{X}$, and let e be a subgradient of $\phi(h) = \Phi(h;\mathcal{A}(\bar{x}))$ taken at $h=0$. For $h\in\mathcal{E}_H$ and all $\alpha>0$ such that $(h,\alpha)\in\mathcal{H}^+$ we have

$$\begin{array}{rcl}\Psi_+(h,\alpha) & \geq & \alpha\Phi(h/\alpha;\mathcal{A}(\bar{x})) - G(\bar{x}) - \upsilon(\bar{x}) + \alpha\ln(2/\epsilon)\\ & \geq & \alpha[\Phi(0;\mathcal{A}(\bar{x})) + \langle e, h/\alpha\rangle] - G(\bar{x}) - \upsilon(\bar{x}) + \alpha\ln(2/\epsilon)\\ & \geq & \langle e,h\rangle - G(\bar{x}) - \upsilon(\bar{x})\end{array}$$

(we have used (3.30)), and therefore $\Psi_+(h,\alpha)$ as a function of α is bounded from below on the set $\{\alpha>0 : h/\alpha\in\mathcal{H}\}$. In addition, this set is nonempty, since $\mathcal{H}$ contains a neighbourhood of the origin. Thus, $\widehat{\Psi}_+$ is real-valued and convex on $\mathcal{E}_H$. Verification of (3.31.b) and of the fact that $\widehat{\Psi}_-(h)$ is real-valued convex function on $\mathcal{E}_H$ is completely similar.

Now, given a feasible solution $(\bar{h},\bar{\varkappa},\widetilde{\rho})$ to (3.32), let us select some $\bar{\rho}>\widetilde{\rho}$. Taking into account the definition of $\widehat{\Psi}_\pm$, we can find $\bar{\alpha}$ and $\bar{\beta}$ such that

$$\begin{array}{l}(\bar{h},\bar{\alpha})\in\mathcal{H}^+ \;\&\; \Psi_+(\bar{h},\bar{\alpha}) + \bar{\alpha}\ln(2/\epsilon) \leq \bar{\rho} - \bar{\varkappa},\\ (\bar{h},\bar{\beta})\in\mathcal{H}^+ \;\&\; \Psi_-(\bar{h},\bar{\beta}) + \bar{\beta}\ln(2/\epsilon) \leq \bar{\rho} + \bar{\varkappa},\end{array}$$

implying that the collection $(\bar{h},\bar{\alpha},\bar{\beta},\bar{\varkappa},\bar{\rho})$ is a feasible solution to (3.29). Invoking Proposition 3.5, we get

$$\mathrm{Prob}_{\omega\sim P}\left\{\omega : |\widehat{g}(\omega) - G(x)| > \bar{\rho} + \upsilon(x)\right\} \leq \epsilon$$

for all $(x\in\mathcal{X}, P\in\mathcal{P})$ satisfying (3.26). Since $\bar{\rho}$ can be selected arbitrarily close to $\widetilde{\rho}$, $\widehat{g}(\cdot)$ indeed is a $(\widetilde{\rho},\epsilon,\upsilon(\cdot))$-accurate estimate. □

3.3.3 Estimation from repeated observations

Assume that in the situation described in Section 3.3.1 we have access to K observations $\omega_1, ..., \omega_K$ sampled, independently of each other, from a probability distribution P, and aim to build the estimate based on these K observations rather than on a single observation. We can immediately reduce this new situation to the previous one, just by redefining the data. Specifically, given initial data

$$\mathcal{H} \subset \mathcal{E}_H,\ \mathcal{M} \subset \mathcal{E}_M,\ \Phi(\cdot;\cdot) : \mathcal{H} \times \mathcal{M} \to \mathbf{R},\ \mathcal{X} \subset \mathcal{E}_X,\ \upsilon(\cdot),\ \mathcal{A}(\cdot),\ G(x) = \langle g, x\rangle + c$$

(see Section 3.3.1) and a positive integer K, let us update part of the data, namely, replace $\mathcal{H} \subset \mathcal{E}_\mathcal{H}$ with

$$\mathcal{H}^K := \underbrace{\mathcal{H} \times ... \times \mathcal{H}}_{K} \subset \mathcal{E}_H^K := \underbrace{\mathcal{E}_H \times ... \times \mathcal{E}_H}_{K},$$

and replace $\Phi(\cdot,\cdot) : \mathcal{H} \times \mathcal{M} \to \mathbf{R}$ with

$$\Phi^K(h^K = (h_1, ..., h_K); \mu) = \sum_{i=1}^{K} \Phi(h_i; \mu) : \mathcal{H}^K \times \mathcal{M} \to \mathbf{R}.$$

It is immediately seen that the updated data satisfy all requirements imposed on the data in Section 3.3.1, and that whenever $x \in \mathcal{X}$ and a Borel probability distribution P on $\mathcal{E}_\mathcal{H}$ are linked by (3.26), x and the distribution P^K of K-element i.i.d. sample $\omega^K = (\omega_1, ..., \omega_K)$ drawn from P are linked by the relation

$$\begin{array}{l} \forall h^K = (h_1, ..., h_K) \in \mathcal{H}^K : \\ \ln\left(\int_{\mathcal{E}_H^K} \mathrm{e}^{\langle h^K, \omega^K\rangle} P^K(d\omega^K)\right) \end{array} \begin{array}{ll} & \\ = & \sum_i \ln\left(\int_{\mathcal{E}_H} \mathrm{e}^{\langle h_i, \omega_i\rangle} P(d\omega_i)\right) \\ \leq & \Phi^K(h^K; \mathcal{A}(x)). \end{array}$$

Applying to our new data the construction from Section 3.3.2, we arrive at "repeated observation" versions of Proposition 3.5 and Corollary 3.6. Note that the resulting convex constraints/objectives are symmetric w.r.t. permutations functions of the components $h_1, ..., h_K$ of h^K, implying that we lose nothing when restricting ourselves with collections h^K with components equal to each other; it is convenient to denote the common value of these components h/K. With this observation in mind, Proposition 3.5 and Corollary 3.6 translate into the following statements (we use the assumptions and the notation from the previous sections):

Proposition 3.7. *Given $\epsilon \in (0, 1)$ and positive integer K, let*

$$\begin{array}{llll} (a) & \Psi_+(h, \alpha) & = & \sup_{x\in\mathcal{X}} [\alpha\Phi(h/\alpha, \mathcal{A}(x)) - G(x) - \upsilon(x)] : \mathcal{H}^+ \to \mathbf{R}, \\ (b) & \Psi_-(h, \beta) & = & \sup_{x\in\mathcal{X}} [\beta\Phi(-h/\beta, \mathcal{A}(x)) + G(x) - \upsilon(x)] : \mathcal{H}^+ \to \mathbf{R}, \end{array}$$

and let $\bar{h}$, $\bar{\alpha}$, $\bar{\beta}$, $\bar{\varkappa}$, $\bar{\rho}$ be a feasible solution to the system of convex constraints

$$\begin{array}{llll} (a_1) & (h, \alpha) & \in & \mathcal{H}^+ \\ (a_2) & (h, \beta) & \in & \mathcal{H}^+ \\ (b_1) & \alpha K^{-1} \ln(\epsilon/2) & \geq & \Psi_+(h, \alpha) - \rho + \varkappa \\ (b_2) & \beta K^{-1} \ln(\epsilon/2) & \geq & \Psi_-(h, \beta) - \rho - \varkappa \end{array} \tag{3.35}$$

in variables h, α, β, ρ, $\varkappa$. Setting

$$\widehat{g}(\omega^K) = \left\langle \bar{h}, \frac{1}{K}\sum\nolimits_{i=1}^K \omega_i \right\rangle + \bar{\varkappa},$$

we obtain an estimate of $G(x)$ via independent K-repeated observations

$$\omega_i \sim P,\ i = 1, ..., K,$$

with the ϵ-risk on $\mathcal{X}$ not exceeding $\bar{\rho}$. In other words, whenever $x \in \mathcal{X}$ and a Borel probability distribution P on $\mathcal{E}_\mathcal{H}$ are linked by (3.26), one has

$$\mathrm{Prob}_{\omega^K \sim P^K}\left\{\omega^K : |\widehat{g}(\omega^K) - G(x)| > \bar{\rho} + \upsilon(x)\right\} \leq \epsilon. \tag{3.36}$$

Corollary 3.8. *In the situation described at the beginning of Section 3.3.1, let Φ satisfy relation (3.30), and let a positive integer K be given. Then*

$$\begin{array}{ll}(a) & \widehat{\Psi}_{+,K}(h) := \inf\limits_{\alpha}\left\{\Psi_+(h,\alpha) + K^{-1}\alpha\ln(2/\epsilon) : \alpha > 0, (h,\alpha) \in \mathcal{H}^+\right\} \\ & \quad = \sup\limits_{x\in\mathcal{X}} \inf\limits_{\alpha>0,(h,\alpha)\in\mathcal{H}^+} \left[\alpha\Phi(h/\alpha, \mathcal{A}(x)) - G(x) - \upsilon(x) + K^{-1}\alpha\ln(2/\epsilon)\right], \\ (b) & \widehat{\Psi}_{-,K}(h) := \inf_{\alpha}\left\{\Psi_-(h,\alpha) + K^{-1}\alpha\ln(2/\epsilon) : \alpha > 0, (h,\alpha) \in \mathcal{H}^+\right\} \\ & \quad = \sup\limits_{x\in\mathcal{X}} \inf\limits_{\alpha>0,(h,\alpha)\in\mathcal{H}^+} \left[\alpha\Phi(-h/\alpha, \mathcal{A}(x)) + G(x) - \upsilon(x) + K^{-1}\alpha\ln(2/\epsilon)\right],\end{array}$$

and functions $\widehat{\Psi}_{\pm,K} : \mathcal{E}_H \to \mathbf{R}$ are convex. Furthermore, let $\bar{h}$, $\bar{\varkappa}$, $\widetilde{\rho}$ be a feasible solution to the system of convex constraints

$$\widehat{\Psi}_{+,K}(h) \leq \rho - \varkappa,\quad \widehat{\Psi}_{-,K}(h) \leq \rho + \varkappa \tag{3.37}$$

in variables h, ρ, $\varkappa$. Then the ϵ-risk of the estimate

$$\widehat{g}(\omega^K) = \left\langle \bar{h}, \frac{1}{K}\sum\nolimits_{i=1}^K \omega_i \right\rangle + \bar{\varkappa},$$

of $G(x)$, $x \in \mathcal{X}$, is at most $\widehat{\Psi}(\bar{h})$, implying that whenever $x \in \mathcal{X}$ and a Borel probability distribution P on $\mathcal{E}_H$ are linked by (3.26), relation (3.36) holds true.

Relation (3.37) clearly holds true when $\bar{h}$ is a candidate solution to the convex optimization problem

$$\mathrm{Opt}_K = \min_h \left\{\widehat{\Psi}_K(h) := \tfrac{1}{2}\left[\widehat{\Psi}_{+,K}(h) + \widehat{\Psi}_{-,K}(h)\right]\right\}, \tag{3.38}$$

$\bar{\rho} = \widehat{\Psi}_K(\bar{h})$, and

$$\bar{\varkappa} = \tfrac{1}{2}\left[\widehat{\Psi}_{-,K}(\bar{h}) - \widehat{\Psi}_{+,K}(\bar{h})\right].$$

As a result, by properly selecting $\bar{h}$ we can make (an upper bound on) the ϵ-risk of the estimate $\widehat{g}(\cdot)$ arbitrarily close to Opt, *and equal to* Opt *when optimization problem (3.38) is solvable.*

From now on, if not explicitly stated otherwise, we deal with K-repeated observations; to get back to single-observation case, it suffices to set $K = 1$.

3.3.4 Application: Estimating linear forms of sub-Gaussianity parameters

Consider the simplest case of the situation from Sections 3.3.1 and 3.3.3, where

- $\mathcal{H} = \mathcal{E}_H = \mathbf{R}^d$, $\mathcal{M} = \mathcal{E}_M = \mathbf{R}^d \times \mathbf{S}^d_+$,

  $$\Phi(h; \mu, M) = h^T\mu + \tfrac{1}{2}h^T M h : \mathbf{R}^d \times (\mathbf{R}^d \times \mathbf{S}^d_+) \to \mathbf{R},$$

  so that $\mathcal{S}[\mathcal{H}, \mathcal{M}, \Phi]$ is the family of all sub-Gaussian distributions on $\mathbf{R}^d$;
- $\mathcal{X} \subset \mathcal{E}_X = \mathbf{R}^{n_x}$ is a nonempty convex compact set;
- $\mathcal{A}(x) = (Ax + a, M(x))$, where A is $d \times n_x$ matrix, and $M(x)$ is a symmetric $d \times d$ matrix affinely depending on x such that $M(x)$ is $\succeq 0$ when $x \in \mathcal{X}$;
- $v(x)$ is a convex continuous function on $\mathcal{X}$;
- $G(x)$ is an affine function on $\mathcal{E}_X$.

In the case in question, (3.30) clearly takes place, and the left-hand sides in constraints (3.37) become

$$\begin{array}{rcl}\widehat{\Psi}_{+,K}(h) &=& \sup\limits_{x\in\mathcal{X}}\inf\limits_{\alpha>0}\left\{h^T[Ax+a] + \frac{1}{2\alpha}h^T M(x)h + K^{-1}\alpha\ln(2/\epsilon) - G(x) - v(x)\right\}\\ &=& \max\limits_{x\in\mathcal{X}}\left\{\sqrt{2K^{-1}\ln(2/\epsilon)[h^T M(x)h]} + h^T[Ax+a] - G(x) - v(x)\right\},\\ \widehat{\Psi}_{-,K}(h) &=& \sup\limits_{x\in\mathcal{X}}\inf\limits_{\alpha>0}\left\{-h^T[Ax+a] + \frac{1}{2\alpha}h^T M(x)h + K^{-1}\alpha\ln(2/\epsilon) + G(x) - v(x)\right\}\\ &=& \max\limits_{x\in\mathcal{X}}\left\{\sqrt{2K^{-1}\ln(2/\epsilon)[h^T M(x)h]} - h^T[Ax+a] + G(x) - v(x)\right\}.\end{array}$$

Thus, system (3.37) reads

$$\begin{array}{rcl}a^T h + \max\limits_{x\in\mathcal{X}}\left[\sqrt{2K^{-1}\ln(2/\epsilon)[h^T M(x)h]} + h^T Ax - G(x) - v(x)\right] &\le& \rho - \varkappa,\\ -a^T h + \max\limits_{x\in\mathcal{X}}\left[\sqrt{2K^{-1}\ln(2/\epsilon)[h^T M(x)h]} - h^T Ax + G(x) - v(x)\right] &\le& \rho + \varkappa.\end{array}$$

We arrive at the following version of Corollary 3.8:

Proposition 3.9. *In the situation described at the beginning of Section 3.3.4, given $\epsilon \in (0,1)$, let $\bar{h}$ be a feasible solution to the convex optimization problem*

$$\mathrm{Opt}_K = \min_{h\in\mathbf{R}^d} \widehat{\Psi}_K(h) \tag{3.39}$$

where

$$\widehat{\Psi}_K(h) := \frac{1}{2}\left[\begin{array}{l}\overbrace{\max\limits_{x\in\mathcal{X}}\left[\sqrt{2K^{-1}\ln(2/\epsilon)[h^T M(x)h]} + h^T Ax - G(x) - v(x)\right] + a^T h}^{\widehat{\Psi}_{+,K}(h)}\\ \underbrace{+\max\limits_{y\in\mathcal{X}}\left[\sqrt{2K^{-1}\ln(2/\epsilon)[h^T M(y)h]} - h^T Ay + G(y) - v(y)\right] - a^T h}_{\widehat{\Psi}_{-,K}(h)}\end{array}\right].$$

Then, setting

$$\bar{\varkappa} = \tfrac{1}{2}\left[\widehat{\Psi}_{-,K}(\bar{h}) - \widehat{\Psi}_{+,K}(\bar{h})\right], \; \bar{\rho} = \widehat{\Psi}_K(\bar{h}),$$

the affine estimate

$$\widehat{g}(\omega^K) = \frac{1}{K}\sum_{i=1}^K \bar{h}^T\omega_i + \bar{\varkappa}$$

has ϵ-risk, taken w.r.t. the data listed at the beginning of this section, at most $\bar{\rho}$.

It is immediately seen that optimization problem (3.39) is solvable, provided that

$$\bigcap_{x\in\mathcal{X}} \mathrm{Ker}(M(x)) = \{0\},$$

and an optimal solution h_* to the problem, taken along with

$$\varkappa_* = \tfrac{1}{2}\left[\widehat{\Psi}_{-,K}(h_*) - \widehat{\Psi}_{+,K}(h_*)\right], \tag{3.40}$$

yields the affine estimate

$$\widehat{g}_*(\omega) = \frac{1}{K}\sum_{i=1}^{K} h_*^T\omega_i + \varkappa_*$$

with ϵ-risk, taken w.r.t. the data listed at the beginning of this section, at most Opt_K.

3.3.4.1 Consistency

Assuming $v(x) \equiv 0$, we can easily answer the natural question "when is the proposed estimation scheme consistent?" meaning that for every $\epsilon \in (0,1)$, it allows us to achieve arbitrarily small ϵ-risk, provided that K is large enough. Specifically, denoting by g^Tx the linear part of $G(x)$: $G(x) = g^Tx + c$, from Proposition 3.9 it is immediately seen that a necessary and sufficient condition for consistency is the existence of $\bar{h} \in \mathbf{R}^d$ such that $\bar{h}^TAx = g^Tx$ for all $x \in \mathcal{X} - \mathcal{X}$, or, equivalently, the condition that g is orthogonal to the intersection of the kernel of A with the linear span of $\mathcal{X} - \mathcal{X}$. Indeed, under this assumption, for every fixed $\epsilon \in (0,1)$ we clearly have $\lim_{K\to\infty} \widehat{\Psi}_K(\bar{h}) = 0$, implying that $\lim_{K\to\infty} \mathrm{Opt}_K = 0$, with $\widehat{\Psi}_K$ and Opt_K given by (3.39). On the other hand, if the condition is violated, then there exist $x', x'' \in \mathcal{X}$ such that $Ax' = Ax''$ and $G(x') \neq G(x'')$; we lose nothing when assuming that $G(x'') > G(x')$. Looking at (3.39), we see that

$$\begin{array}{rcl}\widehat{\Psi}_K(h) &\geq& \frac{1}{2}\Big[\left(\sqrt{2K^{-1}\ln(2/\epsilon)[h^TM(x')h]} + h^TAx' - G(x')\right) + a^Th \\ && + \left(\sqrt{2K^{-1}\ln(2/\epsilon)[h^TM(x'')h]} - h^TAx'' + G(x'')\right) - a^Th\Big] \\ &\geq& G(x'') - G(x'),\end{array}$$

whence Opt_K, for all K, is lower-bounded by $G(x'') - G(x') > 0$.

3.3.4.2 Direct product case

Further simplifications are possible in the *direct product case*, where, in addition to what was assumed at the beginning of Section 3.3.4,

- $\mathcal{E}_X = \mathcal{E}_U \times \mathcal{E}_V$ and $\mathcal{X} = U \times V$, with convex compact sets $U \subset \mathcal{E}_U = \mathbf{R}^{n_u}$ and $V \subset E_V = \mathbf{R}^{n_v}$,
- $\mathcal{A}(x = (u,v)) = [Au + a, M(v)] : U \times V \to \mathbf{R}^d \times \mathbf{S}^d$, with $M(v) \succeq 0$ for $v \in V$,
- $G(x = (u,v)) = g^Tu + c$ depends solely on u, and

• $\upsilon(x=(u,v))=\varrho(u)$ depends solely on u.

It is immediately seen that in the direct product case problem (3.39) reads

$$\mathrm{Opt}_K=\min_{h\in\mathbf{R}^d}\left\{\frac{\phi_U(A^Th-g)+\phi_U(-A^Th+g)}{2}+\max_{v\in V}\sqrt{2K^{-1}\ln(2/\epsilon)h^TM(v)h}\right\}, \tag{3.41}$$

where

$$\phi_U(f)=\max_{u\in U}\left[u^Tf-\varrho(u)\right]. \tag{3.42}$$

Assuming $\bigcap_{v\in V}\mathrm{Ker}(M(v))=\{0\}$, the problem is solvable, and its optimal solution h_* gives rise to the affine estimate

$$\widehat{g}_*(\omega^K)=\frac{1}{K}\sum_i h_*^T\omega_i+\varkappa_*,\ \varkappa_*=\tfrac{1}{2}[\phi_U(-A^Th+g)-\phi_U(A^Th-g)]-a^Th_*+c$$

with ϵ-risk $\leq \mathrm{Opt}_K$.

Near-optimality. In addition to the assumption that we are in the direct product case, assume that $\upsilon(\cdot)\equiv 0$ and, for the sake of simplicity, that $M(v)\succ 0$ whenever $v\in V$. In this case (3.39) reads

$$\begin{array}{rl}\mathrm{Opt}_K= & \min_h\max_{v\in V}\Big\{\Theta(h,v):=\tfrac{1}{2}[\phi_U(A^Th-g)+\phi_U(-A^Th+g)]\\ & \qquad\qquad+\sqrt{2K^{-1}\ln(2/\epsilon)h^TM(v)h}\Big\}.\end{array}$$

Hence, taking into account that $\Theta(h,v)$ clearly is convex in h and concave in v, while V is a convex compact set, by the Sion-Kakutani Theorem we get also

$$\begin{array}{rl}\mathrm{Opt}_K= & \max_{v\in V}\Big[\mathrm{Opt}(v)=\min_h\Big[\tfrac{1}{2}[\phi_U(A^Th-g)+\phi_U(-A^Th+g)]\\ & \qquad\qquad+\sqrt{2K^{-1}\ln(2/\epsilon)h^TM(v)h}\Big]\Big].\end{array} \tag{3.43}$$

Now consider the problem of estimating g^Tu from independent observations ω_i, $i\leq K$, sampled from $\mathcal{N}(Au+a,M(v))$, where unknown u is known to belong to U and $v\in V$ is known. Let $\rho_\epsilon(v)$ be the minimax ϵ-risk of recovery:

$$\rho_\epsilon(v)=\inf_{\widehat{g}(\cdot)}\left\{\rho:\mathrm{Prob}_{\omega^K\sim[\mathcal{N}(Au+a,M(v))]^K}\{\omega^K:|\widehat{g}(\omega^K)-g^Tu|>\rho\}\leq\epsilon\ \forall u\in U\right\},$$

where inf is taken over all Borel functions $\widehat{g}(\cdot):\mathbf{R}^{Kd}\to\mathbf{R}$. Invoking [131, Theorem 3.1], it is immediately seen that whenever $\epsilon<1/4$, one has

$$\rho_\epsilon(v)\geq\left[\frac{2\ln(2/\epsilon)}{\ln\left(\frac{1}{4\epsilon}\right)}\right]^{-1}\mathrm{Opt}(v).$$

Since the family $\mathcal{SG}(U,V)$ of all sub-Gaussian distributions on $\mathbf{R}^d$ with parameters $(Au+a,M(v))$, $u\in U$, $v\in V$, contains all Gaussian distributions $\mathcal{N}(Au+a,M(v))$ induced by $(u,v)\in U\times V$, we arrive at the following conclusion:

Proposition 3.10. *In the just described situation, the minimax optimal ϵ-risk*

$$\mathrm{Risk}_\epsilon^{\mathrm{opt}}(K) = \inf_{\widehat{g}(\cdot)} \mathrm{Risk}_\epsilon(\widehat{g}(\cdot))$$

of recovering $g^T u$ from a K-repeated i.i.d. sub-Gaussian observation with parameters $(Au+a, M(v))$, $(u,v) \in U \times V$, is within a moderate factor of the upper bound Opt_K on the ϵ-risk, taken w.r.t. the same data, of the affine estimate $\widehat{g}_*(\cdot)$ yielded by an optimal solution to (3.41), namely,

$$\mathrm{Opt}_K \leq \frac{2\ln(2/\epsilon)}{\ln\left(\frac{1}{4\epsilon}\right)} \mathrm{Risk}_\epsilon^{\mathrm{opt}}(K).$$

3.3.4.3 Numerical illustration

The numerical illustration we are about to discuss models the situation in which we want to recover a linear form of a signal x known to belong to a given convex compact subset $\mathcal{X}$ via indirect observations Ax affected by sub-Gaussian "relative noise," meaning that the variance of observation is larger the larger is the signal. Specifically, our observation is

$$\omega \sim \mathcal{SG}(Ax, M(x)),$$

where

$$x \in \mathcal{X} = \left\{x \in \mathbf{R}^n : 0 \leq x_j \leq j^{-\alpha}, 1 \leq j \leq n\right\},\ M(x) = \sigma^2 \sum_{j=1}^n x_j \Theta_j \tag{3.44}$$

where $A \in \mathbf{R}^{d\times n}$ and $\Theta_j \in \mathbf{S}^d_+$, $j = 1, ..., n$, are given matrices; the linear form to be estimated is $G(x) = g^T x$. The entities g, A and $\{\Theta_j\}_{j=1}^n$ and reals $\alpha \geq 0$ ("degree of smoothness") and $\sigma > 0$ ("noise intensity") are parameters of the estimation problem we intend to process. The parameters g, A, Θ_j are as follows:

- $g \geq 0$ is selected at random and then normalized to have

$$\max_{x\in\mathcal{X}} g^T x = \max_{x,y\in\mathcal{X}} g^T[x-y] = 2;$$

- we deal with the case of $n > d$ ("deficient observations"); the d nonzero singular values of A were set to $\theta^{-\frac{i-1}{d-1}}$, where "condition number" $\theta \geq 1$ is a parameter; the orthonormal systems U and V of the first d left and, respectively, right singular vectors of A were drawn at random from rotationally invariant distributions;
- the positive semidefinite $d\times d$ matrices Θ_j are orthogonal projectors on randomly selected subspaces in $\mathbf{R}^d$ of dimension $\lfloor d/2 \rfloor$;
- in all our experiments, we consider the single-observation case $K = 1$ and use $v(\cdot) \equiv 0$.

Note that $\mathcal{X}$ possesses the $\geq$-largest point $\bar{x}$, whence $M(x) \preceq M(\bar{x})$ whenever $x \in \mathcal{X}$; as a result, sub-Gaussian distributions with matrix parameter $M(x)$, $x \in \mathcal{X}$, can be thought also to have matrix parameter $M(\bar{x})$. One of the goals of the considered experiment is to understand how much we might lose were we replacing $M(\cdot)$ with $\widehat{M}(x) \equiv M(\bar{x})$, that is, were we ignoring the fact that small signals result in low-noise observations.

In our experiment we use $d = 32$, $m = 48$, $\alpha = 2$, $\theta = 2$, and $\sigma = 0.01$. With these parameters, we generated at random, as described above, 10 collections $\{g, A, \Theta_j, j \leq d\}$, thus arriving at 10 estimation problems. For each problem, we apply the outlined machinery to build an estimate of $g^T x$ affine in ω as yielded by the optimal solution to (3.39), and compute the upper bound Opt on the ($\epsilon = 0.01$)-risk of this estimate. In fact, for each problem, we build two estimates and two risk bounds: the first for the problem "as is," and the second for the aforementioned "direct product envelope" of the problem, where the mapping $x \mapsto M(x)$ is replaced with conservative $x \mapsto \widehat{M}(x) := M(\bar{x})$. The results are as follows:

min	median	mean	max
0.138	0.190	0.212	0.299
0.150	0.210	0.227	0.320

Upper bounds on 0.01-risk, data over 10 estimation problems
$[d = 32, m = 48, \alpha = 2, \theta = 2, \sigma = 0.01]$
First row: $\omega \sim \mathcal{SG}(Ax, M(x))$; second row: $\omega \sim \mathcal{SG}(Ax, M(\bar{x}))$

Note the significant "noise amplification" in the estimate (about 20 times the observation noise level σ) and high risk variability across the experiments. Seemingly, both these phenomena stem from the fact that we have highly deficient observations ($n/d = 1.5$) combined with a random orientation of the 16-dimensional kernel of A.

3.4 ESTIMATING QUADRATIC FORMS VIA QUADRATIC LIFTING

In the situation of Section 3.3.1, passing from "original" observations (3.25) to their quadratic lifting, we can use the machinery just developed to estimate quadratic, rather than linear, forms of the underlying parameters. We investigate the related possibilities in the cases of Gaussian and sub-Gaussian observations. The results of this section form an essential extension of the results of [39, 81] where a similar approach to estimating quadratic functionals of the mean of a Gaussian vector was used.

3.4.1 Estimating quadratic forms, Gaussian case

3.4.1.1 Preliminaries

Consider the situation where we are given

- a nonempty bounded set U in $\mathbf{R}^m$;
- a nonempty convex compact subset $\mathcal{V}$ of the positive semidefinite cone $\mathbf{S}^d_+$;
- a matrix $\Theta_* \succ 0$ such that $\Theta_* \succeq \Theta$ for all $\Theta \in \mathcal{V}$;
- an affine mapping $u \mapsto A[u; 1] : \mathbf{R}^m \to \Omega = \mathbf{R}^d$, where A is a given $d \times (m + 1)$ matrix;
- a convex continuous function $\varrho(\cdot)$ on $\mathbf{S}^{m+1}_+$.

A pair $(u \in U, \Theta \in \mathcal{V})$ specifies Gaussian random vector $\zeta \sim \mathcal{N}(A[u; 1], \Theta)$ and thus specifies probability distribution $P[u,\Theta]$ of $(\zeta,\zeta\zeta^T)$. Let $\mathcal{Q}(U,\mathcal{V})$ be the family of probability distributions on $\Omega = \mathbf{R}^d \times \mathbf{S}^d$ stemming this way from Gaussian distributions with parameters from $U \times \mathcal{V}$. Our goal is to cover the family $\mathcal{Q}(U,\mathcal{V})$ by a family of the type $\mathcal{S}[\mathcal{N},\mathcal{M},\Phi]$.

It is convenient to represent a linear form on $\Omega = \mathbf{R}^d \times \mathbf{S}^d$ as

$$h^T z + \tfrac{1}{2}\mathrm{Tr}(HZ),$$

where $(h,H) \in \mathbf{R}^d \times \mathbf{S}^d$ is the "vector of coefficients" of the form, and $(z,Z) \in \mathbf{R}^d \times \mathbf{S}^d$ is the argument of the form.

We assume that for some $\delta \in [0,2]$ it holds

$$\|\Theta^{1/2}\Theta_*^{-1/2} - I_d\| \leq \delta \ \forall \Theta \in \mathcal{V}, \tag{3.45}$$

where $\|\cdot\|$ is the spectral norm (cf. (2.129)). Finally, we set

$$b = [0;...;0;1] \in \mathbf{R}^{m+1},\ B = \left[\begin{array}{c} A \\ b^T \end{array}\right]$$

and

$$\mathcal{Z}^+ = \{W \in \mathbf{S}^{m+1}_+ : W_{m+1,m+1} = 1\}.$$

The statement below is nothing but a straightforward reformulation of Proposition 2.43.i:

Proposition 3.11. *In the just described situation, let us select $\gamma \in (0,1)$ and set*

$$\begin{array}{rcl} \mathcal{H} &=& \mathcal{H}_\gamma := \{(h,H) \in \mathbf{R}^d \times \mathbf{S}^d : -\gamma\Theta_*^{-1} \preceq H \preceq \gamma\Theta_*^{-1}\}, \\ \mathcal{M}^+ &=& \mathcal{V} \times \mathcal{Z}^+, \\ \Phi(h,H;\Theta,Z) &=& -\frac{1}{2}\ln \mathrm{Det}(I - \Theta_*^{1/2}H\Theta_*^{1/2}) + \frac{1}{2}\mathrm{Tr}([\Theta - \Theta_*]H) \\ && + \frac{\delta(2+\delta)}{2(1-\|\Theta_*^{1/2}H\Theta_*^{1/2}\|)}\|\Theta_*^{1/2}H\Theta_*^{1/2}\|_F^2 + \Gamma(h,H;Z) : \mathcal{H} \times \mathcal{M}^+ \to \mathbf{R}, \end{array}$$

where $\|\cdot\|$ is the spectral, $\|\cdot\|_F$ is the Frobenius norm, and

$$\begin{array}{rcl} \Gamma(h,H;Z) &=& \frac{1}{2}\mathrm{Tr}\left(Z[bh^TA + A^Thb^T + A^THA + B^T[H,h]^T[\Theta_*^{-1} - H]^{-1}[H,h]B]\right) \\ &=& \frac{1}{2}\mathrm{Tr}\left(ZB^T\left[\left[\begin{array}{c|c} H & h \\ \hline h^T & \end{array}\right] + [H,h]^T\left[\Theta_*^{-1} - H\right]^{-1}[H,h]\right]B\right). \end{array}$$

Then $\mathcal{H},\mathcal{M}^+,\Phi$ is a regular data, and for every $(u,\Theta) \in \mathbf{R}^m \times \mathcal{V}$ it holds

$$\forall (h,H) \in \mathcal{H} : \ln\left(\mathbf{E}_{\zeta\sim\mathcal{N}(A[u;1],\Theta)}\left\{\mathrm{e}^{h^T\zeta + \frac{1}{2}\zeta^TH\zeta}\right\}\right) \leq \Phi(h,H;\Theta,[u;1][u;1]^T).$$

Besides this, function $\Phi(h,H;\Theta,Z)$ is coercive in the convex argument: whenever $(\Theta,Z) \in \mathcal{M}$ and $(h_i,H_i) \in \mathcal{H}$ and $\|(h_i,H_i)\| \to \infty$ as $i \to \infty$, we have $\Phi(h_i,H_i;\Theta,Z) \to \infty$, $i \to \infty$.

3.4.1.2 Estimating quadratic form: Situation and goal

Let us assume that we are given a sample $\zeta^K = (\zeta_1,...,\zeta_K)$ of identically distributed observations

$$\zeta_i \sim \mathcal{N}(A[u;1], M(v)),\ 1 \leq i \leq K \tag{3.46}$$

independent across i, where

- (u, v) is an unknown "signal" known to belong to a given set $U \times V$, where
  - $U \subset \mathbf{R}^m$ is a compact set, and
  - $V \subset \mathbf{R}^k$ is a compact convex set;
- A is a given $d \times (m+1)$ matrix, and $v \mapsto M(v) : \mathbf{R}^k \to \mathbf{S}^d$ is an affine mapping such that $M(v) \succeq 0$ whenever $v \in V$.

We are also given a convex calibrating function $\varrho(Z) : \mathbf{S}^{m+1}_+ \to \mathbf{R}$ and "functional of interest"

$$F(u, v) = [u; 1]^T Q [u; 1] + q^T v, \tag{3.47}$$

where Q and q are a known $(m+1) \times (m+1)$ symmetric matrix and a k-dimensional vector, respectively. Our goal is to estimate the value $F(u, v)$, for unknown (u, v) known to belong to $U \times V$. Given a tolerance $\epsilon \in (0, 1)$, we quantify the quality of a candidate estimate $\widehat{g}(\zeta^K)$ of $F(u, v)$ by the smallest ρ such that for all $(u, v) \in U \times V$ it holds

$$\mathrm{Prob}_{\zeta^K \sim [\mathcal{N}(A[u;1], M(v))]^K} \left\{ |\widehat{g}(\zeta^K) - F(u, v)| > \rho + \varrho([u; 1][u; 1]^T) \right\} \leq \epsilon.$$

3.4.1.3 *Construction and result*

Let

$$\mathcal{V} = \{M(v) : v \in V\},$$

so that $\mathcal{V}$ is a convex compact subset of the positive semidefinite cone $\mathbf{S}^d_+$. Let us select some

1. matrix $\Theta_* \succ 0$ such that $\Theta_* \succeq \Theta$, for all $\Theta \in \mathcal{V}$;
2. convex compact subset $\mathcal{Z}$ of the set $\mathcal{Z}^+ = \{Z \in \mathbf{S}^{m+1}_+ : Z_{m+1,m+1} = 1\}$ such that $[u; 1][u; 1]^T \in \mathcal{Z}$ for all $u \in U$;
3. real $\gamma \in (0, 1)$ and a nonnegative real δ such that (3.45) takes place.

We further set (cf. Proposition 3.11)

$$\begin{array}{rcl}
B & = & \left[\begin{array}{c} A \\ [0, ..., 0, 1] \end{array}\right] \in \mathbf{R}^{(d+1) \times (m+1)}, \\
\mathcal{H} & = & \mathcal{H}_\gamma := \{(h, H) \in \mathbf{R}^d \times \mathbf{S}^d : -\gamma \Theta_*^{-1} \preceq H \preceq \gamma \Theta_*^{-1}\}, \\
\mathcal{M} & = & \mathcal{V} \times \mathcal{Z}, \\
\Phi(h, H; \Theta, Z) & = & -\frac{1}{2} \ln \mathrm{Det}(I - \Theta_*^{1/2} H \Theta_*^{1/2}) + \frac{1}{2} \mathrm{Tr}([\Theta - \Theta_*] H) \\
& & + \frac{\delta(2+\delta)}{2(1 - \|\Theta_*^{1/2} H \Theta_*^{1/2}\|)} \|\Theta_*^{1/2} H \Theta_*^{1/2}\|_F^2 + \Gamma(h, H; Z) : \mathcal{H} \times \mathcal{M} \to \mathbf{R}
\end{array} \tag{3.48}$$

where

$$\begin{array}{rcl}
\Gamma(h, H; Z) & = & \frac{1}{2} \mathrm{Tr}\left(Z[bh^T A + A^T h b^T + A^T H A + B^T [H, h]^T [\Theta_*^{-1} - H]^{-1} [H, h] B]\right) \\
& = & \frac{1}{2} \mathrm{Tr}\left(Z B^T \left[\left[\begin{array}{c|c} H & h \\ \hline h^T & \end{array}\right] + [H, h]^T \left[\Theta_*^{-1} - H\right]^{-1} [H, h]\right] B\right)
\end{array}$$

and treat, as observation, the quadratic lifting of observation (3.46), that is, our observation is

$$\omega^K = \{\omega_i = (\zeta_i, \zeta_i \zeta_i^T)\}_{i=1}^K, \text{ with independent } \zeta_i \sim \mathcal{N}(A[u; 1], M(v)). \tag{3.49}$$

Note that by Proposition 3.11 function $\Phi(h,H;\Theta,Z):\ \mathcal{H}\times\mathcal{M}\to\mathbf{R}$ is a continuous convex-concave function which is coercive in convex argument and is such that

$$\begin{array}{l}\forall(u\in U,v\in V,(h,H)\in\mathcal{H}):\\ \ln\left(\mathbf{E}_{\zeta\sim\mathcal{N}(A[u;1],M(v))}\left\{\mathrm{e}^{\frac{1}{2}\zeta^TH\zeta+h^T\zeta}\right\}\right)\leq\Phi(h,H;M(v),[u;1][u;1]^T).\end{array}\tag{3.50}$$

We are about to demonstrate that when estimating the functional of interest (3.47) at a point $(u,v)\in U\times V$ via observation (3.49), we are in the situation considered in Section 3.3 and can utilize the corresponding machinery. Indeed, let us specify the following data introduced in Section 3.3.1:

- $\mathcal{H}=\{f=(h,H)\in\mathcal{H}\}\subset\mathcal{E}_H=\mathbf{R}^d\times\mathbf{S}^d$, with $\mathcal{H}$ defined in (3.48), and the inner product on $\mathcal{E}_H$ defined as

$$\langle(h,H),(h',H')\rangle=h^Th'+\frac{1}{2}\mathrm{Tr}(HH'),$$

  $\mathcal{E}_M=\mathbf{S}^d\times\mathbf{S}^{m+1}$, and $\mathcal{M}$, Φ defined as in (3.48);
- $\mathcal{E}_X=\mathbf{R}^k\times\mathbf{S}^{m+1}$, $\mathcal{X}=V\times\mathcal{Z}$;
- $\mathcal{A}(x=(v,Z))=(M(v),Z)$; note that $\mathcal{A}$ is an affine mapping from $\mathcal{E}_X$ into $\mathcal{E}_M$ which maps $\mathcal{X}$ into $\mathcal{M}$, as required in Section 3.3.1. Observe that when $u\in U$ and $v\in V$, the common distribution $P=P_{u,v}$ of i.i.d. observations ω_i defined by (3.49) satisfies the relation

$$\begin{array}{rcl}\forall(f=(h,H)\in\mathcal{H}):&&\\ \ln\left(\mathbf{E}_{\omega\sim P}\left\{\mathrm{e}^{\langle f,\omega\rangle}\right\}\right)&=&\ln\left(\mathbf{E}_{\zeta\sim\mathcal{N}(A[u;1],M(v))}\left\{\mathrm{e}^{h^T\zeta+\frac{1}{2}\zeta^TH\zeta}\right\}\right)\\ &\leq&\Phi(h,H;M(v),[u;1][u;1]^T);\end{array}\tag{3.51}$$

  see (3.50);
- $v(x=(v,Z))=\varrho(Z):\mathcal{X}\to\mathbf{R}$;
- we define affine functional $G(x)$ on $\mathcal{E}_X$ by the relation

$$\langle g,x:=(v,Z)\rangle=q^Tv+\mathrm{Tr}(QZ);$$

  see (3.47). As a result, for $x=(v,[u;1][u;1]^T)$ with $v\in V$ and $u\in U$ we have

$$F(u,v)=G(x).$$

Applying Corollary 3.8 to the data just specified (which is legitimate, because our Φ clearly satisfies (3.30)), we arrive at the result as follows:

Proposition 3.12. *In the situation just described, let us set*

$$
\begin{aligned}
&\widehat{\Psi}_{+,K}(h,H)\\
&:= \inf_{\alpha}\Bigg\{ \max_{(v,Z)\in V\times\mathcal{Z}} \left[\alpha\Phi(h/\alpha,H/\alpha;M(v),Z) - G(v,Z) - \varrho(Z) + K^{-1}\alpha\ln(2/\epsilon)\right] :\\
&\qquad\qquad\qquad\qquad\qquad\qquad\qquad\qquad \alpha>0, -\gamma\alpha\Theta_*^{-1}\preceq H\preceq\gamma\alpha\Theta_*^{-1}\Bigg\}\\
&= \max_{(v,Z)\in V\times\mathcal{Z}} \inf_{\substack{\alpha>0,\\ -\gamma\alpha\Theta_*^{-1}\preceq H\preceq\gamma\alpha\Theta_*^{-1}}} \Bigg[\alpha\Phi(h/\alpha,H/\alpha;M(v),Z) - G(v,Z) - \varrho(Z)\\
&\qquad\qquad\qquad\qquad\qquad\qquad\qquad\qquad\qquad\qquad + K^{-1}\alpha\ln(2/\epsilon)\Bigg],\\
&\widehat{\Psi}_{-,K}(h,H)\\
&:= \inf_{\alpha}\Bigg\{ \max_{(v,Z)\in V\times\mathcal{Z}} \left[\alpha\Phi(-h/\alpha,-H/\alpha;M(v),Z) + G(v,Z) - \varrho(Z) + K^{-1}\alpha\ln(2/\epsilon)\right] :\\
&\qquad\qquad\qquad\qquad\qquad\qquad\qquad\qquad \alpha>0, -\gamma\alpha\Theta_*^{-1}\preceq H\preceq\gamma\alpha\Theta_*^{-1}\Bigg\}\\
&= \max_{(v,Z)\in V\times\mathcal{Z}} \inf_{\substack{\alpha>0,\\ -\gamma\alpha\Theta_*^{-1}\preceq H\preceq\gamma\alpha\Theta_*^{-1}}} \Bigg[\alpha\Phi(-h/\alpha,-H/\alpha;M(v),Z) + G(v,Z) - \varrho(Z)\\
&\qquad\qquad\qquad\qquad\qquad\qquad\qquad\qquad\qquad\qquad + K^{-1}\alpha\ln(2/\epsilon)\Bigg],
\end{aligned} \tag{3.52}
$$

so that functions $\widehat{\Psi}_{\pm,K}(h,H) : \mathbf{R}^d\times\mathbf{S}^d\to\mathbf{R}$ *are convex. Furthermore, whenever* $\bar{h},\bar{H},\bar{\rho},\bar{\varkappa}$ *form a feasible solution to the system of convex constraints*

$$
\widehat{\Psi}_{+,K}(h,H)\le\rho-\varkappa,\ \ \widehat{\Psi}_{-,K}(h,H)\le\rho+\varkappa \tag{3.53}
$$

in variables $(h,H)\in\mathbf{R}^d\times\mathbf{S}^d$, $\rho\in\mathbf{R}$, $\varkappa\in\mathbf{R}$, *setting*

$$
\widehat{g}(\zeta^K := (\zeta_1,...,\zeta_K)) = \frac{1}{K}\sum_{i=1}^K \left[h^T\zeta_i + \frac{1}{2}\zeta_i^T H\zeta_i\right] + \bar{\varkappa}, \tag{3.54}
$$

we get an estimate of the functional of interest $F(u,v) = [u;1]^TQ[u;1] + q^Tv$ *via* K *independent observations*

$$
\zeta_i\sim\mathcal{N}(A[u;1],M(v)),\ i=1,...,K,
$$

with the following property:

$$
\begin{array}{l}
\forall(u,v)\in U\times V:\\
\mathrm{Prob}_{\zeta^K\sim[\mathcal{N}(A[u;1],M(v))]^K}\left\{|F(u,v)-\widehat{g}(\zeta^K)| > \bar{\rho}+\varrho([u;1][u;1]^T)\right\}\le\epsilon.
\end{array} \tag{3.55}
$$

Proof. Under the premise of the proposition, let us fix $u\in U$, $v\in V$, so that $x := (v, Z := [u;1][u;1]^T)\in\mathcal{X}$. Denoting, as above, by $P = P_{u,v}$ the distribution of $\omega := (\zeta,\zeta\zeta^T)$ with $\zeta\sim\mathcal{N}(A[u;1],M(v))$, and invoking (3.51), we see that for the (x,P) just defined, relation (3.26) takes place. Applying Corollary 3.8, we conclude that

$$
\mathrm{Prob}_{\zeta^K\sim[\mathcal{N}(A[u;1],M(v))]^K}\left\{|\widehat{g}(\zeta^K)-G(x)| > \bar{\rho}+\varrho([u;1][u;1]^T)\right\}\le\epsilon.
$$

It remains to note that by construction for the $x = (v, Z)$ in question it holds

$$G(x) = q^T v + \mathrm{Tr}(QZ) = q^T v + \mathrm{Tr}(Q[u;1][u;1]^T) = q^T v + [u;1]^T Q[u,1] = F(u,v).$$

□

An immediate consequence of Proposition 3.12 is as follows:

Corollary 3.13. *Under the premise and in the notation of Proposition 3.12, let* $(h, H) \in \mathbf{R}^d \times \mathbf{S}^d$. *Setting*

$$\begin{array}{rcl}\rho & = & \frac{1}{2}\left[\widehat{\Psi}_{+,K}(h,H) + \widehat{\Psi}_{-,K}(h,H)\right],\\ \varkappa & = & \frac{1}{2}\left[\widehat{\Psi}_{-,K}(h,H) - \widehat{\Psi}_{+,K}(h,H)\right],\end{array} \tag{3.56}$$

the ϵ-risk of estimate (3.54) *does not exceed* ρ.

Indeed, with ρ and $\varkappa$ given by (3.56), $h, H, \rho, \varkappa$ satisfy (3.53).

3.4.1.4 Consistency

We are about to present a simple sufficient condition for the estimator defined in Proposition 3.12 to be consistent in the sense of Section 3.3.4.1. Specifically, in the situation and with the notation from Sections 3.4.1.1 and 3.4.1.3 assume that

A.1. $\varrho(\cdot) \equiv 0$;
A.2. $V = \{\bar{v}\}$ is a singleton and $M(v) \succ 0$, which allows us to set $\Theta_* = M(\bar{v})$, to satisfy (3.45) with $\delta = 0$, and to assume w.l.o.g. that

$$F(u,v) = [u;1]^T Q[u;1],\ G(Z) = \mathrm{Tr}(QZ);$$

A.3. the first m columns of the $d \times (m+1)$ matrix A are linearly independent.

By A.3, the columns of $(d+1) \times (m+1)$ matrix B (see (3.48)) are linearly independent, so that we can find $(m+1) \times (d+1)$ matrix C such that $CB = I_{m+1}$. Let us define $(\bar{h}, \bar{H}) \in \mathbf{R}^d \times \mathbf{S}^d$ from the relation

$$\left[\begin{array}{c|c}\bar{H} & \bar{h}\\ \hline \bar{h}^T & \end{array}\right] = 2(C^T QC)^o, \tag{3.57}$$

where for $(d+1) \times (d+1)$ matrix S, S^o is the matrix obtained from S by zeroing our the entry in the cell $(d+1, d+1)$.

The consistency of our estimation machinery is given by the following simple statement:

Proposition 3.14. *In the situation just described and under assumptions A.1–3, given $\epsilon \in (0,1)$, consider the estimate*

$$\widehat{g}_K(\zeta^K) = \frac{1}{K}\sum_{k=1}^K [\bar{h}^T \zeta_k + \tfrac{1}{2}\zeta^T \bar{H}\zeta_k] + \varkappa_K,$$

where

$$\varkappa_K = \tfrac{1}{2}\left[\widehat{\Psi}_{-,K}(\bar{h},\bar{H}) - \widehat{\Psi}_{+,K}(\bar{h},\bar{H})\right]$$

and $\widehat{\Psi}_{\pm,K}$ are given by (3.52). *Then the ϵ-risk of $\widehat{g}_K(\cdot)$ goes to 0 as $K \to \infty$.*

For proof, see Section 3.6.4.

3.4.1.5 A modification

In the situation described at the beginning of Section 3.4.1.2, let a set $W \subset U \times V$ be given, and assume we are interested in estimating the value of $F(u, v)$, as defined in (3.47), at points $(u, v) \in W$ only. When reducing the "domain of interest" $U \times V$ to W, we hopefully can reduce the attainable ϵ-risk of recovery. Let us assume that we can point out a convex compact set $\mathcal{W} \subset V \times \mathcal{Z}$ such that

$$(u, v) \in W \Rightarrow (v, [u; 1][u; 1]^T) \in \mathcal{W}$$

A straightforward inspection justifies the following:

Remark 3.15. In the situation just described, the conclusion of Proposition 3.12 remains valid when the set $U \times V$ participating in *(3.55)* is reduced to W, and the set $V \times \mathcal{Z}$ participating in relations *(3.52)* is reduced to $\mathcal{W}$. This modification enlarges the feasible set of *(3.53)* and thus reduces the risk bound $\bar{\rho}$.

3.4.2 Estimating quadratic form, sub-Gaussian case

3.4.2.1 Situation

In the rest of this section we are interested in the situation as follows: we are given K i.i.d. observations

$$\zeta_i \sim \mathcal{SG}(A[u; 1], M(v)), \; i = 1, ..., K \tag{3.58}$$

(i.e., ζ_i are sub-Gaussian random vectors with parameters $A[u; 1] \in \mathbf{R}^d$ and $M(v) \in \mathcal{S}^d_+$), where

- (u, v) is an unknown "signal" known to belong to a given set $U \times V$, where
  - $U \subset \mathbf{R}^m$ is a compact set, and
  - $V \subset \mathbf{R}^k$ is a compact convex set;
- A is a given $d \times (m + 1)$ matrix, and $v \mapsto M(v) : \mathbf{R}^k \to \mathbf{S}^d$ is an affine mapping such that $M(v) \succeq 0$ whenever $v \in V$.

We are also given a convex calibrating function $\varrho(Z) : \mathbf{S}^{m+1}_+ \to \mathbf{R}$ and "functional of interest"

$$F(u, v) = [u; 1]^T Q[u; 1] + q^T v, \tag{3.59}$$

where Q and q are a known $(m+1)\times(m+1)$ symmetric matrix and a k-dimensional vector, respectively. Our goal is to recover $F(u, v)$, for unknown (u, v) known to belong to $U \times V$, via observation (3.58).

Note that the only difference between our present setting and that considered in Section 3.4.1.1 is that now we allow for sub-Gaussian, and not necessary Gaussian, observations.

3.4.2.2 Construction and result

Let

$$\mathcal{V} = \{M(v) : v \in V\},$$

so that $\mathcal{V}$ is a convex compact subset of the positive semidefinite cone $\mathbf{S}^d_+$. Let us select some

1. matrix $\Theta_* \succ 0$ such that $\Theta_* \succeq \Theta$, for all $\Theta \in \mathcal{V}$;
2. convex compact subset $\mathcal{Z}$ of the set $\mathcal{Z}^+ = \{Z \in \mathbf{S}^{m+1}_+ : Z_{m+1,m+1} = 1\}$ such that $[u;1][u;1]^T \in \mathcal{Z}$ for all $u \in U$;
3. reals $\gamma, \gamma^+ \in (0,1)$ with $\gamma < \gamma^+$ (say, $\gamma = 0.99, \gamma^+ = 0.999$).

Preliminaries. Given the data of the above description and $\delta \in [0,2]$, we set (cf. Proposition 3.11)

$$\begin{array}{rcl}
\mathcal{H} &=& \mathcal{H}_\gamma := \{(h,H) \in \mathbf{R}^d \times \mathbf{S}^d : -\gamma\Theta_*^{-1} \preceq H \preceq \gamma\Theta_*^{-1}\},\\
B &=& \left[\begin{array}{c} A \\ {[0,...,0,1]} \end{array}\right] \in \mathbf{R}^{(d+1)\times(m+1)},\\
\mathcal{M} &=& \mathcal{V} \times \mathcal{Z},\\
\Psi(h,H,G;Z) &=& -\frac{1}{2}\ln\mathrm{Det}(I - \Theta_*^{1/2} G \Theta_*^{1/2})\\
&& +\frac{1}{2}\mathrm{Tr}\left(ZB^T\left[\left[\begin{array}{c|c} H & h \\ \hline h^T & \end{array}\right] + [H,h]^T[\Theta_*^{-1} - G]^{-1}[H,h]\right]B\right):\\
&& (\mathcal{H} \times \{G : 0 \preceq G \preceq \gamma^+\Theta_*^{-1}\}) \times \mathcal{Z} \to \mathbf{R},
\end{array} \tag{3.60}$$

where

$$\begin{array}{rcl}
\Psi_\delta(h,H,G;\Theta,Z) &=& -\frac{1}{2}\ln\mathrm{Det}(I - \Theta_*^{1/2} G \Theta_*^{1/2}) + \frac{1}{2}\mathrm{Tr}([\Theta - \Theta_*]G)\\
&& + \frac{\delta(2+\delta)}{2(1 - \|\Theta_*^{1/2} G \Theta_*^{1/2}\|)}\|\Theta_*^{1/2} G \Theta_*^{1/2}\|_F^2\\
&& +\frac{1}{2}\mathrm{Tr}\left(ZB^T\left[\left[\begin{array}{c|c} H & h \\ \hline h^T & \end{array}\right] + [H,h]^T[\Theta_*^{-1} - G]^{-1}[H,h]\right]B\right):\\
&& (\mathcal{H} \times \{G : 0 \preceq G \preceq \gamma^+\Theta_*^{-1}\}) \times (\{0 \preceq \Theta \preceq \Theta_*\} \times \mathcal{Z}) \to \mathbf{R},\\
\Phi(h,H;Z) &=& \min\limits_G \left\{\Psi(h,H,G;Z) : 0 \preceq G \preceq \gamma^+\Theta_*^{-1}, G \succeq H\right\} : \mathcal{H} \times \mathcal{Z} \to \mathbf{R},\\
\Phi_\delta(h,H;\Theta,Z) &=& \min\limits_G \left\{\Psi_\delta(h,H,G;\Theta,Z) : 0 \preceq G \preceq \gamma^+\Theta_*^{-1}, G \succeq H\right\} :\\
&& \mathcal{H} \times (\{0 \preceq \Theta \preceq \Theta_*\} \times \mathcal{Z}) \to \mathbf{R}.
\end{array}$$

The following statement is a straightforward reformulation of Proposition 2.46.i:

Proposition 3.16. *In the situation described in Sections 3.4.2.1 and 3.4.2.2 we have*

(i) Φ is well-defined real-valued continuous function on the domain $\mathcal{H} \times \mathcal{Z}$; the function is convex in $(h,H) \in \mathcal{H}$, concave in $Z \in \mathcal{Z}$, and $\Phi(0;Z) \geq 0$. Furthermore, let $(h,H) \in \mathcal{H}$, $u \in U$, $v \in V$, and let $\zeta \sim \mathcal{SG}(A[u;1], M(v))$. Then

$$\ln\left(\mathbf{E}_\zeta\left\{\exp\{h^T\zeta + \tfrac{1}{2}\zeta^T H \zeta\}\right\}\right) \leq \Phi(h,H;[u;1][u;1]^T). \tag{3.61}$$

(ii) Assume that

$$\forall \Theta \in \mathcal{V} : \|\Theta^{1/2}\Theta_*^{-1/2} - I_d\| \leq \delta. \tag{3.62}$$

Then $\Phi_\delta(h,H;\Theta,Z)$ is a well-defined real-valued continuous function on the domain $\mathcal{H} \times (\mathcal{V} \times \mathcal{Z})$; it is convex in $(h,H) \in \mathcal{H}$, concave in $(\Theta, Z) \in \mathcal{V} \times \mathcal{Z}$, and $\Phi_\delta(0;\Theta,Z) \geq 0$. Furthermore, let $(h,H) \in \mathcal{H}$, $u \in U$, $v \in V$, and let $\zeta \sim \mathcal{SG}(A[u;1], M(v))$. Then

$$\ln\left(\mathbf{E}_\zeta\left\{\exp\{h^T\zeta + \tfrac{1}{2}\zeta^T H \zeta\}\right\}\right) \leq \Phi_\delta(h,H;M(v),[u;1][u;1]^T). \tag{3.63}$$

The estimate. Our construction of the estimate is completely similar to the case of Gaussian observations. Specifically, let us pass from observations (3.58) to their quadratic lifts, so that our observations become

$$\omega_i = (\zeta_i, \zeta_i\zeta_i^T),\ 1 \leq i \leq K,\ \ \zeta_i \sim \mathcal{SG}(A[u;1], M(v)) \text{ are i.i.d.} \tag{3.64}$$

As in the Gaussian case, we find ourselves in the situation considered in Section 3.3.3 and can use the corresponding constructions. Indeed, let us specify the data introduced in Section 3.3.1 and participating in the constructions of Section 3.3 as follows:

- $\mathcal{H} = \{f = (h, H) \in \mathcal{H}\} \subset \mathcal{E}_H = \mathbf{R}^d \times \mathbf{S}^d$, with $\mathcal{H}$ defined in (3.60), and the inner product on $\mathcal{E}_H$ defined as

$$\langle (h, H), (h', H') \rangle = h^T h' + \frac{1}{2}\mathrm{Tr}(HH'),$$

  $\mathcal{E}_M = \mathbf{S}^d \times \mathbf{S}^{m+1}$, and $\mathcal{M}$, Φ defined as in (3.60);
- $\mathcal{E}_X = \mathbf{R}^k \times \mathbf{S}^{m+1}$, $\mathcal{X} = V \times \mathcal{Z}$;
- $\mathcal{A}(x = (v, Z)) = (M(v), Z)$; note that $\mathcal{A}$ is an affine mapping from $\mathcal{E}_X$ into $\mathcal{E}_M$ mapping $\mathcal{X}$ into $\mathcal{M}$, as required in Section 3.3. Observe that when $u \in U$ and $v \in V$, the common distribution $P = P_{u,v}$ of i.i.d. observations ω_i defined by (3.64) satisfies the relation

$$\begin{array}{l} \forall (f = (h, H) \in \mathcal{H}): \\ \ln\left(\mathbf{E}_{\omega\sim P}\left\{e^{\langle f, \omega\rangle}\right\}\right) \ = \ \ln\left(\mathbf{E}_{\zeta\sim\mathcal{SG}(A[u;1],M(v))}\left\{e^{h^T\zeta + \frac{1}{2}\zeta^T H \zeta}\right\}\right) \\ \qquad\qquad\qquad\qquad\ \leq \ \Phi(h, H; [u;1][u;1]^T); \end{array} \tag{3.65}$$

  see (3.61). Moreover, in the case of (3.62), we have also

$$\begin{array}{l} \forall (f = (h, H) \in \mathcal{H}): \\ \ln\left(\mathbf{E}_{\omega\sim P}\left\{e^{\langle f, \omega\rangle}\right\}\right) \ = \ \ln\left(\mathbf{E}_{\zeta\sim\mathcal{SG}(A[u;1],M(v))}\left\{e^{h^T\zeta + \frac{1}{2}\zeta^T H \zeta}\right\}\right) \\ \qquad\qquad\qquad\qquad\ \leq \ \Phi_\delta(h, H; M(v), [u;1][u;1]^T); \end{array} \tag{3.66}$$

  see (3.63);
- we set $v(x = (v, Z)) = \varrho(Z)$;
- we define affine functional $G(x)$ on $\mathcal{E}_X$ by the relation

$$G(x := (v, Z)) = q^T v + \mathrm{Tr}(QZ);$$

  see (3.59). As a result, for $x = (v, [u;1][u;1]^T)$ with $v \in V$ and $u \in U$ we have

$$F(u, v) = G(x).$$

The result. Applying to the data just specified Corollary 3.8 (which is legitimate, because our Φ clearly satisfies (3.30)), we arrive at the result as follows:

Proposition 3.17. *In the situation described in Sections 3.4.2.1 and 3.4.2.2 let us* *set*

$$\begin{array}{l}\widehat{\Psi}_{+,K}(h,H) := \inf_{\alpha}\left\{\max_{(v,Z)\in V\times\mathcal{Z}}\left[\alpha\Phi(h/\alpha,H/\alpha;Z)-G(v,Z)-\varrho(Z)+\alpha K^{-1}\ln(2/\epsilon)\right]:\right.\\ \qquad\qquad\qquad\qquad\qquad\qquad\left.\alpha>0,-\gamma\alpha\Theta_*^{-1}\preceq H\preceq\gamma\alpha\Theta_*^{-1}\right\}\\ \quad=\max_{(v,Z)\in V\times\mathcal{Z}}\inf_{\substack{\alpha>0,\\ -\gamma\alpha\Theta_*^{-1}\preceq H\preceq\gamma\alpha\Theta_*^{-1}}}\left[\alpha\Phi(h/\alpha,H/\alpha;Z)-G(v,Z)-\varrho(Z)+\alpha K^{-1}\ln(2/\epsilon)\right],\\ \widehat{\Psi}_{-,K}(h,H) := \inf_{\alpha}\left\{\max_{(v,Z)\in V\times\mathcal{Z}}\left[\alpha\Phi(-h/\alpha,-H/\alpha;Z)+G(v,Z)-\varrho(Z)+\alpha K^{-1}\ln(2/\epsilon)\right]:\right.\\ \qquad\qquad\qquad\qquad\qquad\qquad\left.\alpha>0,-\gamma\alpha\Theta_*^{-1}\preceq H\preceq\gamma\alpha\Theta_*^{-1}\right\}\\ \quad=\max_{(v,Z)\in V\times\mathcal{Z}}\inf_{\substack{\alpha>0,\\ -\gamma\alpha\Theta_*^{-1}\preceq H\preceq\gamma\alpha\Theta_*^{-1}}}\left[\alpha\Phi(-h/\alpha,-H/\alpha;Z)+G(v,Z)-\varrho(Z)+\alpha K^{-1}\ln(2/\epsilon)\right].\end{array}\tag{3.67}$$

Thus, functions $\widehat{\Psi}_{\pm,K}(h,H):\mathbf{R}^d\times\mathbf{S}^d\to\mathbf{R}$ are convex. Furthermore, whenever $\bar{h},\bar{H},\bar{\rho},\bar{\varkappa}$ form a feasible solution to the system of convex constraints

$$\widehat{\Psi}_{+,K}(h,H)\leq\rho-\varkappa,\ \widehat{\Psi}_{-,K}(h,H)\leq\rho+\varkappa\tag{3.68}$$

in variables $(h,H)\in\mathbf{R}^d\times\mathbf{S}^d$, $\rho\in\mathbf{R}$, $\varkappa\in\mathbf{R}$, the estimate

$$\widehat{g}(\zeta^K)=\frac{1}{K}\sum_{i=1}^K\left[h^T\zeta_i+\frac{1}{2}\zeta_i^TH\zeta_i\right]+\bar{\varkappa},$$

of $F(u,v)=[u;1]^TQ[u;1]+q^Tv$ via i.i.d. observations

$$\zeta_i\sim\mathcal{SG}(A[u;1],M(v)),\ 1\leq i\leq K,$$

satisfies for all $(u,v)\in U\times V$:

$$\mathrm{Prob}_{\zeta^K\sim[\mathcal{SG}(A[u;1],M(v))]^K}\left\{|F(u,v)-\widehat{g}(\zeta^K)|>\bar{\rho}+\varrho([u;1][u;1]^T)\right\}\leq\epsilon.$$

Proof. Under the premise of the proposition, let us fix $u\in U$, $v\in V$, and let $x=(v,Z:=[u;1][u;1]^T)$. Denoting by P the distribution of $\omega:=(\zeta,\zeta\zeta^T)$ with $\zeta\sim\mathcal{SG}(A[u;1],M(v))$, and invoking (3.65), we see that for the (x,P) just defined relation (3.26) takes place. Applying Corollary 3.8, we conclude that

$$\mathrm{Prob}_{\zeta^K\sim[\mathcal{N}(A[u;1],M(v))]^K}\left\{|\widehat{g}(\zeta^K)-G(x)|>\bar{\rho}+\varrho([u;1][u;1]^T)\right\}\leq\epsilon.$$

It remains to note that by construction for the $x=(v,Z)$ in question it holds

$$G(x)=q^Tv+\mathrm{Tr}(QZ)=q^Tv+[u;1]^TQ[u,1]=F(u,v).\qquad\square$$

Remark 3.18. In the situation described in Sections 3.4.2.1 and 3.4.2.2 let $\delta\in[0,2]$ be such that

$$\|\Theta^{1/2}\Theta_*^{-1/2}-I_d\|\leq\delta\ \forall\Theta\in\mathcal{V}.$$

Then the conclusion of Proposition 3.17 remains valid when the function Φ in *(3.67)* is replaced with the function Φ_δ, that is, when $\widehat{\Psi}_{\pm,K}$ are defined as

$$\begin{aligned}
\widehat{\Psi}_{+,K}(h,H) &:= \inf_{\alpha}\Big\{ \max_{(v,Z)\in V\times\mathcal{Z}} \left[\alpha\Phi_\delta(h/\alpha,H/\alpha;M(v),Z) - G(v,Z) - \varrho(Z) + \alpha K^{-1}\ln(2/\epsilon)\right] : \\
&\qquad\qquad \alpha>0, -\gamma\alpha\Theta_*^{-1}\preceq H\preceq\gamma\alpha\Theta_*^{-1}\Big\} \\
&= \max_{(v,Z)\in V\times\mathcal{Z}} \inf_{\substack{\alpha>0,\\ -\gamma\alpha\Theta_*^{-1}\preceq H\preceq\gamma\alpha\Theta_*^{-1}}} \left[\alpha\Phi_\delta(h/\alpha,H/\alpha;M(v),Z) - G(v,Z) - \varrho(Z) + \alpha K^{-1}\ln(2/\epsilon)\right],
\end{aligned}$$

$$\begin{aligned}
\widehat{\Psi}_{-,K}(h,H) &:= \inf_{\alpha}\Big\{ \max_{(v,Z)\in V\times\mathcal{Z}} \left[\alpha\Phi_\delta(-h/\alpha,-H/\alpha;M(v),Z) + G(v,Z) - \varrho(Z) + \alpha K^{-1}\ln(2/\epsilon)\right] : \\
&\qquad\qquad \alpha>0, -\gamma\alpha\Theta_*^{-1}\preceq H\preceq\gamma\alpha\Theta_*^{-1}\Big\} \\
&= \max_{(v,Z)\in V\times\mathcal{Z}} \inf_{\substack{\alpha>0,\\ -\gamma\alpha\Theta_*^{-1}\preceq H\preceq\gamma\alpha\Theta_*^{-1}}} \left[\alpha\Phi_\delta(-h/\alpha,-H/\alpha;M(v),Z) + G(v,Z) - \varrho(Z) + \alpha K^{-1}\ln(2/\epsilon)\right].
\end{aligned}$$

To justify Remark 3.18, it suffices to replace relation (3.65) in the proof of Proposition 3.17 with (3.66). Note that what is better in terms of the risk of the resulting estimate—Proposition 3.17 "as is" or its modification presented in Remark 3.18—depends on the situation, so that it makes sense to keep in mind both options.

3.4.2.3 Numerical illustration, direct observations

The problem. Our initial illustration is deliberately selected to be extremely simple: given direct noisy observations

$$\zeta = u + \xi$$

of unknown signal $u\in\mathbf{R}^m$ known to belong to a given set U, we want to recover the "energy" u^Tu of u. We are interested in an estimate of u^TU quadratic in ζ with as small as possible an ϵ-risk on U; here $\epsilon\in(0,1)$ is a given design parameter. The details of our setup are as follows:

- U is the "spherical layer" $U=\{u\in\mathbf{R}^m : r^2\leq u^Tu\leq R^2\}$, where r and R, $0\leq r<R<\infty$, are given. As a result, the "main ingredient" of constructions from Sections 3.4.1.3 and 3.4.2.2—the convex compact subset $\mathcal{Z}$ of the set $\{Z\in\mathbf{S}^{m+1}_+ : Z_{m+1,m+1}=1\}$ containing all matrices $[u;1][u;1]^T$, $u\in U$—can be specified as

$$\mathcal{Z}=\{Z\in\mathbf{S}^{m+1}_+ : Z_{m+1,m+1}=1, 1+r^2\leq \mathrm{Tr}(Z)\leq 1+R^2\};$$

- ξ is either $\sim\mathcal{N}(0,\Theta)$ (Gaussian case), or $\sim\mathcal{SG}(0,\Theta)$ (sub-Gaussian case), with matrix Θ known to be diagonal with diagonal entries equal to each other satisfying $\theta\sigma^2\leq\Theta_{ii}\leq\sigma^2$, $1\leq i\leq d=m$, with known $\theta\in[0,1]$ and $\sigma^2>0$;
- the calibrating function $\varrho(Z)$ is $\varrho(Z)=\varsigma(\sum_{i=1}^m Z_{ii})$, where ς is a convex continuous real-valued function on $\mathbf{R}_+$. Note that with this selection, the claim that ϵ-risk of an estimate $\widehat{g}(\cdot)$ is $\leq\rho$ means that whenever $u\in U$, one has

$$\mathrm{Prob}\{|\widehat{g}(u+\xi)-u^Tu|>\rho+\varsigma(u^Tu)\}\leq\epsilon. \tag{3.69}$$

Processing the problem. It is easily seen that in the situation in question the apparatus in Sections 3.4.1 and 3.4.2 translates into the following:

1. We lose nothing when restricting ourselves with estimates of the form

$$\widehat{g}(\zeta) = \tfrac{1}{2}\eta\zeta^T\zeta + \varkappa, \tag{3.70}$$

with properly selected scalars η and $\varkappa$;
2. In Gaussian case, η and $\varkappa$ are yielded by the convex optimization problem with only three variables α_+, α_-, and η, namely the problem

$$\min_{\alpha_\pm,\eta}\left\{\widehat{\Psi}(\alpha_+,\alpha_-,\eta) = \tfrac{1}{2}\left[\widehat{\Psi}_+(\alpha_+,\eta) + \widehat{\Psi}_-(\alpha_-,\eta)\right] : \sigma^2|\eta| < \alpha_\pm\right\} \tag{3.71}$$

where

$$\begin{array}{rcl}\widehat{\Psi}_+(\alpha_+,\eta) &=& -\frac{d\alpha_+}{2}\ln(1-\sigma^2\eta/\alpha_+) + \frac{d}{2}\sigma^2(1-\theta)\max[-\eta,0] + \frac{d\delta(2+\delta)\sigma^4\eta^2}{2(\alpha_+-\sigma^2|\eta|)}\\ && + \max\limits_{r^2\le t\le R^2}\left[\left[\frac{\alpha_+\eta}{2(\alpha_+-\sigma^2\eta)} - 1\right]t - \varsigma(t)\right] + \alpha_+\ln(2/\epsilon)\\ \widehat{\Psi}_-(\alpha_+,\eta) &=& -\frac{d\alpha_-}{2}\ln(1+\sigma^2\eta/\alpha_-) + \frac{d}{2}\sigma^2(1-\theta)\max[\eta,0] + \frac{d\delta(2+\delta)\sigma^4\eta^2}{2(\alpha_--\sigma^2|\eta|)}\\ && + \max\limits_{r^2\le t\le R^2}\left[\left[-\frac{\alpha_-\eta}{2(\alpha_-+\sigma^2\eta)} + 1\right]t - \varsigma(t)\right] + \alpha_-\ln(2/\epsilon),\end{array}$$

with $\delta = 1-\sqrt{\theta}$. Now, the η-component of a feasible solution to (3.71) augmented by the quantity

$$\varkappa = \tfrac{1}{2}\left[\widehat{\Psi}_-(\alpha_-,\eta) - \widehat{\Psi}_+(\alpha_+,\eta)\right]$$

yields estimate (3.70) with ϵ-risk on U not exceeding $\widehat{\Psi}(\alpha_+,\alpha_-,\eta)$;
3. In the sub-Gaussian case, η and $\varkappa$ are yielded by the convex optimization problem with five variables, $\alpha_\pm, g_\pm$, and η, namely, the problem

$$\begin{array}{l}\min\limits_{\alpha_\pm,g_\pm,\eta}\left\{\widehat{\Psi}(\alpha_\pm,g_\pm,\eta) = \tfrac{1}{2}\left[\widehat{\Psi}_+(\alpha_+,g_+,\eta) + \widehat{\Psi}_-(\alpha_-,g_-,\eta)\right] :\right.\\ \qquad\left. 0\le\sigma^2 g_\pm < \alpha_\pm,\ -\alpha_+ < \sigma^2\eta < \alpha_-,\ \eta\le g_+,\ -\eta\le g_-\right\},\end{array} \tag{3.72}$$

where

$$\begin{array}{rcl}\widehat{\Psi}_+(\alpha_+,g_+,\eta) &=& -\frac{d\alpha_+}{2}\ln(1-\sigma^2 g_+/\alpha_+)\\ && +\alpha_+\ln(2/\epsilon) + \max\limits_{r^2\le t\le R^2}\left[\left[\frac{\sigma^2\eta^2}{2(\alpha_+-\sigma^2 g_+)} + \frac{1}{2}\eta - 1\right]t - \varsigma(t)\right]\\ \widehat{\Psi}_-(\alpha_-,g_-,\eta) &=& -\frac{d\alpha_-}{2}\ln(1-\sigma^2 g_-/\alpha_-)\\ && +\alpha_-\ln(2/\epsilon) + \max\limits_{r^2\le t\le R^2}\left[\left[\frac{\sigma^2\eta^2}{2(\alpha_--\sigma^2 g_-)} - \frac{1}{2}\eta + 1\right]t - \varsigma(t)\right]\end{array}$$

The η-component of a feasible solution to (3.72) augmented by the quantity

$$\varkappa = \tfrac{1}{2}\left[\widehat{\Psi}_-(\alpha_-,g_-,\eta) - \widehat{\Psi}_+(\alpha_+,g_+,\eta)\right]$$

yields estimate (3.70) with ϵ-risk on U not exceeding $\widehat{\Psi}(\alpha_\pm,g_\pm,\eta)$.

Note that the Gaussian case of our "energy estimation" problem is well studied in the literature (see, among others, [19, 43, 81, 87, 90, 97, 120, 124, 147, 160]), mainly in the case $\xi\sim\mathcal{N}(0,\sigma^2 I_m)$ of white Gaussian noise with exactly known variance σ^2. Available results investigate analytically the interplay between the dimension m of signal, noise intensity σ^2 and the parameters R, r and offer estimates which are provably optimal, up to absolute constant factors. A nice property of the proposed

d	r	R	θ	Relative 0.01-risk, Gaussian case	Relative 0.01-risk, sub-Gaussian case	Optimality ratio
64	0	16	1	0.34808	0.44469	1.22
64	0	16	0.5	0.43313	0.44469	1.48
64	0	128	1	0.04962	0.05181	1.28
64	0	128	0.5	0.05064	0.05181	1.34
64	8	80	1	0.07827	0.08376	1.28
64	8	80	0.5	0.08095	0.08376	1.34
256	0	32	1	0.19503	0.30457	1.28
256	0	32	0.5	0.26813	0.30457	1.41
256	0	512	1	0.01264	0.01314	1.28
256	0	512	0.5	0.01289	0.01314	1.34
256	16	160	1	0.03996	0.04501	1.28
256	16	160	0.5	0.04255	0.04501	1.34
1024	0	64	1	0.10272	0.21923	1.28
1024	0	64	0.5	0.17032	0.21923	1.34
1024	0	2048	1	0.00317	0.00330	1.28
1024	0	2048	0.5	0.00324	0.00330	1.34
1024	32	320	1	0.02019	0.02516	1.28
1024	32	320	0.5	0.02273	0.02516	1.41

Table 3.3: Estimating the signal energy from direct observations.

approach is that (3.71) automatically takes care of the parameters and results in estimates with seemingly near-optimal performance, as witnessed by the numerical experiments we are about to present.

Numerical results. In the first series of experiments we use the trivial calibrating function: $\varsigma(\cdot) \equiv 0$.

A typical sample of numerical results is presented in Table 3.3. To avoid large numbers, we display in the table *relative* 0.01-risk of the estimates, that is, the plain risk as given by (3.71) divided by R^2; keeping this in mind, one will not be surprised that when extending the range $[r, R]$ of allowed norms of the observed signal, all other components of the setup being fixed, the relative risk can decrease (the actual risk, of course, can only increase). Note that in all our experiments σ is set to 1.

Along with the values of the relative 0.01-risk, we present also the values of "optimality ratios"—the ratios of the upper risk bounds given by (3.71) in the Gaussian case, to (lower bounds on) the best 0.01-risks $\mathrm{Risk}^*_{0.01}$ possible under the circumstances, defined as the infimum of the 0.01-risk over all estimates recovering $\|u\|_2^2$ via single observation $\omega = u + \zeta$. These lower bounds are obtained as follows. Let us select some values $r_1 < r_2$ in the allowed range $[r, R]$ of $\|u\|_2$, along with two values, σ_1, σ_2, in the allowed range $[\theta\sigma, \sigma] = [\theta, 1]$ of values of diagonal entries in diagonal matrices Θ, and consider two distributions of observations P_1 and P_2 as follows: P_χ is the distribution of the random vector $x + \zeta$, where x and ξ are independent, x is uniformly distributed on the sphere $\|x\|_2 = r_\chi$, and $\zeta \sim \mathcal{N}(0, \sigma_\chi^2 I_d)$. It is immediately seen that whenever the two simple hypotheses $\omega \sim P_1$ and $\omega \sim P_2$ can*not* be decided upon via a single observation by a test with total risk (the sum, over the two hypotheses in question, of probabilities for the test to reject the hypothesis when it is true) $\leq 2\epsilon$, the quantity $\delta = \frac{1}{2}(r_2^2 - r_1^2)$ is a lower bound on the optimal ϵ-risk, Risk^*_ϵ. In other words, denoting by $p_\chi(\cdot)$ the density of P_χ, we have

$$0.02 < \int_{\mathbf{R}^d} \min[p_1(\omega), p_2(\omega)]d\omega \Rightarrow \mathrm{Risk}^*_{0.01} \geq \tfrac{1}{2}(r_2^2 - r_1^2).$$

Now, the densities p_χ are spherically symmetric, whence, denoting by $q_\chi(\cdot)$ the univariate density of the energy $\omega^T\omega$ of observation $\omega \sim P_\chi$, we have

$$\int_{\mathbf{R}^d} \min[p_1(\omega), p_2(\omega)]d\omega = \int_0^\infty \min[q_1(s), q_2(s)]ds,$$

so that

$$0.02 < \int_0^\infty \min[q_1(s), q_2(s)]ds \Rightarrow \mathrm{Risk}^*_{0.01} \geq \tfrac{1}{2}(r_2^2 - r_1^2). \tag{3.73}$$

On closer inspection, q_χ is the convolution of two univariate densities representable by explicit computation-friendly formulas, implying that given r_1, r_2, σ_1 and σ_2, we can check numerically whether the premise in (3.73) indeed takes place, and whenever the latter is the case, the quantity $(r_2^2 - r_1^2)/2$ is a lower bound on $\mathrm{Risk}^*_{0.01}$. In our experiments, we implement a simple search strategy (not described here) aimed at crudely maximizing this bound in r_1, r_2, σ_1, and σ_2 and use the resulting lower bounds on $\mathrm{Risk}^*_{0.01}$ to compute the optimality ratios presented in the table.[10]

We believe that quite moderate values of the optimality ratios presented in the table (these results are typical for a much larger series of experiments we have conducted) witness quite good performance of our machinery.

Optimizing the relative risk. The "relative risk" displayed in Table 3.3 is the 0.01-risk of recovery of u^Tu, corresponding to the trivial calibrating function, divided by the largest value R^2 of this risk allowed by the inclusion $u \in U$. When R is large, low relative risk can correspond to rather high "actual" risk. For example, when $d := \dim u = 1024$, $\theta = 1$, and $U = \{u \in \mathbf{R}^d : \|u\|_2 \leq 1.e6\}$, the 0.01-risk becomes as large as $\rho \approx 6.5e6$. For "relatively small" signals, like $u^Tu \approx 10^4$, recovering u^Tu within accuracy ρ is of no interest. In order to allow for "large" domains U it makes sense to pass from the trivial calibrating function to a nontrivial one, e.g., $\varsigma(t) = \alpha t$, with small positive α. With this calibrating function, (3.69) reads

$$\mathrm{Prob}\left\{|\widehat{g}(u+\xi) - u^Tu| > \rho + \alpha u^Tu\right\} \leq \epsilon.$$

It turns out that (quite reasonable when U is large) "relative" risk quantification results in risk values essentially smaller than those of "absolute" risk. Here is some instructive numerical data:

r	R	0.01-Risk, $\alpha = 0$	0.01-Risk, $\alpha = 0.01$	0.01-Risk, $\alpha = 0.1$
0	$1.e7$	$6.51e7/6.51e7$	$1.33e3/1.58e3$	$474/642$
$1.e2$	$1.e7$	$6.51e7/6.51e7$	$1.33e3/1.58e3$	$-123/92.3$
$1.1e3$	$1.e7$	$6.51e7/6.51e7$	$-4.73e3/-4.48e3$	$-1.14e5/-1.14e5$

$U = \{u \in \mathbf{R}^{1024} : r \leq \|u\|_2 \leq R\}$, $\theta = 1/2$
Left/Right: risks in Gaussian/sub-Gaussian cases

[10]The reader should not be surprised by the "narrow numerical spectrum" of optimality ratios displayed in Table 3.3: our lower bounding scheme was restricted to identify actual optimality ratios among the candidates on the grid 1.05^i, $i = 1, 2, \ldots$.

3.4.2.4 Numerical illustration, indirect observations

The problem. Let us consider the estimation problem as follows. Our observations are

$$\zeta = Bu + \xi, \tag{3.74}$$

where

- B is a given $d \times m$ matrix, with $m > d$ ("deficient observations"),
- $u \in \mathbf{R}^m$ is a signal known to belong to a compact set U,
- $\xi \sim \mathcal{N}(0, \Theta)$ (Gaussian case) or $\xi \sim \mathcal{SG}(0, \Theta)$ (sub-Gaussian case) is the observation noise; Θ is a positive semidefinite $d \times d$ matrix known to belong to a given convex compact set $\mathcal{V} \subset \mathbf{S}^d_+$.

Our goal is to estimate the energy

$$F(u) = \tfrac{1}{m}\|u\|_2^2$$

of the signal given observation (3.74).

In our experiment, the data is specified as follows:

1. We think of $u \in \mathbf{R}^m$ as of discretization of a smooth function $x(t)$ of continuous argument $t \in [0; 1]$: $u_i = x(\frac{i}{m})$, $1 \le i \le m$. We set $U = \{u : \|Su\|_2 \le 1\}$, where $u \mapsto Su$ is the finite-difference approximation of the mapping $x(\cdot) \mapsto (x(0), x'(0), x''(\cdot))$, so that U is a natural discrete-time analog of the Sobolev-type ball $\{x : [x(0)]^2 + [x'(0)]^2 + \int_0^1 [x''(t)]^2 dt \le 1\}$.
2. $d \times m$ matrix B is of the form UDV^T, where U and V are randomly selected $d \times d$ and $m \times m$ orthogonal matrices, and the d diagonal entries in diagonal $d \times m$ matrix D are of the form $\theta^{-\frac{i-1}{d-1}}$, $1 \le i \le d$.
3. The set $\mathcal{V}$ of admissible matrices Θ is the set of all diagonal $d \times d$ matrices with diagonal entries varying in $[0, \sigma^2]$.

Both σ and θ are components of the experiment setup.

Processing the problem. The described estimation problem clearly is covered by the setups considered in Sections 3.4.1 (Gaussian case) and 3.4.2 (sub-Gaussian case); in terms of these setups, it suffices to specify Θ_* as $\sigma^2 I_d$, $M(v)$ as the identity mapping of $\mathcal{V}$ onto itself, the mapping $u \mapsto A[u; 1]$ as the mapping $u \mapsto Bu$, and the set $\mathcal{Z}$ (which should be a convex compact subset of the set $\{Z \in \mathbf{S}^{d+1}_+ : Z_{d+1,d+1} = 0\}$ containing all matrices of the form $[u; 1][u; 1]^T$, $u \in U$) as the set

$$\mathcal{Z} = \{Z \in \mathbf{S}^{d+1}_+ : Z_{d+1,d+1} = 1, \operatorname{Tr}\left(Z \operatorname{Diag}\{S^T S, 0\}\right) \le 1\}.$$

As suggested by Propositions 3.12 (Gaussian case) and 3.17 (sub-Gaussian case), the linear in "lifted observation" $\omega = (\zeta, \zeta\zeta^T)$ estimates of $F(u) = \frac{1}{m}\|u\|_2^2$ stem from the optimal solution (h_*, H_*) to the convex optimization problem

$$\mathrm{Opt} = \min_{h,H} \tfrac{1}{2}\left[\widehat{\Psi}_+(h, H) + \widehat{\Psi}_-(h, H)\right], \tag{3.75}$$

with $\widehat{\Psi}_\pm(\cdot)$ given by (3.52) in the Gaussian, and by (3.67) in the sub-Gaussian cases, with the number K of observations in (3.52) and (3.67) set to 1. The resulting

d, m	Opt, Gaussian case	Opt, sub-Gaussian case	LwBnd
8, 12	0.1362(+65%)	0.1382(+67%)	0.0825
16, 24	0.1614(+53%)	0.1640(+55%)	0.1058
32, 48	0.0687(+46%)	0.0692(+48%)	0.0469

Table 3.4: Upper bound (Opt) on the 0.01-risk of estimate (3.76), (3.75) vs. lower bound (LwBnd) on the 0.01-risk attainable under the circumstances. In the experiments, $\sigma = 0.025$ and $\theta = 10$. Data in parentheses: excess of Opt over LwBnd.

estimate is

$$\zeta \mapsto h_*^T\zeta + \tfrac{1}{2}\zeta^T H_*\zeta + \varkappa,\ \varkappa = \tfrac{1}{2}\left[\widehat{\Psi}_-(h_*, H_*) - \widehat{\Psi}_+(h_*, H_*)\right] \tag{3.76}$$

and the ϵ-risk of the estimate is (upper-bounded by) Opt.

Problem (3.75) is a well-structured convex-concave saddle point problem and as such is beyond the "immediate scope" of the standard Convex Programming software toolbox primarily aimed at solving well-structured convex minimization (or maximization) problems. However, applying conic duality, one can easily eliminate in (3.52) and (3.67) the inner maxima over v, Z to end up with a reformulation which can be solved numerically by CVX [108], and this is how we process (3.75) in our experiments.

Numerical results. In the experiments to be reported, we use the trivial calibrating function: $\varrho(\cdot) \equiv 0$.

We present some typical numerical results in Table 3.4. To qualify the performance of our approach, we present, along with the upper risk bounds for the computed estimates, simple lower bounds on ϵ-risk. The origin of the lower bounds is as follows. Assume we have at our disposal a signal $w \in U$, and let $t(w) = \|Bw\|_2$, $\rho = 2\sigma\mathrm{ErfcInv}(\epsilon)$, where ErfcInv is the inverse error function as defined in (1.26). Setting $\theta(w) = \max[1 - \rho/t(w), 0]$, observe that $w' := \theta(w)w \in U$ and $\|Bw - Bw'\|_2 \leq \rho$, which, due to the origin of ρ, implies that there is no way to decide via observation $Bu + \xi$, $\xi \sim \mathcal{N}(0, \sigma^2)$, with risk $< \epsilon$ on the two simple hypotheses $u = w$ and $u = w'$. As an immediate consequence, the quantity $\phi(w) := \frac{1}{2}[\|w\|_2^2 - \|w'\|_2^2] = \|w\|_2^2[1 - \theta^2(w)]/2$ is a lower bound on the ϵ-risk, on U, of any estimate of $\|u\|_2^2$. We can now try to maximize the resulting lower risk bound over U, thus arriving at the lower risk bound

$$\mathrm{LwBnd} = \max_{w \in U}\left\{\tfrac{1}{2}\|w\|_2^2(1 - \theta^2(w))\right\}.$$

On closer inspection, the latter problem is not a convex one, which does not prevent building a suboptimal solution to this problem, and this is how the lower risk bounds in Table 3.4 are built (we omit the details). We see that the ϵ-risks of our estimates are within a moderate factor of the optimal ones.

Figure 3.4 shows empirical error distributions of the estimates built in the three experiments reported in Table 3.4. When simulating the observations and estimates, we used $\mathcal{N}(0, \sigma^2 I_d)$ noise and selected signals in U by maximizing over U randomly selected linear forms. Finally, we note that already with fixed design parameters d, m, θ and σ we deal with a family of estimation problems rather than with a single problem, the reason being that our U is an ellipsoid with half-axes es-

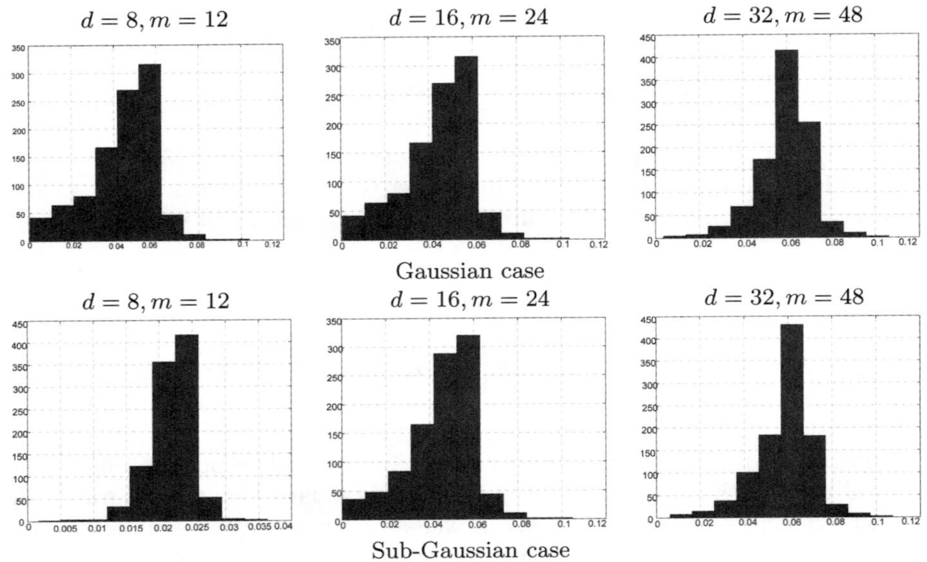

Figure 3.4: Histograms of recovery errors in experiments, 1,000 simulations per experiment.

sentially different from each other. In this situation, attainable risks heavily depend on how the right singular vectors of A are oriented with respect to the directions of the half-axes of U, so that the risks of our estimates vary significantly from instance to instance. Note also that the "sub-Gaussian experiments" were conducted on exactly the same data as "Gaussian experiments" of the same sizes d and m.

3.5 EXERCISES FOR CHAPTER 3

Exercise 3.1.

In the situation of Section 3.3.4, design of a "good" estimate is reduced to solving convex optimization problem (3.39). Note that the objective in this problem is, in a sense, "implicit"—the design variable is h, and the objective is obtained from an explicit convex-concave function of h and (x, y) by maximization over (x, y). There exist solvers able to process problems of this type efficiently. However, commonly used off-the-shelf solvers, like `cvx`, cannot handle problems of this type. The goal of the exercise to follow is to reformulate (3.39) as a semidefinite program, thus making it amenable for `cvx`.

On an immediate inspection, the situation we are interested in is as follows. We are given

- a nonempty convex compact set $X \subset \mathbf{R}^n$ along with affine function $M(x)$ taking values in $\mathbf{S}^d$ and such that $M(x) \succeq 0$ when $x \in X$, and
- an affine function $F(h) : \mathbf{R}^d \to \mathbf{R}^n$.

Given $\gamma > 0$, this data gives rise to the convex function

$$\Psi(h) = \max_{x\in X}\left\{F^T(h)x + \gamma\sqrt{h^T M(x)h}\right\},$$

and we want to find a "nice" representation of this function, specifically, we want to represent the inequality $\tau \geq \Psi(h)$ by a bunch of LMIs in variables τ, h, and perhaps additional variables.

To achieve our goal, we assume in the sequel that the set

$$X^+ = \{(x, M) : x \in X, M = M(x)\}$$

can be described by a system of linear and semidefinite constraints in variables x, M, and additional variables ξ, namely,

$$X^+ = \left\{(x,M) : \exists\xi : \left\{\begin{array}{ll} (a) & s_i - a_i^T x - b_i^T\xi - \mathrm{Tr}(C_i M) \geq 0,\ i \leq I \\ (b) & S - \mathcal{A}(x) - \mathcal{B}(\xi) - \mathcal{C}(M) \succeq 0 \\ (c) & M \succeq 0 \end{array}\right.\right\}.$$

Here $s_i \in \mathbf{R}$, $S \in \mathbf{S}^N$ are some constants, and $\mathcal{A}(\cdot), \mathcal{B}(\cdot), \mathcal{C}(\cdot)$ are (homogeneous) linear functions taking values in $\mathbf{S}^N$. We assume that this system of constraints is essentially strictly feasible, meaning that there exists a feasible solution at which the semidefinite constraints (b) and (c) are satisfied strictly (i.e., the left-hand sides of the LMIs are positive definite).

Here comes the exercise:

1) Check that $\Psi(h)$ is the optimal value in the semidefinite program

$$\Psi(h) = \max_{x,M,\xi,t}\left\{F^T(h)x + \gamma t : \left\{\begin{array}{ll} s_i - a_i^T x - b_i^T\xi - \mathrm{Tr}(C_i M) \geq 0, i \leq I & (a) \\ S - \mathcal{A}(x) - \mathcal{B}(\xi) - \mathcal{C}(M) \succeq 0 & (b) \\ M \succeq 0 & (c) \\ \left[\begin{array}{c|c} h^T M h & t \\ \hline t & 1 \end{array}\right] \succeq 0 & (d) \end{array}\right.\right\}. \tag{P}$$

2) Passing from (P) to the semidefinite dual of (P), build explicit semidefinite representation of Ψ, that is, an explicit system $\mathcal{S}$ of LMIs in variables h, τ, and additional variables u such that

$$\{\tau \geq \Psi(h)\} \Leftrightarrow \{\exists u : (\tau, h, u) \text{ satisfies } \mathcal{S}\}.$$

Exercise 3.2.

Let us consider the situation as follows. Given an $m \times n$ "sensing matrix" A which is stochastic with columns from the probabilistic simplex

$$\Delta_m = \left\{v \in \mathbf{R}^m : v \geq 0, \sum_i v_i = 1\right\}$$

and a nonempty closed subset U of Δ_n, we observe an M-element, $M > 1$, i.i.d. sample $\zeta^M = (\zeta_1, ..., \zeta_M)$ with ζ_k drawn from the discrete distribution Au_*, where u_* is an unknown probabilistic vector ("signal") known to belong to U. We handle the discrete distribution Au, $u \in \Delta_n$, as a distribution on the vertices $e_1, ..., e_m$ of Δ_m, so that possible values of ζ_k are basic orths $e_1, ..., e_m$ in $\mathbf{R}^m$. Our goal is to recover the value $F(u_*)$ of a given quadratic form

$$F(u) = u^T Qu + 2q^T u.$$

Observe that for $u \in \Delta_n$, we have $u = [uu^T]\mathbf{1}_n$, where $\mathbf{1}_k$ is the all-ones vector in $\mathbf{R}^k$. This observation allows us to rewrite $F(u)$ as a homogeneous quadratic form:

$$F(u) = u^T \bar{Q} u, \ \bar{Q} = Q + [q\mathbf{1}_n^T + \mathbf{1}_n q^T]. \tag{3.77}$$

The goal of the exercise is to follow the approach developed in Section 3.4.1 for the Gaussian case in order to build an estimate $\widehat{g}(\zeta^M)$ of $F(u)$. To this end, consider the following construction.

Let

$$\mathcal{J}_M = \{(i,j) : 1 \leq i < j \leq M\}, J_M = \mathrm{Card}(\mathcal{J}_M).$$

For $\zeta^M = (\zeta_1, ..., \zeta_M)$ with $\zeta_k \in \{e_1, ..., e_m\}$, $1 \leq k \leq M$, let

$$\omega_{ij}[\zeta^M] = \tfrac{1}{2}[\zeta_i \zeta_j^T + \zeta_j \zeta_i^T], \ (i,j) \in \mathcal{J}_M.$$

The estimates we are interested in are of the form

$$\widehat{g}(\zeta^M) = \mathrm{Tr}\Bigg(h \underbrace{\left[\frac{1}{J_M} \sum\nolimits_{(i,j)\in\mathcal{J}_M} \omega_{ij}[\zeta^M]\right]}_{\omega[\zeta^M]}\Bigg) + \kappa$$

where $h \in \mathbf{S}^m$ and $\kappa \in \mathbf{R}$ are the parameters of the estimate.

Now comes the exercise:

1) Verify that when the ζ_k's stem from signal $u \in U$, the expectation of $\omega[\zeta^M]$ is a linear image $Az[u]A^T$ of the matrix $z[u] = uu^T \in \mathbf{S}^n$: denoting by P_u^M the distribution of ζ^M, we have

$$\mathbf{E}_{\zeta^M \sim P_u^M}\{\omega[\zeta^M]\} = Az[u]A^T. \tag{3.78}$$

Check that when setting

$$\mathcal{Z}_k = \{\omega \in \mathbf{S}^k : \omega \succeq 0, \omega \geq 0, \mathbf{1}_k^T \omega \mathbf{1}_k = 1\},$$

where $x \geq 0$ for a matrix x means that x is entrywise nonnegative, the image of $\mathcal{Z}_n$ under the mapping $z \mapsto AzA^T$ is contained in $\mathcal{Z}_m$.

2) Let $\Delta^k = \{z \in \mathbf{S}^k : z \geq 0, \mathbf{1}_n^T z \mathbf{1}_n = 1\}$, so that $\mathcal{Z}_k$ is the set of all positive semidefinite matrices from Δ^k. For $\mu \in \Delta^m$, let P_μ be the distribution of the random matrix w taking values in $\mathbf{S}^m$ as follows: the possible values of w are matrices of the form $e^{ij} = \frac{1}{2}[e_i e_j^T + e_j e_i^T]$, $1 \leq i \leq j \leq m$; for every $i \leq m$, w takes value e^{ii} with probability μ_{ii}, and for every i, j with $i < j$, w takes value e^{ij} with probability $2\mu_{ij}$. Let us set

$$\Phi_1(h; \mu) = \ln\left(\sum_{i,j=1}^m \mu_{ij} \exp\{h_{ij}\}\right) : \mathbf{S}^m \times \Delta^m \to \mathbf{R},$$

so that Φ_1 is a continuous convex-concave function on $\mathbf{S}^m \times \Delta^m$.

2.1. Prove that

$$\forall (h \in \mathbf{S}^m, \mu \in \mathcal{Z}_m) : \ln\left(\mathbf{E}_{w\sim P_\mu}\left\{\exp\{\mathrm{Tr}(hw)\}\right\}\right) = \Phi_1(h;\mu).$$

2.2. Derive from 2.1 that setting

$$K = K(M) = \lfloor M/2 \rfloor,\ \ \Phi_M(h;\mu) = K\Phi_1(h/K;\mu) : \mathbf{S}^m \times \Delta^m \to \mathbf{R},$$

Φ_M is a continuous convex-concave function on $\mathbf{S}^m \times \Delta^m$ such $\Phi_K(0;\mu) = 0$ for all $\mu \in \mathcal{Z}_m$, and whenever $u \in U$, the following holds true:

Let $P_{u,M}$ be the distribution of $\omega = \omega[\zeta^M]$, $\zeta^M \sim P_u^M$. Then for all $u \in U$, $h \in \mathbf{S}^m$,

$$\ln\left(\mathbf{E}_{\omega\sim P_{u,M}}\left\{\exp\{\mathrm{Tr}(h\omega)\}\right\}\right) \le \Phi_M(h; Az[u]A^T),\ z[u] = uu^T. \tag{3.79}$$

3) Combine the above observations with Corollary 3.6 to arrive at the following result:

Proposition 3.19. *In the situation in question, let $\mathcal{Z}$ be a convex compact subset of $\mathcal{Z}_n$ such that $uu^T \in \mathcal{Z}$ for all $u \in U$. Given $\epsilon \in (0,1)$, let*

$$\begin{array}{rcl}
\Psi_+(h,\alpha) &=& \max\limits_{z\in\mathcal{Z}}\left[\alpha\Phi_M(h/\alpha, AzA^T) - \mathrm{Tr}(\bar{Q}z)\right] : \mathbf{S}^m \times \{\alpha > 0\} \to \mathbf{R},\\
\Psi_-(h,\alpha) &=& \max\limits_{z\in\mathcal{Z}}\left[\alpha\Phi_M(-h/\alpha, AzA^T) + \mathrm{Tr}(\bar{Q}z)\right] : \mathbf{S}^m \times \{\alpha > 0\} \to \mathbf{R}\\
\widehat{\Psi}_+(h) &:=& \inf_{\alpha>0}\left[\Psi_+(h,\alpha) + \alpha\ln(2/\epsilon)\right]\\
&=& \max\limits_{z\in\mathcal{Z}}\inf\limits_{\alpha>0}\left[\alpha\Phi_M(h/\alpha, AzA^T) - \mathrm{Tr}(\bar{Q}z) + \alpha\ln(2/\epsilon)\right]\\
&=& \max\limits_{z\in\mathcal{Z}}\inf\limits_{\beta>0}\left[\beta\Phi_1(h/\beta, AzA^T) - \mathrm{Tr}(\bar{Q}z) + \frac{\beta}{K}\ln(2/\epsilon)\right] \quad [\beta = K\alpha],\\
\widehat{\Psi}_-(h) &:=& \inf\limits_{\alpha>0}\left[\Psi_-(h,\alpha) + \alpha\ln(2/\epsilon)\right]\\
&=& \max\limits_{z\in\mathcal{Z}}\inf\limits_{\alpha>0}\left[\alpha\Phi_M(-h/\alpha, AzA^T) + \mathrm{Tr}(\bar{Q}z) + \alpha\ln(2/\epsilon)\right]\\
&=& \max\limits_{z\in\mathcal{Z}}\inf\limits_{\beta>0}\left[\beta\Phi_1(-h/\beta, AzA^T) + \mathrm{Tr}(\bar{Q}z) + \frac{\beta}{K}\ln(2/\epsilon)\right] \quad [\beta = K\alpha].
\end{array}$$

The functions $\widehat{\Psi}_\pm$ are real-valued and convex on $\mathbf{S}^m$, and every candidate solution h to the convex optimization problem

$$\mathrm{Opt} = \min_h\left\{\widehat{\Psi}(h) := \tfrac{1}{2}\left[\widehat{\Psi}_+(h) + \widehat{\Psi}_-(h)\right]\right\} \tag{3.80}$$

induces the estimate

$$\widehat{g}_h(\zeta^M) = \mathrm{Tr}(h\omega[\zeta^M]) + \kappa(h),\ \ \kappa(h) = \tfrac{1}{2}[\widehat{\Psi}_-(h) - \widehat{\Psi}_+(h)]$$

of the functional of interest (3.77) *via observation ζ^M with ϵ-risk on U not exceeding $\rho = \widehat{\Psi}(h)$:*

$$\forall (u \in U) : \mathrm{Prob}_{\zeta^M\sim P_u^M}\{|F(u) - \widehat{g}_h(\zeta^M)| > \rho\} \le \epsilon.$$

4) Consider an alternative way to estimate $F(u)$, namely, as follows. Let $u \in U$. Given a pair of independent observations ζ_1, ζ_2 drawn from distribution Au, let us convert them into the symmetric matrix $\omega_{1,2}[\zeta^2] = \frac{1}{2}[\zeta_1\zeta_2^T + \zeta_2\zeta_1^T]$. The distribution $P_{u,2}$ of this matrix is exactly the distribution $P_{\mu(z[u])}$—see item B—where $\mu(z) = AzA^T : \Delta^n \to \Delta^m$. Now, given $M = 2K$ observations $\zeta^{2K} = (\zeta_1, ..., \zeta_{2K})$ stemming from signal u, we can split them into K consecutive pairs giving rise

to K observations $\omega^K=(\omega_1,...,\omega_K)$, $\omega_k=\omega[[\zeta_{2k-1};\zeta_{2k}]]$, drawn independently of each other from probability distribution $P_{\mu(z[u])}$, and the functional of interest (3.77) is a linear function $\mathrm{Tr}(\bar{Q}z[u])$ of $z[u]$. Assume that we are given a set $\mathcal{Z}$ as in the premise of Proposition 3.19. Observe that we are in the situation as follows:

> Given K i.i.d. observations $\omega^K=(\omega_1,...,\omega_K)$ with $\omega_k\sim P_{\mu(z)}$, where z is an unknown signal known to belong to $\mathcal{Z}$, we want to recover the value at z of linear function $G(v)=\mathrm{Tr}(\bar{Q}v)$ of $v\in\mathbf{S}^n$. Besides this, we know that P_μ, for every $\mu\in\Delta^m$, satisfies the relation
>
> $$\forall(h\in\mathbf{S}^m):\ln\left(\mathbf{E}_{\omega\sim P_\mu}\{\exp\{\mathrm{Tr}(h\omega)\}\}\right)\leq\Phi_1(h;\mu).$$

This situation fits the setting of Section 3.3.3, with the data specified as

$$\begin{array}{c}\mathcal{H}=\mathcal{E}_H=\mathbf{S}^m,\mathcal{M}=\Delta^m\subset\mathcal{E}_M=\mathbf{S}^m,\Phi=\Phi_1,\\ \mathcal{X}:=\mathcal{Z}\subset\mathcal{E}_X=\mathbf{S}^n,\mathcal{A}(z)=AzA^T.\end{array}$$

Therefore, we can use the apparatus developed in that section to upper-bound the ϵ-risk of the affine estimate

$$\mathrm{Tr}\left(h\frac{1}{K}\sum_{k=1}^K\omega_k\right)+\kappa$$

of $F(u):=G(z[u])=u^T\bar{Q}u$ and to build the best, in terms of the upper risk bound, estimate; see Corollary 3.8. On closer inspection (carry it out!), the associated with the above data functions $\widehat{\Psi}_\pm$ arising in (3.38) are exactly the functions $\widehat{\Psi}_\pm$ specified in Proposition 3.19 for $M=2K$. Thus, the approach to estimating $F(u)$ via observations ζ^{2K} stemming from $u\in U$ results in a family of estimates

$$\widetilde{g}_h(\zeta^{2K})=\mathrm{Tr}\left(h\frac{1}{K}\sum_{k=1}^K\omega[[\zeta_{2k-1};\zeta_{2k}]]\right)+\kappa(h),\ h\in\mathbf{S}^m.$$

The resulting upper bound on the ϵ-risk of estimate $\widetilde{g}_h$ is $\widehat{\Psi}(h)$, where $\widehat{\Psi}(\cdot)$ is associated with $M=2K$ according to Proposition 3.19. In other words, this is exactly the upper bound on the ϵ-risk of the estimate $\widehat{g}_h$ offered by the proposition. Note, however, that the estimates $\widetilde{g}_h$ and $\widehat{g}_h$ are not identical:

$$\begin{array}{rcl}\widetilde{g}_h(\zeta^{2K})&=&\mathrm{Tr}\left(h\frac{1}{K}\sum_{k=1}^K\omega_{2k-1,2k}[\zeta^{2K}]\right)+\kappa(h),\\ \widehat{g}_h(\zeta^{2K})&=&\mathrm{Tr}\left(h\frac{1}{K(2K-1)}\sum_{1\leq i<j\leq 2K}\omega_{ij}[\zeta^{2K}]\right)+\kappa(h).\end{array}$$

Now goes the question:

• Which of the estimates $\widetilde{g}_h$ and $\widehat{g}_h$ would you prefer? That is, which one of these estimates, in your opinion, exhibits better practical performance?

To check your intuition, compare the estimate performance by simulation. Consider the following story underlying the recommended simulation model:

> *"Tomorrow, tomorrow not today, all the lazy people say." Is it profitable to be lazy?* Imagine you are supposed to carry out a job, and should decide

> whether to do it today or tomorrow. The reward for the job is drawn at random "by nature," with unknown to you time-invariant distribution u on an n-element set $\{r_1, ..., r_n\}$, with $r_1 \le r_2 \le ... \le r_n$. Given $2K$ historical observations of the rewards, what would be better—to complete the job today or tomorrow? In other words, is the probability for tomorrow's reward to be at least the reward of today greater than 0.5? What is this probability? How do we estimate it from historical data?

State the above problem as that of estimating a quadratic functional $u^T \bar{Q} u$ of distribution u from direct observations ($m = n, A = I_n$). Pick $u \in \Delta_n$ at random and run simulations to check which of the estimates $\widehat{g}_h$ and $\widetilde{g}_h$ works better. To avoid the necessity of solving optimization problem (3.80), you can use $h = \bar{Q}$, resulting in an unbiased estimate of $u^T \bar{Q} u$.

Exercise 3.3.

What follows is a variation of Exercise 3.2. Consider the situation as follows. We observe K realizations η_k, $k \le K$, of a discrete random variable with p possible values, and $L \ge K$ realizations ζ_ℓ, $\ell \le L$, of a discrete random variable with q possible values. All realizations are independent of each other; η_k's are drawn from distribution Pu, and the ζ_ℓ's from distribution Qv, where $P \in \mathbf{R}^{p\times r}$, $Q \in \mathbf{R}^{q\times s}$ are given stochastic "sensing matrices," and u, v are unknown "signals" known to belong to given subsets U, V of probabilistic simplexes Δ_r, Δ_s. Our goal is to recover from observations $\{\eta_k, \zeta_\ell\}$ the value at u, v of a given bilinear function

$$F(u, v) = u^T F v = \mathrm{Tr}(F[uv^T]^T). \tag{3.81}$$

> A "covering story" could be as follows. Imagine that there are two possible actions, say, administering to a patient drug A or drug B. Let u be the probability distribution of a (quantified) outcome of the first action, and v be a similar distribution for the second action. Observing what happens when the first action is utilized K, and the second L times, we could ask ourselves what the probability is of the outcome of the first action being better than the outcome of the second one. This amounts to computing the probability π of the event "$\eta > \zeta$," where η, ζ are discrete real-valued random variables independent of each other with distributions u, v, and π is a linear function of the "joint distribution" uv^T of η, ζ. This story gives rise to the aforementioned estimation problem with the unit sensing matrices P and Q. Assuming that there are "measurement errors"—instead of observing an action's outcome "as is," we observe a realization of a random variable with distribution depending, in a prescribed fashion, on the outcome—we arrive at problems where P and Q can be general type stochastic matrices.

As always, we encode the p possible values of η_k by the basic orths $e_1, ..., e_p$ in $\mathbf{R}^p$, and the q possible values of ζ by the basic orths $f_1, ..., f_q$ in $\mathbf{R}^q$.

We focus on estimates of the form

$$\widehat{g}_{h,\kappa}(\eta^K, \zeta^L) = \left[\frac{1}{K}\sum_k \eta_k\right]^T h \left[\frac{1}{L}\sum_\ell \zeta_\ell\right] + \kappa \qquad [h \in \mathbf{R}^{p\times q}, \kappa \in \mathbf{R}].$$

This is what you are supposed to do:

1) (cf. item 2 in Exercise 3.2) Denoting by Δ_{mn} the set of nonnegative $m \times n$ matrices with unit sum of all entries (i.e., the set of all probability distributions on $\{1, ..., m\} \times \{1, ..., n\}$) and assuming $L \geq K$, let us set

$$\mathcal{A}(z) = PzQ^T : \mathbf{R}^{r\times s} \to \mathbf{R}^{p\times q}$$

and

$$\begin{array}{rcl}\Phi(h;\mu) & = & \ln\left(\sum_{i=1}^{p}\sum_{j=1}^{q}\mu_{ij}\exp\{h_{ij}\}\right) : \mathbf{R}^{p\times q}\times\Delta_{pq}\to\mathbf{R},\\ \Phi_K(h;\mu) & = & K\Phi(h/K;\mu) : \mathbf{R}^{p\times q}\times\Delta_{pq}\to\mathbf{R}.\end{array}$$

Verify that $\mathcal{A}$ maps Δ_{rs} into Δ_{pq}, Φ and Φ_K are continuous convex-concave functions on their domains, and that for every $u \in \Delta_r$, $v \in \Delta_s$, the following holds true:

(!) When $\eta^K = (\eta_1, ..., \eta_K)$, $\zeta^L = (\zeta_1, ..., \zeta_K)$ with mutually independent $\eta_1, ..., \zeta_L$ such that $\eta_k \sim Pu$, $\eta_\ell \sim Qv$ for all k, ℓ, we have

$$\ln\left(\mathbf{E}_{\eta,\zeta}\left\{\exp\left\{\left[\frac{1}{K}\sum_k \eta_k\right]^T h\left[\frac{1}{L}\sum_\ell \zeta_\ell\right]\right\}\right\}\right) \leq \Phi_K(h;\mathcal{A}(uv^T)). \quad (3.82)$$

2) Combine (!) with Corollary 3.6 to arrive at the following analog of Proposition 3.19:

Proposition 3.20. *In the situation in question, let $\mathcal{Z}$ be a convex compact subset of Δ_{rs} such that $uv^T \in \mathcal{Z}$ for all $u \in U$, $v \in V$. Given $\epsilon \in (0,1)$, let*

$$\begin{array}{rcl}\Psi_+(h,\alpha) & = & \max\limits_{z\in\mathcal{Z}}\left[\alpha\Phi_K(h/\alpha, PzQ^T) - \mathrm{Tr}(Fz^T)\right] : \mathbf{R}^{p\times q}\times\{\alpha>0\}\to\mathbf{R},\\ \Psi_-(h,\alpha) & = & \max\limits_{z\in\mathcal{Z}}\left[\alpha\Phi_K(-h/\alpha, PzQ^T) + \mathrm{Tr}(Fz^T)\right] : \mathbf{R}^{p\times q}\times\{\alpha>0\}\to\mathbf{R},\\ \widehat{\Psi}_+(h) & := & \inf_{\alpha>0}\left[\Psi_+(h,\alpha)+\alpha\ln(2/\epsilon)\right]\\ & = & \max\limits_{z\in\mathcal{Z}}\inf\limits_{\alpha>0}\left[\alpha\Phi_K(h/\alpha, PzQ^T) - \mathrm{Tr}(Fz^T) + \alpha\ln(2/\epsilon)\right]\\ & = & \max\limits_{z\in\mathcal{Z}}\inf\limits_{\beta>0}\left[\beta\Phi(h/\beta, PzQ^T) - \mathrm{Tr}(Fz^T) + \frac{\beta}{K}\ln(2/\epsilon)\right] \quad [\beta = K\alpha],\\ \widehat{\Psi}_-(h) & := & \inf\limits_{\alpha>0}\left[\Psi_-(h,\alpha)+\alpha\ln(2/\epsilon)\right]\\ & = & \max\limits_{z\in\mathcal{Z}}\inf\limits_{\alpha>0}\left[\alpha\Phi_K(-h/\alpha, PzQ^T) + \mathrm{Tr}(Fz^T) + \alpha\ln(2/\epsilon)\right]\\ & = & \max\limits_{z\in\mathcal{Z}}\inf\limits_{\beta>0}\left[\beta\Phi(-h/\beta, PzQ^T) + \mathrm{Tr}(Fz^T) + \frac{\beta}{K}\ln(2/\epsilon)\right] \quad [\beta = K\alpha].\end{array}$$

The functions $\widehat{\Psi}_\pm$ are real-valued and convex on $\mathbf{R}^{p\times q}$, and every candidate solution h to the convex optimization problem

$$\mathrm{Opt} = \min_h\left\{\widehat{\Psi}(h) := \tfrac{1}{2}\left[\widehat{\Psi}_+(h) + \widehat{\Psi}_-(h)\right]\right\}$$

induces the estimate

$$\widehat{g}_h(\eta^K, \zeta^L) = \mathrm{Tr}\left(h\left[\left[\frac{1}{K}\sum_k \eta_k\right]\left[\frac{1}{L}\sum_\ell \zeta_\ell\right]^T\right]^T\right) + \kappa(h), \quad \kappa(h) = \tfrac{1}{2}[\widehat{\Psi}_-(h) - \widehat{\Psi}_+(h)]$$

of the functional of interest (3.81) *via observation η^K, ζ^L with ϵ-risk on $U \times V$ not exceeding $\rho = \widehat{\Psi}(h)$:*

$$\forall (u \in U, v \in V): \ \mathrm{Prob}\{|F(u,v) - \widehat{g}_h(\eta^K, \zeta^L)| > \rho\} \leq \epsilon,$$

the probability being taken w.r.t. the distribution of observations η^K, ζ^L *stemming from signals* u, v.

Exercise 3.4.

[recovering mixture weights] The problem to be addressed in this exercise is as follows. We are given K probability distributions $P_1, ..., P_K$ on observation space Ω, and let these distributions have densities $p_k(\cdot)$ w.r.t. some reference measure Π on Ω; we assume that $\sum_k p_k(\cdot)$ is positive on Ω. We are given also N independent observations

$$\omega_t \sim P_\mu,\ t = 1, ..., N,$$

drawn from distribution

$$P_\mu = \sum_{k=1}^{K} \mu_k P_k,$$

where μ is an unknown "signal" known to belong to the probabilistic simplex $\mathbf{\Delta}_K = \{\mu \in \mathbf{R}^K : \mu \geq 0, \sum_k \mu_k = 1\}$. Given $\omega^N = (\omega_1, ..., \omega_N)$, we want to recover the linear image $G\mu$ of μ, where $G \in \mathbf{R}^{\nu \times K}$ is given.

We intend to measure the risk of a candidate estimate $\widehat{G}(\omega^N) : \Omega \times ... \times \Omega \to \mathbf{R}^\nu$ by the quantity

$$\mathrm{Risk}[\widehat{G}(\cdot)] = \sup_{\mu \in \mathbf{\Delta}} \left[\mathbf{E}_{\omega^N \sim P_\mu \times ... \times P_\mu} \left\{\|\widehat{G}(\omega^N) - G\mu\|_2^2\right\}\right]^{1/2}.$$

3.4.A. Recovering linear form. Let us start with the case when $G = g^T$ is a $1 \times K$ matrix.

3.4.A.1. Preliminaries. To motivate the construction to follow, consider the case when Ω is a finite set (obtained, e.g., by "fine discretization" of the "true" observation space). In this situation our problem becomes an estimation problem in Discrete o.s.: *given a stationary* N*-repeated observation stemming from a discrete probability distribution* P_μ *affinely parameterized by signal* $\mu \in \mathbf{\Delta}_K$, *we want to recover a linear form of* μ. It is shown in Section 3.1—see Remark 3.2—that in this case a nearly optimal, in terms of its ϵ-risk, estimate is of the form

$$\widehat{g}(\omega^N) = \frac{1}{N}\sum_{t=1}^{N} \phi(\omega_t) \tag{3.83}$$

with properly selected ϕ. The difficulty with this approach is that as far as computations are concerned, optimal design of ϕ requires solving a convex optimization problem of design dimension of order of the cardinality of Ω, and this cardinality could be huge, as is the case when Ω is a discretization of a domain in $\mathbf{R}^d$ with d in the range of tens. To circumvent this problem, we are to simplify the outlined approach: from the construction of Section 3.1 we inherit the simple structure (3.83) of the estimator; taking this structure for granted, we are to develop an alternative design of ϕ. With this new design, we have no theoretical guarantees for the resulting estimates to be near-optimal; we sacrifice these guarantees in order to reduce dramatically the computational effort of building the estimates.

3.4.A.2. Generic estimate. Let us select somehow L functions $F_\ell(\cdot)$ on Ω such that

$$\int F_\ell^2(\omega)p_k(\omega)\Pi(d\omega) < \infty,\ 1 \le \ell \le L, 1 \le k \le K. \tag{3.84}$$

With $\lambda \in \mathbf{R}^L$, consider estimate of the form

$$\widehat{g}_\lambda(\omega^N) = \frac{1}{N}\sum_{t=1}^N \Phi_\lambda(\omega_t),\ \Phi_\lambda(\omega) = \sum_\ell \lambda_\ell F_\ell(\omega). \tag{3.85}$$

1) Prove that

$$\begin{array}{rcl}\mathrm{Risk}[\widehat{g}_\lambda] & \le & \overline{\mathrm{Risk}}(\lambda) \\ & := & \max\limits_{k\le K}\Big[\frac{1}{N}\int\left[\sum_\ell \lambda_\ell F_\ell(\omega)\right]^2 p_k(\omega)\Pi(d\omega) \\ & & + \left[\int\left[\sum_\ell \lambda_\ell F_\ell(\omega)\right]p_k(\omega)\Pi(d\omega) - g^T e_k\right]^2\Big]^{1/2} \\ & = & \max\limits_{k\le K}\left[\frac{1}{N}\lambda^T W_k\lambda + [e_k^T[M\lambda - g]]^2\right]^{1/2},\end{array} \tag{3.86}$$

where

$$\begin{array}{rcl} M & = & \left[M_{k\ell} := \int F_\ell(\omega)p_k(\omega)\Pi(d\omega)\right]_{\substack{k\le K\\ \ell\le L}}, \\ W_k & = & \left[[W_k]_{\ell\ell'} := \int F_\ell(\omega)F_{\ell'}(\omega)p_k(\omega)\Pi(d\omega)\right]_{\substack{\ell\le L\\ \ell'\le L}},\ 1 \le k \le K,\end{array}$$

and $e_1, ..., e_K$ are the standard basic orths in $\mathbf{R}^K$.

Note that $\overline{\mathrm{Risk}}(\lambda)$ is a convex function of λ; this function is easy to compute, provided the matrices M and W_k, $k \le K$, are available. Assuming this is the case, we can solve the convex optimization problem

$$\mathrm{Opt} = \min_{\lambda\in\mathbf{R}^K} \overline{\mathrm{Risk}}(\lambda) \tag{3.87}$$

and use the estimate (3.85) associated with the optimal solution to this problem; the risk of this estimate will be upper-bounded by Opt.

3.4.A.3. Implementation. When implementing the generic estimate we arrive at the "Measurement Design" question: how do we select the value of L and functions F_ℓ, $1 \le \ell \le L$, resulting in small (upper bound Opt on the) risk of the estimate (3.85) yielded by an optimal solution to (3.87)? We are about to consider three related options—*naive*, *basic*, and *Maximum Likelihood* (ML).

The naive option is to take $F_\ell = p_\ell$, $1 \le \ell \le L = K$, assuming that this selection meets (3.84). For the sake of definiteness, consider the "Gaussian case," where $\Omega = \mathbf{R}^d$, Π is the Lebesgue measure, and p_k is Gaussian distribution with parameters ν_k, Σ_k:

$$p_k(\omega) = (2\pi)^{-d/2}\mathrm{Det}(\Sigma_k)^{-1/2}\exp\left\{-\tfrac{1}{2}(\omega-\nu_k)^T\Sigma_k^{-1}(\omega-\nu_k)\right\}.$$

In this case, the Naive option leads to easily computable matrices M and W_k appearing in (3.86).

2) Check that in the Gaussian case, when setting

$$\Sigma_{k\ell} = [\Sigma_k^{-1} + \Sigma_\ell^{-1}]^{-1}, \ \Sigma_{k\ell m} = [\Sigma_k^{-1} + \Sigma_\ell^{-1} + \Sigma_m^{-1}]^{-1}, \ \chi_k = \Sigma_k^{-1}\nu_k,$$

$$\alpha_{k\ell} = \sqrt{\frac{\mathrm{Det}(\Sigma_{k\ell})}{(2\pi)^d \mathrm{Det}(\Sigma_k)\mathrm{Det}(\Sigma_\ell)}}, \ \beta_{k\ell m} = (2\pi)^{-d}\sqrt{\frac{\mathrm{Det}(\Sigma_{k\ell m})}{\mathrm{Det}(\Sigma_k)\mathrm{Det}(\Sigma_\ell)\mathrm{Det}(\Sigma_m)}},$$

we have

$$\begin{array}{rcl} M_{k\ell} & := & \int p_\ell(\omega)p_k(\omega)\Pi(d\omega) \\ & = & \alpha_{k\ell}\exp\left\{\frac{1}{2}\left[[\chi_k+\chi_\ell]^T\Sigma_{k\ell}[\chi_k+\chi_\ell] - \chi_k^T\Sigma_k\chi_k - \chi_\ell^T\Sigma_\ell\chi_\ell\right]\right\}, \\ {[W_k]_{\ell m}} & := & \int p_\ell(\omega)p_m(\omega)p_k(\omega)\Pi(d\omega) \\ & = & \beta_{k\ell m}\exp\left\{\frac{1}{2}\left[[\chi_k+\chi_\ell+\chi_m]^T\Sigma_{k\ell m}[\chi_k+\chi_\ell+\chi_m]\right.\right. \\ & & \left.\left. -\chi_k^T\Sigma_k\chi_k - \chi_\ell^T\Sigma_\ell\chi_\ell - \chi_m^T\Sigma_m\chi_m\right]\right\}. \end{array}$$

Basic option. Though simple, the Naive option does not make much sense: when replacing the reference measure Π with another measure Π' which has positive density $\theta(\cdot)$ w.r.t. Π, the densities p_k are updated according to $p_k(\cdot) \mapsto p'_k(\cdot) = \theta^{-1}(\cdot)p(\cdot)$, so that selecting $F'_\ell = p'_\ell$, the matrices M and W_k become M' and W'_k with

$$\begin{array}{rcl} M'_{k\ell} & = & \int \frac{p_k(\omega)p_\ell(\omega)}{\theta^2(\omega)}\Pi'(d\omega) = \int \frac{p_k(\omega)p_\ell(\omega)}{\theta(\omega)}\Pi(d\omega), \\ {[W'_k]_{\ell m}} & = & \int \frac{p_k(\omega)p_\ell(\omega)p_m(\omega)}{\theta^3(\omega)}\Pi'(d\omega) = \int \frac{p_k(\omega)p_\ell(\omega)}{\theta^2(\omega)}\Pi(d\omega). \end{array}$$

We see that in general $M \neq M'$ and $W_k \neq W'_k$, which makes the Naive option rather unnatural. In the alternative *Basic* option we set

$$L = K, \ F_\ell(\omega) = \pi(\omega) := \frac{p_\ell(\omega)}{\sum_k p_k(\omega)}.$$

The motivation is that the functions F_ℓ are invariant when replacing Π with Π', so that here $M = M'$ and $W_k = W'_k$. Besides this, there are statistical arguments in favor of the Basic option, namely, as follows. Let Π_* be the measure with the density $\sum_k p_k(\cdot)$ w.r.t. Π; taken w.r.t. Π_*, the densities of P_k are exactly the above $\pi_k(\cdot)$, and $\sum_k \pi_k(\omega) \equiv 1$. Now, (3.86) says that the risk of estimate $\widehat{g}_\lambda$ can be upper-bounded by the function $\overline{\mathrm{Risk}}(\lambda)$ defined in (3.86), and this function, in turn, can be upper-bounded by the function

$$\begin{array}{rl} \mathrm{Risk}^+(\lambda) := & \Big[\frac{1}{N}\sum_k \int \left[\sum_\ell \lambda_\ell F_\ell(\omega)\right]^2 p_k(\omega)\Pi(d\omega) \\ & \qquad + \max_k \left[\int \left[\sum_k \lambda_\ell F_\ell(\omega)\right] p_k(\omega)\Pi(d\omega) - g^T e_k\right]^2\Big]^{1/2} \\ & = \Big[\frac{1}{N}\int \left[\sum_\ell \lambda_\ell F_\ell(\omega)\right]^2 \Pi_*(d\omega) \\ & \qquad + \max_k \left[\int \left[\sum_k \lambda_\ell F_\ell(\omega)\right] \pi_k(\omega)\Pi_*(d\omega) - g^T e_k\right]^2\Big]^{1/2} \\ & \leq K\overline{\mathrm{Risk}}(\lambda) \end{array}$$

(we have said that the maximum of K nonnegative quantities is at most their sum, and the latter is at most K times the maximum of the quantities). Consequently, the risk of the estimate (3.85) stemming from an optimal solution to (3.87) can be upper-bounded by the quantity

$$\mathrm{Opt}^+ := \min_\lambda \mathrm{Risk}^+(\lambda) \quad [\geq \mathrm{Opt} := \max_\lambda \overline{\mathrm{Risk}}(\lambda)].$$

And here comes the punchline:

3.1) Prove that both the quantities Opt defined in (3.87) and the above Opt^+ depend only on the linear span of the functions F_ℓ, $\ell = 1, ..., L$, not on how the functions F_ℓ are selected in this span.

3.2) Prove that the selection $F_\ell = \pi_\ell$, $1 \leq \ell \leq L = K$, minimizes Opt^+ among all possible selections L, $\{F_\ell\}_{\ell=1}^L$ satisfying (3.84).

Conclude that the selection $F_\ell = \pi_\ell$, $1 \leq \ell \leq L = K$, while not necessarily optimal in terms of Opt, definitely is meaningful: this selection optimizes the natural upper bound Opt^+ on Opt. Observe that $\mathrm{Opt}^+ \leq K\mathrm{Opt}$, so that optimizing instead of Opt the upper bound Opt^+, although rough, is not completely meaningless.

A downside of the Basic option is that it seems problematic to get closed form expressions for the associated matrices M and W_k; see (3.86). For example, in the Gaussian case, the Naive choice of F_ℓ's allows us to represent M and W_k in an explicit closed form; in contrast to this, when selecting $F_\ell = \pi_\ell$, $\ell \leq L = K$, seemingly the only way to get M and W_k is to use Monte-Carlo simulations. This being said, we indeed can use Monte-Carlo simulations to compute M and W_k, provided we can sample from distributions $P_1, ..., P_K$. In this respect, it should be stressed that with $F_\ell \equiv \pi_\ell$, the entries in M and W_k are expectations, w.r.t. $P_1, ..., P_K$, of functions of ω *bounded in magnitude by 1*, and thus well-suited for Monte-Carlo simulation.

Maximum Likelihood option. This choice of $\{F_\ell\}_{\ell \leq L}$ follows straightforwardly the idea of discretization we started with in this exercise. Specifically, we split Ω into L cells $\Omega_1, ..., \Omega_L$ in such a way that the intersection of any two different cells is of Π-measure zero, and treat as our observations not the actual observations ω_t, but the indexes of the cells to which the ω_t's belong. With our estimation scheme, this is the same as selecting F_ℓ as the characteristic function of Ω_ℓ, $\ell \leq L$. Assuming that for distinct k, k' the densities p_k, $p_{k'}$ differ from each other Π-almost surely, the simplest discretization independent of how the reference measure is selected is the Maximum Likelihood discretization

$$\Omega_\ell = \{\omega : \max_k p_k(\omega) = p_\ell(\omega)\},\ 1 \leq \ell \leq L = K;$$

with the ML option, we take, as F_ℓ's, the characteristic functions of the sets Ω_ℓ, $1 \leq \ell \leq L = K$, just defined. As with the Basic option, the matrices M and W_k associated with the ML option can be found by Monte-Carlo simulation.

We have discussed three simple options for selecting F_ℓ's. In applications, one can compute the upper risk bounds Opt—see (3.87)—associated with each option, and use the option with the best—the smallest—risk bound ("smart" choice of F_ℓ's). Alternatively, one can take as $\{F_\ell, \ell \leq L\}$ the union of the three collections yielded by the above options (and, perhaps, further extend this union). Note that the larger is the collection of the F_ℓ's, the smaller is the associated Opt, so that the only price for combining different selections is in increasing the computational cost of solving (3.87).

3.4.A.4. Illustration. In the experimental part of this exercise your are expected to

4.1) Run numerical experiments to compare the estimates yielded by the above three options (Naive, Basic, ML). Recommended setup:

- $d = 8$, $K = 90$;
- Gaussian case with the covariance matrices Σ_k of P_k selected at random,

$$S_k = \texttt{rand}(d,d),\ \Sigma_k = \frac{S_k S_k^T}{\|S_k\|^2} \qquad [\|\cdot\|\text{: spectral norm}]$$

  and the expectations ν_k of P_k selected at random from $\mathcal{N}(0, \sigma^2 I_d)$, with $\sigma = 0.1$;
- values of N: $\{10^s,\ s = 0, 1, ..., 5\}$;
- linear form to be recovered: $g^T\mu \equiv \mu_1$.

4.2†). Utilize the Cramer-Rao lower risk bound (see Proposition 4.37, Exercise 4.22) to upper-bound the level of conservatism $\frac{\text{Opt}}{\text{Risk}_*}$ of the estimates built in item 4.1. Here Risk_* is the minimax risk in our estimation problem:

$$\text{Risk}_* = \inf_{\widehat{g}(\cdot)} \text{Risk}[\widehat{g}(\omega^N)] = \inf_{\widehat{g}(\cdot)} \sup_{\mu\in\mathbf{\Delta}} \left[\mathbf{E}_{\omega^N\sim P_\mu\times\ldots\times P_\mu}\left\{|\widehat{g}(\omega^N) - g^T\mu|^2\right\}\right]^{1/2},$$

where inf is taken over all estimates.

3.4.B. Recovering linear images. Now consider the case when G is a general $\nu \times K$ matrix. The analog of the estimate $\widehat{g}_\lambda(\cdot)$ is now as follows: with somehow chosen $F_1, ..., F_L$ satisfying (3.84), we select a $\nu \times L$ matrix $\Lambda = [\lambda_{i\ell}]$, set

$$\Phi_\Lambda(\omega) = [\textstyle\sum_\ell \lambda_{1\ell}F_\ell(\omega); \sum_\ell \lambda_{2\ell}F_\ell(\omega); ...; \sum_\ell \lambda_{\nu\ell}F_\ell(\omega)],$$

and estimate $G\mu$ by

$$\widehat{G}_\Lambda(\omega^N) = \frac{1}{N}\sum_{t=1}^N \Phi_\lambda(\omega_t).$$

5) Prove the following counterpart of the results of item 3.4.A:

Proposition 3.21. *The risk of the proposed estimator can be upper-bounded as follows:*

$$\begin{array}{rcl}
\text{Risk}[\widehat{G}_\Lambda] & := & \max_{\mu\in\mathbf{\Delta}_K}\left[\mathbf{E}_{\omega^N\sim P_\mu\times\ldots\times P_\mu}\left\{\|\widehat{G}(\omega^N) - G\mu\|_2^2\right\}\right]^{1/2} \\
& \leq & \overline{\text{Risk}}(\Lambda) := \max_{k\leq K}\overline{\Psi}(\Lambda, e_k), \\
\overline{\Psi}(\Lambda,\mu) & = & \left[\frac{1}{N}\sum_{k=1}^K \mu_k \mathbf{E}_{\omega\sim P_k}\left\{\|\Phi_\Lambda(\omega)\|_2^2\right\} + \|[\psi_\Lambda - G]\mu\|_2^2\right]^{1/2} \\
& = & \left[\|[\psi_\Lambda - G]\mu\|_2^2 + \frac{1}{N}\sum_{k=1}^K \mu_k \int [\sum_{i\leq\nu}[\sum_\ell \lambda_{i\ell}F_\ell(\omega)]^2]P_k(d\omega)\right]^{1/2},
\end{array}$$

where

$$\text{Col}_k[\psi_\Lambda] = \mathbf{E}_{\omega\sim P_k(\cdot)}\Phi_\Lambda(\omega) = \left[\begin{array}{c} \int[\sum_\ell \lambda_{1\ell}F_\ell(\omega)]P_k(d\omega) \\ \cdots \\ \int[\sum_\ell \lambda_{\nu\ell}F_\ell(\omega)]P_k(d\omega) \end{array}\right],\ 1 \leq k \leq K$$

and $e_1, ..., e_K$ are the standard basic orths in $\mathbf{R}^K$.

Note that exactly the same reasoning as in the case of the scalar $G\mu \equiv g^T\mu$ demonstrates that a reasonable way to select L and F_ℓ, $\ell = 1, ..., L$, is to set $L = K$ and $F_\ell(\cdot) = \pi_\ell(\cdot)$, $1 \leq \ell \leq L$.

3.6 PROOFS

3.6.1 Proof of Proposition 3.3

1°. Observe that $\mathrm{Opt}_{ij}(K)$ is the saddle point value in the convex-concave saddle point problem:

$$\begin{array}{rcl}\mathrm{Opt}_{ij}(K) & = & \inf\limits_{\alpha>0,\phi\in\mathcal{F}} \max\limits_{x\in X_i, y\in X_j} \left[\frac{1}{2}K\alpha\left[\Phi_{\mathcal{O}}(\phi/\alpha; A_i(x)) + \Phi_{\mathcal{O}}(-\phi/\alpha; A_j(y))\right]\right. \\ & & \left. +\frac{1}{2}g^T[y-x] + \alpha\ln(2I/\epsilon)\right].\end{array}$$

The domain of the maximization variable is compact and the cost function is continuous on its domain, whence, by the Sion-Kakutani Theorem, we also have

$$\begin{array}{rcl}\mathrm{Opt}_{ij}(K) & = & \max\limits_{x\in X_i, y\in X_j} \Theta_{ij}(x,y), \\ \Theta_{ij}(x,y) & = & \inf\limits_{\alpha>0,\phi\in\mathcal{F}} \left[\frac{1}{2}K\alpha\left[\Phi_{\mathcal{O}}(\phi/\alpha; A_i(x)) + \Phi_{\mathcal{O}}(-\phi/\alpha; A_j(y))\right]\right. \\ & & \left. +\alpha\ln(2I/\epsilon)\right] + \frac{1}{2}g^T[y-x].\end{array} \tag{3.88}$$

Note that

$$\begin{array}{rcl}\Theta_{ij}(x,y) & = & \inf\limits_{\alpha>0,\psi\in\mathcal{F}} \left[\frac{1}{2}K\alpha\left[\Phi_{\mathcal{O}}(\psi; A_i(x)) + \Phi_{\mathcal{O}}(-\psi; A_j(y))\right] + \alpha\ln(2I/\epsilon)\right] \\ & & +\frac{1}{2}g^T[y-x] \\ & = & \inf\limits_{\alpha>0} \left[\frac{1}{2}\alpha K \inf\limits_{\psi\in\mathcal{F}} \left[\Phi_{\mathcal{O}}(\psi; A_i(x)) + \Phi_{\mathcal{O}}(-\psi; A_j(y))\right] + \alpha\ln(2I/\epsilon)\right] \\ & & +\frac{1}{2}g^T[y-x].\end{array}$$

Given $x \in X_i$, $y \in X_j$ and setting $\mu = A_i(x)$, $\nu = A_j(y)$, we obtain

$$\begin{array}{l}\inf\limits_{\psi\in\mathcal{F}}\left[\Phi_{\mathcal{O}}(\psi; A_i(x)) + \Phi_{\mathcal{O}}(-\psi; A_j(y))\right] \\ = \inf\limits_{\psi\in\mathcal{F}} \left[\ln\left(\int \exp\{\psi(\omega)\}p_\mu(\omega)\Pi(d\omega)\right) + \ln\left(\int \exp\{-\psi(\omega)\}p_\nu(\omega)\Pi(d\omega)\right)\right].\end{array}$$

Since $\mathcal{O}$ is a good o.s., the function $\bar{\psi}(\omega) = \frac{1}{2}\ln(p_\nu(\omega)/p_\mu(\omega))$ belongs to $\mathcal{F}$, and

$$\begin{array}{l}\inf\limits_{\psi\in\mathcal{F}} \left[\ln\left(\int \exp\{\psi(\omega)\}p_\mu(\omega)\Pi(d\omega)\right) + \ln\left(\int \exp\{-\psi(\omega)\}p_\nu(\omega)\Pi(d\omega)\right)\right] \\ = \inf\limits_{\delta\in\mathcal{F}} \left[\ln\left(\int \exp\{\bar{\psi}(\omega)+\delta(\omega)\}p_\mu(\omega)\Pi(d\omega)\right) + \ln\left(\int \exp\{-\bar{\psi}(\omega)-\delta(\omega)\}p_\nu(\omega)\Pi(d\omega)\right)\right] \\ = \inf\limits_{\delta\in\mathcal{F}} \underbrace{\left[\ln\left(\int \exp\{\delta(\omega)\}\sqrt{p_\mu(\omega)p_\nu(\omega)}\Pi(d\omega)\right) + \ln\left(\int \exp\{-\delta(\omega)\}\sqrt{p_\mu(\omega)p_\nu(\omega)}\Pi(d\omega)\right)\right]}_{f(\delta)}.\end{array}$$

Observe that $f(\delta)$ clearly is a convex and even function of $\delta \in \mathcal{F}$; as such, it attains its minimum over $\delta \in \mathcal{F}$ when $\delta = 0$. The bottom line is that

$$\inf_{\psi \in \mathcal{F}} [\Phi_{\mathcal{O}}(\psi; A_i(x)) + \Phi_{\mathcal{O}}(-\psi; A_j(y))] = 2\ln\left(\int \sqrt{p_{A_i(x)}(\omega) p_{A_j(y)}(\omega)} \Pi(d\omega)\right), \tag{3.89}$$

and

$$\begin{array}{rcl}\Theta_{ij}(x,y) & = & \inf_{\alpha>0} \alpha \left[K \ln\left(\int \sqrt{p_{A_i(x)}(\omega) p_{A_j(y)}(\omega)} \Pi(d\omega)\right) + \ln(2I/\epsilon)\right] + \frac{1}{2} g^T[y-x] \\ & = & \left\{\begin{array}{ll} \frac{1}{2} g^T[y-x], & K \ln\left(\int \sqrt{p_{A_i(x)}(\omega) p_{A_j(y)}(\omega)} \Pi(d\omega)\right) + \ln(2I/\epsilon) \geq 0, \\ -\infty, & \text{otherwise.} \end{array}\right.\end{array}$$

This combines with (3.88) to imply that

$$\begin{array}{rcl}\mathrm{Opt}_{ij}(K) & = & \max_{x,y} \left\{ \frac{1}{2} g^T[y-x] : x \in X_i, y \in X_j, \right. \\ & & \left. \left[\int \sqrt{p_{A_i(x)}(\omega) p_{A_j(y)}(\omega)} \Pi(d\omega)\right]^K \geq \frac{\epsilon}{2I} \right\}.\end{array} \tag{3.90}$$

2°. We claim that under the premise of the proposition, for all i, j, $1 \leq i, j \leq I$, one has

$$\mathrm{Opt}_{ij}(K) \leq \mathrm{Risk}^*_\epsilon(\overline{K}),$$

implying the validity of (3.13). Indeed, assume that for some pair i, j the opposite inequality holds true,

$$\mathrm{Opt}_{ij}(K) > \mathrm{Risk}^*_\epsilon(\overline{K}),$$

and let us lead this assumption to a contradiction. Under our assumption optimization problem in (3.90) has a feasible solution $(\bar{x}, \bar{y})$ such that

$$r := \tfrac{1}{2} g^T[\bar{y} - \bar{x}] > \mathrm{Risk}^*_\epsilon(\overline{K}), \tag{3.91}$$

implying, due to the origin of $\mathrm{Risk}^*_\epsilon(\overline{K})$, that there exists an estimate $\widetilde{g}(\omega^{\overline{K}})$ such that for $\mu = A_i(\bar{x})$, $\nu = A_j(\bar{y})$ it holds

$$\begin{array}{rcl}\mathrm{Prob}_{\omega^{\overline{K}} \sim p_\nu^{\overline{K}}} \left\{ \widetilde{g}(\omega^{\overline{K}}) \leq \frac{1}{2} g^T[\bar{x} + \bar{y}] \right\} & \leq & \mathrm{Prob}_{\omega^{\overline{K}} \sim p_\nu^{\overline{K}}} \left\{ |\widetilde{g}(\omega^{\overline{K}}) - g^T\bar{y}| \geq r \right\} \leq \epsilon \\ \mathrm{Prob}_{\omega^{\overline{K}} \sim p_\mu^{\overline{K}}} \left\{ \widetilde{g}(\omega^{\overline{K}}) \geq \frac{1}{2} g^T[\bar{x} + \bar{y}] \right\} & \leq & \mathrm{Prob}_{\omega^{\overline{K}} \sim p_\mu^{\overline{K}}} \left\{ |\widetilde{g}(\omega^{\overline{K}}) - g^T\bar{x}| \geq r \right\} \leq \epsilon.\end{array}$$

In other words, we can decide on two simple hypotheses stating that observation $\omega^{\overline{K}}$ obeys distribution $p_\mu^{\overline{K}}$ or $p_\nu^{\overline{K}}$, with risk $\leq \epsilon$. Consequently, setting $\Pi^K = \underbrace{\Pi \times \ldots \times \Pi}_{K}$ and $p_\theta^K(\omega^K) = \prod_{k=1}^K p_\theta(\omega_k)$, we have

$$\int \min\left[p_\mu^{\overline{K}}(\omega^{\overline{K}}), p_\nu^{\overline{K}}(\omega^{\overline{K}})\right] \Pi^{\overline{K}}(d\omega^{\overline{K}}) \leq 2\epsilon.$$

Hence,

$$\begin{array}{rl}\left[\int \sqrt{p_\mu(\omega)p_\nu(\omega)}\Pi(d\omega)\right]^{\overline{K}} = \int \sqrt{p_\mu^{\overline{K}}(\omega^{\overline{K}})p_\nu^{\overline{K}}(\omega^{\overline{K}})}\Pi^{\overline{K}}(d\omega^{\overline{K}}) & \\ = & \int \sqrt{\min\left[p_\mu^{\overline{K}}(\omega^{\overline{K}}), p_\nu^{\overline{K}}(\omega^{\overline{K}})\right]}\sqrt{\max\left[p_\mu^{\overline{K}}(\omega^{\overline{K}}), p_\nu^{\overline{K}}(\omega^{\overline{K}})\right]}\Pi^{\overline{K}}(d\omega^{\overline{K}}) \\ \le & \left(\int \min\left[p_\mu^{\overline{K}}(\omega^{\overline{K}}), p_\nu^{\overline{K}}(\omega^{\overline{K}})\right]\Pi^{\overline{K}}(d\omega^{\overline{K}})\right)^{\frac{1}{2}}\left(\int \max\left[p_\mu^{\overline{K}}(\omega^{\overline{K}}), p_\nu^{\overline{K}}(\omega^{\overline{K}})\right]\Pi^{\overline{K}}(d\omega^{\overline{K}})\right)^{\frac{1}{2}} \\ = & \left(\int \min\left[p_\mu^{\overline{K}}(\omega^{\overline{K}}), p_\nu^{\overline{K}}(\omega^{\overline{K}})\right]\Pi^{\overline{K}}(d\omega^{\overline{K}})\right)^{\frac{1}{2}} \\ & \times\left(\int \left[p_\mu^{\overline{K}}(\omega^{\overline{K}}) + p_\nu^{\overline{K}}(\omega^{\overline{K}}) - \min\left[p_\mu^{\overline{K}}(\omega^{\overline{K}}), p_\nu^{\overline{K}}(\omega^{\overline{K}})\right]\right]\Pi^{\overline{K}}(d\omega^{\overline{K}})\right)^{\frac{1}{2}} \\ = & \left(\int \min\left[p_\mu^{\overline{K}}(\omega^{\overline{K}}), p_\nu^{\overline{K}}(\omega^{\overline{K}})\right]\Pi^{\overline{K}}(d\omega^{\overline{K}})\right)^{\frac{1}{2}}\left(2 - \int \min\left[p_\mu^{\overline{K}}(\omega^{\overline{K}}), p_\nu^{\overline{K}}(\omega^{\overline{K}})\right]\Pi^{\overline{K}}(d\omega^{\overline{K}})\right)^{\frac{1}{2}} \\ \le & 2\sqrt{\epsilon(1-\epsilon)}.\end{array}$$

Therefore, for K satisfying (3.12) we have

$$\left[\int \sqrt{p_\mu(\omega)p_\nu(\omega)}\Pi(d\omega)\right]^K \le [2\sqrt{\epsilon(1-\epsilon)}]^{K/\overline{K}} < \frac{\epsilon}{2I},$$

which is the desired contradiction (recall that $\mu = A_i(\bar{x})$, $\nu = A_j(\bar{y})$ and $(\bar{x}, \bar{y})$, is feasible for (3.90)).

3°. Now let us prove that under the premise of the proposition, (3.14) takes place. To this end, let us set

$$w_{ij}(s) = \max_{x\in X_j, y\in X_j}\left\{\tfrac{1}{2}g^T[y-x] : \underbrace{\overline{K}\ln\left(\int \sqrt{p_{A_i(x)}(\omega)p_{A_j(y)}(\omega)}\Pi(d\omega)\right)}_{H(x,y)} + s \ge 0\right\}. \tag{3.92}$$

As we have seen in item 1°—see (3.89)—one has

$$H(x,y) = \inf_{\psi\in\mathcal{F}} \tfrac{1}{2}\left[\Phi_{\mathcal{O}}(\psi; A_i(x)) + \Phi_{\mathcal{O}}(-\psi, A_j(y))\right],$$

that is, $H(x,y)$ is the infimum of a parametric family of concave functions of $(x,y) \in X_i \times X_j$ and as such is concave. Besides this, the optimization problem in (3.92) is feasible whenever $s \ge 0$, a feasible solution being $y = x = x_{ij}$. At this feasible solution we have $g^T[y-x] = 0$, implying that $w_{ij}(s) \ge 0$ for $s \ge 0$. Observe also that from concavity of $H(x,y)$ it follows that $w_{ij}(s)$ is concave on the ray $\{s \ge 0\}$. Finally, we claim that

$$w_{ij}(\bar{s}) \le \mathrm{Risk}^*_\epsilon(\overline{K}),\ \bar{s} = -\ln(2\sqrt{\epsilon(1-\epsilon)}). \tag{3.93}$$

Indeed, $w_{ij}(s)$ is nonnegative, concave, and bounded (since X_i, X_j are compact) on $\mathbf{R}_+$, implying that $w_{ij}(s)$ is continuous on $\{s > 0\}$. Assuming, on the contrary to what we need to prove, that $w_{ij}(\bar{s}) > \mathrm{Risk}^*_\epsilon(\overline{K})$, there exists $s' \in (0, \bar{s})$ such that $w_{ij}(s') > \mathrm{Risk}^*_\epsilon(\overline{K})$ and thus there exist $\bar{x} \in X_i$, $\bar{y} \in X_j$ such that $(\bar{x}, \bar{y})$ is feasible for the optimization problem specifying $w_{ij}(s')$ and (3.91) takes place. We have seen in item 2° that the latter relation implies that for $\mu = A_i(\bar{x})$, $\nu = A_j(\bar{y})$ it holds

$$\left[\int \sqrt{p_\mu(\omega)p_\nu(\omega)}\Pi(d\omega)\right]^{\overline{K}} \le 2\sqrt{\epsilon(1-\epsilon)},$$

that is,

$$\overline{K}\ln\left(\int\sqrt{p_\mu(\omega)p_\nu(\omega)}\Pi(d\omega)\right)+\bar{s}\leq 0.$$

Hence,

$$\overline{K}\ln\left(\int\sqrt{p_\mu(\omega)p_\nu(\omega)}\Pi(d\omega)\right)+s'<0,$$

contradicting the feasibility of $(\bar{x},\bar{y})$ to the optimization problem specifying $w_{ij}(s')$.

It remains to note that (3.93) combines with concavity of $w_{ij}(\cdot)$ and the relation $w_{ij}(0)\geq 0$ to imply that

$$w_{ij}(\ln(2I/\epsilon))\leq\vartheta w_{ij}(\bar{s})\leq\vartheta\mathrm{Risk}^*_\epsilon(\overline{K})$$

where

$$\vartheta=\ln(2I/\epsilon)/\bar{s}=\frac{2\ln(2I/\epsilon)}{\ln([4\epsilon(1-\epsilon)]^{-1})}.$$

Invoking (3.90), we conclude that

$$\mathrm{Opt}_{ij}(\overline{K})=w_{ij}(\ln(2I/\epsilon))\leq\vartheta\mathrm{Risk}^*_\epsilon(\overline{K})\,\forall i,j.$$

Finally, from (3.90) it immediately follows that $\mathrm{Opt}_{ij}(K)$ is nonincreasing in K (as K grows, the feasible set of the optimization problem in (3.90) shrinks), so that for $K\geq\overline{K}$ we have

$$\mathrm{Opt}(K)\leq\mathrm{Opt}(\overline{K})=\max_{i,j}\mathrm{Opt}_{ij}(\overline{K})\leq\vartheta\mathrm{Risk}^*_\epsilon(\overline{K}),$$

and (3.14) follows. □

3.6.2 Verifying 1-convexity of the conditional quantile

Let r be a nonvanishing probability distribution on S, and let

$$F_m(r)=\sum_{i=1}^m r_i,\ 1\leq m\leq M,$$

so that $0<F_1(r)<F_2(r)<...<F_M(r)=1$. Denoting by $\mathcal{P}$ the set of all nonvanishing probability distributions on S, observe that for every $p\in\mathcal{P}$ $\chi_\alpha[r]$ is a piecewise linear function of $\alpha\in[0,1]$ with breakpoints $0,F_1(r),F_2(r),F_3(r),...,F_M(r)$, the values of the function at these breakpoints being $s_1,s_1,s_2,s_3,...,s_M$. In particular, this function is equal to s_1 on $[0,F_1(r)]$ and is strictly increasing on $[F_1(r),1]$. Now let $s\in\mathbf{R}$, and let

$$\mathcal{P}^{\leq}_{\alpha}[s]=\{r\in\mathcal{P}:\chi_\alpha[r]\leq s\},\ \ \mathcal{P}^{\geq}_{\alpha}[s]=\{r\in\mathcal{P}:\chi_\alpha[r]\geq s\}.$$

Observe that the just introduced sets are cut off $\mathcal{P}$ by nonstrict linear inequalities, specifically,

- when $s<s_1$, we have $\mathcal{P}^{\leq}_{\alpha}[s]=\emptyset$, $\mathcal{P}^{\geq}_{\alpha}[s]=\mathcal{P}$;
- when $s=s_1$, we have $\mathcal{P}^{\leq}_{\alpha}[s]=\{r\in\mathcal{P}:F_1(r)\geq\alpha\}$, $\mathcal{P}^{\geq}_{\alpha}[s]=\mathcal{P}$;
- when $s>s_M$, we have $\mathcal{P}^{\leq}_{\alpha}[s]=\mathcal{P}$, $\mathcal{P}^{\geq}_{\alpha}[s]=\emptyset$;
- when $s_1<s\leq s_M$, for every $r\in\mathcal{P}$ the equation $\chi_\gamma[r]=s$ in variable $\gamma\in[0,1]$

has exactly one solution $\gamma(r)$ which can be found as follows: we specify $k = k^s \in \{1, ..., M-1\}$ such that $s_k < s \leq s_{k+1}$ and set

$$\gamma(r) = \frac{(s_{k+1}-s)F_k(r) + (s-s_k)F_{k+1}(r)}{s_{k+1}-s_k}.$$

Since $\chi_\alpha[r]$ is strictly increasing in α when $\alpha \in [F_1(p), 1]$, for $s \in (s_1, s_M]$ we have

$$\mathcal{P}_{\alpha}^{\leq}[s] = \{r \in \mathcal{P} : \alpha \leq \gamma(r)\} = \left\{r \in \mathcal{P} : \frac{(s_{k+1}-s)F_k(r) + (s-s_k)F_{k+1}(r)}{s_{k+1}-s_k} \geq \alpha\right\},$$
$$\mathcal{P}_{\alpha}^{\geq}[s] = \{r \in \mathcal{P} : \alpha \geq \gamma(r)\} = \left\{r \in \mathcal{P} : \frac{(s_{k+1}-s)F_k(r) + (s-s_k)F_{k+1}(r)}{s_{k+1}-s_k} \leq \alpha\right\}.$$

As an immediate consequence of this description, given $\alpha \in [0,1]$ and $\tau \in T$ and setting

$$G_{\tau,\mu}(p) = \sum_{\iota=1}^{\mu} p(\iota, \tau),\ 1 \leq \mu \leq M,$$

and

$$\mathcal{X}^{s,\leq} = \{p(\cdot,\cdot) \in \mathcal{X} : \chi_\alpha[p_\tau] \leq s\},\quad \mathcal{X}^{s,\geq} = \{p(\cdot,\cdot) \in \mathcal{X} : \chi_\alpha[p_\tau] \geq s\},$$

we get

$$\begin{array}{rcl}
s < s_1 & \Rightarrow & \mathcal{X}^{s,\leq} = \emptyset,\ \mathcal{X}^{s,\geq} = \mathcal{X},\\
s = s_1 & \Rightarrow & \mathcal{X}^{s,\leq} = \{p \in \mathcal{X} : G_{\tau,1}(p) \leq s_1 G_{\tau,M}(p)\},\ \mathcal{X}^{s,\geq} = \mathcal{X},\\
s > s_M & \Rightarrow & \mathcal{X}^{s,\leq} = \mathcal{X},\ \mathcal{X}^{s,\geq} = \emptyset,\\
s_1 < s \leq s_M & \Rightarrow & \left\{\begin{array}{l} \mathcal{X}^{s,\leq} = \left\{p \in \mathcal{X} : \frac{(s_{k+1}-s)G_{\tau,k}(r) + (s-s_k)G_{\tau,k+1}(r)}{s_{k+1}-s_k} \geq \alpha G_{\tau,M}(p)\right\},\\ \mathcal{X}^{s,\geq} = \left\{p \in \mathcal{X} : \frac{(s_{k+1}-s)G_{\tau,k}(r) + (s-s_k)G_{\tau,k+1}(r)}{s_{k+1}-s_k} \leq \alpha G_{\tau,M}(p)\right\},\end{array}\right.\\
& & k = k_s : s_k < s \leq s_{k+1},
\end{array}$$

implying 1-convexity of the conditional quantile on $\mathcal{X}$ (recall that $G_{\tau,\mu}(p)$ are linear in p). □

3.6.3 Proof of Proposition 3.4

3.6.3.1 Proof of Proposition 3.4.i

We call step ℓ *essential* if at this step rule 2d is invoked.

1°. Let $x \in X$ be the true signal underlying the observation $\bar{\omega}^K$, so that $\bar{\omega}_1, ..., \bar{\omega}_K$ are drawn from the distribution $p_{A(x)}$ independently of each other. Consider the "ideal" estimate given by exactly the same rules as the Bisection procedure in Section 3.2.4.2 (in the sequel, we refer to the latter as the "true" one), with tests $\mathcal{T}^K_{\Delta_{\ell,\text{rg}},\text{r}}(\cdot)$, $\mathcal{T}^K_{\Delta_{\ell,\text{lf}},\text{l}}(\cdot)$ in rule 2d replaced with the "ideal tests"

$$\widehat{T}_{\Delta_{\ell,\text{rg}},\text{r}} = \widehat{T}_{\Delta_{\ell,\text{lf}},\text{l}} = \left\{\begin{array}{ll} \texttt{right}, & f(x) > c_\ell,\\ \texttt{left}, & f(x) \leq c_\ell.\end{array}\right.$$

Marking by * the entities produced by the resulting *fully deterministic* procedure, we arrive at the sequence of nested segments $\Delta^*_\ell = [a^*_\ell, b^*_\ell]$, $0 \leq \ell \leq L^* \leq L$, along with subsegments $\Delta^*_{\ell,\mathrm{rg}} = [c^*_\ell, v^*_\ell]$, $\Delta^*_{\ell,\mathrm{lf}} = [u^*_\ell, c^*_\ell]$ of $\Delta^*_{\ell-1}$, defined for all *-essential values of ℓ, and the output segment $\bar{\Delta}^*$ claimed to contain $f(x)$. Note that the ideal procedure cannot terminate due to arriving at a disagreement, and that $f(x)$, as is immediately seen, is contained in all segments Δ^*_ℓ, $0 \le \ell \le L^*$, just as $f(x) \in \bar{\Delta}^*$.

Let $\mathcal{L}^*$ be the set of all *-essential values of ℓ. For $\ell \in \mathcal{L}^*$, let the event $\mathcal{E}_\ell[x]$ parameterized by x be defined as follows:

$$\mathcal{E}_\ell[x] = \left\{ \begin{array}{ll} \left\{\omega^K : \mathcal{T}^K_{\Delta^*_{\ell,\mathrm{rg}},\mathrm{r}}(\omega^K) = \texttt{right} \text{ or } \mathcal{T}^K_{\Delta^*_{\ell,\mathrm{lf}},\mathrm{l}}(\omega^K) = \texttt{right}\right\}, & f(x) \le u^*_\ell, \\ \left\{\omega^K : \mathcal{T}^K_{\Delta^*_{\ell,\mathrm{rg}},\mathrm{r}}(\omega^K) = \texttt{right}\right\}, & u^*_\ell < f(x) \le c^*_\ell, \\ \left\{\omega^K : \mathcal{T}^K_{\Delta^*_{\ell,\mathrm{lf}},\mathrm{l}}(\omega^K) = \texttt{left}\right\}, & c^*_\ell < f(x) < v^*_\ell, \\ \left\{\omega^K : \mathcal{T}^K_{\Delta^*_{\ell,\mathrm{rg}},\mathrm{r}}(\omega^K) = \texttt{left} \text{ or } \mathcal{T}^K_{\Delta^*_{\ell,\mathrm{lf}},\mathrm{l}}(\omega^K) = \texttt{left}\right\}, & f(x) \ge v^*_\ell. \end{array} \right. \tag{3.94}$$

2°. Observe that by construction and in view of Proposition 2.27 we have

$$\forall \ell \in \mathcal{L}^* : \mathrm{Prob}_{\omega^K \sim p_{A(x)} \times \ldots \times p_{A(x)}} \{\mathcal{E}_\ell[x]\} \le 2\delta. \tag{3.95}$$

Indeed, let $\ell \in \mathcal{L}^*$.

- When $f(x) \le u^*_\ell$, we have $x \in X$ and $f(x) \le u^*_\ell \le c^*_\ell$, implying that $\mathcal{E}_\ell[x]$ takes place only when either the left test $\mathcal{T}^K_{\Delta^*_{\ell,\mathrm{lf}},\mathrm{l}}$ or the right test $\mathcal{T}^K_{\Delta^*_{\ell,\mathrm{rg}},\mathrm{r}}$, or both, accept wrong—right—hypotheses from the pairs of right and left hypotheses. Since the corresponding intervals ($[u^*_\ell, c^*_\ell]$ for the left side test, $[c^*_\ell, v^*_\ell]$ for the right side one) are δ-good left and right, respectively, the risks of the tests do not exceed δ, and the $p_{A(x)}$-probability of the event $\mathcal{E}_\ell[x]$ is at most 2δ;
- when $u^*_\ell < f(x) \le c^*_\ell$, the event $\mathcal{E}_\ell[x]$ takes place only when the right side test $\mathcal{T}^K_{\Delta^*_{\ell,\mathrm{rg}},\mathrm{r}}$ accepts the wrong—right—hypothesis from the pair; as above, this can happen with $p_{A(x)}$-probability at most δ;
- when $c_\ell < f(x) \le v_\ell$, the event $\mathcal{E}_\ell[x]$ takes place only if the left test $\mathcal{T}^K_{\Delta^*_{\ell,\mathrm{lf}},\mathrm{l}}$ accepts the wrong—left—hypothesis from the pair to which it was applied, which again happens with $p_{A(x)}$-probability $\le \delta$;
- finally, when $f(x) > v_\ell$, the event $\mathcal{E}_\ell[x]$ takes place only when either the left side test $\mathcal{T}^K_{\Delta^*_{\ell,\mathrm{lf}},\mathrm{l}}$ or the right side test $\mathcal{T}^K_{\Delta^*_{\ell,\mathrm{rg}},\mathrm{r}}$, or both, accept wrong—left—hypotheses from the pairs; as above, this can happen with $p_{A(x)}$-probability at most 2δ.

3°. Let $\bar{L} = \bar{L}(\bar{\omega}^K)$ be the last step of the true estimating procedure as run on the observation $\bar{\omega}^K$. We claim that the following holds true:

(!) *Let $\mathcal{E} := \bigcup_{\ell \in \mathcal{L}^*} \mathcal{E}_\ell[x]$, so that the $p_{A(x)}$-probability of the event $\mathcal{E}$, the observations stemming from x, is at most*

$$2\delta L = \epsilon$$

(see (3.17), (3.95)). Assume that $\bar{\omega}^K \notin \mathcal{E}$. Then $\bar{L}(\bar{\omega}^K) \le L^*$, and only two cases are possible:

A. *The true estimating procedure does not terminate due to arriving at a disagreement. In this case $L^* = \bar{L}(\bar{\omega}^K)$ and the trajectories of the ideal and* *the true procedures are identical* (same localizers and essential steps, same output segments, etc.), and, in particular, $f(x) \in \bar{\Delta}$, *or*

B. *The true estimating procedure terminates due to arriving at a disagreement. Then $\Delta_\ell = \Delta_\ell^*$ for $\ell < \bar{L}$, and $f(x) \in \bar{\Delta}$.*

In view of **A** *and* **B** *the $p_{A(x)}$-probability of the event $f(x) \in \bar{\Delta}$ is at least $1 - \epsilon$, as claimed in Proposition 3.4.*

To prove **(!)**, note that the actions at step ℓ in ideal and true procedures depend solely on $\Delta_{\ell-1}$ and on the outcome of rule 2d. Taking into account that $\Delta_0 = \Delta_0^*$, all we need to verify is the following claim:

> **(!!)** *Let $\bar{\omega}^K \notin \mathcal{E}$, and let $\ell \leq L^*$ be such that $\Delta_{\ell-1} = \Delta_{\ell-1}^*$, whence also $u_\ell = u_\ell^*, c_\ell = c_\ell^*$, and $v_\ell = v_\ell^*$. Assume that ℓ is essential (given that $\Delta_{\ell-1} = \Delta_{\ell-1}^*$, this may happen if and only if ℓ is *-essential as well). Then either*
>
> **C.** *At step ℓ the true procedure terminates due to disagreement, in which case $f(x) \in \bar{\Delta}$, or*
>
> **D.** *At step ℓ there was no disagreement, in which case Δ_ℓ as given by* (3.16) *is identical to Δ_ℓ^* as given by the ideal counterpart of* (3.16) *in the case of $\Delta_{\ell-1}^* = \Delta_{\ell-1}$, that is, by the rule*
>
> $$\Delta_\ell^* = \begin{cases} [c_\ell, b_{\ell-1}], & f(x) > c_\ell, \\ [a_{\ell-1}, c_\ell], & f(x) \leq c_\ell. \end{cases} \tag{3.96}$$

To verify **(!!)**, let $\bar{\omega}^K$ and ℓ satisfy the premise of **(!!)**. Note that due to $\Delta_{\ell-1} = \Delta_{\ell-1}^*$ we have $u_\ell = u_\ell^*$, $c_\ell = c_\ell^*$, and $v_\ell = v_\ell^*$, and thus also $\Delta_{\ell,\mathrm{lf}}^* = \Delta_{\ell,\mathrm{lf}}$, $\Delta_{\ell,\mathrm{rg}}^* = \Delta_{\ell,\mathrm{rg}}$. Consider first the case when the true estimation procedure terminates by disagreement at step ℓ, so that $\mathcal{T}^K_{\Delta_{\ell,\mathrm{lf}}^*,\mathrm{l}}(\bar{\omega}^K) \neq \mathcal{T}^K_{\Delta_{\ell,\mathrm{rg}}^*,\mathrm{r}}(\bar{\omega}^K)$. When assuming that $f(x) < u_\ell = u_\ell^*$, the relation $\bar{\omega}^K \notin \mathcal{E}_\ell[x]$ combines with (3.94) to imply that $\mathcal{T}^K_{\Delta_{\ell,\mathrm{rg}}^*,\mathrm{r}}(\bar{\omega}^K) = \mathcal{T}^K_{\Delta_{\ell,\mathrm{lf}}^*,\mathrm{l}}(\bar{\omega}^K) = \mathtt{left}$, which under disagreement is impossible. Assuming $f(x) > v_\ell = v_\ell^*$, the same argument results in $\mathcal{T}^K_{\Delta_{\ell,\mathrm{rg}}^*,\mathrm{r}}(\bar{\omega}^K) = \mathcal{T}^K_{\Delta_{\ell,\mathrm{lf}}^*,\mathrm{l}}(\bar{\omega}^K) = \mathtt{right}$, which again is impossible. We conclude that in the case in question $u_\ell \leq f(x) \leq v_\ell$, i.e., $f(x) \in \bar{\Delta}$, as claimed in **C**. **C** is proved.

Now, suppose that there was a consensus at step ℓ in the true estimating procedure. Because $\bar{\omega}^K \notin \mathcal{E}_\ell[x]$ this can happen in the following four cases:

1. $\mathcal{T}^K_{\Delta_{\ell,\mathrm{rg}}^*,\mathrm{r}}(\bar{\omega}^K) = \mathtt{left}$ and $f(x) \leq u_\ell = u_\ell^*$,
2. $\mathcal{T}^K_{\Delta_{\ell,\mathrm{rg}}^*,\mathrm{r}}(\bar{\omega}^K) = \mathtt{left}$ and $u_\ell < f(x) \leq c_\ell = c_\ell^*$,
3. $\mathcal{T}^K_{\Delta_{\ell,\mathrm{lf}}^*,\mathrm{l}}(\bar{\omega}^K) = \mathtt{right}$ and $c_\ell < f(x) < v_\ell = v_\ell^*$,
4. $\mathcal{T}^K_{\Delta_{\ell,\mathrm{lf}}^*,\mathrm{l}}(\bar{\omega}^K) = \mathtt{right}$ and $v_\ell \leq f(x)$.

Due to consensus at step ℓ, in situations 1 and 2 (3.16) says that $\Delta_\ell = [a_{\ell-1}, c_\ell]$, which combines with (3.96) and $v_\ell = v_\ell^*$ to imply that $\Delta_\ell = \Delta_\ell^*$. Similarly, in situations 3 and 4, due to consensus at step ℓ, (3.16) implies that $\Delta_\ell = [c_\ell, b_{\ell-1}]$, which combines with $u_\ell = u_\ell^*$ and (3.96) to imply that $\Delta_\ell = \Delta_\ell^*$. **D** is proved. □

3.6.3.2 Proof of Proposition 3.4.ii

There is nothing to prove when $\frac{b_0-a_0}{2} \leq \widehat{\rho}$, since in this case the estimate $\frac{a_0+b_0}{2}$ which does not use observations at all is $(\widehat{\rho}, 0)$-reliable. From now on we assume that $b_0 - a_0 > 2\widehat{\rho}$, implying that L is a positive integer.

1°. Observe, first, that if a and b are such that a is lower-feasible, b is upper-feasible, and $b - a > 2\rho$, then for every $i \leq I_{b,\geq}$ and $j \leq I_{a,\leq}$ there exists a test, based on $\overline{K}$ observations, which decides upon the hypotheses H_1, H_2, stating that the observations are drawn from $p_{A(x)}$ with $x \in Z_i^{b,\geq}$ (H_1) or with $x \in Z_j^{a,\leq}$ (H_2) with risk at most ϵ. Indeed, it suffices to consider the test which accepts H_1 and rejects H_2 when $\widehat{f}(\omega^{\overline{K}}) \geq \frac{a+b}{2}$ and accepts H_2 and rejects H_1 otherwise.

2°. With parameters of Bisection chosen according to (3.19), by already proved Proposition 3.4.i, we have

> **E.** *For every $x \in X$, the $p_{A(x)}$-probability of the event $f(x) \in \bar{\Delta}$, $\bar{\Delta}$ being the output segment of our Bisection, is at least $1 - \epsilon$.*

3°. We claim also that

F.1. *Every segment $\Delta = [a, b]$ with $b - a > 2\rho$ and lower-feasible a is δ-good (right),*
F.2. *Every segment $\Delta = [a, b]$ with $b - a > 2\rho$ and upper-feasible b is δ-good (left),*
F.3. *Every $\varkappa$-maximal δ-good (left or right) segment has length at most $2\rho + \varkappa = \widehat{\rho}$. As a result, for every essential step ℓ, the lengths of the segments $\Delta_{\ell,\mathrm{rg}}$ and $\Delta_{\ell,\mathrm{lf}}$ do not exceed $\widehat{\rho}$.*

Let us verify F.1 (verification of F.2 is completely similar, and F.3 is an immediate consequence of the definitions and F.1-2). Let $[a, b]$ satisfy the premise of F.1. It may happen that b is upper-infeasible, whence $\Delta = [a, b]$ is 0-good (right), and we are done. Now let b be upper-feasible. As we have already seen, whenever $i \leq I_{b,\geq}$ and $j \leq I_{a,\leq}$, the hypotheses stating that ω_k are sampled from $p_{A(x)}$ for some $x \in Z_i^{b,\geq}$ and for some $x \in Z_j^{a,\leq}$, respectively, can be decided upon with risk $\leq \epsilon$, implying, as in the proof of Proposition 2.25, that

$$\epsilon_{ij\Delta} \leq [2\sqrt{\epsilon(1-\epsilon)}]^{1/\overline{K}}.$$

Hence, taking into account that the column and the row sizes of $E_{\Delta,\mathrm{r}}$ do not exceed NI,

$$\sigma_{\Delta,\mathrm{r}} \leq NI \max_{i,j} \epsilon_{ij\Delta}^K \leq NI[2\sqrt{\epsilon(1-\epsilon)}]^{K/\overline{K}} \leq \frac{\epsilon}{2L} = \delta$$

(we have used (3.19)), that is, Δ indeed is δ-good (right).

4°. Let us fix $x \in X$ and consider a trajectory of Bisection, the observation being drawn from $p_{A(x)}$. The output $\bar{\Delta}$ of the procedure is given by one of the following options:

1. At some step ℓ of Bisection, the process terminated according to rules in 2b or 2c. In the first case, the segment $[c_\ell, b_{\ell-1}]$ has lower-feasible left endpoint and is not δ-good (right), implying by F.1 that the length of this segment (which is half the length of $\bar{\Delta} = \Delta_{\ell-1}$) is $\leq 2\rho$, so that the length $|\bar{\Delta}|$ of $\bar{\Delta}$ is at most

$4\rho \le 2\widehat{\rho}$. The same conclusion, by a completely similar argument, holds true if the process terminated at step ℓ according to rule 2c.

2. At some step ℓ of Bisection, the process terminated due to disagreement. In this case, by F.3, we have $|\bar{\Delta}| \le 2\widehat{\rho}$.
3. Bisection terminated at step L, and $\bar{\Delta} = \Delta_L$. In this case, termination clauses in rules 2b, 2c, and 2d were never invoked, clearly implying that $|\Delta_s| \le |\Delta_{s-1}|/2$, $1 \le s \le L$, and thus $|\bar{\Delta}| = |\Delta_L| \le 2^{-L}|\Delta_0| \le 2\widehat{\rho}$ (see (3.19)).

Thus, we have $|\bar{\Delta}| \le 2\widehat{\rho}$, implying that whenever the signal $x \in X$ underlying observations and the output segment $\bar{\Delta}$ are such that $f(x) \in \bar{\Delta}$, the error of the Bisection estimate (which is the midpoint of $\bar{\Delta}$) is at most $\widehat{\rho}$. Invoking **E**, we conclude that the Bisection estimate is $(\widehat{\rho}, \epsilon)$-reliable. □

3.6.4 Proof of Proposition 3.14

Let us fix $\epsilon \in (0,1)$. Setting

$$\rho_K = \tfrac{1}{2}\left[\widehat{\Psi}_{+,K}(\bar{h},\bar{H}) + \widehat{\Psi}_{-,K}(\bar{h},\bar{H})\right]$$

and invoking Corollary 3.13, all we need to prove is that in the case of A.1-3 one has

$$\limsup_{K\to\infty}\left[\widehat{\Psi}_{+,K}(\bar{h},\bar{H}) + \widehat{\Psi}_{-,K}(\bar{h},\bar{H})\right] \le 0. \tag{3.97}$$

To this end, note that in our current situation, (3.48) and (3.52) simplify to

$$\begin{array}{l}
\Phi(h,H;Z) = -\frac{1}{2}\ln\mathrm{Det}(I - \Theta_*^{1/2}H\Theta_*^{1/2}) \\
\quad +\frac{1}{2}\mathrm{Tr}\Big(Z\underbrace{\Big(B^T\Big[\Big[\begin{array}{c|c} H & h \\ \hline h^T & \end{array}\Big] + [H,h]^T\left[\Theta_*^{-1} - H\right]^{-1}[H,h]\Big]B\Big)}_{\mathcal{Q}(h,H)}\Big), \\
\widehat{\Psi}_{+,K}(h,H) = \inf\limits_{\alpha}\Big\{\max\limits_{Z\in\mathcal{Z}}\left[\alpha\Phi(h/\alpha,H/\alpha;Z) - \mathrm{Tr}(QZ) + K^{-1}\alpha\ln(2/\epsilon)\right] : \\
\qquad\qquad\qquad\qquad \alpha > 0, -\gamma\alpha\Theta_*^{-1} \preceq H \preceq \gamma\alpha\Theta_*^{-1}\Big\}, \\
\widehat{\Psi}_{-,K}(h,H) = \inf\limits_{\alpha}\Big\{\max\limits_{Z\in\mathcal{Z}}\left[\alpha\Phi(-h/\alpha,-H/\alpha;Z) + \mathrm{Tr}(QZ) + K^{-1}\alpha\ln(2/\epsilon)\right] : \\
\qquad\qquad\qquad\qquad \alpha > 0, -\gamma\alpha\Theta_*^{-1} \preceq H \preceq \gamma\alpha\Theta_*^{-1}\Big\}.
\end{array}$$

Hence

$$\begin{array}{l}
\left[\widehat{\Psi}_{+,K}(\bar{h},\bar{H}) + \widehat{\Psi}_{-,K}(\bar{h},\bar{H})\right] \le \inf\limits_{\alpha}\Big\{\max\limits_{Z_1,Z_2\in\mathcal{Z}}\Big[\alpha\Phi(\bar{h}/\alpha,\bar{H}/\alpha;Z_1) - \mathrm{Tr}(QZ_1) \\
\qquad +\Phi(-\bar{h}/\alpha,-\bar{H}/\alpha;Z_1) + \mathrm{Tr}(QZ_2) + 2K^{-1}\alpha\ln(2/\epsilon)\Big] : \\
\qquad\qquad \alpha > 0, -\gamma\alpha\Theta_*^{-1} \preceq \bar{H} \preceq \gamma\alpha\Theta_*^{-1}\Big\} \\
= \inf\limits_{\alpha}\max\limits_{Z_1,Z_2\in\mathcal{Z}}\Big\{-\frac{1}{2}\alpha\ln\mathrm{Det}\left(I - [\Theta_*^{1/2}\bar{H}\Theta_*^{1/2}]^2/\alpha^2\right) + 2K^{-1}\alpha\ln(2/\epsilon) \\
\quad +\mathrm{Tr}(Q[Z_2 - Z_1]) + \frac{1}{2}\left[\alpha\mathrm{Tr}\left(Z_1\mathcal{Q}(\bar{h}/\alpha,\bar{H}/\alpha)\right) + \alpha\mathrm{Tr}\left(Z_2\mathcal{Q}(-\bar{h}/\alpha,-\bar{H}/\alpha)\right)\right] : \\
\qquad\qquad \alpha > 0, -\gamma\alpha\Theta_*^{-1} \preceq \bar{H} \preceq \gamma\alpha\Theta_*^{-1}\Big\}
\end{array}$$

$$
\begin{array}{r}
= \inf\limits_{\alpha} \max\limits_{Z_1,Z_2\in\mathcal{Z}} \Big\{ -\frac{1}{2}\alpha \ln \mathrm{Det}\left(I - [\Theta_*^{1/2}\bar{H}\Theta_*^{1/2}]^2/\alpha^2\right) + 2K^{-1}\alpha\ln(2/\epsilon) \\
+\frac{1}{2}\mathrm{Tr}\left(Z_1B^T[\bar{H},\bar{h}]^T[\alpha\Theta_*^{-1}-\bar{H}]^{-1}[\bar{H},\bar{h}]B\right) \\
+\frac{1}{2}\mathrm{Tr}\left(Z_2B^T[\bar{H},\bar{h}]^T[\alpha\Theta_*^{-1}+\bar{H}]^{-1}[\bar{H},\bar{h}]B\right) \\
\underbrace{+\,\mathrm{Tr}(Q[Z_2-Z_1]) + \frac{1}{2}\mathrm{Tr}([Z_1-Z_2]B^T\left[\begin{array}{c|c}\bar{H} & \bar{h}\\ \hline \bar{h}^T & \end{array}\right]B)}_{T(Z_1,Z_2)} : \\
\alpha>0, -\gamma\alpha\Theta_*^{-1}\preceq \bar{H}\preceq\gamma\alpha\Theta_*^{-1}\Big\}\,.
\end{array} \tag{3.98}
$$

By (3.57) we have $\frac{1}{2}B^T\left[\begin{array}{c|c}\bar{H} & \bar{h}\\ \hline \bar{h}^T & \end{array}\right]B = B^T[C^TQC+J]B$, where the only nonzero entry, if any, in the $(d+1)\times(d+1)$ matrix J is in the cell $(d+1,d+1)$. By definition of B—see (3.48)—the only nonzero element, if any, in $\bar{J}=B^TJB$ is in the cell $(m+1,m+1)$, and we conclude that

$$
\frac{1}{2}B^T\left[\begin{array}{c|c}\bar{H} & \bar{h}\\ \hline \bar{h}^T & \end{array}\right]B = (CB)^TQ(CB)+\bar{J} = Q+\bar{J}
$$

(recall that $CB = I_{m+1}$). Now, when $Z_1, Z_2\in\mathcal{Z}$, the entries of Z_1, Z_2 in the cell $(m+1,m+1)$ both are equal to 1, whence

$$
\frac{1}{2}\mathrm{Tr}([Z_1-Z_2]B^T\left[\begin{array}{c|c}\bar{H} & \bar{h}\\ \hline \bar{h}^T & \end{array}\right]B) = \mathrm{Tr}([Z_1-Z_2]Q)+\mathrm{Tr}([Z_1-Z_2]\bar{J}) = \mathrm{Tr}([Z_1-Z_2]Q),
$$

implying that the quantity $T(Z_1,Z_2)$ in (3.98) is zero, provided $Z_1,Z_2\in\mathcal{Z}$. Consequently, (3.98) becomes

$$
\begin{array}{r}
\left[\widehat{\Psi}_{+,K}(\bar{h},\bar{H})+\widehat{\Psi}_{-,K}(\bar{h},\bar{H})\right] \leq \inf\limits_{\alpha}\max\limits_{Z_1,Z_2\in\mathcal{Z}}\Big\{ -\frac{1}{2}\alpha\ln\mathrm{Det}\left(I-[\Theta_*^{1/2}\bar{H}\Theta_*^{1/2}]^2/\alpha^2\right) \\
+2K^{-1}\alpha\ln(2/\epsilon) + \frac{1}{2}\mathrm{Tr}\left(Z_1B^T[\bar{H},h][\alpha\Theta_*^{-1}-\bar{H}]^{-1}[\bar{H},\bar{h}]^TB\right) \\
+\frac{1}{2}\mathrm{Tr}\left(Z_2B^T[\bar{H},\bar{h}]^T[\alpha\Theta_*^{-1}+\bar{H}]^{-1}[\bar{H},\bar{h}]B\right) : \alpha>0, -\gamma\alpha\Theta_*^{-1}\preceq\bar{H}\preceq\gamma\alpha\Theta_*^{-1}\Big\}.
\end{array} \tag{3.99}
$$

Now, for an appropriately selected real c independent of K, for α allowed by (3.99), and all $Z_1,Z_2\in\mathcal{Z}$ we have (recall that $\mathcal{Z}$ is bounded)

$$
\begin{array}{l}
\frac{1}{2}\mathrm{Tr}\left(Z_1B^T[\bar{H},\bar{h}]^T[\alpha\Theta_*^{-1}-\bar{H}]^{-1}[\bar{H},\bar{h}]B\right) \\
\quad +\frac{1}{2}\mathrm{Tr}\left(Z_2B^T[\bar{H},\bar{h}]^T[\alpha\Theta_*^{-1}+\bar{H}]^{-1}[\bar{H},\bar{h}]B\right) \leq c/\alpha,
\end{array}
$$

along with

$$
-\frac{1}{2}\alpha\ln\mathrm{Det}\left(I-[\Theta_*^{1/2}\bar{H}\Theta_*^{1/2}]^2/\alpha^2\right)\leq c/\alpha.
$$

Therefore, given $\delta>0$, we can find $\alpha=\alpha_\delta>0$ large enough to ensure that

$$
-\gamma\alpha_\delta\Theta_*^{-1}\preceq\bar{H}\preceq\gamma\alpha_\delta\Theta_*^{-1} \text{ and } 2c/\alpha_\delta\leq\delta,
$$

which combines with (3.99) to imply that

$$
\left[\widehat{\Psi}_{+,K}(\bar{h},\bar{H})+\widehat{\Psi}_{-,K}(\bar{h},\bar{H})\right]\leq\delta+2K^{-1}\alpha_\delta\ln(2/\epsilon),
$$

and (3.97) follows. □

Chapter Four

Signal Recovery by Linear Estimation

OVERVIEW

In this chapter we consider several variations of one of the most basic problems of high-dimensional statistics—*signal recovery*. In its simplest form the problem is as follows: given positive definite $m \times m$ matrix Γ, $m \times n$ matrix A, $\nu \times n$ matrix B, and indirect noisy observation

$$\omega = Ax + \xi \qquad [\xi \sim \mathcal{N}(0, \Gamma)] \tag{4.1}$$

of unknown "signal" x known to belong to a given convex compact subset $\mathcal{X}$ of $\mathbf{R}^n$, we want to recover the vector $Bx \in \mathbf{R}^\nu$ of x. We focus first on the case where the quality of a candidate recovery $\omega \mapsto \widehat{x}(\omega)$ is quantified by its worst-case, over $x \in \mathcal{X}$, expected $\|\cdot\|_2^2$-error, that is, by the risk

$$\mathrm{Risk}[\widehat{x}(\cdot)|\mathcal{X}] = \sup_{x \in \mathcal{X}} \sqrt{\mathbf{E}_{\xi \sim \mathcal{N}(0,\Gamma)} \{\|\widehat{x}(Ax + \xi) - Bx\|_2^2\}}. \tag{4.2}$$

The simplest and the most studied type of recovery is an affine one: $\widehat{x}(\omega) = H^T\omega + h$; assuming $\mathcal{X}$ to be symmetric w.r.t. the origin, we lose nothing when passing from affine estimates to linear ones—those of the form $\widehat{x}_H(\omega) = H^T\omega$. An advantage of linear estimates is that under favorable circumstances (e.g., when $\mathcal{X}$ is an ellipsoid), minimizing risk over linear estimates is an efficiently solvable problem, and there exists a huge body of literature on optimal in terms of their risk linear estimates (see, e.g., [6, 57, 82, 155, 156, 197, 206, 207] and references therein). Moreover, in the case of signal recovery from direct observations in white Gaussian noise (the case of $B = A = I_n$, $\Gamma = \sigma^2 I_n$), there is huge body of results on near-optimality of properly selected linear estimates among *all* possible recovery routines; see, e.g., [79, 88, 106, 124, 198, 230, 239] and references therein. A typical result of this type states that when recovering $x \in \mathcal{X}$ from direct observation $\omega = x + \sigma\xi$, $\xi \sim \mathcal{N}(0, I_m)$, where $\mathcal{X}$ is an ellipsoid of the form

$$\{x \in \mathbf{R}^n : \sum_j j^{2\alpha} x_j^2 \le L^2\},$$

or the box

$$\{x \in \mathbf{R}^n : j^\alpha |x_j| \le L, j \le n\},$$

with fixed $L < \infty$ and $\alpha > 0$, the ratio of the risk of a properly selected linear estimate to the *minimax risk*

$$\mathrm{Risk}_{\mathrm{opt}}[\mathcal{X}] := \inf_{\widehat{x}(\cdot)} \mathrm{Risk}[\widehat{x}|\mathcal{X}] \tag{4.3}$$

(the infimum is taken over all estimates, not necessarily linear) remains bounded, or even tends to 1, as $\sigma \to +0$, and this happens *uniformly in* n, α and L being fixed.

Similar "near-optimality" results are known for the "diagonal" case, where $\mathcal{X}$ is an ellipsoid/box and A, B, Γ are diagonal matrices. To the best of our knowledge, the only "general" (that is, not imposing severe restrictions on how the geometries of $\mathcal{X}$, A, B, Γ are linked to each other) result on optimality of linear estimates is due to D. Donoho, who proved [64], that *when recovering a linear form* (i.e., in the case of one-dimensional Bx), the best risk over all linear estimates is within the factor 1.2 of the minimax risk.

The primary goal of this chapter is to establish rather general results on near-optimality of properly built linear estimates as compared to all possible estimates. Results of this type are bound to impose some restrictions on $\mathcal{X}$, since there are cases (e.g., the case of a high-dimensional $\|\cdot\|_1$-ball $\mathcal{X}$) where linear estimates are *by far* nonoptimal. Our restrictions on $\mathcal{X}$ reduce to the existence of a special type representation of $\mathcal{X}$ and are satisfied, e.g., when $\mathcal{X}$ is the intersection of $K < \infty$ ellipsoids/elliptic cylinders,

$$\mathcal{X} = \{x \in \mathbf{R}^n : x^T R_k x \leq 1, 1 \leq k \leq K\} \qquad [R_k \succeq 0, \textstyle\sum_k R_k \succ 0] \tag{4.4}$$

in particular, $\mathcal{X}$ can be a symmetric w.r.t. the origin compact polytope given by $2K$ linear inequalities $-1 \leq r_k^T x \leq 1$, $1 \leq k \leq K$, or, equivalently, $\mathcal{X} = \{x : x^T \underbrace{(r_k r_k^T)}_{R_k} x, 1 \leq k \leq K\}$. Another instructive example is a set of the form $\mathcal{X} = \{x : \|Sx\|_p \leq L\}$, where $p \geq 2$ and S is a matrix with trivial kernel. It should be stressed that while imposing some restrictions on $\mathcal{X}$, *we require nothing from A, B, and Γ, aside from positive definiteness of the latter matrix.* Our main result (Proposition 4.5) states, in particular, that with $\mathcal{X}$ given by (4.4) and with arbitrary A and B, the risk of properly selected linear estimate $\widehat{x}_{H_*}$ with both H_* and the risk efficiently computable, satisfies the bound

$$\mathrm{Risk}[\widehat{x}_{H_*}|\mathcal{X}] \leq O(1)\sqrt{\ln(K+1)}\mathrm{Risk}_{\mathrm{opt}}[\mathcal{X}], \tag{$*$}$$

where $\mathrm{Risk}_{\mathrm{opt}}[\mathcal{X}]$ is the minimax risk, and $O(1)$ is an absolute constant. Note that the outlined result is an "operational" one—the risk of *provably nearly optimal* estimate and the estimate itself are given by efficient computation. This is in sharp contrast with traditional results of nonparametric statistics, where near-optimal estimates and their risks are given in a "closed analytical form," at the price of severe restrictions on the structure of the "data" $\mathcal{X}$, A, B, Γ. This being said, it should be stressed that one of the crucial components in our construction is quite classical—this is the idea, going back to M.S. Pinsker [198], of bounding from below the minimax risk via the Bayesian risk associated with a properly selected Gaussian prior.[1]

The main body of the chapter originates from [138, 137] and is organized as follows.

- Section 4.1 presents basic results on Conic Programming and Conic Duality—the

[1] [88, 198] address the problem of $\|\cdot\|_2$-recovery of a signal x from direct observations ($A = B = I$) in the case when $\mathcal{X}$ is a high-dimensional ellipsoid with "regularly decreasing half-axes," like $\mathcal{X} = \{x \in \mathbf{R}^n : \sum_j j^{2\alpha} x_j^2 \leq L^2\}$ with $\alpha > 0$. In this case Pinsker's construction shows that as $\sigma \to +0$, the risk of a properly built linear estimate is, uniformly in n, $(1 + o(1))$ times the minimax risk. This is much stronger than $(*)$, and it seems to be unlikely that a similarly strong result holds true in the general case underlying $(*)$.

principal optimization tools utilized in all subsequent constructions and proofs.

- Section 4.2 contains problem formulation (Section 4.2.1), construction of the linear estimate we deal with (Section 4.2.2) and the central result on near-optimality of this estimate (Section 4.2.2.2). We discuss also the "expressive abilities" of the family of sets (we call them *ellitopes*) to which our main result applies.
- In Section 4.3 we extend the results of the previous section from ellitopes to their "matrix analogs"—*spectratopes* in the role of signal sets, passing simultaneously from the norm $\|\cdot\|_2$ in which the recovery error is measured to arbitrary *spectratopic* norms, those for which the unit ball of the conjugate norm is a spectratope. In addition, we allow for observation noise to have nonzero mean and to be non-Gaussian.
- Section 4.4 adjusts our preceding results on linear estimation to the case where the signals to be recovered possess stochastic components.
- Finally, Section 4.5 deals with "uncertain-but-bounded" observation noise, that is, noise selected "by nature," perhaps in an adversarial fashion, from a given bounded set.

4.1 PRELIMINARIES: EXECUTIVE SUMMARY ON CONIC PROGRAMMING

4.1.1 Cones

A *cone* in Euclidean space E is a nonempty set K which is closed w.r.t. taking *conic* combinations of its elements, that is, linear combinations with nonnegative coefficients. Equivalently: $K \subset E$ is a cone if K is nonempty, and

- $x, y \in K \Rightarrow x + y \in K$;
- $x \in K, \lambda \geq 0 \Rightarrow \lambda x \in K$.

It is immediately seen that a cone is a convex set. We call a cone K *regular* if it is closed, *pointed* (that is, does not contain lines passing through the origin, or, equivalently, $K \bigcap [-K] = \{0\}$) and possesses a nonempty interior.

Given a cone $K \subset E$, we can associate with it its *dual cone* K^* defined as

$$K^* = \{y \in E : \langle y, x\rangle \geq 0 \,\forall x \in K\} \qquad [\langle\cdot,\cdot\rangle \text{ is inner product on } E].$$

It is immediately seen that K^* is a closed cone, and $K \subset (K^*)^*$. It is well known that

- if K is a closed cone, it holds $K = (K^*)^*$;
- K is a regular cone if and only if K^* is so.

Examples of regular cones "useful in applications" are as follows:

1. *Nonnegative orthants* $\mathbf{R}^d_+ = \{x \in \mathbf{R}^d : x \geq 0\}$;
2. *Lorentz cones* $\mathbf{L}^d_+ = \{x \in \mathbf{R}^d : x_d \geq \sqrt{\sum_{i=1}^{d-1} x_i^2}\}$;
3. *Semidefinite cones* $\mathbf{S}^d_+$ comprised of positive semidefinite symmetric $d \times d$ matrices. Semidefinite cone $\mathbf{S}^d_+$ lives in the space $\mathbf{S}^d$ of symmetric matrices equipped

with the Frobenius inner product

$$\langle A, B\rangle = \mathrm{Tr}(AB^T) = \mathrm{Tr}(AB) = \sum_{i,j=1}^{d} A_{ij}B_{ij}, \quad A, B \in \mathbf{S}^d.$$

All cones listed so far are self-dual.

4. Let $\|\cdot\|$ be a norm on $\mathbf{R}^n$. The set $\{[x;t] \in \mathbf{R}^n \times \mathbf{R} : t \geq \|x\|\}$ is a regular cone, and the dual cone is $\{[y;\tau] : \|y\|_* \leq \tau\}$, where

$$\|y\|_* = \max_x \{x^T y : \|x\| \leq 1\}$$

is the norm on $\mathbf{R}^n$ conjugate to $\|\cdot\|$.

An additional example of a regular cone useful for the sequel is the *conic hull* of a convex compact set defined as follows. Let $\mathcal{T}$ be a convex compact set with a nonempty interior in Euclidean space E. We can associate with $\mathcal{T}$ its *closed conic hull*

$$\mathbf{T} = \mathrm{cl}\underbrace{\{[t;\tau] \in E^+ = E \times \mathbf{R} : \tau > 0, t/\tau \in \mathcal{T}\}}_{K^o(\mathcal{T})}.$$

It is immediately seen that $\mathbf{T}$ is a regular cone, and that to get this cone, one should add to the convex set $K^o(\mathcal{T})$ the origin of E^+. It is also clear that one can "see $\mathcal{T}$ in $\mathbf{T}$:"—$\mathcal{T}$ is nothing but the cross-section of the cone $\mathbf{T}$ by the hyperplane $\tau = 1$ in $E^+ = \{[t;\tau]\}$:

$$\mathcal{T} = \{t \in E : [t;1] \in \mathbf{T}\}.$$

It is easily seen that the cone $\mathbf{T}_*$ dual to $\mathbf{T}$ is given by

$$\mathbf{T}_* = \{[g;s] \in \mathbf{E}^+ : s \geq \phi_{\mathcal{T}}(-g)\},$$

where

$$\phi_{\mathcal{T}}(g) = \max_{t\in\mathcal{T}} \langle g, t\rangle$$

is the support function of $\mathcal{T}$.

4.1.2 Conic problems and their duals

Given regular cones $K_i \subset E_i$, $1 \leq i \leq m$, consider an optimization problem of the form

$$\mathrm{Opt}(P) = \min\left\{\langle c, x\rangle : \begin{array}{l} A_i x - b_i \in K_i,\ i = 1, ..., m \\ Rx = r \end{array}\right\}, \tag{P}$$

where $x \mapsto A_i x - b_i$ are affine mappings acting from some Euclidean space E to the spaces E_i where the cones K_i live. A problem in this form is called a *conic problem on the cones* $K_1, ..., K_m$; the constraints $A_i x - b_i \in K_i$ on x are called *conic constraints.* We call a conic problem (P) *strictly feasible* if it admits a *strictly feasible* solution $\bar{x}$, meaning that $\bar{x}$ satisfies the equality constraints and satisfies *strictly* the conic constraints, i.e., $A_i\bar{x} - b_i \in \mathrm{int}\, K_i$.

One can associate with conic problem (P) its *dual*, which also is a conic problem. The origin of the dual problem is the desire to obtain lower bounds on the optimal value $\mathrm{Opt}(P)$ of the *primal* problem (P) in a systematic way—*by linear aggregation* *of constraints.* Linear aggregation of constraints works as follows: let us equip every conic constraint $A_i x - b_i \in K_i$ with aggregation weight, called *Lagrange multiplier*, y_i restricted to reside in the cone K_i^* dual to K_i. Similarly, we equip the system $Rx = r$ of equality constraints in (P) with Lagrange multiplier z—a vector of the same dimension as r. Now let x be a feasible solution to the conic problem, and let $y_i \in K_i^*$, $i \le m$, and z be Lagrange multipliers. By the definition of the dual cone and due to $A_i x - b_i \in K_i$, $y_i \in K_i^*$ we have

$$\langle y_i, A_i x\rangle \ge \langle y_i, b_i\rangle, 1 \le i \le m$$

and of course

$$z^T Rx \ge r^T z.$$

Summing up all resulting inequalities, we arrive at the scalar linear inequality

$$\left\langle R^* z + \sum\nolimits_i A_i^* y_i, x \right\rangle \ge r^T z + \sum\nolimits_i \langle b_i, y_i\rangle \tag{!}$$

where A_i^* are the conjugates to A_i: $\langle y, A_i x\rangle_{E_i} \equiv \langle A_i^* y, x\rangle_E$, and R^* is the conjugate of R. By its origin, (!) is a consequence of the system of constraints in (P) and as such is satisfied everywhere on the feasible domain of the problem. If we are lucky to get the objective of (P) as the linear function of x in the left hand side of (!), that is, if

$$R^* z + \sum_i A_i^* y_i = c,$$

(!) imposes a lower bound on the objective of the primal conic problem (P) everywhere on the feasible domain of the primal problem, and the *conic dual* of (P) is the problem

$$\mathrm{Opt}(D) = \max_{y_i, z} \left\{ r^T z + \sum_i \langle b_i, y_i\rangle : \begin{array}{l} y_i \in K_i^*, 1 \le i \le m \\ R^* z + \sum_{i=1}^m A_i^* y_i = c \end{array} \right\} \tag{D}$$

of maximizing this lower bound on $\mathrm{Opt}(P)$.

The relations between the primal and the dual conic problems are the subject of the standard *Conic Duality Theorem* as follows:

Theorem 4.1. *[Conic Duality Theorem] Consider conic problem (P) (where all K_i are regular cones) along with its dual problem (D). Then*

1. *Duality is symmetric: the dual problem (D) is conic, and the conic dual of (D) is (equivalent to) (P);*
2. *Weak duality: It always holds* $\mathrm{Opt}(D) \le \mathrm{Opt}(P)$
3. *Strong duality: If one of the problems (P), (D) is strictly feasible and bounded,*[2] *then the other problem in the pair is solvable, and the optimal values of the problems are equal to each other. In particular, if both (P) and (D) are strictly feasible, then both problems are solvable with equal optimal values.*

Remark 4.2. While the Conic Duality Theorem in the form just presented meets all our subsequent needs, it makes sense to note that in fact the Strong Duality part of

[2]For a minimization problem, boundedness means that the objective is bounded from below on the feasible set, for a maximization problem, that it is bounded from above on the feasible set.

the theorem can be strengthened by replacing strict feasibility with "essential strict feasibility" defined as follows: a conic problem in the form of (P) (or, which is the same, form of (D)) is called essentially strictly feasible if it admits a feasible solution $\bar{x}$ *which satisfies strictly the* non-polyhedral *conic constraints, that is,* $A_i\bar{x} - b_i \in \operatorname{int} K_i$ *for all* i *for which the cone* K_i *is* not *polyhedral—is* not *given by a finite list of homogeneous linear inequality constraints.*

The proof of the Conic Duality Theorem can be found in numerous sources, e.g., in [187, Section 7.1.3].

4.1.3 Schur Complement Lemma

The following simple fact is extremely useful:

Lemma 4.3. *[Schur Complement Lemma] A symmetric block matrix*

$$A = \left[\begin{array}{c|c} P & Q^T \\ \hline Q & R \end{array}\right]$$

with $R \succ 0$ *is positive (semi)definite if and only if the matrix* $P - Q^TR^{-1}Q$ *is so.*

Proof. With u, v of the same sizes as P, R, we have

$$\min_v \, [u;v]^T A\,[u;v] = u^T[P - Q^TR^{-1}Q]u$$

(direct computation utilizing the fact that $R \succ 0$). It follows that the quadratic form associated with A is nonnegative everywhere if and only if the quadratic form with the matrix $[P - Q^TR^{-1}Q]$ is nonnegative everywhere (since the latter quadratic form is obtained from the former one by partial minimization). □

4.2 NEAR-OPTIMAL LINEAR ESTIMATION FROM GAUSSIAN OBSERVATIONS

4.2.1 Situation and goal

Given an $m \times n$ matrix A, a $\nu \times n$ matrix B, and an $m \times m$ matrix $\Gamma \succ 0$, consider the problem of estimating the linear image Bx of an unknown signal x known to belong to a given set $\mathcal{X} \subset \mathbf{R}^n$ via noisy observation

$$\omega = Ax + \xi, \;\; \xi \sim \mathcal{N}(0, \Gamma), \tag{4.5}$$

where ξ is the observation noise. A candidate estimate in this case is a (Borel) function $\widehat{x}(\cdot) : \mathbf{R}^m \to \mathbf{R}^\nu$, and the performance of such an estimate in what follows will be quantified by the *Euclidean risk* $\operatorname{Risk}[\widehat{x}|\mathcal{X}]$ defined by (4.2).

4.2.1.1 Ellitopes

From now on we assume that $\mathcal{X} \subset \mathbf{R}^n$ is a set given by

$$\begin{array}{r}\mathcal{X} = \big\{x \in \mathbf{R}^n : \exists (y \in \mathbf{R}^{\bar{n}}, t \in \mathcal{T}) : \\ x = Py, \; y^TR_ky \le t_k, \; 1 \le k \le K\big\},\end{array} \tag{4.6}$$

where

- P is an $n \times \bar{n}$ matrix,
- $R_k \succeq 0$ are $\bar{n} \times \bar{n}$ matrices with $\sum_k R_k \succ 0$,
- $\mathcal{T}$ is a nonempty computationally tractable convex compact subset of $\mathbf{R}^K_+$ intersecting the interior of $\mathbf{R}^K_+$ and such that $\mathcal{T}$ is monotone, meaning that the relations $0 \leq \tau \leq t$ and $t \in \mathcal{T}$ imply that $\tau \in \mathcal{T}$.[3] Note that under our assumptions $\operatorname{int} \mathcal{T} \neq \emptyset$.

In the sequel, we refer to a set of the form (4.6) with data $[P, \{R_k, 1 \leq k \leq K\}, \mathcal{T}]$ satisfying the assumptions just formulated as an *ellitope*, and to (4.6) as an *ellitopic representation* of $\mathcal{X}$. Here are instructive examples of ellitopes (in all these examples, P is the identity mapping; in the sequel, we call ellitopes of this type *basic*):

- when $K = 1$, $\mathcal{T} = [0, 1]$, and $R_1 \succ 0$, $\mathcal{X}$ is the ellipsoid $\{x : x^T R_1 x \leq 1\}$;
- when $K \geq 1$, $\mathcal{T} = \{t \in \mathbf{R}^K : 0 \leq t_k \leq 1, k \leq K\}$, and $\mathcal{X}$ is the intersection of

$$\bigcap_{1 \leq k \leq K} \{x : x^T R_k x \leq 1\}$$

  ellipsoids/elliptic cylinders centered at the origin. In particular, when U is a $K \times n$ matrix of rank n with rows u_k^T, $1 \leq k \leq K$, and $R_k = u_k u_k^T$, $\mathcal{X}$ is the symmetric w.r.t. the origin polytope $\{x : \|Ux\|_\infty \leq 1\}$;
- when U, u_k and R_k are as in the latter example and $\mathcal{T} = \{t \in \mathbf{R}^K_+ : \sum_k t_k^{p/2} \leq 1\}$ for some $p \geq 2$, we get $\mathcal{X} = \{x : \|Ux\|_p \leq 1\}$.

It should be added that the family of ellitope-representable sets is quite rich: this family admits a "calculus," so that more ellitopes can be constructed by taking intersections, direct products, linear images (direct and inverse) or arithmetic sums of ellitopes given by the above examples. In fact, the property of being an ellitope is preserved by nearly all basic operations with sets preserving convexity and symmetry w.r.t. the origin (a regrettable exception is taking the convex hull of a finite union); see Section 4.6;.

As another example of an ellitope instructive in the context of nonparametric statistics, consider the situation where our signals x are discretizations of functions of continuous argument running through a compact d-dimensional domain D, and the functions f we are interested in are those satisfying a Sobolev-type smoothness constraint – an upper bound on the $L_p(D)$-norm of $\mathcal{L}f$, where $\mathcal{L}$ is a linear differential operator with constant coefficients. After discretization, this restriction can be modeled as $\|Lx\|_p \leq 1$, with properly selected matrix L. As we already know from the above example, when $p \geq 2$, the set $\mathcal{X} = \{x : \|Lx\|_p \leq 1\}$ is an ellitope, and as such is captured by our machinery. Note also that by the outlined calculus, imposing on the functions f in question *several Sobolev-type smoothness constraints* with parameters $p \geq 2$, still results in a set of signals which is an ellitope.

[3]The latter relation is "for free"—given a nonempty convex compact set $\mathcal{T} \subset \mathbf{R}^K_+$, the right-hand side of (4.6) remains intact when passing from $\mathcal{T}$ to its "monotone hull" $\{\tau \in \mathbf{R}^K_+ : \exists t \in \mathcal{T} : \tau \leq t\}$ which already is a monotone convex compact set.

4.2.1.2 Estimates and their risks

In the outlined situation, a candidate estimate is a Borel function $\widehat{x}(\cdot) : \mathbf{R}^m \to \mathbf{R}^\nu$; given observation (4.5), we recover $w = Bx$ as $\widehat{x}(\omega)$. In the sequel, we quantify the quality of an estimate by its worst-case, over $x \in \mathcal{X}$, expected $\|\cdot\|_2^2$ recovery error

$$\mathrm{Risk}[\widehat{x}|\mathcal{X}] = \sup_{x\in\mathcal{X}} \left[\mathbf{E}_{\xi\sim\mathcal{N}(0,\Gamma)}\left\{\|\widehat{x}(Ax+\xi) - Bx\|_2^2\right\}\right]^{1/2},$$

and define the optimal, or the *minimax*, risk as

$$\mathrm{Risk}_{\mathrm{opt}}[\mathcal{X}] = \inf_{\widehat{x}(\cdot)} \mathrm{Risk}[\widehat{x}|\mathcal{X}], \tag{4.7}$$

where inf is taken over all Borel candidate estimates.

4.2.1.3 Main goal

The main goal of what follows is to demonstrate that an estimate *linear in* ω

$$\widehat{x}_H(\omega) = H^T\omega \tag{4.8}$$

with a properly selected efficiently computable matrix H is near-optimal in terms of its risk.

Our first observation is that when $\mathcal{X}$ is the ellitope (4.6), replacing matrices A and B with AP and BP, respectively, we pass from the initial estimation problem of interest to the *transformed problem*, where the signal set is

$$\bar{X} = \{y \in \mathbf{R}^{\bar{n}} : \exists t \in \mathcal{T} :\ y^T R_k y \le t_k,\ 1 \le k \le K\},$$

and we want to recover $[BP]y$, $y \in \bar{X}$, via observation

$$\omega = [AP]y + \xi.$$

It is obvious that the considered families of estimates (the family of all linear estimates and the family of all estimates), like the risks of the estimates, remain intact under this transformation; in particular,

$$\mathrm{Risk}[\widehat{x}|\mathcal{X}] = \sup_{y\in\bar{X}} \left[\mathbf{E}_\xi\{\|\widehat{x}([AP]\,y+\xi) - [BP]\,y\|_2^2\}\right]^{1/2}.$$

Therefore, to save notation, from now on, unless explicitly stated otherwise, we assume that matrix P is identity, so that $\mathcal{X}$ is the basic ellitope

$$\mathcal{X} = \left\{x \in \mathbf{R}^n :\ \exists t \in \mathcal{T},\ x^T R_k x \le t_k,\ 1 \le k \le K\right\}. \tag{4.9}$$

We assume in the sequel that $B \neq 0$, since otherwise one has $Bx = 0$ for all $x \in \mathcal{X}$, and the estimation problem is trivial.

4.2.2 Building a linear estimate

We start with building a "presumably good" linear estimate. Restricting ourselves to linear estimates (4.8), we may be interested in the estimate with the smallest risk, that is, the estimate associated with a $\nu \times m$ matrix H which is an optimal solution to the optimization problem

$$\min_H \left\{R(H) := \mathrm{Risk}^2[\widehat{x}_H|\mathcal{X}]\right\}.$$

We have

$$\begin{array}{rcl} R(H) & = & \max_{x\in\mathcal{X}} \mathbf{E}_\xi\{\|H^T\omega - Bx\|_2^2\} = \mathbf{E}_\xi\{\|H^T\xi\|_2^2\} + \max_{x\in\mathcal{X}} \|H^TAx - Bx\|_2^2 \\ & = & \mathrm{Tr}(H^T\Gamma H) + \max_{x\in\mathcal{X}} x^T(H^TA-B)^T(H^TA-B)x. \end{array}$$

This function, while convex, can be hard to compute. For this reason, we use a linear estimate yielded by minimizing an *efficiently computable convex upper bound* on $R(H)$ which is built as follows. Let $\phi_\mathcal{T}$ be the support function of $\mathcal{T}$:

$$\phi_\mathcal{T}(\lambda) = \max_{t\in\mathcal{T}} \lambda^T t : \mathbf{R}^K \to \mathbf{R}.$$

Observe that whenever $\lambda \in \mathbf{R}^K_+$ and H are such that

$$[B - H^TA]^T[B - H^TA] \preceq \sum_k \lambda_k R_k, \tag{4.10}$$

for $x \in \mathcal{X}$ it holds

$$\|Bx - H^TAx\|_2^2 \le \phi_\mathcal{T}(\lambda). \tag{4.11}$$

Indeed, in the case of (4.10) and with $x \in \mathcal{X}$, there exists $t \in \mathcal{T}$ such that $x^TR_kx \le t_k$ for all t, and consequently vector $\bar{t}$ with the entries $\bar{t}_k = x^TR_kx$ also belongs to $\mathcal{T}$, whence

$$\|Bx - H^TAx\|_2^2 = x^T[B-H^TA]^T[B-H^TA]x \le \sum_k \lambda_k x^TR_kx = \lambda^T\bar{t} \le \phi_\mathcal{T}(\lambda),$$

which combines with (4.9) to imply (4.11).

From (4.11) it follows that if H and $\lambda \ge 0$ are linked by (4.10), then

$$\begin{array}{rcl} \mathrm{Risk}^2[\widehat{x}_H|\mathcal{X}] & = & \max_{x\in\mathcal{X}} \mathbf{E}\left\{\|Bx - H^T(Ax+\xi)\|_2^2\right\} \\ & = & \mathrm{Tr}(H^T\Gamma H) + \max_{x\in\mathcal{X}} \|[B-H^TA]x\|_2^2 \\ & \le & \mathrm{Tr}(H^T\Gamma H) + \phi_\mathcal{T}(\lambda). \end{array}$$

We see that the efficiently computable convex function

$$\widehat{R}(H) = \inf_\lambda \left\{ \mathrm{Tr}(H^T\Gamma H) + \phi_\mathcal{T}(\lambda) : (B-H^TA)^T(B-H^TA) \preceq \sum_k \lambda_k R_k, \lambda \ge 0 \right\}$$

(which clearly is well defined due to compactness of $\mathcal{T}$ combined with $\sum_k R_k \succ 0$) is an upper bound on $R(H)$.[4] Note that by Schur Complement Lemma the matrix

[4]It is well known that when $K = 1$ (i.e., $\mathcal{X}$ is an ellipsoid), the above bounding scheme is exact: $R(\cdot) \equiv \widehat{R}(\cdot)$. For more complicated $\mathcal{X}$'s, $\widehat{R}(\cdot)$ could be larger than $R(\cdot)$, although the ratio $\widehat{R}(\cdot)/R(\cdot)$ is bounded by $O(\log(K))$; see Section 4.2.3.

inequality $(B-H^TA)^T(B-H^TA) \preceq \sum_k \lambda_k R_k$ is equivalent to the matrix inequality

$$\left[\begin{array}{c|c}\sum_k \lambda_k R_k & B^T - A^T H \\ \hline B - H^T A & I_\nu\end{array}\right] \succeq 0$$

linear in H, λ. We have arrived at the following result:

Proposition 4.4. *In the situation of this section, the risk of the "presumably good" linear estimate $\widehat{x}_{H_*}(\omega) = H_*^T\omega$ yielded by an optimal solution (H_*, λ_*) to the (clearly solvable) convex optimization problem*

$$\begin{array}{rcl}\mathrm{Opt} &=& \min\limits_{H,\lambda}\left\{\mathrm{Tr}(H^T\Gamma H) + \phi_{\mathcal{T}}(\lambda) : (B-H^TA)^T(B-H^TA) \preceq \sum\limits_k \lambda_k R_k, \lambda \geq 0\right\} \\ &=& \min\limits_{H,\lambda}\left\{\mathrm{Tr}(H^T\Gamma H) + \phi_{\mathcal{T}}(\lambda) : \left[\begin{array}{c|c}\sum_k \lambda_k R_k & B^T - A^T H \\ \hline B - H^T A & I_\nu\end{array}\right] \succeq 0, \lambda \geq 0\right\}\end{array} \quad (4.12)$$

is upper-bounded by $\sqrt{\mathrm{Opt}}$.

4.2.2.1 Illustration: Recovering temperature distribution

Situation: A square steel plate was somewhat heated at time 0 and left to cool, the temperature along the perimeter of the plate being all the time kept zero. At time t_1, we measure the temperatures at m points of the plate, and want to recover the distribution of the temperature along the plate at a given time t_0, $0 < t_0 < t_1$.

Physics, after suitable discretization of spatial variables, offers the following model of the situation. We represent the distribution of temperature at time t as $(2N-1) \times (2N-1)$ matrix $U(t) = [u_{ij}(t)]_{i,j=1}^{2N-1}$, where $u_{ij}(t)$ is the temperature, at time t, at the point

$$P_{ij} = (p_i, p_j),\ p_k = k/N - 1, \quad 1 \leq i, j \leq 2N-1$$

of the plate (in our model, this plate occupies the square $S = \{(p,q) : |p| \leq 1, |q| \leq 1\}$). Here positive integer N is responsible for spatial discretization.

For $1 \leq k \leq 2N-1$, let us specify functions $\phi_k(s)$ on the segment $-1 \leq s \leq 1$ as follows:

$$\phi_{2\ell-1}(s) = c_{2\ell-1}\cos(\omega_{2\ell-1}s),\ \phi_{2\ell}(s) = c_{2\ell}\sin(\omega_{2\ell}s),\ \omega_{2\ell-1} = (\ell - 1/2)\pi,\ \omega_{2\ell} = \ell\pi,$$

where c_k are readily given by the normalization condition $\sum_{i=1}^{2N-1}\phi_k^2(p_i) = 1$; note that $\phi_k(\pm 1) = 0$. It is immediately seen that the matrices

$$\Phi^{k\ell} = [\phi_k(p_i)\phi_\ell(p_j)]_{i,j=1}^{2N-1},\ 1 \leq k, \ell \leq 2N-1$$

form an orthonormal basis in the space of $(2N-1) \times (2N-1)$ matrices, so that we can write

$$U(t) = \sum_{k,\ell \leq 2N-1} x_{k\ell}(t)\Phi^{k\ell}.$$

The advantage of representing temperature fields in the basis $\{\Phi^{k\ell}\}_{k,\ell \leq 2N-1}$ stems from the fact that in this basis the heat equation governing evolution of the temperature distribution in time becomes extremely simple, just

$$\frac{d}{dt}x_{k\ell}(t) = -(\omega_k^2 + \omega_\ell^2)x_{k\ell}(t) \Rightarrow x_{k\ell}(t) = \exp\{-(\omega_k^2 + \omega_\ell^2)t\}x_{k\ell}.$$ [5]

Now we can convert the situation into the one considered in our general estimation scheme, namely, as follows:

- We select some discretization parameter N and treat $x = \{x_{k\ell}(0), 1 \leq k, \ell \leq 2N-1\}$ as the signal underlying our observations.
  In every potential application, we can safely upper-bound the magnitudes of the initial temperatures and thus the magnitude of x, say, by a constraint of the form

$$\sum_{k,\ell} x_{k\ell}^2(0) \leq R^2$$

  with properly selected R, which allows us to specify the domain $\mathcal{X}$ of the signal as the Euclidean ball:

$$\mathcal{X} = \{x \in \mathbf{R}^{(2N-1)\times(2N-1)} : \|x\|_2^2 \leq R^2\}. \tag{4.13}$$

- Let the measurements of the temperature at time t_1 be taken along the points $P_{i(\nu),j(\nu)}$, $1 \leq \nu \leq m$,[6] and let them be affected by a $\mathcal{N}(0, \sigma^2 I_m)$-noise, so that our observation is

$$\omega = A(x) + \xi,\ \xi \sim \mathcal{N}(0, \sigma^2 I_m).$$

  Here $x \mapsto A(x)$ is the linear mapping from $\mathbf{R}^{(2N-1)\times(2N-1)}$ into $\mathbf{R}^m$ given by

$$[A(x)]_\nu = \sum_{k,\ell=1}^{2N-1} \mathrm{e}^{-(\omega_k^2+\omega_\ell^2)t_1} \phi_k(p_{i(\nu)})\phi_\ell(p_{j(\nu)})x_{k\ell}(0). \tag{4.14}$$

- We want to recover the temperatures at time t_0 taken along some grid, say, the square $(2K-1) \times (2K-1)$ grid $\{Q_{ij} = (r_i, r_j), 1 \leq i, j \leq 2K-1\}$, where $r_i = i/K - 1$, $1 \leq i \leq 2K-1$. In other words, we want to recover $B(x)$, where the linear mapping $x \mapsto B(x)$ from $\mathbf{R}^{(2N-1)\times(2N-1)}$ into $\mathbf{R}^{(2K-1)\times(2K-1)}$ is given by

$$[B(x)]_{ij} = \sum_{k,\ell=1}^{2N-1} \mathrm{e}^{-(\omega_k^2+\omega_\ell^2)t_0} \phi_k(r_i)\phi_\ell(r_j)x_{k\ell}(0).$$

---

[5] The explanation is simple: the functions $\phi_{k\ell}(p,q) = \phi_k(p)\phi_\ell(q)$, $k, \ell = 1, 2, \ldots$, form an orthogonal basis in $L_2(S)$ and vanish on the boundary of S, and the heat equation

$$\frac{\partial}{\partial t}u(t;p,q) = \left[\frac{\partial^2}{\partial p^2} + \frac{\partial^2}{\partial q^2}\right] u(t;p,q)$$

governing evolution of the temperature field $u(t;p,q)$, $(p,q) \in S$, with time t, in terms of the coefficients $x_{k\ell}(t)$ of the temperature field in the orthogonal basis $\{\phi_{k\ell}(p,q)\}_{k,\ell}$ becomes

$$\frac{d}{dt}x_{k\ell}(t) = -(\omega_k^2 + \omega_\ell^2)x_{k\ell}(t).$$

In our discretization, we truncate the expansion of $u(t;p,q)$, keeping only the terms with $k, \ell \leq 2N-1$, and restrict the spatial variables to reside in the grid $\{P_{ij}, 1 \leq i, j \leq 2N-1\}$.

[6] The construction can be easily extended to allow for measurement points outside of the grid $\{P_{ij}\}$.

Ill-posedness. Our problem is a typical example of an *ill-posed inverse problem*, where one wants to recover a past state of a dynamical system converging exponentially fast to equilibrium and thus "forgetting rapidly" its past. More specifically, in our situation ill-posedness stems from the fact that, as is clearly seen from (4.14), contributions of "high frequency" (i.e., with large $\omega_k^2 + \omega_\ell^2$) components $x_{k\ell}(0)$ of the signal to $A(x)$ decrease exponentially fast, with high decay rate, as t_1 grows. As a result, high frequency components $x_{k\ell}(0)$ are impossible to recover from noisy observations of $A(x)$, unless the corresponding time instant t_1 is very small. As a kind of compensation, contributions of high frequency components $x_{k\ell}(0)$ to $B(x)$ are also very small, provided that t_0 is not too small, implying that there is no necessity to recover well high frequency components, unless they are huge. Our linear estimate, roughly speaking, seeks for the best trade-off between these two opposite phenomena, utilizing (4.13) as the source of upper bounds on the magnitudes of high frequency components of the signal.

Numerical results. In the experiment to be reported, we used $N = 32$, $m = 100$, $K = 6$, $t_0 = 0.01$, $t_1 = 0.03$ (i.e., temperature is measured at time 0.03 at 100 points selected at random on a 63×63 square grid, and we want to recover the temperatures at time 0.01 along an 11×11 square grid). We used $R = 15$, that is,

$$\mathcal{X} = \{[x_{k\ell}]_{k,\ell=1}^{63} : \sum_{k,\ell} x_{k\ell}^2 \leq 225\},$$

and $\sigma = 0.001$.

Under the circumstances, the risk of the best linear estimate turns out to be 0.3968. Figure 4.1 shows a sample temperature distribution $B(x) = U_*(t_0)$ at time t_0 resulting from a randomly selected signal $x \in \mathcal{X}$ along with the recovery $\widehat{U}(t_0)$ of U_* by the optimal linear estimate and the naive "least squares" recovery $\widetilde{U}(t_0)$ of U_*. The latter is defined as $B(x_*)$, where x_* is the least squares recovery of the signal underlying observation ω:

$$x = x_*(\omega) := \operatorname*{argmin}_x \|A(x) - \omega\|_2.$$

Notice the dramatic difference in performances of the "naive least squares" and the optimal linear estimate.

4.2.2.2 Near-optimality of $\widehat{x}_{H_*}$

Proposition 4.5. *The efficiently computable linear estimate $\widehat{x}_{H_*}(\omega) = H_*^T\omega$ yielded by an optimal solution to the optimization problem* (4.12) *is nearly optimal in terms of its risk:*

$$\mathrm{Risk}[\widehat{x}_{H_*}|\mathcal{X}] \leq \sqrt{\mathrm{Opt}} \leq 64\sqrt{45\ln 2(\ln K + 5\ln 2)}\,\mathrm{Risk}_{\mathrm{opt}}[\mathcal{X}], \tag{4.15}$$

where the minimax optimal risk $\mathrm{Risk}_{\mathrm{opt}}[\mathcal{X}]$ *is given by* (4.7).

For proof, see Section 4.8.5. Note that the "nonoptimality factor" in (4.15) depends *logarithmically* on K and is completely independent on what A, B, Γ are and the "details" R_k, $\mathcal{T}$—see (4.9)—specifying ellitope $\mathcal{X}$.

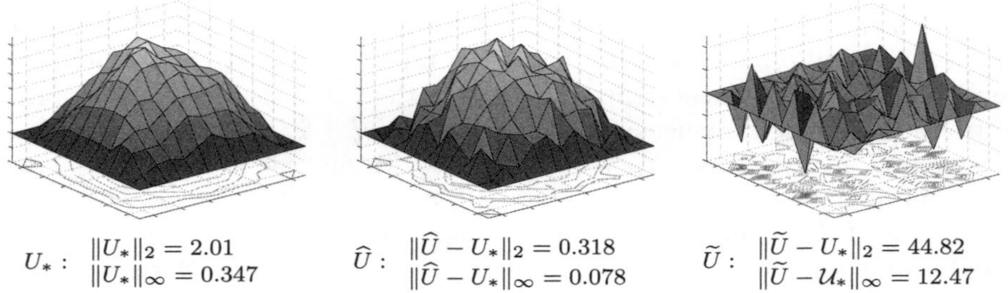

U_* : $\|U_*\|_2 = 2.01$
$\|U_*\|_\infty = 0.347$
$\widehat{U}$: $\|\widehat{U} - U_*\|_2 = 0.318$
$\|\widehat{U} - U_*\|_\infty = 0.078$
$\widetilde{U}$: $\|\widetilde{U} - U_*\|_2 = 44.82$
$\|\widetilde{U} - \mathcal{U}_*\|_\infty = 12.47$

Figure 4.1: True distribution of temperature $U_* = B(x)$ at time $t_0 = 0.01$ (left) along with its recovery $\widehat{U}$ via the optimal linear estimate (center) and the "naive" recovery $\widetilde{U}$ (right).

4.2.2.3 Relaxing the symmetry requirement

Sets $\mathcal{X}$ of the form (4.6)—we called them ellitopes—are symmetric w.r.t. the origin convex compact sets of special structure. This structure is rather flexible, but the symmetry is "built in." We are about to demonstrate that, to some extent, the symmetry requirement can be somewhat relaxed. Specifically, assume instead of (4.6) that the convex compact set $\mathcal{X}$ known to contain the signals x underlying observations (4.5) can be "sandwiched" by two ellitopes known to us and similar to each other, with coefficient $\alpha \geq 1$:

$$\underbrace{\{x \in \mathbf{R}^n : \exists (y \in \mathbf{R}^{\bar{n}}, t \in \mathcal{T}) : x = Py \ \&\ y^T R_k y \leq t_k,\ 1 \leq k \leq K\}}_{\underline{\mathcal{X}}} \subset \mathcal{X} \subset \alpha \underline{\mathcal{X}},$$

with R_k and $\mathcal{T}$ possessing the properties postulated in Section 4.2.1.1. Let Opt and H_* be the optimal value and optimal solution of the optimization problem (4.12) associated with the data R_1, ..., R_K, $\mathcal{T}$ and matrices $\bar{A} = AP$, $\bar{B} = BP$ in the role of A, B, respectively. It is immediately seen that the risk $\mathrm{Risk}[\widehat{x}_{H_*}|\mathcal{X}]$ of the linear estimate $\widehat{x}_{H_*}(\omega)$ is at most $\alpha\sqrt{\mathrm{Opt}}$. On the other hand, we have $\mathrm{Risk_{opt}}[\underline{\mathcal{X}}] \leq \mathrm{Risk_{opt}}[\mathcal{X}]$, and by Proposition 4.5 also $\sqrt{\mathrm{Opt}} \leq O(1)\sqrt{\ln(2K)}\mathrm{Risk_{opt}}[\underline{\mathcal{X}}]$. Taken together, these relations imply that

$$\mathrm{Risk}[\widehat{x}_{H^*}|\mathcal{X}] \leq O(1)\alpha\sqrt{\ln(2K)}\mathrm{Risk_{opt}}[\mathcal{X}]. \tag{4.16}$$

In other words, as far as the "level of nonoptimality" of efficiently computable linear estimates is concerned, signal sets $\mathcal{X}$ which can be approximated by ellitopes within a factor α *of order of 1* are nearly as good as the ellitopes. To give an example: it is known that whenever the intersection $\mathcal{X}$ of K elliptic cylinders $\{x : (x - c_k)^T R_k (x - c_k) \leq 1\}$, $R_k \succeq 0$, concentric or not, is bounded and has a nonempty interior, $\mathcal{X}$ can be approximated by an ellipsoid within the factor

$\alpha = K + 2\sqrt{K}$.[7] Assuming w.l.o.g. that the approximating ellipsoid is centered at the origin, the level of nonoptimality of a linear estimate is bounded by (4.16) with $O(1)K$ in the role of α.

4.2.2.4 Comments

Note that bound (4.16) rapidly deteriorates when α grows, and this phenomenon to some extent "reflects the reality." For example, a perfect simplex $\mathcal{X}$ inscribed into the unit sphere in $\mathbf{R}^n$ is in-between two Euclidean balls centered at the origin with the ratio of radii equal to n (i.e. $\alpha = n$). It is immediately seen that with $A = B = I$, $\Gamma = \sigma^2 I$, in the range $\sigma \leq n\sigma^2 \leq 1$ of values of n and σ, we have

$$\mathrm{Risk}_{\mathrm{opt}}[\mathcal{X}] \approx \sqrt{\sigma},\ \mathrm{Risk}_{\mathrm{opt}}[\widehat{x}_{H_*}|\mathcal{X}] = O(1)\sqrt{n\sigma},$$

with $\approx$ meaning "up to logarithmic in n/σ factor." In other words, for large $n\sigma$ linear estimates indeed are significantly (albeit not to the full extent of (4.16)) outperformed by nonlinear ones.

Another situation "bad for linear estimates" suggested by (4.15) is the one where the description (4.6) of $\mathcal{X}$, albeit possible, requires a very large value of K. Here again (4.15) reflects to some extent the reality: when $\mathcal{X}$ is the unit $\|\cdot\|_1$ ball in $\mathbf{R}^n$, (4.6) takes place with $K = 2^{n-1}$; consequently, the factor at $\mathrm{Risk}_{\mathrm{opt}}[\mathcal{X}]$ in the right-hand side of (4.15) becomes at least $\sqrt{n}$. On the other hand, with $A = B = I$, $\Gamma = \sigma^2 I$, in the range $\sigma \leq n\sigma^2 \leq 1$ of values of n, σ, the risks $\mathrm{Risk}_{\mathrm{opt}}[\mathcal{X}]$, $\mathrm{Risk}_{\mathrm{opt}}[\widehat{x}_{H_*}|\mathcal{X}]$ are basically the same as in the case of $\mathcal{X}$ being the perfect simplex inscribed into the unit sphere in $\mathbf{R}^n$, and linear estimates indeed are "heavily nonoptimal" when $n\sigma$ is large.

4.2.2.5 How near is "near-optimal": Numerical illustration

The "nonoptimality factor" θ in the upper bound $\sqrt{\mathrm{Opt}} \leq \theta \mathrm{Risk}_{\mathrm{opt}}[\mathcal{X}]$ from Proposition 4.5, while logarithmic, seems to be unpleasantly large. On closer inspection, one can get numerically less conservative bounds on non-optimality factors. Here are some illustrations. In the six experiments to be reported, we used $n = m = \nu = 32$ and $\Gamma = \sigma^2 I_m$. In the first triple of experiments, $\mathcal{X}$ was the ellipsoid

$$X = \{x \in \mathbf{R}^{32} : \sum_{j=1}^{32} j^2 x_j^2 \leq 1\},$$

that is, P was the identity, $K = 1$, $R_1 = \sum_{j=1}^{32} j^2 e_j e_j^T$ (e_j are basic orths), and $\mathcal{T} = [0, 1]$. In the second triple of experiments, $\mathcal{X}$ was the box circumscribed around the above ellipsoid:

$$\begin{array}{c} X = \{x \in \mathbf{R}^{32} : j|x_j| \leq 1,\, 1 \leq j \leq 32\} \\ \left[P = I, K = 32, R_k = k^2 e_k e_k^T, k \leq K, \mathcal{T} = [0,1]^K\right]. \end{array}$$

[7] Namely, setting $F(x) = -\sum_{k=1}^K \ln(1 - (x - c_k)^T R_k (x - c_k)) : \mathrm{int}\,\mathcal{X} \to \mathbf{R}$ and denoting by $\bar{x}$ the *analytic center* $\mathrm{argmin}_{x \in \mathrm{int}\,\mathcal{X}} F(x)$, one has

$$\{x : (x - \bar{x})^T F''(\bar{x})(x - \bar{x}) \leq 1\} \subset \mathcal{X} \subset \{x : (x - \bar{x})^T F''(\bar{x})(x - \bar{x}) \leq [K + 2\sqrt{K}]^2\}.$$

X	σ	$\sqrt{\text{Opt}}$	LwB	$\sqrt{\text{Opt}}/\text{LwB}$
ellipsoid	0.0100	0.288	0.153	1.88
ellipsoid	0.0010	0.103	0.060	1.71
ellipsoid	0.0001	0.019	0.018	1.06
box	0.0100	0.698	0.231	3.02
box	0.0010	0.163	0.082	2.00
box	0.0001	0.021	0.020	1.06

Table 4.1: Performance of linear estimates (4.8), (4.12), $m = n = 32$, $B = I$.

In these experiments, B was the identity matrix, and A was a randomly rotated matrix common for all experiments, with singular values λ_j, $1 \leq j \leq 32$, forming a geometric progression, with $\lambda_1 = 1$ and $\lambda_{32} = 0.01$. Experiments in a triple differed by the values of σ (0.01,0.001,0.0001).

The results of the experiments are presented in Table 4.1, where, as above, $\sqrt{\text{Opt}}$ is the upper bound given by (4.12) on the risk $\text{Risk}[\widehat{x}_{H_*}|X]$ of recovering $Bx = x$, $x \in X$, by the linear estimate yielded by (4.8) and (4.12), and LwB is the lower bound on $\text{Risk}_{\text{opt}}[X]$ computed via the techniques outlined in Exercise 4.22 (we skip the details). Whatever might be your attitude to the "reality" as reflected by the data in Table 4.1, this reality is much better than the theoretical upper bound on θ appearing in (4.15).

4.2.3 Byproduct on semidefinite relaxation

We are about to present a byproduct, important in its own right, of the reasoning underlying Proposition 4.5. This byproduct is not directly related to Statistics; it relates to the quality of the standard semidefinite relaxation. Specifically, given a quadratic from $x^T Cx$ and an ellitope $\mathcal{X}$ represented by (4.6), consider the problem

$$\text{Opt}_* = \max_{x \in \mathcal{X}} x^T Cx = \max_y \left\{ y^T P^T CPy : \exists t \in \mathcal{T} : y^T R_k y \leq t_k,\, k \leq K \right\}. \quad (4.17)$$

This problem can be NP-hard (this is already so when $\mathcal{X}$ is the unit box and C a general-type positive semidefinite matrix); however, Opt admits an efficiently computable upper bound given by *semidefinite relaxation* as follows: whenever $\lambda \geq 0$ is such that

$$P^T CP \preceq \sum_{k=1}^K \lambda_k R_k,$$

for $y \in \bar{X} := \{y : \exists t \in \mathcal{T} : y^T R_k y \leq t_k,\, k \leq K\}$ we clearly have

$$[Py]^T CPy \leq \sum_k \lambda_k y^T R_k y \leq \phi_{\mathcal{T}}(\lambda)$$

where the last $\leq$ is due to the fact that the vector with the entries $y^T R_k y$, $1 \leq k \leq K$, belongs to $\mathcal{T}$. As a result, the efficiently computable quantity

$$\text{Opt} = \min_\lambda \left\{ \phi_{\mathcal{T}}(\lambda) : \lambda \geq 0, P^T CP \preceq \sum_k \lambda_k R_k \right\} \quad (4.18)$$

is an upper bound on Opt_*. We have the following

Proposition 4.6. *Let C be a symmetric $n \times n$ matrix and $\mathcal{X}$ be given by ellitopic representation* (4.6)*, and let* Opt_* *and* Opt *be given by* (4.17) *and* (4.18)*. Then*

$$\frac{\mathrm{Opt}}{3\ln(\sqrt{3}K)} \leq \mathrm{Opt}_* \leq \mathrm{Opt}. \tag{4.19}$$

For proof, see Section 4.8.2.

4.3 FROM ELLITOPES TO SPECTRATOPES

So far, the domains of signals we dealt with were ellitopes. In this section we demonstrate that basically all our constructions and results can be extended onto a much wider family of signal domains, namely, *spectratopes.*

4.3.1 Spectratopes: Definition and examples

We call a set $\mathcal{X} \subset \mathbf{R}^n$ a *basic spectratope* if it admits a *simple spectratopic representation*—representation of the form

$$\mathcal{X} = \left\{x \in \mathbf{R}^n : \exists t \in \mathcal{T} : R_k^2[x] \preceq t_k I_{d_k}, 1 \leq k \leq K\right\} \tag{4.20}$$

where

S.1. $R_k[x] = \sum_{i=1}^n x_i R^{ki}$ are symmetric $d_k \times d_k$ matrices linearly depending on $x \in \mathbf{R}^n$ (i,e., "matrix coefficients" R^{ki} belong to $\mathbf{S}^n$).

S.2. $\mathcal{T} \in \mathbf{R}^K_+$ is the set with the same properties as in the definition of an ellitope, that is, $\mathcal{T}$ is a convex compact subset of $\mathbf{R}^K_+$ which contains a positive vector and is monotone:
$$0 \leq t' \leq t \in \mathcal{T} \Rightarrow t' \in \mathcal{T}.$$

S.3. Whenever $x \neq 0$, it holds $R_k[x] \neq 0$ for at least one $k \leq K$.

An immediate observation is as follows:

Remark 4.7. By the Schur Complement Lemma, the set *(4.20)* given by data satisfying S.1-2 can be represented as

$$\mathcal{X} = \left\{x \in \mathbf{R}^n : \exists t \in \mathcal{T} : \left[\begin{array}{c|c} t_k I_{d_k} & R_k[x] \\ \hline R_k[x] & I_{d_k} \end{array}\right] \succeq 0,\, k \leq K\right\}.$$

By the latter representation, $\mathcal{X}$ is nonempty, closed, convex, symmetric w.r.t. the origin, and contains a neighbourhood of the origin. This set is bounded if and only if the data, in addition to S.1–2, satisfies S.3.

A spectratope $\mathcal{X} \subset \mathbf{R}^\nu$ is a set represented as a linear image of a basic spectratope:

$$\mathcal{X} = \{x \in \mathbf{R}^\nu : \exists (y \in \mathbf{R}^n, t \in \mathcal{T}) : x = Py, R_k^2[y] \preceq t_k I_{d_k}, 1 \leq k \leq K\}, \tag{4.21}$$

where P is a $\nu \times n$ matrix, and $R_k[\cdot]$, $\mathcal{T}$ are as in S.1–3.

We associate with a basic spectratope (4.20), S.1–3, the following entities:

1. The *size*

$$D = \sum_{k=1}^{K} d_k;$$

2. Linear mappings

$$Q \mapsto \mathcal{R}_k[Q] = \sum_{i,j} Q_{ij} R^{ki} R^{kj} : \mathbf{S}^n \to \mathbf{S}^{d_k}.$$

As is immediately seen, we have

$$\mathcal{R}_k[xx^T] \equiv R_k^2[x], \tag{4.22}$$

implying that $\mathcal{R}_k[Q] \succeq 0$ whenever $Q \succeq 0$, whence $\mathcal{R}_k[\cdot]$ is $\succeq$-monotone:

$$Q' \succeq Q \Rightarrow \mathcal{R}_k[Q'] \succeq \mathcal{R}_k[Q]. \tag{4.23}$$

Besides this, we have

$$Q \succeq 0 \Rightarrow \mathbf{E}_{\xi \sim \mathcal{N}(0,Q)}\{R_k^2[\xi]\} = \mathbf{E}_{\xi \sim \mathcal{N}(0,Q)}\{\mathcal{R}_k[\xi\xi^T]\} = \mathcal{R}_k[Q], \tag{4.24}$$

where the first equality is given by (4.22).

3. Linear mappings $\Lambda_k \mapsto \mathcal{R}_k^*[\Lambda_k] : \mathbf{S}^{d_k} \to \mathbf{S}^n$ given by

$$[\mathcal{R}_k^*[\Lambda_k]]_{ij} = \tfrac{1}{2}\mathrm{Tr}(\Lambda_k[R^{ki}R^{kj} + R^{kj}R^{ki}]),\ 1 \le i,j \le n. \tag{4.25}$$

It is immediately seen that $\mathcal{R}_k^*[\cdot]$ is the conjugate of $\mathcal{R}_k[\cdot]$:

$$\langle \Lambda_k, \mathcal{R}_k[Q] \rangle_F = \mathrm{Tr}(\Lambda_k \mathcal{R}_k[Q]) = \mathrm{Tr}(\mathcal{R}_k^*[\Lambda_k]Q) = \langle \mathcal{R}_k^*[\Lambda_k], Q \rangle_F, \tag{4.26}$$

where $\langle A, B \rangle_F = \mathrm{Tr}(AB)$ is the Frobenius inner product of symmetric matrices. Besides this, we have

$$\Lambda_k \succeq 0 \Rightarrow \mathcal{R}_k^*[\Lambda_k] \succeq 0. \tag{4.27}$$

Indeed, $\mathcal{R}_k^*[\Lambda_k]$ is linear in Λ_k, so that it suffices to verify (4.27) for dyadic matrices $\Lambda_k = ff^T$; for such a Λ_k, (4.25) reads

$$(\mathcal{R}_k^*[ff^T])_{ij} = [R^{ki}f]^T[R^{kj}f],$$

that is, $\mathcal{R}_k^*[ff^T]$ is a Gram matrix and as such is $\succeq 0$. Another way to arrive at (4.27) is to note that when $\Lambda_k \succeq 0$ and $Q = xx^T$, the first quantity in (4.26) is nonnegative by (4.22), and therefore (4.26) states that $x^T\mathcal{R}_k^*[\Lambda_k]x \ge 0$ for every x, implying $\mathcal{R}_k^*[\Lambda_k] \succeq 0$.

4. The linear space $\Lambda^K = \mathbf{S}^{d_1} \times \ldots \times \mathbf{S}^{d_K}$ of all ordered collections $\Lambda = \{\Lambda_k \in \mathbf{S}^{d_k}\}_{k \le K}$ along with the linear mapping

$$\Lambda \mapsto \lambda[\Lambda] := [\mathrm{Tr}(\Lambda_1); \ldots; \mathrm{Tr}(\Lambda_K)] : \Lambda^K \to \mathbf{R}^K.$$

4.3.1.1 Examples of spectratopes

Example: Ellitopes. Every ellitope

$$\mathcal{X} = \{x \in \mathbf{R}^\nu : \exists (y \in \mathbf{R}^n, t \in \mathcal{T}) : x = Py, y^T R_k y \leq t_k, \, k \leq K\}$$
$$[R_k \succeq 0, \sum_k R_k \succ 0]$$

is a spectratope as well. Indeed, let $R_k = \sum_{j=1}^{p_k} r_{kj} r_{kj}^T$, $p_k = \mathrm{Rank}(R_k)$, be a dyadic representation of the positive semidefinite matrix R_k, so that

$$y^T R_k y = \sum_j (r_{kj}^T y)^2 \; \forall y,$$

and let

$$\widehat{\mathcal{T}} = \{\{t_{kj} \geq 0, 1 \leq j \leq p_k, 1 \leq k \leq K\} : \exists t \in \mathcal{T} : \sum_j t_{kj} \leq t_k\},$$
$$R_{kj}[y] = r_{kj}^T y \in \mathbf{S}^1 = \mathbf{R}.$$

We clearly have

$$\mathcal{X} = \{x \in \mathbf{R}^\nu : \exists (\{t_{kj}\} \in \widehat{\mathcal{T}}, y) : x = Py, R_{kj}^2[y] \preceq t_{kj} I_1 \, \forall k, j\},$$

and the right-hand side is a legitimate spectratopic representation of $\mathcal{X}$.

Example: "Matrix box." Let L be a positive definite $d \times d$ matrix. Then the "matrix box"

$$\begin{array}{rcl}\mathcal{X} & = & \{X \in \mathbf{S}^d : -L \preceq X \preceq L\} = \{X \in \mathbf{S}^d : -I_d \preceq L^{-1/2} X L^{-1/2} \preceq I_d\} \\ & = & \{X \in \mathbf{S}^d : R^2[X] := [L^{-1/2} X L^{-1/2}]^2 \preceq I_d\}\end{array}$$

is a basic spectratope (augment $R_1[\cdot] := R[\cdot]$ with $K = 1$, $\mathcal{T} = [0, 1]$). As a result, a *bounded* set $\mathcal{X} \subset \mathbf{R}^\nu$ given by a system of "two-sided" Linear Matrix Inequalities, specifically,

$$\mathcal{X} = \{x \in \mathbf{R}^\nu : \exists t \in \mathcal{T} : -\sqrt{t_k} L_k \preceq S_k[x] \preceq \sqrt{t_k} L_k, k \leq K\}$$

where $S_k[x]$ are symmetric $d_k \times d_k$ matrices linearly depending on x, $L_k \succ 0$, and $\mathcal{T}$ satisfies S.2, is a basic spectratope:

$$\mathcal{X} = \{x \in \mathbf{R}^\nu : \exists t \in \mathcal{T} : R_k^2[x] \leq t_k I_{d_k}, \, k \leq K\} \qquad [R_k[x] = L_k^{-1/2} S_k[x] L_k^{-1/2}].$$

Like ellitopes, spectratopes admit fully algorithmic calculus; see Section 4.6.

4.3.2 Semidefinite relaxation on spectratopes

Now let us extend Proposition 4.6 to our current situation. The extension reads as follows:

Proposition 4.8. *Let C be a symmetric $n \times n$ matrix and $\mathcal{X}$ be given by spectratopic representation*

$$\mathcal{X} = \{x \in \mathbf{R}^n : \exists y \in \mathbf{R}^\mu, t \in \mathcal{T} : x = Py, R_k^2[y] \preceq t_k I_{d_k}, k \leq K\}, \tag{4.28}$$

let

$$\mathrm{Opt}_* = \max_{x\in\mathcal{X}} x^T Cx,$$

and let

$$\begin{array}{c}\mathrm{Opt} = \min\limits_{\Lambda=\{\Lambda_k\}_{k\leq K}} \left\{\phi_{\mathcal{T}}(\lambda[\Lambda]) : \Lambda_k \succeq 0, P^TCP \preceq \sum_k \mathcal{R}_k^*[\Lambda_k]\right\}\\ {[\lambda[\Lambda] = [\mathrm{Tr}(\Lambda_1); ...; \mathrm{Tr}(\Lambda_K)]]}\,.\end{array} \tag{4.29}$$

Then (4.29) *is solvable, and*

$$\mathrm{Opt}_* \leq \mathrm{Opt} \leq 2\max[\ln(2D), 1]\mathrm{Opt}_*,\ D = \sum_k d_k. \tag{4.30}$$

Let us verify the easy and instructive part of the proposition, namely, the left inequality in (4.30); the remaining claims will be proved in Section 4.8.3.

The left inequality in (4.30) is readily given by the following

Lemma 4.9. *Let $\mathcal{X}$ be spectratope* (4.28) *and $Q \in \mathbf{S}^n$. Whenever $\Lambda_k \in \mathbf{S}_+^{d_k}$ satisfy*

$$P^TQP \preceq \sum_k \mathcal{R}_k^*[\Lambda_k],$$

for all $x \in \mathcal{X}$ we have

$$x^TQx \leq \phi_{\mathcal{T}}(\lambda[\Lambda]),\ \lambda[\Lambda] = [\mathrm{Tr}(\Lambda_1); ...; \mathrm{Tr}(\Lambda_K)].$$

Proof of the lemma: Let $x \in \mathcal{X}$, so that for some $t \in \mathcal{T}$ and y it holds

$$x = Py,\ R_k^2[y] \preceq t_k I_{d_k}\ \forall k \leq K.$$

Consequently,

$$\begin{array}{rl}x^TQx & = y^TP^TQPy \leq y^T \sum_k \mathcal{R}_k^*[\Lambda_k]y = \sum_k \mathrm{Tr}(\mathcal{R}_k^*[\Lambda_k][yy^T])\\ & = \sum_k \mathrm{Tr}(\Lambda_k \mathcal{R}_k[yy^T])\ [\text{by } (4.26)]\\ & = \sum_k \mathrm{Tr}(\Lambda_k R_k^2[y])\ [\text{by } (4.22)]\\ & \leq \sum_k t_k \mathrm{Tr}(\Lambda_k I_{d_k})\ [\text{since } \Lambda_k \succeq 0 \text{ and } R_k^2[y] \preceq t_k I_{d_k}]\\ & \leq \phi_{\mathcal{T}}(\lambda[\Lambda]).\end{array}$$

□

4.3.3 Linear estimates beyond ellitopic signal sets and $\|\cdot\|_2$-risk

In Section 4.2, we have developed a computationally efficient scheme for building "presumably good" linear estimates of the linear image Bx of unknown signal x known to belong to a given ellitope $\mathcal{X}$ in the case when the (squared) risk is defined as the worst, w.r.t. $x \in \mathcal{X}$, expected squared Euclidean norm $\|\cdot\|_2^2$ of the recovery error. We are about to extend these results to the case when $\mathcal{X}$ is a spectratope, and the norm used to measure the recovery error, while not being completely arbitrary, is not necessarily $\|\cdot\|_2$. Besides this, in what follows we also relax our assumptions on observation noise.

4.3.3.1 Situation and goal

We consider the problem of recovering the image $Bx \in \mathbf{R}^\nu$ of a signal $x \in \mathbf{R}^n$ known to belong to a given spectratope

$$\mathcal{X} = \{x \in \mathbf{R}^n : \exists t \in \mathcal{T} : R_k^2[x] \preceq t_k I_{d_k}, \ 1 \leq k \leq K\}$$

from noisy observation

$$\omega = Ax + \xi, \tag{4.31}$$

where A is a known $m \times n$ matrix, and ξ is random observation noise.

Observation noise. In typical signal processing applications, the distribution of noise is fixed and is a part of the data of the estimation problem. In order to cover some applications (e.g., the one in Section 4.3.3.7), we allow for "ambiguous" noise distributions; all we know is that this distribution belongs to a family $\mathcal{P}$ of Borel probability distributions on $\mathbf{R}^m$ associated with a given convex compact subset Π of the interior of the cone $\mathbf{S}_+^m$ of positive semidefinite $m \times m$ matrices, "association" meaning that the matrix of second moments of every distribution $P \in \mathcal{P}$ is $\succeq$-dominated by a matrix from Π:

$$P \in \mathcal{P} \Rightarrow \exists Q \in \Pi : \mathrm{Var}[P] := \mathbf{E}_{\xi \sim P}\{\xi\xi^T\} \preceq Q. \tag{4.32}$$

The actual distribution of noise in (4.31) is selected from $\mathcal{P}$ by nature (and may, e.g., depend on x).

In the sequel, for a probability distribution P on $\mathbf{R}^m$ we write $P \lhd \Pi$ to express the fact that the matrix of second moments of P is $\succeq$-dominated by a matrix from Π:

$$\{P \lhd \Pi\} \Leftrightarrow \{\exists \Theta \in \Pi : \mathrm{Var}[P] \preceq \Theta\}.$$

Quantifying risk. Given Π and a norm $\|\cdot\|$ on $\mathbf{R}^\nu$, we quantify the quality of a candidate estimate $\widehat{x}(\cdot) : \mathbf{R}^m \to \mathbf{R}^\nu$ by its $(\Pi, \|\cdot\|)$-risk on $\mathcal{X}$ defined as

$$\mathrm{Risk}_{\Pi,\|\cdot\|}[\widehat{x}|\mathcal{X}] = \sup_{x \in \mathcal{X}, P \lhd \Pi} \mathbf{E}_{\xi \sim P}\left\{\|\widehat{x}(Ax+\xi) - Bx\|\right\}.$$

Goal. As before, our focus is on *linear estimates*—estimates of the form

$$\widehat{x}_H(\omega) = H^T\omega$$

given by $m \times \nu$ matrices H. Our goal is to demonstrate that under some restrictions on the signal domain $\mathcal{X}$, a "presumably good" linear estimate yielded by an optimal solution to an efficiently solvable convex optimization problem is near-optimal in terms of its risk among *all* estimates, linear and nonlinear alike.

4.3.3.2 Assumptions

Preliminaries: Conjugate norms. Recall that a norm $\|\cdot\|$ on a Euclidean space $\mathcal{E}$, e.g., on $\mathbf{R}^k$, gives rise to its *conjugate* norm

$$\|y\|_* = \max_x\{\langle y, x\rangle : \|x\| \leq 1\},$$

where $\langle\cdot,\cdot\rangle$ is the inner product in $\mathcal{E}$. Equivalently, $\|\cdot\|_*$ is the smallest norm such that

$$\langle x, y\rangle \leq \|x\|\|y\|_* \ \forall x, y. \tag{4.33}$$

It is well known that taken twice, norm conjugation recovers the initial norm: $(\|\cdot\|_*)_*$ is exactly $\|\cdot\|$; in other words,

$$\|x\| = \max_y\{\langle x, y\rangle : \|y\|_* \leq 1\}.$$

The standard examples are the conjugates to the standard ℓ_p-norms on $\mathcal{E} = \mathbf{R}^k$, $p \in [1,\infty]$: it turns out that

$$(\|\cdot\|_p)_* = \|\cdot\|_{p*},$$

where $p_* \in [1,\infty]$ is linked to $p \in [1,\infty]$ by the symmetric relation

$$\frac{1}{p} + \frac{1}{p_*} = 1,$$

so that $1_* = \infty$, $\infty_* = 1$, $2_* = 2$. The corresponding version of inequality (4.33) is called *Hölder inequality*—an extension of the Cauchy-Schwartz inequality dealing with the case $\|\cdot\| = \|\cdot\|_* = \|\cdot\|_2$.

Assumptions. From now on we make the following assumptions:

Assumption A: *The unit ball $\mathcal{B}_*$ of the norm $\|\cdot\|_*$* conjugate *to the norm $\|\cdot\|$ in the formulation of our estimation problem is a spectratope:*

$$\begin{array}{c}\mathcal{B}_* = \{z \in \mathbf{R}^\nu : \exists y \in \mathcal{Y} : z = My\},\\ \mathcal{Y} := \{y \in \mathbf{R}^q : \exists r \in \mathcal{R} : S_\ell^2[y] \preceq r_\ell I_{f_\ell},\ 1 \leq \ell \leq L\},\end{array} \tag{4.34}$$

where the right-hand side data are as required in a spectratopic representation.

Note that Assumption **A** is satisfied when $\|\cdot\| = \|\cdot\|_p$ with $p \in [1,2]$: in this case,

$$\mathcal{B}_* = \{u \in \mathbf{R}^\nu : \|u\|_{p_*} \leq 1\},\ p_* = \frac{p}{p-1} \in [2,\infty],$$

so that $\mathcal{B}_*$ is an ellitope—see Section 4.2.1.1—and thus is a spectratope. Another potentially useful example of norm $\|\cdot\|$ which obeys Assumption **A** is the *nuclear norm* $\|V\|_{\mathrm{Sh},1}$ on the space $\mathbf{R}^\nu = \mathbf{R}^{p\times q}$ of $p\times q$ matrices—the sum of singular values of a matrix V. In this case the conjugate norm is the spectral norm $\|\cdot\| = \|\cdot\|_{2,2}$ on $\mathbf{R}^\nu = \mathbf{R}^{p\times q}$, and the unit ball of the latter norm is a spectratope:

$$\begin{array}{c}\{X \in \mathbf{R}^{p\times q} : \|X\| \leq 1\} = \{X : \exists t \in \mathcal{T} = [0,1] : R^2[X] \preceq tI_{p+q}\},\\ R[X] = \left[\begin{array}{c|c} & X^T \\ \hline X & \end{array}\right].\end{array}$$

Besides Assumption **A**, we make

Assumption B: *The signal set $\mathcal{X}$ is a basic spectratope:*

$$\mathcal{X} = \{x \in \mathbf{R}^n : \exists t \in \mathcal{T} : R_k^2[x] \preceq t_k I_{d_k},\ 1 \leq k \leq K\},$$

where the right-hand side data are as required in a spectratopic representation.

Note: Similarly to what we have observed in Section 4.2.1.3 in the case of ellitopes, the situation where the signal set is a general type spectratope can be straightforwardly reduced to the one where $\mathcal{X}$ is a *basic* spectratope.

In addition we make the following regularity assumption:

Assumption R: *All matrices from* Π *are positive definite.*

4.3.3.3 Building linear estimate

Let $H \in \mathbf{R}^{m\times\nu}$. We clearly have

$$\begin{array}{rcl}\mathrm{Risk}_{\Pi,\|\cdot\|}[\widehat{x}_H(\cdot)|\mathcal{X}] &=& \sup\limits_{x\in X, P\lhd\Pi} \mathbf{E}_{\xi\sim P}\left\{\|[B-H^TA]x - H^T\xi\|\right\}\\ &\leq& \sup_{x\in X}\|[B-H^TA]x\| + \sup_{P\lhd\Pi}\mathbf{E}_{\xi\sim P}\left\{\|H^T\xi\|\right\}\\ &=& \|B-H^TA\|_{\mathcal{X},\|\cdot\|} + \Psi_\Pi(H),\end{array} \tag{4.35}$$

where

$$\begin{array}{rcl}\|V\|_{\mathcal{X},\|\cdot\|} &=& \max_x\left\{\|Vx\| : x\in\mathcal{X}\right\} : \mathbf{R}^{\nu\times n}\to\mathbf{R},\\ \Psi_\Pi(H) &=& \sup\limits_{P\lhd\Pi}\mathbf{E}_{\xi\sim P}\left\{\|H^T\xi\|\right\}.\end{array}$$

As in Section 4.2.2, we need to derive efficiently computable convex upper bounds on the norm $\|\cdot\|_{\mathcal{X},\|\cdot\|}$ and the function Ψ_Π, which by themselves, while being convex, can be difficult to compute.

4.3.3.4 Upper-bounding $\|\cdot\|_{\mathcal{X},\|\cdot\|}$

With Assumptions **A**, **B** in force, consider the spectratope

$$\begin{array}{rcl}\mathcal{Z} &:=& \mathcal{X}\times\mathcal{Y} = \{[x;y]\in\mathbf{R}^n\times\mathbf{R}^q : \exists s=[t;r]\in\mathcal{T}\times\mathcal{R}:\\ && \qquad R_k^2[x]\preceq t_kI_{d_k},\, 1\leq k\leq K,\, S_\ell^2[y]\preceq r_\ell I_{f_\ell},\, 1\leq\ell\leq L\}\\ &=& \{w=[x;y]\in\mathbf{R}^n\times\mathbf{R}^q : \exists s=[t;r]\in\mathcal{S}=\mathcal{T}\times\mathcal{R} : U_i^2[w]\preceq s_iI_{g_i},\\ && \qquad 1\leq i\leq I=K+L\}\end{array}$$

with $U_i[\cdot]$ readily given by $R_k[\cdot]$ and $S_\ell[\cdot]$. Given a $\nu\times n$ matrix V and setting

$$W[V] = \frac{1}{2}\left[\begin{array}{c|c} & V^TM\\ \hline M^TV & \end{array}\right]$$

we have

$$\|V\|_{\mathcal{X},\|\cdot\|} = \max_{x\in\mathcal{X}}\|Vx\| = \max_{x\in\mathcal{X},z\in\mathcal{B}_*} z^TVx = \max_{x\in\mathcal{X},y\in\mathcal{Y}} y^TM^TVx = \max_{w\in\mathcal{Z}} w^TW[V]w.$$

Applying Proposition 4.8, we arrive at the following

Corollary 4.10. *In the situation just defined, the efficiently computable convex* *function*

$$\begin{array}{rcl}\|V\|^+_{\mathcal{X},\|\cdot\|} & = & \min\limits_{\Lambda,\Upsilon}\Big\{\phi_{\mathcal{T}}(\lambda[\Lambda])+\phi_{\mathcal{R}}(\lambda[\Upsilon]):\\ & & \Lambda=\{\Lambda_k\in\mathbf{S}^{d_k}_+\}_{k\le K},\Upsilon=\{\Upsilon_\ell\in\mathbf{S}^{f_\ell}_+\}_{\ell\le L},\\ & & \left[\begin{array}{c|c}\sum_k\mathcal{R}^*_k[\Lambda_k] & \frac{1}{2}V^TM\\ \hline \frac{1}{2}M^TV & \sum_\ell\mathcal{S}^*_\ell[\Upsilon_\ell]\end{array}\right]\succeq 0\Big\}\end{array}\tag{4.36}$$

$$\left[\begin{array}{l}\phi_{\mathcal{T}}(\lambda)=\max\limits_{t\in\mathcal{T}}\lambda^Tt,\ \phi_{\mathcal{R}}(\lambda)=\max\limits_{r\in\mathcal{R}}\lambda^Tr,\ \lambda[\{\Xi_1,...,\Xi_N\}]=[\mathrm{Tr}(\Xi_1);...;\mathrm{Tr}(\Xi_N)],\\ {[\mathcal{R}^*_k[\Lambda_k]]_{ij}}=\frac{1}{2}\mathrm{Tr}(\Lambda_k[R^{ki}_kR^{kj}_k+R^{kj}_kR^{ki}_k]),\ \textit{where } R_k[x]=\sum_i x_iR^{ki},\\ {[\mathcal{S}^*_\ell[\Upsilon_\ell]]_{ij}}=\frac{1}{2}\mathrm{Tr}(\Upsilon_\ell[S^{\ell i}_\ell S^{\ell j}_\ell+S^{\ell j}_\ell S^{\ell i}_\ell]),\ \textit{where } S_\ell[y]=\sum_i y_iS^{\ell i}\end{array}\right]$$

is a norm on $\mathbf{R}^{\nu\times n}$*, and this norm is a tight upper bound on* $\|\cdot\|_{\mathcal{X},\|\cdot\|}$*, namely,*

$$\begin{array}{c}\forall V\in\mathbf{R}^{\nu\times n}:\|V\|_{\mathcal{X},\|\cdot\|}\le\|V\|^+_{\mathcal{X},\|\cdot\|}\le 2\max[\ln(2\mathcal{D}),1]\|V\|_{\mathcal{X},\|\cdot\|},\\ \mathcal{D}=\sum_k d_k+\sum_\ell f_\ell.\end{array}$$

4.3.3.5 Upper-bounding $\Psi_\Pi(\cdot)$

The next step is to derive an efficiently computable convex upper bound on the function Ψ_Π stemming from a norm obeying Assumption **B**. The underlying observation is as follows:

Lemma 4.11. *Let* V *be an* $m\times\nu$ *matrix,* $Q\in\mathbf{S}^m_+$*, and* P *be a probability distribution on* $\mathbf{R}^m$ *with* $\mathrm{Var}[P]\preceq Q$*. Let, further,* $\|\cdot\|$ *be a norm on* $\mathbf{R}^\nu$ *with the unit ball* $\mathcal{B}_*$ *of the conjugate norm* $\|\cdot\|_*$ *given by* (4.34)*. Finally, let* $\Upsilon=\{\Upsilon_\ell\in\mathbf{S}^{f_\ell}_+\}_{\ell\le L}$ *and a matrix* $\Theta\in\mathbf{S}^m$ *satisfy the constraint*

$$\left[\begin{array}{c|c}\Theta & \frac{1}{2}VM\\ \hline \frac{1}{2}M^TV^T & \sum_\ell\mathcal{S}^*_\ell[\Upsilon_\ell]\end{array}\right]\succeq 0\tag{4.37}$$

(for notation, see (4.34), (4.36)*). Then*

$$\mathbf{E}_{\eta\sim P}\{\|V^T\eta\|\}\le\mathrm{Tr}(Q\Theta)+\phi_{\mathcal{R}}(\lambda[\Upsilon]).\tag{4.38}$$

Proof is immediate. In the case of (4.37), we have

$$\begin{array}{rcl}\|V^T\xi\| & = & \max\limits_{z\in\mathcal{B}_*}z^TV^T\xi=\max\limits_{y\in\mathcal{Y}}y^TM^TV^T\xi\\ & \le & \max\limits_{y\in\mathcal{Y}}\left[\xi^T\Theta\xi+\sum_\ell y^T\mathcal{S}^*_\ell[\Upsilon_\ell]y\right]\quad\text{[by (4.37)]}\\ & = & \max\limits_{y\in\mathcal{Y}}\left[\xi^T\Theta\xi+\sum_\ell\mathrm{Tr}(\mathcal{S}^*_\ell[\Upsilon_\ell]yy^T)\right]\\ & = & \max\limits_{y\in\mathcal{Y}}\left[\xi^T\Theta\xi+\sum_\ell\mathrm{Tr}(\Upsilon_\ell S^2_\ell[y])\right]\quad\text{[by (4.22) and (4.26)]}\\ & = & \xi^T\Theta\xi+\max\limits_{y,r}\left\{\sum_\ell\mathrm{Tr}(\Upsilon_\ell S^2_\ell[y]):\ S^2_\ell[y]\preceq r_\ell I_{f_\ell},\ell\le L,r\in\mathcal{R}\right\}\\ & & \hfill\text{[by (4.34)]}\\ & \le & \xi^T\Theta\xi+\max\limits_{r\in\mathcal{R}}\sum_\ell\mathrm{Tr}(\Upsilon_\ell)r_\ell\ \text{[by }\Upsilon_\ell\succeq 0]\\ & \le & \xi^T\Theta\xi+\phi_{\mathcal{R}}(\lambda[\Upsilon]).\end{array}$$

Taking the expectation of both sides of the resulting inequality w.r.t. distribution P of ξ and taking into account that $\mathrm{Tr}(\mathrm{Var}[P]\Theta)\le\mathrm{Tr}(Q\Theta)$ due to $\Theta\succeq 0$ (by (4.37)) and $\mathrm{Var}[P]\preceq Q$, we get (4.38). □

Note that when $P = \mathcal{N}(0,Q)$, the smallest upper bound on $\mathbf{E}_{\eta\sim P}\{\|V^T\eta\|\}$ which can be extracted from Lemma 4.11 (this bound is efficiently computable) is tight; see Lemma 4.17 below.

An immediate consequence of the bound in Lemma (4.11) is:

Corollary 4.12. *Let*

$$\Gamma(\Theta) = \max_{Q\in\Pi} \mathrm{Tr}(Q\Theta) \tag{4.39}$$

and

$$\begin{array}{rcl}\overline{\Psi}_\Pi(H) & = & \min\limits_{\{\Upsilon_\ell\}_{\ell\le L},\Theta\in\mathbf{S}^m}\left\{\Gamma(\Theta)+\phi_{\mathcal{R}}(\lambda[\Upsilon]) : \Upsilon_\ell\succeq 0\,\forall\ell,\right.\\ && \left.\left[\begin{array}{c|c}\Theta & \frac{1}{2}HM\\ \hline \frac{1}{2}M^TH^T & \sum_\ell \mathcal{S}_\ell^*[\Upsilon_\ell]\end{array}\right]\succeq 0\right\}.\end{array} \tag{4.40}$$

Then $\overline{\Psi}_\Pi(\cdot) : \mathbf{R}^{m\times\nu}\to\mathbf{R}$ *is an efficiently computable convex upper bound on* $\Psi_\Pi(\cdot)$.

Indeed, given Lemma 4.11, the only non-evident part of the corollary is that $\overline{\Psi}_\Pi(\cdot)$ is a well-defined real-valued function, which is readily given by Lemma 4.44 stating, in particular, that the optimization problem in (4.40) is feasible, combined with the fact that the objective is coercive on the feasible set (i.e., is not bounded from above along every *un*bounded sequence of feasible solutions).

Remark 4.13. When $\Upsilon = \{\Upsilon_\ell\}_{\ell\le L}$, Θ is a feasible solution to the right-hand side problem in (4.40) and $s>0$, the pair $\Upsilon' = \{s\Upsilon_\ell\}_{\ell\le L}$, $\Theta' = s^{-1}\Theta$ also is a feasible solution. Since $\phi_{\mathcal{R}}(\cdot)$ and $\Gamma(\cdot)$ are positive homogeneous of degree 1, we conclude that $\overline{\Psi}_\Pi$ is in fact the infimum of the function

$$2\sqrt{\Gamma(\Theta)\phi_{\mathcal{R}}(\lambda[\Upsilon])} = \inf_{s>0}\left[s^{-1}\Gamma(\Theta)+s\phi_{\mathcal{R}}(\lambda[\Upsilon])\right]$$

over Υ,Θ satisfying the constraints of the problem *(4.40)*.

In addition, for every feasible solution $\Upsilon = \{\Upsilon_\ell\}_{\ell\le L}$, Θ to *(4.40)* with $\mathcal{M}[\Upsilon] := \sum_\ell \mathcal{S}_\ell^*[\Upsilon_\ell]\succ 0$, the pair Υ, $\widehat{\Theta} = \frac{1}{4}HM\mathcal{M}^{-1}[\Upsilon]M^TH^T$ is feasible for the problem as well, and $0\preceq\widehat{\Theta}\preceq\Theta$ (Schur Complement Lemma), so that $\Gamma(\widehat{\Theta})\le\Gamma(\Theta)$. As a result,

$$\overline{\Psi}_\Pi(H) = \inf_{\Upsilon}\left\{\begin{array}{c}\frac{1}{4}\Gamma(HM\mathcal{M}^{-1}[\Upsilon]M^TH^T)+\phi_{\mathcal{R}}(\lambda[\Upsilon]) :\\ \Upsilon = \{\Upsilon_\ell\in\mathbf{S}_+^{f_\ell}\}_{\ell\le L}, \mathcal{M}[\Upsilon]\succ 0\end{array}\right\}. \tag{4.41}$$

Illustration. Suppose that $\|u\| = \|u\|_p$ with $p\in[1,2]$, and let us apply the just described scheme for upper-bounding Ψ_Π, assuming $\{Q\}\subset\Pi\subset\{S\in\mathbf{S}_+^m : S\preceq Q\}$ for some given $Q\succ 0$, so that $\Gamma(\Theta) = \mathrm{Tr}(Q\Theta)$, $\Theta\succeq 0$. The unit ball of the norm conjugate to $\|\cdot\|$, that is, the norm $\|\cdot\|_q$, $q=\frac{p}{p-1}\in[2,\infty]$, is the basic spectratope (in fact, ellitope)

$$\begin{array}{c}\mathcal{B}_* = \{y\in\mathbf{R}^\mu : \exists r\in\mathcal{R} := \{\mathbf{R}_+^\nu : \|r\|_{q/2}\le 1\} : S_\ell^2[y]\le r_\ell,\ 1\le\ell\le L=\nu\},\\ S_\ell[y] = y_\ell.\end{array}$$

As a result, Υ's from Remark 4.13 are collections of ν positive semidefinite 1×1 matrices, and we can identify them with ν-dimensional nonnegative vectors υ, resulting in $\lambda[\Upsilon] = \upsilon$ and $\mathcal{M}[\Upsilon] = \mathrm{Diag}\{\upsilon\}$. Furthermore, for *nonnegative* υ we clearly have $\phi_{\mathcal{R}}(\upsilon) = \|\upsilon\|_{p/(2-p)}$, so the optimization problem in (4.41) now reads

$$\overline{\Psi}_\Pi(H) = \inf_{\upsilon \in \mathbf{R}^\nu} \left\{ \tfrac{1}{4}\mathrm{Tr}(V\mathrm{Diag}^{-1}\{\upsilon\}V^T) + \|\upsilon\|_{p/(2-p)} : \upsilon > 0 \right\} \qquad [V = Q^{1/2}H],$$

and when setting $a_\ell = \|\mathrm{Col}_\ell[V]\|_2$, (4.41) becomes

$$\overline{\Psi}_\Pi(H) = \inf_{\upsilon > 0} \left\{ \frac{1}{4}\sum_\ell \frac{a_\ell^2}{\upsilon_\ell} + \|\upsilon\|_{p/(2-p)} \right\}.$$

This results in $\overline{\Psi}_\Pi(H) = \|[a_1; ...; a_\mu]\|_p$. Recalling what a_ℓ and V are, we end up with

$$\begin{array}{l}\forall P, \mathrm{Var}[P] \preceq Q : \\ \mathbf{E}_{\xi \sim P}\{\|H^T\xi\|\} \le \overline{\Psi}_\Pi(H) := \left\| [\|\mathrm{Row}_1[H^TQ^{1/2}]\|_2; \ldots; \|\mathrm{Row}_\nu[H^TQ^{1/2}]\|_2] \right\|_p.\end{array}$$

This result is quite transparent and could be easily obtained straightforwardly. Indeed, when $\mathrm{Var}[P] \preceq Q$, and $\xi \sim P$, the vector $\zeta = H^T\xi$ clearly satisfies $\mathbf{E}\{\zeta_i^2\} \le \sigma_i^2 := \|\mathrm{Row}_i[H^TQ^{1/2}]\|_2^2$, implying, due to $p \in [1,2]$, that $\mathbf{E}\{\sum_i |\zeta_i|^p\} \le \sum_i \sigma_i^p$, whence $\mathbf{E}\{\|\zeta\|_p\} \le \|[\sigma_1; ...; \sigma_\nu]\|_p$.

4.3.3.6 Putting things together

An immediate outcome of Corollaries 4.10 and 4.12 is the following recipe for building a "presumably good" linear estimate:

Proposition 4.14. *In the situation of Section 4.3.3.1 and under Assumptions* ***A****,* ***B****, and* ***R*** *(see Section 4.3.3.2) consider the convex optimization problem (for notation, see* (4.36) *and* (4.39)*)*

$$\mathrm{Opt} = \min_{H,\Lambda,\Upsilon,\Upsilon',\Theta} \left\{ \begin{array}{l} \phi_{\mathcal{T}}(\lambda[\Lambda]) + \phi_{\mathcal{R}}(\lambda[\Upsilon]) + \phi_{\mathcal{R}}(\lambda[\Upsilon']) + \Gamma(\Theta) : \\ \Lambda = \{\Lambda_k \succeq 0, k \le K\}, \Upsilon = \{\Upsilon_\ell \succeq 0, \ell \le L\}, \Upsilon' = \{\Upsilon'_\ell \succeq 0, \ell \le L\}, \\ \left[\begin{array}{c|c} \sum_k \mathcal{R}_k^*[\Lambda_k] & \frac{1}{2}[B^T - A^TH]M \\ \hline \frac{1}{2}M^T[B - H^TA] & \sum_\ell \mathcal{S}_\ell^*[\Upsilon_\ell] \end{array}\right] \succeq 0, \\ \left[\begin{array}{c|c} \Theta & \frac{1}{2}HM \\ \hline \frac{1}{2}M^TH^T & \sum_\ell \mathcal{S}_\ell^*[\Upsilon'_\ell] \end{array}\right] \succeq 0. \end{array} \right. \tag{4.42}$$

The problem is solvable, and the H-component H_* of its optimal solution yields linear estimate $\widehat{x}_{H_*}(\omega) = H_*^T\omega$ such that

$$\mathrm{Risk}_{\Pi,\|\cdot\|}[\widehat{x}_{H_*}(\cdot)|\mathcal{X}] \le \mathrm{Opt}. \tag{4.43}$$

Note that the only claim in Proposition 4.14 which is not an immediate consequence of Corollaries 4.10 and 4.12 is that problem (4.42) is solvable; this fact is readily given by the feasibility of the problem (by Lemma 4.44) and the coerciveness of the objective on the feasible set (recall that $\Gamma(\Theta)$ is coercive on $\mathbf{S}_+^m$ due to $\Pi \subset \mathrm{int}\,\mathbf{S}_+^m$ and that $y \mapsto My$ is an onto mapping, since $\mathcal{B}_*$ is full-dimensional).

4.3.3.7 Illustration: Covariance matrix estimation

Suppose that we observe a sample

$$\eta^T = \{\eta_k = A\xi_k\}_{k\leq T} \tag{4.44}$$

where A is a given $m\times n$ matrix, and $\xi_1,...,\xi_T$ are sampled, independently of each other, from a zero mean Gaussian distribution with unknown covariance matrix ϑ known to satisfy

$$\gamma\vartheta_* \preceq \vartheta \preceq \vartheta_*, \tag{4.45}$$

where $\gamma\geq 0$ and $\vartheta_*\succ 0$ are given. Our goal is to recover ϑ, and the norm on $\mathbf{S}^n$ in which the recovery error is measured satisfies Assumption **A**.

Processing the problem. We can process the problem just outlined as follows.

1. We represent the set $\{\vartheta\in\mathbf{S}^n_+ : \gamma\vartheta_*\preceq\vartheta\preceq\vartheta_*\}$ as the image of the matrix box

$$\mathcal{V}=\{v\in\mathbf{S}^n : \|v\|_{2,2}\leq 1\} \qquad [\|\cdot\|_{2,2}\text{: spectral norm}]$$

under affine mapping; specifically, we set

$$\vartheta_0=\frac{1+\gamma}{2}\vartheta_*,\ \sigma=\frac{1-\gamma}{2}$$

and treat the matrix

$$v=\sigma^{-1}\vartheta_*^{-1/2}(\vartheta-\vartheta_0)\vartheta_*^{-1/2} \qquad \left[\Leftrightarrow \vartheta=\vartheta_0+\sigma\vartheta_*^{1/2}v\vartheta_*^{1/2}\right]$$

as the signal underlying our observations. Note that our a priori information on ϑ reduces to $v\in\mathcal{V}$.

2. We pass from observations η_k to "lifted" observations $\eta_k\eta_k^T\in\mathbf{S}^m$, so that

$$\mathbf{E}\{\eta_k\eta_k^T\}=\mathbf{E}\{A\xi_k\xi_k^TA^T\}=A\vartheta A^T=A\underbrace{\left(\vartheta_0+\sigma A\vartheta_*^{1/2}v\vartheta_*^{1/2}\right)}_{\vartheta[v]}A^T,$$

and treat as "actual" observations the matrices

$$\omega_k=\eta_k\eta_k^T-A\vartheta_0A^T.$$

We have[8]

$$\omega_k=\mathcal{A}v+\zeta_k \text{ with } \mathcal{A}v=\sigma A\vartheta_*^{1/2}v\vartheta_*^{1/2}A^T \text{ and } \zeta_k=\eta_k\eta_k^T-A\vartheta[v]A^T. \tag{4.46}$$

Observe that random matrices $\zeta_1,...,\zeta_T$ are i.i.d. with zero mean and covariance mapping $\mathcal{Q}[v]$ (that of random matrix-valued variable $\zeta=\eta\eta^T-\mathbf{E}\{\eta\eta^T\}$, $\eta\sim$

[8]In our current considerations, we need to operate with linear mappings acting from $\mathbf{S}^p$ to $\mathbf{S}^q$. We treat $\mathbf{S}^k$ as Euclidean space equipped with the Frobenius inner product $\langle u,v\rangle=\mathrm{Tr}(uv)$ and denote linear mappings from $\mathbf{S}^p$ into $\mathbf{S}^q$ by capital calligraphic letters, like $\mathcal{A}$, $\mathcal{Q}$, etc. Thus, $\mathcal{A}$ in (4.46) denotes the linear mapping which, on closer inspection, maps matrix $v\in\mathbf{S}^n$ into the matrix $\mathcal{A}v=A[\vartheta[v]-\vartheta[0]]A^T$.

$\mathcal{N}(0, A\vartheta[v]A^T))$.

3. Let us $\succeq$-upper-bound the covariance mapping of ζ. Observe that $\mathcal{Q}[v]$ is a symmetric linear mapping of $\mathbf{S}^m$ into itself given by

$$\langle h, \mathcal{Q}[v]h\rangle = \mathbf{E}\{\langle h, \zeta\rangle^2\} = \mathbf{E}\{\langle h, \eta\eta^T\rangle^2\} - \langle h, \mathbf{E}\{\eta\eta^T\}\rangle^2, \;\; h \in \mathbf{S}^m.$$

Given $v \in \mathcal{V}$, let us set $\theta = \vartheta[v]$, so that $0 \preceq \theta \preceq \theta_*$, and let $\mathcal{H}(h) = \theta^{1/2}A^ThA\theta^{1/2}$. We have

$$\begin{array}{rcl}\langle h, \mathcal{Q}[v]h\rangle &=& \mathbf{E}_{\xi\sim\mathcal{N}(0,\theta)}\{\mathrm{Tr}^2(hA\xi\xi^TA^T)\} - \mathrm{Tr}^2(h\mathbf{E}_{\xi\sim\mathcal{N}(0,\theta)}\{A\xi\xi^TA^T\})\\ &=& \mathbf{E}_{\chi\sim\mathcal{N}(0,I_n)}\{\mathrm{Tr}^2(hA\theta^{1/2}\chi\chi^T\theta^{1/2}A^T))\} - \mathrm{Tr}^2(hA\theta A^T)\\ &=& \mathbf{E}_{\chi\sim\mathcal{N}(0,I_n)}\{(\chi^T\mathcal{H}(h)\chi)^2\} - \mathrm{Tr}^2(\mathcal{H}(h)).\end{array}$$

We have $\mathcal{H}(h) = U\mathrm{Diag}\{\lambda\}U^T$ with orthogonal U, so that

$$\begin{array}{l}\mathbf{E}_{\chi\sim\mathcal{N}(0,I_n)}\{(\chi^T\mathcal{H}(h)\chi)^2\} - \mathrm{Tr}^2(\mathcal{H}(h))\\ \quad = \mathbf{E}_{\bar\chi:=U^T\chi\sim\mathcal{N}(0,I_n)}\{(\bar\chi^T\mathrm{Diag}\{\lambda\}\bar\chi)^2\} - (\sum_i\lambda_i)^2\\ \quad = \mathbf{E}_{\bar\chi\sim\mathcal{N}(0,I_n)}\{(\sum_i\lambda_i\bar\chi_i^2)^2\} - (\sum_i\lambda_i)^2 = \sum_{i\neq j}\lambda_i\lambda_j + 3\sum_i\lambda_i^2 - (\sum_i\lambda_i)^2\\ \quad = 2\sum_i\lambda_i^2 = 2\mathrm{Tr}([\mathcal{H}(h)]^2).\end{array}$$

Thus,

$$\begin{array}{rcl}\langle h, \mathcal{Q}[v]h\rangle &=& 2\mathrm{Tr}([\mathcal{H}(h)]^2) = 2\mathrm{Tr}(\theta^{1/2}A^ThA\theta A^ThA\theta^{1/2})\\ &\leq& 2\mathrm{Tr}(\theta^{1/2}A^ThA\theta_*A^ThA\theta^{1/2}) \text{ [since } 0 \preceq \theta \preceq \theta_*]\\ &=& 2\mathrm{Tr}(\theta_*^{1/2}A^ThA\theta A^ThA\theta_*^{1/2}) \leq 2\mathrm{Tr}(\theta_*^{1/2}A^ThA\theta_*A^ThA\theta_*^{1/2})\\ &=& 2\mathrm{Tr}(\theta_*A^ThA\theta_*A^ThA).\end{array}$$

We conclude that

$$\forall v \in \mathcal{V} : \mathcal{Q}[v] \preceq \mathcal{Q}, \; \langle e, \mathcal{Q}h\rangle = 2\mathrm{Tr}(\vartheta_*A^ThA\vartheta_*A^TeA), \, e, h \in \mathbf{S}^m. \tag{4.47}$$

4. To continue, we need to set some additional notation to be used when operating with Euclidean spaces $\mathbf{S}^p$, $p = 1, 2, ...$

- We denote $\bar p = \frac{p(p+1)}{2} = \dim \mathbf{S}^p$, $\mathcal{I}_p = \{(i,j) : 1 \leq i \leq j \leq p\}$, and for $(i,j) \in \mathcal{I}_p$ set
$$e_p^{ij} = \left\{\begin{array}{ll} e_ie_i^T, & i = j\\ \frac{1}{\sqrt2}[e_ie_j^T + e_je_i^T], & i < j\end{array}\right.,$$
where the e_i are standard basic orths in $\mathbf{R}^p$. Note that $\{e_p^{ij} : (i,j) \in \mathcal{I}_p\}$ is the standard orthonormal basis in $\mathbf{S}^p$. Given $v \in \mathbf{S}^p$, we denote by $\mathrm{X}^p(v)$ the vector of coordinates of v in this basis:
$$\mathrm{X}^p_{ij}(v) = \mathrm{Tr}(ve_p^{ij}) = \left\{\begin{array}{ll} v_{ii}, & i = j\\ \sqrt2 v_{ij}, & i < j\end{array}\right., \; (i,j) \in \mathcal{I}_p.$$
Similarly, for $x \in \mathbf{R}^{\bar p}$, we index the entries in x by pairs ij, $(i,j) \in \mathcal{I}_p$, and set $\mathrm{V}^p(x) = \sum_{(i,j)\in\mathcal{I}_p} x_{ij}e_p^{ij}$, so that $v \mapsto \mathrm{X}^p(v)$ and $x \mapsto \mathrm{V}^p(x)$ are linear norm-preserving maps inverse to each other identifying the Euclidean spaces $\mathbf{S}^p$ and $\mathbf{R}^{\bar p}$ (recall that the inner products on these spaces are, respectively, the Frobenius and the standard one).

- Recall that $\mathcal{V}$ is the matrix box $\{v \in \mathbf{S}^n : v^2 \preceq I_n\} = \{v \in \mathbf{S}^n : \exists t \in \mathcal{T} := [0,1] : v^2 \preceq tI_n\}$. We denote by $\mathcal{X}$ the image of $\mathcal{V}$ under the mapping X^n:

$$\mathcal{X} = \{x \in \mathbf{R}^{\bar{n}} : \exists t \in \mathcal{T} : R^2[x] \preceq tI_n\},\ R[x] = \sum_{(i,j)\in\mathcal{I}_n} x_{ij}e_n^{ij},\ \ \bar{n} = \tfrac{1}{2}n(n+1).$$

  Note that $\mathcal{X}$ is a basic spectratope of size n.

Now we can assume that the signal underlying our observations is $x \in \mathcal{X}$, and the observations themselves are

$$w_k = \mathrm{X}^m(\omega_k) = \underbrace{\mathrm{X}^m(\mathcal{A}\mathrm{V}^n(x))}_{=:\overline{A}x} + z_k,\ \ z_k = \mathrm{X}^m(\zeta_k).$$

Note that $z_k \in \mathbf{R}^{\bar{m}}$, $1 \le k \le T$, are zero mean i.i.d. random vectors with covariance matrix $Q[x]$ satisfying, in view of (4.47), the relation

$$Q[x] \preceq Q,\ \text{where } Q_{ij,k\ell} = 2\mathrm{Tr}(\vartheta_* A^T e_m^{ij} A \vartheta_* A^T e_m^{k\ell} A),\ (i,j) \in \mathcal{I}_m, (k,\ell) \in \mathcal{I}_m.$$

Our goal is to estimate $\vartheta[v] - \vartheta[0]$, or, which is the same, to recover

$$\overline{B}x := \mathrm{X}^n(\vartheta[\mathrm{V}^n(x)] - \vartheta[0]).$$

We assume that the norm in which the estimation error is measured is "transferred" from $\mathbf{S}^n$ to $\mathbf{R}^{\bar{n}}$; we denote the resulting norm on $\mathbf{R}^{\bar{n}}$ by $\|\cdot\|$ and assume that the unit ball $\mathcal{B}_*$ of the conjugate norm $\|\cdot\|_*$ is given by spectratopic representation:

$$\begin{array}{l}\{u \in \mathbf{R}^{\bar{n}} : \|u\|_* \le 1\} = \{u \in \mathbf{R}^{\bar{n}} : \exists y \in \mathcal{Y} : u = My\},\\ \mathcal{Y} := \{y \in \mathbf{R}^q : \exists r \in \mathcal{R} : S_\ell^2[y] \preceq r_\ell I_{f_\ell},\ 1 \le \ell \le L\}.\end{array} \tag{4.48}$$

The formulated description of the estimation problem fits the premises of Proposition 4.14, specifically:

- the signal x underlying our observation $w^{(T)} = [w_1; ...; w_T]$ is known to belong to basic spectratope $\mathcal{X} \in \mathbf{R}^{\bar{n}}$, and the observation itself is of the form

$$w^{(T)} = \overline{A}^{(T)}x + z^{(T)},\ \overline{A}^{(T)} = \underbrace{[\overline{A}; ...; \overline{A}]}_{T},\ z^{(T)} = [z_1; ...; z_T];$$

- the noise $z^{(T)}$ is zero mean, and its covariance matrix is $\preceq Q_T := \mathrm{Diag}\underbrace{\{Q, ..., Q\}}_{T}$, which allows us to set $\Pi = \{Q_T\}$;
- our goal is to recover $\overline{B}x$, and the norm $\|\cdot\|$ in which the recovery error is measured satisfies (4.48).

Proposition 4.14 supplies the linear estimate

$$\widehat{x}(w^{(T)}) = \sum_{k=1}^{T} H_{*k}^T w_k$$

of $\overline{B}x$ with $H_* = [H_{*1}; ...; H_{*T}]$ stemming from the optimal solution to the convex optimization problem

$$\begin{array}{rcl}\mathrm{Opt} & = & \min\limits_{H=[H_1;...;H_T],\Lambda,\Upsilon} \Big\{ \mathrm{Tr}(\Lambda) + \phi_{\mathcal{R}}(\lambda[\Upsilon]) + \overline{\Psi}_{\{Q_T\}}(H_1, ..., H_T) : \\ & & \Lambda \in \mathbf{S}^n_+, \Upsilon = \{\Upsilon_\ell \succeq 0, \ell \leq L\}, \\ & & \left[\begin{array}{c|c} \mathcal{R}^*[\Lambda] & \frac{1}{2}[\overline{B}^T - \overline{A}^T \sum_k H_k]M \\ \hline \frac{1}{2}M^T[\overline{B} - [\sum_k H_k]^T \overline{A}] & \sum_\ell \mathcal{S}_\ell^*[\Upsilon_\ell] \end{array}\right] \succeq 0 \Big\}\end{array} \tag{4.49}$$

where

$$\mathcal{R}^*[\Lambda] \in \mathbf{S}^{\bar{n}} : (\mathcal{R}^*[\Lambda])_{ij,k\ell} = \mathrm{Tr}(\Lambda e_n^{ij} e_n^{k\ell}), \, (i,j) \in \mathcal{I}_n, \, (k,\ell) \in \mathcal{I}_n,$$

and (cf. (4.40))

$$\begin{array}{l}\overline{\Psi}_{\{Q_T\}}(H_1, ..., H_T) = \min\limits_{\Upsilon',\Theta} \Big\{ \mathrm{Tr}(Q_T\Theta) + \phi_{\mathcal{R}}(\lambda[\Upsilon']) : \Theta \in \mathbf{S}^{mT}, \Upsilon' = \{\Upsilon'_\ell \succeq 0, \ell \leq L\}, \\ \qquad \left[\begin{array}{c|c} \Theta & \frac{1}{2}[H_1M; ...; H_TM] \\ \hline \frac{1}{2}[M^TH_1^T, ..., M^TH_T^T] & \sum_\ell \mathcal{S}_\ell^*[\Upsilon'_\ell] \end{array}\right] \succeq 0 \Big\}.\end{array}$$

5. Evidently, the function $\overline{\Psi}_{\{Q_T\}}([H_1, ..., H_T])$ remains intact when permuting $H_1, ..., H_T$; with this in mind, it is clear that permuting $H_1, ..., H_T$ and keeping intact Λ and Υ is a symmetry of (4.49)—such a transformation maps the feasible set onto itself and preserves the value of the objective. Since (4.49) is convex and solvable, it follows that there exists an optimal solution to the problem with $H_1 = ... = H_T = H$. On the other hand,

$$\begin{array}{l}\overline{\Psi}_{\{Q_T\}}(H, ..., H) \\ = \min\limits_{\Upsilon',\Theta} \Big\{ \mathrm{Tr}(Q_T\Theta) + \phi_{\mathcal{R}}(\lambda[\Upsilon']) : \Theta \in \mathbf{S}^{mT}, \Upsilon' = \{\Upsilon'_\ell \succeq 0, \ell \leq L\} \\ \qquad \left[\begin{array}{c|c} \Theta & \frac{1}{2}[HM; ...; HM] \\ \hline \frac{1}{2}[M^TH^T, ..., M^TH^T] & \sum_\ell \mathcal{S}_\ell^*[\Upsilon'_\ell] \end{array}\right] \succeq 0 \Big\} \\ = \inf\limits_{\Upsilon',\Theta} \Big\{ \mathrm{Tr}(Q_T\Theta) + \phi_{\mathcal{R}}(\lambda[\Upsilon']) : \Theta \in \mathbf{S}^{mT}, \Upsilon' = \{\Upsilon'_\ell \succ 0, \ell \leq L\}, \\ \qquad \left[\begin{array}{c|c} \Theta & \frac{1}{2}[HM; ...; HM] \\ \hline \frac{1}{2}[M^TH^T, ..., M^TH^T] & \sum_\ell \mathcal{S}_\ell^*[\Upsilon'_\ell] \end{array}\right] \succeq 0 \Big\} \\ = \inf\limits_{\Upsilon',\Theta} \Big\{ \mathrm{Tr}(Q_T\Theta) + \phi_{\mathcal{R}}(\lambda[\Upsilon']) : \Theta \in \mathbf{S}^{mT}, \Upsilon' = \{\Upsilon'_\ell \succ 0, \ell \leq L\}, \\ \qquad \Theta \succeq \frac{1}{4}[HM; ...; HM] \left[\sum_\ell \mathcal{S}_\ell^*[\Upsilon'_\ell]\right]^{-1} [HM; ...; HM]^T \Big\} \\ = \inf\limits_{\Upsilon'} \Big\{ \phi_{\mathcal{R}}(\lambda[\Upsilon']) + \frac{T}{4}\mathrm{Tr}\left(QHM \left[\sum_\ell \mathcal{S}_\ell^*[\Upsilon'_\ell]\right]^{-1} M^TH^T \right) : \Upsilon' = \{\Upsilon'_\ell \succ 0, \ell \leq L\} \Big\}\end{array}$$

due to $Q_T = \mathrm{Diag}\{Q, ..., Q\}$, and we arrive at

$$\begin{array}{l}\overline{\Psi}_{\{Q_T\}}(H, ..., H) = \min\limits_{\Upsilon',G} \Big\{ T\mathrm{Tr}(QG) + \phi_{\mathcal{R}}(\lambda[\Upsilon']) : \Upsilon' = \{\Upsilon'_\ell \succeq 0, \ell \leq L\}, \\ \qquad G \in \mathbf{S}^m, \left[\begin{array}{c|c} G & \frac{1}{2}HM \\ \hline \frac{1}{2}M^TH^T & \sum_\ell \mathcal{S}_\ell^*[\Upsilon'_\ell] \end{array}\right] \succeq 0 \Big\}\end{array} \tag{4.50}$$

(we have used the Schur Complement Lemma combined with the fact that $\sum_\ell \mathcal{S}_\ell^*[\Upsilon'_\ell] \succ 0$ whenever $\Upsilon'_\ell \succ 0$ for all ℓ; see Lemma 4.44).

In view of the above observations, when replacing variables H and G with $\overline{H} = TH$ and $\overline{G} = T^2G$, respectively, problem (4.49), (4.50) becomes

$$\begin{array}{rcl}\mathrm{Opt} & = & \min\limits_{\overline{H},\overline{G},\Lambda,\Upsilon,\Upsilon'} \left\{ \mathrm{Tr}(\Lambda) + \phi_{\mathcal{R}}(\lambda[\Upsilon]) + \phi_{\mathcal{R}}(\lambda[\Upsilon']) + \frac{1}{T}\mathrm{Tr}(Q\overline{G}) : \right. \\ & & \left. \begin{array}{l} \Lambda \in \mathbf{S}^n_+, \Upsilon = \{\Upsilon_\ell \succeq 0, \ell \leq L\}, \Upsilon' = \{\Upsilon'_\ell \succeq 0, \ell \leq L\}, \\ \left[\begin{array}{c|c} \mathcal{R}^*[\Lambda] & \frac{1}{2}[\overline{B}^T - \overline{A}^T\overline{H}]M \\ \hline \frac{1}{2}M^T[\overline{B} - \overline{H}^T\overline{A}] & \sum_\ell \mathcal{S}^*_\ell[\Upsilon_\ell] \end{array}\right] \succeq 0, \\ \left[\begin{array}{c|c} \overline{G} & \frac{1}{2}\overline{H}M \\ \hline \frac{1}{2}M^T\overline{H}^T & \sum_\ell \mathcal{S}^*_\ell[\Upsilon'_\ell] \end{array}\right] \succeq 0 \end{array} \right\}, \end{array} \tag{4.51}$$

and the estimate

$$\widehat{x}(w^T) = \frac{1}{T}\overline{H}^T \sum_{k=1}^T w_k$$

brought about by an optimal solution to (4.51) satisfies $\mathrm{Risk}_{\Pi,\|\cdot\|}[\widehat{x}|\mathcal{X}] \leq \mathrm{Opt}$ where $\Pi = \{Q_T\}$.

4.3.3.8 Estimation from repeated observations

Consider the special case of the situation from Section 4.3.3.1 where observation ω in (4.31) is a T-element sample $\omega = [\bar{\omega}_1; ...; \bar{\omega}_T]$ with components

$$\bar{\omega}_t = \bar{A}x + \xi_t, \; t = 1, ..., T$$

and ξ_t are i.i.d. observation noises with *zero mean* distribution $\bar{P}$ satisfying $\bar{P} \lhd \bar{\Pi}$ for some convex compact set $\bar{\Pi} \subset \mathrm{int}\,\mathbf{S}^{\bar{m}}_+$. In other words, we are in the situation where

$$\begin{array}{l} A = [\underbrace{\bar{A}; ...; \bar{A}}_{T}] \in \mathbf{R}^{m\times n} \text{ for some } \bar{A} \in \mathbf{R}^{\bar{m}\times n} \text{ and } m = T\bar{m}, \\ \Pi = \{Q = \mathrm{Diag}\{\underbrace{\bar{Q}, ..., \bar{Q}}_{T}\}, \bar{Q} \in \bar{\Pi}\}. \end{array}$$

The same argument as used in item 5 of Section 4.3.3.7 above justifies the following

Proposition 4.15. *In the situation in question and under Assumptions **A**, **B**, and **R** the linear estimate of Bx yielded by an optimal solution to problem (4.42) can be found as follows. Consider the convex optimization problem*

$$\begin{array}{l}\overline{\mathrm{Opt}} = \min\limits_{\bar{H},\Lambda,\Upsilon,\Upsilon',\bar{\Theta}} \left\{ \phi_{\mathcal{T}}(\lambda[\Lambda]) + \phi_{\mathcal{R}}(\lambda[\Upsilon]) + \phi_{\mathcal{R}}(\lambda[\Upsilon']) + \frac{1}{T}\overline{\Gamma}(\bar{\Theta}) : \right. \\ \left. \begin{array}{l} \Lambda = \{\Lambda_k \succeq 0, k \leq K\}, \; \Upsilon = \{\Upsilon_\ell \succeq 0, \ell \leq L\}, \; \Upsilon' = \{\Upsilon'_\ell \succeq 0, \ell \leq L\}, \\ \left[\begin{array}{c|c} \sum_k \mathcal{R}^*_k[\Lambda_k] & \frac{1}{2}[B^T - A^T\bar{H}]M \\ \hline \frac{1}{2}M^T[B - \bar{H}^TA] & \sum_\ell \mathcal{S}^*_\ell[\Upsilon_\ell] \end{array}\right] \succeq 0, \\ \left[\begin{array}{c|c} \bar{\Theta} & \frac{1}{2}\bar{H}M \\ \hline \frac{1}{2}M^T\bar{H}^T & \sum_\ell \mathcal{S}^*_\ell[\Upsilon'_\ell] \end{array}\right] \succeq 0 \end{array} \right\} \end{array} \tag{4.52}$$

where

$$\overline{\Gamma}(\bar{\Theta}) = \max_{\bar{Q}\in\bar{\Pi}} \mathrm{Tr}(\bar{Q}\bar{\Theta}).$$

The problem is solvable, and the estimate in question is yielded by the $\bar{H}$-component

$\bar{H}_*$ of the optimal solution according to

$$\widehat{x}([\bar{\omega}_1;...;\bar{\omega}_T]) = \frac{1}{T}\bar{H}_*^T \sum_{t=1}^{T} \bar{\omega}_t.$$

The upper bound provided by Proposition 4.14 on the risk $\mathrm{Risk}_{\Pi,\|\cdot\|}[\widehat{x}(\cdot)|\mathcal{X}]$ *of this estimate is equal to* $\overline{\mathrm{Opt}}$.

The advantage of this result as compared to what is stated under the circumstances by Proposition 4.14 is that the sizes of optimization problem (4.52) are independent of T.

4.3.3.9 Near-optimality in the Gaussian case

The risk of the linear estimate $\widehat{x}_{H_*}(\cdot)$ constructed in (4.42) can be compared to the minimax optimal risk of recovering Bx, $x \in \mathcal{X}$, from observations corrupted by zero mean Gaussian noise with covariance matrix from Π. Formally, the minimax risk is defined as

$$\mathrm{RiskOpt}_{\Pi,\|\cdot\|}[\mathcal{X}] = \sup_{Q\in\Pi}\inf_{\widehat{x}(\cdot)}\left[\sup_{x\in\mathcal{X}} \mathbf{E}_{\xi\sim\mathcal{N}(0,Q)}\{\|Bx - \widehat{x}(Ax+\xi)\|\}\right] \tag{4.53}$$

where the infimum is taken over all estimates.

Proposition 4.16. *Under the premise and in the notation of Proposition 4.14, we have*

$$\mathrm{RiskOpt}_{\Pi,\|\cdot\|}[\mathcal{X}] \geq \frac{\mathrm{Opt}}{64\sqrt{(2\ln F + 10\ln 2)(2\ln D + 10\ln 2)}}, \tag{4.54}$$

where

$$D = \sum_k d_k,\ F = \sum_\ell f_\ell. \tag{4.55}$$

Thus, the upper bound Opt *on the risk* $\mathrm{Risk}_{\Pi,\|\cdot\|}[\widehat{x}_{H_*}|\mathcal{X}]$ *of the presumably good linear estimate* $\widehat{x}_{H_*}$ *yielded by an optimal solution to optimization problem* (4.42) *is within logarithmic in the sizes of spectratopes* $\mathcal{X}$ *and* $\mathcal{B}_*$ *factor of the Gaussian minimax risk* $\mathrm{RiskOpt}_{\Pi,\|\cdot\|}[\mathcal{X}]$.

For the proof, see Section 4.8.5. The key component of the proof is the following fact important in its own right (for proof, see Section 4.8.4):

Lemma 4.17. *Let Y be an $N\times\nu$ matrix, let $\|\cdot\|$ be a norm on $\mathbf{R}^\nu$ such that the unit ball $\mathcal{B}_*$ of the conjugate norm is the spectratope* (4.34)*, and let $\zeta\sim\mathcal{N}(0,Q)$ for some positive semidefinite $N\times N$ matrix Q. Then the best upper bound on $\psi_Q(Y) := \mathbf{E}\{\|Y^T\zeta\|\}$ yielded by Lemma 4.11, that is, the optimal value* $\mathrm{Opt}[Q]$ *in the convex optimization problem (cf.* (4.40)*)*

$$\begin{array}{rcl}\mathrm{Opt}[Q] & = & \min\limits_{\Theta,\Upsilon}\left\{\phi_{\mathcal{R}}(\lambda[\Upsilon]) + \mathrm{Tr}(Q\Theta) : \Upsilon = \{\Upsilon_\ell \succeq 0, 1\leq \ell\leq L\},\right.\\ && \left.\Theta\in\mathbf{S}^N,\ \left[\begin{array}{c|c}\Theta & \frac{1}{2}YM\\ \hline \frac{1}{2}M^TY^T & \sum_\ell S_\ell^*[\Upsilon_\ell]\end{array}\right]\succeq 0\right\}\end{array} \tag{4.56}$$

(for notation, see Lemma 4.11 and (4.36)), satisfies the identity

$$\begin{array}{l}\forall(Q\succeq 0):\\ \mathrm{Opt}[Q]=\overline{\mathrm{Opt}}[Q]:=\min\limits_{G,\Upsilon=\{\Upsilon_\ell,\ell\leq L\}}\Big\{\phi_{\mathcal{R}}(\lambda[\Upsilon])+\mathrm{Tr}(G):\Upsilon_\ell\succeq 0,\\ \qquad\left[\begin{array}{c|c}G & \frac{1}{2}Q^{1/2}YM\\ \hline \frac{1}{2}M^TY^TQ^{1/2} & \sum_\ell \mathcal{S}_\ell^*[\Upsilon_\ell]\end{array}\right]\succeq 0\Big\},\end{array}\tag{4.57}$$

and is a tight bound on $\psi_Q(Y)$, namely,

$$\psi_Q(Y)\leq \mathrm{Opt}[Q]\leq 22\sqrt{2\ln F+10\ln 2}\,\psi_Q(Y),\tag{4.58}$$

where $F=\sum_\ell f_\ell$ is the size of the spectratope (4.34).

Besides this, for all $\varkappa\geq 1$ one has

$$\mathrm{Prob}_\zeta\left\{\|Y^T\zeta\|\geq\frac{\mathrm{Opt}[Q]}{4\varkappa}\right\}\geq\beta_\varkappa:=1-\frac{\mathrm{e}^{3/8}}{2}-2F\mathrm{e}^{-\varkappa^2/2}.\tag{4.59}$$

In particular, when selecting $\varkappa=\sqrt{2\ln F+10\ln 2}$, we obtain

$$\mathrm{Prob}_\zeta\left\{\|Y^T\zeta\|\geq\frac{\mathrm{Opt}[Q]}{4\sqrt{2\ln F+10\ln 2}}\right\}\geq 0.2100>\tfrac{3}{16}.\tag{4.60}$$

4.4 LINEAR ESTIMATES OF STOCHASTIC SIGNALS

In the recovery problem considered so far in this chapter, the signal x underlying observation $\omega=Ax+\xi$ was "deterministic uncertain but bounded"—all the a priori information on x was that $x\in\mathcal{X}$ for a given signal set $\mathcal{X}$. There is a well-known alternative model, where the signal x has a random component, specifically,

$$x=[\eta;u]$$

where the "stochastic component" η is random with (partly) known probability distribution P_η, and the "deterministic component" u is known to belong to a given set $\mathcal{X}$. As a typical example, consider a linear dynamical system given by

$$\begin{array}{rcl}y_{t+1}&=&P_ty_t+\eta_t+u_t,\\ \omega_t&=&C_ty_t+\xi_t,\quad 1\leq t\leq T,\end{array}\tag{4.61}$$

where y_t, η_t, and u_t are, respectively, the state, the random "process noise," and the deterministic "uncertain but bounded" disturbance affecting the system at time t, ω_t is the output (it is what we observe at time t), and ξ_t is the observation noise. We assume that the matrices P_t, C_t are known in advance. Note that the trajectory

$$y=[y_1;...;y_T]$$

of the states depends not only on the trajectories of process noises η_t and disturbances u_t, but also on the initial state y_1, which can be modeled as a realization of either the initial noise η_0, or the initial disturbance u_0. When $u_t\equiv 0$, $y_1=\eta_0$

and the random vectors $\{\eta_t, 0 \leq t \leq T, \xi_t, 1 \leq t \leq T\}$ are zero mean Gaussian independent of each other, (4.61) is the model underlying the celebrated *Kalman filter* [143, 144, 171, 172].

Now, given model (4.61), we can use the equations of the model to represent the trajectory of the states as a linear image of the trajectory of noises $\eta = \{\eta_t\}$ and the trajectory of disturbances $u = \{u_t\}$,

$$y = P\eta + Qu$$

(recall that the initial state is either the component η_0 of η, or the component u_0 of u), and our "full observation" becomes

$$\omega = [\omega_1; ...; \omega_T] = A[\eta; u] + \xi, \;\; \xi = [\xi_1, ..., \xi_T].$$

A typical statistical problem associated with the outlined situation is to estimate the linear image $B[\eta; u]$ of the "signal" $x = [\eta; u]$ underlying the observation. For example, when speaking about (4.61), the goal could be to recover y_{T+1} ("forecast").

We arrive at the following estimation problem:

> Given noisy observation
> $$\omega = Ax + \xi \in \mathbf{R}^m$$
> of signal $x = [\eta; u]$ with random component $\eta \in \mathbf{R}^p$ and deterministic component u known to belong to a given set $\mathcal{X} \subset \mathbf{R}^q$, we want to recover the image $Bx \in \mathbf{R}^\nu$ of the signal. Here A and B are given matrices, η is independent of ξ, and we have a priori (perhaps, incomplete) information on the probability distribution P_η of η, specifically, we know that $P_\eta \in \mathcal{P}_\eta$ for a given family $\mathcal{P}_\eta$ of probability distributions. Similarly, we assume that what we know about the noise ξ is that its distribution belongs to a given family $\mathcal{P}_\xi$ of distributions on the observation space.

Given a norm $\|\cdot\|$ on the image space of B, it makes sense to specify the risk of a candidate estimate $\widehat{x}(\omega)$ by taking the expectation of the norm $\|\widehat{x}(A[\eta; u] + \xi) - B[\eta; u]\|$ of the error over *both* ξ and η and then taking the supremum of the result over the allowed distributions of η, ξ and over $u \in \mathcal{X}$:

$$\mathrm{Risk}_{\|\cdot\|}[\widehat{x}] = \sup_{u \in \mathcal{X}} \sup_{P_\xi \in \mathcal{P}_\xi, P_\eta \in \mathcal{P}_\eta} \mathbf{E}_{[\xi;\eta] \sim P_\xi \times P_\eta} \left\{ \|\widehat{x}(A[\eta; u] + \xi) - B[\eta; u]\| \right\}.$$

When $\|\cdot\| = \|\cdot\|_2$ and all distributions from $\mathcal{P}_\xi$ and $\mathcal{P}_\eta$ are with zero means and finite covariance matrices, it is technically more convenient to operate with the *Euclidean risk*

$$\mathrm{Risk}_{\mathrm{Eucl}}[\widehat{x}] = \left[\sup_{u \in \mathcal{X}} \sup_{P_\xi \in \mathcal{P}_\xi, P_\eta \in \mathcal{P}_\eta} \mathbf{E}_{[\xi;\eta] \sim P_\xi \times P_\eta} \left\{ \|\widehat{x}(A[\eta; u] + \xi) - B[\eta; u]\|_2^2 \right\} \right]^{1/2}.$$

Our next goal is to show that as far as the design of "presumably good" *linear estimates* $\widehat{x}(\omega) = H^T\omega$ is concerned, the techniques developed so far can be straightforwardly extended to the case of signals with random component.

4.4.1 Minimizing Euclidean risk

For the time being, assume that $\mathcal{P}_\xi$ is comprised of all probability distributions P on $\mathbf{R}^m$ with zero mean and covariance matrices $\mathrm{Cov}[P] = \mathbf{E}_{\xi\sim P}\{\xi\xi^T\}$ running through a computationally tractable convex compact subset $\mathcal{Q}_\xi \subset \mathrm{int}\,\mathbf{S}^m_+$, and $\mathcal{P}_\eta$ is comprised of all probability distributions P on $\mathbf{R}^p$ with zero mean and covariance matrices running through a computationally tractable convex compact subset $\mathcal{Q}_\eta \subset \mathrm{int}\,\mathbf{S}^p_+$. Let, in addition, $\mathcal{X}$ be a basic spectratope:

$$\mathcal{X} = \{x \in \mathbf{R}^q : \exists t \in \mathcal{T} : R_k^2[x] \preceq t_k I_{d_k},\ k \leq K\}$$

with our standard restrictions on $\mathcal{T}$ and $R_k[\cdot]$. Let us derive an efficiently solvable convex optimization problem "responsible" for a presumably good, in terms of its Euclidean risk, linear estimate.

For a linear estimate $H^T\omega$, $u \in \mathcal{X}$, $P_\xi \in \mathcal{P}_\xi$, $P_\eta \in \mathcal{P}_\eta$, denoting by Q_ξ and Q_η the covariance matrices of P_ξ and P_η, and partitioning A as $A = [A_\eta, A_u]$ and $B = [B_\eta, B_u]$ according to the partition $x = [\eta; u]$, we have

$$\begin{array}{l}
\mathbf{E}_{[\xi;\eta]\sim P_\xi\times P_\eta}\left\{\|H^T(A[\eta;u]+\xi) - B[\eta;u]\|_2^2\right\}\\
\quad = \mathbf{E}_{[\xi;\eta]\sim P_\xi\times P_\eta}\left\{\|[H^TA_\eta - B_\eta]\eta + H^T\xi + [H^TA_u - B_u]u\|_2^2\right\}\\
\quad = u^T[B_u - H^TA_u]^T[B_u - H^TA_u]u + \mathbf{E}_{\xi\sim P_\xi}\left\{\mathrm{Tr}(H^T\xi\xi^TH)\right\}\\
\quad\quad +\mathbf{E}_{\eta\sim P_\eta}\left\{\mathrm{Tr}([B_\eta - H^TA_\eta]\eta\eta^T[B_\eta - H^TA_\eta]^T)\right\}\\
\quad = u^T[B_u - H^TA_u]^T[B_u - H^TA_u]u + \mathrm{Tr}(H^TQ_\xi H)\\
\quad\quad +\mathrm{Tr}([B_\eta - H^TA_\eta]Q_\eta[B_\eta - H^TA_\eta]^T).
\end{array}$$

Hence, the squared Euclidean risk of the linear estimate $\widehat{x}_H(\omega) = H^T\omega$ is

$$\begin{array}{rcl}
\mathrm{Risk}^2_{\mathrm{Eucl}}[\widehat{x}_H] &=& \Phi(H) + \Psi_\xi(H) + \Psi_\eta(H),\\
\Phi(H) &=& \max\limits_{u\in\mathcal{X}} u^T[B_u - H^TA_u]^T[B_u - H^TA_u]u,\\
\Psi_\xi(H) &=& \max\limits_{Q\in\mathcal{Q}_\xi} \mathrm{Tr}(H^TQH),\\
\Psi_\eta(H) &=& \max_{Q\in\mathcal{Q}_\eta} \mathrm{Tr}([B_\eta - H^TA_\eta]Q[B_\eta - H^TA_\eta]^T).
\end{array}$$

Functions Ψ_ξ and Ψ_η are convex and efficiently computable, function $\Phi(H)$, by Proposition 4.8, admits an efficiently computable convex upper bound

$$\overline{\Phi}(H) = \min_\Lambda \left\{ \phi_{\mathcal{T}}(\lambda[\Lambda]) : \Lambda = \{\Lambda_k \succeq 0, k \leq K\}, \right.$$
$$\left. [B_u - H^TA_u]^T[B_u - H^TA_u] \preceq \textstyle\sum_k \mathcal{R}_k^*[\Lambda_k] \right\}$$

which is tight within the factor $2\max[\ln(2\sum_k d_k), 1]$ (see Proposition 4.8). Thus, the efficiently solvable convex problem yielding a presumably good linear estimate is

$$\mathrm{Opt} = \min_H \left[\overline{\Phi}(H) + \Psi_\xi(H) + \Psi_\eta(H)\right];$$

the Euclidean risk of the linear estimate $H_*^T\omega$ yielded by the optimal solution to the problem is upper-bounded by $\sqrt{\mathrm{Opt}}$ and is within factor $\sqrt{2\max[\ln(2\sum_k d_k), 1]}$ of the minimal Euclidean risk achievable with linear estimates.

4.4.2 Minimizing $\|\cdot\|$-risk

Now let $\mathcal{P}_\xi$ be comprised of all probability distributions P on $\mathbf{R}^m$ with matrices of second moments $\mathrm{Var}[P] = \mathbf{E}_{\xi\sim P}\{\xi\xi^T\}$ running through a computationally tractable convex compact subset $\mathcal{Q}_\xi \subset \mathrm{int}\,\mathbf{S}^m_+$, and $\mathcal{P}_\eta$ be comprised of all probability distributions P on $\mathbf{R}^p$ with matrices of second moments $\mathrm{Var}[P]$ running through a computationally tractable convex compact subset $\mathcal{Q}_\eta \subset \mathrm{int}\,\mathbf{S}^p_+$. Let, as above, $\mathcal{X}$ be a basic spectratope,

$$\mathcal{X} = \{u \in \mathbf{R}^n : \exists t \in \mathcal{T} : R_k^2[u] \preceq t_k I_{d_k},\ k \le K\},$$

and let $\|\cdot\|$ be such that the unit ball $\mathcal{B}_*$ of the conjugate norm $\|\cdot\|_*$ is a spectratope:

$$\mathcal{B}_* = \{y : \|y\|_* \le 1\} = \left\{y \in \mathbf{R}^\nu : \exists (r \in \mathcal{R}, z \in \mathbf{R}^N) : y = Mz, S_\ell^2[z] \preceq r_\ell I_{f_\ell}, \ell \le L\right\},$$

with our standard restrictions on $\mathcal{T}, \mathcal{R}, R_k[\cdot]$ and $S_\ell[\cdot]$. Here the efficiently solvable convex optimization problem "responsible" for a presumably good, in terms of its risk $\mathrm{Risk}_{\|\cdot\|}$, linear estimate can be built as follows.

For a linear estimate $H^T\omega$, $u \in \mathcal{X}$, $P_\xi \in \mathcal{P}_\xi$, $P_\eta \in \mathcal{P}_\eta$, denoting by Q_ξ and Q_η the matrices of second moments of P_ξ and P_η, and partitioning A as $A = [A_\eta, A_u]$ and $B = [B_\eta, B_u]$ according to the partition $x = [\eta; u]$, we have

$$\begin{array}{l}
\mathbf{E}_{[\xi;\eta]\sim P_\xi\times P_\eta}\left\{\|H^T(A[\eta;u]+\xi) - B[\eta;u]\|\right\} \\
\quad = \ \mathbf{E}_{[\xi;\eta]\sim P_\xi\times P_\eta}\left\{\|[H^TA_\eta - B_\eta]\eta + H^T\xi + [H^TA_u - B_u]u\|\right\} \\
\quad \le \ \|[B_u - H^TA_u]u\| + \mathbf{E}_{\xi\sim P_\xi}\left\{\|H^T\xi\|\right\} + \mathbf{E}_{\eta\sim P_\eta}\left\{\|[B_\eta - H^TA_\eta]\eta\|\right\}.
\end{array}$$

It follows that for a linear estimate $\widehat{x}_H(\omega) = H^T\omega$ one has

$$\begin{array}{rcl}
\mathrm{Risk}_{\|\cdot\|}[\widehat{x}_H] & \le & \Phi(H) + \Psi_\xi(H) + \Psi_\eta(H),\\
\Phi(H) & = & \max_{u\in\mathcal{X}} \|[B_u - H^TA_u]u\|,\\
\Psi_\xi(H) & = & \sup_{P_\xi\in\mathcal{P}_\xi} \mathbf{E}_{\xi\sim P_\xi}\{\|H^T\xi\|\},\\
\Psi_\eta(H) & = & \sup_{P_\eta\in\mathcal{P}_\eta} \mathbf{E}_{\xi\sim P_\xi}\{\|[B_\eta - H^TA_\eta]\eta\|\}.
\end{array}$$

As was shown in Section 4.3.3.3, the functions Φ, Ψ_ξ, Ψ_η admit efficiently computable upper bounds as follows (for notation, see Section 4.3.3.3):

$$\begin{array}{l}
\Phi(H) \le \overline{\Phi}(H) := \min\limits_{\Lambda,\Upsilon}\Bigg\{\phi_{\mathcal{T}}(\lambda[\Lambda]) + \phi_{\mathcal{R}}(\lambda[\Upsilon]) : \\
\qquad\qquad \Lambda = \{\Lambda_k \succeq 0, k \le K\}, \Upsilon = \{\Upsilon_\ell \succeq 0, \ell \le L\} \\
\qquad\qquad \left[\begin{array}{c|c} \sum_k \mathcal{R}_k^*[\Lambda_k] & \frac{1}{2}[B_u^T - A_u^TH]M \\ \hline \frac{1}{2}M^T[B_u - H^TAu] & \sum_\ell \mathcal{S}_\ell[\Upsilon_\ell]\end{array}\right] \succeq 0 \Bigg\};\\
\Psi_\xi(H) \le \overline{\Psi}_\xi(H) := \min\limits_{\Upsilon,G}\Bigg\{\phi_{\mathcal{R}}(\lambda[\Upsilon]) + \max_{Q\in\mathcal{Q}_\xi}\mathrm{Tr}(GQ) : \Upsilon = \{\Upsilon_\ell \succeq 0, \ell \le L\} \\
\qquad\qquad \left[\begin{array}{c|c} G & \frac{1}{2}HM \\ \hline \frac{1}{2}M^TH^T & \sum_\ell \mathcal{S}_\ell[\Upsilon_\ell]\end{array}\right] \succeq 0 \Bigg\},\\
\Psi_\eta(H) \le \overline{\Psi}_\eta(H) := \min\limits_{\Upsilon,G}\Bigg\{\phi_{\mathcal{R}}(\lambda[\Upsilon]) + \max_{Q\in\mathcal{Q}_\eta}\mathrm{Tr}(GQ) : \Upsilon = \{\Upsilon_\ell \succeq 0, \ell \le L\},\\
\qquad\qquad \left[\begin{array}{c|c} G & \frac{1}{2}[B_\eta^T - A_\eta^TH]M \\ \hline \frac{1}{2}M^T[B_\eta - H^TA_\eta] & \sum_\ell \mathcal{S}_\ell[\Upsilon_\ell]\end{array}\right] \succeq 0 \Bigg\},
\end{array}$$

and these bounds are reasonably tight (for details on tightness, see Proposition 4.8

and Lemma 4.17). As a result, to get a presumably good linear estimate, one needs to solve the efficiently solvable convex optimization problem

$$\mathrm{Opt} = \min_H \left[\overline{\Phi}(H) + \overline{\Psi}_\xi(H) + \overline{\Psi}_\eta(H)\right].$$

The linear estimate $\widehat{x}_{H_*} = H_*^T \omega$ yielded by an optimal solution H_* to this problem admits the risk bound

$$\mathrm{Risk}_{\|\cdot\|}[\widehat{x}_{H_*}] \leq \mathrm{Opt}.$$

Note that the above derivation did not use independence of ξ and η.

4.5 LINEAR ESTIMATION UNDER UNCERTAIN-BUT-BOUNDED NOISE

So far, the main subject of our interest was recovering (linear images of) signals via indirect observations of these signals corrupted by random noise. In this section, we focus on alternative observation schemes – those with "uncertain-but-bounded" and "mixed" noise.

4.5.1 Uncertain-but-bounded noise

Consider the estimation problem where, given observation

$$\omega = Ax + \eta \tag{4.62}$$

of unknown signal x known to belong to a given signal set $\mathcal{X}$, one wants to recover linear image Bx of x. Here A and B are given $m \times n$ and $\nu \times n$ matrices. The situation looks exactly as before, the difference with our previous considerations is that now we do not assume the observation noise to be random—all we assume about η is that it belongs to a given compact set $\mathcal{H}$ ("uncertain-but-bounded observation noise"). In the situation in question, a natural definition of the risk on $\mathcal{X}$ of a candidate estimate $\omega \mapsto \widehat{x}(\omega)$ is

$$\mathrm{Risk}_{\mathcal{H},\|\cdot\|}[\widehat{x}|\mathcal{X}] = \sup_{x\in X, \eta\in\mathcal{H}} \|Bx - \widehat{x}(Ax+\eta)\|$$

("$\mathcal{H}$-risk").

We are about to prove that when $\mathcal{X}$, $\mathcal{H}$, and the unit ball $\mathcal{B}_*$ of the norm $\|\cdot\|_*$ conjugate to $\|\cdot\|$ are spectratopes, which we assume from now on, an efficiently computable linear estimate is near-optimal in terms of its $\mathcal{H}$-risk.

Our initial observation is that in this case the model (4.62) reduces straightforwardly to the model without observation noise. Indeed, let $\mathcal{Y} = \mathcal{X} \times \mathcal{H}$; then $\mathcal{Y}$ is a spectratope, and we lose nothing when assuming that the signal underlying observation ω is $y = [x; \eta] \in \mathcal{Y}$:

$$\omega = Ax + \eta = \bar{A}y,\ \bar{A} = [A, I_m],$$

while the entity to be recovered is

$$Bx = \bar{B}y, \ \bar{B} = [B, 0_{\nu \times m}].$$

With these conventions, the $\mathcal{H}$-risk of a candidate estimate $\widehat{x}(\cdot) : \mathbf{R}^m \to \mathbf{R}^\nu$ becomes the quantity

$$\mathrm{Risk}_{\|\cdot\|}[\widehat{x}|\mathcal{X} \times \mathcal{H}] = \sup_{y=[x;\eta]\in\mathcal{X}\times\mathcal{H}} \|\bar{B}y - \widehat{x}(\bar{A}y)\|,$$

and we indeed arrive at the situation where the observation noise is identically zero.

To avoid messy notation, let us assume that the outlined reduction has been carried out in advance, so that

(!) *The problem of interest is to recover the linear image $Bx \in \mathbf{R}^\nu$ of an unknown signal x known to belong to a given spectratope $\mathcal{X}$ (which, as always, we can assume w.l.o.g. to be basic) from (noiseless) observation*

$$\omega = Ax \in \mathbf{R}^m.$$

The risk of a candidate estimate is defined as

$$\mathrm{Risk}_{\|\cdot\|}[\widehat{x}|\mathcal{X}] = \sup_{x\in\mathcal{X}} \|Bx - \widehat{x}(Ax)\|,$$

where $\|\cdot\|$ is a given norm with a spectratope $\mathcal{B}_*$—see (4.34)—as the unit ball of the conjugate norm:

$$\begin{array}{rcl}\mathcal{X} & = & \{x \in \mathbf{R}^n : \exists t \in \mathcal{T} : R_k^2[x] \preceq t_k I_{d_k}, k \leq K\}, \\ \mathcal{B}_* & = & \{z \in \mathbf{R}^\nu : \exists y \in \mathcal{Y} : z = My\}, \\ & & \mathcal{Y} := \{y \in \mathbf{R}^q : \exists r \in \mathcal{R} : S_\ell^2[y] \preceq r_\ell I_{f_\ell}, 1 \leq \ell \leq L\},\end{array} \tag{4.63}$$

with our standard restrictions on $\mathcal{T}, \mathcal{R}$ and $R_k[\cdot]$, $S_\ell[\cdot]$.

4.5.1.1 Building a linear estimate

Let us build a presumably good linear estimate. For a linear estimate $\widehat{x}_H(\omega) = H^T\omega$, we have

$$\begin{array}{rcl}\mathrm{Risk}_{\|\cdot\|}[\widehat{x}_H|\mathcal{X}] & = & \max_{x\in\mathcal{X}} \|(B - H^TA)x\| \\ & = & \max_{[u;x]\in\mathcal{B}_*\times\mathcal{X}} [u;x]^T \left[\begin{array}{c|c} & \frac{1}{2}(B - H^TA) \\ \hline \frac{1}{2}(B - H^TA)^T & \end{array}\right][u;x].\end{array}$$

Applying Proposition 4.8, we arrive at the following:

Proposition 4.18. *In the situation of this section, consider the convex optimization problem*

$$\begin{array}{rcl}\mathrm{Opt}_\# & = & \min_{H,\Upsilon=\{\Upsilon_\ell\},\Lambda=\{\Lambda_k\}} \left\{ \phi_\mathcal{R}(\lambda[\Upsilon]) + \phi_\mathcal{T}(\lambda[\Lambda]) : \Upsilon_\ell \succeq 0, \ \Lambda_k \succeq 0, \ \forall(\ell,k) \right. \\ & & \left. \left[\begin{array}{c|c} \sum_k \mathcal{R}_k^*[\Lambda_k] & \frac{1}{2}[B - H^TA]^TM \\ \hline \frac{1}{2}M^T[B - H^TA] & \sum_\ell \mathcal{S}_\ell^*[\Upsilon_\ell] \end{array}\right] \succeq 0 \right\},\end{array} \tag{4.64}$$

where $\mathcal{R}_k^*[\cdot]$, $\mathcal{S}_\ell^*[\cdot]$ are induced by $R_k[\cdot]$, $S_\ell[\cdot]$, respectively, as explained in Section 4.3.1. The problem is solvable, and the risk of the linear estimate $\widehat{x}_{H_*}(\cdot)$ yielded by the H-component of an optimal solution does not exceed $\mathrm{Opt}_\#$.

For proof, see Section 4.8.6.1.

4.5.1.2 Near-optimality

Proposition 4.19. *The linear estimate $\widehat{x}_{H_*}$ yielded by Proposition 4.18 is near-optimal in terms of its risk:*

$$\mathrm{Risk}_{\|\cdot\|}[\widehat{x}_{H_*}|\mathcal{X}] \leq \mathrm{Opt}_\# \leq O(1)\ln(D)\mathrm{Risk}_{\mathrm{opt}}[\mathcal{X}], \quad D = \sum_k d_k + \sum_\ell f_\ell, \tag{4.65}$$

where $\mathrm{Risk}_{\mathrm{opt}}[\mathcal{X}]$ *is the minimax optimal risk:*

$$\mathrm{Risk}_{\mathrm{opt}}[\mathcal{X}] = \inf_{\widehat{x}} \mathrm{Risk}_{\|\cdot\|}[\widehat{x}|\mathcal{X}]$$

with inf *taken w.r.t. all Borel estimates.*

Remark 4.20. When $\mathcal{X}$ and $\mathcal{B}_*$ are ellitopes rather than spectratopes,

$$\begin{array}{rcl}\mathcal{X} & = & \{x \in \mathbf{R}^n : \exists t \in \mathcal{T} : x^T R_k x \leq t_k, k \leq K\},\\ \mathcal{B}_* & := & \{u \in \mathbf{R}^\nu : \|u\|_* \leq 1\}\\ & = & \{u \in \mathbf{R}^\nu : \exists r \in \mathcal{R}, z : u = Mz, z^T S_\ell z \leq r_\ell, \ell \leq L\}\\ & & [R_k \succeq 0, \sum_k R_k \succ 0, S_\ell \succeq 0, \sum_\ell S_\ell \succ 0],\end{array}$$

problem *(4.64)* becomes

$$\begin{array}{rcl}\mathrm{Opt}_\# & = & \min\limits_{H,\lambda,\mu} \left\{ \phi_{\mathcal{R}}(\mu) + \phi_{\mathcal{T}}(\lambda) : \lambda \geq 0, \mu \geq 0, \right.\\ & & \left. \left[\begin{array}{c|c} \sum_k \lambda_k R_k & \frac{1}{2}[B - H^T A]^T M \\ \hline \frac{1}{2} M^T [B - H^T A] & \sum_\ell \mu_\ell S_\ell \end{array}\right] \succeq 0 \right\},\end{array}$$

and *(4.65)* can be strengthened to

$$\mathrm{Risk}_{\|\cdot\|}[\widehat{x}_{H_*}|\mathcal{X}] \leq \mathrm{Opt}_\# \leq O(1)\ln(K + L)\mathrm{Risk}_{\mathrm{opt}}[\mathcal{X}].$$

For proofs, see Section 4.8.6.

4.5.1.3 Nonlinear estimation

The uncertain-but-bounded model of observation error makes it easy to point out an efficiently computable near-optimal *nonlinear* estimate. Indeed, in the situation described at the beginning of Section 4.5.1, let us assume that the range of observation error η is

$$\mathcal{H} = \{\eta \in \mathbf{R}^m : \|\eta\|_{(m)} \leq \sigma\},$$

where $\|\cdot\|_{(m)}$ and $\sigma > 0$ are a given norm on $\mathbf{R}^m$ and a given error bound, and let us measure the recovery error by a given norm $\|\cdot\|_{(\nu)}$ on $\mathbf{R}^\nu$. We can immediately point out a (nonlinear) estimate optimal within factor 2 in terms of its $\mathcal{H}$-risk, namely, estimate $\widehat{x}_*$, as follows:

Given ω, we solve the feasibility problem

$$\text{find } x \in \mathcal{X} : \|Ax - \omega\|_{(m)} \leq \sigma. \qquad (F[\omega])$$

Let x_ω be a feasible solution; we set $\widehat{x}_*(\omega) = Bx_\omega$.

Note that the estimate is well-defined, since $(F[\omega])$ clearly is solvable, with one of the feasible solutions being the true signal underlying observation ω. When $\mathcal{X}$ is a computationally tractable convex compact set, and $\|\cdot\|_{(m)}$ is an efficiently computable norm, a feasible solution to $(F[\omega])$ can be found in a computationally efficient fashion. Let us make the following immediate observation:

Proposition 4.21. *The estimate $\widehat{x}_*$ is optimal within factor 2:*

$$\begin{array}{rcl}\mathrm{Risk}_{\mathcal{H}}[\widehat{x}_*|\mathcal{X}] & \leq & \mathrm{Opt}_* := \sup\limits_{x,y} \left\{\|Bx - By\|_{(\nu)} : x, y \in \mathcal{X}, \|A(x-y)\|_{(m)} \leq 2\sigma\right\} \\ & \leq & 2\mathrm{Risk}_{\mathrm{opt},\mathcal{H}} \end{array} \qquad (4.66)$$

where $\mathrm{Risk}_{\mathrm{opt},\mathcal{H}}$ *is the infimum of $\mathcal{H}$-risk over all estimates.*

The proof of the proposition is the subject of Exercise 4.28.

4.5.1.4 Quantifying risk

Note that Proposition 4.21 does not impose restrictions on $\mathcal{X}$ and the norms $\|\cdot\|_{(m)}$, $\|\cdot\|_{(\nu)}$.

The only—but essential—shortcoming of the estimate $\widehat{x}_*$ is that we do not know, in general, what its $\mathcal{H}$-risk is. From (4.66) it follows that this risk is tightly (namely, within factor 2) upper-bounded by Opt_*, but this quantity, being the maximum of a convex function over some domain, can be difficult to compute. Aside from a handful of special cases where this difficulty does not arise, there is a generic situation when Opt_* can be tightly upper-bounded by efficient computation. This is the situation where $\mathcal{X}$ is the spectratope defined in (4.63), $\|\cdot\|_{(m)}$ is such that the unit ball of this norm is a basic spectratope,

$$B_{(m)} := \{u : \|u\|_{(m)} \leq 1\} = \{u \in \mathbf{R}^m : \exists p \in \mathcal{P} : Q_j^2[u] \preceq p_j I_{e_j}, 1 \leq j \leq J\},$$

and the unit ball of the norm $\|\cdot\|_{(\nu),*}$ *conjugate* to the norm $\|\cdot\|_{(\nu)}$ is a spectratope,

$$\begin{array}{rcl}B^*_{(\nu)} & := & \{v \in \mathbf{R}^\nu : \|v\|_{(\nu),*} \leq 1\} \\ & = & \{v : \exists (w \in \mathbf{R}^N, r \in \mathcal{R}) : v = Mw, S_\ell^2[w] \preceq r_\ell I_{f_\ell}, 1 \leq \ell \leq L\},\end{array}$$

with the usual restrictions on $\mathcal{P}$, $\mathcal{R}$, $Q_j[\cdot]$, and $S_\ell[\cdot]$.

Proposition 4.22. *In the situation in question, consider the convex optimization problem*

$$\begin{array}{rcl}\mathrm{Opt} & = & \min\limits_{\substack{\Lambda=\{\Lambda_k,k\leq K\},\\ \Upsilon=\{\Upsilon_\ell,\ell\leq L\},\\ \Sigma=\{\Sigma_j,j\leq J\}}} \left\{\phi_{\mathcal{T}}(\lambda[\Lambda]) + \phi_{\mathcal{R}}(\lambda[\Upsilon]) + \sigma^2\phi_{\mathcal{P}}(\lambda[\Sigma]) + \phi_{\mathcal{R}}(\lambda[\Sigma]) : \right. \\ & & \Lambda_k \succeq 0, \Upsilon_\ell \succeq 0, \Sigma_j \succeq 0 \,\forall (k,\ell,j), \\ & & \left. \left[\begin{array}{c|c} \sum_\ell \mathcal{S}_\ell^*[\Upsilon_\ell] & M^T B \\ \hline B^T M & \sum_k \mathcal{R}_k^*[\Lambda_k] + A^T[\sum_j \mathcal{Q}_j^*[\Sigma_j]]A \end{array}\right] \succeq 0 \right\}\end{array} \qquad (4.67)$$

where $\mathcal{R}_k^*[\cdot]$ *is associated with mapping* $x \mapsto R_k[x]$ *according to* (4.25), $\mathcal{S}_\ell^*[\cdot]$ *and* $\mathcal{Q}_j^*[\cdot]$ *are associated in the same fashion with mappings* $w \mapsto \mathcal{S}_\ell[w]$ *and* $u \mapsto Q_j[u]$, *respectively, and* $\phi_{\mathcal{T}}$, $\phi_{\mathcal{R}}$, *and* $\phi_{\mathcal{P}}$ *are the support functions of the corresponding sets* $\mathcal{T}$, $\mathcal{R}$, *and* $\mathcal{P}$.

The optimal value in (4.67) *is an efficiently computable upper bound on the quantity* $\mathrm{Opt}_\#$ *defined in* (4.66), *and this bound is tight within factor*

$$2\max[\ln(2D), 1], \;\; D = \sum_k d_k + \sum_\ell f_\ell + \sum_j e_j.$$

Proof of the proposition is the subject of Exercise 4.29.

4.5.2 Mixed noise

So far, we have considered separately the cases of random and uncertain-but-bounded observation noises in (4.31). Note that both these observation schemes are covered by the following "mixed" scheme:

$$\omega = Ax + \xi + \eta,$$

where, as above, A is a given $m \times n$ matrix, x is an unknown deterministic signal known to belong to a given signal set $\mathcal{X}$, ξ is random noise with distribution known to belong to a family $\mathcal{P}$ of Borel probability distributions on $\mathbf{R}^m$ satisfying (4.32) for a given convex compact set $\Pi \subset \mathrm{int}\,\mathbf{S}^m_+$, and η is an "uncertain-but-bounded" observation error known to belong to a given set $\mathcal{H}$. As before, our goal is to estimate $Bx \in \mathbf{R}^\nu$ via observation ω. In our present setting, given a norm $\|\cdot\|$ on $\mathbf{R}^\nu$, we can quantify the performance of a candidate estimate $\omega \mapsto \widehat{x}(\omega) : \mathbf{R}^m \to \mathbf{R}^\nu$ by its risk

$$\mathrm{Risk}_{\Pi,\mathcal{H},\|\cdot\|}[\widehat{x}|\mathcal{X}] = \sup_{x\in\mathcal{X}, P\lhd\Pi, \eta\in\mathcal{H}} \mathbf{E}_{\xi\sim P}\{\|Bx - \widehat{x}(Ax+\xi+\eta)\|\}.$$

Observe that the estimation problem associated with the "mixed" observation scheme straightforwardly reduces to a similar problem for the random observation scheme, by the same trick we have used in Section 4.5 to eliminate the observation noise. Indeed, let us treat $x^+ = [x;\eta] \in \mathcal{X}^+ := \mathcal{X} \times \mathcal{H}$ and $\mathcal{X}^+$ as the new signal/signal set underlying our observation, and set $\bar{A}x^+ = Ax + \eta$, $\bar{B}x^+ = Bx$. With these conventions, the "mixed" observation scheme reduces to

$$\omega = \bar{A}x^+ + \xi,$$

and for every candidate estimate $\widehat{x}(\cdot)$ it clearly holds

$$\mathrm{Risk}_{\Pi,\mathcal{H},\|\cdot\|}[\widehat{x}|\mathcal{X}] = \mathrm{Risk}_{\Pi,\|\cdot\|}[\widehat{x}|\mathcal{X}^+],$$

so that we find ourselves in the situation of Section 4.3.3.1. Assuming that $\mathcal{X}$ and $\mathcal{H}$ are spectratopes, so is $\mathcal{X}^+$, meaning that all results of Section 4.3.3 on building presumably good linear estimates and their near-optimality are applicable to our present setup.

4.6 CALCULUS OF ELLITOPES/SPECTRATOPES

We present here the rules of the calculus of ellitopes/spectratopes. We formulate these rules for ellitopes; the "spectratopic versions" of the rules are straightforward modifications of their "ellitopic versions."

- Intersection $\mathcal{X} = \bigcap_{i=1}^{I} \mathcal{X}_i$ of ellitopes

$$\mathcal{X}_i = \{x \in \mathbf{R}^n : \exists (y^i \in \mathbf{R}^{n_i}, t^i \in \mathcal{T}_i) : x = P_i y^i \ \&\ [y^i]^T R_{ik} y^i \leq t^i_k, 1 \leq k \leq K_i\}$$

is an ellitope. Indeed, this is evident when $\mathcal{X} = \{0\}$. Assuming $\mathcal{X} \neq \{0\}$, we have

$$\begin{array}{rcl} \mathcal{X} & = & \{x \in \mathbf{R}^n : \exists (y = [y^1; ...; y^I] \in \mathcal{Y}, t = (t^1, ..., t^I) \in \mathcal{T} = \mathcal{T}_1 \times ... \times \mathcal{T}_I) : \\ & & x = Py := P_1 y^1 \ \&\ \underbrace{[y^i]^T R_{ik} y^i}_{y^T R^+_{ik} y} \leq t^i_k, 1 \leq k \leq K_i, 1 \leq i \leq I\}, \\ \mathcal{Y} & = & \{[y^1; ...; y^I] \in \mathbf{R}^{n_1 + ... + n_I} : P_i y^i = P_1 y^1, \ 2 \leq i \leq I\} \end{array}$$

(note that $\mathcal{Y}$ can be identified with $\mathbf{R}^{\bar{n}}$ with a properly selected $\bar{n} > 0$).

- The direct product $\mathcal{X} = \prod_{i=1}^{I} \mathcal{X}_i$ of ellitopes

$$\begin{array}{rl} \mathcal{X}_i = & \{x^i \in \mathbf{R}^{n_i} : \exists (y^i \in \mathbf{R}^{\bar{n}_i}, t^i \in \mathcal{T}_i) : \\ & x^i = P_i y^i, \ 1 \leq i \leq I \ \&\ [y^i]^T R_{ik} y^i \leq t^i_k, 1 \leq k \leq K_i\} \end{array}$$

is an ellitope:

$$\mathcal{X} = \left\{ [x^1; ...; x^I] \in \mathbf{R}^{n_1} \times ... \times \mathbf{R}^{n_I} : \exists \left(\begin{array}{c} y = [y^1; ...; y^I] \in \mathbf{R}^{\bar{n}_1 + ... \bar{n}_I} \\ t = (t^1, ..., t^I) \in \mathcal{T} = \mathcal{T}_1 \times ... \times \mathcal{T}_I \end{array} \right) \right. \\ \left. x = Py := [P_1 y^1; ...; P_I y^I], \ \underbrace{[y^i]^T R_{ik} y^i}_{y^T R^+_{ik} y} \leq t^i_k, 1 \leq k \leq K_i, 1 \leq i \leq I \right\}.$$

- The linear image $\mathcal{Z} = \{Rx : x \in \mathcal{X}\}$, $R \in \mathbf{R}^{p \times n}$, of an ellitope

$$\mathcal{X} = \{x \in \mathbf{R}^n : \exists (y \in \mathbf{R}^{\bar{n}}, t \in \mathcal{T}) : x = Py \ \&\ y^T R_k y \leq t_k, \ 1 \leq k \leq K\}$$

is an ellitope:

$$\mathcal{Z} = \{z \in \mathbf{R}^p : \exists (y \in \mathbf{R}^{\bar{n}}, t \in \mathcal{T}) : z = [RP]y \ \&\ y^T R_k y \leq t_k, \ 1 \leq k \leq K\}.$$

- The inverse linear image $\mathcal{Z} = \{z \in \mathbf{R}^q : Rz \in \mathcal{X}\}$, $R \in \mathbf{R}^{n \times q}$, of an ellitope $\mathcal{X} = \{x \in \mathbf{R}^n : \exists (y \in \mathbf{R}^{\bar{n}}, t \in \mathcal{T}) : x = Py \ \&\ y^T R_k y \leq t_k, 1 \leq k \leq K\}$ under linear mapping $z \mapsto Rz : \mathbf{R}^q \to \mathbf{R}^n$ is an ellitope, *provided that the mapping is an embedding:* $\operatorname{Ker} R = \{0\}$. Indeed, setting $E = \{y \in \mathbf{R}^{\bar{n}} : Py \in \operatorname{Im} R\}$, we get a linear subspace in $\mathbf{R}^{\bar{n}}$. If $E = \{0\}$, $\mathcal{Z} = \{0\}$ is an ellitope; if $E \neq \{0\}$, we have

$$\begin{array}{rcl} \mathcal{Z} & = & \{z \in \mathbf{R}^q : \exists (y \in E, t \in \mathcal{T}) : z = \bar{P}y \ \&\ y^T R_k y \leq t_k, \ 1 \leq k \leq K\}, \\ \bar{P} & = & \Pi P, \ \text{ where } \Pi : \operatorname{Im} R \to \mathbf{R}^q \text{ is the inverse of } z \mapsto Rz : \mathbf{R}^q \to \operatorname{Im} R \end{array}$$

(E can be identified with some $\mathbf{R}^k$, and Π is well-defined since R is an embedding).

- The arithmetic sum $\mathcal{X} = \left\{x = \sum_{i=1}^I x^i : x^i \in \mathcal{X}_i,\ 1 \le i \le I\right\}$ of ellitopes $\mathcal{X}_i$ is an ellitope, with representation readily given by those of $\mathcal{X}_1, ..., \mathcal{X}_I$.
  Indeed, $\mathcal{X}$ is the image of $\mathcal{X}_1 \times ... \times \mathcal{X}_I$ under the linear mapping $[x^1; ...; x^I] \mapsto x^1 + ... + x^I$, and taking direct products and images under linear mappings preserves ellitopes.
- "$\mathcal{S}$-product." Let $\mathcal{X}_i = \{x^i \in \mathbf{R}^{n_i} : \exists (y^i \in \mathbf{R}^{\bar{n}_i}, t^i \in \mathcal{T}_i) : x^i = P_i y^i,\ 1 \le i \le I\ \&\ [y^i]^T R_{ik} y^i \le t^i_k, 1 \le k \le K_i\}$ be ellitopes, and let $\mathcal{S}$ be a convex compact set in $\mathbf{R}^I_+$ which intersects the interior of $\mathbf{R}^I_+$ and is monotone: $0 \le s' \le s \in \mathcal{S}$ implies $s' \in \mathcal{S}$. We associate with $\mathcal{S}$ the set

$$\mathcal{S}^{1/2} = \left\{s \in \mathbf{R}^I_+ : [s_1^2; ...; s_I^2] \in \mathcal{S}\right\}$$

of entrywise square roots of points from $\mathcal{S}$; clearly, $\mathcal{S}^{1/2}$ is a convex compact set.
$\mathcal{X}_i$ and $\mathcal{S}$ specify the *$\mathcal{S}$-product* of the sets $\mathcal{X}_i$, $i \le I$, defined as the set

$$\mathcal{Z} = \left\{z = [z^1; ...; z^I] : \exists (s \in \mathcal{S}^{1/2}, x^i \in \mathcal{X}_i, i \le I) : z^i = s_i x^i, 1 \le i \le I\right\},$$

or, equivalently,

$$\begin{array}{rcl}\mathcal{Z} & = & \Big\{z = [z^1; ...; z^I] : \exists (r = [r^1; ...; r^I] \in \mathcal{R}, y^1, ..., y^I) : \\ & & \qquad z_i = P_i y_i \,\forall i \le I, [y^i]^T R_{ik} y^i \le r^i_k \,\forall (i \le I, k \le K_i)\Big\}, \\ & & \mathcal{R} = \{[r^1; ...; r^I] \ge 0 : \exists (s \in \mathcal{S}^{1/2}, t^i \in \mathcal{T}_i) : r^i = s_i^2 t^i \,\forall i \le I\}.\end{array}$$

We claim that $\mathcal{Z}$ is an ellitope. All we need to verify to this end is that the set $\mathcal{R}$ is as it should be in an ellitopic representation, that is, that $\mathcal{R}$ is a compact and monotone subset of $\mathbf{R}_+^{K_1+...+K_I}$ containing a strictly positive vector (all this is evident), and that $\mathcal{R}$ is convex. To verify convexity, let $\mathbf{T}_i = \mathrm{cl}\{[t^i; \tau_i] : \tau_i > 0, t^i/\tau_i \in \mathcal{T}_i\}$ be the conic hulls of $\mathcal{T}_i$'s. We clearly have

$$\begin{array}{l}\mathcal{R} = \{[r^1; ...; r^I] : \exists s \in \mathcal{S}^{1/2} : [r^i; s_i^2] \in \mathbf{T}_i,\ i \le I\} \\ \quad = \{[r^1; ...; r^I] : \exists \sigma \in \mathcal{S} : [r^i; \sigma_i] \in \mathbf{T}_i,\ i \le I\},\end{array}$$

where the concluding equality is due to the origin of $\mathcal{S}^{1/2}$. The concluding set in the above chain clearly is convex, and we are done.
As an example, consider the situation where the ellitopes $\mathcal{X}_i$ possess nonempty interiors and thus can be thought of as unit balls of norms $\|\cdot\|_{(i)}$ on the respective spaces $\mathbf{R}^{n_i}$, and let $\mathcal{S} = \{s \in \mathbf{R}^I_+ : \|s\|_{p/2} \le 1\}$, where $p \ge 2$. In this situation, $\mathcal{S}^{1/2} = \{s \in \mathbf{R}^I_+ : \|s\|_p \le 1\}$, whence $\mathcal{Z}$ is the unit ball of the "block p-norm"

$$\|[z^1; ...; z^I]\| = \|\,\left[\|z^1\|_{(1)}; ...; \|z^I\|_{(I)}\right]\|_p.$$

Note also that the usual direct product of I ellitopes is their $\mathcal{S}$-product, with $\mathcal{S} = [0,1]^I$.
- "$\mathcal{S}$-weighted sum." Let $\mathcal{X}_i \subset \mathbf{R}^n$ be ellitopes, $1 \le i \le I$, and let $\mathcal{S} \subset \mathbf{R}^I_+$, $\mathcal{S}^{1/2}$ be the same as in the previous rule. Then the *$\mathcal{S}$-weighted sum* of the sets $\mathcal{X}_i$,

defined as

$$\mathcal{X} = \{x : \exists (s \in \mathcal{S}^{1/2}, x^i \in \mathcal{X}_i, i \leq I) : x = \sum_i s_i x^i\},$$

is an ellitope. Indeed, the set in question is the image of the $\mathcal{S}$-product of $\mathcal{X}_i$ under the linear mapping $[z^1; ...; z^I] \mapsto z^1 + ... + z^I$, and taking $\mathcal{S}$-products and linear images preserves the property of being an ellitope.

It should be stressed that the outlined "calculus rules" are fully algorithmic: representation (4.6) of the result of an operation is readily given by the representations (4.6) of the operands.

4.7 EXERCISES FOR CHAPTER 4

4.7.1 Linear estimates vs. Maximum Likelihood

Exercise 4.1.

Consider the problem posed at the beginning of Chapter 4: *Given observation*

$$\omega = Ax + \sigma\xi, \ \xi \sim \mathcal{N}(0, I)$$

of unknown signal x known to belong to a given signal set $\mathcal{X} \subset \mathbf{R}^n$, we want to recover Bx.

Let us consider the case where matrix A is square and invertible, B is the identity, and $\mathcal{X}$ is a computationally tractable convex compact set. As far as computational aspects are concerned, the situation is well suited for utilizing the "magic wand" of Statistics—the *Maximum Likelihood* (ML) estimate where the recovery of x is

$$\widehat{x}_{\mathrm{ML}}(\omega) = \operatorname*{argmin}_{y \in \mathcal{X}} \|\omega - Ay\|_2 \qquad \text{(ML)}$$

—the signal which maximizes, over $y \in \mathcal{X}$, the likelihood (the probability density) of getting the observation we actually got. Indeed, with computationally tractable $\mathcal{X}$, (ML) is an explicit convex, and therefore efficiently solvable, optimization problem. Given the exclusive role played by the ML estimate in Statistics, perhaps the first question about our estimation problem is: *how good is the ML estimate?*

The goal of this exercise is to show that *in the situation we are interested in, the ML estimate can be "heavily nonoptimal," and this may happen even when the techniques we develop in Chapter 4 do result in an efficiently computable near-optimal linear estimate.*

To justify the claim, investigate the risk (4.2) of the ML estimate in the case where

$$\mathcal{X} = \left\{ x \in \mathbf{R}^n : x_1^2 + \epsilon^{-2} \sum_{i=2}^n x_i^2 \leq 1 \right\} \ \& \ A = \mathrm{Diag}\{1, \epsilon^{-1}, ..., \epsilon^{-1}\},$$

ϵ and σ are small, and n is large, so that $\sigma^2(n-1) \geq 2$. Accompany your theoretical analysis by numerical experiments—compare the empirical risks of the ML estimate with theoretical and empirical risks of the linear estimate optimal under the circumstances.

Recommended setup: n runs through $\{256, 1024, 2048\}$, $\epsilon = \sigma$ runs through $\{0.01; 0.05; 0.1\}$, and signal x is generated as

$$x = [\cos(\phi); \sin(\phi)\epsilon\zeta],$$

where $\phi \sim \text{Uniform}[0, 2\pi]$ and random vector ζ is independent of ϕ and is distributed uniformly on the unit sphere in $\mathbf{R}^{n-1}$.

4.7.2 Measurement Design in Signal Recovery

Exercise 4.2.

[Measurement Design in Gaussian o.s.] As a preamble to the exercise, please read the story about possible "physics" of Gaussian o.s. from Section 2.7.3.3. The summary of the story is as follows:

We consider the Measurement Design version of signal recovery in Gaussian o.s., specifically, we are allowed to use observations

$$\omega = A_q x + \sigma\xi \qquad [\xi \sim \mathcal{N}(0, I_m)]$$

where

$$A_q = \text{Diag}\{\sqrt{q_1}, \sqrt{q_2}, ..., \sqrt{q_m}\}A,$$

with a given $A \in \mathbf{R}^{m\times n}$ and vector q which we can select in a given convex compact set $\mathcal{Q} \subset \mathbf{R}^m_+$. The signal x underlying the observation is known to belong to a given ellitope $\mathcal{X}$. Your goal is to select $q \in \mathcal{Q}$ and a linear recovery $\omega \mapsto G^T\omega$ of the image Bx of $x \in \mathcal{X}$, with given B, resulting in the smallest worst-case, over $x \in \mathcal{X}$, expected $\|\cdot\|_2^2$ recovery risk. Modify, according to this goal, problem (4.12). Is it possible to end up with a tractable problem? Work out in full detail the case when $\mathcal{Q} = \{q \in \mathbf{R}^m_+ : \sum_i q_i = m\}$.

Exercise 4.3.

[follow-up to Exercise 4.2] A translucent bar of length $n = 32$ is comprised of 32 consecutive segments of length 1 each, with density ρ_i of i-th segment known to belong to the interval $[\mu - \delta_i, \mu + \delta_i]$.

Sample translucent bar

The bar is lit from the left end; when light passes through a segment with density ρ, the light's intensity is reduced by factor $e^{-\alpha\rho}$. The light intensity at the left endpoint of the bar is 1. You can scan the segments one by one from left to right and measure light intensity ℓ_i at the right endpoint of the i-th segment during time q_i; the result z_i of the measurement is $\ell_i e^{\sigma\xi_i/\sqrt{q_i}}$, where $\xi_i \sim \mathcal{N}(0, 1)$ are independent across i. The total time budget is n, and you are interested in recovering the $m = n/2$-dimensional vector of densities of the right m segments. Build an optimization problem responsible for near-optimal linear recovery with and without Measurement Design (in the latter case, we assume that each segment is observed during unit time) and compare the resulting near-optimal risks.
Recommended data:

$$\alpha = 0.01,\ \delta_i = 1.2 + \cos(4\pi(i-1)/n),\ \mu = 1.1 \max_i \delta_i,\ \sigma = 0.001.$$

Exercise 4.4.

Let X be a basic ellitope in $\mathbf{R}^n$:

$$X = \{x \in \mathbf{R}^n : \exists t \in \mathcal{T} : x^T S_k x \leq t_k,\ 1 \leq k \leq K\}$$

with our usual restrictions on S_k and $\mathcal{T}$. Let, further, m be a given positive integer, and $x \mapsto Bx : \mathbf{R}^n \to \mathbf{R}^\nu$ be a given linear mapping. Consider the Measurement Design problem where you are looking for a linear recovery $\omega \mapsto \widehat{x}_H(\omega) := H^T\omega$ of Bx, $x \in X$, from observation

$$\omega = Ax + \sigma\xi \qquad\qquad [\sigma > 0 \text{ is given and } \xi \sim \mathcal{N}(0, I_m)]$$

in which the $m \times n$ sensing matrix A is under your control—it is allowed to be any $m \times n$ matrix of spectral norm not exceeding 1. You are interested in selecting H and A in order to minimize the worst-case, over $x \in X$, expected $\|\cdot\|_2^2$ recovery error. Similarly to (4.12), this problem can be posed as

$$\begin{array}{rl}\mathrm{Opt} = & \min_{H,\lambda,A}\left\{\sigma^2\mathrm{Tr}(H^TH) + \phi_{\mathcal{T}}(\lambda) : \right.\\ & \left.\left[\begin{array}{c|c}\sum_k \lambda_k S_k & B^T - A^TH \\ \hline B - H^TA & I_\nu\end{array}\right] \succeq 0,\ \|A\| \leq 1,\ \lambda \geq 0\right\},\end{array} \tag{4.68}$$

where $\|\cdot\|$ stands for the spectral norm. The objective in this problem is the (upper bound on the) squared risk $\mathrm{Risk}^2[\widehat{x}_H|X]$, the sensing matrix being A. The problem is nonconvex, since the matrix participating in the semidefinite constraint is bilinear in H and A.

A natural way to handle an optimization problem with objective and/or constraints bilinear in the decision variables u, v is to use "alternating minimization," where one alternates optimization in v for u fixed and optimization in u for v fixed, the value of the variable fixed in a round being the result of optimization w.r.t. this variable in the previous round. Alternating minimizations are carried out until the value of the objective (which in the outlined process definitely improves from round to round) stops to improve (or nearly so). Since the algorithm does not necessarily converge to the globally optimal solution to the problem of interest, it makes sense to run the algorithm several times from different, say, randomly selected, starting points.

Now comes the exercise.

1. Implement Alternating Minimization as applied to (4.68). You may restrict your experimentation to the case where the sizes m, n, ν are quite moderate, in the range of tens, and X is either the box $\{x : j^{2\gamma}x_j^2 \leq 1, 1 \leq j \leq n\}$, or the ellipsoid $\{x : \sum_{j=1}^n j^{2\gamma}x_j^2 \leq 1\}$, where γ is a nonnegative parameter (try $\gamma = 0, 1, 2, 3$). As for B, you can generate it at random, or enforce B to have prescribed singular values, say, $\sigma_j = j^{-\theta}$, $1 \leq j \leq \nu$, and a randomly selected system of singular vectors.
2. Identify cases where a globally optimal solution to (4.68) is easy to find and use this information in order to understand how reliable Alternating Minimization is in the application in question, reliability meaning the ability to identify near-optimal, in terms of the objective, solutions.

   If you are not satisfied with Alternating Minimization "as is," try to improve it.

3. Modify (4.68) and your experiment to cover the cases where the constraint $\|A\| \leq 1$ on the sensing matrix is replaced with one of the following:
   - $\|\mathrm{Row}_i[A]\|_2 \leq 1,\ 1 \leq i \leq m$,
   - $|A_{ij}| \leq 1$ for all i, j

   (note that these two types of restrictions mimic what happens if you are interested in recovering (the linear image of) the vector of parameters in a linear regression model from noisy observations of the model's outputs at the m points which you are allowed to select in the unit ball or unit box).
4. [Embedded Exercise] Recall that a $\nu \times n$ matrix G admits *singular value decomposition* $G = UDV^T$ with orthogonal matrices $U \in \mathbf{R}^{\nu\times\nu}$ and $V \in \mathbf{R}^{n\times n}$ and diagonal $\nu \times n$ matrix D with nonnegative and nonincreasing diagonal entries.[9] These entries are uniquely defined by G and are called *singular values* $\sigma_i(G)$, $1 \leq i \leq \min[\nu, n]$. Singular values admit characterization similar to variational characterization of eigenvalues of a symmetric matrix; see, e.g., [15, Section A.7.3]:

   Theorem 4.23. *[VCSV—Variational Characterization of Singular Values] For a $\nu \times n$ matrix G it holds*

   $$\sigma_i(G) = \min_{E\in\mathcal{E}_i} \max_{e\in E, \|e\|_2=1} \|Ge\|_2,\ 1 \leq i \leq \min[\nu, n], \tag{4.69}$$

   where $\mathcal{E}_i$ is the family of all subspaces in $\mathbf{R}^n$ of codimension $i-1$.

   Corollary 4.24. *[SVI—Singular Value Interlacement] Let G and G' be $\nu \times n$ matrices, and let $k = \mathrm{Rank}(G - G')$. Then*

   $$\sigma_i(G) \geq \sigma_{i+k}(G'),\ 1 \leq i \leq \min[\nu, n],$$

   where, by definition, singular values of a $\nu \times n$ matrix with indexes $> \min[\nu, n]$ are zeros.

   We denote by $\sigma(G)$ the vector of singular values of G arranged in nonincreasing order. The function $\|G\|_{\mathrm{Sh},p} = \|\sigma(G)\|_p$ is called the *Shatten p-norm* of matrix G; this indeed is a norm on the space of $\nu \times n$ matrices, and the conjugate norm is $\|\cdot\|_{\mathrm{Sh},q}$, with $\frac{1}{p} + \frac{1}{q} = 1$. An easy and important consequence of Corollary 4.24 is the following fact:

   Corollary 4.25. *Given a $\nu \times n$ matrix G, an integer k, $0 \leq k \leq \min[\nu, n]$, and $p \in [1, \infty]$, (one of) the best approximation of G in the Shatten p-norm among matrices of rank $\leq k$ is obtained from G by zeroing out all but k largest singular values, that is, the matrix $G^k = \sum_{i=1}^k \sigma_i(G)\mathrm{Col}_i[U]\mathrm{Col}_i^T[V]$, where $G = UDV^T$ is the singular value decomposition of G.*

   Prove Theorem 4.23 and Corollaries 4.24 and 4.25.
5. Consider the Measurement Design problem (4.68) in the case when X is an ellipsoid:

   $$X = \left\{x \in \mathbf{R}^n : \sum\nolimits_{j=1}^n x_j^2/a_j^2 \leq 1\right\},$$

[9] We say that a rectangular matrix D is diagonal if all entries D_{ij} in D with $i \neq j$ are zeros.

A is an $m \times n$ matrix of spectral norm not exceeding 1, and there is no noise in observations: $\sigma = 0$. Find an optimal solution to this problem. Think how this result can be used to get a (hopefully) good starting point for Alternating Minimization in the case when X is an ellipsoid and σ is small.

4.7.3 Around semidefinite relaxation

Exercise 4.5.

Let $\mathcal{X}$ be an ellitope:

$$\mathcal{X} = \{x \in \mathbf{R}^n : \exists (y \in \mathbf{R}^N, t \in \mathcal{T}) : x = Py, y^T S_k y \le t_k, k \le K\}$$

with our standard restrictions on $\mathcal{T}$ and S_k. Representing $S_k = \sum_{j=1}^{r_k} s_{kj} s_{kj}^T$, we can pass from the initial ellitopic representation of $\mathcal{X}$ to the spectratopic representation of the same set:

$$\begin{array}{c}\mathcal{X} = \{x \in \mathbf{R}^n : \exists (y \in \mathbf{R}^N, t^+ \in \mathcal{T}^+) : x = Py, [s_{kj}^T x]^2 \preceq t_{kj}^+ I_1, 1 \le k \le K, 1 \le j \le r_k\} \\ \left[\mathcal{T}^+ = \{t^+ = \{t_{kj}^+ \ge 0\} : \exists t \in \mathcal{T} : \sum_{j=1}^{r_k} t_{kj}^+ \le t_k,\ 1 \le k \le K\}\right].\end{array}$$

If now C is a symmetric $n \times n$ matrix and $\mathrm{Opt} = \max_{x \in \mathcal{X}} x^T C x$, we have

$$\begin{array}{rcl}\mathrm{Opt}_* & \le & \mathrm{Opt}_e := \min\limits_{\lambda = \{\lambda_k \in \mathbf{R}_+\}} \left\{\phi_{\mathcal{T}}(\lambda) : P^T C P \preceq \sum_k \lambda_k S_k\right\} \\ \mathrm{Opt}_* & \le & \mathrm{Opt}_s := \min\limits_{\Lambda = \{\Lambda_{kj} \in \mathbf{R}_+\}} \left\{\phi_{\mathcal{T}^+}(\Lambda) : P^T C P \preceq \sum_{k,j} \Lambda_{kj} s_{kj} s_{kj}^T\right\}\end{array}$$

where the first relation is yielded by the ellitopic representation of $\mathcal{X}$ and Proposition 4.6, and the second, on closer inspection (carry this inspection out!), by the spectratopic representation of $\mathcal{X}$ and Proposition 4.8.

Prove that $\mathrm{Opt}_e = \mathrm{Opt}_s$.

Exercise 4.6.

Proposition 4.6 provides us with an upper bound on the quality of the semidefinite relaxation as applied to the problem of upper-bounding the maximum of a homogeneous quadratic form over an ellitope. Extend the construction to the case where an inhomogeneous quadratic form is maximized over a shifted ellitope, so that the quantity to upper-bound is

$$\begin{array}{c}\mathrm{Opt} = \max\limits_{x \in X} \left[f(x) := x^T A x + 2b^T x + c\right], \\ X = \{x : \exists (y, t \in \mathcal{T}) : x = Py + p, y^T S_k y \le t_k, 1 \le k \le K\}\end{array}$$

with our standard assumptions on S_k and $\mathcal{T}$.

Note: X is centered at p, and a natural upper bound on Opt is

$$\mathrm{Opt} \le f(p) + \widehat{\mathrm{Opt}},$$

where $\widehat{\mathrm{Opt}}$ is an upper bound on the quantity

$$\overline{\mathrm{Opt}} = \max_{x \in X} [f(x) - f(p)].$$

What you are interested in upper-bounding is the ratio $\widehat{\mathrm{Opt}}/\overline{\mathrm{Opt}}$.

Exercise 4.7.

[estimating Kolmogorov widths of spectratopes/ellitopes]

4.7.A. Preliminaries: Kolmogorov and Gelfand widths. Let $\mathcal{X}$ be a convex compact set in $\mathbf{R}^n$, and let $\|\cdot\|$ be a norm on $\mathbf{R}^n$. Given a linear subspace E in $\mathbf{R}^n$, let

$$\mathrm{dist}_{\|\cdot\|}(x,E) = \min_{z\in E} \|x-z\| : \mathbf{R}^n \to \mathbf{R}_+$$

be the $\|\cdot\|$-distance from x to E. The quantity

$$\mathrm{dist}_{\|\cdot\|}(\mathcal{X},E) = \max_{x\in\mathcal{X}} \mathrm{dist}_{\|\cdot\|}(x,E)$$

can be viewed as the worst-case $\|\cdot\|$-accuracy to which vectors from $\mathcal{X}$ can be approximated by vectors from E. Given positive integer $m \le n$ and denoting by $\mathcal{E}_m$ the family of all linear subspaces in $\mathbf{R}^m$ of dimension m, the quantity

$$\delta_m(\mathcal{X},\|\cdot\|) = \min_{E\in\mathcal{E}_m} \mathrm{dist}_{\|\cdot\|}(\mathcal{X},E)$$

can be viewed as the best achievable quality of approximation, measured in $\|\cdot\|$, of vectors from $\mathcal{X}$ by vectors from an m-dimensional linear subspace of $\mathbf{R}^n$. This quantity is called the *m-th Kolmogorov width* of $\mathcal{X}$ w.r.t. $\|\cdot\|$.

Observe that one has

$$\begin{array}{rl} \mathrm{dist}_{\|\cdot\|}(x,E) = & \max_\xi\{\xi^T x : \|\xi\|_* \le 1, \xi \in E^\perp\}, \\ \mathrm{dist}_{\|\cdot\|}(\mathcal{X},E) = & \max\limits_{\substack{x\in\mathcal{X},\\ \|\xi\|_*\le 1,\xi\in E^\perp}} \xi^T x, \end{array} \tag{4.70}$$

where $E^\perp$ is the orthogonal complement to E.

1) Prove (4.70).
<u>Hint:</u> Represent $\mathrm{dist}_{\|\cdot\|}(x,E)$ as the optimal value in a conic problem on the cone $\mathbf{K} = \{[x;t] : t \ge \|x\|\}$ and use the Conic Duality Theorem.

Now consider the case when $\mathcal{X}$ is the unit ball of some norm $\|\cdot\|_{\mathcal{X}}$. In this case (4.70) combines with the definition of Kolmogorov width to imply that

$$\begin{array}{rcl} \delta_m(\mathcal{X},\|\cdot\|) & = & \min\limits_{E\in\mathcal{E}_m} \mathrm{dist}_{\|\cdot\|}(x,E) = \min\limits_{E\in\mathcal{E}_m} \max\limits_{x\in\mathcal{X}} \max\limits_{y\in E^\perp,\|y\|_*\le 1} y^T x \\ & = & \min\limits_{E\in\mathcal{E}_m} \max\limits_{y\in E^\perp,\|y\|_*\le 1} \max\limits_{x:\|x\|_{\mathcal{X}}\le 1} y^T x \\ & = & \min\limits_{F\in\mathcal{E}_{n-m}} \max\limits_{y\in F,\|y\|_*\le 1} \|y\|_{\mathcal{X},*}, \end{array} \tag{4.71}$$

where $\|\cdot\|_{\mathcal{X},*}$ is the norm conjugate to $\|\cdot\|_{\mathcal{X}}$. Note that when $\mathcal{Y}$ is a convex compact set in $\mathbf{R}^n$ and $|\cdot|$ is a norm on $\mathbf{R}^n$, the quantity

$$d^m(\mathcal{Y},|\cdot|) = \min_{F\in\mathcal{E}_{n-m}} \max_{y\in\mathcal{Y}\cap F} |y|$$

has a name—it is called the m-th *Gelfand width* of $\mathcal{Y}$ taken w.r.t. $|\cdot|$. The "duality relation" (4.71) states that

When $\mathcal{X}$, $\mathcal{Y}$ are the unit balls of respective norms $\|\cdot\|_{\mathcal{X}}$, $\|\cdot\|_{\mathcal{Y}}$, for every $m < n$ the m-th Kolmogorov width of $\mathcal{X}$ taken w.r.t. $\|\cdot\|_{\mathcal{Y},*}$ is the same as

the m-th Gelfand width of $\mathcal{Y}$ taken w.r.t. $\|\cdot\|_{\mathcal{X},*}$.

The goal of the remaining part of the exercise is to use our results on the quality of semidefinite relaxation on ellitopes/spectratopes to infer efficiently computable upper bounds on Kolmogorov widths of a given set $\mathcal{X} \subset \mathbf{R}^n$. In the sequel we assume that

- $\mathcal{X}$ is a spectratope:

$$\mathcal{X} = \{x \in \mathbf{R}^n : \exists (t \in \mathcal{T}, u) : x = Pu, R_k^2[u] \preceq t_k I_{d_k}, k \leq K\};$$

- The unit ball $\mathcal{B}_*$ of the norm conjugate to $\|\cdot\|$ is a spectratope:

$$\mathcal{B}_* = \{y : \|y\|_* \leq 1\} = \{y \in \mathbf{R}^n : \exists (r \in \mathcal{R}, z) : y = Mz, S_\ell^2[z] \preceq r_\ell I_{f_\ell}, \ell \leq L\}.$$

with our usual restrictions on $\mathcal{T}, \mathcal{R}$ and $R_k[\cdot]$ and $S_\ell[\cdot]$.

4.7.B. Simple case: $\|\cdot\| = \|\cdot\|_2$. We start with the *simple case* where $\|\cdot\| = \|\cdot\|_2$, so that $\mathcal{B}_*$ is the ellitope $\{y : y^T y \leq 1\}$.

Let $D = \sum_k d_k$ be the size of the spectratope $\mathcal{X}$, and let

$$\varkappa = 2\max[\ln(2D), 1].$$

Given integer $m < n$, consider the convex optimization problem

$$\begin{array}{rcl}\mathrm{Opt}(m) & = & \min_{\Lambda=\{\Lambda_k, k\leq K\}, Y} \left\{ \phi_{\mathcal{T}}(\lambda[\Lambda]) : \Lambda_k \succeq 0 \forall k, 0 \preceq Y \preceq I_n, \right. \\ & & \left. \sum_k \mathcal{S}_k^*[\Lambda_k] \succeq P^T Y P, \, \mathrm{Tr}(Y) = n - m \right\}. \end{array} \qquad (P_m)$$

2) Prove the following:

Proposition 4.26. *Whenever $1 \leq \mu \leq m < n$, one has*

$$\mathrm{Opt}(m) \leq \varkappa \delta_m^2(\mathcal{X}, \|\cdot\|_2) \; \& \; \delta_m^2(\mathcal{X}, \|\cdot\|_2) \leq \frac{m+1}{m+1-\mu} \mathrm{Opt}(\mu). \qquad (4.72)$$

Moreover, the above upper bounds on $\delta_m(\mathcal{X}, \|\cdot\|_2)$ are "constructive," meaning that an optimal solution to (P_μ), $\mu \leq m$, can be straightforwardly converted into a linear subspace $E^{m,\mu}$ of dimension m such that

$$\mathrm{dist}_{\|\cdot\|_2}(\mathcal{X}, E^{m,\mu}) \leq \sqrt{\frac{m+1}{m+1-\mu} \mathrm{Opt}(\mu)}.$$

Finally, $\mathrm{Opt}(\mu)$ *is nonincreasing in $\mu < n$.*

4.7.C. General case. Now consider the case when both $\mathcal{X}$ and the unit ball $\mathcal{B}_*$ of the norm conjugate to $\|\cdot\|$ are spectratopes. As we are about to see, this case is essentially more difficult than the case of $\|\cdot\| = \|\cdot\|_2$, but something still can be done.

3) Prove the following statement:

(!) *Given* $m < n$, *let* Y *be an orthoprojector of* $\mathbf{R}^n$ *of rank* $n - m$, *and let collections* $\Lambda = \{\Lambda_k \succeq 0, k \leq K\}$ *and* $\Upsilon = \{\Upsilon_\ell \succeq 0, \ell \leq L\}$ *satisfy the relation*

$$\left[\begin{array}{c|c} \sum_k \mathcal{R}_k^*[\Lambda_k] & \frac{1}{2}P^TYM \\ \hline \frac{1}{2}M^TYP & \sum_\ell \mathcal{S}_\ell^*[\Upsilon_\ell] \end{array}\right] \succeq 0. \tag{4.73}$$

Then

$$\mathrm{dist}_{\|\cdot\|}(\mathcal{X}, \mathrm{Ker}\, Y) \leq \phi_{\mathcal{T}}(\lambda[\Lambda]) + \phi_{\mathcal{R}}(\lambda[\Upsilon]). \tag{4.74}$$

As a result,

$$\begin{array}{rcl} \delta_m(\mathcal{X}, \|\cdot\|) & \leq & \mathrm{dist}_{\|\cdot\|}(\mathcal{X}, \mathrm{Ker}\, Y) \\ & \leq & \mathrm{Opt} := \min\limits_{\substack{\Lambda=\{\Lambda_k, k\leq K\}, \\ \Upsilon=\{\Upsilon_\ell, \ell\leq L\}}} \left\{ \phi_{\mathcal{T}}(\lambda[\Lambda]) + \phi_{\mathcal{R}}(\lambda[\Upsilon]) : \right. \\ & & \Lambda_k \succeq 0\,\forall k, \Upsilon_\ell \succeq 0\,\forall \ell, \\ & & \left. \left[\begin{array}{c|c} \sum_k \mathcal{R}_k^*[\Lambda_k] & \frac{1}{2}P^TYM \\ \hline \frac{1}{2}M^TYP & \sum_\ell \mathcal{S}_\ell^*[\Upsilon_\ell] \end{array}\right] \succeq 0 \right\}. \end{array} \tag{4.75}$$

4) Prove the following statement:
(!!) *Let* m, n, Y *be as in* (!). *Then*

$$\begin{array}{l} \delta_m(\mathcal{X}, \|\cdot\|) \leq \mathrm{dist}_{\|\cdot\|}(\mathcal{X}, \mathrm{Ker}\, Y) \\ \quad \leq \widehat{\mathrm{Opt}} := \min\limits_{\substack{\nu, \Lambda=\{\Lambda_k, k\leq K\}, \\ \Upsilon=\{\Upsilon_\ell, \ell\leq L\}}} \left\{ \phi_{\mathcal{T}}(\lambda[\Lambda]) + \phi_{\mathcal{R}}(\lambda[\Upsilon]) : \right. \\ \quad \nu \geq 0, \Lambda_k \succeq 0\,\forall k, \Upsilon_\ell \succeq 0\,\forall \ell, \\ \quad \left. \left[\begin{array}{c|c} \sum_k \mathcal{R}_k^*[\Lambda_k] & \frac{1}{2}P^TM \\ \hline \frac{1}{2}M^TP & \sum_\ell \mathcal{S}_\ell^*[\Upsilon_\ell] + \nu M^T(I-Y)M \end{array}\right] \succeq 0 \right\}, \end{array} \tag{4.76}$$

and $\widehat{\mathrm{Opt}} \leq \mathrm{Opt}$, *with* Opt *given by* (4.75).

Statements (!) and (!!) suggest the following policy for upper-bounding the Kolmogorov width $\delta_m(\mathcal{X}, \|\cdot\|)$:

A. First, we select an integer μ, $1 \leq \mu < n$, and solve the convex optimization problem

$$\begin{array}{l} \min\limits_{\Lambda, \Upsilon, Y} \left\{ \phi_{\mathcal{T}}(\lambda[\Lambda]) + \phi_{\mathcal{R}}(\lambda[\Upsilon]) : 0 \preceq Y \preceq I, \mathrm{Tr}(Y) = n - \mu, \right. \\ \qquad \Lambda = \{\Lambda_k \succeq 0, k \leq K\}, \Upsilon = \{\Upsilon_\ell \succeq 0, \ell \leq L\}, \\ \qquad \left. \left[\begin{array}{c|c} \sum_k \mathcal{R}_k^*[\Lambda_k] & \frac{1}{2}P^TYM \\ \hline \frac{1}{2}M^TYP & \sum_\ell \mathcal{S}_\ell^*[\Upsilon_\ell] \end{array}\right] \succeq 0 \right\}. \end{array} \tag{P^μ}$$

B. Next, we take the Y-component Y^μ of the optimal solution to (P^μ) and "round" it to a orthoprojector Y of rank $n - m$ in the same fashion as in the case of $\|\cdot\| = \|\cdot\|_2$, that is, keep the eigenvectors of Y^μ intact and replace the m smallest eigenvalues with zeros, and all remaining eigenvalues with ones.

C. Finally, we solve the convex optimization problem

$$\begin{array}{rl}\mathrm{Opt}_{m,\mu} = \min\limits_{\Lambda,\Upsilon,\nu} \Big\{ \phi_{\mathcal{T}}(\lambda[\Lambda]) + \phi_{\mathcal{R}}(\lambda[\Upsilon]) : & \\ \nu \geq 0, \Lambda = \{\Lambda_k \succeq 0, k \leq K\}, \Upsilon = \{\Upsilon_\ell \succeq 0, \ell \leq L\}, & \\ \left[\begin{array}{c|c} \sum_k \mathcal{R}_k^*[\Lambda_k] & \frac{1}{2}P^T M \\ \hline \frac{1}{2}M^T P & \sum_\ell \mathcal{S}_\ell^*[\Upsilon_\ell] + \nu M^T(I-Y)M \end{array}\right] \succeq 0 \Big\}. & \end{array} \qquad (P^{m,\mu})$$

By (!!), $\mathrm{Opt}_{m,\mu}$ is an upper bound on the Kolmogorov width $\delta_m(\mathcal{X}, \|\cdot\|)$ (and in fact also on $\mathrm{dist}_{\|\cdot\|}(\mathcal{X}, \mathrm{Ker}\, Y)$).

Observe all the complications we encounter when passing from the simple case $\|\cdot\| = \|\cdot\|_2$ to the case of general norm $\|\cdot\|$ with a spectratope as the unit ball of the conjugate norm. Note that Proposition 4.26 gives both a lower bound $\sqrt{\mathrm{Opt}(m)/\varkappa}$ on the m-th Kolmogorov width of $\mathcal{X}$ w.r.t. $\|\cdot\|_2$, and a family of upper bounds $\sqrt{\frac{m+1}{m+1-\mu}\mathrm{Opt}(\mu)}$, $1 \leq \mu \leq m$, on this width. As a result, we can approximate $\mathcal{X}$ by m-dimensional subspaces in the Euclidean norm in a "nearly optimal" fashion. Indeed, if for some ϵ and k it holds $\delta_k(\mathcal{X}, \|\cdot\|_2) \leq \epsilon$, then $\mathrm{Opt}(k) \leq \varkappa\epsilon^2$ by Proposition 4.26 as applied with $m = k$. On the other hand, assuming $k < n/2$, the same proposition when applied with $m = 2k$ and $\mu = k$ says that

$$\mathrm{dist}_{\|\cdot\|_2}(\mathcal{X}, E^{m,k}) \leq \sqrt{\frac{2k+1}{k+1}\mathrm{Opt}(k)} \leq \sqrt{2\mathrm{Opt}(k)} \leq \sqrt{2\varkappa}\,\epsilon.$$

Thus, if $\mathcal{X}$ can be approximated by a k-dimensional subspace within $\|\cdot\|_2$-accuracy ϵ, we can efficiently get approximation of "nearly the same quality" ($\sqrt{2\varkappa}\epsilon$ instead of ϵ; recall that $\varkappa$ is just logarithmic in D) and "nearly the same dimension" ($2k$ instead of k).

Neither of these options is preserved when passing from the Euclidean norm to a general one: in the latter case, we do not have lower bounds on Kolmogorov widths, and have no understanding of how tight our upper bounds are.

Now, two concluding questions:

5) Why in step A of the above bounding scheme do we utilize statement (!) rather than the less conservative (since $\widehat{\mathrm{Opt}} \leq \mathrm{Opt}$) statement (!!)?
6) Implement the scheme numerically and run experiments.
   Recommended setup:

   - Given $\sigma > 0$ and positive integers n and κ, let f be a function of continuous argument $t \in [0,1]$ satisfying the smoothness restriction $|f^{(k)}(t)| \leq \sigma^k$, $0 \leq t \leq 1$, $k = 0, 1, 2, ..., \kappa$. Specify $\mathcal{X}$ as the set of n-dimensional vectors x obtained by restricting f onto the n-point equidistant grid $\{t_i = i/n\}_{i=1}^n$. To this end, translate the description on f into a bunch of two-sided linear constraints on x:

     $$|d_{(k)}^T[x_i; x_{i+1}; ...; x_{i+k}]| \leq \sigma^k, \ 1 \leq i \leq n-k, \ 0 \leq k \leq \kappa,$$

     where $d_{(k)} \in \mathbf{R}^{k+1}$ is the vector of coefficients of finite-difference approximation, with resolution $1/n$, of the k-th derivative:

     $$d_{(0)} = 1, \ d_{(1)} = n[-1;1], \ d_{(2)} = n^2[1;-2;1], \\ d_{(3)} = n^3[-1;3;-3;1], \ d_{(4)} = n^4[1;-4;6;-4;1], ...$$

- Recommended parameters: $n = 32$, $m = 8$, $\kappa = 5$, $\sigma \in \{0.25, 0.5; 1, 2, 4\}$.
- Run experiments with $\|\cdot\| = \|\cdot\|_1$ and $\|\cdot\| = \|\cdot\|_2$.

Exercise 4.8.

[more on semidefinite relaxation] The goal of this exercise is to extend SDP relaxation beyond ellitopes/spectratopes.

SDP relaxation is aimed at upper-bounding the quantity

$$\mathrm{Opt}_{\mathcal{X}}(B) = \max_{x\in\mathcal{X}} x^T Bx, \qquad [B \in \mathbf{S}^n]$$

where $\mathcal{X} \subset \mathbf{R}^n$ is a given set (which we from now on assume to be nonempty convex compact). To this end we look for a computationally tractable convex compact set $\mathcal{U} \subset \mathbf{S}^n$ such that for every $x \in \mathcal{X}$ it holds $xx^T \in \mathcal{U}$; in this case, we refer to $\mathcal{U}$ as to a set *matching* $\mathcal{X}$ (equivalent wording: "$\mathcal{U}$ matches $\mathcal{X}$"). Given such a set $\mathcal{U}$, the optimal value in the convex optimization problem

$$\overline{\mathrm{Opt}}_{\mathcal{U}}(B) = \max_{U\in\mathcal{U}} \mathrm{Tr}(BU) \tag{4.77}$$

is an efficiently computable convex upper bound on $\mathrm{Opt}_{\mathcal{X}}(B)$.

Given $\mathcal{U}$ matching $\mathcal{X}$, we can pass from $\mathcal{U}$ to the conic hull of $\mathcal{U}$–to the set

$$\mathbf{U}[\mathcal{U}] = \mathrm{cl}\{(U,\mu) \in \mathbf{S}^n \times \mathbf{R}_+ : \mu > 0, U/\mu \in \mathcal{U}\}$$

which, as is immediately seen, is a closed convex cone contained in $\mathbf{S}^n \times \mathbf{R}_+$. The only point (U,μ) in this cone with $\mu = 0$ has $U = 0$ (since $\mathcal{U}$ is compact), and

$$\mathcal{U} = \{U : (U,1) \in \mathbf{U}\} = \{U : \exists \mu \le 1 : (U,\mu) \in \mathbf{U}\},$$

so that the definition of $\overline{\mathrm{Opt}}_{\mathcal{U}}$ can be rewritten equivalently as

$$\overline{\mathrm{Opt}}_{\mathcal{U}}(B) = \min_{U,\mu} \{\mathrm{Tr}(BU) : (U,\mu) \in \mathbf{U}, \mu \le 1\}.$$

The question, of course, is where to take a set $\mathcal{U}$ matching $\mathcal{X}$, and the answer depends on what we know about $\mathcal{X}$. For example, when $\mathcal{X}$ is a basic ellitope

$$\mathcal{X} = \{x \in \mathbf{R}^n : \exists t \in \mathcal{T} : x^T S_k x \le t_k, k \le K\}$$

with our usual restrictions on $\mathcal{T}$ and S_k, it is immediately seen that

$$x \in \mathcal{X} \Rightarrow xx^T \in \mathcal{U} = \{U \in \mathbf{S}^n : U \succeq 0, \exists t \in \mathcal{T} : \mathrm{Tr}(US_k) \le t_k, k \le K\}.$$

Similarly, when $\mathcal{X}$ is a basic spectratope

$$\mathcal{X} = \{x \in \mathbf{R}^n : \exists t \in \mathcal{T} : S_k^2[x] \preceq t_k I_{d_k}, k \le K\}$$

with our usual restrictions on $\mathcal{T}$ and $S_k[\cdot]$, it is immediately seen that

$$x \in \mathcal{X} \Rightarrow xx^T \in \mathcal{U} = \{U \in \mathbf{S}^n : U \succeq 0, \exists t \in \mathcal{T} : \mathcal{S}_k[U] \preceq t_k I_{d_k}, k \le K\}.$$

One can verify that the semidefinite relaxation bounds on the maximum of a quadratic form on an ellitope/spectratope $\mathcal{X}$ derived in Sections 4.2.3 (for ellitopes) and 4.3.2 (for spectratopes) are nothing but the bounds (4.77) associated with the $\mathcal{U}$ just defined.

4.8.A Matching via absolute norms. There are other ways to specify a set matching $\mathcal{X}$. The seemingly simplest of them is as follows. Let $p(\cdot)$ be an absolute norm on $\mathbf{R}^n$ (recall that this is a norm $p(x)$ which depends solely on $\mathrm{abs}[x]$, where $\mathrm{abs}[x]$ is the vector comprised of the magnitudes of entries in x). We can convert $p(\cdot)$ into the norm $p^+(\cdot)$ on the space $\mathbf{S}^n$ as follows:

$$p^+(U) = p([p(\mathrm{Col}_1[U]); ...; p(\mathrm{Col}_n[U])]) \qquad [U \in \mathbf{S}^n].$$

1.1) Prove that p^+ indeed is a norm on $\mathbf{S}^n$, and $p^+(xx^T) = p^2(x)$. Denoting by $q(\cdot)$ the norm conjugate to $p(\cdot)$, what is the relation between the norm $(p^+)_*(\cdot)$ conjugate to $p^+(\cdot)$ and the norm $q^+(\cdot)$?

1.2) Derive from 1.1 that whenever $p(\cdot)$ is an absolute norm such that $\mathcal{X}$ is contained in the unit ball $\mathcal{B}_{p(\cdot)} = \{x : p(x) \leq 1\}$ of the norm p, the set

$$\mathcal{U}_{p(\cdot)} = \{U \in \mathbf{S}^n : U \succeq 0, p^+(U) \leq 1\}$$

is matching $\mathcal{X}$. If, in addition,

$$\mathcal{X} \subset \{x : p(x) \leq 1, Px = 0\}, \tag{4.78}$$

then the set

$$\mathcal{U}_{p(\cdot),P} = \{U \in \mathbf{S}^n : U \succeq 0, p^+(U) \leq 1, PU = 0\}$$

is matching $\mathcal{X}$.

Assume that in addition to $p(\cdot)$, we have at our disposal a computationally tractable closed convex set $\mathcal{D}$ such that whenever $p(x) \leq 1$, the vector $[x]^2 := [x_1^2; ...; x_n^2]$ belongs to $\mathcal{D}$; in the sequel we call such a $\mathcal{D}$ *square-dominating* $p(\cdot)$. For example, when $p(\cdot) = \|\cdot\|_r$, we can take

$$\mathcal{D} = \left\{ \begin{array}{ll} \{y \in \mathbf{R}^n_+ : \sum_i y_1 \leq 1\}, & r \leq 2 \\ \{y \in \mathbf{R}^n_+ : \|y\|_{r/2} \leq 1\}, & r > 2 \end{array} \right. .$$

Prove that in this situation the above construction can be refined: whenever $\mathcal{X}$ satisfies (4.78), the set

$$\mathcal{U}^{\mathcal{D}}_{p(\cdot),P} = \{U \in \mathbf{S}^n : U \succeq 0, p^+(U) \leq 1, PU = 0, \mathrm{dg}(U) \in \mathcal{D}\}$$
$$[\mathrm{dg}(U) = [U_{11}; U_{22}; ...; U_{nn}]]$$

matches $\mathcal{X}$.

Note: in the sequel, we suppress P in the notation $\mathcal{U}_{p(\cdot),P}$ and $\mathcal{U}^{\mathcal{D}}_{p(\cdot),P}$ when $P = 0$; thus, $\mathcal{U}_{p(\cdot)}$ is the same as $\mathcal{U}_{p(\cdot),0}$.

1.3) Check that when $p(\cdot) = \|\cdot\|_r$ with $r \in [1, \infty]$, one has

$$p^+(U) = \|U\|_r := \left\{ \begin{array}{ll} \left(\sum_{i,j} |U_{ij}|^r\right)^{1/r}, & 1 \leq r < \infty, \\ \max_{i,j} |U_{ij}|, & r = \infty \end{array} \right. .$$

1.4) Let $\mathcal{X} = \{x \in \mathbf{R}^n : \|x\|_1 \leq 1\}$ and $p(x) = \|x\|_1$, so that $\mathcal{X} \subset \{x : p(x) \leq 1\}$, and

$$\mathrm{Conv}\{[x]^2 : x \in \mathcal{X}\} \subset \mathcal{D} = \left\{y \in \mathbf{R}^n_+ : \sum\nolimits_i y_i = 1\right\}. \tag{4.79}$$

What are the bounds $\overline{\mathrm{Opt}}_{\mathcal{U}_{p(\cdot)}}(B)$ and $\overline{\mathrm{Opt}}_{\mathcal{U}^{\mathcal{D}}_{p(\cdot)}}(B)$? Is it true that the former (the latter) of the bounds is precise? Is it true that the former (the latter) bound is precise when $B \succeq 0$?

1.5) Let $\mathcal{X} = \{x \in \mathbf{R}^n : \|x\|_2 \leq 1\}$ and $p(x) = \|x\|_2$, so that $\mathcal{X} \subset \{x : p(x) \leq 1\}$ and (4.79) holds true. What are the bounds $\overline{\mathrm{Opt}}_{\mathcal{U}_{p(\cdot)}}(B)$ and $\overline{\mathrm{Opt}}_{\mathcal{U}^{\mathcal{D}}_{p(\cdot)}}(B)$? Is the former (the latter) bound precise?

1.6) Let $\mathcal{X} \subset \mathbf{R}^n_+$ be closed, convex, bounded, and with a nonempty interior. Verify that the set

$$\mathcal{X}^+ = \{x \in \mathbf{R}^n : \exists y \in \mathcal{X} : \mathrm{abs}[x] \leq y\}$$

is the unit ball of an absolute norm $p_{\mathcal{X}}$, and this is the largest absolute norm $p(\cdot)$ such that $\mathcal{X} \subset \{x : p(x) \leq 1\}$. Derive from this observation that the norm $p_{\mathcal{X}}(\cdot)$ is the best (i.e., resulting in the least conservative bounding scheme) among absolute norms which allow us to upper-bound $\mathrm{Opt}_{\mathcal{X}}(B)$ via the construction from item 1.2.

4.8.B "Calculus of matchings." Observe that the matching we have introduced admits a kind of "calculus." Specifically, consider the situation as follows: for $1 \leq \ell \leq L$, we are given

- nonempty convex compact sets $\mathcal{X}_\ell \subset \mathbf{R}^{n_\ell}$, $0 \in \mathcal{X}_\ell$, along with matching $\mathcal{X}_\ell$ convex compact sets $\mathcal{U}_\ell \subset \mathbf{S}^{n_\ell}$ giving rise to the closed convex cones

$$\mathbf{U}_\ell = \mathrm{cl}\{(U_\ell, \mu_\ell) \in \mathbf{S}^{n_\ell} \times \mathbf{R}_+ : \mu_\ell > 0, \mu_\ell^{-1} U_\ell \in \mathcal{U}_\ell\}.$$

  We denote by $\vartheta_\ell(\cdot)$ the Minkowski functions of $\mathcal{X}_\ell$:

$$\vartheta_\ell(y^\ell) = \inf\{t : t > 0, t^{-1} y^\ell \in \mathcal{X}_\ell\} : \mathbf{R}^{n_\ell} \to \mathbf{R} \cup \{+\infty\};$$

  note that $\mathcal{X}_\ell = \{y^\ell : \vartheta_\ell(y^\ell) \leq 1\}$;
- $n_\ell \times n$ matrices A_ℓ such that $\sum_\ell A_\ell^T A_\ell \succ 0$.

On top of that, we are given a monotone convex set $\mathcal{T} \subset \mathbf{R}^L_+$ intersecting the interior of $\mathbf{R}^L_+$.

These data specify the convex set

$$\mathcal{X} = \{x \in \mathbf{R}^n : \exists t \in \mathcal{T} : \vartheta_\ell^2(A_\ell x) \leq t_\ell, \ell \leq L\}. \tag{$*$}$$

2.1) Prove the following:

Lemma 4.27. *In the situation in question, the set*

$$\mathcal{U} = \left\{U \in \mathbf{S}^n : U \succeq 0 \ \&\ \exists t \in \mathcal{T} : (A_\ell U A_\ell^T, t_\ell) \in \mathbf{U}_\ell, \ell \leq L\right\}$$

is a closed and bounded convex set which matches $\mathcal{X}$. As a result, the efficiently *computable quantity*

$$\overline{\mathrm{Opt}}_{\mathcal{U}}(B) = \max_U \{\mathrm{Tr}(BU) : U \in \mathcal{U}\}$$

is an upper bound on

$$\mathrm{Opt}_{\mathcal{X}}(B) = \max_{x\in\mathcal{X}} x^T Bx.$$

2.2) Prove that if $\mathcal{X} \subset \mathbf{R}^n$ is a nonempty convex compact set, P is an $m \times n$ matrix, and $\mathcal{U}$ matches $\mathcal{X}$, then the set $\mathcal{V} = \{V = PUP^T : U \in \mathcal{U}\}$ matches $\mathcal{Y} = \{y : \exists x \in \mathcal{X} : y = Px\}$.

2.3) Prove that if $\mathcal{X} \subset \mathbf{R}^n$ is a nonempty convex compact set, P is an $n \times m$ matrix of rank m, and $\mathcal{U}$ matches $\mathcal{X}$, then the set $\mathcal{V} = \{V \succeq 0 : PVP^T \in \mathcal{U}\}$ matches $\mathcal{Y} = \{y : Py \in \mathcal{X}\}$.

2.4) Consider the "direct product" case where $\mathcal{X} = \mathcal{X}_1 \times ... \times \mathcal{X}_L$. When specifying A_ℓ as the matrix which "cuts" the ℓ-th block $A_\ell x = x^\ell$ of a block vector $x = [x^1; ...; x^L] \in \mathbf{R}^{n_1} \times ... \times \mathbf{R}^{n_L}$ and setting $\mathcal{T} = [0, 1]^L$, we cover this situation by the setup under consideration. In the direct product case, the construction from item 2.1 is as follows: given the sets $\mathcal{U}_\ell$ matching $\mathcal{X}_\ell$, we build the set

$$\mathcal{U} = \{U = [U^{\ell\ell'} \in \mathbf{R}^{n_\ell \times n_{\ell'}}]_{\ell,\ell' \leq L} \in \mathbf{S}^{n_1+...+n_L} : U \succeq 0, U^{\ell\ell} \in \mathcal{U}_\ell, \ell \leq L\}$$

and claim that this set matches $\mathcal{X}$.

Could we be less conservative? While we do not know how to be less conservative in general, we do know how to be less conservative in the special case when the $\mathcal{U}_\ell$ are built via absolute norms. Namely, let $p_\ell(\cdot) : \mathbf{R}^{n_\ell} \to \mathbf{R}_+$, $\ell \leq L$, be absolute norms, let sets $\mathcal{D}_\ell$ be square-dominating $p_\ell(\cdot)$,

$$\mathcal{X}^\ell \subset \widehat{X}_\ell = \{x^\ell \in \mathbf{R}^{n_\ell} : P_\ell x_\ell = 0, p_\ell(x^\ell) \leq 1\},$$

and let

$$\mathcal{U}_\ell = \{U \in \mathbf{S}^{n_\ell} : U \succeq 0, P_\ell U = 0, p_\ell^+(U) \leq 1, \mathrm{dg}(U) \in \mathcal{D}_\ell\}.$$

In this case the above construction results in

$$\mathcal{U} = \left\{ U = [U^{\ell\ell'} \in \mathbf{R}^{n_\ell \times n_{\ell'}}]_{\ell,\ell' \leq L} \in \mathbf{S}_+^{n_1+...+n_L} : U \succeq 0, \begin{array}{l} P_\ell U^{\ell\ell} = 0 \\ p_\ell^+(U^{\ell\ell}) \leq 1 \\ \mathrm{dg}(U^{\ell\ell}) \in \mathcal{D}_\ell \end{array}, \ell \leq L \right\}.$$

Now let

$$p([x^1; ...; x^L]) = \max[p_1(x^1), ..., p_L(x^L)] : \mathbf{R}^{n_1} \times ... \times \mathbf{R}^{n_L} \to \mathbf{R},$$

so that p is an absolute norm and

$$\mathcal{X} \subset \{x = [x^1; ...; x^L] : p(x) \leq 1, P_\ell x^\ell = 0, \ell \leq L\}.$$

Prove that in fact the set

$$\overline{\mathcal{U}} = \left\{ U = [U^{\ell\ell'} \in \mathbf{R}^{n_\ell \times n_{\ell'}}]_{\ell,\ell' \leq L} \in \mathbf{S}_+^{n_1+...+n_L} : U \succeq 0, \begin{array}{l} P_\ell U^{\ell\ell} = 0 \\ \mathrm{dg}(U^{\ell\ell}) \in \mathcal{D}_\ell \\ p^+(U) \leq 1 \end{array}, \ell \leq L \right\}$$

matches $\mathcal{X}$, and that we always have $\overline{\mathcal{U}} \subset \mathcal{U}$. Verify that in general this inclusion is strict.

4.8.C Illustration: Nullspace property revisited. Recall the sparsity-oriented signal recovery via ℓ_1 minimization from Chapter 1: Given an $m \times n$ sensing matrix A and (noiseless) observation $y = Aw$ of unknown signal w known to have at most s nonzero entries, we recover w as

$$\widehat{w} \in \operatorname*{Argmin}_z \{\|z\|_1 : Az = y\}.$$

We called matrix A s-good if whenever $y = Aw$ with s-sparse w, the only optimal solution to the right-hand side optimization problem is w. The (difficult to verify!) necessary and sufficient condition for s-goodness is the Nullspace property:

$$\mathrm{Opt} := \max_z \left\{\|z\|_{(s)} : z \in \operatorname{Ker} A, \|z\|_1 \leq 1\right\} < 1/2,$$

where $\|z\|_{(k)}$ is the sum of the k largest entries in the vector abs$[z]$. A verifiable sufficient condition for s-goodness is

$$\widehat{\mathrm{Opt}} := \min_H \max_j \|\mathrm{Col}_j[I - H^T A]\|_{(s)} < \tfrac{1}{2}, \tag{4.80}$$

the reason being that, as is immediately seen, $\widehat{\mathrm{Opt}}$ is an upper bound on Opt (see Proposition 1.9 with $q = 1$).

An immediate observation is that Opt is nothing but the maximum of quadratic form over an appropriate convex compact set. Specifically, let

$$\mathcal{X} = \{[u; v] \in \mathbf{R}^n \times \mathbf{R}^n : Au = 0, \|u\|_1 \leq 1, \textstyle\sum_i |v_i| \leq s, \|v\|_\infty \leq 1\},$$
$$B = \left[\begin{array}{c|c} & \frac{1}{2}I_n \\ \hline \frac{1}{2}I_n & \end{array}\right].$$

Then

$$\begin{array}{rcl}
\mathrm{Opt}_{\mathcal{X}}(B) & = & \max\limits_{[u;v]\in\mathcal{X}} [u; v]^T B[u; v] \\
& = & \max\limits_{u,v} \left\{u^T v : Au = 0, \|u\|_1 \leq 1, \sum_i |v_i| \leq s, \|v\|_\infty \leq 1\right\} \\
& \underbrace{=}_{(a)} & \max_u \left\{\|u\|_{(s)} : Au = 0, \|u\|_1 \leq 1\right\} \\
& = & \mathrm{Opt},
\end{array}$$

where (a) is due to the well-known fact (prove it!) that *whenever s is a positive integer $\leq n$, the extreme points of the set*

$$V = \{v \in \mathbf{R}^n : \sum_i |v_i| \leq s, \|v\|_\infty \leq 1\}$$

are exactly the vectors with at most s nonzero entries, the nonzero entries being ± 1; as a result

$$\forall (z \in \mathbf{R}^n) : \max_{v \in V} z^T v = \|z\|_{(s)}.$$

Now, V is the unit ball of the absolute norm

$$r(v) = \min\{t : \|v\|_1 \leq st, \|v\|_\infty \leq t\},$$

so that $\mathcal{X}$ is contained in the unit ball $\mathcal{B}$ of the absolute norm on $\mathbf{R}^{2n}$ specified as

$$p([u;v]) = \max\{\|u\|_1, r(v)\} \qquad [u, v \in \mathbf{R}^n],$$

i.e.,

$$\mathcal{X} = \{[u;v] : p([u,v]) \leq 1, Au = 0\}.$$

As a result, whenever $x = [u;v] \in \mathcal{X}$, the matrix

$$U = xx^T = \left[\begin{array}{c|c} U^{11} = uu^T & U^{12} = uv^T \\ \hline U^{21} = vu^T & U^{22} = vv^T \end{array}\right]$$

satisfies the condition $p^+(U) \leq 1$ (see item 1.2 above). In addition, this matrix clearly satisfies the condition

$$A[U^{11}, U^{12}] = 0.$$

It follows that the set

$$\mathcal{U} = \{U = \left[\begin{array}{c|c} U^{11} & U^{12} \\ \hline U^{21} & U^{22} \end{array}\right] \in \mathbf{S}^{2n} : U \succeq 0, p^+(U) \leq 1, AU^{11} = 0, AU^{12} = 0\}$$

(which clearly is a nonempty convex compact set) matches $\mathcal{X}$. As a result, the efficiently computable quantity

$$\begin{array}{rcl} \overline{\text{Opt}} & = & \max\limits_{U \in \mathcal{U}} \text{Tr}(BU) \\ & = & \max\limits_{U} \left\{ \text{Tr}(U^{12}) : U = \left[\begin{array}{c|c} U^{11} & U^{12} \\ \hline U^{21} & U^{22} \end{array}\right] \succeq 0, p^+(U) \leq 1, AU^{11} = 0, AU^{12} = 0 \right\} \end{array} \tag{4.81}$$

is an upper bound on Opt. As a result, the verifiable condition

$$\overline{\text{Opt}} < 1/2$$

is sufficient for s-goodness of A.

Now comes the concluding part of the exercise:

3.1) Prove that $\overline{\text{Opt}} \leq \widehat{\text{Opt}}$, so that (4.81) is less conservative than (4.80).
<u>Hint:</u> Apply Conic Duality to verify that

$$\widehat{\text{Opt}} = \max_V \left\{ \text{Tr}(V) : V \in \mathbf{R}^{n \times n}, AV = 0, \sum_{i=1}^n r(\text{Col}_i[V^T]) \leq 1 \right\}. \tag{4.82}$$

3.2) Run simulations with randomly generated Gaussian matrices A and play with different values of s to compare $\widehat{\text{Opt}}$ and $\overline{\text{Opt}}$. To save time, you can use toy sizes m, n, say, $m = 18, n = 24$.

4.7.4 Around Propositions 4.4 and 4.14

4.7.4.1 Optimizing linear estimates on convex hulls of unions of spectratopes

Exercise 4.9.

Let

- $\mathcal{X}_1, ..., \mathcal{X}_J$ be spectratopes in $\mathbf{R}^n$:

$$\mathcal{X}_j = \{x \in \mathbf{R}^n : \exists (y \in \mathbf{R}^{N_j}, t \in \mathcal{T}_j) : x = P_j y, R_{kj}^2[y] \preceq t_k I_{d_{kj}}, k \leq K_j\}, \ 1 \leq j \leq J,$$
$$\left[R_{kj}[y] = \sum_{i=1}^{N_j} y_i R^{kji}\right],$$

- $A \in \mathbf{R}^{m \times n}$ and $B \in \mathbf{R}^{\nu \times n}$ be given matrices,
- $\|\cdot\|$ be a norm on $\mathbf{R}^\nu$ such that the unit ball $\mathcal{B}_*$ of the conjugate norm $\|\cdot\|_*$ is a spectratope:

$$\begin{array}{rcl}\mathcal{B}_* & := & \{u : \|u\|_* \leq 1\} \\ & = & \{u \in \mathbf{R}^\nu : \exists (z \in \mathbf{R}^N, r \in \mathcal{R}) : u = Mz, S_\ell^2[z] \preceq r_\ell I_{f_\ell}, \ell \leq L\} \\ & & \left[S_\ell[z] = \sum_{i=1}^N z_i S^{\ell i}\right],\end{array}$$

- Π be a convex compact subset of the interior of the positive semidefinite cone $\mathbf{S}_+^m$,

with our standard restrictions on $R_{kj}[\cdot]$, $S_\ell[\cdot]$, $\mathcal{T}_j$ and $\mathcal{R}$. Let, further,

$$\mathcal{X} = \mathrm{Conv}\left(\bigcup_j \mathcal{X}_j\right)$$

be the convex hull of the union of spectratopes $\mathcal{X}_j$. Consider the situation where, given observation

$$\omega = Ax + \xi$$

of unknown signal x known to belong to $\mathcal{X}$, we want to recover Bx. We assume that the matrix of second moments of noise is $\succeq$-dominated by a matrix from Π, and quantify the performance of a candidate estimate $\widehat{x}(\cdot)$ by its $\|\cdot\|$-*risk*

$$\mathrm{Risk}_{\Pi,\|\cdot\|}[\widehat{x}|\mathcal{X}] = \sup_{x \in \mathcal{X}} \sup_{P: P \lhd \Pi} \mathbf{E}_{\xi \sim P}\left\{\|Bx - \widehat{x}(Ax + \xi)\|\right\}$$

where $P \lhd \Pi$ means that the matrix $\mathrm{Var}[P] = \mathbf{E}_{\xi \sim P}\{\xi\xi^T\}$ of second moments of distribution P is $\succeq$-dominated by a matrix from Π.

Prove the following:

Proposition 4.28. *In the situation in question, consider the convex optimization problem*

$$\mathrm{Opt} = \min_{H, \Theta, \Lambda^j, \Upsilon^j, \Upsilon'} \left\{ \max_j \left[\phi_{\mathcal{T}_j}(\lambda[\Lambda^j]) + \phi_{\mathcal{R}}(\lambda[\Upsilon^j])\right] + \phi_{\mathcal{R}}(\lambda[\Upsilon']) + \Gamma_\Pi(\Theta) : \right.$$

$$\left.\begin{array}{r}\Lambda^j = \{\Lambda^j_k \succeq 0, j \leq K_j\}, j \leq J, \\ \Upsilon^j = \{\Upsilon^j_\ell \succeq 0, \ell \leq L\}, j \leq J, \Upsilon' = \{\Upsilon'_\ell \succeq 0, \ell \leq L\} \\ \left[\begin{array}{c|c}\sum_k \mathcal{R}^*_{kj}[\Lambda^j_k] & \frac{1}{2}P_j^T[B^T - A^T H]M \\ \hline \frac{1}{2}M^T[B - H^T A]P_j & \sum_\ell \mathcal{S}^*_\ell[\Upsilon^j_\ell]\end{array}\right] \succeq 0, j \leq J, \\ \left[\begin{array}{c|c}\Theta & \frac{1}{2}HM \\ \hline \frac{1}{2}M^T H^T & \sum_\ell \mathcal{S}^*_\ell[\Upsilon'_\ell]\end{array}\right] \succeq 0 \Big\},\end{array}\right\} \tag{4.83}$$

where, as usual,

$$\begin{array}{c}\phi_{\mathcal{T}_j}(\lambda) = \max_{t\in\mathcal{T}_j} t^T\lambda,\ \phi_{\mathcal{R}}(\lambda) = \max_{r\in\mathcal{R}} r^T\lambda, \\ \Gamma_\Pi(\Theta) = \max_{Q\in\Pi} \mathrm{Tr}(Q\Theta),\ \lambda[U_1, ..., U_s] = [\mathrm{Tr}(U_1); ...; \mathrm{Tr}(U_S)], \\ \mathcal{S}^*_\ell[\cdot] : \mathbf{S}^{f_\ell} \to \mathbf{S}^N : \mathcal{S}^*_\ell[U] = \left[\mathrm{Tr}(S^{\ell p} U S^{\ell q})\right]_{p,q\leq N}, \\ \mathcal{R}^*_{kj}[\cdot] : \mathbf{S}^{d_{kj}} \to \mathbf{S}^{N_j} : \mathcal{R}^*_{kj}[U] = \left[\mathrm{Tr}(R^{kjp} U R^{kjq})\right]_{p,q\leq N_j}.\end{array}$$

Problem (4.83) is solvable, and H-component H_* of its optimal solution gives rise to linear estimate $\widehat{x}_{H_*}(\omega) = H_*^T\omega$ such that

$$\mathrm{Risk}_{\Pi,\|\cdot\|}[\widehat{x}_{H_*}|\mathcal{X}] \leq \mathrm{Opt}. \tag{4.84}$$

Moreover, the estimate $\widehat{x}_{H_*}$ is near-optimal among linear estimates:

$$\begin{array}{c}\mathrm{Opt} \leq O(1)\ln(D + F)\mathrm{RiskOpt}_{\mathit{lin}} \\ \left[D = \max_j \sum_{k\leq K_j} d_{kj},\ F = \sum_{\ell\leq L} f_\ell\right]\end{array} \tag{4.85}$$

where

$$\mathrm{RiskOpt}_{\mathit{lin}} = \inf_H \sup_{x\in\mathcal{X}, Q\in\Pi} \mathbf{E}_{\xi\sim\mathcal{N}(0,Q)}\left\{\|Bx - H^T(Ax + \xi)\|\right\}$$

is the best risk attainable by linear estimates in the current setting under zero mean Gaussian observation noise.

It should be stressed that the convex hull of a union of spectratopes is not necessarily a spectratope, and that Proposition 4.28 states that the linear estimate stemming from (4.83) is near-optimal only among linear, not among all estimates (the latter might indeed not be the case).

4.7.4.2 Recovering nonlinear vector-valued functions

Exercise 4.10.

Consider the situation as follows: We are given a noisy observation

$$\omega = Ax + \xi_x \qquad [A \in \mathbf{R}^{m\times n}]$$

of the linear image Ax of an unknown signal x known to belong to a given spectratope $\mathcal{X} \subset \mathbf{R}^n$; here ξ_x is the observation noise with distribution P_x which can depend on x. As in Section 4.3.3, we assume that we are given a computationally tractable convex compact set $\Pi \subset \mathrm{int}\,\mathbf{S}^m_+$ such that for every $x \in \mathcal{X}$, $\mathrm{Var}[P_x] \preceq \Theta$ for some $\Theta \in \Pi$; cf. (4.32). We want to recover the value $f(x)$ of a given vector-valued function $f : \mathcal{X} \to \mathbf{R}^\nu$, and we measure the recovery error in a given norm $|\cdot|$ on $\mathbf{R}^\nu$.

4.10.A. Preliminaries and the Main observation. Let $\|\cdot\|$ be a norm on $\mathbf{R}^n$, and $g(\cdot):\mathcal{X}\to\mathbf{R}^\nu$ be a function. Recall that the function is called *Lipschitz continuous on $\mathcal{X}$ w.r.t. the pair of norms $\|\cdot\|$ on the argument and $|\cdot|$ on the image spaces,* if there exist $L<\infty$ such that

$$|g(x)-g(y)|\leq L\|x-y\|\ \forall(x,y\in\mathcal{X});$$

every L with this property is called a Lipschitz constant of g. It is well known that in our finite-dimensional situation, the property of g to be Lipschitz continuous is independent of how the norms $\|\cdot\|$, $|\cdot|$ are selected; this selection affects only the value(s) of Lipschitz constant(s).

Assume from now on that the function of interest f is Lipschitz continuous on $\mathcal{X}$. Let us call a norm $\|\cdot\|$ on $\mathbf{R}^n$ *appropriate* for f if f is Lipschitz continuous *with constant 1* on $\mathcal{X}$ w.r.t. $\|\cdot\|$, $|\cdot|$. Our immediate observation is as follows:

Observation 4.29. In the situation in question, let $\|\cdot\|$ be appropriate for f. Then recovering $f(x)$ is not more difficult than recovering x in the norm $\|\cdot\|$: every estimate $\widehat{x}(\omega)$ of x via ω such that $\widehat{x}(\cdot)\in\mathcal{X}$ induces the "plug-in" estimate

$$\widehat{f}(\omega)=f(\widehat{x}(\omega))$$

of $f(x)$, and the $\|\cdot\|$-risk

$$\mathrm{Risk}_{\|\cdot\|}[\widehat{x}|\mathcal{X}]=\sup_{x\in\mathcal{X}}\mathbf{E}_{\xi\sim P_x}\left\{\|\widehat{x}(Ax+\xi)-x\|\right\}$$

of estimate $\widehat{x}$ upper-bounds the $|\cdot|$-risk

$$\mathrm{Risk}_{|\cdot|}[\widehat{f}|\mathcal{X}]=\sup_{x\in\mathcal{X}}\mathbf{E}_{\xi\sim P_x}\left\{|\widehat{f}(Ax+\xi)-f(x)|\right\}$$

of the estimate $\widehat{f}$ induced by $\widehat{x}$:

$$\mathrm{Risk}_{|\cdot|}[\widehat{f}|\mathcal{X}]\leq\mathrm{Risk}_{\|\cdot\|}[\widehat{x}|\mathcal{X}].$$

When f is defined and Lipschitz continuous with constant 1 w.r.t. $\|\cdot\|,|\cdot|$ on the entire $\mathbf{R}^n$, this conclusion remains valid without the assumption that $\widehat{x}$ is $\mathcal{X}$-valued.

4.10.B. Consequences. Observation 4.29 suggests the following simple approach to solving the estimation problem we started with: assuming that we have at our disposal a norm $\|\cdot\|$ on $\mathbf{R}^n$ such that

- $\|\cdot\|$ is appropriate for f, and
- $\|\cdot\|$ is *good*, meaning that the unit ball $\mathcal{B}_*$ of the norm $\|\cdot\|_*$ conjugate to $\|\cdot\|$ is a spectratope given by explicit spectratopic representation,

we use the machinery of linear estimation developed in Section 4.3.3 to build a near-optimal, in terms of its $\|\cdot\|$-risk, linear estimate of x via ω, and convert this estimate into an estimate of $f(x)$. By the above observation, the $|\cdot|$- risk of the resulting estimate is upper-bounded by the $\|\cdot\|$-risk of the underlying linear estimate. The construction just outlined needs a correction: in general, the linear estimate $\widetilde{x}(\cdot)$ yielded by Proposition 4.14 (same as any nontrivial—not identically zero—*linear* estimate) is *not* guaranteed to take values in $\mathcal{X}$, which is, in general, required for

Observation 4.29 to be applicable. This correction is easy: it is enough to convert $\widetilde{x}$ into the estimate $\widehat{x}$ defined by

$$\widehat{x}(\omega) \in \operatorname*{Argmin}_{u\in\mathcal{X}} \|u - \widetilde{x}(\omega)\|.$$

This transformation preserves efficient computability of the estimate, and ensures that the corrected estimate takes its values in $\mathcal{X}$; at the same time, "correction" $\widetilde{x} \mapsto \widehat{x}$ nearly preserves the $\|\cdot\|$-risk:

$$\mathrm{Risk}_{\|\cdot\|}[\widehat{x}|\mathcal{X}] \le 2\mathrm{Risk}_{\|\cdot\|}[\widetilde{x}|\mathcal{X}]. \qquad (*)$$

Note that when $\|\cdot\|$ is a (general-type) Euclidean norm: $\|x\|^2 = x^TQx$ for some $Q \succ 0$, factor 2 on the right-hand side can be discarded.

1) Justify $(*)$.

4.10.C. How to select $\|\cdot\|$. When implementing the outlined approach, the major question is how to select a norm $\|\cdot\|$ appropriate for f. The best choice would be to select the smallest among the norms appropriate for f (such a norm does exist under mild assumptions), because the smaller the $\|\cdot\|$, the smaller the $\|\cdot\|$-risk of an estimate of x. This ideal can be achieved in rare cases only: first, it could be difficult to identify the smallest among the norms appropriate for f; second, our approach requires for $\|\cdot\|$ to have an explicitly given spectratope as the unit ball of the conjugate norm. Let us look at a couple of "favorable cases," where the difficulties just outlined can be (partially) overcome.

Example: A norm-induced f. Let us start with the case, important in its own right, when f is a scalar functional which itself is a norm, and this norm has a spectratope as the unit ball of the conjugate norm, as is the case when $f(\cdot) = \|\cdot\|_r$, $r \in [1, 2]$, or when $f(\cdot)$ is the nuclear norm. In this case the smallest of the norms appropriate for f clearly is f itself, and none of the outlined difficulties arises. As an extension, when $f(x)$ is obtained from a good norm $\|\cdot\|$ by operations preserving Lipschitz continuity and constant, such as $f(x) = \|x - c\|$, or $f(x) = \sum_i a_i\|x - c_i\|$, $\sum_i |a_i| \le 1$, or

$$f(x) = \sup_{c\in C} / \inf \|x - c\|,$$

or even something like

$$f(x) = \sup_{\alpha\in\mathcal{A}} / \inf \left\{ \sup_{c\in C_\alpha} / \inf \|x - c\| \right\}.$$

In such a case, it seems natural to use this norm in our construction, although now this, perhaps, is not the smallest of the norms appropriate for f.

Now let us consider the general case. Note that *in principle* the smallest of the norms appropriate for a given Lipschitz continuous f admits a description. Specifically, assume that $\mathcal{X}$ has a nonempty interior (this is w.l.o.g.—we can always replace $\mathbf{R}^n$ with the linear span of $\mathcal{X}$). A well-known fact of Analysis (Rademacher Theorem) states that in this situation (more generally, when $\mathcal{X}$ is convex with a nonempty interior), a Lipschitz continuous f is differentiable almost everywhere in $\mathcal{X}^o = \operatorname{int}\mathcal{X}$, and f is Lipschitz continuous with constant 1 w.r.t. a norm $\|\cdot\|$ if and only if

$$\|f'(x)\|_{\|\cdot\|\to|\cdot|} \leq 1$$

whenever $x \in \mathcal{X}^o$ is such that the derivative (a.k.a. Jacobian) of f at x exists; here $\|Q\|_{\|\cdot\|\to|\cdot|}$ is the matrix norm of a $\nu \times n$ matrix Q induced by the norms $\|\cdot\|$ on $\mathbf{R}^n$ and $|\cdot|$ on $\mathbf{R}^\nu$:

$$\|Q\|_{\|\cdot\|\to|\cdot|} := \max_{\|x\|\leq 1} |Qx| = \max_{\substack{\|x\|\leq 1\\ |y|_*\leq 1}} y^T Qx = \max_{\substack{|y|_*\leq 1\\ [\|x\|_*]_*\leq 1}} x^T Q^T y = \|Q^T\|_{|\cdot|_*\to\|\cdot\|_*},$$

where $\|\cdot\|_*$, $|\cdot|_*$ are the conjugates of $\|\cdot\|$, $|\cdot|$.

2) Prove that a norm $\|\cdot\|$ is appropriate for f if and only if the unit ball of the *conjugate* to $\|\cdot\|$ norm contains the set

$$\mathcal{B}_{f,*} = \mathrm{cl\,Conv}\{z : \exists (x \in \mathcal{X}_o, y, |y|_* \leq 1) : z = [f'(x)]^T y\},$$

where $\mathcal{X}_o$ is the set of all $x \in \mathcal{X}^o$ where $f'(x)$ exists. Geometrically, $\mathcal{B}_{f,*}$ is the closed convex hull of the union of all images of the unit ball $\mathcal{B}_*$ of $|\cdot|_*$ under the linear mappings $y \mapsto [f'(x)]^T y$ stemming from $x \in \mathcal{X}_o$.
Equivalently: $\|\cdot\|$ is appropriate for f if and only if

$$\|u\| \geq \|u\|_f := \max_{z\in\mathcal{B}_{f,*}} z^T u. \qquad (!)$$

Check that $\|u\|_f$ is a norm, provided that $\mathcal{B}_{f,*}$ (this set by construction is a convex compact set symmetric w.r.t. the origin) possesses a nonempty interior; whenever this is the case, $\|u\|_f$ is the smallest of the norms appropriate for f. Derive from the above that the norms $\|\cdot\|$ we can use in our approach are the norms on $\mathbf{R}^n$ for which the unit ball of the conjugate norm is a spectratope containing $\mathcal{B}_{f,*}$.

Example. Consider the case of componentwise quadratic f:

$$f(x) = \left[\tfrac{1}{2}x^T Q_1 x; \tfrac{1}{2}x^T Q_2 x; ...; \tfrac{1}{2}x^T Q_\nu x\right] \qquad [Q_i \in \mathbf{S}^n]$$

and $|u| = \|u\|_q$ with $q \in [1,2]$.[10] In this case

$$\mathcal{B}_* = \{u \in \mathbf{R}^\nu : \|u\|_p \leq 1\},\ p = \frac{q}{q-1} \in [2,\infty[,\ \text{and}\ f'(x) = \left[x^T Q_1; x^T Q_2; ...; x^T Q_\nu\right].$$

Setting $\mathcal{S} = \{s \in \mathbf{R}^\nu_+ : \|s\|_{p/2} \leq 1\}$ and

$$\mathcal{S}^{1/2} = \{s \in \mathbf{R}^\nu_+ : [s_1^2; ...; s_\nu^2] \in \mathcal{S}\} = \{s \in \mathbf{R}^\nu_+ : \|s\|_p \leq 1\},$$

the set

$$\mathcal{Z} = \{[f'(x)]^T u : x \in \mathcal{X}, u \in \mathcal{B}_*\}$$

[10] To save notation, we assume that the linear parts in the components of f_i are trivial—just zeros. In this respect, note that we always can subtract from f any linear mapping and reduce our estimation problem to two distinct problems of estimating separately the values at the signal x of the modified f and the linear mapping we have subtracted (we know how to solve the latter problem reasonably well).

is contained in the set

$$\mathcal{Y} = \left\{ y \in \mathbf{R}^n : \exists (s \in \mathcal{S}^{1/2}, x^i \in \mathcal{X}, i \le \nu) : y = \sum_i s_i Q_i x_i \right\}.$$

The set $\mathcal{Y}$ is a spectratope with spectratopic representation readily given by that of $\mathcal{X}$. Indeed, $\mathcal{Y}$ is nothing but the $\mathcal{S}$-sum of the spectratopes $Q_i\mathcal{X}$, $i = 1, ..., \nu$; see Section 4.10. As a result, we can use the spectratope $\mathcal{Y}$ (when $\operatorname{int}\mathcal{Y} \neq \emptyset$) or the arithmetic sum of $\mathcal{Y}$ with a small Euclidean ball (when $\operatorname{int}\mathcal{Y} = \emptyset$) as the unit ball of the norm conjugate to $\|\cdot\|$, thus ensuring that $\|\cdot\|$ is appropriate for f. We then can use $\|\cdot\|$ in order to build an estimate of $f(\cdot)$.

3.1) For illustration, work out the problem of recovering the value of a scalar quadratic form

$$f(x) = x^T M x,\ M = \operatorname{Diag}\{i^\alpha, i = 1, ..., n\} \qquad [\nu = 1, |\cdot| \text{ is the absolute value}]$$

from noisy observation

$$\omega = Ax + \sigma\eta,\ A = \operatorname{Diag}\{i^\beta, i = 1, ..., n\},\ \eta \sim \mathcal{N}(0, I_n) \tag{4.86}$$

of a signal x known to belong to the ellipsoid

$$\mathcal{X} = \{x \in \mathbf{R}^n : \|Px\|_2 \le 1\},\ P = \operatorname{Diag}\{i^\gamma, i = 1, ..., n\},$$

where α, β, γ are given reals satisfying

$$\alpha - \gamma - \beta < -1/2.$$

You could start with the simplest unbiased estimate

$$\widetilde{x}(\omega) = [1^{-\beta}\omega_1; 2^{-\beta}\omega_2; ...; n^{-\beta}\omega_n]$$

of x.

3.2) Work out the problem of recovering the norm

$$f(x) = \|Mx\|_p,\ M = \operatorname{Diag}\{i^\alpha, i = 1, ..., n\},\ p \in [1, 2],$$

from observation (4.86) with

$$\mathcal{X} = \{x : \|Px\|_r \le 1\},\ P = \operatorname{Diag}\{i^\gamma, i = 1, ..., n\},\ r \in [2, \infty].$$

4.7.4.3 Suboptimal linear estimation

Exercise 4.11.

[recovery of large-scale signals] Consider the problem of estimating the image $Bx \in \mathbf{R}^\nu$ of signal $x \in \mathcal{X}$ from observation

$$\omega = Ax + \sigma\xi \in \mathbf{R}^m$$

in the simplest case where $\mathcal{X} = \{x \in \mathbf{R}^n : x^T S x \le 1\}$ is an ellipsoid (so that $S \succ 0$), the recovery error is measured in $\|\cdot\|_2$, and $\xi \sim \mathcal{N}(0, I_m)$. In this case,

Problem (4.12) to solve when building "presumably good linear estimate" reduces to

$$\mathrm{Opt} = \min_{H,\lambda} \left\{ \lambda + \sigma^2 \|H\|_F^2 : \left[\begin{array}{c|c} \lambda S & B^T - A^T H \\ \hline B - H^T A & I_\nu \end{array} \right] \succeq 0 \right\}, \tag{4.87}$$

where $\|\cdot\|_F$ is the Frobenius norm of a matrix. An optimal solution H_* to this problem results in the linear estimate $\widehat{x}_{H_*}(\omega) = H_*^T \omega$ satisfying the risk bound

$$\mathrm{Risk}[\widehat{x}_{H_*}|\mathcal{X}] := \max_{x\in\mathcal{X}} \sqrt{\mathbf{E}\{\|Bx - H_*^T(Ax+\sigma\xi)\|_2^2\}} \leq \sqrt{\mathrm{Opt}}.$$

Now, (4.87) is an efficiently solvable convex optimization problem. However, when the sizes m, n of the problem are large, solving the problem by standard optimization techniques could become prohibitively time-consuming. The goal of what follows is to develop a relatively cheap computational technique for finding a good enough *sub*optimal solution to (4.87). In the sequel, we assume that $A \neq 0$; otherwise (4.87) is trivial.

1) Prove that problem (4.87) can be reduced to a similar problem with $S = I_n$ and diagonal positive semidefinite matrix A, the reduction requiring several singular value decompositions and multiplications of matrices of the same sizes as those of A, B, and S.

2) By item 1, we can assume from the very beginning that $S = I$ and $A = \mathrm{Diag}\{\alpha_1, ..., \alpha_n\}$ with $0 \leq \alpha_1 \leq \alpha_2 \leq ... \leq \alpha_n$. Passing in (4.87) from variables λ, H to variables $\tau = \sqrt{\lambda}, G = H^T$, the problem becomes

$$\mathrm{Opt} = \min_{G,\tau} \left\{ \tau^2 + \sigma^2 \|G\|_F^2 : \|B - GA\| \leq \tau \right\}, \tag{4.88}$$

where $\|\cdot\|$ is the spectral norm. Now consider the construction as follows:

- Consider a partition $\{1, ..., n\} = I_0 \cup I_1 \cup ... \cup I_K$ of the index set $\{1, ..., n\}$ into consecutive segments in such a way that
(a) I_0 is the set of those i, if any, for which $\alpha_i = 0$, and $I_k \neq \emptyset$ when $k \geq 1$,
(b) for $k \geq 1$ the ratios α_j/α_i, $i, j \in I_k$, do not exceed $\theta > 1$ (θ is the parameter of our construction), while
(c) for $1 \leq k < k' \leq K$, the ratios α_j/α_i, $i \in I_k$, $j \in I_{k'}$, are $> \theta$.
The recipe for building the partition is self-evident, and we clearly have

$$K \leq \ln(\overline{\alpha}/\underline{\alpha})/\ln(\theta) + 1,$$

where $\overline{\alpha}$ is the largest of α_i, and $\underline{\alpha}$ is the smallest of those α_i which are positive.
- For $1 \leq k \leq K$, we denote by i_k the first index in I_k, set $\alpha^k = \alpha_{i_k}$, $n_k = \mathrm{Card}\, I_k$, and define A_k as the $n_k \times n_k$ diagonal matrix with diagonal entries α_i, $i \in I_k$.

Now, given a $\nu \times n$ matrix C, let us specify C_k, $0 \leq k \leq K$, as the $\nu \times n_k$ submatrix of C comprised of columns with indexes from I_k, and consider the following parametric optimization problems:

$$\begin{array}{rcll} \mathrm{Opt}_k^*(\tau) & = & \min_{G_k\in\mathbf{R}^{\nu\times n_k}} \left\{ \|G_k\|_F^2 : \|B_k - G_k A_k\| \leq \tau \right\} & (P_k^*[\tau]) \\ \mathrm{Opt}_k(\tau) & = & \min_{G_k\in\mathbf{R}^{\nu\times n_k}} \left\{ \|G_k\|_F^2 : \|B_k - \alpha^k G_k\| \leq \tau \right\} & (P_k[\tau]) \end{array}$$

where $\tau \geq 0$ is the parameter, and $1 \leq k \leq K$.

Justify the following simple observations:

2.1) G_k is feasible for $(P_k[\tau])$ if and only if the matrix

$$G_k^* = \alpha^k G_k A_k^{-1}$$

is feasible for $(P_k^*[\tau])$, and $\|G_k^*\|_F \leq \|G_k\|_F \leq \theta\|G_k^*\|_F$, implying that

$$\text{Opt}_k^*(\tau) \leq \text{Opt}_k(\tau) \leq \theta^2 \text{Opt}_k^*(\tau);$$

2.2) Problems $(P_k[\tau])$ are easy to solve: if $B_k = U_k D_k V_k^T$ is the singular value decomposition of B_k and $\sigma_{k\ell}$, $1 \leq \ell \leq \nu_k := \min[\nu, n_k]$, are diagonal entries of D_k, then an optimal solution to $(P_k[\tau])$ is

$$\widehat{G}_k[\tau] = [\alpha^k]^{-1} U_k D_k[\tau] V_k^T,$$

where $D_k[\tau]$ is the diagonal matrix obtained from D_k by truncating the diagonal entries $\sigma_{k\ell} \mapsto [\sigma_{k\ell} - \tau]_+$ (from now on, $a_+ = \max[a, 0]$, $a \in \mathbf{R}$). The optimal value in $(P_k[\tau])$ is

$$\text{Opt}_k(\tau) = [\alpha^k]^{-2} \sum_{\ell=1}^{\nu_k} [\sigma_{k\ell} - \tau]_+^2.$$

2.3) If (τ, G) is a feasible solution to (4.88) then $\tau \geq \underline{\tau} := \|B_0\|$, and the matrices G_k, $1 \leq k \leq K$, are feasible solutions to problems $(P_k^*[\tau])$, implying that

$$\sum_k \text{Opt}_k^*(\tau) \leq \|G\|_F^2.$$

And vice versa: if $\tau \geq \underline{\tau}$, G_k, $1 \leq k \leq K$, are feasible solutions to problems $(P_k^*[\tau])$, and

$$K_+ = \left\{ \begin{array}{ll} K, & I_0 = \emptyset \\ K+1, & I_0 \neq \emptyset \end{array} \right.,$$

then the matrix $G = [0_{\nu \times n_0}, G_1, ..., G_K]$ and $\tau_+ = \sqrt{K_+}\tau$ form a feasible solution to (4.88).

Extract from these observations that if τ_* is an optimal solution to the convex optimization problem

$$\min_\tau \left\{ \theta^2 \tau^2 + \sigma^2 \sum_{k=1}^K \text{Opt}_k(\tau) : \tau \geq \underline{\tau} \right\} \tag{4.89}$$

and $G_{k,*}$ are optimal solutions to the problems $(P_k[\tau_*])$, then the pair

$$\widehat{\tau} = \sqrt{K_+}\tau_*, \ \widehat{G} = [0_{\nu \times n_0}, G_{1,*}^*, ..., G_{K,*}^*] \qquad [G_{k,*}^* = \alpha^k G_{k,*} A_k^{-1}]$$

is a feasible solution to (4.88), and the value of the objective of the latter problem at this feasible solution is within the factor $\max[K_+, \theta^2]$ of the true optimal value Opt of this problem. As a result, $\widehat{G}$ gives rise to a linear estimate with risk on $\mathcal{X}$ which is within the factor $\max[\sqrt{K_+}, \theta]$ of the risk $\sqrt{\text{Opt}}$ of the "presumably good" linear estimate yielded by an optimal solution to (4.87).

Notice that

- After carrying out singular value decompositions of matrices B_k, $1 \le k \le K$, specifying τ_* and $G_{k,*}$ requires solving univariate convex minimization problem with an easy-to-compute objective, so that the problem can be easily solved, e.g., by bisection;
- The computationally cheap suboptimal solution we end up with is not that bad, since K is "moderate"—just logarithmic in the condition number $\overline{\alpha}/\underline{\alpha}$ of A.

Your next task is as follows:

3) To get an idea of the performance of the proposed synthesis of "suboptimal" linear estimation, run numerical experiments as follows:
   - select some n and generate at random the $n \times n$ data matrices S, A, B;
   - for "moderate" values of n compute both the linear estimate yielded by the optimal solution to (4.12)[11] and the suboptimal estimate as yielded by the above construction. Compare their risk bounds and the associated CPU times. For "large" n, where solving (4.12) becomes prohibitively time-consuming, compute only a suboptimal estimate in order to get an impression of how the corresponding CPU time grows with n.

   Recommended setup:
   - range of n: 50, 100 ("moderate" values), 1000, 2000 ("large" values)
   - range of σ: $\{1.0, 0.01, 0.0001\}$
   - generation of S, A, B: generate the matrices at random according to

   $$S = U_S \mathrm{Diag}\{1, 2, ..., n\} U_S^T,\ A = U_A \mathrm{Diag}\{\mu_1, ..., \mu_n\} V_A^T,$$
   $$B = U_B \mathrm{Diag}\{\mu_1, ..., \mu_n\} V_B^T,$$

   where U_S, U_A, V_A, U_B, V_B are random orthogonal $n \times n$ matrices, and the μ_i form a geometric progression with $\mu_1 = 0.01$ and $\mu_n = 1$.

   You could run the above construction for several values of θ and select the best, in terms of its risk bound, of the resulting suboptimal estimates.

4.11.A. Simple case. There is a trivial case where (4.88) is really easy; this is the case where the right orthogonal factors in the singular value decompositions of A and B are the same, that is, when

$$B = WFV^T,\ A = UDV^T$$

with orthogonal $n \times n$ matrices W, U, V and diagonal F, D. This very special case is in fact of some importance—it covers the *denoising* situation where $B = A$, so that our goal is to denoise our observation of Ax given a priori information $x \in \mathcal{X}$

[11]When $\mathcal{X}$ is an ellipsoid, semidefinite relaxation bound on the maximum of a quadratic form over $x \in \mathcal{X}$ is exact, so that we are in the case when an optimal solution to (4.12) yields the best, in terms of risk on $\mathcal{X}$, linear estimate.

on x. In this situation, setting $W^T H^T U = G$, problem (4.88) becomes

$$\mathrm{Opt} = \min_G \left\{ \|F - GD\|^2 + \sigma^2 \|G\|_F^2 \right\}. \tag{4.90}$$

Now goes the concluding part of the exercise:

4) Prove that in the situation in question an optimal solution G_* to (4.90) can be selected to be diagonal, with diagonal entries γ_i, $1 \leq i \leq n$, yielded by the optimal solution to the optimization problem

$$\mathrm{Opt} = \min_\gamma \left\{ f(G) := \max_{i \leq n} (\phi_i - \gamma_i \delta_i)^2 + \sigma^2 \sum_{i=1}^n \gamma_i^2 \right\} \qquad [\phi_i = F_{ii}, \delta_i = D_{ii}].$$

Exercise 4.12.

[image reconstruction—follow-up to Exercise 4.11] A grayscale image can be represented by an $m \times n$ matrix $x = [x_{pq}]_{\substack{0 \leq p < m, \\ 0 \leq q < n}}$ with entries in the range $[-\overline{x}, \overline{x}]$, with $\overline{x} = 255/2$.[12] Taking a picture can be modeled as observing in noise the 2D convolution $x \star \kappa$ of image x with known *blurring kernel* $\kappa = [\kappa_{uv}]_{\substack{0 \leq u \leq 2\mu, \\ 0 \leq v \leq 2\nu}}$, so that the observation is the random matrix

$$\omega = \Big[\omega_{rs} = \underbrace{\sum\nolimits_{\substack{0 \leq u \leq 2\mu, 0 \leq v \leq 2\nu \\ 0 \leq p < m, 0 \leq q < n: \\ u+p=r, v+q=s}} x_{pq} \kappa_{uv}}_{[x \star \kappa]_{rs}} + \sigma \xi_{rs}\Big]_{\substack{0 \leq r < m+2\mu, \\ 0 \leq s < n+2\nu}},$$

where mutually independent random variables $\xi_{rs} \sim \mathcal{N}(0,1)$ form the observation noise.[13] Our goal is to build a presumably good linear estimate of x via ω, the recovery error being measured in $\|\cdot\|_2$. To apply the techniques developed in Section 4.2.2, we need to cover the set of signals x allowed by our a priori assumptions with an ellitope $\mathcal{X}$, and then solve the associated optimization problem (4.12). The difficulty, however, is that this problem is really high-dimensional—with 256×256 images (a rather poor resolution!), the matrix H we are looking for is of the size $\dim \omega \times \dim x = ((256 + 2\mu)(256 + 2\nu)) \times 256^2 \geq 4.295 \times 10^9$. It is difficult to store such a matrix in the memory of a typical computer, let alone speaking about optimizing w.r.t. such a matrix. For this reason, in what follows we develop a "practically," and not just theoretically, efficiently computable estimate.

4.12.A. The construction. Our key observation is that when passing from representations of x and ω "as they are" to their Discrete Fourier Transforms, the situation simplifies dramatically. Specifically, for matrices y, x of the same sizes, let $y \bullet z$ be the entrywise product of y and z: $[y \bullet z]_{pq} = y_{pq} z_{pq}$. Setting

$$\alpha = 2\mu + m, \ \beta = 2\nu + n,$$

let $F_{\alpha,\beta}$ be the 2D discrete Fourier Transform—a linear mapping from the space

[12] The actual grayscale image is a matrix with entries, representing the pixels' light intensities, in the range $[0, 255]$. It is convenient for us to represent this actual image as the shift, by $\overline{x}$, of a matrix with entries in $[-\overline{x}, \overline{x}]$.

[13] Be careful: everywhere in this exercise indexing of elements of 2D arrays starts from 0, and not from 1!

$\mathbf{C}^{\alpha\times\beta}$ onto itself given by

$$[F_{\alpha,\beta}y]_{rs} = \frac{1}{\sqrt{\alpha\beta}} \sum_{\substack{0\leq p<\alpha,\\ 0\leq q<\beta}} y_{pq} \exp\left\{-\frac{2\pi i r}{\alpha} - \frac{2\pi i s}{\beta}\right\},$$

where i is the imaginary unit. It is well known that it is a unitary transformation which is easy to compute (it can be computed in $O(\alpha\beta\ln(\alpha\beta))$ arithmetic operations) which "nearly diagonalizes" the convolution: whenever $x \in \mathbf{R}^{m\times n}$, setting

$$x^+ = \left[\begin{array}{c|c} x & 0_{m\times 2\nu} \\ \hline 0_{2\mu\times n} & 0_{2\mu\times 2\nu} \end{array}\right] \in \mathbf{R}^{\alpha\times\beta},$$

we have

$$F_{\alpha,\beta}(x \star \kappa) = \chi \bullet [F_{\alpha,\beta}x^+]$$

with easy-to-compute χ.[14] Now, let δ be another $(2\mu+1)\times(2\nu+1)$ kernel, with the only nonzero entry, equal to 1, in the position (μ,ν) (recall that indices are enumerated starting from 0); then

$$F_{\alpha,\beta}(x \star \delta) = \theta \bullet [F_{\alpha,\beta}x^+]$$

with easy-to-compute θ. Now consider the auxiliary estimation problem as follows:

Given $R > 0$ and noisy observation

$$\widehat{\omega} = \chi \bullet \widehat{x} + \sigma \underbrace{F_{\alpha,\beta}\xi}_{\eta} \qquad [\xi = [\xi_{rs}] \text{ with independent } \xi_{rs} \sim \mathcal{N}(0,1)],$$

of signal $\widehat{x} \in \mathbf{C}^{\alpha\times\beta}$ known to satisfy $\|\widehat{x}\|_2 \leq R$, we want to recover the matrix $\theta \bullet \widehat{x}$, the error being measured in the Frobenius norm $\|\cdot\|_2$.

Treating signals $\widehat{x}$ and noises η as long vectors rather than matrices and taking into account that $F_{\alpha,\beta}$ is a unitary transformation, we see that our auxiliary problem is nothing but the problem of recovery, in $\|\cdot\|_2$-norm, of the image Θz of signal z known to belong to the Euclidean ball $\mathcal{Z}_R$ of radius R centered at the origin in $\mathbf{C}^{\alpha\beta}$, from noisy observation

$$\zeta = Az + \sigma\eta.$$

Here Θ and A are *diagonal* matrices with complex entries, and η is random complex-valued noise with zero mean and unit covariance matrix. Exactly the same argument as in the real case demonstrates that as far as linear estimates $\widehat{z} = H\zeta$ are concerned, we lose nothing when restricting ourselves with diagonal matrices $H = \mathrm{Diag}\{h\}$, and the best, in terms of its worst-case, over $z \in \mathcal{Z}_R$, expected $\|\cdot\|_2^2$ error, estimate corresponds to h solving the optimization problem

$$R^2 \max_{\ell\leq\alpha\beta} |\Theta_{\ell\ell} - h_\ell A_{\ell\ell}|^2 + \sigma^2 \sum_{\ell\leq\alpha\beta} |h_\ell|^2.$$

Coming back to the initial setting of our auxiliary estimation problem, we conclude

[14]Here $\chi = \sqrt{\alpha\beta}F_{\alpha,\beta}\kappa^+$, where κ^+ is the $\alpha\times\beta$ matrix with κ as its $(2\mu+1)\times(2\nu+1)$ upper-left block and zeros outside this block.

that the best linear recovery of $\theta \bullet \widehat{x}$ via $\widehat{\omega}$ is given by

$$\widehat{z} = h \bullet \widehat{\omega},$$

where h is an optimal solution to the optimization problem

$$\text{Opt} = \min_{h \in \mathbf{C}^{\alpha \times \beta}} \left\{ R^2 \max_{r,s} |\theta_{rs} - h_{rs}\chi_{rs}|^2 + \sigma^2 \sum_{r,s} |h_{rs}|^2 \right\}, \tag{4.91}$$

and the $\|\cdot\|_2$-risk

$$\text{Risk}_R[\widehat{z}] = \max_{\|\widehat{x}\|_2 \le R} \mathbf{E}\left\{\|\theta \bullet \widehat{x} - h \bullet [\chi \bullet \widehat{x} + \sigma\eta]\|_2\right\}$$

of this estimate does not exceed $\sqrt{\text{Opt}}$.

Now comes your first task:

1.1) Prove that the above h induces the estimate

$$\widehat{w}(\omega) = F_{\alpha,\beta}^{-1}\left[h \bullet [F_{\alpha,\beta}\omega]\right]$$

of $x \star \delta$, $x \in \mathcal{X}_R = \{x \in \mathbf{R}^{m \times n} : \|x\|_2 \le R\}$, via observation $\omega = x \star \kappa + \sigma\xi$, with risk

$$\text{Risk}[\widehat{w}|R] = \max_{x \in \mathbf{R}^{m \times n} : \|x\|_2 \le R} \mathbf{E}\left\{\|x \star \delta - \widehat{w}(x \star \kappa + \sigma\xi)\|_2\right\}$$

not exceeding $\sqrt{\text{Opt}}$. Note that x itself is nothing but a block in $x \star \delta$; observe also that in order for $\mathcal{X}_R$ to cover all images we are interested in, it suffices to take $R = \sqrt{mn}\overline{x}$.

1.2) Prove that finding an optimal solution to (4.91) is easy—the problem is in fact one-dimensional!

1.3) What are the sources, if any, of the conservatism of the estimate $\widehat{w}$ we have built as compared to the linear estimate given by an optimal solution to (4.12)?

1.4) Think how to incorporate in the above construction a small number L (say, five to 10) of additional a priori constraints on x of the form

$$\|x \star \kappa_\ell\|_2 \le R_\ell,$$

where $\kappa_\ell \in \mathbf{R}^{(2\mu+1)\times(2\nu+1)}$, along with a priori upper bounds u_{rs} on the magnitudes of Fourier coefficients of x^+:

$$|[F_{\alpha\beta}x^+]_{rs}| \le u_{rs},\ 0 \le r < \alpha, 0 \le s < \beta.$$

4.12.B. Mimicking Total Variation constraints. For an $m \times n$ image $x \in \mathbf{R}^{m \times n}$, its (anisotropic) total variation is defined as the ℓ_1 norm of the "discrete gradient field" of x:

$$\text{TV}(x) = \underbrace{\sum_{p=0}^{m-1}\sum_{q=0}^{n} |x_{p+1,q} - x_{p,q}|}_{\text{TV}_a(x)} + \underbrace{\sum_{p=0}^{m}\sum_{q=0}^{n-1} |x_{p,q+1} - x_{p,q}|}_{\text{TV}_b(x)}.$$

A well-established experimental fact is that for naturally arising images, their total variation is essentially less than what could be expected given the magnitudes of entries in x and the sizes m, n of the image. As a result, it is tempting to incorporate a priori upper bounds on the total variation of the image into an image reconstruction procedure. Unfortunately, while an upper bound on total variation is a convex constraint on the image, incorporating this constraint into our construction would completely destroy its "practical computability." What we can do, is to *speculate* that bounds on $\mathrm{TV}_{a,b}(x)$ can be somewhat mimicked by bounds on the energy of two convolutions: one with kernel $\kappa_a \in \mathbf{R}^{(2\mu+1)\times(2\nu+1)}$ with the only nonzero entries

$$[\kappa_a]_{\mu,\nu} = -1, [\kappa_a]_{\mu+1,\nu} = 1,$$

and the other one with kernel $\kappa_b \in \mathbf{R}^{(2\mu+1)\times(2\nu+1)}$ with the only nonzero entries

$$[\kappa_b]_{\mu,\nu} = -1, [\kappa_b]_{\mu,\nu+1} = 1$$

(recall that the indices start from 0, and not from 1). Note that $x \star \kappa_a$ and $x \star \kappa_b$ are "discrete partial derivatives" of $x \star \delta$.

For a small library of the grayscale $m \times n$ images x we dealt with, an experiment shows that, in addition to the energy constraint $\|x\|_2 \leq R = \sqrt{mn\overline{x}}$, the images satisfy the constraints

$$\|x \star \kappa_a\|_2 \leq \gamma_2 R, \ \|x \star \kappa_b\|_2 \leq \gamma_2 R \qquad (*)$$

with small γ_2, e.g., $\gamma_2 = 0.25$. In addition, it turns out that the ∞-norms of the Fourier transforms of $x \star \kappa_a$ and $x \star \kappa_b$ for these images are much less than one could expect looking at the energy of the transform's argument. Specifically, for all images x from the library it holds

$$\begin{array}{l}\|F_{\alpha\beta}[x \star \kappa_a]\|_\infty \leq \gamma_\infty R,\\ \|F_{\alpha\beta}[x \star \kappa_b]\|_\infty \leq \gamma_\infty R,\end{array}, \quad \|\{z_{rs}\}_{r,s}\|_\infty = \max_{r,s} |z_{rs}| \qquad (**)$$

with $\gamma_\infty = 0.01$.[15] Now, relations $(**)$ read

$$\max[|\omega^a_{rs}|, |\omega^b_{rs}|] |[F_{\alpha\beta} x^+]_{rs}| \leq \gamma_\infty R \, \forall r, s$$

with easy-to-compute ω^a, ω^b, and in addition $|[F_{\alpha\beta} x^+]_{rs}| \leq R$ due to $\|F_{\alpha\beta} x^+\|_2 = \|x^+\|_2 \leq R$. We arrive at the bounds

$$|[F_{\alpha\beta} x^+]_{rs}| \leq \min\left[1, 1/|\omega^a_{rs}|, 1/|\omega^b_{rs}|\right] R \, \forall r, s$$

on the magnitudes of entries in $F_{\alpha\beta} x^+$, and can utilize item 1.4 to incorporate these bounds, along with relations $(*)$.

Here is your next task:

2) Write software implementing the outlined deblurring and denoising image reconstruction routine and run numerical experiments.

[15] Note that from $(*)$ it follows that $(**)$ holds with $\gamma_\infty = \gamma_2$, while with our empirical γ's, γ_∞ is 25 times smaller than γ_2.

Recommended kernel κ: set $\mu = \rfloor m/32 \lfloor$, $\nu = \rfloor n/32 \lfloor$, start with

$$\kappa_{uv} = \frac{1}{(2\mu+1)(2\nu+1)} + \left\{ \begin{array}{ll} \Delta, & u = \mu, v = \nu \\ 0, & \text{otherwise} \end{array} \right. , 0 \le u \le 2\mu, 0 \le v \le 2\nu,$$

and then normalize this kernel to make the sum of entries equal to 1. In this description, $\Delta \ge 0$ is a control parameter responsible for the well-posedness of the auxiliary estimation problem we end up with: the smaller is Δ, the smaller is $\min_{r,s} |\chi_{rs}|$ (note that when decreasing the magnitudes of χ_{rs}, we increase the optimal value in (4.91)).
We recommend comparing what happens when $\Delta = 0$ with what happens when $\Delta = 0.25$, and also comparing the estimates accounting and not accounting for the constraints $(*)$ and $(**)$.
On top of that, you can compare your results with what is given by "ℓ_1-minimization recovery," described as follows:

As we remember from item 4.12.A, our problem of interest can be equivalently reformulated as recovering the image Θz of a signal $z \in \mathbf{C}^{\alpha\beta}$ from noisy observation $\widehat{\omega} = Az + \sigma\eta$, where Θ and A are diagonal matrices, and η is the zero mean complex Gaussian noise with unit covariance matrix. In other words, the entries η_ℓ in η are real two-dimensional Gaussian vectors independent of each other with zero mean and the covariance matrix $\frac{1}{2}I_2$. Given a reasonable "reliability tolerance" ϵ, say, $\epsilon = 0.1$, we can easily point out the smallest "confidence radius" ρ such that for $\zeta \sim \mathcal{N}(0, \frac{1}{2}I_2)$ it holds $\text{Prob}\{\|\zeta\|_2 > \rho\} \le \frac{\epsilon}{\alpha\beta}$, implying that for every ℓ it holds

$$\text{Prob}_\eta \left\{ |\widehat{\omega}_\ell - A_\ell z_\ell| > \sigma\rho \right\} \le \frac{\epsilon}{\alpha\beta},$$

and therefore

$$\text{Prob}_\eta \left\{ \|\widehat{\omega} - Az\|_\infty > \sigma\rho \right\} \le \epsilon.$$

We can now easily find the smallest, in $\|\cdot\|_1$, vector $\widehat{z} = \widehat{z}(\omega)$ which is "compatible with our observation," that is, satisfies the constraint

$$\|\widehat{\omega} - A\widehat{z}\|_\infty \le \sigma\rho,$$

and take $\Theta\widehat{z}$ as the estimate of the "entity of interest" Θz (cf. Regular ℓ_1 recovery from Section 1.2.3).
Note that this recovery needs no a priori information on z.

Exercise 4.13.

[classical periodic nonparametric deconvolution] In classical univariate nonparametric regression, one is interested in recovering a function $f(t)$ of continuous argument $t \in [0, 1]$ from noisy observations $\omega_i = f(i/n) + \sigma\eta_i$, $0 \le i \le n$, where $\eta_i \sim \mathcal{N}(0, 1)$ are observation noises independent across i. Usually, a priori restrictions on f are *smoothness assumptions*—existence of $\varkappa$ continuous derivatives satisfying a priori upper bounds

$$\left(\int_0^1 |f^{(k)}(t)|^{p_k} dt \right)^{1/p_k} \le L_k, \; 0 \le k \le \varkappa,$$

on their L_{p_k}-norms. The risk of an estimate is defined as the supremum of expected L_r-norm of the recovery error over f's of given smoothness, and the primary emphasis of classical studies here is on how the minimax optimal (i.e., the best, over all estimates) risk goes to 0 as the number of observations n goes to infinity, what the near-optimal estimates are, etc. Many of these studies deal with the *periodic case*—one where f can be extended onto the entire real axis as a $\varkappa$ times continuously differentiable periodic function with period 1 or, which is the same, when f is treated as a smooth function on the circumference of length 1 rather than on the unit segment $[0, 1]$. While being slightly simpler for analysis than the general case, the periodic case turned out to be highly instructive: what was first established for the latter, usually extends straightforwardly to the former.

What you are about to do in this exercise is apply our machinery of building linear estimates to the outlined recovery of smooth univariate periodic regression functions.

4.13.A. Setup. What follows is aimed at handling restrictions of smooth functions on the unit (i.e., of unit length) circumference C onto an equidistant n-point grid Γ_n on the circumference. These restrictions form the usual n-dimensional coordinate space $\mathbf{R}^n$; it is convenient to index the entries in $f \in \mathbf{R}^n$ starting from 0 rather than from 1. We equip $\mathbf{R}^n$ with two linear operators:

- *Cyclic shift* (in the sequel, just *shift*) Δ:

$$\Delta \cdot [f_0; f_1; ... : f_{n-2}; f_{n-1}] = [f_{n-1}; f_0; f_1; ...; f_{n-2}],$$

  and
- *Derivative* D:

$$D = n[I - \Delta].$$

Treating $f \in \mathbf{R}^n$ as a restriction of a function F on C onto Γ_n, Df is the finite-difference version of the first order derivative of the function, and the norms

$$|f|_p = n^{-1/p}\|f\|_p,\ p \in [1, \infty],$$

are the discrete versions of L_p-norms of F.

Next, we associate with $\chi \in \mathbf{R}^n$ the operator $\sum_{i=0}^{n-1} \chi_i \Delta^i$; the image of $f \in \mathbf{R}^n$ under this operator is denoted by $\chi \star f$ and is called (cyclic) *convolution* of χ and f.

The problem we focus on is as follows:

Given are:

- *smoothness data* represented by a nonnegative integer $\varkappa$ and two collections: $\{L_\iota > 0, 0 \leq \iota \leq \varkappa\}$, $\{p_\iota \in [2, \infty], 0 \leq \iota \leq \varkappa\}$. The smoothness data specify the set

$$\mathcal{F} = \{f \in \mathbf{R}^n : |f|_{p_\iota} \leq L_\iota, 0 \leq \iota \leq \varkappa\}$$

  of signals we are interested in (this is the discrete analog of *periodic Sobolev ball*—the set of $\varkappa$ times continuously differentiable functions on C with derivatives of orders up to $\varkappa$ bounded, in integral p_ι-norms, by given quantities L_ι);

- two vectors $\alpha \in \mathbf{R}^n$ (*sensing kernel*) and $\beta \in \mathbf{R}^n$ (*decoding kernel*);
- positive integer σ (noise intensity) and a real $q \in [1,2]$.

These data define the estimation problem as follows: given noisy observation

$$\omega = \alpha \star f + \sigma\eta$$

of unknown signal f known to belong to $\mathcal{F}$, where $\eta \in \mathbf{R}^n$ is a random observation noise, we want to recover $\beta\star f$ in norm $|\cdot|_q$. Our only assumption on the noise is that

$$\mathrm{Var}[\eta] := \mathbf{E}\left\{\eta\eta^T\right\} \preceq I_n.$$

The risk of a candidate estimate $\widehat{f}$ is defined as

$$\mathrm{Risk}_r[\widehat{f}|\mathcal{F}] = \sup_{\substack{f\in\mathcal{F},\\ \eta:\mathrm{Var}[\eta]\preceq I_n}} \mathbf{E}_\eta\left\{|\beta\star f - \widehat{f}(\alpha\star f+\sigma\eta)|_q\right\}.$$

Here is the exercise:

1) Check that the situation in question fits the framework of Section 4.3.3 and figure out to what, under the circumstances, reduces the optimization problem (4.42) responsible for the presumably good linear estimate $\widehat{f}_H(\omega) = H^T\omega$.
2) Prove that in the case in question the linear estimate yielded by an appropriate optimal solution to (4.42) is just the cyclic convolution

$$\widehat{f}(\omega) = h\star\omega$$

and work out a computationally cheap way to identify h.
3) Implement your findings in software and run simulations. You could, in particular, consider the denoising problem—the problem where $\alpha\star x \equiv \beta\star x \equiv x$ and $\eta \sim \mathcal{N}(0, I_n)$—and compare numerically the computed risks of your estimates with the classical results on the limits of performance in recovering smooth univariate regression functions. According to those results, in the situation in question and under the natural assumption that the L_ι are nondecreasing in ι, the minimax optimal risk, up to a factor depending solely on $\varkappa$, is $(\sigma^2/n)^{\frac{\varkappa}{2\varkappa+1}} L_\varkappa^{\frac{1}{2\varkappa+1}}$.

4.7.4.4 Probabilities of large deviations in linear estimation under sub-Gaussian noise

Exercise 4.14.

The goal of the exercise is to derive bounds for probabilities of large deviations for estimates built in Proposition 4.14.

1) Prove the following fact:

Lemma 4.30. *Let $\Theta, Q \in \mathbf{S}^m_+$, with $Q \succ 0$, and let ξ be sub-Gaussian random vector with sub-Gaussianity parameters (μ, S), where μ and S satisfy $\mu\mu^T + S \preceq Q$. Setting $\rho = \mathrm{Tr}(\Theta Q)$, we have*

$$\mathbf{E}_\xi\left\{\exp\{\tfrac{1}{8\rho}\xi^T\Theta\xi\}\right\} \leq \sqrt{2}\exp\{1/4\}. \tag{4.92}$$

As a result, for $t > 0$ it holds

$$\text{Prob}\left\{\sqrt{\xi^T\Theta\xi} \geq t\sqrt{\rho}\right\} \leq \sqrt{2}\exp\{1/4\}\exp\{-t^2/8\},\ t \geq 0. \tag{4.93}$$

Hint: Use the same trick as in the proof of Lemma 2.53.

2) Recall that (proof of) Proposition 4.14 states that in the situation of Section 4.3.3.1 and under Assumptions **A**, **B**, **R**, for every feasible solution $(H, \Lambda, \Upsilon, \Upsilon', \Theta)$ to the optimization problem[16]

$$\begin{array}{l}\text{Opt} = \min\limits_{H,\Lambda,\Upsilon,\Upsilon',\Theta} \left\{ \underbrace{\phi_{\mathcal{T}}(\lambda[\Lambda]) + \phi_{\mathcal{R}}(\lambda[\Upsilon])}_{\mathcal{A}=\mathcal{A}(\Lambda,\Upsilon)} + \underbrace{\phi_{\mathcal{R}}(\lambda[\Upsilon']) + \Gamma_\Pi(\Theta)}_{\mathcal{B}=\mathcal{B}(\Theta,\Upsilon')} : \right.\\ \left.\begin{array}{r}\Lambda = \{\Lambda_k \succeq 0, k \leq K\},\ \Upsilon = \{\Upsilon_\ell \succeq 0, \ell \leq L\},\ \Upsilon' = \{\Upsilon'_\ell \succeq 0, \ell \leq L\},\\ \left[\begin{array}{c|c}\sum_k \mathcal{R}_k^*[\Lambda_k] & \frac{1}{2}[B^T - A^T H]M \\ \hline \frac{1}{2}M^T[B - H^T A] & \sum_\ell \mathcal{S}_\ell^*[\Upsilon_\ell]\end{array}\right] \succeq 0,\\ \left[\begin{array}{c|c}\Theta & \frac{1}{2}HM \\ \hline \frac{1}{2}M^T H^T & \sum_\ell \mathcal{S}_\ell^*[\Upsilon'_\ell]\end{array}\right] \succeq 0\end{array}\right\},\end{array} \tag{4.94}$$

one has

$$\max_{x\in\mathcal{X}} \|[B - H^T A]x\| \leq \mathcal{A}\ \ \&\ \ \max_{P:\text{Var}[P]\lhd\Pi} \mathbf{E}_{\xi\sim P}\left\{\|H^T\xi\|\right\} \leq \mathcal{B}, \tag{4.95}$$

implying that the linear estimate $\widehat{x}_H(\omega) = H^T\omega$ satisfies the risk bound

$$\text{Risk}_{\Pi,\|\cdot\|}[\widehat{x}_H(\cdot)|\mathcal{X}] \leq \mathcal{A} + \mathcal{B}. \tag{4.96}$$

Prove the following:

Proposition 4.31. *Let $H, \Lambda, \Upsilon, \Upsilon', \Theta$ be a feasible solution to (4.94), and let $\widehat{x}_H(\omega) = H^T\omega$. Let, further, P be a sub-Gaussian probability distribution on $\mathbf{R}^m$, with parameters (μ, S) satisfying*

$$\mu\mu^T + S \lhd \Pi,$$

and, finally, let $x \in \mathcal{X}$. Then

(i) One has

$$\begin{array}{rcl} & & \mathbf{E}_{\xi\sim P}\left\{\|Bx - \widehat{x}_H(Ax + \xi)\|\right\} \leq \mathcal{A}_* + \mathcal{B}_*,\\ \mathcal{A}_* & = & \mathcal{A}_*(\Lambda,\Upsilon) := 2\sqrt{\phi_{\mathcal{T}}(\lambda[\Lambda])\phi_{\mathcal{R}}(\lambda[\Upsilon])} \leq \mathcal{A}(\Lambda,\Upsilon) := \phi_{\mathcal{T}}(\lambda[\Lambda]) + \phi_{\mathcal{R}}(\lambda[\Upsilon])\\ \mathcal{B}_* & = & \mathcal{B}_*(\Theta,\Upsilon') := 2\sqrt{\Gamma_\Pi(\Theta)\phi_{\mathcal{R}}(\lambda[\Upsilon'])} \leq \mathcal{B}(\Theta,\Upsilon') := \Gamma_\Pi(\Theta) + \phi_{\mathcal{R}}(\lambda[\Upsilon'].)\end{array}$$

(ii) For every $\epsilon \in (0, 1)$ one has

$$\text{Prob}_{\xi\sim P}\left\{\xi : \|Bx - \widehat{x}_H(Ax + \xi)\| > \mathcal{A}_* + \theta_\epsilon\mathcal{B}_*\right\} \leq \epsilon \tag{4.97}$$

where $\theta_\epsilon = 2\sqrt{2\ln(\sqrt{2}\mathrm{e}^{1/4}/\epsilon)}$.

[16]For notation, see Section 4.3.3.1, (4.36), and (4.39). For the reader's convenience, we recall part of this notation: for a probability distribution P on $\mathbf{R}^m$, $\text{Var}[P] = \mathbf{E}_{\xi\sim P}\{\xi^T\xi\}$, Π is a convex compact subset of $\text{int}\,\mathbf{S}^m_+$, $Q \lhd \Pi$ means that $Q \preceq Q'$ for some $Q' \in \Pi$, and $\Gamma_\Pi(\Theta) = \max_{Q\in\Pi}\text{Tr}(\Theta Q)$.

3) Suppose we are given observation $\omega = Ax + \xi$ of unknown signal x known to belong to a given spectratope $\mathcal{X} \subset \mathbf{R}^n$ and want to recover the signal. We quantify the error of a recovery $\widehat{x}$ by $\max_{k\leq K} \|B_k(\widehat{x}-x)\|_{(k)}$, where $B_k \in \mathbf{R}^{\nu_k \times n}$ are given matrices, and $\|\cdot\|_{(k)}$ are given norms on $\mathbf{R}^{\nu_k}$ (for example, x can represent a discretization of a continuous-time signal, and $B_k x$ can be finite-difference approximations of the signal's derivatives). We also assume, as in item 2, that observation noise ξ is independent of signal x and is sub-Gaussian with sub-Gaussianity parameters μ, S satisfying $\mu\mu^T + S \preceq Q$, for some given matrix $Q \succ 0$. Finally, we suppose that the unit balls of the norms conjugate to the norms $\|\cdot\|_{(k)}$ are spectratopes. In this situation, Proposition 4.14 provides us with K efficiently computable linear estimates $\widehat{x}_k(\omega) = H_k^T\omega : \mathbf{R}^{\dim \omega} \to \mathbf{R}^{\nu_k}$ along with upper bounds Opt_k on their risks $\max_{x\in\mathcal{X}} \mathbf{E}\left\{\|B_k x - \widehat{x}_k(Ax+\xi)\|_{(k)}\right\}$. Think about how, given reliability tolerance $\epsilon \in (0,1)$, to aggregate these linear estimates into a single estimate $\widehat{x}(\omega) : \mathbf{R}^{\dim \omega} \to \mathbf{R}^n$ such that for every $x \in \mathcal{X}$, the probability of the event

$$\|B_k(\widehat{x}(Ax+\xi) - x)\|_{(k)} \leq \theta \mathrm{Opt}_k,\ 1 \leq k \leq K, \tag{!}$$

is at least $1-\epsilon$, for some moderate (namely, logarithmic in K and $1/\epsilon$) "assembling price" θ.

Exercise 4.15.

Prove that if ξ is uniformly distributed on the unit sphere $\{x : \|x\|_2 = 1\}$ in $\mathbf{R}^n$, then ξ is sub-Gaussian with parameters $(0, \frac{1}{n}I_n)$.

4.7.4.5 Linear recovery under signal-dependent noise

Exercise 4.16.

Consider the situation as follows: we observe a realization ω of an m-dimensional random vector

$$\omega = Ax + \xi_x,$$

where

- x is an unknown signal belonging to a given signal set $\mathcal{X}$, specifically, spectratope (which, as usual, we can assume to be basic):

$$\mathcal{X} = \{x \in \mathbf{R}^n : \exists t \in \mathcal{T} : R_k^2[x] \preceq t_k I_{d_k}, k \leq K\}$$

  with standard restrictions on $\mathcal{T}$ and $R_k[\cdot]$;
- ξ_x is the observation noise with distribution which can depend on x; all we know is that

$$\mathrm{Var}[\xi_x] := \mathbf{E}\{\xi_x\xi_x^T\} \preceq \mathcal{C}[x],$$

  where the entries of symmetric matrix $\mathcal{C}[x]$ are quadratic in x. We assume in the sequel that signals x belong to the subset

$$\mathcal{X}_\mathcal{C} = \{x \in \mathcal{X} : \mathcal{C}[x] \succeq 0\}$$

  of $\mathcal{X}$;
- Our goal is to recover Bx, with given $B \in \mathbf{R}^{\nu\times n}$, in a given norm $\|\cdot\|$ such that

the unit ball $\mathcal{B}_*$ of the conjugate norm is a spectratope:

$$\mathcal{B}_* = \{u : \|u\|_* \leq 1\} = M\mathcal{V}, \mathcal{V} = \{v : \exists r \in \mathcal{R} : S_\ell^2[v] \preceq r_\ell I_{f_\ell}, \ell \leq L\}.$$

We quantify the performance of a candidate estimate $\widehat{x}(\omega) : \mathbf{R}^m \to \mathbf{R}^\nu$ by the risk

$$\mathrm{Risk}_{\|\cdot\|}[\widehat{x}|\mathcal{X}_{\mathcal{C}}] = \sup_{x \in \mathcal{X}_{\mathcal{C}}} \sup_{\xi_x : \mathrm{Var}[\xi_x] \preceq \mathcal{C}[x]} \mathbf{E}\{\|Bx - \widehat{x}(Ax + \xi_x)\|\}.$$

1) Utilize semidefinite relaxation in order to build, in a computationally efficient fashion, a "presumably good" linear estimate, specifically, prove the following:

Proposition 4.32. *In the situation in question, for $G \in \mathbf{S}^m$ let us define $\alpha_0[G] \in \mathbf{R}$, $\alpha_1[G] \in \mathbf{R}^n$, $\alpha_2[G] \in \mathbf{S}^n$ from the identity*

$$\mathrm{Tr}(\mathcal{C}[x]G) = \alpha_0[G] + \alpha_1^T[G]x + x^T\alpha_2[G]x \;\forall (x \in \mathbf{R}^n, G \in \mathbf{S}^m),$$

so that $\alpha_\chi[G]$ are affine in G. Consider the convex optimization problem

$$\mathrm{Opt} = \min_{H,\mu,D,\Lambda,\Upsilon,\Upsilon',G} \left\{ \begin{array}{l} \mu + \phi_{\mathcal{T}}(\lambda[\Lambda]) + \phi_{\mathcal{R}}(\lambda[\Upsilon]) + \phi_{\mathcal{R}}(\lambda[\Upsilon']) : \\ \Lambda = \{\Lambda_k \in \mathbf{S}_+^{d_k}, k \leq K\}, \Upsilon = \{\Upsilon_\ell \in \mathbf{S}_+^{f_\ell}, \ell \leq L\}, \Upsilon' = \{\Upsilon'_\ell \in \mathbf{S}_+^{f_\ell}, \ell \leq L\}, D \in \mathbf{S}_+^m, \\ \left[\begin{array}{c|c|c} \alpha_0[G] & \frac{1}{2}\alpha_1^T[G] & \\ \hline \frac{1}{2}\alpha_1[G] & \alpha_2[G] & \frac{1}{2}[B^T - A^TH]M \\ \hline & \frac{1}{2}M^T[B - H^TA] & \end{array}\right] \\ \qquad \preceq \left[\begin{array}{c|c|c} \mu - \alpha_0[D] & -\frac{1}{2}\alpha_1^T[D] & \\ \hline -\frac{1}{2}\alpha_1[D] & \sum_k \mathcal{R}_k^*[\Lambda_k] - \alpha_2[D] & \\ \hline & & \sum_\ell \mathcal{S}_\ell^*[\Upsilon_\ell] \end{array}\right] \\ \qquad \left[\begin{array}{c|c} G & \frac{1}{2}HM \\ \hline \frac{1}{2}M^TH^T & \sum_\ell \mathcal{S}_\ell^*[\Upsilon'_\ell] \end{array}\right] \succeq 0, \end{array} \right\}$$

$$\left[\begin{array}{c} [\mathcal{R}_k^*[\Lambda_k]]_{ij} = \mathrm{Tr}(\Lambda_k \frac{1}{2}[R^{ki}R^{kj} + R^{kj}R^{ki}]),\; R_k[x] = \sum_j x_j R^{kj}, \\ [\mathcal{S}_\ell^*[\Upsilon_\ell]]_{ij} = \mathrm{Tr}(\Upsilon_\ell \frac{1}{2}[S^{\ell i}S^{\ell j} + S^{\ell j}S^{\ell i}]),\; S_\ell[v] = \sum_j v_j S^{\ell j}, \\ \lambda[\{Z_i, i \leq I\}] = [\mathrm{Tr}(Z_1); ...; \mathrm{Tr}(Z_I)],\; \phi_A(q) = \max_{s \in A} q^T s. \end{array}\right]$$

Whenever $H, \mu, D, \Lambda, \Upsilon, \Upsilon'$ and G are feasible for the problem, one has

$$\mathrm{Risk}_{\|\cdot\|}[\widehat{x}^H|\mathcal{X}_{\mathcal{C}}] \leq \mu + \phi_{\mathcal{T}}(\lambda[\Upsilon]) + \phi_{\mathcal{R}}(\lambda[\Upsilon]) + \phi_{\mathcal{R}}(\lambda[\Upsilon'])$$
$$\text{where } \widehat{x}^H(\omega) = H^T\omega.$$

2) Work out the following special case of the situation above dealing with Poisson Imaging (see Section 2.4.3.2): your observation is an m-dimensional random vector with independent Poisson entries, the vector of parameters of the corresponding Poisson distributions being Py; here P is an $m \times n$ entrywise nonnegative matrix, and the unknown signal y is known to belong to a given box $Y = \{y \in \mathbf{R}^n : \underline{a} \leq y \leq \overline{a}\}$, where $0 \leq \underline{a} < \overline{a}$. You want to recover y in $\|\cdot\|_p$-norm with given $p \in [1, 2]$.

4.7.5 Signal recovery in Discrete and Poisson observation schemes

Exercise 4.17.

The goal of what follows is to "transfer" the constructions of linear estimates to the case of multiple indirect observations of discrete random variables. Specifically, we are interested in the situation where

- Our observation is a K-element sample $\omega^K = (\omega_1, .., \omega_K)$ with independent identically distributed components ω_k taking values in an m-element set. As always, we encode the points from this m-element set by the standard basic orths $e_1, ..., e_m$ in $\mathbf{R}^m$.
- The (common for all k) probability distribution of ω_k is Ax, where x is an unknown "signal"—n-dimensional probabilistic vector known to belong to a closed convex subset $\mathcal{X}$ of the n-dimensional probabilistic simplex $\mathbf{\Delta}_n = \{x \in \mathbf{R}^n : x \geq 0, \sum_i x_i = 1\}$, and A is a given $m \times n$ column-stochastic matrix (i.e., entrywise nonnegative matrix with unit column sums).
- Our goal is to recover Bx, where B is a given $\nu \times n$ matrix, and we quantify a candidate estimate $\widehat{x}(\omega^K) : \mathbf{R}^{mK} \to \mathbf{R}^\nu$ by its *risk*

$$\mathrm{Risk}_{\|\cdot\|}[\widehat{x}|\mathcal{X}] = \sup_{x \in \mathcal{X}} \mathbf{E}_{\omega^K \sim [Ax] \times ... \times [Ax]} \left\{ \|Bx - \widehat{x}(\omega^K)\| \right\},$$

where $\|\cdot\|$ is a given norm on $\mathbf{R}^\nu$.

We use *linear* estimates—estimates of the form

$$\widehat{x}_H(\omega^K) = H^T \underbrace{\left[\frac{1}{K} \sum_{k=1}^K \omega_k \right]}_{\widehat{\omega}_K[\omega^K]}, \tag{4.98}$$

where $H \in \mathbf{R}^{m \times \nu}$.

1) In the main body of Chapter 4, $\mathcal{X}$ always was assumed to be symmetric w.r.t. the origin, which easily implies that we gain nothing when passing from linear estimates to affine ones (sums of linear estimates and constants). Now we are in the case where $\mathcal{X}$ can be "heavily asymmetric," which, in general, can make "genuinely affine" estimates preferable. Show that in the case in question, we still lose nothing when restricting ourselves to linear, rather than affine, estimates.

4.17.A. Observation scheme revisited. When observation ω^K stems from a signal $x \in \mathbf{\Delta}_n$, we have

$$\widehat{\omega}_K[\omega^K] = Ax + \xi_x,$$

where

$$\xi_x = \frac{1}{K} \sum_{k=1}^K [\omega_k - Ax]$$

is the average of K independent identically distributed zero mean random vectors with common covariance matrix $Q[x]$.

2) Check that

$$Q[x] = \mathrm{Diag}\{Ax\} - [Ax][Ax]^T,$$

and derive from this fact that the covariance matrix of ξ_x is

$$Q_K[x] = \frac{1}{K} Q[x].$$

Setting

$$\Pi = \Pi_{\mathcal{X}} = \left\{ Q = \frac{1}{K}\mathrm{Diag}\{Ax\} : x \in \mathcal{X} \right\},$$

check that $\Pi_{\mathcal{X}}$ is a convex compact subset of the positive semidefinite cone $\mathbf{S}^m_+$, and that whenever $x \in \mathcal{X}$, one has $Q[x] \preceq Q$ for some $Q \in \Pi$.

4.17.B. Upper-bounding risk of a linear estimate. We can upper-bound the risk of a linear estimate $\widehat{x}_H$ as follows:

$$\begin{array}{rcl}\mathrm{Risk}_{\|\cdot\|}[\widehat{x}_H|\mathcal{X}] & = & \sup_{x\in\mathcal{X}} \mathbf{E}_{\omega^K\sim[Ax]\times\ldots\times[Ax]} \left\{\|Bx - H^T\widehat{\omega}_K[\omega^K]\|\right\} \\ & = & \sup_{x\in\mathcal{X}} \mathbf{E}_{\xi_x} \left\{\|[Bx - H^T A]x - H^T\xi_x\|\right\} \\ & \leq & \underbrace{\sup_{x\in\mathcal{X}} \|[B - H^T A]x\|}_{\Phi(H)} + \underbrace{\sup_{\xi:\mathrm{Cov}[\xi]\in\Pi_{\mathcal{X}}} \mathbf{E}_\xi \left\{\|H^T\xi\|\right\}}_{\Psi^{\mathcal{X}}(H)}.\end{array}$$

As in the main body of Chapter 4, we intend to build a "presumably good" linear estimate by minimizing over H the sum of efficiently computable upper bounds $\overline{\Phi}(H)$ on $\Phi(H)$ and $\overline{\Psi}^{\mathcal{X}}(H)$ on $\Psi^{\mathcal{X}}(H)$.

Assuming from now on that the unit ball $\mathcal{B}_*$ of the norm conjugate to $\|\cdot\|$ is a spectratope,

$$\mathcal{B}_* := \{u : \|u\|_* \leq 1\} = \{u : \exists r \in \mathcal{R}, y : u = My, S_\ell^2[y] \preceq r_\ell I_{f_\ell}, \ell \leq L\}$$

with our usual restrictions on $\mathcal{R}$ and S_ℓ, we can take as $\overline{\Psi}^{\mathcal{X}}(\cdot)$ the function (4.40).

For the sake of simplicity, we from now on assume that $\mathcal{X}$ is cut off $\mathbf{\Delta}_n$ by linear inequalities:

$$\mathcal{X} = \{x \in \mathbf{\Delta}_n : Gx \leq g, \; Ex = e\} \qquad [G \in \mathbf{R}^{p\times n}, E \in \mathbf{R}^{q\times n}].$$

Observe that replacing G with $G - g\mathbf{1}_n^T$ and E with $E - e\mathbf{1}_n^T$, we reduce the situation to that where all linear constraints are homogeneous, that is,

$$\mathcal{X} = \{x \in \mathbf{\Delta}_n : Gx \leq 0, \; Ex = 0\},$$

and this is what we assume from now on. Setting

$$F = [G; E; -E] \in \mathbf{R}^{(p+2q)\times n},$$

we have also

$$\mathcal{X} = \{x \in \mathbf{\Delta}_n : Fx \leq 0\}.$$

Suppose that $\mathcal{X}$ is nonempty. Finally, in addition to what was already assumed about the norm $\|\cdot\|$, let us also suppose that this norm is *absolute*, that is, $\|u\|$ depends only on the vector of *magnitudes* of entries in u. From this assumption it immediately follows that if $0 \leq u \leq u'$, then $\|u\| \leq \|u'\|$ (why?).

Our next task is to efficiently upper-bound $\Phi(\cdot)$.

4.17.C. Bounding Φ, simple case. We start with the *simple case* where there are no linear constraints (formally, G and E are zero matrices); in this case bounding Φ is straightforward:

3) Prove that in the simple case Φ is convex and efficiently computable "as is":

$$\Phi(H) = \max_{i\leq n} \|(B - H^T A)g_i\|,$$

where $g_1, ..., g_n$ are the standard basic orths in $\mathbf{R}^n$.

4.17.D. Lagrange upper bound on Φ.

4) Observing that when $\mu \in \mathbf{R}^{p+2q}_+$, the function

$$\|(B - H^T A)x\| - \mu^T F x$$

of x is convex in $x \in \mathbf{\Delta}_n$ and overestimates $\|(B - H^T A)x\|$ everywhere on $\mathcal{X}$, conclude that the efficiently computable convex function

$$\Phi_L(H) = \min_{\mu} \max_{i\leq n} \{\|(B - H^T A)g_i\| - \mu^T F g_i : \mu \geq 0\}$$

upper-bounds $\Phi(H)$. In the sequel, we call this function the *Lagrange* upper bound on Φ.

4.17.E. Basic upper bound on Φ. For vectors u and v of the same dimension, say, k, let $\mathrm{Max}[u, v]$ stand for the entrywise maximum of u, v,

$$[\mathrm{Max}[u, v]]_i = \max[u_i, v_i],$$

and let

$$[u]_+ = \mathrm{Max}[u, 0_k],$$

where 0_k is the k-dimensional zero vector.

5.1) Let $\Lambda_+ \geq 0$ and $\Lambda_- \geq 0$ be $\nu \times (p + 2q)$ matrices, $\Lambda \geq 0$ meaning that matrix Λ is entrywise nonnegative. Prove that whenever $x \in \mathcal{X}$, one has

$$\begin{array}{l}\|(B - H^T A)x\| \leq \mathcal{B}(x, H, \Lambda_+, \Lambda_-) \\ \quad := \min_t \left\{\|t\| : t \geq \mathrm{Max}\left[[(B - H^T A)x - \Lambda_+ F x]_+, [-(B - H^T A)x - \Lambda_- F x]_+\right]\right\}\end{array}$$

and that $\mathcal{B}(x, H, \Lambda_+, \Lambda_-)$ is convex in x.

5.2) Derive from 5.1 that whenever $\Lambda_\pm$ are as in 5.1, one has

$$\Phi(H) \leq \mathcal{B}^+(H, \Lambda_+, \Lambda_-) := \max_{i\leq n} \mathcal{B}(g_i, H, \Lambda_+, \Lambda_-),$$

where, as in item 3, $g_1, ..., g_n$ are the standard basic orths in $\mathbf{R}^n$. Conclude that

$$\Phi(H) \leq \Phi_B(H) = \inf_{\Lambda_\pm} \left\{\mathcal{B}^+(H, \Lambda_+, \Lambda_-) : \Lambda_\pm \in \mathbf{R}_+^{\nu\times(p+2q)}\right\}$$

and that Φ_B is convex and real-valued. In the sequel we refer to $\Phi_B(\cdot)$ as the *Basic* upper bound on $\Phi(\cdot)$.

4.17.F. Sherali-Adams upper bound on Φ. Let us apply the approach we used in Chapter 1, Section 1.3.2, when deriving verifiable sufficient conditions for

s-goodness; see p. 21. Specifically, setting

$$W = \left[\begin{array}{c|c} G & I \\ \hline E & \end{array}\right],$$

let us introduce the slack variable $z \in \mathbf{R}^p$ and rewrite the description of $\mathcal{X}$ as

$$\mathcal{X} = \{x \in \mathbf{\Delta}_n : \exists z \geq 0 : W[x; z] = 0\},$$

so that $\mathcal{X}$ is the projection of the polyhedral set

$$\mathcal{X}^+ = \{[x; z] : x \in \mathbf{\Delta}_n, z \geq 0, W[x; z] = 0\}$$

on the x-space. Projection of $\mathcal{X}^+$ on the z-space is a nonempty (since $\mathcal{X}$ is so) and clearly bounded subset of the nonnegative orthant $\mathbf{R}^p_+$, and we can in many ways cover Z by the simplex

$$\Delta[\alpha] = \{z \in \mathbf{R}^p : z \geq 0, \sum_i \alpha_i z_i \leq 1\},$$

where all α_i are positive.

6.1) Let $\alpha > 0$ be such that $Z \subset \Delta[\alpha]$. Prove that

$$\mathcal{X}^+ = \{[x; z] : W[x; z] = 0, [x; z] \in \mathrm{Conv}\{v_{ij} = [g_i; h_j], 1 \leq i \leq n, 0 \leq j \leq p\}\}, \qquad (!)$$

where the g_i are the standard basic orths in $\mathbf{R}^n$, $h_0 = 0 \in \mathbf{R}^p$, and $\alpha_j h_j$, $1 \leq j \leq p$, are the standard basic orths in $\mathbf{R}^p$.

6.2) Derive from 5.1 that the efficiently computable convex function

$$\Phi_{SA}(H) = \inf_C \max_{i,j} \left\{ \|(B - H^T A) g_i + C^T W v_{ij}\| : C \in \mathbf{R}^{(p+q)\times\nu} \right\}$$

is an upper bound on $\Phi(H)$. In the sequel, we refer to $\Phi_{SA}(H)$ as to the *Sherali-Adams* bound [214].

4.17.G. Combined bound. We can combine the above bounds, specifically, as follows:

7) Prove that the efficiently computable convex function

$$\Phi_{LBS}(H) = \inf_{(\Lambda_\pm, C_\pm, \mu, \mu_+) \in \mathcal{R}} \max_{i,j} \mathcal{G}_{ij}(H, \Lambda_\pm, C_\pm, \mu, \mu_+), \qquad (\#)$$

where

$$\begin{array}{l} \mathcal{G}_{ij}(H, \Lambda_\pm, C_\pm, \mu, \mu_+) := -\mu^T F g_i + \mu_+^T W v_{ij} + \min\limits_t \Big\{ \|t\| : \\ t \geq \mathrm{Max}\left[[(B - H^T A - \Lambda_+ F) g_i + C_+^T W v_{ij}]_+, [(-B + H^T A - \Lambda_- F) g_i + C_-^T W v_{ij}]_+\right] \Big\}, \\ \mathcal{R} = \{(\Lambda_\pm, C_\pm, \mu, \mu_+) : \Lambda_\pm \in \mathbf{R}_+^{\nu\times(p+2q)}, C_\pm \in \mathbf{R}^{(p+q)\times\nu}, \mu \in \mathbf{R}_+^{p+2q}, \mu_+ \in \mathbf{R}^{p+q}\} \end{array}$$

is an upper bound on $\Phi(H)$, and that this *Combined* bound is at least as good as any of the Lagrange, Basic, or Sherali-Adams bounds.

4.17.H. How to select α? A shortcoming of the Sherali-Adams and the combined upper bounds on Φ is the presence of a "degree of freedom"—on the positive vector α. Intuitively, we would like to select α to make the simplex $\Delta[\alpha] \supset Z$ to be "as small as possible." It is unclear, however, what "as small as possible" is in our context, not to speak of how to select the required α after we agree on how we measure the "size" of $\Delta[\alpha]$. It turns out, however, that we can efficiently select α resulting in the *smallest volume* $\Delta[\alpha]$.

8) Prove that minimizing the volume of $\Delta[\alpha] \supset Z$ in α reduces to solving the following convex optimization problem:

$$\inf_{\alpha,u,v}\left\{-\sum_{s=1}^{p}\ln(\alpha_s) : 0 \leq \alpha \leq -v, E^T u + G^T v \leq \mathbf{1}_n\right\}. \qquad (*)$$

9) Run numerical experiments to evaluate the quality of the above bounds. It makes sense to generate problems where we know in advance the actual value of Φ, e.g., to take

$$\mathcal{X} = \{x \in \mathbf{\Delta}_n : x \geq a\} \qquad (a)$$

with $a \geq 0$ such that $\sum_i a_i \leq 1$. In this case, we can easily list the extreme points of $\mathcal{X}$ (how?) and thus can easily compute $\Phi(H)$.
In your experiments, you can use the matrices stemming from "presumably good" linear estimates yielded by the optimization problems

$$\begin{array}{rcl}\mathrm{Opt} &=& \min\limits_{H,\Upsilon,\Theta}\Big\{\overline{\Phi}(H) + \phi_{\mathcal{R}}(\lambda[\Upsilon]) + \Gamma_{\mathcal{X}}(\Theta) : \Upsilon = \{\Upsilon_\ell \succeq 0, \ell \leq L,\} \\ && \left[\begin{array}{c|c}\Theta & \frac{1}{2}HM \\ \hline \frac{1}{2}M^T H^T & \sum_\ell \mathcal{S}_\ell^*[\Upsilon_\ell]\end{array}\right] \succeq 0\Big\}\end{array} \qquad (4.99)$$

where

$$\Gamma_{\mathcal{X}}(\Theta) = \frac{1}{K}\max_{x\in\mathcal{X}} \mathrm{Tr}(\mathrm{Diag}\{Ax\}\Theta),$$

(see Corollary 4.12), with the actual Φ (which is available for our $\mathcal{X}$), or the upper bounds on Φ (Lagrange, Basic, Sherali-Adams, and Combined) in the role of $\overline{\Phi}$. Note that it may make sense to test seven bounds rather than just four. Indeed, with additional constraints on the optimization variables in (#), we can get, besides "pure" Lagrange, Basic, and Sherali-Adams bounds and their "three-component combination" (Combined bound), pairwise combinations of the pure bounds as well. For example, to combine Lagrange and Sherali-Adams bounds, it suffices to add to (#) the constraints $\Lambda_\pm = 0$.

Exercise 4.18.

The exercise to follow deals with recovering discrete probability distributions in the *Wasserstein norm.*

The Wasserstein distance between probability distributions is extremely popular today in Statistics; it is defined as follows.[17] Consider discrete random variables taking values in finite observation space $\Omega = \{1, 2, ..., n\}$ which is equipped with

[17]The distance we consider stems from the Wasserstein 1-distance between discrete probability distributions. This is a particular case of the general Wasserstein p-distance between (not necessarily discrete) probability distributions.

the metric $\{d_{ij} : 1 \le i,j \le n\}$ satisfying the standard axioms.[18] As always, we identify probability distributions on Ω with n-dimensional probabilistic vectors $p = [p_1; ...; p_n]$, where p_i is the probability mass assigned by p to $i \in \Omega$. The Wasserstein distance between probability distributions p and q is defined as

$$W(p,q) = \min_{x=[x_{ij}]} \left\{ \sum_{ij} d_{ij}x_{ij} : x_{ij} \ge 0, \sum_j x_{ij} = p_i, \sum_i x_{ij} = q_j \ \forall 1 \le i,j \le n \right\}. \quad (4.100)$$

In other words, one may think of p and q as two distributions of unit mass on the points of Ω, and consider the mass transport problem of redistributing the mass assigned to points by distribution p to get the distribution q. Denoting by x_{ij} the mass moved from point i to point j, constraints $\sum_j x_{ij} = p_i$ say that the total mass taken from point i is exactly p_i, constraints $\sum_i x_{ij} = q_j$ say that as the result of transportation, the mass at point j will be exactly q_j, and the constraints $x_{ij} \ge 0$ reflect the fact that transport of a negative mass is forbidden. Assuming that the cost of transporting a mass μ from point i to point j is $d_{ij}\mu$, the Wasserstein distance $W(p,q)$ between p and q is the cost of the cheapest transportation plan which converts p into q. As compared to other natural distances between discrete probability distributions, like $\|p-q\|_1$, the advantage of the Wasserstein distance is that it allows us to model the situation (indeed arising in some applications) where the effect, measured in terms of intended application, of changing probability masses of points from Ω is small when the probability mass of a point is redistributed among *close* points.[19]

Now comes the first part of the exercise:

1) Let p, q be two probability distributions. Prove that

$$W(p,q) = \max_{f \in \mathbf{R}^n} \left\{ \sum_i f_i(p_i - q_i) : |f_i - f_j| \le d_{ij} \ \forall i,j \right\}. \quad (4.101)$$

Treating vector $f \in \mathbf{R}^n$ as a function on Ω, the value of the function at a point $i \in \Omega$ being f_i, (4.101) admits a very transparent interpretation: the Wasserstein distance $W(p,q)$ between probability distributions p and q is the maximum of inner products of $p - q$ and functions f on Ω which are Lipschitz continuous w.r.t. the metric d, with constant 1. When shifting f by a constant, the inner product remains intact (since $p - q$ is a vector with zero sum of entries). Therefore, denoting by

$$D = \max_{i,j} d_{ij}$$

the d-diameter of Ω, we have

$$W(p,q) = \max_f \left\{ f^T(p-q) : |f_i - f_j| \le d_{ij}, \ |f_i| \le D/2 \ \forall i,j \right\}, \quad (4.102)$$

[18] Namely, positivity: $d_{ij} = d_{ji} \ge 0$, with $d_{ij} = 0$ if and only if $i = j$; and the triangle inequality: $d_{ik} \le d_{ij} + d_{jk}$ for all triples i, j, k.

[19] In fact, the Wasserstein distance shares this property with some other distances between distributions used in Probability Theory, such as Skorohod, or Prokhorov, or Ky Fan distances. What makes the Wasserstein distance so "special" is its representation (4.100) as the optimal value of a Linear Programming problem, responsible for efficient computational handling of this distance.

the reason being that every function f on Ω which is Lipschitz continuous, with constant 1, w.r.t. metric d can be shifted by a constant to ensure $\|f\|_\infty \leq D/2$ (look what happens when the shift ensures that $\min_i f_i = -D/2$).

Representation (4.102) shows that the Wasserstein distance is generated by a norm on $\mathbf{R}^n$: for all probability distributions on Ω one has

$$W(p,q) = \|p-q\|_W,$$

where $\|\cdot\|_W$ is the *Wasserstein norm* on $\mathbf{R}^n$ given by

$$\begin{array}{rcl}\|x\|_W &=& \max\limits_{f\in\mathcal{B}_*} f^Tx,\\ \mathcal{B}_* &=& \left\{u\in\mathbf{R}^n : u^TS_{ij}u\leq 1, 1\leq i\leq j\leq n\right\},\\ S_{ij} &=& \left\{\begin{array}{ll} d_{ij}^{-2}[e_i-e_j][e_i-e_j]^T, & 1\leq i<j\leq n,\\ 4D^{-2}e_ie_i^T, & 1\leq i=j\leq n,\end{array}\right.\end{array} \tag{4.103}$$

where $e_1,...,e_n$ are the standard basic orths in $\mathbf{R}^n$.

2) Let us equip n-element set $\Omega=\{1,...,d\}$ with the metric $d_{ij}=\left\{\begin{array}{ll}2, & i\neq j\\ 0, & i=j\end{array}\right.$. What is the associated Wasserstein norm?

Note that the set $\mathcal{B}_*$ in (4.103) is the unit ball of the norm conjugate to $\|\cdot\|_W$, and as we see, this set is a basic ellitope. As a result, the estimation machinery developed in Chapter 4 is well suited for recovering discrete probability distributions in the Wasserstein norm. This observation motivates the concluding part of the exercise:

3) Consider the situation as follows: Given an $m\times n$ column-stochastic matrix A and a $\nu\times n$ column-stochastic matrix B, we observe K samples ω_k, $1\leq k\leq K$, independent of each other, drawn from the discrete probability distribution $Ax\in\mathbf{\Delta}_m$ (as always, $\mathbf{\Delta}_\nu\subset\mathbf{R}^\nu$ is the probabilistic simplex in $\mathbf{R}^\nu$), $x\in\mathbf{\Delta}_n$ being an unknown "signal" underlying the observations; realizations of ω_k are identified with respective vertices $f_1,...,f_m$ of $\mathbf{\Delta}_m$. Our goal is to use the observations to estimate the distribution $Bx\in\mathbf{\Delta}_\nu$. We are given a metric d on the set $\Omega_\nu=\{1,2,...,\nu\}$ of indices of entries in Bx, and measure the recovery error in the Wasserstein norm $\|\cdot\|_W$ associated with d.
Build an explicit convex optimization problem responsible for a "presumably good" linear recovery of the form

$$\widehat{x}_H = \frac{1}{K}H^T\sum_{k=1}^K\omega_k.$$

Exercise 4.19.

[follow-up to Exercise 4.17] In Exercise 4.17, we have built a "presumably good" linear estimate $\widehat{x}_{H_*}(\cdot)$—see (4.98)—yielded by the H-component H_* of an optimal solution to problem (4.99). The optimal value Opt in this problem is an upper bound on the risk $\mathrm{Risk}_{\|\cdot\|}[\widehat{x}_{H_*}|\mathcal{X}]$ (here and in what follows we use the same notation and impose the same assumptions as in Exercise 4.17). Recall that $\mathrm{Risk}_{\|\cdot\|}$ is the worst, w.r.t. signals $x\in\mathcal{X}$ underlying our observations, expected norm of the recovery error. It makes sense also to provide upper bounds on the probabilities of deviations of the error's magnitude from its expected value, and this is the problem we consider here; cf. Exercise 4.14.

1) Prove the following

Lemma 4.33. *Let $Q \in \mathbf{S}^m_+$, let K be a positive integer, and let $p \in \mathbf{\Delta}_m$. Let, further, $\omega^K = (\omega_1, ..., \omega_K)$ be i.i.d. random vectors, with ω_k taking the value e_j ($e_1, ..., e_m$ are the standard basic orths in $\mathbf{R}^m$) with probability p_j. Finally, let $\xi_k = \omega_k - \mathbf{E}\{\omega_k\} = \omega_k - p$, and $\widehat{\xi} = \frac{1}{K}\sum_{k=1}^K \xi_k$. Then for every $\epsilon \in (0,1)$ it holds*

$$\mathrm{Prob}\left\{\|\widehat{\xi}\|_2^2 \leq \frac{12\ln(2m/\epsilon)}{K}\right\} \geq 1-\epsilon.$$

Hint: use the classical
Bernstein inequality: *Let $X_1, ..., X_K$ be independent zero mean random variables taking values in $[-M, M]$, and let $\sigma_k^2 = \mathbf{E}\{X_k^2\}$. Then for every $t \geq 0$ one has*

$$\mathrm{Prob}\left\{\sum\nolimits_{k=1}^K X_k \geq t\right\} \leq \exp\left\{-\frac{t^2}{2[\sum_k \sigma_k^2 + \frac{1}{3}Mt]}\right\}.$$

2) Consider the situation described in Exercise 4.17 with $\mathcal{X} = \mathbf{\Delta}_n$, specifically,

- Our observation is a sample $\omega^K = (\omega_1, ..., \omega_K)$ with i.i.d. components $\omega_k \sim Ax$, where $X \in \mathbf{\Delta}_n$ is an unknown n-dimensional probabilistic vector, A is an $m \times n$ stochastic matrix (nonnegative matrix with unit column sums), and $\omega \sim Ax$ means that ω is a random vector taking value e_i (e_i are standard basic orths in $\mathbf{R}^m$) with probability $[Ax]_i$, $1 \leq i \leq m$.
- Our goal is to recover Bx in a given norm $\|\cdot\|$; here B is a given $\nu \times n$ matrix.
- We assume that the unit ball $\mathcal{B}_*$ of the norm $\|\cdot\|_*$ conjugate to $\|\cdot\|$ is a spectratope:

$$\mathcal{B}_* = \{u = My, y \in \mathcal{Y}\},\ \mathcal{Y} = \{y \in \mathbf{R}^N : \exists r \in \mathcal{R} : S_\ell^2[y] \preceq r_\ell I_{f_\ell}, \ell \leq L\}.$$

Our goal is to build a presumably good linear estimate

$$\widehat{x}_H(\omega^K) = H^T\widehat{\omega}[\omega^K],\ \widehat{\omega}[\omega^K] = \frac{1}{K}\sum_k \omega_k.$$

Prove the following

Proposition 4.34. *Let H, Θ, Υ be a feasible solution to the convex optimization problem*

$$\begin{array}{l}\min_{H,\Theta,\Upsilon}\left\{\Phi(H) + \phi_{\mathcal{R}}(\lambda[\Upsilon]) + \Gamma(\Theta)/K : \Upsilon = \{\Upsilon_\ell \succeq 0, \ell \leq L\},\right.\\ \qquad\qquad\left.\left[\begin{array}{c|c}\Theta & \frac{1}{2}HM\\ \hline \frac{1}{2}M^TH^T & \sum_\ell S_\ell^*[\Upsilon_\ell]\end{array}\right] \succeq 0\right\}\end{array} \tag{4.104}$$

where

$$\Phi(H) = \max_{j \leq n}\|\mathrm{Col}_j[B - H^TA]\|,\ \ \Gamma(\Theta) = \max_{x \in \mathbf{\Delta}_n}\mathrm{Tr}(\mathrm{Diag}\{Ax\}\Theta).$$

Then

(i) For every $x \in \mathbf{\Delta}_n$ it holds

$$\begin{array}{rcl}\mathbf{E}_{\omega^K}\left\{\|Bx-\widehat{x}_H(\omega^K)\|\right\} & \leq & \Phi(H)+2K^{-1/2}\sqrt{\phi_{\mathcal{R}}(\lambda[\Upsilon])\Gamma(\Theta)} \\ \Big[& \leq & \Phi(H)+\phi_{\mathcal{R}}(\lambda[\Upsilon])+\Phi(H)+\Gamma(\Theta)/K\Big].\end{array} \tag{4.105}$$

(ii) Let $\epsilon \in (0,1)$. For every $x \in \mathbf{\Delta}_n$ with

$$\gamma = 2\sqrt{3\ln(2m/\epsilon)}$$

one has

$$\begin{array}{l}\mathrm{Prob}_{\omega^K}\left\{\|Bx-\widehat{x}_H(\omega^K)\| \leq \Phi(H)+2\gamma K^{-1/2}\sqrt{\phi_{\mathcal{R}}(\lambda[\Upsilon])\|\Theta\|_{Sh,\infty}}\right\} \\ \geq 1-\epsilon.\end{array} \tag{4.106}$$

3) Look what happens when $\nu = m = n$, A and B are the unit matrices, and $H = I$, i.e., we want to understand how good is the recovery of a discrete probability distribution by empirical distribution of a K-element i.i.d. sample drawn from the original distribution. Take, as $\|\cdot\|$, the norm $\|\cdot\|_p$ with $p \in [1,2]$, and show that for every $x \in \mathbf{\Delta}_n$ and every $\epsilon \in (0,1)$ one has

$$\begin{array}{l}\forall(x \in \mathbf{\Delta}_n): \\ \mathbf{E}\left\{\|x-\widehat{x}_I(\omega^K)\|_p\right\} \leq n^{\frac{1}{p}-\frac{1}{2}}K^{-\frac{1}{2}}, \\ \mathrm{Prob}\left\{\|x-\widehat{x}_I(\omega^K)\|_p \leq 2\sqrt{3\ln(2n/\epsilon)}n^{\frac{1}{p}-\frac{1}{2}}K^{-\frac{1}{2}}\right\} \geq 1-\epsilon.\end{array}$$

Exercise 4.20.

[follow-up to Exercise 4.17] Consider the situation as follows. A retailer sells n items by offering customers, via internet, bundles of $m < n$ items, so that an offer is an m-element subset B of the set $S = \{1, ..., n\}$ of the items. A customer has personal preferences represented by a subset P of S—customer's *preference set.* We assume that if an offer B intersects with the preference set P of a customer, the latter buys an item drawn at random from the uniform distribution on $B \cap P$, and if $B \cap P = \emptyset$, the customer declines the offer. In the pilot stage we are interested in, the seller learns the market by making offers to K customers. Specifically, the seller draws the k-th customer, $k \leq K$, at random from the uniform distribution on the population of customers, and makes the selected customer an offer drawn at random from the uniform distribution on the set $\mathcal{S}_{m,n}$ of all m-item offers. What is observed in the k-th experiment is the item, if any, bought by the customer, and we want to make statistical inferences from these observations.

The outlined observation scheme can be formalized as follows. Let $\mathcal{S}$ be the set of all subsets of the n-element set, so that $\mathcal{S}$ is of cardinality $N = 2^n$. The population of customers induces a probability distribution p on $\mathcal{S}$: for $P \in \mathcal{S}$, p_P is the fraction of customers with the preference set being P; we refer to p as to the *preference distribution.* An outcome of a single experiment can be represented by a pair (ι, B), where $B \in \mathcal{S}_{m,n}$ is the offer used in the experiment, and ι is either 0 ("nothing is bought", $P \cap B = \emptyset$), or a point from $P \cap B$, the item which was bought, when $P \cap B \neq \emptyset$. Note that A_P is a probability distribution on the $(M = (m+1)\binom{n}{m})$-element set $\Omega = \{(\iota, B)\}$ of possible outcomes. As a result, our observation scheme is fully specified by an $M \times N$ column-stochastic matrix A known to us with the columns A_P indexed by $P \in \mathcal{S}$. When a customer is drawn at random from the uniform distribution on the population of customers, the distribution of the outcome clearly is Ap, where p is the (unknown) preference distribution. Our inferences should be based on the K-element sample $\omega^K = (\omega_1, ..., \omega_K)$, with $\omega_1, .., \omega_K$ drawn, independently of each other, from the distribution Ap.

Now we can pose various inference problems, e.g., that of estimating p. We, however, intend to focus on a simpler problem—one of recovering Ap. In terms of our story, this makes sense: when we know Ap, we know, e.g., what the probability is for every offer to be "successful" (something indeed is bought) and/or to result in a specific profit, etc. With this knowledge at hand, the seller can pass from a "blind" offering policy (drawing an offer at random from the uniform distribution on the set $\mathcal{S}_{m,n}$) to something more rewarding.

Now comes the exercise:

1. Use the results of Exercise 4.17 to build a "presumably good" linear estimate

$$\widehat{x}_H(\omega^K) = H^T \left[\frac{1}{K} \sum_{k=1}^{K} \omega_k \right]$$

   of Ap (as always, we encode observations ω, which are elements of the M-element set Ω, by standard basic orths in $\mathbf{R}^M$). As the norm $\|\cdot\|$ quantifying the recovery error, use $\|\cdot\|_1$ and/or $\|\cdot\|_2$. In order to avoid computational difficulties, use small m and n (e.g., $m = 3$ and $n = 5$). Compare your results with those for the "straightforward" estimate $\frac{1}{K}\sum_{k=1}^{K} \omega_k$ (the empirical distribution of $\omega \sim Ap$).
2. Assuming that the "presumably good" linear estimate outperforms the straightforward one, how could this phenomenon be explained? Note that we have no nontrivial a priori information on p!

Exercise 4.21.

[Poisson Imaging] The *Poisson Imaging Problem* is to recover an unknown signal observed via the Poisson observation scheme. More specifically, assume that our observation is a realization of random vector $\omega \in \mathbf{R}^m_+$ with Poisson entries $\omega_i = \text{Poisson}([Ax]_i)$ independent of each other. Here A is a given entrywise nonnegative $m \times n$ matrix, and x is an unknown signal known to belong to a given compact convex subset $\mathcal{X}$ of $\mathbf{R}^n_+$. Our goal is to recover in a given norm $\|\cdot\|$ the linear image Bx of x, where B is a given $\nu \times n$ matrix.

We assume in the sequel that $\mathcal{X}$ is a subset cut off the n-dimensional probabilistic simplex $\mathbf{\Delta}_n$ by a collection of linear equality and inequality constraints. The assumption $\mathcal{X} \subset \mathbf{\Delta}_n$ is not too restrictive. Indeed, assume that we know in advance a linear inequality $\sum_i \alpha_i x_i \leq 1$ with positive coefficients which is valid on $\mathcal{X}$.[20] Introducing slack variable s given by $\sum_i \alpha_i x_i + s = 1$ and passing from signal x to the new signal $[\alpha_1 x_1; ...; \alpha_n x_n; s]$, after a straightforward modification of matrices A and B, we arrive at the situation where $\mathcal{X}$ is a subset of the probabilistic simplex.

Our goal in the sequel is to build a presumably good linear estimate $\widehat{x}_H(\omega) = H^T\omega$ of Bx. As in Exercise 4.17, we start with upper-bounding the risk of a linear

[20] For example, in PET—see Section 2.4.3.2—where x is the density of a radioactive tracer injected into the patient taking the PET procedure, we know in advance the total amount $\sum_i v_i x_i$ of the tracer, v_i being the volume of voxels.

estimate. When representing

$$\omega = Ax + \xi_x,$$

we arrive at zero mean observation noise ξ_x with entries $[\xi_x]_i = \omega_i - [Ax]_i$ independent of each other and covariance matrix $\mathrm{Diag}\{Ax\}$. We now can upper-bound the risk of a linear estimate $\widehat{x}_H(\cdot)$ in the same way as in Exercise 4.17. Specifically, denoting by $\Pi_{\mathcal{X}}$ the set of all diagonal matrices $\mathrm{Diag}\{Ax\}$, $x \in \mathcal{X}$, and by $P_{i,x}$ the Poisson distribution with parameter $[Ax]_i$, we have

$$\begin{array}{rcl}\mathrm{Risk}_{\|\cdot\|}[\widehat{x}_H|\mathcal{X}] &=& \sup_{x\in\mathcal{X}} \mathbf{E}_{\omega\sim P_{1,x}\times\ldots\times P_{m,x}}\left\{\|Bx - H^T\omega\|\right\}\\ &=& \sup_{x\in\mathcal{X}} \mathbf{E}_{\xi_x}\left\{\|[Bx - H^T A]x - H^T\xi_x\|\right\}\\ &\leq& \underbrace{\sup_{x\in\mathcal{X}} \|[B - H^T A]x\|}_{\Phi(H)} + \underbrace{\sup_{\xi:\mathrm{Cov}[\xi]\in\Pi_{\mathcal{X}}} \mathbf{E}_{\xi}\left\{\|H^T\xi\|\right\}}_{\Psi^{\mathcal{X}}(H)}.\end{array}$$

In order to build a presumably good linear estimate, it suffices to build efficiently computable upper bounds $\overline{\Phi}(H)$ on $\Phi(H)$ and $\overline{\Psi}^{\mathcal{X}}(H)$ on $\Psi^{\mathcal{X}}(H)$ convex in H, and then take as H an optimal solution to the convex optimization problem

$$\mathrm{Opt} = \min_H \left[\overline{\Phi}(H) + \overline{\Psi}^{\mathcal{X}}(H)\right].$$

As in Exercise 4.17, assume from now on that $\|\cdot\|$ is an absolute norm, and the unit ball $\mathcal{B}_*$ of the conjugate norm is a spectratope:

$$\mathcal{B}_* := \{u : \|u\|_* \leq 1\} = \{u : \exists r \in \mathcal{R}, y : u = My, S_\ell^2[y] \preceq r_\ell I_{f_\ell}, \ell \leq L\}.$$

Observe that

- In order to build $\overline{\Phi}$, we can use exactly the same techniques as those developed in Exercise 4.17. Indeed, as far as building $\overline{\Phi}$ is concerned, the only difference with the situation of Exercise 4.17 is that in the latter, A was column-stochastic matrix, while now A is just an entrywise nonnegative matrix. Note, however, that when upper-bounding Φ in Exercise 4.17, we never used the fact that A is column-stochastic.
- In order to upper-bound $\Psi^{\mathcal{X}}$, we can use the bound (4.40) of Exercise 4.17.

The bottom line is that in order to build a presumably good linear estimate, we need to solve the convex optimization problem

$$\mathrm{Opt} = \min_{H,\Upsilon,\Theta} \left\{\overline{\Phi}(H) + \phi_{\mathcal{R}}(\lambda[\Upsilon]) + \Gamma_{\mathcal{X}}(\Theta) : \Upsilon = \{\Upsilon_\ell \succeq 0, \ell \leq L\} \atop \left[\begin{array}{c|c}\Theta & \frac{1}{2}HM\\ \hline \frac{1}{2}M^T H^T & \sum_\ell S_\ell^*[\Upsilon_\ell]\end{array}\right] \succeq 0\right\} \tag{P}$$

where

$$\Gamma_{\mathcal{X}}(\Theta) = \max_{x\in\mathcal{X}} \mathrm{Tr}(\mathrm{Diag}\{Ax\}\Theta)$$

(cf. problem (4.99)) with $\overline{\Phi}$ yielded by any construction from Exercise 4.17, e.g., the least conservative Combined upper bound on Φ.

What in our present situation differs significantly from the situation of Exercise 4.17, are the bounds on probabilities of large deviations (for Discrete o.s., established in Exercise 4.19). The goal of what follows is to establish these bounds for

Poisson Imaging.

Here is what you are supposed to do:

1. Let $\omega \in \mathbf{R}^m$ be a random vector with independent entries $\omega_i \sim \text{Poisson}(\mu_i)$, and let $\mu = [\mu_1; ...; \mu_m]$. Prove that whenever $h \in \mathbf{R}^m$, $\gamma > 0$, and $\delta \geq 0$, one has

$$\ln\left(\text{Prob}\{h^T\omega > h^T\mu + \delta\}\right) \leq \sum_i [\exp\{\gamma h_i\} - 1]\mu_i - \gamma h^T\mu - \gamma\delta. \tag{4.107}$$

2. Taking for granted (or see, e.g., [178]) that $e^x - x - 1 \leq \frac{x^2}{2(1-x/3)}$ when $|x| < 3$, prove that in the situation of item 1 one has for $t > 0$:

$$0 \leq \gamma < \frac{3}{\|h\|_\infty} \Rightarrow \ln\left(\text{Prob}\{h^T\omega > h^T\mu + t\}\right) \leq \frac{\gamma^2 \sum_i h_i^2 \mu_i}{2(1 - \gamma\|h\|_\infty/3)} - \gamma t. \tag{4.108}$$

Derive from the latter fact that

$$\text{Prob}\left\{h^T\omega > h^T\mu + \delta\right\} \leq \exp\left\{-\frac{\delta^2}{2[\sum_i h_i^2\mu_i + \|h\|_\infty\delta/3]}\right\}, \tag{4.109}$$

and conclude that

$$\text{Prob}\left\{|h^T\omega - h^T\mu| > \delta\right\} \leq 2\exp\left\{-\frac{\delta^2}{2[\sum_i h_i^2\mu_i + \|h\|_\infty\delta/3]}\right\}. \tag{4.110}$$

3. Extract from (4.110) the following

Proposition 4.35. *In the situation and under the assumptions of Exercise 4.21, let* Opt *be the optimal value, and* H, Υ, Θ *be a feasible solution to problem (P). Whenever* $x \in \mathcal{X}$ *and* $\epsilon \in (0,1)$*, denoting by* P_x *the distribution of observations stemming from* x *(i.e., the distribution of random vector* ω *with independent entries* $\omega_i \sim \text{Poisson}([Ax]_i)$*), one has*

$$\begin{array}{rcl}\mathbf{E}_{\omega\sim P_x}\{\|Bx - \widehat{x}_H(\omega)\|\} & \leq & \overline{\Phi}(H) + 2\sqrt{\phi_{\mathcal{R}}(\lambda[\Upsilon])\text{Tr}(\text{Diag}(Ax\}\Theta)} \\ & \leq & \overline{\Phi}(H) + \phi_{\mathcal{R}}(\lambda[\Upsilon]) + \Gamma_{\mathcal{X}}(\Theta)\end{array} \tag{4.111}$$

and

$$\begin{array}{l}\text{Prob}_{\omega\sim P_x}\Big\{\|Bx - \widehat{x}_H(\omega)\| \leq \overline{\Phi}(H) \\ \quad +4\sqrt{\frac{2}{9}\ln^2(2m/\epsilon)\text{Tr}(\Theta) + \ln(2m/\epsilon)\text{Tr}(\text{Diag}\{Ax\}\Theta)}\sqrt{\phi_{\mathcal{R}}(\lambda[\Upsilon])}\Big\} \geq 1 - \epsilon.\end{array} \tag{4.112}$$

Note that in the case of $[Ax]_i \geq 1$ *for all* $x \in \mathcal{X}$ *and all* i *we have* $\text{Tr}(\Theta) \leq \text{Tr}(\text{Diag}\{Ax\}\Theta)$*, so that in this case the* P_x*-probability of the event*

$$\left\{\omega : \|Bx - \widehat{x}_H(\omega)\| \leq \overline{\Phi}(H) + O(1)\ln(2m/\epsilon)\sqrt{\phi_{\mathcal{R}}(\lambda[\Upsilon])\Gamma_{\mathcal{X}}(\Theta)}\right\}$$

is at least $1 - \epsilon$.

4.7.6 Numerical lower-bounding minimax risk

Exercise 4.22.

4.22.A. Motivation. From the theoretical viewpoint, the results on near-optimality of presumably good linear estimates stated in Propositions 4.5 and 4.16 seem to be quite strong and general. This being said, for a practically oriented user the "nonoptimality factors" arising in these propositions can be too large to make any practical sense. This drawback of our theoretical results is not too crucial—what matters in applications, is whether the risk of a proposed estimate is appropriate for the application in question, and not by how much it could be improved were we smart enough to build the "ideal" estimate; results of the latter type from a practical viewpoint offer no more than some "moral support." Nevertheless, the "moral support" has its value, and it makes sense to strengthen it by improving the lower risk bounds as compared to those underlying Propositions 4.5 and 4.16. In this respect, an appealing idea is to pass from lower risk bounds yielded by theoretical considerations to *computation-based* ones. The goal of this exercise is to develop some methodology yielding computation-based lower risk bounds. We start with the main ingredient of this methodology—the classical *Cramer-Rao* bound.

4.22.B. Cramer-Rao bound. Consider the situation as follows: we are given

- an observation space Ω equipped with reference measure Π, basic examples being (A) $\Omega = \mathbf{R}^m$ with Lebesgue measure Π, and (B) (finite or countable) discrete set Ω with counting measure Π;
- a convex compact set $\Theta \subset \mathbf{R}^k$ and a family $\mathcal{P} = \{p(\omega,\theta) : \theta \in \Theta\}$ of probability densities, taken w.r.t. Π.

Our goal is, given an observation $\omega \sim p(\cdot,\theta)$ stemming from unknown θ known to belong to Θ, to recover θ. We quantify the risk of a candidate estimate $\widehat{\theta}$ as

$$\mathrm{Risk}[\widehat{\theta}|\Theta] = \sup_{\theta\in\Theta}\left(\mathbf{E}_{\omega\sim p(\cdot,\theta)}\left\{\|\widehat{\theta}(\omega)-\theta\|_2^2\right\}\right)^{1/2}, \tag{4.113}$$

and define the "ideal" minimax risk as

$$\mathrm{Risk}_{\mathrm{opt}} = \inf_{\widehat{\theta}} \mathrm{Risk}[\widehat{\theta}],$$

the infimum being taken w.r.t. all estimates, or, which is the same, all *bounded* estimates (indeed, passing from a candidate estimate $\widehat{\theta}$ to the projected estimate $\widehat{\theta}_\Theta(\omega) = \mathrm{argmin}_{\theta\in\Theta}\|\widehat{\theta}(\omega)-\theta\|_2$ will only reduce the estimate risk).

The Cramer-Rao inequality [58, 205], which we intend to use,[21] is a certain relation between the covariance matrix of a bounded estimate and its bias; this relation is valid under mild regularity assumptions on the family $\mathcal{P}$, specifically, as follows:

1) $p(\omega,\theta) > 0$ for all $\omega \in \Omega, \theta \in U$, and $p(\omega,\theta)$ is differentiable in θ, with $\nabla_\theta p(\omega,\theta)$ continuous in $\theta \in \Theta$;
2) The *Fisher Information matrix*

$$\mathcal{I}(\theta) = \int_\Omega \frac{\nabla_\theta p(\omega,\theta)[\nabla_\theta p(\omega,\theta)]^T}{p(\omega,\theta)}\Pi(d\omega)$$

[21] As a matter of fact, the classical Cramer-Rao inequality dealing with unbiased estimates is not sufficient for our purposes "as is." What we need to build is a "bias enabled" version of this inequality. Such an inequality may be developed using Bayesian argument [99, 233].

is well-defined for all $\theta \in \Theta$;

3) There exists function $M(\omega) \geq 0$ such that $\int_\Omega M(\omega)\Pi(d\omega) < \infty$ and

$$\|\nabla_\theta p(\omega,\theta)\|_2 \leq M(\omega)\ \forall \omega \in \Omega, \theta \in \Theta.$$

The derivation of the Cramer-Rao bound is as follows. Let $\widehat{\theta}(\omega)$ be a bounded estimate, and let

$$\phi(\theta) = [\phi_1(\theta); ...; \phi_k(\theta)] = \int_\Omega \widehat{\theta}(\omega) p(\omega,\theta)\Pi(d\omega)$$

be the expected value of the estimate. By item 3, $\phi(\theta)$ is differentiable on Θ, with the Jacobian $\phi'(\theta) = \left[\frac{\partial \phi_i(\theta)}{\partial \theta_j}\right]_{i,j\leq k}$ given by

$$\phi'(\theta)h = \int_\Omega \widehat{\theta}(\omega) h^T \nabla_\theta p(\omega,\theta)\Pi(d\omega),\ h \in \mathbf{R}^k.$$

Besides this, recalling that $\int_\Omega p(\omega,\theta)\Pi(d\omega) \equiv 1$ and invoking item 3, we have $\int_\Omega h^T \nabla_\theta p(\omega,\theta)\Pi(d\omega) = 0$, whence, in view of the previous identity,

$$\phi'(\theta)h = \int_\Omega [\widehat{\theta}(\omega) - \phi(\theta)] h^T \nabla_\theta p(\omega,\theta)\Pi(d\omega),\ h \in \mathbf{R}^k.$$

Therefore, for all $g, h \in \mathbf{R}^k$ we have

$$\begin{array}{rcl}
[g^T\phi'(\theta)h]^2 &=& \left[\int_\omega [g^T(\widehat{\theta} - \phi(\theta)][h^T\nabla_\theta p(\omega,\theta)/p(\omega,\theta)]p(\omega,\theta)\Pi(d\omega)\right]^2 \\
&\leq& \left[\int_\Omega g^T[\widehat{\theta} - \phi(\theta)][\widehat{\theta} - \phi(\theta)]^T g p(\omega,\theta)\Pi(d\omega)\right] \\
&& \quad \times \left[\int_\Omega [h^T\nabla_\theta p(\omega,\theta)/p(\omega,\theta)]^2 p(\omega,\theta)\Pi(d\omega)\right] \\
&& \text{[by the Cauchy Inequality]} \\
&=& \left[g^T \mathrm{Cov}_{\widehat{\theta}}(\theta) g\right]\left[h^T\mathcal{I}(\theta)h\right],
\end{array}$$

where $\mathrm{Cov}_{\widehat{\theta}}(\theta)$ is the covariance matrix $\mathbf{E}_{\omega\sim p(\cdot,\theta)}\left\{[\widehat{\theta}(\omega) - \phi(\theta)][\widehat{\theta}(\omega) - \phi(\theta)]^T\right\}$ of $\widehat{\theta}(\omega)$ induced by $\omega \sim p(\cdot,\theta)$. We have arrived at the inequality

$$\left[g^T \mathrm{Cov}_{\widehat{\theta}}(\theta) g\right]\left[h^T\mathcal{I}(\theta)h\right] \geq [g^T\phi'(\theta)h]^2\ \forall (g,h \in \mathbf{R}^k, \theta \in \Theta). \qquad (*)$$

For $\theta \in \Theta$ fixed, let $\mathcal{J}$ be a positive definite matrix such that $\mathcal{J} \succeq \mathcal{I}(\theta)$, whence by $(*)$ it holds

$$\left[g^T \mathrm{Cov}_{\widehat{\theta}}(\theta) g\right]\left[h^T\mathcal{J}h\right] \geq [g^T\phi'(\theta)h]^2\ \forall (g,h \in \mathbf{R}^k). \qquad (**)$$

For g fixed, the maximum of the right-hand side quantity in $(**)$ over h satisfying $h^T\mathcal{J}h \leq 1$ is $g^T\phi'(\theta)\mathcal{J}^{-1}[\phi'(\theta]^T g$, and we arrive at the *Cramer-Rao inequality*

$$\forall (\theta \in \Theta, \mathcal{J} \succeq \mathcal{I}(\theta), \mathcal{J} \succ 0) : \mathrm{Cov}_{\widehat{\theta}}(\theta) \succeq \phi'(\theta)\mathcal{J}^{-1}[\phi'(\theta]^T \qquad (4.114)$$

$$\left[\mathrm{Cov}_{\widehat{\theta}}(\theta) = \mathbf{E}_{\omega\sim p(\cdot,\theta)}\left\{[\widehat{\theta} - \phi(\theta)][\widehat{\theta} - \phi(\theta)]^T\right\},\ \phi(\theta) = \mathbf{E}_{\omega\sim p(\cdot,\theta)}\left\{\widehat{\theta}(\omega)\right\}\right]$$

which holds true for every bounded estimate $\widehat{\theta}(\cdot)$. Note also that for every $\theta \in \Theta$

and every bounded estimate x we have

$$\begin{aligned}\text{Risk}^2[\widehat{\theta}] &\geq \mathbf{E}_{\omega\sim p(\cdot,\theta)}\left\{\|\widehat{\theta}(\omega)-\theta\|_2^2\right\} = \mathbf{E}_{\omega\sim p(\cdot,\theta)}\left\{\|[\widehat{\theta}(\omega)-\phi(\theta)]+[\phi(\theta)-\theta]\|_2^2\right\}\\ &= \mathbf{E}_{\omega\sim p(\cdot,\theta)}\left\{\|\widehat{\theta}(\omega)-\phi(\theta)\|_2^2\right\} + \|\phi(\theta)-\theta)\|_2^2 \\ &\qquad -2\underbrace{\mathbf{E}_{\omega\sim p(\cdot,\theta)}\left[[\widehat{\theta}(\omega)-\phi(\theta)]^T[\phi(\theta)-\theta)]\right\}}_{=0}\\ &= \text{Tr}(\text{Cov}_{\widehat{\theta}}(\theta)) + \|\phi(\theta)-\theta\|_2^2.\end{aligned}$$

Hence, in view of (4.114), for every bounded estimate $\widehat{\theta}$ it holds

$$\begin{array}{l}\forall(\mathcal{J}\succ 0:\mathcal{J}\succeq\mathcal{I}(\theta)\,\forall\theta\in\Theta):\\ \text{Risk}^2[\widehat{\theta}]\geq\sup_{\theta\in\Theta}\left[\text{Tr}(\phi'(\theta)\mathcal{J}^{-1}[\phi'(\theta)]^T)+\|\phi(\theta)-\theta\|_2^2\right]\\ \qquad\left[\phi(\theta)=\mathbf{E}_{\omega\sim p(\cdot,\theta)}\{\widehat{\theta}(\omega)\}\right].\end{array} \tag{4.115}$$

The fact that we considered the risk of estimating "the entire" θ rather than a given vector-valued function $f(\theta):\Theta\to\mathbf{R}^\nu$ plays no special role, and in fact the Cramer-Rao inequality admits the following modification yielded by a completely similar reasoning:

Proposition 4.36. *In the situation described in item 4.22.B and under assumptions 1)–3) of this item, let $f(\cdot):\Theta\to\mathbf{R}^\nu$ be a bounded Borel function, and let $\widehat{f}(\omega)$ be a bounded estimate of $f(\omega)$ via observation $\omega\sim p(\cdot,\theta)$. Then, setting for $\theta\in\Theta$*

$$\begin{array}{rcl}\phi(\theta) &=& \mathbf{E}_{\omega\sim p(\cdot,\theta)}\left\{\widehat{f}(\theta)\right\},\\ \text{Cov}_{\widehat{f}}(\theta) &=& \mathbf{E}_{\omega\sim p(\cdot,\theta)}\left\{[\widehat{f}(\omega)-\phi(\theta)][\widehat{f}(\omega)-\phi(\theta)]^T\right\},\end{array}$$

one has

$$\forall(\theta\in\Theta,\mathcal{J}\succeq\mathcal{I}(\theta),\mathcal{J}\succ 0):\text{Cov}_{\widehat{f}}(\theta)\succeq\phi'(\theta)\mathcal{J}^{-1}[\phi'(\theta)]^T.$$

As a result, for

$$\text{Risk}[\widehat{f}]=\sup_{\theta\in\Theta}\left[\mathbf{E}_{\omega\sim p(\cdot,\theta)}\left\{\|\widehat{f}(\omega)-f(\theta)\|_2^2\right\}\right]^{1/2}$$

it holds

$$\begin{array}{l}\forall(\mathcal{J}\succ 0:\mathcal{J}\succeq\mathcal{I}(\theta)\,\forall\theta\in\Theta):\\ \text{Risk}^2[\widehat{f}]\geq\sup_{\theta\in\Theta}\left[\text{Tr}(\phi'(\theta)\mathcal{J}^{-1}[\phi'(\theta)]^T)+\|\phi(\theta)-f(\theta)\|_2^2\right]\end{array}$$

Now comes the first part of the exercise:

1) Derive from (4.115) the following

Proposition 4.37. *In the situation of item 4.22.B, let*

- *$\Theta\subset\mathbf{R}^k$ be a $\|\cdot\|_2$-ball of radius $r>0$,*
- *the family $\mathcal{P}$ be such that $\mathcal{I}(\theta)\preceq\mathcal{J}$ for some $\mathcal{J}\succ 0$ and all $\theta\in\Theta$.*

Then the minimax optimal risk satisfies the bound

$$\mathrm{Risk}_{\mathrm{opt}} \geq \frac{rk}{r\sqrt{\mathrm{Tr}(\mathcal{J})}+k}. \tag{4.116}$$

In particular, when $\mathcal{J} = \alpha^{-1}I_k$, we have

$$\mathrm{Risk}_{\mathrm{opt}} \geq \frac{r\sqrt{\alpha k}}{r+\sqrt{\alpha k}}. \tag{4.117}$$

Hint. Assuming w.l.o.g. that Θ is centered at the origin, and given a bounded estimate $\widehat{\theta}$ with risk $\mathfrak{R}$, let $\phi(\theta)$ be associated with the estimate via (4.115). Select $\gamma \in (0,1)$ and consider two cases: (a): there exists $\theta \in \partial\Theta$ such that $\|\phi(\theta) - \theta\|_2 > \gamma r$, and (b): $\|\phi(\theta)-\theta\|_2 \leq \gamma r$ for all $\theta \in \partial\Theta$. In the case of (a), lower-bound $\mathfrak{R}$ by $\max_{\theta\in\Theta}\|\phi(\theta)-\theta\|_2$; see (4.115). In the case of (b), lower-bound $\mathfrak{R}^2$ by $\max_{\theta\in\Theta}\mathrm{Tr}(\phi'(\theta)\mathcal{J}^{-1}[\phi'(\theta)]^T)$—see (4.115)—and use the Gauss Divergence theorem to lower-bound the latter quantity in terms of the flux of the vector field $\phi(\cdot)$ over $\partial\Theta$.

When implementing the above program, you might find useful the following fact (prove it!):

Lemma 4.38. *Let Φ be an $n\times n$ matrix, and $\mathcal{J}$ be a positive definite $n \times n$ matrix. Then*

$$\mathrm{Tr}(\Phi\mathcal{J}^{-1}\Phi^T) \geq \frac{\mathrm{Tr}^2(\Phi)}{\mathrm{Tr}(\mathcal{J})}.$$

4.22.C. Application to signal recovery. Proposition 4.37 allows us to build computation-based lower risk bounds in the signal recovery problem considered in Section 4.2, in particular, the problem where one wants to recover the linear image Bx of an unknown signal x known to belong to a given ellitope

$$\mathcal{X} = \{x \in \mathbf{R}^n : \exists t \in \mathcal{T} : x^T S_\ell x \leq t_\ell, \ell \leq L\}$$

(with our usual restriction on S_ℓ and $\mathcal{T}$) via observation

$$\omega = Ax + \sigma\xi,\ \xi \sim \mathcal{N}(0, I_m),$$

and the risk of a candidate estimate, as in Section 4.2, is defined according to (4.113).[22] It is convenient to assume that the matrix B (which in our general setup can be an arbitrary $\nu \times n$ matrix) is a *nonsingular* $n \times n$ matrix.[23] Under this

[22]In fact, the approach to be developed can be applied to signal recovery problems involving Discrete/Poisson observation schemes and norms different from $\|\cdot\|_2$ used to measure the recovery error, signal-dependent noises, etc.

[23]This assumption is nonrestrictive. Indeed, when $B \in \mathbf{R}^{\nu\times n}$ with $\nu < n$, we can add to B $n-\nu$ zero rows, which keeps our estimation problem intact. When $\nu \geq n$, we can add to B a small perturbation to ensure $\mathrm{Ker}\, B = \{0\}$, which, for small enough perturbation, again keeps our estimation problem basically intact. It remains to note that when $\mathrm{Ker}\, B = \{0\}$ we can replace $\mathbf{R}^\nu$ with the image space of B, which again does not affect the estimation problem we are interested in.

assumption, setting

$$\mathcal{Y} = B^{-1}\mathcal{X} = \{y \in \mathbf{R}^n : \exists t \in \mathcal{T} : y^T[B^{-1}]^T S_\ell B^{-1} y \leq t_\ell, \ell \leq L\}$$

and $\bar{A} = AB^{-1}$, we lose nothing when replacing the sensing matrix A with $\bar{A}$ and treating as our signal $y \in \mathcal{Y}$ rather than $\mathcal{X}$. Note that in our new situation A is replaced with $\bar{A}$, $\mathcal{X}$ with $\mathcal{Y}$, and B is the unit matrix I_n. For the sake of simplicity, we assume from now on that A (and therefore $\bar{A}$) has trivial kernel. Finally, let $\tilde{S}_\ell \succeq S_\ell$ be close to S_k positive definite matrices, e.g., $\tilde{S}_\ell = S_\ell + 10^{-100} I_n$. Setting $\bar{S}_\ell = [B^{-1}]^T \tilde{S}_\ell B^{-1}$ and

$$\bar{\mathcal{Y}} = \{y \in \mathbf{R}^n : \exists t \in \mathcal{T} : y^T \bar{S}_\ell y \leq t_\ell, \ell \leq L\},$$

we get $\bar{S}_\ell \succ 0$ and $\bar{\mathcal{Y}} \subset \mathcal{Y}$. Therefore, any lower bound on the $\|\cdot\|_2$-risk of recovery $y \in \bar{\mathcal{Y}}$ via observation $\omega = AB^{-1}y + \sigma\xi$, $\xi \sim \mathcal{N}(0, I_m)$, automatically is a lower bound on the minimax risk Risk_{opt} corresponding to our original problem of interest.

Now assume that we can point out a k-dimensional linear subspace E in $\mathbf{R}^n$ and positive reals r, γ such that

(i) the $\|\cdot\|_2$-ball $\Theta = \{\theta \in E : \|\theta\|_2 \leq r\}$ is contained in $\bar{\mathcal{Y}}$;

(ii) The restriction $\bar{A}_E$ of $\bar{A}$ onto E satisfies the relation

$$\text{Tr}(\bar{A}_E^* \bar{A}_E) \leq \gamma$$

($\bar{A}_E^* : \mathbf{R}^m \to E$ is the conjugate of the linear map $\bar{A}_E : E \to \mathbf{R}^m$).

Consider the auxiliary estimation problem obtained from the (reformulated) problem of interest by replacing the signal set $\bar{\mathcal{Y}}$ with Θ. Since $\Theta \subset \bar{\mathcal{Y}}$, the minimax risk in the auxiliary problem is a lower bound on the minimax risk Risk_{opt} we are interested in. On the other hand, the auxiliary problem is nothing but the problem of recovering parameter $\theta \in \Theta$ from observation $\omega \sim \mathcal{N}(\bar{A}\theta, \sigma^2 I)$, which is just a special case of the problem considered in item 4.22.B. As it is immediately seen, the Fisher Information matrix in this problem is independent of θ and is $\sigma^{-2}\bar{A}_E^*\bar{A}_E$:

$$e^T \mathcal{I}(\theta) e = \sigma^{-2} e^T \bar{A}_E^* \bar{A}_E e, \; e \in E.$$

Invoking Proposition 4.37, we arrive at the lower bound on the minimax risk in the auxiliary problem (and thus in the problem of interest as well):

$$\text{Risk}_{\text{opt}} \geq \frac{r\sigma k}{r\sqrt{\gamma} + \sigma k}. \tag{4.118}$$

The resulting risk bound depends on r, k, γ and is larger the smaller γ is and the larger k and r are.

Lower-bounding Risk_{opt}. In order to make the bounding scheme just outlined give its best, we need a mechanism which allows us to generate k-dimensional "disks" $\Theta \subset \bar{\mathcal{Y}}$ along with associated quantities r, γ. In order to design such a mechanism, it is convenient to represent k-dimensional linear subspaces of $\mathbf{R}^n$ as the image spaces of orthogonal $n \times n$ projectors P of rank k. Such a projector P gives rise to the disk Θ_P of the radius $r = r_P$ contained in $\bar{\mathcal{Y}}$, where r_P is the largest ρ such that the set $\{y \in \text{Im}P : y^T P y \leq \rho^2\}$ is contained in $\bar{\mathcal{Y}}$ ("condition

$\mathcal{C}(r)$"), and we can equip the disk with γ satisfying (ii) if and only if

$$\mathrm{Tr}(P\bar{A}^T\bar{A}P) \leq \gamma,$$

or, which is the same (recall that P is an orthogonal projector)

$$\mathrm{Tr}(\bar{A}P\bar{A}^T) \leq \gamma \tag{4.119}$$

("condition $\mathcal{D}(\gamma)$"). Now, when P is a nonzero orthogonal projector, the simplest sufficient condition for the validity of $\mathcal{C}(r)$ is the existence of $t \in \mathcal{T}$ such that

$$\forall (y \in \mathbf{R}^n, \ell \leq L) : y^T P\bar{S}_\ell Py \leq t_\ell r^{-2} y^T Py,$$

or, which is the same,

$$\exists s : r^2 s \in \mathcal{T} \ \&\ P\bar{S}_\ell P \preceq s_\ell P,\ \ell \leq L. \tag{4.120}$$

Let us rewrite (4.119) and (4.120) as a system of *linear* matrix inequalities. This is what you are supposed to do:

2.1) Prove the following simple fact:

Observation 4.39. Let Q be a positive definite, R be a nonzero positive semidefinite matrix, and let s be a real. Then

$$RQR \preceq sR$$

if and only if

$$sQ^{-1} \succeq R.$$

2.2) Extract from the above observation the conclusion as follows. Let $\mathbf{T}$ be the conic hull of $\mathcal{T}$:

$$\mathbf{T} = \mathrm{cl}\{[s;\tau] : \tau > 0, s/\tau \in \mathcal{T}\} = \{[s;\tau] : \tau > 0, s/\tau \in \mathcal{T}\} \cup \{0\}.$$

Consider the system of constraints

$$\begin{array}{c} [s;\tau] \in \mathbf{T} \ \&\ s_\ell \bar{S}_\ell^{-1} \succeq P,\ \ell \leq L \ \&\ \mathrm{Tr}(\bar{A}P\bar{A}^T) \leq \gamma, \\ P \text{ is an orthogonal projector of rank } k \geq 1 \end{array} \tag{\#}$$

in variables $[s;\tau]$, k, γ, and P. Every feasible solution to this system gives rise to a k-dimensional Euclidean subspace $E \subset \mathbf{R}^n$ (the image space of P) such that the Euclidean ball Θ in E centered at the origin of radius

$$r = 1/\sqrt{\tau}$$

taken along with γ satisfies conditions (i)–(ii). Consequently, such a feasible solution yields the lower bound

$$\mathrm{Risk}_{\mathrm{opt}} \geq \psi_{\sigma,k}(\gamma,\tau) := \frac{\sigma k}{\sqrt{\gamma} + \sigma\sqrt{\tau k}}$$

on the minimax risk in the problem of interest.

Ideally, to utilize item 2.2 to lower-bound $\mathrm{Risk}_{\mathrm{opt}}$, we should look through $k =$

$1,...,n$ and maximize for every k the lower risk bound $\psi_{\sigma,k}(\gamma,\tau)$ under constraints (#), thus arriving at the problem

$$\min_{[s;\tau],\gamma,P}\left\{\begin{array}{l}\frac{\sigma}{\psi_{\sigma,k}(\gamma,\tau)}=\sqrt{\gamma}/k+\sigma\sqrt{\tau}:\\ {[s;\tau]}\in\mathbf{T}\ \&\ s_\ell\bar{S}_\ell^{-1}\succeq P,\ell\leq L\ \&\ \mathrm{Tr}(\bar{A}P\bar{A}^T)\leq\gamma,\\ P\text{ is an orthogonal projector of rank }k.\end{array}\right\} \qquad (P_k)$$

This problem seems to be computationally intractable, since the constraints of (P_k) include the nonconvex restriction on P to be a projector of rank k. A natural convex relaxation of this constraint is

$$0\preceq P\preceq I_n,\ \mathrm{Tr}(P)=k.$$

The (minor) remaining difficulty is that the objective in (P) is nonconvex. Note, however, that to minimize $\sqrt{\gamma}/k+\sigma\sqrt{\tau}$ is basically the same as to minimize the convex function $\gamma/k^2+\sigma^2\tau$ which is a tight "proxy" of the squared objective of (P_k). We arrive at a convex "proxy" of (P_k)—the problem

$$\min_{[s;\tau],\gamma,P}\left\{\gamma/k^2+\sigma^2\tau:\ \begin{array}{l}[s;\tau]\in\mathbf{T},0\preceq P\preceq I_n,\mathrm{Tr}(P)=k\\ s_\ell\bar{S}_\ell^{-1}\succeq P,\ell\leq L,\mathrm{Tr}(\bar{A}P\bar{A}^T)\leq\gamma\end{array}\right\}, \qquad (P[k])$$

$k=1,...,n$. Problem $(P[k])$ clearly is solvable, and the P-component $P^{(k)}$ of its optimal solution gives rise to a collection of orthogonal projectors $P^{(k)}_\kappa$, $\kappa=1,...,n$ obtained from $P^{(k)}$ by "rounding"—to get $P^{(k)}_\kappa$, we replace the κ leading eigenvalues of $P^{(k)}$ with ones, and the remaining eigenvalues with zeros, while keeping the eigenvectors intact. We can now for every $\kappa=1,...,n$ fix the P-variable in (P_k) as $P^{(k)}_\kappa$ and solve the resulting problem in the remaining variables $[s;\tau]$ and γ, which is easy—with P fixed, the problem clearly reduces to minimizing τ under the convex constraints

$$s_\ell\bar{S}_\ell^{-1}\succeq P,\ell\leq L,\ [s;\tau]\in\mathbf{T}$$

on $[s;\tau]$. As a result, for every $k\in\{1,...,n\}$, we get n lower bounds on $\mathrm{Risk}_{\rm opt}$, that is, a total of n^2 lower risk bounds, of which we select the best—the largest.

Now comes the next part of the exercise:

3) Implement the outlined program numerically and compare the lower bound on the minimax risk with the upper risk bounds of presumably good linear estimates yielded by Proposition 4.4.
Recommended setup:
   - Sizes: $m=n=\nu=16$.
   - A, B: $B=I_n$, $A=\mathrm{Diag}\{a_1,...,a_n\}$ with $a_i=i^{-\alpha}$ and α running through $\{0,1,2\}$.
   - $\mathcal{X}=\{x\in\mathbf{R}^n: x^TS_\ell x\leq 1,\ell\leq L\}$ (i.e., $\mathcal{T}=[0,1]^L$) with randomly generated S_ℓ.
   - Range of L: $\{1,4,16\}$. For L in this range, you can generate S_ℓ, $\ell\leq L$, as $S_\ell=R_\ell R_\ell^T$ with $R_\ell=\mathtt{randn}(n,p)$, where $p=\rfloor n/L\lfloor$.
   - Range of σ: $\{1.0,0.1,0.01,0.001,0.0001\}$.

Exercise 4.23.

[follow-up to Exercise 4.22]

1) Prove the following version of Proposition 4.37:

Proposition 4.40. *In the situation of item 4.22.B and under Assumptions 1)–3) from this item, let*

- $\|\cdot\|$ *be a norm on* $\mathbf{R}^k$ *such that*

$$\|\theta\|_2 \leq \kappa\|\theta\| \ \forall \theta \in \mathbf{R}^k,$$

- $\Theta \subset \mathbf{R}^k$ *be a* $\|\cdot\|$*-ball of radius* $r > 0$*,*
- *the family* $\mathcal{P}$ *be such that* $\mathcal{I}(\theta) \preceq \mathcal{J}$ *for some* $\mathcal{J} \succ 0$ *and all* $\theta \in \Theta$*.*

Then the minimax optimal risk

$$\mathrm{Risk}_{\mathrm{opt},\|\cdot\|} = \inf_{\widehat{\theta}(\cdot)} \left(\sup_{\theta\in\Theta} \mathbf{E}_{\omega\sim p(\cdot,\theta)} \left\{ \|\theta - \widehat{\theta}(\omega)\|^2 \right\} \right)^{1/2}$$

of recovering parameter $\theta \in \Theta$ *from observation* $\omega \sim p(\cdot,\theta)$ *in the norm* $\|\cdot\|$ *satisfies the bound*

$$\mathrm{Risk}_{\mathrm{opt},\|\cdot\|} \geq \frac{rk}{r\kappa\sqrt{\mathrm{Tr}(\mathcal{J})} + k}. \tag{4.121}$$

In particular, when $\mathcal{J} = \alpha^{-1} I_k$*, we get*

$$\mathrm{Risk}_{\mathrm{opt},\|\cdot\|} \geq \frac{r\sqrt{\alpha k}}{r\kappa + \sqrt{\alpha k}}. \tag{4.122}$$

2) Apply Proposition 4.40 to get lower bounds on the minimax $\|\cdot\|$-risk in the following estimation problems:

2.1) Given indirect observation $\omega = A\theta + \sigma\xi$, $\xi \sim \mathcal{N}(0, I_m)$ of unknown vector θ known to belong to $\Theta = \{\theta \in \mathbf{R}^k : \|\theta\|_p \leq r\}$ with given A, $\mathrm{Ker}\, A = \{0\}$, $p \in [2,\infty]$, $r > 0$, we want to recover θ in $\|\cdot\|_p$.

2.2) Given indirect observation $\omega = L\theta R + \sigma\xi$, where θ is unknown $\mu \times \nu$ matrix known to belong to the Shatten norm ball $\Theta \in \mathbf{R}^{\mu\times\nu} : \|\theta\|_{\mathrm{Sh},p} \leq r$, we want to recover θ in $\|\cdot\|_{\mathrm{Sh},p}$. Here $L \in \mathbf{R}^{m\times\mu}$, $\mathrm{Ker}\, L = \{0\}$ and $R \in \mathbf{R}^{\nu\times n}$, $\mathrm{Ker}\, R^T = \{0\}$ are given matrices, $p \in [2,\infty]$, and ξ is a random Gaussian $m \times n$ matrix (i.e., the entries in ξ are $\mathcal{N}(0,1)$ random variables independent of each other).

2.3) Given a K-repeated observation $\omega^K = (\omega_1, ..., \omega_K)$ with i.i.d. components $\omega_t \sim \mathcal{N}(0,\theta)$, $1 \leq t \leq K$, with unknown $\theta \in \mathbf{S}^n$ known to belong to the matrix box $\Theta = \{\theta : \beta_- I_n \preceq \theta \preceq \beta_+ I_n\}$ with given $0 < \beta_- < \beta_+ < \infty$, we want to recover θ in the spectral norm.

Exercise 4.24.

[More on Cramer-Rao risk bound] Let us fix $\mu \in (1,\infty)$ and a norm $\|\cdot\|$ on $\mathbf{R}^k$, and let $\|\cdot\|_*$ be the norm conjugate to $\|\cdot\|$, and $\mu_* = \frac{\mu}{\mu-1}$. Assume that we are in the situation of item 4.22.B and under assumptions 1) and 3) from this item; as for assumption 2) we now replace it with the assumption that the quantity

$$\mathcal{I}_{\|\cdot\|_*,\mu_*}(\theta) := \left[\mathbf{E}_{\omega\sim p(\cdot,\theta)} \{\|\nabla_\theta \ln(p(\omega,\theta))\|_*^{\mu_*}\}\right]^{1/\mu_*}$$

is well-defined and bounded on Θ; in the sequel, we set

$$\mathcal{I}_{\|\cdot\|_*,\mu_*} = \sup_{\theta\in\Theta} \mathcal{I}_{\|\cdot\|_*,\mu_*}(\theta).$$

1) Prove the following variant of the Cramer-Rao risk hound:

Proposition 4.41. *In the situation described at the beginning of item 4.22.D, let $\Theta \subset \mathbf{R}^k$ be a $\|\cdot\|$-ball of radius r. Then the minimax $\|\cdot\|$-risk of recovering $\theta \in \Theta$ via observation $\omega \sim p(\cdot,\theta)$ can be lower-bounded as*

$$\begin{array}{l}\mathrm{Risk}_{\mathrm{opt},\|\cdot\|}[\Theta] := \inf\limits_{\widehat{\theta}(\cdot)} \sup\limits_{\theta\in\Theta} \left[\mathbf{E}_{\omega\sim p(\cdot,\theta)}\left\{\|\widehat{\theta}(\omega)-\theta\|^\mu\right\}\right]^{1/\mu} \geq \frac{rk}{r\mathcal{I}_{\|\cdot\|_*,\mu_*}+k},\\ \mathcal{I}_{\|\cdot\|_*,\mu_*} = \max\limits_{\theta\in\Theta}\left[\mathcal{I}_{\|\cdot\|_*,\mu_*}(\theta) := \left[\mathbf{E}_{\omega\sim p(\cdot,\theta)}\left\{\|\nabla_\theta \ln(p(\omega,\theta))\|_*^{\mu_*}\right\}\right]^{1/\mu_*}\right].\end{array} \tag{4.123}$$

Example I: Gaussian case, estimating shift. Let $\mu = 2$, and let $p(\omega,\theta) = \mathcal{N}(A\theta, \sigma^2 I_m)$ with $A \in \mathbf{R}^{m\times k}$. Then

$$\begin{array}{rcl}\nabla_\theta \ln(p(\omega,\theta)) = \sigma^{-2}A^T(\omega - A\theta) \Rightarrow & & \\ \int \|\nabla_\theta \ln(p(\omega,\theta))\|_*^2 p(\omega,\theta)d\omega & = & \sigma^{-4}\int \|A^T(\omega-A\theta)\|_*^2 p(\omega,\theta)d\omega \\ & = & \sigma^{-4}\frac{1}{[\sqrt{2\pi}\sigma]^m}\int \|A^T\omega\|_*^2 \exp\{-\frac{\omega^T\omega}{2\sigma^2}\}d\omega \\ & = & \sigma^{-4}\frac{1}{[2\pi]^{m/2}}\int \|A^T\sigma\xi\|_*^2 \exp\{-\xi^T\xi/2\}d\xi \\ & = & \sigma^{-2}\frac{1}{[2\pi]^{m/2}}\int \|A^T\xi\|_*^2 \exp\{-\xi^T\xi/2\}d\xi\end{array}$$

whence

$$\mathcal{I}_{\|\cdot\|_*,2} = \sigma^{-1}\underbrace{\left[\mathbf{E}_{\xi\sim\mathcal{N}(0,I_m)}\left\{\|A^T\xi\|_*^2\right\}\right]^{1/2}}_{\gamma_{\|\cdot\|}(A)}.$$

Consequently, assuming Θ to be a $\|\cdot\|$-ball of radius r in $\mathbf{R}^k$, lower bound (4.123) becomes

$$\mathrm{Risk}_{\mathrm{opt},\|\cdot\|}[\Theta] \geq \frac{rk}{r\mathcal{I}_{\|\cdot\|_*}+k} = \frac{rk}{r\sigma^{-1}\gamma_{\|\cdot\|}(A)+k} = \frac{r\sigma k}{r\gamma_{\|\cdot\|}(A)+\sigma k}. \tag{4.124}$$

The case of direct observations. To see "how it works," consider the case $m = k$, $A = I_k$ of direct observations, and let $\Theta = \{\theta \in \mathbf{R}^k : \|\theta\| \leq r\}$. Then

- We have $\gamma_{\|\cdot\|_1}(I_k) \leq O(1)\sqrt{\ln(k)}$, whence the $\|\cdot\|_1$-risk bound is

$$\mathrm{Risk}_{\mathrm{opt},\|\cdot\|_1}[\Theta] \geq O(1)\frac{r\sigma k}{r\sqrt{\ln(k)}+\sigma k} \qquad [\Theta = \{\theta\in\mathbf{R}^k : \|\theta - a\|_1 \leq r\}].$$

- We have $\gamma_{\|\cdot\|_2}(I_k) = \sqrt{k}$, whence the $\|\cdot\|_2$-risk bound is

$$\mathrm{Risk}_{\mathrm{opt},\|\cdot\|_2}[\Theta] \geq \frac{r\sigma\sqrt{k}}{r+\sigma\sqrt{k}} \qquad [\Theta = \{\theta\in\mathbf{R}^k : \|\theta - a\|_2 \leq r\}].$$

- We have $\gamma_{\|\cdot\|_\infty}(I_k) \leq O(1)k$, whence the $\|\cdot\|_\infty$-risk bound is

$$\mathrm{Risk}_{\mathrm{opt},\|\cdot\|_\infty}[\Theta] \geq O(1)\frac{r\sigma}{r+\sigma} \qquad [\Theta = \{\theta \in \mathbf{R}^k : \|\theta - a\|_\infty \leq r\}].$$

In fact, the above examples are essentially covered by the following:

Observation 4.42. Let $\|\cdot\|$ be a norm on $\mathbf{R}^k$, and let

$$\Theta = \{\theta \in \mathbf{R}^k : \|\theta\| \leq r\}.$$

Consider the problem of recovering signal $\theta \in \Theta$ via observation $\omega \sim \mathcal{N}(\theta, \sigma^2 I_k)$. Let

$$\mathrm{Risk}_{\|\cdot\|}[\widehat{\theta}|\Theta] = \sup_{\theta\in\Theta}\left(\mathbf{E}_{\omega\sim\mathcal{N}(\theta,\sigma^2 I)}\left\{\|\widehat{\theta}(\omega) - \theta\|^2\right\}\right)^{1/2}$$

be the $\|\cdot\|$-risk of an estimate $\widehat{\theta}(\cdot)$, and let

$$\mathrm{Risk}_{\mathrm{opt},\|\cdot\|}[\Theta] = \inf_{\widehat{\theta}(\cdot)} \mathrm{Risk}_{\|\cdot\|}[\widehat{\theta}|\Theta]$$

be the associated minimax risk.

Assume that the norm $\|\cdot\|$ is absolute and symmetric w.r.t. permutations of the coordinates. Then

$$\mathrm{Risk}_{\mathrm{opt},\|\cdot\|}[\Theta] \geq \frac{r\sigma k}{2\sqrt{\ln(ek)}r\alpha_* + \sigma k}, \quad \alpha_* = \|[1;...;1]\|_*. \tag{4.125}$$

Here is the concluding part of the exercise:

2) Prove the observation and compare the lower risk bound it yields with the $\|\cdot\|$-risk of the "plug-in" estimate $\widehat{\chi}(\omega) \equiv \omega$.

Example II: Gaussian case, estimating covariance. Let $\mu = 2$, let K be a positive integer, and let our observation ω be a collection of K i.i.d. samples $\omega_t \sim \mathcal{N}(0, \theta)$, $1 \leq t \leq K$, with unknown θ known to belong to a given convex compact subset Θ of the interior of the positive semidefinite cone $\mathbf{S}^n_+$. Given ω_1,...,ω_K, we want to recover θ in the Shatten norm $\|\cdot\|_{\mathrm{Sh},s}$ with $s \in [1, \infty]$. Our estimation problem is covered by the setup of Exercise 4.22 with $\mathcal{P}$ comprised of the product probability densities $p(\omega, \theta) = \prod_{t=1}^K g(\omega_t, \theta)$, $\theta \in \Theta$, where $g(\cdot, \theta)$ is the density of $\mathcal{N}(0, \theta)$. We have

$$\begin{array}{rcl} \nabla_\theta \ln(p(\omega,\theta)) & = & \frac{1}{2}\sum_t \nabla_\theta \ln(g(\omega_t,\theta)) = \frac{1}{2}\sum_t \left[\theta^{-1}\omega_t\omega_t^T\theta^{-1} - \theta^{-1}\right] \\ & = & \frac{1}{2}\theta^{-1/2}\left[\sum_t \left[[\theta^{-1/2}\omega_t][\theta^{-1/2}\omega_t]^T - I_n\right]\right]\theta^{-1/2}. \end{array} \tag{4.126}$$

With some effort [149] it can be proved that when

$$K \geq n,$$

which we assume from now on, for random vectors $\xi_1, ..., \xi_K$ independent across t sampled from the standard Gaussian distribution $\mathcal{N}(0, I_n)$ for every $u \in [1, \infty]$ one has

$$\left[\mathbf{E}\left\{\left\|\sum_{t=1}^{K}[\xi_t\xi_t^T - I_n]\right\|_{\mathrm{Sh},u}^2\right\}\right]^{1/2} \leq Cn^{\frac{1}{2}+\frac{1}{u}}\sqrt{K} \tag{4.127}$$

with appropriate *absolute constant* C. Consequently, for $\theta \in \Theta$ and all $u \in [1,\infty]$ we have

$$\begin{array}{rl}
\mathbf{E}_{\omega\sim p(\cdot,\theta)} & \left\{\|\nabla_\theta \ln(p(\omega,\theta))\|_{\mathrm{Sh},u}^2\right\} \\
= & \frac{1}{4}\mathbf{E}_{\omega\sim p(\cdot,\theta)}\left\{\|\theta^{-1/2}\left[\sum_t\left[[\theta^{-1/2}\omega_t][\theta^{-1/2}\omega_t]^T - I_n\right]\right]\theta^{-1/2}\|_{\mathrm{Sh},u}^2\right\} \\
& \hfill [\text{by } (4.126)] \\
= & \frac{1}{4}\mathbf{E}_{\xi\sim p(\cdot,I_n)}\left\{\|\theta^{-1/2}\left[\sum_t\left[\xi_t\xi_t^T - I_n\right]\right]\theta^{-1/2}\|_{\mathrm{Sh},u}^2\right\} \quad [\text{setting } \theta^{-1/2}\omega_t = \xi_t] \\
\leq & \frac{1}{4}\|\theta^{-1/2}\|_{\mathrm{Sh},\infty}^4\mathbf{E}_{\xi\sim p(\cdot,I_n)}\left\{\|\sum_t\left[\xi_t\xi_t^T - I_n\right]\|_{\mathrm{Sh},u}^2\right\} \\
& \hfill [\text{since } \|AB\|_{\mathrm{Sh},u} \leq \|A\|_{\mathrm{Sh},\infty}\|B\|_{\mathrm{Sh},u}] \\
\leq & \frac{1}{4}\|\theta^{-1/2}\|_{\mathrm{Sh},\infty}^4\left[Cn^{\frac{1}{2}+\frac{1}{u}}\sqrt{K}\right]^2 \quad [\text{by } (4.127)]
\end{array}$$

and we arrive at

$$\left[\mathbf{E}_{\omega\sim p(\cdot,\theta)}\left\{\|\nabla_\theta \ln(p(\omega,\theta))\|_{\mathrm{Sh},u}^2\right\}\right]^{1/2} \leq \frac{C}{2}\|\theta^{-1}\|_{\mathrm{Sh},\infty}n^{\frac{1}{2}+\frac{1}{u}}\sqrt{K}. \tag{4.128}$$

Now assume that Θ is $\|\cdot\|_{\mathrm{Sh},s}$-ball of radius $r < 1$ centered at I_n:

$$\Theta = \{\theta \in \mathbf{S}^n : \|\theta - I_n\|_{\mathrm{Sh},s} \leq r\}. \tag{4.129}$$

In this case the estimation problem from Example II is the scope of Proposition 4.41, and the quantity $I_{\|\cdot\|_*,2}$ as defined in (4.123) can be upper-bounded as follows:

$$\begin{array}{rcl}
I_{\|\cdot\|_*,2} & = & \max\limits_{\theta\in\Theta}\left[\mathbf{E}_{\omega\sim p(\cdot,\theta)}\left\{\|\nabla_\theta \ln(p(\omega,\theta))\|_{\mathrm{Sh},s_*}^2\right\}\right]^{1/2} \\
& \leq & O(1)n^{\frac{1}{2}+\frac{1}{s_*}}\sqrt{K}\max_{\theta\in\Theta}\|\theta^{-1}\|_{\mathrm{Sh},\infty} \quad [\text{see } (4.128)] \\
& \leq & O(1)\frac{n^{\frac{1}{2}+\frac{1}{s_*}}\sqrt{K}}{1-r}.
\end{array}$$

We can now use Proposition 4.41 to lower-bound the minimax $\|\cdot\|_{\mathrm{Sh},s}$-risk, thus arriving at

$$\mathrm{Risk}_{\mathrm{opt},\|\cdot\|_{\mathrm{Sh},s}}[\Theta] \geq O(1)\frac{n(1-r)r}{\sqrt{K}n^{\frac{1}{2}-\frac{1}{s}}r + n(1-r)} \tag{4.130}$$

(note that we are in the case of $k = \dim\theta = \frac{n(n+1)}{2}$).

Let us compare this lower risk bound with the $\|\cdot\|_{\mathrm{Sh},s}$-risk of the "plug-in" estimate

$$\widehat{\theta}(\omega) = \frac{1}{K}\sum_{t=1}^{K}\omega_t\omega_t^T.$$

Assuming $\theta \in \Theta$, we have

$$\begin{array}{l}
\mathbf{E}_{\omega\sim p(\cdot,\theta)}\left\{\|K[\widehat{\theta}(\omega)-\theta]\|_{\mathrm{Sh},s}^2\right\} = \mathbf{E}_{\omega\sim p(\cdot,\theta)}\left\{\|\sum_t[\omega_t\omega_t^T-\theta]\|_{\mathrm{Sh},s}^2\right\}\\
= \mathbf{E}_{\omega\sim p(\cdot,\theta)}\left\{\|\theta^{1/2}\left[\sum_t[[\theta^{-1/2}\omega_t][\theta^{-1/2}\omega_t]^T-I_n]\right]\theta^{1/2}\|_{\mathrm{Sh},s}^2\right\}\\
= \mathbf{E}_{\xi\sim p(\cdot,I_n)}\left\{\|\theta^{1/2}\left[\sum_t[\xi_t\xi_t^T-I_n]\right]\theta^{1/2}\|_{\mathrm{Sh},s}^2\right\}\\
\leq \|\theta^{1/2}\|_{\mathrm{Sh},\infty}^4\mathbf{E}_{\xi\sim p(\cdot,I_n)}\left\{\|\sum_t[\xi_t\xi_t^T-I_n]\|_{\mathrm{Sh},s}^2\right\}\\
\leq \|\theta^{1/2}\|_{\mathrm{Sh},\infty}^4\left[Cn^{\frac{1}{2}+\frac{1}{s}}\sqrt{K}\right]^2, \qquad \text{[see (4.127)]}
\end{array}$$

and we arrive at

$$\mathrm{Risk}_{\|\cdot\|_{\mathrm{Sh},s}}[\widehat{\theta}|\Theta] \leq O(1)\max_{\theta\in\Theta}\|\theta\|_{\mathrm{Sh},\infty}\frac{n^{\frac{1}{2}+\frac{1}{s}}}{\sqrt{K}}.$$

In the case of (4.129), the latter bound becomes

$$\mathrm{Risk}_{\|\cdot\|_{\mathrm{Sh},s}}[\widehat{\theta}|\Theta] \leq O(1)\max_{\theta\in\Theta}\|\theta\|_{\mathrm{Sh},\infty}\frac{n^{\frac{1}{2}+\frac{1}{s}}}{\sqrt{K}}. \tag{4.131}$$

For the sake of simplicity, assume that r in (4.129) is 1/2 (what actually matters below is that $r \in (0,1)$ is bounded away from 0 and from 1). In this case the lower bound (4.130) on the minimax $\|\cdot\|_{\mathrm{Sh},s}$-risk reads

$$\mathrm{Risk}_{\mathrm{opt},\|\cdot\|_{\mathrm{Sh},s}}[\Theta] \geq O(1)\min\left[\frac{n^{\frac{1}{2}+\frac{1}{s}}}{\sqrt{K}}, 1\right].$$

When K is "large": $K \geq n^{1+\frac{2}{s}}$, this lower bound matches, within an absolute constant factor, the upper bound (4.131) on the risk of the plug-in estimate, so that the latter estimate is near-optimal. When $K < n^{1+\frac{2}{s}}$, the lower risk bound becomes $O(1)$, so that here a nearly optimal estimate is the trivial estimate $\widehat{\theta}(\omega) \equiv I_n$.

4.7.7 Around $\mathcal{S}$-Lemma

$\mathcal{S}$-Lemma is a classical result of extreme importance in Semidefinite Optimization. Basically, the lemma states that when the ellitope $\mathcal{X}$ in Proposition 4.6 is an ellipsoid, (4.19) can be strengthened to $\mathrm{Opt} = \mathrm{Opt}_*$. In fact, $\mathcal{S}$-Lemma is even stronger:

Lemma 4.43. *[$\mathcal{S}$-Lemma] Consider two quadratic forms $f(x) = x^TAx + 2a^Tx + \alpha$ and $g(x) = x^TBx + 2b^Tx + \beta$ such that $g(\bar{x}) < 0$ for some $\bar{x}$. Then the implication*

$$g(x) \leq 0 \Rightarrow f(x) \leq 0$$

takes place if and only if for some $\lambda \geq 0$ it holds $f(x) \leq \lambda g(x)$ for all x, or, which is the same, if and only if Linear Matrix Inequality

$$\left[\begin{array}{c|c}\lambda B - A & \lambda b - a\\ \hline \lambda b^T - a^T & \lambda\beta - \alpha\end{array}\right] \succeq 0$$

in scalar variable λ has a nonnegative solution.

Proof of $\mathcal{S}$-Lemma can be found, e.g., in [15, Section 3.5.2].

The goal of subsequent exercises is to get "tight" tractable outer approximations of sets obtained from ellitopes by quadratic lifting. We fix an ellitope

$$X = \{x \in \mathbf{R}^n : \exists t \in \mathcal{T} : x^T S_k x \leq t_k, 1 \leq k \leq K\} \tag{4.132}$$

where, as always, S_k are positive semidefinite matrices with positive definite sum, and $\mathcal{T}$ is a computationally tractable convex compact subset in $\mathbf{R}^k_+$ such that $t \in \mathcal{T}$ implies $t' \in \mathcal{T}$ whenever $0 \leq t' \leq t$ and $\mathcal{T}$ contains a positive vector.

Exercise 4.25.

Let us associate with ellitope X given by (4.132) the sets

$$\begin{array}{rcl} \mathcal{X} & = & \mathrm{Conv}\{xx^T : x \in X\}, \\ \widehat{\mathcal{X}} & = & \{Y \in \mathbf{S}^n : Y \succeq 0, \exists t \in \mathcal{T} : \mathrm{Tr}(S_k Y) \leq t_k, 1 \leq k \leq K\}, \end{array}$$

so that $\mathcal{X}$, $\widehat{\mathcal{X}}$ are convex compact sets containing the origin, and $\widehat{\mathcal{X}}$ is computationally tractable along with $\mathcal{T}$. Prove that

1. When $K = 1$, we have $\mathcal{X} = \widehat{\mathcal{X}}$;
2. We always have $\mathcal{X} \subset \widehat{\mathcal{X}} \subset 3\ln(\sqrt{3}K)\mathcal{X}$.

Exercise 4.26.

For $x \in \mathbf{R}^n$ let $Z(x) = [x; 1][x; 1]^T$, $Z^o[x] = \left[\begin{array}{c|c} xx^T & x \\ \hline x^T & \end{array}\right]$. Let

$$C = \left[\begin{array}{c|c} & \\ \hline & 1 \end{array}\right],$$

and let us associate with ellitope X given by (4.132) the sets

$$\begin{array}{rcl} \mathcal{X}^+ & = & \mathrm{Conv}\{Z^o[x] : x \in X\}, \\ \widehat{\mathcal{X}}^+ & = & \left\{Y = \left[\begin{array}{c|c} U & u \\ \hline u^T & \end{array}\right] \in \mathbf{S}^{n+1} : Y + C \succeq 0, \exists t \in \mathcal{T} : \mathrm{Tr}(S_k U) \leq t_k, 1 \leq k \leq K\right\}, \end{array}$$

so that $\mathcal{X}^+$, $\widehat{\mathcal{X}}^+$ are convex compact sets containing the origin, and $\widehat{\mathcal{X}}^+$ is computationally tractable along with $\mathcal{T}$. Prove that

1. When $K = 1$, we have $\mathcal{X}^+ = \widehat{\mathcal{X}}^+$;
2. We always have $\mathcal{X}^+ \subset \widehat{\mathcal{X}}^+ \subset 3\ln(\sqrt{3}(K+1))\mathcal{X}^+$.

4.7.8 Miscellaneous exercises

Exercise 4.27.

Let $X \subset \mathbf{R}^n$ be a convex compact set, let $b \in \mathbf{R}^n$, and let A be an $m \times n$ matrix. Consider the problem of affine recovery $\omega \mapsto h^T\omega + c$ of the linear function $Bx = b^T x$ of $x \in X$ from indirect observation

$$\omega = Ax + \sigma\xi, \ \xi \sim \mathcal{N}(0, I_m).$$

Given tolerance $\epsilon \in (0,1)$, we are interested in minimizing the worst-case, over $x \in X$, width of $(1-\epsilon)$ confidence interval, that is, the smallest ρ such that

$$\text{Prob}\{\xi : b^T x - f^T(Ax+\sigma\xi) > \rho\} \le \epsilon/2 \ \& \ \text{Prob}\{\xi : b^T x - f^T(Ax+\sigma\xi) < \rho\} \le \epsilon/2 \ \forall x \in X.$$

Pose the problem as a convex optimization problem and consider in detail the case where X is the box $\{x \in \mathbf{R}^n : a_j|x_j| \le 1,\, 1 \le j \le n\}$, where $a_j > 0$ for all j.

Exercise 4.28.

Prove Proposition 4.21.

Exercise 4.29.

Prove Proposition 4.22.

4.8 PROOFS

4.8.1 Preliminaries

4.8.1.1 Technical lemma

Lemma 4.44. *Given basic spectratope*

$$\mathcal{X} = \{x \in \mathbf{R}^n : \exists t \in \mathcal{T} : R_k^2[x] \preceq t_k I_{d_k}, 1 \le k \le K\} \tag{4.133}$$

and a positive definite $n \times n$ matrix Q and setting $\Lambda_k = \mathcal{R}_k[Q]$ (for notation, see Section 4.3.1), we get a collection of positive semidefinite matrices, and $\sum_k \mathcal{R}_k^*[\Lambda_k]$ is positive definite.

As a corollaries,

(i) whenever M_k, $k \le K$, are positive definite matrices, the matrix $\sum_k \mathcal{R}_k^*[M_k]$ is positive definite;

(ii) the set $\mathcal{Q}_T = \{Q \succeq 0 : \mathcal{R}_k[Q] \preceq T I_{d_k}, k \le K\}$ is bounded for every T.

Proof. Let us prove the first claim. Assuming the opposite, we would be able to find a nonzero vector y such that $\sum_k y^T \mathcal{R}_k^*[\Lambda_k] y \le 0$, whence

$$0 \ge \sum_k y^T \mathcal{R}_k^*[\Lambda_k] y = \sum_k \text{Tr}(\mathcal{R}_k^*[\Lambda_k][yy^T]) = \sum_k \text{Tr}(\Lambda_k \mathcal{R}_k[yy^T])$$

(we have used (4.26), (4.22)). Since $\Lambda_k = \mathcal{R}_k[Q] \succeq 0$ due to $Q \succeq 0$—see (4.23)—it follows that $\text{Tr}(\Lambda_k \mathcal{R}_k[yy^T]) = 0$ for all k. Now, the linear mapping $\mathcal{R}_k[\cdot]$ is $\succeq$-monotone, and Q is positive definite, implying that $Q \succeq r_k yy^T$ for some $r_k > 0$, whence $\Lambda_k \succeq r_k \mathcal{R}_k[yy^T]$, and therefore $\text{Tr}(\Lambda_k \mathcal{R}_k[yy^T]) = 0$ implies that $\text{Tr}(\mathcal{R}_k^2[yy^T]) = 0$, that is, $\mathcal{R}_k[yy^T] = R_k^2[y] = 0$. Since $R_k[\cdot]$ takes values in $\mathbf{S}^{d_k}$, we get $R_k[y] = 0$ for al k, which is impossible due to $y \ne 0$ and property S.3; see Section 4.3.1.

To verify (i), note that when M_k are positive definite, we can find $\gamma > 0$ such that $\Lambda_k \preceq \gamma M_k$ for all $k \le K$; invoking (4.27), we conclude that $\mathcal{R}_k^*[\Lambda_k] \preceq \gamma \mathcal{R}_k^*[M_k]$, whence $\sum_k \mathcal{R}_k^*[M_k]$ is positive definite along with $\sum_k \mathcal{R}_k^*[\Lambda_k]$.

To verify (ii), assume, on the contrary to what should be proved, that $\mathcal{Q}_T$ is unbounded. Since $\mathcal{Q}_T$ is closed and convex, it must possess a nonzero recessive direction, that is, there should exist nonzero positive semidefinite matrix D such that $\mathcal{R}_k[D] \preceq 0$ for all k. Selecting positive definite matrices M_k, the matrices $\mathcal{R}_k^*[M_k]$ are positive semidefinite (see Section 4.3.1), and their sum S is positive definite by (i). We have

$$0 \geq \sum_k \mathrm{Tr}(\mathcal{R}_k[D]M_k) = \sum_k \mathrm{Tr}(D\mathcal{R}_k^*[M_k]) = \mathrm{Tr}(DS),$$

where the first inequality is due to $M_k \succeq 0$, and the first equality is due to (4.26). The resulting inequality is impossible due to $0 \neq D \succeq 0$ and $S \succ 0$, which is the desired contradiction. □

4.8.1.2 Noncommutative Khintchine Inequality

We will use a deep result from Functional Analysis ("Noncommutative Khintchine Inequality") due to Lust-Piquard [175], Pisier [199] and Buchholz [34]; see [228, Theorem 4.6.1]:

Theorem 4.45. *Let $Q_i \in \mathbf{S}^n$, $1 \leq i \leq I$, and let ξ_i, $i = 1, ..., I$, be independent Rademacher (± 1 with probabilities $1/2$) or $\mathcal{N}(0,1)$ random variables. Then for all $t \geq 0$ one has*

$$\mathrm{Prob}\left\{\left\|\sum_{i=1}^I \xi_i Q_i\right\| \geq t\right\} \leq 2n\exp\left\{-\frac{t^2}{2v_Q}\right\}$$

where $\|\cdot\|$ is the spectral norm, and $v_Q = \left\|\sum_{i=1}^I Q_i^2\right\|$.

We need the following immediate consequence of the theorem:

Lemma 4.46. *Given spectratope (4.20), let $Q \in \mathbf{S}^n_+$ be such that*

$$\mathcal{R}_k[Q] \preceq \rho t_k I_{d_k},\ 1 \leq k \leq K, \tag{4.134}$$

for some $t \in \mathcal{T}$ and some $\rho \in (0,1]$. Then

$$\mathrm{Prob}_{\xi\sim\mathcal{N}(0,Q)}\{\xi \notin \mathcal{X}\} \leq \min\left[2D\mathrm{e}^{-\frac{1}{2\rho}}, 1\right],\ D := \sum_{k=1}^K d_k.$$

Proof. When setting $\xi = Q^{1/2}\eta$, $\eta \sim \mathcal{N}(0, I_n)$, we have

$$R_k[\xi] = R_k[Q^{1/2}\eta] =: \sum_{i=1}^n \eta_i \bar{R}^{ki} = \bar{R}_k[\eta]$$

with

$$\sum_i [\bar{R}^{ki}]^2 = \mathbf{E}_{\eta\sim\mathcal{N}(0,I_n)}\left\{\bar{R}_k^2[\eta]\right\} = \mathbf{E}_{\xi\sim\mathcal{N}(0,Q)}\left\{R_k^2[\xi]\right\} = \mathcal{R}_k[Q] \preceq \rho t_k I_{d_k}$$

due to (4.24). Hence, by Theorem 4.45

$$\mathrm{Prob}_{\xi\sim\mathcal{N}(0,Q)}\{\|R_k[\xi]\|^2 \geq t_k\} = \mathrm{Prob}_{\eta\sim\mathcal{N}(0,I_n)}\{\|\bar{R}_k[\eta]\|^2 \geq t_k\} \leq 2d_k\mathrm{e}^{-\frac{1}{2\rho}}.$$

We conclude that

$$\mathrm{Prob}_{\xi\sim\mathcal{N}(0,Q)}\{\xi \notin \mathcal{X}\} \leq \mathrm{Prob}_{\xi\sim\mathcal{N}(0,Q)}\{\exists k : \|R_k[\xi]\|^2 > t_k\} \leq 2D\mathrm{e}^{-\frac{1}{2\rho}}. \qquad \square$$

The ellitopic version of Lemma 4.46 is as follows:

Lemma 4.47. *Given ellitope (4.9), let $Q \in \mathbf{S}^n_+$ be such that*

$$\mathrm{Tr}(R_k Q) \leq \rho t_k,\ 1 \leq k \leq K, \tag{4.135}$$

for some $t \in \mathcal{T}$ and some $\rho \in (0,1]$. Then

$$\mathrm{Prob}_{\xi\sim\mathcal{N}(0,Q)}\{\xi \notin \mathcal{X}\} \leq 2K \exp\left\{-\frac{1}{3\rho}\right\}.$$

Proof. Observe that if $P \in \mathbf{S}^n_+$ satisfies $\mathrm{Tr}(R) \leq 1$, we have

$$\mathbf{E}_{\eta\sim\mathcal{N}(0,I_n)}\left\{\exp\left\{\tfrac{1}{3}\eta^T P\eta\right\}\right\} \leq \sqrt{3}. \tag{4.136}$$

Indeed, we lose nothing when assuming that $P = \mathrm{Diag}\{\lambda_1, ..., \lambda_n\}$ with $\lambda_i \geq 0$, $\sum_i \lambda_i \leq 1$. In this case

$$\mathbf{E}_{\eta\sim\mathcal{N}(0,I_n)}\left\{\exp\{\tfrac{1}{3}\eta^T P\eta\}\right\} = f(\lambda) := \mathbf{E}_{\eta\sim\mathcal{N}(0,I_n)}\left\{\exp\{\tfrac{1}{3}\sum_i \lambda_i \eta_i^2\}\right\}.$$

Function f is convex, so that its maximum on the simplex $\{\lambda \geq 0 : \sum_i \lambda_i \leq 1\}$ is achieved at a vertex, that is,

$$f(\lambda) \leq \mathbf{E}_{\eta\sim\mathcal{N}(0,1)}\left\{\exp\{\tfrac{1}{3}\eta^2\}\right\} = \sqrt{3};$$

(4.136) is proved. Note that (4.136) implies that

$$\mathrm{Prob}_{\eta\sim\mathcal{N}(0,I_n)}\left\{\eta : \eta^T P\eta > s\right\} < \sqrt{3}\exp\{-s/3\},\ s \geq 0. \tag{4.137}$$

Now let Q and t satisfy the Lemma's premise. Setting $\xi = Q^{1/2}\eta$, $\eta \sim \mathcal{N}(0, I_n)$, for $k \leq K$ such that $t_k > 0$ we have

$$\xi^T R_k \xi = \rho t_k \eta^T P_k \eta,\ P_k := [\rho t_k]^{-1} Q^{1/2} R_k Q^{1/2} \succeq 0\ \&\ \mathrm{Tr}(P_k) = [\rho t_k]^{-1}\mathrm{Tr}(QR_k) \leq 1,$$

so that

$$\begin{array}{rcl}\mathrm{Prob}_{\xi\sim\mathcal{N}(0,Q)}\left\{\xi : \xi^T R_k \xi > s\rho t_k\right\} & = & \mathrm{Prob}_{\eta\sim\mathcal{N}(0,I_n)}\left\{\eta^T P_k \eta > s\right\} \\ & < & \sqrt{3}\exp\{-s/3\},\end{array} \tag{4.138}$$

where the inequality is due to (4.137). Relation (4.138) was established for k with $t_k > 0$; it is trivially true when $t_k = 0$, since in this case $Q^{1/2}R_kQ^{1/2} = 0$ due to $\mathrm{Tr}(QR_k) \leq 0$ and $R_k, Q \in \mathbf{S}^n_+$. Setting $s = 1/\rho$, we get from (4.138) that

$$\mathrm{Prob}_{x\sim\mathcal{N}(0,Q)}\left\{\xi^T R_k \xi > t_k\right\} \leq \sqrt{3}\exp\{-\frac{1}{3\rho}\},\ k \leq K,$$

and (4.137) follows due to the union bound. $\square$

4.8.1.3 Anderson's Lemma

Below we use a simple-looking, but by far nontrivial, fact.

Anderson's Lemma [4]. *Let f be a nonnegative even ($f(x) \equiv f(-x)$) summable function on $\mathbf{R}^N$ such that the level sets $\{x : f(x) \geq t\}$ are convex for all t and let $X \subset \mathbf{R}^N$ be a closed convex set symmetric w.r.t. the origin. Then for every $y \in \mathbf{R}^N$*

$$\int_{X+ty} f(z)dz$$

is a nonincreasing function of $t \geq 0$. In particular, if ζ is a zero mean N-dimensional Gaussian random vector, then for every $y \in \mathbf{R}^N$

$$\mathrm{Prob}\{\zeta \notin y + X\} \geq \mathrm{Prob}\{\zeta \notin X\}.$$

Hence, for every norm $\|\cdot\|$ on $\mathbf{R}^N$ it holds

$$\mathrm{Prob}\{\zeta : \|\zeta - y\| > \rho\} \geq \mathrm{Prob}\{\zeta : \|\zeta\| > \rho\}\ \forall (y \in \mathbf{R}^N, \rho \geq 0).$$

4.8.2 Proof of Proposition 4.6

1°. We need the following:

Lemma 4.48. *Let S be a positive semidefinite $N \times N$ matrix with trace ≤ 1 and ξ be an N-dimensional Rademacher random vector (i.e., the entries in ξ are independent and take values ± 1 with probabilities $1/2$). Then*

$$\mathbf{E}\left\{\exp\left\{\tfrac{1}{3}\zeta^T S\zeta\right\}\right\} \leq \sqrt{3},$$

implying that

$$\mathrm{Prob}\{\xi^T S\xi > s\} \leq \sqrt{3}\exp\{-s/3\},\ s \geq 0.$$

Proof. Let $S = \sum_i \sigma_i h^i [h^i]^T$ be the eigenvalue decomposition of S, so that $[h^i]^T h^i = 1$, $\sigma_i \geq 0$, and $\sum_i \sigma_i \leq 1$. The function

$$F(\sigma_1, ..., \sigma_{\bar{n}}) = \mathbf{E}\left\{\mathrm{e}^{\frac{1}{3}\sum_i \sigma_i \xi^T h^i [h^i]^T \xi}\right\}$$

is convex on the simplex $\{\sigma \geq 0, \sum_i \sigma_i \leq 1\}$ and thus attains its maximum over the simplex at a vertex, implying that for some $f = h^i$, $f^T f = 1$, it holds

$$\mathbf{E}\{\mathrm{e}^{\frac{1}{3}\xi^T S\xi}\} \leq \mathbf{E}\{\mathrm{e}^{\frac{1}{3}(f^T\xi)^2}\}.$$

Let $\zeta \sim \mathcal{N}(0,1)$ be independent of ξ. We have

$$\begin{array}{l}
\mathbf{E}_\xi\left\{\exp\{\frac{1}{3}(f^T\xi)^2\}\right\} = \mathbf{E}_\xi\left\{\mathbf{E}_\zeta\left\{\exp\{[\sqrt{2/3}f^T\xi]\zeta\}\right\}\right\}\\
= \mathbf{E}_\zeta\left\{\mathbf{E}_\xi\left\{\exp\{[\sqrt{2/3}f^T\xi]\zeta\}\right\}\right\} = \mathbf{E}_\zeta\left\{\prod_{j=1}^N \mathbf{E}_\xi\left\{\exp\{\sqrt{2/3}\zeta f_j\xi_j\}\right\}\right\}\\
= \mathbf{E}_\zeta\left\{\prod_{j=1}^N \cosh(\sqrt{2/3}\zeta f_j)\right\} \leq \mathbf{E}_\zeta\left\{\prod_{j=1}^N \exp\{\frac{1}{3}\zeta^2 f_j^2\}\right\}\\
= \mathbf{E}_\zeta\left\{\exp\{\frac{1}{3}\zeta^2\}\right\} = \sqrt{3}.
\end{array}$$

□

2°. The right inequality in (4.19) has been justified in Section 4.2.3. To prove the left inequality in (4.19), let $\mathbf{T}$ be the closed conic hull of $\mathcal{T}$ (see Section 4.1.1), and let us consider the conic problem

$$\mathrm{Opt}_\# = \max_{Q,t}\left\{\mathrm{Tr}(P^TCPQ) : Q \succeq 0, \mathrm{Tr}(QR_k) \leq t_k\ \forall k \leq K, [t;1] \in \mathbf{T}\right\}. \quad (4.139)$$

We claim that

$$\mathrm{Opt} = \mathrm{Opt}_\#. \quad (4.140)$$

Indeed, (4.139) clearly is a strictly feasible and bounded conic problem, so that its optimal value is equal to the optimal value of its conic dual (Conic Duality Theorem). Taking into account that the cone $\mathbf{T}_*$ dual to $\mathbf{T}$ is $\{[g;s] : s \geq \phi_\mathcal{T}(-g)\}$—see Section 4.1.1—we therefore get

$$\begin{array}{rcl}
\mathrm{Opt}_\# &=& \min\limits_{\lambda,[g;s],L}\left\{s : \begin{array}{l}\mathrm{Tr}([\sum_k \lambda_k R_k - L]Q) - \sum_k[\lambda_k + g_k]t_k = \mathrm{Tr}(P^TCPQ)\ \forall (Q,t),\\ \lambda \geq 0, L \succeq 0, s \geq \phi_\mathcal{T}(-g)\end{array}\right\}\\
&=& \min\limits_{\lambda,[g;s],L}\left\{s : \begin{array}{l}\sum_k \lambda_k R_k - L = P^TCP,\ g = -\lambda,\\ \lambda \geq 0, L \succeq 0, s \geq \phi_\mathcal{T}(-g)\end{array}\right\}\\
&=& \min\limits_\lambda\left\{\phi_\mathcal{T}(\lambda) : \sum_k \lambda_k R_k \succeq P^TCP, \lambda \geq 0\right\} = \mathrm{Opt},
\end{array}$$

as claimed.

3°. With Lemma 4.48 and (4.140) at our disposal, we can now complete the proof of Proposition 4.6 by adjusting the technique from [191]. Specifically, problem (4.139) clearly is solvable; let Q_*, t^* be an optimal solution to the problem. Next, let us set $R_* = Q_*^{1/2}$, $\bar{C} = R_*P^TCPR_*$, let $\bar{C} = UDU^T$ be the eigenvalue decomposition of $\bar{C}$, and let $\bar{R}_k = U^TR_*R_kR_*U$. Observe that

$$\begin{array}{rcl}
\mathrm{Tr}(\bar{C}) &=& \mathrm{Tr}(R_*P^TCPR_*) = \mathrm{Tr}(Q_*P^TCP) = \mathrm{Opt}_\# = \mathrm{Opt},\\
\mathrm{Tr}(\bar{R}_k) &=& \mathrm{Tr}(R_*R_kR_*) = \mathrm{Tr}(Q_*R_k) \leq t_k^*.
\end{array}$$

Now let ξ be a Rademacher random vector. For k with $t_k^* > 0$, applying Lemma 4.48 to matrices $\bar{R}_k/t_k^*$, we get for $s > 0$

$$\mathrm{Prob}\{\xi^T\bar{R}_k\xi > st_k^*\} \leq \sqrt{3}\exp\{-s/3\}; \quad (4.141)$$

if k is such that $t_k^* = 0$, we have $\mathrm{Tr}(\bar{R}_k) = 0$, that is, $\bar{R}_k = 0$ (since $\bar{R}_k \succeq 0$), and (4.141) holds true as well. Now let

$$s_* = 3\ln(\sqrt{3}K),$$

so that $\sqrt{3}\exp\{-s/3\} < 1/K$ when $s > s_*$. The latter relation combines with (4.141) to imply that for every $s > s_*$ there exists a realization $\bar{\xi}$ of ξ such that

$$\bar{\xi}^T \bar{R}_k \bar{\xi} \leq s t_k^* \,\forall k.$$

Let us set $\bar{y} = \frac{1}{\sqrt{s}} R_* U \bar{\xi}$. Then

$$\bar{y}^T R_k \bar{y} = s^{-1} \bar{\xi}^T U^T R_* R_k R_* U \bar{\xi} = s^{-1} \bar{\xi}^T \bar{R}_k \bar{\xi} \leq t_k^* \;\; \forall k$$

implying that $\bar{x} := P\bar{y} \in \mathcal{X}$, and

$$\bar{x}^T C \bar{x} = s^{-1} \bar{\xi}^T U^T \underbrace{R_* P^T C P R_*}_{\bar{C}} U \bar{\xi} = s^{-1} \bar{\xi}^T D \bar{\xi} = s^{-1} \mathrm{Tr}(D) = s^{-1} \mathrm{Tr}(\bar{C}) = s^{-1} \mathrm{Opt}.$$

Thus, $\mathrm{Opt}_* := \max_{x \in \mathcal{X}} x^T C x \geq s^{-1}\mathrm{Opt}$ whenever $s > s_*$, which implies the left inequality in (4.19). □

4.8.3 Proof of Proposition 4.8

The proof follows the lines of the proof of Proposition 4.6. First, passing from C to the matrix $\bar{C} = P^T C P$, the situation clearly reduces to the one where $P = I$. To save notation, in the rest of the proof we assume that P is the identity.

Second, from Lemma 4.44 and the fact that the level sets of $\phi_{\mathcal{T}}(\cdot)$ on the nonnegative orthant are bounded (since $\mathcal{T}$ contains a positive vector) it immediately follows that problem (4.29) is feasible with bounded level sets of the objective, so that the problem is solvable. The left inequality in (4.30) was proved in Section 4.3.2. Thus, all we need is to prove the right inequality in (4.30).

1°. Let $\mathbf{T}$ be the closed conic hull of $\mathcal{T}$ (see Section 4.1.1). Consider the conic problem

$$\mathrm{Opt}_\# = \max_{Q,t} \left\{ \mathrm{Tr}(CQ) : Q \succeq 0, \mathcal{R}_k[Q] \preceq t_k I_{d_k} \,\forall k \leq K, [t;1] \in \mathbf{T} \right\}. \tag{4.142}$$

This problem clearly is strictly feasible; by Lemma 4.44, the feasible set of the problem is bounded, so the problem is solvable. We claim that

$$\mathrm{Opt}_\# = \mathrm{Opt}.$$

Indeed, (4.142) is a strictly feasible and bounded conic problem, so that its optimal value is equal to the one in its conic dual, that is,

$$\begin{array}{rcl}
\mathrm{Opt}_\# & = & \min\limits_{\Lambda=\{\Lambda_k\}_{k\leq K},[g;s],L} \left\{ s : \begin{array}{r} \mathrm{Tr}([\sum_k \mathcal{R}_k^*[\Lambda_k] - L]Q) - \sum_k [\mathrm{Tr}(\Lambda_k) + g_k] t_k \\ = \mathrm{Tr}(CQ) \;\forall (Q,t), \\ \Lambda_k \succeq 0 \,\forall k, L \succeq 0, s \geq \phi_{\mathcal{T}}(-g) \end{array} \right\} \\
& = & \min\limits_{\Lambda,[g;s],L} \left\{ s : \begin{array}{l} \sum_k \mathcal{R}_k^*[\Lambda_k] - L = C, g = -\lambda[\Lambda], \\ \Lambda_k \succeq 0 \,\forall k, L \succeq 0, s \geq \phi_{\mathcal{T}}(-g) \end{array} \right\} \\
& = & \min\limits_{\Lambda} \left\{ \phi_{\mathcal{T}}(\lambda[\Lambda]) : \sum_k \mathcal{R}_k^*[\Lambda_k] \succeq C, \Lambda_k \succeq 0 \,\forall k \right\} = \mathrm{Opt},
\end{array}$$

as claimed.

2°. Problem (4.142), as we already know, is solvable; let Q_*, t^* be an optimal solution to the problem. Next, let us set $R_* = Q_*^{1/2}$, $\widehat{C} = R_*CR_*$, and let $\widehat{C} = UDU^T$ be the eigenvalue decomposition of $\widehat{C}$, so that the matrix $D = U^TR_*CR_*U$ is diagonal, and the trace of this matrix is $\mathrm{Tr}(R_*CR_*) = \mathrm{Tr}(CQ_*) = \mathrm{Opt}_\# = \mathrm{Opt}$. Now let $V = R_*U$, and let $\xi = V\eta$, where η is n-dimensional random Rademacher vector (independent entries taking values ± 1 with probabilities 1/2). We have

$$\xi^TC\xi = \eta^T[V^TCV]\eta = \eta^T[U^TR_*CR_*U]\eta = \eta^TD\eta \equiv \mathrm{Tr}(D) = \mathrm{Opt} \tag{4.143}$$

(recall that D is diagonal) and

$$\mathbf{E}_\xi\{\xi\xi^T\} = \mathbf{E}_\eta\{V\eta\eta^TV^T\} = VV^T = R_*UU^TR_* = R_*^2 = Q_*.$$

From the latter relation,

$$\begin{array}{rcl}\mathbf{E}_\xi\left\{R_k^2[\xi]\right\} & = & \mathbf{E}_\xi\left\{\mathcal{R}_k[\xi\xi^T]\right\} = \mathcal{R}_k[\mathbf{E}_\xi\{\xi\xi^T\}] \\ & = & \mathcal{R}_k[Q_*] \preceq t_k^* I_{d_k}, 1 \le k \le K.\end{array} \tag{4.144}$$

On the other hand, with properly selected symmetric matrices $\bar{R}^{kj}$ we have

$$R_k[Vy] = \sum_i y_i \bar{R}^{ki}$$

identically in $y \in \mathbf{R}^n$, whence

$$\mathbf{E}_\xi\left\{R_k^2[\xi]\right\} = \mathbf{E}_\eta\left\{R_k^2[V\eta]\right\} = \mathbf{E}_\eta\left\{\left[\sum\nolimits_i \eta_i \bar{R}^{ki}\right]^2\right\} = \sum_{i,j}\mathbf{E}_\eta\{\eta_i\eta_j\}\bar{R}^{ki}\bar{R}^{kj} = \sum_i [\bar{R}^{ki}]^2.$$

This combines with (4.144) to imply that

$$\sum_i [\bar{R}^{ki}]^2 \preceq t_k^* I_{d_k}, \ 1 \le k \le K. \tag{4.145}$$

3°. Let us fix $k \le K$. Assuming $t_k^* > 0$ and applying Theorem 4.45, we derive from (4.145) that

$$\mathrm{Prob}\{\eta : \|\bar{R}_k[\eta]\|^2 > t_k^*/\rho\} < 2d_k \mathrm{e}^{-\frac{1}{2\rho}},$$

and recalling the relation between ξ and η, we arrive at

$$\mathrm{Prob}\{\xi : \|R_k[\xi]\|^2 > t_k^*/\rho\} < 2d_k \mathrm{e}^{-\frac{1}{2\rho}} \ \ \forall \rho \in (0,1]. \tag{4.146}$$

Note that when $t_k^* = 0$ (4.145) implies $\bar{R}^{ki} = 0$ for all i, so that $R_k[\xi] \equiv \bar{R}_k[\eta] \equiv 0$, and (4.146) holds for those k as well.

Now let us set $\rho = \frac{1}{2\max[\ln(2D),1]}$. For this ρ, the sum over $k \le K$ of the right-hand sides in inequalities (4.146) is ≤ 1, implying that there exists a realization $\bar{\xi}$ of ξ such that

$$\|R_k[\bar{\xi}]\|^2 \le t_k^*/\rho, \ \forall k,$$

or, equivalently,

$$\bar{x} := \rho^{1/2}\bar{\xi} \in \mathcal{X}$$

(recall that $P = I$), implying that

$$\mathrm{Opt}_* := \max_{x\in\mathcal{X}} x^TCx \ge \bar{x}^TC\bar{x} = \rho\xi^TC\xi = \rho\mathrm{Opt}$$

(the concluding equality is due to (4.143)), and we arrive at the right inequality in (4.30). □

4.8.4 Proof of Lemma 4.17

1°. Let us verify (4.57). When $Q \succ 0$, passing from variables (Θ, Υ) in problem (4.56) to the variables $(G = Q^{1/2}\Theta Q^{1/2}, \Upsilon)$, the problem becomes exactly the optimization problem in (4.57), implying that $\mathrm{Opt}[Q] = \overline{\mathrm{Opt}}[Q]$ when $Q \succ 0$. As is easily seen, both sides in this equality are continuous in $Q \succeq 0$, and (4.57) follows.

2°. Let us prove (4.59). Setting $\zeta = Q^{1/2}\eta$ with $\eta \sim \mathcal{N}(0, I_N)$ and $Z = Q^{1/2}Y$, to justify (4.59) we have to show that when $\varkappa \geq 1$ one has

$$\bar{\delta} = \frac{\mathrm{Opt}[Q]}{4\varkappa} \Rightarrow \mathrm{Prob}_\eta\{\|Z^T\eta\| \geq \bar{\delta}\} \geq \beta_\varkappa := 1 - \frac{\mathrm{e}^{3/8}}{2} - 2F\mathrm{e}^{-\varkappa^2/2}, \tag{4.147}$$

where (cf. (4.57))

$$[\overline{\mathrm{Opt}}[Q] =]\ \mathrm{Opt}[Q] := \min_{\Theta, \Upsilon = \{\Upsilon_\ell, \ell \leq L\}} \left\{ \phi_{\mathcal{R}}(\lambda[\Upsilon]) + \mathrm{Tr}(\Theta) : \Upsilon_\ell \succeq 0, \left[\begin{array}{c|c} \Theta & \frac{1}{2}ZM \\ \hline \frac{1}{2}M^T Z^T & \sum_\ell \mathcal{S}_\ell^*[\Upsilon_\ell] \end{array}\right] \succeq 0 \right\}. \tag{4.148}$$

Justification of (4.147) is as follows.

2.1°. Let us represent $\mathrm{Opt}[Q]$ as the optimal value of a conic problem. Setting

$$\mathbf{K} = \mathrm{cl}\{[r; s] : s > 0, r/s \in \mathcal{R}\},$$

we ensure that

$$\mathcal{R} = \{r : [r; 1] \in \mathbf{K}\},\ \mathbf{K}_* = \{[g; s] : s \geq \phi_{\mathcal{R}}(-g)\},$$

where $\mathbf{K}_*$ is the cone dual to $\mathbf{K}$. Consequently, (4.148) reads

$$\mathrm{Opt}[Q] = \min_{\Theta, \Upsilon, \theta} \left\{ \theta + \mathrm{Tr}(\Theta) : \begin{array}{ll} \Upsilon_\ell \succeq 0, 1 \leq \ell \leq L & (a) \\ \left[\begin{array}{c|c} \Theta & \frac{1}{2}ZM \\ \hline \frac{1}{2}M^T Z^T & \sum_\ell \mathcal{S}_\ell^*[\Upsilon_\ell] \end{array}\right] \succeq 0 & (b) \\ [-\lambda[\Upsilon]; \theta] \in \mathbf{K}_* & (c) \end{array} \right\}. \tag{P}$$

2.2°. Now let us prove that there exists a matrix $W \in \mathbf{S}^q_+$ and $r \in \mathcal{R}$ such that

$$\mathcal{S}_\ell[W] \preceq r_\ell I_{f_\ell},\ \ell \leq L, \tag{4.149}$$

and

$$\mathrm{Opt}[Q] \leq \sum_i \sigma_i(ZMW^{1/2}), \tag{4.150}$$

where $\sigma_1(\cdot) \geq \sigma_2(\cdot) \geq ...$ are singular values.

To get the announced result, let us pass from problem (P) to its conic dual. Applying Lemma 4.44 we conclude that (P) is strictly feasible; in addition, (P)

clearly is bounded, so that the dual to (P) problem (D) is solvable with optimal value $\mathrm{Opt}[Q]$. Let us build (D). Denoting by $\Lambda_\ell \succeq 0, \ell \leq L$, $\left[\begin{array}{c|c} G & -R \\ \hline -R^T & W \end{array}\right] \succeq 0$, $[r;\tau] \in \mathbf{K}$ the Lagrange multipliers for the respective constraints in (P), and aggregating these constraints, the multipliers being the aggregation weights, we arrive at the following aggregated constraint:

$$\mathrm{Tr}(\Theta G) + \mathrm{Tr}(W \textstyle\sum_\ell \mathcal{S}_\ell^*[\Upsilon_\ell]) + \sum_\ell \mathrm{Tr}(\Lambda_\ell \Upsilon_\ell) - \sum_\ell r_\ell \mathrm{Tr}(\Upsilon_\ell) + \theta\tau \geq \mathrm{Tr}(ZMR^T).$$

To get the dual problem, we impose on the Lagrange multipliers, in addition to the initial conic constraints like $\Lambda_\ell \succeq 0$, $1 \leq \ell \leq L$, the restriction that the left-hand side in the aggregated constraint, identically in Θ, Υ_ℓ, and θ, is equal to the objective of (P), that is,

$$G = I,\ \mathcal{S}_\ell[W] + \Lambda_\ell - r_\ell I_{f_\ell} = 0,\ 1\ \leq \ell \leq L,\ \tau = 1,$$

and maximize, under the resulting restrictions, the right-hand side of the aggregated constraint. After immediate simplifications, we arrive at

$$\mathrm{Opt}[Q] = \max_{W,R,r} \left\{\mathrm{Tr}(ZMR^T) : W \succeq R^T R, r \in \mathcal{R}, \mathcal{S}_\ell[W] \preceq r_\ell I_{f_\ell}, 1 \leq \ell \leq L\right\}$$

(note that $r \in \mathcal{R}$ is equivalent to $[r;1] \in \mathbf{K}$, and $W \succeq R^T R$ is the same as $\left[\begin{array}{c|c} I & -R \\ \hline -R^T & W \end{array}\right] \succeq 0$). Now, to say that $R^T R \preceq W$ is exactly the same as to say that $R = SW^{1/2}$ with the spectral norm $\|S\|_{2,2}$ of S not exceeding 1, so that

$$\mathrm{Opt}[Q] = \max_{W,S,r} \Big\{ \underbrace{\mathrm{Tr}(ZM[SW^{1/2}]^T)}_{=\mathrm{Tr}([ZMW^{1/2}]S^T)} : W \succeq 0, \|S\|_{2,2} \leq 1, r \in \mathcal{R}, \mathcal{S}_\ell[W] \preceq r_\ell I_{f_\ell}, \ell \leq L \Big\}.$$

We can immediately eliminate the S-variable, using the well-known fact that for a $p \times q$ matrix J it holds

$$\max_{S \in \mathbf{R}^{p\times q}, \|S\|_{2,2} \leq 1} \mathrm{Tr}(JS^T) = \|J\|_{\mathrm{Sh},1},$$

where $\|J\|_{\mathrm{Sh},1}$ is the nuclear norm (the sum of singular values) of J. We arrive at

$$\mathrm{Opt}[Q] = \max_{W,r} \left\{ \|ZMW^{1/2}\|_{\mathrm{Sh},1} : r \in \mathcal{R}, W \succeq 0, \mathcal{S}_\ell[W] \preceq r_\ell I_{f_\ell}, \ell \leq L \right\}.$$

The resulting problem clearly is solvable, and its optimal solution W ensures the target relations (4.149) and (4.150).

2.3^o. Given W satisfying (4.149) and (4.150), let $UJV = W^{1/2}M^T Z^T$ be the singular value decomposition of $W^{1/2}M^T Z^T$, so that U and V are, respectively, $q\times q$ and $N\times N$ orthogonal matrices, J is $q\times N$ matrix with diagonal $\sigma = [\sigma_1;...;\sigma_p]$, $p = \min[q,N]$, and zero off-diagonal entries; the diagonal entries σ_i, $1 \leq i \leq p$ are the singular values of $W^{1/2}M^T Z^T$, or, which is the same, of $ZMW^{1/2}$. Therefore, by (4.150) we have

$$\sum_i \sigma_i \geq \mathrm{Opt}[Q]. \tag{4.151}$$

Now consider the following construction. Let $\eta \sim \mathcal{N}(0, I_N)$; we denote by υ the vector comprised of the first p entries in $V\eta$; note that $\upsilon \sim \mathcal{N}(0, I_p)$, since V is orthogonal. We then augment, if necessary, υ by $q-p$ $\mathcal{N}(0,1)$ random variables independent of each other and of η to obtain a q-dimensional random vector $\upsilon' \sim \mathcal{N}(0, I_q)$, and set $\chi = U\upsilon'$. Because U is orthogonal we also have $\chi \sim \mathcal{N}(0, I_q)$. Observe that

$$\chi^T W^{1/2} M^T Z^T \eta = \chi^T UJV\eta = [\upsilon']^T J\upsilon = \sum_{i=1}^{p} \sigma_i \upsilon_i^2. \tag{4.152}$$

To continue we need two simple observations.

(i) *One has*

$$\alpha := \mathrm{Prob}\left\{\sum_{i=1}^{p} \sigma_i \upsilon_i^2 < \tfrac{1}{4}\sum_{i=1}^{p} \sigma_i\right\} \leq \frac{\mathrm{e}^{3/8}}{2} \; [= 0.7275...]. \tag{4.153}$$

The claim is evident when $\sigma := \sum_i \sigma_i = 0$. Now let $\sigma > 0$, and let us apply the Cramer bounding scheme. Namely, given $\gamma > 0$, consider the random variable

$$\omega = \exp\left\{\tfrac{1}{4}\gamma \sum_i \sigma_i - \gamma \sum_i \sigma_i \upsilon_i^2\right\}.$$

Note that $\omega > 0$ a.s., and is > 1 when $\sum_{i=1}^p \sigma_i \upsilon_i^2 < \frac{1}{4}\sum_{i=1}^p \sigma_i$, so that $\alpha \leq \mathbf{E}\{\omega\}$, or, equivalently, thanks to $\upsilon \sim \mathcal{N}(0, I_p)$,

$$\begin{array}{rcl}\ln(\alpha) & \leq & \ln(\mathbf{E}\{\omega\}) = \frac{1}{4}\gamma \sum_i \sigma_i + \sum_i \ln\left(\mathbf{E}\{\exp\{-\gamma\sigma_i \upsilon_i^2\}\}\right) \\ & \leq & \frac{1}{4}\gamma\sigma - \frac{1}{2}\sum_i \ln(1 + 2\gamma\sigma_i).\end{array}$$

Function $-\sum_i \ln(1 + 2\gamma\sigma_i)$ is convex in $[\sigma_1; ...; \sigma_p] \geq 0$; therefore, its maximum over the simplex $\{\sigma_i \geq 0, i \leq p, \sum_i \sigma_i = \sigma\}$ is attained at a vertex, and we get

$$\ln(\alpha) \leq \tfrac{1}{4}\gamma\sigma - \tfrac{1}{2}\ln(1 + 2\gamma\sigma).$$

Minimizing the right-hand side in $\gamma > 0$, we arrive at (4.153).

(ii) *Whenever $\varkappa \geq 1$, one has*

$$\mathrm{Prob}\{\|MW^{1/2}\chi\|_* > \varkappa\} \leq 2F\exp\{-\varkappa^2/2\}, \tag{4.154}$$

with F given by (4.55).

Indeed, setting $\rho = 1/\varkappa^2 \leq 1$ and $\omega = \sqrt{\rho}W^{1/2}\chi$, we get $\omega \sim \mathcal{N}(0, \rho W)$. Let us apply Lemma 4.46 to $Q = \rho W$, $\mathcal{R}$ in the role of $\mathcal{T}$, L in the role of K, and $\mathcal{S}_\ell[\cdot]$ in the role of $\mathcal{R}_k[\cdot]$. Denoting

$$\mathcal{Y} := \{y : \exists r \in \mathcal{R} : S_\ell^2[y] \preceq r_\ell I_{f_\ell}, \ell \leq L\},$$

we have $\mathcal{S}_\ell[Q] = \rho\mathcal{S}_\ell[W] \preceq \rho r_\ell I_{f_\ell}$, $\ell \leq L$, with $r \in \mathcal{R}$ (see (4.149)), so we are under the premise of Lemma 4.46 (with $\mathcal{Y}$ in the role of $\mathcal{X}$ and thus with F in the role of D). Applying the lemma, we conclude that

$$\mathrm{Prob}\left\{\chi : \varkappa^{-1}W^{1/2}\chi \notin \mathcal{Y}\right\} \leq 2F\exp\{-1/(2\rho)\} = 2F\exp\{-\varkappa^2/2\}.$$

Recalling that $\mathcal{B}_* = M\mathcal{Y}$, we see that $\text{Prob}\{\chi : \varkappa^{-1} MW^{1/2}\chi \not\in \mathcal{B}_*\}$ is indeed upper-bounded by the right-hand side of (4.154), and (4.154) follows.

2.4^o. Now, for $\varkappa \geq 1$, let

$$E_\varkappa = \left\{ (\chi, \eta) : \|MW^{1/2}\chi\|_* \leq \varkappa, \sum_i \sigma_i v_i^2 \geq \tfrac{1}{4} \sum_i \sigma_i \right\},$$

and let $E_\varkappa^+ = \{\eta : \exists \chi : (\chi, \eta) \in E_\varkappa\}$. For $\eta \in E_\varkappa^+$ there exists χ such that $(\chi, \eta) \in E_\varkappa$, leading to

$$\varkappa\|Z^T\eta\| \geq \|MW^{1/2}\chi\|_* \|Z^T\eta\| \geq \chi^T W^{1/2} M^T Z^T \eta = \sum_i \sigma_i v_i^2 \geq \tfrac{1}{4} \sum_i \sigma_i \geq \tfrac{1}{4}\text{Opt}[Q]$$

(we have used (4.152) and (4.151)). Thus,

$$\eta \in E_\varkappa^+ \Rightarrow \|Z^T\eta\| \geq \frac{\text{Opt}[Q]}{4\varkappa}.$$

On the other hand, due to (4.153) and (4.154), for our random (χ, η) it holds

$$\text{Prob}\{E_\varkappa\} \geq 1 - \frac{\mathrm{e}^{3/8}}{2} - 2F\mathrm{e}^{-\varkappa^2/2} = \beta_\varkappa,$$

and the marginal distribution of η is $\mathcal{N}(0, I_N)$, implying that

$$\text{Prob}_{\eta \sim \mathcal{N}(0, I_N)}\{\eta \in E_\varkappa^+\} \geq \beta_\varkappa.$$

(4.147) is proved.

3°. As was explained in the beginning of item 2°, (4.147) is exactly the same as (4.59). The latter relation clearly implies (4.60) which, in turn, implies the right inequality in (4.58). □

4.8.5 Proofs of Propositions 4.5, 4.16 and 4.19

Below, we focus on the proof of Proposition 4.16; Propositions 4.5 and 4.19 will be derived from it in Sections 4.8.5.2, 4.8.6.2, respectively.

4.8.5.1 Proof of Proposition 4.16

In what follows, we use the assumptions and the notation of Proposition 4.16.

1°. Let

$$\Phi(H, \Lambda, \Upsilon, \Upsilon', \Theta; Q) = \phi_{\mathcal{T}}(\lambda[\Lambda]) + \phi_{\mathcal{R}}(\lambda[\Upsilon]) + \phi_{\mathcal{R}}(\lambda[\Upsilon']) + \text{Tr}(Q\Theta) : \mathcal{M} \times \Pi \to \mathbf{R},$$

where

$$\mathcal{M} = \left\{(H,\Lambda,\Upsilon,\Upsilon',\Theta): \begin{array}{l}\Lambda=\{\Lambda_k\succeq 0,k\leq K\},\\ \Upsilon=\{\Upsilon_\ell\succeq 0,\ell\leq L\},\ \Upsilon'=\{\Upsilon'_\ell\succeq 0,\ell\leq L\},\\ \left[\begin{array}{c|c}\sum_k \mathcal{R}_k^*[\Lambda_k] & \frac{1}{2}[B^T-A^TH]M\\ \hline \frac{1}{2}M^T[B-H^TA] & \sum_\ell \mathcal{S}_\ell^*[\Upsilon_\ell]\end{array}\right]\succeq 0,\\ \left[\begin{array}{c|c}\Theta & \frac{1}{2}HM\\ \hline \frac{1}{2}M^TH^T & \sum_\ell \mathcal{S}_\ell^*[\Upsilon'_\ell]\end{array}\right]\succeq 0.\end{array}\right\}$$

Looking at (4.42), we see immediately that the optimal value Opt in (4.42) is nothing but

$$\mathrm{Opt} = \min_{(H,\Lambda,\Upsilon,\Upsilon',\Theta)\in\mathcal{M}} \left[\overline{\Phi}(H,\Lambda,\Upsilon,\Upsilon',\Theta) := \max_{Q\in\Pi}\Phi(H,\Lambda,\Upsilon,\Upsilon',\Theta;Q)\right]. \tag{4.155}$$

Note that sets $\mathcal{M}$ and Π are closed and convex, Π is compact, and Φ is a continuous convex-concave function on $\mathcal{M}\times\Pi$. In view of these observations, the fact that $\Pi\subset \operatorname{int}\mathbf{S}^m_+$ combines with the Sion-Kakutani Theorem to imply that Φ possesses saddle point $(H_*,\Lambda_*,\Upsilon_*,\Upsilon'_*,\Theta_*;Q_*)$ (min in $(H,\Lambda,\Upsilon,\Upsilon',\Theta)$, max in Q) on $\mathcal{M}\times\Pi$, whence Opt is the saddle point value of Φ by (4.155). We conclude that for properly selected $Q_*\in\Pi$ it holds

$$\begin{array}{rl}\mathrm{Opt} = & \min\limits_{(H,\Lambda,\Upsilon,\Upsilon',\Theta)\in\mathcal{M}} \Phi(H,\Lambda,\Upsilon,\Upsilon',\Theta;Q_*)\\
= & \min\limits_{H,\Lambda,\Upsilon,\Upsilon',\Theta}\Big\{\phi_{\mathcal{T}}(\lambda[\Lambda])+\phi_{\mathcal{R}}(\lambda[\Upsilon])+\phi_{\mathcal{R}}(\lambda[\Upsilon'])+\mathrm{Tr}(Q_*\Theta):\\
& \quad \Lambda=\{\Lambda_k\succeq 0,k\leq K\},\ \Upsilon=\{\Upsilon_\ell\succeq 0,\ell\leq L\},\ \Upsilon'=\{\Upsilon'_\ell\succeq 0,\ell\leq L\},\\
& \quad \left[\begin{array}{c|c}\sum_k \mathcal{R}_k^*[\Lambda_k] & \frac{1}{2}[B^T-A^TH]M\\ \hline \frac{1}{2}M^T[B-H^TA] & \sum_\ell \mathcal{S}_\ell^*[\Upsilon_\ell]\end{array}\right]\succeq 0,\\
& \quad \left[\begin{array}{c|c}\Theta & \frac{1}{2}HM\\ \hline \frac{1}{2}M^TH^T & \sum_\ell \mathcal{S}_\ell^*[\Upsilon'_\ell]\end{array}\right]\succeq 0\Big\}\\
= & \min\limits_{H,\Lambda,\Upsilon,\Upsilon',G}\Big\{\phi_{\mathcal{T}}(\lambda[\Lambda])+\phi_{\mathcal{R}}(\lambda[\Upsilon])+\phi_{\mathcal{R}}(\lambda[\Upsilon'])+\mathrm{Tr}(G):\\
& \quad \Lambda=\{\Lambda_k\succeq 0,k\leq K\},\ \Upsilon=\{\Upsilon_\ell\succeq 0,\ell\leq L\},\ \Upsilon'=\{\Upsilon'_\ell\succeq 0,\ell\leq L\},\\
& \quad \left[\begin{array}{c|c}\sum_k \mathcal{R}_k^*[\Lambda_k] & \frac{1}{2}[B^T-A^TH]M\\ \hline \frac{1}{2}M^T[B-H^TA] & \sum_\ell \mathcal{S}_\ell^*[\Upsilon_\ell]\end{array}\right]\succeq 0,\\
& \quad \left[\begin{array}{c|c}G & \frac{1}{2}Q_*^{1/2}HM\\ \hline \frac{1}{2}M^TH^TQ_*^{1/2} & \sum_\ell \mathcal{S}_\ell^*[\Upsilon'_\ell]\end{array}\right]\succeq 0\Big\}\\
= & \min\limits_{H,\Lambda,\Upsilon}\Big\{\phi_{\mathcal{T}}(\lambda[\Lambda])+\phi_{\mathcal{R}}(\lambda[\Upsilon])+\overline{\Psi}(H):\\
& \quad \Lambda=\{\Lambda_k\succeq 0,k\leq K\},\ \Upsilon=\{\Upsilon_\ell\succeq 0,\ell\leq L\},\\
& \quad \left[\begin{array}{c|c}\sum_k \mathcal{R}_k^*[\Lambda_k] & \frac{1}{2}[B^T-A^TH]M\\ \hline \frac{1}{2}M^T[B-H^TA] & \sum_\ell \mathcal{S}_\ell^*[\Upsilon_\ell]\end{array}\right]\succeq 0\Big\}\end{array} \tag{4.156}$$

where

$$\overline{\Psi}(H) := \min_{G,\Upsilon'}\Big\{\phi_{\mathcal{R}}(\lambda[\Upsilon'])+\mathrm{Tr}(G):\Upsilon'=\{\Upsilon'_\ell\succeq 0,\ell\leq L\},\ \left[\begin{array}{c|c}G & \frac{1}{2}Q_*^{1/2}HM\\ \hline \frac{1}{2}M^TH^TQ_*^{1/2} & \sum_\ell \mathcal{S}_\ell^*[\Upsilon'_\ell]\end{array}\right]\succeq 0\Big\},$$

and Opt is given by (4.42), and the equalities are due to (4.56) and (4.57).

From now on we assume that the noise ξ in observation (4.31) is $\xi \sim \mathcal{N}(0, Q_*)$. We also assume that $B \neq 0$, since otherwise the conclusion of Proposition 4.16 is evident.

2°. ϵ-risk. In Proposition 4.16, we are speaking about $\|\cdot\|$-risk of an estimate—the maximal, over signals $x \in \mathcal{X}$, expected norm $\|\cdot\|$ of the error of recovering Bx; what we need to prove is that the minimax optimal risk $\mathrm{RiskOpt}_{\Pi,\|\cdot\|}[\mathcal{X}]$ as given by (4.53) can be lower-bounded by a quantity "of order of" Opt. To this end, of course, it suffices to build such a lower bound for the quantity

$$\mathrm{RiskOpt}_{\|\cdot\|} := \inf_{\widehat{x}(\cdot)} \left[\sup_{x \in \mathcal{X}} \mathbf{E}_{\xi \sim \mathcal{N}(0,Q_*)} \{\|Bx - \widehat{x}(Ax + \xi)\|\} \right],$$

since this quantity is a lower bound on $\mathrm{RiskOpt}_{\Pi,\|\cdot\|}$. Technically, it is more convenient to work with the *ϵ-risk* defined in terms of "$\|\cdot\|$-confidence intervals" rather than in terms of the expected norm of the error. Specifically, in the sequel we will heavily use the *minimax ϵ-risk* defined as

$$\mathrm{RiskOpt}_\epsilon = \inf_{\widehat{x},\rho} \left\{ \rho : \mathrm{Prob}_{\xi \sim \mathcal{N}(0,Q_*)} \{\|Bx - \widehat{x}(Ax + \xi)\| > \rho\} \leq \epsilon \;\forall x \in \mathcal{X} \right\},$$

where $\widehat{x}$ in the infimum runs through the set of all Borel estimates. When $\epsilon \in (0,1)$ is once and forever fixed (in the sequel, we use $\epsilon = \frac{1}{8}$) we can use ϵ-risk to lower-bound $\mathrm{RiskOpt}_{\|\cdot\|}$, since by evident reasons

$$\mathrm{RiskOpt}_{\|\cdot\|} \geq \epsilon \cdot \mathrm{RiskOpt}_\epsilon. \tag{4.157}$$

Consequently, all we need in order to prove Proposition 4.16 is to lower-bound $\mathrm{RiskOpt}_{\frac{1}{8}}$ by a "not too small" multiple of Opt, and this is our current objective.

3°. Let W be a positive semidefinite $n \times n$ matrix, let $\eta \sim \mathcal{N}(0, W)$ be random signal, and let $\xi \sim \mathcal{N}(0, Q_*)$ be independent of η; vectors (η, ξ) induce random vector

$$\omega = A\eta + \xi \sim \mathcal{N}(0, AWA^T + Q_*).$$

Consider the Bayesian version of the estimation problem where given ω we are interested in recovering $B\eta$. Recall that, because $[\omega; B\eta]$ is zero mean Gaussian, the conditional expectation $\mathbf{E}_{|\omega}\{B\eta\}$ of $B\eta$ given ω is linear in ω: $\mathbf{E}_{|\omega}\{B\eta\} = \bar{H}^T\omega$ for some $\bar{H}$ depending on W only.[24] Therefore, denoting by $P_{|\omega}$ the conditional probability distribution given ω, for any $\rho > 0$ and estimate $\widehat{x}(\cdot)$ one has

$$\begin{array}{l}\mathrm{Prob}_{\eta,\xi}\{\|B\eta - \widehat{x}(A\eta + \xi)\| \geq \rho\} = \mathbf{E}_\omega\left\{\mathrm{Prob}_{|\omega}\{\|B\eta - \widehat{x}(\omega)\| \geq \rho\}\right\} \\ \quad \geq \mathbf{E}_\omega\left\{\mathrm{Prob}_{|\omega}\{\|B\eta - \mathbf{E}_{|\omega}\{B\eta\}\| \geq \rho\}\right\} = \mathrm{Prob}_{\eta,\xi}\{\|B\eta - \bar{H}^T(A\eta + \xi)\| \geq \rho\},\end{array}$$

with the inequality given by the Anderson Lemma as applied to the shift of the Gaussian distribution $P_{|\omega}$ by its mean. Applying the Anderson Lemma again we

[24]We have used the following standard fact [172]: *let $\zeta = [\omega; \eta] \sim \mathcal{N}(0, S)$, the covariance matrix of the marginal distribution of ω being nonsingular. Then the conditional distribution of η given ω is Gaussian with the mean linearly depending on ω and covariance matrix independent of ω.*

get

$$\begin{array}{rcl}\mathrm{Prob}_{\eta,\xi}\{\|B\eta-\bar{H}^T(A\eta+\xi)\|\geq\rho\} &=& \mathbf{E}_\xi\{\mathrm{Prob}_\eta\{\|(B-\bar{H}^TA)\eta-\bar{H}^T\xi\|\geq\rho\}\}\\ &\geq& \mathrm{Prob}_\eta\{\|(B-\bar{H}^TA)\eta\|\geq\rho\},\end{array}$$

and, by "symmetric" reasoning,

$$\mathrm{Prob}_{\eta,\xi}\{\|B\eta-\bar{H}^T(A\eta+\xi)\|\geq\rho\}\geq\mathrm{Prob}_\xi\{\|\bar{H}^T\xi\|\geq\rho\}.$$

We conclude that for any $\widehat{x}(\cdot)$

$$\begin{array}{l}\mathrm{Prob}_{\eta,\xi}\{\|B\eta-\widehat{x}(\omega)\|\geq\rho\}\\ \quad\geq\max\left\{\mathrm{Prob}_\eta\{\|(B-\bar{H}^TA)\eta\|\geq\rho\},\,\mathrm{Prob}_\xi\{\|\bar{H}^T\xi\|\geq\rho\}\right\}.\end{array}\tag{4.158}$$

4°. Let H be an $m\times\nu$ matrix. Applying Lemma 4.17 to $N=m$, $Y=\bar{H}$, $Q=Q_*$, we get from (4.59)

$$\mathrm{Prob}_{\xi\sim\mathcal{N}(0,Q_*)}\{\|\bar{H}^T\xi\|\geq[4\varkappa]^{-1}\overline{\Psi}(\bar{H})\}\geq\beta_\varkappa\ \forall\varkappa\geq1,\tag{4.159}$$

where $\overline{\Psi}(H)$ is defined by (4.156). Similarly, applying Lemma 4.17 to $N=n$, $Y=(B-\bar{H}^TA)^T$, $Q=W$, we obtain

$$\mathrm{Prob}_{\eta\sim\mathcal{N}(0,W)}\{\|(B-\bar{H}^TA)\eta\|\geq[4\varkappa]^{-1}\overline{\Phi}(W,\bar{H})\}\geq\beta_\varkappa\ \forall\kappa\geq1,\tag{4.160}$$

where

$$\begin{array}{rcl}\overline{\Phi}(W,H) &=& \min\limits_{\Upsilon=\{\Upsilon_\ell,\ell\leq L\},\Theta}\left\{\mathrm{Tr}\,(W\Theta)+\phi_{\mathcal{R}}(\lambda[\Upsilon]):\ \Upsilon_\ell\succeq0\,\forall\ell,\right.\\ && \left.\left[\begin{array}{c|c}\Theta & \frac{1}{2}[B^T-A^TH]M\\ \hline \frac{1}{2}M^T[B-H^TA] & \sum_\ell\mathcal{S}_\ell^*[\Upsilon_\ell]\end{array}\right]\succeq0\right\}.\end{array}\tag{4.161}$$

Let us put $\rho(W,\bar{H})=[8\varkappa]^{-1}[\overline{\Psi}(\bar{H})+\overline{\Phi}(W,\bar{H})]$; when combining (4.160) with (4.159) we conclude that

$$\max\left\{\mathrm{Prob}_\eta\{\|(B-\bar{H}^TA)\eta\|\geq\rho(W,\bar{H})\},\,\mathrm{Prob}_\xi\{\|\bar{H}^T\xi\|\geq\rho(W,\bar{H})\}\right\}\geq\beta_\varkappa,$$

and the same inequality holds if $\rho(W,\bar{H})$ is replaced with the smaller quantity

$$\bar{\rho}(W)=[8\varkappa]^{-1}\inf_H[\overline{\Psi}(H)+\overline{\Phi}(W,H)].$$

Now, the latter bound combines with (4.158) to imply the following result:

Lemma 4.49. *Let W be a positive semidefinite $n\times n$ matrix, and $\varkappa\geq1$. Then for any estimate $\widehat{x}(\cdot)$ of $B\eta$ given observation $\omega=A\eta+\xi$, where $\eta\sim\mathcal{N}(0,W)$ is independent of $\xi\sim\mathcal{N}(0,Q_*)$, one has*

$$\mathrm{Prob}_{\eta,\xi}\left\{\|B\eta-\widehat{x}(\omega)\|\geq[8\varkappa]^{-1}\inf_H[\overline{\Psi}(H)+\overline{\Phi}(W,H)]\right\}\geq\beta_\varkappa=1-\frac{\mathrm{e}^{3/8}}{2}-2F\mathrm{e}^{-\varkappa^2/2}$$

where $\overline{\Psi}(H)$ and $\overline{\Phi}(W,H)$ are defined, respectively, by (4.156) and (4.161).

In particular, for

$$\varkappa = \bar{\varkappa} := \sqrt{2\ln F + 10\ln 2} \qquad (4.162)$$

it holds

$$\mathrm{Prob}_{\eta,\xi}\{\|B\eta - \widehat{x}(\omega)\| \geq [8\bar{\varkappa}]^{-1}\inf_H[\overline{\Psi}(H) + \overline{\Phi}(W,H)]\} > \tfrac{3}{16}.$$

5°. For $0 < \kappa \leq 1$, let us set

$$\begin{array}{ll}(a) & \mathcal{W}_\kappa = \{W \in \mathbf{S}^n_+ : \exists t \in \mathcal{T} : \mathcal{R}_k[W] \preceq \kappa t_k I_{d_k}, 1 \leq k \leq K\},\\ (b) & \mathcal{Z} = \left\{(\Upsilon = \{\Upsilon_\ell, \ell \leq L\}, \Theta, H) : \begin{array}{c}\Upsilon_\ell \succeq 0\,\forall \ell,\\ \left[\begin{array}{c|c}\Theta & \frac{1}{2}[B^T - A^T H]M\\ \hline \frac{1}{2}M^T[B - H^T A] & \sum_\ell \mathcal{S}_\ell^*[\Upsilon_\ell]\end{array}\right] \succeq 0\end{array}\right\}.\end{array}$$

Note that $\mathcal{W}_\kappa$ is a nonempty convex and compact (by Lemma 4.44) set such that $\mathcal{W}_\kappa = \kappa\mathcal{W}_1$, and $\mathcal{Z}$ is a nonempty closed convex set. Consider the parametric saddle point problem

$$\mathrm{Opt}(\kappa) = \max_{W \in \mathcal{W}_\kappa} \inf_{(\Upsilon,\Theta,H)\in\mathcal{Z}} \left[E(W;\Upsilon,\Theta,H) := \mathrm{Tr}(W\Theta) + \phi_{\mathcal{R}}(\lambda[\Upsilon]) + \overline{\Psi}(H)\right]. \qquad (4.163)$$

This problem is convex-concave; utilizing the fact that $\mathcal{W}_\kappa$ is compact and contains positive definite matrices, it is immediately seen that the Sion-Kakutani theorem ensures the existence of a saddle point whenever $\kappa \in (0,1]$. We claim that

$$0 < \kappa \leq 1 \Rightarrow \mathrm{Opt}(\kappa) \geq \sqrt{\kappa}\,\mathrm{Opt}(1). \qquad (4.164)$$

Indeed, $\mathcal{Z}$ is invariant w.r.t. scalings

$$(\Upsilon = \{\Upsilon_\ell, \ell \leq L\}, \Theta, H) \mapsto (\theta\Upsilon := \{\theta\Upsilon_\ell, \ell \leq L\}, \theta^{-1}\Theta, H), \qquad [\theta > 0].$$

When taking into account that $\phi_{\mathcal{R}}(\lambda[\theta\Upsilon]) = \theta\phi_{\mathcal{R}}(\lambda[\Upsilon])$, we get

$$\begin{array}{rcl}\underline{E}(W) & := & \inf\limits_{(\Upsilon,\Theta,H)\in\mathcal{Z}} E(W;\Upsilon,\Theta,H) = \inf\limits_{(\Upsilon,\Theta,H)\in\mathcal{Z}}\inf\limits_{\theta>0} E(W;\theta\Upsilon,\theta^{-1}\Theta,H)\\ & = & \inf\limits_{(\Upsilon,\Theta,H)\in\mathcal{Z}}\left[2\sqrt{\mathrm{Tr}(W\Theta)\phi_{\mathcal{R}}(\lambda[\Upsilon])} + \overline{\Psi}(H)\right].\end{array}$$

Because $\overline{\Psi}$ is nonnegative we conclude that whenever $W \succeq 0$ and $\kappa \in (0,1]$, one has

$$\underline{E}(\kappa W) \geq \sqrt{\kappa}\,\underline{E}(W).$$

This combines with $\mathcal{W}_\kappa = \kappa\mathcal{W}_1$ to imply that

$$\mathrm{Opt}(\kappa) = \max_{W\in\mathcal{W}_\kappa}\underline{E}(W) = \max_{W\in\mathcal{W}_1}\underline{E}(\kappa W) \geq \sqrt{\kappa}\max_{W\in\mathcal{W}_1}\underline{E}(W) = \sqrt{\kappa}\,\mathrm{Opt}(1),$$

and (4.164) follows.

6°. We claim that

$$\mathrm{Opt}(1) = \mathrm{Opt}, \qquad (4.165)$$

where Opt is given by (4.42) (and, as we have seen, by (4.156) as well). Note that (4.165) combines with (4.164) to imply that

$$0 < \kappa \leq 1 \Rightarrow \mathrm{Opt}(\kappa) \geq \sqrt{\kappa}\,\mathrm{Opt}. \qquad (4.166)$$

Verification of (4.165) is given by the following computation. By the Sion-Kakutani Theorem,

$$\begin{array}{rcl}\mathrm{Opt}(1) &=& \max\limits_{W\in\mathcal{W}_1}\inf\limits_{(\Upsilon,\Theta,H)\in\mathcal{Z}}\left\{\mathrm{Tr}(W\Theta)+\phi_{\mathcal{R}}(\lambda[\Upsilon])+\overline{\Psi}(H)\right\}\\
&=& \inf\limits_{(\Upsilon,\Theta,H)\in\mathcal{Z}}\max\limits_{W\in\mathcal{W}_1}\left\{\mathrm{Tr}(W\Theta)+\phi_{\mathcal{R}}(\lambda[\Upsilon])+\overline{\Psi}(H)\right\}\\
&=& \inf\limits_{(\Upsilon,\Theta,H)\in\mathcal{Z}}\left\{\overline{\Psi}(H)+\phi_{\mathcal{R}}(\lambda[\Upsilon])+\max\limits_{W}\left\{\mathrm{Tr}(\Theta W):\right.\right.\\
&& \left.\left. W\succeq 0,\exists t\in\mathcal{T}:\mathcal{R}_k[W]\preceq t_kI_{d_k},k\leq K\right\}\right\}\\
&=& \inf\limits_{(\Upsilon,\Theta,H)\in\mathcal{Z}}\left\{\overline{\Psi}(H)+\phi_{\mathcal{R}}(\lambda[\Upsilon])+\max\limits_{W,t}\left\{\mathrm{Tr}(\Theta W):\right.\right.\\
&& \left.\left. W\succeq 0,[t;1]\in\mathbf{T},\mathcal{R}_k[W]\preceq t_kI_{d_k},k\leq K\right\}\right\},\end{array}$$

where $\mathbf{T}$ is the closed conic hull of $\mathcal{T}$. On the other hand, using Conic Duality combined with the fact that $\mathbf{T}_*=\{[g;s]:s\geq\phi_{\mathcal{T}}(-g)\}$ we obtain

$$\begin{array}{l}\max\limits_{W,t}\{\mathrm{Tr}(\Theta W):\ W\succeq 0,[t;1]\in\mathbf{T},\,\mathcal{R}_k[W]\preceq t_kI_{d_k},\,k\leq K\}\\
=\min\limits_{Z,[g;s],\Lambda=\{\Lambda_k\}}\left\{s:\begin{array}{l}Z\succeq 0,[g;s]\in\mathbf{T}_*,\,\Lambda_k\succeq 0,k\leq K,\\ -\mathrm{Tr}(ZW)-g^Tt+\sum_k\mathrm{Tr}(\mathcal{R}_k^*[\Lambda_k]W)\\ \qquad -\sum_k t_k\mathrm{Tr}(\Lambda_k)=\Theta,\\ \forall(W\in\mathbf{S}^n,t\in\mathbf{R}^K)\end{array}\right\}\\
=\min\limits_{Z,[g;s],\Lambda=\{\Lambda_k\}}\left\{s:\begin{array}{l}Z\succeq 0,\,s\geq\phi_{\mathcal{T}}(-g),\,\Lambda_k\succeq 0,\,k\leq K,\\ \Theta=\sum_k\mathcal{R}_k^*[\Lambda_k]-Z,\,g=-\lambda[\Lambda]\end{array}\right\}\\
=\min\limits_{\Lambda}\left\{\phi_{\mathcal{T}}(\lambda[\Lambda]):\,\Lambda=\{\Lambda_k\succeq 0,k\leq K\},\,\Theta\preceq\sum\limits_k\mathcal{R}_k^*[\Lambda_k]\right\},\end{array}$$

and we arrive at

$$\begin{array}{rl}\mathrm{Opt}(1)=\inf\limits_{\Upsilon,\Theta,H,\Lambda} & \left\{\overline{\Psi}(H)+\phi_{\mathcal{R}}(\lambda[\Upsilon])+\phi_{\mathcal{T}}(\lambda[\Lambda]):\right.\\
& \left.\begin{array}{l}\Upsilon=\{\Upsilon_\ell\succeq 0,\ell\leq L\},\Lambda=\{\Lambda_k\succeq 0,k\leq K\},\\ \Theta\preceq\sum_k\mathcal{R}_k^*[\Lambda_k],\\ \left[\begin{array}{c|c}\Theta & \frac{1}{2}[B^T-A^TH]M\\ \hline \frac{1}{2}M^T[B-H^TA] & \sum_\ell\mathcal{S}_\ell^*[\Upsilon_\ell]\end{array}\right]\succeq 0\end{array}\right\}\\
=\inf\limits_{\Upsilon,H,\Lambda} & \left\{\overline{\Psi}(H)+\phi_{\mathcal{R}}(\lambda[\Upsilon])+\phi_{\mathcal{T}}(\lambda[\Lambda]):\right.\\
& \left.\begin{array}{l}\Upsilon=\{\Upsilon_\ell\succeq 0,\,\ell\leq L\},\Lambda=\{\Lambda_k\succeq 0,k\leq K\},\\ \left[\begin{array}{c|c}\sum_k\mathcal{R}_k^*[\Lambda_k] & \frac{1}{2}[B^T-A^TH]M\\ \hline \frac{1}{2}M^T[B-H^TA] & \sum_\ell\mathcal{S}_\ell^*[\Upsilon_\ell]\end{array}\right]\succeq 0\end{array}\right\}\\
=\mathrm{Opt} & [\text{see (4.156)}].\end{array}$$

7°. Now we can complete the proof. For $\kappa\in(0,1]$, let W_κ be the W-component of a saddle point solution to the saddle point problem (4.163). Then, by (4.166),

$$\begin{array}{rcl}\sqrt{\kappa}\mathrm{Opt} \leq \mathrm{Opt}(\kappa) & = & \inf\limits_{(\Upsilon,\Theta,H)\in\mathcal{Z}} \left\{\mathrm{Tr}(W_\kappa\Theta) + \phi_{\mathcal{R}}(\lambda[\Upsilon]) + \overline{\Psi}(H)\right\} \\ & = & \inf\limits_{H} \left\{\overline{\Phi}(W_\kappa, H) + \overline{\Psi}(H)\right\}.\end{array} \tag{4.167}$$

On the other hand, when applying Lemma 4.46 to $Q = W_\kappa$ and $\rho = \kappa$, we obtain, in view of relations $0 < \kappa \leq 1$, $W_\kappa \in \mathcal{W}_\kappa$,

$$\delta(\kappa) := \mathrm{Prob}_{\zeta\sim\mathcal{N}(0,I_n)}\{W_\kappa^{1/2}\zeta \notin \mathcal{X}\} \leq 2D\mathrm{e}^{-\frac{1}{2\kappa}}, \tag{4.168}$$

with D given by (4.55). In particular, when setting

$$\bar{\kappa} = \frac{1}{2\ln D + 10\ln 2} \tag{4.169}$$

we obtain $\delta_{\bar\kappa} \leq 1/16$. Therefore,

$$\mathrm{Prob}_{\eta\sim\mathcal{N}(0,W_{\bar\kappa})}\{\eta \notin \mathcal{X}\} \leq \tfrac{1}{16}. \tag{4.170}$$

Now let

$$\varrho_* := \frac{\mathrm{Opt}}{8\sqrt{(2\ln F + 10\ln 2)(2\ln D + 10\ln 2)}}. \tag{4.171}$$

All we need in order to achieve our goal of justifying (4.54) is to show that

$$\mathrm{RiskOpt}_{\frac{1}{8}} \geq \varrho_*, \tag{4.172}$$

since given the latter relation, (4.54) will be immediately given by (4.157) as applied with $\epsilon = \frac{1}{8}$.

To prove (4.172), assume, on the contrary to what should be proved, that the $\frac{1}{8}$-risk is $< \varrho_*$, and let $\bar{x}(\cdot)$ be an estimate with $\frac{1}{8}$-risk $\varrho' < \varrho_*$. We can utilize $\bar{x}$ to estimate $B\eta$, in the Bayesian problem of recovering $B\eta$ from observation $\omega = A\eta+\xi$, $(\eta,\xi) \sim \mathcal{N}(0,\Sigma)$ with $\Sigma = \mathrm{Diag}\{W_{\bar\kappa}, Q_*\}$. From (4.170) we conclude that

$$\begin{array}{l}\mathrm{Prob}_{(\eta,\xi)\sim\mathcal{N}(0,\Sigma)}\{\|B\eta - \bar{x}(A\eta+\xi)\| > \varrho'\} \\ \quad\leq\ \mathrm{Prob}_{(\eta,\xi)\sim\mathcal{N}(0,\Sigma)}\{\|B\eta - \bar{x}(A\eta+\xi)\| > \varrho',\ \eta \in \mathcal{X}\} \\ \qquad + \mathrm{Prob}_{\eta\sim\mathcal{N}(0,W_{\bar\kappa})}\{\eta \notin \mathcal{X}\} \leq \frac{1}{8} + \frac{1}{16} = \frac{3}{16}.\end{array} \tag{4.173}$$

On the other hand, by (4.167) we have

$$\inf_H [\overline{\Phi}(W_{\bar\kappa}, H) + \overline{\Psi}(H)] = \mathrm{Opt}(\bar\kappa) \geq \sqrt{\bar\kappa}\mathrm{Opt} = [8\bar{\varkappa}]\varrho_*$$

with $\bar\varkappa$ given by (4.162). Thus, by Lemma 4.49, for any estimate $\hat{x}(\cdot)$ of $B\eta$ via observation $\omega = Ax + \xi$ it holds

$$\mathrm{Prob}_{\eta,\xi}\{\|B\eta - \widehat{x}(A\eta+\xi)\| \geq \varrho_*\} \geq \beta_{\bar\varkappa} > 3/16;$$

in particular, this relation should hold true for $\widehat{x}(\cdot) \equiv \bar{x}(\cdot)$, but the latter is impossible: the $\frac{3}{16}$-risk of $\bar{x}$ is $\leq \varrho' < \varrho_*$; see (4.173). □

4.8.5.2 Proof of Proposition 4.5

We shall extract Proposition 4.5 from the following result, meaningful by its own right (it can be considered as an "ellitopic refinement" of Proposition 4.16):

Proposition 4.50. *Consider the recovery of the linear image $Bx \in \mathbf{R}^\nu$ of unknown signal x known to belong to a given signal set $\mathcal{X} \subset \mathbf{R}^n$ from noisy observation*

$$\omega = Ax + \xi \in \mathbf{R}^m \qquad [\xi \sim \mathcal{N}(0,\Gamma),\ \Gamma \succ 0],$$

the recovery error being measured in norm $\|\cdot\|$ on $\mathbf{R}^\nu$. Assume that $\mathcal{X}$ and the unit ball $\mathcal{B}_*$ of the norm $\|\cdot\|_*$ conjugate to $\|\cdot\|$ are ellitopes:

$$\begin{array}{rcl}\mathcal{X} &=& \{x \in \mathbf{R}^n : \exists t \in \mathcal{T} : x^T R_k x \leq t_k,\ k \leq K\},\\ \mathcal{B}_* &=& \{y \in \mathbf{R}^\nu : \exists (r \in \mathcal{R}, y) : u = My, y^T S_\ell y \leq r_\ell,\ \ell \leq L\},\end{array} \tag{4.174}$$

with our standard restrictions on $\mathcal{T}$, $\mathcal{R}$, R_k and S_ℓ (as always, we lose nothing when assuming that the ellitope $\mathcal{X}$ is basic).

Consider the optimization problem

$$\begin{array}{rcl}\mathrm{Opt}_\# &=& \min\limits_{\Theta,H,\lambda,\mu,\mu'} \left\{ \phi_{\mathcal{T}}(\lambda) + \phi_{\mathcal{R}}(\mu) + \phi_{\mathcal{R}}(\mu') + \mathrm{Tr}(\Gamma\Theta) : \right.\\ && \lambda \geq 0, \mu \geq 0, \mu' \geq 0,\\ && \left[\begin{array}{c|c}\sum_k \lambda_k R_k & \frac{1}{2}[B - H^T A]^T M \\ \hline \frac{1}{2}M^T[B - H^T A] & \sum_\ell \mu_\ell S_\ell\end{array}\right] \succeq 0,\\ && \left.\left[\begin{array}{c|c}\Theta & \frac{1}{2}HM \\ \hline \frac{1}{2}M^T H^T & \sum_\ell \mu'_\ell S_\ell\end{array}\right] \succeq 0 \right\}.\end{array} \tag{4.175}$$

The problem is solvable, and the linear estimate $\widehat{x}_{H_*}(\omega) = H_*^T\omega$ yielded by the H-component of an optimal solution to the problem satisfies the risk bound

$$\mathrm{Risk}_{\Gamma,\|\cdot\|}[\widehat{x}_{H_*}|\mathcal{X}] := \max_{x\in\mathcal{X}} \mathbf{E}_{\xi\sim\mathcal{N}(0,\Gamma)} \{\|Bx - \widehat{x}_{H_*}(Ax+\xi)\|\} \leq \mathrm{Opt}_\#.$$

Furthermore, the estimate $\widehat{x}_{H_*}(\cdot)$ is near-optimal:

$$\mathrm{Opt}_\# \leq 64\sqrt{(3\ln K + 15\ln 2)(3\ln L + 15\ln 2)}\,\mathrm{RiskOpt}, \tag{4.176}$$

where RiskOpt *is the minimax optimal risk*

$$\mathrm{RiskOpt} = \inf_{\widehat{x}} \sup_{x\in\mathcal{X}} \mathbf{E}_{\xi\sim\mathcal{N}(0,\Gamma)} \{\|Bx - \widehat{x}(Ax+\xi)\|\},$$

the infimum being taken w.r.t. all estimates.

Proposition 4.50 ⇒ Proposition 4.5: Clearly, the situation considered in Proposition 4.5 is a particular case of the setting of Proposition 4.50, namely, the case where $\mathcal{B}_*$ is the standard Euclidean ball, $\mathcal{B}_* = \{u \in \mathbf{R}^\nu : u^T u \leq 1\}$. In this case, problem (4.175) reads

$$\begin{array}{rl}
\mathrm{Opt}_{\#} = \min\limits_{\Theta,H,\lambda,\mu,\mu'} & \Big\{\phi_{\mathcal{T}}(\lambda)+\mu+\mu'+\mathrm{Tr}(\Gamma\Theta): \\
& \quad \lambda\geq 0,\mu\geq 0,\mu'\geq 0, \\
& \quad \left[\begin{array}{c|c}\sum_k\lambda_kR_k & \frac{1}{2}[B-H^TA]^T\\ \hline \frac{1}{2}[B-H^TA] & \mu I_\nu\end{array}\right]\succeq 0, \\
& \quad \left[\begin{array}{c|c}\Theta & \frac{1}{2}H\\ \hline \frac{1}{2}H^T & \mu' I_\nu\end{array}\right]\succeq 0 \Big\} \\
= \min\limits_{\Theta,H,\lambda,\mu,\mu'} & \Big\{\phi_{\mathcal{T}}(\lambda)+\mu+\mu'+\mathrm{Tr}(\Gamma\Theta): \\
& \quad \lambda\geq 0,\mu\geq 0,\mu'\geq 0, \\
& \quad \mu\left[\sum_k\lambda_kR_k\right]\succeq\frac{1}{4}[B-H^TA]^T[B-H^TA], \\
& \quad \mu'\Theta\succeq\frac{1}{4}HH^T \Big\} \\
& \qquad\qquad [\text{Schur Complement Lemma}] \\
= \min\limits_{\chi,H} & \Big\{\sqrt{\phi_{\mathcal{T}}(\chi)}+\sqrt{\mathrm{Tr}(H\Gamma H^T)}: \\
& \quad \chi\geq 0, \left[\begin{array}{c|c}\sum_k\chi'_kR_k & [B-H^TA]^T\\ \hline [B-H^TA] & I_\nu\end{array}\right]\succeq 0\Big\} \\
& \quad [\text{by eliminating } \mu,\mu'; \text{ note that } \phi_{\mathcal{T}}(\cdot) \\
& \quad \text{is positively homogeneous of degree 1}].
\end{array}$$

Comparing the resulting representation of $\mathrm{Opt}_{\#}$ with (4.12), we see that the upper bound $\sqrt{\mathrm{Opt}}$ on the risk of the linear estimate $\widehat{x}_{H_*}$ appearing in (4.15) is $\leq \mathrm{Opt}_{\#}$. Combining this observation with (4.176) and the evident relation

$$\begin{array}{rcl}
\mathrm{RiskOpt} & = & \inf_{\widehat{x}}\sup_{x\in\mathcal{X}}\mathbf{E}_{x\sim\mathcal{N}(0,\Gamma)}\left\{\|Bx-\widehat{x}(Ax+\xi)\|_2\right\} \\
& \leq & \inf_{\widehat{x}}\sqrt{\sup_{x\in\mathcal{X}}\mathbf{E}_{x\sim\mathcal{N}(0,\Gamma)}\left\{\|Bx-\widehat{x}(Ax+\xi)\|_2^2\right\}} = \mathrm{Risk}_{\mathrm{opt}}
\end{array}$$

(recall that we are in the case of $\|\cdot\| = \|\cdot\|_2$), we arrive at (4.15) and thus justify Proposition 4.5. □

Proof of Proposition 4.50. It is immediately seen that problem (4.175) is nothing but problem (4.42) in the case when the spectratopes $\mathcal{X},\mathcal{B}_*$ and the set Π participating in Proposition 4.14 are, respectively, the ellitopes given by (4.174), and the singleton $\{\Gamma\}$. Thus, Proposition 4.50 is, essentially, a particular case of Proposition 4.16. The only refinement in Proposition 4.50 as compared to Proposition 4.16 is the form of the logarithmic "nonoptimality" factor in (4.176); a similar factor in Proposition 4.16 is expressed in terms of spectratopic sizes D, F of $\mathcal{X}$ and $\mathcal{B}_*$ (the total ranks of matrices R_k, $k\leq K$, and S_ℓ, $\ell\leq L$, in the case of (4.174)), while in (4.176) the nonoptimality factor is expressed in terms of ellitopic sizes K, L of $\mathcal{X}$ and $\mathcal{B}_*$. Strictly speaking, to arrive at this (slight—the sizes in question are under logs) refinement, we were supposed to reproduce, with minimal modifications, the reasoning of items 2°–7° of Section 4.8.5.1, with Γ in the role of Q_*, and slightly refine Lemma 4.17 underlying this reasoning. Instead of carrying out this plan literally, we detail "local modifications" to be made in the proof of Proposition 4.16 in order to prove Proposition 4.50. Here are these modifications:

A. The collections of matrices $\Lambda=\{\Lambda_k\succeq 0,k\leq K\}$, $\Upsilon=\{\Upsilon_\ell\succeq 0,\ell\leq L\}$ should be substituted by collections of nonnegative reals $\lambda\in\mathbf{R}^K_+$ or $\mu\in\mathbf{R}^L_+$, and vectors

$\lambda[\Lambda]$, $\lambda[\Upsilon]$—with vectors λ or μ. Expressions like $\mathcal{R}_k[W]$, $\mathcal{R}_k^*[\Lambda_k]$, and $\mathcal{S}_\ell^*[\Upsilon_\ell]$ should be replaced, respectively, with $\mathrm{Tr}(R_kW)$, λ_kR_k, and $\mu_\ell S_\ell$. Finally, Q_* should be replaced with Γ, and scalar matrices, like $t_kI_{d_k}$, should be replaced with the corresponding reals, like t_k.

B. The role of Lemma 4.17 is now played by

Lemma 4.51. *Let Y be an $N\times\nu$ matrix, let $\|\cdot\|$ be a norm on $\mathbf{R}^\nu$ such that the unit ball $\mathcal{B}_*$ of the conjugate norm is the ellitope*

$$\mathcal{B}_* = \{y\in\mathbf{R}^\nu : \exists(r\in\mathcal{R}, y) : u = My, y^TS_\ell y\leq r_\ell, \ell\leq L\}, \tag{4.174}$$

and let $\zeta\sim\mathcal{N}(0,Q)$ for some positive semidefinite $N\times N$ matrix Q. Then the best upper bound on $\psi_Q(Y) := \mathbf{E}\{\|Y^T\zeta\|\}$ yielded by Lemma 4.11, that is, the optimal value $\mathrm{Opt}[Q]$ *in the convex optimization problem (cf. (4.40))*

$$\mathrm{Opt}[Q] \;=\; \min_{\Theta,\mu}\left\{\phi_{\mathcal{R}}(\mu)+\mathrm{Tr}(Q\Theta) : \mu\geq 0, \left[\begin{array}{c|c}\Theta & \frac{1}{2}YM\\ \hline \frac{1}{2}M^TY^T & \sum_\ell \mu_\ell R_\ell\end{array}\right]\succeq 0\right\}$$

satisfies for all $Q\succeq 0$ the identity

$$\begin{array}{rl}\mathrm{Opt}[Q]=\overline{\mathrm{Opt}}[Q] := & \min\limits_{G,\mu}\left\{\phi_{\mathcal{R}}(\mu)+\mathrm{Tr}(G) : \right.\\ & \left.\mu\geq 0, \left[\begin{array}{c|c} G & \frac{1}{2}Q^{1/2}YM\\ \hline \frac{1}{2}M^TY^TQ^{1/2} & \sum_\ell\mu_\ell R_\ell\end{array}\right]\succeq 0\right\},\end{array} \tag{4.177}$$

and is a tight bound on $\psi_Q(Y)$. Namely,

$$\psi_Q(Y)\leq \mathrm{Opt}[Q]\leq 22\sqrt{3\ln L+15\ln 2}\,\psi_Q(Y),$$

where L is the size of the ellitope $\mathcal{B}_*$; see (4.174). *Furthermore, for all $\varkappa\geq 1$ one has*

$$\mathrm{Prob}_\zeta\left\{\|Y^T\zeta\|\geq\frac{\mathrm{Opt}[Q]}{4\varkappa}\right\}\geq\beta_\varkappa := 1-\frac{\mathrm{e}^{3/8}}{2}-2L\mathrm{e}^{-\varkappa^2/3}. \tag{4.178}$$

In particular, when selecting $\varkappa=\sqrt{3\ln L+15\ln 2}$, we obtain

$$\mathrm{Prob}_\zeta\left\{\|Y^T\zeta\|\geq\frac{\mathrm{Opt}[Q]}{4\sqrt{3\ln L+15\ln 2}}\right\}\geq\beta_\varkappa = 0.2100>\tfrac{3}{16}.$$

Proof of Lemma 4.51 follows the lines of the proof of Lemma 4.17, with Lemma 4.47 substituting Lemma 4.46.

1°. Relation (4.177) can be verified exactly in the same fashion as in the case of Lemma 4.17.

2°. Let us set $\zeta = Q^{1/2}\eta$ with $\eta\sim\mathcal{N}(0,I_N)$ and $Z=Q^{1/2}Y$. Observe that to prove (4.178) is the same as to show that when $\varkappa\geq 1$ one has

$$\bar{\delta}=\frac{\mathrm{Opt}[Q]}{4\varkappa}\Rightarrow \mathrm{Prob}_\eta\{\|Z^T\eta\|\geq\bar{\delta}\}\geq\beta_\varkappa := 1-\frac{\mathrm{e}^{3/8}}{2}-2L\mathrm{e}^{-\varkappa^2/3}, \tag{4.179}$$

where

$$[\overline{\mathrm{Opt}}[Q] =] \quad \mathrm{Opt}[Q] \quad := \quad \min_{\Theta,\mu}\left\{\phi_{\mathcal{R}}(\mu) + \mathrm{Tr}(\Theta) : \mu \geq 0, \left[\begin{array}{c|c}\Theta & \frac{1}{2}ZM \\ \hline \frac{1}{2}M^T Z^T & \sum_\ell \mu_\ell R_\ell\end{array}\right] \succeq 0\right\}. \tag{4.180}$$

Justification of (4.179) goes as follows.

2.1^o. Let us represent Opt$[Q]$ as the optimal value of a conic problem. Setting

$$\mathbf{K} = \mathrm{cl}\{[r;s] : s > 0, r/s \in \mathcal{R}\},$$

we ensure that

$$\mathcal{R} = \{r : [r;1] \in \mathbf{K}\},\ \mathbf{K}_* = \{[g;s] : s \geq \phi_{\mathcal{R}}(-g)\},$$

where $\mathbf{K}_*$ is the cone dual to $\mathbf{K}$. Consequently, (4.180) reads

$$\mathrm{Opt}[Q] = \min_{\Theta,\Upsilon,\theta}\left\{\theta + \mathrm{Tr}(\Theta) : \begin{array}{lr} \mu \geq 0 & (a) \\ \left[\begin{array}{c|c}\Theta & \frac{1}{2}ZM \\ \hline \frac{1}{2}M^T Z^T & \sum_\ell \mu_\ell S_\ell\end{array}\right] \succeq 0 & (b) \\ [-\mu;\theta] \in \mathbf{K}_* & (c)\end{array}\right\}. \tag{$P_\mathcal{E}$}$$

2.2^o. Now let us prove that there exist matrix $W \in \mathbf{S}^q_+$ and $r \in \mathcal{R}$ such that

$$\mathrm{Tr}(WS_\ell) \leq r_\ell,\ \ell \leq L, \tag{4.181}$$

and

$$\mathrm{Opt}[Q] \leq \sum_i \sigma_i(ZMW^{1/2}), \tag{4.182}$$

where $\sigma_1(\cdot) \geq \sigma_2(\cdot) \geq ...$ are singular values.

To get the announced result, let us pass from problem (P) to its conic dual. $(P_\mathcal{E})$ clearly is strictly feasible and bounded, so that the dual to $(P_\mathcal{E})$ problem $(D_\mathcal{E})$ is solvable with optimal value Opt$[Q]$. Denoting by $\lambda_\ell \geq 0, \ell \leq L$, $\left[\begin{array}{c|c} G & -R \\ \hline -R^T & W\end{array}\right] \succeq 0$, $[r;\tau] \in \mathbf{K}$, the Lagrange multipliers for the respective constraints in $(P_\mathcal{E})$, and aggregating these constraints, the multipliers being the aggregation weights, we arrive at the aggregated constraint:

$$\mathrm{Tr}(\Theta G) + \mathrm{Tr}(W \textstyle\sum_\ell \mu_\ell S_\ell) + \sum_\ell \lambda_\ell \mu_\ell - \sum_\ell r_\ell \mu_\ell + \theta\tau \geq \mathrm{Tr}(ZMR^T).$$

To get the dual problem, we impose on the Lagrange multipliers, in addition to the initial constraints, the restriction that the left-hand side in the aggregated constraint is equal to the objective of (P), identically in Θ, μ_ℓ, and θ, that is,

$$G = I,\ \mathrm{Tr}(WS_\ell) + \lambda_\ell - r_\ell = 0,\ 1 \leq \ell \leq L,\ \tau = 1,$$

and maximize the right-hand side of the aggregated constraint. After immediate simplifications, we arrive at

$$\mathrm{Opt}[Q] = \max_{W,R,r}\left\{\mathrm{Tr}(ZMR^T) : W \succeq R^T R, r \in \mathcal{R}, \mathrm{Tr}(WS_\ell) \leq r_\ell,\ 1 \leq \ell \leq L\right\}$$

(note that $r \in \mathcal{R}$ is equivalent to $[r;1] \in \mathbf{K}$, and $W \succeq R^T R$ is the same as $\left[\begin{array}{c|c} I & -R \\ \hline -R^T & W\end{array}\right] \succeq 0$). Exactly as in the proof of Lemma 4.17, the above representation of Opt$[Q]$ implies that

$$\mathrm{Opt}[Q] = \max_{W,r}\left\{\|ZMW^{1/2}\|_{\mathrm{Sh},1} : r \in \mathcal{R}, W \succeq 0, \mathrm{Tr}(WS_\ell) \leq r_\ell,\ \ell \leq L\right\}.$$

The resulting problem clearly is solvable, and its optimal solution W ensures the target relations (4.181) and (4.182).

2.3^o. Given W satisfying (4.181) and (4.182), we proceed exactly as in item 2.3^o of the proof of Lemma 4.17, thus arriving at three random vectors (χ, υ, η) with marginal distributions $\mathcal{N}(0, I_q)$, $\mathcal{N}(0, I_q)$, and $\mathcal{N}(0, I_N)$, respectively, such that

$$\chi^T W^{1/2} M^T Z^T \eta = \sum_{i=1}^{p} \sigma_i \upsilon_i^2, \tag{4.183}$$

where $p = \min[q, N]$ and $\sigma_i = \sigma_i(ZMW^{1/2})$. As in item 3°.i of the proof of Lemma 4.17, we have *(i)*

$$\alpha := \mathrm{Prob}\left\{\sum_{i=1}^{p} \sigma_i \upsilon_i^2 < \tfrac{1}{4}\sum_{i=1}^{p} \sigma_i\right\} \leq \frac{\mathrm{e}^{3/8}}{2} \; [= 0.7275...]. \tag{4.184}$$

The role of item 3°.ii in the aforementioned proof is now played by

(ii) Whenever $\varkappa \geq 1$, one has

$$\mathrm{Prob}\{\|MW^{1/2}\chi\|_* > \varkappa\} \leq 2L\exp\{-\varkappa^2/3\}, \tag{4.185}$$

with L as defined in (4.174).

Indeed, setting $\rho = 1/\varkappa^2 \leq 1$ and $\omega = \sqrt{\rho}W^{1/2}\chi$, we get $\omega \sim \mathcal{N}(0, \rho W)$. Let us apply Lemma 4.47 to $Q = \rho W$, $\mathcal{R}$ in the role of $\mathcal{T}$, with L in the role of K, and S_ℓ's in the role of R_k's. Denoting

$$\mathcal{Y} := \{y : \exists r \in \mathcal{R} : y^T S_\ell y \preceq r_\ell, \ell \leq L\},$$

we have $\mathrm{Tr}(QS_\ell) = \rho\mathrm{Tr}(WS_\ell) = \rho\mathrm{Tr}(WS_\ell) \leq \rho r_\ell$, $\ell \leq L$, with $r \in \mathcal{R}$ (see (4.181)), so we are under the premise of Lemma 4.47 (with $\mathcal{Y}$ in the role of $\mathcal{X}$ and therefore with L in the role of K). Applying the lemma, we conclude that

$$\mathrm{Prob}\left\{\chi : \varkappa^{-1}W^{1/2}\chi \notin \mathcal{Y}\right\} \leq 2L\exp\{-1/(3\rho)\} = 2L\exp\{-\varkappa^2/3\}.$$

Recalling that $\mathcal{B}_* = M\mathcal{Y}$, we see that $\mathrm{Prob}\{\chi : \varkappa^{-1}MW^{1/2}\chi \notin \mathcal{B}_*\}$ is indeed upper-bounded by the right-hand side of (4.185), and (4.185) follows.

With (i) and (ii) at our disposal, we complete the proof of Lemma 4.51 in exactly the same way as in items 2.4^o and 3° of the proof of Lemma 4.17. □

C. As a result of substituting Lemma 4.17 with Lemma 4.51, the counterpart of Lemma 4.49 used in item 4° of the proof of Proposition 4.16 now reads as follows:

Lemma 4.52. *Let W be a positive semidefinite $n \times n$ matrix, and $\varkappa \geq 1$. Then for any estimate $\widehat{x}(\cdot)$ of $B\eta$ given observation $\omega = A\eta + \xi$ with $\eta \sim \mathcal{N}(0, W)$ and $\xi \sim \mathcal{N}(0, \Gamma)$ independent of each other, one has*

$$\mathrm{Prob}_{\eta,\xi}\left\{\|B\eta - \widehat{x}(\omega)\| \geq [8\varkappa]^{-1}\inf_H[\overline{\Psi}(H) + \overline{\Phi}(W, H)]\right\} \geq \beta_\varkappa = 1 - \frac{\mathrm{e}^{3/8}}{2} - 2L\mathrm{e}^{-\varkappa^2/3}$$

where $\overline{\Psi}(H)$ and $\overline{\Phi}(W, H)$ are defined, respectively, by (4.156) (where Q_* should be set to Γ) and (4.161).

In particular, for

$$\varkappa = \bar{\varkappa} := \sqrt{3\ln K + 15\ln 2}$$

the latter probability is $> 3/16$.

D. We substitute the reference to Lemma 4.46 in item 7° of the proof with Lemma 4.47, resulting in replacing

- definition of $\delta(\kappa)$ in (4.168) with

$$\delta(\kappa) := \mathrm{Prob}_{\zeta\sim\mathcal{N}(0,I_n)}\{W_\kappa^{1/2}\zeta \not\in \mathcal{X}\} \leq 3K\mathrm{e}^{-\frac{1}{3\kappa}},$$

- definition (4.169) of $\bar{\kappa}$ with

$$\bar{\kappa} = \frac{1}{3\ln K + 15\ln 2},$$

- and, finally, definition (4.171) of ρ_* with

$$\varrho_* := \frac{\mathrm{Opt}}{8\sqrt{(3\ln L + 15\ln 2)(3\ln K + 15\ln 2)}}.$$

4.8.6 Proofs of Propositions 4.18 and 4.19, and justification of Remark 4.20

4.8.6.1 Proof of Proposition 4.18

The only claim of the proposition which is not an immediate consequence of Proposition 4.8 is that problem (4.64) is solvable; let us justify this claim. Let $F = \mathrm{Im}A$. Clearly, feasibility of a candidate solution (H, Λ, Υ) to the problem depends solely on the restriction of the linear mapping $z \mapsto H^Tz$ onto F, so that adding to the constraints of the problem the requirement that the restriction of this linear mapping on the orthogonal complement of F in $\mathbf{R}^m$ is identically zero, we get an equivalent problem. It is immediately seen that in the resulting problem, the feasible solutions with the value of the objective $\leq a$ for every $a \in \mathbf{R}$ form a compact set, so that the latter problem (and thus the original one) indeed is solvable. □

4.8.6.2 Proof of Proposition 4.19

We are about to derive Proposition 4.19 from Proposition 4.16. Observe that in the situation of the latter Proposition, setting formally $\Pi = \{0\}$, problem (4.42) becomes problem (4.64), so that Proposition 4.19 looks like the special case $\Pi = \{0\}$ of Proposition 4.16. However, the premise of the latter proposition forbids specializing Π as $\{0\}$—this would violate the regularity assumption **R** which is part of the premise. The difficulty, however, can be easily resolved. Assume w.l.o.g. that the image space of A is the entire $\mathbf{R}^m$ (otherwise we could from the very beginning replace $\mathbf{R}^m$ with the image space of A), and let us pass from our current noiseless recovery problem of interest **(!)**—see Section 4.5.1—to its "noisy modification," the differences with **(!)** being

- noisy observation $\omega = Ax + \sigma\xi$, $\sigma > 0$, $\xi \sim \mathcal{N}(0, I_m)$;
- risk quantification of a candidate estimate $\widehat{x}(\cdot)$ according to

$$\mathrm{Risk}^\sigma_{\|\cdot\|}[\widehat{x}(Ax+\sigma\xi)|\mathcal{X}] = \sup_{x\in\mathcal{X}} \mathbf{E}_{\xi\sim\mathcal{N}(0,I_m)}\left\{\|Bx - \widehat{x}(Ax+\sigma\xi)\|\right\},$$

the corresponding minimax optimal risk being

$$\mathrm{RiskOpt}^\sigma_{\|\cdot\|}[\mathcal{X}] = \inf_{\widehat{x}(\cdot)} \mathrm{Risk}^\sigma_{\|\cdot\|}[\widehat{x}(Ax+\sigma\xi)|\mathcal{X}].$$

Proposition 4.16 does apply to the modified problem—it suffices to specify Π as $\{\sigma^2 I_m\}$. According to this proposition, the quantity

$$\begin{array}{rcl}\mathrm{Opt}[\sigma] & = & \min\limits_{H,\Lambda,\Upsilon,\Upsilon',\Theta}\left\{\phi_{\mathcal{T}}(\lambda[\Lambda])+\phi_{\mathcal{R}}(\lambda[\Upsilon])+\phi_{\mathcal{R}}(\lambda[\Upsilon'])+\sigma^2\mathrm{Tr}(\Theta):\right.\\ && \Lambda=\{\Lambda_k\succeq 0,k\leq K\},\ \Upsilon=\{\Upsilon_\ell\succeq 0,\ell\leq L\},\ \Upsilon'=\{\Upsilon'_\ell\succeq 0,\ell\leq L\},\\ && \left[\begin{array}{c|c}\sum_k \mathcal{R}_k^*[\Lambda_k] & \frac{1}{2}[B^T-A^TH]M\\ \hline \frac{1}{2}M^T[B-H^TA] & \sum_\ell \mathcal{S}_\ell^*[\Upsilon_\ell]\end{array}\right]\succeq 0,\\ && \left.\left[\begin{array}{c|c}\Theta & \frac{1}{2}HM\\ \hline \frac{1}{2}M^TH^T & \sum_\ell \mathcal{S}_\ell^*[\Upsilon'_\ell]\end{array}\right]\succeq 0\right\}\end{array}$$

satisfies the relation

$$\mathrm{Opt}[\sigma]\leq O(1)\ln(D)\mathrm{RiskOpt}^{\sigma}_{\|\cdot\|}[\mathcal{X}] \tag{4.186}$$

with D defined in (4.65). Looking at problem (4.64) we immediately conclude that $\mathrm{Opt}_{\#}\leq\mathrm{Opt}[\sigma]$. Thus, all we need in order to extract the target relation (4.65) from (4.186) is to prove that the minimax optimal risk $\mathrm{Risk}_{\mathrm{opt}}[\mathcal{X}]$ defined in Proposition 4.19 satisfies the relation

$$\liminf_{\sigma\to+0}\mathrm{RiskOpt}^{\sigma}_{\|\cdot\|}[\mathcal{X}]\leq\mathrm{Risk}_{\mathrm{opt}}[\mathcal{X}]. \tag{4.187}$$

To prove this relation, let us fix $r>\mathrm{Risk}_{\mathrm{opt}}[\mathcal{X}]$, so that for some Borel estimate $\widehat{x}(\cdot)$ it holds

$$\sup_{x\in\mathcal{X}}\|Bx-\widehat{x}(Ax)\|<r. \tag{4.188}$$

Were we able to ensure that $\widehat{x}(\cdot)$ is bounded and continuous, we would be done, since in this case, due to compactness of $\mathcal{X}$, it clearly holds

$$\begin{array}{rl}\liminf_{\sigma\to+0}\mathrm{RiskOpt}^{\sigma}_{\|\cdot\|}[\mathcal{X}] & \\ \leq & \liminf_{\sigma\to+0}\sup_{x\in\mathcal{X}}\mathbf{E}_{\xi\sim\mathcal{N}(0,I_m)}\{\|Bx-\widehat{x}(Ax+\sigma\xi)\|\}\\ \leq & \sup_{x\in\mathcal{X}}\|Bx-\widehat{x}(Ax)\|<r,\end{array}$$

and since $r>\mathrm{Risk}_{\mathrm{opt}}[\mathcal{X}]$ is arbitrary, (4.187) would follow. Thus, all we need to do is to verify that given Borel estimate $\widehat{x}(\cdot)$ satisfying (4.188), we can update it into a bounded and continuous estimate satisfying the same relation. Verification is as follows:

1. Setting $\beta=\max_{x\in\mathcal{X}}\|Bx\|$ and replacing estimate $\widehat{x}$ with its truncation

$$\widetilde{x}(\omega)=\left\{\begin{array}{ll}\widehat{x}(\omega), & \|\widehat{x}(\omega)\|\leq 2\beta\\ 0, & \text{otherwise}\end{array}\right.$$

for any $x\in\mathcal{X}$ we only reduce the norm of the recovery error. At the same time, $\widetilde{x}$ is Borel and bounded. Thus, we lose nothing when assuming in the rest of the proof that $\widehat{x}(\cdot)$ is Borel and bounded.

2. For $\epsilon>0$, let $\widehat{x}_\epsilon(\omega)=(1+\epsilon)\widehat{x}(\omega/(1+\epsilon))$ and let $\mathcal{X}_\epsilon=(1+\epsilon)\mathcal{X}$. Observe that

$$\begin{array}{l}\sup_{x\in\mathcal{X}_\epsilon}\|Bx-\widehat{x}_\epsilon(Ax)\|=\sup_{y\in\mathcal{X}}\|B[1+\epsilon]y-\widehat{x}_\epsilon(A[1+\epsilon]y)\|\\ \quad=\ \sup_{y\in\mathcal{X}}\|B[1+\epsilon]y-[1+\epsilon]\widehat{x}(Ay)\|=[1+\epsilon]\sup_{y\in\mathcal{X}}\|By-\widehat{x}(Ay)\|,\end{array}$$

implying, in view of (4.188), that for small enough positive ϵ we have

$$\bar{r} := \sup_{x\in\mathcal{X}_\epsilon} \|Bx - \widehat{x}_\epsilon(Ax)\| < r. \tag{4.189}$$

3. Finally, let $A^\dagger$ be the pseudoinverse of A, so that $AA^\dagger z = z$ for every $z \in \mathbf{R}^m$ (recall that the image space of A is the entire $\mathbf{R}^m$). Given $\rho > 0$, let $\theta_\rho(\cdot)$ be a nonnegative smooth function on $\mathbf{R}^m$ with integral 1 such that θ_ρ vanishes outside of the ball of radius ρ centered at the origin, and let

$$\widehat{x}_{\epsilon,\rho}(\omega) = \int_{\mathbf{R}^m} \widehat{x}_\epsilon(\omega - z)\theta_\rho(z)dz$$

be the convolution of $\widehat{x}_\epsilon$ and θ_ρ. Since $\widehat{x}_\epsilon(\cdot)$ is Borel and bounded, this convolution is a well-defined smooth function on $\mathbf{R}^m$. Because $\mathcal{X}$ contains a neighbourhood of the origin, for all small enough $\rho > 0$, all z from the support of θ_ρ and all $x \in \mathcal{X}$ the point $x - A^\dagger z$ belongs to $\mathcal{X}_\epsilon$. For such ρ and any $x \in \mathcal{X}$ we have

$$\begin{array}{rcl}\|Bx - \widehat{x}_\epsilon(Ax - z)\| &=& \|Bx - \widehat{x}_\epsilon(A[x - A^\dagger z])\| \\ &\leq& \|BA^\dagger z\| + \|B[x - A^\dagger z] - \widehat{x}_\epsilon(A[x - A^\dagger z])\| \\ &\leq& C\rho + \bar{r}\end{array}$$

with properly selected constant C independent of ρ (we have used (4.189); note that for our ρ and x we have $x - A^\dagger z \in \mathcal{X}_\epsilon$). We conclude that for properly selected $r' < r$, $\rho > 0$ and all $x \in \mathcal{X}$ we have

$$\|Bx - \widehat{x}_\epsilon(Ax - z)\| \leq r' \,\forall (z \in \operatorname{supp}\theta_\rho),$$

implying, by construction of $\widehat{x}_{\epsilon,\rho}$, that

$$\forall (x \in \mathcal{X}) : \|Bx - \widehat{x}_{\epsilon,\rho}(Ax)\| \leq r' < r.$$

The resulting estimate $\widehat{x}_{\epsilon,\rho}$ is the continuous and bounded estimate satisfying (4.188) we were looking for. □

4.8.6.3 Justification of Remark 4.20

Justification of Remark is given by repeating word by word the proof of Proposition 4.19, with Proposition 4.50 in the role of Proposition 4.16.

Chapter Five

Signal Recovery Beyond Linear Estimates

OVERVIEW

In this chapter, as in Chapter 4, we focus on signal recovery. In contrast to the previous chapter, on our agenda now are

- a special kind of nonlinear estimation—*polyhedral estimate* (Section 5.1), an alternative to linear estimates which were our subject in Chapter 4. We demonstrate that as applied to the same estimation problem as in Chapter 4—recovery of an unknown signal via noisy observation of a linear image of the signal, polyhedral estimation possesses the same attractive properties as linear estimation, that is, efficient computability and near-optimality, provided the signal set is an ellitope/spectratope. Besides this, we show that properly built polyhedral estimates are near-optimal in several special cases where linear estimates could be heavily suboptimal.
- recovering signals from noisy observations of *nonlinear* images of the signal. Specifically, we consider signal recovery in *generalized linear models*, where the expected value of an observation is a known *nonlinear* transformation of the signal we want to recover, in contrast to observation model (4.1) where this expectation is linear in the signal.

5.1 POLYHEDRAL ESTIMATION

5.1.1 Motivation

The estimation problem we were considering so far is as follows:

> *We want to recover the image $Bx \in \mathbf{R}^\nu$ of unknown signal x known to belong to signal set $\mathcal{X} \subset \mathbf{R}^n$ from a noisy observation*
>
> $$\omega = Ax + \xi_x \in \mathbf{R}^m,$$
>
> *where ξ_x is observation noise (index $_x$ in ξ_x indicates that the distribution P_x of the observation noise may depend on x). Here $\mathcal{X}$ is a given nonempty convex compact set, and A and B are given $m \times n$ and $\nu \times n$ matrices; in addition, we are given a norm $\|\cdot\|$ on $\mathbf{R}^\nu$ in which the recovery error is measured.*

We have seen that if $\mathcal{X}$ is an ellitope/spectratope then, under reasonable assumptions on observation noise and $\|\cdot\|$, an appropriate efficiently computable estimate *linear in ω* is near-optimal. Note that the ellitopic/spectratopic structure of $\mathcal{X}$ is crucial here. What follows is motivated by the desire to build an alternative estimation scheme which works beyond the ellitopic/spectratopic case, where linear estimates can become "heavily nonoptimal."

Motivating example. Consider the simply-looking problem of recovering $Bx = x$ in the $\|\cdot\|_2$-norm from *direct* observations ($Ax = x$) corrupted by the standard Gaussian noise $\xi \sim \mathcal{N}(0, \sigma^2 I)$, and let $\mathcal{X}$ be the unit $\|\cdot\|_1$-ball:

$$\mathcal{X} = \{x \in \mathbf{R}^n : \sum_i |x_i| \leq 1\}.$$

In this situation, one can easily build the optimal, in terms of the worst-case, over $x \in \mathcal{X}$, expected squared risk, linear estimate $\widehat{x}_H(\omega) = H^T\omega$:

$$\begin{array}{rcl}\mathrm{Risk}^2[\widehat{x}_H|\mathcal{X}] & := & \max_{x\in\mathcal{X}} \mathbf{E}\left\{\|\widehat{x}_H(\omega) - Bx\|_2^2\right\} \\ & = & \max_{x\in\mathcal{X}} \left\{\|[I - H^T]x\|_2^2 + \sigma^2 \mathrm{Tr}(HH^T)\right\} \\ & = & \max_{i\leq n} \|\mathrm{Col}_i[I - H^T]\|_2^2 + \sigma^2 \mathrm{Tr}(HH^T).\end{array}$$

Clearly, the optimal H is just a scalar matrix hI, the optimal h is the minimizer of the univariate quadratic function $(1-h)^2 + \sigma^2 nh^2$, and the best squared risk attainable with linear estimates is

$$R^2 = \min_h \left[(1-h)^2 + \sigma^2 nh^2\right] = \frac{n\sigma^2}{1 + n\sigma^2}.$$

On the other hand, consider a *nonlinear* estimate $\widehat{x}(\omega)$ as follows. Given observation ω, specify $\widehat{x}(\omega)$ as an optimal solution to the optimization problem

$$\mathrm{Opt}(\omega) = \min_{y\in\mathcal{X}} \|y - \omega\|_\infty.$$

Note that for every $\rho > 0$ the probability that the true signal satisfies $\|x - \omega\|_\infty \leq \rho\sigma$ ("event $\mathcal{E}$") is at least $1 - 2n\exp\{-\rho^2/2\}$, and if this event happens, then both x and $\widehat{x}$ belong to the box $\{y : \|y - \omega\|_\infty \leq \rho\sigma\}$, implying that $\|x - \widehat{x}\|_\infty \leq 2\rho\sigma$. In addition, we always have $\|x - \widehat{x}\|_2 \leq \|x - \widehat{x}\|_1 \leq 2$, since $x \in \mathcal{X}$ and $\widehat{x} \in \mathcal{X}$. We therefore have

$$\|x - \widehat{x}\|_2 \leq \sqrt{\|x - \widehat{x}\|_\infty \|x - \widehat{x}\|_1} \leq \left\{\begin{array}{ll} 2\sqrt{\rho\sigma}, & \omega \in \mathcal{E}, \\ 2, & \omega \notin \mathcal{E}, \end{array}\right.$$

whence

$$\mathbf{E}\left\{\|\widehat{x} - x\|_2^2\right\} \leq 4\rho\sigma + 8n\exp\{-\rho^2/2\}. \qquad (*)$$

Assuming $\sigma \leq 2n\exp\{-1/2\}$ and specifying ρ as $\sqrt{2\ln(2n/\sigma)}$, we get $\rho \geq 1$ and $2n\exp\{-\rho^2/2\} \leq \sigma$, implying that the right hand side in $(*)$ is at most $8\rho\sigma$. In other words, for our nonlinear estimate $\widehat{x}(\omega)$ it holds

$$\mathrm{Risk}^2[\widehat{x}|\mathcal{X}] \leq 8\sqrt{\ln(2n/\sigma)}\sigma.$$

When $n\sigma^2$ is of order of 1, the latter bound on the squared risk is of order of $\sigma\sqrt{\ln(1/\sigma)}$, while the best squared risk achievable with linear estimates under the circumstances is of order of 1. We conclude that when σ is small and n is large (specifically, is of order of $1/\sigma^2$), the best linear estimate is *by far* inferior compared to our nonlinear estimate—the ratio of the corresponding squared risks is as large as $\frac{O(1)}{\sigma\sqrt{\ln(1/\sigma)}}$, the factor which is "by far" worse than the nonoptimality factor in the case of ellitope/spectratope $\mathcal{X}$.

The construction of the nonlinear estimate $\widehat{x}$ which we have built[1] admits a natural extension yielding what we shall call *polyhedral estimate*, and our present goal is to design and to analyse presumably good estimates of this type.

5.1.2 Generic polyhedral estimate

A generic polyhedral estimate is as follows:

> Given the data $A \in \mathbf{R}^{m\times n}, B \in \mathbf{R}^{\nu\times n}, \mathcal{X} \subset \mathbf{R}^n$ of our recovery problem (where $\mathcal{X}$ is a computationally tractable convex compact set) and a "reliability tolerance" $\epsilon \in (0,1)$, we specify somehow positive integer N along with N linear forms $h_\ell^T z$ on the space $\mathbf{R}^m$ where observations live. These forms define linear forms $g_\ell^T x := h_\ell^T Ax$ on the space of signals $\mathbf{R}^n$. Assuming that the observation noise ξ_x is zero mean for every $x \in \mathcal{X}$, the "plug-in" estimates $h_\ell^T\omega$ are unbiased estimates of the forms $g_i^T x$. Assume that vectors h_ℓ are selected in such a way that
>
> $$\forall (x\in\mathcal{X}): \mathrm{Prob}\{|h_\ell^T\xi_x|>1\}\leq \epsilon/N \;\; \forall \ell. \tag{5.1}$$
>
> In this situation, setting $H=[h_1,...,h_N]$ (in the sequel, H is referred to as *contrast matrix*), we can ensure that whatever be the signal $x\in\mathcal{X}$ underlying our observation $\omega = Ax+\xi_x$, the observable vector $H^T\omega$ satisfies the relation
>
> $$\mathrm{Prob}\left\{\|H^T\omega - H^TAx\|_\infty > 1\right\}\leq\epsilon. \tag{5.2}$$
>
> With the polyhedral estimation scheme, we act *as if* all information about x contained in our observation ω were represented by $H^T\omega$, and we estimate Bx by $B\bar{x}$, where $\bar{x}=\bar{x}(\omega)$ is any vector from $\mathcal{X}$ compatible with this information, specifically, such that $\bar{x}$ solves the feasibility problem
>
> $$\text{find } \bar{x}\in\mathcal{X} \text{ such that } \|H^T\omega - H^TA\bar{x}\|_\infty \leq 1.$$
>
> Note that this feasibility problem with positive probability can be unsolvable; all we know in this respect is that the latter probability is $\leq 1-\epsilon$, since by construction the true signal x underlying observation ω is with probability $1-\epsilon$ a feasible solution. In other words, such $\bar{x}$ is not always well-defined. To circumvent this difficulty, let us define $\bar{x}$ as
>
> $$\bar{x}\in\underset{u}{\mathrm{Argmin}}\left\{\|H^T\omega - H^TAu\|_\infty : u\in\mathcal{X}\right\}, \tag{5.3}$$
>
> so that $\bar{x}$ always is well-defined and belongs to $\mathcal{X}$, and estimate Bx by $B\bar{x}$. Thus,
>
> > *a polyhedral estimate is specified by an $m\times N$ contrast matrix $H=[h_1,...,h_N]$ with columns h_ℓ satisfying* (5.1) *and is as follows: given observation ω, we build $\bar{x}=\bar{x}(\omega)\in\mathcal{X}$ according to* (5.3) *and estimate Bx by $\widehat{x}^H(\omega)=B\bar{x}(\omega)$.*

The rationale behind polyhedral estimation scheme is the desire to reduce complex

[1] In fact, this estimate is nearly optimal under the circumstances in a meaningful range of values of n and σ.

estimating problems to those of estimating linear forms. To the best of our knowledge, this approach was first used in [192] (see also [185, Chapter 2]) in connection with recovering from direct observations (restrictions on regular grids of) multivariate functions from Sobolev balls. Recently, the ideas underlying the results of [192] have been taken up in the MIND estimator of [109], then applied to multiple testing in [203]. What follows is based on [139].

$(\epsilon, \|\cdot\|)$-risk. Given a desired "reliability tolerance" $\epsilon \in (0,1)$, it is convenient to quantify the performance of polyhedral estimate by its $(\epsilon, \|\cdot\|)$-*risk*

$$\mathrm{Risk}_{\epsilon,\|\cdot\|}[\widehat{x}(\cdot)|\mathcal{X}] = \inf\left\{\rho : \mathrm{Prob}\left\{\|Bx - \widehat{x}(Ax+\xi_x)\| > \rho\right\} \leq \epsilon\,\forall x \in \mathcal{X}\right\}, \tag{5.4}$$

that is, the worst, over $x \in \mathcal{X}$, size of "$(1-\epsilon)$-reliable $\|\cdot\|$-confidence interval" associated with the estimate $\widehat{x}(\cdot)$.

An immediate observation is as follows:

Proposition 5.1. *In the situation in question, denoting by $\mathcal{X}_{\mathrm{s}} = \frac{1}{2}(\mathcal{X}-\mathcal{X})$ the symmetrization of $\mathcal{X}$, given a contrast matrix $H = [h_1, ..., h_N]$ with columns satisfying* (5.1), *the quantity*

$$\mathfrak{R}[H] = \max_z \left\{\|Bz\| : \|H^T Az\|_\infty \leq 2, z \in 2\mathcal{X}_{\mathrm{s}}\right\} \tag{5.5}$$

is an upper bound on the $(\epsilon, \|\cdot\|)$-risk of the polyhedral estimate $\widehat{x}^H(\cdot)$:

$$\mathrm{Risk}_{\epsilon,\|\cdot\|}[\widehat{x}^H|\mathcal{X}] \leq \mathfrak{R}[H].$$

Proof is immediate. Let us fix $x \in \mathcal{X}$, and let $\mathcal{E}$ be the set of all realizations of ξ_x such that $\|H^T\xi_x\|_\infty \leq 1$, so that $P_x(\mathcal{E}) \geq 1-\epsilon$ by (5.2). Let us fix a realization $\xi \in \mathcal{E}$ of the observation noise, and let $\omega = Ax+\xi$, $\bar{x} = \bar{x}(Ax+\xi)$. Then $u = x$ is a feasible solution to the optimization problem (5.3) with the value of the objective ≤ 1, implying that the value of this objective at the optimal solution $\bar{x}$ to the problem is ≤ 1 as well, so that $\|H^T A[x-\bar{x}]\|_\infty \leq 2$. Besides this, $z = x - \bar{x} \in 2\mathcal{X}_{\mathrm{s}}$. We see that z is a feasible solution to (5.5), whence $\|B[x-\bar{x}]\| = \|Bx - \widehat{x}^H(\omega)\| \leq \mathfrak{R}[H]$. It remains to note that the latter relation holds true whenever $\omega = Ax + \xi$ with $\xi \in \mathcal{E}$, and the P_x-probability of the latter inclusion is at least $1-\epsilon$, whatever be $x \in \mathcal{X}$. □

What is ahead. In what follows our focus will be on the following questions pertinent to the design of polyhedral estimates:

1. Given the data of our estimation problem and a tolerance $\delta \in (0,1)$, how to find a set $\mathcal{H}_\delta$ of vectors $h \in \mathbf{R}^m$ satisfying the relation

   $$\forall (x \in \mathcal{X}) : \mathrm{Prob}\left\{|h^T\xi_x| > 1\right\} \leq \delta. \tag{5.6}$$

   With our approach, after the number N of columns in a contrast matrix has been selected, we choose the columns of H from $\mathcal{H}_\delta$, with $\delta = \epsilon/N$, ϵ being a given reliability tolerance of the estimate we are designing. Thus, the problem of constructing sets $\mathcal{H}_\delta$ arises, the larger $\mathcal{H}_\delta$, the better.
2. The upper bound $\mathfrak{R}[H]$ on the $(\epsilon, \|\cdot\|)$-risk of the polyhedral estimate $\widehat{x}^H$ is, in general, difficult to compute—this is the maximum of a convex function over a computationally tractable convex set. Thus, similarly to the case of linear

estimates, we need techniques for computationally efficient upper bounding of $\mathfrak{R}[\cdot]$.

3. With "raw materials" (sets $\mathcal{H}_\delta$) and efficiently computable upper bounds on the risk of candidate polyhedral estimates at our disposal, how do we design the best in terms of (the upper bound on) its risk polyhedral estimate?

We are about to consider these questions one by one.

5.1.3 Specifying sets $\mathcal{H}_\delta$ for basic observation schemes

To specify reasonable sets $\mathcal{H}_\delta$ we need to make some assumptions on the distributions of observation noises we want to handle. In the sequel we restrict ourselves to three special cases as follows:

- *sub-Gaussian case:* For every $x \in \mathcal{X}$, the observation noise ξ_x is sub-Gaussian with parameters $(0, \sigma^2 I_m)$, where $\sigma > 0$, i.e. $\xi_x \sim \mathcal{SG}(0, \sigma^2 I_m)$.
- *Discrete case:* $\mathcal{X}$ is a convex compact subset of the probabilistic simplex $\mathbf{\Delta}_n = \{x \in \mathbf{R}^n : x \geq 0, \sum_i x_i = 1\}$, A is a column-stochastic matrix, and

$$\omega = \frac{1}{K} \sum_{k=1}^{K} \zeta_k$$

with random vectors ζ_k independent across $k \leq K$, ζ_k taking value e_i with probability $[Ax]_i$, $i = 1, ..., m$, e_i being the basic orths in $\mathbf{R}^m$.
- *Poisson case:* $\mathcal{X}$ is a convex compact subset of the nonnegative orthant $\mathbf{R}^n_+$, A is entrywise nonnegative, and the observation ω stemming from $x \in \mathcal{X}$ is a random vector with entries $\omega_i \sim \text{Poisson}([Ax]_i)$ independent across i.

The associated sets $\mathcal{H}_\delta$ can be built as follows.

5.1.3.1 *Sub-Gaussian case*

When $h \in \mathbf{R}^n$ is deterministic and ξ is sub-Gaussian with parameters $0, \sigma^2 I_m$, we have

$$\text{Prob}\{|h^T \xi| > 1\} \leq 2 \exp\left\{-\frac{1}{2\sigma^2 \|h\|_2^2}\right\}.$$

Indeed, when $h \neq 0$ and $\gamma > 0$, we have

$$\text{Prob}\{h^T \xi > 1\} \leq \exp\{-\gamma\} \mathbf{E}\left\{\exp\{\gamma h^T \xi\}\right\} \leq \exp\{\tfrac{1}{2}\sigma^2\gamma^2\|h\|_2^2 - \gamma\}.$$

Minimizing the resulting bound in $\gamma > 0$, we get $\text{Prob}\{h^T \xi > 1\} \leq \exp\left\{-\frac{1}{2\|h\|_2^2\sigma^2}\right\}$; the same reasoning as applied to $-h$ in the role of h results in $\text{Prob}\{h^T \xi < -1\} \leq \exp\left\{-\frac{1}{2\|h\|_2^2\sigma^2}\right\}$.

Consequently

$$\pi_G(h) := \underbrace{\sigma\sqrt{2\ln(2/\delta)}}_{\vartheta_G} \|h\|_2 \leq 1 \Rightarrow \text{Prob}\{|h^T \xi| > 1\} \leq \delta,$$

and we can set

$$\mathcal{H}_\delta = \mathcal{H}_\delta^G := \{h : \pi_G(h) \leq 1\}.$$

5.1.3.2 Discrete case

Given $x \in \mathcal{X}$, setting $\mu = Ax$ and $\eta_k = \zeta_k - \mu$, we get

$$\omega = Ax + \underbrace{\frac{1}{K}\sum_{k=1}^{K}\eta_k}_{\xi_x}.$$

Given $h \in \mathbf{R}^m$,

$$h^T\xi_x = \frac{1}{K}\sum_k \underbrace{h^T\eta_k}_{\chi_k}.$$

Random variables $\chi_1, ..., \chi_K$ are independent zero mean and clearly satisfy

$$\mathbf{E}\left\{\chi_k^2\right\} \leq \sum_i [Ax]_i h_i^2, \; |\chi_k| \leq 2\|h\|_\infty \;\; \forall k.$$

When applying Bernstein's inequality[2] we get (cf. Exercise 4.19)

$$\begin{array}{rcl}\mathrm{Prob}\{|h^T\xi_x| > 1\} & = & \mathrm{Prob}\{|\sum_k \chi_k| > K\} \\ & \leq & 2\exp\left\{-\frac{K}{2\sum_i [Ax]_i h_i^2 + \frac{4}{3}\|h\|_\infty}\right\}.\end{array} \tag{5.7}$$

Setting

$$\begin{array}{rcl}\pi_D(h) & = & \sqrt{\vartheta_D^2 \max_{x\in\mathcal{X}} \sum_i [Ax]_i h_i^2 + \varrho_D^2 \|h\|_\infty^2}, \\ & & \vartheta_D = 2\sqrt{\frac{\ln(2/\delta)}{K}}, \; \varrho_D = \frac{8\ln(2/\delta)}{3K},\end{array}$$

after a completely straightforward computation, we conclude from (5.7) that

$$\pi_D(h) \leq 1 \Rightarrow \mathrm{Prob}\{|h^T\xi_x| > 1\} \leq \delta, \; \forall x \in \mathcal{X}.$$

Thus, in the Discrete case we can set

$$\mathcal{H}_\delta = \mathcal{H}_\delta^D := \{h : \pi_D(h) \leq 1\}.$$

5.1.3.3 Poisson case

In the Poisson case, for $x \in \mathcal{X}$, setting $\mu = Ax$, we have

$$\omega = Ax + \xi_x, \; \xi_x = \omega - \mu.$$

It turns out that for every $h \in \mathbf{R}^m$ one has

$$\forall t \geq 0 : \mathrm{Prob}\left\{|h^T\xi_x| \geq t\right\} \leq 2\exp\left\{-\frac{t^2}{2[\sum_i h_i^2\mu_i + \frac{1}{3}\|h\|_\infty t]}\right\} \tag{5.8}$$

[2]The classical Bernstein inequality states that if $X_1, ..., X_K$ are independent zero mean scalar random variables with finite variances σ_k^2 such that $|X_k| \leq M$ a.s., then for every $t > 0$ one has

$$\mathrm{Prob}\{X_1 + ... + X_k > t\} \leq \exp\left\{-\frac{t^2}{2[\sum_k \sigma_k^2 + \frac{1}{3}Mt]}\right\}.$$

(for verification, see Exercise 4.21 or Section 5.4.1). As a result, we conclude via a straightforward computation that setting

$$\begin{aligned}\pi_P(h) &= \sqrt{\vartheta_P^2 \max_{x\in\mathcal{X}} \sum_i [Ax]_i h_i^2 + \varrho_P^2 \|h\|_\infty^2},\\ &\vartheta_P = 2\sqrt{\ln(2/\delta)},\ \varrho_P = \tfrac{4}{3}\ln(2/\delta),\end{aligned}$$

we ensure that

$$\pi_P(h) \le 1 \Rightarrow \mathrm{Prob}\{|h^T\xi_x| > 1\} \le \delta,\ \forall x \in \mathcal{X}.$$

Thus, in the Poisson case we can set

$$\mathcal{H}_\delta = \mathcal{H}_\delta^P := \{h : \pi_P(h) \le 1\}.$$

5.1.4 Efficient upper-bounding of $\mathfrak{R}[H]$ and contrast design, I.

The scheme for upper-bounding $\mathfrak{R}[H]$ to be presented in this section (an alternative, completely different, scheme will be presented in Section 5.1.5) is inspired by our motivating example. Note that there is a special case of (5.5) where $\mathfrak{R}[H]$ is easy to compute—the case where $\|\cdot\|$ is the uniform norm $\|\cdot\|_\infty$, whence

$$\mathfrak{R}[H] = \widehat{\mathfrak{R}}[H] := 2\max_{i\le\nu}\max_x\left\{\mathrm{Row}_i^T[B]x : x \in \mathcal{X}_\mathrm{s}, \|H^TAx\|_\infty \le 1\right\}$$

is the maximum of ν efficiently computable convex functions. It turns out that when $\|\cdot\| = \|\cdot\|_\infty$, it is not only easy to compute $\mathfrak{R}[H]$, but to optimize this risk bound in H as well.[3] These observations underlie the forthcoming developments in this section: under appropriate assumptions, we bound the risk of a polyhedral estimate with contrast matrix H via the efficiently computable quantity $\widehat{\mathfrak{R}}[H]$ and then show that the resulting risk bounds can be efficiently optimized w.r.t. H. We shall also see that in some "simple for analytical analysis" situations, like that of the example, the resulting estimates are nearly minimax optimal.

5.1.4.1 Assumptions

We stay within the setup introduced in Section 5.1.1 which we augment with the following assumptions:

A.1. $\|\cdot\| = \|\cdot\|_r$ with $r \in [1,\infty]$.

A.2. We have at our disposal a sequence $\gamma = \{\gamma_i > 0, i \le \nu\}$ and $\rho \in [1,\infty]$ such that the image of $\mathcal{X}_\mathrm{s}$ under the mapping $x \mapsto Bx$ is contained in the "scaled $\|\cdot\|_\rho$-ball"

$$\mathcal{Y} = \{y \in \mathbf{R}^\nu : \|\mathrm{Diag}\{\gamma\}y\|_\rho \le 1\}. \tag{5.9}$$

5.1.4.2 Simple observation

Let B_ℓ^T be the ℓ-th row in B, $1 \le \ell \le \nu$. Let us make the following observation:

[3]On closer inspection, in the situation considered in the motivating example the $\|\cdot\|_\infty$-optimal contrast matrix H is proportional to the unit matrix, and the quantity $\widehat{\mathfrak{R}}[H]$ can be easily translated into an upper bound on, say, the $\|\cdot\|_2$-risk of the associated polyhedral estimate.

Proposition 5.2. *In the situation described in Section 5.1.1, let us assume that Assumptions* **A.1-2** *hold. Let* $\epsilon \in (0,1)$ *and let a positive real* $N \geq \nu$ *be given; let also* $\pi(\cdot)$ *be a norm on* $\mathbf{R}^m$ *such that*

$$\forall (h : \pi(h) \leq 1, x \in \mathcal{X}) : \mathrm{Prob}\{|h^T \xi_x| > 1\} \leq \epsilon / N.$$

Next, let a matrix $H = [H_1, ..., H_\nu]$ *with* $H_\ell \in \mathbf{R}^{m \times m_\ell}$, $m_\ell \geq 1$, *and positive reals* ς_ℓ, $\ell \leq \nu$, *satisfy the relations*

$$\begin{array}{ll} (a) & \pi(\mathrm{Col}_j[H]) \leq 1,\ 1 \leq j \leq N; \\ (b) & \max_x \left\{B_\ell^T x : x \in \mathcal{X}_{\mathrm{s}}, \|H_\ell^T A x\|_\infty \leq 1\right\} \leq \varsigma_\ell,\ 1 \leq \ell \leq \nu. \end{array} \tag{5.10}$$

Then the quantity $\mathfrak{R}[H]$ *as defined in* (5.5) *can be upper-bounded as follows:*

$$\begin{array}{rcl} \mathfrak{R}[H] \leq \Psi(\varsigma) & := & 2 \max_w \{\|[w_1/\gamma_1; ...; w_\nu/\gamma_\nu]\|_r : \\ & & \|w\|_\rho \leq 1,\ 0 \leq w_\ell \leq \gamma_\ell \varsigma_\ell,\ \ell \leq \nu\}, \end{array} \tag{5.11}$$

which combines with Proposition 5.1 to imply that

$$\mathrm{Risk}_{\epsilon, \|\cdot\|}[\widehat{x}^H | \mathcal{X}] \leq \Psi(\varsigma). \tag{5.12}$$

Function Ψ *is nondecreasing on the nonnegative orthant and is easy to compute.*

Proof. Let $z = 2\bar{z}$ be a feasible solution to (5.5), thus $\bar{z} \in \mathcal{X}_{\mathrm{s}}$ and $\|H^T A \bar{z}\|_\infty \leq 1$. Let $y = B\bar{z}$, so that $y \in \mathcal{Y}$ (see (5.9)) due to $\bar{z} \in \mathcal{X}_{\mathrm{s}}$ and **A.2**. Then $\|\mathrm{Diag}\{\gamma\} y\|_p \leq 1$. Besides this, by (5.10.b) relations $\bar{z} \in \mathcal{X}_{\mathrm{s}}$ and $\|H^T A \bar{z}\|_\infty \leq 1$ combine with the symmetry of $\mathcal{X}_{\mathrm{s}}$ w.r.t. the origin to imply that

$$|y_\ell| = |B_\ell^T \bar{z}| \leq \varsigma_\ell,\ \ell \leq \nu.$$

Taking into account that $\|\cdot\| = \|\cdot\|_r$ by **A.1**, we see that

$$\begin{array}{rcl} \mathfrak{R}[H] & = & \max_z \left\{\|Bz\|_r : z \in 2\mathcal{X}_{\mathrm{s}}, \|H^T A z\|_\infty \leq 2\right\} \\ & \leq & 2 \max_y \left\{\|y\|_r : |y_\ell| \leq \varsigma_\ell, \ell \leq \nu,\ \&\ \|\mathrm{Diag}\{\gamma\} y\|_\rho \leq 1\right\} \\ & = & 2 \max_w \left\{\|[w_1/\gamma_1; ...; w_\nu/\gamma_\nu]\|_r : \|w\|_\rho \leq 1, 0 \leq w_\ell \leq \gamma_\ell \varsigma_\ell, \ell \leq \nu\right\}, \end{array}$$

as stated in (5.11).

It is evident that Ψ is nondecreasing on the nonnegative orthant. Computing Ψ can be carried out as follows:

1. When $r = \infty$, we need to compute $\max_{\ell \leq \nu} \max_w \{w_\ell / \gamma_\ell : \|w\|_\rho \leq 1, 0 \leq w_j \leq \gamma_j \varsigma_j, j \leq \nu\}$ so that evaluating Ψ reduces to solving ν simple convex optimization problems;
2. When $\rho = \infty$, we clearly have $\Psi(\varsigma) = \|[\bar{w}_1/\gamma_1; ...; \bar{w}_\nu/\gamma_\nu]\|_r$, $\bar{w}_\ell = \min[1, \gamma_\ell \varsigma_\ell]$;
3. When $1 \leq r, \rho < \infty$, passing from variables w_ℓ to variables $u_\ell = w_\ell^\rho$, we get

$$\Psi^r(\varsigma) = 2^r \max_u \left\{\sum_\ell \gamma_\ell^{-r} u_\ell^{r/\rho} : \sum_\ell u_\ell \leq 1, 0 \leq u_\ell \leq (\gamma_\ell \varsigma_\ell)^\rho\right\}.$$

When $r \leq \rho$, the optimization problem on the right-hand side is the easily solvable problem of maximizing a simple concave function over a simple convex compact set. When $\infty > r > \rho$, this problem can be solved by Dynamic

Programming. □

Comment. When we want to recover Bx in $\|\cdot\|_\infty$ (i.e., we are in the case of $r=\infty$), under the premise of Proposition 5.2 we clearly have $\Psi(\varsigma)\leq\max_\ell\varsigma_\ell$, resulting in the bound

$$\mathrm{Risk}_{\epsilon,\|\cdot\|_\infty}[\widehat{x}^H|\mathcal{X}]\leq 2\max_{\ell\leq\nu}\varsigma_\ell.$$

Note that this bound in fact does not require Assumption **A.2** (since it is satisfied for any ρ with large enough γ_i's).

5.1.4.3 Specifying contrasts

Risk bound (5.12) allows for an easy design of contrast matrices. Recalling that Ψ is monotone on the nonnegative orthant, all we need is to select h_ℓ's satisfying (5.10) and resulting in the smallest possible ς_ℓ's, which is what we are about to do now.

Preliminaries. Given a vector $b\in\mathbf{R}^m$ and a norm $s(\cdot)$ on $\mathbf{R}^m$, consider convex-concave saddle point problem

$$\mathrm{Opt}=\inf_{g\in\mathbf{R}^m}\max_{x\in\mathcal{X}_\mathrm{s}}\left\{\phi(g,x):=[b-A^Tg]^Tx+s(g)\right\}\tag{SP}$$

along with the induced primal and dual problems

$$\begin{array}{rcl}\mathrm{Opt}(P)&=&\inf_{g\in\mathbf{R}^m}\left[\overline{\phi}(g):=\max_{x\in\mathcal{X}_\mathrm{s}}\phi(g,x)\right]\\&=&\inf_{g\in\mathbf{R}^m}\left[s(g)+\max_{x\in\mathcal{X}_\mathrm{s}}[b-A^Tg]^Tx\right],\end{array}\tag{P}$$

and

$$\begin{array}{rcl}\mathrm{Opt}(D)&=&\max_{x\in\mathcal{X}_\mathrm{s}}\left[\underline{\phi}(g):=\inf_{g\in\mathbf{R}^m}\phi(g,x)\right]\\&=&\max_{x\in\mathcal{X}_\mathrm{s}}\left[\inf_{g\in\mathbf{R}^m}\left[b^Tx-[Ax]^Tg+s(g)\right]\right]\\&=&\max_x\left[b^Tx:x\in\mathcal{X}_\mathrm{s},q(Ax)\leq1\right]\end{array}\tag{D}$$

where $q(\cdot)$ is the norm conjugate to $s(\cdot)$ (we have used the evident fact that $\inf_{g\in\mathbf{R}^m}[f^Tg+s(g)]$ is either $-\infty$ or 0 depending on whether $q(f)>1$ or $q(f)\leq1$). Since $\mathcal{X}_\mathrm{s}$ is compact, we have $\mathrm{Opt}(P)=\mathrm{Opt}(D)=\mathrm{Opt}$ by the Sion-Kakutani Theorem. Besides this, (D) is solvable (evident) and (P) is solvable as well, since $\overline{\phi}(g)$ is continuous due to the compactness of $\mathcal{X}_\mathrm{s}$ and $\overline{\phi}(g)\geq s(g)$, so that $\overline{\phi}(\cdot)$ has bounded level sets. Let $\bar{g}$ be an optimal solution to (P), let $\bar{x}$ be an optimal solution to (D), and let $\bar{h}$ be the $s(\cdot)$-unit normalization of $\bar{g}$, so that $s(\bar{h})=1$ and $\bar{g}=s(\bar{g})\bar{h}$. Now let us make the following observation:

Observation 5.3. In the situation in question, we have

$$\max_x\left\{|b^Tx|:x\in\mathcal{X}_\mathrm{s},|\bar{h}^TAx|\leq1\right\}\leq\mathrm{Opt}.\tag{5.13}$$

In addition, for any matrix $G=[g^1,...,g^M]\in\mathbf{R}^{m\times M}$ with $s(g^j)\leq1$, $j\leq M$, one has

$$\begin{array}{l}\max_x\left\{|b^Tx|:x\in\mathcal{X}_\mathrm{s},\|G^TAx\|_\infty\leq1\right\}\\=\ \max_x\left\{b^Tx:x\in\mathcal{X}_\mathrm{s},\|G^TAx\|_\infty\leq1\right\}\geq\mathrm{Opt}.\end{array}\tag{5.14}$$

Proof. Let x be a feasible solution to the problem in (5.13). Replacing, if necessary, x with $-x$, we can assume that $|b^T x| = b^T x$. We now have

$$\begin{array}{rcl} |b^T x| & = & b^T x = [\bar{g}^T A x - s(\bar{g})] + \underbrace{[b - A^T \bar{g}]^T x + s(\bar{g})}_{\leq \overline{\phi}(\bar{g}) = \mathrm{Opt}(P)} \\ & \leq & \mathrm{Opt}(P) + [s(\bar{g}) \bar{h}^T A x - s(\bar{g})] \leq \mathrm{Opt}(P) + s(\bar{g}) \underbrace{|\bar{h}^T A x|}_{\leq 1} - s(\bar{g}) \\ & \leq & \mathrm{Opt}(P) = \mathrm{Opt}, \end{array}$$

as claimed in (5.13). Now, the equality in (5.14) is due to the symmetry of $\mathcal{X}_s$ w.r.t. the origin. To verify the inequality in (5.14), note that $\bar{x}$ satisfies the relations $\bar{x} \in \mathcal{X}_s$ and $q(A\bar{x}) \leq 1$, implying, due to the fact that the columns of G are of $s(\cdot)$-norm ≤ 1, that $\bar{x}$ is a feasible solution to the optimization problems in (5.14). As a result, the second quantity in (5.14) is at least $b^T \bar{x} = \mathrm{Opt}(D) = \mathrm{Opt}$, and (5.14) follows. □

Comment. Note that problem (P) has a very transparent origin. In the situation of Section 5.1.1, assume that our goal is, to estimate, given observation $\omega = Ax + \xi_x$, the value at $x \in \mathcal{X}$ of the linear function $b^T x$, and we want to use for this purpose an estimate $\widehat{g}(\omega) = g^T \omega + \gamma$ affine in ω. Given $\epsilon \in (0,1)$, how do we construct a presumably good in terms of its ϵ-risk estimate? Let us show that a meaningful answer is yielded by the optimal solution to (P). Indeed, we have

$$b^T x - \widehat{g}(Ax + \xi_x) = [b - A^T g]^T x - \gamma - g^T \xi_x.$$

Assume that we have at our disposal a norm $s(\cdot)$ on $\mathbf{R}^m$ such that

$$\forall (h \in \mathbf{R}^m, s(h) \leq 1, x \in \mathcal{X}) : \mathrm{Prob}\{\xi_x : |h^T \xi_x| > 1\} \leq \epsilon,$$

or, which is the same,

$$\forall (g \in \mathbf{R}^m, x \in \mathcal{X}) : \mathrm{Prob}\{\xi_x : |g^T \xi_x| > s(g)\} \leq \epsilon.$$

Then we can safely upper-bound the ϵ-risk of a candidate estimate $\widehat{g}(\cdot)$ by the quantity

$$\rho = \underbrace{\max_{x \in \mathcal{X}} |[b - A^T g]^T x - \gamma|}_{\text{bias } B(g,\gamma)} + s(g).$$

Observe that for g fixed, the minimal, over γ, bias is

$$M(g) := \max_{x \in \mathcal{X}_s} [b - A^T g] x.$$

Postponing verification of this claim, here is the conclusion:

> *in the present setting, problem (P) is nothing but the problem of building the best in terms of the upper bound ρ on the ϵ-risk affine estimate of linear function $b^T x$.*
>
> It remains to justify the above claim, which is immediate: on one hand, for all $u \in \mathcal{X}, v \in \mathcal{X}$ we have
>
> $$B(g,\gamma) \geq [b - A^T g]^T u - \gamma, \quad B(g,\gamma) \geq -[b - A^T g]^T v + \gamma$$

implying that

$$B(g,\gamma) \geq \tfrac{1}{2}[b - A^T g]^T [u - v] \;\forall (u \in \mathcal{X}, v \in \mathcal{X}),$$

just as $B(g,\gamma) \geq M(g)$. On the other hand, let

$$M_+(g) = \max_{x\in\mathcal{X}} [b - A^T g]^T x, \;\; M_-(g) = -\min_{x\in\mathcal{X}} [b - A^T g]^T x,$$

so that $M(g) = \frac{1}{2}[M_+(g) + M_-(g)]$. Setting $\bar{\gamma} = \frac{1}{2}[M_+(g) - M_-(g)]$, we have

$$\begin{array}{rcl}\max_{x\in\mathcal{X}} \left[[b - A^T g]^T x - \bar{\gamma}\right] & = M_+(g) - \bar{\gamma} = \frac{1}{2}[M_+(g) + M_-(g)] = M(g), \\ \min_{x\in\mathcal{X}} \left[[b - A^T g]^T x - \bar{\gamma}\right] & = -M_-(g) - \bar{\gamma} = -\frac{1}{2}[M_+(g) + M_-(g)] = -M(g).\end{array}$$

That is, $B(g,\bar{\gamma}) = M(g)$. Combining these observations, we arrive at $\min_{\gamma} B(g,\gamma) = M(g)$, as claimed. □

Contrast design. Proposition 5.2 and Observation 5.3 allow for a straightforward solution of the associated contrast design problem, at least in the case of sub-Gaussian, Discrete, and Poisson observation schemes. Indeed, in these cases, when designing a contrast matrix with N columns, with our approach we are supposed to select its columns in the respective sets $\mathcal{H}_{\epsilon/N}$; see Section 5.1.3. Note that these sets, while shrinking as N grows, are "nearly independent" of N, since the norms π_G, π_D, π_P in the description of the respective sets $\mathcal{H}^G_\delta$, $\mathcal{H}^D_\delta$, $\mathcal{H}^P_\delta$ depend on $1/\delta$ via factors logarithmic in $1/\delta$. It follows that we lose nearly nothing when assuming that $N \geq \nu$. Let us act as follows:

We set $N = \nu$, specify $\bar{\pi}(\cdot)$ as the norm (π_G, or π_D, or π_P) associated with the observation scheme (sub-Gaussian, or Discrete, or Poisson) in question and $\delta = \epsilon/\nu$. We solve ν convex optimization problems

$$\begin{array}{rcl}\mathrm{Opt}_\ell & = & \min_{g\in\mathbf{R}^m} \left[\overline{\phi}_\ell(g) := \max_{x\in\mathcal{X}_s} \phi_\ell(g,x)\right], \\ \phi_\ell(g,x) & = & [B_\ell - A^T g]^T x + \bar{\pi}(g).\end{array} \qquad (P_\ell)$$

Next, we convert optimal solution g_ℓ to (P_ℓ) into vector $h_\ell \in \mathbf{R}^m$ by representing $g_\ell = \bar{\pi}(g_\ell)h_\ell$ with $\bar{\pi}(h_\ell) = 1$, and set $H_\ell = h_\ell$. As a result, we obtain an $m \times \nu$ contrast matrix $H = [h_1, ..., h_\nu]$ which, taken along with $N = \nu$, quantities

$$\varsigma_\ell = \mathrm{Opt}_\ell, \; 1 \leq \ell \leq \nu, \qquad (5.15)$$

and with $\pi(\cdot) \equiv \bar{\pi}(\cdot)$, in view of the first claim in Observation 5.3 as applied with $s(\cdot) \equiv \bar{\pi}(\cdot)$, satisfies the premise of Proposition 5.2.

Consequently, by Proposition 5.2 we have

$$\mathrm{Risk}_{\epsilon,\|\cdot\|}[\widehat{x}^H|\mathcal{X}] \leq \Psi([\mathrm{Opt}_1; ...; \mathrm{Opt}_\nu]). \qquad (5.16)$$

Comment. Optimality of the outlined contrast design for the sub-Gaussian, or Discrete, or Poisson observation scheme stems, within the framework set by Proposition 5.2, from the second claim of Observation 5.3, which states that when $N \geq \nu$ and the columns of the $m \times N$ contrast matrix $H = [H_1, ..., H_\nu]$ belong to the set $\mathcal{H}_{\epsilon/N}$ associated with the observation scheme in question—i.e., the norm $\pi(\cdot)$ in the proposition is the norm π_G, or π_D, or π_P associated with $\delta = \epsilon/N$—the quantities ς_ℓ participating in (5.10.b) cannot be less than Opt_ℓ.

Indeed, the norm $\pi(\cdot)$ from Proposition 5.2 is $\geq$ the norm $\bar{\pi}(\cdot)$ participating in (P_ℓ) (because the value ϵ/N in the definition of $\pi(\cdot)$ is at most $\frac{\epsilon}{\nu}$), implying, by (5.10.a), that the columns of matrix H obeying the premise of the proposition satisfy the relation $\bar{\pi}(\mathrm{Col}_j[H]) \leq 1$. Invoking the second part of Observation 5.3 with $s(\cdot) \equiv \bar{\pi}(\cdot)$, $b = B_\ell$, and $G = H_\ell$, and taking (5.10.b) into account, we conclude that $\varsigma_\ell \geq \mathrm{Opt}_\ell$ for all ℓ, as claimed.

Since the bound on the risk of a polyhedral estimate offered by Proposition 5.2 is better the lesser are the ς_ℓ's, we see that as far as this bound is concerned, the outlined design procedure is the best possible, provided $N \geq \nu$.

An attractive feature of the contrast design we have just presented is that it is completely independent of the entities participating in assumptions **A.1-2**—these entities affect theoretical risk bounds of the resulting polyhedral estimate, but not the estimate itself.

5.1.4.4 Illustration: Diagonal case

Let us consider the *diagonal case* of our estimation problem, where

- $\mathcal{X} = \{x \in \mathbf{R}^n : \|Dx\|_\rho \leq 1\}$, where D is a diagonal matrix with positive diagonal entries $D_{\ell\ell} =: d_\ell$,
- $m = \nu = n$, and A and B are diagonal matrices with diagonal entries $0 < A_{\ell\ell} =: a_\ell$, $0 < B_{\ell\ell} =: b_\ell$,
- $\|\cdot\| = \|\cdot\|_r$,
- We are in the sub-Gaussian case, that is, observation noise ξ_x is $(0, \sigma^2 I_n)$-sub-Gaussian for every $x \in \mathcal{X}$.

Let us implement the approach developed in Sections 5.1.4.1–5.1.4.3.

1. Given reliability tolerance ϵ, we set

$$\delta = \epsilon/n, \;\; \vartheta_G := \sigma\sqrt{2\ln(2/\delta)} = \sigma\sqrt{2\ln(2n/\epsilon)}, \tag{5.17}$$

   and

$$\mathcal{H} = \mathcal{H}_\delta^G = \{h \in \mathbf{R}^n : \pi_G(h) := \vartheta_G\|h\|_2 \leq 1\}.$$

2. We solve $\nu = n$ convex optimization problems (P_ℓ) associated with $\bar{\pi}(\cdot) \equiv \pi_G(\cdot)$, which is immediate: the resulting contrast matrix is $H = \vartheta_G^{-1} I_n$, and

$$\mathrm{Opt}_\ell = \varsigma_\ell := b_\ell \min[\vartheta_G/a_\ell, 1/d_\ell]. \tag{5.18}$$

Risk analysis. The $(\epsilon, \|\cdot\|)$-risk of the resulting polyhedral estimate $\widehat{x}(\cdot)$ can be bounded by Proposition 5.2. Note that setting $\gamma_\ell = d_\ell/b_\ell$, $1 \leq \ell \leq n$, we meet assumptions **A.1-2**, and the above choice of $H, N = n$, and ς_ℓ satisfies the premise of Proposition 5.2. By this proposition,

$$\begin{array}{rcl}\mathrm{Risk}_{\epsilon,\|\cdot\|_r}[\widehat{x}^H|\mathcal{X}] \leq \Psi & := & 2\max_w \{\|[w_1/\gamma_1; ...; w_n/\gamma_n]\|_r : \\ & & \|w\|_\rho \leq 1,\, 0 \leq w_\ell \leq \gamma_\ell\varsigma_\ell\}.\end{array} \tag{5.19}$$

Let us work out what happens in the *simple case* where

$$\begin{array}{ll}(a) & 1 \le \rho \le r < \infty,\\ (b) & a_\ell/d_\ell \text{ and } b_\ell/a_\ell \text{ are nonincreasing in } \ell.\end{array} \tag{5.20}$$

Proposition 5.4. *In the simple case just defined, let $\mathfrak{n} = n$ when*

$$\sum_{\ell=1}^{n} (\vartheta_G d_\ell/a_\ell)^\rho \le 1;$$

otherwise let $\mathfrak{n}$ be the smallest integer such that

$$\sum_{\ell=1}^{\mathfrak{n}} (\vartheta_G d_\ell/a_\ell)^\rho > 1,$$

with ϑ_G given by (5.17). *Then for the contrast matrix $H = \vartheta_G^{-1} I_n$ one has*

$$\mathrm{Risk}_{\epsilon,\|\cdot\|_r}[\widehat{x}^H|\mathcal{X}] \le \Psi \le 2\left[\sum\nolimits_{\ell=1}^{\mathfrak{n}} (\vartheta_G b_\ell/a_\ell)^r\right]^{1/r}.$$

Proof. Consider the optimization problem specifying Ψ in (5.19). Setting $\theta = r/\rho \ge 1$, let us pass in this problem from variables w_ℓ to variables $z_\ell = w_\ell^\rho$, so that

$$\Psi^r = 2^r \max_z \left\{ \sum_\ell z_\ell^\theta (b_\ell/d_\ell)^r : \sum_\ell z_\ell \le 1,\, 0 \le z_\ell \le (d_\ell \varsigma_\ell / b_\ell)^\rho \right\} \le 2^r \Gamma,$$

where

$$\Gamma = \max_z \left\{ \sum_\ell z_\ell^\theta (b_\ell/d_\ell)^r : \sum_\ell z_\ell \le 1,\, 0 \le z_\ell \le \chi_\ell := (\vartheta_G d_\ell/a_\ell)^\rho \right\}$$

(we have used (5.18)). Note that Γ is the optimal value in the problem of maximizing a convex (since $\theta \ge 1$) function $\sum_\ell z_\ell^\theta (b_\ell/d_\ell)^r$ over a bounded polyhedral set, so that the maximum is attained at an extreme point $\bar{z}$ of the feasible set. By the standard characterization of extreme points, the (clearly nonempty) set I of positive entries in $\bar{z}$ is as follows. Let us denote by I' the set of indexes $\ell \in I$ such that $\bar{z}_\ell$ is on its upper bound $\bar{z}_\ell = \chi_\ell$; note that the cardinality $|I'|$ of I' is at least $|I| - 1$. Since $\sum_{\ell \in I'} \bar{z}_\ell = \sum_{\ell \in I'} \chi_\ell \le 1$ and χ_ℓ are nondecreasing in ℓ by (5.20.b), we conclude that

$$\sum_{\ell=1}^{|I'|} \chi_\ell \le 1,$$

implying that $|I'| < \mathfrak{n}$ provided that $\mathfrak{n} < n$, so that in this case $|I| \le \mathfrak{n}$; and of course $|I| \le \mathfrak{n}$ when $\mathfrak{n} = n$. Next, we have

$$\Gamma = \sum_{\ell \in I} \bar{z}_\ell^\theta (b_\ell/d_\ell)^r \le \sum_{\ell \in I} \chi_\ell^\theta (b_\ell/d_\ell)^r = \sum_{\ell \in I} (\vartheta_G b_\ell/a_\ell)^r,$$

and since b_ℓ/a_ℓ is nonincreasing in ℓ and $|I| \le \mathfrak{n}$, the latter quantity is at most $\sum_{\ell=1}^{\mathfrak{n}} (\vartheta_G b_\ell/a_\ell)^r$. □

Application. Consider the "standard case" [72, 74] where

$$0 < \sqrt{\ln(2n/\epsilon)}\sigma \leq 1,\ a_\ell = \ell^{-\alpha},\ b_\ell = \ell^{-\beta},\ d_\ell = \ell^{\varkappa}$$

with $\beta \geq \alpha \geq 0$, $\varkappa \geq 0$ and $(\beta - \alpha)r < 1$. In this case for large n, namely,

$$n \geq c\vartheta_G^{-\frac{1}{\alpha+\varkappa+1/\rho}} \qquad [\vartheta_G = \sigma\sqrt{2\ln(2n/\epsilon)}] \tag{5.21}$$

(here and in what follows, the factors denoted by c and C depend solely on $\alpha, \beta, \varkappa, r, \rho$) we get

$$\mathfrak{n} \leq C\vartheta_G^{-\frac{1}{\alpha+\varkappa+1/\rho}},$$

resulting in

$$\mathrm{Risk}_{\epsilon,\|\cdot\|_r}[\widehat{x}|\mathcal{X}] \leq C\vartheta_G^{\frac{\beta+\varkappa+1/\rho-1/r}{\alpha+\varkappa+1/\rho}}. \tag{5.22}$$

Setting $x = D^{-1}y$, $\bar{\alpha} = \alpha + \varkappa$, $\bar{\beta} = \beta + \varkappa$ and treating y, rather than x, as the signal underlying the observation, we obtain the estimation problem which is similar to the original one in which $\alpha, \beta, \varkappa$ and $\mathcal{X}$ are replaced, respectively, with $\bar{\alpha}$, $\bar{\beta}$, $\bar{\varkappa} = 0$, and $\mathcal{Y} = \{y : \|y\|_\rho \leq 1\}$, and A, B replaced with $\bar{A} = \mathrm{Diag}\{\ell^{-\bar{\alpha}}, \ell \leq n\}$, $\bar{B} = \mathrm{Diag}\{\ell^{-\bar{\beta}}, \ell \leq n\}$. When n is large enough, namely, $n \geq \sigma^{-\frac{1}{\bar{\alpha}+1/\rho}}$, $\mathcal{Y}$ contains the "coordinate box"

$$\overline{\mathcal{Y}} = \{x : |x_\ell| \leq \mathfrak{m}^{-1/\rho}, \mathfrak{m}/2 \leq \ell \leq \mathfrak{m}, x_\ell = 0 \text{ otherwise}\}$$

of dimension $\geq \mathfrak{m}/2$, where

$$\mathfrak{m} \geq c\sigma^{-\frac{1}{\bar{\alpha}+1/\rho}}.$$

Observe that for all $y \in \overline{\mathcal{Y}}$, $\|\bar{A}y\|_2 \leq C\mathfrak{m}^{-\bar{\alpha}}\|y\|_2$, and $\|\bar{B}y\|_r \geq c\mathfrak{m}^{-\bar{\beta}}\|y\|_r$. This observation, when combined with the Fano inequality, implies (cf. [79]) that for $\epsilon \ll 1$ the minimax optimal w.r.t. the family of all Borel estimates $(\epsilon, \|\cdot\|_r)$-risk on the signal set $\overline{\mathcal{X}} = D^{-1}\overline{\mathcal{Y}} \subset \mathcal{X}$ is at least

$$c\sigma^{\frac{\bar{\beta}+1/\rho-1/r}{\bar{\alpha}+1/\rho}}.$$

In other words, in this situation, the upper bound (5.22) on the risk of the polyhedral estimate is within a factor logarithmic in n/ϵ from the minimax risk. In particular, without surprise, in the case of $\beta = 0$ the polyhedral estimates attain well-known optimal rates [72, 109].

5.1.5 Efficient upper-bounding of $\mathfrak{R}[H]$ and contrast design, II.

5.1.5.1 *Outline*

In this section we develop and alternative approach to the design of polyhedral estimates which resembles in many aspects the approach to building linear estimates from Chapter 4. Recall that the principal technique underlying the design of a presumably good linear estimate $\widehat{x}_H(\omega) = H^T\omega$ was upper-bounding of maximal risk of the estimate—the maximum of a quadratic form, depending on H as a parameter, over the signal set $\mathcal{X}$, and we were looking for a bounding scheme allowing us to efficiently optimize the bound in H.

The design of a presumably good polyhedral estimate also reduces to minimizing the optimal value in a parametric maximization problem (5.5) over the contrast matrix H. However, while the design of a presumably good linear estimate reduces to *unconstrained minimization,* to conceive a polyhedral estimate we need to minimize bound $\mathcal{R}[H]$ on the estimation risk under the restriction on the contrast matrix H—the columns h_ℓ of this matrix should satisfy condition (5.1). In other words, in the case of polyhedral estimate the "design parameter" affects the constraints of the optimization problem rather than the objective.

Our strategy can be outlined as follows. Let us denote by

$$\mathcal{B}_* = \{u \in \mathbf{R}^\nu : \|u\|_* \leq 1\}$$

the unit ball of the norm $\|\cdot\|_*$ *conjugate* to the norm $\|\cdot\|$ in the formulation of the estimation problem in Section 5.1.2. Assume that we have at our disposal a technique for bounding quadratic forms on the set $\mathcal{B}_* \times \mathcal{X}_s$, in other words, we have an efficiently computable convex function $\mathcal{M}(M)$ on $\mathbf{S}^{\nu+n}$ such that

$$\mathcal{M}(M) \geq \max_{[u;z]\in\mathcal{B}_*\times\mathcal{X}_s} [u;z]^T M[u;z] \;\forall M \in \mathbf{S}^{\nu+n}. \tag{5.23}$$

Note that the upper bound $\mathfrak{R}[H]$, as defined in (5.5), on the risk of a candidate polyhedral estimate $\widehat{x}^H$ is nothing but

$$\mathfrak{R}[H] = 2\max_{[u;z]}\left\{[u;z]^T \underbrace{\left[\begin{array}{c|c} & \frac{1}{2}B \\ \hline \frac{1}{2}B^T & \end{array}\right]}_{B_+}[u;z] : \begin{array}{l} u \in \mathcal{B}_*, z \in \mathcal{X}_s, \\ z^T A^T h_\ell h_\ell^T Az \leq 1, \ell \leq N \end{array}\right\}. \tag{5.24}$$

Given $\lambda \in \mathbf{R}^N_+$, the constraints $z^T A^T h_\ell h_\ell^T Az \leq 1$ in (5.24) can be aggregated to yield the quadratic constraint

$$z^T A^T \Theta_\lambda Az \leq \mu_\lambda,\; \Theta_\lambda = H\mathrm{Diag}\{\lambda\}H^T,\; \mu_\lambda = \sum_\ell \lambda_\ell.$$

Observe that for every $\lambda \geq 0$ we have

$$\mathfrak{R}[H] \leq 2\mathcal{M}\Bigg(\underbrace{\left[\begin{array}{c|c} & \frac{1}{2}B \\ \hline \frac{1}{2}B^T & -A^T\Theta_\lambda A \end{array}\right]}_{B_+[\Theta_\lambda]}\Bigg) + 2\mu_\lambda. \tag{5.25}$$

Indeed, let $[u;z]$ be a feasible solution to the optimization problem (5.24) specifying $\mathfrak{R}[H]$. Then

$$[u;z]^T B_+[u;z] = [u;z]^T B_+[\Theta_\lambda][u;z] + z^T A^T \Theta_\lambda Az;$$

the first term on the right-hand side is $\leq \mathcal{M}(B_+[\Theta_\lambda])$ since $[u;z] \in \mathcal{B}_* \times \mathcal{X}_s$, and the second term on the right-hand side, as we have already seen, is $\leq \mu_\lambda$, and (5.25) follows.

Now assume that we have at our disposal a computationally tractable cone

$$\mathbf{H} \subset \mathbf{S}^N_+ \times \mathbf{R}_+$$

satisfying the following assumption:

C. *Whenever $(\Theta, \mu) \in \mathbf{H}$, we can efficiently find an $n \times N$ matrix $H = [h_1, ..., h_N]$ and a nonnegative vector $\lambda \in \mathbf{R}^N_+$ such that*

$$\begin{array}{ll} (a) & h_\ell \text{ satisfies (5.1)}, 1 \leq \ell \leq N, \\ (b) & \Theta = H\mathrm{Diag}\{\lambda\}H^T, \\ (c) & \sum_i \lambda_i \leq \mu. \end{array} \tag{5.26}$$

The following simple observation is crucial to what follows:

Proposition 5.5. *Consider the estimation problem posed in Section 5.1.1, and let efficiently computable convex function $\mathcal{M}$ and computationally tractable closed convex cone $\mathbf{H}$ satisfy* (5.23) *and Assumption* **C**, *respectively. Consider the convex optimization problem*

$$\begin{array}{c} \mathrm{Opt} = \min_{\tau,\Theta,\mu} \{2\tau + 2\mu : (\Theta, \mu) \in \mathbf{H}, \mathcal{M}(B_+[\Theta]) \leq \tau\} \\ \left[B_+[\Theta] = \left[\begin{array}{c|c} & \frac{1}{2}B \\ \hline \frac{1}{2}B^T & -A^T\Theta A \end{array}\right]\right]. \end{array} \tag{5.27}$$

Given a feasible solution (τ, Θ, μ) to this problem, by **C** *we can efficiently convert it to (H, λ) such that $H = [h_1, ..., h_N]$ with h_ℓ satisfying* (5.1) *and $\lambda \geq 0$ with $\sum_\ell \lambda_\ell \leq \mu$. We have*

$$\mathfrak{R}[H] \leq 2\tau + 2\mu,$$

whence the $(\epsilon, \|\cdot\|)$-risk of the polyhedral estimate $\widehat{x}^H$ satisfies the bound

$$\mathrm{Risk}_{\epsilon,\|\cdot\|}[\widehat{x}^H|\mathcal{X}] \leq 2\tau + 2\mu. \tag{5.28}$$

Consequently, we can efficiently construct polyhedral estimates with $(\epsilon, \|\cdot\|)$-risk arbitrarily close to Opt *(and with risk exactly* Opt, *provided problem* (5.27) *is solvable).*

Proof is readily given by the reasoning preceding the proposition. Indeed, with $\tau, \Theta, \mu, H, \lambda$ as in the premise of the proposition, the columns h_ℓ of H satisfy (5.1) by **C**, implying, by Proposition 5.1, that $\mathrm{Risk}_{\epsilon,\|\cdot\|}[\widehat{x}^H|\mathcal{X}] \leq \mathfrak{R}[H]$. Besides this, **C** says that for our H, λ it holds $\Theta = \Theta_\lambda$ and $\mu_\lambda \leq \mu$, so that (5.25) combines with the constraints of (5.27) to imply that $\mathfrak{R}[H] \leq 2\tau + 2\mu$, and (5.28) follows by Proposition 5.1. □

The approach to the design of polyhedral estimates we develop in this section amounts to reducing the construction of the estimate (i.e., construction of the contrast matrix H) to finding (nearly) optimal solutions to (5.27). Implementing this approach requires devising techniques for constructing cones $\mathbf{H}$ satisfying **C** along with efficiently computable functions $\mathcal{M}(\cdot)$ satisfying (5.23). These tasks are the subjects of the sections to follow.

5.1.5.2 Specifying cones H

We specify cones $\mathbf{H}$ in the case when the number N of columns in the candidate contrast matrices is m and under the following assumption on the given reliability tolerance ϵ and observation scheme in question:

D. *There is a computationally tractable convex compact subset $Z \subset \mathbf{R}^m_+$*

intersecting $\mathrm{int}\,\mathbf{R}^m_+$ *such that the norm* $\pi(\cdot)$

$$\pi(h) = \sqrt{\max_{z\in Z} \sum_i z_i h_i^2}$$

induced by Z *satisfies the relation*

$$\pi(h) \leq 1 \Rightarrow \mathrm{Prob}\{|h^T\xi_x| > 1\} \leq \epsilon/m \;\forall x \in \mathcal{X}.$$

Note that condition **D** is satisfied for sub-Gaussian, Discrete, and Poisson observation schemes: according to the results of Section 5.1.3,

- in the sub-Gaussian case, it suffices to take

$$Z = \{2\sigma^2 \ln(2m/\epsilon)[1;...;1]\};$$

- in the Discrete case, it suffices to take

$$Z = \frac{4\ln(2m/\epsilon)}{K} A\mathcal{X} + \frac{64\ln^2(2m/\epsilon)}{9K^2}\mathbf{\Delta}_m,$$

  where

$$A\mathcal{X} = \{Ax : x \in \mathcal{X}\},\ \mathbf{\Delta}_m = \{y \in \mathbf{R}^m : y \geq 0, \sum_i y_i = 1\}.$$

- in the Poisson case, it suffices to take

$$Z = 2\ln(2m/\epsilon)A\mathcal{X} + \tfrac{16}{9}\ln^2(2m/\epsilon)\mathbf{\Delta}_m,$$

  with $A\mathcal{X}$ and $\mathbf{\Delta}_m$ as above.

Note that in all these cases Z only "marginally"—logarithmically—depends on ϵ and m.

Under Assumption **D**, the cone **H** can be built as follows:

- When $\mathcal{Z}$ is a singleton, $\mathcal{Z} = \{\bar{z}\}$, so that $\pi(\cdot)$ is a scaled Euclidean norm, we set

$$\mathbf{H} = \left\{(\Theta, \mu) \in \mathbf{S}^m_+ \times \mathbf{R}_+ : \mu \geq \sum_i \bar{z}_i\Theta_{ii}\right\}.$$

  Given $(\Theta, \mu) \in \mathbf{H}$, the $m \times m$ matrix H and $\lambda \in \mathbf{R}^m_+$ are built as follows: setting $S = \mathrm{Diag}\{\sqrt{\bar{z}_1}, ..., \sqrt{\bar{z}_m}\}$, we compute the eigenvalue decomposition of the matrix $S\Theta S$:

$$S\Theta S = U\mathrm{Diag}\{\lambda\}U^T,$$

  where U is orthonormal, and set $H = S^{-1}U$, thus ensuring $\Theta = H\mathrm{Diag}\{\lambda\}H^T$. Since $\mu \geq \sum_i \bar{z}_i\Theta_{ii}$, we have $\sum_i \lambda_i = \mathrm{Tr}(S\Theta S) \leq \mu$. Finally, a column h of H is of the form $S^{-1}f$ with $\|\cdot\|_2$-unit vector f, implying that

$$\pi(h) = \sqrt{\sum_i \bar{z}_i[S^{-1}f]_i^2} = \sqrt{\sum_i f_i^2} = 1,$$

  so that h satisfies (5.1) by **D**.

- When Z is not a singleton, we set

$$\begin{array}{rcl}\phi(r) & = & \max_{z\in Z} z^T r,\\ \varkappa & = & 6\ln(2\sqrt{3}m^2),\\ \mathbf{H} & = & \{(\Theta,\mu)\in \mathbf{S}^m_+\times \mathbf{R}_+ : \mu\geq \varkappa\phi(\mathrm{dg}(\Theta))\},\end{array} \tag{5.29}$$

where $\mathrm{dg}(Q)$ is the diagonal of a (square) matrix Q. Note that $\phi(r) > 0$ whenever $r \geq 0$, $r \neq 0$, since Z contains a positive vector.
The justification of this construction and the efficient (randomized) algorithm for converting a pair $(\Theta,\mu)\in\mathbf{H}$ into (H,λ) satisfying, when taken along with (Θ,μ), the requirements of $\mathbf{C}$ are given by the following:

Lemma 5.6. *Let norm $\pi(\cdot)$ satisfy **D**.*
(i) *Whenever H is an $m\times m$ matrix with columns h_ℓ satisfying $\pi(h_\ell)\leq 1$ and $\lambda\in\mathbf{R}^m_+$, we have*

$$\left(\Theta_\lambda = H\mathrm{Diag}\{\lambda\}H^T,\ \mu=\varkappa\sum_i\lambda_i\right)\in\mathbf{H}.$$

(ii) *Given $(\Theta,\mu)\in\mathbf{H}$ with $\Theta\neq 0$, we find decomposition $\Theta=QQ^T$ with $m\times m$ matrix Q, and fix an orthonormal $m\times m$ matrix V with magnitudes of entries not exceeding $\sqrt{2/m}$ (e.g., the orthonormal scaling of the matrix of the cosine transform). When $\mu>0$, we set $\lambda=\frac{\mu}{m}[1;...;1]\in\mathbf{R}^m$ and consider the random matrix*

$$H_\chi=\sqrt{\frac{m}{\mu}}Q\mathrm{Diag}\{\chi\}V,$$

where χ is the m-dimensional Rademacher random vector. We have

$$H_\chi\mathrm{Diag}\{\lambda\}H_\chi^T\equiv\Theta,\ \lambda\geq 0,\sum_i\lambda_i=\mu. \tag{5.30}$$

Moreover, the probability of the event

$$\pi(\mathrm{Col}_\ell[H_\chi])\leq 1\,\forall \ell\leq m \tag{5.31}$$

is at least 1/2. Thus, generating independent samples of χ and terminating with $H=H_\chi$ when the latter matrix satisfies (5.31)*, we with probability 1 terminate with (H,λ) satisfying **C**, and the probability for the outlined procedure to terminate in the course of the first $M=1,2,...$ steps is at least $1-2^{-M}$.*
When $\mu=0$, we have $\Theta=0$ (since $\mu=0$ implies $\phi(\mathrm{dg}(\Theta))=0$, which with $\Theta\succeq 0$ is possible only when $\Theta=0$); thus, when $\mu=0$, we set $H=0_{m\times m}$ and $\lambda=0_{m\times 1}$.

Note that the lemma states, essentially, that the cone $\mathbf{H}$ is a tight, up to a factor logarithmic in m, inner approximation of the set

$$\left\{(\Theta,\mu):\exists(\lambda\in\mathbf{R}^m_+,H\in\mathbf{R}^{m\times m}):\begin{array}{c}\Theta=H\mathrm{Diag}\{\lambda\}H^T,\\ \pi(\mathrm{Col}_\ell[H])\leq 1,\ \ell\leq m,\\ \mu\geq\sum_\ell\lambda_\ell\end{array}\right\}.$$

For proof, see Section 5.4.2.

5.1.5.3 Specifying functions $\mathcal{M}$

In this section we focus on computationally efficient upper-bounding of maxima of quadratic forms over convex compact sets symmetric w.r.t. the origin by semidefinite relaxation, our goal being to specify a "presumably good" efficiently computable convex function $\mathcal{M}(\cdot)$ satisfying (5.23).

Cones compatible with convex sets. Given a nonempty convex compact set $\mathcal{Y} \subset \mathbf{R}^N$, we say that a cone $\mathbf{Y}$ is *compatible* with $\mathcal{Y}$ if

- $\mathbf{Y}$ is a closed convex computationally tractable cone contained in $\mathbf{S}^N_+ \times \mathbf{R}_+$
- one has

$$\forall (V,\tau) \in \mathbf{Y} : \max_{y\in\mathcal{Y}} y^T V y \leq \tau \tag{5.32}$$

- $\mathbf{Y}$ contains a pair (V,τ) with $V \succ 0$
- relations $(V,\tau) \in \mathbf{Y}$ and $\tau' \geq \tau$ imply that $(V,\tau') \in \mathbf{Y}$.[4]

We call a cone $\mathbf{Y}$ *sharp* if $\mathbf{Y}$ is a closed convex cone contained in $\mathbf{S}^N_+ \times \mathbf{R}_+$ and such that the only pair $(V,\tau) \in \mathbf{Y}$ with $\tau = 0$ is the pair $(0,0)$, or, equivalently, a sequence $\{(V_i,\tau_i) \in \mathbf{Y}, i \geq 1\}$ is bounded if and only if the sequence $\{\tau_i, i \geq 1\}$ is bounded.

Note that whenever the linear span of $\mathcal{Y}$ is the entire $\mathbf{R}^N$, every cone compatible with $\mathcal{Y}$ is sharp.

Observe that *if $\mathcal{Y} \subset \mathbf{R}^N$ is a nonempty convex compact set and $\mathbf{Y}$ is a cone compatible with a shift $\mathcal{Y} - a$ of $\mathcal{Y}$, then $\mathbf{Y}$ is compatible with $\mathcal{Y}_s$.*

Indeed, when shifting a set $\mathcal{Y}$, its symmetrization $\frac{1}{2}[\mathcal{Y} - \mathcal{Y}]$ remains intact, so that we can assume that $\mathbf{Y}$ is compatible with $\mathcal{Y}$. Now let $(V,\tau) \in \mathbf{Y}$ and $y, y' \in \mathcal{Y}$. We have

$$[y-y']^T V [y-y'] + \underbrace{[y+y']^T V [y+y']}_{\geq 0} = 2[y^T V y + [y']^T V y'] \leq 4\tau,$$

whence for $z = \frac{1}{2}[y-y']$ it holds $z^T V z \leq \tau$. Since every $z \in \mathcal{Y}_s$ is of the form $\frac{1}{2}[y-y']$ with $y, y' \in \mathcal{Y}$, the claim follows.

Note that the claim can be "nearly inverted": *if $0 \in \mathcal{Y}$ and $\mathbf{Y}$ is compatible with $\mathcal{Y}_s$, then the "widening" of $\mathbf{Y}$—the cone*

$$\mathbf{Y}^+ = \{(V,\tau) : (V,\tau/4) \in \mathbf{Y}\}$$

—is compatible with $\mathcal{Y}$ (evident, since when $0 \in \mathcal{Y}$, every vector from $\mathcal{Y}$ is proportional, with coefficient 2, to a vector from $\mathcal{Y}_s$).

Constructing functions $\mathcal{M}$. The role of compatibility in our context becomes clear from the following observation:

Proposition 5.7. *In the situation described in Section 5.1.1, assume that we have at our disposal cones $\mathbf{X}$ and $\mathbf{U}$ compatible, respectively, with $\mathcal{X}_s$ and with the unit*

[4]The latter requirement is "for free"—passing from a computationally tractable closed convex cone $\mathbf{Y} \subset \mathbf{S}^N_+ \times \mathbf{R}_+$ satisfying (5.32) to the cone $\mathbf{Y}^+ = \{(V,\tau) : \exists \bar{\tau} \leq \tau : (V,\bar{\tau}) \in \mathbf{Y}\}$, we get a cone larger than $\mathbf{Y}$ and still compatible with $\mathcal{Y}$. It will be clear from the sequel that in our context, the larger is a cone compatible with $\mathcal{Y}$, the better.

ball

$$\mathcal{B}_* = \{v \in \mathbf{R}^\nu : \|u\|_* \leq 1\}$$

of the norm $\|\cdot\|_*$ *conjugate to the norm* $\|\cdot\|$. *Given* $M \in \mathbf{S}^{\nu+n}$, *let us set*

$$\mathcal{M}(M) = \inf_{X,t,U,s} \{t + s : (X,t) \in \mathbf{X}, (U,s) \in \mathbf{U}, \mathrm{Diag}\{U,X\} \succeq M\}. \tag{5.33}$$

Then $\mathcal{M}$ *is a real-valued efficiently computable convex function on* $\mathbf{S}^{\nu+n}$ *such that* (5.23) *takes place: for every* $M \in \mathbf{S}^{n+\nu}$ *it holds*

$$\mathcal{M}(M) \geq \max_{[u;z]\in\mathcal{B}_*\times\mathcal{X}_\mathrm{s}} [u;z]^T M[u;z].$$

In addition, when $\mathbf{X}$ *and* $\mathbf{U}$ *are sharp, the infimum in* (5.33) *is achieved.*

Proof is immediate. Given that the objective of the optimization problem specifying $\mathcal{M}(M)$ is nonnegative on the feasible set, the fact that $\mathcal{M}$ is real-valued is equivalent to problem's feasibility, and the latter is readily given by the fact that $\mathbf{X}$ is a cone containing a pair (X,t) with $X \succ 0$ and similarly for $\mathbf{U}$. Convexity of $\mathcal{M}$ is evident. To verify (5.23), let (X,t,U,s) form a feasible solution to the optimization problem in (5.33). When $[u;z] \in \mathcal{B}_* \times \mathcal{X}_\mathrm{s}$ we have

$$[u;z]^T M[u;z] \leq u^T U u + z^T X z \leq s + t,$$

where the first inequality is due to the $\succeq$-constraint in (5.33), and the second is due to the fact that $\mathbf{U}$ is compatible with $\mathcal{B}_*$, and $\mathbf{X}$ is compatible with $\mathcal{X}_\mathrm{s}$. Since the resulting inequality holds true for all feasible solutions to the optimization problem in (5.33), (5.23) follows. Finally, when $\mathbf{X}$ and $\mathbf{U}$ are sharp, (5.33) is a feasible conic problem with bounded level sets of the objective and as such is solvable. □

5.1.5.4 Putting things together

The following statement combining the results of Propositions 5.7 and 5.5 summarizes our second approach to the design of the polyhedral estimate.

Proposition 5.8. *In the situation of Section 5.1.1, assume that we have at our disposal cones* $\mathbf{X}$ *and* $\mathbf{U}$ *compatible, respectively, with* $\mathcal{X}_\mathrm{s}$ *and with the unit ball* $\mathcal{B}_*$ *of the norm conjugate to* $\|\cdot\|$. *Given reliability tolerance* $\epsilon \in (0,1)$ *along with a positive integer* N *and a computationally tractable cone* $\mathbf{H}$ *satisfying Assumption* ***C***, *consider the (clearly feasible) convex optimization problem*

$$\mathrm{Opt} = \min_{\Theta,\mu,X,t,U,s} \left\{ \begin{array}{l} f(t,s,\mu) := 2(t+s+\mu) : \\ (\Theta,\mu) \in \mathbf{H}, (X,t) \in \mathbf{X}, (U,s) \in \mathbf{U}, \\ \left[\begin{array}{c|c} U & \frac{1}{2}B \\ \hline \frac{1}{2}B^T & A^T\Theta A + X \end{array}\right] \succeq 0 \end{array} \right\}. \tag{5.34}$$

Let Θ, μ, X, t, U, s *be a feasible solution to* (5.34). *Invoking* ***C***, *we can convert, in a computationally efficient manner,* (Θ,μ) *into* (H,λ) *such that the columns of the* $m \times N$ *contrast matrix* H *satisfy* (5.1), $\Theta = H\mathrm{Diag}\{\lambda\}H^T$, *and* $\mu \geq \sum_\ell \lambda_\ell$. *The*

$(\epsilon, \|\cdot\|)$-risk of the polyhedral estimate $\widehat{x}^H$ satisfies the bound

$$\mathrm{Risk}_{\epsilon,\|\cdot\|}[\widehat{x}^H|\mathcal{X}] \leq f(t,s,\mu). \tag{5.35}$$

In particular, we can build, in a computationally efficient manner, polyhedral estimates with risks arbitrarily close to Opt *(and with risk* Opt*, provided that* (5.34) *is solvable).*

Proof. Let Θ, μ, X, t, U, s form a feasible solution to (5.34). By the semidefinite constraint in (5.34) we have

$$0 \preceq \left[\begin{array}{c|c} U & -\frac{1}{2}B \\ \hline -\frac{1}{2}B^T & A^T\Theta A + X \end{array}\right] = \mathrm{Diag}\{U, X\} - \underbrace{\left[\begin{array}{c|c} & \frac{1}{2}B \\ \hline \frac{1}{2}B^T & -A^T\Theta A \end{array}\right]}_{=:M},$$

whence for the function $\mathcal{M}$ defined in (5.33) one has

$$\mathcal{M}(M) \leq t + s.$$

Since $\mathcal{M}$, by Proposition 5.7, satisfies (5.23), invoking Proposition 5.5 we arrive at

$$\mathfrak{R}[H] \leq 2(\mu + \mathcal{M}(M)) \leq f(t,s,\mu).$$

By Proposition 5.1 this implies the target relation (5.35). □

5.1.5.5 Compatibility: Basic examples and calculus

Our approach to the design of polyhedral estimates utilizing the recipe described in Proposition 5.8 relies upon our ability to equip convex "sets of interest" (in our context, these are the symmetrization $\mathcal{X}_s$ of the signal set and the unit ball $\mathcal{B}_*$ of the norm conjugate to the norm $\|\cdot\|$) with compatible cones.[5] Below, we discuss two principal sources of such cones, namely (a) spectratopes/ellitopes, and (b) absolute norms. More examples of compatible cones can be constructed using a "compatibility calculus." Namely, let us assume that we are given a finite collection of convex sets (operands) and apply to them some basic operation, such as taking the intersection, or arithmetic sum, direct or inverse linear image, or convex hull of the union. It turns out that cones compatible with the results of such operations can be easily (in a fully algorithmic fashion) obtained from the cones compatible with the operands; see Section 5.1.8 for principal calculus rules.

In view of Proposition 5.8, the larger are the cones $\mathbf{X}$ and $\mathbf{U}$ compatible with $\mathcal{X}_s$ and $\mathcal{B}_*$, the better—the wider is the optimization domain in (5.34) and, consequently, the less is (the best) risk bound achievable with the recipe presented in the proposition. Given convex compact set $\mathcal{Y} \in \mathbf{R}^N$, the "ideal"—the largest—candidate to the role of the cone compatible with $\mathcal{Y}$ would be

$$\mathbf{Y}^* = \{(V,\tau) \in \mathbf{S}^N_+ \times \mathbf{R}_+ : \tau \geq \max_{y\in\mathcal{Y}} y^T V y\}.$$

However, this cone is typically intractable, therefore, we look for "as large as pos-

[5]Recall that we already know how to specify the second element of the construction, the cone $\mathbf{H}$.

sible" *tractable* inner approximations of $\mathbf{Y}^*$.

5.1.5.5.A. Cones compatible with ellitopes/spectratopes are readily given by semidefinite relaxation. Specifically, when

$$\mathcal{Y} = \{y \in \mathbf{R}^N : \exists (r \in \mathcal{R}, z \in \mathbf{R}^K) : y = Mz, R_\ell^2[z] \preceq r_\ell I_{d_\ell}, \ell \leq L\}$$
$$\left[R_\ell[z] = \sum\nolimits_j z_j R^{\ell j},\ R^{\ell j} \in \mathbf{S}^{d_\ell}\right]$$

with our standard restrictions on $\mathcal{R}$, invoking Proposition 4.8 it is immediately seen that the set

$$\mathbf{Y} = \big\{(V,\tau) \in \mathbf{S}^N_+ \times \mathbf{R}_+ : \exists \Lambda = \{\Lambda_\ell \in \mathbf{S}^{d_\ell}_+, \ell \leq L\} : \phi_{\mathcal{R}}(\lambda[\Lambda]) \leq \tau \\ M^T V M \preceq \sum_\ell \mathcal{R}^*[\Lambda_\ell]\big\} \tag{5.36}$$

is a closed convex cone which is compatible with $\mathcal{Y}$; here, as usual,

$$[\mathcal{R}^*_\ell[\Lambda_\ell]]_{ij} = \mathrm{Tr}(R^{\ell i}\Lambda_\ell R^{\ell j}),\ \lambda[\Lambda] = [\mathrm{Tr}(\Lambda_1); ...; \mathrm{Tr}(\Lambda_L)], \phi_{\mathcal{R}}(\lambda) = \max_{r \in \mathcal{R}} r^T \lambda.$$

Similarly, when $\mathcal{Y}$ is an ellitope:

$$\mathcal{Y} = \{y \in \mathbf{R}^N : \exists (r \in \mathcal{R}, z \in \mathbf{R}^K) : y = Mz, z^T R_\ell z \leq r_\ell,\ \ell \leq L\}$$

with our standard restrictions on R_ℓ, invoking Proposition 4.6, the set

$$\mathbf{Y} = \{(V,\tau) \in \mathbf{S}^N_+ \times \mathbf{R}_+ : \exists \lambda \in \mathbf{R}^L_+ : M^T V M \preceq \sum_\ell \lambda_\ell R_\ell, \phi_{\mathcal{R}}(\lambda) \leq \tau\} \tag{5.37}$$

is a closed convex cone which is compatible with $\mathcal{Y}$. In both cases, $\mathbf{Y}$ is sharp, provided that the image space of M is the entire $\mathbf{R}^N$.

Note that in both these cases $\mathbf{Y}$ is a reasonably tight inner approximation of $\mathbf{Y}^*$: whenever $(V,\tau) \in \mathbf{Y}^*$, we have $(V,\theta\tau) \in \mathbf{Y}$, with a moderate θ (specifically, $\theta = O(1)\ln\left(2\sum_\ell d_\ell\right)$ in the spectratopic, and $\theta = O(1)\ln(2L)$ in the ellitopic case; see Propositions 4.8, 4.6, respectively).

5.1.5.5.B. Compatibility via absolute norms.
Preliminaries. Recall that a norm $p(\cdot)$ on $\mathbf{R}^N$ is called *absolute* if $p(x)$ is a function of the vector $\mathrm{abs}[x] := [|x_1|; ...; |x_N|]$ of the magnitudes of entries in x. It is well known that an absolute norm p is monotone on $\mathbf{R}^N_+$, so that $\mathrm{abs}[x] \leq \mathrm{abs}[x']$ implies that $p(x) \leq p(x')$, and that the norm

$$p_*(x) = \max_{y: p(y) \leq 1} x^T y$$

conjugate to $p(\cdot)$ is absolute along with p.

Let us say that an absolute norm $r(\cdot)$ *fits* an absolute norm $p(\cdot)$ on $\mathbf{R}^N$ if for every vector x with $p(x) \leq 1$ the entrywise square $[x]^2 = [x_1^2; ...; x_N^2]$ of x satisfies $r([x]^2) \leq 1$. For example, the largest norm $r(\cdot)$ which fits the absolute norm $p(\cdot) = \|\cdot\|_s$, $s \in [1,\infty]$, is

$$r(\cdot) = \left\{\begin{array}{ll} \|\cdot\|_1, & 1 \leq s \leq 2 \\ \|\cdot\|_{s/2}, & s \geq 2 \end{array}\right. .$$

An immediate observation is that an absolute norm $p(\cdot)$ on $\mathbf{R}^N$ can be "lifted" to a norm on $\mathbf{S}^N$, specifically, the norm

$$p^+(Y) = p([p(\mathrm{Col}_1[Y]); ...; p(\mathrm{Col}_N[Y])]) : \mathbf{S}^N \to \mathbf{R}_+, \tag{5.38}$$

where $\mathrm{Col}_j[Y]$ is j-th column in Y. It is immediately seen that when p is an absolute norm, the right-hand side in (5.38) indeed is a norm on $\mathbf{S}^N$ satisfying the identity

$$p^+(xx^T) = p^2(x),\ x \in \mathbf{R}^N. \tag{5.39}$$

Absolute norms and compatibility. Our interest in absolute norms is motivated by the following immediate observation:

Observation 5.9. Let $p(\cdot)$ be an absolute norm on $\mathbf{R}^N$, and $r(\cdot)$ be another absolute norm which fits $p(\cdot)$, both norms being computationally tractable. These norms give rise to the computationally tractable and sharp closed convex cone

$$\mathbf{P} = \mathbf{P}_{p(\cdot),r(\cdot)} = \left\{ (V,\tau) \in \mathbf{S}^N_+ \times \mathbf{R}_+ : \exists (W \in \mathbf{S}^N, w \in \mathbf{R}^N_+) : \begin{array}{l} V \preceq W + \mathrm{Diag}\{w\}, \\ [p^+]_*(W) + r_*(w) \le \tau \end{array} \right\} \tag{5.40}$$

where $[p^+]_*(\cdot)$ is the norm on $\mathbf{S}^N$ conjugate to the norm $p^+(\cdot)$, and $r_*(\cdot)$ is the norm on $\mathbf{R}^N$ conjugate to the norm $r(\cdot)$, and this cone is compatible with the unit ball of the norm $p(\cdot)$ (and thus with any convex compact subset of this ball).

Verification is immediate. The fact that $\mathbf{P}$ is a computationally tractable and closed convex cone is evident. Now let $(V,\tau) \in \mathbf{P}$, so that $V \succeq 0$ and $V \preceq W + \mathrm{Diag}\{w\}$ with $[p^+]_*(W) + r_*(w) \le \tau$. For x with $p(x) \le 1$ we have

$$\begin{array}{rcl} x^TVx & \le & x^T[W + \mathrm{Diag}\{w\}]x = \mathrm{Tr}(W[xx^T]) + w^T[x]^2 \\ & \le & p^+(xx^T)[p^+]_*(W) + r([x]^2)r_*(w) = p^2(x)[p^+]_*(W) + r_*(w) \\ & \le & [p^+]_*(W) + r_*(w) \le \tau \end{array}$$

(we have used (5.40)), whence $x^TVx \le \tau$ for all x with $p(x) \le 1$. □

Let us look at the proposed construction in the case where $p(\cdot) = \|\cdot\|_s$, $s \in [1, \infty]$, and let $r(\cdot) = \|\cdot\|_{\bar{s}}$, $\bar{s} = \max[s/2, 1]$. Setting $s_* = \frac{s}{s-1}$, $\bar{s}_* = \frac{\bar{s}}{\bar{s}-1}$, we clearly have

$$[p^+]_*(W) = \|W\|_{s_*} := \left\{ \begin{array}{ll} \left(\sum_{i,j} |W_{ij}|^{s_*}\right)^{1/s_*}, & s_* < \infty \\ \max_{i,j} |W_{ij}|, & s_* = \infty \end{array} \right., \ r_*(w) = \|w\|_{\bar{s}_*}, \tag{5.41}$$

resulting in

$$\mathbf{P}^s := \mathbf{P}_{\|\cdot\|_s, \|\cdot\|_{\bar{s}}} = \left\{ (V,\tau) : V \in \mathbf{S}^N_+, \exists (W \in \mathbf{S}^N, w \in \mathbf{R}^N_+) : \begin{array}{l} V \preceq W + \mathrm{Diag}\{w\}, \\ \|W\|_{s_*} + \|w\|_{\bar{s}_*} \le \tau \end{array} \right\}. \tag{5.42}$$

By Observation 5.9, $\mathbf{P}^s$ is compatible with the unit ball of $\|\cdot\|_s$-norm on $\mathbf{R}^N$ (and therefore with every closed convex subset of this ball).

When $s = 1$, that is, $s_* = \bar{s}_* = \infty$, (5.42) results in

$$\begin{array}{rcl}\mathbf{P}^1 & = & \left\{(V,\tau) : V \succeq 0,\ \exists (W \in \mathbf{S}^N, w \in \mathbf{R}^N_+) : \begin{array}{l} V \preceq W + \mathrm{Diag}\{w\}, \\ \|W\|_\infty + \|w\|_\infty \leq \tau \end{array}\right\} \\ & = & \{(V,\tau) : V \succeq 0, \|V\|_\infty \leq \tau\},\end{array} \tag{5.43}$$

and it is easily seen that the situation is a good as it could be, namely,

$$\mathbf{P}^1 = \{(V,\tau) : V \succeq 0,\ \max_{\|x\|_1 \leq 1} x^T V x \leq \tau\}.$$

It can be shown (see Section 5.4.3) that when $s \in [2,\infty]$, and so $\bar{s}_* = \frac{s}{s-2}$, (5.42) results in

$$\mathbf{P}^s = \{(V,\tau) : V \succeq 0, \exists (w \in \mathbf{R}^N_+) : V \preceq \mathrm{Diag}\{w\} \ \&\ \|w\|_{\frac{s}{s-2}} \leq \tau\}. \tag{5.44}$$

Note that

$$\mathbf{P}^2 = \{(V,\tau) : V \succeq 0, \|V\|_{2,2} \leq \tau\},$$

and this is *exactly* the largest cone compatible with the unit Euclidean ball.

When $s \geq 2$, the unit ball $\mathcal{Y}$ of the norm $\|\cdot\|_s$ is an ellitope:

$$\{y \in \mathbf{R}^N : \|y\|_s \leq 1\} = \{y \in \mathbf{R}^N : \exists (t \geq 0, \|t\|_{\bar{s}} \leq 1) : y^T R_\ell y := y_\ell^2 \leq t_\ell,\ \ell \leq L = N\},$$

so that one of the cones compatible with $\mathcal{Y}$ is given by (5.37) with the identity matrix in the role of M. As it is immediately seen, the latter cone is nothing but the cone (5.44).

5.1.5.6 *Near-optimality of polyhedral estimate in the spectratopic sub-Gaussian case*

As an instructive application of the approach developed so far, consider the special case of the estimation problem stated in Section 5.1.1, where

1. The signal set $\mathcal{X}$ and the unit ball $\mathcal{B}_*$ of the norm conjugate to $\|\cdot\|$ are spectratopes:

$$\begin{array}{rcl}\mathcal{X} & = & \{x \in \mathbf{R}^n : \exists t \in \mathcal{T} : R_k^2[x] \preceq t_k I_{d_k},\ 1 \leq k \leq K\}, \\ \mathcal{B}_* & = & \{z \in \mathbf{R}^\nu : \exists y \in \mathcal{Y} : z = My\}, \\ & & \mathcal{Y} := \{y \in \mathbf{R}^q : \exists r \in \mathcal{R} : S_\ell^2[y] \preceq r_\ell I_{f_\ell},\ 1 \leq \ell \leq L\},\end{array}$$

   (cf. Assumptions **A**, **B** in Section 4.3.3.2; as always, we lose nothing assuming spectratope $\mathcal{X}$ to be basic).
2. For every $x \in \mathcal{X}$, observation noise ξ_x is sub-Gaussian, i.e., $\xi_x \sim \mathcal{SG}(0, \sigma^2 I_m)$.

We are about to show that in the present situation, *the polyhedral estimate constructed in Sections 5.1.5.2–5.1.5.4, i.e., yielded by the efficiently computable (high accuracy near-) optimal solution to the optimization problem* (5.34), *is near-optimal in the minimax sense.*

Given reliability tolerance $\epsilon \in (0,1)$, the recipe for constructing the $m \times m$ contrast matrix H as presented in Proposition 5.8 is as follows:

- Set

$$Z = \{\vartheta^2[1;...;1]\},\ \vartheta = \sigma\kappa,\ \kappa = \sqrt{2\ln(2m/\epsilon)},$$

and utilize the construction from Section 5.1.5.2, thus arriving at the cone

$$\mathbf{H} = \{(\Theta, \mu) \in \mathbf{S}^m_+ \times \mathbf{R}_+ : \sigma^2\kappa^2 \mathrm{Tr}(\Theta) \leq \mu\}$$

satisfying the requirements of Assumption **C**.

- Specify the cones **X** and **U** compatible with $\mathcal{X}_s = \mathcal{X}$, and $\mathcal{B}_*$, respectively, according to (5.36).

The resulting problem (5.34), after immediate straightforward simplifications, reads

$$\mathrm{Opt} = \min_{\Theta, U, \Lambda, \Upsilon} \left\{ \begin{array}{l} 2\left[\phi_{\mathcal{R}}(\lambda[\Upsilon]) + \phi_{\mathcal{T}}(\lambda[\Lambda]) + \sigma^2\kappa^2\mathrm{Tr}(\Theta)\right] : \\ \Theta \succeq 0, U \succeq 0, \Lambda = \{\Lambda_k \succeq 0, k \leq K\}, \\ \Upsilon = \{\Upsilon_\ell \succeq 0, \ell \leq L\}, M^TUM \preceq \sum_\ell \mathcal{S}^*_\ell[\Upsilon_\ell], \\ \left[\begin{array}{c|c} U & \frac{1}{2}B \\ \hline \frac{1}{2}B^T & A^T\Theta A + \sum_k \mathcal{R}^*_k[\Lambda_k] \end{array}\right] \succeq 0 \end{array} \right\} \tag{5.45}$$

where, as always,

$$\begin{array}{rl} [\mathcal{R}^*_k[\Lambda_k]]_{ij} = \mathrm{Tr}(R^{ki}\Lambda_k R^{kj}) & [R_k[x] = \sum_i x_i R^{ki}], \\ [\mathcal{S}^*_\ell[\Upsilon_\ell]]_{ij} = \mathrm{Tr}(S^{\ell i}\Upsilon_\ell S^{\ell j}) & [S_\ell[u] = \sum_i u_i S^{\ell i}], \end{array}$$

and

$$\lambda[\Lambda] = [\mathrm{Tr}(\Lambda_1); ...; \mathrm{Tr}(\Lambda_K)],\ \lambda[\Upsilon] = [\mathrm{Tr}(\Upsilon_1); ...; \mathrm{Tr}(\Upsilon_L)],\ \phi_W(f) = \max_{w \in W} w^T f.$$

Let now

$$\mathrm{RiskOpt}_\epsilon = \inf_{\widehat{x}(\cdot)} \sup_{x \in \mathcal{X}} \inf\left\{\rho : \mathrm{Prob}_{\xi \sim \mathcal{N}(0,\sigma^2 I)}\{\|Bx - \widehat{x}(Ax + \xi)\| > \rho\} \leq \epsilon \forall x \in \mathcal{X}\right\},$$

be the minimax optimal $(\epsilon, \|\cdot\|)$-risk of estimating Bx in the *Gaussian* observation scheme where $\xi_x \sim \mathcal{N}(0, \sigma^2 I_m)$ independently of $x \in X$.

Proposition 5.10. *When $\epsilon \leq 1/8$, the polyhedral estimate $\widehat{x}^H$ yielded by a feasible near-optimal, in terms of the objective, solution to problem* (5.45) *is minimax optimal within the logarithmic factor, namely*

$$\begin{array}{rcl} \mathrm{Risk}_{\epsilon,\|\cdot\|}[\widehat{x}^H|\mathcal{X}] & \leq & O(1)\sqrt{\ln\left(\sum_k d_k\right)\ln\left(\sum_\ell f_\ell\right)\ln(2m/\epsilon)}\ \mathrm{RiskOpt}_{\frac{1}{8}} \\ & \leq & O(1)\sqrt{\ln\left(\sum_k d_k\right)\ln\left(\sum_\ell f_\ell\right)\ln(2m/\epsilon)}\ \mathrm{RiskOpt}_\epsilon \end{array}$$

where $O(1)$ is an absolute constant.

See Section 5.4.4 for the proof.

Discussion. It is worth mentioning that the approach described in Section 5.1.4 is complementary to the approach developed in this section. In fact, it is easily seen that the bound Opt for the risk of the polyhedral estimate stemming from (5.34) is suboptimal in the simple situation described in the motivating example from Section 5.1.1. Indeed, let $\mathcal{X}$ be the unit $\|\cdot\|_1$-ball, $\|\cdot\| = \|\cdot\|_2$, and let us consider the problem of estimating $x \in \mathcal{X}$ from the direct observation $\omega = x + \xi$ with Gaussian observation noise $\xi \sim \mathcal{N}(0, \sigma^2 I)$. We equip the ball $\mathcal{B}_* = \{u \in \mathbf{R}^n : \|u\|_2 \leq 1\}$

with the cone

$$\mathbf{U} = \mathbf{P}^2 = \{(U,\tau) : U \succeq 0, \|U\|_{2,2} \leq \tau\}$$

and $\mathcal{X}$ with the cone

$$\mathbf{X} = \mathbf{P}^1 = \{(X,t) : X \succeq 0, \|X\|_\infty \leq t\},$$

(note that both cones are the largest w.r.t. inclusion cones compatible with the respective sets). The corresponding problem (5.34) reads

$$\begin{array}{rcl} \mathrm{Opt} & = & \min\limits_{\Theta,X,U} \left\{ 2\left(\kappa^2\sigma^2\mathrm{Tr}(\Theta) + \max\limits_i X_{ii} + \|U\|_{2,2}\right) : \begin{array}{c} \Theta \succeq 0,\ X \succeq 0,\ U \succeq 0, \\ \left[\begin{array}{c|c} U & \frac{1}{2}I_n \\ \hline \frac{1}{2}I_n & \Theta + X \end{array}\right] \succeq 0 \end{array} \right\} \\ & = & \min\limits_{\Theta,X,U} \left\{ 2\left(\kappa^2\sigma^2\mathrm{Tr}(\Theta) + \max\limits_i X_{ii} + \tau\right) : \begin{array}{c} \Theta \succeq 0,\ X \succeq 0,\ U \succeq 0, \\ \left[\begin{array}{c|c} \tau I_n & \frac{1}{2}I_n \\ \hline \frac{1}{2}I_n & \Theta + X \end{array}\right] \succeq 0 \end{array} \right\}. \end{array} \tag{5.46}$$

Observe that every $n \times n$ matrix of the form $Q = EP$, where E is diagonal with diagonal entries ± 1, and P is a permutation matrix, induces a symmetry $(\Theta, X, \tau) \mapsto (Q\Theta Q^T, QXQ^T, \tau)$ of the second optimization problem in (5.46), that is, a transformation which maps the feasible set onto itself and keeps the objective intact. Since the problem is convex and solvable, we conclude that it has an optimal solution which remains intact under the symmetries in question, i.e., solution with scalar matrices $\Theta = \theta I_n$ and $X = uI_n$. As a result,

$$\mathrm{Opt} = \min_{\theta\geq 0, u\geq 0, \tau} \left\{2(\kappa^2\sigma^2 n\theta + u + \tau) : \tau(\theta + u) \geq \tfrac{1}{4}\right\} = 2\min\left[\kappa\sigma\sqrt{n}, 1\right]. \tag{5.47}$$

A similar derivation shows that the value Opt remains intact if we replace the set $\mathcal{X} = \{x : \|x\|_1 \leq 1\}$ with $\mathcal{X} = \{x : \|x\|_s \leq 1\}$, $s \in [1,2]$, and the cone $\mathbf{X} = \mathbf{P}^1$ with $\mathbf{X} = \mathbf{P}^s$; see (5.42). Since the Θ-component of an optimal solution to (5.46) can be selected to be scalar, the contrast matrix H we end up with can be selected to be the unit matrix. An unpleasant observation is that when $s < 2$, the quantity Opt given by (5.47) "heavily overestimates" the actual risk of the polyhedral estimate with $H = I_n$. Indeed, the analysis of this estimate in Section 5.1.4 results in the risk bound (up to a factor logarithmic in n) $\min[\sigma^{1-s/2}, \sigma\sqrt{n}]$, which can be much less than $\mathrm{Opt} = 2\min[\kappa\sigma\sqrt{n}, 1]$, e.g., in the case of large n, and $\sigma\sqrt{n} = O(1)$.

5.1.6 Assembling estimates: Contrast aggregation

The good news is that whenever the approaches to the design of polyhedral estimates presented in Sections 5.1.4 and 5.1.5 are applicable, they can be utilized simultaneously. The underlying observation is that

> (!) *In the problem setting described in Section 5.1.2, a collection of K candidate polyhedral estimates can be assembled into a single polyhedral estimate with the (upper bound on the) risk, as given by Proposition 5.1, being nearly the minimum of the risks of estimates we aggregate.*

Indeed, given an observation scheme (that is, collection of probability distributions P_x of noises ξ_x, $x \in \mathcal{X}$), assume we have at our disposal norms $\pi_\delta(\cdot) : \mathbf{R}^m \to \mathbf{R}$ parameterized by $\delta \in (0,1)$ such that $\pi_\delta(h)$, for every h, is larger the lesser δ is,

and

$$\forall (x \in \mathcal{X}, \delta \in (0,1), h \in \mathbf{R}^m) : \pi_\delta(h) \leq 1 \Rightarrow \mathrm{Prob}_{\xi \sim P_x}\{\xi : |h^T \xi| > 1\} \leq \delta.$$

Assume also (as is indeed the case in all our constructions) that we ensure (5.1) by imposing on the columns h_ℓ of an $m \times N$ contrast matrix H the restrictions $\pi_{\epsilon/N}(h_\ell) \leq 1$.

Now suppose that given risk tolerance $\epsilon \in (0,1)$, we have generated K candidate contrast matrices $H_k \in \mathbf{R}^{m \times N_k}$ such that

$$\pi_{\epsilon/N_k}(\mathrm{Col}_j[H_k]) \leq 1,\; j \leq N_k,$$

so that the $(\epsilon, \|\cdot\|)$-risk of the polyhedral estimate yielded by the contrast matrix H_k does not exceed

$$\mathfrak{R}_k = \max_x \left\{ \|Bx\| : x \in 2\mathcal{X}_s, \|H_k^T Ax\|_\infty \leq 2 \right\}.$$

Let us combine the contrast matrices $H_1, ..., H_K$ into a single contrast matrix H with $N = N_1 + ... + N_K$ columns by normalizing the columns of the concatenated matrix $[H_1, ..., H_K]$ to have $\pi_{\epsilon/N}$-norms equal to 1, so that

$$H = [\bar{H}_1, ..., \bar{H}_K],\; \mathrm{Col}_j[\bar{H}_k] = \theta_{jk}\mathrm{Col}_j[H_k]\; \forall (k \leq K, j \leq N_k)$$

with

$$\theta_{jk} = \frac{1}{\pi_{\epsilon/N}(\mathrm{Col}_j[H_k])} \geq \vartheta_k := \min_{h \neq 0} \frac{\pi_{\epsilon/N_k}(h)}{\pi_{\epsilon/N}(h)},$$

where the concluding $\geq$ is due to $\pi_{\epsilon/N_k}(\mathrm{Col}_j[H_k]) \leq 1$. We claim that in terms of $(\epsilon, \|\cdot\|)$-risk, the polyhedral estimate yielded by H is "almost as good" as the best of the polyhedral estimates yielded by the contrast matrices $H_1, ..., H_K$, specifically,[6]

$$\mathfrak{R}[H] := \max_x \left\{ \|Bx\| : x \in 2\mathcal{X}_s, \|H^T Ax\|_\infty \leq 2 \right\} \leq \min_k \vartheta_k^{-1} \mathfrak{R}_k.$$

The justification is readily given by the following observation: when $\vartheta \in (0,1)$, we have

$$\mathfrak{R}_{k,\vartheta} := \max_x \left\{ \|Bx\| : x \in 2\mathcal{X}_s, \|H_k^T Ax\|_\infty \leq 2/\vartheta \right\} \leq \mathfrak{R}_k/\vartheta.$$

Indeed, when x is a feasible solution to the maximization problem specifying $\mathfrak{R}_{k,\vartheta}$, ϑx is a feasible solution to the problem specifying $\mathfrak{R}_k$, implying that $\vartheta\|Bx\| \leq \mathfrak{R}_k$. It remains to note that we clearly have $\mathfrak{R}[H] \leq \min_k \mathfrak{R}_{k,\vartheta_k}$.

The bottom line is that the aggregation just described of contrast matrices $H_1, ..., H_K$ into a single contrast matrix H results in a polyhedral estimate which in terms of upper bound $\mathfrak{R}[\cdot]$ on its $(\epsilon, \|\cdot\|)$-risk is, up to factor $\bar{\vartheta} = \max_k \vartheta_k^{-1}$, not worse than the best of the K estimates yielded by the original contrast matrices. Consequently, if $\pi_\delta(\cdot)$ grows slowly as δ decreases, the "price" $\bar{\vartheta}$ of assembling the original estimates is quite moderate. For example, in our basic cases (sub-Gaussian, Discrete, and Poisson), $\bar{\vartheta}$ is logarithmic in $\max_k N_k^{-1}(N_1 + ... + N_K)$, and $\bar{\vartheta} = 1 + o(1)$ as $\epsilon \to +0$ for $K, N_1, ..., N_K$ fixed.

[6] This is the precise "quantitative expression" of the observation (!).

5.1.7 Numerical illustration

We are about to illustrate the numerical performance of polyhedral estimates by comparing it to the performance of a "presumably good" linear estimate. Our setup is deliberately simple: the signal set $\mathcal{X}$ is just the unit box $\{x \in \mathbf{R}^n : \|x\|_\infty \leq 1\}$, $B \in \mathbf{R}^{n\times n}$ is "numerical double integration": for a $\delta > 0$,

$$B_{ij} = \left\{ \begin{array}{ll} \delta^2(i-j+1), & j \leq i \\ 0, & j > i \end{array} \right.,$$

so that x, modulo boundary effects, is the second order finite difference derivative of $w = Bx$,

$$x_i = \frac{w_i - 2w_{i-1} + w_{i-2}}{\delta^2},\ 2 < i \leq n;$$

and Ax is comprised of m randomly selected entries of Bx. The observation is

$$\omega = Ax + \xi,\ \xi \sim \mathcal{N}(0, \sigma^2 I_m)$$

and the recovery norm is $\|\cdot\|_2$. In other words, we want to recover a restriction of a twice differentiable function of one variable on the n-point regular grid on the segment $\Delta = [0, n\delta]$ from noisy observations of this restriction taken along m randomly selected points of the grid. A priori information on the function is that the magnitude of its second order derivative does not exceed 1.

Note that in the considered situation both linear estimate $\widehat{x}_H$ yielded by Proposition 4.14 and polyhedral estimate $\widehat{x}^H$ yielded by Proposition 5.7, are near-optimal in the minimax sense in terms of their $\|\cdot\|_2$- or $(\epsilon, \|\cdot\|_2)$-risk.

In the experiments reported in Figure 5.1, we used $n = 64$, $m = 32$, and $\delta = 4/n$ (i.e., $\Delta = [0, 4]$); the reliability parameter for the polyhedral estimate was set to $\epsilon = 0.1$. For different noise levels $\sigma = \{0.1, 0.01, 0.001, 0.0001\}$ we generate 20 random signals x from $\mathcal{X}$ and record the $\|\cdot\|_2$-recovery errors of the linear and the polyhedral estimates. In addition to testing the nearly optimal polyhedral estimate *PolyI* yielded by Proposition 5.8 as applied in the framework of item 5.1.5.5.A, we also record the performance of the polyhedral estimate *PolyII* yielded by the construction from Section 5.1.4. The observed $\|\cdot\|_2$-recovery errors of the three estimates are plotted in Figure 5.1.

All three estimates exhibit similar empirical performance in these simulations. However, when the noise level becomes small, polyhedral estimates seem to outperform the linear one. In addition, the estimate *PolyII* seems to "work" better than or, at the very worst, similarly to *PolyI* in spite of the fact that in the situation in question the estimate *PolyI,* in contrast to *PolyII,* is provably near-optimal.

5.1.8 Calculus of compatibility

The principal rules of the calculus of compatibility are as follows (verification of the rules is straightforward and is therefore skipped):

1. [passing to a subset] When $\mathcal{Y}' \subset \mathcal{Y}$ are convex compact subsets of $\mathbf{R}^N$ and a cone $\mathbf{Y}$ is compatible with $\mathcal{Y}$, the cone is compatible with $\mathcal{Y}'$ as well.

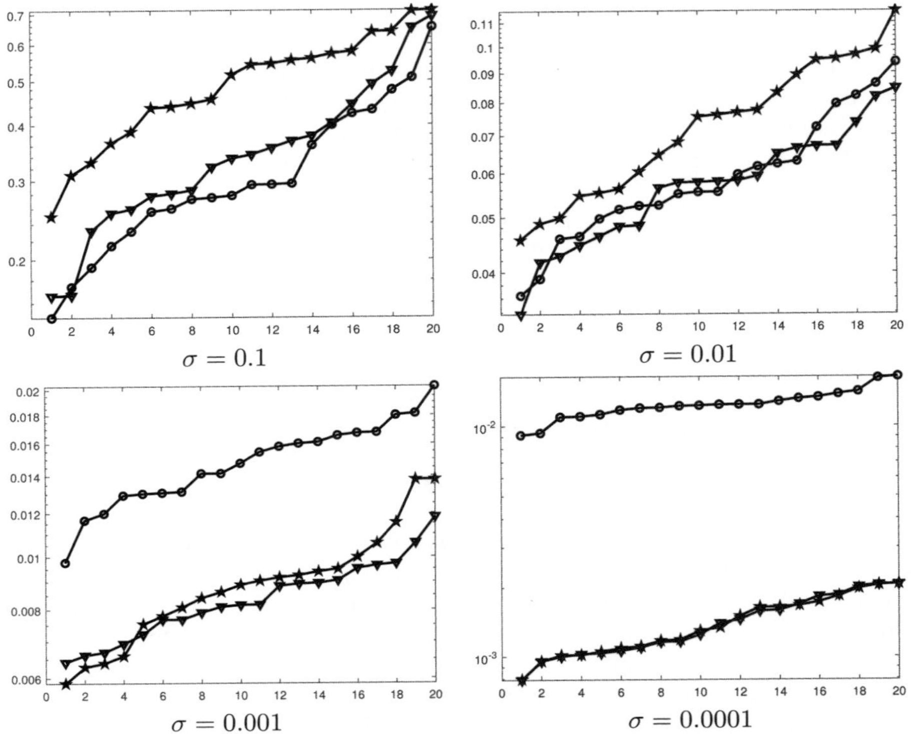

Figure 5.1: Recovery errors for the near-optimal linear estimate (circles) and for polyhedral estimates yielded by Proposition 5.8 (*PolyI*, pentagrams) and by the construction from Section 5.1.4 (*PolyII*, triangles), 20 simulations per each value of σ.

2. [finite intersection] Let cones $\mathbf{Y}^j$ be compatible with convex compact sets $\mathcal{Y}_j \subset \mathbf{R}^N$, $j = 1, ..., J$. Then the cone

$$\mathbf{Y} = \mathrm{cl}\{(V,\tau) \in \mathbf{S}^N_+ \times \mathbf{R}_+ : \exists ((V_j,\tau_j) \in \mathbf{Y}^j, j \leq J) : V \preceq \sum_j V_j, \sum_j \tau_j \leq \tau\}$$

is compatible with $\mathcal{Y} = \bigcap_j \mathcal{Y}_j$. The closure operation can be skipped when all cones $\mathbf{Y}^j$ are sharp, in which case $\mathbf{Y}$ is sharp as well.

3. [convex hulls of finite union] Let cones $\mathbf{Y}^j$ be compatible with convex compact sets $\mathcal{Y}_j \subset \mathbf{R}^N$, $j = 1, ..., J$, and let there exist (V, τ) such that $V \succ 0$ and

$$(V,\tau) \in \mathbf{Y} := \bigcap_j \mathbf{Y}^j.$$

Then $\mathbf{Y}$ is compatible with $\mathcal{Y} = \mathrm{Conv}\{\bigcup_j \mathcal{Y}_j\}$ and, in addition, is sharp provided that at least one of the $\mathbf{Y}^j$ is sharp.

4. [direct product] Let cones $\mathbf{Y}^j$ be compatible with convex compact sets $\mathcal{Y}_j \subset \mathbf{R}^{N_j}$, $j = 1, ..., J$. Then the cone

$$\mathbf{Y} = \{(V,\tau) \in \mathbf{S}^{N_1+...+N_J}_+ \times \mathbf{R}_+ : \exists (V_j,\tau_j) \in \mathbf{Y}^j : V \preceq \mathrm{Diag}\{V_1,...,V_J\} \ \&\ \tau \geq \sum_j \tau_j\}$$

is compatible with $\mathcal{Y} = \mathcal{Y}_1 \times ... \times \mathcal{Y}_J$. This cone is sharp, provided that all the $\mathbf{Y}^j$ are so.

5. [linear image] Let cone $\mathbf{Y}$ be compatible with convex compact set $\mathcal{Y} \subset \mathbf{R}^N$, let A be a $K \times N$ matrix, and let $\mathcal{Z} = A\mathcal{Y}$. The cone
$$\mathbf{Z} = \mathrm{cl}\{(V, \tau) \in \mathbf{S}^K_+ \times \mathbf{R}_+ : \exists U \succeq A^T V A : (U, \tau) \in \mathbf{Y}\}$$
is compatible with $\mathcal{Z}$. The closure operation can be skipped whenever $\mathbf{Y}$ is either sharp, or *complete*, completeness meaning that $(V, \tau) \in \mathbf{Y}$ and $0 \preceq V' \preceq V$ imply that $(V', \tau) \in \mathbf{Y}$. The cone $\mathbf{Z}$ is sharp, provided $\mathbf{Y}$ is so and the rank of A is K.
6. [inverse linear image] Let cone $\mathbf{Y}$ be compatible with convex compact set $\mathcal{Y} \subset \mathbf{R}^N$, let A be an $N \times K$ matrix with trivial kernel, and let $\mathcal{Z} = A^{-1}\mathcal{Y} := \{z \in \mathbf{R}^K : Az \in \mathcal{Y}\}$. The cone
$$\mathbf{Z} = \mathrm{cl}\{(V, \tau) \in \mathbf{S}^K_+ \times \mathbf{R}_+ : \exists U : A^T U A \succeq V \;\&\; (U, \tau) \in \mathbf{Y}\}$$
is compatible with $\mathcal{Z}$. The closure operations can be skipped whenever $\mathbf{Y}$ is sharp, in which case $\mathbf{Z}$ is sharp as well.
7. [arithmetic summation] Let cones $\mathbf{Y}^j$ be compatible with convex compact sets $\mathcal{Y}_j \subset \mathbf{R}^N$, $j = 1, ..., J$. Then the arithmetic sum $\mathcal{Y} = \mathcal{Y}_1 + ... + \mathcal{Y}_J$ of the sets $\mathcal{Y}_j$ can be equipped with a compatible cone readily given by the cones $\mathbf{Y}^j$; this cone is sharp, provided all the $\mathbf{Y}^j$ are so.
Indeed, the arithmetic sum of $\mathcal{Y}_j$ is the linear image of the direct product of the $\mathcal{Y}_j$'s under the mapping $[y^1; ...; y^J] \mapsto y^1 + ... + y^J$, and it remains to combine rules 4 and 5; note the cone yielded by rule 4 is complete, so that when applying rule 5, the closure operation can be skipped.

5.2 RECOVERING SIGNALS FROM NONLINEAR OBSERVATIONS BY STOCHASTIC OPTIMIZATION

The "common denominator" of all estimation problems considered so far in this chapter is that what we observed was obtained by adding noise to the *linear* image of the unknown signal to be recovered. In this section we consider the problem of signal estimation in the case where the observation is obtained by adding noise to a *nonlinear* transformation of the signal.

5.2.1 Problem setting

A motivating example for what follows is provided by the *logistic regression* model, where

- the unknown signal to be recovered is a vector x known to belong to a given signal set $\mathcal{X} \subset \mathbf{R}^n$, which we assume to be a nonempty convex compact set;
- our observation
$$\omega^K = \{\omega_k = (\eta_k, y_k), 1 \leq k \leq K\}$$
stemming from a signal x is as follows:
  - the *regressors* $\eta_1, ..., \eta_K$ are i.i.d. realizations of an n-dimensional random

vector η with distribution Q independent of x and such that Q possesses a finite and positive definite matrix $\mathbf{E}_{\eta\sim Q}\{\eta\eta^T\}$ of second moments;

– the *labels* y_k are generated as follows: y_k is the Bernoulli random variable independent of the "history" $\eta_1, ..., \eta_{k-1}, y_1, ..., y_{k-1}$, and the conditional, given η_k, probability for y_k to be 1 is $\phi(\eta_k^T x)$, where

$$\phi(s) = \frac{\exp\{s\}}{1 + \exp\{s\}}.$$

In this model, the standard (and very well-studied) approach to estimating the signal x underlying the observations is to use the Maximum Likelihood (ML) estimate: the *logarithm* of the conditional, given η_k, $1 \leq k \leq K$, probability of getting the observed labels as a function of a candidate signal z is

$$\begin{array}{rcl} \ell(z, \omega^K) & = & \sum_{k=1}^{K} \left[y_k \ln \left(\phi(\eta_k^T z) \right) + (1 - y_k) \ln \left(1 - \phi(\eta_k^T z) \right) \right] \\ & = & \left[\sum_k y_k \eta_k \right]^T z - \sum_k \ln \left(1 + \exp\{\eta_k^T z\} \right), \end{array} \tag{5.48}$$

and the ML estimate of the "true" signal x underlying our observation ω^K is obtained by maximizing the log-likelihood $\ell(z, \omega^K)$ over $z \in \mathcal{X}$,

$$\widehat{x}_{\rm ML}(\omega^K) \in \operatorname*{Argmax}_{z \in \mathcal{X}} \ell(z, \omega^K), \tag{5.49}$$

which is a convex optimization problem.

The problem we intend to consider (referred to as the *generalized linear model* (GLM) in Statistics) can be viewed as a natural generalization of the logistic regression just presented and is as follows:

Our observation depends on unknown signal x known to belong to a given convex compact set $\mathcal{X} \subset \mathbf{R}^n$ and is

$$\omega^K = \{\omega_k = (\eta_k, y_k), 1 \leq k \leq K\} \tag{5.50}$$

with ω_k, $1 \leq k \leq K$, which are i.i.d. realizations of a random pair (η, y) with the distribution P_x such that

- the *regressor* η is a random $n \times m$ matrix with some probability distribution Q *independent of* x;
- the *label* y is an m-dimensional random vector such that the conditional distribution of y given η induced by P_x has the expectation $f(\eta^T x)$:

$$\mathbf{E}^x_{|\eta}\{y\} = f(\eta^T x), \tag{5.51}$$

where $\mathbf{E}^x_{|\eta}\{y\}$ is the conditional expectation of y given η stemming from the distribution P_x of $\omega = (\eta, y)$, and $f(\cdot) : \mathbf{R}^m \to \mathbf{R}^m$ ("link function") is a given mapping.

Note that the logistic regression model corresponds to the case where $m = 1$,

$f(s) = \frac{\exp\{s\}}{1+\exp\{s\}}$, and y takes values 0,1, with the conditional probability of taking value 1 given η equal to $f(\eta^T x)$.

Another example is provided by the model

$$y = f(\eta^T x) + \xi,$$

where ξ is a random vector with zero mean independent of η, say, $\xi \sim \mathcal{N}(0, \sigma^2 I_m)$. Note that in the latter case the ML estimate of the signal x underlying the observations is

$$\widehat{x}_{\text{ML}}(\omega^K) \in \underset{z\in\mathcal{X}}{\text{Argmin}} \sum_k \|y_k - f(\eta_k^T z)\|_2^2. \tag{5.52}$$

In contrast to what happens with logistic regression, now the optimization problem—"Nonlinear Least Squares"—responsible for the ML estimate typically is nonconvex and can be computationally difficult.

Following [140], we intend to impose on the data of the estimation problem we have just described (namely, on $\mathcal{X}$, $f(\cdot)$, and the distributions P_x, $x \in \mathcal{X}$, of the pair (η, y)) assumptions which allow us to reduce our estimation problem to a problem with convex structure—a *strongly monotone variational inequality represented by a stochastic oracle.* At the end of the day, this will lead to a consistent estimate of the signal, with explicit "finite sample" accuracy guarantees.

5.2.2 Assumptions

Preliminaries: Monotone vector fields. A monotone vector field on $\mathbf{R}^m$ is a single-valued everywhere defined mapping $g(\cdot) : \mathbf{R}^m \to \mathbf{R}^m$ which possesses the *monotonicity property*

$$[g(z) - g(z')]^T [z - z'] \geq 0 \;\forall z, z' \in \mathbf{R}^m.$$

We say that such a field is *monotone with modulus $\varkappa \geq 0$ on a closed convex set $Z \subset \mathbf{R}^m$* if

$$[g(z) - g(z')]^T [z - z'] \geq \varkappa \|z - z'\|_2^2, \forall z\, z' \in Z,$$

and say that g is *strongly monotone* on Z if the modulus of monotonicity of g on Z is positive. It is immediately seen that for a monotone vector field which is continuously differentiable on a closed convex set Z with a nonempty interior, the necessary and sufficient condition for being monotone with modulus $\varkappa$ on the set is

$$d^T f'(z) d \geq \varkappa d^T d \;\forall (d \in \mathbf{R}^n, z \in Z). \tag{5.53}$$

Basic examples of monotone vector fields are:

- gradient fields $\nabla\phi(x)$ of continuously differentiable convex functions of m variables or, more generally, the vector fields $[\nabla_x \phi(x, y); -\nabla_y \phi(x, y)]$ stemming from continuously differentiable functions $\phi(x, y)$ which are convex in x and concave in y;
- "diagonal" vector fields $f(x) = [f_1(x_1); f_2(x_2); ...; f_m(x_m)]$ with monotonically nondecreasing univariate components $f_i(\cdot)$. If, in addition, the $f_i(\cdot)$ are continuously differentiable with positive first order derivatives, then the associated field f is strongly monotone on every compact convex subset of $\mathbf{R}^m$, the monotonicity modulus depending on the subset.

Monotone vector fields on $\mathbf{R}^n$ admit simple calculus which includes, in particular, the following two rules:

I. [affine substitution of argument]: If $f(\cdot)$ is a monotone vector field on $\mathbf{R}^m$ and A is an $n \times m$ matrix, the vector field

$$g(x) = Af(A^T x + a)$$

is monotone on $\mathbf{R}^n$; if, in addition, f is monotone with modulus $\varkappa \geq 0$ on a closed convex set $Z \subset \mathbf{R}^m$ and $X \subset \mathbf{R}^n$ is closed, convex, and such that $A^T x + a \in Z$ whenever $x \in X$, g is monotone with modulus $\sigma^2 \varkappa$ on X, where σ is the n-th singular value of A (i.e., the largest γ such that $\|A^T x\|_2 \geq \gamma \|x\|_2$ for all x).

II. [summation]: If S is a Polish space, $f(x, s) : \mathbf{R}^m \times S \to \mathbf{R}^m$ is a Borel vector-valued function which is monotone in x for every $s \in S$, and $\mu(ds)$ is a Borel probability measure on S such that the vector field

$$F(x) = \int_S f(x, s)\mu(ds)$$

is well-defined for all x, then $F(\cdot)$ is monotone. If, in addition, X is a closed convex set in $\mathbf{R}^m$ and $f(\cdot, s)$ is monotone on X with Borel in s modulus $\varkappa(s)$ for every $s \in S$, then F is monotone on X with modulus $\int_S \varkappa(s)\mu(ds)$.

Assumptions. In what follows, we make the following assumptions on the ingredients of the estimation problem posed in Section 5.2.1:

- **A.1**. The vector field $f(\cdot)$ is continuous and monotone, and the vector field

$$F(z) = \mathbf{E}_{\eta \sim Q} \left\{ \eta f(\eta^T z) \right\}$$

  is well-defined (and therefore is monotone along with f by **I**, **II**);
- **A.2**. The signal set $\mathcal{X}$ is a nonempty convex compact set, and the vector field F is monotone with positive modulus $\varkappa$ on $\mathcal{X}$;
- **A.3**. For properly selected $M < \infty$ and every $x \in \mathcal{X}$ it holds

$$\mathbf{E}_{(\eta,y) \sim P_x} \left\{ \|\eta y\|_2^2 \right\} \leq M^2. \tag{5.54}$$

A simple *sufficient* condition for the validity of Assumptions **A.1-3** with properly selected $M < \infty$ and $\varkappa > 0$ is as follows:

- The distribution Q of η has finite moments of all orders, and $\mathbf{E}_{\eta \sim Q}\{\eta \eta^T\} \succ 0$;
- f is continuously differentiable, and $d^T f'(z) d > 0$ for all $d \neq 0$ and all z. Besides this, f is of polynomial growth: for some constants $C \geq 0$ and $p \geq 0$ and all z one has $\|f(z)\|_2 \leq C(1 + \|z\|_2^p)$.

Verification of sufficiency is straightforward.

The principal observation underlying the construction we are about to discuss is as follows.

Proposition 5.11. *With Assumptions* **A.1–3** *in force, let us associate with a pair*

$(\eta, y) \in \mathbf{R}^{n\times m} \times \mathbf{R}^m$ *the vector field*

$$G_{(\eta,y)}(z) = \eta f(\eta^T z) - \eta y : \mathbf{R}^n \to \mathbf{R}^n. \tag{5.55}$$

Then for every $x \in \mathcal{X}$ we have

$$\begin{array}{rcll} \mathbf{E}_{(\eta,y)\sim P_x}\left\{G_{(\eta,y)}(z)\right\} & = & F(z) - F(x)\ \forall z \in \mathbf{R}^n & (a)\\ \|F(z)\|_2 & \leq & M\,\forall z \in \mathcal{X} & (b)\\ \mathbf{E}_{(\eta,y)\sim P_x}\left\{\|G_{(\eta,y)}(z)\|_2^2\right\} & \leq & 4M^2\ \forall z \in \mathcal{X}. & (c) \end{array} \tag{5.56}$$

Proof is immediate. Indeed, let $x \in \mathcal{X}$. Then

$$\mathbf{E}_{(\eta,y)\sim P_x}\{\eta y\} = \mathbf{E}_{\eta\sim Q}\left\{\mathbf{E}^x_{|\eta}\{\eta y\}\right\} = \mathbf{E}_\eta\left\{\eta f(\eta^T x)\right\} = F(x)$$

(we have used (5.51) and the definition of F), whence,

$$\begin{array}{rcl} \mathbf{E}_{(\eta,y)\sim P_x}\left\{G_{(\eta,y)}(z)\right\} & = & \mathbf{E}_{(\eta,y)\sim P_x}\left\{\eta f(\eta^T z) - \eta y\right\} = \mathbf{E}_{(\eta,y)\sim P_x}\left\{\eta f(\eta^T z)\right\} - F(x)\\ & = & \mathbf{E}_{\eta\sim Q}\left\{\eta f(\eta^T z)\right\} - F(x) = F(z) - F(x), \end{array}$$

as stated in (5.56.a). Besides this, for $x, z \in \mathcal{X}$, taking into account that the marginal distribution of η induced by P_z is Q, we have

$$\begin{array}{rcll} \mathbf{E}_{(\eta,y)\sim P_x}\{\|\eta f(\eta^T z)\|_2^2\} & = & \mathbf{E}_{\eta\sim Q}\left\{\|\eta f(\eta^T z)\|_2^2\right\} & \\ & = & \mathbf{E}_{\eta\sim Q}\left\{\|\mathbf{E}_{y\sim P^z_{|\eta}}\{\eta y\}\|_2^2\right\} & [\text{since } \mathbf{E}_{y\sim P^z_{|\eta}}\{y\} = f(\eta^T z)]\\ & \leq & \mathbf{E}_{\eta\sim Q}\left\{\mathbf{E}_{y\sim P^z_{|\eta}}\left\{\|\eta y\|_2^2\right\}\right\} & [\text{by Jensen's inequality}]\\ & = & \mathbf{E}_{(\eta,y)\sim P_z}\left\{\|\eta y\|_2^2\right\} \leq M^2 & [\text{by } \mathbf{A.3} \text{ due to } z \in \mathcal{X}]. \end{array}$$

This combines with the relation $\mathbf{E}_{(\eta,y)\sim P_x}\{\|\eta y\|_2^2\} \leq M^2$ given by **A.3** due to $x \in \mathcal{X}$ to imply (5.56.b) and (5.56.c). □

Consequences. Our goal is to recover the signal $x \in \mathcal{X}$ underlying observations (5.50), and under assumptions **A.1–3**, x is a root of the monotone vector field

$$G(z) = F(z) - F(x),\ F(z) = \mathbf{E}_{\eta\sim Q}\left\{\eta f(\eta^T z)\right\}; \tag{5.57}$$

we know that this root belongs to $\mathcal{X}$, and this root is unique because $G(\cdot)$ is strongly monotone on $\mathcal{X}$ along with $F(\cdot)$. Now, the problem of finding a root, known to belong to a given convex compact set $\mathcal{X}$, of a vector field G which is strongly monotone on this set is known to be computationally tractable, provided we have access to an "oracle" which, given on input a point $z \in \mathcal{X}$, returns the value $G(z)$ of the field at the point. The latter is not exactly the case in the situation we are interested in: the field G is the expectation of a random field:

$$G(z) = \mathbf{E}_{(\eta,y)\sim P_x}\left\{\eta f(\eta^T z) - \eta y\right\},$$

and we do not know a priori what the distribution is over which the expectation is taken. However, we can sample from this distribution—the samples are exactly the observations (5.50), and we can use these samples to approximate G and use this approximation to approximate the signal x.[7] Two standard implementations of this idea are *Sample Average Approximation* (SAA) and *Stochastic Approximation* (SA). We are about to consider these two techniques as applied to the situation we are in.

5.2.3 Estimating via Sample Average Approximation

The idea underlying SAA is quite transparent: given observations (5.50), let us approximate the field of interest G with its empirical counterpart

$$G_{\omega^K}(z) = \frac{1}{K}\sum_{k=1}^{K}\left[\eta_k f(\eta_k^T z) - \eta_k y_k\right].$$

By the Law of Large Numbers, as $K \to \infty$, the empirical field G_{ω^K} converges to the field of interest G, so that under mild regularity assumptions, when K is large, G_{ω^K}, with overwhelming probability, will be close to G uniformly on $\mathcal{X}$. Due to strong monotonicity of G, this would imply that a set of "near-zeros" of G_{ω^K} on $\mathcal{X}$ will be close to the zero x of G, which is nothing but the signal we want to recover. The only question is how we can consistently define a "near-zero" of G_{ω^K} on $\mathcal{X}$.[8] A convenient notion of a "near-zero" in our context is provided by the concept of a *weak solution* to a variational inequality with a monotone operator, defined as follows (we restrict the general definition to the situation of interest):

> Let $\mathcal{X} \subset \mathbf{R}^n$ be a nonempty convex compact set, and $H(z) : \mathcal{X} \to \mathbf{R}^n$ be a monotone (i.e., $[H(z) - H(z')]^T[z - z'] \geq 0$ for all $z, z' \in \mathcal{X}$) vector field. A vector $z_* \in \mathcal{X}$ is called a *weak solution* to the variational inequality (VI) associated with $H, \mathcal{X}$ when
>
> $$H^T(z)[z - z_*] \geq 0 \;\forall z \in \mathcal{X}.$$

Let $\mathcal{X} \subset \mathbf{R}^n$ be a nonempty convex compact set and H be monotone on $\mathcal{X}$. It is well known that

- The VI associated with $H, \mathcal{X}$ (let us denote it by VI$(H, \mathcal{X})$) always has a weak solution. It is clear that if $\bar{z} \in \mathcal{X}$ is a root of H, then $\bar{z}$ is a weak solution to VI$(H, \mathcal{X})$.[9]
- When H is continuous on $\mathcal{X}$, every weak solution $\bar{z}$ to VI$(H, \mathcal{X})$ is also a *strong solution*, meaning that

$$H^T(\bar{z})(z - \bar{z}) \geq 0 \;\forall z \in \mathcal{X}. \tag{5.58}$$

Indeed, (5.58) clearly holds true when $z = \bar{z}$. Assuming $z \neq \bar{z}$ and setting $z_t = \bar{z}+t(z-\bar{z})$, $0 < t \leq 1$, we have $H^T(z_t)(z_t-\bar{z}) \geq 0$ (since $\bar{z}$ is a weak solution),

[7] The observation expressed by Proposition 5.11, however simple, and the resulting course of actions seem to be new. In retrospect, one can recognize unperceived ad hoc utilization of this approach in Perceptron and Isotron algorithms, see [1, 2, 29, 62, 116, 141, 142, 210] and references therein.

[8] Note that we in general cannot define a "near-zero" of G_{ω^K} on $\mathcal{X}$ as a root of G_{ω^K} on this set—while G does have a root belonging to $\mathcal{X}$, nobody told us that the same holds true for G_{ω^K}.

[9] Indeed, when $\bar{z} \in \mathcal{X}$ and $H(\bar{z}) = 0$, monotonicity of H implies that $H^T(z)[z - \bar{z}] = [H(z) - H(\bar{z})]^T[z - \bar{z}] \geq 0$ for all $z \in \mathcal{X}$, that is, $\bar{z}$ is a weak solution to the VI.

whence $H^T(z_t)(z-\bar{z}) \geq 0$ (since $z-\bar{z}$ is a positive multiple of $z_t-\bar{z}$). Passing to limit as $t \to +0$ and invoking the continuity of H, we get $H^T(\bar{z})(z-\bar{z}) \geq 0$, as claimed.

- When H is the gradient field of a continuously differentiable convex function on $\mathcal{X}$ (such a field indeed is monotone), weak (or strong, which in the case of continuous H is the same) solutions to VI$(H, \mathcal{X})$ are exactly the minimizers of the function on $\mathcal{X}$.

  Note also that a strong solution to VI$(H, \mathcal{X})$ with monotone H always is a weak one: if $\bar{z} \in \mathcal{X}$ satisfies $H^T(\bar{z})(z-\bar{z}) \geq 0$ for all $z \in \mathcal{X}$, then $H(z)^T(z-\bar{z}) \geq 0$ for all $z \in \mathcal{X}$, since by monotonicity $H^T(z)(z-\bar{z}) \geq H^T(\bar{z})(z-\bar{z})$.

In the sequel, we utilize the following simple and well-known fact:

Lemma 5.12. *Let $\mathcal{X}$ be a convex compact set, and H be a monotone vector field on $\mathcal{X}$ with monotonicity modulus $\varkappa > 0$, i.e.*

$$\forall z, z' \in X \; [H(z) - H(z')]^T [z - z'] \geq \varkappa \|z - z'\|_2^2.$$

Further, let $\bar{z}$ be a weak solution to VI$(H, \mathcal{X})$. Then the weak solution to VI$(H, \mathcal{X})$ is unique. Besides this,

$$H^T(z)[z - \bar{z}] \geq \varkappa \|z - \bar{z}\|_2^2 \;\forall z \in \mathcal{X}. \tag{5.59}$$

Proof: Under the premise of lemma, let $z \in \mathcal{X}$ and let $\bar{z}$ be a weak solution to VI$(H, \mathcal{X})$ (recall that it does exist). Setting $z_t = \bar{z} + t(z - \bar{z})$, for $t \in (0, 1)$ we have

$$H^T(z)[z - z_t] \geq H^T(z_t)[z - z_t] + \varkappa \|z - z_t\|_2^2 \geq \varkappa \|z - z_t\|_2^2,$$

where the first $\geq$ is due to strong monotonicity of H, and the second $\geq$ is due to the fact that $H^T(z_t)[z - z_t]$ is proportional, with positive coefficient, to $H^T(z_t)[z_t - \bar{z}]$, and the latter quantity is nonnegative since $\bar{z}$ is a weak solution to the VI in question. We end up with $H^T(z)(z - z_t) \geq \varkappa \|z - z_t\|_2^2$; passing to limit as $t \to +0$, we arrive at (5.59). To prove uniqueness of a weak solution, assume that besides the weak solution $\bar{z}$ there exists a weak solution $\widetilde{z}$ distinct from $\bar{z}$, and let us set $z' = \frac{1}{2}[\bar{z} + \widetilde{z}]$. Since both $\bar{z}$ and $\widetilde{z}$ are weak solutions, both the quantities $H^T(z')[z' - \bar{z}]$ and $H^T(z')[z' - \widetilde{z}]$ should be nonnegative, and because the sum of these quantities is 0, both of them are zero. Thus, when applying (5.59) to $z = z'$, we get $z' = \bar{z}$, whence $\widetilde{z} = \bar{z}$ as well. □

Now let us come back to the estimation problem under consideration. Let Assumptions **A.1-3** hold, so that vector fields $G_{(\eta_k, y_k)}(z)$ defined in (5.55), and therefore vector field $G_{\omega^K}(z)$ are continuous and monotone. When using the SAA, we compute a weak solution $\widehat{x}(\omega^K)$ to VI$(G_{\omega^K}, \mathcal{X})$ and treat it as the SAA estimate of signal x underlying observations (5.50). Since the vector field $G_{\omega^K}(\cdot)$ is monotone with efficiently computable values, provided that so is f, computing (a high accuracy approximation to) a weak solution to VI$(G_{\omega^K}, \mathcal{X})$ is a computationally tractable problem (see, e.g., [189]). Moreover, utilizing the techniques from [30, 204, 220, 212, 213], under mild regularity assumptions additional to **A.1–3** one can get a non-asymptotical upper bound on, say, the expected $\|\cdot\|_2^2$-error of the SAA estimate as a function of the sample size K and find out the rate at which this bound converges to 0 as $K \to \infty$; this analysis, however, goes beyond our scope.

Let us specify the SAA estimate in the logistic regression model. In this case we have $f(u) = (1 + \mathrm{e}^{-u})^{-1}$, and

$$\begin{aligned} G_{(\eta_k, y_k)}(z) &= \left[\frac{\exp\{\eta_k^T z\}}{1+\exp\{\eta_k^T z\}} - y_k\right]\eta_k, \\ G_{\omega^K}(z) &= \frac{1}{K}\sum_{k=1}^K \left[\frac{\exp\{\eta_k^T z\}}{1+\exp\{\eta_k^T z\}} - y_k\right]\eta_k \\ &= \frac{1}{K}\nabla_z\left[\sum_k \left(\ln\left(1+\exp\{\eta_k^T z\}\right) - y_k\eta_k^T z\right)\right]. \end{aligned}$$

In other words, $G_{\omega^K}(z)$ is proportional, with negative coefficient $-1/K$, to the gradient field of the log-likelihood $\ell(z, \omega^K)$; see (5.48). As a result, in the case in question weak solutions to $\mathrm{VI}(G_{\omega^K}, \mathcal{X})$ are exactly the maximizers of the log-likelihood $\ell(z, \omega^K)$ over $z \in \mathcal{X}$, that is, *for the logistic regression the SAA estimate is nothing but the Maximum Likelihood estimate* $\widehat{x}_{\mathrm{ML}}(\omega^K)$ *as defined in* (5.49).[10] On the other hand, in the "nonlinear least squares" example described in Section 5.2.1 with (for the sake of simplicity, scalar) monotone $f(\cdot)$ the vector field $G_{\omega^K}(\cdot)$ is given by

$$G_{\omega^K}(z) = \frac{1}{K}\sum_{k=1}^K \left[f(\eta_k^T z) - y_k\right]\eta_k$$

which is different (provided that f is nonlinear) from the gradient field

$$2\sum_{k=1}^K f'(\eta_k^T z)\left[f(\eta_k^T z) - y_k\right]\eta_k$$

of the minus log-likelihood appearing in (5.52). As a result, in this case the ML estimate (5.52) is, in general, different from the SAA estimate (and, in contrast to the ML, the SAA estimate is easy to compute).

[10]This phenomenon is specific to the logistic regression model. The equality of the SAA and the ML estimates in this case is due to the fact that the logistic sigmoid $f(s) = \exp\{s\}/(1+\exp\{s\})$ "happens" to satisfy the identity $f'(s) = f(s)(1 - f(s))$. When replacing the logistic sigmoid with $f(s) = \phi(s)/(1 + \phi(s))$ with differentiable monotonically nondecreasing positive $\phi(\cdot)$, the SAA estimate becomes the weak solution to $\mathrm{VI}(\Phi, \mathcal{X})$ with

$$\Phi(z) = \sum_k \left[\frac{\phi(\eta_k^T z)}{1+\phi(\eta_k^T z)} - y_k\right]\eta_k.$$

On the other hand, the gradient field of the *minus* log-likelihood

$$-\sum_k \left[y_k \ln(f(\eta_k^T z)) + (1 - y_k)\ln(1 - f(\eta_k^T z))\right]$$

we need to minimize when computing the ML estimate is

$$\Psi(z) = \sum_k \frac{\phi'(\eta_k^T z)}{\phi(\eta_k^T z)}\left[\frac{\phi(\eta_k^T z)}{1+\phi(\eta_k^T z)} - y_k\right]\eta_k.$$

When $k > 1$ and ϕ is not an exponent, Φ and Ψ are "essentially different," so that the SAA estimate typically will differ from the ML one.

5.2.4 Stochastic Approximation estimate

The *Stochastic Approximation* (SA) estimate stems from a simple algorithm—*Subgradient Descent*—for solving variational inequality VI$(G, \mathcal{X})$. Were the values of the vector field $G(\cdot)$ available, one could approximate a root $x \in \mathcal{X}$ of this VI using the recurrence

$$z_k = \mathrm{Proj}_{\mathcal{X}}[z_{k-1} - \gamma_k G(z_{k-1})],\ k = 1, 2, ..., K,$$

where

- $\mathrm{Proj}_{\mathcal{X}}[z]$ is the metric projection of $\mathbf{R}^n$ onto $\mathcal{X}$:

$$\mathrm{Proj}_{\mathcal{X}}[z] = \operatorname*{argmin}_{u \in \mathcal{X}} \|z - u\|_2;$$

- $\gamma_k > 0$ are given stepsizes;
- the initial point z_0 is an arbitrary point of $\mathcal{X}$.

It is well known that under Assumptions **A.1-3** this recurrence with properly selected stepsizes and started at a point from $\mathcal{X}$ allows to approximate the root of G (in fact, the unique weak solution to VI$(G, \mathcal{X})$) to any desired accuracy, provided K is large enough. However, we are in the situation when the actual values of G are not available; the standard way to cope with this difficulty is to replace in the above recurrence the "unobservable" values $G(z_{k-1})$ of G with their unbiased random estimates $G_{(\eta_k, y_k)}(z_{k-1})$. This modification gives rise to *Stochastic Approximation* (coming back to [146])—the recurrence

$$z_k = \mathrm{Proj}_{\mathcal{X}}[z_{k-1} - \gamma_k G_{(\eta_k, y_k)}(z_{k-1})],\ 1 \le k \le K, \tag{5.60}$$

where z_0 is a once and forever chosen point from $\mathcal{X}$, and $\gamma_k > 0$ are deterministic stepsizes.

The next item on our agenda is the (well-known) convergence analysis of SA under assumptions **A.1–3**. To this end observe that the z_k are deterministic functions of the initial fragments $\omega^k = \{\omega_t, 1 \le t \le k\} \sim \underbrace{P_x \times ... \times P_x}_{P_x^k}$ of our sequence of observations $\omega^K = \{\omega_k = (\eta_k, y_k), 1 \le k \le K\}$: $z_k = Z_k(\omega^k)$. Let us set

$$D_k(\omega^k) = \tfrac{1}{2}\|Z_k(\omega^k) - x\|_2^2 = \tfrac{1}{2}\|z_k - x\|_2^2, \quad d_k = \mathbf{E}_{\omega^k \sim P_x^k}\{D_k(\omega^k)\},$$

where $x \in \mathcal{X}$ is the signal underlying observations (5.50). Note that, as is well known, the metric projection onto a closed convex set $\mathcal{X}$ is contracting:

$$\forall (z \in \mathbf{R}^n, u \in \mathcal{X}) : \ \|\mathrm{Proj}_{\mathcal{X}}[z] - u\|_2 \le \|z - u\|_2.$$

Consequently, for $1 \le k \le K$ it holds

$$\begin{array}{rcl} D_k(\omega^k) & = & \tfrac{1}{2}\|\mathrm{Proj}_{\mathcal{X}}[z_{k-1} - \gamma_k G_{\omega_k}(z_{k-1})] - x\|_2^2 \\ & \le & \tfrac{1}{2}\|z_{k-1} - \gamma_k G_{\omega_k}(z_{k-1}) - x\|_2^2 \\ & = & \tfrac{1}{2}\|z_{k-1} - x\|_2^2 - \gamma_k G_{\omega_k}(z_{k-1})^T(z_{k-1} - x) + \tfrac{1}{2}\gamma_k^2\|G_{\omega_k}(z_{k-1})\|_2^2. \end{array}$$

Taking expectations w.r.t. $\omega^k \sim P_x^k$ on both sides of the resulting inequality and keeping in mind relations (5.56) along with the fact that $z_{k-1} \in \mathcal{X}$, we get

$$d_k \leq d_{k-1} - \gamma_k \mathbf{E}_{\omega^{k-1}\sim P_x^{k-1}} \left\{ G(z_{k-1})^T (z_{k-1} - x) \right\} + 2\gamma_k^2 M^2. \tag{5.61}$$

Recalling that we are in the case where G is strongly monotone on $\mathcal{X}$ with modulus $\varkappa > 0$, x is the weak solution VI$(G, \mathcal{X})$, and z_{k-1} takes values in $\mathcal{X}$, invoking (5.59), the expectation in (5.61) is at least $2\varkappa d_k$, and we arrive at the relation

$$d_k \leq (1 - 2\varkappa\gamma_k) d_{k-1} + 2\gamma_k^2 M^2. \tag{5.62}$$

We put

$$S = \frac{2M^2}{\varkappa^2}, \quad \gamma_k = \frac{1}{\varkappa(k+1)}. \tag{5.63}$$

Let us verify by induction in k that for $k = 0, 1, ..., K$ it holds

$$d_k \leq (k+1)^{-1} S. \tag{$*_k$}$$

Base $k = 0$**.** Let D stand for the $\|\cdot\|_2$-diameter of $\mathcal{X}$, and $z_\pm \in \mathcal{X}$ be such that $\|z_+ - z_-\|_2 = D$. By (5.56) we have $\|F(z)\|_2 \leq M$ for all $z \in \mathcal{X}$, and by strong monotonicity of $G(\cdot)$ on $\mathcal{X}$ we have

$$[G(z_+) - G(z_-)]^T [z_+ - z_-] = [F(z_+) - F(z_-)][z_+ - z_-] \geq \varkappa \|z_+ - z_-\|_2^2 = \varkappa D^2.$$

By the Cauchy inequality, the left-hand side in the concluding $\geq$ is at most $2MD$, and we get

$$D \leq \frac{2M}{\varkappa},$$

whence $S \geq D^2/2$. On the other hand, due to the origin of d_0 we have $d_0 \leq D^2/2$. Thus, $(*_0)$ holds true.

Inductive step $(*_{k-1}) \Rightarrow (*_k)$**.** Now assume that $(*_{k-1})$ holds true for some k, $1 \leq k \leq K$, and let us prove that $(*_k)$ holds true as well. Observe that $\varkappa\gamma_k = (k+1)^{-1} \leq 1/2$, so that

$$\begin{array}{rcl} d_k & \leq & d_{k-1}(1 - 2\varkappa\gamma_k) + 2\gamma_k^2 M^2 \text{ [by (5.62)]} \\ & \leq & \frac{S}{k}(1 - 2\varkappa\gamma_k) + 2\gamma_k^2 M^2 \text{ [by } (*_{k-1}) \text{ and due to } \varkappa\gamma_k \leq 1/2] \\ & = & \frac{S}{k}\left(1 - \frac{2}{k+1}\right) + \frac{S}{(k+1)^2} = \frac{S}{k+1}\left(\frac{k-1}{k} + \frac{1}{k+1}\right) \leq \frac{S}{k+1}, \end{array}$$

so that $(*_k)$ hods true. Induction is complete.

Recalling that $d_k = \frac{1}{2}\mathbf{E}\{\|z_k - x\|_2^2\}$, we arrive at the following:

Proposition 5.13. *Under Assumptions* ***A.1–3*** *and with the stepsizes*

$$\gamma_k = \frac{1}{\varkappa(k+1)}, \; k = 1, 2, \dots, \tag{5.64}$$

for every signal $x \in \mathcal{X}$ *the sequence of estimates* $\widehat{x}_k(\omega^k) = z_k$ *given by the SA recurrence* (5.60) *and* $\omega_k = (\eta_k, y_k)$ *defined in* (5.50) *obeys the error bound*

$$\mathbf{E}_{\omega^k \sim P_x^k}\left\{\|\widehat{x}_k(\omega^k) - x\|_2^2\right\} \leq \frac{4M^2}{\varkappa^2(k+1)},\ k = 0, 1, \dots, \tag{5.65}$$

P_x *being the distribution of* (η, y) *stemming from signal* x.

5.2.5 Numerical illustration

To illustrate the above developments, we present here the results of some numerical experiments. Our deliberately simplistic setup is as follows:

- $\mathcal{X} = \{x \in \mathbf{R}^n : \|x\|_2 \leq 1\}$;
- the distribution Q of η is $\mathcal{N}(0, I_n)$;
- f is the monotone vector field on $\mathbf{R}$ given by one of the following four options:
  A. $f(s) = \exp\{s\}/(1 + \exp\{s\})$ ("logistic sigmoid");
  B. $f(s) = s$ ("linear regression");
  C. $f(s) = \max[s, 0]$ ("hinge function");
  D. $f(s) = \min[1, \max[s, 0]]$ ("ramp sigmoid").
- the conditional distribution of y given η induced by P_x is
  - Bernoulli distribution with probability $f(\eta^T x)$ of outcome 1 in the case of A (i.e., A corresponds to the logistic model),
  - Gaussian distribution $\mathcal{N}(f(\eta^T x), I_n)$ in cases B–D.

Note that when $m = 1$ and $\eta \sim \mathcal{N}(0, I_n)$, one can easily compute the field $F(z)$. Indeed, we have $\forall z \in \mathbf{R}^n \backslash \{0\}$:

$$\eta = \frac{zz^T}{\|z\|_2^2}\eta + \underbrace{\left(I - \frac{zz^T}{\|z\|_2^2}\right)\eta}_{\eta_\perp},$$

and due to the independence of $\eta^T z$ and $\eta_\perp$,

$$F(z) = \mathbf{E}_{\eta\sim\mathcal{N}(0,I)}\{\eta f(\eta^T z)\} = \mathbf{E}_{\eta\sim\mathcal{N}(0,I)}\left\{\frac{zz^T\eta}{\|z\|_2^2} f(\eta^T z)\right\} = \frac{z}{\|z\|_2}\mathbf{E}_{\zeta\sim\mathcal{N}(0,1)}\{\zeta f(\|z\|_2\zeta)\},$$

so that $F(z)$ is proportional to $z/\|z\|_2$ with proportionality coefficient

$$h(\|z\|_2) = \mathbf{E}_{\zeta\sim\mathcal{N}(0,1)}\{\zeta f(\|z\|_2\zeta)\}.$$

In Figure 5.2 we present the plots of the function $h(t)$ for the situations A–D and of the moduli of strong monotonicity of the corresponding mappings F on the $\|\cdot\|_2$-ball of radius R centered at the origin, as functions of R. The dimension n in all experiments was set to 100, and the number of observations K was $400, 1,000, 4,000, 10,000$, and $40,000$. For each combination of parameters we ran 10 simulations for signals x underlying observations (5.50) drawn randomly from the uniform distribution on the unit sphere (the boundary of $\mathcal{X}$).

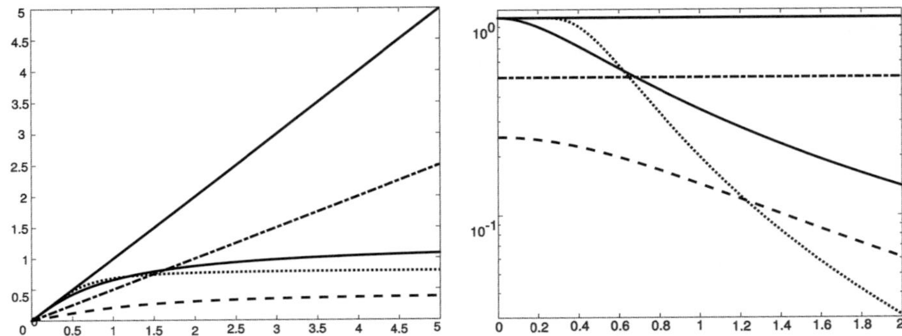

Figure 5.2: Left: functions h; right: moduli of strong monotonicity of the operators $F(\cdot)$ in $\{z : \|z\|_2 \leq R\}$ as functions of R. Dashed lines – case A (logistic sigmoid), solid lines – case B (linear regression), dash-dotted lines – case C (hinge function), dotted line – case D (ramp sigmoid).

In each experiment, we computed the SAA and the SA estimates (note that in the cases A and B the SAA estimate is the Maximum Likelihood estimate as well). The SA stepsizes γ_k were selected according to (5.64) with "empirically tuned" $\varkappa$.[11] Namely, given observations $\omega_k = (\eta_k, y_k)$, $k \leq K$—see (5.50)—we used them to build the SA estimate in two stages:

— at the *tuning stage*, we generate a random "training signal" $x' \in \mathcal{X}$ and then generate labels y'_k as if x' were the actual signal. For instance, in the case of A, y'_k is assigned value 1 with probability $f(\eta_k^T x')$ and value 0 with complementary probability. After the "training signal" and associated labels are generated, we run on the resulting artificial observations SA with different values of $\varkappa$, compute the accuracy of the resulting estimates, and select the value of $\varkappa$ resulting in the best recovery;

— at the *execution stage*, we run SA on the actual data with stepsizes (5.64) specified by the $\varkappa$ found at the tuning stage.

The results of some numerical experiments are presented in Figure 5.3.

Note that the CPU time for SA includes both tuning and execution stages. The conclusion from these experiments is that as far as estimation quality is concerned, the SAA estimate marginally outperforms the SA, while being significantly more time consuming. Note also that the dependence of recovery errors on K observed in our experiments is consistent with the convergence rate $O(1/\sqrt{K})$ established by Proposition 5.13.

Comparison with Nonlinear Least Squares. Observe that in the case $m = 1$ of scalar monotone field f, the SAA estimate yielded by our approach as applied to observation ω^K is the minimizer of the convex function

$$H_{\omega^K}(z) = \frac{1}{K} \sum_{k=1}^{k} \left[v(\eta_k^T z) - y_k \eta_k^T z \right], \; v(r) = \int_0^t f(s) ds,$$

[11] We could get (lower bounds on) the moduli of strong monotonicity of the vector fields $F(\cdot)$ we are interested in analytically, but this would be boring and conservative.

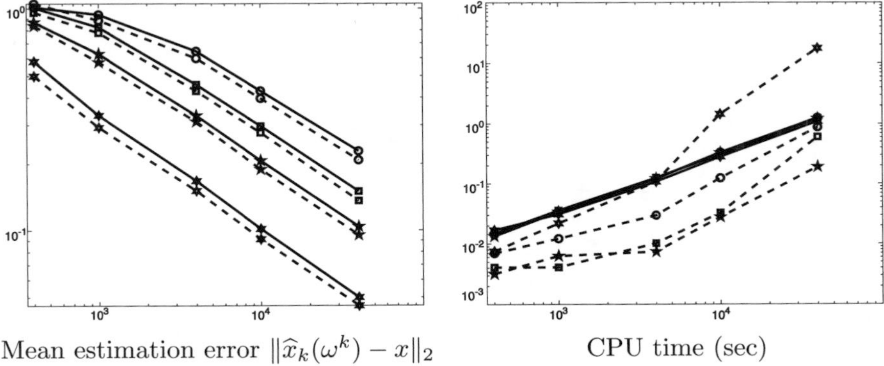

Mean estimation error $\|\widehat{x}_k(\omega^k) - x\|_2$ CPU time (sec)

Figure 5.3: Mean errors and CPU times for SA (solid lines) and SAA estimates (dashed lines) as functions of the number of observations K. o – case A (logistic link), x – case B (linear link), + – case C (hinge function), □ – case D (ramp sigmoid).

on the signal set $\mathcal{X}$. When f is the logistic sigmoid, $H_{\omega^K}(\cdot)$ is exactly the convex loss function leading to the ML estimate in the logistic regression model. As we have already mentioned, this is not the case for a general GLM. Consider, e.g., the situation where the regressors and the signals are reals, the distribution of regressor η is $\mathcal{N}(0,1)$, and the conditional distribution of y given η is $\mathcal{N}(f(\eta x), \sigma^2)$, with $f(s) = \arctan(s)$. In this situation the ML estimate stemming from observation ω^K is the minimizer on $\mathcal{X}$ of the function

$$M_{\omega^K}(z) = \frac{1}{K}\sum_{k=1}^{k} [y_k - \arctan(\eta_k z)]^2 . \tag{5.66}$$

The latter function is typically nonconvex and can be multi-extremal. For example, when running simulations[12] we from time to time observe the situation similar to that presented in Figure 5.4.

Of course, in our toy situation of scalar x the existence of several local minima of $M_{\omega^K}(\cdot)$ is not an issue—we can easily compute the ML estimate by a brute force search along a dense grid. What to do in the multidimensional case—this is another question. We could also add that in the simulations which led to Figure 5.4 both the SAA and the ML estimates exhibited nearly the same performance in terms of the estimation error: in 1,000 experiments, the median of the observed recovery errors was 0.969 for the ML, and 0.932 for the SAA estimate. When increasing the number of observations to 1,000, the empirical median (taken over 1,000 simulations) of recovery errors became 0.079 for the ML, and 0.085 for the SAA estimate.

[12] In these simulations, the "true" signal x underlying observations was drawn from $\mathcal{N}(0,1)$, the number K of observations also was random with uniform distribution on $\{1, ..., 20\}$, and $\mathcal{X} = [-20, 20]$, $\sigma = 3$ were used.

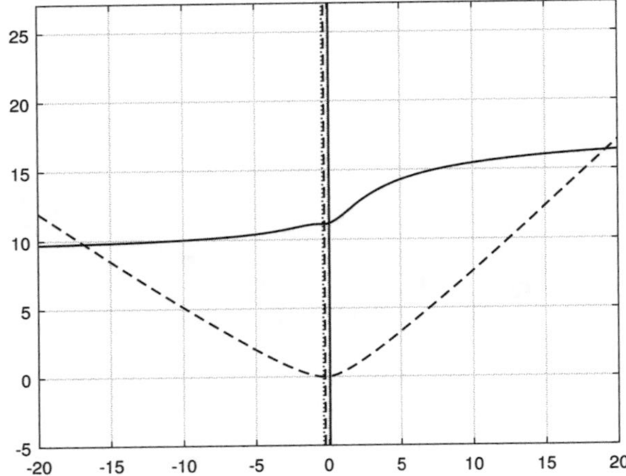

Figure 5.4: Solid curve: $M_{\omega^K}(z)$, dashed curve: $H_{\omega^K}(z)$. True signal x (solid vertical line): $+0.081$; SAA estimate (unique minimizer of H_{ω^K}, dashed vertical line): -0.252; ML estimate (global minimizer of M_{ω^K} on $[-20, 20]$): -20.00, closest to x local minimizer of M_{ω^K} (dotted vertical line): -0.363.

5.2.6 "Single-observation" case

Let us look at the special case of our estimation problem where the sequence $\eta_1, ..., \eta_K$ of regressors in (5.50) is deterministic. At first glance, this situation goes beyond our setup, where the regressors should be i.i.d. drawn from some distribution Q. However, we can circumvent this "contradiction" by saying that we are now in the *single-observation case* with the regressor being the matrix $[\eta_1, ..., \eta_K]$ and Q being a degenerate distribution supported at a singleton. Specifically, consider the case where our observation is

$$\omega = (\eta, y) \in \mathbf{R}^{n \times mK} \times \mathbf{R}^{mK} \tag{5.67}$$

(m, n, K are given positive integers), and the distribution P_x of observation stemming from a signal $x \in \mathbf{R}^n$ is as follows:

- η is a given deterministic matrix independent of x;
- y is random, and the distribution of y induced by P_x is with mean $\phi(\eta^T x)$, where $\phi : \mathbf{R}^{mK} \to \mathbf{R}^{mK}$ is a given mapping.

As an instructive example connecting our current setup with the previous one, consider the case where $\eta = [\eta_1, ..., \eta_K]$ with $n \times m$ deterministic "individual regressors" η_k, and $y = [y_1; ...; y_K]$ with random "individual labels" $y_k \in \mathbf{R}^m$ conditionally independent, given x, across k, and such that the expectations of y_k induced by x are $f(\eta_k^T x)$ for some $f : \mathbf{R}^m \to \mathbf{R}^m$. We set $\phi([u_1; ...; u_K]) = [f(u_1); ...; f(u_K)]$. The resulting "single observation" model is a natural analogy of the K-observation model considered so far, the only difference being that the individual regressors now form a fixed deterministic sequence rather than being a sample of realizations of some random matrix.

As before, our goal is to use observation (5.67) to recover the (unknown) signal x underlying, as explained above, the distribution of the observation. Formally, we are now in the case $K = 1$ of our previous recovery problem where Q is supported on a singleton $\{\eta\}$ and can use the constructions developed so far. Specifically,

- The field $F(z)$ associated with our problem (it used to be $\mathbf{E}_{\eta\sim Q}\{\eta f(\eta^T z)\}$) is

$$F(z) = \eta\phi(\eta^T z),$$

and the vector field $G(z) = F(z) - F(x)$, x being the signal underlying observation (5.67), is

$$G(z) = \mathbf{E}_{(\eta,y)\sim P_x}\{F(z) - \eta y\}$$

(cf. (5.57)). As before, the signal to recover is a zero of the latter field. Note that now the vector field $F(z)$ is observable, and the vector field G still is the expectation, over P_x, of an observable vector field:

$$G(z) = \mathbf{E}_{(\eta,y)\sim P_x}\{\underbrace{\eta\phi(\eta^T z) - \eta y}_{G_y(z)}\};$$

cf. Lemma 5.11.

- Assumptions **A.1–2** now read

A.1′ The vector field $\phi(\cdot) : \mathbf{R}^{mK} \to \mathbf{R}^{mK}$ is continuous and monotone, so that $F(\cdot)$ is continuous and monotone as well,

A.2′ $\mathcal{X}$ is a nonempty compact convex set, and F is strongly monotone, with modulus $\varkappa > 0$, on $\mathcal{X}$.

A simple sufficient condition for the validity of the above monotonicity assumptions is positive definiteness of the matrix $\eta\eta^T$ plus strong monotonicity of ϕ on every bounded set.

- For our present purposes, it is convenient to reformulate assumption **A.3** in the following equivalent form:

A.3′ For properly selected $\sigma \geq 0$ and every $x \in \mathcal{X}$ it holds

$$\mathbf{E}_{(\eta,y)\sim P_x}\{\|\eta[y - \phi(\eta^T x)]\|_2^2\} \leq \sigma^2.$$

In the present setting, the SAA $\widehat{x}(y)$ is the unique weak solution to $\mathrm{VI}(G_y, \mathcal{X})$, and we can easily quantify the quality of this estimate:

Proposition 5.14. *In the situation in question, let Assumptions* **A.1′–3′** *hold. Then for every $x \in \mathcal{X}$ induced by x and every realization (η, y) of observation* (5.67) *one has*

$$\|\widehat{x}(y) - x\|_2 \leq \varkappa^{-1}\|\underbrace{\eta[y - \phi(\eta^T x)]}_{\Delta(x,y)}\|_2, \tag{5.68}$$

whence also

$$\mathbf{E}_{(\eta,y)\sim P_x}\{\|\widehat{x}(y) - x\|_2^2\} \leq \sigma^2/\varkappa^2. \tag{5.69}$$

Proof. Let $x \in \mathcal{X}$ be the signal underlying observation (5.67), and $G(z) = F(z) - F(x)$ be the associated vector field G. We have

$$G_y(z) = F(z) - \eta y = F(z) - F(x) + [F(x) - \eta y] = G(z) - \eta[y - \phi(\eta^T x)] = G(z) - \Delta(x, y).$$

For y fixed, $\bar{z} = \widehat{x}(y)$ is the weak, and therefore the strong (since $G_y(\cdot)$ is continuous), solution to VI$(G_y, \mathcal{X})$, implying, due to $x \in \mathcal{X}$, that

$$0 \leq G_y^T(\bar{z})[x - \bar{z}] = G^T(\bar{z})[x - \bar{z}] - \Delta^T(x, y)[x - \bar{z}],$$

whence

$$-G^T(\bar{z})[x - \bar{z}] \leq -\Delta^T(x, y)[x - \bar{z}].$$

Besides this, $G(x) = 0$, whence $G^T(x)[x - \bar{z}] = 0$, and we arrive at

$$[G(x) - G(\bar{z})]^T[x - \bar{z}] \leq -\Delta^T(x, y)[x - \bar{z}],$$

whence also

$$\varkappa\|x - \bar{z}\|_2^2 \leq -\Delta^T(x, y)[x - \bar{z}]$$

(recall that G, along with F, is strongly monotone with modulus $\varkappa$ on $\mathcal{X}$ and $x, \bar{z} \in \mathcal{X}$). Applying the Cauchy inequality, we arrive at (5.68). □

Example. Consider the case where $m = 1$, ϕ is strongly monotone, with modulus $\varkappa_\phi > 0$, on the entire $\mathbf{R}^K$, and η in (5.67) is drawn from a "Gaussian ensemble"—the columns η_k of the $n \times K$ matrix η are independent $\mathcal{N}(0, I_n)$-random vectors. Assume also that the observation noise is Gaussian:

$$y = \phi(\eta^T x) + \lambda\xi, \ \ \xi \sim \mathcal{N}(0, I_K).$$

It is well known that as $K/n \to \infty$, the minimal singular value of the $n \times n$ matrix $\eta\eta^T$ is at least $O(1)K$ with overwhelming probability, implying that when $K/n \gg 1$, the typical modulus of strong monotonicity of $F(\cdot)$ is $\varkappa \geq O(1)K\varkappa_\phi$. Furthermore, in our situation, as $K/n \to \infty$, the Frobenius norm of η with overwhelming probability is at most $O(1)\sqrt{nK}$. In other words, when K/n is large, a "typical" recovery problem from the ensemble just described satisfies the premise of Proposition 5.14 with $\varkappa = O(1)K\varkappa_\phi$ and $\sigma^2 = O(\lambda^2 nK)$. As a result, (5.69) reads

$$\mathbf{E}_{(\eta,y)\sim P_x}\{\|\widehat{x}(y) - x\|_2^2\} \leq O(1)\frac{\lambda^2 n}{\varkappa_\phi^2 K}. \qquad [K \gg n]$$

It is well known that in the standard case of linear regression, where $\phi(x) = \varkappa_\phi x$, the resulting bound is near-optimal, provided $\mathcal{X}$ is large enough.

Numerical illustration: In the situation described in the example above, we set $m = 1$, $n = 100$, and use

$$\phi(u) = \arctan[u] := [\arctan(u_1); ...; \arctan(u_K)] : \mathbf{R}^K \to \mathbf{R}^K;$$

the set $\mathcal{X}$ is the unit ball $\{x \in \mathbf{R}^n : \|x\|_2 \leq 1\}$. In a particular experiment, η is chosen at random from the Gaussian ensemble as described above, and signal $x \in \mathcal{X}$ underlying observation (5.67) is drawn at random; the observation noise $y - \phi(\eta^T x)$ is $\mathcal{N}(0, \lambda^2 I_K)$. Some typical results (10 simulations for each combination of the samples size and noise variance λ^2) are presented in Figure 5.5.

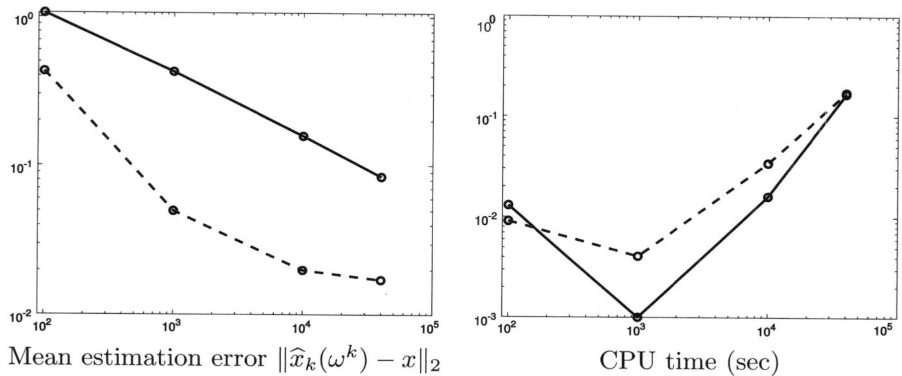

Mean estimation error $\|\widehat{x}_k(\omega^k) - x\|_2$ CPU time (sec)

Figure 5.5: Mean errors and CPU times for standard deviation $\lambda = 1$ (solid line) and $\lambda = 0.1$ (dashed line).

5.3 EXERCISES FOR CHAPTER 5

5.3.1 Estimation by Stochastic Optimization

Exercise 5.1.

Consider the following "multinomial" version of the logistic regression problem from Section 5.2.1:
For $k = 1, ..., K$, we observe pairs

$$(\zeta_k, \ell_k) \in \mathbf{R}^n \times \{0, 1, ..., m\} \tag{5.70}$$

drawn independently of each other from a probability distribution P_x parameterized by an unknown signal $x = [x^1; ...; x^m] \in \mathbf{R}^n \times ... \times \mathbf{R}^n$ as follows:

- The probability distribution of *regressor* ζ induced by the distribution S_x of (ζ, ℓ) is a once forever fixed, independent of x, distribution R on $\mathbf{R}^n$ with finite second order moments and positive definite matrix $Z = \mathbf{E}_{\zeta\sim R}\{\zeta\zeta^T\}$ of second order moments;
- The conditional distribution of *label* ℓ given ζ induced by the distribution S_x of (ζ, ℓ) is the distribution of the discrete random variable taking value $\iota \in \{0, 1, ..., m\}$ with probability

$$p_\iota = \left\{ \begin{array}{ll} \frac{\exp\{\zeta^T x^\iota\}}{1+\sum_{i=1}^m \exp\{\zeta^T x^i\}}, & 1 \leq \iota \leq m, \\ \frac{1}{1+\sum_{i=1}^m \exp\{\zeta^T x^i\}}, & \iota = 0. \end{array} \right. \qquad [x = [x^1; ...; x^m]]$$

Given a nonempty convex compact set $\mathcal{X} \in \mathbf{R}^{mn}$ known to contain the (unknown) signal x underlying observations (5.70), we want to recover x. Note that the recovery problem associated with the standard logistic regression model is the case $m = 1$ of the problem just defined.

Your task is to process the above recovery problem via the approach developed in Section 5.2 and to compare the resulting SAA estimate with the Maximum Likelihood estimate.

Exercise 5.2.

Let

$$H(x) : \mathbf{R}^n \to \mathbf{R}^n$$

be a vector field strongly monotone and Lipschitz continuous on the entire space:

$$\forall (x, x' \in \mathbf{R}^n): \quad \begin{array}{l} [H(x) - H(x')]^T [x - x'] \geq \varkappa \|x - x'\|_2, \\ \|H(x) - H(x')\|_2 \leq L \|x - x'\|_2 \end{array} \tag{5.71}$$

for some $\varkappa > 0$ and $L < \infty$.

1.1) Prove that for every $x \in \mathbf{R}^n$, the vector equation

$$H(z) = x$$

in variable $z \in \mathbf{R}^n$ has a unique solution (which we denote by $H^{-1}(x)$), and that for every $x, y \in \mathbf{R}^n$ one has

$$\|H^{-1}(x) - y\|_2 \leq \varkappa^{-1} \|x - H(y)\|_2. \tag{5.72}$$

1.2) Prove that the vector field

$$x \mapsto H^{-1}(x)$$

is strongly monotone with modulus

$$\varkappa_* = \varkappa / L^2$$

and Lipschitz continuous, with constant $1/\varkappa$ w.r.t. $\|\cdot\|_2$, on the entire $\mathbf{R}^n$.

Let us interpret $-H(\cdot)$ as a field of "reaction forces" applied to a particle: when the particle is in a position $y \in \mathbf{R}^n$ the reaction force applied to it is $-H(y)$. Next, let us interpret $x \in \mathbf{R}^n$ as an external force applied to the particle. An equilibrium y is a point in space where the reaction force $-H(y)$ compensates the external force, that is, $H(y) = x$, or, which for our H is the same, $y = H^{-1}(x)$. Note that with this interpretation, strong monotonicity of H makes perfect sense, implying that the equilibrium in question is stable: when the particle is moved from the equilibrium $y = H^{-1}(x)$ to a position $y + \Delta$, the total force acting at the particle becomes $f = x - H(y + \Delta)$, so that

$$f^T \Delta = [x - H(y + \Delta)]^T \Delta = [H(y) - H(y + \Delta)]^T [\Delta] \leq -\varkappa \Delta^2,$$

that is, the force is oriented "against" the displacement Δ and "wants" to return the particle to the equilibrium position.

Now imagine that we can observe in noise equilibrium $H^{-1}(x)$ of the particle, the external force x being unknown, and want to recover x from our observation. For the sake of simplicity, let the observation noise be zero mean Gaussian, so that our observation is

$$\omega = H^{-1}(x) + \sigma \xi, \ \xi \sim \mathcal{N}(0, I_n).$$

2) Verify that the recovery problem we have posed is a special case of the "single observation" recovery problem from Section 5.2.6, with $\mathbf{R}^n$ in the role of $\mathcal{X}$, [13]

---

[13] In Section 5.2.6, $\mathcal{X}$ was assumed to be closed, convex, *and bounded*; a straightforward inspection shows that when the vector field ϕ is strongly monotone, with some positive modulus,

and that the SAA estimate $\widehat{x}(\omega)$ from that section under the circumstances is just the root of the equation

$$H^{-1}(\cdot) = \omega,$$

that is,

$$\widehat{x}(\omega) = H(\omega).$$

Prove also that

$$\mathbf{E}\{\|\widehat{x}(\omega) - x\|_2^2\} \leq n\sigma^2 L^2. \tag{5.73}$$

Note that in the situation in question the ML estimate should be the minimizer of the function

$$f(z) = \|\omega - H^{-1}(z)\|_2^2,$$

and this minimizer is nothing but $\widehat{x}(\omega)$.

Exercise 5.3.

[identification of parameters of a linear dynamic system] Consider the problem as follows:

A deterministic sequence $x = \{x_t : t \geq -d+1\}$ satisfies the linear finite-difference equation

$$\sum_{i=0}^{d} \alpha_i x_{t-i} = y_t,\ t = 1, 2, ... \tag{5.74}$$

of given order d and is bounded:

$$|x_t| \leq M_x < \infty,\ \forall t \geq -d+1,$$

implying that the sequence $\{y_t\}$ also is bounded,

$$|y_t| \leq M_y < \infty, \forall t \geq 1.$$

The vector $\alpha = [\alpha_0; ...; \alpha_d]$ is unknown; all we know is that this vector belongs to a given closed convex set $\mathcal{X} \subset \mathbf{R}^{d+1}$. We have at our disposal observations

$$\omega_t = x_t + \sigma_x \xi_t,\ -d+1 \leq t \leq K, \tag{5.75}$$

of the terms in the sequence, with $\xi_t \sim \mathcal{N}(0,1)$ independent across t, with some given σ_x, and observations

$$\zeta_t = y_t + \sigma_y \eta_t \tag{5.76}$$

with $\eta_t \sim \mathcal{N}(0,1)$ independent across t and independent of $\{\xi_\tau\}_\tau$. Our goal is to recover from these observations the vector α.

Strategy. To get the rationale underlying the construction to follow, let us start with the case when there is no observation noise at all: $\sigma_x = \sigma_y = 0$. In this case

on the entire space, and η has trivial kernel, all constructions and results of Section 5.2.6 can be extended to the case of an arbitrary closed convex $\mathcal{X}$.

we could act as follows: let us denote

$$x^t = [x_t; x_{t-1}; x_{t-2}; ...; x_{t-d}],\ 1 \le t \le K,$$

and rewrite (5.74) as

$$[x^t]^T \alpha = y_t,\ 1 \le t \le K.$$

When setting

$$A_K = \frac{1}{K}\sum_{t=1}^K x^t[x^t]^T,\ a_K = \frac{1}{K}\sum_{t=1}^K y_t x^t,$$

we get

$$A_K \alpha = a_K. \tag{5.77}$$

Assuming that K is large and trajectory x is "rich enough" to ensure that A_K is nonsingular, we could identify α by solving the linear system (5.77).

Now, when the observation noise is present, we could try to use the noisy observations of x^t and y_t we have at our disposal in order to build empirical approximations to A_K and a_K which are good for large K, and identify α by solving the "empirical counterpart" of (5.77). The straightforward way would be to define ω^t as an "observable version" of x^t,

$$\omega^t = [\omega_t; \omega_{t-1}; ...; \omega_{t-d}] = x^t + \sigma_x \underbrace{[\xi_t; \xi_{t-1}; ...; \xi_{t-d}]}_{\xi^t},$$

and to replace A_K, a_K with

$$\widetilde{A}_t = \frac{1}{K}\sum_{t=1}^K \omega^t[\omega^t]^T,\ \widetilde{a}_K = \sum_{t=1}^K \zeta_t \omega^t.$$

As far as empirical approximation of a_K is concerned, this approach works: we have

$$\widetilde{a}_K = a_K + \delta a_K,\ \delta a_K = \frac{1}{K}\sum_{t=1}^K \underbrace{\left[\sigma_x y_t \xi^t + \sigma_y \eta_t x^t + \sigma_x \sigma_y \eta_t \xi^t\right]}_{\delta_t}.$$

Since the sequence $\{y_t\}$ is bounded, the random error δa_K of approximation $\widetilde{a}_K$ of a_K is small for large K with overwhelming probability. Indeed, δa_K is the average of K zero mean random vectors δ_t (recall that ξ^t and η_t are independent and have zero mean) with[14]

$$\mathbf{E}\{\|\delta_t\|_2^2\} \le 3(d+1)\left[\sigma_x^2 M_y^2 + \sigma_y^2 M_x^2 + \sigma_x^2 \sigma_y^2\right],$$

and δ_t is independent of δ_s whenever $|t-s| > d+1$, implying that

$$\mathbf{E}\{\|\delta a_K\|_2^2\} \le \frac{3(d+1)(2d+1)\left[\sigma_x^2 M_y^2 + \sigma_y^2 M_x^2 + \sigma_x^2 \sigma_y^2\right]}{K}. \tag{5.78}$$

[14]We use the elementary inequality $\|\sum_{t=1}^p a_t\|_2^2 \le p\sum_{t=1}^p \|a_t\|_2^2$.

The quality of approximating A_K with $\widetilde{A}_K$ is essentially worse: setting

$$\delta A_K = \widetilde{A}_K - A_K = \frac{1}{K}\sum_{t=1}^{K} \underbrace{[\sigma_x^2 \xi^t [\xi^t]^T + \sigma_x \xi^T [x^t]^T + \sigma_x x^t [\xi^t]^T]}_{\Delta_t}$$

we see that δA_K is the average of K random matrices Δ_t with *nonzero* mean, namely, the mean $\sigma_x^2 I_{d+1}$, and as such ΔA_K is "large" independently of how large K is. There is, however, a simple way to overcome this difficulty—*splitting observations* ω_t.[15]

Splitting observations. Let θ be a random n-dimensional vector with unknown mean μ and known covariance matrix, namely, $\sigma^2 I_n$, and let $\chi \sim \mathcal{N}(0, I_n)$ be independent of θ. Finally, let $\kappa > 0$ be a deterministic real.

1) Prove that setting

$$\theta' = \theta + \sigma\kappa\chi,\ \theta'' = \theta - \sigma\kappa^{-1}\chi,$$

we get two random vectors with mean μ and covariance matrices $\sigma^2(1+\kappa^2)I_n$ and $\sigma^2(1+1/\kappa^2)I_n$, respectively, and these vectors are uncorrelated:

$$\mathbf{E}\{[\theta' - \mu][\theta'' - \mu]^T\} = 0.$$

In view of item 1, let us do as follows: given observations $\{\omega_t\}$ and $\{\zeta_t\}$, let us generate i.i.d. sequence $\{\chi_t \sim \mathcal{N}(0,1), t \geq -d+1\}$, so that the sequences $\{\xi_t\}$, $\{\eta_t\}$, and $\{\chi_t\}$ are i.i.d. and independent of each other, and let us set

$$u_t = \omega_t + \sigma_x \chi_t,\ v_t = \omega_t - \sigma_x \chi_t.$$

Note that given the sequence $\{\omega_t\}$ of actual observations, sequences $\{u_t\}$ and $\{v_t\}$ are observable as well, and that the sequence $\{(u_t, v_t)\}$ is i.i.d.. Moreover, for all t,

$$\mathbf{E}\{u_t\} = \mathbf{E}\{v_t\} = x_t,\ \mathbf{E}\{[u_t - x_t]^2\} = 2\sigma_x^2,\ \mathbf{E}\{[v_t - x_t]^2\} = 2\sigma_x^2,$$

and for all t and s

$$\mathbf{E}\{[u_t - x_t][v_s - x_s]\} = 0.$$

Now, let us put

$$u^t = [u_t; u_{t-1}; ...; u_{t-d}],\ \ v^t = [v_t; v_{t-1}; ...; v_{t-d}],$$

and let

$$\widehat{A}_K = \frac{1}{K}\sum_{t=1}^{K} u^t [v^t]^T.$$

2) Prove that $\widehat{A}_K$ is a good empirical approximation of A_K:

$$\mathbf{E}\{\widehat{A}_K\} = A_K,\ \ \mathbf{E}\{\|\widehat{A}_K - A_K\|_F^2\} \leq \frac{12[d+1]^2[2d+3]\left[M_x^2 + \sigma_x^2\right]\sigma_x^2}{K}, \tag{5.79}$$

[15]The model (5.74)–(5.76) is referred to as Errors in Variables model [85] in the statistical literature or Output Error model in the literature on System Identification [173, 218]. In general, statistical inference for such models is difficult—for instance, parameter estimation problem in such models is ill-posed. The estimate we develop in this exercise can be seen as a special application of the general Instrumental Variables methodology [7, 219, 241].

the expectation being taken over the distribution of observation noises $\{\xi_t\}$ and auxiliary random sequence $\{\chi_t\}$.

Conclusion. We see that as $K \to \infty$, the differences of typical realizations of $\widehat{A}_K - A_K$ and $\widetilde{a}_K - a_K$ approach 0. It follows that if the sequence $\{x_t\}$ is "rich enough" to ensure that the minimal eigenvalue of A_K for large K stays bounded away from 0, the estimate

$$\widehat{\alpha}_K \in \operatorname*{Argmin}_{\beta \in \mathbf{R}^{d+1}} \|\widehat{A}_K \beta - \widetilde{a}_K\|_2^2$$

will converge in probability to the desired vector α, and we can even say something reasonable about the rate of convergence. To account for a priori information $\alpha \in \mathcal{X}$, we can modify the estimate by setting

$$\widehat{\alpha}_K \in \operatorname*{Argmin}_{\beta \in \mathcal{X}} \|\widehat{A}_K \beta - \widetilde{a}_K\|_2^2.$$

Note that the assumption that noises affecting observations of x_t's and y_t's are zero mean *Gaussian* random variables independent of each other with known dispersions is not that important; we could survive the situation where samples $\{\omega_t - x_t, t > -d\}$, $\{\zeta_t - y_t, t \geq 1\}$ are zero mean i.i.d., and independent of each other, *with a priori known variance of* $\omega_t - x_t$. Under this and mild additional assumptions (like finiteness of the fourth moments of $\omega_t - x_t$ and $\zeta_t - y_t$), the obtained results would be similar to those for the Gaussian case.

Now comes the concluding part of the exercise:

3) To evaluate numerically the performance of the proposed identification scheme, run experiments as follows:
   - Given an even value of d and $\rho \in (0, 1]$, select $d/2$ complex numbers λ_i at random on the circle $\{z \in \mathbf{C} : |z| = \rho\}$, and build a real polynomial of degree d with roots λ_i, λ_i^* (* here stands for complex conjugation). Build a finite-difference equation (5.77) with this polynomial as the characteristic polynomial.
   - Generate i.i.d. $\mathcal{N}(0, 1)$ "inputs" $\{y_t, t = 1, 2, ...\}$, select at random initial conditions $x_{-d+1}, x_{-d+2}, ..., x_0$ for the trajectory $\{x_t\}$ of states (5.77), and simulate the trajectory along with observations ω_t of x_t and ζ_t of y_t, with σ_x, σ_y being the experiment's parameters.
   - Look at the performance of the estimate $\widehat{\alpha}_K$ on the simulated data.

Exercise 5.4.

[more on generalized linear models] Consider a generalized linear model as follows: we observe i.i.d. random pairs

$$\omega_k = (y_k, \zeta_k) \in \mathbf{R} \times \mathbf{R}^{\nu \times \mu},\ k = 1, ..., K,$$

where the conditional expectation of the scalar label y_k given ζ_k is $\psi(\zeta_k^T z)$, z being an unknown signal underlying the observations. What we know is that z belongs to a given convex compact set $\mathcal{Z} \subset \mathbf{R}^n$. Our goal is to recover z.

Note that while the estimation problem we have just posed looks similar to those treated in Section 5.2, it cannot be straightforwardly handled via techniques developed in that section unless $\mu = 1$. Indeed, these techniques in the case of $\mu > 1$ require ψ to be a monotone vector field on $\mathbf{R}^\mu$, while our ψ is just a scalar function on $\mathbf{R}^\mu$. The goal of the exercise is to show that *when*

$$\psi(w) = \sum_{q\in\mathcal{Q}} c_q w^q \equiv \sum_{q\in\mathcal{Q}} c_q w_1^{q_1}...w_\mu^{q_\mu} \qquad [c_q \neq 0, q \in \mathcal{Q} \subset \mathbf{Z}_+^\mu]$$

is an algebraic polynomial (which we assume from now on), one can use lifting to reduce the situation to that considered in Section 5.2.

The construction is straightforward. Let us associate with algebraic monomial

$$z^p := z_1^{p_1} z_2^{p_2} ... z_\nu^{p_\nu}$$

with ν variables[16] a real variable x_p. For example, monomial $z_1 z_2$ is associated with $x_{1,1,0,...,0}$, $z_1^2 z_\nu^3$ is associated with $x_{2,0,...,0,3}$, etc. For $q \in \mathcal{Q}$, the contribution of the monomial $c_q w^q$ into $\psi(\zeta^T z)$ is

$$c_q[\mathrm{Col}_1^T[\zeta]z]^{q_1}[\mathrm{Col}_2^T[\zeta]z]^{q_2}...[\mathrm{Col}_\mu^T[\zeta]z]^{q_\mu} = \sum_{p\in\mathcal{P}_q} h_{pq}(\zeta) z_1^{p_1} z_2^{p_2}...z_\nu^{p_\nu},$$

where $\mathcal{P}_q$ is a properly built set of multi-indices $p = (p_1, ..., p_\nu)$, and $h_{pq}(\zeta)$ are easily computable functions of ζ. Consequently,

$$\psi(\zeta^T z) = \sum_{q\in\mathcal{Q}} \sum_{p\in\mathcal{P}_q} h_{pq}(\zeta) z^p = \sum_{p\in\mathcal{P}} H_p(\zeta) z^p,$$

with properly selected finite set $\mathcal{P}$ and readily given functions $H_p(\zeta)$, $p \in \mathcal{P}$. We can always take, as $\mathcal{P}$, the set of all ν-entry multi-indices with the sum of entries not exceeding d, where d is the total degree of the polynomial ψ. This being said, the structure of ψ and/or the common structure, if any, of regressors ζ_k can enforce some of the functions $H_p(\cdot)$ to be identically zero. When this is the case, it makes sense to eliminate the corresponding "redundant" multi-index p from $\mathcal{P}$.

Next, consider the mapping $x[z]$ which maps a vector $z \in \mathbf{R}^\nu$ into a vector with entries $x_p[z] = z^p$ indexed by $p \in \mathcal{P}$, and let us associate with our estimation problem its "lifting" with observations

$$\overline{\omega}_k = (y_k, \eta_k = \{H_p(\zeta_k), p \in \mathcal{P}\}).$$

I.e., new observations are deterministic transformations of the actual observations; observe that the new observations still are i.i.d., and the conditional expectation of y_k given η_k is nothing but

$$\sum_{p\in\mathcal{P}} [\eta_k]_p x_p[z].$$

In our new situation, the "signal" underlying observations is a vector from $\mathbf{R}^N$, $N = \mathrm{Card}(\mathcal{P})$, the regressors are vectors from the same $\mathbf{R}^N$, and *regression is linear*—the conditional expectation of the label y_k given regressor η_k is a linear function $\eta_k^T x$ of our new signal. Given a convex compact localizer $\mathcal{Z}$ for the "true signal" z, we can in many ways find a convex compact localizer $\mathcal{X}$ for $x = x[z]$.

[16]Note that factors in the monomial are ordered according to the indices of the variables.

Thus, we find ourselves in the simplest possible case of the situation considered in Section 5.2 (one with scalar $\phi(s) \equiv s$), and can apply the estimation procedures developed in this section. Note that in the "lifted" problem the SAA estimate $\widehat{x}(\cdot)$ of the lifted signal $x = x[z]$ is nothing but the standard Least Squares:

$$\begin{array}{rcl}\widehat{x}(\varpi^K) & \in & \mathrm{Argmin}_{x\in\mathcal{X}}\left\{\frac{1}{2}x^T\left[\sum_{k=1}^K \eta_k\eta_k^T\right]x - \left[\sum_{k=1}^K y_k\eta_k\right]^T x\right\} \\ & = & \mathrm{Argmin}_{x\in\mathcal{X}}\left\{\sum_k (y_k - \eta_k^T x)^2\right\}.\end{array} \tag{5.80}$$

Of course, there is no free lunch, and there are some potential complications:

- It may happen that the matrix $\mathcal{H} = \mathbf{E}_{\eta\sim Q}\{\eta\eta^T\}$ (Q is the common distribution of "artificial" regressors η_k induced by the common distribution of the actual regressors ζ_k) is *not* positive definite, which would make it impossible to recover well the signal $x[z]$ underlying our transformed observations, however large be K.
- Even when $\mathcal{H}$ is positive definite, so that $x[z]$ can be recovered well, provided K is large, we still need to recover z from $x[z]$, that is, to solve a system of polynomial equations, which can be difficult; besides, this system can have more than one solution.
- Even when the above difficulties can be somehow avoided, "lifting" $z \to x[z]$ typically increases significantly the number of parameters to be identified, which, in turn, deteriorates "finite time" accuracy bounds.

Note also that when $\mathcal{H}$ is not positive definite, this still is not the end of the world. Indeed, $\mathcal{H}$ is positive semidefinite; assuming that it has a nontrivial kernel L which we can identify, a realization η_k of our artificial regressor is orthogonal to L with probability 1, implying that replacing artificial signal x with its orthogonal projection onto $L^\perp$, we almost surely keep the value of the objective in (5.80) intact. Thus, we lose nothing when restricting the optimization domain in (5.80) to the orthogonal projection of $\mathcal{X}$ onto $L^\perp$. Since the restriction of $\mathcal{H}$ onto $L^\perp$ is positive definite, with this approach, for large enough values of K we will still get a good approximation of the projection of $x[z]$ onto $L^\perp$. With luck, this approximation, taken together with the fact that the "artificial signal" we are looking for is not an arbitrary vector from $\mathcal{X}$—it is of the form $x[z]$ for some $z \in \mathcal{Z}$—will allow us to get a good approximation of z. Here is the first part of the exercise:

1) Carry out the outlined approach in the situation where
   - The common distribution Π of regressors ζ_k has density w.r.t. the Lebesgue measure on $\mathbf{R}^{\nu\times\mu}$ and possesses finite moments of all orders
   - $\psi(w)$ is a quadratic form, either (case A) homogeneous,

   $$\psi(w) = w^T S w \qquad [S \neq 0],$$

   or (case B) inhomogeneous,

   $$\psi(w) = w^T S w + s^T w \qquad [S \neq 0, s \neq 0].$$

   - The labels are linked to the regressors and to the true signal z by the relation

   $$y_k = \psi(\eta_k^T z) + \chi_k,$$

where the $\chi_k \sim \mathcal{N}(0,1)$ are mutually independent and independent from the regressors.

Now comes the concluding part of the exercise, where you are supposed to apply the approach we have developed to the situation as follows:

> You are given a DC electric circuit comprised of resistors, that is, *connected* oriented graph with m nodes and n arcs $\gamma_j = (s_j, t_j)$, $1 \leq j \leq n$, where $1 \leq s_j, t_j \leq m$ and $s_j \neq t_j$ for all j; arcs γ_j are assigned with resistances $R_j > 0$ known to us. At instant $k = 1, 2, ..., K$, "nature" specifies "external currents" (charge flows from the "environment" into the circuit) $s_1, ..., s_m$ at the nodes; these external currents specify currents in the arcs and voltages at the nodes, and consequently, the power dissipated by the circuit.
>
> Note that nature cannot be completely free in generating the external currents: their total should be zero. As a result, all that matters is the vector $s = [s_1; ...; s_{m-1}]$ of external currents at the first $m-1$ nodes, due to $s_m \equiv -[s_1 + ... + s_{m-1}]$. We assume that the mechanism of generating the vector of external currents at instant k—let this vector be denoted by $s^k \in \mathbf{R}^{m-1}$—is as follows. There are somewhere $m-1$ sources producing currents $z_1, ..., z_{m-1}$. At time k nature selects a one-to-one correspondence $i \mapsto \pi_k(i)$, $i = 1, ..., m-1$, between these sources and the first $m-1$ nodes of the circuit, and "forwards" current $z_{\pi_k(i)}$ to node i:
>
> $$s_i^k = z_{\pi_k(i)},\ 1 \leq i \leq m-1.$$
>
> For the sake of definiteness, assume that the permutations π_k of $1, ..., m-1$, $k = 1, ..., K$, are i.i.d. drawn from the uniform distribution on the set of $(m-1)!$ permutations of $m-1$ elements.
>
> Assume that at time instants $k = 1, ..., K$ we observe the permutations π_k and noisy measurements of the power dissipated at this instant by the circuit; given those observations, we want to recover the vector z.

Here is your task:

2) Assuming the noises in the dissipated power measurements to be independent of each other and of π_k zero mean Gaussian noises with variance σ^2, apply to the estimation problem in question the approach developed in item 1 of the exercise and run numerical experiments.

Exercise 5.5.

[shift estimation] Consider the situation as follows: given a continuous vector field $f(u) : \mathbf{R}^m \to \mathbf{R}^m$ which is strongly monotone on bounded subsets of $\mathbf{R}^m$ and a convex compact set $\mathcal{S} \subset \mathbf{R}^m$, we observe in noise vectors $f(p - s)$, where $p \in \mathbf{R}^m$ is an observation point known to us, and $s \in \mathbf{R}^m$ is a shift unknown to us known to belong to $\mathcal{S}$. Precisely, assume that our observations are

$$y_k = f(p_k - s) + \xi_k,\ k = 1, ..., K,$$

where $p_1, ..., p_K$ is a deterministic sequence known to us, and $\xi_1, ..., \xi_K$ are $\mathcal{N}(0, \gamma^2 I_m)$ observation noises independent of each other. Our goal is to recover from observations $y_1, ..., y_K$ the shift s.

1. Pose the problem as a single-observation version of the estimation problem from Section 5.2
2. Assuming f to be strongly monotone, with modulus $\varkappa > 0$, on the entire space, what is the error bound for the SAA estimate?
3. Run simulations in the case of $m = 2$, $\mathcal{S} = \{u \in \mathbf{R}^2 : \|u\|_2 \leq 1\}$ and

$$f(u) = \left[\begin{array}{c} 2u_1 + \sin(u_1) + 5u_2 \\ 2u_2 - \sin(u_2) - 5u_1 \end{array}\right].$$

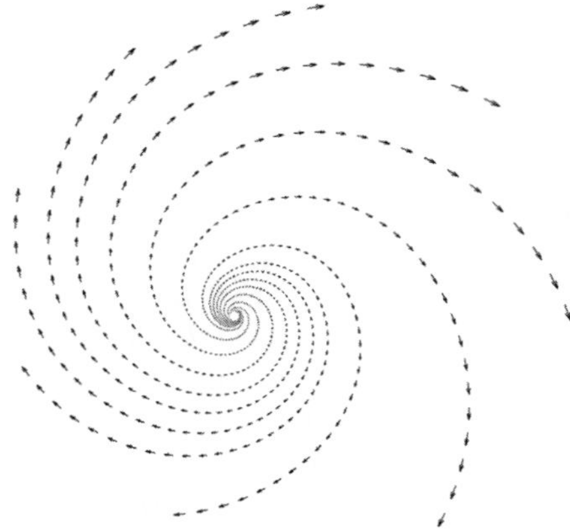

Note: Field $f(\cdot)$ is not potential; this is the monotone vector field associated with the strongly convex-concave function

$$\psi(u) = u_1^2 - \cos(u_1) - u_2^2 - \cos(u_2) + 5u_1u_2,$$

so that $f(u) = [\frac{\partial}{\partial u_1}\phi(u); -\frac{\partial}{\partial u_2}\phi(u)]$. Compare the actual recovery errors with their theoretical upper bounds.

4. Think what can be done when our observations are

$$y_k = f(Ap_k - s) + \xi_k,\ 1 \leq k \leq K$$

with known p_k, noises $\xi_k \sim \mathcal{N}(0, \gamma^2 I_2)$ independent across k, and unknown A and s which we want to recover.

5.4 PROOFS

5.4.1 Proof of (5.8)

Let $h \in \mathbf{R}^m$, and let ω be a random vector with entries $\omega_i \sim \text{Poisson}(\mu_i)$ independent across i. Taking into account that the ω_i are independent across i, we have

$$\begin{array}{l}\mathbf{E}\left\{\exp\{\gamma h^T\omega\}\right\} = \prod_i \mathbf{E}\left\{\gamma h_i\omega_i\right\} = \prod_i \exp\{[\exp\{\gamma h_i\} - 1]\mu_i\} \\ = \exp\{\sum_i [\exp\{\gamma h_i\} - 1]\mu_i\},\end{array}$$

whence by the Chebyshev inequality for $\gamma \geq 0$ it holds

$$\begin{array}{rl} \mathrm{Prob}\{h^T\omega > h^T\mu + t\} &= \mathrm{Prob}\{\gamma h^T\omega > \gamma h^T\mu + \gamma t\} \\ &\leq \mathbf{E}\left\{\exp\{\gamma h^T\omega\}\right\}\exp\{-\gamma h^T\mu - \gamma t\} \\ &\leq \exp\left\{\sum_i [\exp\{\gamma h_i\} - 1]\mu_i - \gamma h^T\mu - \gamma t\right\}. \end{array} \tag{5.81}$$

Now, for $|s| < 3$, one has $e^s \leq 1 + s + \frac{s^2}{2(1-s/3)}$ (see., e.g., [178]), which combines with (5.81) to imply that

$$0 \leq \gamma < \frac{3}{\|h\|_\infty} \Rightarrow \ln\left(\mathrm{Prob}\{h^T\omega > h^T\mu + t\}\right) \leq \frac{\gamma^2 \sum_i h_i^2\mu_i}{2(1 - \gamma\|h\|_\infty/3)} - \gamma t.$$

Minimizing the right hand side in this inequality in $\gamma \in [0, \frac{3}{\|h\|_\infty})$, we get

$$\mathrm{Prob}\left\{h^T\omega > h^T\mu + t\right\} \leq \exp\left\{-\frac{t^2}{2[\sum_i h_i^2\mu_i + \|h\|_\infty t/3]}\right\}.$$

This inequality combines with the same inequality applied to $-h$ in the role of h to imply (5.8). □

5.4.2 Proof of Lemma 5.6

(i): When $\pi(\mathrm{Col}_\ell[H])) \leq 1$ for all ℓ and $\lambda \geq 0$, denoting by $[h]^2$ the vector comprised of squares of the entries in h, we have

$$\begin{array}{rcl} \phi(\mathrm{dg}(H\mathrm{Diag}\{\lambda\}H^T)) &=& \phi(\sum_\ell \lambda_\ell[\mathrm{Col}_\ell[H]]^2) \leq \sum_\ell \lambda_\ell \phi([\mathrm{Col}_\ell[H]]^2) \\ &=& \sum_\ell \lambda_\ell \pi^2(\mathrm{Col}_\ell[H]) \leq \sum_\ell \lambda_\ell, \end{array}$$

implying that $(H^T\mathrm{Diag}\{\lambda\}H^T, \varkappa\sum_\ell \lambda_\ell)$ belongs to $\mathbf{H}$.

(ii): Let Θ, μ, Q, V be as stated in (ii); there is nothing to prove when $\mu = 0$; thus assume that $\mu > 0$. Let $d = \mathrm{dg}(\Theta)$, so that

$$d_i = \sum_j Q_{ij}^2 \ \&\ \varkappa\phi(d) \leq \mu \tag{5.82}$$

(the second relation is due to $(\Theta, \mu) \in \mathbf{H}$). (5.30) is evident. We have

$$[H_\chi]_{ij} = \sqrt{m/\mu}[G_\chi]_{ij},\ G_\chi = Q\mathrm{Diag}\{\chi\}V = \left[\sum_{k=1}^m Q_{ik}\chi_k V_{kj}\right]_{i,j}.$$

We claim that for every i it holds

$$\forall \gamma > 0 : \mathrm{Prob}\left\{[G_\chi]_{ij}^2 > 3\gamma d_i/m\right\} \leq \sqrt{3}\exp\{-\gamma/2\}. \tag{5.83}$$

Indeed, let us fix i. There is nothing to prove when $d_i = 0$, since in this case $Q_{ij} = 0$ for all j and therefore $[G_\chi]_{ij} \equiv 0$. When $d_i > 0$, by homogeneity in Q it suffices to verify (5.83) when $d_i/m = 1/3$. Assuming that this is the case, let $\eta \sim \mathcal{N}(0,1)$ be independent of χ. We have

$$\begin{array}{l}\mathbf{E}_\eta\left\{\mathbf{E}_\chi\{\exp\{\eta[G_\chi]_{ij}\}\}\right\} = \mathbf{E}_\eta\left\{\prod_k \cosh(\eta Q_{ik}V_{kj})\right\} \leq \mathbf{E}_\eta\left\{\prod_k \exp\{\frac{1}{2}\eta^2 Q_{ik}^2 V_{kj}^2\}\right\} \\ = \mathbf{E}_\eta\{\exp\{\frac{1}{2}\eta^2 \underbrace{\sum_k Q_{ik}^2 V_{kj}^2}_{\leq 2d_i/m}\}\} \leq \mathbf{E}_\eta\left\{\eta^2 d_i/m\right\} = \mathbf{E}_\eta\left\{\exp\{\eta^2/3\}\right\} = \sqrt{3},\end{array}$$

and

$$\mathbf{E}_\chi\left\{\mathbf{E}_\eta\{\exp\{\eta[G_\chi]_{ij}\}\}\right\} = \mathbf{E}_\chi\left\{\exp\{\tfrac{1}{2}[G_\chi]_{ij}^2\}\right\},$$

implying that

$$\mathbf{E}_\chi\left\{\exp\{\tfrac{1}{2}[G_\chi]_{ij}^2\}\right\} \leq \sqrt{3}.$$

Therefore in the case of $d_i/m = 1/3$ for all $s > 0$ it holds

$$\mathrm{Prob}\{\chi : [G_\chi]_{ij}^2 > s\} \leq \sqrt{3}\exp\{-s/2\},$$

and (5.83) follows. Recalling the relation between H and G, we get from (5.83) that

$$\forall \gamma > 0 : \mathrm{Prob}\{\chi : [H_\chi]_{ij}^2 > 3\gamma d_i/\mu\} \leq \sqrt{3}\exp\{-\gamma/2\}.$$

By the latter inequality, with $\varkappa$ given by (5.29) the probability of the event

$$\forall i,j : [H_\chi]_{ij}^2 \leq \varkappa \frac{d_i}{\mu}$$

is at least $1/2$. Let this event take place; in this case we have $[\mathrm{Col}_\ell[H_\chi]]^2 \leq \varkappa d/\mu$, whence, by definition of the norm $\pi(\cdot)$, $\pi^2(\mathrm{Col}_\ell[H_\chi]) \leq \varkappa\phi(d)/\mu \leq 1$ (see the second relation in (5.82)). Thus, the probability of the event (5.31) is at least $1/2$. □

5.4.3 Verification of (5.44)

Given $s \in [2,\infty]$ and setting $\bar{s} = s/2$, $s_* = \frac{s}{s-1}$, $\bar{s}_* = \frac{\bar{s}}{\bar{s}-1}$, we want to prove that

$$\begin{array}{l}\{(V,\tau) \in \mathbf{S}_+^N \times \mathbf{R}_+ : \exists (W \in \mathbf{S}^N, w \in \mathbf{R}_+^N) : V \preceq W + \mathrm{Diag}\{w\} \ \&\ \|W\|_{s_*} + \|w\|_{\bar{s}_*} \leq \tau\} \\ = \{(V,\tau) \in \mathbf{S}_+^N \times \mathbf{R}_+ : \exists w \in \mathbf{R}_+^N : V \preceq \mathrm{Diag}\{w\}, \|w\|_{\bar{s}_*} \leq \tau\}.\end{array}$$

To this end it clearly suffices to check that whenever $W \in \mathbf{S}^N$, there exists $w \in \mathbf{R}^N$ satisfying

$$W \preceq \mathrm{Diag}\{w\},\ \|w\|_{\bar{s}_*} \leq \|W\|_{s_*}.$$

The latter is equivalent to saying that for any $W \in \mathbf{S}^N$ such that $\|W\|_{s_*} \leq 1$, the conic optimization problem

$$\mathrm{Opt} = \min_{t,w}\{t : t \geq \|w\|_{\bar{s}_*}, \mathrm{Diag}\{w\} \succeq W\} \tag{5.84}$$

is solvable (which is evident) with optimal value ≤ 1. To see that the latter indeed is the case, note that the problem clearly is strictly feasible, whence its optimal value is the same as the optimal value in the conic problem

$$\begin{array}{c}\mathrm{Opt} = \max_P\left\{\mathrm{Tr}(PW) : P \succeq 0, \|\mathrm{dg}\{P\}\|_{\bar{s}_*/(\bar{s}_*-1)} \leq 1\right\} \\ [\mathrm{dg}\{P\} = [P_{11}; P_{22}; \ldots; P_{NN}]]\end{array}$$

dual to (5.84). Since $\mathrm{Tr}(PW) \leq \|P\|_{s_*/(s_*-1)}\|W\|_{s_*} \leq \|P\|_{s_*/(s_*-1)}$, recalling what s_* and $\bar{s}_*$ are, our task boils down to verifying that when a matrix $P \succeq 0$ satisfies $\|\mathrm{dg}\{P)\|_{s/2} \leq 1$, one also has $\|P\|_s \leq 1$. This is immediate: since P is positive semidefinite, we have $|P_{ij}| \leq P_{ii}^{1/2}P_{jj}^{1/2}$, whence, assuming $s < \infty$,

$$\|P\|_s^s = \sum_{i,j}|P_{ij}|^s \leq \sum_{i,j} P_{ii}^{s/2}P_{jj}^{s/2} = \left(\sum_i P_{ii}^{s/2}\right)^2 \leq 1.$$

When $s = \infty$, the same argument leads to

$$\|P\|_\infty = \max_{i,j}|P_{ij}| = \max_i |P_{ii}| = \|\mathrm{dg}\{P\}\|_\infty. \qquad \square$$

5.4.4 Proof of Proposition 5.10

1°. Let us consider the optimization problem (4.42) (where one should set $\mathcal{Q} = \sigma^2 I_m$), which under the circumstances is responsible for building a nearly optimal *linear* estimate of Bx yielded by Proposition 4.14, namely,

$$\mathrm{Opt}_* = \min_{\Theta,H,\Lambda,\Upsilon',\Upsilon''}\left\{\begin{array}{l}\phi_{\mathcal{T}}(\lambda[\Lambda]) + \phi_{\mathcal{R}}(\lambda[\Upsilon']) + \phi_{\mathcal{R}}(\lambda[\Upsilon'']) + \sigma^2\mathrm{Tr}(\Theta) : \\ \Lambda = \{\Lambda_k \succeq 0, k \leq K\}, \Upsilon' = \{\Upsilon'_\ell \succeq 0, \ell \leq L\}, \\ \Upsilon'' = \{\Upsilon''_\ell \succeq 0, \ell \leq L\}, \left[\begin{array}{c|c}\sum_\ell \mathcal{S}_\ell^*[\Upsilon''_\ell] & \frac{1}{2}M^TH^T \\ \hline \frac{1}{2}HM & \Theta\end{array}\right] \succeq 0, \\ \left[\begin{array}{c|c}\sum_\ell \mathcal{S}_\ell^*[\Upsilon'_\ell] & \frac{1}{2}M^T[B - H^TA] \\ \hline \frac{1}{2}[B-H^TA]^TM & \sum_k \mathcal{R}_k^*[\Lambda_k]\end{array}\right] \succeq 0,\end{array}\right\} \tag{5.85}$$

Let us show that the optimal value Opt of (5.45) satisfies

$$\mathrm{Opt} \leq 2\kappa\mathrm{Opt}_* = 2\sqrt{2\ln(2m/\epsilon)}\mathrm{Opt}_*. \tag{5.86}$$

To this end, observe that the matrices

$$Q := \left[\begin{array}{c|c} U & \frac{1}{2}B \\ \hline \frac{1}{2}B^T & A^T\Theta A + \sum_k \mathcal{R}_k^*[\Lambda_k]\end{array}\right]$$

and

$$\left[\begin{array}{c|c} M^TUM & \frac{1}{2}M^TB \\ \hline \frac{1}{2}B^TM & A^T\Theta A + \sum_k \mathcal{R}_k^*[\Lambda_k]\end{array}\right] = \left[\begin{array}{c|c} M^T & \\ \hline & I_n\end{array}\right] Q \left[\begin{array}{c|c} M & \\ \hline & I_n\end{array}\right]$$

simultaneously are/are not positive semidefinite due to the fact that the image space of M contains the full-dimensional set $\mathcal{B}_*$ and thus is the entire $\mathbf{R}^\nu$, so that the image space of $\left[\begin{array}{c|c} M & \\ \hline & I_n\end{array}\right]$ is the entire $\mathbf{R}^\nu \times \mathbf{R}^n$. Therefore,

$$\mathrm{Opt} = \min_{\Theta,U,\Lambda,\Upsilon}\left\{\begin{array}{l} 2\left[\phi_{\mathcal{R}}(\lambda[\Upsilon]) + \phi_{\mathcal{T}}(\lambda[\Lambda]) + \sigma^2\kappa^2\mathrm{Tr}(\Theta)\right] : \\ \Theta \succeq 0, U \succeq 0, \Lambda = \{\Lambda_k \succeq 0, k \leq K\}, \Upsilon = \{\Upsilon_\ell \succeq 0, \ell \leq L\}, \\ \left[\begin{array}{c|c} M^TUM & \frac{1}{2}M^TB \\ \hline \frac{1}{2}B^TM & A^T\Theta A + \sum_k \mathcal{R}_k^*[\Lambda_k]\end{array}\right] \succeq 0, \ M^TUM \preceq \sum_\ell \mathcal{S}_\ell^*[\Upsilon_\ell]\end{array}\right\}.$$

Further, note that if a collection $\Theta, U, \{\Lambda_k\}, \{\Upsilon_\ell\}$ is a feasible solution to the latter problem and $\theta > 0$, the scaled collection $\theta\Theta, \theta^{-1}U, \{\theta\Lambda_k\}, \{\theta^{-1}\Upsilon_\ell\}$ is also a feasible solution. When optimizing with respect to the scaling, we get

$$\begin{array}{rcl}\mathrm{Opt} &=& \inf\limits_{\Theta,U,\Lambda,\Upsilon}\left\{4\sqrt{\phi_{\mathcal{R}}(\lambda[\Upsilon])\left[\phi_{\mathcal{T}}(\lambda[\Lambda] + \sigma^2\kappa^2\mathrm{Tr}(\Theta)\right]} : \right.\\ && \Theta \succeq 0, U \succeq 0, \Lambda = \{\Lambda_k \succeq 0, k \le K\}, \Upsilon = \{\Upsilon_\ell \succeq 0, \ell \le L\},\\ && \left.\left[\begin{array}{c|c} M^TUM & \frac{1}{2}M^TB \\ \hline \frac{1}{2}B^TM & A^T\Theta A + \sum_k \mathcal{R}_k^*[\Lambda_k]\end{array}\right] \succeq 0,\ M^TUM \preceq \sum_\ell \mathcal{S}_\ell^*[\Upsilon_\ell]\right\}\\ &\le& 2\kappa\mathrm{Opt}_+,\end{array} \tag{5.87}$$

where (note that $\kappa > 1$)

$$\begin{array}{rcl}\mathrm{Opt}_+ &=& \inf\limits_{\Theta,U,\Lambda,\Upsilon}\left\{2\sqrt{\phi_{\mathcal{R}}(\lambda[\Upsilon])\left[\phi_{\mathcal{T}}(\lambda[\Lambda]) + \sigma^2\mathrm{Tr}(\Theta)\right]} : \right.\\ && \left.\begin{array}{l}\Theta \succeq 0, U \succeq 0, \Lambda = \{\Lambda_k \succeq 0, k \le K\},\\ \Upsilon = \{\Upsilon_\ell \succeq 0, \ell \le L\}, M^TUM \preceq \sum_\ell \mathcal{S}_\ell^*[\Upsilon_\ell],\\ \left[\begin{array}{c|c} M^TUM & \frac{1}{2}M^TB \\ \hline \frac{1}{2}B^TM & A^T\Theta A + \sum_k \mathcal{R}_k^*[\Lambda_k]\end{array}\right] \succeq 0\end{array}\right\}.\end{array} \tag{5.88}$$

On the other hand, when strengthening the constraint $\Lambda_k \succeq 0$ of (5.85) to $\Lambda_k \succ 0$, we still have

$$\begin{array}{l}\mathrm{Opt}_* = \inf\limits_{\Theta,H,\Lambda,\Upsilon',\Upsilon''}\left\{\phi_{\mathcal{T}}(\lambda[\Lambda]) + \phi_{\mathcal{R}}(\lambda[\Upsilon']) + \phi_{\mathcal{R}}(\lambda[\Upsilon'']) + \sigma^2\mathrm{Tr}(\Theta) : \right.\\ \left.\begin{array}{l}\Lambda = \{\Lambda_k \succ 0, k \le K\}, \Upsilon' = \{\Upsilon'_\ell \succeq 0, \ell \le L\},\\ \Upsilon'' = \{\Upsilon''_\ell \succeq 0, \ell \le L\},\ \left[\begin{array}{c|c} \sum_\ell \mathcal{S}_\ell^*[\Upsilon''_\ell] & \frac{1}{2}M^TH^T \\ \hline \frac{1}{2}HM & \Theta\end{array}\right] \succeq 0,\\ \left[\begin{array}{c|c} \sum_\ell \mathcal{S}_\ell^*[\Upsilon'_\ell] & \frac{1}{2}M^T[B - H^TA] \\ \hline \frac{1}{2}[B - H^TA]^TM & \sum_k \mathcal{R}_k^*[\Lambda_k]\end{array}\right] \succeq 0\end{array}\right\}.\end{array} \tag{5.89}$$

Now let $\Theta, H, \Lambda, \Upsilon', \Upsilon''$ be a feasible solution to the latter problem. By the second semidefinite constraint in (5.89) we have

$$\left[\begin{array}{c|c} \sum_\ell \mathcal{S}_\ell^*[\Upsilon''_\ell] & \frac{1}{2}M^TH^TA \\ \hline \frac{1}{2}A^THM & A^T\Theta A\end{array}\right] = \left[\begin{array}{c|c} I & \\ \hline & A\end{array}\right]^T \left[\begin{array}{c|c} \sum_\ell \mathcal{S}_\ell^*[\Upsilon''_\ell] & \frac{1}{2}M^TH^T \\ \hline \frac{1}{2}HM & \Theta\end{array}\right] \left[\begin{array}{c|c} I & \\ \hline & A\end{array}\right] \succeq 0,$$

which combines with the first semidefinite constraint in (5.89) to imply that

$$\left[\begin{array}{c|c} \sum_\ell \mathcal{S}_\ell^*[\Upsilon'_\ell + \Upsilon''_\ell] & \frac{1}{2}M^TB \\ \hline \frac{1}{2}B^TM & A^T\Theta A + \sum_k \mathcal{R}_k^*[\Lambda_k]\end{array}\right] \succeq 0.$$

Next, by the Schur Complement Lemma (which is applicable due to

$$A^T\Theta A + \sum_k \mathcal{R}_k^*[\Lambda_k] \succeq \sum_k \mathcal{R}_k^*[\Lambda_k] \succ 0,$$

where the concluding $\succ$ is due to Lemma 4.44 combined with $\Lambda_k \succ 0$), this relation implies that for

$$\Upsilon_\ell = \Upsilon'_\ell + \Upsilon''_\ell,$$

we have

$$\sum_\ell \mathcal{S}_\ell^*[\Upsilon_\ell] \succeq M^T \underbrace{\left[\tfrac{1}{4}B[A^T\Theta A + \sum_k \mathcal{R}_k^*[\Lambda_k]]^{-1}B^T\right]}_{U} M.$$

Using the Schur Complement Lemma again, for the $U \succeq 0$ just defined we obtain

$$\left[\begin{array}{c|c} M^TUM & \frac{1}{2}M^TB \\ \hline \frac{1}{2}B^TM & A^T\Theta A + \sum_k \mathcal{R}_k^*[\Lambda_k] \end{array}\right] \succeq 0,$$

and in addition, by the definition of U,

$$M^TUM \preceq \sum_\ell \mathcal{S}_\ell^*[\Upsilon_\ell].$$

We conclude that

$$(\Theta, U, \Lambda, \Upsilon := \{\Upsilon_\ell = \Upsilon_\ell' + \Upsilon_\ell'', \ell \leq L\})$$

is a feasible solution to optimization problem (5.88) specifying Opt_+. The value of the objective of the latter problem at this feasible solution is

$$\begin{array}{l} 2\sqrt{\phi_\mathcal{R}(\lambda[\Upsilon'] + \lambda[\Upsilon''])\,[\phi_\mathcal{T}(\lambda[\Lambda]) + \sigma^2\mathrm{Tr}(\Theta)]} \\ \quad \leq \quad \phi_\mathcal{R}(\lambda[\Upsilon'] + \lambda[\Upsilon'']) + \phi_\mathcal{T}(\lambda[\Lambda]) + \sigma^2\mathrm{Tr}(\Theta) \\ \quad \leq \quad \phi_\mathcal{R}(\lambda[\Upsilon']) + \phi_\mathcal{R}(\lambda[\Upsilon'']) + \phi_\mathcal{T}(\lambda[\Lambda]) + \sigma^2\mathrm{Tr}(\Theta), \end{array}$$

the concluding quantity in the chain being the value of the objective of problem (5.89) at the feasible solution $\Theta, H, \Lambda, \Upsilon', \Upsilon''$ to this problem. Since the resulting inequality holds true for every feasible solution to (5.89), we conclude that $\mathrm{Opt}_+ \leq \mathrm{Opt}_*$, and we arrive at (5.86) due to (5.87).

2°. Now, from Proposition 4.16 we conclude that Opt_* is within a logarithmic factor of the minimax optimal $(\frac{1}{8}, \|\cdot\|)$-risk corresponding to the case of Gaussian noise $\xi_x \sim \mathcal{N}(0, \sigma^2 I_m)$ for all x:

$$\mathrm{Opt}_* \leq \theta_* \mathrm{RiskOpt}_{1/8},$$

where

$$\theta_* = 8\sqrt{(2\ln F + 10\ln 2)\,(2\ln D + 10\ln 2)}, \quad F = \sum_\ell f_\ell,\ D = \sum_k d_k.$$

Since the minimax optimal $(\epsilon, \|\cdot\|)$-risk clearly only grows when ϵ decreases, we conclude that for $\epsilon \leq 1/8$ a feasible near-optimal solution to (5.45) is minimax optimal within the factor $2\theta^*\kappa$. □

Solutions to Selected Exercises

6.1 SOLUTIONS FOR CHAPTER 1

Exercise 1.1. The k-th Hadamard matrix, $\mathcal{H}_k$ (here k is a nonnegative integer) is the $n_k \times n_k$ matrix, $n_k = 2^k$, given by the recurrence

$$\mathcal{H}_0 = [1]; \mathcal{H}_{k+1} = \left[\begin{array}{c|c} \mathcal{H}_k & \mathcal{H}_k \\ \hline \mathcal{H}_k & -\mathcal{H}_k \end{array}\right].$$

In the sequel, we assume that $k > 0$. Now comes the exercise:

1. Check that $\mathcal{H}_k$ is a symmetric matrix with entries ± 1, and columns of the matrix are mutually orthogonal, so that $\mathcal{H}_k/\sqrt{n_k}$ is an orthogonal matrix.
2. Check that when $k > 0$, $\mathcal{H}_k$ has just two distinct eigenvalues, $\sqrt{n_k}$ and $-\sqrt{n_k}$, each of multiplicity $m_k := 2^{k-1} = n_k/2$.
3. Prove that whenever f is an eigenvector of $\mathcal{H}_k$, one has

$$\|f\|_\infty \le \|f\|_1/\sqrt{n_k}.$$

Derive from this observation the conclusion as follows:

Let $a_1, ..., a_{m_k} \in \mathbf{R}^{n_k}$ be unit vectors orthogonal to each other which are eigenvectors of $\mathcal{H}_k$ with eigenvalues $\sqrt{n_k}$ (by the above, the dimension of the eigenspace of $\mathcal{H}_k$ associated with the eigenvalue $\sqrt{n_k}$ is m_k, so that the required $a_1, ..., a_{m_k}$ do exist), and let A be the $m_k \times n_k$ matrix with the rows $a_1^T, ..., a_{m_k}^T$. For every $x \in \operatorname{Ker} A$ it holds

$$\|x\|_\infty \le \frac{1}{\sqrt{n_k}}\|x\|_1,$$

whence A satisfies the nullspace property whenever the sparsity s satisfies $2s < \sqrt{n_k} = \sqrt{2m_k}$. Moreover, there exists (and can be found efficiently) an $m_k \times n_k$ contrast matrix $H = H_k$ such that for every $s < \frac{1}{2}\sqrt{n_k}$, the pair $(H_k, \|\cdot\|_\infty)$ satisfies the condition $\mathbf{Q}_\infty(s, \kappa_s = \underbrace{s/\sqrt{n_k}}_{<1/2})$ associated with A, and the $\|\cdot\|_2$-norms of columns of H_k do not exceed $\sqrt{2\frac{\sqrt{n_k}+1}{\sqrt{n_k}}}$.

Note that the above conclusion yields a sequence of individual $(m_k = 2^{k-1}) \times (n_k = 2^k)$ sensing matrices, $k = 1, 2, ...$, with "size ratio" $n_k/m_k = 2$, which make an efficiently verifiable condition for s-goodness, say, $\mathbf{Q}_\infty(s, \frac{1}{3})$ satisfiable in basically the entire range of values of s allowed by Proposition 1.13. It would be interesting to get similar "fully constructive" results for other size ratios, like $m : n = 1 : 4$, $m : n = 1 : 8$, etc.

Solution: The facts that $\mathcal{H}_k$ is symmetric, and $\mathcal{H}_k/\sqrt{n_k}$ is orthogonal, are given by straightforward induction in k. Since $\mathcal{H}_k/\sqrt{n_k}$ is both symmetric and orthogonal, all eigenvalues of this matrix are equal to either 1, or -1, and since the trace of the matrix is 0 (recall that $k > 0$), there are m_k eigenvalues equal to $+1$ and m_k

eigenvalues equal to -1, which for $\mathcal{H}_k$ means eigenvalues $\pm\sqrt{n_k}$ of multiplicity m_k each.

Now let f be an eigenvector of $\mathcal{H}_k$, so that $\mathcal{H}_k f = \lambda f$ with $|\lambda| = \sqrt{n_k}$. Since the entries in $\mathcal{H}_k$ are ± 1, we have $\|\mathcal{H}_k f\|_\infty \leq \|f\|_1$, whence $\|f\|_\infty = \|\mathcal{H}_k f\|_\infty / |\lambda|$, and thus $\|f\|_\infty \leq \|f\|_1/\sqrt{n_k}$, as claimed.

Now let $x \in \operatorname{Ker} A$. By construction, $\operatorname{Ker} A$ is the orthogonal complement to the eigenspace of $\mathcal{H}_k$ corresponding to the eigenvalue $\sqrt{n_k}$. From what we know about the eigenstructure of $\mathcal{H}_k$, this orthogonal complement is the eigenspace of $\mathcal{H}_k$ corresponding to the eigenvalue $-\sqrt{n_k}$, whence $\mathcal{H}_k x = -\sqrt{n_k}x$, and thus

$$\|x\|_\infty \leq \|x\|_1/\sqrt{n_k}.$$

Given $i \leq n_k$, consider the LP program

$$\mathrm{Opt}_i = \max_x \{x_i : x \in \operatorname{Ker} A, \|x\|_1 \leq 1\}; \qquad (*)$$

as we have seen, $\mathrm{Opt}_i \leq 1/\sqrt{n_k}$. On the other hand, Opt_i is the optimal value in the dual to $(*)$ LP program, which is

$$\min_{t,h,g} \left\{t : e_i = A^T h + g, \|g\|_\infty \leq t\right\},$$

where e_i is the i-th basic orth in $\mathbf{R}^{n_k}$. Denoting by h_i the h-component of an optimal solution to the dual problem, we get

$$\|A^T h_i - e_i\|_\infty \leq 1/\sqrt{n_k},$$

whence $A^T h_i = e_i + g_i$, $\|g_i\|_\infty \leq 1/\sqrt{n_k}$, and thus

$$\|A^T h_i\|_2^2 = \|e_i + g_i\|_2^2 = \|e_i\|_2^2 + \|g_i\|_2^2 + 2e_i^T g_i \leq 2 + 2/\sqrt{n_k}.$$

By construction, $AA^T = I_{m_k}$, whence $\|A^T h_i\|_2^2 = \|h_i\|_2^2$; thus, $\|h_i\|_2 \leq \sqrt{2\frac{\sqrt{n_k}+1}{\sqrt{n_k}}}$.

Now let H_k be the $m_k \times n_k$ matrix with the columns $h_1, ..., h_{n_k}$. The columns of H_k do satisfy the norm bound from the claim we are proving, and we have

$$|[I - H_k^T A]_{ij}| = |[e_i^T - h_i^T A]_j| \leq \|e_i - A^T h_i\|_\infty \leq \mathrm{Opt}_i \leq 1/\sqrt{n_k}$$

for all i, j, implying that H_k satisfies the condition (1.34) with $\kappa = s/\sqrt{n_k} = \kappa_s$, and thus, by Proposition 1.10, $(H_k, \|\cdot\|_\infty)$ indeed satisfies $\mathbf{Q}_\infty(s, \kappa_s)$.

Similar "fully constructive" results for other size ratios can be extracted from [61]. □

Exercise 1.2. [Follow-up to Exercise 1.1] Exercise 1.1 provides us with an explicitly given $(m = 512) \times (n = 1024)$ sensing matrix $\bar{A}$ such that the efficiently verifiable condition $\mathbf{Q}_\infty(15, \frac{15}{32})$ is satisfiable; in particular, $\bar{A}$ is 15-good. With all we know about the limits of performance of verifiable sufficient conditions for goodness, how should we evaluate this specific sensing matrix? Could we point out a sensing matrix of the same size which is provably s-good for a value of s larger (or "much larger") than 15?

We do not know the answer, and you are requested to explore some possibilities, including (but not reducing to—you are welcome to investigate more options!) the following ones.

1. Generate at random a sample of $m \times n$ sensing matrices A, compute their mutual incoherences and look at how large the goodness levels certified by these incoherences are. What

happens when the matrices are Gaussian (independent $\mathcal{N}(0,1)$ entries) and Rademacher ones (independent entries taking values ± 1 with probabilities $1/2$)?

2. Generate at random a sample of $m \times n$ matrices with independent $\mathcal{N}(0, 1/m)$ entries. Proposition 1.7 suggests that a sampled matrix A has good chances to satisfy $\mathrm{RIP}(\delta, k)$ with some $\delta < 1/3$ and some k, and thus to be s-good (and even more than this; see Proposition 1.8) for every $s \leq k/2$. Of course, given A we cannot check whether the matrix indeed satisfies $\mathrm{RIP}(\delta, k)$ with given δ, k; what we can try to do is to certify that $\mathrm{RIP}(\delta, k)$ does _not_ take place. To this end, it suffices to select at random, say, 200 $m \times k$ submatrices $\tilde{A}$ of A and compute the eigenvalues of $\tilde{A}^T\tilde{A}$; if A possesses $\mathrm{RIP}(\delta, k)$, all these eigenvalues should belong to the segment $[1-\delta, 1+\delta]$, and if in reality this does not happen, A definitely is not $\mathrm{RIP}(\delta, k)$.

Solution:

1. Here are the levels of goodness as justified by mutual incoherence for randomly generated 512×1024 matrices:

Generation	Mutual incoherence	Justified goodness level
Rademacher	0.191406	3
Gaussian	0.199985	3

The mutual incoherences and associated goodness levels in the table are the best, over 768 randomly generated matrices.

2. Here are the *largest* values of s for which in a series of 128 randomly generated 512×1024 matrices there *could be* one satisfying $\mathrm{RIP}(\delta, 2s)$ with $\delta < 1/3$:

Generation	s
Rademacher	6
Gaussian	5

We would say that the above numerical results are rather discouraging. We do know a 15-good individual 512×1024 matrix, and this fact is given by our "heavily conservative and severely restricted in its scope" verifiable sufficient condition for goodness. Moreover, it seems to be immediate to build an $m \times n$ sensing matrix which is s-good for s "nearly" as large as $\sqrt{m}$. For example, it is immediately seen (check it!) that if ξ, η are two independent random m-dimensional Rademacher vectors, then

$$\forall (\alpha > 0) : \mathrm{Prob}\{|\xi^T\eta| > \alpha m\} \leq 2\exp\{-\alpha^2/2\},$$

implying that the mutual incoherence for an $m \times n$ Rademacher matrix is with probability $\geq 1/2$ upper-bounded by the quantity

$$\widehat{\mu} = \sqrt{2\ln(2n^2)/m}.$$

Thus, mutual incoherence of a random, say, Rademacher matrix justifies its "nearly" (up to logarithmic in n factor) $\sqrt{m}$-goodness. This being said, numerical data above demonstrate that the "logarithmic toll" in question in reality may be quite heavy. Similarly, beautiful theoretical results on the RIP property of random Rademacher and Gaussian matrices seem to be poorly suited for "real life" purposes, even when we are ready to skip full-scale verification of RIP. As our numerical results show, in the situation where our "conservative and severely restricted in scope" sufficient condition for goodness allows us to justify goodness 15, the "theoretically perfect"

RIP-based approach cannot justify something like 7-goodness.

To conclude, we remark that our pessimism applies to *what we can provably say about the level of goodness*, and not with this level per se. For example, RIP$(\delta, 2s)$ with $\delta < 1/3$ is just a sufficient condition for s-goodness, and the fact that for the sizes m, n we were considering this sufficient condition is typically *not* satisfied with "meaningful" values of s does *not* prevent the random matrices we were testing from being s-good with "nice" values of s. It could also happen that the sizes m, n we are speaking about are too small for meaningful Compressed Sensing applications, and what happens in this range of sizes is of no actual interest.

Exercise 1.3. Let us start with a preamble. Consider a finite Abelian group; the only thing which matters for us is that such a group G is specified by a collection of a $k \geq 1$ of positive integers $\nu_1, ..., \nu_k$ and is comprised of all collections $\omega = (\omega_1, ..., \omega_k)$ where every ω_i is an integer from the range $\{0, 1, ..., \nu_k - 1\}$; the group operation, denoted by $\oplus$, is

$$(\omega_1, ..., \omega_k) \oplus (\omega_1', ..., \omega_k') = ((\omega_1 + \omega_1') \bmod \nu_1, ..., (\omega_k + \omega_k') \bmod \nu_k),$$

where $a \bmod b$ is the remainder, taking values in $\{0, 1, ..., b-1\}$, in the division of an integer a by positive integer b; say, $5 \bmod 3 = 2$, and $6 \bmod 3 = 0$. Clearly, the cardinality of the above group G is $n_k = \nu_1 \nu_2 ... \nu_k$. A *character* of group G is a homomorphism acting from G into the multiplicative group of complex numbers of modulus 1, or, in simple words, a complex-valued function $\chi(\omega)$ on G such that $|\chi(\omega)| = 1$ for all $\omega \in G$ and $\chi(\omega \oplus \omega') = \chi(\omega)\chi(\omega')$ for all $\omega, \omega' \in G$. Note that characters themselves form a group w.r.t. pointwise multiplication; clearly, all characters of our G are functions of the form

$$\chi((\omega_1, ..., \omega_k)) = \mu_1^{\omega_1} ... \mu_k^{\omega_k},$$

where μ_i are restricted to be roots of degree ν_i from 1: $\mu_i^{\nu_i} = 1$. It is immediately seen that the group G_* of characters of G is of the same cardinality $n_k = \nu_1 ... \nu_k$ as G. We can associate with G the matrix $\mathcal{F}$ of size $n_k \times n_k$; the columns in the matrix are indexed by the elements ω of G, the rows by the characters $\chi \in G_*$ of G, and the element in cell (χ, ω) is $\chi(\omega)$. The standard example here corresponds to $k = 1$, in which case $\mathcal{F}$ clearly is the $\nu_1 \times \nu_1$ matrix of the Discrete Fourier Transform.

Now comes the exercise:

1. Verify that the above $\mathcal{F}$ is, up to factor $\sqrt{n_k}$, a unitary matrix: denoting by $\overline{a}$ the complex conjugate of a complex number a, $\sum_{\omega \in G} \chi(\omega)\overline{\chi'}(\omega)$ is n_k or 0 depending on whether $\chi = \chi'$ or $\chi \neq \chi'$.
2. Let $\bar{\omega}, \bar{\omega}'$ be two elements of G. Prove that there exists a permutation Π of elements of G which maps $\bar{\omega}$ into $\bar{\omega}'$ and is such that

$$\mathrm{Col}_{\Pi(\omega)}[\mathcal{F}] = D\mathrm{Col}_\omega[\mathcal{F}] \;\forall \omega \in G,$$

   where D is a diagonal matrix with diagonal entries $\chi(\bar{\omega}')/\chi(\bar{\omega})$, $\chi \in G_*$.
3. Consider the special case of the above construction where $\nu_1 = \nu_2 = ... = \nu_k = 2$. Verify that in this case $\mathcal{F}$, up to permutation of rows and permutation of columns (these permutations depend on how we assign the elements of G and of G_* their serial numbers), is exactly the Hadamard matrix $\mathcal{H}_k$.
4. Extract from the above the following fact: let m, k be positive integers such that $m \leq n_k := 2^k$, and let sensing matrix A be obtained from $\mathcal{H}_k$ by selecting m distinct rows. Assume we want to find an $m \times n_k$ contrast matrix H such that the pair $(H, \|\cdot\|_\infty)$ satisfies the condition $\mathbf{Q}_\infty(s, \kappa)$ with as small κ as possible; by Proposition 1.10, to this end we should solve n LP programs

$$\mathrm{Opt}_i = \min_h \|e^i - A^T h\|_\infty,$$

where e^i is i-th basic orth in $\mathbf{R}^n$. Prove that with A coming from $\mathcal{H}_k$, all these problems have the same optimal value, and optimal solutions to all of the problems are readily given by the optimal solution to just one of them.

Solution: Item 1: Let $\chi(\omega_1,...,\omega_k) = \prod_i \alpha_i^{\omega_i}$, $\chi'(\omega_1,...,\omega_k) = \prod_i \beta_i^{\omega_i}$, where α_i, β_i are some roots of degree ν_i of 1. Let us set $\gamma_i = \alpha_i/\beta_i$, so that γ_i also are roots of degree ν_i of 1. We have

$$\sum_{\omega\in G} \chi(\omega)\overline{\chi'}(\omega) = \prod_i \left[\sum_{\omega_i=0}^{\nu_i-1} (\alpha_i/\beta_i)^{\omega_i}\right],$$

and if γ is a root of degree ν of 1, then clearly $\sum_{s=0}^{\nu-1}\gamma^s = \begin{cases} \nu, & \gamma=1 \\ \frac{1-\gamma^\nu}{1-\gamma}=0, & \gamma\neq 1\end{cases}$, so that $\sum_{\omega\in G}\chi(\omega)\overline{\chi'}(\omega)$ is $n_k = \prod_i \nu_i$ when all γ_i are equal to 1 (i.e., when $\chi = \chi'$) and is 0 otherwise. □

Item 2: Since G is a group, there exists $\delta \in G$ such that $\bar{\omega}' = \delta \oplus \bar{\omega}$, and the mapping $\omega \mapsto \Pi(\omega) := \delta \oplus \omega$ is a permutation of elements of G which maps $\bar{\omega}$ onto $\bar{\omega}'$. We clearly have

$$[\mathcal{F}]_{\chi,\delta\oplus\omega} = \chi(\delta\oplus\omega) = \chi(\delta)\chi(\omega),$$

implying that

$$\mathrm{Col}_{\delta\oplus\omega}[\mathcal{F}] = \mathrm{Diag}\{\chi(\delta) : \chi \in G_*\}\mathrm{Col}_\omega[\mathcal{F}]\ \forall \omega \in G,$$

as required. □

Item 3: When $\nu_1 = \nu_2 = ... = \nu_k = 2$, the elements of G are just binary words $\omega_1|\omega_2|...|\omega_k$ of length k (ω_i is the i-th bit in the word), and $\oplus$ is the letterwise summation of these words modulo 2. Characters of G also can be encoded by binary words $\zeta_1|\zeta_2|...|\zeta_k$ of length k—such a character is identified by ordered collection $\mu_1,...,\mu_k$ of roots ± 1 of degree 2 of 1, and we can represent such a collection by binary word $\zeta_1|...|\zeta_k$ according to the rule $\mu_i = (-1)^{\zeta_i}$. With these conventions, the value of a character $\zeta = \zeta_1|...|\zeta_k$ at a point $\omega = \omega_1|...|\omega_k \in G$ is $\zeta(\omega) = \prod_{i=1}^k (-1)^{\zeta_i\omega_i}$; for example, $0|0(0|0) = 1$, $0|1(1|0) = 1$, $0|1(0|1) = -1$. Now, we can treat a k-bit binary word $s = s_1|...|s_k$ as the binary representation of the integer $b(s) = s_1 + 2s_2 + 4s_3 + ... + 2^{k-1}s_k$, thus arriving at a one-to-one correspondence between the set B_k of all binary words of length k and the set $\{0,1,...,2^k-1\}$; we treat $b(s)$ as the serial number of $s \in B_k$. Now consider the matrix $\mathcal{F} = \mathcal{F}_k$ corresponding to $\nu_1 = ... = \nu_k = 2$, and let us order its rows and columns, originally indexed by elements of B_k, in the order given by serial numbers of these elements (so that the rows and columns are indexed starting with 0, and not with 1). When $k = 1$, we get $\mathcal{F}_1 = \begin{bmatrix} 0(0) & 0(1) \\ 1(0) & 1(1)\end{bmatrix} = \begin{bmatrix} 1 & 1 \\ 1 & -1\end{bmatrix} = \mathcal{H}_1$. Besides this, our ordering of words from B_{k+1} is as follows: we first write down the words $s|0$ with $s \in B_k$, in the order in which the s's appear in B_k, and then all words $s|1$ with $s \in B_k$, again in the order in which the s's appear in B_k. It follows that with our ordering of rows and columns in $\mathcal{F}_{k+1}$, we have

$$\mathcal{F}_{k+1} = \left[\begin{array}{c|c} [\zeta|0(\omega|0)]_{\zeta,\omega\in B_k} & [\zeta|0(\omega|1)]_{\zeta,\omega\in B_k} \\ \hline [\zeta|1(\omega|0)]_{\zeta,\omega\in B_k} & [\zeta|1(\omega|1)]_{\zeta,\omega\in B_k}\end{array}\right] = \begin{bmatrix} \mathcal{F}_k & \mathcal{F}_k \\ \mathcal{F}_k & -\mathcal{F}_k\end{bmatrix}.$$

We see that $\mathcal{F}_1 = \mathcal{H}_1$, and the recurrence linking matrices $\mathcal{F}_1, \mathcal{F}_2, ...$ is exactly the recurrence linking $\mathcal{H}_1, \mathcal{H}_2, ...$, whence $\mathcal{F}_k = \mathcal{H}_k$ for all $k = 1, 2, ...$ $\square$

Item 4: Indexing rows and columns of $\mathcal{H}_k$ starting with 0, and denoting by $s(i) = s_1(i)|s_2(i)|...|s_k(i)$ the binary representation of an integer $i \in \{0, 1, ..., 2^k - 1\}$ (this representation is augmented by zeros from the right to get a binary word of length k), we get from item 2 that for every $i \in I = \{0, 1, ..., 2^k - 1\}$ there exists a permutation $j \mapsto \Pi_i(j)$ of elements of I which maps 0 onto i and is such that $\mathrm{Col}_{\Pi_i(j)}[\mathcal{H}_k] = D_i \mathrm{Col}_j[\mathcal{H}_k]$, for all $j \in I$, where D_i is the diagonal matrix. Since $i = \Pi_i(0)$ and $\mathrm{Col}_0[\mathcal{H}_k]$ is the all-ones vector, we have $D_i = \mathrm{Diag}\{\mathrm{Col}_i[\mathcal{H}_k]\}$. As a result, $\mathrm{Col}_{\Pi_i(j)}[A] = E_i \mathrm{Col}_j[A]$, $j \in I$, where E_i is diagonal $m \times m$ matrix cut from D_i by the rows of $\mathcal{H}_k$ participating in A and the columns with the same indexes; note that all diagonal entries in E_i are ± 1. It follows that for every $h \in \mathbf{R}^k$ and for every j we have

$$[A^T h]_{\Pi_i(j)} = \mathrm{Col}^T_{\Pi_i(j)}[A]h = (E_i \mathrm{Col}_j[A])^T h = \mathrm{Col}^T_j[A](E_i h).$$

Equivalently, denoting by Π_i^{-1} the permutation inverse to Π_i,

$$[A^T h]_j = \mathrm{Col}^T_{\Pi_i^{-1}(j)}[A](E_i h).$$

It follows that

$$[e^i]_j - [A^T h]_j = [e^i]_j - \mathrm{Col}^T_{\Pi_i^{-1}(j)}[A](E_i h);$$

noting that $\Pi_i(0) = i$, that is, $\Pi_i^{-1}(i) = 0$, so that $[e^i]_j = [e^0]_{\Pi_i^{-1}(j)}$, we conclude that for all $j \in I$ it holds

$$[e^i]_j - [A^T h]_j = [e^0]_{\Pi_i^{-1}(j)} - \mathrm{Col}^T_{\Pi_i^{-1}(j)}[A](E_i h).$$

Hence

$$\|e^i - A^T h\|_\infty = \|e^0 - A^T(E_i h)\|_\infty$$

for all $i \in I$ and all $h \in \mathbf{R}^m$. We see that all problems $\min_h \|e^i - A^T h\|_\infty$ are equivalent to the problem $\min_h \|e^0 - A^T h\|_\infty$, and the optimal solution h^0 to the latter problem induces optimal solutions $h^i = E_i^{-1} h^0 = E_i h^0$ to the former problems. $\square$

Exercise 1.4. Proposition 1.13 states that the verifiable condition $\mathbf{Q}_\infty(s, \kappa)$ can certify s-goodness of an "essentially nonsquare" (with $m \leq n/2$) $m \times n$ sensing matrix A only when s is small as compared to m, namely, $s \leq \sqrt{2m}$. The exercise to follow is aimed at investigating what happens when an $m \times n$ "low" (with $m < n$) sensing matrix A is "nearly square," meaning that $m^o = n - m$ is small compared to n. Specifically, you should prove that for properly selected individual $(n - m^o) \times n$ matrices A the condition $\mathbf{Q}_\infty(s, \kappa)$ with $\kappa < 1/2$ is satisfiable when s is as large as $O(1)n/\sqrt{m^o}$.

1. Let $n = 2^k p$ with positive integer p and integer $k \geq 1$, and let $m^o = 2^{k-1}$. Given a $2m^o$-dimensional vector u, let u^+ be the n-dimensional vector built as follows: we split indexes from $\{1, ..., n = 2^k p\}$ into 2^k consecutive groups $I_1, ..., I_{2^k}$, p elements per group, and all entries of u^+ with indexes from I_i are equal to i-th entry, u_i, of vector u. Now let U be the linear subspace in $\mathbf{R}^{2^k}$ comprised of all eigenvectors, with eigenvalue $\sqrt{2^k}$, of the Hadamard matrix $\mathcal{H}_k$—see Exercise 1.1—so that the dimension of U is $2^{k-1} = m^o$, and let L be given by
$$L = \{u^+ : u \in U\} \subset \mathbf{R}^n.$$

Clearly, L is a linear subspace in $\mathbf{R}^n$ of dimension m^o. Prove that

$$\forall x \in L : \|x\|_\infty \leq \frac{\sqrt{2m^o}}{n}\|x\|_1.$$

Conclude that if A is an $(n - m^o) \times n$ sensing matrix with $\operatorname{Ker} A = L$, then the verifiable sufficient condition $\mathbf{Q}_\infty(s, \kappa)$ does certify s-goodness of A whenever

$$1 \leq s < \frac{n}{2\sqrt{2m^o}}.$$

Solution: Let $x \in L$, so that $x = u^+$ for some $u \in U$. By the result of Exercise 1.1, we have $\|u\|_\infty \leq \|u\|_1/\sqrt{2^k}$, whence, due to the construction of L,

$$\|x\|_\infty = \|u\|_\infty \leq \|u\|_1/\sqrt{2^k} = (\|x\|_1/p)/\sqrt{2^k} = \frac{\sqrt{2^k}}{n}\|x\|_1 = \frac{\sqrt{2m^o}}{n}\|x\|_1.$$

It remains to use Proposition 1.10; see (1.33). □

2. Let L be an m^o-dimensional subspace in $\mathbf{R}^n$. Prove that L contains a nonzero vector x with

$$\|x\|_\infty \geq \frac{\sqrt{m^o}}{n}\|x\|_1,$$

so that the condition $\mathbf{Q}_\infty(s, \kappa)$ cannot certify s-goodness of an $(n-m^o) \times n$ sensing matrix A whenever $s > O(1)n/\sqrt{m^o}$, for properly selected absolute constant $O(1)$.

Solution: Let the columns in $n \times m^o$ matrix F form an orthonormal basis in L. The squared Frobenius norm of F is m^o, implying that there exists a row in F of Euclidean norm $\geq \sqrt{m^o/n}$, whence there exists a unit m^o-dimensional vector d such that $z = Fd$ has an entry of magnitude $\geq \sqrt{m^o/n}$: $\|z\|_\infty \geq \sqrt{m^o/n}$. Now, by construction $z \in L$ and $\|z\|_2 = \|d\|_2 = 1$, whence $\|z\|_1 \leq \sqrt{n}$, and we get

$$\|z\|_\infty \geq \sqrt{m^o/n} = \sqrt{m^o/n}\|z\|_2 = \frac{\sqrt{m^o}}{n}[\sqrt{n}\|z\|_2] \geq \frac{\sqrt{m^o}}{n}\|z\|_1,$$

as claimed. □

Exercise 1.5. Utilize the results of Exercise 1.3 in a numerical experiment as follows.

- select n as an integer power 2^k of 2, say, set $n = 2^{10} = 1024$
- select a "representative" sequence M of values of m, $1 \leq m < n$, including values of m close to n and "much smaller" than n, say,

$$M = \{2, 5, 8, 16, 32, 64, 128, 256, 512, 7, 896, 960, 992, 1008, 1016, 1020, 1022, 1023\}$$

- for every $m \in M$,
  - generate at random an $m \times n$ submatrix A of the $n \times n$ Hadamard matrix $\mathcal{H}_k$ and utilize the result of item 4 of Exercise 1.3 in order to find the largest s such that s-goodness of A can be certified via the condition $\mathbf{Q}_\infty(\cdot, \cdot)$; call $s(m)$ the resulting value of s.
  - generate a moderate sample of Gaussian $m \times n$ sensing matrices A_i with independent $\mathcal{N}(0, 1/m)$ entries and use the construction from Exercise 1.2 to upper-bound the largest s for which a matrix from the sample satisfies $\mathrm{RIP}(1/3, 2s)$; call $\widehat{s}(m)$ the largest, over your A_i's, of the resulting upper bounds.

The goal of the exercise is to compare the computed values of $s(m)$ and $\widehat{s}(m)$; in other words, we again want to understand how "theoretically perfect" RIP compares to "conservative restricted scope" condition $\mathbf{Q}_\infty$.

Solution: Here are the results of our experiment, $n = 1024$:

m	1023	1022	1020	1016	1008	992	960	896
$\widehat{s}(m)$	11	10	10	10	10	10	10	9
$s(m)$	511	256	191	139	100	70	48	32
m	768	512	256	128	64	32	16	8
$\widehat{s}(m)$	7	5	2	0	0	0	0	0
$s(m)$	20	11	5	3	2	1	1	0

And here are the results for $n = 2048$:

m	2047	2046	2044	2040	2032	2016	1984	1920	1792
$\widehat{s}(m)$	22	21	22	22	22	21	21	20	19
$s(m)$	1023	511	384	279	199	143	100	69	46
m	1536	1024	512	256	128	64	32	16	8
$\widehat{s}(m)$	16	10	5	2	1	0	0	0	0
$s(m)$	29	15	8	4	3	2	1	1	0

We believe the results speak for themselves.

6.2 SOLUTIONS FOR CHAPTER 2

6.2.1 Two-point lower risk bound

Exercise 2.1. Let p and q be two probability distributions distinct from each other on d-element observation space $\Omega = \{1, ..., d\}$, and consider two simple hypotheses on the distribution of observation $\omega \in \Omega$, $H_1 : \omega \sim p$, and $H_2 : \omega \sim q$.

1. Is it true that there always exists a simple deterministic test deciding on H_1, H_2 with risk $< 1/2$?
2. Is it true that there always exists a simple randomized test deciding on H_1, H_2 with risk $< 1/2$?
3. Is it true that when quasi-stationary K-repeated observations are allowed, one can decide on H_1, H_2 with any small risk, provided K is large enough?

Solution: The answer to the first question is negative. E.g., when $d = 2$, $p = [0.8; 0.2]$, and $q = [0.9; 0.1]$, no deterministic test can decide on the hypotheses with risk < 0.8 (look at what happens when our observation is $\omega = 1$). The answer to the second question is positive. Indeed, assuming w.l.o.g. that $p + q > 0$, consider randomized test which, given observations $\omega \in \Omega$, accepts H_1 with probability $p_\omega/(p_\omega + q_\omega)$, and accepts H_2 whenever it does not accept H_1. As is immediately seen, both partial risks of this randomized test are equal to each other and equal to

$$\sum_\omega \frac{p_\omega q_\omega}{p_\omega + q_\omega} = \sum_\omega \underbrace{\left[\frac{\sqrt{p_\omega q_\omega}}{p_\omega + q_\omega}\right]}_{\leq \frac{1}{2}} \sqrt{p_\omega q_\omega} \leq \tfrac{1}{2} \sum_\omega \sqrt{p_\omega q_\omega}.$$

Thus, there always exists a randomized test with risk upper-bounded by half of the Hellinger affinity of p and q, and since $p \neq q$, this affinity is < 1 due to

$$2 \sum_\omega \sqrt{p_\omega q \omega} = 2 - \sum_\omega (\sqrt{p_\omega} - \sqrt{q_\omega})^2.$$

A positive answer to the second question implies a positive answer to the third one—it suffices to pass to the majority version of the single-observation randomized test just defined.

6.2.2 Around Euclidean Separation

Exercise 2.2. Justify the "immediate observation" in Section 2.2.2.3.B.

Solution: Function (2.18) clearly is the marginal univariate density of the random variable $\sqrt{Z}\eta$ with Z underlying the Gaussian mixture in question and random vector $\eta \sim \mathcal{N}(0, I_d)$ independent of Z. All we need to prove is that if e is a unit vector in $\mathbf{R}^d$ and $\xi \sim \mathcal{N}(0, \Theta)$ is independent of Z, with $\Theta \preceq I_d$, then for every $\delta \geq 0$ we have $\mathrm{Prob}\{e^T\sqrt{Z}\xi \geq \delta\} \leq \mathrm{Prob}\{e^T\sqrt{Z}\eta \geq \delta\}$ $[= \int_\delta^\infty \gamma_Z(s)ds]$. Indeed, we lose nothing when assuming that $\xi = \Theta^{1/2}\eta$, so that, denoting by $\Phi(s)$ the probability for the $\mathcal{N}(0,1)$ random variable to be $\geq s$, we have

$$\begin{array}{l}
\mathrm{Prob}\{e^T\sqrt{Z}\xi > \delta\} = \mathrm{Prob}\{e^T\sqrt{Z}\Theta^{1/2}\eta \geq \delta\} \\
= \int_{t>0} \mathrm{Prob}\{e^T\Theta^{1/2}\eta \geq \delta t^{-1/2}\}P_Z(dt) = \int_{t>0} \Phi(\delta t^{-1/2}/\sqrt{e^T\Theta e})P_Z(dt) \\
\leq \int_{t>0} \Phi(\delta t^{-1/2})P_Z(dt) \text{ [since } \Theta \preceq I_d] \\
= \int_{t>0} \mathrm{Prob}\{e^T\eta \geq \delta t^{-1/2}\}P_Z(dt) = \mathrm{Prob}\{e^T\sqrt{Z}\eta \geq \delta\} = \int_\delta^\infty \gamma_Z(s)ds,
\end{array}$$

as claimed.

Exercise 2.3.

1) Prove Proposition 2.9.
   Hint: You can find useful the following simple observation (prove it if you indeed use it):

   Let $f(\omega)$, $g(\omega)$ be probability densities taken w.r.t. a reference measure P on an observation space Ω, and let $\epsilon \in (0, 1/2]$ be such that

$$2\bar{\epsilon} := \int_\Omega \min[f(\omega), g(\omega)]P(d\omega) \leq 2\epsilon.$$

   Then

$$\int_\Omega \sqrt{f(\omega)g(\omega)}P(d\omega) \leq 2\sqrt{\epsilon(1-\epsilon)}.$$

Solution: Let us prove the claim in the hint. Setting

$$\underline{h}(\omega) = \min[f(\omega), g(\omega)], \ \ \overline{h}(\omega) = \max[f(\omega), g(\omega)],$$

we have

$$\begin{array}{rl}
\int_\Omega \sqrt{f(\omega)g(\omega)}P(d\omega) & = \int_\Omega \sqrt{\underline{h}(\omega)\overline{h}(\omega)}P(d\omega) \leq \left[\int_\Omega \underline{h}(\omega)P(d\omega)\right]^{1/2}\left[\int_\Omega \overline{h}(\omega)P(d\omega)\right]^{1/2} \\
& = \left[\int_\Omega \underline{h}(\omega)P(d\omega)\right]^{1/2}\left[\int_\Omega [f(\omega)+g(\omega)-\underline{h}(\omega)]P(d\omega)\right]^{1/2} \\
& = \sqrt{2\bar{\epsilon}}\sqrt{2-2\bar{\epsilon}} = 2\sqrt{\bar{\epsilon}(1-\bar{\epsilon})} \leq 2\sqrt{\epsilon(1-\epsilon)},
\end{array}$$

where the first $\leq$ is due to the Cauchy inequality, and the last $\leq$ is because $0 \leq \bar{\epsilon} \leq \epsilon \leq 1/2$. □

We are ready to prove Proposition 2.9. Let $\mathcal{T}_K^{\mathrm{maj}}$ be the K-observation majority test in question. Observe that (2.25) combines with (2.27) and (2.24) to imply that the quantity ϵ_* as given by (2.23) is $\leq \frac{1}{2} - \beta\mathrm{Opt}$, so that by Proposition 2.6 we have

$$\begin{array}{rcl}\mathrm{Risk}(\mathcal{T}_K^{\mathrm{maj}}|\mathcal{H}_1,\mathcal{H}_2) & \leq & \sum\limits_{K\geq k\geq K/2} \binom{K}{k}(1/2-\beta\mathrm{Opt})^k(1/2+\beta\mathrm{Opt})^{K-k}\\ & \leq & \sum\limits_{K\geq k\geq K/2} 2^{-K}\binom{K}{k}(1-4\beta^2\mathrm{Opt}^2)^k\\ & \leq & (1-4\beta^2\mathrm{Opt}^2)^{K/2} \leq \exp\{-2K\beta^2\mathrm{Opt}^2\}.\end{array}$$

In particular, when $\epsilon \in (0,1/4)$, we have

$$K \geq \frac{\ln(1/\epsilon)}{2\beta^2\mathrm{Opt}^2} \Rightarrow \mathrm{Risk}(\mathcal{T}_K^{\mathrm{maj}}|\mathcal{H}_1,\mathcal{H}_2) \leq \epsilon,$$

as claimed in (2.28).

To prove (2.29), let $\mathcal{T}_K$ be a test using a K-repeated stationary observation with risk $\leq \epsilon$. Let x_*^1, x_*^2 form an optimal solution to (2.11), so that $\|x_*^1 - x_*^2\|_2 = 2\mathrm{Opt}$. Consider two simple hypotheses stating that observations $\omega_1,...,\omega_K$ are i.i.d. drawn from the distribution of $x_*^1+\xi$ or of $x_*^2+\xi$, with $\xi \sim q(\cdot)$. Test $\mathcal{T}_K$ decides on these hypotheses with risk $\leq \epsilon$; consequently, assuming w.l.o.g. that $x_*^1 + x_2^* = 0$, so that $x_1^* = e$, $x_2^* = -e$ with $\|e\|_2 = \mathrm{Opt}$, we have by the two-point lower risk bound (see Proposition 2.2)

$$\int \min\Bigg[\underbrace{\prod_{k=1}^K q(\xi_k+e)}_{q_+(\xi^K)}, \underbrace{\prod_{k=1}^K q(\xi_k-e)}_{q_-(\xi^K)}\Bigg] \underbrace{d\xi_1...d\xi_K}_{d\xi^K} \leq 2\epsilon.$$

Applying the claim in the hint, we conclude that

$$\begin{array}{l}\left[\int_{\mathbf{R}^n} \sqrt{q(\xi+e)q(\xi-e)}d\xi\right]^K = \int \sqrt{q_+(\xi^K)q_-(\xi^K)}d\xi^K\\ \quad = \int \sqrt{\min[q_+(\xi^K),q_-(\xi^K)]\max[q_+(\xi^K),q_-(\xi^K)]}d\xi^K \leq 2\sqrt{\epsilon(1-\epsilon)}.\end{array} \tag{6.1}$$

Since $\|e\|_2 = \mathrm{Opt} \leq D$, (2.26) combines with (6.1) to imply that

$$\exp\{-K\alpha\mathrm{Opt}^2\} \leq 2\sqrt{\epsilon(1-\epsilon)} \leq \sqrt{4\epsilon},$$

and (2.29) follows. □

2) Justify the illustration in Section 2.2.3.2.C.

Solution: All we need to verify is that in the situation in question, denoting by $q(\cdot)$ the density of $\mathcal{N}(0,\frac{1}{2}I_n)$, we have $q \in \mathcal{P}_\gamma$. Indeed, taking this fact for granted we ensure the validity of (2.26) with $\alpha = 1$ and $D = \infty$ (check it!). Next, it is immediately seen that the function γ—see (2.30)—satisfies the relation

$$\gamma(s) \geq \tfrac{1}{7},\ \ 0 \leq s \leq 1,$$

implying that (2.25) holds true with $D = 1$ and $\beta = \frac{1}{7}$, as claimed.

It remains to verify that $q \in \mathcal{P}_\gamma$, which reduces to verifying that the marginal univariate density $\bar\gamma(s) = \frac{1}{\sqrt{\pi}}\exp\{-s^2\}$ of $q(\cdot)$ satisfies the relation $\int_\delta^\infty \bar\gamma(s)ds \leq \int_\delta^\infty \gamma(s)ds$, $\delta \geq 0$, or, which is the same since both γ and $\bar\gamma$ are even probability densities on the axis, that

$$\forall(\delta \geq 0) : \int_0^\delta (\bar{\gamma}(s) - \gamma(s))ds \geq 0.$$

The latter relation is an immediate consequence of the fact that the ratio $\bar{\gamma}(s)/\gamma(s)$ is a strictly decreasing function of $s \geq 0$ (check it!) combined with $\int_0^\infty(\bar{\gamma}(s) - \gamma(s))ds = 0$. Indeed, since $\bar{\gamma} \not\equiv \gamma$, the function $\Delta(s) = \bar{\gamma}(s) - \gamma(s)$ of $s \geq 0$ is not identically zero; since $\int_0^\infty \Delta(s)ds = 0$, Δ takes on $\mathbf{R}_+$ both positive and negative values and therefore, being continuous, has zeros on $\mathbf{R}_+$. Since $\bar{\gamma}$ and γ are positive and $\bar{\gamma}(s)/\gamma(s)$ is strictly decreasing on $\mathbf{R}_+$, $\Delta(s)$ has exactly one zero $\bar{s}$ on $\mathbf{R}_+$, this zero is positive, and Δ is ≥ 0 on $[0, \bar{s}]$ and is ≤ 0 on $[\bar{s}, \infty)$. As a result, when $0 \leq \delta \leq \bar{s}$, we have $\int_0^\delta \Delta(s)ds \geq 0$, and when $\delta > \bar{s}$, we have $\int_0^\delta \Delta(s)ds \geq \int_0^\infty \Delta(s)ds = 0$. □

6.2.3 Hypothesis testing via ℓ_1 separation

Let d be a positive integer, and the observation space Ω be the finite set $\{1, ..., d\}$ equipped with the counting reference measure.[1] Probability distributions on Ω can be identified with points p of d-dimensional *probabilistic simplex*

$$\mathbf{\Delta}_d = \{p \in \mathbf{R}^d : p \geq 0, \sum_i p_i = 1\};$$

the i-th entry p_i in $p \in \mathbf{\Delta}_d$ is the probability for the random variable distributed according to p to take value $i \in \{1, ..., d\}$. With this interpretation, p is the probability density taken w.r.t. the counting measure on Ω.

Assume B and W are two nonintersecting nonempty closed convex subsets of $\mathbf{\Delta}_d$; we interpret B and W as the sets of black and white probability distributions on Ω, and our goal is to find an optimal, in terms of its total risk, test deciding on the hypotheses

$$H_1 : p \in B,\ H_2 : p \in W$$

via a single observation $\omega \sim p$.

Warning: Everywhere in this section, "test" means "simple test."

Exercise 2.4. Our first goal is to find an optimal, in terms of its total risk, test deciding on the hypotheses H_1, H_2 via a *single* observation $\omega \sim p \in B \cup W$. To this end we consider the convex optimization problem

$$\text{Opt} = \min_{p \in B, q \in W} \left[f(p, q) := \sum_{i=1}^d |p_i - q_i| \right] \tag{2.154}$$

and let (p^*, q^*) be an optimal solution to this problem (it clearly exists).

1. Extract from optimality conditions that there exist reals $\rho_i \in [-1, 1]$, $1 \leq i \leq n$, such that

$$\rho_i = \begin{cases} 1, & p_i^* > q_i^* \\ -1, & p_i^* < q_i^* \end{cases} \tag{2.155}$$

[1] Counting measure is the measure on a discrete (finite or countable) set Ω which assigns every point of Ω with mass 1, so that the measure of a subset of Ω is the cardinality of the subset when it is finite and is $+\infty$ otherwise.

and
$$\rho^T(p-p^*) \geq 0 \forall p \in B \;\&\; \rho^T(q-q^*) \leq 0 \;\forall q \in W. \tag{2.156}$$

Solution: We are minimizing convex real-valued function $f(p,q) = \|p-q\|_1$ of $z = [p;q]$ over a nonempty convex compact set, so that there exists a subgradient $g = [f'_p; f'_q]$ of this function at the minimizer $[p^*; q^*]$ such that

$$g^T([p;q] - [p^*;q^*]) \geq 0 \;\forall [p;q] \in B \times W,$$

or, equivalently,

$$[f'_p]^T(p-p^*) \geq 0 \forall p \in B, \;[f'_q]^T(q-q^*) \geq 0 \forall q \in W.$$

Taking into account the expression for f, we see that $f'_p = -f'_q$ is a vector with entries from $[-1,1]$, and these entries satisfy (2.155).

2. Extract from the previous item that the test $\mathcal{T}$ which, given an observation $\omega \in \{1,...,d\}$, accepts H_1 with probability $\pi_\omega = (1+\rho_\omega)/2$ and accepts H_2 with complementary probability, has its total risk equal to
$$\sum_{\omega\in\Omega} \min[p^*_\omega, q^*_\omega], \tag{2.157}$$
and thus is minimax optimal in terms of the total risk.

Solution: Let $\omega \sim p \in B$. Then the p-probability to accept H_1 is

$$\sum_{\omega=1}^{d} \tfrac{1}{2}(1+\rho_\omega)p_\omega = \tfrac{1}{2}[1+\rho^T p] \geq \tfrac{1}{2}[1+\rho^T p^*]$$

(we have used (2.156)), and consequently

$$\mathrm{Risk}_1(\mathcal{T}|H_1,H_2) \leq 1 - \tfrac{1}{2}[1+\rho^T p^*] = \tfrac{1}{2}[1-\rho^T p^*].$$

Similarly, when $p \in W$, the probability to accept H_2 is

$$\sum_{\omega=1}^{d} \tfrac{1}{2}(1-\rho_\omega)p_\omega = \tfrac{1}{2}[1-\rho^T p] \geq \tfrac{1}{2}[1-\rho^T q^*]$$

(we have used (2.156)), and thus

$$\mathrm{Risk}_2(\mathcal{T}|H_1,H_2) \leq 1 - \tfrac{1}{2}[1-\rho^T q^*] = \tfrac{1}{2}[1+\rho^T q^*].$$

As a result,

$$\begin{array}{rcl}\mathrm{Risk}_1(\mathcal{T}|H_1,H_2) + \mathrm{Risk}_2(\mathcal{T}|H_1,H_2) & \leq & 1 - \tfrac{1}{2}\rho^T[p^* - q^*] \\ & = & 1 - \tfrac{1}{2}\sum_{\omega=1}^{n} |p^*_i - q^*_i|,\end{array}$$

where the concluding equality follows from (2.155), and we arrive at

$$\begin{array}{rcl}\mathrm{Risk}_1(\mathcal{T}|H_1,H_2) + \mathrm{Risk}_2(\mathcal{T}|H_1,H_2) & \leq & \tfrac{1}{2}\sum_{\omega\in\Omega}[p^*_\omega + q^*_\omega - |p^*_\omega - q^*_\omega|] \\ & = & \sum_{\omega\in\Omega}\min[p^*_\omega, q^*_\omega].\end{array}$$

Strict inequality in the resulting relation is impossible by Proposition 2.2, so that the total risk of our test indeed is as stated in (2.157).

Comments. Exercise 2.4 describes an efficiently computable and *optimal in terms of the worst-case total risk* simple test deciding on a pair of "convex" composite hypotheses on the distribution of a discrete random variable. While it seems an attractive result, we believe *by itself* this result is useless, since typically in the testing problem in question a *single* observation by far is not enough for a reasonable inference; such an inference requires observing *several* independent realizations $\omega_1, ..., \omega_K$ of the random variable in question. And the construction presented in Exercise 2.4 says nothing on how to adjust the test to the case of repeated observation. Of course, when $\omega^K = (\omega_1, ..., \omega_K)$ is a K-element i.i.d. sample drawn from a probability distribution p on $\Omega = \{1, ..., d\}$, ω^K can be thought of as a single observation of a discrete random variable taking values in the set $\Omega^K = \underbrace{\Omega \times ... \times \Omega}_{K}$, the probability distribution p^K of ω^K being readily given by p. So, why not to apply the construction from Exercise 2.4 to ω^K in the role of ω? On a close inspection, this idea fails. One of the reasons for this failure is that the cardinality of Ω^K (which, among other factors, is responsible for the computational complexity of implementing the test in Exercise 2.4) blows up exponentially as K grows. Another, even more serious, complication is that p^K depends on p nonlinearly, so that the family of distributions p^K of ω^K induced by a convex family of distributions p of ω—convexity meaning that the p's in question fill a *convex* subset of the probabilistic simplex—is not convex; and convexity of the sets B, W in the context of Exercise 2.4 is crucial. Thus, passing from a single realization of a discrete random variable to the sample of $K > 1$ independent realizations of the variable results in severe structural and quantitative complications "killing," at least at first glance, the approach undertaken in Exercise 2.4.

In spite of the above pessimistic conclusions, the single-observation test from Exercise 2.4 admits a meaningful multi-observation modification, which is the subject of our next exercise.

Exercise 2.5. There is a straightforward way to use the optimal, in terms of its total risk, single-observation test built in Exercise 2.4 in the "multi-observation" environment. Specifically, following the notation from Exercise 2.4, let $\rho \in \mathbf{R}^d, p^*, q^*$ be the entities built in this exercise, so that $p^* \in B$, $q^* \in W$, all entries in ρ belong to $[-1, 1]$, and

$$\begin{array}{c}\{\rho^T p \geq \alpha := \rho^T p^* \;\forall p \in B\} \;\&\; \{\rho^T q \leq \beta := \rho^T q^* \,\forall q \in W\}\\ \&\; \alpha - \beta = \rho^T[p^* - q^*] = \|p^* - q^*\|_1.\end{array}$$

Given an i.i.d. sample $\omega^K = (\omega_1, ..., \omega_K)$ with $\omega_t \sim p$, where $p \in B \cup W$, we could try to decide on the hypotheses $H_1 : p \in B$, $H_2 : p \in W$ as follows. Let us set $\zeta_t = \rho_{\omega_t}$. For large K, the observable, given ω^K, quantity $\zeta^K := \frac{1}{K}\sum_{t=1}^K \zeta_t$, by the Law of Large Numbers, will be with overwhelming probability close to $\mathbf{E}_{\omega\sim p}\{\rho_\omega\} = \rho^T p$, and the latter quantity is $\geq \alpha$ when $p \in B$ and is $\leq \beta < \alpha$ when $p \in W$. Consequently, selecting a "comparison level" $\ell \in (\beta, \alpha)$, we can decide on the hypotheses $p \in B$ vs. $p \in W$ by computing ζ^K, comparing the result to ℓ, accepting the hypothesis $p \in B$ when $\zeta^K \geq \ell$, and accepting the alternative $p \in W$ otherwise. The goal of this exercise is to quantify the above qualitative considerations. To this end let us fix $\ell \in (\beta, \alpha)$ and K and ask ourselves the following questions:

A. For $p \in B$, how do we upper-bound the probability $\mathrm{Prob}_{p_K}\{\zeta^K \leq \ell\}$?
B. For $p \in W$, how do we upper-bound the probability $\mathrm{Prob}_{p_K}\{\zeta^K \geq \ell\}$?

Here p_K is the probability distribution of the i.i.d. sample $\omega^K = (\omega_1, ..., \omega_K)$ with $\omega_t \sim p$.

The simplest way to answer these questions is to use Bernstein's bounding scheme. Specifically, to answer question A, let us select $\gamma \geq 0$ and observe that for every probability distribution

p on $\{1,2,...,d\}$ it holds

$$\underbrace{\mathrm{Prob}_{p^K}\left\{\zeta^K \leq \ell\right\}}_{\pi_{K,-}[p]}\exp\{-\gamma\ell\} \leq \mathbf{E}_{p^K}\left\{\exp\{-\gamma\zeta^K\}\right\} = \left[\sum_{i=1}^d p_i \exp\left\{-\frac{1}{K}\gamma\rho_i\right\}\right]^K,$$

whence

$$\ln(\pi_{K,-}[p]) \leq K\ln\left(\sum_{i=1}^d p_i \exp\left\{-\frac{1}{K}\gamma\rho_i\right\}\right) + \gamma\ell,$$

implying, via substitution $\gamma = \mu K$, that

$$\forall \mu \geq 0 : \ln(\pi_{K,-}[p]) \leq K\psi_-(\mu,p),\ \psi_-(\mu,p) = \ln\left(\sum_{i=1}^d p_i \exp\{-\mu\rho_i\}\right) + \mu\ell.$$

Similarly, setting $\pi_{K,+}[p] = \mathrm{Prob}_{p^K}\left\{\zeta^K \geq \ell\right\}$, we get

$$\forall \nu \geq 0 : \ln(\pi_{K,+}[p]) \leq K\psi_+(\nu,p),\ \psi_+(\nu,p) = \ln\left(\sum_{i=1}^d p_i \exp\{\nu\rho_i\}\right) - \nu\ell.$$

Now comes the exercise:

1. Extract from the above observations that

$$\mathrm{Risk}(\mathcal{T}^{K,\ell}|H_1,H_2) \leq \exp\{K\varkappa\},\ \varkappa = \max\left[\max_{p\in B}\inf_{\mu\geq 0}\psi_-(\mu,p), \max_{q\in W}\inf_{\nu\geq 0}\psi_+(\nu,q)\right],$$

   where $\mathcal{T}^{K,\ell}$ is the K-observation test which accepts the hypothesis $H_1 : p \in B$ when $\zeta^K \geq \ell$ and accepts the hypothesis $H_2 : p \in W$ otherwise.
2. Verify that $\psi_-(\mu,p)$ is convex in μ and concave in p, and similarly for $\psi_+(\nu,q)$, so that

$$\max_{p\in B}\inf_{\mu\geq 0}\psi_-(\mu,p) = \inf_{\mu\geq 0}\max_{p\in B}\psi_-(\mu,p),\ \max_{q\in W}\inf_{\nu\geq 0}\psi_+(\nu,q) = \inf_{\nu\geq 0}\max_{q\in W}\psi_+(\nu,q).$$

   Thus, computing $\varkappa$ reduces to minimizing on the nonnegative ray the convex functions $\phi_-(\mu) = \max_{p\in B}\psi_+(\mu,p)$ and $\phi_+(\nu) = \max_{q\in W}\psi_+(\nu,q)$.
3. Prove that when $\ell = \frac{1}{2}[\alpha+\beta]$, one has

$$\varkappa \leq -\frac{1}{12}\Delta^2,\ \Delta = \alpha - \beta = \|p^* - q^*\|_1.$$

Note that the above test and the quantity $\varkappa$ responsible for the upper bound on its risk depend, as on a parameter, on the "acceptance level" $\ell \in (\beta,\alpha)$. The simplest way to select a reasonable value of ℓ is to minimize $\varkappa$ over an equidistant grid $\Gamma \subset (\beta,\alpha)$, of small cardinality, of values of ℓ.

Solution: All claims in items 1 and 2 are self-evident. Item 3: One can easily verify that $\exp\{s\} \leq 1+s+\frac{3}{4}s^2$ when $s \in [-1,1]$. Thus, taking into account that $|\rho_i| \leq 1$ for all i, when $\mu \in [0,1]$, for every probabilistic (i.e., nonnegative with unit sum of entries) vector p it holds

$$\begin{array}{rcl}\ln\left(\sum_i p_i\exp\{-\mu\rho_i\}\right) & \leq & \ln\left(\sum_i p_i[1-\mu\rho_i+\frac{3}{4}\mu^2\rho_i^2]\right)\\ & \leq & \ln\left(1-\mu p^T\rho+\frac{3}{4}\mu^2\right) \leq -\mu p^T\rho + \frac{3}{4}\mu^2,\end{array}$$

implying that $\psi_-(\mu,p) \leq -\mu p^T\rho + \frac{3}{4}\mu^2 + \mu\ell$ when $\mu \in [0,1]$. This, in turn, implies that

$$\begin{array}{rcl}\max\limits_{p\in B}\inf\limits_{\mu\geq 0}\psi_-(\mu,p) & = & \inf_{\mu\geq 0}\max\limits_{p\in B}\psi_-(\mu,p) \leq \min\limits_{0\leq\mu\leq 1}\max\limits_{p\in B}\left[-\mu p^T\rho + \frac{3}{4}\mu^2 + \mu\ell\right]\\ & \leq & \min\limits_{0\leq\mu\leq 1}\left[-\mu\alpha + \mu\ell + \frac{3}{4}\mu^2\right]\end{array}$$

(recall that $p^T\rho \geq \alpha$ when $p \in B$). The bottom line is that

$$\max_{p\in B}\inf_{\mu\geq 0}\psi_-(\mu,p) \leq \min_{0\leq\mu\leq 1}[-\mu\alpha+\mu\ell+\tfrac{3}{4}\mu^2] = \min_{0\leq\mu\leq 1}\left[-\tfrac{1}{2}\mu\Delta+\tfrac{3}{4}\mu^2\right] = -\tfrac{1}{12}\Delta^2$$

(note that by the origin of α, β it holds $|\alpha|\leq 1$, $|\beta|\leq 1$, whence $0\leq\Delta\leq 2$). "Symmetric" reasoning shows that

$$\max_{q\in W}\inf_{\nu\geq 0}\psi_+(\nu,q) \leq -\tfrac{1}{12}\Delta^2,$$

and the claim follows.

Now, let us consider an alternative way to pass from a "good" single-observation test to its multi-observation version. Our "building block" now is the minimum risk randomized single-observation test,[2] and its multi-observation modification is just the majority version of this building block. Our first observation is that building the minimum risk single-observation test reduces to solving a *convex* optimization problem.

Exercise 2.6. Let, as above, B and W be nonempty nonintersecting closed convex subsets of probabilistic simplex $\mathbf{\Delta}_d$. Show that the problem of finding the best, in terms of its risk, randomized single-observation test deciding on $H_1 : p \in B$ vs. $H_2 : p \in W$ via observation $\omega \sim p$ reduces to solving a convex optimization problem. Write down this problem as an explicit LO program when B and W are polyhedral sets given by polyhedral representations:

$$\begin{array}{rcl} B &=& \{p : \exists u : P_Bp + Q_Bu \leq a_B\},\\ W &=& \{p : \exists u : P_Wp + Q_Wu \leq a_W\}. \end{array}$$

Solution: We can parameterize a randomized single-observation test by a vector $\pi \in B_d = [0,1]^d$; π_ω, $\omega \in \Omega = \{1,...,d\}$, is the probability for the test to accept H_1 given observation ω. Denoting the test associated with parameter vector π by $\mathcal{T}_\pi$, we clearly have

$$\mathrm{Risk}(\mathcal{T}_\pi|H_1,H_2) = \max\left[f_B(\pi) = \max_{p\in B}[\mathbf{e}-\pi]^Tp,\, f_W(\pi) = \max_{q\in W}\pi^Tq\right],$$

where $\mathbf{e}$ is the all-ones vector. Functions f_B and f_W are convex, so that minimizing risk is a convex programming problem.

With B, W given by the above polyhedral representations, we have

$$\begin{array}{rcl} f_B(\pi) &=& \max_{p,u}\left\{[\mathbf{e}-\pi]^Tp : P_Bp+Q_Nu \leq a_B\right\}\\ &=& \min_{\lambda_B}\left\{a_B^T\lambda_B : \lambda_B \geq 0, P_B^T\lambda_B = \mathbf{e}-\pi, Q_B^T\lambda_B = 0\right\} \text{ [LP Duality]} \end{array}$$

and similarly

$$f_W(\pi) = \min_{\lambda_W}\left\{a_W^T\lambda_W : \lambda_W \geq 0, P_W^T\lambda_W = \pi, Q_W^T\lambda_W = 0\right\}.$$

[2]This test can differ from the test built in Exercise 2.4—the latter test is optimal in terms of the sum, rather than the maximum, of its partial risks.

Thus, the risk minimization problem can be set as

$$\min_{t,\pi,\lambda_B,\lambda_W} \left\{ t : \begin{array}{l} \lambda_B \geq 0, P_B^T\lambda_B = \mathbf{e} - \pi, Q_B^T\lambda_B = 0, a_B^T\lambda_B \leq t \\ \lambda_W \geq 0, P_W^T\lambda_W = \pi, Q_W^T\lambda_W = 0, a_W^T\lambda_W \leq t \\ 0 \leq \pi \leq \mathbf{e} \end{array} \right\}.$$

We see that the "ideal building block"—the minimum-risk single-observation test–can be built efficiently. What is at this point unclear, is whether this block is of any use for majority modifications, that is, is the risk of this test $< 1/2$. This is what we need for the majority version of the minimum-risk single-observation test to be consistent.

Exercise 2.7. Extract from Exercise 2.4 that in the situation of this section, denoting by Δ the optimal value in the optimization problem (2.154), one has

1. The risk of any single-observation test, deterministic or randomized, is $\geq \frac{1}{2} - \frac{\Delta}{4}$
2. There exists a single-observation randomized test with risk $\leq \frac{1}{2} - \frac{\Delta}{8}$, and thus the risk of the minimum risk single-observation test given by Exercise 2.6 does not exceed $\frac{1}{2} - \frac{\Delta}{8} < 1/2$ as well.

Pay attention to the fact that $\Delta > 0$ (since, by assumption, B and W do not intersect).

Solution: 1) With p^*, q^* defined in Exercise 2.4, the best possible total risk of testing simple hypothesis $\omega \sim p^*$ vs. simple alternative $\omega \sim q^*$ is

$$\sum_\omega \min[p^*_\omega, q^*_\omega] = \tfrac{1}{2}\sum_\omega [p^*_\omega + q^*_\omega - |p^*_\omega - q^*_\omega|] = 1 - \tfrac{\Delta}{2}.$$

The best possible risk of testing the same simple hypotheses cannot be less than half of the best total risk, and we arrive at the result announced in the first item of the exercise.

2) By the result stated in Exercise 2.4, there exists a vector ρ, $\|\rho\|_\infty \leq 1$, such that

$$\begin{array}{c} \forall (p \in B, q \in W) : \rho^T p \geq \alpha := \rho^T p^*, \ \rho^T q \leq \beta := \rho^T q^*, \\ \alpha - \beta = \rho^T p^* - \rho^T q^* = \Delta. \end{array} \tag{6.2}$$

Let us set

$$\pi = \left(\tfrac{1}{2} - \tfrac{\alpha+\beta}{8}\right)\mathbf{e} + \tfrac{1}{4}\rho,$$

where $\mathbf{e}$ is the all-ones vector. Note that $\|\rho\|_\infty \leq 1$, that is, $-\mathbf{e} \leq \rho \leq \mathbf{e}$, whence $|\alpha| \leq 1$, $|\beta| \leq 1$ and therefore

$$\pi \geq (\tfrac{1}{2} - \tfrac{1}{4})\mathbf{e} - \tfrac{1}{4}\mathbf{e} \geq 0, \ \pi \leq (\tfrac{1}{2} + \tfrac{1}{4})\mathbf{e} + \tfrac{1}{4}\mathbf{e} = \mathbf{e},$$

so that the randomized single-observation test $\mathcal{T}_\pi$ is well-defined. When $p \in B$, we have in view of (6.2)

$$[\mathbf{e} - \pi]^T p = \left[\tfrac{1}{2} + \tfrac{\alpha+\beta}{8}\right]\mathbf{e}^T p - \tfrac{1}{4}\rho^T p \leq \tfrac{1}{2} + \tfrac{\alpha+\beta}{8} - \tfrac{1}{4}\alpha = \tfrac{1}{2} - \tfrac{\alpha-\beta}{8} = \tfrac{1}{2} - \tfrac{\Delta}{8},$$

and when $q \in W$, we have

$$\pi^T q = \left[\tfrac{1}{2} - \tfrac{\alpha+\beta}{8}\right]\mathbf{e}^T q + \tfrac{1}{4}\rho^T q \leq \tfrac{1}{2} - \tfrac{\alpha+\beta}{8} + \tfrac{1}{4}\beta = \tfrac{1}{2} - \tfrac{\Delta}{8}.$$

Taken together, the last two inequalities imply that

$$\mathrm{Risk}(\mathcal{T}_\pi | H_1, H_2) \leq \tfrac{1}{2} - \tfrac{\Delta}{8},$$

as required in the second item of the exercise.
The bottom line is that in the situation of this section, given a target value ϵ of risk and assuming stationary repeated observations are allowed, we have (at least) three options to meet the risk specifications:

1. To start with the optimal, in terms of its total risk, single-observation detector as explained in Exercise 2.4, and then to pass to its multi-observation version built in Exercise 2.5;
2. To use the majority version of the minimum-risk randomized single-observation test built in Exercise 2.6;
3. To use the test based on the minimum risk detector for B, W, as explained in the main body of Chapter 2.

In all cases, we have to specify the number K of observations which guarantees that the risk of the resulting multi-observation test is at most a given target ϵ. A bound on K can be easily obtained by utilizing the results on the risk of a detector-based test in a Discrete o.s. from the main body of Chapter 2 along with risk-related results of Exercises 2.5, 2.6, and 2.7.

Exercise 2.8. Run numerical experimentation to see if one of the three options above always dominates the others (that is, requires smaller sample of observations to ensure the same risk).

Solution: The simplest way to understand what is going on is to consider the case of $d=2$, so that $\mathbf{\Delta}_d$ is just the segment $[e_1,e_2]$ (e_1,e_2 are the basic orths in $\mathbf{R}^2$), to take an n-point equidistant grid $\{r_i=e_1+\lambda_i(e_2-e_1), 1\leq i\leq n\}$ on this segment, with $\lambda_i=\frac{2i-1}{2n}$, and to look at $n(n-1)/2$ pairs of indexes (i,j), $1\leq i<j\leq n$. Such a pair (i,j) specifies two nonoverlapping segments $B=B_{ij}:=\{\lambda e_1+(1-\lambda)e_2: 0\leq\lambda\leq\lambda_i\}$ and $W=W_{ij}:=\{\lambda e_1+(1-\lambda)e_2:\lambda_j\leq\lambda\leq 1\}$. We then compute the sample sizes $K_o=K_d(i,j)$, $K_m=K_m(i,j)$, and $K_d=K_d(i,j)$, as given by the first, the second, and the third of the above options, associated with B_{ij} and W_{ij}, and compare these quantities to each other. Here are the results for $\epsilon=0.01$ and $n=16$:

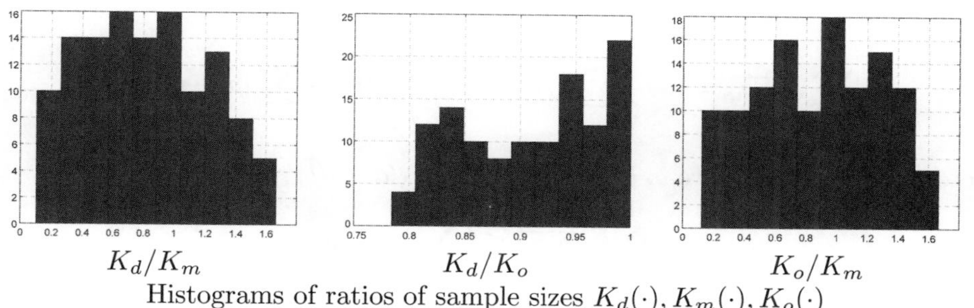

K_d/K_m $\quad$ K_d/K_o $\quad$ K_o/K_m

Histograms of ratios of sample sizes $K_d(\cdot), K_m(\cdot), K_o(\cdot)$

	K_d	K_m	K_o
K_d		0.7391	1.0000
K_m	0.2609		0.3043
K_o	0.0435	0.6957	

Fractions of pairs i,j, $1\leq i<j\leq 15$ with $K_{row}(i,j)\leq K_{column}(i,j)$

We see that the first option—detector-based test—always outperforms the third option, and for $\approx 70\%$ of our pairs i,j outperforms the second one. This should not be too surprising—the test based on the minimum risk detector is provably near-optimal!

Let us now focus on a theoretical comparison of the detector-based test and the majority version of the minimum-risk single-observation test (options 1 and 2 above) in the general situation described at the beginning of Section 2.10.3. Given $\epsilon \in (0,1)$, the corresponding sample sizes K_d and K_m are completely determined by the relevant "measure of closeness" between B and W. Specifically,

- For K_d, the closeness measure is

$$\rho_d(B,W) = 1 - \max_{p\in B, q\in W} \sum_\omega \sqrt{p_\omega q_\omega}; \tag{2.158}$$

  $1-\rho_d(B,W)$ is the minimal risk of a detector for B,W, and for $\rho_d(B,W)$ and ϵ small, we have $K_d \approx \ln(1/\epsilon)/\rho_d(B,W)$ (why?).
- Given ϵ, K_m is fully specified by the minimal risk ρ of simple randomized single-observation test $\mathcal{T}$ deciding on the hypotheses associated with B, W. By Exercise 2.7, we have $\rho = \frac{1}{2} - \delta$, where δ is within absolute constant factor of the optimal value $\Delta = \min_{p\in B, q\in W}\|p-q\|_1$ of (2.154). The risk bound for the K-observation majority version of $\mathcal{T}$ is the probability to get at least $K/2$ heads in K independent tosses of a coin with probability to get heads in a single toss equal to $\rho = 1/2 - \delta$. When ρ is not close to 0 and ϵ is small, the $(1-\epsilon)$-quantile of the number of heads in our K coin tosses is $K\rho + O(1)\sqrt{K\ln(1/\epsilon)} = K/2 - \delta K + O(1)\sqrt{K\ln(1/\epsilon)}$ (why?). K_m is the smallest K for which this quantile is $< K/2$, so that K_m is of the order of $\ln(1/\epsilon)/\delta^2$, or, which is the same, of the order of $\ln(1/\epsilon)/\Delta^2$. We see that the "responsible for K_m" closeness between B and W is

$$\rho_m(B,W) = \Delta^2 = \left[\min_{p\in B, q\in W}\|p-q\|_1\right]^2,$$

  and K_m is of the order of $\ln(1/\epsilon)/\rho_m(B,W)$.

The goal of the next exercise is to compare ρ_b and ρ_m.

Exercise 2.9. Prove that in the situation of this section one has

$$\tfrac{1}{8}\rho_m(B,W) \le \rho_d(B,W) \le \tfrac{1}{2}\sqrt{\rho_m(B,W)}. \tag{2.159}$$

<u>Solution:</u> Let $p\in B$, $q \in W$. Then

$$\sum_\omega \sqrt{p_\omega q_\omega} = 1 - \tfrac{1}{2}\sum_\omega (\sqrt{p_\omega} - \sqrt{q_\omega})^2 \ge 1 - \tfrac{1}{2}\sum_\omega |p_\omega - q_\omega|$$

(note that for nonnegative a, b we always have $|\sqrt{a}-\sqrt{b}| \le \sqrt{|a-b|}$), whence

$$1 - \sum_\omega \sqrt{p_\omega q_\omega} \le \tfrac{1}{2}\sum_\omega |p_\omega - q_\omega|$$

and therefore

$$\rho_d(B,W) = \min_{p\in B, q\in W}\left[1 - \sum_\omega \sqrt{p_\omega q_\omega}\right] \le \tfrac{1}{2}\min_{p\in B, q\in W}\|p-q\|_1 = \tfrac{1}{2}\Delta = \tfrac{1}{2}\sqrt{\rho_m(b,W)}.$$

On the other hand, for $p \in B$, $q\in W$ we have

$$\begin{array}{rcl}
\sum_\omega |p_\omega - q_\omega| &=& \sum_\omega |\sqrt{p_\omega} - \sqrt{q_\omega}|(\sqrt{p_\omega}+\sqrt{q_\omega})\\
&\le& \left(\sum_\omega(\sqrt{p_\omega}-\sqrt{q_\omega})^2\right)^{1/2}\left(\sum_\omega(\sqrt{p_\omega}+\sqrt{q_\omega})^2\right)^{1/2}\\
&\le& \left(\sum_\omega(\sqrt{p_\omega}-\sqrt{q_\omega})^2\right)^{1/2}\left(\sum_\omega[2p_\omega+2q_\omega]\right)^{1/2}\\
&\le& 2\left(\sum_\omega(\sqrt{p_\omega}-\sqrt{q_\omega})^2\right)^{1/2} = 2\sqrt{2}\left(1-\sum_\omega\sqrt{p_\omega q_\omega}\right)^{1/2},
\end{array}$$

whence

$$\forall (p\in B, q\in W): 1-\sum_\omega \sqrt{p_\omega q_\omega} \geq \frac{1}{8}\|p-q\|_1^2.$$

Taking infima of the both sides of the resulting inequality over $p \in B$, $q \in W$, we get $\rho_d(B,W) \geq \frac{1}{8}\rho_m(B,W)$. □

Relation (2.159) suggests that while K_d never is "much larger" than K_m (this we know in advance: in repeated versions of Discrete o.s., a properly built detector-based test provably is nearly optimal), K_m might be much larger than K_d. This indeed is the case:

Exercise 2.10. Given $\delta \in (0,1/2)$, let $B = \{[\delta;0;1-\delta]\}$ and $W = \{[0;\delta;1-\delta]\}$. Verify that in this case, the numbers of observations K_d and K_m resulting in a given risk $\epsilon \ll 1$ of multi-observation tests, as functions of δ are proportional to $1/\delta$ and $1/\delta^2$, respectively. Compare the numbers when $\epsilon = 0.01$ and $\delta \in \{0.01; 0.05; 0.1\}$.

Solution: We clearly have $\rho_d(B,W) = \delta$, see (2.158), implying that

$$K_d = \mathrm{Ceil}(\ln(1/\epsilon)/\ln(1/(1-\delta/2))) \approx 2\ln(1/\epsilon)/\delta.$$

At the same time, the ℓ_1-distance Δ between B and W is 2δ, whence $K_m = O(1)\ln(1/\epsilon)/\delta^2$. Here are the numbers for $\epsilon = 0.01$:

δ	K_d	K_m
0.10	44	557
0.05	90	2201
0.01	459	54315

6.2.4 Miscellaneous exercises

Exercise 2.11. Prove that the conclusion in Proposition 2.18 remains true when the test $\mathcal{T}$ in the premise of the proposition is randomized.

Solution: Assume that simple randomized test $\mathcal{T}$ given observation ω accepts H_1 with probability $\pi(\omega) \in [0,1]$. Let us associate with a probability distribution P on Ω the probability distribution P^+ on $\Omega \times [0,1]$, where P^+ is the distribution of independent pair (ω,θ) with $\omega \sim P$ and θ uniformly distributed on $[0,1]$. Next, let us augment observation $\omega \in \Omega$ sampled from a distribution P on Ω with θ sampled, independently of ω, from the uniform distribution on $[0,1]$. With this association, families of distributions $\mathcal{P}_\chi$, $\chi = 1,2$, become associated with families $\mathcal{P}_\chi^+$ of probability distributions on Ω^+, and hypotheses H_χ, $\chi = 1,2$, on the distribution P underlying observation ω give rise to hypotheses H_χ^+, on the distribution P^+ underlying the augmented observation $\omega^+ = (\omega,\theta)$. The randomized test $\mathcal{T}$ deciding on H_1, H_2 via observation ω gives rise to a deterministic simple test $\mathcal{T}_+$ deciding on H_1^+, H_2^+ via observation $\omega^+ = (\omega,\theta)$ as follows: given (ω,θ), $\mathcal{T}_+$ accepts H_1^+ when $\theta \leq \pi(\omega)$, and accepts H_2^+ otherwise.

It is immediately seen that the risks

$$\begin{array}{rcl}\mathrm{Risk}_1(\mathcal{T}|H_1,H_2) &=& \sup_{P\in\mathcal{P}_1}\int_\Omega (1-\pi(\omega))P(d\omega),\\ \mathrm{Risk}_2(\mathcal{T}|H_1,H_2) &=& \sup_{P\in\mathcal{P}_2}\int_\Omega \pi(\omega)P(d\omega)\end{array}$$

of test $\mathcal{T}$ deciding on H_1, H_2 via observation ω are exactly the same as the risks $\mathrm{Risk}_1(\mathcal{T}_+|H_1^+, H_2^+)$, $\mathrm{Risk}_2(\mathcal{T}_+|H_1^+, H_2^+)$ of test $\mathcal{T}_+$ deciding on hypotheses H_1^+, H_2^+ via observation ω^+. Under the premise of Proposition 2.18, these risks are $\leq \epsilon \leq 1/2$, and since $\mathcal{T}_+$ is a deterministic test, the already proved version of the proposition says that there exists a detector $\phi_+(\omega^+) = \phi_+(\omega, \theta)$ such that

$$\begin{array}{l}\forall (P \in \mathcal{P}_1) : \int_\Omega \left[\int_0^1 \exp\{-\phi_+(\omega,\theta)\}d\theta\right] P(d\omega) \leq \epsilon^+ := 2\sqrt{\epsilon(1-\epsilon)},\\ \forall (P \in \mathcal{P}_2) : \int_\Omega \left[\int_0^1 \exp\{\phi_+(\omega,\theta)\}d\theta\right] P(d\omega) \leq \epsilon^+.\end{array} \tag{6.3}$$

After setting $\phi(\omega) = \int_0^1 \phi_+(\omega,\theta)d\theta$ and invoking Jensen's inequality, (6.3) implies that

$$\forall (P \in \mathcal{P}_1) : \int_\Omega \exp\{-\phi(\omega)\}P(d\omega) \leq \epsilon^+,\ \forall (P \in \mathcal{P}_2) : \int_\Omega \exp\{\phi(\omega)\}P(d\omega) \leq \epsilon^+,$$

and ϕ is the detector yielding simple test $\mathcal{T}_\phi$ with risks $\leq \epsilon^+$; see Proposition 2.14. □

Exercise 2.12. Let $p_1(\omega), p_2(\omega)$ be two positive probability densities, taken w.r.t. a reference measure Π on an observation space Ω, and let $\mathcal{P}_\chi = \{p_\chi\}$, $\chi = 1, 2$. Find the optimal, in terms of its risk, balanced detector for $\mathcal{P}_\chi$, $\chi = 1, 2$.

Solution: An optimal detector is $\phi_*(\omega) = \frac{1}{2}\ln(p_1(\omega)/p_2(\omega))$, and the corresponding risk is the Hellinger affinity $\epsilon_\star := \int_\Omega \sqrt{p_1(\omega)p_2(\omega)}\Pi(d\omega)$ of the distributions. Indeed, the fact that the risk of the detector just defined is $\epsilon_\star$, is evident. On the other hand, if ϕ is a balanced detector with risk ϵ, we have

$$2\epsilon \geq \int_\Omega \underbrace{\left(e^{-\phi(\omega)}p_1(\omega) + e^{\phi(\omega)}p_2(\omega)\right)}_{\geq \sqrt{p_1(\omega)p_2(\omega)}} \Pi(d\omega) \geq 2\epsilon_\star \Rightarrow \epsilon \geq \epsilon_\star. \qquad \square$$

Exercise 2.13. Recall that the exponential distribution with parameter $\mu > 0$ on $\Omega = \mathbf{R}_+$ is the distribution with the density $p_\mu(\omega) = \mu e^{-\mu\omega}$, $\omega \geq 0$. Given positive reals $\alpha < \beta$, consider two families of exponential distributions, $\mathcal{P}_1 = \{p_\mu : 0 < \mu \leq \alpha\}$ and $\mathcal{P}_2 = \{p_\mu : \mu \geq \beta\}$. Build the optimal, in terms of its risk, balanced detector for $\mathcal{P}_1$, $\mathcal{P}_2$. What happens with the risk of the detector you have built when the families $\mathcal{P}_\chi$, $\chi = 1, 2$, are replaced with their convex hulls?

Solution: In light of Exercise 2.12, the risk of a balanced detector cannot be less than the Hellinger affinity of the probability densities $p(\omega) = \alpha e^{-\alpha\omega}$, $q(\omega) = \beta e^{-\beta\omega}$, that is, less than

$$\epsilon_\star := \frac{2\sqrt{\alpha\beta}}{\alpha + \beta}.$$

An educated guess is that the detector $\phi_*(\omega) = \frac{1}{2}\ln(p(\omega)/q(\omega))$ for $\mathcal{P}_1$, $\mathcal{P}_2$ is balanced and has risk $\epsilon_\star$. All we need to verify in order to justify the guess is that

$$\begin{array}{l}\int_\Omega \sqrt{q(\omega)/p(\omega)}p_\mu(\omega)d\omega \leq \epsilon_\star,\ 0 < \mu \leq \alpha,\\ \int_\Omega \sqrt{p(\omega)/q(\omega)}p_\mu(\omega)d\omega \leq \epsilon_\star,\ \mu \geq \beta,\end{array}$$

or, equivalently, that

$$\begin{array}{l}\frac{2\sqrt{\beta/\alpha}\mu}{\beta-\alpha+2\mu} \leq \frac{2\sqrt{\alpha\beta}}{\alpha+\beta},\ 0<\mu\leq\alpha\\ \frac{2\sqrt{\alpha/\beta}\mu}{\alpha-\beta+2\mu} \leq \frac{2\sqrt{\alpha\beta}}{\alpha+\beta},\ \mu\geq\beta.\end{array}$$

The above inequalities are evident due to the fact that when $0<\alpha<\beta$, the function $\frac{\mu}{\beta-\alpha+2\mu}$ is nondecreasing in $\mu\in(0,\alpha]$, and the function $\frac{\mu}{\alpha-\beta+2\mu}$ is nonincreasing in $\mu\in[\beta,\infty)$.

Finally, risks of a detector remain unchanged when the families of distributions in question are replaced with their convex hulls. □

Exercise 2.14. [Follow-up to Exercise 2.13] Assume that the "lifetime" ζ of a lightbulb is a realization of a random variable with exponential distribution (i.e., the density $p_\mu(\zeta)=\mu e^{-\mu\zeta}$, $\zeta\geq 0$; in particular, the expected lifespan of a lightbulb in this model is $1/\mu$).[3] Given a lot of lightbulbs, you should decide whether they were produced under normal conditions (resulting in $\mu\leq\alpha=1$) or under abnormal ones (resulting in $\mu\geq\beta=1.5$). To this end, you can select at random K lightbulbs and test them. How many lightbulbs should you test in order to make a 0.99-reliable conclusion? Answer this question in the situations when the observation ω in a test is

1. the lifespan of a lightbulb (i.e., $\omega\sim p_\mu(\cdot)$)
2. the minimum $\omega=\min[\zeta,\delta]$ of the lifespan $\zeta\sim p_\mu(\cdot)$ and the allowed duration $\delta>0$ of your test (i.e., if the lightbulb you are testing does not "die" on time horizon δ, you terminate the test)
3. $\omega=\chi_{\zeta<\delta}$, that is, $\omega=1$ when $\zeta<\delta$, and $\omega=0$ otherwise; here, as above, $\zeta\sim p_\mu(\cdot)$ is the random lifespan of a lightbulb, and $\delta>0$ is the allowed test duration (i.e., you observe whether or not a lightbulb "dies" on time horizon δ, but do not register the lifespan when it is $<\delta$).

Consider the values $0.25, 0.5, 1, 2, 4$ of δ.

Solution: In item 1, we deal with testing hypotheses via stationary K-repeated observations in the situation where the hypotheses admit a balanced detector with risk $\epsilon_\star=\frac{2\sqrt{\alpha\beta}}{\alpha+\beta}=\frac{2\sqrt{1\cdot 1.5}}{1+1.5}\approx 0.9798$. This detector, for every K, induces a test $\mathcal{T}^K$ deciding on our hypotheses via K-repeated observation with risk $\epsilon_\star^K$. The smallest K for which the latter quantity is ≤ 0.01 (we want 0.99-reliable inference) is $K=226$.

In item 2, the observation space is the segment $\Delta=[0,\delta]$, and a natural reference measure is the Lebesgue measure on $[0,\delta)$ augmented by unit mass assigned to the point δ, to account for the case that the lightbulb you are testing did not "die" on the testing time horizon (that is, your observation $\min[\delta,\omega]$ is δ). The density of observation w.r.t. the reference measure just defined is

$$\widehat{p}_\mu(\omega)=\left\{\begin{array}{ll}\mu e^{-\mu\omega}, & 0\leq\omega<\delta\\ e^{-\mu\delta}, & \omega=\delta\end{array}\right.$$

[3] In Reliability, probability distribution of the lifespan ζ of an organism or a technical device is characterized by the *failure rate* $\lambda(t)=\lim_{\delta\to+0}\frac{\mathrm{Prob}\{t\leq\zeta\leq t+\delta\}}{\delta\cdot\mathrm{Prob}\{\zeta\geq t\}}$ (so that for small δ, $\lambda(t)\delta$ is the conditional probability to "die" in the time interval $[t,t+\delta]$ provided the organism or device is still 'alive" at time t). The exponential distribution corresponds to the case of failure rate independent of t; in applications, this indeed is often the case except for "very small" and "very large" values of t.

The hypotheses to be decided upon state that the distribution of observation belongs to $\widehat{\mathcal{P}}_\chi$, $\chi = 1, 2$, with

$$\widehat{\mathcal{P}}_1 = \{\widehat{p}_\mu : 0 < \mu \leq \alpha = 1\}, \ \widehat{\mathcal{P}}_2 = \{\widehat{p}_\mu : \mu \geq \beta = 1.5\}.$$

There is numerical evidence that specifying $\phi_*(\omega)$ as the optimal detector for the pair of "extreme" distributions from our families, that is,

$$\phi_*(\omega) = \tfrac{1}{2}\ln(\widehat{p}_\alpha(\omega)/\widehat{p}_\beta(\omega)),$$

we get a detector with risk on the families $\widehat{\mathcal{P}}_1$, $\widehat{\mathcal{P}}_2$, equal to the risk of the detector on the pair of distributions $\widehat{p}_\alpha$, $\widehat{p}_\beta$:

$$\begin{array}{l} \forall \mu \in (0,\alpha) : \int_\Delta e^{-\phi_*(\omega)}\widehat{p}_\mu(\omega)\Pi(d\omega) \leq \epsilon_\star := \int_\Delta e^{-\phi_*(\omega)}\widehat{p}_\alpha(\omega)\Pi(d\omega) \\ \qquad\qquad\qquad\qquad\qquad\qquad\qquad\qquad = \int_\Delta \sqrt{\widehat{p}_\alpha(\omega)\widehat{p}_\beta(\omega)}\Pi(d\omega), \\ \forall \mu \geq \beta : \int_\Delta e^{\phi_*(\omega)}\widehat{p}_\mu(\omega)\Pi(d\omega) \leq \epsilon_\star. \end{array}$$

A reasoning similar to that used in the basic setting yields

$$K = \rfloor \ln(1/0.01)/\ln(1/\epsilon_\star)\lfloor.$$

In item 3, our observation takes value 0 with probability $1 - e^{-\mu\delta}$ and value 1 with complementary probability, and our testing problem becomes to decide, via stationary K-repeated observations in Discrete o.s. with $\Omega = \{0, 1\}$, on two families of distributions

$$\widehat{\mathcal{P}}_1 = \{[p; 1-p] : 1 > p \geq u := e^{-\alpha\delta}\}, \ \widehat{\mathcal{P}}_2 = \{[p; 1-p] : 0 < p \leq v := e^{-\beta\delta}\}.$$

The risk of the optimal balanced single-observation test is

$$\epsilon_\star = \sqrt{(1-u)(1-v)} + \sqrt{uv},$$

and the number of observations allowing for 0.99-reliable inference, as yielded by out theory, is $\rfloor \ln(1/0.01)/\ln(1/\epsilon_\star)\lfloor$.

Here are the values of K for items 2 and 3:

observation	$\delta = 0.25$	$\delta = 0.5$	$\delta = 1$	$\delta = 2$	$\delta = 4$
$\min[\zeta, \delta]$	848	489	318	247	230
$\chi_{\zeta<\delta}$	854	504	360	399	1246

Note that when $\delta = 2$ (twice the expected lifetime of a "normal" lightbulb), the observations from item 2 are nearly as informative as those from item 1 (the resulting values of K are 247 and 226, respectively). We see also that, in full accordance with common sense, the observations from item 2 are nearly as informative as those from item 3 when δ is small as compared to the expected lifetime of a lightbulb.

Exercise 2.15. [Follow-up to Exercise 2.14] In the situation of Exercise 2.14, build a sequential test for deciding on null hypothesis "the lifespan of a lightbulb from a given lot is $\zeta \sim p_\mu(\cdot)$ with $\mu \leq 1$" (recall that $p_\mu(z)$ is the exponential density $\mu e^{-\mu z}$ on the ray $\{z \geq 0\}$) vs. the alternative "the lifespan is $\zeta \sim p_\mu(\cdot)$ with $\mu > 1$." In this test, you can select a number K of lightbulbs from the lot, switch them on at time 0 and record the actual lifetimes of the lightbulbs you are testing. As a result at the end of (any) observation interval $\Delta = [0, \delta]$, you observe K independent realizations of r.v. $\min[\zeta, \delta]$, where $\zeta \sim p_\mu(\cdot)$ with some unknown μ. In your sequential test, you are welcome to make conclusions at the endpoints $\delta_1 < \delta_2 < ... < \delta_S$ of several observation intervals.

Note: We deliberately skip details of problem's setting; how you decide on these missing details, is part of your solution to the exercise.

Solution: The construction we are about to describe has several design parameters, specifically

- number K of lightbulbs we test,
- number S and S times $0 < \delta_1 < \delta_2 < ... < \delta_S$ at which we observe what happened so far with the lightbulbs we are testing to decide whether the null hypothesis should be rejected.

We assume we are given *false alarm probability* $\epsilon \in (0,1)$; the probability that our inference rejects the null hypothesis when it is true (i.e., when the actual parameter μ of the lot we are testing is ≤ 1) must be at most ϵ. Since we decide on the hypotheses S times, at time instants $\delta_1, ..., \delta_S$, the simplest way to control the false alarm probability is to split the corresponding bound ϵ into S parts, $\epsilon_1, ..., \epsilon_S$,

$$0 < \epsilon_s,\ s \leq S, \ \text{ with } \sum_{s=1}^{S} \epsilon_s = \epsilon,$$

and to ensure that when deciding on the hypotheses at time instant $\delta = \delta_s$, the probability of false alarm is $\leq \epsilon_s$, and this is so for all $s \leq S$.

Now, given $s \leq S$, we can find the smallest $\beta = \beta_s > 1$ such that the two hypotheses on the distribution of the lifetime of a lightbulb from the lot, one, $H^s_{\leq 1}$, stating that this lifetime is $\sim p_\mu(\cdot)$ with $\mu \leq 1$, and another, $H^s_{\geq\beta}$, stating that the lifetime is $\sim p_\mu(\cdot)$ with $\mu \geq \beta$, admit a detector with the risk $\rho := \sqrt{\epsilon_s \epsilon}$, the observations being what we see at time δ_s, that is, the collection

$$\omega^s = \{\omega_{ks} = \min[\zeta_k, \delta_s] : 1 \leq k \leq K\},$$

where $\zeta_1, ..., \zeta_K$ are drawn independently of each other "by nature" from the "true" lifetime distribution $p_\mu(\cdot)$. Invoking the result of Exercise 2.14, $\beta = \beta_s$ is the minimal solution of the equation

$$\left[\frac{2\sqrt{\beta}}{1+\beta}\left[1 - \mathrm{e}^{-\frac{1+\beta}{2}\delta_s}\right] + \mathrm{e}^{-\frac{1+\beta}{2}\delta_s}\right]^K = \sqrt{\epsilon\epsilon_s} \tag{6.4}$$

on the ray $\beta > 1$. The quantity β_s induces the single-observation balanced detector applicable at time instant δ_s, namely,

$$\phi_s(\omega) = \left\{ \begin{array}{ll} \frac{1}{2}\left[\ln(1/\beta_s) + (\beta_s - 1)\omega\right], & \omega < \delta_s \\ \frac{1}{2}(\beta_s - 1)\delta_s, & \omega = \delta_s \end{array} \right. : [0, \delta_s] \to \mathbf{R}$$

and its K-repeated version

$$\phi_s^{(K)}(\omega^s) = \sum_{k=1}^{K} \phi_s(\omega_{ks}).$$

By construction, the risk of the detector $\phi_s^{(K)}$ on the pair of hypotheses $H^s_{\leq 1}$, $H^s_{\geq \beta_s}$, the observation being $\omega^s = \{\omega_{ks} : 1 \leq k \leq K\}$, is equal to $\sqrt{\epsilon\epsilon_s}$. As a result, the test $\mathcal{T}_s$ which, given observation ω^s,

— accepts $H^s_{\leq 1}$ and rejects $H^s_{\geq \beta_s}$, when $\phi_s^{(K)}(\omega^s) + \frac{1}{2}\ln(\epsilon/\epsilon_s) \geq 0$,
— accepts $H^s_{\geq \beta_s}$ and rejects $H^s_{\leq 1}$, when $\phi_s^{(K)}(\omega^s) + \frac{1}{2}\ln(\epsilon/\epsilon_s) < 0$

possesses the following properties (see Section 2.3.2.2):

- when $H^s_{\leq 1}$ is true, the probability for $\mathcal{T}_s$ to reject the hypothesis is at most ϵ_s;
- when $H^{\bar{s}}_{\geq \beta_s}$ is true, the probability for $\mathcal{T}_s$ to reject the hypothesis is at most ϵ.

Now consider sequential test $\mathcal{T}$ where the inferences are made at time instants $\delta_1, ..., \delta_S$ as follows. At time δ_s, we apply $\mathcal{T}_s$ to observations ω^s (this is what we see at time instant δ_s). If the result is in acceptance of $H^s_{\leq 1}$, we say that at time δ_s the observations support the null hypothesis "the true lifetime parameter μ for our lot is ≤ 1" and pass to time δ_{s+1} (when $s < S$) or terminate (when $s = S$). In the opposite case, that is, when $\mathcal{T}_s$ accepts $H^s_{\geq \beta_s}$, we claim that the alternative to null hypothesis, that is, the hypothesis "the true lifetime parameter μ for our lot is > 1" takes place, and terminate further testing of our K lightbulbs.

By construction, it is immediately seen that

- for $\mathcal{T}$, the probability of false alarm (probability to reject the null hypothesis in one of our S decision making acts) is at most ϵ;
- on the other hand, for every s, if the true lifetime parameter μ is "large," namely, $\mu \geq \beta_s$, then the probability for $\mathcal{T}$ to terminate at the s-th decision making act *or earlier* with the conclusion that the alternative to null hypothesis takes place, is at least $1 - \epsilon$.

Here are numerical illustrations for several setups; we used $\epsilon = 0.01$ and $\epsilon_s = \epsilon/S$, $1 \leq s \leq S = 4$.

s	δ_s	β_s
1	0.100	1.786
2	0.200	1.549
3	0.400	1.398
4	0.800	1.305

$K = 1000$

s	δ_s	β_s
1	0.100	2.183
2	0.200	1.817
3	0.400	1.587
4	0.800	1.448

$K = 500$

s	δ_s	β_s
1	0.100	2.620
2	0.200	2.106
3	0.400	1.790
4	0.800	1.601

$K = 300$

s	δ_s	β_s
1	0.100	2.620
2	0.200	4.345
3	0.400	3.224
4	0.800	2.183

$K = 100$

Exercise 2.17. Work out the following extension of the Opinion Poll Design problem. You are given two finite sets, $\Omega_1 = \{1, ..., I\}$ and $\Omega_2 = \{1, ..., M\}$, along with L nonempty closed convex subsets Y_ℓ of the set

$$\mathbf{\Delta}_{IM} = \left\{ [y_{im} > 0]_{i,m} : \sum_{i=1}^{I} \sum_{m=1}^{M} y_{im} = 1 \right\}$$

of all nonvanishing probability distributions on $\Omega = \Omega_1 \times \Omega_2 = \{(i, m) : 1 \leq i \leq I, 1 \leq m \leq M\}$. Sets Y_ℓ are such that all distributions from Y_ℓ have a common marginal distribution $\theta^\ell > 0$ of i:

$$\sum_{m=1}^{M} y_{im} = \theta_i^\ell, \; 1 \leq i \leq I, \; \forall y \in Y_\ell, \; 1 \leq \ell \leq L.$$

Your observations $\omega_1, \omega_2, ...$ are sampled, independently of each other, from a distribution partly selected "by nature," and partly by you. Specifically, nature selects $\ell \leq L$ and a distribution $y \in Y_\ell$, and you select a positive I-dimensional probabilistic vector q from a given convex compact subset $\mathcal{Q}$ of the positive part of the I-dimensional probabilistic simplex. Let $y_{|i}$ be the conditional distribution of $m \in \Omega_2$, i being given, induced by y, so that $y_{|i}$ is the M-dimensional probabilistic vector with entries

$$[y_{|i}]_m = \frac{y_{im}}{\sum_{\mu \leq M} y_{i\mu}} = \frac{y_{im}}{\theta_i^\ell}.$$

In order to generate $\omega_t = (i_t, m_t) \in \Omega$, you draw i_t at random from the distribution q, and then nature draws m_t at random from the distribution $y_{|i_t}$.

Given closeness relation $\mathcal{C}$, your goal is to decide, up to closeness $\mathcal{C}$, on the hypotheses $H_1, ..., H_L$, with H_ℓ stating that the distribution y selected by nature belongs to Y_ℓ. Given "observation budget" (a number K of observations ω_k you can use), you want to find a probabilistic vector q which results in the test with as small a $\mathcal{C}$-risk as possible. Pose this Measurement Design problem as an efficiently solvable convex optimization problem.

Solution: Observation ω induced by your selection of q and the choice of y by nature takes value $(i, m) \in \Omega$ with probability

$$q_i y_{im} / \theta_i^\ell.$$

When y runs through Y_ℓ, the "signal" $x(y) \in \mathbf{R}^{I \times M} : x_{im}(y) = y_{im}/\theta_i^\ell$ runs through the convex compact subset

$$X_\ell = \{x \in \mathbf{R}^{I \times M} : \exists y \in Y_\ell : x_{im} = y_{im}/\theta_i^\ell, 1 \leq i \leq I, 1 \leq m \leq M\}$$

of the convex compact set

$$\mathcal{X} = \{x \in \mathbf{R}_+^{I \times M} : \sum_{m=1}^{M} x_{im} = 1 \, \forall i\},$$

and you lose nothing by assuming that H_ℓ states that nature selects ℓ and signal $x \in X_\ell$, you select $q \in \mathcal{Q}$, and your observations ω_t are drawn at random and independently of each other from the probability distribution

$$\mu = A_q x : \mu_{im} = [A_q x]_{im} \equiv q_i x_{im}$$

on Ω; and of course $A_q x$ is a probability distribution on Ω whenever $x \in \mathcal{X}$ and q is a probabilistic vector. Consequently, you are in the Simple case (2.102) of Discrete o.s., and (2.101) is convex and tractable.

Exercise 2.18. [probabilities of deviations from the mean] The goal of what follows is to present the most straightforward application of simple families of distributions—bounds on probabilities of deviations of random vectors from their means. Let $\mathcal{H} \subset \Omega = \mathbf{R}^d, \mathcal{M}, \Phi$ be regular data such that $0 \in \operatorname{int} \mathcal{H}$, $\mathcal{M}$ is compact, $\Phi(0; \mu) = 0 \, \forall \mu \in \mathcal{M}$, and $\Phi(h; \mu)$ is differentiable at $h = 0$ for every $\mu \in \mathcal{M}$. Let, further, $\bar{P} \in \mathcal{S}[\mathcal{H}, \mathcal{M}, \Phi]$ and let $\bar{\mu} \in \mathcal{M}$ be a parameter of $\bar{P}$. Prove that

1. $\bar{P}$ possesses expectation $e[\bar{P}]$, and

$$e[\bar{P}] = \nabla_h \Phi(0; \bar{\mu}).$$

Solution: Denoting by ω_i the coordinates of ω, and taking into account that $0 \in \operatorname{int} \mathcal{H}$, we see that $\int_{\mathbf{R}^d} [\mathrm{e}^{t\omega_i} + \mathrm{e}^{-t\omega_i}] \bar{P}(d\omega) < \infty$ for all i and all small positive t, whence $e[\bar{P}]$ exists. Now, for every $h \in \mathbf{R}^d$ and all small positive t we have

$$\int_{\mathbf{R}^d} [1 \pm th^T\omega]\bar{P}(d\omega) \leq \int_{\mathbf{R}^d} \mathrm{e}^{\pm th^T\omega}\bar{P}(d\omega) \leq \exp\{\Phi(\pm th;\bar{\mu})\},$$

so that

$$\pm \int_{\mathbf{R}^d} h^T\omega\bar{P}(d\omega) \leq t^{-1}\left[\mathrm{e}^{\Phi(\pm th;\bar{\mu})} - 1\right] \to_{t\to+0} \pm h^T\nabla_h\Phi(0;\bar{\mu}).$$

The resulting relation holds true for all h, implying that $\int_{\mathbf{R}^d}\omega\bar{P}(d\omega) = \nabla_h\Phi(0;\bar{\mu})$.

2. For every linear form $e^T\omega$ on Ω it holds

$$\begin{array}{rcl}\pi & := & \bar{P}\{\omega : e^T(\omega - e[\bar{P}]) \geq 1\} \\ & \leq & \exp\left\{\inf\limits_{t\geq 0: te\in\mathcal{H}}\left[\Phi(te;\bar{\mu}) - te^T\nabla_h\Phi(0;\bar{\mu}) - t\right]\right\}.\end{array} \tag{2.160}$$

What are the consequences of (2.160) for sub-Gaussian distributions?

Solution: Whenever $t \geq 0$ is such that $te \in \mathcal{H}$, we have

$$\begin{array}{rcl}\ln(\pi) & \leq & \ln\left(\int_{\mathbf{R}^d}\mathrm{e}^{te^T(\omega - e[\bar{P}])-t}\bar{P}(d\omega)\right) \\ & \leq & \Phi(te,\bar{\mu}) - te^Te[\bar{P}] - t = \Phi(te,\bar{\mu}) - te^T\nabla_h\Phi(0;\bar{\mu}) - t,\end{array}$$

where the concluding equality is due to item 1, and (2.160) follows.

Specifying $\mathcal{H} = \mathbf{R}^d$, $\mathcal{M} = \mathbf{R}^d \times \mathbf{S}^d$, $\Phi(h;\mu,\theta) = h^T\mu + \frac{1}{2}h^T\theta h$, the family $\mathcal{S}[\mathcal{H},\mathcal{M},\Phi]$ is the family of sub-Gaussian distributions on $\mathbf{R}^d$, and for a distribution $\bar{P}$ of this type, a parameter $\bar{\mu}$ w.r.t. $\mathcal{S}$ is a pair $e[\bar{P}], \bar{\theta}$ of sub-Gaussianity parameters of $\bar{P}$. Consequently, (2.160) reads

$$\begin{array}{rcl}\mathrm{Prob}_{\omega\sim\bar{P}}\{e^T(\omega - e[\bar{p}]) \geq 1\} & \leq & \exp\left\{\inf_{t\geq 0}\left\{te^Te[\bar{P}] + \frac{1}{2}t^2e^T\bar{\theta}e - te^Te[\bar{p}] - t\right\}\right\} \\ & = & \exp\{-\frac{1}{2e^T\bar{\theta}e}\},\end{array}$$

or, in a more convenient form,

$$\mathrm{Prob}_{\omega\sim\bar{P}}\{e^T(\omega - e[\bar{P}]) > s\sqrt{e^T\bar{\theta}e}\} \leq \exp\{-s^2/2\}\ \forall s \geq 0.$$

□

Exercise 2.19. [testing convex hypotheses on mixtures] Consider the situation as follows. For given positive integers K, L and for $\chi = 1, 2$, given are

- nonempty convex compact *signal sets* $U_\chi \subset \mathbf{R}^{n_\chi}$
- regular data $\mathcal{H}^\chi_{k\ell} \subset \mathbf{R}^{d_k}, \mathcal{M}^\chi_{k\ell}, \Phi^\chi_{k\ell}$, and affine mappings

$$u_\chi \mapsto A^\chi_{k\ell}[u_\chi; 1] : \mathbf{R}^{n_\chi} \to \mathbf{R}^{d_k}$$

  such that

$$u_\chi \in U_\chi \Rightarrow A^\chi_{k\ell}[u_\chi; 1] \in \mathcal{M}^\chi_{k\ell},$$

  $1 \leq k \leq K$, $1 \leq \ell \leq L$,
- probabilistic vectors $\mu^k = [\mu^k_1; ...; \mu^k_L]$, $1 \leq k \leq K$.

We can associate with the outlined data families of probability distributions $\mathcal{P}_\chi$ on the observation space $\Omega = \mathbf{R}^{d_1} \times ... \times \mathbf{R}^{d_K}$ as follows. For $\chi = 1, 2$, $\mathcal{P}_\chi$ is comprised of all probability distributions P of random vectors $\omega^K = [\omega_1; ...; \omega_K] \in \Omega$ generated as follows:
We select

- a signal $u_\chi \in U_\chi$,
- a collection of probability distributions $P_{k\ell} \in \mathcal{S}[\mathcal{H}^\chi_{k\ell}, \mathcal{M}^\chi_{k\ell}, \Phi^\chi_{k\ell}]$, $1 \leq k \leq K$, $1 \leq \ell \leq L$, in such a way that $A^\chi_{k\ell}[u_\chi; 1]$ is a parameter of $P_{k\ell}$:

$$\forall h \in \mathcal{H}^\chi_{k\ell} : \ln\left(\mathbf{E}_{\omega_k \sim P_{k\ell}}\{e^{h^T\omega_k}\}\right) \leq \Phi^\chi_{k\ell}(h_k; A^\chi_{k\ell}[u_\chi; 1]);$$

- we generate the components ω_k, $k = 1, ..., K$, *independently across* k, from μ^k-mixture $\Pi[\{P_{k\ell}\}_{\ell=1}^L, \mu]$ of distributions $P_{k\ell}$, $\ell = 1, ..., L$, that is, draw at random, from distribution μ^k on $\{1, ..., L\}$, index ℓ, and then draw ω_k from the distribution $P_{k\ell}$.

Prove that when setting

$$\begin{array}{rcl}\mathcal{H}_\chi & = & \{h = [h_1; ...; h_K] \in \mathbf{R}^{d=d_1+...+d_K} : h_k \in \bigcap_{\ell=1}^L \mathcal{H}^\chi_{k\ell}, 1 \leq k \leq K\},\\ \mathcal{M}_\chi & = & \{0\} \subset \mathbf{R},\\ \Phi_\chi(h; \mu) & = & \sum_{k=1}^K \ln\left(\sum_{\ell=1}^L \mu_\ell^k \exp\left\{\max\limits_{u_\chi \in U_\chi} \Phi^\chi_{k\ell}(h_k; A^\chi_{k\ell}[u_\chi; 1])\right\}\right) : \mathcal{H}_\chi \times \mathcal{M}_\chi \to \mathbf{R},\end{array}$$

we obtain the regular data such that

$$\mathcal{P}_\chi \subset \mathcal{S}[\mathcal{H}_\chi, \mathcal{M}_\chi, \Phi_\chi].$$

Explain how to use this observation to compute via Convex Programming an affine detector and its risk for the families of distributions $\mathcal{P}_1$ and $\mathcal{P}_2$.

Solution: The claim is an immediate consequence of the "calculus rules" in items 2.8.1.3.B ("mixtures") and 2.8.1.3.A ("direct products"). Affine detectors in question are of the form

$$\phi_h(\omega^K) = \sum_{k=1}^K h_k^T\omega_k + \tfrac{1}{2}\left[\Phi_1(-h; 0) - \Phi_2(h; 0)\right]$$

with $h \in \mathcal{H}_1 \cap \mathcal{H}_2$, and

$$\mathrm{Risk}(\phi_h|\mathcal{P}_1, \mathcal{P}_2) \leq \tfrac{1}{2}[\Phi_1(h) + \Phi_2(h)],$$

see Proposition 2.39.

Exercise 2.20 [mixture of sub-Gaussian distributions] Let P_ℓ be sub-Gaussian distributions on $\mathbf{R}^d$ with sub-Gaussianity parameters θ_ℓ, Θ, $1 \leq \ell \leq L$, with a common Θ-parameter, and let $\nu = [\nu_1; ...; \nu_L]$ be a probabilistic vector. Consider the ν-mixture $P = \Pi[P^L, \nu]$ of distributions P_ℓ, so that $\omega \sim P$ is generated as follows: we draw at random from distribution ν index ℓ and then draw ω at random from distribution P_ℓ. Prove that P is sub-Gaussian with sub-Gaussianity parameters $\bar\theta = \sum_\ell \nu_\ell\theta_\ell$ and $\bar\Theta$, with (any) $\bar\Theta$ chosen to satisfy

$$\bar\Theta \succeq \Theta + \frac{6}{5}[\theta_\ell - \bar\theta][\theta_\ell - \bar\theta]^T\ \forall \ell,$$

in particular, according to any of the following rules:

1. $\bar\Theta = \Theta + \left(\frac{6}{5}\max_\ell \|\theta_\ell - \bar\theta\|_2^2\right) I_d$,
2. $\bar\Theta = \Theta + \frac{6}{5}\sum_\ell(\theta_\ell - \bar\theta)(\theta_\ell - \bar\theta)^T$,
3. $\bar\Theta = \Theta + \frac{6}{5}\sum_\ell \theta_\ell\theta_\ell^T$, provided that $\nu_1 = ... = \nu_L = 1/L$,

Solution: Setting $\delta_\ell = \theta_\ell - \bar{\theta}$, we have

$$\begin{array}{rcl}
\ln\left(\mathbf{E}_{\omega\sim P}\{\mathrm{e}^{h^T\omega}\}\right) & = & \ln\left(\sum_\ell \nu_\ell \mathbf{E}_{\omega\sim P_\ell}\{\mathrm{e}^{h^T\omega}\}\right) \leq \ln\left(\sum_\ell \nu_\ell \mathrm{e}^{h^T\theta_\ell+\frac{1}{2}h^T\Theta h}\right)\\
& = & h^T\bar{\theta} + \frac{1}{2}h^T\Theta h + \ln\left(\sum_\ell \nu_\ell \mathrm{e}^{h^T\delta_\ell}\right)\\
& \leq & h^T\bar{\theta} + \frac{1}{2}h^T\Theta h + \ln\left(\sum_\ell \nu_\ell\left[h^T\delta_\ell + \mathrm{e}^{\frac{3}{5}(h^T\delta_\ell)^2}\right]\right)\\
& & [\text{due to } \exp\{a\} \leq a + \exp\{\frac{3}{5}a^2\} \text{ for all } a]\\
& = & h^T\bar{\theta} + \frac{1}{2}h^T\Theta h + \ln\left(\sum_\ell \nu_\ell \mathrm{e}^{\frac{3}{5}(h^T\delta_\ell)^2}\right)\\
& & [\text{due to } \sum_\ell \nu_\ell\delta_\ell = 0]\\
& \leq & h^T\bar{\theta} + \frac{1}{2}h^T\Theta h + \frac{3}{5}\max_\ell(h^T\delta_\ell)^2\\
& = & h^T\bar{\theta} + \frac{1}{2}h^T\Theta h + \frac{1}{2}\left[\frac{6}{5}\max_\ell(h^T\delta_\ell)^2\right].
\end{array}$$

□

Exercise 2.21. The goal of this exercise is to give a simple sufficient condition for quadratic lifting "to work" in the Gaussian case. Namely, let $\mathcal{A}_\chi$, U_χ, $\mathcal{V}_\chi$, $\mathcal{G}_\chi$, $\chi = 1,2$, be as in Section 2.9.3, with the only difference that now we do *not* assume the compact sets $\mathcal{U}_\chi$ to be convex, and let $\mathcal{Z}_\chi$ be convex compact subsets of the sets $\mathcal{Z}^{n_\chi}$, see item i.2. in Proposition 2.43, such that

$$[u_\chi; 1][u_\chi; 1]^T \in \mathcal{Z}_\chi \;\forall u_\chi \in U_\chi,\; \chi = 1,2.$$

Augmenting the above data with $\Theta_\chi^{(*)}$, δ_χ such that $\mathcal{V} = \mathcal{V}_\chi$, $\Theta_* = \Theta_*^{(\chi)}$, $\delta = \delta_\chi$ satisfy (2.129), $\chi = 1,2$, and invoking Proposition 2.43.ii, we get at our disposal a quadratic detector ϕ_{lift} such that

$$\text{Risk}[\phi_{\text{lift}}|\mathcal{G}_1, \mathcal{G}_2] \leq \exp\{\text{SadVal}_{\text{lift}}\},$$

with $\text{SadVal}_{\text{lift}}$ given by (2.134). A natural question is, when $\text{SadVal}_{\text{lift}}$ is negative, meaning that our quadratic detector indeed "is working," that is, its risk is < 1. When repeated observations are allowed, tests based upon this detector are consistent—they are able to decide on the hypotheses $H_\chi : P \in \mathcal{G}_\chi$, $\chi = 1,2$, on the distribution of observation $\zeta \sim P$ with any desired risk $\epsilon \in (0,1)$? With our computation-oriented ideology, this is not too important a question, since we can answer it via efficient computation. This being said, there is no harm in a "theoretical" answer which could provide us with an additional insight. The goal of the exercise is to justify a simple result on the subject. Here is the exercise:

In the situation in the question, assume that $\mathcal{V}_1 = \mathcal{V}_2 = \{\Theta_*\}$, which allows us to set $\Theta_*^{(\chi)} = \Theta_*$, $\delta_\chi = 0$, $\chi = 1,2$. Prove that in this case a necessary and sufficient condition for $\text{SadVal}_{\text{lift}}$ to be negative is that the convex compact sets

$$\mathcal{U}_\chi = \{B_\chi Z B_\chi^T : Z \in \mathcal{Z}_\chi\} \subset \mathbf{S}_+^{d+1},\; \chi = 1,2$$

do not intersect with each other.

Solution: Substituting the descriptions (2.130) of $\Phi_{\mathcal{A}_\chi,\mathcal{Z}_\chi}$ into (2.134), under the exercise premise we get

$$\begin{array}{rcl}
\mathrm{SadVal_{lift}} &=& \min\limits_{(h,H)\in\mathcal{H}_1\cap\mathcal{H}_2}\Big[\underbrace{-\tfrac{1}{2}\left[\ln\mathrm{Det}(I_d+\Theta_*^{1/2}H\Theta_*^{1/2})+\ln\mathrm{Det}(I_d-\Theta_*^{1/2}H\Theta_*^{1/2})\right]}_{A(H)\geq 0}\\
&&+\max\limits_{W_1\in\mathcal{U}_1}\mathrm{Tr}\big(\underbrace{W_1}_{\succeq 0}\big[-\left[\begin{array}{c|c}H&h\\ \hline h^T&\end{array}\right]+\underbrace{[H,h]^T[\Theta_*^{-1}+H]^{-1}[H,h]}_{B_+(h,H)\succeq 0}\big]\big)\\
&&+\max\limits_{W_2\in\mathcal{U}_2}\mathrm{Tr}\big(\underbrace{W_2}_{\succeq 0}\big[\left[\begin{array}{c|c}H&h\\ \hline h^T&\end{array}\right]+\underbrace{[H,h]^T[\Theta_*^{-1}-H]^{-1}[H,h]}_{B_-(h,H)\succeq 0}\big]\big)\Big]\\
&\geq& \min\limits_{(h,H)\in\mathcal{H}_1\cap\mathcal{H}_2}\max\limits_{W_1\in\mathcal{U}_1,W_2\in\mathcal{U}_2}\mathrm{Tr}\big([W_2-W_1]\left[\begin{array}{c|c}H&h\\ \hline h^T&\end{array}\right]\big),
\end{array} \tag{6.5}$$

and the concluding quantity in the chain clearly is nonnegative when $\mathcal{U}_1\cap\mathcal{U}_2\neq\emptyset$. Now assume that $\mathcal{U}_1\cap\mathcal{U}_2=\emptyset$, and let us verify that $\mathrm{SadVal_{lift}}<0$. Indeed, $\mathcal{U}_1\cap\mathcal{U}_2=\emptyset$, so that the nonempty convex compact sets $\mathcal{U}_1$, $\mathcal{U}_2$ can be strictly separated: there exist $\bar{h}\in\mathbf{R}^d$, $\bar{H}\in\mathbf{S}^d$, $\eta\in\mathbf{R}$, and $\alpha>0$ such that

$$\mathrm{Tr}\left([W_2-W_1]\left[\begin{array}{c|c}\bar{H}&\bar{h}\\ \hline \bar{h}^T&\eta\end{array}\right]\right)\leq-\alpha\ \forall(W_1\in\mathcal{U}_1,W_2\in\mathcal{U}_2). \tag{6.6}$$

From the structure of matrices B_χ, $\chi=1,2$—see (2.131)—it follows that $[W_1]_{d+1,d+1}$ $=[W_2]_{d+1,d+1}=1$ whenever $W_1\in\mathcal{U}_1$ and $W_2\in\mathcal{U}_2$, whence (6.6) remains intact when keeping $\bar{h}$, $\bar{H}$ as they are and setting $\eta=0$. Now, looking at (6.5) it is clear that for all small enough $t>0$ we have $(t\bar{h},t\bar{H})\in\mathcal{H}_1\cap\mathcal{H}_2$ and

$$A(t\bar{H})\leq Ct^2,\ \|B_\pm(t\bar{h},t\bar{H})\|\leq Ct^2,$$

with C independent of t. Hence, due to boundedness of $\mathcal{U}_\chi$, $\chi=1,2$, (6.5) implies that for all small enough positive t and some constant $\bar{C}$ independent of t it holds

$$\mathrm{SadVal_{lift}}\leq\bar{C}t^2+t\max_{W_1\in\mathcal{U}_1,W_2\in\mathcal{U}_2}\mathrm{Tr}\left([W_2-W_2]\left[\begin{array}{c|c}\bar{H}&\bar{h}\\ \hline \bar{h}^T&\eta\end{array}\right]\right)\leq\bar{C}t^2-\alpha t$$

(we have used (6.6)), implying that $\mathrm{SadVal_{lift}}<0$ due to $\alpha>0$. □

Exercise 2.22. Prove that if X is a nonempty convex compact set in $\mathbf{R}^d$, then the function $\widehat{\Phi}(h;\mu)$ given by (2.114) is real-valued and continuous on $\mathbf{R}^d\times X$ and is convex in h and concave in μ.

Solution: Let $\bar{x}\in\mathrm{rint}\,X$, $\bar{X}=X-\bar{x}$, and E be the linear span of $\bar{X}$. Let $\widetilde{\Phi}(h;\mu)$ be associated with $\bar{X}$ in exactly the same way as $\widehat{\Phi}$ is associated with X:

$$\widetilde{\Phi}(h;\mu)=\inf_g\left[\widetilde{\Psi}(h,g;\mu):=(h-g)^T\mu+\tfrac{1}{8}[\phi_{\bar{X}}(h-g)+\phi_{\bar{X}}(-h+g)]^2+\phi_{\bar{X}}(g)\right].$$

1°. We clearly have $\phi_{\bar{X}}(u)=\phi_X(u)-\bar{x}^Tu$, implying that

$$\begin{array}{rcl}
\widetilde{\Psi}(h,g;\mu)&=&(h-g)^T\mu+\tfrac{1}{8}[\phi_X(h-g)+\phi_X(-h+g)]^2+\phi_X(g)-g^T\bar{x}\\
&=&(h-g)^T[\mu+\bar{x}]+\tfrac{1}{8}[\phi_X(h-g)+\phi_X(-h+g)]^2+\phi_X(g)-h^T\bar{x}\\
&=&\widehat{\Psi}(h,g;\mu+\bar{x})-h^T\bar{x},
\end{array}$$

implying that

$$\widetilde{\Phi}(h;\mu)=\widehat{\Phi}(h;\mu+\bar{x})-h^T\bar{x}.$$

We conclude that verifying continuity and convexity-concavity of $\widehat{\Phi}(h;\mu)$ on $\mathbf{R}^d\times X$ is equivalent to verifying continuity and convexity-concavity of $\widetilde{\Phi}(h;\mu)$ on $\mathbf{R}^d\times\bar{X}$, and this is what we do next.

2°. When $\mu\in\bar{X}$, the functions $\phi_{\bar{X}}(g)$ and $\widetilde{\Psi}(h,g;\mu)$ clearly depend solely on the orthogonal projections g_E and h_E of g and h on E:

$$\widetilde{\Psi}(h,g;\mu)=[h_E-g_E]^T\mu+\tfrac{1}{8}[\phi_{\bar{X}}(h_E-g_E)+\phi_{\bar{X}}(-h_E+g_E)]^2+\phi_{\bar{X}}(g_E).$$

Now, $\bar{X}$ contains a Euclidean ball of positive radius centered at the origin in the linear space E, implying that $[\phi_{\bar{X}}(e)+\phi_{\bar{X}}(-e)]^2\geq\alpha\|e\|_2^2$ for properly selected $\alpha>0$ and all $e\in E$, and that $\phi_{\bar{X}}(g)\geq 0$ for all g. As an immediate consequence, given a nonempty convex compact subset H of $\mathbf{R}^d$, we can find a nonempty convex compact set $G=G[H]\subset E$ such that

$$\begin{array}{l}\forall(h\in H,\mu\in\bar{X}):\\ \inf_g\widetilde{\Psi}(h,g;\mu)=\min_{e\in G[H]}\left[[h_E-e]^T\mu+\tfrac{1}{8}[\phi_{\bar{X}}(h_E-e)+\phi_{\bar{X}}(-h_E+e)]^2+\phi_{\bar{X}}(e)\right].\end{array}$$

Since $\bar{X}$ and $G[H]$ are nonempty convex compact sets, the right hand side in this representation is real-valued and continuous in (h,μ) on $H\times\bar{X}$, implying that $\widetilde{\Phi}(h;\mu)$ is continuous on $\mathbf{R}^d\times\bar{X}$. Convexity-concavity of $\widetilde{\Phi}$ is evident. □

Exercise 2.23. The goal of what follows is to refine the change detection procedure (let us refer to it as the "basic" one) developed in Section 2.9.5.1. The idea is pretty simple. With the notation from Section 2.9.5.1, in the basic procedure, when testing the null hypothesis H_0 vs. signal hypothesis H_t^ρ, we look at the difference $\zeta_t=\omega_t-\omega_1$ and try to decide whether the energy of the deterministic component x_t-x_1 of ζ_t is 0, as is the case under H_0, or is $\geq\rho^2$, as is the case under H_t^ρ. Note that if $\sigma\in[\underline{\sigma},\overline{\sigma}]$ is the actual intensity of the observation noise, then the noise component of ζ_t is $\mathcal{N}(0,2\sigma^2I_d)$; other things being equal, the larger is the noise in ζ_t, the larger should be ρ to allow for a reliable, with a given reliability level, decision. Now note that under the hypothesis H_t^ρ, we have $x_1=...=x_{t-1}$, so that the deterministic component of the difference $\zeta_t=\omega_t-\omega_1$ is exactly the same as for the difference $\widetilde{\zeta}_t=\omega_t-\frac{1}{t-1}\sum_{s=1}^{t-1}\omega_s$, while the noise component in $\widetilde{\zeta}_t$ is $\mathcal{N}(0,\sigma_t^2I_d)$ with $\sigma_t^2=\sigma^2+\frac{1}{t-1}\sigma^2=\frac{t}{t-1}\sigma^2$. Thus, the intensity of noise in $\widetilde{\zeta}_t$ is at most the same as in ζ_t, and this intensity, in contrast to that for ζ_t, decreases as t grows. Here goes the exercise:

Let reliability tolerances $\epsilon,\varepsilon\in(0,1)$ be given, and let our goal be to design a system of inferences $\mathcal{T}_t$, $t=2,3,...,K$, which, when used in the same fashion as tests $\mathcal{T}_t^\kappa$ were used in the basic procedure, results in false alarm probability at most ϵ and in probability to miss a change of energy $\geq\rho^2$ at most ε. Needless to say, we want to achieve this goal with as small ρ as possible. Think how to utilize the above observation to refine the basic procedure eventually reducing (and provably not increasing) the required value of ρ. Implement the basic and the refined change detection procedures and compare their quality (the resulting values of ρ) e.g., on the data used in the experiment reported in Section 2.9.5.1.

Solution: Given $\rho>0$, let us act as follows. For every t, $2\leq t\leq K$, we use Proposition 2.43.ii, in the same fashion as in the Basic procedure, to decide on two hypotheses, $G_{1,t}$ and $G_{2,t}^\rho$, on the distribution of observation

$$\widetilde{\zeta}_t=\omega_t-\frac{1}{t-1}\sum_{s=1}^{t-1}\omega_s=x_t-x_1+\xi^t.$$

Both hypotheses state that the noise component ξ^t of the observation is $\mathcal{N}(0, \frac{t\sigma^2}{t-1}I_d)$, with $\sigma \in [\underline{\sigma}, \overline{\sigma}]$. On top of that, $G_{1,t}$ states that the deterministic component $x_t - x_1$ of the observation is 0, while $G_{2,t}^\rho$ states that the energy of this component is $\geq \rho^2$. Let $\phi_t^\rho(\cdot)$ and $\kappa_t = \kappa_t(\rho)$ be a detector quadratic in $\widetilde{\zeta}_t$ and (the upper bound on) its risk yielded by Proposition 2.43.ii. Passing from this detector to its shift

$$\bar{\phi}_t^\rho(\zeta) = \phi_t^\rho(\zeta) + \ln(\varepsilon/\kappa_t),$$

the simple test $\mathcal{T}_t^\rho$ which, given observation $\widetilde{\zeta}_t$, accepts $G_{1,t}$ when $\bar{\phi}_t^\rho(\widetilde{\zeta}_t) \geq 0$ and accepts $G_{2,t}^\rho$ otherwise, satisfies the relations (cf. (2.151))

$$\mathrm{Risk}_1(\bar{\phi}_t^\rho | G_{1,t}, G_{2,t}^\rho) \leq \frac{\kappa_t^2(\rho)}{\varepsilon}, \quad \mathrm{Risk}_2(\bar{\phi}_t^\rho | G_{1,t}, G_{2,t}^\rho) \leq \varepsilon.$$

Same as in Section 2.9.5.1, the system of tests $\mathcal{T}_t^\rho$, $2 \leq t \leq K$, gives rise to a change detection procedure Π_ρ with the probability to miss a change of energy $\geq \rho^2$ at most ε, and the probability of false alarm at most

$$\epsilon(\rho) = \sum_{t=2}^K \kappa_t^2(\rho)/\varepsilon.$$

It is immediately seen that $\epsilon(\rho) \to 0$ as $\rho \to \infty$, $\epsilon(\rho) > \epsilon$ for small ρ, and $\epsilon(\rho)$ is nonincreasing function of ρ. Applying Bisection, we can find the largest $\rho = \rho_*$ for which $\epsilon(\rho) \leq \epsilon$, and then use the procedure Π_{ρ_*} in actual change detection, thus ensuring the reliability specifications.

Note that with the above refinement, we sacrifice Proposition 2.48.

As for numerical results for the data from the experiment reported in Section 2.9.5.1 ($\dim x = 256^2$, $K = 9$, $\sqrt{2}\underline{\sigma} = \overline{\sigma} = 10$, $\epsilon = \varepsilon = 0.01$), they are as follows: the "resolution" ρ of the Basic procedure is 2,717.4, and that of the refined one 2,702.4. Not too impressive progress!

6.3 SOLUTIONS FOR CHAPTER 3

Exercise 3.1. In the situation of Section 3.3.4, design of a "good" estimate is reduced to solving convex optimization problem (3.39). Note that the objective in this problem is, in a sense, "implicit"—the design variable is h, and the objective is obtained from an explicit convex-concave function of h and (x, y) by maximization over (x, y). There exist solvers capable of processing problems of this type efficiently. However, commonly used off-the-shelf solvers, like cvx, cannot handle problems of this type. The goal of the exercise to follow is to reformulate (3.39) as a semidefinite program, thus making it amenable for cvx.

On an immediate inspection, the situation we are interested in is as follows. We are given

- a nonempty convex compact set $X \subset \mathbf{R}^n$ along with affine function $M(x)$ taking values in $\mathbf{S}^d$ and such that $M(x) \succeq 0$ when $x \in X$, and
- affine function $F(h) : \mathbf{R}^d \to \mathbf{R}^n$.

Given $\gamma > 0$, this data gives rise to the convex function

$$\Psi(h) = \max_{x \in X} \left\{ F^T(h)x + \gamma\sqrt{h^T M(x) h} \right\},$$

and we want to find a "nice" representation of this function, specifically, want to represent the inequality $\tau \geq \Psi(h)$ by a bunch of LMI's in variables τ, h, and perhaps additional variables.

To achieve our goal, we assume in the sequel that the set

$$X^+ = \{(x, M) : x \in X, M = M(x)\}$$

can be described by a system of linear and semidefinite constraints in variables x, M, and additional variables ξ, namely,

$$X^+ = \left\{(x, M) : \exists \xi : \left\{\begin{array}{ll} (a) & s_i - a_i^T x - b_i^T \xi - \mathrm{Tr}(C_i M) \geq 0,\ i \leq I \\ (b) & S - \mathcal{A}(x) - \mathcal{B}(\xi) - \mathcal{C}(M) \succeq 0 \\ (c) & M \succeq 0 \end{array}\right.\right\}.$$

Here $s_i \in \mathbf{R}$, $S \in \mathbf{S}^N$ are some constants, and $\mathcal{A}(\cdot), \mathcal{B}(\cdot), \mathcal{C}(\cdot)$ are (homogeneous) linear functions taking values in $\mathbf{S}^N$. We assume that this system of constraints is essentially strictly feasible, meaning that there exists a feasible solution at which the semidefinite constraints $(b), (c)$ are satisfied strictly (i.e., the left-hand sides of the LMI's are positive definite).

Now comes the exercise:

1) Check that $\Psi(h)$ is the optimal value in the semidefinite program

$$\Psi(h) = \max_{x,M,\xi,t} \left\{ F^T(h)x + \gamma t : \left\{\begin{array}{ll} s_i - a_i^T x - b_i^T \xi - \mathrm{Tr}(C_i M) \geq 0, i \leq I & (a) \\ S - \mathcal{A}(x) - \mathcal{B}(\xi) - \mathcal{C}(M) \succeq 0 & (b) \\ M \succeq 0 & (c) \\ \left[\begin{array}{c|c} h^T M h & t \\ \hline t & 1 \end{array}\right] \succeq 0 & (d) \end{array}\right.\right\}. \qquad (P)$$

2) Passing from (P) to the semidefinite dual of (P), build an explicit semidefinite representation of Ψ, that is, an explicit system $\mathcal{S}$ of LMIs in variables h, τ and additional variables u such that

$$\{\tau \geq \Psi(h)\} \Leftrightarrow \{\exists u : (\tau, h, u) \text{ satisfies } \mathcal{S}\}.$$

Solution: 1: The last LMI in (P) represents equivalently the constraint $t \leq \sqrt{h^T M(x) h}$, so that the optimal value in (P) indeed is $\Psi(h)$.

2: From our assumptions it immediately follows that (P) is feasible and bounded; in addition, in the case of $h \neq 0$ (which we assume for the time being) (P) admits a feasible solution at which all semidefinite constraints are satisfied strictly. Invoking the Refined Conic Duality Theorem (see Remark 4.2), $\Psi(h)$ is the optimal value in the solvable semidefinite dual of (P). To build this dual, let $\lambda_i \geq 0$ be Lagrange multipliers for (a), $W \succeq 0$ be the Lagrange multiplier for (b), $Z \succeq 0$ be the Lagrange multiplier for (c), and $\left[\begin{array}{c|c} u & -v \\ \hline -v & w \end{array}\right] \succeq 0$ be Lagrange multiplier for (d). Taking the weighted sum of the constraints in (P) with the weights given by the multipliers, we get the following consequence of the constraints of (P):

$$\begin{array}{l} \lambda^T s + \mathrm{Tr}(WS) + w - \left[\sum_i \lambda_i a_i + \mathcal{A}^*(W)\right]^T x \\ \quad -\mathrm{Tr}\left(\left[\sum_i \lambda_i C_i + \mathcal{C}^*(W) - uhh^T - Z\right] M\right) - 2vt - \left[\sum_i \lambda_i b_i + \mathcal{B}^*(W)\right]^T \xi \geq 0 \end{array}$$

where for $\mathcal{D}(y) = \sum_i y_i D_i : \mathbf{R}^k \to \mathbf{S}^N$ the mapping $\mathcal{D}^*(\cdot) : \mathbf{S}^N \to \mathbf{R}^k$ is given by

$$\mathcal{D}^*(Y) = [\mathrm{Tr}(D_1 Y); ...; \mathrm{Tr}(D_k Y)] \qquad \left[\Leftrightarrow y^T \mathcal{D}^*(Y) \equiv \mathrm{Tr}(\mathcal{D}(y)Y)\right].$$

To get the dual problem, we should impose on Lagrange multipliers, besides the above restrictions, also the constraint that the part of the left-hand side in the aggregated constraint homogeneous in x, M, ξ, t is identically equal to the minus objective in (P), so that the dual problem reads

$$\begin{array}{rcl}
\Psi(h) & = & \min\limits_{\lambda,W,Z,u,v,w}\left\{\lambda^Ts+\mathrm{Tr}(WS)+w:\left\{\begin{array}{l}\sum_i\lambda_ia_i+\mathcal{A}^*(W)=F(h)\\ uhh^T+Z=\sum_i\lambda_iC_i+\mathcal{C}^*(W)\\ 2v=\gamma\\ \sum_i\lambda_ib_i+\mathcal{B}^*(W)=0\\ uw\geq v^2\\ \lambda\geq0,W\succeq0,Z\succeq0,u\geq0,w\geq0\end{array}\right.\right\}\\
& = & \min\limits_{\lambda,W,Z,u,w}\left\{\lambda^Ts+\mathrm{Tr}(WS)+w:\left\{\begin{array}{l}\sum_i\lambda_ia_i+\mathcal{A}^*(W)=F(h)\\ uhh^T+Z=\sum_i\lambda_iC_i+\mathcal{C}^*(W)\\ \sum_i\lambda_ib_i+\mathcal{B}^*(W)=0\\ uw\geq\frac{\gamma^2}{4}\\ \lambda\geq0,W\succeq0,Z\succeq0,w\geq0\end{array}\right.\right\}\\
& = & \min\limits_{\lambda,W,w}\left\{\lambda^Ts+\mathrm{Tr}(WS)+w:\left\{\begin{array}{l}\sum_i\lambda_ia_i+\mathcal{A}^*(W)=F(h)\\ \frac{\gamma^2}{4w}hh^T\preceq\sum_i\lambda_iC_i+\mathcal{C}^*(W)\\ \sum_i\lambda_ib_i+\mathcal{B}^*(W)=0\\ \lambda\geq0,W\succeq0,w\geq0\end{array}\right.\right\}
\end{array}$$

that is,

$$\Psi(h)=\min_{\lambda,W,w}\left\{\lambda^Ts+\mathrm{Tr}(WS)+w:\left\{\begin{array}{l}\sum_i\lambda_ia_i+\mathcal{A}^*(W)=F(h)\\ \left[\begin{array}{c|c}\sum_i\lambda_iC_i+\mathcal{C}^*(W)&\frac{\gamma}{2}h\\ \hline \frac{\gamma}{2}h^T&w\end{array}\right]\succeq0\\ \sum_i\lambda_ib_i+\mathcal{B}^*(W)=0,\\ \lambda\geq0,W\succeq0,w\geq0\end{array}\right.\right\}. \quad (6.7)$$

(6.7) was obtained under the assumption that $h\neq0$, but in fact the relation holds true when $h=0$ as well, due to

$$\begin{array}{rcl}
\Psi(0) & = & \max_{x,M,\xi}\left\{F^T(0)x:\left\{\begin{array}{ll}s_i-a_i^Tx-b_i^T\xi-\mathrm{Tr}(C_iM)\geq0,i\leq I & (a)\\ S-\mathcal{A}(x)-\mathcal{B}(\xi)-\mathcal{C}(M)\succeq0 & (b)\\ M\succeq0 & (c)\end{array}\right.\right\}\\
\Rightarrow & & \text{[by conic duality]}\\
& = & \min\limits_{\lambda,W,Z}\left\{\lambda^Ts+\mathrm{Tr}(WS):\left\{\begin{array}{l}\sum_i\lambda_ia_i+\mathcal{A}^*(W)=F(0)\\ Z=\sum_i\lambda_iC_i+\mathcal{C}^*(W)\\ \sum_i\lambda_ib_i+\mathcal{B}^*(W)=0,\\ \lambda\geq0,W\succeq0,Z\succeq0\end{array}\right.\right\}\\
& = & \min\limits_{\lambda,W}\left\{\lambda^Ts+\mathrm{Tr}(WS):\left\{\begin{array}{l}\sum_i\lambda_ia_i+\mathcal{A}^*(W)=F(0)\\ \sum_i\lambda_iC_i+\mathcal{C}^*(W)\succeq0\\ \sum_i\lambda_ib_i+\mathcal{B}^*(W)=0,\lambda\geq0,W\succeq0\end{array}\right.\right\},
\end{array}$$

which is exactly the same as (6.7) with $h=0$.

We conclude that

$$\{\tau\geq\Psi(h)\}\Leftrightarrow\exists(\lambda,W,w):\lambda^Ts+\mathrm{Tr}(WS)+w\leq\tau\,\&\left\{\begin{array}{l}\sum_i\lambda_ia_i+\mathcal{A}^*(W)=F(h)\\ \left[\begin{array}{c|c}\sum_i\lambda_iC_i+\mathcal{C}^*(W)&\frac{\gamma}{2}h\\ \hline \frac{\gamma}{2}h^T&w\end{array}\right]\succeq0\\ \sum_i\lambda_ib_i+\mathcal{B}^*(W)=0,\\ \lambda\geq0,W\succeq0\end{array}\right.,$$

which is the desired representation of Ψ. □

Exercise 3.2. Let us consider the situation as follows. Given an $m\times n$ "sensing matrix" A which is stochastic—with columns from the probabilistic simplex

$$\Delta_m=\left\{v\in\mathbf{R}^m:v\geq0,\sum_i v_i=1\right\}$$

and a nonempty closed subset U of Δ_n, we observe an M-element, $M > 1$, i.i.d. sample $\zeta^M = (\zeta_1, ..., \zeta_M)$ with ζ_k drawn from the discrete distribution Au_*, where u_* is an unknown probabilistic vector ("signal") known to belong to U. We handle the discrete distribution Au, $u \in \Delta_n$, as a distribution on the vertices $e_1, ..., e_m$ of Δ_m, so that possible values of ζ_k are basic orths $e_1, ..., e_m$ in $\mathbf{R}^m$. Our goal is to recover the value $F(u_*)$ of a given quadratic form

$$F(u) = u^T Q u + 2q^T u.$$

Observe that for $u \in \Delta_n$, we have $u = [uu^T]\mathbf{1}_n$, where $\mathbf{1}_k$ is the all-ones vector in $\mathbf{R}^k$. This observation allows us to rewrite $F(u)$ as a homogeneous quadratic form:

$$F(u) = u^T \bar{Q} u, \ \bar{Q} = Q + [q\mathbf{1}_n^T + \mathbf{1}_n q^T]. \tag{3.77}$$

The goal of the exercise is to follow the approach developed in Section 3.4.1 for the Gaussian case in order to build an estimate $\widehat{g}(\zeta^M)$ of $F(u)$. To this end, consider the following construction.

Let

$$\mathcal{J}_M = \{(i,j) : 1 \le i < j \le M\}, J_M = \mathrm{Card}(\mathcal{J}_M).$$

For $\zeta^M = (\zeta_1, ..., \zeta_M)$ with $\zeta_k \in \{e_1, ..., e_m\}$, $1 \le k \le M$, let

$$\omega_{ij}[\zeta^M] = \tfrac{1}{2}[\zeta_i\zeta_j^T + \zeta_j\zeta_i^T], \ (i,j) \in \mathcal{J}_M.$$

The estimates we are interested in are of the form

$$\widehat{g}(\zeta^M) = \mathrm{Tr}\Big(h \underbrace{\left[\frac{1}{J_M}\sum\nolimits_{(i,j)\in\mathcal{J}_M} \omega_{ij}[\zeta^M]\right]}_{\omega[\zeta^M]}\Big) + \kappa$$

where $h \in \mathbf{S}^m$ and $\kappa \in \mathbf{R}$ are the parameters of the estimate.

Now comes the exercise:

1) Verify that when the ζ_k's stem from signal $u \in U$, the expectation of $\omega[\zeta^M]$ is a linear image $Az[u]A^T$ of the matrix $z[u] = uu^T \in \mathbf{S}^n$: denoting by P_u^M the distribution of ζ^M, we have

$$\mathbf{E}_{\zeta^M \sim P_u^M}\{\omega[\zeta^M]\} = Az[u]A^T. \tag{3.78}$$

Check that when setting

$$\mathcal{Z}_k = \{\omega \in \mathbf{S}^k : \omega \succeq 0, \omega \ge 0, \mathbf{1}_k^T \omega \mathbf{1}_k = 1\},$$

where $x \ge 0$ for a matrix x means that x is entrywise nonnegative, the image of $\mathcal{Z}_n$ under the mapping $z \mapsto AzA^T$ is contained in $\mathcal{Z}_m$.

Solution: Let us fix signal $u \in \mathbf{\Delta}_n$ underlying observations $\zeta_1, ..., \zeta_M$. Since ζ_1, ..., ζ_M are independent, and ζ_k takes value e_i with probability $[Au]_i$, $i = 1, ..., m$, all matrices $\omega_{ij}[\zeta^M]$, $(i,j) \in \mathcal{J}_M$, have common expectation, namely, $[Au][Au]^T$, and (3.78) follows. The fact that $AzA^T \in \mathcal{Z}_m$ when $z \in \mathcal{Z}_n$ immediately follows from $A \ge 0$ and $A^T\mathbf{1}_m = \mathbf{1}_n$.

2) Let $\Delta^k = \{z \in \mathbf{S}^k : z \ge 0, \mathbf{1}_n^T z \mathbf{1}_n = 1\}$, so that $\mathcal{Z}_k$ is the set of all positive semidefinite matrices from Δ^k. For $\mu \in \Delta^m$, let P_μ be the distribution of the random matrix w taking values in $\mathbf{S}^m$ as follows: the possible values of w are matrices of the form $e^{ij} = \frac{1}{2}[e_ie_j^T + e_je_i^T]$, $1 \le i \le j \le m$; for every $i \le m$, w takes value e^{ii} with probability μ_{ii}, and for every i, j with $i < j$, w takes value e^{ij} with probability $2\mu_{ij}$. Let us set

$$\Phi_1(h; \mu) = \ln\left(\sum_{i,j=1}^{m} \mu_{ij}\exp\{h_{ij}\}\right) : \mathbf{S}^m \times \Delta^m \to \mathbf{R},$$

so that Φ_1 is a continuous convex-concave function on $\mathbf{S}^m \times \Delta^m$.

2.1. Prove that

$$\forall (h \in \mathbf{S}^m, \mu \in \mathcal{Z}_m) : \ln\left(\mathbf{E}_{w\sim P_\mu}\left\{\exp\{\mathrm{Tr}(hw)\}\right\}\right) = \Phi_1(h;\mu).$$

2.2. Derive from 2.1 that setting

$$K = K(M) = \lfloor M/2 \rfloor,\ \Phi_M(h;\mu) = K\Phi_1(h/K;\mu) : \mathbf{S}^m \times \Delta^m \to \mathbf{R},$$

Φ_M is a continuous convex-concave function on $\mathbf{S}^m \times \Delta^m$ such that $\Phi_K(0;\mu) = 0$ for all $\mu \in \mathcal{Z}_m$, and whenever $u \in U$, the following holds true:

Let $P_{u,M}$ be the distribution of $\omega = \omega[\zeta^M]$, $\zeta^M \sim P_u^M$. Then for all $u \in U$, $h \in \mathbf{S}^m$,

$$\ln\left(\mathbf{E}_{\omega\sim P_{u,M}}\left\{\exp\{\mathrm{Tr}(h\omega)\}\right\}\right) \leq \Phi_M(h; Az[u]A^T),\ z[u] = uu^T. \tag{3.79}$$

Solution: 2.1 is evident. Let us prove 2.2. Continuity and convexity-concavity of Φ_M and the relation $\Phi_M(0;\mu) = 0$, $\mu \in \mathcal{Z}_m$, are evident, so that all we need is to verify (3.79).

Let us fix $u \in U$ and $h \in \mathbf{S}^m$, and let $\mu = Az[u]A^T$, so that $\mu \in \mathcal{Z}_m \subset \Delta^m$. Let $\mathcal{S}_M$ be the set of all permutations σ of $1, ..., M$ such that $\sigma(2k-1) < \sigma(2k)$ for $k = 1, ..., K$, and let

$$\omega^\sigma[\zeta^M] = \frac{1}{K}\sum_{k=1}^K \omega_{\sigma(2k-1)\sigma(2k)}[\zeta^M],\ \sigma \in \mathcal{S}_M.$$

We claim that

$$\omega[\zeta^M] = \frac{1}{\mathrm{Card}(\mathcal{S}_M)} \sum_{\sigma\in\mathcal{S}_M} \omega^\sigma[\zeta^M]. \tag{6.8}$$

Indeed, we clearly have

$$\sum_{\sigma\in\mathcal{S}_M}\sum_{k=1}^K \omega_{\sigma(2k-1)\sigma(2k)}[\zeta^M] = N \sum_{(i,j)\in\mathcal{J}_M} \omega_{ij}[\zeta^M],$$

where N is the number of permutations $\sigma \in \mathcal{S}_M$ such that a particular pair $(i,j) \in \mathcal{J}_M$ is met among the pairs $(\sigma(2k-1), \sigma(2k))$, $1 \leq k \leq K$. Comparing the total number of ω_{ij}-terms in the left- and the right-hand sides of the latter equality, we get $\mathrm{Card}(\mathcal{S}_K)K = NJ_M$, which combines with the equality itself to imply that

$$\frac{1}{J_M}\sum_{(i,j)\in\mathcal{J}_M} \omega_{ij}[\zeta^M] = \frac{1}{\mathrm{Card}(\mathcal{S}_M)}\sum_{\sigma\in\mathcal{S}_M}\frac{1}{K}\sum_{k=1}^K \omega_{\sigma(2k-1)\sigma(2k)}[\zeta^M],$$

which is exactly (6.8) (see what $\omega^\sigma[\cdot]$ is).

Let σ_{id} be the identity permutation of $1, ..., M$; it clearly belongs to $\mathcal{S}_M$. We have

$$\begin{array}{l}
\mathbf{E}_{\omega\sim P_{u,M}}\left\{\exp\{\mathrm{Tr}(h\omega)\}\right\}=\mathbf{E}_{\zeta^M\sim P_u^M}\left\{\exp\{\mathrm{Tr}(h\omega[\zeta^M])\}\right\}\\
=\mathbf{E}_{\zeta^M\sim P_u^M}\left\{\exp\left\{\frac{1}{\mathrm{Card}(\mathcal{S}_M)}\sum_{\sigma\in\mathcal{S}_M}\mathrm{Tr}(h\omega^\sigma[\zeta^M])\right\}\right\}\ \text{[by (6.8)]}\\
\le\prod\limits_{\sigma\in\mathcal{S}_M}\left[\mathbf{E}_{\zeta^M\sim P_u^M}\left\{\exp\{\mathrm{Tr}(h\omega^\sigma[\zeta^M])\}\right\}\right]^{1/\mathrm{Card}(\mathcal{S}_M)}\ \text{[by Hölder's inequality]}\\
=\mathbf{E}_{\zeta^M\sim P_u^M}\left\{\exp\{\mathrm{Tr}(h\omega^{\sigma_{\mathrm{id}}}[\zeta^M])\}\right\}\\
\text{[since the distribution of } \omega^\sigma[\zeta^M],\ \zeta^M\sim P_u^M, \text{ is independent of } \sigma\in\mathcal{S}_M]\\
=\mathbf{E}_{\zeta^M\sim P_u^M}\left\{\prod_{k=1}^K\exp\{\frac{1}{K}\mathrm{Tr}(h[\zeta_{2k-1}\zeta_{2k}^T+\zeta_{2k}\zeta_{2k-1}^T]/2)\}\right\}\\
\ \text{[by definition of } \omega^{\sigma_{\mathrm{id}}}[\cdot]]\\
=\left[\mathbf{E}_{\zeta^2\sim P_u^2}\left\{\exp\{\mathrm{Tr}((h/K)[\zeta_1\zeta_2^T+\zeta_2\zeta_1^T]/2)\}\right\}\right]^K\ \text{[since } \zeta_1,...,\zeta_M \text{ are i.i.d.]}\\
=\left[\mathbf{E}_{w\sim P_\mu}\left\{\exp\{\mathrm{Tr}((h/K)w)\}\right\}\right]^K\\
\text{[since the distribution of } \frac{1}{2}[\zeta_1\zeta_2^T+\zeta_2\zeta_1^T],\ \zeta^2\sim P_u^2, \text{ clearly is } P_\mu \text{ with } \mu=Auu^TA^T].
\end{array}$$

The resulting inequality combines with (3.78) to imply that

$$\ln\left(\mathbf{E}_{\omega\sim P_{u,M}}\left\{\exp\{\mathrm{Tr}(h\omega)\}\right\}\right)\le K\Phi_1(h/K;Az[u]A^T),$$

and (3.79) follows. □

3) Combine the above observations with Corollary 3.6 to arrive at the following result:

Proposition 3.19 *In the situation in question, let $\mathcal{Z}$ be a convex compact subset of $\mathcal{Z}_n$ such that $uu^T\in\mathcal{Z}$ for all $u\in U$. Given $\epsilon\in(0,1)$, let*

$$\begin{array}{rcl}
\Psi_+(h,\alpha)&=&\max\limits_{z\in\mathcal{Z}}\left[\alpha\Phi_M(h/\alpha,AzA^T)-\mathrm{Tr}(\bar{Q}z)\right]:\mathbf{S}^m\times\{\alpha>0\}\to\mathbf{R},\\
\Psi_-(h,\alpha)&=&\max\limits_{z\in\mathcal{Z}}\left[\alpha\Phi_M(-h/\alpha,AzA^T)+\mathrm{Tr}(\bar{Q}z)\right]:\mathbf{S}^m\times\{\alpha>0\}\to\mathbf{R}\\
\widehat{\Psi}_+(h)&:=&\inf_{\alpha>0}\left[\Psi_+(h,\alpha)+\alpha\ln(2/\epsilon)\right]\\
&=&\max\limits_{z\in\mathcal{Z}}\inf\limits_{\alpha>0}\left[\alpha\Phi_M(h/\alpha,AzA^T)-\mathrm{Tr}(\bar{Q}z)+\alpha\ln(2/\epsilon)\right]\\
&=&\max\limits_{z\in\mathcal{Z}}\inf\limits_{\beta>0}\left[\beta\Phi_1(h/\beta,AzA^T)-\mathrm{Tr}(\bar{Q}z)+\frac{\beta}{K}\ln(2/\epsilon)\right]\quad[\beta=K\alpha],\\
\widehat{\Psi}_-(h)&:=&\inf\limits_{\alpha>0}\left[\Psi_-(h,\alpha)+\alpha\ln(2/\epsilon)\right]\\
&=&\max\limits_{z\in\mathcal{Z}}\inf\limits_{\alpha>0}\left[\alpha\Phi_M(-h/\alpha,AzA^T)+\mathrm{Tr}(\bar{Q}z)+\alpha\ln(2/\epsilon)\right]\\
&=&\max\limits_{z\in\mathcal{Z}}\inf\limits_{\beta>0}\left[\beta\Phi_1(-h/\beta,AzA^T)+\mathrm{Tr}(\bar{Q}z)+\frac{\beta}{K}\ln(2/\epsilon)\right]\quad[\beta=K\alpha].
\end{array}$$

The functions $\widehat{\Psi}_\pm$ are real valued and convex on $\mathbf{S}^m$, and every candidate solution h to the convex optimization problem

$$\mathrm{Opt}=\min_h\left\{\widehat{\Psi}(h):=\tfrac{1}{2}\left[\widehat{\Psi}_+(h)+\widehat{\Psi}_-(h)\right]\right\},\tag{3.80}$$

induces the estimate

$$\widehat{g}_h(\zeta^M)=\mathrm{Tr}(h\omega[\zeta^M])+\kappa(h),\ \ \kappa(h)=\tfrac{1}{2}[\widehat{\Psi}_-(h)-\widehat{\Psi}_+(h)]$$

of the functional of interest (3.77) via observation ζ^M with ϵ-risk on U not exceeding $\rho=\widehat{\Psi}(h)$:

$$\forall(u\in U):\mathrm{Prob}_{\zeta^M\sim P_u^M}\{|F(u)-\widehat{g}_h(\zeta^M)|>\rho\}\le\epsilon.$$

Solution: Verification is straightforward.

4) Consider an alternative way to estimate $F(u)$, namely, as follows. Let $u \in U$. Given a pair of independent observations ζ_1, ζ_2 drawn from distribution Au, let us convert them into the symmetric matrix $\omega_{1,2}[\zeta^2] = \frac{1}{2}[\zeta_1\zeta_2^T + \zeta_2\zeta_1^T]$. The distribution $P_{u,2}$ of this matrix is exactly the distribution $P_{\mu(z[u])}$—see item B—where $\mu(z) = AzA^T : \Delta^n \to \Delta^m$. Now, given $M = 2K$ observations $\zeta^{2K} = (\zeta_1, ..., \zeta_{2K})$ stemming from signal u, we can split them into K consecutive pairs giving rise to K observations $\omega^K = (\omega_1, ..., \omega_K)$, $\omega_k = \omega[[\zeta_{2k-1}; \zeta_{2k}]]$ drawn independently of each other from probability distribution $P_{\mu(z[u])}$, and the functional of interest (3.77) is a linear function $\mathrm{Tr}(\bar{Q}z[u])$ of $z[u]$. Assume that we are given a set $\mathcal{Z}$ as in the premise of Proposition 3.19. Observe that we are in the situation as follows:

> Given K independent identically distributed observations $\omega^K = (\omega_1, ..., \omega_K)$ with $\omega_k \sim P_{\mu(z)}$, where z is an unknown signal known to belong to $\mathcal{Z}$, we want to recover the value at z of linear function $G(v) = \mathrm{Tr}(\bar{Q}v)$ of $v \in \mathbf{S}^n$. Besides this, we know that P_μ, for every $\mu \in \Delta^m$, satisfies the relation
>
> $$\forall (h \in \mathbf{S}^m) : \ln\left(\mathbf{E}_{\omega \sim P_\mu}\{\exp\{\mathrm{Tr}(h\omega)\}\}\right) \leq \Phi_1(h; \mu).$$

This situation fits the setting of Section 3.3.3, with the data specified as

$$\begin{array}{c}\mathcal{H} = \mathcal{E}_H = \mathbf{S}^m, \mathcal{M} = \Delta^m \subset \mathcal{E}_M = \mathbf{S}^m, \Phi = \Phi_1, \\ \mathcal{X} := \mathcal{Z} \subset \mathcal{E}_X = \mathbf{S}^n, \mathcal{A}(z) = AzA^T.\end{array}$$

Therefore, we can use the apparatus developed in that section to upper-bound the ϵ-risk of the affine estimate

$$\mathrm{Tr}\left(h\frac{1}{K}\sum_{k=1}^{K}\omega_k\right) + \kappa$$

of $F(u) := G(z[u]) = u^T\bar{Q}u$ and to build the best, in terms of the upper risk bound, estimate; see Corollary 3.8. On closer inspection (carry it out!), the associated with the above data functions $\widehat{\Psi}_\pm$ arising in (3.38) are exactly the functions $\widehat{\Psi}_\pm$ specified in Proposition 3.19 for $M = 2K$. Thus, the approach just outlined to estimating $F(u)$ via stemming from $u \in U$ observations ζ^{2K} results in a family of estimates

$$\widetilde{g}_h(\zeta^{2K}) = \mathrm{Tr}\left(h\frac{1}{K}\sum_{k=1}^{K}\omega[[\zeta_{2k-1}; \zeta_{2k}]]\right) + \kappa(h),\ h \in \mathbf{S}^m.$$

The resulting upper bound on the ϵ-risk of estimate $\widetilde{g}_h$ is $\widehat{\Psi}(h)$, where $\widehat{\Psi}(\cdot)$ is associated with $M = 2K$ according to Proposition 3.19. In other words, this is exactly the upper bound on the ϵ-risk of the estimate $\widehat{g}_h$ offered by the proposition. Note, however, that the estimates $\widetilde{g}_h$ and $\widehat{g}_h$ are not identical:

$$\begin{array}{rcl}\widetilde{g}_h(\zeta^{2K}) & = & \mathrm{Tr}\left(h\frac{1}{K}\sum_{k=1}^{K}\omega_{2k-1,2k}[\zeta^{2K}]\right) + \kappa(h), \\ \widehat{g}_h(\zeta^{2K}) & = & \mathrm{Tr}\left(h\frac{1}{K(2K-1)}\sum_{1\leq i<j\leq 2K}\omega_{ij}[\zeta^{2K}]\right) + \kappa(h).\end{array}$$

Now comes the question:

• Which of the estimates $\widetilde{g}_h$ and $\widehat{g}_h$ would you prefer? That is, which one of these estimates, in your opinion, exhibits better practical performance?

To check your intuition, compare the estimate performance by simulation. Consider the following story underlying the recommended simulation model:

> *"Tomorrow, tomorrow not today, all the lazy people say." Is it profitable to be lazy?* Imagine you are supposed to carry out a job, and should decide whether to do it today or tomorrow. The reward for the job is drawn at random "by nature," with unknown to you time-invariant distribution u on an n-element set $\{r_1, ..., r_n\}$, with $r_1 \leq r_2 \leq ... \leq r_n$. Given $2K$ historical observations of the rewards, what would be better—to complete the job today or tomorrow? In other words, is the probability for tomorrow's reward to be at least the reward of today greater than 0.5? What is this probability? How do we estimate it from historical data?

State the above problem as that of estimating a quadratic functional $u^T\bar{Q}u$ of distribution u from direct observations ($m = n, A = I_n$). Pick $u \in \Delta_n$ at random and run simulations to check which of the estimates $\widehat{g}_h$ and $\widetilde{g}_h$ works better. To avoid the necessity of solving optimization problem (3.80), you can use $h = \bar{Q}$, resulting in an unbiased estimate of $u^T\bar{Q}u$.

Solution: In estimate $\widehat{g}_h$, one averages $K(2K-1)$ identically distributed random terms, and in $\widetilde{g}_h$ just K random terms with the same distribution. Therefore we can expect that the magnitude of the stochastic component in $\widetilde{g}_h$ (the difference of the estimate and its expectation) will typically be greater than the magnitude of the stochastic component in $\widehat{g}_h$. Given u, the expected values of both estimates are the same, and thus we can expect that the deviation of $\widehat{g}_h(\zeta^{2K})$ from $F(u)$ typically will be less than a similar deviation for $\widetilde{g}_h$. We hardly can quantify this phenomenon theoretically, the obstacle being dependence of the random terms comprising $\widehat{g}_h(\zeta^{2K})$.[4] With this in mind, the fact that theoretical upper bounds on the risks of both estimates are the same already is good (and not completely trivial) news.

Now about "tomorrow and today." The probability for tomorrow's reward to be at least that of today is

$$\sum_{i=1}^{n}\sum_{j=i}^{n} u_i u_j = \frac{1}{2}\sum_{i,j} u_i u_j + \frac{1}{2}\sum_i u_i^2 = 1 + \frac{1}{2}\sum_i u_i^2 = u^T\bar{Q}u,$$

where $\bar{Q}$ is the $n \times n$ symmetric matrix with diagonal entries equal to 1 and off-diagonal entries equal to $1/2$. We see that the probability in question always is $> 1/2$—it makes sense to be lazy! Note, however, that the probability of today's reward being at least that of tomorrow's is exactly the same $u^T\bar{Q}u$, so that lazy people have scientific reasons to be lazy, and industrious people have equally scientific reasons to stay so ....

Now, M historical data can be treated as a stationary M-repeated observation ζ^M of a discrete random variable with distribution u, so that we find ourselves in the situation considered in the exercise, with $m = n$, $A = I_n$, and the matrix $\bar{Q}$ just defined, and we can apply the estimate we have just derived.

Our numerical experiments (where we used $n = 16$ and $\mathcal{Z} = \mathcal{Z}_n$) fully support the claim that "in reality" estimate $\widehat{g}_h$ outperforms its competitor $\widetilde{g}_h$. Here is a

[4] In statistical literature aggregates $\widehat{g}_h$ and $\widetilde{g}_h$ are referred to as *U-statistics of order 2*. There exists a rich and technically advanced theory of U-statistics (cf. [100, 151, 195] and references therein); yet, available results usually impose restrictions on the choice of matrix h which we wish to avoid. This being said, in the simple situation considered in this exercise, a theoretical comparison of estimates $\widehat{g}_h$ and $\widetilde{g}_h$ could be carried out using, for instance, results of [119], but we were just too lazy to do it and leave this task to an interested reader.

typical result:

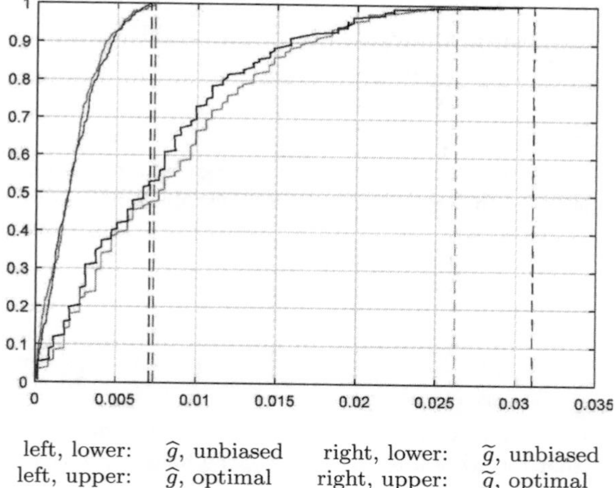

left, lower:	$\widehat{g}$, unbiased	right, lower:	$\widetilde{g}$, unbiased
left, upper:	$\widehat{g}$, optimal	right, upper:	$\widetilde{g}$, optimal

What you see are empirical cumulative distribution functions, built via 200 simulations of ζ^{2K} with $K = 256$, of magnitudes $|\widehat{g}(\zeta^{2K}) - u^T\bar{Q}u|$ of recovery errors. In each simulation, we compute $\widehat{g}_h$- and $\widetilde{g}_h$-estimates, for both $h = \bar{Q}$ ("unbiased estimates"), and for h selected as the optimal solution to (3.80) ("optimal estimates"). We see that while there is no significant difference between unbiased and optimal estimates of the same type, the $\widehat{g}_h$-estimates significantly outperform their $\widetilde{g}_h$ counterparts.

Exercise 3.3. What follows is a variation of Exercise 3.2. Consider the situation as follows. We observe K realizations η_k, $k \leq K$, of a discrete random variable with p possible values, and $L \geq K$ realizations ζ_ℓ, $\ell \leq L$, of a discrete random variable with q possible values. All realizations are independent of each other; the η_k's are drawn from distribution Pu, and the ζ_ℓ's from distribution Qv, where $P \in \mathbf{R}^{p\times r}$, $Q \in \mathbf{R}^{q\times s}$ are given stochastic "sensing matrices," and u, v are unknown "signals" known to belong to given subsets U, V of probabilistic simplexes Δ_r, Δ_s. Our goal is to recover from observations $\{\eta_k, \zeta_\ell\}$ the value at u, v of a given bilinear function

$$F(u,v) = u^T F v = \mathrm{Tr}(F[uv^T]^T). \tag{3.81}$$

A "covering story" could be as follows. Imagine that there are two possible actions, say, administering to a patient drug A or drug B. Let u be the probability distribution of a (quantified) outcome of the first action, and v be a similar distribution for the second action. Observing what happens when the first action is utilized K, and the second L times, we could ask ourselves what the probability is of the outcome of the first action being better than the outcome of the second one. This amounts to computing the probability π of the event "$\eta > \zeta$," where η, ζ are discrete real-valued random variables independent of each other with distributions u, v, and π is a linear function of the "joint distribution" uv^T of η, ζ. This story gives rise to the aforementioned estimation problem with the unit sensing matrices P and Q. Assuming that there are "measurement errors"—instead of observing the action's outcome "as is," we observe a realization of a random variable with distribution depending, in a prescribed fashion, on the outcome—we arrive at problems where P and Q can be general type stochastic matrices.

As always, we encode the p possible values of η_k by the basic orths $e_1,...,e_p$ in $\mathbf{R}^p$, and the q possible values of ζ by the basic orths $f_1,...,f_q$ in $\mathbf{R}^q$.

We focus on estimates of the form

$$\widehat{g}_{h,\kappa}(\eta^K,\zeta^L)=\left[\frac{1}{K}\sum_k \eta_k\right]^T h\left[\frac{1}{L}\sum_\ell \zeta_\ell\right]+\kappa \qquad [h\in\mathbf{R}^{p\times q},\kappa\in\mathbf{R}].$$

This is what you are supposed to do:

1) (cf. item 2 in Exercise 3.2) Denoting by Δ_{mn} the set of nonnegative $m\times n$ matrices with unit sum of all entries (i.e., the set of all probability distributions on $\{1,...,m\}\times\{1,...,n\}$) and assuming $L\geq K$, let us set

$$\mathcal{A}(z)=PzQ^T:\mathbf{R}^{r\times s}\to\mathbf{R}^{p\times q}$$

and

$$\begin{array}{rcl}\Phi(h;\mu)&=&\ln\left(\sum_{i=1}^p\sum_{j=1}^q\mu_{ij}\exp\{h_{ij}\}\right):\mathbf{R}^{p\times q}\times\Delta_{pq}\to\mathbf{R},\\ \Phi_K(h;\mu)&=&K\Phi(h/K;\mu):\mathbf{R}^{p\times q}\times\Delta_{pq}\to\mathbf{R}.\end{array}$$

Verify that $\mathcal{A}$ maps Δ_{rs} into Δ_{pq}, Φ and Φ_K are continuous convex-concave functions on their domains, and that for every $u\in\Delta_r$, $v\in\Delta_s$, the following holds true:

(!) When $\eta^K=(\eta_1,...,\eta_K)$, $\zeta^L=(\zeta_1,...,\zeta_K)$ with mutually independent $\eta_1,...,\zeta_L$ such that $\eta_k\sim Pu$, $\eta_\ell\sim Qv$ for all k, ℓ, we have

$$\ln\left(\mathbf{E}_{\eta,\zeta}\left\{\exp\left\{\left[\frac{1}{K}\sum_k\eta_k\right]^T h\left[\frac{1}{L}\sum_\ell\zeta_\ell\right]\right\}\right\}\right)\leq\Phi_K(h;\mathcal{A}(uv^T)). \tag{3.82}$$

<u>Solution:</u> The only nontrivial claim is (!). To verify this claim, let us start with the immediate observation that when $\mu\in\mathbf{\Delta}_{pq}$ and $h\in\mathbf{R}^{p\times q}$, treating μ as the probability distribution of a random $p\times q$ matrix w which takes values $e_if_j^T$ with probabilities μ_{ij}, $1\leq i\leq p$, $1\leq j\leq q$, we have

$$\ln\left(\mathbf{E}_{w\sim\mu}\{\exp\{\mathrm{Tr}(hw^T)\}\}\right)\leq\Phi(h,\mu),$$

whence, in particular, with η_k and ζ_ℓ stemming, as explained above, from u,v, we have

$$\ln\left(\mathbf{E}\{\exp\{\mathrm{Tr}(\zeta_\ell^T h\eta_k)\}\}\right)\leq\Phi(h;\mathcal{A}(uv^T))\ \forall(h\in\mathbf{R}^{p\times q},k\leq K,\ell\leq L). \tag{$*$}$$

Now let $\mathcal{S}$ be the set of those mappings $k\mapsto\sigma(k)$ of $\{1,...,K\}$ into $\{1,...,L\}$ which are embeddings (i.e., $\sigma(k)\neq\sigma(k')$ when $k\neq k'$). We clearly have

$$\sum_{\sigma\in\mathcal{S}}\sum_{k=1}^K\eta_k\zeta_{\sigma(k)}^T=N\sum_{k=1}^K\sum_{\ell=1}^L\eta_k\zeta_\ell^T$$

with properly selected N; counting the numbers of $\eta\zeta^T$-terms in the left- and in the right-hand side, we get

$$N=\frac{\mathrm{Card}(\mathcal{S})}{L},$$

whence

$$\begin{array}{rcl}\left[\frac{1}{K}\sum_{k=1}^K\eta_k\right]\left[\frac{1}{L}\sum_{\ell=1}^L\zeta_\ell\right]^T&=&\frac{1}{NKL}\left[N\sum_{k=1}^K\sum_{\ell=1}^L\eta_k\zeta_\ell^T\right]\\ &=&\frac{1}{\mathrm{Card}(\mathcal{S})}\sum_{\sigma\in\mathcal{S}}\left[\frac{1}{K}\sum_{k=1}^K\eta_k\zeta_{\sigma(k)}^T\right],\end{array}$$

and therefore

$$\begin{array}{l}\ln\left(\mathbf{E}\left\{\exp\left\{\mathrm{Tr}\left(h\left[\left[\frac{1}{K}\sum_{k=1}^K\eta_k\right]\left[\frac{1}{L}\sum_{\ell=1}^L\zeta_\ell\right]^T\right]^T\right)\right\}\right\}\right)\\ \quad=\frac{1}{\mathrm{Card}(\mathcal{S})}\sum_{\sigma\in\mathcal{S}}\ln\left(\mathbf{E}\left\{\exp\left\{\mathrm{Tr}\left(h\left[\frac{1}{K}\sum_{k=1}^K\eta_k\zeta_{\sigma(k)}^T\right]^T\right)\right\}\right\}\right)\\ \quad[\text{by Hölder inequality}]\\ \quad=K\ln\left(\mathbf{E}\left\{\exp\left\{\mathrm{Tr}(K^{-1}h[\eta_1\zeta_1^T]^T)\right\}\right\}\right)\\ \quad[\text{since } \eta_1,\eta_2,...,\eta_K,\zeta_1,...,\zeta_L \text{ are independent}]\\ \quad\leq K\Phi(h/K;Puv^TQ^T)\ [\text{since } \eta_1\zeta_1^T\sim Puv^TQ^T],\end{array}$$

and (!) follows. □

2) Combine (!) with Corollary 3.6 to arrive at the following analog of Proposition 3.19:

Proposition 3.20 *In the situation in question, let $\mathcal{Z}$ be a convex compact subset of Δ_{rs} such that $uv^T\in\mathcal{Z}$ for all $u\in U$, $v\in V$. Given $\epsilon\in(0,1)$, let*

$$\begin{array}{rcl}\Psi_+(h,\alpha)&=&\max\limits_{z\in\mathcal{Z}}\left[\alpha\Phi_K(h/\alpha,PzQ^T)-\mathrm{Tr}(Fz^T)\right]:\mathbf{R}^{p\times q}\times\{\alpha>0\}\to\mathbf{R},\\ \Psi_-(h,\alpha)&=&\max\limits_{z\in\mathcal{Z}}\left[\alpha\Phi_K(-h/\alpha,PzQ^T)+\mathrm{Tr}(Fz^T)\right]:\mathbf{R}^{p\times q}\times\{\alpha>0\}\to\mathbf{R}\\ \widehat{\Psi}_+(h)&:=&\inf_{\alpha>0}\left[\Psi_+(h,\alpha)+\alpha\ln(2/\epsilon)\right]\\ &=&\max\limits_{z\in\mathcal{Z}}\inf\limits_{\alpha>0}\left[\alpha\Phi_K(h/\alpha,PzQ^T)-\mathrm{Tr}(Fz^T)+\alpha\ln(2/\epsilon)\right]\\ &=&\max\limits_{z\in\mathcal{Z}}\inf\limits_{\beta>0}\left[\beta\Phi(h/\beta,PzQ^T)-\mathrm{Tr}(Fz^T)+\frac{\beta}{K}\ln(2/\epsilon)\right]\quad[\beta=K\alpha],\\ \widehat{\Psi}_-(h)&:=&\inf\limits_{\alpha>0}\left[\Psi_-(h,\alpha)+\alpha\ln(2/\epsilon)\right]\\ &=&\max\limits_{z\in\mathcal{Z}}\inf\limits_{\alpha>0}\left[\alpha\Phi_K(-h/\alpha,PzQ^T)+\mathrm{Tr}(Fz^T)+\alpha\ln(2/\epsilon)\right]\\ &=&\max\limits_{z\in\mathcal{Z}}\inf\limits_{\beta>0}\left[\beta\Phi(-h/\beta,PzQ^T)+\mathrm{Tr}(Fz^T)+\frac{\beta}{K}\ln(2/\epsilon)\right]\quad[\beta=K\alpha].\end{array}$$

The functions $\widehat{\Psi}_\pm$ are real-valued and convex on $\mathbf{R}^{p\times q}$, and every candidate solution h to the convex optimization problem

$$\mathrm{Opt}=\min_h\left\{\widehat{\Psi}(h):=\tfrac{1}{2}\left[\widehat{\Psi}_+(h)+\widehat{\Psi}_-(h)\right]\right\}$$

induces the estimate

$$\widehat{g}_h(\eta^K,\zeta^L)=\mathrm{Tr}\left(h\left[\left[\frac{1}{K}\sum_k\eta_k\right]\left[\frac{1}{L}\sum_\ell\zeta_\ell\right]^T\right]^T\right)+\kappa(h),\quad\kappa(h)=\tfrac{1}{2}[\widehat{\Psi}_-(h)-\widehat{\Psi}_+(h)]$$

of the functional of interest (3.81) *via observation η^K,ζ^L with ϵ-risk on $U\times V$ not exceeding $\rho=\widehat{\Psi}(h)$:*

$$\forall(u\in U,v\in V):\ \mathrm{Prob}\{|F(u,v)-\widehat{g}_h(\eta^K,\zeta^L)|>\rho\}\leq\epsilon,$$

the probability being taken w.r.t. the distribution of observations η^K,ζ^L stemming from signals u, v.

Solution: Verification is straightforward.

Exercise 3.4 [recovering mixture weights] The problem to be addressed in this exercise is as follows. We are given K probability distributions $P_1,...,P_K$ on observation space Ω, and let these distributions have densities $p_k(\cdot)$ w.r.t. some reference measure Π on Ω; we assume that $\sum_k p_k(\cdot)$ is positive on Ω. We are given also N independent observations

$$\omega_t\sim P_\mu,\ t=1,...,N,$$

drawn from distribution

$$P_\mu = \sum_{k=1}^{K} \mu_k P_k,$$

where μ is an unknown "signal" known to belong to the probabilistic simplex $\mathbf{\Delta}_K = \{\mu \in \mathbf{R}^K : \mu \geq 0, \sum_k \mu_k = 1\}$. Given $\omega^N = (\omega_1, ..., \omega_N)$, we want to recover the linear image $G\mu$ of μ, where $G \in \mathbf{R}^{\nu \times K}$ is given.

We intend to measure the risk of a candidate estimate $\widehat{G}(\omega^N) : \Omega \times ... \times \Omega \to \mathbf{R}^\nu$ by the quantity

$$\mathrm{Risk}[\widehat{G}(\cdot)] = \sup_{\mu \in \mathbf{\Delta}} \left[\mathbf{E}_{\omega^N \sim P_\mu \times ... \times P_\mu} \left\{\|\widehat{G}(\omega^N) - G\mu\|_2^2\right\}\right]^{1/2}.$$

3.4.A. Recovering linear form. Let us start with the case when $G = g^T$ is a $1 \times K$ matrix.

3.4.A.1. Preliminaries. To motivate the construction to follow, consider the case when Ω is a finite set (obtained, e.g., by "fine discretization" of the "true" observation space). In this situation our problem becomes an estimation problem in Discrete o.s.: *given a stationary N-repeated observation stemming from discrete probability distribution P_μ affinely parameterized by signal $\mu \in \mathbf{\Delta}_K$, we want to recover a linear form of μ.* It is shown in Section 3.1—see Remark 3.2—that in this case a nearly optimal, in terms of its ϵ-risk, estimate is of the form

$$\widehat{g}(\omega^N) = \frac{1}{N} \sum_{t=1}^{N} \phi(\omega_t) \tag{3.83}$$

with properly selected ϕ. The difficulty with this approach is that as far as computations are concerned, an optimal design of ϕ requires solving a convex optimization problem of design dimension of order of the cardinality of Ω, and this cardinality could be huge, as is the case when Ω is a discretization of a domain in $\mathbf{R}^d$ with d in the range of tens. To circumvent this problem, we are to simplify the outlined approach: from the construction of Section 3.1 we inherit the simple structure (3.83) of the estimator; taking this structure for granted, we are to develop an alternative design of ϕ. With this new design, we have no theoretical guarantees for the resulting estimates to be near-optimal; we sacrifice these guarantees in order to reduce dramatically the computational effort of building the estimates.

3.4.A.2. Generic estimate. Let us select somehow L functions $F_\ell(\cdot)$ on Ω such that

$$\int F_\ell^2(\omega) p_k(\omega) \Pi(d\omega) < \infty, \ 1 \leq \ell \leq L, 1 \leq k \leq K. \tag{3.84}$$

With $\lambda \in \mathbf{R}^L$, consider an estimate of the form

$$\widehat{g}_\lambda(\omega^N) = \frac{1}{N} \sum_{t=1}^{N} \Phi_\lambda(\omega_t), \ \Phi_\lambda(\omega) = \sum_\ell \lambda_\ell F_\ell(\omega). \tag{3.85}$$

1) Prove that

$$\begin{array}{rcl}
\mathrm{Risk}[\widehat{g}_\lambda] & \leq & \overline{\mathrm{Risk}}(\lambda) \\
& := & \max\limits_{k \leq K} \Big[\frac{1}{N} \int \left[\sum_\ell \lambda_\ell F_\ell(\omega)\right]^2 p_k(\omega) \Pi(d\omega) \\
& & + \left[\int \left[\sum_\ell \lambda_\ell F_\ell(\omega)\right] p_k(\omega) \Pi(d\omega) - g^T e_k\right]^2 \Big]^{1/2} \\
& = & \max\limits_{k \leq K} \left[\frac{1}{N} \lambda^T W_k \lambda + [e_k^T [M\lambda - g]]^2\right]^{1/2},
\end{array} \tag{3.86}$$

where

$$\begin{array}{rcl} M & = & \left[M_{k\ell} := \int F_\ell(\omega)p_k(\omega)\Pi(d\omega)\right]_{\substack{k\le K\\ \ell\le L}},\\ W_k & = & \left[[W_k]_{\ell\ell'} := \int F_\ell(\omega)F_{\ell'}(\omega)p_k(\omega)\Pi(d\omega)\right]_{\substack{\ell\le L\\ \ell'\le L}},\ 1\le k\le K, \end{array}$$

and $e_1, ..., e_K$ are the standard basic orths in $\mathbf{R}^K$.

Note that $\overline{\text{Risk}}(\lambda)$ is a convex function of λ; this function is easy to compute, provided the matrices M and W_k, $k \le K$, are available. Assuming this is the case, we can solve the convex optimization problem

$$\text{Opt} = \min_{\lambda\in\mathbf{R}^K} \overline{\text{Risk}}(\lambda) \tag{3.87}$$

and use the estimate (3.85) associated with optimal solution to this problem; the risk of this estimate will be upper-bounded by Opt.

Solution: Let

$$\Psi(\lambda,\mu) = \mathbf{E}_{\omega^N\sim P_\mu\times...\times P_\mu}\left\{[\widehat{g}_\lambda(\omega^N) - g^T\mu]^2\right\}.$$

Let us associate with λ vector

$$\phi_\lambda \in \mathbf{R}^K : [\phi_\lambda]_k = \int \Phi_\lambda(\omega)p_k(\omega)\Pi(d\omega),\ k\le K;$$

note that ϕ_λ is linear in λ and that we have

$$\mathbf{E}_{\omega\sim P_\mu(\cdot)}\{\Phi_\lambda(\omega)\} = \phi_\lambda^T\mu.$$

Hence,

$$\widehat{g}_\lambda(\omega^N) - g^T\mu = \frac{1}{N}\sum_{t=1}^N[\Phi_\lambda(\omega_t) - \phi_\lambda^T\mu] + [\phi_\lambda - g]^T\mu,$$

and random variables $[\Phi_\lambda(\omega_t) - \phi_\lambda^T\mu]$, $t = 1, ..., N$, with $\omega_1, ..., \omega_N$ sampled, independently from each other, from P_μ are zero mean i.i.d., implying that

$$\begin{array}{rcl} \Psi(\lambda,\mu) & = & \mathbf{E}_{\omega^N\sim P_\mu\times...\times P_\mu}\left\{[\widehat{g}_\lambda(\omega^N) - g^T\mu]^2\right\}\\ & = & \frac{1}{N}\mathbf{E}_{\omega\sim P_\mu}\left\{[\Phi_\lambda(\omega) - \phi_\lambda^T\mu]^2\right\} + ([\phi_\lambda - g]^T\mu)^2. \end{array}$$

That is,

$$\begin{array}{l} \Psi(\lambda,\mu) = \frac{1}{N}\sum_{k=1}^K \mu_k\mathbf{E}_{\omega\sim P_k}\left\{[\Phi_\lambda(\omega) - \phi_\lambda^T\mu]^2\right\} + ([\phi_\lambda - g]^T\mu)^2\\ \quad = \frac{1}{N}\sum_{k=1}^K \mu_k\left[\mathbf{E}_{\omega\sim P_k}\left\{\Phi_\lambda^2(\omega)\right\} - 2[\phi_\lambda^T\mu]\mathbf{E}_{\omega\sim P_k}\left\{\Phi_\lambda(\omega)\right\} + [\phi_\lambda^T\mu]^2\right]\\ \qquad\qquad\qquad\qquad\qquad\qquad\qquad\qquad + ([\phi_\lambda - g]^T\mu)^2\\ \quad = \frac{1}{N}\left[\sum_{k=1}^K \mu_k\mathbf{E}_{\omega\sim P_k}\left\{\Phi_\lambda^2(\omega)\right\} - 2[\phi_\lambda^T\mu]\overbrace{\sum_k \mu_k\mathbf{E}_{\omega\sim P_k}\left\{\Phi_\lambda(\omega)\right\}}^{\phi_\lambda^T\mu} + [\phi_\lambda^T\mu]^2\right]\\ \qquad\qquad\qquad\qquad\qquad\qquad\qquad\qquad + ([\phi_\lambda - g]^T\mu)^2\\ \quad = \frac{1}{N}\sum_{k=1}^K \mu_k\mathbf{E}_{\omega\sim P_k}\left\{\Phi_\lambda^2(\omega)\right\} - \frac{1}{N}[\phi_\lambda^T\mu]^2 + ([\phi_\lambda - g]^T\mu)^2\\ \quad \le \overline{\Psi}(\lambda,\mu) := \frac{1}{N}\sum_{k=1}^K \mu_k\mathbf{E}_{\omega\sim P_k}\left\{\Phi_\lambda^2(\omega)\right\} + ([\phi_\lambda - g]^T\mu)^2. \end{array}$$

Note that $\Phi_\lambda(\cdot)$ and ϕ_λ are linear in λ, implying that $\overline{\Psi}(\lambda,\mu)$ is convex in λ. Besides this, we see that this function is convex in μ as well. We conclude that

$$\mathrm{Risk}^2[\widehat{g}_\lambda] := \max_{\mu\in\mathbf{\Delta}_K} \Psi(\lambda,\mu) \leq \max_{\mu\in\mathbf{\Delta}_K} \overline{\Psi}(\lambda,\mu) = \overline{\mathrm{Risk}}^2(\lambda) := \max_k \overline{\Psi}(\lambda, e_k).$$

In other words,

$$\begin{array}{rcl}\mathrm{Risk}(\widehat{g}_\lambda) \leq \overline{\mathrm{Risk}}(\lambda) & := & \max_{k\leq K} \left[\overline{\Psi}(\lambda,e_k)\right]^{1/2} \\ & = & \max_{k\leq K}\left[\frac{1}{N}\lambda^T W_k\lambda + [e_k^T[M\lambda - g]]^2\right]^{1/2},\end{array}$$

$$\left[\begin{array}{c} M = \left[M_{k\ell} := \int F_\ell(\omega)p_k(\omega)\Pi(d\omega)\right]_{\substack{k\leq K\\ \ell\leq L}}, \\ W_k = \left[[W_k]_{\ell\ell'} := \int F_\ell(\omega)F_{\ell'}(\omega)p_k(\omega)\Pi(d\omega)\right]_{\substack{\ell\leq L\\ \ell'\leq L}}, \ 1\leq k\leq K\end{array}\right]$$

as required in (3.86) □

3.4.A.3. Implementation. When implementing the generic estimate we arrive at the "Measurement Design" question: how do we select the value of L and functions F_ℓ, $1 \leq \ell \leq L$, resulting in a small (upper bound Opt on the) risk of the estimate (3.85) yielded by an optimal solution to (3.87)? We are about to consider three related options—*naive*, *basic*, and *Maximum Likelihood* (ML).

The naive option is to take $F_\ell = p_\ell$, $1 \leq \ell \leq L = K$, assuming that this selection meets (3.84). For the sake of definiteness, consider the "Gaussian case," where $\Omega = \mathbf{R}^d$, Π is the Lebesgue measure, and p_k is Gaussian distribution with parameters ν_k, Σ_k:

$$p_k(\omega) = (2\pi)^{-d/2}\mathrm{Det}(\Sigma_k)^{-1/2}\exp\left\{-\tfrac{1}{2}(\omega-\nu_k)^T\Sigma_k^{-1}(\omega-\nu_k)\right\}.$$

In this case, the Naive option leads to easily computable matrices M and W_k appearing in (3.86).

2) Check that in the Gaussian case, when setting

$$\Sigma_{k\ell} = [\Sigma_k^{-1}+\Sigma_\ell^{-1}]^{-1},\ \ \Sigma_{k\ell m} = [\Sigma_k^{-1}+\Sigma_\ell^{-1}+\Sigma_m^{-1}]^{-1},\ \ \chi_k = \Sigma_k^{-1}\nu_k,$$

$$\alpha_{k\ell} = \sqrt{\frac{\mathrm{Det}(\Sigma_{k\ell})}{(2\pi)^d\mathrm{Det}(\Sigma_k)\mathrm{Det}(\Sigma_\ell)}},\ \ \beta_{k\ell m} = (2\pi)^{-d}\sqrt{\frac{\mathrm{Det}(\Sigma_{k\ell m})}{\mathrm{Det}(\Sigma_k)\mathrm{Det}(\Sigma_\ell)\mathrm{Det}(\Sigma_m)}},$$

we have

$$\begin{array}{rcl} M_{k\ell} & := & \int p_\ell(\omega)p_k(\omega)\Pi(d\omega) \\ & = & \alpha_{k\ell}\exp\left\{\frac{1}{2}\left[[\chi_k+\chi_\ell]^T\Sigma_{k\ell}[\chi_k+\chi_\ell] - \chi_k^T\Sigma_k\chi_k - \chi_\ell^T\Sigma_\ell\chi_\ell\right]\right\}, \\ {[W_k]_{\ell m}} & := & \int p_\ell(\omega)p_m(\omega)p_k(\omega)\Pi(d\omega) \\ & = & \beta_{k\ell m}\exp\left\{\frac{1}{2}\left[[\chi_k+\chi_\ell+\chi_m]^T\Sigma_{k\ell m}[\chi_k+\chi_\ell+\chi_m]\right.\right. \\ & & \left.\left. -\chi_k^T\Sigma_k\chi_k - \chi_\ell^T\Sigma_\ell\chi_\ell - \chi_m^T\Sigma_m\chi_m\right]\right\}. \end{array}$$

Solution: Verification is straightforward.

Basic option. Though simple, the Naive option does not make much sense: when replacing the reference measure Π with another measure Π' which has positive density $\theta(\cdot)$ w.r.t. Π, the densities p_k are updated according to $p_k(\cdot) \mapsto p_k'(\cdot) = \theta^{-1}(\cdot)p(\cdot)$, so that selecting $F_\ell' = p_\ell'$, the matrices M and W_k become M' and W_k' with

$$\begin{array}{rcl} M_{k\ell}' & = & \int \frac{p_k(\omega)p_\ell(\omega)}{\theta^2(\omega)}\Pi'(d\omega) = \int \frac{p_k(\omega)p_\ell(\omega)}{\theta(\omega)}\Pi(d\omega), \\ {[W_k']_{\ell m}} & = & \int \frac{p_k(\omega)p_\ell(\omega)p_m(\omega)}{\theta^3(\omega)}\Pi'(d\omega) = \int \frac{p_k(\omega)p_\ell(\omega)}{\theta^2(\omega)}\Pi(d\omega). \end{array}$$

We see that in general $M \neq M'$ and $W_k \neq W_k'$, which makes the Naive option rather unnatural. In the alternative *Basic* option we set

$$L = K,\ F_\ell(\omega) = \pi(\omega) := \frac{p_\ell(\omega)}{\sum_k p_k(\omega)}.$$

The motivation is that the functions F_ℓ are invariant when replacing Π with Π', so that here $M = M'$ and $W_k = W_k'$. Besides this, there are statistical arguments in favor of the Basic option, namely, as follows. Let Π_* be the measure with the density $\sum_k p_k(\cdot)$ w.r.t. Π; taken w.r.t. Π_*, the densities of P_k are exactly the above $\pi_k(\cdot)$, and $\sum_k \pi_k(\omega) \equiv 1$. Now, (3.86) says that the risk of estimate $\widehat{g}_\lambda$ can be upper-bounded by the function $\overline{\mathrm{Risk}}(\lambda)$ defined in (3.86), and this function, in turn, can be upper-bounded by the function

$$\begin{array}{rl}\mathrm{Risk}^+(\lambda) := & \Big[\frac{1}{N}\sum_k \int \left[\sum_\ell \lambda_\ell F_\ell(\omega)\right]^2 p_k(\omega)\Pi(d\omega) \\ & \qquad + \max_k \left[\int \left[\sum_k \lambda_\ell F_\ell(\omega)\right] p_k(\omega)\Pi(d\omega) - g^T e_k\right]^2\Big]^{1/2} \\ & = \Big[\frac{1}{N}\int \left[\sum_\ell \lambda_\ell F_\ell(\omega)\right]^2 \Pi_*(d\omega) \\ & \qquad + \max_k \left[\int \left[\sum_k \lambda_\ell F_\ell(\omega)\right] \pi_k(\omega)\Pi_*(d\omega) - g^T e_k\right]^2\Big]^{1/2} \\ & \le K\overline{\mathrm{Risk}}(\lambda)\end{array}$$

(we have said that the maximum of K nonnegative quantities is at most their sum, and the latter is at most K times the maximum of the quantities). Consequently, the risk of the estimate (3.85) stemming from an optimal solution to (3.87) can be upper-bounded by the quantity

$$\mathrm{Opt}^+ := \min_\lambda \mathrm{Risk}^+(\lambda) \quad [\ge \mathrm{Opt} := \max_\lambda \overline{\mathrm{Risk}}(\lambda)].$$

And here comes the punchline:

3.1) Prove that both the quantities Opt defined in (3.87) and the above Opt^+ depend only on the linear span of the functions F_ℓ, $\ell = 1, ..., L$, not on how the functions F_ℓ are selected in this span.

3.2) Prove that the selection $F_\ell = \pi_\ell$, $1 \le \ell \le L = K$, minimizes Opt^+ among all possible selections L, $\{F_\ell\}_{\ell=1}^L$ satisfying (3.84).

Conclude that the selection $F_\ell = \pi_\ell$, $1 \le \ell \le L = K$, while not necessary optimal in terms of Opt, definitely is meaningful: this selection optimizes the natural upper bound Opt^+ on Opt. Observe that $\mathrm{Opt}^+ \le K\mathrm{Opt}$, so that optimizing instead of Opt the upper bound Opt^+, although rough, is not completely meaningless.

<u>Solution:</u> 3.1: By construction, both $\overline{\mathrm{Risk}}(\lambda)$ and $\mathrm{Risk}^+(\lambda)$ depend on λ solely through the function $\Phi_\lambda(\cdot) = \sum_\ell \lambda_\ell F_\ell(\cdot)$; when λ runs through $\mathbf{R}^L$, this function runs exactly through the linear span $\mathcal{L}$ of $F_1(\cdot), ..., F_L(\cdot)$. Consequently,

$$\begin{array}{rcl}\mathrm{Opt}^2 & = & \min\limits_{\Phi\in\mathcal{L}} \max\limits_k \left[\frac{1}{N}\int \Phi^2(\omega)p_k(\omega)\Pi(d\omega) + \left[\int \Phi(\omega)p_k(\omega)\Pi(d\omega) - g^T e_k\right]^2\right], \\ {[\mathrm{Opt}^+]^2} & = & \min\limits_{\Phi\in\mathcal{L}} \left[\frac{1}{N}\int \Phi^2(\omega)\left[\sum\limits_k p_k(\omega)\right]\Pi(d\omega) + \max\limits_k \left[\int \Phi(\omega)p_k(\omega)\Pi(d\omega) - g^T e_k\right]^2\right],\end{array} \tag{6.9}$$

so that Opt and Opt^+ depend solely on $\mathcal{L}$. □

3.2: We have

$$\begin{array}{rcl}\mathrm{Risk}^+(\lambda) & = & \Big[\frac{1}{N}\int \left[\sum_\ell \lambda_\ell F_\ell(\omega)\right]^2 \Pi_*(d\omega) \\ & & \qquad + \max_k \left[\int \left[\sum_k \lambda_\ell F_\ell(\omega)\right] \pi_k(\omega)\Pi_*(d\omega) - g^T e_k\right]^2\Big]^{1/2}.\end{array}$$

Thus, when replacing F_ℓ with their orthogonal, in $L_2[\Omega,\Pi_*]$, projections on the linear span of $\pi_k(\cdot)$, $1\leq k\leq K$, we can only reduce $\mathrm{Risk}^+(\lambda)$ at every point λ, implying that Opt^+ as given by (6.9) cannot be smaller than what this relation yields when $\mathcal{L}$ is the linear span of $\pi_1(\cdot),...,\pi_K(\cdot)$, i.e., when setting $F_\ell=\pi_\ell$, $1\leq\ell\leq L=K$. □

A downside of the Basic option is that it seems problematic to get closed form expressions for the associated matrices M and W_k; see (3.86). For example, in the Gaussian case, the Naive choice of F_ℓ's allows us to represent M and W_k in an explicit closed form; in contrast to this, when selecting $F_\ell=\pi_\ell$, $\ell\leq L=K$, seemingly the only way to get M and W_k is to use Monte-Carlo simulations. This being said, we indeed can use Monte-Carlo simulations to compute M and W_k, provided we can sample from distributions $P_1,...,P_K$. In this respect, it should be stressed that with $F_\ell\equiv\pi_\ell$, the entries in M and W_k are expectations, w.r.t. $P_1,...,P_K$, of functions of ω *bounded in magnitude by 1*, and thus well suited for Monte-Carlo simulation.

Maximum Likelihood option. This choice of $\{F_\ell\}_{\ell\leq L}$ follows straightforwardly the idea of discretization we started with in this exercise. Specifically, we split Ω into L cells $\Omega_1,...,\Omega_L$ in such a way that the intersection of any two different cells is of Π-measure zero, and treat as our observations not the actual observations ω_t, but the indexes of the cells to which the ω_t's belong. With our estimation scheme, this is the same as selecting F_ℓ as the characteristic function of Ω_ℓ, $\ell\leq L$. Assuming that for distinct k, k' the densities p_k, $p_{k'}$ differ from each other Π-almost surely, the simplest discretization independent of how the reference measure is selected is the Maximum Likelihood discretization

$$\Omega_\ell=\{\omega:\max_k p_k(\omega)=p_\ell(\omega)\},\ 1\leq\ell\leq L=K;$$

with the ML option, we take, as F_ℓ's, the characteristic functions of the sets Ω_ℓ just defined, $1\leq\ell\leq L=K$. As with the Basic option, the matrices M and W_k associated with the ML option can be found by Monte-Carlo simulation.

We have discussed three simple options for selecting the F_ℓ's. In applications, one can compute the upper risk bounds Opt—see (3.87)—associated with each option, and use the option with the best—the smallest—risk bound ("smart" choice of F_ℓ's). Alternatively, one can take as $\{F_\ell,\ell\leq L\}$ the union of the three collections yielded by the above options (and, perhaps, further extend this union). Note that the larger is the collection of the F_ℓ's, the smaller is the associated Opt, so that the only price for combining different selections is in increasing the computational cost of solving (3.87).

3.4.A.4. Illustration. In the experimental part of this exercise you are expected to

4.1) Run numerical experiments to compare the estimates yielded by the above three options (Naive, Basic, ML). Recommended setup:
   - $d=8$, $K=90$;
   - Gaussian case with the covariance matrices Σ_k of P_k selected at random,

$$S_k=\mathtt{rand}(d,d),\ \Sigma_k=\frac{S_kS_k^T}{\|S_k\|^2}\qquad[\|\cdot\|\text{: spectral norm}]$$

   and the expectations ν_k of P_k selected at random from $\mathcal{N}(0,\sigma^2I_d)$, with $\sigma=0.1$;
   - values of N: $\{10^s,\ s=0,1,...,5\}$;
   - linear form to be recovered: $g^T\mu\equiv\mu_1$.

4.2†). Utilize the Cramer-Rao lower risk bound (see Proposition 4.37, Exercise 4.22) to upper-bound the level of conservatism $\frac{\mathrm{Opt}}{\mathrm{Risk}_*}$ of the estimates built in item 4.1. Here Risk_* is the minimax risk in our estimation problem:

$$\mathrm{Risk}_*=\inf_{\widehat{g}(\cdot)}\mathrm{Risk}[\widehat{g}(\omega^N)]=\inf_{\widehat{g}(\cdot)}\sup_{\mu\in\mathbf{\Delta}}\left[\mathbf{E}_{\omega^N\sim P_\mu\times...\times P_\mu}\left\{|\widehat{g}(\omega^N)-g^T\mu|^2\right\}\right]^{1/2},$$

where inf is taken over all estimates.

Solution: 4.1: In our implementation of Basic and ML options, M and W_k were computed by a 30,000-sample Monte-Carlo simulation. Here are the results of a typical experiment following the above setup.

Risk bounds for the three estimates we are comparing (that is, the respective values Opt_N, Opt_B, Opt_{ML} of Opt as given by (3.87)) are as follows:

N	Opt_N	Opt_B	Opt_{ML}
1	0.8639	0.7071	0.7071
10	0.4767	0.3179	0.3206
10^2	0.1698	0.1099	0.1116
10^3	0.0553	0.0355	0.0361
10^4	0.0177	0.0113	0.0115
10^5	0.0056	0.0036	0.0036

We see that Basic and ML choices of the F_ℓ's result in similar risk bounds and in this respect are significantly (although not dramatically) better than the Naive choice.

Empirical risks of the estimate yielded by the "smart" choice of F_ℓ's, as observed in 100 experiments with randomly selected μ's, 100 experiments per every tested value of N, are as follows :

		empirical risk		
N	$\mathrm{Opt_B}$	mean	median	max
1	0.7071	0.4945	0.4956	0.5055
10	0.3179	0.2966	0.2910	0.3863
10^2	0.1099	0.1097	0.1071	0.1468
10^3	0.0355	0.0350	0.0352	0.0450
10^4	0.0113	0.0106	0.0107	0.0134

Level of conservatism of the estimate yielded by the smart choice of F_ℓ's in our experiment was as follows:

N	1	10	10^2	10^3	10^4	10^5
$\mathrm{Opt}_B/\mathrm{Risk}_* \leq$	3.35	3.23	2.59	2.34	2.26	2.23

Our scheme for lower-bounding the minimax risk is as follows. Let us select two points $[\underline{\mu}, \overline{\mu}]$ in $\mathbf{\Delta}_K$, and let $\|\overline{\mu} - \underline{\mu}\|_2 = 2r$, so that $\overline{\mu} = \underline{\mu} + 2rf$ with $\|\cdot\|_2$-unit vector f. From the Cramer-Rao inequality it is easy to derive that the minimax risk corresponding to the stationary N-repeated observations can be lower-bounded as follows:

$$\mathrm{Risk}_* \geq \underline{\mathrm{Risk}} = \frac{r|g^T e|}{r\sqrt{JN}+1}, \quad J = \max_{0 \leq s \leq 1} \left[J(s) := \int \frac{\left[\sum_k f_k p_k(\omega)\right]^2}{\sum_k \mu_k(s) p_k(\omega)} \Pi(d\omega) \right],$$

where $f_1, ..., f_K$ are the coordinates of f, and $\mu_1(s), ..., \mu_K(s)$ are the coordinates of the vector $(1-s)\underline{\mu} + s\overline{\mu}$. Note that $J(s)$ is convex in $s \in [0,1]$, so that $J = \max[J(0), J(1)]$, and therefore we can compute J by Monte-Carlo simulation.

Lower bounds on Risk$_*$ underlying the above numbers were obtained in the fashion just described via a simple mechanism for selecting $\underline{\mu}$ and $\overline{\mu}$ as follows:

- We specify $\mu_{\min}$ and $\mu_{\max}$ as a minimizer and a maximizer of the linear form $g^T\mu$ on $\mathbf{\Delta}_K$, and set $f = [\mu_{\max} - \mu_{\min}]/\|\mu_{\max} - \mu_{\min}\|_2$;
- We build an $m = 21$-element equidistant grid $\mu^0 = \mu_{\min}, \mu^1, ..., \mu^{20} = \mu_{\max}$ on $[\mu_{\min}, \mu_{\max}]$ and compute by Monte-Carlo simulation the quantities

$$J_i = \int \frac{\left[\sum_k f_k p_k(\omega)\right]^2}{\sum_k \mu^i_k p_k(\omega)} \Pi(d\omega),\ 0 \leq i \leq m-1;$$

- Finally, we apply the above scheme for lower-bounding the minimax risk to each selection $\underline{\mu} = \mu^i$, $\overline{\mu} = \mu^j$, $0 \leq i < j < m$, with $J = \max[J_i, J_j]$ (for the selection associated with i, j, the quantities J_i and J_j are exactly the above quantities $J(0)$ and $J(1)$). From the $m(m-1)/2$ resulting lower bounds we select the best—the largest—one, and this is the lower hound on the minimax risk which we use to build the above table.

While one can exhibit different attitudes to the claim that an estimation procedure is minimax optimal within a factor like 2.5, one thing is for sure: such a factor is much smaller than similar factors yielded by typical theoretical results on near-optimality of statistical inferences.

3.4.B. Recovering linear images. Now consider the case when G is a general $\nu \times K$ matrix. The analog of the estimate $\widehat{g}_\lambda(\cdot)$ is now as follows: with somehow chosen $F_1, ..., F_L$ satisfying (3.84), we select a $\nu \times L$ matrix $\Lambda = [\lambda_{i\ell}]$, set

$$\Phi_\Lambda(\omega) = [\sum\nolimits_\ell \lambda_{1\ell}F_\ell(\omega); \sum\nolimits_\ell \lambda_{2\ell}F_\ell(\omega); ...; \sum\nolimits_\ell \lambda_{\nu\ell}F_\ell(\omega)],$$

and estimate $G\mu$ by

$$\widehat{G}_\Lambda(\omega^N) = \frac{1}{N}\sum_{t=1}^N \Phi_\lambda(\omega_t).$$

5) Prove the following analogy of the results of item 3.4.A:

Proposition 3.21 *The risk of the proposed estimator can be upper-bounded as follows:*

$$\begin{array}{rcl}\mathrm{Risk}[\widehat{G}_\Lambda] & := & \max_{\mu\in\mathbf{\Delta}_K}\left[\mathbf{E}_{\omega^N\sim P_\mu\times...\times P_\mu}\left\{\|\widehat{G}(\omega^N) - G\mu\|_2^2\right\}\right]^{1/2}\\ & \leq & \overline{\mathrm{Risk}}(\Lambda) := \max_{k\leq K}\overline{\Psi}(\Lambda, e_k),\\ \overline{\Psi}(\Lambda,\mu) & = & \left[\frac{1}{N}\sum_{k=1}^K \mu_k \mathbf{E}_{\omega\sim P_k}\left\{\|\Phi_\Lambda(\omega)\|_2^2\right\} + \|[\psi_\Lambda - G]\mu\|_2^2\right]^{1/2}\\ & = & \left[\|[\psi_\Lambda - G]\mu\|_2^2 + \frac{1}{N}\sum_{k=1}^K \mu_k \int [\sum_{i\leq\nu}[\sum_\ell \lambda_{i\ell}F_\ell(\omega)]^2] P_k(d\omega)\right]^{1/2},\end{array}$$

where

$$\mathrm{Col}_k[\psi_\Lambda] = \mathbf{E}_{\omega\sim P_k(\cdot)}\Phi_\Lambda(\omega) = \left[\begin{array}{c}\int[\sum_\ell \lambda_{1\ell}F_\ell(\omega)]P_k(d\omega)\\ \cdots\\ \int[\sum_\ell \lambda_{\nu\ell}F_\ell(\omega)]P_k(d\omega)\end{array}\right],\ 1 \leq k \leq K$$

and $e_1, ..., e_K$ are the standard basic orths in $\mathbf{R}^K$.

Note that exactly the same reasoning as in the case of the scalar $G\mu \equiv g^T\mu$ demonstrates that a reasonable way to select L and F_ℓ, $\ell = 1, ..., L$, is to set $L = K$ and $F_\ell(\cdot) = \pi_\ell(\cdot)$, $1 \leq \ell \leq L$.

Solution: The proof is given by straightforward modification of the reasoning which led to (3.86).

6.4 SOLUTIONS FOR CHAPTER 4

6.4.1 Linear Estimates vs. Maximum Likelihood

Exercise 4.1. Consider the problem posed at the beginning of Chapter 4: *Given observation*

$$\omega = Ax + \sigma\xi,\ \xi \sim \mathcal{N}(0, I)$$

of an unknown signal x known to belong to a given signal set $\mathcal{X} \subset \mathbf{R}^n$, we want to recover Bx.

Let us consider the case where matrix A is square and invertible, B is the identity, and $\mathcal{X}$ is a computationally tractable convex compact set. As far as computational aspects are concerned, the situation is well suited for utilizing the "magic wand" of Statistics—the *Maximum Likelihood* (ML) estimate where the recovery of x is

$$\widehat{x}_{\mathrm{ML}}(\omega) = \operatorname*{argmin}_{y\in\mathcal{X}} \|\omega - Ay\|_2 \tag{ML}$$

—the signal which maximizes, over $y \in \mathcal{X}$, the likelihood (the probability density) to get the observation we actually got. Indeed, with computationally tractable $\mathcal{X}$, (ML) is an explicit convex, and therefore efficiently solvable, optimization problem. Given the exclusive role played by the ML estimate in Statistics, perhaps the first question about our estimation problem is: *how good is the ML estimate?*

The goal of this exercise is to show that *in the situation we are interested in, the ML estimate can be "heavily nonoptimal," and this may happen even when the techniques we develop in Chapter 4 do result in an efficiently computable near-optimal linear estimate.*

To justify the claim, investigate the risk (4.2) of the ML estimate in the case where

$$\mathcal{X} = \{x \in \mathbf{R}^n : x_1^2 + \epsilon^{-2}\sum_{i=2}^n x_i^2 \le 1\}\ \&\ A = \mathrm{Diag}\{1, \epsilon^{-1}, ..., \epsilon^{-1}\},$$

ϵ and σ are small, and n is large, so that $\sigma^2(n-1) \ge 2$. Accompany your theoretical analysis by numerical experiments—compare the empirical risks of the ML estimate with theoretical and empirical risks of the linear estimate optimal under the circumstances.

Recommended setup: n runs through $\{256, 1024, 2048\}$, $\epsilon = \sigma$ runs through $\{0.01; 0.05; 0.1\}$, and signal x is generated as

$$x = [\cos(\phi); \sin(\phi)\epsilon\zeta],$$

where $\phi \sim$ Uniform$[0, 2\pi]$ and random vector ζ is independent of ϕ and is distributed uniformly on the unit sphere in $\mathbf{R}^{n-1}$.

Solution: We have $\{Ay : y \in \mathcal{X}\} = \{u : \|u\|_2 \le 1\}$, so that

$$\widehat{x}_{\mathrm{ML}}(\omega) = A^{-1} \operatorname*{argmin}_{\|u\|_2\le 1} \|\omega - u\|_2.$$

Let us look what happens when the true signal x is the first basic orth e_1 (which does belong to $\mathcal{X}$). In this case,

$$\widehat{x}(Ax + \sigma\xi) = A^{-1} \operatorname*{argmin}_{\|u\|_2\le 1} \|e_1 + \sigma\xi - u\|_2.$$

Now, it is easily seen that when $\eta \sim \mathcal{N}(0, I_k)$ and $\delta \in (0,1)$, we have

$$\mathrm{Prob}\left\{\|\eta\|_2^2 < (1-\delta)k\right\} \leq \exp\left\{\frac{k}{2}[\delta + \ln(1-\delta)]\right\}. \qquad (!)$$

In particular,

$$\mathrm{Prob}_{\eta\sim\mathcal{N}(0,I_k)}\left\{\|\eta\|_2^2 < \frac{3k}{4}\right\} \leq \exp\left\{\frac{k}{2}[1/4 + \ln(3/4)]\right\} \leq 0.9813,$$

implying that when $\sigma^2(n-1) \geq 2$ and $\xi \sim \mathcal{N}(0, \sigma^2 I_n)$, the probability of the event

$$\mathcal{E} = \Big\{\xi : \|e_1 + \sigma\xi\|_2^2 \geq \underbrace{\tfrac{3}{4}\sigma^2(n-1)}_{\geq \frac{3}{2}} \;\&\; \xi_1 \leq 0\Big\}$$

is at least some positive absolute constant C (one can take $C = (1 - 0.9813)/2$). When $\mathcal{E}$ takes place, we have $\|e_1 + \sigma\xi\|_2 \geq \sqrt{3/2} > 1$, whence

$$z := \underset{\|u\|_2 \leq 1}{\mathrm{argmin}}\, \|e_1 + \sigma\xi - u\|_2 = [e_1 + \sigma\xi]/\|e_1 + \sigma\xi\|_2,$$

implying that the first entry in z is $\leq \sqrt{2/3}$ and therefore the norm of the ML recovery error $e_1 - z$ is at least $1 - \sqrt{2/3}$. We see that under the circumstances the risk (4.2) of the ML estimate is at least a positive absolute constant $(1 - \sqrt{2/3})C$. On the other hand, the risk of the simplest linear estimate

$$\widehat{x} = \omega_1 = x_1 + \sigma\xi_1$$

is at most

$$\sqrt{\max_{x\in\mathcal{X}} \mathbf{E}_{\xi\sim\mathcal{N}}\left\{\sigma^2\xi_1^2 + \sum_{i=2}^n x_i^2\right\}} \leq \sqrt{\sigma^2 + \epsilon^2},$$

and we see that when σ, ϵ are small and $\sigma^2(n-1) \geq 2$, the risk of the ML estimate can be worse, by a large factor, than the risk of a simple linear estimate. It should be added that we are in the situation when $\mathcal{X}$ is an ellipsoid centered at the origin; in this case, first, it is easy to compute the minimal risk linear estimate, and, second, the latter estimate is near-optimal among all estimates, linear and nonlinear alike, see Proposition 4.5.

Here are the results of our numerical experiments, 1000 simulations per every pair (n, σ) with $\epsilon = \sigma$:

σ	$n = 256$			$n = 1024$			$n = 2048$		
0.01	0.010	0.010	0.019	0.010	0.010	0.035	0.011	0.011	0.065
0.05	0.061	0.060	0.241	0.064	0.064	0.340	0.066	0.066	0.424
0.10	0.129	0.128	0.434	0.132	0.128	0.511	0.133	0.136	0.570

For each combination of parameters, from left to right: upper risk bound for the linear estimate, empirical risk of the linear estimate, empirical risk of the ML estimate

Note that when $\epsilon \leq 1$, the risk of the trivial estimate $\widehat{x}(\omega) \equiv 0$ does not exceed 1, and that this trivial risk bound is not that far from the empirical risk of the ML estimate when $\sigma \geq 0.05$.

To make the solution self-contained, here is the demonstration of (!) via Cramer bounding:

$$\begin{array}{l}\forall \gamma > 0: \ \mathrm{Prob}\{\|\eta\|_2^2 \leq (1-\delta)k\}\exp\{-\gamma(1-\delta)k\} \leq \mathbf{E}\{\exp\{-\gamma\|\eta\|_2^2\}\} = (1+2\gamma)^{-\frac{k}{2}} \\ \Rightarrow \ \ln\left(\mathrm{Prob}\{\|\eta\|_2^2 \leq (1-\delta)k\}\right) \leq \inf\limits_{\gamma>0}\left[\gamma(1-\delta)k - \frac{k}{2}\ln(1+2\gamma)\right] = \frac{k}{2}\left[\delta + \ln(1-\delta)\right].\end{array}$$

6.4.2 Measurement Design in Signal Recovery

Exercise 4.2. [Measurement Design in Gaussian o.s.] As a preamble to the exercise, please read the story about the possible "physics" of Gaussian o.s. from Section 2.7.3.3. The summary of the story is as follows:

We consider the Measurement Design version of signal recovery in Gaussian o.s.; specifically, we are allowed to use observations

$$\omega = A_q x + \sigma\xi \qquad [\xi \sim \mathcal{N}(0, I_m)]$$

where

$$A_q = \mathrm{Diag}\{\sqrt{q_1}, \sqrt{q_2}, ..., \sqrt{q_m}\}A,$$

with a given $A \in \mathbf{R}^{m\times n}$ and vector q which we can select in a given convex compact set $\mathcal{Q} \subset \mathbf{R}^m_+$. The signal x underlying the observation is known to belong to a given ellitope $\mathcal{X}$. Your goal is to select $q \in \mathcal{Q}$ and a linear recovery $\omega \mapsto G^T\omega$ of the image Bx of $x \in \mathcal{X}$, with given B, resulting in the smallest worst-case, over $x \in \mathcal{X}$, expected $\|\cdot\|_2^2$ recovery risk. Modify, according to this goal, problem (4.12). Is it possible to end up with a tractable problem? Work out in full details the case when $\mathcal{Q} = \{q \in \mathbf{R}^m_+ : \sum_i q_i = m\}$.

Solution: For the same reasons as in the case of problem (4.12), we lose nothing by assuming that

$$\mathcal{X} = \{x \in \mathbf{R}^n : \exists t \in \mathcal{T} : x^T S_k x \leq t_k, \ 1 \leq k \leq K\},$$

(we use the same notation and assumptions as everywhere in this chapter). Let us find out the squared risk of a candidate estimate $G^T\omega$. We are in the situation when the i-th entry in ω is $\sqrt{q_i}[Ax]_i + \sigma\xi_i$, where $\xi_1, ..., \xi_m$ are independent $\mathcal{N}(0,1)$ random variables. Consequently, passing from the matrix variable G in linear recovery to variable $H = \mathrm{Diag}\{q_1^{1/2}, ..., q_m^{1/2}\}G$, we get

$$\begin{array}{l}\mathbf{E}_{\xi_1,...,\xi_m}\left\{\|G^T\omega - Bx\|_2^2\right\} \\ \quad = \ \mathbf{E}_{\xi_1,...,\xi_m}\left\{\|\sigma H^T \mathrm{Diag}\{q_1^{-1/2}, ..., q_m^{-1/2}\}[\xi_1; ...; \xi_m] + [H^T A - B]x\|_2^2\right\} \\ \quad = \ \sigma^2 \sum_{i=1}^m \|\mathrm{Row}_i[H]\|_2^2/q_i + \|[H^T A - B]x\|_2^2,\end{array}$$

where $\mathrm{Row}_i[H]$ is the transpose of the i-th row of H. Similarly to the case of (4.12), a natural upper bound on the worst-case, over $x \in \mathcal{X}$, squared $\|\cdot\|_2^2$-risk of recovery is

$$\begin{array}{rcl}\widehat{R}(H,q) & = & \sigma^2 \sum_{i=1}^m \|\mathrm{Row}_i[H]\|_2^2/q_i \\ & & + \min_\lambda \left\{\phi_{\mathcal{T}}(\lambda) : \lambda \geq 0, \left[\begin{array}{c|c}\sum_k \lambda_k S_k & B^T - A^T H \\ \hline B - H^T A & I_\nu\end{array}\right] \succeq 0\right\},\end{array}$$

and the Measurement Design analog of (4.12) is the optimization problem

$$\mathrm{Opt} = \min_{H,q,\lambda} \left\{ \sigma^2 \sum_i q_i^{-1} \mathrm{Row}_i^T[H]\mathrm{Row}_i[H] + \phi_{\mathcal{T}}(\lambda) : \right.$$
$$\left. \lambda \geq 0,\ q \in \mathcal{Q}, \left[\begin{array}{c|c} \sum_k \lambda_k S_k & B^T - A^T H \\ \hline B - H^T A & I_\nu \end{array}\right] \succeq 0 \right\},$$

where $\frac{a^Ta}{0}$ is $+\infty$ or 0 depending on whether a is nonzero or not. The problem is convex, since the function a^Ta/s is convex in (a,s) in the domain $s \geq 0$; indeed, the epigraph of this function is given by the Linear Matrix Inequality,

$$\tau \geq \frac{a^Ta}{s} \Leftrightarrow \left[\begin{array}{c|c} \tau & a^T \\ \hline a & sI \end{array}\right] \succeq 0$$

(Schur Complement Lemma) and thus is convex.

Finally, when $\mathcal{Q} = \{q \geq 0 : \sum_i q_i = m\}$, we can carry out partial minimization in q analytically:

$$\min_{q \in \mathcal{Q}} \sum_i \frac{\mathrm{Row}_i^T[H]\mathrm{Row}_i[H]}{q_i}$$

is achieved when

$$q_i = \frac{m\|\mathrm{Row}_i[H]\|_2}{\sum_j \|\mathrm{Row}_j[H]\|_2},\ 1 \leq i \leq m,$$

and is equal to

$$\frac{1}{m}\left(\sum_i \|\mathrm{Row}_i[H]\|_2\right)^2;$$

as it should be, this quantity is $\leq$ its counterpart $\mathrm{Tr}(H^TH) = \sum_i \|\mathrm{Row}_i[H]\|_2^2$ appearing in (4.12).

Exercise 4.3. [follow-up to Exercise 4.2] A translucent bar of length $n = 32$ is comprised of 32 consecutive segments of length 1 each, with density ρ_i of i-th segment known to belong to the interval $[\mu - \delta_i, \mu + \delta_i]$.

Sample translucent bar

The bar is lit from the left end; when light passes through a segment with density ρ, the light's intensity is reduced by factor $\mathrm{e}^{-\alpha\rho}$. The light intensity at the left endpoint of the bar is 1. You can scan the segments one by one from left to right and measure light intensity ℓ_i at the right endpoint of the i-th segment during time q_i; the result z_i of the measurement is $\ell_i \mathrm{e}^{\sigma\xi_i/\sqrt{q_i}}$, where the $\xi_i \sim \mathcal{N}(0,1)$ are independent across i. The total time budget is n, and you are interested in recovering the $m = n/2$-dimensional vector of densities of the right m segments. Build an optimization problem responsible for near-optimal linear recovery with and without Measurement Design (in the latter case, we assume that each segment is observed during unit time) and compare the resulting near-optimal risks.

Recommended data:

$$\alpha = 0.01,\ \delta_i = 1.2 + \cos(4\pi(i-1)/n),\ \mu = 1.1 \max_i \delta_i,\ \sigma = 0.001.$$

Solution: Passing from measurements to their logarithms, we arrive at observations

$$\bar{\omega}_i = -\alpha[\rho_1 + ... + \rho_i] + \sigma\xi_i/\sqrt{q_i}.$$

Passing from densities ρ_i to the differences $x_i = \rho_i - \mu$ and from observations $\bar{\omega}_i$ to observations $\omega_i = \bar{\omega}_i + \alpha\mu i$, we arrive at the Gaussian o.s.

$$\omega_i = -\alpha[x_1 + ... + x_i] + \sigma\xi_i/\sqrt{q_i}, \ 1 \le i \le n.$$

We are in the situation of Exercise 4.2 with square sensing matrix A proportional, with coefficient $-\alpha$, to lower-triangular $n \times n$ matrix with all ones in the lower-triangular part, with the "lower half" of the $n \times n$ unit matrix in the role of B, and with the box $\{x : |x_i| \le \delta_i, \ 1 \le i \le n\}$ in the role of $\mathcal{X}$; in other words, $\mathcal{X}$ is the ellitope given by

$$\left\{x : \exists t \in [0,1]^n : x^T S_k x := \delta_k^{-2} x_k^2 \le t_k, \ 1 \le k \le K := n\right\}.$$

Problem (4.12) in this case reads

$$\mathrm{Opt} = \min_{H,\lambda} \left\{\sigma^2 \mathrm{Tr}(H^T H) + \textstyle\sum_k \lambda_k : \lambda \ge 0, \left[\begin{array}{c|c} \mathrm{Diag}\{\lambda_i/\delta_i^2, 1 \le i \le n\} & B^T - A^T H \\ \hline B - H^T A & I_{n/2} \end{array}\right] \succeq 0\right\}.$$

This problem is different from the one of Measurement Design: the term $\sigma^2\mathrm{Tr}(H^T H)$ in the objective of the latter is replaced with the term $\frac{\sigma^2}{n}\left(\sum_{i=1}^n \|\mathrm{Row}_i[H]\|_2\right)^2$, and the optimal q_i's are obtained from the optimal solution (H_*, λ_*) via the relation

$$q_i^* = \frac{n\|\mathrm{Row}_i[H_*]\|_2}{\sum_j \|\mathrm{Row}_j[H_*]\|_2}, \ 1 \le i \le n.$$

Solving the problems, we find that with no Measurement Design, the risk of the near-optimal linear estimate is ≈ 0.5656; Measurement Design reduces this risk to ≈ 0.4103. The optimal q's are as shown on the plot below:

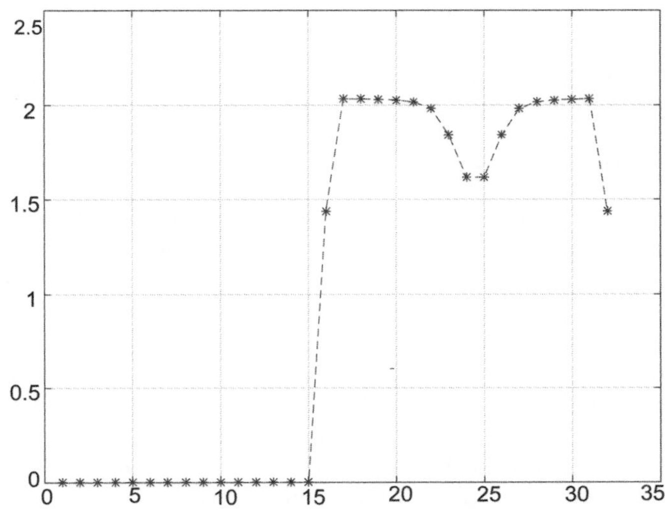

Exercise 4.4. Let X be a basic ellitope in $\mathbf{R}^n$:

$$X = \{x \in \mathbf{R}^n : \exists t \in \mathcal{T} : x^T S_k x \le t_k,\ 1 \le k \le K\}$$

with our usual restrictions on S_k and $\mathcal{T}$. Let, further, m be a given positive integer, and $x \mapsto Bx : \mathbf{R}^n \to \mathbf{R}^\nu$ be a given linear mapping. Consider the Measurement Design problem where you are looking for a linear recovery $\omega \mapsto \widehat{x}_H(\omega) := H^T\omega$ of Bx, $x \in X$, from observation

$$\omega = Ax + \sigma\xi \qquad [\sigma > 0 \text{ is given and } \xi \sim \mathcal{N}(0, I_m)]$$

in which the $m \times n$ sensing matrix A is under your control—it is allowed to be any $m \times n$ matrix of spectral norm not exceeding 1. You are interested in selecting H and A in order to minimize the worst-case, over $x \in X$, expected $\|\cdot\|_2^2$ recovery error. Similarly to (4.12), this problem can be posed as

$$\begin{array}{rl}\mathrm{Opt} = & \min_{H,\lambda,A} \Big\{ \sigma^2 \mathrm{Tr}(H^T H) + \phi_{\mathcal{T}}(\lambda) : \\ & \left[\begin{array}{c|c} \sum_k \lambda_k S_k & B^T - A^T H \\ \hline B - H^T A & I_\nu \end{array}\right] \succeq 0,\ \|A\| \le 1,\ \lambda \ge 0 \Big\},\end{array} \tag{4.68}$$

where $\|\cdot\|$ stands for the spectral norm. The objective in this problem is the (upper bound on the) squared risk $\mathrm{Risk}^2[\widehat{x}_H|X]$, the sensing matrix being A. The problem is nonconvex, since the matrix participating in the semidefinite constraint is bilinear in H and A.

A natural way to handle an optimization problem with objective and/or constraints bilinear in the decision variables u, v is to use "alternating minimization," where one alternates optimization in v for u fixed and optimization in u for v fixed, the value of the variable fixed in a round being the result of optimization w.r.t. this variable in the previous round. Alternating minimizations are carried out until the value of the objective (which in the outlined process definitely improves from round to round) stops improving (or nearly so). Since the algorithm does not necessarily converge to the globally optimal solution to the problem of interest, it makes sense to run the algorithm several times from different, say, randomly selected, starting points.

Now comes the exercise.

1. Implement Alternating Minimization as applied to (4.68). You may restrict your experimentation to the case where the sizes m, n, ν are quite moderate, in the range of tens, and X is either the box $\{x : j^{2\gamma} x_j^2 \le 1, 1 \le j \le n\}$, or the ellipsoid $\{x : \sum_{j=1}^n j^{2\gamma} x_j^2 \le 1\}$, where γ is a nonnegative parameter (try $\gamma = 0, 1, 2, 3$). As for B, you can generate it at random, or enforce B to have prescribed singular values, say, $\sigma_j = j^{-\theta}$, $1 \le j \le \nu$, and a randomly selected system of singular vectors.
2. Identify cases where a globally optimal solution to (4.68) is easy to find and use this information in order to understand how reliable Alternating Minimization is in the application in question, reliability meaning the ability to identify near-optimal, in terms of the objective, solutions.

   If you are not satisfied with Alternating Minimization "as is," try to improve it.

Solution: When σ is small and $m \ge \nu$, we clearly can recover Bx with high accuracy. To this end, we can select A as the projector on the orthogonal complement to the kernel of B; note that when $\sigma = 0$, this selection of A allows us to recover Bx *exactly.* This being said, in our experiments Alternating minimization "as is" typically was unable to take full advantage of the just outlined favorable circumstances. Seemingly the simplest way to cure this drawback is to start with singular value decomposition of B and to select the initial A for alternating minimization as the projector on the linear span $\mathcal{L}$ of the $\mu = \min[m, \mathrm{Rank}(B)]$ leading right singular vectors of B ($\mathcal{L}$ can be thought of as the best approximation of $\mathrm{Ker}^\perp(B)$ by a linear subspace of dimension $\le m$).

3. Modify (4.68) and your experiment to cover the cases where the constraint $\|A\| \leq 1$ on the sensing matrix is replaced with one of the following:
   - $\|\mathrm{Row}_i[A]\|_2 \leq 1,\ 1 \leq i \leq m$,
   - $|A_{ij}| \leq 1$ for all i, j

   (note that these two types of restrictions mimic what happens if you are interested in recovering (the linear image of) the vector of parameters in a linear regression model from noisy observations of the model's outputs at m points which you are allowed to select in the unit ball or unit box).
4. [Embedded Exercise] Recall that a $\nu \times n$ matrix G admits *singular value decomposition* $G = UDV^T$ with orthogonal matrices $U \in \mathbf{R}^{\nu\times\nu}$ and $V \in \mathbf{R}^{n\times n}$ and diagonal $\nu \times n$ matrix D with nonnegative and nonincreasing diagonal entries.[5] These entries are uniquely defined by G and are called *singular values* $\sigma_i(G)$, $1 \leq i \leq \min[\nu, n]$. Singular values admit characterization similar to variational characterization of eigenvalues of a symmetric matrix; see, e.g., [15, Section A.7.3]:

   Theorem 4.23 *[VCSV—Variational Characterization of Singular Values] For a $\nu \times n$ matrix G it holds*
   $$\sigma_i(G) = \min_{E\in\mathcal{E}_i} \max_{e\in E, \|e\|_2=1} \|Ge\|_2,\ 1 \leq i \leq \min[\nu, n],$$
   where $\mathcal{E}_i$ is the family of all subspaces in $\mathbf{R}^n$ of codimension $i-1$.

   Corollary 4.24 *[SVI—Singular Value Interlacement] Let G and G' be $\nu \times n$ matrices, and let $k = \mathrm{Rank}(G - G')$. Then*
   $$\sigma_i(G) \geq \sigma_{i+k}(G'),\ 1 \leq i \leq \min[\nu, n], \tag{4.69}$$
   where, by definition, singular values of a $\nu \times n$ matrix with indexes $> \min[\nu, n]$ are zeros.

   We denote by $\sigma(G)$ the vector of singular values of G arranged in nonincreasing order. The function $\|G\|_{\mathsf{Sh},p} = \|\sigma(G)\|_p$ is called the *Shatten p-norm* of matrix G; this indeed is a norm on the space of $\nu \times n$ matrices, and the conjugate norm is $\|\cdot\|_{\mathsf{Sh},q}$, with $\frac{1}{p} + \frac{1}{q} = 1$. An easy and important consequence of Corollary 4.24 is the following fact:

   Corollary 4.25 *Given a $\nu \times n$ matrix G, an integer k, $0 \leq k \leq \min[\nu, n]$, and $p \in [1, \infty]$, (one of) the best approximations of G in the Shatten p-norm among matrices of rank $\leq k$ is obtained from G by zeroing our all but k largest singular values, that is, the matrix $G^k = \sum_{i=1}^k \sigma_i(G)\mathrm{Col}_i[U]\mathrm{Col}_i^T[V]$, where $G = UDV^T$ is the singular value decomposition of G.*

   Prove Theorem 4.23 and Corollaries 4.24 and 4.25.

Solution: Let us start with VCSV. Let us denote the right hand sides in (4.69) by $\sigma'_i(G)$. Observe that when we multiply G from the left and from the right by orthogonal matrices, the quantities $\sigma_i(G)$ and $\sigma'_i(G)$ remain intact, so that we lose nothing when considering the case when G is a $\nu \times n$ matrix with the only nonzero entries in diagonal cells, these entries being nonnegative. Given $i \leq \min[\nu, n]$, let E_i be the linear subspace of $\mathbf{R}^n$ spanned by the first i standard basic orths in $\mathbf{R}^n$. The dimension of this space is i, and for every unit vector e from E_i we have $\|Ge\|_2 \geq \sigma_i(G)$. When E is a linear subspace of $\mathbf{R}^N$ of codimension $i-1$, the intersection of E and E_i is a nontrivial linear subspace of $\mathbf{R}^n$, so that it contains a unit vector e, and since $e \in E_i$, we have $\|Ge\|_2 \geq \sigma_i(g)$. Consequently, $\max_{e\in E:\|e\|_2=1} \|Ge\|_2 \geq \sigma_i(G)$, implying that $\sigma'_i(G) \geq \sigma_i(G)$. On the other hand, the linear span E^i of all but the first $i-1$ standard basic orths in $\mathbf{R}^n$ has codimension $i-1$, and we clearly have $\max_{e\in E^i:\|e\|_2=1} \|Ge\|_2 \leq \sigma_i(G)$, implying that $\sigma'_i(G) \leq \sigma_i(G)$. The bottom line is that $\sigma^i(G) = \sigma'_i(G)$ for all i. □

[5] We say that a rectangular matrix D is diagonal if all entries D_{ij} in D with $i \neq j$ are zeros.

Now let us prove SVI. Let us fix $i \leq \min[\nu, n]$. The inequality $\sigma_i(G) \geq \sigma_{i+k}(G')$ is evident when $i + k > \min[\nu, n]$, since in this case $\sigma_{i+k}(G') = 0$, while $\sigma_i(G)$ is nonnegative. Now consider the case when $i + k \leq \min[\nu, n]$. Let $\Delta = G - G'$, so that the kernel F of Δ is a linear subspace of $\mathbf{R}^n$ of codimension at most $k = \mathrm{Rank}(\Delta)$. For every linear subspace E of $\mathbf{R}^n$ of codimension $i-1$, the linear subspace $F \cap E$ of $\mathbf{R}^n$ is of codimension at most $i - 1 + k$, implying by VCSV that $F \cap E$ contains a unit vector e with $\|G'e\|_2 \geq \sigma_{i+k}(G')$. Since $e \in F$, we have $Ge = G'e$, so that E contains a unit vector e with $\|Ge\|_2 \geq \sigma_{i+k}(G')$. Since this is so for every linear subspace E in $\mathbf{R}^n$ of codimension $\leq i - 1$, VCSV says that $\sigma_i(G) \geq \sigma_{i+k}(G')$. □

Corollary 4.25: The Shatten p-norm of $G - G^k$ clearly is the p-norm of the $\mu = \min[\nu, n]$-dimensional vector $\bar{\sigma}$ obtained from $\sigma(G)$ by zeroing out the first k entries. On the other hand, let H be a $\nu \times n$ matrix of rank $\leq k$; by SVI, the singular values $\sigma_i(G-H)$ satisfy the inequalities $\sigma_i(G-H) \geq \sigma_{i+k}(G)$, implying that $\|G - H\|_{\mathrm{Sh},p} = \|\sigma(G - H)\|_p \geq \|\bar{\sigma}\|_p = \|G - G^k\|_{\mathrm{Sh},p}$. Thus, G^k is one of the matrices of rank $\leq k$ closest to G in $\|\cdot\|_{\mathrm{Sh},p}$-norm. □

5. Consider the Measurement Design problem (4.68) in the case when X is an ellipsoid:

$$X = \left\{x \in \mathbf{R}^n : \sum\nolimits_{j=1}^n x_i^2/a_j^2 \leq 1\right\},$$

A is an $m \times n$ matrix of spectral norm not exceeding 1, and there is no noise in the observations: $\sigma = 0$. Find an optimal solution to this problem. Think how this result can be used to get a (hopefully) good starting point for Alternating Minimization in the case when X is an ellipsoid and σ is small.

Solution: Without loss of generality, we can assume that $m \leq \nu \leq n$ (why?). When $\sigma = 0$, (4.68) reduces to minimizing the quantity

$$\max_{x \in X} \|[H^T A - B]x\|_2 = \max_{y \in \mathbf{R}^n, \|y\|_2 \leq 1} \|[H^T AD - BD]y\|_2, \;\; D = \mathrm{Diag}\{a_1, ..., a_n\}$$

over H and A with an $m \times n$ matrix A with spectral norm ≤ 1. Passing from variables H and A to a single matrix variable $G = H^T AD$ and setting $\bar{B} = BD$, observe that the only constraint on G in the case of $m \leq \nu \leq n$ is that $\mathrm{Rank}(G) \leq m$. Thus, all we need is to find the best, in the spectral norm (i.e., in the Shatten ∞-norm) approximation of the matrix $\bar{B}$ by a matrix of rank $\leq m$. The solution G_* to this problem is readily given by Corollary 4.25. After G_* is found, it suffices to represent the $\nu \times n$ matrix $G_* D^{-1}$ (which has rank $\leq m$ along with G_*) as the product of a $\nu \times m$ matrix H^T and an $m \times n$ matrix A of norm not exceeding 1. The decomposition is readily given by the singular value decomposition UDV^T of $G_* D^{-1}$: since this matrix is of rank $\leq m$, the number of its nonzero singular values is at most m, so that the SVD of the matrix is $G_* D^{-1} = \sum_{i=1}^m \sigma_i \mathrm{Col}_i[U]\mathrm{Col}_i^T[V]$, and we can set $A = [\mathrm{Col}_1^T[V]; ...; \mathrm{Col}_m^T[V]]$ and $H^T = [\sigma_1 \mathrm{Col}_1[U], ..., \sigma_m \mathrm{Col}_m[U]]$.

6.4.3 Around semidefinite relaxation

Exercise 4.5. Let $\mathcal{X}$ be an ellitope:

$$\mathcal{X} = \{x \in \mathbf{R}^n : \exists (y \in \mathbf{R}^N, t \in \mathcal{T}) : x = Py, y^T S_k y \leq t_k, k \leq K\}$$

with our standard restrictions on $\mathcal{T}$ and S_k. Representing $S_k = \sum_{j=1}^{r_k} s_{kj}s_{kj}^T$, we can pass from initial ellitopic representation of $\mathcal{X}$ to the spectratopic representation of the same set:

$$\mathcal{X} = \{x \in \mathbf{R}^n : \exists (y \in \mathbf{R}^N, t^+ \in \mathcal{T}^+) : x = Py, [s_{kj}^T x]^2 \preceq t_{kj}^+ I_1, 1 \le k \le K, 1 \le j \le r_k\}$$
$$\left[\mathcal{T}^+ = \{t^+ = \{t_{kj}^+ \ge 0\} : \exists t \in \mathcal{T} : \textstyle\sum_{j=1}^{r_k} t_{kj}^+ \le t_k, \, 1 \le k \le K\}\right].$$

If now C is a symmetric $n \times n$ matrix and $\mathrm{Opt} = \max_{x \in \mathcal{X}} x^T C x$, we have

$$\begin{array}{rcl}\mathrm{Opt}_* & \le & \mathrm{Opt}_e := \min\limits_{\lambda = \{\lambda_k \in \mathbf{R}_+\}} \left\{\phi_{\mathcal{T}}(\lambda) : P^T C P \preceq \sum_k \lambda_k S_k\right\} \\ \mathrm{Opt}_* & \le & \mathrm{Opt}_s := \min\limits_{\Lambda = \{\Lambda_{kj} \in \mathbf{R}_+\}} \left\{\phi_{\mathcal{T}^+}(\Lambda) : P^T C P \preceq \sum_{k,j} \Lambda_{kj} s_{kj} s_{kj}^T\right\}\end{array}$$

where the first relation is yielded by ellitopic representation of $\mathcal{X}$ and Proposition 4.6, and the second, on closer inspection (carry this inspection out!) by the spectratopic representation of $\mathcal{X}$ and Proposition 4.8.

Prove that $\mathrm{Opt}_e = \mathrm{Opt}_s$.

<u>Solution:</u> For $\Lambda = \{\Lambda_{kj} \ge 0, 1 \le k \le K, 1 \le j \le r_k\}$, we have

$$\begin{array}{rcl}\phi_{\mathcal{T}^+}(\Lambda) & = & \max\limits_{t_{kj}} \left\{\sum_{k,j} t_{kj} \Lambda_{kj} : t_{kj} \ge 0, \exists t \in \mathcal{T} : \sum_j t_{kj} \le t_k \, \forall k\right\} \\ & = & \max\limits_{t \in \mathcal{T}} \sum_{k=1}^K t_k \max_j \Lambda_{kj} = \phi_{\mathcal{T}}(\underbrace{[\lambda_1(\Lambda); ...; \lambda_K(\Lambda)]}_{\lambda(\Lambda)}), \ \lambda_k(\Lambda) = \max\limits_j \Lambda_{kj}.\end{array}$$

It follows that if Λ is a feasible solution to the optimization problem specifying Opt_s and $\overline{\Lambda}$ is obtained from Λ by replacing all Λ_{kj}, $1 \le j \le r_k$, with $\lambda_k(\Lambda)$, then $\overline{\Lambda}$ is another feasible solution to the same problem and with the same value of the objective, namely, the value equal to $\phi_{\mathcal{T}}(\lambda(\Lambda))$. We conclude that

$$\begin{array}{rcl}\mathrm{Opt}_s & = & \min\limits_{\Lambda = \{\Lambda_{kj} = \lambda_k \ge 0\}} \left\{\phi_{\mathcal{T}}(\lambda) : P^T C P \preceq \sum_{k,j} \Lambda_{kj} s_{kj} s_{kj}^T\right\} \\ & = & \min\limits_{\lambda = \{\lambda_k \ge 0\}} \left\{\phi_{\mathcal{T}}(\lambda) : P^T C P \preceq \sum_k \lambda_k \underbrace{\left[\sum_j s_{kj} s_{kj}^T\right]}_{S_k}\right\} = \mathrm{Opt}_e.\end{array}$$

Exercise 4.6. Proposition 4.6 provides us with an upper bound on the quality of semidefinite relaxation as applied to the problem of upper-bounding the maximum of a homogeneous quadratic form over an ellitope. Extend the construction to the case when an inhomogeneous quadratic form is maximized over a shifted ellitope, so that the quantity to upper-bound is

$$\begin{array}{c}\mathrm{Opt} = \max\limits_{x \in X} \left[f(x) := x^T A x + 2b^T x + c\right], \\ X = \{x : \exists (y, t \in \mathcal{T}) : x = Py + p, y^T S_k y \le t_k, 1 \le k \le K\}\end{array}$$

with our standard assumptions on S_k and $\mathcal{T}$.

<u>Note:</u> X is centered at p, and a natural upper bound on Opt is

$$\mathrm{Opt} \le f(p) + \widehat{\mathrm{Opt}},$$

where $\widehat{\mathrm{Opt}}$ is an upper bound on the quantity

$$\overline{\mathrm{Opt}} = \max_{x \in X} \left[f(x) - f(p)\right].$$

What you are interested in upper-bounding is the ratio $\widehat{\mathrm{Opt}}/\overline{\mathrm{Opt}}$.

Solution: Passing from variables x to variables $u = x - p$, we get

$$\begin{array}{rcl}\overline{\mathrm{Opt}} & = & \max_{u,y,t}\left\{u^TAu + 2e^Tu : u = Py, y^TS_ky \le t_k, k \le K, t \in \mathcal{T}\right\}\\ & = & \max_{y,t}\left\{y^T[P^TAP]y + 2[Pe]^Ty : y^TS_ky \le t_k, k \le K, t \in \mathcal{T}\right\}\\ & \underbrace{=}_{(*)} & \max_{y,t}\left\{g(y,s) := y^T[P^TAP]y + 2s[Pe]^Ty : \begin{array}{l} y^TS_ky \le t_k, k \le K\\ s^2 \le 1, t \in \mathcal{T}\end{array}\right\}\end{array}$$

where $(*)$ is due to the symmetry w.r.t. the origin of the set $\{y : \exists t \in \mathcal{T} : y^TS_ky \le t_k, k \le K\}$. The bottom line is that $\overline{\mathrm{Opt}}$ is the maximum of the homogeneous quadratic form

$$g(y,s) := [y;s]^T\underbrace{\left[\begin{array}{c|c} P^TAP & Pe\\ \hline e^TP^T & \end{array}\right]}_{C^+}[y;s]$$

over the ellitope

$$\begin{array}{l}X^+ = \{z := [y;s] : \exists t^+ \in \mathcal{T}^+ : z^TS_k^+z \le t_k^+, 1 \le k \le K^+ = K+1\}\\ \left[\mathcal{T}^+ = [t;\tau] : t \in \mathcal{T}, 0 \le \tau \le 1,\ S_k^+ = \left\{\begin{array}{ll}\left[\begin{array}{c|c} S_k & \\ \hline & \end{array}\right], & k \le K,\\ \left[\begin{array}{c|c} & \\ \hline & 1\end{array}\right], & k = K+1\end{array}\right.\right].\end{array}$$

Applying Proposition 4.6, the efficiently computable quantity

$$\widehat{\mathrm{Opt}} = \min_\lambda\left\{\phi_{\mathcal{T}^+}(\lambda) : \lambda \ge 0, C^+ \preceq \sum_{k=1}^{K+1}\lambda_kS_k^+\right\}$$

satisfies

$$\overline{\mathrm{Opt}} \le \widehat{\mathrm{Opt}} \le 3\ln(\sqrt{3}(K+1))\overline{\mathrm{Opt}}.$$

Exercise 4.7. [estimating Kolmogorov widths of sperctratopes/ellitopes]

4.7.A. Preliminaries: Kolmogorov and Gelfand widths. Let $\mathcal{X}$ be a convex compact set in $\mathbf{R}^n$, and let $\|\cdot\|$ be a norm on $\mathbf{R}^n$. Given a linear subspace E in $\mathbf{R}^n$, let

$$\mathrm{dist}_{\|\cdot\|}(x,E) = \min_{z\in E}\|x - z\| : \mathbf{R}^n \to \mathbf{R}_+$$

be the $\|\cdot\|$-distance from x to E. The quantity

$$\mathrm{dist}_{\|\cdot\|}(\mathcal{X},E) = \max_{x\in\mathcal{X}}\mathrm{dist}_{\|\cdot\|}(x,E)$$

can be viewed as the worst-case $\|\cdot\|$-accuracy to which vectors from $\mathcal{X}$ can be approximated by vectors from E. Given positive integer $m \le n$ and denoting by $\mathcal{E}_m$ the family of all linear subspaces in $\mathbf{R}^m$ of dimension m, the quantity

$$\delta_m(\mathcal{X},\|\cdot\|) = \min_{E\in\mathcal{E}_m}\mathrm{dist}_{\|\cdot\|}(\mathcal{X},E)$$

can be viewed as the best achievable quality of approximation, measured in $\|\cdot\|$, of vectors from $\mathcal{X}$ by vectors from an m-dimensional linear subspace of $\mathbf{R}^n$. This quantity is called the *m-th Kolmogorov width* of $\mathcal{X}$ w.r.t. $\|\cdot\|$.

Observe that one has

$$\begin{array}{l}\mathrm{dist}_{\|\cdot\|}(x,E)=\max_\xi\{\xi^Tx:\|\xi\|_*\leq 1,\xi\in E^\perp\},\\ \mathrm{dist}_{\|\cdot\|}(\mathcal{X},E)=\max\limits_{\substack{x\in\mathcal{X},\\ \|\xi\|_*\leq 1,\xi\in E^\perp}}\xi^Tx\end{array}\tag{4.70}$$

where $E^\perp$ is the orthogonal complement to E.

1) Prove (4.70).
Hint: Represent $\mathrm{dist}_{\|\cdot\|}(x,E)$ as the optimal value in a conic problem on the cone $\mathbf{K}=\{[x;t]:t\geq\|x\|\}$ and use the Conic Duality Theorem.

Solution: Representing $E=\mathrm{Ker}\,A$ for a properly selected matrix A, we clearly have

$$\mathrm{dist}_{\|\cdot\|}(x,E)=\min_{[z;t]}\{t:Az=0,[x-z;t]\in\mathbf{K}\},\tag{$*$}$$

and the conic problem just specified clearly is strictly feasible and bounded. It is immediately seen that the cone $\mathbf{K}_*$ dual to $\mathbf{K}$ is $\{[y;s]:s\geq\|y\|_*\}$, so that the problem dual to $(*)$ reads

$$\begin{array}{l}\max\limits_{\lambda,y,s}\left\{-y^Tx:\|y\|_*\leq s,-y^Tz+ts+\lambda^TAz=t\forall(z,t)\right\}\\ \quad\Leftrightarrow\max\limits_{\lambda,y}\left\{-y^Tx:\|y\|_*\leq 1,y=A^T\lambda\right\}\\ \quad\Leftrightarrow\max\limits_{y}\{y^Tx:y\in\mathrm{Im}A^T,\|y\|_*\leq 1\}.\end{array}$$

To justify the first equality in (4.70), it remains to note that the optimal value in the dual is equal to the one in the primal, and that $E=\mathrm{Ker}\,A$ implies $E^\perp=\mathrm{Im}A^T$. The second equality in (4.70) is an immediate consequence of the first one. □

Now consider the case when $\mathcal{X}$ is the unit ball of some norm $\|\cdot\|_{\mathcal{X}}$. In this case (4.70) combines with the definition of Kolmogorov width to imply that

$$\begin{array}{rcl}\delta_m(\mathcal{X},\|\cdot\|)&=&\min\limits_{E\in\mathcal{E}_m}\mathrm{dist}_{\|\cdot\|}(x,E)=\min\limits_{E\in\mathcal{E}_m}\max\limits_{x\in\mathcal{X}}\max\limits_{y\in E^\perp,\|y\|_*\leq 1}y^Tx\\ &=&=\min\limits_{E\in\mathcal{E}_m}\max\limits_{y\in E^\perp,\|y\|_*\leq 1}\max\limits_{x:\|x\|_{\mathcal{X}}\leq 1}y^Tx\\ &=&\min\limits_{F\in\mathcal{E}_{n-m}}\max\limits_{y\in F,\|y\|_*\leq 1}\|y\|_{\mathcal{X},*},\end{array}\tag{4.71}$$

where $\|\cdot\|_{\mathcal{X},*}$ is the norm conjugate to $\|\cdot\|_{\mathcal{X}}$. Note that when $\mathcal{Y}$ is a convex compact set in $\mathbf{R}^n$ and $|\cdot|$ is a norm on $\mathbf{R}^n$, the quantity

$$d^m(\mathcal{Y},|\cdot|)=\min_{F\in\mathcal{E}_{n-m}}\max_{y\in\mathcal{Y}\cap F}|y|$$

has a name—it is called the m-th *Gelfand width* of $\mathcal{Y}$ taken w.r.t. $|\cdot|$. The "duality relation" (4.71) states that

When $\mathcal{X}$, $\mathcal{Y}$ are the unit balls of respective norms $\|\cdot\|_{\mathcal{X}}$, $\|\cdot\|_{\mathcal{Y}}$, for every $m<n$ the m-th Kolmogorov width of $\mathcal{X}$ taken w.r.t. $\|\cdot\|_{\mathcal{Y},*}$ is the same as m-th Gelfand width of $\mathcal{Y}$ taken w.r.t. $\|\cdot\|_{\mathcal{X},*}$.

The goal of the remaining part of the exercise is to use our results on the quality of semidefinite relaxation on ellitopes/spectratopes to infer efficiently computable upper bounds on Kolmogorov widths of a given set $\mathcal{X}\subset\mathbf{R}^n$. In the sequel we assume that

- $\mathcal{X}$ is a spectratope,

$$\mathcal{X}=\{x\in\mathbf{R}^n:\exists(t\in\mathcal{T},u):x=Pu,R_k^2[u]\preceq t_kI_{d_k},k\leq K\};$$

- The unit ball $\mathcal{B}_*$ of the norm conjugate to $\|\cdot\|$ is a spectratope:

$$\mathcal{B}_* = \{y : \|y\|_* \leq 1\} = \{y \in \mathbf{R}^n : \exists (r \in \mathcal{R}, z) : y = Mz, S_\ell^2[z] \preceq r_\ell I_{f_\ell}, \ell \leq L\}.$$

with our usual restrictions on $\mathcal{T}, \mathcal{R}$ and $R_k[\cdot]$ and $S_\ell[\cdot]$.

4.7.B. Simple case: $\|\cdot\| = \|\cdot\|_2$. We start with the *simple case* where $\|\cdot\| = \|\cdot\|_2$, so that $\mathcal{B}_*$ is the ellitope $\{y : y^T y \leq 1\}$.

Let $D = \sum_k d_k$ be the size of the spectratope $\mathcal{X}$, and let

$$\varkappa = 2\max[\ln(2D), 1].$$

Given integer $m < n$, consider the convex optimization problem

$$\begin{array}{rcl}\mathrm{Opt}(m) & = & \min_{\Lambda=\{\Lambda_k, k\leq K\}, Y} \Big\{\phi_{\mathcal{T}}(\lambda[\Lambda]) : \Lambda_k \succeq 0 \forall k, 0 \preceq Y \preceq I_n, \\ && \sum_k \mathcal{S}_k^*[\Lambda_k] \succeq P^T Y P, \ \mathrm{Tr}(Y) = n - m\Big\}.\end{array} \qquad (P_m)$$

2) Prove the following

Proposition 4.26 *Whenever* $1 \leq \mu \leq m < n$, *one has*

$$\mathrm{Opt}(m) \leq \varkappa \delta_m^2(\mathcal{X}, \|\cdot\|_2) \ \& \ \delta_m^2(\mathcal{X}, \|\cdot\|_2) \leq \frac{m+1}{m+1-\mu}\mathrm{Opt}(\mu). \qquad (4.72)$$

Moreover, the above upper bounds on $\delta_m(\mathcal{X}, \|\cdot\|_2)$ *are "constructive," meaning that an optimal solution to* (P_μ), $\mu \leq m$, *can be straightforwardly converted into a linear subspace* $E^{m,\mu}$ *of dimension* m *such that*

$$\mathrm{dist}_{\|\cdot\|_2}(\mathcal{X}, E^{m,\mu}) \leq \sqrt{\frac{m+1}{m+1-\mu}\mathrm{Opt}(\mu)}.$$

Finally, $\mathrm{Opt}(\mu)$ *is nonincreasing in* $\mu < n$.

Solution: Let E_m be an m-dimensional linear subspace of $\mathbf{R}^n$ such that $\delta_m(\mathcal{X}, \|\cdot\|_2) = \max_{x\in\mathcal{X}} \mathrm{dist}_{\|\cdot\|_2}(x, E_m)$, and let Y_m be the projector onto the orthogonal complement of E_m. The maximum of the quadratic form $x^T Y_m x$ over $\mathcal{X}$ is nothing but $\delta_m^2(\mathcal{X}, \|\cdot\|_2)$, implying by Proposition 4.8 that there exists a collection $\Lambda = \{\Lambda_k \succeq 0, k \leq K\}$ such that $\phi_{\mathcal{T}}(\lambda[\lambda]) \leq \varkappa \delta_m^2(\mathcal{X}, \|\cdot\|_2)$ and $\sum_k \mathcal{S}_k^*[\Lambda_k] \succeq P^T Y_m P$. We see that (Λ, Y_m) is a feasible solution of (P_m) with the value of the objective at most $\varkappa\delta_m^2(\mathcal{X}, \|\cdot\|_2)$, which proves the first inequality in (4.72).

To prove the second inequality, let (Λ^*, Y_*) be an optimal solution to (P_μ), and let $\nu_1 \leq \nu_2 \leq \ldots \leq \nu_n$ be the eigenvalues of Y_*, so that $0 \leq \nu_s \leq 1$, $s \leq n$. Denoting by $E^{m,\mu}$ the linear span of the first m eigenvectors of Y_*, we have dim $E^{m,\mu} = m$ and

$$\begin{array}{l} n - \mu = \mathrm{Tr}(Y_*) = \sum_i \nu_i \leq (m+1)\nu_{m+1} + (n-m-1) \Rightarrow \nu_{m+1} \geq \frac{m+1-\mu}{m+1} \\ \Rightarrow x^T Y_* x \geq \frac{m+1-\mu}{m+1}\mathrm{dist}^2_{\|\cdot\|_2}(x, E^{m,\mu}) \Rightarrow \mathrm{dist}^2_{\|\cdot\|_2}(x, E^{m,\mu}) \leq \frac{m+1}{m+1-\mu} x^T Y_* x \\ \Rightarrow \delta_m^2(\mathcal{X}, \|\cdot\|_2) \leq \max_{x\in\mathcal{X}} \mathrm{dist}^2_{\|\cdot\|_2}(x, E^{m,\mu}) \leq \frac{m+1}{m+1-\mu} \max_{x\in\mathcal{X}} x^T Y_* x. \end{array}$$

On the other hand, (Λ^*, Y_*) is feasible for (P_μ), implying by Proposition 4.8 that $\max\limits_{x\in\mathcal{X}} x^T Y_* x \leq \phi_{\mathcal{T}}(\lambda[\Lambda^*]) = \mathrm{Opt}(\mu)$, and we arrive at the second inequality in (4.72).

Finally, if (Λ, Y) is feasible for (P_k) and $k \leq k' < n$, then (Λ, Y'), with $Y' = \frac{n-k'}{n-k} Y$, is feasible for $(P_{k'})$, implying that $\mathrm{Opt}(k) \geq \mathrm{Opt}(k')$. □

Remark 6.15. Inspecting the above proof, we arrive at the bound

$$\delta_m^2(\mathcal{X}, \|\cdot\|_2) \leq \mathrm{dist}^2_{\|\cdot\|_2}(\mathcal{X}, E^{m,\mu}) \leq \nu_{m+1}^{-1}\mathrm{Opt}(\mu)$$

where ν_{m+1} is the $(m+1)$-st smallest eigenvalue of the Y-component of an optimal solution to (P_μ). This bound holds true whenever $\mu < n$, and when $\mu \leq m$, it is at least as good as the upper bound on Kolmogorov width stated in Proposition 4.26.

Remark 6.16. When $\mathcal{X}$ is an ellitope,

$$\mathcal{X} = \{x \in \mathbf{R}^n : \exists (t \in \mathcal{T}, y) : x = Py, y^T S_k y \leq t_k, k \leq K\},$$

setting

$$\mathrm{Opt}(m) = \min_{\lambda, Y}\left\{\sum_k \lambda_k : \lambda \geq 0, \sum_k \lambda_k S_k \succeq P^T Y P, 0 \preceq Y \preceq I_n, \mathrm{Tr}(Y) = n - m\right\},$$

the same reasoning as in the proof of the lemma with Proposition 4.6 in the role of Proposition 4.8 results in the validity of *(4.72)* with $\varkappa = 3\ln(\sqrt{3}K)$.

4.7.C. General case. Now consider the case when both $\mathcal{X}$ and the unit ball $\mathcal{B}_*$ of the norm conjugate to $\|\cdot\|$ are spectratopes. As we are about to see, this case is essentially more difficult than the case of $\|\cdot\| = \|\cdot\|_2$, but something still can be done.

3) Prove the following statement:
(!) *Given $m < n$, let Y be a projector of $\mathbf{R}^n$ of rank $n - m$, and let collections $\Lambda = \{\Lambda_k \succeq 0, k \leq K\}$ and $\Upsilon = \{\Upsilon_\ell \succeq 0, \ell \leq L\}$ satisfy the relation*

$$\left[\begin{array}{c|c} \sum_k \mathcal{R}_k^*[\Lambda_k] & \frac{1}{2}P^T Y M \\ \hline \frac{1}{2}M^T Y P & \sum_\ell \mathcal{S}_\ell^*[\Upsilon_\ell] \end{array}\right] \succeq 0. \tag{4.73}$$

Then

$$\mathrm{dist}_{\|\cdot\|}(\mathcal{X}, \mathrm{Ker}\, Y) \leq \phi_{\mathcal{T}}(\lambda[\Lambda]) + \phi_{\mathcal{R}}(\lambda[\Upsilon]). \tag{4.74}$$

As a result,

$$\begin{array}{rcl} \delta_m(\mathcal{X}, \|\cdot\|) & \leq & \mathrm{dist}_{\|\cdot\|}(\mathcal{X}, \mathrm{Ker}\, Y) \\ & \leq & \mathrm{Opt} := \min\limits_{\substack{\Lambda = \{\Lambda_k, k \leq K\}, \\ \Upsilon = \{\Upsilon_\ell, \ell \leq L\}}} \left\{ \phi_{\mathcal{T}}(\lambda[\Lambda]) + \phi_{\mathcal{R}}(\lambda[\Upsilon]) : \right. \\ & & \Lambda_k \succeq 0\,\forall k, \Upsilon_\ell \succeq 0\,\forall \ell, \\ & & \left. \left[\begin{array}{c|c} \sum_k \mathcal{R}_k^*[\Lambda_k] & \frac{1}{2}P^T Y M \\ \hline \frac{1}{2}M^T Y P & \sum_\ell \mathcal{S}_\ell^*[\Upsilon_\ell] \end{array}\right] \succeq 0 \right\}. \end{array} \tag{4.75}$$

<u>Solution:</u> Let $F = \mathrm{Im} Y$, and let $\mathcal{Y} = \{y \in F : \|y\|_* \leq 1\}$. Let us fix $y \in \mathcal{Y}$, so that $y = Yy$ due to $y \in F$ and $y = Mz$ with z satisfying for some $r \in \mathcal{R}$ relations $S_\ell^2[z] \preceq r_\ell I_{f_\ell}$, $\ell \leq L$. Now let $x \in \mathcal{X}$, so that $x = Pu$ with u satisfying for some $t \in \mathcal{T}$ relations $R_k^2[u] \leq t_k I_{d_k}$, $k \leq K$. We have

$$\begin{array}{rcl} x^T y & = & x^T Y y = u^T P^T Y M z \\ & \leq & u^T \left[\sum_k \mathcal{R}_k^*[\Lambda_k]\right] u + z^T \left[\sum_\ell \mathcal{S}_\ell^*[\Upsilon_\ell]\right] z \text{ [by (4.73)]} \\ & = & \sum_k \mathrm{Tr}(\mathcal{R}_k^*[\Lambda_k][uu^T]) + \sum_\ell \mathrm{Tr}(\mathcal{S}_\ell^*[\Upsilon_\ell][zz^T]) \\ & = & \sum_k \mathrm{Tr}(\mathcal{R}_k[uu^T]\Lambda_k) + \sum_\ell \mathrm{Tr}(\mathcal{S}_\ell[zz^T]\Upsilon_\ell) \\ & = & \sum_k \mathrm{Tr}(R_k^2[u]\Lambda_k) + \sum_\ell \mathrm{Tr}(S_\ell^2[z]\Upsilon_\ell) \\ & \leq & \sum_k t_k \mathrm{Tr}(\Lambda_k) + \sum_\ell r_\ell \mathrm{Tr}(\Upsilon_\ell) \text{ [due to the origin of } u \text{ and } z] \\ & \leq & \theta := \phi_{\mathcal{T}}(\lambda[\Lambda]) + \phi_{\mathcal{R}}(\lambda[\Upsilon]), \end{array}$$

whence

$$\max_{y\in\mathcal{Y}}\max_{x\in\mathcal{X}} y^Tx \leq \theta.$$

Recalling that $\mathcal{Y}=\{y\in F: \|y\|_*\leq 1\}$ and invoking (4.70), we see that

$$\mathrm{dist}_{\|\cdot\|}(\mathcal{X},F^\perp)\leq\theta,$$

noting that $F^\perp=\mathrm{Ker}\,Y$, (4.74) follows. □

4) Prove the following statement:
(!!) *Let* m,n,Y *be as in* (!). *Then*

$$\begin{array}{l}\delta_m(\mathcal{X},\|\cdot\|)\leq\mathrm{dist}_{\|\cdot\|}(\mathcal{X},\mathrm{Ker}\,Y)\\ \quad\leq\ \widehat{\mathrm{Opt}}:=\min\limits_{\substack{\nu,\Lambda=\{\Lambda_k,k\leq K\},\\ \Upsilon=\{\Upsilon_\ell,\ell\leq L\}}}\Bigg\{\phi_{\mathcal{T}}(\lambda[\Lambda])+\phi_{\mathcal{R}}(\lambda[\Upsilon]):\\ \quad\nu\geq0,\Lambda_k\succeq0\,\forall k,\Upsilon_\ell\succeq0\,\forall\ell,\\ \quad\left[\begin{array}{c|c}\sum_k\mathcal{R}^*_k[\Lambda_k] & \frac{1}{2}P^TM\\ \hline \frac{1}{2}M^TP & \sum_\ell\mathcal{S}^*_\ell[\Upsilon_\ell]+\nu M^T(I-Y)M\end{array}\right]\succeq0\Bigg\},\end{array}\tag{4.76}$$

and $\widehat{\mathrm{Opt}}\leq\mathrm{Opt}$, *with* Opt *given by* (4.75).

Solution: Same as in the proof of (!), let $F=\mathrm{Im}Y$ and

$$\mathcal{Y}=\{y\in F:\|y\|_*\leq1\}=\{y\in F:\exists(r\in\mathcal{R},z):y=Mz,S^2_\ell[z]\preceq r_\ell I_{f_\ell},\ell\leq L\}.$$

Given $\epsilon>0$, consider the spectratope

$$\begin{array}{r}\mathcal{Y}_\epsilon=\{y:\exists([r;\rho]\in\mathcal{R}^+:=\mathcal{R}\times[0,\epsilon],z):y=Mz,S^2_\ell[z]\preceq r_\ell I_{f_\ell},\ell\leq L,\\ z^TM^T(I-Y)Mz\leq\rho\}.\end{array}$$

Observe that for evident reasons we have $\mathcal{Y}\subset\mathcal{Y}_\epsilon$. Now let ν,Λ,Υ be a feasible solution to the optimization problem in (4.76). We have

$$\begin{array}{rcl}\max\limits_{y\in\mathcal{Y},x\in\mathcal{X}}y^Tx&\leq&\max\limits_{y\in\mathcal{Y}_\epsilon,x\in\mathcal{X}}y^Tx\\ &=&\max\limits_{u,t,z,[r;\rho]}\left\{u^TP^TMz:\begin{array}{c}t\in\mathcal{T},R^2_k[u]\preceq t_kI_{d_k}\,\forall k\\ r\in\mathcal{R},\rho\in[0,\epsilon],S^2_\ell[z]\preceq r_\ell I_{f_\ell}\,\forall\ell,\\ z^TM^T(I-Y)Mz\leq\rho\end{array}\right\}\\ &\leq&\phi_{\mathcal{T}}(\lambda[\Lambda])+\phi_{\mathcal{R}^+}([\lambda[\Upsilon];\nu])=\phi_{\mathcal{T}}(\lambda[\Lambda])+\phi_{\mathcal{R}}(\lambda[\Upsilon])+\epsilon\nu,\end{array}$$

where the concluding $\leq$ is by Proposition 4.8. Recalling that by (4.70) and the construction of F and $\mathcal{Y}$ it holds

$$\max_{y\in\mathcal{Y},x\in\mathcal{X}}y^Tx=\max_{x\in\mathcal{X},\|y\|_*\leq1,y\in F}y^Tx=\mathrm{dist}_{\|\cdot\|}(\mathcal{X},F^\perp)=\mathrm{dist}_{\|\cdot\|}(\mathcal{X},\mathrm{Ker}\,Y).$$

We see that

$$\mathrm{dist}_{\|\cdot\|}(\mathcal{X},\mathrm{Ker}\,Y)\leq\phi_{\mathcal{T}}(\lambda[\Lambda])+\phi_{\mathcal{R}}(\lambda[\Upsilon])+\epsilon\nu$$

whenever (Λ,Υ,ν) is feasible for the optimization problem in (4.76). Since $\epsilon>0$ can be made arbitrarily small, (4.76) follows.

It remains to verify that $\widehat{\mathrm{Opt}}\leq\mathrm{Opt}$. Indeed, let $\Lambda^o=\{\Lambda^o_k\in\mathbf{S}^{d_k}\}$ be a once and forever fixed collection of positive *definite* matrices, so that the matrix $Q_o=\sum_k\mathcal{R}^*_k[\Lambda^o_k]$ is positive definite by Lemma 4.44. Since Y is a projector, we have $\left[\begin{array}{c|c}\frac{1}{4}I & \frac{1}{2}(I-Y)\\ \hline \frac{1}{2}(I-Y) & I-Y\end{array}\right]\succeq 0$, whence also

$$\begin{array}{l}\left[\begin{array}{c|c}\frac{1}{4}P^TP & \frac{1}{2}P^T(I-Y)M\\ \hline \frac{1}{2}M^T(I-Y)P & M^T(I-Y)M\end{array}\right]\\ \quad=\ \mathrm{Diag}\{P^T,M^T\}\left[\begin{array}{c|c}\frac{1}{4}I & \frac{1}{2}(I-Y)\\ \hline \frac{1}{2}(I-Y) & I-Y\end{array}\right]\mathrm{Diag}\{P,M\}\succeq 0,\end{array}$$

so that for properly selected $a_o>0$ it holds

$$\left[\begin{array}{c|c}\sum_k\mathcal{R}_k^*[a_o\Lambda_k^o] & \frac{1}{2}P^T(I-Y)M\\ \hline \frac{1}{2}M^T(I-Y)P & M^T(I-Y)M\end{array}\right]\succeq 0,$$

whence

$$\left[\begin{array}{c|c}\sum_k\mathcal{R}_k^*[\epsilon a_o\Lambda_k^o] & \frac{1}{2}P^T(I-Y)M\\ \hline \frac{1}{2}M^T(I-Y)P & \epsilon^{-1}M^T(I-Y)M\end{array}\right]\succeq 0\ \forall\epsilon>0.$$

It follows that whenever (Λ,Υ) is feasible for the optimization problem in (4.74) and $\epsilon>0$, the collection $(\Lambda_\epsilon=\{\Lambda_k+\epsilon a_o\Lambda_k^o\},\Upsilon,\nu=\epsilon^{-1})$ is feasible for the optimization problem in (4.76); since $\epsilon>0$ can be arbitrarily small and the objective of the problem in (4.76) is independent of ν, we arrive at the desired relation $\widehat{\mathrm{Opt}}\leq\mathrm{Opt}$. □

Statements (!), (!!) suggest the following policy for upper-bounding the Kolmogorov width $\delta_m(\mathcal{X},\|\cdot\|)$:

A. First, we select an integer μ, $1\leq\mu<n$, and solve the convex optimization problem

$$\min_{\Lambda,\Upsilon,Y}\left\{\begin{array}{l}\phi_{\mathcal{T}}(\lambda[\Lambda])+\phi_{\mathcal{R}}(\lambda[\Upsilon]):\ 0\preceq Y\preceq I,\mathrm{Tr}(Y)=n-\mu\\ \qquad\Lambda=\{\Lambda_k\succeq 0,k\leq K\},\Upsilon=\{\Upsilon_\ell\succeq 0,\ell\leq L\},\\ \qquad\left[\begin{array}{c|c}\sum_k\mathcal{R}_k^*[\Lambda_k] & \frac{1}{2}P^TYM\\ \hline \frac{1}{2}M^TYP & \sum_\ell\mathcal{S}_\ell^*[\Upsilon_\ell]\end{array}\right]\succeq 0\end{array}\right\}\qquad(P^\mu)$$

B. Next, we take the Y-component Y^μ of the optimal solution to (P^μ) and "round" it to a projector Y of rank $n-m$ in the same fashion as in the case of $\|\cdot\|=\|\cdot\|_2$, that is, keep the eigenvectors of Y^μ intact and replace m smallest eigenvalues with zeros, and all remaining eigenvalues with ones.

C. Finally, we solve the convex optimization problem

$$\mathrm{Opt}_{m,\mu}=\min_{\Lambda,\Upsilon,\nu}\left\{\begin{array}{l}\phi_{\mathcal{T}}(\lambda[\Lambda])+\phi_{\mathcal{R}}(\lambda[\Upsilon]):\\ \nu\geq 0,\Lambda=\{\Lambda_k\succeq 0,k\leq K\},\Upsilon=\{\Upsilon_\ell\succeq 0,\ell\leq L\},\\ \left[\begin{array}{c|c}\sum_k\mathcal{R}_k^*[\Lambda_k] & \frac{1}{2}P^TM\\ \hline \frac{1}{2}M^TP & \sum_\ell\mathcal{S}_\ell^*[\Upsilon_\ell]+\nu M^T(I-Y)M\end{array}\right]\succeq 0\end{array}\right\}\qquad(P^{m,\mu})$$

By (!!), $\mathrm{Opt}_{m,\mu}$ is an upper bound on the Kolmogorov width $\delta_m(\mathcal{X},\|\cdot\|)$ (and in fact also on $\mathrm{dist}_{\|\cdot\|}(\mathcal{X},\mathrm{Ker}\,Y)$).

Observe all the complications we encounter when passing from the simple case $\|\cdot\|=\|\cdot\|_2$ to the case of a general norm $\|\cdot\|$ with a spectratope as the unit ball of the conjugate norm. Note that Proposition 4.26 gives both a lower bound $\sqrt{\mathrm{Opt}(m)/\varkappa}$ on the m-th Kolmogorov width of $\mathcal{X}$ w.r.t. $\|\cdot\|_2$, and a family of upper bounds $\sqrt{\frac{m+1}{m+1-\mu}\mathrm{Opt}(\mu)}$, $1\leq\mu\leq m$, on this width. As a result, we can approximate $\mathcal{X}$ by m-dimensional subspaces in the Euclidean norm in a "nearly optimal" fashion. Indeed, if for some ϵ and k it holds $\delta_k(\mathcal{X},\|\cdot\|_2)\leq\epsilon$, then $\mathrm{Opt}(k)\leq\varkappa\epsilon^2$ by Proposition 4.26 as applied with $m=k$. On the other hand, assuming $k<n/2$, the same proposition when applied with $m=2k$ and $\mu=k$ says that

$$\mathrm{dist}_{\|\cdot\|_2}(\mathcal{X},E^{m,k})\leq\sqrt{\frac{2k+1}{k+1}\mathrm{Opt}(k)}\leq\sqrt{2\mathrm{Opt}(k)}\leq\sqrt{2\varkappa}\,\epsilon.$$

Thus, if $\mathcal{X}$ may somehow be approximated by a k-dimensional subspace within $\|\cdot\|_2$-accuracy ϵ, we can efficiently get an approximation of "nearly the same quality" ($\sqrt{2\varkappa}\epsilon$ instead of ϵ; recall that $\varkappa$ is just logarithmic in D) and "nearly the same dimension" ($2k$ instead of k).

Neither of these options is preserved when passing from the Euclidean norm to a general one: in the latter case, we do not have lower bounds on Kolmogorov widths, and have no understanding of how tight our upper bounds are.

Now, two concluding questions:

5) Why in step A of the above bounding scheme do we utilize statement (!) rather than less conservative (since $\widehat{\text{Opt}} \leq \text{Opt}$) statement (!!)?
6) Implement the scheme numerically and run experiments.
   Recommended setup:
   - Given $\sigma > 0$ and positive integers n and κ, let f be a function of continuous argument $t \in [0,1]$ satisfying the smoothness restriction $|f^{(k)}(t)| \leq \sigma^k$, $0 \leq t \leq 1$, $k = 0,1,2,...,\kappa$. Specify $\mathcal{X}$ as the set of n-dimensional vectors x obtained by restricting f onto the n-point equidistant grid $\{t_i = i/n\}_{i=1}^n$. To this end,translate the description on f into a bunch of two-sided linear constraints on x:
   $$|d_{(k)}^T[x_i; x_{i+1}; ...; x_{i+k}]| \leq \sigma^k, \; 1 \leq i \leq n-k, \; 0 \leq k \leq \kappa,$$
   where $d_{(k)} \in \mathbf{R}^{k+1}$ is the vector of coefficients of finite-difference approximation, with resolution $1/n$, of the k-th derivative:
   $$\begin{array}{c} d_{(0)} = 1, \; d_{(1)} = n[-1;1], \; d_{(2)} = n^2[1;-2;1], \\ d_{(3)} = n^3[-1;3;-3;1], \; d_{(4)} = n^4[1;-4;6;-4;1], ... \end{array}$$
   - Recommended parameters: $n = 32$, $m = 8$, $\kappa = 5$, $\sigma \in \{0.25, 0.5; 1, 2, 4\}$.
   - Run experiments with $\|\cdot\| = \|\cdot\|_1$ and $\|\cdot\| = \|\cdot\|_2$.

Solution: Were we able to optimize in (4.76) over (Λ, Υ) *and over projector* Y *of rank* $n - m$, this is definitely what we would do. The problem is that projectors form a nonconvex set, so that we should relax the constraint "Y is a projector." Now, whatever be a convex outer approximation $\mathcal{P}$ of the set of all projectors of a given positive rank $n - m$, it *must* contain the matrix $Y_* = \frac{n-m}{n} I$, since this matrix is the average of a properly selected sequence of projectors of rank $n - m$. As a result, setting $Y = Y_*$ and selecting small enough positive definite Λ_k, $\Upsilon_\ell = 0$, and large enough ν, we get a feasible solution to the relaxed version of (4.76), and the value of the objective at this feasible solution can be made as close to 0 as we want. Thus, all natural convex relaxations of (4.76) are, in a sense, trivial and do not provide any clue to what a good projector Y is. In contrast to this, the optimization problem in (4.76) combines with our "rounding" to yield a "qualified guess" on what may be a good Y.

Here are some numerical results:

σ	$\text{dist}_{\|\cdot\|_1}(\mathcal{X}, E)$	$\text{dist}_{\|\cdot\|_2}(\mathcal{X}, F)$
0.25	0.0030	0.0015
0.50	0.0104	0.0036
1.00	0.0025	0.0118
2.00	0.0197	0.0105
4.00	0.0462	0.0493

As recommended, we used $n = 32$ and $m = 8$; E and F stand for the resulting eight-dimensional subspaces obtained when minimizing $\|\cdot\|_1$ and $\|\cdot\|_2$ width, respectively.

Notice the "counterintuitive" phenomena exhibited by the presented results: the actual $\|\cdot\|_p$-widths should decrease in p and increase in σ (since $\|\cdot\|_p$ decreases as p grows, and $\mathcal{X}$ grows as σ grows). We see from the table that some of these relations are violated by our bounds. This is unpleasant, but not surprising: what we compute are conservative upper bounds on the actual widths, and there is no guarantee that these bounds preserve inequalities between the widths.

Exercise 4.8. [more on semidefinite relaxation] The goal of this exercise is to extend SDP relaxation beyond ellitopes/spectratopes.

SDP relaxation is aimed at upper-bounding the quantity

$$\mathrm{Opt}_{\mathcal{X}}(B) = \max_{x\in\mathcal{X}} x^T Bx, \qquad [B \in \mathbf{S}^n]$$

where $\mathcal{X} \subset \mathbf{R}^n$ is a given set (which we from now on assume to be nonempty convex compact). To this end we look for a computationally tractable convex compact set $\mathcal{U} \subset \mathbf{S}^n$ such that for every $x \in \mathcal{X}$ it holds $xx^T \in \mathcal{U}$; in this case, we refer to $\mathcal{U}$ as to a set *matching* $\mathcal{X}$ (equivalent wording: "$\mathcal{U}$ matches $\mathcal{X}$"). Given such a set $\mathcal{U}$, the optimal value in the convex optimization problem

$$\overline{\mathrm{Opt}}_{\mathcal{U}}(B) = \max_{U\in\mathcal{U}} \mathrm{Tr}(BU) \tag{4.77}$$

is an efficiently computable convex upper bound on $\mathrm{Opt}_{\mathcal{X}}(B)$.

Given $\mathcal{U}$ matching $\mathcal{X}$, we can pass from $\mathcal{U}$ to the conic hull of $\mathcal{U}$, that is, to the set

$$\mathbf{U}[\mathcal{U}] = \mathrm{cl}\{(U,\mu) \in \mathbf{S}^n \times \mathbf{R}_+ : \mu > 0, U/\mu \in \mathcal{U}\}$$

which, as it is immediately seen, is a closed convex cone contained in $\mathbf{S}^n \times \mathbf{R}_+$. The only point (U,μ) in this cone with $\mu = 0$ has $U = 0$ (since $\mathcal{U}$ is compact), and

$$\mathcal{U} = \{U : (U,1) \in \mathbf{U}\} = \{U : \exists \mu \leq 1 : (U,\mu) \in \mathbf{U}\},$$

so that the definition of $\overline{\mathrm{Opt}}_{\mathcal{U}}$ can be rewritten equivalently as

$$\overline{\mathrm{Opt}}_{\mathcal{U}}(B) = \min_{U,\mu} \{\mathrm{Tr}(BU) : (U,\mu) \in \mathbf{U}, \mu \leq 1\}.$$

The question, of course, is where to take a set $\mathcal{U}$ matching $\mathcal{X}$, and the answer depends on what we know about $\mathcal{X}$. For example, when $\mathcal{X}$ is a basic ellitope:

$$\mathcal{X} = \{x \in \mathbf{R}^n : \exists t \in \mathcal{T} : x^T S_k x \leq t_k, k \leq K\}$$

with our usual restrictions on $\mathcal{T}$ and S_k, it is immediately seen that

$$x \in \mathcal{X} \Rightarrow xx^T \in \mathcal{U} = \{U \in \mathbf{S}^n : U \succeq 0, \exists t \in \mathcal{T} : \mathrm{Tr}(US_k) \leq t_k, k \leq K\}.$$

Similarly, when $\mathcal{X}$ is a basic spectratope,

$$\mathcal{X} = \{x \in \mathbf{R}^n : \exists t \in \mathcal{T} : S_k^2[x] \preceq t_k I_{d_k}, k \leq K\}$$

with our usual restrictions on $\mathcal{T}$ and $S_k[\cdot]$, it is immediately seen that

$$x \in \mathcal{X} \Rightarrow xx^T \in \mathcal{U} = \{U \in \mathbf{S}^n : U \succeq 0, \exists t \in \mathcal{T} : \mathcal{S}_k[U] \preceq t_k I_{d_k}, k \leq K\}.$$

One can verify that the semidefinite relaxation bounds on the maximum of a quadratic form on an ellitope/spectratope $\mathcal{X}$ derived in Sections 4.2.3 (for ellitopes) and 4.3.2 (for spectratopes) are nothing but the bounds (4.77) associated with the $\mathcal{U}$ just defined.

4.8.A Matching via absolute norms. There are other ways to specify a set matching $\mathcal{X}$. The seemingly simplest of them is as follows. Let $p(\cdot)$ be an absolute norm on $\mathbf{R}^n$ (recall that this is a norm $p(x)$ which depends solely on $\mathrm{abs}[x]$, where $\mathrm{abs}[x]$ is the vector comprised of the magnitudes of entries in x). We can convert $p(\cdot)$ into the norm $p^+(\cdot)$ on the space $\mathbf{S}^n$ as follows:

$$p^+(U) = p([p(\mathrm{Col}_1[U]); ...; p(\mathrm{Col}_n[U])]) \qquad [U \in \mathbf{S}^n]$$

1.1) Prove that p^+ indeed is a norm on $\mathbf{S}^n$, and $p^+(xx^T) = p^2(x)$. Denoting by $q(\cdot)$ the norm conjugate to $p(\cdot)$, what is the relation between the norm $(p^+)_*(\cdot)$ conjugate to $p^+(\cdot)$ and the norm $q^+(\cdot)$?

1.2) Derive from 1.1 that whenever $p(\cdot)$ is an absolute norm such that $\mathcal{X}$ is contained in the unit ball $\mathcal{B}_{p(\cdot)} = \{x : p(x) \leq 1\}$ of the norm p, the set

$$\mathcal{U}_{p(\cdot)} = \{U \in \mathbf{S}^n : U \succeq 0, p^+(U) \leq 1\}$$

is matching $\mathcal{X}$. If, in addition,

$$\mathcal{X} \subset \{x : p(x) \leq 1, Px = 0\}, \tag{4.78}$$

then the set

$$\mathcal{U}_{p(\cdot),P} = \{U \in \mathbf{S}^n : U \succeq 0, p^+(U) \leq 1, PU = 0\}$$

is matching $\mathcal{X}$.

Assume that in addition to $p(\cdot)$, we have at our disposal a computationally tractable closed convex set $\mathcal{D}$ such that whenever $p(x) \leq 1$, the vector $[x]^2 := [x_1^2; ...; x_n^2]$ belongs to $\mathcal{D}$; in the sequel we call such a $\mathcal{D}$ *square-dominating* $p(\cdot)$. For example, when $p(\cdot) = \|\cdot\|_r$, we can take

$$\mathcal{D} = \left\{ \begin{array}{ll} \{y \in \mathbf{R}^n_+ : \sum_i y_1 \leq 1\}, & r \leq 2 \\ \{y \in \mathbf{R}^n_+ : \|y\|_{r/2} \leq 1\}, & r > 2 \end{array} \right. .$$

Prove that in this situation the above construction can be refined: whenever $\mathcal{X}$ satisfies (4.78), the set

$$\mathcal{U}^{\mathcal{D}}_{p(\cdot),P} = \{U \in \mathbf{S}^n : U \succeq 0, p^+(U) \leq 1, PU = 0, \mathrm{dg}(U) \in \mathcal{D}\}$$
$$[\mathrm{dg}(U) = [U_{11}; U_{22}; ...; U_{nn}]]$$

matches $\mathcal{X}$.

Note: in the sequel, we suppress P in the notation $\mathcal{U}_{p(\cdot),P}$ and $\mathcal{U}^{\mathcal{D}}_{p(\cdot),P}$ when $P = 0$; thus, $\mathcal{U}_{p(\cdot)}$ is the same as $\mathcal{U}_{p(\cdot),0}$.

1.3) Check that when $p(\cdot) = \|\cdot\|_r$ with $r \in [1, \infty]$, one has

$$p^+(U) = \|U\|_r := \left\{ \begin{array}{ll} \left(\sum_{i,j} |U_{ij}|^r\right)^{1/r}, & 1 \leq r < \infty, \\ \max_{i,j} |U_{ij}|, & r = \infty \end{array} \right. .$$

1.4) Let $\mathcal{X} = \{x \in \mathbf{R}^n : \|x\|_1 \leq 1\}$ and $p(x) = \|x\|_1$, so that $\mathcal{X} \subset \{x : p(x) \leq 1\}$, and

$$\mathrm{Conv}\{[x]^2 : x \in \mathcal{X}\} \subset \mathcal{D} = \left\{y \in \mathbf{R}^n_+ : \sum_i y_i = 1\right\}. \tag{4.79}$$

What are the bounds $\overline{\mathrm{Opt}}_{\mathcal{U}_{p(\cdot)}}(B)$ and $\overline{\mathrm{Opt}}_{\mathcal{U}^{\mathcal{D}}_{p(\cdot)}}(B)$? Is it true that the former (the latter) of the bounds is precise? Is it true that the former (the latter) bound is precise when $B \succeq 0$?

1.5) Let $\mathcal{X} = \{x \in \mathbf{R}^n : \|x\|_2 \leq 1\}$ and $p(x) = \|x\|_2$, so that $\mathcal{X} \subset \{x : p(x) \leq 1\}$ and (4.79) holds true. What are the bounds $\overline{\mathrm{Opt}}_{\mathcal{U}_{p(\cdot)}}(B)$ and $\overline{\mathrm{Opt}}_{\mathcal{U}^{\mathcal{D}}_{p(\cdot)}}(B)$? Is the former (or the latter) bound precise?

1.6) Let $\mathcal{X} \subset \mathbf{R}^n_+$ be closed, convex, bounded, and with a nonempty interior. Verify that the set

$$\mathcal{X}^+ = \{x \in \mathbf{R}^n : \exists y \in \mathcal{X} : \mathrm{abs}[x] \leq y\}$$

is the unit ball of an absolute norm $p_{\mathcal{X}}$, and this is the largest absolute norm $p(\cdot)$ such that $\mathcal{X} \subset \{x : p(x) \leq 1\}$. Derive from this observation that the norm $p_{\mathcal{X}}(\cdot)$ is the best (i.e., resulting in the least conservative bounding scheme) among absolute norms which allow us to upper-bound $\mathrm{Opt}_{\mathcal{X}}(B)$ via the construction from item 1.2.

Solution: 1.1: Verification that p^+ is a norm is completely straightforward. For example, here is the derivation of the Triangle inequality:

$$\begin{array}{rcl} p^+(U+V) & = & p([\underbrace{p(\mathrm{Col}_1[U]+\mathrm{Col}_1[V])}_{\leq p(\mathrm{Col}_1[U])+p(\mathrm{Col}_1[V])};...;\underbrace{p(\mathrm{Col}_n[U]+\mathrm{Col}_n[V])}_{\leq p(\mathrm{Col}_n[U])+p(\mathrm{Col}_n[V])}]) \\ & \underbrace{\leq}_{(a)} & p([p(\mathrm{Col}_1[U])+p(\mathrm{Col}_1[V]);...;p(\mathrm{Col}_n[U])+p(\mathrm{Col}_n[V])]) \\ & \leq & p([p(\mathrm{Col}_1[U]);...;p(\mathrm{Col}_n[U])])+p([p(\mathrm{Col}_1[V]);...;p(\mathrm{Col}_n[V])]) \\ & = & p^+(U)+p^+(V), \end{array}$$

where (a) is due to the fact that every absolute norm is entrywise monotone on the nonnegative orthant. Next, we have

$$p^+(xx^T) = p([|x_1|p(x);|x_2|p(x);...;|x_n|p(x)]) = p(x)p(\mathrm{abs}[x]) = p^2(x).$$

Finally, we have $(p^+)_*(\cdot) \leq q^+(\cdot)$, *and the inequality may be strict.* Indeed, to verify the inequality, let $V \in \mathbf{S}^n$. Then for every $U \in \mathbf{S}^n$, denoting by u the vector comprised of $p(\cdot)$-norms of the columns in U, and by v the vector comprised of $q(\cdot)$-norms of the columns of V, it holds (recall that U and V are symmetric)

$$\mathrm{Tr}(VU) = \sum_i \mathrm{Col}_i^T[V]\mathrm{Col}_i[U] \leq \sum_i v_i u_i \leq q(v)p(u) = q^+(V)p^+(U),$$

implying that

$$(p^+)_*(V) = \max_{U\in\mathbf{S}^n:p^+(U)\leq 1} \mathrm{Tr}(VU) \leq q^+(V).$$

To check that in general the inequality is strict, it suffices to compare $(p^+)_*$ and q^+ numerically for, say, $n=5$ and $p(x)=\|x\|_{(2)}$, where $\|x\|_{(k)}$ is the sum of the k largest magnitudes of entries in x. Of course, sometimes $(p^+)_* \equiv q^+$; this is so, e.g., when $p(\cdot) = \|\cdot\|_r$ with $r \in [1,\infty]$. □

1.2 is immediate: when $\mathcal{X} \subset \{x : p(x) \leq 1\}$, we have

$$\begin{array}{c} x \in \mathcal{X} \Rightarrow p(x) \leq 1 \Rightarrow p^+(xx^T) = p^2(x) \leq 1 \\ \Rightarrow xx^T \in \mathcal{U}_{p(\cdot)} = \{U \in \mathbf{S}^n : U \succeq 0, p^+(U) \leq 1\}. \end{array}$$

Besides this, $\mathrm{dg}[xx^T] = [x]^2$, and if $Px = 0$ for all $x \in \mathcal{X}$, then also $P[xx^T] = 0$ for all $x \in \mathcal{X}$. □

1.3: Evident.

1.4: Observing that $p^+(U) = \sum_{i,j}|U_{ij}|$, we see that whenever $U \succeq 0$ and $p^+(U) \leq 1$, we have also $\mathrm{dg}(U) \in \mathcal{D}$, implying that $\mathcal{U}_{\|\cdot\|_1} = \mathcal{U}^{\mathcal{D}}_{\|\cdot\|_1}$; thus, the bounds in question are identical. We have

$$\begin{array}{rcl} \overline{\mathrm{Opt}}_{\mathcal{U}_{\|\cdot\|_1}}(B) & = & \max\limits_U \left\{\mathrm{Tr}(BU) : U \succeq 0, \sum_{i,j}|U_{ij}| \leq 1\right\} \\ & \leq & \max\limits_U \left\{\sum_{i,j} B_{ij}U_{ij} : \sum_{i,j}|U_{ij}| \leq 1\right\} \\ & = & \max_{i,j}|B_{ij}|. \end{array}$$

The bound $\overline{\mathrm{Opt}}_{\mathcal{U}_{\|\cdot\|_1}}(B)$ in general is strictly greater than $\mathrm{Opt}_{\mathcal{X}}(B) = \max\limits_{\|x\|_1\leq 1} x^TBx$. However, the bound is equal to $\mathrm{Opt}_{\mathcal{X}}(B)$ when $B \succeq 0$. Indeed, in the latter case the function x^TBx is convex, and its maximum over $\mathcal{X}$ is achieved at a vertex of

$\mathcal{X} = \{x : \|x\|_1 \leq 1\}$, that is, $\mathrm{Opt}_{\mathcal{X}}(B) = \max_i B_{ii}$. On the other hand, when $B \succeq 0$, we have $|B_{ij}| \leq \sqrt{B_{ii}B_{jj}}$, implying that $\max_{i,j} |B_{ij}| = \max_i B_{ii}$.

1.5: We have

$$\begin{array}{rcl} \mathcal{U}_{\|\cdot\|_2} & = & \{U \in \mathbf{S}^n : U \succeq 0, \|U\|_2 \leq 1\}, \\ \mathcal{U}^{\mathcal{D}}_{\|\cdot\|_2} & = & \{U \in \mathbf{S}^n : U \succeq 0, \|U\|_2 \leq 1, \mathrm{Tr}(U) \leq 1\} \\ & = & \{U \in \mathbf{S}^n : U \succeq 0, \mathrm{Tr}(U) \leq 1\}. \end{array}$$

and

$$\overline{\mathrm{Opt}}_{\mathcal{U}_{\|\cdot\|_2}}(B) = \min_U \{\mathrm{Tr}(BU) : \|U\|_2 \leq 1\}.$$

Let $\lambda(Q)$ be the vector of eigenvalues of $Q \in \mathbf{S}^n$ taken with their multiplicities and arranged in the non-ascending order. Since $\|Q\|_2$ is the Frobenius norm of a matrix Q and thus remains invariant when replacing Q with VQW, with orthogonal V, W, we can assume w.l.o.g. that B is diagonal, which results in $\overline{\mathrm{Opt}}_{\mathcal{U}_{\|\cdot\|_2}}(B) = \|\lambda^+[B]\|_2$, where $\lambda^+[B]$ is the vector comprised of the positive parts $\max[\lambda_i, 0]$ of the eigenvalues λ_i of B. Recalling that under the circumstances we have $\mathrm{Opt}_{\mathcal{X}}(B) = \max[\lambda_1(B), 0]$, we conclude that the bound in question is precise if and only if B has at most one positive eigenvalue, and can be greater than $\mathrm{Opt}_{\mathcal{X}}(B)$ by factor as large as $\sqrt{n}$, as in the case of $B = I_n$.

In contrast to this, the bound $\overline{\mathrm{Opt}}_{\mathcal{U}^{\mathcal{D}}_{\|\cdot\|_2}}(B)$ is precise for all B; indeed, for the same reasons as above, it suffices to verify this claim when B is diagonal, in which case the claim becomes evident. □

1.6: It is clear that $\mathcal{X}^+$ is a closed and bounded convex set symmetric w.r.t. the origin and containing a neighbourhood of the origin. Consequently, $\mathcal{X}^+$ is the unit ball of some norm $p_{\mathcal{X}}(\cdot)$. By construction, if E is a diagonal matrix with diagonal entries ± 1, then $Ex \in \mathcal{X}^+$ if and only if $x \in \mathcal{X}^+$, implying that $p_{\mathcal{X}}$ is an absolute norm. Now, if p is an absolute norm such that $\mathcal{X} \subset \mathcal{B} := \{x : p(x) \leq 1\}$ then $\mathcal{X}^+ \subset \mathcal{B}$. Indeed, when $x \in \mathcal{X}^+$, we have $\mathrm{abs}[x] \leq y$ for some $y \in X$, that is, $p(x) = p(\mathrm{abs}[x]) \leq p(y) \leq 1$, where the first inequality is due to the fact that the absolute norm $p(\cdot)$ is monotone in the nonnegative orthant, and the last inequality is due to $\mathcal{X} \subset \mathcal{B}$. Thus, $\mathcal{X}^+ \subset \mathcal{B}$, implying that $p(\cdot) \leq p_{\mathcal{X}}(\cdot)$ and consequently $\mathcal{U}_{p_{\mathcal{X}}(\cdot)} \subset \mathcal{U}_{p(\cdot)}$. □

4.8.B. "Calculus of matchings." Observe that the matching we have introduced admits a kind of "calculus." Specifically, consider the situation as follows: for $1 \leq \ell \leq L$, we are given

- nonempty convex compact sets $\mathcal{X}_\ell \subset \mathbf{R}^{n_\ell}$, $0 \in \mathcal{X}_\ell$, along with matching $\mathcal{X}_\ell$ convex compact sets $\mathcal{U}_\ell \subset \mathbf{S}^{n_\ell}$ giving rise to the closed convex cones

$$\mathbf{U}_\ell = \mathrm{cl}\{(U_\ell, \mu_\ell) \in \mathbf{S}^{n_\ell} \times \mathbf{R}_+ : \mu_\ell > 0, \mu_\ell^{-1} U_\ell \in \mathcal{U}_\ell\}.$$

  We denote by $\vartheta_\ell(\cdot)$ the Minkowski functions of $\mathcal{X}_\ell$:

$$\vartheta_\ell(y^\ell) = \inf\{t : t > 0, t^{-1} y^\ell \in \mathcal{X}_\ell\} : \mathbf{R}^{n_\ell} \to \mathbf{R} \cup \{+\infty\};$$

  note that $\mathcal{X}_\ell = \{y^\ell : \vartheta_\ell(y^\ell) \leq 1\}$;
- $n_\ell \times n$ matrices A_ℓ such that $\sum_\ell A_\ell^T A_\ell \succ 0$.

On top of that, we are given a monotone convex set $\mathcal{T} \subset \mathbf{R}^L_+$ intersecting the interior of $\mathbf{R}^L_+$.

These data specify the convex set

$$\mathcal{X} = \{x \in \mathbf{R}^n : \exists t \in \mathcal{T} : \vartheta_\ell^2(A_\ell x) \leq t_\ell, \ell \leq L\} \tag{$*$}$$

2.1) Prove the following

Lemma 4.27 *In the situation in question, the set*

$$\mathcal{U} = \{U \in \mathbf{S}^n : U \succeq 0 \ \&\ \exists t \in \mathcal{T} : (A_\ell U A_\ell^T, t_\ell) \in \mathbf{U}_\ell, \ell \le L\}$$

is a closed and bounded convex set which matches $\mathcal{X}$. As a result, the efficiently computable quantity

$$\overline{\mathrm{Opt}}_{\mathcal{U}}(B) = \max_U \{\mathrm{Tr}(BU) : U \in \mathcal{U}\}$$

is an upper bound on

$$\mathrm{Opt}_{\mathcal{X}}(B) = \max_{x \in \mathcal{X}} x^T B x.$$

2.2) Prove that if $\mathcal{X} \subset \mathbf{R}^n$ is a nonempty convex compact set, P is an $m \times n$ matrix, and $\mathcal{U}$ matches $\mathcal{X}$, then the set $\mathcal{V} = \{V = PUP^T : U \in \mathcal{U}\}$ matches $\mathcal{Y} = \{y : \exists x \in \mathcal{X} : y = Px\}$.

2.3) Prove that if $\mathcal{X} \subset \mathbf{R}^n$ is a nonempty convex compact set, P is an $n \times m$ matrix of rank m, and $\mathcal{U}$ matches $\mathcal{X}$, then the set $\mathcal{V} = \{V \succeq 0 : PVP^T \in \mathcal{U}\}$ matches $\mathcal{Y} = \{y : Py \in \mathcal{X}\}$.

2.4) Consider the "direct product" case where $\mathcal{X} = \mathcal{X}_1 \times ... \times \mathcal{X}_L$. When specifying A_ℓ as the matrix which "cuts" the ℓ-th block $A_\ell x = x^\ell$ of a block vector $x = [x^1; ...; x^L] \in \mathbf{R}^{n_1} \times ... \times \mathbf{R}^{n_L}$ and setting $\mathcal{T} = [0,1]^L$, we cover this situation by the setup under consideration. In the direct product case, the construction from item 2.1 is as follows: given the sets $\mathcal{U}_\ell$ matching $\mathcal{X}_\ell$, we build the set

$$\mathcal{U} = \{U = [U^{\ell\ell'} \in \mathbf{R}^{n_\ell \times n_{\ell'}}]_{\ell,\ell' \le L} \in \mathbf{S}^{n_1+\dots+n_L} : U \succeq 0, U^{\ell\ell} \in \mathcal{U}_\ell, \ell \le L\}$$

and claim that this set matches $\mathcal{X}$.

Could we be less conservative? While we do not know how to be less conservative in general, we do know how to be less conservative in the special case when $\mathcal{U}_\ell$ are built via absolute norms. Namely, let $p_\ell(\cdot) : \mathbf{R}^{n_\ell} \to \mathbf{R}_+$, $\ell \le L$, be absolute norms, let sets $\mathcal{D}_\ell$ be square-dominating $p_\ell(\cdot)$,

$$\mathcal{X}^\ell \subset \widehat{\mathcal{X}}_\ell = \{x^\ell \in \mathbf{R}^{n_\ell} : P_\ell x_\ell = 0, p_\ell(x^\ell) \le 1\},$$

and let

$$\mathcal{U}_\ell = \{U \in \mathbf{S}^{n_\ell} : U \succeq 0, P_\ell U = 0, p_\ell^+(U) \le 1, \mathrm{dg}(U) \in \mathcal{D}_\ell\}.$$

In this case the above construction results in

$$\mathcal{U} = \left\{ U = [U^{\ell\ell'} \in \mathbf{R}^{n_\ell \times n_{\ell'}}]_{\ell,\ell' \le L} \in \mathbf{S}_+^{n_1+\dots+n_L} : U \succeq 0, \begin{array}{c} P_\ell U^{\ell\ell} = 0 \\ p_\ell^+(U^{\ell\ell}) \le 1 \\ \mathrm{dg}(U^{\ell\ell}) \in \mathcal{D}_\ell \end{array}, \ell \le L \right\}.$$

Now let

$$p([x^1; ...; x^L]) = \max[p_1(x^1), ..., p_L(x^L)] : \mathbf{R}^{n_1} \times ... \times \mathbf{R}^{n_L} \to \mathbf{R},$$

so that p is an absolute norm and

$$\mathcal{X} \subset \{x = [x^1; ...; x^L] : p(x) \le 1, P_\ell x^\ell = 0, \ell \le L\}.$$

Prove that in fact the set

$$\overline{\mathcal{U}} = \left\{ U = [U^{\ell\ell'} \in \mathbf{R}^{n_\ell \times n_{\ell'}}]_{\ell,\ell' \le L} \in \mathbf{S}_+^{n_1+\dots+n_L} : U \succeq 0, \begin{array}{c} P_\ell U^{\ell\ell} = 0 \\ \mathrm{dg}(U^{\ell\ell}) \in \mathcal{D}_\ell \\ p^+(U) \le 1 \end{array}, \ell \le L \right\}$$

matches $\mathcal{X}$, and that we always have $\overline{\mathcal{U}} \subset \mathcal{U}$. Verify that in general this inclusion is strict.

Solution: 2.1: The fact that $\mathcal{U}$ is closed and convex is evident (recall that $\mathcal{T}$ is compact and the $\mathbf{U}_\ell$ are closed convex cones). To check that the set is bounded, assume the opposite; in this case $\mathcal{U}$ admits a nonzero recessive direction δ, and since $\mathcal{T}$ and the sets $\mathcal{U}_\ell$ are bounded, it is immediately seen that $A_\ell \delta A_\ell^T = 0$ for every ℓ. In addition, we should have $\delta \succeq 0$ due to $\mathcal{U} \subset \mathbf{S}^n_+$. Since $\delta \succeq 0$ is nonzero, we can find a nonzero vector $e \in \mathbf{R}^n$ such that $ee^T \preceq \delta$. Then

$$0 \preceq A_\ell ee^T A_\ell^T \preceq A_\ell \delta A_\ell^T = 0,$$

implying that $A_\ell e = 0$ for all ℓ; but the latter is impossible since $e \neq 0$ and $\sum_\ell A_\ell^T A_\ell \succ 0$.

It remains to verify that $\mathcal{U}$ matches $\mathcal{X}$. Let $x \in \mathcal{X}$, so that for some $t \in \mathcal{T}$ it holds $\vartheta_\ell(A_\ell x) \leq \sqrt{t_\ell}$ for all ℓ. When ℓ is such that $t_\ell = 0$, we have $A_\ell x = 0$ (indeed, $\mathcal{X}_\ell$ is bounded, so that $\vartheta_\ell(\cdot) > 0$ outside of the origin). Let us set $y_\ell = A_\ell x/\sqrt{t_\ell}$ when $t_\ell > 0$ and $y_\ell = 0_{n_\ell}$ when $t_\ell = 0$, so that $\vartheta_\ell(y_\ell) \leq 1$ for all ℓ, that is, $y_\ell \in \mathcal{X}_\ell$ and therefore $y_\ell y_\ell^T \in \mathcal{U}_\ell$, implying that $(t_\ell y_\ell y_\ell^T, t_\ell) \in \mathbf{U}_\ell$ for all ℓ. Taking into account that $t_\ell y_\ell y_\ell^T = A_\ell x x^T A_\ell^T$ for all ℓ, we see that there exists $t \in \mathcal{T}$ such that $(A_\ell x x^T A_\ell^T, t_\ell) \in \mathbf{U}_\ell$ for all ℓ, implying that $xx^T \in \mathcal{U}$. □

2.2: Obviously, $\mathcal{V}$ is a convex compact set; when $y \in \mathcal{Y}$, there exists $x \in \mathcal{X}$ such that $y = Px$, whence $yy^T = P[xx^T]P^T$ and $U := xx^T \in \mathcal{U}$, so that $yy^T \in \mathcal{V}$. □

2.3: The fact that $\mathcal{V}$ is a closed convex set is evident. Boundedness of $\mathcal{V}$ is readily given by $\operatorname{Ker} P = 0$, implying that $\operatorname{Im} P^T = \mathbf{R}^m$, combined with boundedness of $\mathcal{U}$. Finally, if $y \in \mathcal{Y}$, then $Py \in \mathcal{X}$, whence $Pxx^TP^T \in \mathcal{U}$, so that $x^Tx \in \mathcal{V}$. □

2.4: The fact that $\overline{\mathcal{U}}$ matches $\mathcal{X}$ is readily given by item 1.2. The inclusion $\overline{\mathcal{U}} \subset \mathcal{U}$ is completely straightforward. The fact that inclusion $\overline{\mathcal{U}} \subset \mathcal{U}$ in general is strict can be easily verified numerically—take $L = n_1 = n_2 = 2$, zero P_1 and P_2, $p_1(\cdot) = \|\cdot\|_2$, $p_2(\cdot) = \|\cdot\|_2$, generate several 4×4 positive semidefinite matrices, normalize them to have the maximum of Frobenius norms of diagonal 2×2 blocks to be 1 (that is, normalize them to become points of $\mathcal{U}$), and see whether the resulting matrices are of p^+-norm ≤ 1; if they are not, you are done. □

4.8.C Illustration: Nullspace property revisited. Recall the sparsity-oriented signal recovery via ℓ_1 minimization from Chapter 1: Given an $m \times n$ sensing matrix A and (noiseless) observation $y = Aw$ of an unknown signal w known to have at most s nonzero entries, we recover w as

$$\widehat{w} \in \operatorname*{Argmin}_z \{\|z\|_1 : Az = y\}.$$

We called matrix A s-good, if whenever $y = Aw$ with s-sparse w, the only optimal solution to the right-hand side optimization problem is w. The (difficult to verify!) necessary and sufficient condition for s-goodness is the Nullspace property:

$$\mathrm{Opt} := \max_z \left\{\|z\|_{(s)} : z \in \operatorname{Ker} A, \|z\|_1 \leq 1\right\} < 1/2,$$

where $\|z\|_{(k)}$ is the sum of the k largest entries in the vector $\mathrm{abs}[z]$. A verifiable sufficient condition for s-goodness is

$$\widehat{\mathrm{Opt}} := \min_H \max_j \|\mathrm{Col}_j[I - H^TA]\|_{(s)} < \tfrac{1}{2}, \tag{4.80}$$

the reason being that, as is immediately seen, $\widehat{\mathrm{Opt}}$ is an upper bound on Opt (see Proposition 1.9 with $q = 1$).

An immediate observation is that Opt is nothing but the maximum of a quadratic form over an appropriate convex compact set. Specifically, let

$$\mathcal{X} = \{[u;v] \in \mathbf{R}^n \times \mathbf{R}^n : Au = 0, \|u\|_1 \le 1, \sum_i |v_i| \le s, \|v\|_\infty \le 1\},$$
$$B = \left[\begin{array}{c|c} & \frac{1}{2}I_n \\ \hline \frac{1}{2}I_n & \end{array}\right].$$

Then

$$\begin{array}{rcl}\mathrm{Opt}_{\mathcal{X}}(B) & = & \max\limits_{[u;v]\in\mathcal{X}} [u;v]^T B[u;v] \\ & = & \max\limits_{u,v} \left\{u^T v : Au = 0, \|u\|_1 \le 1, \sum_i |v_i| \le s, \|v\|_\infty \le 1\right\} \\ & \underbrace{=}_{(a)} & \max_u \left\{\|u\|_{(s)} : Au = 0, \|u\|_1 \le 1\right\} \\ & = & \mathrm{Opt},\end{array}$$

where (a) is due to the well-known fact (prove it!) that *whenever s is a positive integer $\le n$, the extreme points of the set*

$$V = \{v \in \mathbf{R}^n : \sum_i |v_i| \le s, \|v\|_\infty \le 1\}$$

are exactly the vectors with at most s nonzero entries, the nonzero entries being ± 1; as a result

$$\forall (z \in \mathbf{R}^n) : \max_{v\in V} z^T v = \|z\|_{(s)}.$$

Now, V is the unit ball of the absolute norm

$$r(v) = \min\left\{t : \|v\|_1 \le st, \|v\|_\infty \le t\right\},$$

so that $\mathcal{X}$ is contained in the unit ball $\mathcal{B}$ of the absolute norm on $\mathbf{R}^{2n}$ specified as

$$p([u;v]) = \max\left\{\|u\|_1, r(v)\right\} \qquad [u, v \in \mathbf{R}^n],$$

i.e.,

$$\mathcal{X} = \left\{[u;v] : p([u,v]) \le 1, Au = 0\right\}.$$

As a result, whenever $x = [u;v] \in \mathcal{X}$, the matrix

$$U = xx^T = \left[\begin{array}{c|c} U^{11} = uu^T & U^{12} = uv^T \\ \hline U^{21} = vu^T & U^{22} = vv^T \end{array}\right]$$

satisfies the condition $p^+(U) \le 1$ (see item 1.2 above). In addition, this matrix clearly satisfies the condition

$$A[U^{11}, U^{12}] = 0.$$

It follows that the set

$$\mathcal{U} = \{U = \left[\begin{array}{c|c} U^{11} & U^{12} \\ \hline U^{21} & U^{22} \end{array}\right] \in \mathbf{S}^{2n} : U \succeq 0, p^+(U) \le 1, AU^{11} = 0, AU^{12} = 0\}$$

(which clearly is a nonempty convex compact set) matches $\mathcal{X}$. As a result, the efficiently computable quantity

$$\begin{array}{rcl}\overline{\mathrm{Opt}} & = & \max\limits_{U\in\mathcal{U}} \mathrm{Tr}(BU) \\ & = & \max\limits_{U} \left\{\mathrm{Tr}(U^{12}) : U = \left[\begin{array}{c|c} U^{11} & U^{12} \\ \hline U^{21} & U^{22} \end{array}\right] \succeq 0, p^+(U) \le 1, AU^{11} = 0, AU^{12} = 0\right\}\end{array} \tag{4.81}$$

is an upper bound on Opt. As a result, the verifiable condition

$$\overline{\mathrm{Opt}} < 1/2$$

is sufficient for s-goodness of A.

Now comes the concluding part of the exercise:

3.1) Prove that $\overline{\mathrm{Opt}} \leq \widehat{\mathrm{Opt}}$, so that (4.81) is less conservative than (4.80).
Hint: Apply Conic Duality to verify that

$$\widehat{\mathrm{Opt}} = \max_V \left\{ \mathrm{Tr}(V) : V \in \mathbf{R}^{n\times n}, AV = 0, \sum_{i=1}^n r(\mathrm{Col}_i[V^T]) \leq 1 \right\}. \tag{4.82}$$

3.2) Run simulations with randomly generated Gaussian matrices A and play with different values of s to compare $\widehat{\mathrm{Opt}}$ and $\overline{\mathrm{Opt}}$. To save time, you can use toy sizes m, n, say, $m = 18, n = 24$.

Solution: 3.1: $\widehat{\mathrm{Opt}}$ is the optimal value in the strictly feasible and bounded conic problem

$$\widehat{\mathrm{Opt}} = \min_{H,\tau} \left\{ \tau : [\mathrm{Col}_j[I - H^T A]; \tau] \in \mathbf{K}, j \leq n \right\}, \mathbf{K} = \{[h;\tau] \in \mathbf{R}^n \times \mathbf{R} : \|h\|_{(s)} \leq \tau\}. \tag{P}$$

From the above considerations, the cone $\mathbf{K}_*$ is the cone $\{[v;t] \in \mathbf{R}^n \times \mathbf{R} : r(v) \leq t\}$, so that by the Conic Duality Theorem $\widehat{\mathrm{Opt}}$ is the optimal value in the problem dual to (P). Straightforward computation shows that the dual problem is

$$\begin{array}{l} \max_{\{[v_j;t_j], j\leq n\}} \left\{ \sum_j [v_j]_j : [v_j; t_j] \in \mathbf{K}_*, \sum_j t_j = 1, \sum_j v_j^T H^T \mathrm{Col}_j[A] = 0 \forall H \right\} \\ = \max_{V = [v_1^T; v_2^T; \ldots; v_n^T] \in \mathbf{R}^{n\times n}} \left\{ \mathrm{Tr}(V) : \sum_i r(v_i) \leq 1, AV = 0 \right\}, \end{array}$$

implying the validity of (4.82). It remains to note that when U runs through the feasible set of (4.81), $V = U^{12}$ is contained in the feasible set of (4.82) (see what $p^+(\cdot)$ is), implying that $\overline{\mathrm{Opt}} \leq \widehat{\mathrm{Opt}}$. □

3.2: Here is an instructive simulation run:

s	$\widehat{\mathrm{Opt}}$	$\overline{\mathrm{Opt}}$
1	0.2497	0.2497
2	0.3751	0.3725
3	0.5075	0.4975
4	0.6319	0.6209

We see that (4.81) is slightly better than (4.80). For example, (4.81) certifies that matrix A underlying the experiment is 3-good, while (4.80) only certifies 2-goodness.

6.4.4 Around Propositions 4.4 and 4.14

6.4.4.1 Optimizing linear estimates on convex hulls of unions of spectratopes

Exercise 4.9 Let

- $\mathcal{X}_1, \ldots, \mathcal{X}_J$ be spectratopes in $\mathbf{R}^n$:

$$\mathcal{X}_j = \{x \in \mathbf{R}^n : \exists (y \in \mathbf{R}^{N_j}, t \in \mathcal{T}_j) : x = P_j y, R_{kj}^2[y] \preceq t_k I_{d_{kj}}, k \leq K_j\}, \ 1 \leq j \leq J$$
$$\left[R_{kj}[y] = \textstyle\sum_{i=1}^{N_j} y_i R^{kji} \right]$$

- $A \in \mathbf{R}^{m\times n}$ and $B \in \mathbf{R}^{\nu\times n}$ be given matrices,
- $\|\cdot\|$ be a norm on $\mathbf{R}^\nu$ such that the unit ball $\mathcal{B}_*$ of the conjugate norm $\|\cdot\|_*$ is a spectratope:

$$\begin{array}{rcl} \mathcal{B}_* & := & \{u : \|u\|_* \leq 1\} \\ & = & \{u \in \mathbf{R}^\nu : \exists (z \in \mathbf{R}^N, r \in \mathcal{R}) : u = Mz, S_\ell^2[z] \preceq r_\ell I_{f_\ell}, \ell \leq L\} \\ & & \left[S_\ell[z] = \sum_{i=1}^N z_i S^{\ell i} \right] \end{array}$$

• Π be a convex compact subset of the interior of the positive semidefinite cone $\mathbf{S}^m_+$, with our standard restrictions on $R_{kj}[\cdot]$, $S_\ell[\cdot]$, $\mathcal{T}_j$ and $\mathcal{R}$. Let, further,

$$\mathcal{X} = \mathrm{Conv}\left(\bigcup_j \mathcal{X}_j\right)$$

be the convex hull of the union of spectratopes $\mathcal{X}_j$. Consider the situation where, given observation

$$\omega = Ax + \xi$$

of an unknown signal x known to belong to $\mathcal{X}$, we want to recover Bx. We assume that the matrix of second moments of noise is $\succeq$-dominated by a matrix from Π, and quantify the performance of a candidate estimate $\widehat{x}(\cdot)$ by its $\|\cdot\|$-*risk*

$$\mathrm{Risk}_{\Pi,\|\cdot\|}[\widehat{x}|\mathcal{X}] = \sup_{x\in\mathcal{X}} \sup_{P:P\lhd\Pi} \mathbf{E}_{\xi\sim P}\left\{\|Bx - \widehat{x}(Ax+\xi)\|\right\}$$

where $P\lhd\Pi$ means that the matrix $\mathrm{Var}[P] = \mathbf{E}_{\xi\sim P}\{\xi\xi^T\}$ of second moments of distribution P is $\succeq$-dominated by a matrix from Π.

Prove the following

Proposition 4.28 *In the situation in question, consider the convex optimization problem*

$$\mathrm{Opt} = \min_{H,\Theta,\Lambda^j,\Upsilon^j,\Upsilon'} \left\{ \max_j \left[\phi_{\mathcal{T}_j}(\lambda[\Lambda^j]) + \phi_{\mathcal{R}}(\lambda[\Upsilon^j])\right] + \phi_{\mathcal{R}}(\lambda[\Upsilon']) + \Gamma_\Pi(\Theta) : \right.$$
$$\left.\begin{array}{r}\Lambda^j = \{\Lambda^j_k \succeq 0, j\leq K_j\}, j\leq J,\\ \Upsilon^j = \{\Upsilon^j_\ell \succeq 0, \ell\leq L\}, j\leq J, \Upsilon' = \{\Upsilon'_\ell \succeq 0, \ell\leq L\}\\ \left[\begin{array}{c|c}\sum_k \mathcal{R}^*_{kj}[\Lambda^j_k] & \frac{1}{2}P_j^T[B^T - A^TH]M\\ \hline \frac{1}{2}M^T[B-H^TA]P_j & \sum_\ell \mathcal{S}^*_\ell[\Upsilon^j_\ell]\end{array}\right] \succeq 0, j\leq J,\\ \left[\begin{array}{c|c}\Theta & \frac{1}{2}HM\\ \hline \frac{1}{2}M^TH^T & \sum_\ell \mathcal{S}^*_\ell[\Upsilon'_\ell]\end{array}\right] \succeq 0\end{array}\right\}, \tag{4.83}$$

where, as usual,

$$\begin{array}{c}\phi_{\mathcal{T}_j}(\lambda) = \max\limits_{t\in\mathcal{T}_j} t^T\lambda,\ \phi_{\mathcal{R}}(\lambda) = \max\limits_{r\in\mathcal{R}} r^T\lambda,\\ \Gamma_\Pi(\Theta) = \max\limits_{Q\in\Pi}\mathrm{Tr}(Q\Theta),\ \lambda[U_1,...,U_s] = [\mathrm{Tr}(U_1);...;\mathrm{Tr}(U_S)],\\ \mathcal{S}^*_\ell[\cdot]:\mathbf{S}^{f_\ell}\to\mathbf{S}^N : \mathcal{S}^*_\ell[U] = \left[\mathrm{Tr}(S^{\ell p}US^{\ell q})\right]_{p,q\leq N},\\ \mathcal{R}^*_{kj}[\cdot]:\mathbf{S}^{d_{kj}}\to\mathbf{S}^{N_j} : \mathcal{R}^*_{kj}[U] = \left[\mathrm{Tr}(R^{kjp}UR^{kjq})\right]_{p,q\leq N_j}.\end{array}$$

Problem (4.83) is solvable, and H-component H_* of its optimal solution gives rise to linear estimate $\widehat{x}_{H_*}(\omega) = H_*^T\omega$ such that

$$\mathrm{Risk}_{\Pi,\|\cdot\|}[\widehat{x}_{H_*}|\mathcal{X}] \leq \mathrm{Opt}. \tag{4.84}$$

Moreover, the estimate $\widehat{x}_{H_*}$ is near-optimal among linear estimates:

$$\begin{array}{c}\mathrm{Opt} \leq O(1)\ln(D+F)\mathrm{RiskOpt}_{lin}\\ \left[D = \max_j \sum_{k\leq K_j} d_{kj},\ F = \sum_{\ell\leq L} f_\ell\right]\end{array} \tag{4.85}$$

where

$$\mathrm{RiskOpt}_{lin} = \inf_H \sup_{x\in\mathcal{X},Q\in\Pi} \mathbf{E}_{\xi\sim\mathcal{N}(0,Q)}\left\{\|Bx - H^T(Ax+\xi)\|\right\}$$

is the best risk attainable by linear estimates in the current setting under zero mean Gaussian observation noise.

It should be stressed that the convex hull of unions of spectratopes is not necessarily a spectratope, and that Proposition 4.28 states that the linear estimate stemming from (4.83) is near-optimal only among linear, not among all estimates (the latter might indeed not be the case).

Solution: **1°.** Let

$$\begin{array}{rcl}
\Phi(H) &=& \max\limits_{x\in\mathcal{X}} \|[B-H^TA]x\|,\\
\Phi_j(H) &=& \max\limits_{x\in\mathcal{X}_j} \|[B-H^TA]x\|,\\
\Phi_j^+(H) &=& \min\limits_{\Lambda_k^j,\Upsilon_\ell^j} \left\{\phi_{\mathcal{T}_j}(\lambda[\Lambda^j])+\phi_{\mathcal{R}}(\lambda[\Upsilon^j]):\right.\\
&& \qquad \Lambda_k^j\succeq 0, j\leq K_j, \Upsilon_\ell^j\succeq 0,\ell\leq L\\
&& \qquad \left.\left[\begin{array}{c|c}\sum_k\mathcal{R}_{kj}^*[\Lambda_k^j] & \frac{1}{2}P_j^T[B^T-A^TH]M\\ \hline \frac{1}{2}M^T[B-H^TA]P_j & \sum_\ell\mathcal{S}_\ell^*[\Upsilon_\ell^j]\end{array}\right]\succeq 0\right\},\\
\Psi(H) &=& \max\limits_{P\lhd\Pi}\mathbf{E}_{\xi\sim P}\left\{\|H^T\xi\|\right\},\\
\Psi^+(H) &=& \min\limits_{\Theta,\Upsilon_\ell'}\left\{\phi_{\mathcal{R}}(\lambda[\Upsilon'])+\Gamma_\Pi(\Theta): \begin{array}{c}\Upsilon'_\ell\succeq 0,\ell\leq L\\ \left[\begin{array}{c|c}\Theta & \frac{1}{2}HM\\ \hline \frac{1}{2}M^TH^T & \sum_\ell\mathcal{S}_\ell^*[\Upsilon_\ell']\end{array}\right]\succeq 0\end{array}\right\}\\
\Phi^+(H) &=& \max\limits_j \Phi_j^+(H).
\end{array}$$

Observe that $\Phi_j(H)$ is the maximum of a convex function of x (independent of j and depending on H as parameter) over $x\in\mathcal{X}_j$, while $\Phi(H)$ is the maximum of the same function over $x\in\mathcal{X}$. Since $\mathcal{X}$ is the convex hull of $\cup_j\mathcal{X}_j$, we get

$$\Phi(H)=\max_j\Phi_j(H). \tag{6.10}$$

It is easily seen that (4.83) is a feasible problem with bounded level sets of the objective, so that the problem is solvable, and we clearly have

$$\mathrm{Opt}=\min_H\left\{\Phi^+(H)+\Psi^+(H)\right\}=\Phi^+(H_*)+\Psi^+(H_*). \tag{6.11}$$

Applying Proposition 4.8 to the quadratic form $f_H([u;x])=u^T[B-H^TA]x$ and spectratope $\mathcal{B}_*\times\mathcal{X}_j$, we get

$$\Phi_j(H)\leq\Phi_j^+(H)\leq 2\ln(2(D_j+F))\Phi_j(H),\ D_j=\sum_{k\leq K_j}d_{kj},\ 1\leq j\leq J, \tag{6.12}$$

while by Lemma 4.11 it holds

$$\Psi(H)\leq\Psi^+(H). \tag{6.13}$$

In addition, (6.10) combines with the definition of Φ^+ and (6.12) to imply that

$$\Phi(H)\leq\Phi^+(H)\leq 2\ln(2(D+F))\Phi(H). \tag{6.14}$$

Finally, by the origin of Φ and Ψ we have

$$\mathrm{Risk}_{\Pi,\|\cdot\|}[\widehat{x}_{H_*}|\mathcal{X}]\leq\Phi(H_*)+\Psi(H_*)=\max_j\Phi_j(H_*)+\Psi(H_*)$$

which combines with (6.11), (6.14), and (6.13) to imply (4.84).

2°. Let Ω be the feasible set of (4.83). Consider the function

$$F(\underbrace{H,\Theta,\{\Lambda^j,\Upsilon^j,j\leq J\},\Upsilon'}_{\zeta};Q) = \max_j\Big[\phi_{\mathcal{T}_j}(\lambda[\Lambda^j])+\phi_{\mathcal{R}}(\lambda[\Upsilon^j]) \\ +\phi_{\mathcal{R}}(\lambda[\Upsilon'])+\mathrm{Tr}(Q\Theta)\Big]:\Omega\times\Pi\to\mathbf{R}.$$

This function clearly is continuous convex-concave on its domain, and is coercive in the convex argument ζ (restricted to vary in Ω). By the Sion-Kakutani Theorem, the function possesses a saddle point (ζ_*,Q_*). In other words, there exists $Q_*\in\Pi$ such that

$$\begin{array}{l}\mathrm{Opt}=\min\limits_{H,\Theta,\Lambda^j,\Upsilon^j,\Upsilon'}\Big\{\max\limits_j\left[\phi_{\mathcal{T}_j}(\lambda[\Lambda^j])+\phi_{\mathcal{R}}(\lambda[\Upsilon^j])\right]+\phi_{\mathcal{R}}(\lambda[\Upsilon'])+\mathrm{Tr}(Q_*\Theta):\\ \quad\Lambda^j=\{\Lambda^j_k\succeq 0,k\leq K_j\},j\leq J,\Upsilon^j=\{\Upsilon^j_\ell\succeq0,\ell\leq L\},j\leq J,\Upsilon'=\{\Upsilon'_\ell\succeq 0,\ell\leq L\},\\ \qquad\left[\begin{array}{c|c}\sum_k\mathcal{R}^*_{kj}[\Lambda^j_k] & \frac{1}{2}P_j^T[B^T-A^TH]M\\ \hline \frac{1}{2}M^T[B-H^TA]P_j & \sum_\ell\mathcal{S}^*_\ell[\Upsilon^j_\ell]\end{array}\right]\succeq0,j\leq J,\\ \qquad\qquad\left[\begin{array}{c|c}\Theta & \frac{1}{2}HM\\ \hline \frac{1}{2}M^TH^T & \sum_\ell\mathcal{S}^*_\ell[\Upsilon'_\ell]\end{array}\right]\succeq0\Big\}\\ =\min\limits_H\left\{\Phi^+(H)+\Psi^+_*(H)\right\}\end{array} \tag{6.15}$$

where

$$\Psi^+_*(H):=\min_{\Theta,\Upsilon'}\left\{\phi_{\mathcal{R}}(\lambda[\Upsilon'])+\mathrm{Tr}(Q_*\Theta):\begin{array}{c}\Upsilon'_\ell\succeq0,\ell\leq L,\\ \left[\begin{array}{c|c}\Theta & \frac{1}{2}HM\\ \hline \frac{1}{2}M^TH^T & \sum_\ell\mathcal{S}^*_\ell[\Upsilon'_\ell]\end{array}\right]\succeq0\end{array}\right\}.$$

Now consider the optimization problem

$$\mathrm{Opt}_*=\min_H\left\{\Phi(H)+\Psi_*(H)\right\},\ \Psi_*(H)=\mathbf{E}_{\xi\sim\mathcal{N}(0,Q_*)}\left\{\|H^T\xi\|\right\}.$$

We claim that

$$\mathrm{RiskOpt}_{\mathrm{lin}}\geq\tfrac{1}{2}\mathrm{Opt}_*. \tag{6.16}$$

Indeed, taking into account that $Q_*\in\Pi$, it suffices to verify that for every linear estimate $\widehat{x}(\omega)=H^T\omega$ we have

$$\mathrm{Risk}[H]:=\max_{x\in\mathcal{X}}\mathbf{E}_{\xi\sim\mathcal{N}(0,Q_*)}\left\{\|Bx-H^T(Ax+\xi)\|\right\}\geq\frac{1}{2}\mathrm{Opt}_*.$$

By Jensen's inequality, we have

$$\mathrm{Risk}[H]\geq\max_{x\in\mathcal{X}}\|\mathbf{E}_{\xi\sim\mathcal{N}(0,Q_*)}\{Bx-H^T(Ax+\xi)\}\|=\max_{x\in\mathcal{X}}\|Bx-H^TAx\|=\Phi(H).$$

On the other hand, $0\in\mathcal{X}$, whence

$$\mathrm{Risk}[H]\geq\mathbf{E}_{\xi\sim\mathcal{N}(0,Q_*)}\{\|B\cdot0-H^T(A\cdot0+\xi)\|\}=\mathbf{E}_{\xi\sim\mathcal{N}(0,Q)}\{\|H^T\xi\|\}=\Psi_*(H).$$

We conclude that

$$2\mathrm{Risk}[H]\geq\Phi(H)+\Psi_*(H)\geq\mathrm{Opt}_*,$$

as claimed.

3°. It remains to note that by Lemma 4.17 we have

$$\Psi_*(H) \leq \Psi_*^+(H) \leq O(1)\sqrt{\ln(2F)}\Psi_*(H),$$

which combines with (6.14) and (6.15) to imply that

$$\begin{array}{rcl}\mathrm{Opt} &=& \min_H \{\Phi^+(H) + \Psi_*^+(H)\} \leq O(1)\ln(D+F)\min_H \{\Phi(H) + \Psi_*(H)\}\\ &=& O(1)\ln(D+F)\mathrm{Opt}_* \leq O(1)\ln(D+F)\cdot 2\mathrm{RiskOpt}_{\mathrm{lin}},\end{array}$$

where the concluding inequality is due to (6.16), and (4.85) follows. □

6.4.4.2 Recovering nonlinear vector-valued functions

Exercise 4.10 Consider the situation as follows: We are given a noisy observation

$$\omega = Ax + \xi_x \qquad [A \in \mathbf{R}^{m\times n}]$$

of the linear image Ax of an unknown signal x known to belong to a given spectratope $\mathcal{X} \subset \mathbf{R}^n$; here ξ_x is the observation noise with distribution P_x which can depend on x. Similarly to Section 4.3.3, we assume that we are given a computationally tractable convex compact set $\Pi \subset \mathrm{int}\, \mathbf{S}_+^m$ such that for every $x \in \mathcal{X}$, $\mathrm{Var}[P_x] \preceq \Theta$ for some $\Theta \in \Pi$; cf. (4.32). We want to recover the value $f(x)$ of a given vector-valued function $f : \mathcal{X} \to \mathbf{R}^\nu$, and we measure the recovery error in a given norm $|\cdot|$ on $\mathbf{R}^\nu$.

4.10.A. Preliminaries and Main observation. Let $\|\cdot\|$ be a norm on $\mathbf{R}^n$, and $g(\cdot) : \mathcal{X} \to \mathbf{R}^\nu$ be a function. Recall that the function is called *Lipschitz continuous on $\mathcal{X}$ w.r.t. the pair of norms $\|\cdot\|$ on the argument and $|\cdot|$ on the image spaces,* if there exist $L < \infty$ such that

$$|g(x) - g(y)| \leq L\|x - y\| \;\forall (x, y \in \mathcal{X});$$

every L with this property is called a Lipschitz constant of g. It is well known that in our finite-dimensional situation, the property of g to be Lipschitz continuous is independent of how the norms $\|\cdot\|$, $|\cdot|$ are selected; this selection affects only the value(s) of the Lipschitz constant(s).

Assume from now on that the function of interest f is Lipschitz continuous on $\mathcal{X}$. Let us call a norm $\|\cdot\|$ on $\mathbf{R}^n$ *appropriate* for f if f is Lipschitz continuous *with constant 1* on $\mathcal{X}$ w.r.t. $\|\cdot\|$, $|\cdot|$. Our immediate observation is as follows:

Observation 4.29 *In the situation in question, let $\|\cdot\|$ be appropriate for f. Then recovering $f(x)$ is not more difficult than recovering x in the norm $\|\cdot\|$: every estimate $\widehat{x}(\omega)$ of x via ω such that $\widehat{x}(\cdot) \in \mathcal{X}$ induces the "plug-in" estimate*

$$\widehat{f}(\omega) = f(\widehat{x}(\omega))$$

of $f(x)$, and the $\|\cdot\|$-risk

$$\mathrm{Risk}_{\|\cdot\|}[\widehat{x}|\mathcal{X}] = \sup_{x\in\mathcal{X}} \mathbf{E}_{\xi\sim P_x}\left\{\|\widehat{x}(Ax+\xi) - x\|\right\}$$

of estimate $\widehat{x}$ upper-bounds the $|\cdot|$-risk

$$\mathrm{Risk}_{|\cdot|}[\widehat{f}|\mathcal{X}] = \sup_{x\in\mathcal{X}} \mathbf{E}_{\xi\sim P_x}\left\{\|\widehat{f}(Ax+\xi) - f(x)\|\right\}$$

of the induced by $\widehat{x}$ estimate $\widehat{f}$:

$$\mathrm{Risk}_{|\cdot|}[\widehat{f}|\mathcal{X}] \leq \mathrm{Risk}_{\|\cdot\|}[\widehat{x}|\mathcal{X}].$$

When f is defined and Lipschitz continuous with constant 1 w.r.t. $\|\cdot\|, |\cdot|$ on the entire $\mathbf{R}^n$, this conclusion remains valid without the assumption that $\widehat{x}$ is $\mathcal{X}$-valued.

4.10.B. Consequences. Observation 4.29 suggests the following simple approach to solving the estimation problem we started with: assuming that we have at our disposal a norm $\|\cdot\|$ on $\mathbf{R}^n$ such that

- $\|\cdot\|$ is appropriate for f, and
- $\|\cdot\|$ is *good*, goodness meaning that the unit ball $\mathcal{B}_*$ of the norm $\|\cdot\|_*$ *conjugate* to $\|\cdot\|$ is a spectratope given by an explicit spectratopic representation;

we use the machinery of linear estimation developed in Section 4.3.3 to build a near-optimal, in terms of its $\|\cdot\|$-risk, linear estimate of x via ω, and convert this estimate in an estimate of $f(x)$. By the above observation, the $|\cdot|$- risk of the resulting estimate is upper-bounded by the $\|\cdot\|$-risk of the underlying linear estimate. The construction just outlined needs a correction: in general, the linear estimate $\widetilde{x}(\cdot)$ yielded by Proposition 4.14 (same as any nontrivial—not identically zero—*linear* estimate) is *not* guaranteed to take values in $\mathcal{X}$, which is, in general, required for Observation 4.29 to be applicable. This correction is easy: it is enough to convert $\widetilde{x}$ into the estimate $\widehat{x}$ defined by

$$\widehat{x}(\omega) \in \operatorname*{Argmin}_{u\in\mathcal{X}} \|u - \widetilde{x}(\omega)\|.$$

This transformation preserves efficient computability of the estimate, and ensures that the corrected estimate takes its values in $\mathcal{X}$; at the same time, "correction" $\widetilde{x} \mapsto \widehat{x}$ nearly preserves the $\|\cdot\|$-risk:

$$\mathrm{Risk}_{\|\cdot\|}[\widehat{x}|\mathcal{X}] \leq 2\mathrm{Risk}_{\|\cdot\|}[\widetilde{x}|\mathcal{X}]. \tag{$*$}$$

Note that when $\|\cdot\|$ is a (general-type) Euclidean norm—$\|x\|^2 = x^TQx$ for some $Q \succ 0$—the factor 2 on the right-hand side can be discarded.

1) Justify ($*$).

Solution: When $x \in \mathcal{X}$, we have

$$\begin{array}{l}
\|\widehat{x}(Ax+\xi) - \widetilde{x}(Ax+\xi)\| = \min_{u\in\mathcal{X}} \|u - \widetilde{x}(Ax+\xi)\| \leq \|x - \widetilde{x}(Ax+\xi)\| \\
\Rightarrow \|\widehat{x}(Ax+\xi) - x\| \leq \|\widehat{x}(Ax+\xi) - \widetilde{x}(Ax+\xi)\| + \|\widetilde{x}(Ax+\xi) - x\| \\
\qquad\qquad\qquad\qquad\qquad\qquad\qquad \leq 2\|x - \widetilde{x}(Ax+\xi)\| \\
\Rightarrow (*).
\end{array}$$

When $\|\cdot\|$ is a Euclidean norm, the situation is simpler: as it is well known, in this case projection onto a convex set makes the point closer to all points of the set:

$$\|\widehat{x}(\omega) - x\| \leq \|\widetilde{x} - x\|, \forall x \in \mathcal{X},$$

and the factor 2 does not arise. □

4.10.C. How to select $\|\cdot\|$. When implementing the outlined approach, the major question is how to select a norm $\|\cdot\|$ appropriate for f. The best choice would be to select the smallest among the norms appropriate for f (such a norm does exist under mild assumptions), because the smaller the $\|\cdot\|$, the smaller the $\|\cdot\|$-risk of an estimate of x. This ideal can be achieved in rare cases only: first, it could be difficult to identify the smallest among the norms appropriate for f; second, our approach requires from $\|\cdot\|$ to have an explicitly given spectratope as the unit ball of the conjugate norm. Let us look at a couple of "favorable cases," where the difficulties just outlined can be (partially) overcome.

Example: a norm-induced f. Let us start with the case, important in its own right, when f is a scalar functional which itself is a norm, and this norm has a spectratope as the unit ball of the conjugate norm, as is the case when $f(\cdot) = \|\cdot\|_r$, $r \in [1,2]$, or when $f(\cdot)$ is the nuclear norm. In this case the smallest of the norms appropriate for f clearly is f itself, and none of the outlined difficulties arises. As an extension, when $f(x)$ is obtained from a good norm $\|\cdot\|$ by operations preserving Lipschitz continuity and constant, such as $f(x) = \|x - c\|$, or $f(x) = \sum_i a_i\|x - c_i\|$, $\sum_i |a_i| \le 1$, or

$$f(x) = \sup_{c\in C} / \inf \|x - c\|,$$

or even something like

$$f(x) = \sup_{\alpha\in\mathcal{A}} / \inf \left\{ \sup_{c\in C_\alpha} / \inf \|x - c\| \right\},$$

it seems natural to use this norm in our construction, although now this, perhaps, is not the smallest of the norms appropriate for f.

Now let us consider the general case. Note that *in principle* the smallest of the norms appropriate for a given Lipschitz continuous f admits a description. Specifically, assume that $\mathcal{X}$ has a nonempty interior (this is w.l.o.g.—we can always replace $\mathbf{R}^n$ with the linear span of $\mathcal{X}$). A well-known fact of Analysis (Rademacher Theorem) states that in this situation (more generally, when $\mathcal{X}$ is convex with a nonempty interior), a Lipschitz continuous f is differentiable almost everywhere in $\mathcal{X}^o = \mathrm{int}\,\mathcal{X}$, and f is Lipschitz continuous with constant 1 w.r.t. a norm $\|\cdot\|$ if and only if

$$\|f'(x)\|_{\|\cdot\|\to|\cdot|} \le 1$$

whenever $x \in \mathcal{X}^o$ is such that the derivative (a.k.a. Jacobian) of f at x exists; here $\|Q\|_{\|\cdot\|\to|\cdot|}$ is the matrix norm of a $\nu \times n$ matrix Q induced by the norms $\|\cdot\|$ on $\mathbf{R}^n$ and $|\cdot|$ on $\mathbf{R}^\nu$:

$$\|Q\|_{\|\cdot\|\to|\cdot|} := \max_{\|x\|\le 1} |Qx| = \max_{\substack{\|x\|\le 1\\ |y|_*\le 1}} y^T Q x = \max_{\substack{|y|_*\le 1\\ [\|x\|_*]_*\le 1}} x^T Q^T y = \|Q^T\|_{|\cdot|_*\to\|\cdot\|_*},$$

where $\|\cdot\|_*$, $|\cdot|_*$ are the conjugates of $\|\cdot\|$, $|\cdot|$.

2) Prove that a norm $\|\cdot\|$ is appropriate for f if and only if the unit ball of the norm *conjugate* to $\|\cdot\|$ contains the set

$$\mathcal{B}_{f,*} = \mathrm{cl}\,\mathrm{Conv}\{z : \exists (x \in \mathcal{X}_o, y, |y|_* \le 1) : z = [f'(x)]^T y\},$$

where $\mathcal{X}_o$ is the set of all $x \in \mathcal{X}^o$ where $f'(x)$ exists. Geometrically, $\mathcal{B}_{f,*}$ is the closed convex hull of the union of all images of the unit ball $\mathcal{B}_*$ of $|\cdot|_*$ under the linear mappings $y \mapsto [f'(x)]^T y$ stemming from $x \in \mathcal{X}_o$.
Equivalently: $\|\cdot\|$ is appropriate for f if and only if

$$\|u\| \ge \|u\|_f := \max_{z\in\mathcal{B}_{f,*}} z^T u. \tag{!}$$

Check that $\|u\|_f$ is a norm, provided that $\mathcal{B}_{f,*}$ (this set by construction is a convex compact set symmetric w.r.t. the origin) possesses a nonempty interior; whenever this is the case, $\|u\|_f$ is the smallest of the norms appropriate for f.
Derive from the above that the norms $\|\cdot\|$ we can use in our approach are the norms on $\mathbf{R}^n$ for which the unit ball of the conjugate norm is a spectratope containing $\mathcal{B}_{f,*}$.

Proof. By the above Rademacher Theorem, $\|\cdot\|$ is appropriate for f if and only if

$$[\|f'(x)\|_{\|\cdot\|\to|\cdot|} =]\ \ \|[f'(x)]^T\|_{|\cdot|_*\to\|\cdot\|_*} \le 1\ \forall x \in \mathcal{X}_o,$$

or, equivalently, if and only if

$$\|u\| \le 1 \Rightarrow u^T[f'(x)]^T y \le 1\,\forall (y : |y|_* \le 1, x \in \mathcal{X}_o),$$

or, which is the same, if and only if

$$\|u\| \le 1 \Rightarrow u^T z \le 1 \;\forall z \in \mathcal{B}_{f,*},$$

which is nothing but (!). □

Example. Consider the case of componentwise quadratic f:

$$f(x) = \left[\tfrac{1}{2}x^T Q_1 x; \tfrac{1}{2}x^T Q_2 x; ...; \tfrac{1}{2}x^T Q_\nu x\right] \qquad [Q_i \in \mathbf{S}^n]$$

and $|u| = \|u\|_q$ with $q \in [1,2]$.[6] In this case

$$\mathcal{B}_* = \{u \in \mathbf{R}^\nu : \|u\|_p \le 1\},\; p = \frac{q}{q-1} \in [2,\infty[, \text{ and } f'(x) = \left[x^T Q_1; x^T Q_2; ...; x^T Q_\nu\right].$$

Setting $\mathcal{S} = \{s \in \mathbf{R}^\nu_+ : \|s\|_{p/2} \le 1\}$ and

$$\mathcal{S}^{1/2} = \{s \in \mathbf{R}^\nu_+ : [s_1^2; ...; s_\nu^2] \in \mathcal{S}\} = \{s \in \mathbf{R}^\nu_+ : \|s\|_p \le 1\},$$

the set

$$\mathcal{Z} = \{[f'(x)]^T u : x \in \mathcal{X}, u \in \mathcal{B}_*\}$$

is contained in the set

$$\mathcal{Y} = \left\{ y \in \mathbf{R}^n : \exists (s \in \mathcal{S}^{1/2}, x^i \in \mathcal{X}, i \le \nu) : y = \sum_i s_i Q_i x_i \right\}.$$

Set $\mathcal{Y}$ is a spectratope with spectratopic representation readily given by that of $\mathcal{X}$. Indeed, $\mathcal{Y}$ is nothing but the $\mathcal{S}$-sum of the spectratopes $Q_i\mathcal{X}$, $i = 1, ..., \nu$; see Section 4.10. As a result, we can use the spectratope $\mathcal{Y}$ (when $\operatorname{int}\mathcal{Y} \neq \emptyset$) or the arithmetic sum of $\mathcal{Y}$ with a small Euclidean ball (when $\operatorname{int}\mathcal{Y} = \emptyset$) as the unit ball of the norm conjugate to $\|\cdot\|$, thus ensuring, by 2), that $\|\cdot\|$ is appropriate for f. We then can use $\|\cdot\|$ in order to build an estimate of $f(\cdot)$.

3.1) For illustration, work out the problem of recovering the value of a scalar quadratic form

$$f(x) = x^T M x,\; M = \operatorname{Diag}\{i^\alpha, i = 1, ..., n\} \qquad [\nu = 1, |\cdot| \text{ is the absolute value}]$$

from noisy observation

$$\omega = Ax + \sigma\eta,\; A = \operatorname{Diag}\{i^\beta, i = 1, ..., n\},\; \eta \sim \mathcal{N}(0, I_n) \tag{4.86}$$

of a signal x known to belong to the ellipsoid

$$\mathcal{X} = \{x \in \mathbf{R}^n : \|Px\|_2 \le 1\},\; P = \operatorname{Diag}\{i^\gamma, i = 1, ..., n\},$$

where α, β, γ are given reals satisfying

$$\alpha - \gamma - \beta < -1/2.$$

You could start with the simplest unbiased estimate

$$\widetilde{x}(\omega) = [1^{-\beta}\omega_1; 2^{-\beta}\omega_2; ...; n^{-\beta}\omega_n]$$

of x.

[6] To save notation, we assume that the linear parts in the components of f_i are trivial—just zeros. In this respect, note that we always can subtract from f any linear mapping and reduce our estimation problem to two distinct problems of estimating separately the values at the signal x of the modified f and the linear mapping we have subtracted (we know how to solve the latter problem reasonably well).

3.2) Work out the problem of recovering the norm

$$f(x) = \|Mx\|_p,\ M = \mathrm{Diag}\{i^\alpha, i = 1, ..., n\},\ p \in [1,2],$$

from observation (4.86) with

$$\mathcal{X} = \{x : \|Px\|_r \leq 1\},\ P = \mathrm{Diag}\{i^\gamma, i = 1, ..., n\},\ r \in [2,\infty].$$

Solution: 3.1: We are in the situation where the set $\{y = [f'(x)]^T u : x \in \mathcal{X}, u \in \mathcal{B}_* = [-1,1]\}$ is the set of gradients of the quadratic form f taken at points from $\mathcal{X}$, so that this set is just the ellipsoid $\{y : \sum_{i=1}^n i^{2\gamma-2\alpha} y_i^2 \leq 1\}$. We can take it as our $\mathcal{Y}$, resulting in

$$\|y\|_* = \sqrt{\sum\nolimits_{i=1}^n i^{2\gamma-2\alpha} y_i^2},$$

so that the norm $\|\cdot\|$ is

$$\|z\|^2 = \sqrt{\sum\nolimits_{i=1}^n i^{2\alpha-2\gamma} z_i^2}.$$

With $\widetilde{x}(\omega)$ given by $[\widetilde{x}(\omega)]_i = i^{-\beta}\omega_i$, $i \leq n$, we have

$$\|\widetilde{x}(Ax+\sigma\eta) - x\|^2 = \sum_i i^{2\alpha-2\gamma}\left[i^{-\beta}[i^\beta x_i + \sigma\eta_i] - x_i\right]^2 = \sigma^2 \sum_i i^{2\alpha-2\gamma-2\beta}\eta_i^2,$$

implying that

$$\mathrm{Risk}_{\|\cdot\|}[\widetilde{x}|\mathcal{X}] \leq \max_{x\in\mathcal{X}} \sqrt{\mathbf{E}_{\eta\sim\mathcal{N}(0,I_n)}\{\|\widetilde{x}(Ax+\sigma\eta)-x\|^2\}} \leq \sigma\sqrt{\sum\nolimits_i i^{2\alpha-2\gamma-2\beta}}.$$

When $\delta := 2\alpha - 2\gamma - 2\beta < -1$, this bound results in

$$\mathrm{Risk}_{\|\cdot\|}[\widetilde{x}|\mathcal{X}] \leq C\sigma,$$

with C depending solely on δ. The bottom line is that in the situation in question, our construction results in an estimate of $f(x)$ with "parametric risk" $O(\sigma)$. Note that when restricting the signals to have all entries, except for the first one, equal to 0, the risk of recovering $f(x)$ can clearly be *lower-bounded* by $c\sigma$, with an absolute constant $c > 0$. In other words, in the situation in question, already a completely unsophisticated implementation of the approach we have developed results in a near-optimal estimate.

3.2: We are in the situation where the set $\mathcal{B}_{f,*} = \{y \in \mathbf{R}^n : \|M^{-1}y\|_{p_*} \leq 1\}$, $p_* = \frac{p}{p-1} \in [2,\infty]$, and the optimal norm $\|\cdot\|$ is just $f(\cdot)$. Problem (4.42) yielding the estimate $\widetilde{x}$ reads

$$\begin{array}{rcl}\mathrm{Opt} & = & \min\limits_{H,\lambda,\mu,\mu',\Theta}\left\{\phi_r(\lambda)+\psi_p(\mu)+\psi_p(\mu')+\sigma^2\mathrm{Tr}(\Theta) : \lambda\in\mathbf{R}^n_+, \mu\in\mathbf{R}^n_+, \mu'\in\mathbf{R}^n_+,\right.\\ && \left.\begin{array}{r}\left[\begin{array}{c|c}\mathrm{Diag}\{\lambda_i i^{2\gamma}, i\leq n\} & \frac{1}{2}[I_n - AH]\\ \hline \frac{1}{2}[I_n - H^TA] & \mathrm{Diag}\{\mu_i i^{-2\alpha}, i \leq n\}\end{array}\right]\succeq 0\\ \left[\begin{array}{c|c}\Theta & \frac{1}{2}H\\ \hline \frac{1}{2}H^T & \mathrm{Diag}\{\mu'_i i^{-2\alpha}, i\leq n\}\end{array}\right]\succeq 0\end{array}\right\},\\ && \phi_r(\lambda) = \|\lambda\|_{\frac{r}{r-2}},\ \psi_p(\mu) = \|\mu\|_{\frac{p}{2-p}}\end{array}.$$

It is immediately seen that when seeking for an optimal solution, we can restrict ourselves with diagonal H and Θ, which makes the problem easy to solve in the large scale case. It is also worthy of mentioning that we can use $\widehat{x} \equiv \widetilde{x}$, since f is Lipschitz continuous with constant 1 w.r.t. the norm $\|\cdot\|$ on the entire $\mathbf{R}^n$.

6.4.4.3 Suboptimal linear estimation

Exercise 4.11. [recovery of large-scale signals] Consider the problem of estimating the image $Bx \in \mathbf{R}^\nu$ of signal $x \in \mathcal{X}$ from observation

$$\omega = Ax + \sigma\xi \in \mathbf{R}^m$$

in the simplest case where $\mathcal{X} = \{x \in \mathbf{R}^n : x^T S x \leq 1\}$ is an ellipsoid (so that $S \succ 0$), the recovery error is measured in $\|\cdot\|_2$, and $\xi \sim \mathcal{N}(0, I_m)$. In this case, Problem (4.12) to solve when building "presumably good linear estimate" reduces to

$$\mathrm{Opt} = \min_{H,\lambda} \left\{ \lambda + \sigma^2 \|H\|_F^2 : \left[\begin{array}{c|c} \lambda S & B^T - A^T H \\ \hline B - H^T A & I_\nu \end{array}\right] \succeq 0 \right\}, \tag{4.87}$$

where $\|\cdot\|_F$ is the Frobenius norm of a matrix. An optimal solution H_* to this problem results in the linear estimate $\widehat{x}_{H_*}(\omega) = H_*^T \omega$ satisfying the risk bound

$$\mathrm{Risk}[\widehat{x}_{H_*}|\mathcal{X}] := \max_{x\in\mathcal{X}} \sqrt{\mathbf{E}\{\|Bx - H_*^T(Ax + \sigma\xi)\|_2^2\}} \leq \sqrt{\mathrm{Opt}}.$$

Now, (4.87) is an efficiently solvable convex optimization problem. However, when the sizes m, n of the problem are large, solving the problem by standard optimization techniques could become prohibitively time-consuming. The goal of what follows is to develop a relatively cheap computational technique for finding a good enough *sub*optimal solution to (4.87). In the sequel, we assume that $A \neq 0$; otherwise (4.87) is trivial.

1) Prove that problem (4.87) can be reduced to a similar problem with $S = I_n$ and diagonal positive semidefinite matrix A, the reduction requiring several singular value decompositions and multiplications of matrices of the same sizes as those of A, B and S.

Solution: First, we can assume w.l.o.g. that A is a square matrix (i.e., $m = n$). Indeed, when $m > n$, computing the singular value decomposition $A = UDV^T$ of A, multiplying A from the left by U^T and passing from ω to $U^T\omega$, we reduce the situation to the one where the last $m - n$ rows of A are zero. Eliminating these rows in A and in H, we arrive at a problem equivalent to (4.87) of the same form with the initial value of m replaced with $m = n$. Similarly, when $m < n$, adding to A $n - m$ zero rows, we arrive at a problem equivalent to (4.87) of the same form with $m = n$.
Assuming from now on that A is square, let us reduce the situation to the one where $S = I_n$. To this end we find the Cholesky decomposition $S = DD^T$ with triangular D and observe that by multiplying the LMI constraint in (4.87) by $\mathrm{Diag}\{D^{-1}, I_\nu\}$ from the left and by the transpose of this matrix from the right the problem becomes

$$\begin{gathered}\mathrm{Opt} = \min_{H,\lambda} \left\{ \lambda + \sigma^2 \|H\|_F^2 : \left[\begin{array}{c|c} \lambda I_n & \widetilde{B}^T - \widetilde{A}^T H \\ \hline \widetilde{B} - H^T \widetilde{A} & I_\nu \end{array}\right] \succeq 0 \right\}, \\ \widetilde{B} = B(D^{-1})^T,\ \widetilde{A} = A(D^{-1})^T. \end{gathered} \tag{6.17}$$

Now let us compute singular value decomposition

$$\widetilde{A} = U\Lambda V^T$$

of $\widetilde{A}$. Multiplying the semidefinite constraint in (6.17) by $\mathrm{Diag}\{V^T, I_\nu\}$ from the left and by the transpose of this matrix from the right, we arrive at the equivalent to (6.17) problem

$$\mathrm{Opt} = \min_{H,\lambda}\left\{\lambda + \sigma^2\|H\|_F^2 : \left[\begin{array}{c|c}\lambda I_n & \widehat{B}^T - \Lambda^T U^T H \\ \hline \widehat{B} - H^T U\Lambda & I_\nu\end{array}\right] \succeq 0\right\},$$

with $\widehat{B} = \widetilde{B}V$, and it remains to pass in the latter problem from variable H to variable $\bar{H} = U^T H$. □

2) By item 1, we can assume from the very beginning that $S = I$ and $A = \mathrm{Diag}\{\alpha_1, ..., \alpha_n\}$ with $0 \le \alpha_1 \le \alpha_2 \le ... \le \alpha_n$. Passing in (4.87) from variables λ, H to variables $\tau = \sqrt{\lambda}, G = H^T$, the problem becomes

$$\mathrm{Opt} = \min_{G,\tau}\left\{\tau^2 + \sigma^2\|G\|_F^2 : \|B - GA\| \le \tau\right\}, \tag{4.88}$$

where $\|\cdot\|$ is the spectral norm. Now consider the construction as follows:

- Consider a partition $\{1, ..., n\} = I_0 \cup I_1 \cup ... \cup I_K$ of the index set $\{1, ..., n\}$ into consecutive segments in such a way that
(a) I_0 is the set of those i, if any, for which $\alpha_i = 0$, and $I_k \neq \emptyset$ when $k \ge 1$,
(b) for $k \ge 1$ the ratios α_j/α_i, $i, j \in I_k$, do not exceed $\theta > 1$ (θ is the parameter of our construction), while
(c) for $1 \le k < k' \le K$, the ratios α_j/α_i, $i \in I_k$, $j \in I_{k'}$, are $> \theta$.
The recipe for building the partition is self-evident, and we clearly have
$$K \le \ln(\overline{\alpha}/\underline{\alpha})/\ln(\theta) + 1,$$
where $\overline{\alpha}$ is the largest of the α_i, and $\underline{\alpha}$ is the smallest of those α_i which are positive.
- For $1 \le k \le K$, we denote by i_k the first index in I_k, set $\alpha^k = \alpha_{i_k}$, $n_k = \mathrm{Card}\, I_k$, and define A_k as an $n_k \times n_k$ diagonal matrix with diagonal entries α_i, $i \in I_k$.

Now, given a $\nu \times n$ matrix C, let us specify C_k, $0 \le k \le K$, as the $\nu \times n_k$ submatrix of C comprised of columns with indexes from I_k, and consider the following parametric optimization problems:

$$\begin{array}{rcll}\mathrm{Opt}_k^*(\tau) & = & \min_{G_k \in \mathbf{R}^{\nu\times n_k}}\left\{\|G_k\|_F^2 : \|B_k - G_k A_k\| \le \tau\right\} & (P_k^*[\tau]) \\ \mathrm{Opt}_k(\tau) & = & \min_{G_k \in \mathbf{R}^{\nu\times n_k}}\left\{\|G_k\|_F^2 : \|B_k - \alpha^k G_k\| \le \tau\right\} & (P_k[\tau])\end{array}$$

where $\tau \ge 0$ is the parameter, and $1 \le k \le K$.
Justify the following simple observations:

2.1) G_k is feasible for $(P_k[\tau])$ if and only if the matrix
$$G_k^* = \alpha^k G_k A_k^{-1}$$
is feasible for $(P_k^*[\tau])$, and $\|G_k^*\|_F \le \|G_k\|_F \le \theta\|G_k^*\|_F$, implying that
$$\mathrm{Opt}_k^*(\tau) \le \mathrm{Opt}_k(\tau) \le \theta^2 \mathrm{Opt}_k^*(\tau);$$

2.2) Problems $(P_k[\tau])$ are easy to solve: if $B_k = U_k D_k V_k^T$ is the singular value decomposition of B_k and $\sigma_{k\ell}$, $1 \le \ell \le \nu_k := \min[\nu, n_k]$, are diagonal entries of D_k, then an optimal solution to $(P_k[\tau])$ is
$$\widehat{G}_k[\tau] = [\alpha^k]^{-1} U_k D_k[\tau] V_k^T,$$
where $D_k[\tau]$ is the diagonal matrix obtained from D_k by truncating the diagonal entries $\sigma_{k\ell} \mapsto [\sigma_{k\ell} - \tau]_+$ (from now on, $a_+ = \max[a, 0]$, $a \in \mathbf{R}$). The optimal value in $(P_k[\tau])$ is
$$\mathrm{Opt}_k(\tau) = [\alpha^k]^{-2}\sum_{\ell=1}^{\nu_k}[\sigma_{k\ell} - \tau]_+^2.$$

2.3) If (τ, G) is a feasible solution to (4.88) then $\tau \geq \underline{\tau} := \|B_0\|$, and the matrices G_k, $1 \leq k \leq K$, are feasible solutions to problems $(P_k^*[\tau])$, implying that

$$\sum_k \mathrm{Opt}_k^*(\tau) \leq \|G\|_F^2.$$

And vice versa: if $\tau \geq \underline{\tau}$, G_k, $1 \leq k \leq K$, are feasible solutions to problems $(P_k^*[\tau])$, and

$$K_+ = \begin{cases} K, & I_0 = \emptyset \\ K+1, & I_0 \neq \emptyset \end{cases},$$

then the matrix $G = [0_{\nu \times n_0}, G_1, ..., G_K]$ and $\tau_+ = \sqrt{K_+}\tau$ form a feasible solution to (4.88).

Extract from these observations that if τ_* is an optimal solution to the convex optimization problem

$$\min_\tau \left\{ \theta^2\tau^2 + \sigma^2 \sum_{k=1}^K \mathrm{Opt}_k(\tau) : \tau \geq \underline{\tau} \right\} \tag{4.89}$$

and $G_{k,*}$ are optimal solutions to the problems $(P_k[\tau_*])$, then the pair

$$\widehat{\tau} = \sqrt{K_+}\tau_*,\ \widehat{G} = [0_{\nu \times n_0}, G_{1,*}^*, ..., G_{K,*}^*] \qquad [G_{k,*}^* = \alpha^k G_{k,*} A_k^{-1}]$$

is a feasible solution to (4.88), and the value of the objective of the latter problem at this feasible solution is within factor $\max[K_+, \theta^2]$ of the true optimal value Opt of this problem. As a result, $\widehat{G}$ gives rise to a linear estimate with risk on $\mathcal{X}$ which is within factor $\max[\sqrt{K_+}, \theta]$ of the risk $\sqrt{\mathrm{Opt}}$ of the "presumably good" linear estimate yielded by an optimal solution to (4.87).

Notice that

- After carrying out singular value decompositions of matrices B_k, $1 \leq k \leq K$, specifying τ_* and $G_{k,*}$ requires solving an univariate convex minimization problem with an easy to compute objective, so that the problem can be easily solved, e.g., by bisection;
- The computationally cheap suboptimal solution we end up with is not that bad, since K is "moderate"—just logarithmic in the condition number $\overline{\alpha}/\underline{\alpha}$ of A.

Solution: 2.1 is readily given by the fact that by construction

$$B_k - \alpha^k G_k = B_k - G_k^* A_k$$

combined with $\|G_k^*\|_F \leq \|G_k\|_F$ (because G_k^* is obtained from G_k by multiplying G_k by a matrix with spectral norm ≤ 1) and $\|G_k\|_F \leq \theta\|G_k^*\|_F$ (because G_k is obtained from G_k^* by multiplying G_k^* by a matrix with spectral norm $\leq \theta$).

2.2: Representing the variable G_k in $(P_k[\tau])$ as $G_k = U_k E_k V_k^T$ and taking into account that U_k and V_k are orthogonal, $(P_k[\tau])$ becomes the problem

$$\mathrm{Opt}_k(\tau) = \min_{E_k} \{\|E_k\|_F^2 : \|D_k - \alpha^k E_k\| \leq \tau\}, \tag{$*$}$$

and all we need to verify is that an optimal solution of the latter problem is obtained from D_k by replacing the diagonal entries $\sigma_{k\ell}$ in D_k with $\gamma_{k\ell} = [\alpha^k]^{-1}[\sigma_{k\ell} - \tau]_+$ and keeping the off-diagonal entries zeros. This is immediate—the latter matrix is a feasible solution to $(*)$ with the value of the objective $\sum_\ell \gamma_{k\ell}^2$, while the ℓ-th diagonal entry in a feasible solution to $(*)$ should be $\geq \gamma_{k\ell}$ (since the magnitude of an entry in a matrix does not exceed the spectral norm of the matrix), so that the value of the objective at any feasible solution to the problem is $\geq \sum_\ell \gamma_{k\ell}^2$. □

2.3: Let (τ, G) be a feasible solution to (4.88). Since A is diagonal, we have $[B-GA]_0 = B_0 - G_0A_0 = B_0$, and since the spectral norm of a submatrix is $\leq$ the one of a matrix, we have $\tau \geq \|B_0\| = \underline{\tau}$. By similar reasons, for $k \geq 1$ we have

$$\tau \geq \|[B-GA]_k\| = \|B_k - G_kA_k\|,$$

that is, G_k is feasible for $(P_k^*[\tau])$, $1 \leq k \leq K$. Now let $\tau \geq \underline{\tau}$ and let G_k be feasible for $(P_k^*[\tau])$, $1 \leq k \leq K$. Setting $G = [0_{\nu\times n_0}, G_1, ..., G_K]$, observe that

$$\|[B-GA]_0\| = \|B_0\| = \underline{\tau} \leq \tau$$

and

$$\|[B-GA]_k\| = \|B_k - G_kA_k\| \leq \tau.$$

It follows that for every vector $z = [z_0; ...; z_K] \in \mathbf{R}^n$ with blocks z_k of dimension n_k it holds

$$\|[B-GA]z\|_2 \leq \sum_{k=0}^{K} \|[B_k - G_kA_k]z_k\|_2 \leq \tau \sum_{k=0}^{K} \|z_k\|_2 \leq \tau\sqrt{K_+}\|z\|_2,$$

that is, $\|B - GA\| \leq \sqrt{K_+}\tau$, implying that $(\sqrt{K_+}\tau, G)$ is a feasible solution to (4.88). □

Now let us extract from 2.1–3 the desired conclusions. Let $(\bar{\tau}, \bar{G})$ be an optimal solution to (4.88), so that

$$\mathrm{Opt} = \bar{\tau}^2 + \sigma^2 \sum_{k=0}^{K} \|\bar{G}_k\|_F^2.$$

Replacing $\bar{G}_0$ with $0_{\nu\times n_0}$, we get another feasible solution to the problem and can only reduce the value of the objective, implying that $\bar{G}_0 = 0$, whence

$$\mathrm{Opt} = \bar{\tau}^2 + \sigma^2 \sum_{k=1}^{K} \|\bar{G}_k\|_F^2.$$

By the first part of 2.3, the matrices $\bar{G}_k$ are feasible solutions to $(P_k^*[\bar{\tau}])$, whence

$$\bar{\tau}^2 + \sigma^2 \sum_{k=1}^{K} \mathrm{Opt}_k^*(\bar{\tau}) \leq \bar{\tau}^2 + \sigma^2 \sum_{k=1} \|\bar{G}_k\|_F^2 = \mathrm{Opt}.$$

By 2.1, we have $\bar{G}_k = \alpha^k G_k A_k^{-1}$ for some feasible solutions G_k to problems $(P_k[\bar{\tau}])$ such that $\|G_k\|_F^2 \leq \theta^2 \|\bar{G}_k\|_F^2$, whence

$$\theta^2\bar{\tau}^2 + \sigma^2 \sum_k \mathrm{Opt}_k(\bar{\tau}) \leq \theta^2\bar{\tau}^2 + \sigma^2\theta^2 \sum_k \|G_k\|_F^2 \leq \theta^2 \left[\bar{\tau}^2 + \sigma^2 \sum_k \|\bar{G}_k\|_F^2\right] \leq \theta^2 \mathrm{Opt}.$$

Due to the origin of τ_*, we conclude that

$$\theta^2\tau_*^2 + \sigma^2 \sum_k \|G_{k,*}\|_F^2 = \theta^2\tau_*^2 + \sigma^2 \sum_k \mathrm{Opt}_k(\tau_*) \leq \theta^2\bar{\tau}^2 + \sigma^2 \sum_k \mathrm{Opt}_k(\bar{\theta}) \leq \theta^2 \mathrm{Opt},$$

whence, by 2.1,

$$\theta^2\tau_*^2 + \sigma^2 \sum_k \|G_{k,*}^*\|_F^2 \le \theta^2 \mathrm{Opt}. \tag{!}$$

Now, by 2.1, the matrices $G_{k,*}^*$ are feasible solutions to problems $(P_k^*[\tau_*])$, whence by the second part of 2.3, the resulted pair $(\sqrt{K_+}\tau_*, [0_{\nu\times n_0}, G_{1,*}^*, ..., G_{K,*}^*])$ is a feasible solution to (4.88), as claimed. Next, when $K_+ \le \theta^2$, (!) implies that

$$K_+\tau_*^2 + \sigma^2 \sum_k \|G_{k,*}^*\|_F^2 \le \theta^2 \mathrm{Opt},$$

that is, the value of the objective of problem (4.88) at the resulting feasible solution to the problem is within factor θ^2 of Opt. When $K_+ > \theta^2$, we get from (!)

$$K_+\tau_*^2 + \sigma^2 \sum_k \|G_{k,*}^*\|_F^2 \le K_+\theta^{-2}\left[\theta^2\tau_*^2 + \sigma^2 \sum_k \|G_{k,*}^*\|_F^2\right] \le K_+\mathrm{Opt},$$

where the concluding inequality is due to (!). Thus, in all cases the value of the objective of problem (4.88) at the feasible solution to this problem yielded by our construction is at most $\max[K_+, \theta^2]\mathrm{Opt}$. □

Your next task is a follows:

3) To get an idea of the performance of the proposed synthesis of "suboptimal" linear estimation, run numerical experiments as follows:
   - select somehow n and generate at random the $n \times n$ data matrices S, A, B
   - for "moderate" values of n compute both the linear estimate yielded by the optimal solution to (4.12)[7] and the suboptimal estimate as yielded by the above construction and compare their risk bounds and the associated CPU times. For "large" n, where solving (4.12) becomes prohibitively time consuming, compute only a suboptimal estimate in order to get an impression of how the corresponding CPU time grows with n.

   Recommended setup:
   - range of n: 50, 100 ("moderate" values), $1,000, 2,000$ ("large" values)
   - range of σ: $\{1.0, 0.01, 0.0001\}$
   - generation of S, A, B: generate the matrices at random according to

   $$S = U_S \mathrm{Diag}\{1, 2, ..., n\} U_S^T,\ A = U_A \mathrm{Diag}\{\mu_1, ..., \mu_n\} V_A^T,$$
   $$B = U_B \mathrm{Diag}\{\mu_1, ..., \mu_n\} V_B^T,$$

   where U_S, U_A, V_A, U_B, V_B are random orthogonal $n \times n$ matrices, and μ_i form a geometric progression with $\mu_1 = 0.01$ and $\mu_n = 1$.

   You could run the above construction for several values of θ and select the best, in terms of its risk bound, of the resulting suboptimal estimates.

Solution: Here are some numerical results (θ runs through the range $\{2^{i/2}, 1 \le i \le 4\}$):

[7] When $\mathcal{X}$ is an ellipsoid, the semidefinite relaxation bound on the maximum of a quadratic form over $x \in \mathcal{X}$ is exact, so that we are in the case where an optimal solution to (4.12) yields the best, in terms of its risk on $\mathcal{X}$, linear estimate.

n=50						
σ	RiskO	RiskSO	RiskSO/RiskO	cpuO	cpuSO	cpuO/cpuSO
1.0000	0.37877	0.41538	1.10	14.92	0.02	7.5e2
0.0100	0.13623	0.21085	1.55	10.18	0.01	1.0e3
0.0001	0.00775	0.00778	1.00	14.72	0.02	7.7e2

n=100						
σ	RiskO	RiskSO	RiskSO/RiskO	cpuO	cpuSO	cpuO/cpuSO
1.0000	0.33499	0.37705	1.13	633.29	0.10	6.3e3
0.0100	0.12953	0.17978	1.39	364.68	0.04	9.1e3
0.0001	0.01095	0.01114	1.02	532.94	0.05	1.2e4

n=1000		
σ	RiskSO	cpuSO
1.0000	0.34067	3.84
0.0100	0.08272	3.97
0.0001	0.02631	4.03

n=2000		
σ	RiskSO	cpuSO
1.0000	0.31069	53.72
0.0100	0.07098	53.20
0.0001	0.02714	53.98

RiskO, cpuO: risk bound and CPU (sec) for the optimal linear estimate
RiskSO, cpuSO: risk bound and CPU (sec) for suboptimal linear estimate

No data for RiskO and cpuO for $n = 1000$ and $n = 2000$: (4.12) becomes too computationally expensive

We believe the results speak for themselves.

4.11.A. Simple case. There is a trivial case where (4.88) is really easy; this is the case where the right orthogonal factors in the singular value decompositions of A and B are the same, that is, when

$$B = WFV^T,\ A = UDV^T$$

with orthogonal $n \times n$ matrices W, U, V and diagonal F, D. This, at the first glance, very special case is in fact of some importance—it covers the *denoising* situation where $B = A$, so that our goal is to denoise our observation of Ax given a priori information $x \in \mathcal{X}$ on x. In this situation, setting $W^T H^T U = G$, problem (4.88) becomes

$$\mathrm{Opt} = \min_G \left\{ \|F - GD\|^2 + \sigma^2 \|G\|_F^2 \right\}. \tag{4.90}$$

Now comes the concluding part of the exercise:

4) Prove that in the situation in question an optimal solution G_* to (4.90) can be selected to be diagonal, with diagonal entries γ_i, $1 \le i \le n$, yielded by the optimal solution to the optimization problem

$$\mathrm{Opt} = \min_\gamma \left\{ f(G) := \max_{i \le n} (\phi_i - \gamma_i \delta_i)^2 + \sigma^2 \sum_{i=1}^n \gamma_i^2 \right\} \qquad [\phi_i = F_{ii}, \delta_i = D_{ii}]$$

Solution: Let E be a diagonal $n \times n$ matrix with diagonal entries ± 1. It is immediately seen that for every candidate solution G to (4.90), EGE is another candidate solution with the same value of the objective. Since the problem is convex, taking the average of solutions EGE over all 2^n matrices E of the above type, we get a solution $\bar{G}$ which clearly is diagonal and satisfies $f(\bar{G}) \le f(G)$. Thus, we lose nothing when restricting ourselves to diagonal solutions to (4.90). □

Exercise 4.12. [image reconstruction–follow-up to Exercise 4.11] A grayscale image can be represented by an $m \times n$ matrix $x = [x_{pq}]_{\substack{0 \le p < m, \\ 0 \le q < n}}$ with entries in the range $[-\overline{x}, \overline{x}]$, with

$\overline{x} = 255/2$.[8]. Taking a picture can be modeled as observing in noise the 2D convolution $x \star \kappa$ of an image x with known *blurring kernel* $\kappa = [\kappa_{uv}]_{\substack{0\leq u\leq 2\mu,\\ 0\leq v\leq 2\nu}}$, so that the observation is the random matrix

$$\omega = \Big[\omega_{rs} = \underbrace{\sum\nolimits_{\substack{0\leq u\leq 2\mu, 0\leq v\leq 2\nu\\ 0\leq p<m, 0\leq q<n:\\ u+p=r, v+q=s}} x_{pq}\kappa_{uv}}_{[x\star\kappa]_{rs}} + \sigma\xi_{rs}\Big]_{\substack{0\leq r<m+2\mu,\\ 0\leq s<n+2\nu}},$$

where random variables $\xi_{rs} \sim \mathcal{N}(0,1)$ independent of each other form the observation noise.[9] Our goal is to build a presumably good linear estimate of x via ω, the recovery error being measured in $\|\cdot\|_2$. To apply the techniques developed in Section 4.2.2, we need to cover the set of signals x allowed by our a priori assumptions by an ellitope $\mathcal{X}$, and then solve the associated optimization problem (4.12). The difficulty, however, is that this problem is really high-dimensional—with 256×256 images (a rather poor resolution!), the matrix H we are looking for is of size $\dim\omega \times \dim x = ((256+2\mu)(256+2\nu)) \times 256^2 \geq 4.295 \times 10^9$. It is difficult just to store such a matrix in the memory of a usual computer, not to mention optimizing w.r.t. it. By this reason, in what follows we develop a "practically," and not just theoretically, efficiently computable estimate.

4.12.A. The construction. Our key observation is that when passing from representations of x and ω "as they are" to their Discrete Fourier Transforms, the situation simplifies dramatically. Specifically, for matrices y, x of the same sizes, let $y \bullet z$ be the entrywise product of y and z: $[y \bullet z]_{pq} = y_{pq}z_{pq}$. Setting

$$\alpha = 2\mu + m,\ \beta = 2\nu + n,$$

let $F_{\alpha,\beta}$ be the 2D Discrete Fourier Transform—a linear mapping from the space $\mathbf{C}^{\alpha\times\beta}$ onto itself given by

$$[F_{\alpha,\beta}y]_{rs} = \frac{1}{\sqrt{\alpha\beta}} \sum_{\substack{0\leq p<\alpha,\\ 0\leq q<\beta}} y_{pq}\exp\left\{-\frac{2\pi ir}{\alpha} - \frac{2\pi is}{\beta}\right\},$$

where i is the imaginary unit. It is well known that it is a unitary transformation which is an easy-to-compute (it can be computed in $O(\alpha\beta\ln(\alpha\beta))$ arithmetic operations), and which "nearly diagonalizes" the convolution: whenever $x \in \mathbf{R}^{m\times n}$, setting

$$x^+ = \left[\begin{array}{c|c} x & 0_{m\times 2\nu} \\ \hline 0_{2\mu\times n} & 0_{2\mu\times 2\nu}\end{array}\right] \in \mathbf{R}^{\alpha\times\beta},$$

we have

$$F_{\alpha,\beta}(x \star \kappa) = \chi \bullet [F_{\alpha,\beta}x^+]$$

with easy-to-compute χ.[10] Now, let δ be another $(2\mu+1)\times(2\nu+1)$ kernel, with the only nonzero entry, equal to 1, in the position (μ,ν) (recall that indices are enumerated starting from 0); then

$$F_{\alpha,\beta}(x \star \delta) = \theta \bullet [F_{\alpha,\beta}x^+]$$

with easy-to-compute θ. Now consider the auxiliary estimation problem as follows:

Given $R > 0$ and noisy observation

$$\widehat{\omega} = \chi \bullet \widehat{x} + \sigma \underbrace{F_{\alpha,\beta}\xi}_{\eta} \qquad [\xi = [\xi_{rs}] \text{ with independent } \xi_{rs} \sim \mathcal{N}(0,1)],$$

---

[8]The actual grayscale image is a matrix with entries, representing the pixels' light intensities, in the range $[0, 255]$. It is convenient for us to represent this actual image as the shift, by $\overline{x}$, of a matrix with entries in $[-\overline{x}, \overline{x}]$.

[9]Be careful; everywhere in this exercise indexing of elements of 2D arrays starts from 0, and not from 1!

[10]Here $\chi = \sqrt{\alpha\beta}F_{\alpha,\beta}\kappa^+$, where κ^+ is the $\alpha\times\beta$ matrix with κ as its $(2\mu+1)\times(2\nu+1)$ upper-left block and zeros outside this block.

of signal $\widehat{x} \in \mathbf{C}^{\alpha\times\beta}$ known to satisfy $\|\widehat{x}\|_2 \leq R$, we want to recover the matrix $\theta \bullet \widehat{x}$, the error being measured in the Frobenius norm $\|\cdot\|_2$. .

Treating signals $\widehat{x}$ and noises η as long vectors rather than matrices and taking into account that $F_{\alpha,\beta}$ is a unitary transformation, we see that our auxiliary problem is nothing but the problem of recovery, in $\|\cdot\|_2$-norm, of the image Θz of signal z known to belong to the Euclidean ball $\mathcal{Z}_R$ of radius R centered at the origin in $\mathbf{C}^{\alpha\beta}$, from noisy observation

$$\zeta = Az + \sigma\eta,$$

where Θ and A are *diagonal* matrices with complex entries, and η is random complex-valued noise with zero mean and unit covariance matrix. Exactly the same argument as in the real case demonstrates that as far as linear estimates $\widehat{z} = H\zeta$ are concerned, we lose nothing when restricting ourselves with diagonal matrices $H = \mathrm{Diag}\{h\}$, and the best, in terms of its worst-case, over $z \in \mathcal{Z}_R$, expected $\|\cdot\|_2^2$ error, estimate corresponds to h solving the optimization problem

$$R^2 \max_{\ell\leq\alpha\beta} |\Theta_{\ell\ell} - h_\ell A_{\ell\ell}|^2 + \sigma^2 \sum_{\ell\leq\alpha\beta} |h_\ell|^2.$$

Coming back to the initial setting of our auxiliary estimation problem, we conclude that the best linear recovery of $\theta \bullet \widehat{x}$ via $\widehat{\omega}$ is given by

$$\widehat{z} = h \bullet \widehat{\omega},$$

where h is an optimal solution to the optimization problem

$$\mathrm{Opt} = \min_{h\in\mathbf{C}^{\alpha\times\beta}} \left\{ R^2 \max_{r,s} |\theta_{rs} - h_{rs}\chi_{rs}|^2 + \sigma^2 \sum_{r,s} |h_{rs}|^2 \right\}, \tag{4.91}$$

and the $\|\cdot\|_2$-risk

$$\mathrm{Risk}_R[\widehat{z}] = \max_{\|\widehat{x}\|_2\leq R} \mathbf{E}\left\{\|\theta \bullet \widehat{x} - h \bullet [\chi \bullet \widehat{x} + \sigma\eta]\|_2\right\}$$

of this estimate does not exceed $\sqrt{\mathrm{Opt}}$.

Now comes your first task:

1.1) Prove that the above h induces the estimate

$$\widehat{w}(\omega) = F_{\alpha,\beta}^{-1}\left[h \bullet [F_{\alpha,\beta}\omega]\right]$$

of $x \star \delta$, $x \in \mathcal{X}_R = \{x \in \mathbf{R}^{m\times n} : \|x\|_2 \leq R\}$, via observation $\omega = x \star \kappa + \sigma\xi$, with risk

$$\mathrm{Risk}[\widehat{w}|R] = \max_{x\in\mathbf{R}^{m\times n}:\|x\|_2\leq R} \mathbf{E}\left\{\|x \star \delta - \widehat{w}(x \star \kappa + \sigma\xi)\|_2\right\}$$

not exceeding $\sqrt{\mathrm{Opt}}$. Note that x itself is nothing but a block in $x \star \delta$; observe also that in order for $\mathcal{X}_R$ to cover all images we are interested in, it suffices to take $R = \sqrt{mn\overline{x}}$.

1.2) Prove that finding optimal solution to (4.91) is easy—the problem is in fact one-dimensional!

1.3) What are the sources, if any, of the conservatism of the estimate $\widehat{w}$ we have built as compared to the linear estimate given by an optimal solution to (4.12)?

1.4) Think about how to incorporate in the above construction a small number L (say, just five to 10) of additional a priori constraints on x of the form

$$\|x \star \kappa_\ell\|_2 \leq R_\ell,$$

where $\kappa_\ell \in \mathbf{R}^{(2\mu+1)\times(2\nu+1)}$, along with a priori upper bounds u_{rs} on the magnitudes of the Fourier coefficients of x^+:

$$|[F_{\alpha\beta}x^+]_{rs}| \leq u_{rs},\ 0 \leq r < \alpha, 0 \leq s < \beta.$$

Solution: 1.1: Let us fix $x \in \mathcal{X}_R$, and let $x^+ \in \mathbf{R}^{\alpha\times\beta}$ be the above "zero bordering" of x. Setting $\widehat{x} = F_{\alpha,\beta}x^+$, we have

$$\begin{array}{rl} & \omega = x \star \kappa + \sigma\xi \\ \Rightarrow & F_{\alpha,\beta}\omega = \chi \bullet \widehat{x} + \sigma\eta \text{ [by origin of } \chi], \\ & F_{\alpha,\beta}(x\star\delta) = \theta \bullet \widehat{x} \text{ [by origin of } \theta] \\ \Rightarrow & \widehat{w}(\omega) - x\star\delta = F_{\alpha,\beta}^{-1}\left[h \bullet [\chi\bullet\widehat{x} + \sigma\eta] - \theta\bullet\widehat{x}\right] \\ \Rightarrow & \|\widehat{w}(\omega) - x\star\delta\|_2 = \|[h\bullet\chi - \theta]\bullet\widehat{x} + \sigma h\bullet\eta\|_2 \text{ [since } F_{\alpha,\beta} \text{ is unitary]} \\ \Rightarrow & \mathbf{E}\left\{\|\widehat{w}(\omega) - x\star\delta\|_2^2\right\} = \|[h\bullet\chi - \theta]\bullet\widehat{x}\|_2^2 + \sigma^2\mathbf{E}\left\{\|h\bullet\eta\|_2^2\right\} \\ & \quad \le R^2\max_{r,s}|\theta_{rs} - h_{rs}\chi_{rs}|^2 + \sigma^2\|h\|_2^2 \\ & \quad \text{[since } \|\widehat{x}\|_2 = \|x^+\|_2 = \|x\|_2 \le R \text{ and } \eta \text{ is with zero mean} \\ & \quad \text{and unit covariance matrix]} \\ & \quad \le \text{Opt}. \end{array}$$

□

1.2: Introducing the upper bound ρ on $\max_{r,s}|\theta_{rs} - h_{rs}\chi_{rs}|$, (4.91) reads

$$\min_{\rho}\left\{R^2\rho^2 + \sigma^2\min_h\left\{\sum_{r,s}|h_{rs}|^2 : |\theta_{rs} - h_{rs}\chi_{rs}| \le \rho\right\}\right\}.$$

Inner minimization in h is immediate: assuming that ρ is such that the inner problem is feasible, which is the case if and only if

$$\rho \ge \bar{\rho} = \max_{r,s}\left\{|\theta_{rs}| : \chi_{rs} = 0\,\forall r,s\right\},$$

the best h for this ρ is given by

$$h_{rs} = \begin{cases} 0, & \chi_{rs} = 0 \text{ or } |\theta_{rs}| \le \rho, \\ \theta_{rs}(1-\rho/|\theta_{rs}|)/\chi_{rs}, & \chi_{rs} \ne 0 \;\&\; |\theta_{rs}| > \rho \end{cases},$$

which reduces (4.91) to minimizing an efficiently computable (provided $\alpha\beta$ is not astronomically large) univariate convex function. □

1.3: The most evident source of conservatism is that with our approach, we reduce the problem of interest to the auxiliary estimation problem where *the only restriction on the signal* $\widehat{x} \in \mathbf{C}^{\alpha\times\beta}$ *underlying the observation is the energy restriction* $\|\widehat{x}\|_2 \le R$, while the signals $\widehat{x}$ we are actually interested in satisfy additional constraints—they are of the form $F_{\alpha,\beta}(x\star\delta)$, where x runs through the box $\{x : \|x\|_\infty \le \overline{x}\}$ in $\mathbf{R}^n$. Unfortunately, incorporating these additional constraints would, in general, destroy the diagonal structure (and thus computational simplicity) of our auxiliary problem.

1.4: The simplest way to incorporate the indicated additional constraints is to replace (4.91) with the optimization problem

$$\begin{array}{l}\text{Opt} = \min\limits_{h\in\mathbf{C}^{\alpha\times\beta}}\left\{\sigma^2\sum_{r,s}|h_{rs}|^2\right. \\ \left. + \max\limits_{\widehat{x}\in\mathbf{C}^{\alpha\times\beta}}\left\{\sum_{r,s}|\theta_{rs} - h_{rs}\chi_{rs}|^2|\widehat{x}_{rs}|^2 : \begin{array}{l}\sum_{r,s}|\gamma_{rs\ell}|^2|\widehat{x}_{rs}|^2 \le R_\ell^2, 0 \le \ell \le L \\ |\widehat{x}_{rs}| \le u_{rs},\ 0 \le r < \alpha, 0 \le s < \beta\end{array}\right\}\right\},\end{array} \tag{6.18}$$

where $\gamma_{rs0} = 1$, $R_0 = R$, while for $1 \le \ell \le L$, the $\gamma_{rs\ell}$ are readily given by the requirement

$$F_{\alpha,\beta}(x\star\kappa_\ell) = \gamma_{\cdot,\cdot,\ell}\bullet[F_{\alpha,\beta}x^+];$$

cf. the origin of χ and θ. Note that when $\ell = 0$ and u_{rs} are large, specifically,

$u_{rs} \geq R$, problem (6.18) becomes (4.91). Indeed, recalling that $\gamma_{rs0} = 1$, $R_0 = R$, in the case in question the bounds in (6.18) become redundant, and

$$\max_{\widehat{x}} \left\{ \sum_{r,s} |\theta_{rs} - h_{rs}\chi_{rs}|^2 |\widehat{x}_{rs}|^2 : \sum_{r,s} |\gamma_{rs0}|^2 |\widehat{x}_{rs}|^2 \leq R_0^2, \right\} = R^2 \max_{r,s} |\theta_{rs} - h_{rs}\chi_{rs}|^2.$$

In other words, (4.91) is what (6.18) becomes when the only constraint on x we take into account is $\|x\|_2^2 \leq R^2$, which translates into the constraint $\|\widehat{x}\|_2^2 \leq R^2$ on $\widehat{x} = F_{\alpha,\beta}x^+$.

The simplest way to solve (6.18) is to note that setting $y_{rs} = |\widehat{x}_{rs}|^2$, this problem is equivalent to

$$\begin{array}{ll} & \min\limits_{h} \left\{ \sigma^2 \sum_{r,s} |h_{rs}|^2 + \max\limits_{y} \sum_{r,s} |\theta_s - h_{rs}\chi_{rs}|^2 y_{rs} : \begin{array}{l} \\ \sum_{r,s} \gamma_{rs\ell} y_{rs} \leq R_\ell^2,\ 0 \leq \ell \leq L \\ 0 \leq y_{rs} \leq u_{rs}^2, 0 \leq r \leq \alpha, 0 \leq s < \beta \end{array} \right\} \\ \Leftrightarrow & \min\limits_{h} \left\{ \sigma^2 \sum_{r,s} |h_{rs}|^2 \right. \\ & \left. + \min\limits_{\lambda,\mu} \left[\sum_\ell \lambda_\ell R_\ell^2 + \sum_{r,s} u_{rs}^2 \mu_{rs} : \begin{array}{l} |\theta_{rs} - h_{rs}\chi_{rs}|^2 \leq \sum_\ell \lambda_\ell \gamma_{rs\ell} + \mu_{rs}\ \forall r,s \\ \lambda \geq 0, \mu \geq 0 \end{array} \right] \right\} \\ & \hfill \text{[by LP duality]} \\ \Leftrightarrow & \min\limits_{\lambda \geq 0} \left\{ \sum_\ell \lambda_\ell R_\ell^2 \right. \\ & \left. + \sum_{r,s} \underbrace{\min_{\mu_{rs} \geq 0} \left[u_{rs}^2 \mu_{rs} + \min_{h_{rs}} \left[\sigma^2 |h_{rs}|^2 : |\theta_{rs} - h_{rs}\chi_{rs}|^2 \leq \sum_\ell \gamma_{rs\ell}\lambda_\ell + \mu_{rs} \right] \right]}_{\phi_{rs}(\sum_\ell \gamma_{rs\ell}\lambda_\ell)} \right\}. \end{array}$$

Univariate functions $\phi_{rs}(\cdot)$ are convex and easy to compute—they admit closed form analytic representation; thus, we have reduced (6.18) to a low-dimensional convex problem which can be solved, e.g., by the Ellipsoid algorithm.

4.12.B. Mimicking Total Variation constraints. For an $m \times n$ image $x \in \mathbf{R}^{m \times n}$, its (anisotropic) total variation is defined as the ℓ_1 norm of the "discrete gradient field" of x:

$$\mathrm{TV}(x) = \underbrace{\sum_{p=0}^{m-1} \sum_{q=0}^{n} |x_{p+1,q} - x_{p,q}|}_{\mathrm{TV}_a(x)} + \underbrace{\sum_{p=0}^{m} \sum_{q=0}^{n-1} |x_{p,q+1} - x_{p,q}|}_{\mathrm{TV}_b(x)}.$$

A well-established experimental fact is that for naturally arising images, their total variation is essentially less than what could be expected given the magnitudes of entries in x and the sizes m, n of the image. As a result, it is tempting to incorporate a priori upper bounds on the total variation of the image into an image reconstruction procedure. Unfortunately, while an upper bound on total variation is a convex constraint on the image, incorporating this constraint into our construction would completely destroy its "practical computability." What we can do, is to *speculate* that bounds on $\mathrm{TV}_{a,b}(x)$ can be mimicked to some extent by bounds on the energy of two convolutions: one with kernel $\kappa_a \in \mathbf{R}^{(2\mu+1)\times(2\nu+1)}$ with the only nonzero entries

$$[\kappa_a]_{\mu,\nu} = -1, [\kappa_a]_{\mu+1,\nu} = 1,$$

and the other one with kernel $\kappa_b \in \mathbf{R}^{(2\mu+1)\times(2\nu+1)}$ with the only nonzero entries

$$[\kappa_b]_{\mu,\nu} = -1, [\kappa_b]_{\mu,\nu+1} = 1$$

(recall that the indices start from 0, and not from 1). Note that $x\star\kappa_a$ and $x\star\kappa_b$ are "discrete partial derivatives" of $x\star\delta$.

For a small library of grayscale $m\times n$ images x we dealt with, experiment shows that, in addition to the energy constraint $\|x\|_2\leq R=\sqrt{mn\overline{x}}$, the images satisfy the constraints

$$\|x\star\kappa_a\|_2\leq\gamma_2 R,\ \|x\star\kappa_b\|_2\leq\gamma_2 R \tag{$*$}$$

with small γ_2, e.g., $\gamma_2=0.25$. In addition, it turns out that the ∞-norms of the Fourier transforms of $x\star\kappa_a$ and $x\star\kappa_b$ for these images are much less than one could expect looking at the energy of the transform's argument. Specifically, for all images x from the library it holds

$$\begin{array}{l}\|F_{\alpha\beta}[x\star\kappa_a]\|_\infty\leq\gamma_\infty R,\\ \|F_{\alpha\beta}[x\star\kappa_b]\|_\infty\leq\gamma_\infty R,\end{array},\ \ \|\{z_{rs}\}_{r,s}\|_\infty=\max_{r,s}|z_{rs}| \tag{$**$}$$

with $\gamma_\infty=0.01$.[11] Now, relations $(**)$ read

$$\max[|\omega^a_{rs}|,|\omega^b_{rs}|]|F_{\alpha\beta}x^+]_{rs}|\leq\gamma_\infty R\,\forall r,s$$

with easy-to-compute ω^a, ω^b, and in addition $|[F_{\alpha\beta}x^+]_{rs}|\leq R$ due to $\|F_{\alpha\beta}x^+\|_2=\|x^+\|_2\leq R$. We arrive at the bounds

$$|[F_{\alpha\beta}x^+]_{rs}|\leq\min\left[1,1/|\omega^a_{rs}|,1/|\omega^b_{rs}|\right]R\,\forall r,s.$$

on the magnitudes of entries in $F_{\alpha\beta}x^+$, and can utilize item 1.4 to incorporate these bounds, along with relations $(*)$,

Here is your next task:

2) Write software implementing the outlined deblurring and denoising image reconstruction routine and run numerical experiments.
Recommended kernel κ: set $\mu=\rfloor m/32\lfloor$, $\nu=\rfloor n/32\lfloor$, start with

$$\kappa_{uv}=\frac{1}{(2\mu+1)(2\nu+1)}+\left\{\begin{array}{ll}\Delta, & u=\mu,v=\nu\\ 0, & \text{otherwise}\end{array}\right.,0\leq u\leq 2\mu,0\leq v\leq 2\nu,$$

and then normalize this kernel to make the sum of entries equal to 1. In this description, $\Delta\geq 0$ is a control parameter responsible for well-posedness of the auxiliary estimation problem we end up with: the smaller is Δ, the smaller is $\min_{r,s}|\chi_{rs}|$ (note that when decreasing the magnitudes of χ_{rs}, we increase the optimal value in (4.91)).
We recommend checking what happens when $\Delta=0$ with what happens when $\Delta=0.25$, and comparing the estimates accounting and not accounting for the constraints $(*)$ and $(**)$.
On top of that, you can compare your results with what is given by the "ℓ_1-minimization recovery" described as follows:

As we remember from item 4.12.A, our problem of interest can be equivalently reformulated as recovering the image Θz of a signal $z\in\mathbf{C}^{\alpha\beta}$ from noisy observation $\widehat{\omega}=Az+\sigma\eta$, where Θ and A are diagonal matrices, and η is the zero mean complex Gaussian noise with unit covariance matrix. In other words, the entries η_ℓ in η are real two-dimensional Gaussian vectors independent of each other with zero mean and the covariance matrix $\frac{1}{2}I_2$. Given a reasonable "reliability tolerance" ϵ, say, $\epsilon=0.1$, we can easily point out the smallest "confidence radius" ρ such that for $\zeta\sim\mathcal{N}(0,\frac{1}{2}I_2)$ it holds $\mathrm{Prob}\{\|\zeta\|_2>\rho\}\leq\frac{\epsilon}{\alpha\beta}$, implying that for every ℓ it holds

$$\mathrm{Prob}_\eta\left\{|\widehat{\omega}_\ell-A_\ell z_\ell|>\sigma\rho\right\}\leq\frac{\epsilon}{\alpha\beta},$$

[11] Note that from $(*)$ it follows that $(**)$ holds with $\gamma_\infty=\gamma_2$, while with our empirical γ's, γ_∞ is 25 times smaller than γ_2.

and therefore

$$\text{Prob}_\eta\left\{\|\widehat{\omega} - Az\|_\infty > \sigma\rho\right\} \leq \epsilon.$$

We now can easily find the smallest, in $\|\cdot\|_1$, vector $\widehat{z} = \widehat{z}(\omega)$ which is "compatible with our observation," that is, satisfies the constraint

$$\|\widehat{\omega} - A\widehat{z}\|_\infty \leq \sigma\rho,$$

and take $\Theta\widehat{z}$ as the estimate of the "entity of interest" Θz (cf. Regular ℓ_1 recovery from Section 1.2.3).

Note that this recovery needs no a priori information on z.

Solution: Figure 6.1 displays the recovery of an $(m = 1200) \times (n = 1600)$ image when $\Delta = 0$ (ill-posed estimation problem: in (4.91), all θ_{rs} are of magnitude 1, magnitudes of χ_{rs} are at most 1, and 1.3% of the $|\chi_{rs}|$'s are $\leq 10^{-6}$). When $\Delta = 0.25$, the situation becomes much simpler (see Fig. 6.2), since now the estimation problem is reasonably well-posed: in (4.91), all θ_{rs}, as before, are of magnitude 1, while the range of magnitudes of χ_{rs}'s is $[0.026, 1]$.

The computed values of recovery risk and the observed actual $\|\cdot\|_2$ recovery errors are presented in Table 6.1. When presenting the data, we refer to recoveries yielded by (4.91) (that is, with no constraints (∗), (∗∗) accounted for) as *plain* ones, and to recoveries yielded by (6.18) (i.e., with constraints (∗), (∗∗) accounted for) as *TV recoveries*. In our experiments, TV recoveries gain a clear advantage over the plain ones, especially in the ill-posed case. Note also that ℓ_1 minimization significantly outperforms TV recovery in the ill-posed case and is essentially worse than the latter recovery in the "low σ" well-posed case.

Exercise 4.13 [classical periodic nonparametric deconvolution] In classical univariate nonparametric regression, one is interested in recovering a function $f(t)$ of continuous argument $t \in [0, 1]$ from noisy observations $\omega_i = f(i/n) + \sigma\eta_i$, $0 \leq i \leq n$, where $\eta_i \sim \mathcal{N}(0, 1)$ are observation noises independent across i. Usually, a priory restrictions on f are *smoothness assumptions*—existence of $\varkappa$ continuous derivatives satisfying a priori upper bounds

$$\left(\int_0^1 |f^{(k)}(t)|^{p_k} dt\right)^{1/p_k} \leq L_k,\ 0 \leq k \leq \varkappa,$$

on their L_{p_k}-norms. The risk of an estimate is defined as the supremum, over f's of given smoothness, of the expected L_r-norm of the recovery error, and the primary emphasis of classical studies here is on how the minimax optimal (i.e., the best, over all estimates) risk goes to 0 as the number of observations n goes to infinity, what the near-optimal estimates are, etc. Many of these studies deal with *periodic case*—one where f can be extended onto the entire real axis as a $\varkappa$ times continuously differentiable periodic, with period 1, function, or, which is the same, one where f is treated as a smooth function on the circumference of length 1 rather than on the unit segment $[0, 1]$. While being slightly simpler for analysis than the general case, the periodic case turned out to be highly instructive: what was first established for the latter, usually extends straightforwardly to the former.

What you are about to do in this exercise, is to apply our machinery of building linear estimates to the outlined recovery of smooth univariate periodic regression functions.

4.13.A. Setup. What follows is aimed at handling restrictions of smooth functions on the unit (i.e., of unit length) circumference C onto an equidistant n-point grid Γ_n on the circumference. These restrictions form the usual n-dimensional coordinate space $\mathbf{R}^n$; it is convenient to index the entries in $f \in \mathbf{R}^n$ starting from 0 rather than from 1. We equip $\mathbf{R}^n$ with two linear operators:

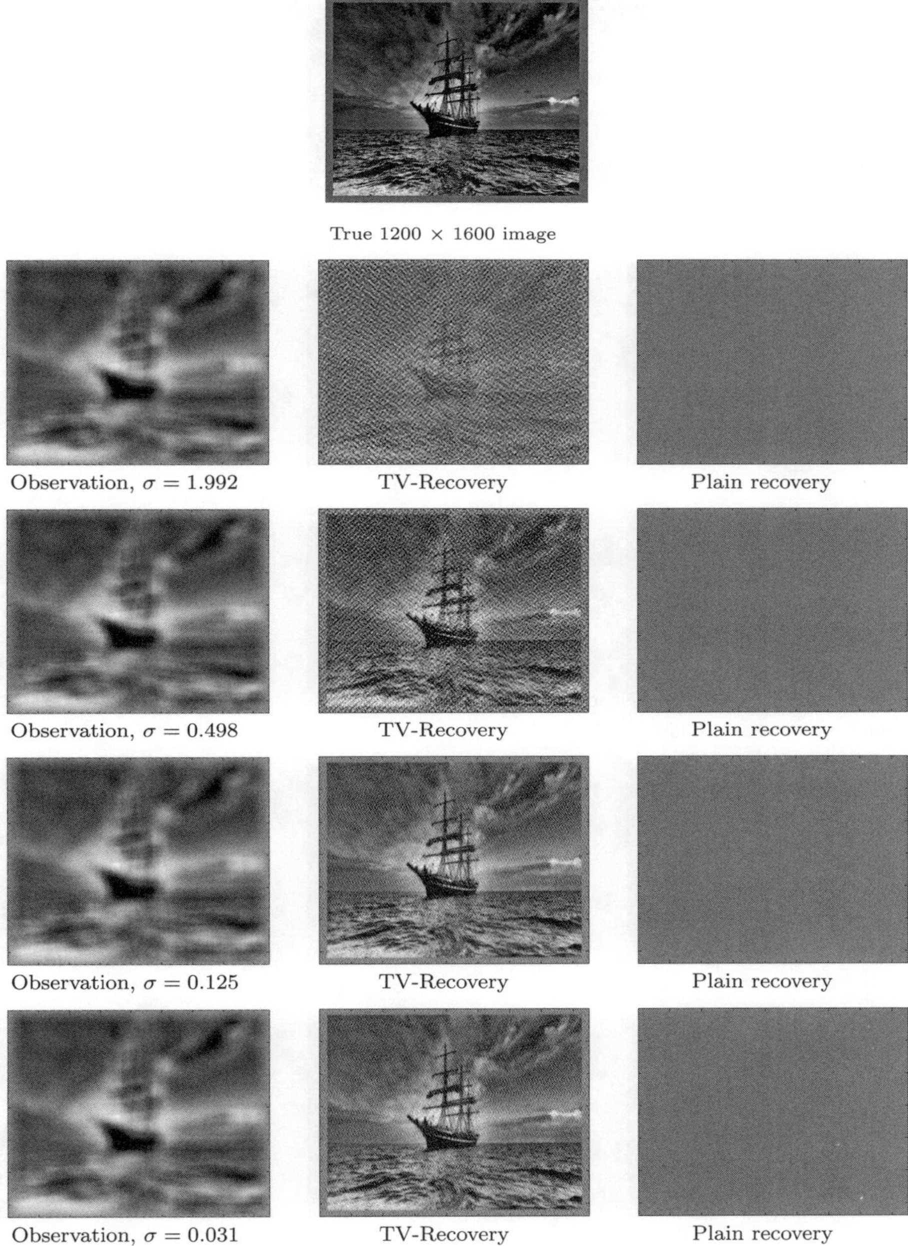

Figure 6.1: Recovery of a 1200×1600 image at different noise levels, ill-posed case $\Delta = 0$.

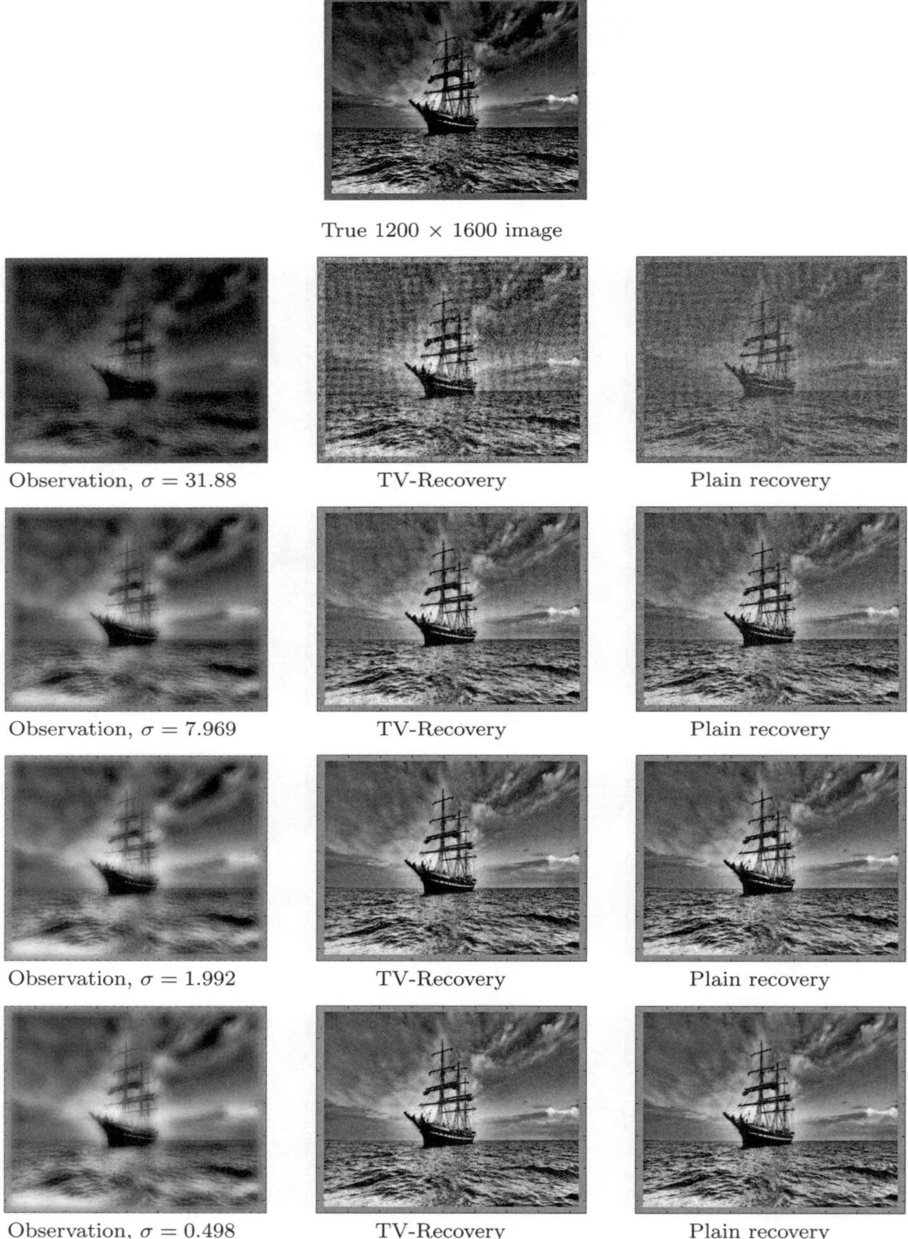

Figure 6.2: Recovery of a 1200×1600 image at different noise levels, well-posed case $\Delta = 0.25$.

σ	TV-recovery	Plain recovery
1.992	122.3/54.08/29.13	127.5/57.05/29.12
0.498	100.2/45.44/26.25	127.5/54.05/26.25
0.125	83.91/36.81/23.27	127.5/54.05/23.26
0.031	70.64/34.02/19.97	127.5/54.05/19.98

$\Delta = 0$

σ	TV-recovery	Plain recovery
31.88	42.41/34.47/33.96	99.65/70.39/33.89
7.969	22.13/17.88/27.43	38.11/36.68/27.44
1.992	9.205/8.771/17.30	9.951/9.921/17.30
0.498	2.481/2.474/7.805	2.497/2.492/7.806

$\Delta = 0.25$

Table 6.1: In a cell, the first number is the normalized risk (square root of the optimal value in (4.91) or (6.18), divided by $\sqrt{mn}$); the second number is the actually observed normalized error of linear estimator ($\|\cdot\|_2$ of recovery error divided by $\sqrt{mn}$); and the third number is the observed normalized error of the ℓ_1 estimator.

- *Cyclic shift* (in the sequel – just *shift*) Δ:

$$\Delta \cdot [f_0; f_1; \dots : f_{n-2}; f_{n-1}] = [f_{n-1}; f_0; f_1; \dots; f_{n-2}],$$

and

- *Derivative* D:

$$D = n[I - \Delta];$$

Treating $f \in \mathbf{R}^n$ as a restriction of a function F on C onto Γ_n, Df is the finite-difference version of the first order derivative of the function, and the norms

$$|f|_p = n^{-1/p}\|f\|_p,\ p \in [1, \infty],$$

are the discrete versions of L_p-norms of F.

Next, we associate with $\chi \in \mathbf{R}^n$ the operator $\sum_{i=0}^{n-1} \chi_i \Delta^i$; the image of $f \in \mathbf{R}^n$ under this operator is denoted by $\chi \star f$ and is called (cyclic) *convolution* of χ and f.

The problem we focus on is as follows.

Given are:

- *smoothness data* represented by a nonnegative integer $\varkappa$ and two collections: $\{L_\iota > 0 : 0 \leq \iota \leq \varkappa\}$, $\{p_\iota \in [2, \infty], 0 \leq \iota \leq \varkappa\}$. The smoothness data specify the set

$$\mathcal{F} = \{f \in \mathbf{R}^n : |f|_{p_\iota} \leq L_\iota, 0 \leq \iota \leq \varkappa\}$$

of signals we are interested in (this is the discrete analog of the *periodic Sobolev ball*—the set of $\varkappa$ times continuously differentiable functions on C with derivatives of orders up to $\varkappa$ bounded, in integral p_ι-norms, by given quantities L_ι);
- two vectors $\alpha \in \mathbf{R}^n$ (*sensing kernel*) and $\beta \in \mathbf{R}^n$ (*decoding kernel*);
- positive integer σ (noise intensity) and a real $q \in [1, 2]$.

These data define the estimation problem as follows: given noisy observation

$$\omega = \alpha \star f + \sigma\eta$$

of an unknown signal f known to belong to $\mathcal{F}$, where $\eta \in \mathbf{R}^n$ is a random observation noise, we want to recover $\beta \star f$ in the norm $|\cdot|_q$.

Our only assumption on the noise is that

$$\mathrm{Var}[\eta] := \mathbf{E}\left\{\eta\eta^T\right\} \preceq I_n.$$

The risk of a candidate estimate $\widehat{f}$ is defined as

$$\mathrm{Risk}_r[\widehat{f}|\mathcal{F}] = \sup_{\substack{f\in\mathcal{F},\\ \eta:\mathrm{Var}[\eta]\preceq I_n}} \mathbf{E}_\eta\left\{|\beta\star f - \widehat{f}(\alpha\star f+\sigma\eta)|_q\right\}.$$

Here is the exercise:

1) Check that the situation in question fits the framework of Section 4.3.3 and figure out to what, under the circumstances, the optimization problem (4.42) responsible for the presumably good linear estimate $\widehat{f}_H(\omega)=H^T\omega$ reduces.
2) Prove that in the case in question the linear estimate yielded by an appropriate optimal solution to (4.42) is just the cyclic convolution

$$\widehat{f}(\omega)=h\star\omega$$

and work out a computationally cheap way to identify h.
3) Implement your findings in software and run simulations. You could, in particular, consider the denoising problem—the problem where $\alpha\star x\equiv\beta\star x\equiv x$ and $\eta\sim\mathcal{N}(0,I_n)$)—and compare numerically the computed risks of your estimates with the classical results on the limits of performance in recovering smooth univariate regression functions. According to those results, in the situation in question and under the natural assumption that the L_ι are nondecreasing in ι, the minimax optimal risk, up to a factor depending solely on $\varkappa$, is $(\sigma^2/n)^{\frac{\varkappa}{2\varkappa+1}}L_\varkappa^{\frac{1}{2\varkappa+1}}$.

Solution: 1: In the notation from Section 4.3.3 and Proposition 4.14, our problem is to recover $Bx:=\beta\star x$ via observation

$$\omega=\underbrace{\alpha\star x}_{Ax}+\underbrace{\sigma\eta}_{\xi}$$

in the situation where

- x is known to belong to the basic spectratope

$$\left[\begin{array}{c}\mathcal{X}=\left\{x\in\mathbf{R}^n:\exists t\in\mathcal{T}:R^2_{k\iota}[x]\le t_{k\iota},0\le k\le n-1,0\le\iota\le\varkappa\right\},\\ R_{k\iota}[x]=\mathrm{Row}_k[D^\iota]\cdot x,\\ \mathcal{T}=\{t=[t_{k\iota}]_{\substack{0\le k\le n-1\\0\le\iota\le\varkappa}}:t\ge0,\|\mathrm{Col}_\iota[t]\|_{p_\iota/2}\le n^{2/p_\iota}L_\iota^2,0\le\iota\le\varkappa\}\end{array}\right],$$

  implying that

$$\begin{array}{rcl}\mathcal{R}^*_{k\iota}[\Lambda_{k\iota}]&=&\Lambda_{k\iota}\mathrm{Row}_k^T[D^\iota]\mathrm{Row}_k[D^\iota],\\ \phi_{\mathcal{T}}\left(\lambda\left[\Lambda:=[\Lambda_{k\iota}\ge0]_{k,\iota}\right]\right)&=&\sum_\iota n^{2/p_\iota}L_\iota^2\|\mathrm{Col}_\iota[\Lambda]\|_{\frac{p_\iota}{p_\iota-2}};\end{array}$$

- the unit ball $\mathcal{B}_*$ of the norm conjugate to the recovery norm $|\cdot|_q=n^{-1/q}\|\cdot\|_q$ is the basic spectratope

$$\begin{array}{rcl}\mathcal{B}_*&=&\{u\in\mathbf{R}^n:n^{1/q}\|u\|_{q/(q-1)}\le1\}\\ &=&\left\{u:\exists r\in\mathcal{R}:S^2_\ell[u]\le r_\ell,\,0\le\ell\le n-1\right\},\\ S_\ell[u]&=&u_\ell\\ \mathcal{R}&=&\{r\ge0:\|r\|_{\frac{q}{2q-2}}\le n^{-2/q}\},\end{array}$$

  so that

$$\begin{array}{rcl}\mathcal{S}^*_\ell[\Upsilon_\ell]&=&\Upsilon_\ell E_\ell,\\ \phi_{\mathcal{R}}\left(\lambda\left[\Upsilon:=\{\Upsilon_\ell\ge0,0\le\ell\le n-1\}\right]\right)&=&n^{-2/q}\|[\Upsilon_0;...;\Upsilon_{n-1}]\|_{\frac{q}{2-q}};\end{array}$$

- $\Pi = \{\sigma^2 I_n\}$.

Consequently, problem (4.42) reads

$$\mathrm{Opt} = \min_{H,\Lambda,\Upsilon,\Upsilon',\Theta} \left\{ \begin{array}{l} \sum_{\iota=0}^{\varkappa} n^{2/p_\iota} L_\iota^2 \|\Lambda^\iota\|_{\frac{p_\iota}{p_\iota-2}} + n^{-2/q}\|\Upsilon\|_{\frac{q}{2-q}} + n^{-2/q}\|\Upsilon'\|_{\frac{q}{2-q}} \\ \qquad\qquad +\sigma^2 \mathrm{Tr}(\Theta) : \\ \Lambda = \{\Lambda^\iota \in \mathbf{R}^n_+\}_{\iota=0}^{\varkappa}, \Upsilon \in \mathbf{R}^n_+, \Upsilon' \in \mathbf{R}^n_+, \\ \left[\begin{array}{c|c} \sum_{k,\iota}[\Lambda^\iota]_k \left[\mathrm{Row}_k^T[D^\iota]\mathrm{Row}_k[D^\iota]\right] & \frac{1}{2}[B^T - A^T H] \\ \hline \frac{1}{2}[B - H^T A] & \mathrm{Diag}\{\Upsilon\} \end{array}\right] \succeq 0, \\ \left[\begin{array}{c|c} \Theta & \frac{1}{2}H \\ \hline \frac{1}{2}H^T & \mathrm{Diag}\{\Upsilon'\} \end{array}\right] \succeq 0 \end{array} \right\} \tag{6.19}$$

and Opt is an upper bound on the risk Risk_q of the linear estimate yielded by an optimal solution to (6.19).

2: Problem (6.19) admits a finite group of symmetries (inherited from the symmetry of our estimation problem w.r.t. to a cyclic shift); formally, when

$$(H, \Lambda = \{\Lambda^\iota, 0 \le \iota \le \varkappa\}, \Upsilon, \Upsilon', \Theta)$$

is a feasible solution to (6.19), so are the collections

$$([\Delta^\ell]^T H \Delta^\ell, \Lambda' = \{[\Delta^\ell]^T \Lambda^\iota, 0 \le \iota \le \varkappa\}, [\Delta^\ell]^T \Upsilon, [\Delta^\ell]^T \Upsilon', [\Delta^\ell]^T \Theta \Delta^\ell),\ \ell = 1, ..., n-1,$$

the value of the objective on all these feasible solutions being the same. Since the problem is convex and I, Δ, Δ^2,..., Δ^{n-1} is a group w.r.t. multiplication, it follows that (6.19) admits an optimal solution which is invariant w.r.t. the above symmetry transformations, that is, such that $D^T H D = H$, $\Lambda^\iota = \lambda_\iota \mathbf{1}_n$, $\Upsilon = \upsilon \mathbf{1}_n$, $\Theta = \Delta^T D \Delta$. Now, linear transformations $x \mapsto Gx$ of $\mathbf{R}^n$ such that $G = \Delta^T G \Delta$ (or, which is the same due to the fact that Δ is orthogonal, $G = \Delta^{-1} G \Delta$, or, which is again the same, the transformations which commute with cyclic shift Δ) are cyclic convolutions, so that there exists an optimal solution where

$$H = \mathcal{H}[h] := \sum_{k=0}^{n-1} h_k \Delta^k, \Lambda = \{\lambda_\iota \mathbf{1}_n, 0 \le \iota \le \varkappa\}, \Upsilon = \upsilon \mathbf{1}_n, \Upsilon' = \upsilon' \mathbf{1}_n.$$

Restricting the feasible solutions to (6.19) by the solutions of the structure just outlined, problem (!), as it is immediately seen, becomes

$$\mathrm{Opt} = \min_{h,\lambda,\upsilon,\upsilon',\Theta} \left\{ \begin{array}{r} n\sum_{\iota=0}^{\varkappa} L_\iota^2 \lambda_\iota + n^{-1}(\upsilon + \upsilon') + \sigma^2 \mathrm{Tr}(\Theta) : \lambda \ge 0, \upsilon \ge 0, \upsilon' \ge 0, \\ \left[\begin{array}{c|c} \sum_{\iota=0}^{\varkappa} \lambda_\iota [D^\iota]^T D^\iota & \frac{1}{2}[\mathcal{H}^T[\beta] - \mathcal{H}^T[\alpha]\mathcal{H}[h]] \\ \hline \frac{1}{2}[\mathcal{H}[\beta] - \mathcal{H}^T[h]\mathcal{H}[\alpha]] & \upsilon I \end{array}\right] \succeq 0, \\ \left[\begin{array}{c|c} \Theta & \frac{1}{2}\mathcal{H}[h] \\ \hline \frac{1}{2}\mathcal{H}^T[h] & \upsilon' I \end{array}\right] \succeq 0 \end{array} \right\}.$$

In this problem, one can immediately carry out partial optimization w.r.t. υ' and Θ (cf. Illustration in Section 4.3.3.5). The resulting problem of optimizing in

n	10	10^2	10^3	10^4	10^5	10^6
Opt/Risk$_*$, $\varkappa=1$, $p_0=p_1=\infty, q=2$	0.1265	0.2726	0.2847	0.2836	0.2836	0.2836
Opt/Risk$_*$, $\varkappa=2$ $p_0=p_1=p_2=2, q=2$	0.0291	0.0530	0.0515	0.0515	0.0515	0.0515

Table 6.2: Ratios Opt/Risk$_*$ in the case of denoising.

remaining variables reads

$$\begin{array}{rcl}\mathrm{Opt} &=& \min\limits_{h\in\mathbf{R}^n,\lambda\in\mathbf{R}^{\varkappa+1},\upsilon}\Big\{n\sum_{\iota=0}^{\varkappa}L_\iota^2\lambda_\iota+n^{-1}\upsilon+\sigma\|h\|_2:\lambda\geq0,\upsilon\geq0,\\ && \left[\begin{array}{c|c}\sum_{\iota=0}^{\varkappa}\lambda_\iota[D^\iota]^TD^\iota & \frac{1}{2}[\mathcal{H}^T[\beta]-\mathcal{H}^T[\alpha]\mathcal{H}[h]]\\ \hline \frac{1}{2}[\mathcal{H}[\beta]-\mathcal{H}^T[h]\mathcal{H}[\alpha]] & \upsilon I\end{array}\right]\succeq0\Big\}.\end{array} \tag{6.20}$$

The latter problem can be immediately simplified by passing from variable h to variable $\zeta=\mathbf{F}_nh$, where $\mathbf{F}_n$ is the matrix of the Discrete Fourier Transform:

$$[\mathbf{F}_nx]_\ell=\sum_{k=0}^{n-1}x_k\exp\{-2\pi ik\ell/n\},\ 0\leq\ell\leq n-1. \qquad [i\text{: imaginary unit}]$$

Multiplying the left-hand side of the LMI constraint in (6.20) by the unitary matrix $U=n^{1/2}\mathrm{Diag}\{\mathbf{F}_n^{-1},\mathbf{F}_n^{-1}\}$ from the right and by the Hermitian conjugate of U from the left, problem (6.20), as it is immediately seen, becomes

$$\begin{array}{rcl}\mathrm{Opt} &=& \min\limits_{z\in\mathbf{C}^n,\lambda\in\mathbf{R}^n,\upsilon\in\mathbf{R}}\Big\{n\sum_{\iota=0}^{\varkappa}L_\iota^2\lambda_\iota+n^{-1}\upsilon+n^{-1/2}\sigma\|z\|_2:\lambda\geq0,\upsilon\geq0,\\ && \left[\begin{array}{c|c}\sum_{\iota=0}^{\varkappa}\lambda_\iota\mathrm{Diag}\{|\delta_0|^{2\iota},...,|\delta_{n-1}|^{2\iota}\} & \frac{1}{2}\mathrm{Diag}\{\overline{\mathbf{F}_n\beta}-\overline{\mathbf{F}_n\alpha}\bullet z\}\\ \hline \frac{1}{2}\mathrm{Diag}\{\mathbf{F}_n\beta-(\mathbf{F}_n\alpha)\bullet\overline{z}\} & \upsilon I_n\end{array}\right]\succeq0\Big\},\\ && [\delta=\mathbf{F}_n\mathrm{Col}_1[D]]\end{array}$$

where $\overline{x}$ is the entrywise complex conjugate of a vector $x\in\mathbf{C}$.[12]. In the resulting problem (cf. item 1.4 in Exercise 4.12), numerical partial minimization in z is easy, and we end up with a low-dimensional problem of minimizing an easy-to-compute function of λ,υ over the nonnegative orthant.

3: In Figure 6.3, we present sample signal recoveries for the cases of denoising ($\beta\star x\equiv\alpha\star x\equiv x$) and deconvolution ($\beta\star x\equiv x$, α is the "Gaussian kernel" shown in Figure 6.3). Table 6.2 deals with the denoising and displays the observed ratios of our upper bound Opt on the risk of a presumably good linear estimate and the theoretical "standard nonparametric risk" $\mathrm{Risk}_*:=(\sigma^2/n)^{\frac{\varkappa}{2\varkappa+1}}L_\varkappa^{\frac{1}{2\varkappa+1}}$ which, as was already explained, coincides, within a constant factor depending on $\varkappa$ only, with the minimax optimal risk.

[12]In principle, we were supposed to further restrict z to be the Discrete Fourier Transform of a real vector; this restriction, however, is redundant, with real vectors α, β, the problem automatically admits an optimal solution with this property.

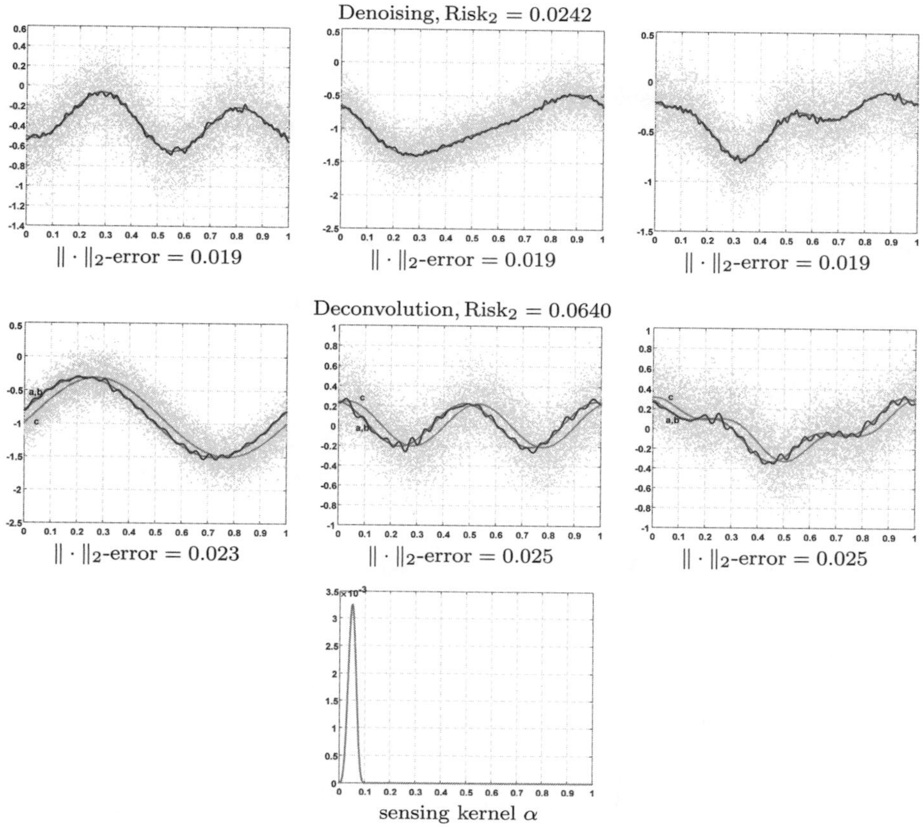

Figure 6.3: Smooth curve: f; "bumpy" curve: recovery; gray cloud: observations. In all experiments, $n = 8192$, $\varkappa = 2$, $p_0 = p_1 = p_2 = 2$, $\sigma = 0.5$, $L_\iota = (10\pi)^\iota, 0 \leq \iota \leq 2$.

6.4.4.4 Probabilities of large deviations in linear estimation under sub-Gaussian noise

Exercise 4.14 The goal of the exercise is to derive bounds for probabilities of large deviations for estimates yielded by Proposition 4.14.

1) Prove the following fact:

Lemma 4.30 *Let $\Theta, Q \in \mathbf{S}^m_+$, with $Q \succ 0$, and let ξ be a sub-Gaussian, with parameters (μ, S), random vector, where μ and S satisfy $\mu\mu^T + S \preceq Q$. Setting $\rho = \mathrm{Tr}(\Theta Q)$, we have*

$$\mathbf{E}_\xi \left\{ \exp\{\tfrac{1}{8\rho}\xi^T\Theta\xi\} \right\} \leq \sqrt{2}\exp\{1/4\}. \tag{4.92}$$

As a result, for $t > 0$ it holds

$$\mathrm{Prob}\left\{ \sqrt{\xi^T\Theta\xi} \geq t\sqrt{\rho} \right\} \leq \sqrt{2}\exp\{1/4\}\exp\{-t^2/8\},\ t \geq 0. \tag{4.93}$$

Hint: Use the same trick as in the proof of Lemma 2.53.

Solution: W.l.o.g., we can assume that $\mu\mu^T + S = Q$. Setting $\eta = Q^{-1/2}\xi$, we get $\xi = Q^{1/2}\eta$ with sub-Gaussian, with parameters $\nu = Q^{-1/2}\mu$, $P = Q^{-1/2}SQ^{-1/2}$, random vector η; note that $\nu\nu^T + P = I_m$. Setting $R = Q^{1/2}\Theta Q^{1/2}$, we have

$$\xi^T\Theta\xi = \eta^T R\eta \ \&\ R \succeq 0, \mathrm{Tr}(R) = \mathrm{Tr}(\Theta Q) = \rho.$$

Next, let $R = U\mathrm{Diag}\{\lambda_1, ..., \lambda_m\}U^T$ be the eigenvalue decomposition of R, and let $\bar{\eta} = U^T\eta$, so that $\bar{\eta}$ is sub-Gaussian with parameters $\bar{\nu} = U^T\nu$, $\bar{P} = U^T PU$. Representing $\bar{\eta} = \bar{\nu} + \zeta$, we get sub-Gaussian, with parameters $0, \bar{P}$, random vector ζ. By construction, we have

$$\xi^T\Theta\xi = \eta^T R\eta = \bar{\eta}^T\mathrm{Diag}\{\lambda_1, ..., \lambda_m\}\bar{\eta} = \sum_i \lambda_i[\bar{\nu}_i + \zeta_i]^2 \ \&\ \lambda_i \geq 0, \sum_i \lambda_i = \rho.$$

Besides this, $\bar{\nu}\bar{\nu}^T + \bar{P} = I_m$, implying that ζ_i are sub-Gaussian with parameters $(0, 1)$ and $\bar{\nu}_i \in [-1, 1]$. Setting $\gamma = \frac{1}{8\rho}$, we have

$$\begin{array}{rcl}\mathbf{E}_\xi\left\{\exp\{\gamma\xi^T\Theta\xi\}\right\} & = & \mathbf{E}_\zeta\left\{\exp\{\gamma\sum_i \lambda_i[\bar{\nu}_i + \zeta_i]^2\}\right\} \leq \exp\{2\gamma\sum_i \lambda_i[\bar{\nu}_i^2 + \zeta_i^2]\} \\ & \leq & \exp\{2\gamma\rho\}\mathbf{E}_\zeta\left\{\exp\{2\gamma\sum_i \lambda_i\zeta_i^2\}\right\}.\end{array}$$

Function $\mathbf{E}_\zeta\left\{\exp\{2\gamma\sum_i \lambda_i\zeta_i^2\}\right\}$ is convex in λ and λ is restricted to reside in the simplex $\{\lambda \geq 0, \sum_i \lambda_i = \rho\}$; the maximum of this function over the simplex is achieved at a vertex of the simplex. Consequently, there exists i_* such that

$$\mathbf{E}_\zeta\left\{\exp\{2\gamma\sum_i \lambda_i\zeta_i^2\}\right\} \leq \mathbf{E}_z\left\{\exp\{2\gamma\rho z^2\}\right\},$$

$z = \zeta_{i_*}$, and we arrive at

$$\begin{array}{rcl}\mathbf{E}_\xi\left\{\exp\{\gamma\xi^T\Theta\xi\}\right\} & \leq & \exp\{2\gamma\rho\}\mathbf{E}_z\left\{\exp\{2\gamma\rho z^2\}\right\} \\ & = & \exp\{1/4\}\mathbf{E}_z\left\{\exp\{z^2/4\}\right\}.\end{array} \tag{6.21}$$

Next, z is sub-Gaussian with parameters $(0, 1)$. Let s be a $\mathcal{N}(0, 1)$ random variable independent of z. Then

$$\begin{array}{rcl}\mathbf{E}_z\left\{\exp\{z^2/4\}\right\} & = & \mathbf{E}_z\left\{\mathbf{E}_s\left\{\exp\{zs/\sqrt{2}\}\right\}\right\} = \mathbf{E}_s\left\{\mathbf{E}_z\left\{\exp\{zs/\sqrt{2}\}\right\}\right\} \\ & \leq & \mathbf{E}_s\left\{\exp\{s^2/4\}\right\} = \sqrt{2},\end{array}$$

where the last $\leq$ is due to the fact that z is $(0, 1)$-sub-Gaussian. The resulting inequality combines with (6.21) to imply (4.92). Relation (4.92) combines with the Chebyshev inequality to imply (4.93). □

2) Recall that (the proof of) Proposition 4.14 states that in the situation of Section 4.3.3.1 and under Assumptions **A**, **B** and **R**, for every feasible solution $(H, \Lambda, \Upsilon, \Upsilon', \Theta)$ to the optimization problem[13]

[13]For notation, see Section 4.3.3.1, (4.36), and (4.39). For the reader's convenience, we recall part of this notation: for a probability distribution P on $\mathbf{R}^m$, $\mathrm{Var}[P] = \mathbf{E}_{\xi\sim P}\{\xi^T\xi\}$, Π is a convex compact subset of $\mathrm{int}\,\mathbf{S}^m_+$, $Q \lhd \Pi$ means that $Q \preceq Q'$ for some $Q' \in \Pi$, and $\Gamma_\Pi(\Theta) = \max_{Q\in\Pi}\mathrm{Tr}(\Theta Q)$.

$$\begin{array}{l}\mathrm{Opt}=\min\limits_{H,\Lambda,\Upsilon,\Upsilon',\Theta}\Big\{\underbrace{\phi_{\mathcal{T}}(\lambda[\Lambda])+\phi_{\mathcal{R}}(\lambda[\Upsilon])}_{\mathcal{A}=\mathcal{A}(\Lambda,\Upsilon)}+\underbrace{\phi_{\mathcal{R}}(\lambda[\Upsilon'])+\Gamma_\Pi(\Theta)}_{\mathcal{B}=\mathcal{B}(\Theta,\Upsilon')}:\\ \left.\begin{array}{r}\Lambda=\{\Lambda_k\succeq 0,k\le K\},\ \Upsilon=\{\Upsilon_\ell\succeq 0,\ell\le L\},\ \Upsilon'=\{\Upsilon'_\ell\succeq 0,\ell\le L\},\\ \left[\begin{array}{c|c}\sum_k\mathcal{R}_k^*[\Lambda_k] & \frac{1}{2}[B^T-A^TH]M\\ \hline \frac{1}{2}M^T[B-H^TA] & \sum_\ell\mathcal{S}_\ell^*[\Upsilon_\ell]\end{array}\right]\succeq 0,\\ \left[\begin{array}{c|c}\Theta & \frac{1}{2}HM\\ \hline \frac{1}{2}M^TH^T & \sum_\ell\mathcal{S}_\ell^*[\Upsilon'_\ell]\end{array}\right]\succeq 0\end{array}\right\}\end{array},\tag{4.94}$$

one has

$$\max_{x\in\mathcal{X}}\|[B-H^TA]x\|\le\mathcal{A}\ \ \&\ \max_{P:\mathrm{Var}[P]\lhd\Pi}\mathbf{E}_{\xi\sim P}\left\{\|H^T\xi\|\right\}\le\mathcal{B},\tag{4.95}$$

implying that the linear estimate $\widehat{x}_H(\omega)=H^T\omega$ satisfies the risk bound

$$\mathrm{Risk}_{\Pi,\|\cdot\|}[\widehat{x}_H(\cdot)|\mathcal{X}]\le\mathcal{A}+\mathcal{B}. \qquad (4.96).$$

Prove the following:

Proposition 4.31 *Let $H,\Lambda,\Upsilon,\Upsilon',\Theta$ be a feasible solution to (4.94), and let $\widehat{x}_H(\omega)=H^T\omega$. Let, further, P be a sub-Gaussian probability distribution on $\mathbf{R}^m$, with parameters (μ,S) satisfying*

$$\mu\mu^T+S\lhd\Pi,$$

and, finally, let $x\in\mathcal{X}$. Then

(i) One has

$$\begin{array}{rcl}&&\mathbf{E}_{\xi\sim P}\{\|Bx-\widehat{x}_H(Ax+\xi)\|\}\le\mathcal{A}_*+\mathcal{B}_*,\\ \mathcal{A}_*&=&\mathcal{A}_*(\Lambda,\Upsilon):=2\sqrt{\phi_{\mathcal{T}}(\lambda[\Lambda])\phi_{\mathcal{R}}(\lambda[\Upsilon])}\le\mathcal{A}(\Lambda,\Upsilon):=\phi_{\mathcal{T}}(\lambda[\Lambda])+\phi_{\mathcal{R}}(\lambda[\Upsilon])\\ \mathcal{B}_*&=&\mathcal{B}_*(\Theta,\Upsilon'):=2\sqrt{\Gamma_\Pi(\Theta)\phi_{\mathcal{R}}(\lambda[\Upsilon'])}\le\mathcal{B}(\Theta,\Upsilon'):=\Gamma_\Pi(\Theta)+\phi_{\mathcal{R}}(\lambda[\Upsilon'])\end{array}$$

(ii) For every $\epsilon\in(0,1)$ one has

$$\mathrm{Prob}_{\xi\sim P}\{\xi:\|Bx-\widehat{x}_H(Ax+\xi)\|>\mathcal{A}_*+\theta_\epsilon\mathcal{B}_*\}\le\epsilon\tag{4.97}$$

where $\theta_\epsilon=2\sqrt{2\ln(\sqrt{2}\mathrm{e}^{1/4}/\epsilon)}$.

Solution: (i): Let $(H,\Lambda,\Upsilon,\Upsilon',\Theta)$ be feasible for (4.42). When $\alpha>0$, the collection $(H,\alpha^{-1}\Lambda,\alpha\Upsilon,\Upsilon',\Theta)$ is feasible for (4.42). Applying to this collection the first inequality in (4.95) and taking into account that $\phi_{\mathcal{T}}$ and $\phi_{\mathcal{R}}$ are positively homogeneous of degree 1, we get

$$\max_{y\in\mathcal{X}}\|[B-H^TA]y\|\le\alpha\phi_{\mathcal{T}}(\lambda[\Lambda])+\alpha^{-1}\phi_{\mathcal{R}}(\lambda[\Upsilon]),$$

which after minimization of the right-hand side in α results in

$$D:=\max_{y\in\mathcal{X}}\|[B-H^TA]y\|\le\mathcal{A}_*.\tag{6.22}$$

Similarly, when $\alpha>0$, the collection $(H,\Lambda,\Upsilon,\alpha^{-1}\Upsilon',\alpha\Theta)$ is feasible for (4.42). Applying to this collection (4.95) and taking into account that $\Gamma_\Pi(\cdot)$ and $\phi_{\mathcal{R}}(\cdot)$ are positively homogeneous of degree 1, we get from the second relation in (4.95) that

$$\mathbf{E}_{\xi\sim P}\left\{\|H^T\xi\|\right\}\le\alpha\Gamma_\Pi(\Theta)+\alpha^{-1}\phi_{\mathcal{R}}(\lambda[\Upsilon']);$$

minimizing the right hand side in $\alpha>0$, we get

$$E:=\mathbf{E}_{\xi\sim P}\left\{\|H^T\xi\|\right\}\le\mathcal{B}_*.\tag{6.23}$$

When $x \in \mathcal{X}$, we clearly have $\mathbf{E}_{\xi\sim P}\{\|Bx - \widehat{x}_H(Ax+\xi)\|\} \leq D+E$, so that (6.22) and (6.23) yield the result stated in (i). □

(ii): Since $\mu\mu^T + S \lhd \Pi$, there exists $Q \in \Pi$ such that $\mu\mu^T + S \preceq Q$; we lose nothing when assuming that $Q = \mu\mu^T + S$. Let us set

$$\varphi := \phi_{\mathcal{R}}(\lambda[\Upsilon']),\ t = \sqrt{8\ln(\sqrt{2}\mathrm{e}^{1/4}/\epsilon)},\ \rho = \mathrm{Tr}(\Theta Q) \leq \Gamma_\Pi(\Theta). \tag{6.24}$$

Note that by Lemma 4.30 we have

$$\mathrm{Prob}_{\xi\sim P}\{\sqrt{\xi^T\Theta\xi} > t\sqrt{\rho}\} \leq \sqrt{2}\mathrm{e}^{1/4}\exp\{-t^2/8\} = \epsilon. \tag{6.25}$$

Now, Θ, Υ', and H satisfy the last semidefinite constraint in (4.42), which, by a slight modification of the computation in the proof of Lemma 4.11, results in the bound

$$\forall \xi : \|H^T\xi\| \leq 2\sqrt{\xi^T\Theta\xi}\sqrt{\varphi}. \tag{6.26}$$

Here is the required computation: for every $\alpha > 0$ we have

$$\begin{array}{l}
\forall y \in \mathcal{Y} : |y^TM^TH^T\xi| = |[\alpha^{1/2}y]^TM^TH^T[\alpha^{-1/2}\xi]| \\
\quad\leq \alpha\sum_\ell y^T\mathcal{S}_\ell^*[\Upsilon'_\ell]y + \alpha^{-1}\xi^T\Theta\xi \text{ [by the last constraint in (4.42)]} \\
\quad\leq \alpha\max_{y\in\mathcal{Y}}\sum_\ell \mathrm{Tr}(\mathcal{S}_\ell^*[\Upsilon'_\ell]yy^T) + \alpha^{-1}\xi^T\Theta\xi \\
\quad= \alpha\max_{y\in\mathcal{Y}}\sum_\ell \mathrm{Tr}(\Upsilon'_\ell\mathcal{S}_\ell[yy^T]) + \alpha^{-1}\xi^T\Theta\xi \\
\quad= \alpha\max_{y\in\mathcal{Y}}\sum_\ell \mathrm{Tr}(\Upsilon'_\ell S_\ell^2[y]) + \alpha^{-1}\xi^T\Theta\xi \\
\quad\leq \alpha\max_{r\in\mathcal{R}}\sum_\ell r_\ell\mathrm{Tr}(\Upsilon'_\ell) + \alpha^{-1}\xi^T\Theta\xi \\
\quad= \alpha\phi_{\mathcal{R}}(\lambda[\Upsilon']) + \alpha^{-1}\xi^T\Theta\xi.
\end{array}$$

For ξ fixed, optimizing the resulting bound in α, we get

$$\forall (y \in \mathcal{Y}) : |y^TM^TH^T\xi| \leq 2\sqrt{[\xi^T\Theta\xi]\varphi}.$$

When maximizing the left hand side in this inequality in $y \in \mathcal{Y}$, we arrive at (6.26).

It remains to note that in view of $x \in \mathcal{X}$ and relations (6.22) and (6.26) one has

$$\begin{array}{rcl}
\|Bx - \widehat{x}_H(Ax+\xi)\| &=& \|[B - H^TA]x - H^T\xi\| \leq \|[B-H^TA]x\| + \|H^T\xi\| \\
&\leq& \mathcal{A}_* + 2\sqrt{\xi^T\Theta\xi}\sqrt{\varphi}.
\end{array}$$

Recalling what φ is and taking into account (6.24) and (6.25), we arrive at (4.97). □

3) Suppose we are given observation $\omega = Ax + \xi$ of an unknown signal x known to belong to a given spectratope $\mathcal{X} \subset \mathbf{R}^n$ and want to recover the signal. We quantify the error of a recovery $\widehat{x}$ by $\max_{k\leq K}\|B_k(\widehat{x} - x)\|_{(k)}$, where $B_k \in \mathbf{R}^{\nu_k\times n}$ are given matrices, and $\|\cdot\|_{(k)}$ are given norms on $\mathbf{R}^{\nu_k}$ (for example, x can represent a discretization of a continuous-time signal, and B_kx can be finite-difference approximations of the signal's derivatives). We also assume, as in item 2, that observation noise ξ is independent of signal x and is sub-Gaussian with sub-Gaussianity parameters μ, S satisfying $\mu\mu^T + S \preceq Q$, for some given matrix $Q \succ 0$. Finally, we suppose that the unit balls of the norms conjugate to the norms $\|\cdot\|_{(k)}$ are spectratopes. In this situation, Proposition 4.14 provides us with K efficiently computable linear estimates $\widehat{x}_k(\omega) = H_k^T\omega : \mathbf{R}^{\dim\,\omega} \to \mathbf{R}^{\nu_k}$ along with upper bounds Opt_k on their risks $\max_{x\in\mathcal{X}}\mathbf{E}\left\{\|B_kx - \widehat{x}_k(Ax+\xi)\|_{(k)}\right\}$.
Think about how, given reliability tolerance $\epsilon \in (0,1)$, to aggregate these linear estimates into a single estimate $\widehat{x}(\omega) : \mathbf{R}^{\dim\,\omega} \to \mathbf{R}^n$ such that for every $x \in \mathcal{X}$, the probability of the event

$$\|B_k(\widehat{x}(Ax+\xi) - x)\|_{(k)} \leq \theta\mathrm{Opt}_k,\ 1 \leq k \leq K, \tag{!}$$

is at least $1-\epsilon$, for some moderate (namely, logarithmic in K and $1/\epsilon$) "assembling price" θ.

Solution: Let $t_* = t_*(\epsilon) := \sqrt{8\ln\left(\sqrt{2}\exp\{1/4\}K/\epsilon\right)}$. By item 2, for every $x \in \mathcal{X}$ and $k \le K$ we have

$$\mathrm{Prob}\{\|B_kx - \widehat{x}_k(Ax+\xi)\|_{(k)} > t_*\mathrm{Opt}_k\} \le \epsilon/K,$$

whence for every $x \in \mathcal{X}$ probability of the event

$$\mathcal{E}_x = \{\xi : \|B_kx - \widehat{x}_k(Ax+\xi)\|_{(k)} \le t_*\mathrm{Opt}_k, 1 \le k \le K\}$$

is at least $1-\epsilon$. Now let us build estimate $\widehat{x}(\cdot)$ as follows: given ω, we compute $x_k = \widehat{x}_k(\omega)$, $1 \le k \le K$, and solve the convex feasibility problem

$$\text{find } \bar{x} : \|B_k\bar{x} - x_k\|_{(k)} \le t_*\mathrm{Opt}_k,\ k \le K.$$

If this problem is feasible, $\widehat{x}(\omega)$ is its feasible solution; otherwise $\widehat{x}(\omega)$ is, say, the origin in the space where signals live. Let us verify that the resulting estimate, augmented with $\theta = 2t_*$, is what we are looking for. Indeed, let $x \in \mathcal{X}$ and ω be such that $\mathcal{E}_x$ takes place (which happens with probability at least $1-\epsilon$). In this case our feasibility problem is solvable, x being one of its feasible solutions, implying that $\bar{x} = \widehat{x}(\omega)$ is another feasible solution to this problem. Since both x and $\bar{x}$ are feasible, we conclude that

$$\|B_k(x-\bar{x})\|_{(k)} \le \|B_kx - x_k\|_{(k)} + \|x_k - B_k\bar{x}\|_{(k)} \le 2t_*\mathrm{Opt}_k\ \ \forall k.$$

A better construction (which does not require a priori knowledge of ϵ and hopefully is less conservative in practice) is to specify $\widehat{x}(\omega)$ as the y-component of the optimal solution to the convex optimization problem

$$\min_{t,y}\left\{t : t \ge 0, \|B_ky - \widehat{x}_k(\omega)\|_{(k)} \le t\mathrm{Opt}_k,\ 1 \le k \le K\right\}.$$

Slightly and straightforwardly modifying the above reasoning, we conclude that this estimate, for every $x \in \mathcal{X}$ and every $\epsilon \in (0,1)$, satisfies the relation

$$\|B_k(x - \widehat{x}(Ax+\xi))\|_{(k)} \le 2t_*(\epsilon)\mathrm{Opt}_k, k \le K$$

with probability at least $1-\epsilon$. □

Exercise 4.15 Prove that if ξ is uniformly distributed on the unit sphere $\{x : \|x\|_2 = 1\}$ in $\mathbf{R}^n$, then ξ is sub-Gaussian with parameters $(0, \frac{1}{n}I_n)$.

Solution: The function $f(\eta) := \mathbf{E}_\xi\left\{\exp\{\xi^T\eta\}\right\}$ clearly depends solely on $\|\eta\|_2$: $f(\eta) = \phi(\|\eta\|_2)$. Now let ζ be an n-dimensional Rademacher vector independent of ξ, and let $t \ge 0$. We have

$$\begin{array}{rcl}\mathbf{E}_{\xi,\zeta}\left\{\exp\{t\xi^T\zeta\}\right\} &=& \mathbf{E}_\zeta\left\{\mathbf{E}_\xi\left\{\exp\{\xi^T[t\zeta]\}\right\}\right\}\\ &=& \mathbf{E}_\zeta\left\{\phi(\|t\zeta\|_2)\right\} = \mathbf{E}_\zeta\left\{\phi(t\sqrt{n})\right\} = \phi(t\sqrt{n}),\end{array}$$

and

$$\begin{array}{rcl}\mathbf{E}_{\xi,\zeta}\left\{\exp\{t\xi^T\eta\}\right\} &=& \mathbf{E}_\xi\left\{\mathbf{E}_\zeta\left\{\exp\{t\sum_i\xi_i\zeta_i\}\right\}\right\} = \mathbf{E}_\xi\left\{\prod_i\cosh(t\xi_i)\right\}\\ &\leq& \mathbf{E}_\xi\left\{\prod_i\exp\{t^2\xi_i^2/2\}\right\} = \exp\{t^2/2\}\end{array}$$

(we have used the fact that $\cosh(a)\leq\exp\{a^2/2\}$—look at the coefficients in the power series expansion of both quantities). We conclude that $\phi(t\sqrt{n})\leq\exp\{t^2/2\}$ for all $t\geq 0$, or, which is the same, $f(\eta)=\phi(\|\eta\|_2)\leq\exp\{\eta^T\eta/(2n)\}$ for all η. □

6.4.4.5 Linear recovery under signal-dependent noise

Exercise 4.16. Consider the situation as follows: we observe a realization ω of an m-dimensional random vector

$$\omega = Ax+\xi_x,$$

where

- x is an unknown signal belonging to a given signal set $\mathcal{X}$, specifically, spectratope (which, as usual, we can assume to be basic):

$$\mathcal{X}=\{x\in\mathbf{R}^n:\exists t\in\mathcal{T}:R_k^2[x]\preceq t_kI_{d_k},k\leq K\}$$

  with standard restrictions on $\mathcal{T}$ and $R_k[\cdot]$;
- ξ_x is the observation noise with distribution which can depend on x; all we know is that

$$\mathrm{Var}[\xi_x]:=\mathbf{E}\{\xi_x\xi_x^T\}\preceq\mathcal{C}[x],$$

  where the entries of symmetric matrix $\mathcal{C}[x]$ are quadratic in x. We assume in the sequel that signals x belong to the subset

$$\mathcal{X}_\mathcal{C}=\{x\in\mathcal{X}:\mathcal{C}[x]\succeq 0\}$$

  of $\mathcal{X}$;
- Our goal is to recover Bx, with given $B\in\mathbf{R}^{\nu\times n}$, in a given norm $\|\cdot\|$ such that the unit ball $\mathcal{B}_*$ of the conjugate norm is a spectratope:

$$\mathcal{B}_*=\{u:\|u\|_*\leq 1\}=M\mathcal{V},\mathcal{V}=\{v:\exists r\in\mathcal{R}:S_\ell^2[v]\preceq r_\ell I_{f_\ell},\ell\leq L\}.$$

We quantify the performance of a candidate estimate $\widehat{x}(\omega):\mathbf{R}^m\to\mathbf{R}^\nu$ by the risk

$$\mathrm{Risk}_{\|\cdot\|}[\widehat{x}|\mathcal{X}_\mathcal{C}]=\sup_{x\in\mathcal{X}_\mathcal{C}}\sup_{\xi_x:\mathrm{Var}[\xi_x]\preceq\mathcal{C}[x]}\mathbf{E}\left\{\|Bx-\widehat{x}(Ax+\xi_x)\|\right\}.$$

1) Utilize semidefinite relaxation in order to build, in a computationally efficient fashion, a "presumably good" linear estimate; specifically, prove the following:

Proposition 4.32 *In the situation in question, for $G\in\mathbf{S}^m$ let us define $\alpha_0[G]\in\mathbf{R}$, $\alpha_1[G]\in\mathbf{R}^n$, $\alpha_2[G]\in\mathbf{S}^n$ from the identity*

$$\mathrm{Tr}(\mathcal{C}[x]G)=\alpha_0[G]+\alpha_1^T[G]x+x^T\alpha_2[G]x\ \forall(x\in\mathbf{R}^n,G\in\mathbf{S}^m),$$

so that the $\alpha_\chi[G]$ are affine in G. Consider the convex optimization problem

$$\mathrm{Opt} = \min_{H,\mu,D,\Lambda,\Upsilon,\Upsilon',G} \left\{ \begin{array}{l} \mu + \phi_{\mathcal{T}}(\lambda[\Lambda]) + \phi_{\mathcal{R}}(\lambda[\Upsilon]) + \phi_{\mathcal{R}}(\lambda[\Upsilon']) : \\ \Lambda = \{\Lambda_k \in \mathbf{S}^{d_k}_+, k \leq K\}, \Upsilon = \{\Upsilon_\ell \in \mathbf{S}^{f_\ell}_+, \ell \leq L\}, \Upsilon' = \{\Upsilon'_\ell \in \mathbf{S}^{f_\ell}_+, \ell \leq L\}, D \in \mathbf{S}^m_+ \\ \left[\begin{array}{c|c|c} \alpha_0[G] & \frac{1}{2}\alpha_1^T[G] & \\ \hline \frac{1}{2}\alpha_1[G] & \alpha_2[G] & \frac{1}{2}[B^T - A^T H]M \\ \hline & \frac{1}{2}M^T[B - H^T A] & \end{array}\right] \\ \qquad \preceq \left[\begin{array}{c|c|c} \mu - \alpha_0[D] & -\frac{1}{2}\alpha_1^T[D] & \\ \hline -\frac{1}{2}\alpha_1[D] & \sum_k \mathcal{R}_k^*[\Lambda_k] - \alpha_2[D] & \\ \hline & & \sum_\ell \mathcal{S}_\ell^*[\Upsilon_\ell] \end{array}\right] \\ \qquad \left[\begin{array}{c|c} G & \frac{1}{2}HM \\ \hline \frac{1}{2}M^T H^T & \sum_\ell \mathcal{S}_\ell^*[\Upsilon'_\ell] \end{array}\right] \succeq 0 \end{array} \right\}$$

$$\left[\begin{array}{c} [\mathcal{R}_k^*[\Lambda_k]]_{ij} = \mathrm{Tr}(\Lambda_k \frac{1}{2}[R^{ki}R^{kj} + R^{kj}R^{ki}]), \text{ where } R_k[x] = \sum_j x_j R^{kj} \\ [\mathcal{S}_\ell^*[\Upsilon_\ell]]_{ij} = \mathrm{Tr}(\Upsilon_\ell \frac{1}{2}[S^{\ell i}S^{\ell j} + S^{\ell j}S^{\ell i}]), \text{ where } S_\ell[v] = \sum_j v_j S^{\ell j} \\ \lambda[\{Z_i, i \leq I\}] = [\mathrm{Tr}(Z_1); ...; \mathrm{Tr}(Z_I)], \ \phi_A(q) = \max_{s\in A} q^T s \end{array}\right].$$

Whenever $H, \mu, D, \Lambda, \Upsilon, \Upsilon'$, and G are feasible for the problem, one has

$$\begin{array}{c} \mathrm{Risk}_{\|\cdot\|}[\widehat{x}^H|\mathcal{X}_C] \leq \mu + \phi_{\mathcal{T}}(\lambda[\Upsilon]) + \phi_{\mathcal{R}}(\lambda[\Upsilon]) + \phi_{\mathcal{R}}(\lambda[\Upsilon']) \\ \text{where } \widehat{x}^H(\omega) = H^T\omega. \end{array}$$

Solution: For a linear estimate $\widehat{x}_H(\omega) = H^T\omega$, $x \in \mathcal{X}_C$, and random noise ξ_x with matrix of second moments $\preceq \mathcal{C}[x]$ we have

$$\begin{array}{l} \mathbf{E}\left\{\|Bx - H^T(Ax + \xi_x)\|\right\} \leq \|[B - H^T A]x\| + \mathbf{E}\left\{\|H^T\xi_x\|\right\} \\ \leq \|[B - H^T A]x\| + \min\limits_{G,\Upsilon'} \left\{ \phi_{\mathcal{R}}(\lambda[\Upsilon']) + \mathrm{Tr}(\mathcal{C}[x]G) : \begin{array}{c} \Upsilon' = \{\Upsilon'_\ell \succeq 0, \ell \leq L\} \\ \left[\begin{array}{c|c} G & \frac{1}{2}HM \\ \hline \frac{1}{2}M^T H^T & \sum_\ell \mathcal{S}_\ell^*[\Upsilon'_\ell] \end{array}\right] \succeq 0 \end{array} \right\} \\ \hfill \text{[by Corollary 4.12]} \\ = \min\limits_{G,\Upsilon'} \left\{ \|[B - H^T A]x\| + \phi_{\mathcal{R}}(\lambda[\Upsilon']) + \mathrm{Tr}(\mathcal{C}[x]G) : \begin{array}{c} \Upsilon' = \{\Upsilon'_\ell \succeq 0, \ell \leq L\} \\ \left[\begin{array}{c|c} G & \frac{1}{2}HM \\ \hline \frac{1}{2}M^T H^T & \sum_\ell \mathcal{S}_\ell^*[\Upsilon'_\ell] \end{array}\right] \succeq 0 \end{array} \right\}. \end{array} \tag{6.27}$$

Consider the quantity

$$F_G(v, x) = v^T M^T[B - H^T A]x + \mathrm{Tr}(\mathcal{C}[x]G).$$

This is a quadratic function of (v, x):

$$F_G(v, x) = \alpha_0[G] + \alpha_1^T[G]x + x^T\alpha_2[G]x + v^T M^T[B - H^T A]x.$$

We can write

$$F_G(v, x) = [1; x; v]^T \underbrace{\left[\begin{array}{c|c|c} \alpha_0[G] & \frac{1}{2}\alpha_1^T[G] & \\ \hline \frac{1}{2}\alpha_1[G] & \alpha_2[G] & \frac{1}{2}[B^T - A^T H]M \\ \hline & \frac{1}{2}M^T[B - H^T A] & \end{array}\right]}_{\mathcal{A}[G,H]} [1; x; v].$$

Now let $\Upsilon = \{\Upsilon_\ell \in \mathbf{S}^{f_\ell}, \ell \leq L\}$, $\Lambda = \{\Lambda_k \in \mathbf{S}^{d_k}, k \leq K\}$, $\mu \in \mathbf{R}$, $D \in \mathbf{S}^m$ be such that

$$\Upsilon_\ell \succeq 0, \ell \leq L, \Lambda_k \succeq 0, k \leq K, D \succeq 0,$$

and, in addition,

$$\mathcal{A}[G,H] \preceq \left[\begin{array}{c|c|c} \mu-\alpha_0[D] & -\frac{1}{2}\alpha_1^T[D] & \\ \hline -\frac{1}{2}\alpha_1[D] & \sum_k \mathcal{R}_k^*[\Lambda_k]-\alpha_2[D] & \\ \hline & & \sum_\ell \mathcal{S}_\ell^*[\Upsilon_\ell] \end{array}\right]. \qquad (!)$$

We claim that in this case for all $x \in \mathcal{X}_C$ it holds

$$\|[B-H^TA]x\| + \mathrm{Tr}(\mathcal{C}[x]G) \le \mu + \phi_{\mathcal{T}}(\lambda[\Lambda]) + \phi_{\mathcal{R}}(\lambda[\Upsilon]).$$

Indeed, we have

$$\begin{array}{rl} \max\limits_{x\in\mathcal{X}_C}\left\{\|[B-H^TA]x\| + \mathrm{Tr}(\mathcal{C}[x]G)\right\} = & \max\limits_{v\in\mathcal{V},x\in\mathcal{X}_C} F_G(v,x) \\ = & \max\limits_{v\in\mathcal{V},x\in\mathcal{X}_C} [1;x;v]^T\mathcal{A}[G,H][1;x;v] \\ \le & \max\limits_{v\in\mathcal{V},x\in\mathcal{X}_C} \left\{[1;x;v]^T\mathcal{A}[G,H][1;x;v] + \mathrm{Tr}(\mathcal{C}[x]D)\right\} \\ & \text{[since } \mathrm{Tr}(\mathcal{C}[x]D)\ge 0 \text{ when } x\in\mathcal{X}_C \text{ due to } D\succeq 0] \\ = & \max\limits_{v\in\mathcal{V},x\in\mathcal{X}_C} [1;x;v]^T\left[\mathcal{A}[G,h] + \left[\begin{array}{c|c|c} \alpha_0[D] & \frac{1}{2}\alpha_1^T[D] & \\ \hline \frac{1}{2}\alpha_1[D] & \alpha_2[D] & \\ \hline & & \end{array}\right]\right][1;x;v] \\ \le & \max\limits_{v\in\mathcal{V},x\in\mathcal{X}} [1;x;v]^T \left[\begin{array}{c|c|c} \mu & & \\ \hline & \sum_k \mathcal{R}_k^*[\Lambda_k] & \\ \hline & & \sum_\ell \mathcal{S}_\ell^*[\Upsilon_\ell] \end{array}\right][1;x;v] \text{ [by (!)]} \\ = & \mu + \max\limits_{x\in\mathcal{X}} x^T[\sum_k \mathcal{R}_k^*[\Lambda_k]]x + \max\limits_{v\in\mathcal{V}} v^T\left[\sum_\ell \mathcal{S}_\ell^*[\Upsilon_\ell]\right]v \\ = & \mu + \max\limits_{x\in\mathcal{X}} \sum_k \mathrm{Tr}(\Lambda_k R_k^2[x]) + \max\limits_{v\in\mathcal{V}} \sum_\ell \mathrm{Tr}(\Upsilon_\ell S_\ell^2[v]) \\ \le & \mu + \phi_{\mathcal{T}}(\lambda[\Lambda]) + \phi_{\mathcal{R}}(\lambda[\Upsilon]) \text{ [cf. proof of Lemma 4.9].} \end{array}$$

Invoking (6.27), we conclude that

$$\mathrm{Risk}_{\|\cdot\|}[\widehat{x}_H|\mathcal{X}_C] \le \min_{\mu,D,\Lambda,\Upsilon,\Upsilon',G}\left\{\begin{array}{l}\mu + \phi_{\mathcal{T}}(\lambda[\Lambda]) + \phi_{\mathcal{R}}(\lambda[\Upsilon]) + \phi_{\mathcal{R}}(\lambda[\Upsilon']) : \\ \Lambda=\{\Lambda_k\in\mathbf{S}_+^{d_k}, k\le K\}, \Upsilon=\{\Upsilon_\ell\in\mathbf{S}_+^{f_\ell}, \ell\le L\}, \Upsilon'=\{\Upsilon'_\ell\in\mathbf{S}_+^{f_\ell},\ell\le L\}, D\in\mathbf{S}_+^m \\ \left[\begin{array}{c|c|c} \alpha_0[G] & \frac{1}{2}\alpha_1^T[G] & \\ \hline \frac{1}{2}\alpha_1[G] & \alpha_2[G] & \frac{1}{2}[B^T-A^TH]M \\ \hline & \frac{1}{2}M^T[B-H^TA] & \end{array}\right] \\ \qquad \preceq \left[\begin{array}{c|c|c} \mu-\alpha_0[D] & -\frac{1}{2}\alpha_1^T[D] & \\ \hline -\frac{1}{2}\alpha_1[D] & \sum_k \mathcal{R}_k^*[\Lambda_k]-\alpha_2[D] & \\ \hline & & \sum_\ell \mathcal{S}_\ell^*[\Upsilon_\ell] \end{array}\right] \\ \qquad\qquad \left[\begin{array}{c|c} G & \frac{1}{2}HM \\ \hline \frac{1}{2}M^TH^T & \sum_\ell \mathcal{S}_\ell^*[\Upsilon'_\ell] \end{array}\right] \succeq 0 \end{array}\right\},$$

and the conclusion of the proposition follows. □

2) Work out the following special case of the above situation dealing with Poisson Imaging (see Section 2.4.3.2): your observation is an m-dimensional random vector with independent Poisson entries, the vector of parameters of the corresponding Poisson distributions being Py; here P is an $\times n$ entrywise nonnegative matrix, and the unknown signal y is known to belong to a given box $Y = \{y \in \mathbf{R}^n : \underline{a} \le y \le \overline{a}\}$, where $0 \le \underline{a} < \overline{a}$. You want to recover y in $\|\cdot\|_p$-norm with given $p \in [1,2]$.

<u>Solution:</u> Setting $y_0 = \frac{1}{2}[\underline{a}+\overline{a}]$, $B = \mathrm{Diag}\{\frac{1}{2}[\overline{a}-\underline{a}]\}$, parameterizing $y \in Y$ as

$$\begin{array}{c} y = y_0 + Bx, \\ x \in \mathcal{X} := [-1;1]^n = \{x\in\mathbf{R}^n : \exists t\in\mathcal{T} := [0,1]^n : S_k^2[x] := x_k^2 \preceq t_k I_1, k\le K := n\}, \end{array}$$

and subtracting from our actual observation the vector Py_0, we reduce the situation to the one where our observation is

$$\omega = Ax + \xi_x,$$

where $A = PB$, and the observation noise ξ_x has independent entries, the i-th entry being the difference between a realization of the Poisson random variable with parameter $[Py_0 + Ax]_i$ and the expectation of this random variable, so that ξ_x has zero mean. Our goal is to recover in $\|\cdot\|_p$ the vector $y_0 + Bx$, or, which is the same, the vector Bx. Immediate computation shows that the covariance matrix of ξ_x is

$$\mathcal{C}[x] = \mathrm{Diag}\{Py_0 + Ax\},$$

and we find ourselves in the situation considered in Proposition 4.32.

Exercise 4.17 The goal of what follows is to "transfer" the constructions of linear estimates to the case of multiple indirect observations of discrete random variables. Specifically, we are interested in the situation where

- Our observation is a K-element sample $\omega^K = (\omega_1, .., \omega_K)$ with independent identically distributed components ω_k taking values in an m-element set. As always, we encode the points from this m-element set by the standard basic orths $e_1, ..., e_m$ in $\mathbf{R}^m$.
- The (common for all k) probability distribution of ω_k is Ax, where x is an unknown "signal"—n-dimensional probabilistic vector known to belong to a closed convex subset $\mathcal{X}$ of n-dimensional probabilistic simplex $\mathbf{\Delta}_n = \{x \in \mathbf{R}^n : x \geq 0, \sum_i x_i = 1\}$—and A is a given $m \times n$ column-stochastic matrix (i.e., entrywise nonnegative matrix with unit column sums).
- Our goal is to recover Bx, where B is a given $\nu \times n$ matrix, and we quantify a candidate estimate $\widehat{x}(\omega^K) : \mathbf{R}^{mK} \to \mathbf{R}^\nu$ by its *risk*

$$\mathrm{Risk}_{\|\cdot\|}[\widehat{x}|\mathcal{X}] = \sup_{x \in \mathcal{X}} \mathbf{E}_{\omega^K \sim [Ax] \times \ldots \times [Ax]} \left\{ \|Bx - \widehat{x}(\omega^K)\| \right\},$$

where $\|\cdot\|$ is a given norm on $\mathbf{R}^\nu$.

We use *linear* estimates—estimates of the form

$$\widehat{x}_H(\omega^K) = H^T \underbrace{\left[\frac{1}{K} \sum_{k=1}^{K} \omega_k \right]}_{\widehat{\omega}_K[\omega^K]}, \tag{4.98}$$

where $H \in \mathbf{R}^{m \times \nu}$.

1) In the main body of Chapter 4, $\mathcal{X}$ always was assumed to be symmetric w.r.t. the origin, which easily implies that we gain nothing when passing from linear estimates to affine ones (sums of linear estimates and constants). Now we are in the case where $\mathcal{X}$ can be "heavily asymmetric," which, in general, can make "genuinely affine" estimates preferable. Show that in the case in question, we still lose nothing when restricting ourselves to linear, rather than affine, estimates.

Solution: Denoting by $\mathbf{1}_k$ the k-dimensional all-ones vector, observe that we are in the situation when $\mathbf{1}_m^T \widehat{\omega}_K[\omega^K] = 1$ for all signals and observations, implying every affine estimate can be written down as a linear one:

$$H^T \widehat{\omega}_K[\omega^K] + h \equiv \bar{H}^T \widehat{\omega}_K[\omega^K], \ \bar{H} = H + \mathbf{1}_m h^T.$$

4.17.A. Observation scheme revisited. When observation ω^K stems from a signal $x \in \mathbf{\Delta}_n$, we have

$$\widehat{\omega}_K[\omega^K] = Ax + \xi_x,$$

where

$$\xi_x = \frac{1}{K}\sum_{k=1}^{K}[\omega_k - Ax]$$

is the average of K independent identically distributed zero mean random vectors with common covariance matrix $Q[x]$.

2) Check that

$$Q[x] = \mathrm{Diag}\{Ax\} - [Ax][Ax]^T,$$

and derive from this fact that the covariance matrix of ξ_x is

$$Q_K[x] = \frac{1}{K}Q[x].$$

Setting

$$\Pi = \Pi_{\mathcal{X}} = \left\{Q = \frac{1}{K}\mathrm{Diag}\{Ax\} : x \in \mathcal{X}\right\},$$

check that $\Pi_{\mathcal{X}}$ is a convex compact subset of the positive semidefinite cone $\mathbf{S}^m_+$, and that whenever $x \in \mathcal{X}$, one has $Q[x] \preceq Q$ for some $Q \in \Pi$.

4.17.B. Upper-bounding risk of a linear estimate. We can upper-bound the risk of a linear estimate $\widehat{x}_H$ as follows:

$$\begin{array}{rcl}\mathrm{Risk}_{\|\cdot\|}[\widehat{x}_H|\mathcal{X}] & = & \sup_{x\in\mathcal{X}} \mathbf{E}_{\omega^K\sim[Ax]\times\ldots\times[Ax]}\left\{\|Bx - H^T\widehat{\omega}_K[\omega^K]\|\right\} \\ & = & \sup_{x\in\mathcal{X}} \mathbf{E}_{\xi_x}\left\{\|[Bx - H^TA]x - H^T\xi_x\|\right\} \\ & \leq & \underbrace{\sup_{x\in\mathcal{X}} \|[B - H^TA]x\|}_{\Phi(H)} + \underbrace{\sup_{\xi:\mathrm{Cov}[\xi]\in\Pi_{\mathcal{X}}} \mathbf{E}_\xi\left\{\|H^T\xi\|\right\}}_{\Psi^{\mathcal{X}}(H)}.\end{array}$$

As in the main body of Chapter 4, we intend to build a "presumably good" linear estimate by minimizing over H the sum of efficiently computable upper bounds $\overline{\Phi}(H)$ on $\Phi(H)$ and $\overline{\Psi}^{\mathcal{X}}(H)$ on $\Psi^{\mathcal{X}}(H)$.

Assuming from now on that the unit ball $\mathcal{B}_*$ of the norm conjugate to $\|\cdot\|$ is a spectratope,

$$\mathcal{B}_* := \{u : \|u\|_* \leq 1\} = \{u : \exists r \in \mathcal{R}, y : u = My, S_\ell^2[y] \preceq r_\ell I_{f_\ell}, \ell \leq L\}$$

with our usual restrictions on $\mathcal{R}$ and S_ℓ, we can take as $\overline{\Psi}^{\mathcal{X}}(\cdot)$ the function (4.40).

For the sake of simplicity, we from now assume that $\mathcal{X}$ is cut off $\mathbf{\Delta}_n$ by linear inequalities:

$$\mathcal{X} = \{x \in \mathbf{\Delta}_n : Gx \leq g, \; Ex = e\} \qquad [G \in \mathbf{R}^{p\times n}, E \in \mathbf{R}^{q\times n}]$$

Observe that replacing G with $G - g\mathbf{1}_n^T$ and E with $E - e\mathbf{1}_n^T$, we reduce the situation to that where all linear constraints are homogeneous, that is,

$$\mathcal{X} = \{x \in \mathbf{\Delta}_n : Gx \leq 0, \; Ex = 0\},$$

and this is what we assume from now on. Setting

$$F = [G; E; -E] \in \mathbf{R}^{(p+2q)\times n},$$

we have also

$$\mathcal{X} = \{x \in \mathbf{\Delta}_n : Fx \leq 0\}.$$

Suppose that $\mathcal{X}$ is nonempty. Finally, in addition to what was already assumed about the norm $\|\cdot\|$, let us also suppose that this norm is *absolute*, that is, $\|u\|$ depends only on the vector of *magnitudes* of entries in u. From this assumption it immediately follows that if $0 \leq u \leq u'$, then $\|u\| \leq \|u'\|$ (why?).

Our next task is to efficiently upper-bound $\Phi(\cdot)$.

4.17.C. Bounding Φ, simple case. We start with the *simple case* where there are no linear constraints (formally, G and E are zero matrices), in this case bounding Φ is straightforward:

3) Prove that in the simple case Φ is convex and efficiently computable "as is:"

$$\Phi(H) = \max_{i \leq n} \|(B - H^T A)g_i\|,$$

where $g_1, ..., g_n$ are the standard basic orths in $\mathbf{R}^n$.

Solution: Recall that in the simple case $\mathcal{X} = \mathbf{\Delta}_n$ is the convex hull of $g_1, ..., g_n$, and that the maximum of a convex function $\|(B - H^T A)x\|$ over $x \in \mathcal{X}$ is achieved at an extreme point of $\mathcal{X}$.

4.17.D. Lagrange upper bound on Φ.

4) Observing that when $\mu \in \mathbf{R}^{p+2q}_+$, the function

$$\|(B - H^T A)x\| - \mu^T F x$$

of x is convex in $x \in \mathbf{\Delta}_n$ and overestimates $\|(B - H^T A)x\|$ everywhere on $\mathcal{X}$, conclude that the efficiently computable convex function

$$\Phi_L(H) = \min_{\mu} \max_{i \leq n} \{\|(B - H^T A)g_i\| - \mu^T F g_i : \mu \geq 0\}$$

upper-bounds $\Phi(H)$. In the sequel, we call this function the *Lagrange* upper bound on Φ.

Solution: $\Phi_L(H)$ is obtained from the convex function of H, μ

$$\Psi(H, \mu) = \max_{i \leq n} \left[\|(B - H^T A)g_i\| - \mu^T F g_i\right]$$

by minimization in $\mu \geq 0$, and the function of this type is convex in H on every convex set where the function does not take value $-\infty$. It remains to note that $\Phi(H) \geq 0$ (since $\mathcal{X}$ is nonempty), and by construction $\Psi(H, \mu) \geq \Phi(H) \geq 0$ whenever $\mu \geq 0$, implying that $\Phi_L(\cdot)$ is nonnegative and thus does not take value $-\infty$. □

4.17.E. Basic upper bound on Φ. For vectors u and v of the same dimension, say, k, let $\mathrm{Max}[u, v]$ stand for the entrywise maximum of u, v:

$$[\mathrm{Max}[u, v]]_i = \max[u_i, v_i],$$

and let

$$[u]_+ = \mathrm{Max}[u, 0_k],$$

where 0_k is the k-dimensional zero vector.

5.1) Let $\Lambda_+ \geq 0$ and $\Lambda_- \geq 0$ be $\nu \times (p+2q)$ matrices, $\Lambda \geq 0$ meaning that matrix Λ is entrywise nonnegative. Prove that whenever $x \in \mathcal{X}$, one has

$$\begin{aligned}\|(B - H^T A)x\| &\leq \mathcal{B}(x, H, \Lambda_+, \Lambda_-)\\ &:= \min_t \left\{\|t\| : t \geq \mathrm{Max}\left[[(B - H^T A)x - \Lambda_+ F x]_+, [-(B - H^T A)x - \Lambda_- F x]_+\right]\right\}\end{aligned}$$

and that $\mathcal{B}(x, H, \Lambda_+, \Lambda_-)$ is convex in x.

5.2) Derive from 5.1 that whenever $\Lambda_\pm$ are as in 5.1, one has

$$\Phi(H) \leq \mathcal{B}^+(H,\Lambda_+,\Lambda_-) := \max_{i\leq n} \mathcal{B}(g_i,H,\Lambda_+,\Lambda_-),$$

where, as in item 3, $g_1,...,g_n$ are the standard basic orths in $\mathbf{R}^n$. Conclude that

$$\Phi(H) \leq \Phi_B(H) = \inf_{\Lambda_\pm} \left\{ \mathcal{B}^+(H,\Lambda_+,\Lambda_-) : \Lambda_\pm \in \mathbf{R}_+^{\nu\times(p+2q)} \right\}$$

and that Φ_B is convex and real-valued. In the sequel we refer to $\Phi_B(\cdot)$ as to *Basic* upper bound on $\Phi(\cdot)$.

Solution: 5.1: Let $x \in \mathcal{X}$, so that $Fx \leq 0$ and therefore $\Lambda_\pm Fx \leq 0$. Since $\|\cdot\|$ is an absolute norm, we have

$$\begin{array}{l}\|(B-H^TA)x\| = \min\limits_t \left\{\|t\| : t \geq \mathrm{Max}\left[[(B-H^TA)x]_+, [-(B-H^TA)x]_+\right]\right\}\\ \quad \leq \min\limits_t \left\{\|t\| : t \geq \mathrm{Max}\left[[(B-H^TA)x - \Lambda_+Fx]_+, [-(B-H^TA)x - \Lambda_-Fx]_+\right]\right\},\end{array}$$

with the inequality due to $x \in \mathcal{X}$, so that $-\Lambda_\pm Fx$ are nonnegative vectors. Thus, we indeed have

$$x \in \mathcal{X}, \Lambda_\pm \geq 0 \Rightarrow \|(B-H^TA)x\| \leq \mathcal{B}(x,H,\Lambda_+,\Lambda_-),$$

and the right hand side in this inequality clearly is convex in x.

5.2: Invoking 5.1, we have

$$\begin{array}{rcl}\Phi(H) & = & \max\limits_{x\in\mathcal{X}} \|(B-H^TA)x\| \leq \max\limits_{x\in\mathcal{X}} \mathcal{B}(x,H,\Lambda_+,\Lambda_-)\\ & \leq & \max\limits_{x\in\mathbf{\Delta}_n} \mathcal{B}(x,H,\Lambda_+,\Lambda_-) = \mathcal{B}^+(H,\Lambda_+\Lambda_-),\end{array}$$

where the last equality is due to the fact that $\mathcal{B}(\cdot)$ is convex in x. The rest of 5.2 is evident. □

4.17.F. Sherali-Adams upper bound on Φ. Let us apply the approach we used in Chapter 1, Section 1.3.2, when deriving verifiable sufficient conditions for s-goodness; see p. 21. Specifically, setting

$$W = \left[\begin{array}{c|c} G & I \\ \hline E & \end{array}\right],$$

let us introduce the slack variable $z \in \mathbf{R}^p$ and rewrite the description of $\mathcal{X}$ as

$$\mathcal{X} = \{x \in \mathbf{\Delta}_n : \exists z \geq 0 : W[x;z] = 0\},$$

so that $\mathcal{X}$ is the projection of the polyhedral set

$$\mathcal{X}^+ = \{[x;z] : x \in \mathbf{\Delta}_n, z \geq 0, W[x;z] = 0\}$$

on the x-space. Projection of $\mathcal{X}^+$ on the z-space is a nonempty (since $\mathcal{X}$ is so) and clearly bounded subset of the nonnegative orthant $\mathbf{R}^p_+$, and we can in many ways cover Z by the simplex

$$\Delta[\alpha] = \{z \in \mathbf{R}^p : z \geq 0, \sum_i \alpha_i z_i \leq 1\},$$

where all α_i are positive.

6.1) Let $\alpha > 0$ be such that $Z \subset \Delta[\alpha]$. Prove that

$$\mathcal{X}^+ = \{[x;z] : W[x;z] = 0, [x;z] \in \mathrm{Conv}\{v_{ij} = [g_i; h_j], 1 \le i \le n, 0 \le j \le p\}\}, \qquad (!)$$

where g_i are the standard basic orhts in $\mathbf{R}^n$, $h_0 = 0 \in \mathbf{R}^p$, and $\alpha_j h_j$, $1 \le j \le p$, are the standard basic orths in $\mathbf{R}^p$.

6.2) Derive from 5.1 that the efficiently computable convex function

$$\Phi_{SA}(H) = \inf_C \max_{i,j} \left\{ \|(B - H^T A)g_i + C^T W v_{ij}\| : C \in \mathbf{R}^{(p+q)\times\nu} \right\}$$

is an upper bound on $\Phi(H)$. In the sequel, we refer to $\Phi_{SA}(H)$ as to *Sherali-Adams* bound [214].

Solution: 6.1: Clearly, whenever $[x;z] \in \mathcal{X}^+$, we have $W[x;z] = 0$, $x \in \mathbf{\Delta}_n$, $z \in Z \subset \Delta[\alpha]$, the latter implying that z is a convex combination of h_j, so that $[x;z]$ is a convex combination of v_{ij}. Thus, $\mathcal{X}^+$ is contained in the right hand side set of (!). Vice versa, if $[x;z]$ belongs to the right hand side set of (!), we have $x \in \mathbf{\Delta}_n$, $z \ge 0$, and $W[x;z] = 0$, implying that $[x;z] \in \mathcal{X}^+$. 6.1 is proved.

6.2: Because $\mathcal{X}$ is the projection of $\mathcal{X}^+$ on the x-space, for every $C \in \mathbf{R}^{(p+q)\times\nu}$ we have

$$\begin{array}{rcl}
\Phi(H) & = & \max_{x\in\mathcal{X}} \|(B - H^T A)x\| = \max_{[x;z]\in\mathcal{X}^+} \|(B - H^T A)x\| \\
& = & \max_{[x;z]\in\mathcal{X}^+} \|(B - H^T A)x + C^T W[x;z]\| \\
& \le & \max_{[x;z]} \left\{ \|(B - H^T A)x + C^T F[x;z]\| : [x;z] \in \mathrm{Conv}\{v_{ij}\} \right\} \ [\text{by } (!)] \\
& = & \max_{i,j} \|(B - H^T A)g_i + C^T W v_{ij}\|, \\
& \Rightarrow & \Phi(H) \le \inf_C \max_{i,j} \|(B - H^T A)g_i + C^T W v_{ij}\| = \Phi_{SA}(H),
\end{array}$$

as claimed in 6.2. □

4.17.G. Combined bound. We can combine the above bounds, specifically, as follows:

7) Prove that the efficiently computable convex function

$$\Phi_{LBS}(H) = \inf_{(\Lambda_\pm, C_\pm, \mu, \mu_+)\in\mathcal{R}} \max_{i,j} \mathcal{G}_{ij}(H, \Lambda_\pm, C_\pm, \mu, \mu_+), \qquad (\#)$$

where

$$\begin{array}{l}
\mathcal{G}_{ij}(H, \Lambda_\pm, C_\pm, \mu, \mu_+) := -\mu^T F g_i + \mu_+^T W v_{ij} + \min_t \Big\{ \|t\| : \\
t \ge \mathrm{Max}\left[[(B - H^T A - \Lambda_+ F)g_i + C_+^T W v_{ij}]_+, [(-B + H^T A - \Lambda_- F)g_i + C_-^T W v_{ij}]_+ \right] \Big\}, \\
\mathcal{R} = \{(\Lambda_\pm, C_\pm, \mu, \mu_+) : \Lambda_\pm \in \mathbf{R}_+^{\nu\times(p+2q)}, C_\pm \in \mathbf{R}^{(p+q)\times\nu}, \mu \in \mathbf{R}_+^{p+2q}, \mu_+ \in \mathbf{R}^{p+q}\}
\end{array}$$

is an upper bound on $\Phi(H)$, and that this *Combined* bound is at least as good as any of the Lagrange, Basic, or Sherali-Adams bounds.

Solution: Whenever $C_\pm \in \mathbf{R}^{(p+q)\times\nu}$, $\mu \in \mathbf{R}_+^{p+2q}$, $\mu_+ \in \mathbf{R}^{p+q}$, and $\Lambda_\pm \in \mathbf{R}_+^{\nu\times(p+2q)}$, for $[x;z] \in \mathcal{X}^+$ one has

$$\begin{array}{l}
\|(B-H^TA)x\| = \min\limits_t \{\|t\| : t \geq \mathrm{Max}[[(B-H^TA)x]_+, [(-B+H^TA)x]_+]\}\\
\quad \leq \min\limits_t \Big\{\|t\| :\\
\quad t \geq \mathrm{Max}\Big([(B-H^TA)x - \Lambda_+Fx + C_+^TW[x;z]]_+, [(-B+H^TA)x - \Lambda_-Fx + C_-^TW[x;z]]_+\Big)\Big\}\\
\hfill [\text{due to } \Lambda_\pm Fx \leq 0 \text{ and } C_\pm^TW[x;z]=0]\\
\quad \leq \min\limits_t \Big\{\|t\| :\\
\quad t \geq \mathrm{Max}\Big([(B-H^TA)x - \Lambda_+Fx + C_+^TW[x;z]]_+, [(-B+H^TA)x - \Lambda_-Fx + C_-^TW[x;z]]_+\Big)\\
\hfill -\mu^TFx + \mu_+^TW[x;z]\Big\}\\
\hfill [\text{due to } \mu \geq 0,\ Fx \leq 0 \text{ and } W[x;z]=0]\\
\quad =: \mathcal{F}(H, \Lambda_\pm, C_\pm, \mu, \mu_+; [x;z]);
\end{array}$$

note that $\mathcal{F}(H, \Lambda_\pm, C_\pm, \mu, \mu_+; [x;z])$ clearly is convex in $[x;z]$. Recalling that $\mathcal{X}^+$ is contained in $\mathrm{Conv}\{v_{ij}, 1 \leq i \leq n, 0 \leq j \leq p\}$, we conclude that

$$\begin{array}{rcl}
\max\limits_{x\in\mathcal{X}} \|(B-H^TA)x\| &=& \max\limits_{[x;z]\in\mathcal{X}^+} \|(B-H^TA)x\|\\
&\leq& \max\limits_{[x;z]\in\mathcal{X}^+} \mathcal{F}(H, \Lambda_\pm, C_\pm, \mu, \mu_+; [x;z])\\
&\leq& \max_{i,j} \mathcal{F}(H, \Lambda_\pm, C_\pm, \mu, \mu_+; v_{ij})\\
&=& \max_{i,j} \mathcal{G}_{ij}(H, \Lambda_\pm, C_\pm, \mu, \mu_+)
\end{array}$$

The concluding quantity clearly is convex in $(H, \Lambda_\pm, C_\pm, \mu, \mu_+)$, implying that $\Phi_{LBS}(H)$ is a convex in H upper bound on $\Phi(H)$.

Finally, to see that the Combined bound is not worse than each of the bounds we merge, it suffices to note that the latter bounds can be obtained from the right-hand side in (#) by imposing additional constraints on the minimization variables, specifically,
— for Lagrange bound—the constraints $\mathcal{C}_\pm = 0$, $\Lambda_\pm = 0$;
— for Basic bound—the constraints $C_\pm = 0$, $\mu = 0$, $\mu_+ = 0$;
— for Sherali-Adams bound—the constraints $C_+ = -C_-$, $\Lambda_\pm = 0$, $\mu = 0$, $\mu_+ = 0$.
□

4.17.H. How to select α? A shortcoming of the Sherali-Adams and the combined upper bounds on Φ is the presence of a "degree of freedom"—the positive vector α. Intuitively, we would like to select α to make the simplex $\Delta[\alpha] \supset Z$ to be "as small as possible." It is unclear, however, what "as small as possible" is in our context, not to mention how to select the required α after we agree on how we measure the "size" of $\Delta[\alpha]$. It turns out, however, that we can efficiently select α resulting in the *smallest volume* $\Delta[\alpha]$.

8) Prove that minimizing the volume of $\Delta[\alpha] \supset Z$ in α reduces to solving the following convex optimization problem:

$$\inf_{\alpha,u,v}\left\{-\sum_{s=1}^p \ln(\alpha_s) : 0 \leq \alpha \leq -v, E^Tu + G^Tv \leq \mathbf{1}_n\right\}. \qquad (*)$$

Solution: $\alpha \geq 0$ satisfies $Z \subset \Delta[\alpha]$ if and only if the linear inequality

$$\alpha^T\zeta \leq 1$$

in variable ζ is a consequence of the feasible system of linear constraints

$$\begin{array}{ll}(a) & x \geq 0 \\ (b) & \mathbf{1}_n^T x = 1 \\ (c) & Ex = 0 \\ (d) & Gx + \zeta = 0 \\ (e) & \zeta \geq 0\end{array}$$

in variables x, ζ. By the Inhomogeneous Farkas Lemma, the latter is the case if and only if there exist Lagrange multipliers $\lambda_a \geq 0$, λ_b, λ_c, λ_d, $\lambda_e \geq 0$ such that

$$\lambda_a + \lambda_b \mathbf{1}_n + E^T \lambda_c + G^T \lambda_d = 0, \; \lambda_d + \lambda_e = -\alpha, \lambda_b \geq -1,$$

which boils down to the existence of $\lambda_c = u$ and $\lambda_d = v$ such that $\alpha \leq -v$ and $E^T u + G^T v \leq \mathbf{1}_n$. The lengths of edges of $\Delta[\alpha]$ incident to the vertex 0 of $\Delta[\alpha]$ are inversely proportional to the corresponding entries in α, and the volume of $\Delta[\alpha]$ is proportional to the product of these edges, so that the minimization of this volume reduces to the convex problem $(*)$. □

9) Run numerical experiments to evaluate the quality of the above bounds. It makes sense to generate problems where we know in advance the actual value of Φ, e.g., to take

$$\mathcal{X} = \{x \in \mathbf{\Delta}_n : x \geq a\} \qquad (a)$$

with $a \geq 0$ such that $\sum_i a_i \leq 1$. In this case, we can easily list the extreme point of $\mathcal{X}$ (how?) and thus can easily compute $\Phi(H)$.
In your experiments, you can use the matrices stemming from "presumably good" linear estimates yielded by the optimization problems

$$\begin{array}{rcl}\mathrm{Opt} & = & \min\limits_{H,\Upsilon,\Theta} \Big\{\overline{\Phi}(H) + \phi_{\mathcal{R}}(\lambda[\Upsilon]) + \Gamma_{\mathcal{X}}(\Theta) : \Upsilon = \{\Upsilon_\ell \succeq 0, \ell \leq L,\} \\ & & \left[\begin{array}{c|c}\Theta & \frac{1}{2}HM \\ \hline \frac{1}{2}M^T H^T & \sum_\ell S_\ell^*[\Upsilon_\ell]\end{array}\right] \succeq 0 \Big\}\end{array} \qquad (4.99)$$

where

$$\Gamma_{\mathcal{X}}(\Theta) = \frac{1}{K} \max_{x \in \mathcal{X}} \mathrm{Tr}(\mathrm{Diag}\{Ax\}\Theta)$$

(see Corollary 4.12) with the actual Φ (which is available for our $\mathcal{X}$), or the upper bounds on Φ (Lagrange, Basic, Sherali-Adams, and Combined) in the role of $\overline{\Phi}$. Note that it may make sense to test seven bounds rather than just four. Indeed, with additional constraints on the optimization variables in (#), we can get, besides "pure" Lagrange, Basic, and Sherali-Adams bounds and their "three-component combination" (Combined bound), pairwise combinations of the pure bounds as well. For example, to combine the Lagrange and Sherali-Adams bound, it suffices to add to (#) the constraints $\Lambda_\pm = 0$.

Solution: In our experiments,

- the sizes of instances were $n = \nu = m = p = 10$, $q = 0$ (no constraints $Ex = 0$), $K = 10000$;
- we used $B = I_{10}$, $\|\cdot\| = \|\cdot\|_1$, and generated 10×10 stochastic matrices A at random, by normalizing the columns in a `rand(10,10)` matrix to have unit column sums;
- we selected a in (a) at random, by generating a `rand(10,1)` vector and then scaling it to have a `rand(1,1)` sum of entries;
- α was selected as explained in item 8.

Here are the optimal values of (P) for 10 randomly selected data sets (corresponding to 10 vertical stripes on the plot):

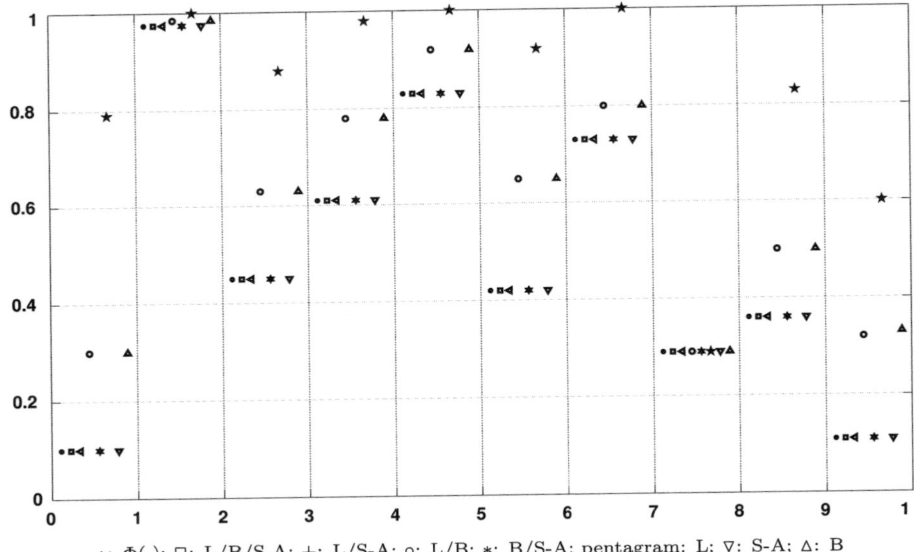

$*$: $\Phi(\cdot)$; $\square$: L/B/S-A; $+$: L/S-A; $\circ$: L/B; $*$: B/S-A; pentagram: L; ∇: S-A; Δ: B

We see that the "winners," besides the Combined bound, are pairwise combinations of Lagrange and Sherali-Adams, and of Basic and Sherali-Adams bounds, as well as the "pure" Sherali-Adams bound, and under the circumstances these bounds are *equal* to the actual Φ within machine accuracy.

Note that excellent performance exhibited in the above experiments by four of our seven estimates is seemingly due to the special structure of the signal set used in these experiments. This is what happens when we modify $\mathcal{X}$ by imposing upper and lower bounds on x-probabilities of some prescribed events rather than imposing lower bounds on the entries in x:

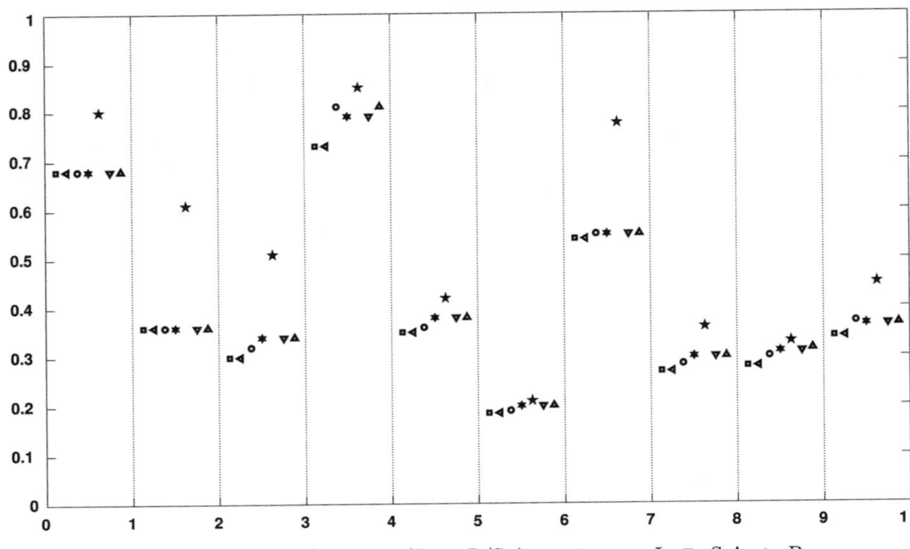

$\square$: L/B/S-A; $+$: L/S-A; $\circ$: L/B; $*$: B/S-A; pentagram: L; ∇: S-A; Δ: B

Now the winner, besides the Combined bound, is the pairwise combination of Lagrange and Sherali-Adams bounds.

Exercise 4.18. The exercise to follow deals with recovering discrete probability distributions in the *Wasserstein norm.*

The Wasserstein distance between probability distributions is extremely popular in Statistics today; it is defined as follows.[14] Consider discrete random variables taking values in finite observation space $\Omega = \{1, 2, ..., n\}$ which is equipped with the metric $\{d_{ij} : 1 \leq i, j \leq n\}$ satisfying the standard axioms.[15] As always, we identify probability distributions on Ω with n-dimensional probabilistic vectors $p = [p_1; ...; p_n]$, where p_i is the probability mass assigned by p to $i \in \Omega$. The Wasserstein distance between probability distributions p and q is defined as

$$W(p, q) = \min_{x=[x_{ij}]} \left\{ \sum_i d_{ij} x_{ij} : x_{ij} \geq 0, \sum_j x_{ij} = p_i, \sum_i x_{ij} = q_j \; \forall 1 \leq i, j \leq n \right\}. \tag{4.100}$$

In other words, one may think of p and q as two distributions of unit mass on the points of Ω, and consider the mass transport problem of redistributing the mass assigned to points by distribution p to get the distribution q. Denoting by x_{ij} the mass moved from point i to point j, constraints $\sum_j x_{ij} = p_i$ say that the total mass taken from point i is exactly p_i, constraints $\sum_i x_{ij} = q_j$ say that as the result of transportation, the mass at point j will be exactly q_j, and the constraints $x_{ij} \geq 0$ reflect the fact that transport of a negative mass is forbidden. Assuming that the cost of transporting a mass μ from point i to point j is $d_{ij}\mu$, the Wasserstein distance $W(p, q)$ between p and q is the cost of the cheapest transportation plan which converts p into q. As compared to other natural distances between discrete probability distributions, like $\|p - q\|_1$, the advantage of the Wasserstein distance is that it allows us to model the situation (indeed arising in some applications) where the effect, measured in terms of the intended application, of changing probability masses of points from Ω is small when the probability mass of a point is redistributed among *close* points.[16]

Now comes the first part of the exercise:

1) Let p, q be two probability distributions. Prove that

$$W(p, q) = \max_{f \in \mathbf{R}^n} \left\{ \sum_i f_i(p_i - q_i) : |f_i - f_j| \leq d_{ij} \; \forall i, j \right\} \tag{4.101}$$

Solution: By definition, $W(p, q)$ is the optimal value in a (clearly solvable) Linear Programming program. By the LP Duality Theorem, passing to the dual problem we have

$$W(p, q) = \max_{f,g} \left\{ f^T p - g^T q : f_i - g_j \leq d_{ij}, \; 1 \leq i, j \leq n \right\}. \tag{$*$}$$

All we need to prove is that we can find an optimal solution to $(*)$ where $f = g$. By the continuity argument, it suffices to prove this claim in the case when $p > 0$,

[14] The distance we consider stems from the Wasserstein 1-distance between discrete probability distributions. This is a particular case of the general Wasserstein p-distance between (not necessarily discrete) probability distributions.

[15] Namely, positivity and symmetry: $d_{ij} = d_{ji} \geq 0$, with $d_{ij} = 0$ if and only if $i = j$; and the triangle inequality: $d_{ik} \leq d_{ij} + d_{jk}$ for all triples i, j, k.

[16] In fact, the Wasserstein distance shares this property with some other distances between distributions used in Probability Theory, such as Skorohod, or Prokhorov, or Ky Fan distances. What makes the Wasserstein distance so "special" is its representation (4.100) as the optimal value of a Linear Programming problem, responsible for efficient computational handling of this distance.

$q > 0$. Assuming that the latter is the case, let $\bar{f}$, $\bar{g}$ be an optimal solution to $(*)$. We claim that for all i, j it holds

$$\bar{g}_i + d_{ij} \geq \bar{g}_j. \qquad (**)$$

Indeed, assume that for some pair i_*, j_* the inequality opposite to $(**)$ holds true:

$$\bar{g}_{i_*} + d_{i_*j_*} < \bar{g}_{j_*},$$

implying that $i_* \neq j_*$, and let g' be obtained from $\bar{g}$ by updating a single entry, specifically, by setting $g'_{j_*} = \bar{g}_{i_*} + d_{i_*j_*}$. Let us verify that $\bar{f}$, g' is a feasible solution to $(*)$. Indeed, all we need to verify is that for all i it holds

$$\bar{f}_i - g'_{j_*} \leq d_{ij_*},$$

or, which is the same, that for all i it holds

$$\bar{f}_i - \bar{g}_{i_*} - d_{i_*j_*} \leq d_{ij_*},$$

which is nothing but

$$\bar{f}_i - \bar{g}_{i_*} \leq d_{ij_*} + d_{i_*j_*}$$

By the triangle inequality, the right-hand side in the latter relation is $\geq d_{ii_*}$, so that the inequality indeed is valid due to $\bar{f}_i - \bar{g}_{i_*} \leq d_{ii_*}$.

Thus, $\bar{f}, g'$ is a feasible solution to $(*)$; when passing from $\bar{f}, \bar{g}$ to $\bar{f}, g'$ we keep $\bar{f}$ and all entries in $\bar{g}$, except for the j_*-th one, intact, and strictly decrease the j_*-th entry in $\bar{g}$, thus strictly increasing the value of the objective in $(*)$ due to $q_{j_*} > 0$. The latter is impossible, since $\bar{f}, \bar{g}$ is an optimal solution to $(*)$. Thus, $(**)$ indeed holds true.

It remains to note that by feasibility of $\bar{f}, \bar{g}$ for $(*)$ we have $\bar{f}_i \leq \bar{g}_i$ for all i due to $d_{ii} = 0$, and by $(**)$ and due to $d_{ij} = d_{ji}$ the pair $\bar{g}, \bar{g}$ is feasible for $(*)$. When passing from the optimal solution $\bar{f}, \bar{g}$ to $(*)$ to the feasible solution $\bar{g}, \bar{g}$, we entrywise increase (perhaps nonstrictly) $\bar{f}$ due to the already proved relation $\bar{f} \leq \bar{g}$ and consequently increase (perhaps nonstrictly) the objective of $(*)$. Since actual increase is impossible ($\bar{f}, \bar{g}$ form an optimal solution to the problem!) and $p > 0$, we conclude that $\bar{f} = \bar{g}$, as claimed. □

Treating vector $f \in \mathbf{R}^n$ as a function on Ω, the value of the function at a point $i \in \Omega$ being f_i, (4.101) admits a very transparent interpretation: the Wasserstein distance $W(p, q)$ between probability distributions p, q is the maximum of inner products of $p - q$ and Lipschitz continuous, with constant 1 w.r.t. the metric d, functions f on Ω. When shifting f by a constant, the inner product remains intact (since $p - q$ is a vector with zero sum of entries). Therefore, denoting by

$$D = \max_{i,j} d_{ij}$$

the d-diameter of Ω, we have

$$W(p, q) = \max_f \left\{ f^T(p - q) : |f_i - f_j| \leq d_{ij},\ |f_i| \leq D/2\,\forall i, j \right\}, \qquad (4.102)$$

the reason being that every Lipschitz continuous, with constant 1 w.r.t. metric d, function f on Ω can be shifted by a constant to ensure $\|f\|_\infty \leq D/2$ (look what happens when the shift ensures that $\min_i f_i = -D/2$).

Representation (4.102) shows that the Wasserstein distance is generated by a norm on $\mathbf{R}^n$: for all probability distributions on Ω one has

$$W(p, q) = \|p - q\|_W,$$

where $\|\cdot\|_W$ is the *Wasserstein norm* on $\mathbf{R}^n$ given by

$$\begin{array}{rcl} \|x\|_W &=& \max\limits_{f\in\mathcal{B}_*} f^T x,\\ \mathcal{B}_* &=& \left\{u\in\mathbf{R}^n : u^T S_{ij} u \leq 1, 1\leq i\leq j\leq n\right\},\\ S_{ij} &=& \left\{\begin{array}{ll} d_{ij}^{-2}[e_i-e_j][e_i-e_j]^T, & 1\leq i<j\leq n,\\ 4D^{-2}e_ie_i^T, & 1\leq i=j\leq n,\end{array}\right.\end{array} \tag{4.103}$$

where $e_1,...,e_n$ are the standard basic orths in $\mathbf{R}^n$.

2) Let us equip n-element set $\Omega=\{1,...,d\}$ with the metric $d_{ij}=\left\{\begin{array}{ll}2, & i\neq j\\ 0, & i=j\end{array}\right.$. What is the associated Wasserstein norm?

Solution: This is just the $\|\cdot\|_1$-norm.

Note that the set $\mathcal{B}_*$ in (4.103) is the unit ball of the norm conjugate to $\|\cdot\|_W$, and as we see, this set is a basic ellitope. As a result, the estimation machinery developed in Chapter 4 is well suited for recovering discrete probability distributions in the Wasserstein norm. This observation motivates the concluding part of the exercise:

3) Consider the situation as follows: Given an $\times n$ column-stochastic matrix A and a $\nu\times n$ column-stochastic matrix B, we observe K samples ω_k, $1\leq k\leq K$, independent of each other and drawn from discrete probability distribution $Ax\in\mathbf{\Delta}_m$ (as always, $\mathbf{\Delta}_\nu\subset\mathbf{R}^\nu$ is the probabilistic simplex in $\mathbf{R}^\nu$), $x\in\mathbf{\Delta}_n$ being an unknown "signal" underlying observations; realizations of ω_k are identified with respective vertices $f_1,...,f_m$ of $\mathbf{\Delta}_m$. Our goal is to use the observations to estimate the distribution $Bx\in\mathbf{\Delta}_\nu$. We are given a metric d on the set $\Omega_\nu=\{1,2,...,\nu\}$ of indices of entries in Bx, and measure the recovery error in the Wasserstein norm $\|\cdot\|_W$ associated with d.
Build an explicit convex optimization problem responsible for "presumably good" linear recovery of the form

$$\widehat{x}_H=\frac{1}{K}H^T\sum_{k=1}^K\omega_k.$$

Solution: The estimate in question is given by the H-component of the optimal solution to the convex optimization problem

$$\begin{array}{rcl}\mathrm{Opt} &=& \min\limits_{H,G,\lambda}\left\{\max\limits_{j\leq n}\|\mathrm{Col}_j[B-H^TA]\|_W+K^{-1}\max\limits_{i\leq n}[A^T\mathrm{dg}(G)]_i+\sum\limits_{1\leq i\leq j\leq\nu}\lambda_{ij}:\right.\\ && \left.\lambda_{ij}\geq 0,\left[\begin{array}{c|c}G & \frac{1}{2}H\\ \hline \frac{1}{2}H^T & \sum_{1\leq i\leq j\leq\nu}\lambda_{ij}S_{ij}\end{array}\right]\succeq 0\right\},\end{array}$$

where $\mathrm{dg}(G)$ is the diagonal of G, and S_{ij} are the matrices specified in (4.103), with ν in the role of n; cf. Simple Case in Exercise 4.17. The risk $\mathrm{Risk}_{\|\cdot\|_W}[\widehat{x}|\mathbf{\Delta}_n]$ of this estimate does not exceed Opt.

Exercise 4.19. [follow-up to Exercise 4.17] In Exercise 4.17, we have built a "presumably good" linear estimate $\widehat{x}_{H_*}(\cdot)$—see (4.98)—yielded by the H-component H_* of an optimal solution to problem (4.99). The optimal value Opt in this problem is an upper bound on the risk $\mathrm{Risk}_{\|\cdot\|}[\widehat{x}_{H_*}|\mathcal{X}]$ (here and in what follows we use the same notation and impose the same assumptions as in Exercise 4.17). Recall that $\mathrm{Risk}_{\|\cdot\|}$ is the worst, w.r.t. signals $x\in\mathcal{X}$ underlying our observations, expected norm of the recovery error. It makes sense also to provide upper bounds on the probabilities of deviations of error's magnitude from its expected value, and this is the problem we consider here, cf. Exercise 4.14.

1) Prove the following

Lemma 4.33 *Let $Q \in \mathbf{S}^m_+$, let K be a positive integer, and let $p \in \mathbf{\Delta}_m$. Let, further, $\omega^K = (\omega_1, ..., \omega_K)$ be i.i.d. random vectors, with ω_k taking the value e_j ($e_1, ..., e_m$ are the standard basic orths in $\mathbf{R}^m$) with probability p_j. Finally, let $\xi_k = \omega_k - \mathbf{E}\{\omega_k\} = \omega_k - p$, and $\widehat{\xi} = \frac{1}{K}\sum_{k=1}^K \xi_k$. Then for every $\epsilon \in (0,1)$ it holds*

$$\mathrm{Prob}\left\{\|\widehat{\xi}\|_2^2 \leq \frac{12\ln(2m/\epsilon)}{K}\right\} \geq 1-\epsilon.$$

Hint: use the classical
Bernstein inequality: *Let $X_1, ..., X_K$ be independent zero mean random variables taking values in $[-M, M]$, and let $\sigma_k^2 = \mathbf{E}\{X_k^2\}$. Then for every $t \geq 0$ one has*

$$\mathrm{Prob}\left\{\sum\nolimits_{k=1}^K X_k \geq t\right\} \leq \exp\left\{-\frac{t^2}{2[\sum_k \sigma_k^2 + \frac{1}{3}Mt]}\right\}.$$

<u>Solution:</u> Let us fix $p \in \mathbf{\Delta}_m$, and let ξ_k, $1 \leq k \leq K$, be the random vectors associated with p, as explained in the premise of Lemma. For $i \leq m$ fixed, denote by X_{ki} the i-th entry in ξ_k, so that X_{ki} are random reals taking values in the segment $[-p_i, 1-p_i] \subset [-1,1]$ with second moments $p_i(1-p_i) \leq p_i$ and independent across k. Applying the Bernstein inequality, we conclude that for every $t_i \geq 0$ it holds

$$\begin{array}{rcl}\mathrm{Prob}\{[\widehat{\xi}]_i > t_i\} & = & \mathrm{Prob}\left\{\sum_k X_{ki} > Kt_i\right\} \leq \exp\left\{-\frac{K^2t_i^2}{2[Kp_i + \frac{1}{3}Kt_i]}\right\} \\ & = & \exp\left\{-\frac{Kt_i^2}{2[p_i+\frac{1}{3}t_i]}\right\},\end{array}$$

and similarly $\mathrm{Prob}\{[\widehat{\xi}]_i < -t_i\} \leq \exp\left\{-\frac{Kt_i^2}{2[p_i+\frac{1}{3}t_i]}\right\}$, implying that

$$\mathrm{Prob}\{|[\widehat{\xi}]_i| > t_i\} \leq 2\exp\left\{-\frac{Kt_i^2}{2[p_i+\frac{1}{3}t_i]}\right\}.$$

Now let us specify t_i in such a way that the right hand side in the above inequality is equal to ϵ/m. Denoting $\alpha = \frac{2\ln(2m/\epsilon)}{K}$, we get

$$t_i = \frac{\alpha}{6} + \sqrt{\frac{\alpha^2}{36} + \alpha p_i} \leq \frac{\alpha}{3} + \sqrt{\alpha p_i}.$$

With our choice of t_i we have

$$\forall i \leq m : \mathrm{Prob}\{|[\widehat{\xi}]_i| > t_i\} \leq \frac{\epsilon}{m},$$

so that the probability of the event $\mathcal{E} = \{\widehat{\xi} : |[\widehat{\xi}]_i| \leq t_i \,\forall i\}$ is at least $1-\epsilon$. Now let

$$I = \{i : p_i \leq \frac{\alpha}{9}\}.$$

Note that $t_i \leq 2\sqrt{\alpha p_i}$ when $i \notin I$ and $t_i \leq 2\alpha/3$ when $i \in I$. Note also that by the origin of $\widehat{\xi}$ we have $\|\widehat{\xi}\|_1 \leq 2$. Consequently, when $\mathcal{E}$ takes place we have

$$\begin{array}{rcl}\sum_{i=1}^m [\widehat{\xi}]_i^2 & = & \sum_{i\in I}[\widehat{\xi}]_i^2 + \sum_{i\notin I}[\widehat{\xi}]_i^2 \leq \sum_{i\in I}|[\widehat{\xi}]_i|t_i + \sum_{i\notin I} t_i^2 \\ & \leq & 2\max_{i\in I} t_i + \sum_{i\notin I} t_i^2 \leq \frac{4\alpha}{3} + 4\sum_{i\notin I}\alpha p_i \leq 6\alpha.\end{array}$$

Substituting the expression for α, we arrive at the desired result. □

2) Consider the situation described in Exercise 4.17 with $\mathcal{X} = \mathbf{\Delta}_n$, specifically,

- Our observation is a sample $\omega^K = (\omega_1, ..., \omega_K)$ with i.i.d. components $\omega_k \sim Ax$, where $X \in \mathbf{\Delta}_n$ is an unknown n-dimensional probabilistic vector, A is an $m \times n$ stochastic matrix (nonnegative matrix with unit column sums), and $\omega \sim Ax$ means that ω is a random vector taking value e_i (e_i are standard basic orths in $\mathbf{R}^m$) with probability $[Ax]_i$, $1 \leq i \leq m$;
- Our goal is to recover Bx in a given norm $\|\cdot\|$; here B is a given $\nu \times n$ matrix;
- We assume that the unit ball $\mathcal{B}_*$ of the norm $\|\cdot\|_*$ conjugate to $\|\cdot\|$ is a spectratope,

$$\mathcal{B}_* = \{u = My, y \in \mathcal{Y}\},\ \mathcal{Y} = \{y \in \mathbf{R}^N : \exists r \in \mathcal{R} : S_\ell^2[y] \preceq r_\ell I_{f_\ell}, \ell \leq L\}.$$

Our goal is to build a presumably good linear estimate

$$\widehat{x}_H(\omega^K) = H^T \widehat{\omega}[\omega^K],\ \widehat{\omega}[\omega^K] = \frac{1}{K}\sum_k \omega_k.$$

Prove the following

Proposition 4.34 *Let H, Θ, Υ be a feasible solution to the convex optimization problem*

$$\min_{H,\Theta,\Upsilon} \left\{\Phi(H) + \phi_{\mathcal{R}}(\lambda[\Upsilon]) + \Gamma(\Theta)/K : \Upsilon = \{\Upsilon_\ell \succeq 0, \ell \leq L\} \atop \left[\begin{array}{c|c} \Theta & \frac{1}{2}HM \\ \hline \frac{1}{2}M^T H^T & \sum_\ell \mathcal{S}_\ell^*[\Upsilon_\ell] \end{array}\right] \succeq 0 \right\} \tag{4.104}$$

where

$$\Phi(H) = \max_{j \leq n} \|\mathrm{Col}_j[B - H^T A]\|,\ \ \Gamma(\Theta) = \max_{x \in \mathbf{\Delta}_n} \mathrm{Tr}(\mathrm{Diag}\{Ax\}\Theta).$$

Then

(i) For every $x \in \mathbf{\Delta}_n$ it holds

$$\begin{array}{rcl} \mathbf{E}_{\omega^K}\left\{\|Bx - \widehat{x}_H(\omega^K)\|\right\} & \leq & \Phi(H) + 2K^{-1/2}\sqrt{\phi_{\mathcal{R}}(\lambda[\Upsilon])\Gamma(\Theta)} \\ & \Big[\leq & \Phi(H) + \phi_{\mathcal{R}}(\lambda[\Upsilon]) + \Phi(H) + \Gamma(\Theta)/K \Big] \end{array} \tag{4.105}$$

(ii) Let $\epsilon \in (0,1)$. For every $x \in \mathbf{\Delta}_n$ with

$$\gamma = 2\sqrt{3\ln(2m/\epsilon)}$$

one has

$$\begin{array}{l} \mathrm{Prob}_{\omega^K}\left\{\|Bx - \widehat{x}_H(\omega^K)\| \leq \Phi(H) + 2\gamma K^{-1/2}\sqrt{\phi_{\mathcal{R}}(\lambda[\Upsilon])\|\Theta\|_{Sh,\infty}}\right\} \\ \geq 1 - \epsilon. \end{array} \tag{4.106}$$

Solution: (i): Let us fix $x \in \mathcal{X}$, let $p = Ax$, so that $p \in \mathbf{\Delta}_m$, and let P be the distribution of i.i.d. sample $\omega^K = (\omega_1, ..., \omega_K)$ with $\omega_k \sim p$. Setting $\xi_k = \omega_k - p$, $\widehat{\xi} = \frac{1}{K}\sum_k \xi_k$, observe that $\xi_1, ..., \xi_K$ are i.i.d. zero mean with covariance matrix $\mathrm{Diag}\{p\} - pp^T$, whence $\widehat{\xi}$ is zero mean with covariance matrix $Q_x = \frac{1}{K}[\mathrm{Diag}\{p\} - pp^T] \preceq \frac{1}{K}\mathrm{Diag}\{p\}$. Besides this, $\widehat{\omega}[\omega^K] = p + \widehat{\xi}$. We now have

$$\begin{array}{rcl} \|Bx - \widehat{x}_H(\omega^K)\| & = & \|Bx - H^T(p + \widehat{\xi})\| = \|Bx - H^T(Ax + \widehat{\xi})\| \\ & \leq & \|[B - H^T A]x\| + \|H^T\widehat{\xi}\| \\ & \leq & \Phi(H) + \|H^T\widehat{\xi}\| \text{ [recall that } x \in \mathbf{\Delta}_n]. \end{array} \tag{6.28}$$

Next, from the last semidefinite constraint in (4.104) by exactly the same argument as in the proof of Proposition 4.31 (see derivation of (6.26) in Solution to Exercise 4.14) one has

$$\|H^T\widehat{\xi}\| \leq 2\sqrt{\phi_{\mathcal{R}}(\lambda[\Upsilon])[\widehat{\xi}^T\Theta\widehat{\xi}]}. \tag{6.29}$$

Taking expectations and invoking Cauchy's inequality, we get

$$\mathbf{E}\left\{\|H^T\widehat{\xi}\|\right\} \leq 2\sqrt{\phi_{\mathcal{R}}(\lambda[\Upsilon])\mathbf{E}\{\widehat{\xi}^T\Theta\widehat{\xi}\}} \leq 2K^{-1/2}\sqrt{\phi_{\mathcal{R}}(\lambda[\Upsilon])\mathrm{Tr}(\mathrm{Diag}\{Ax\}\Theta)},$$

where the concluding inequality is due to $\Theta \succeq 0$ and $\widehat{\xi}$ being zero mean with covariance matrix $\preceq K^{-1}\mathrm{Diag}\{Ax\}$. Thus,

$$\mathbf{E}\left\{\|H^T\widehat{\xi}\|\right\} \leq 2K^{-1/2}\sqrt{\phi_{\mathcal{R}}(\lambda[\Upsilon])\Gamma(\Theta)},$$

which combines with (6.28) to yield the first inequality in (4.105). The second inequality in (4.105) is evident. □

(ii): In the notation from the proof of (i), by (6.28), (6.29) we have

$$\|Bx - \widehat{x}_H(\omega^K)\| \leq \Phi(H) + 2\sqrt{\phi_{\mathcal{R}}(\lambda[\Upsilon])[\widehat{\xi}^T\Theta\widehat{\xi}]}$$

whence

$$\|Bx - \widehat{x}_H(\omega^K)\| \leq \Phi(H) + 2\|\widehat{\xi}\|_2\sqrt{\phi_{\mathcal{R}}(\lambda[\Upsilon])\|\Theta\|_{\mathrm{Sh},\infty}}.$$

It remains to apply Lemma 4.33. □

3) Look what happens when $\nu = m = n$, A and B are the unit matrices, and $H = I$, i.e., we want to understand how good is the recovery of a discrete probability distribution by empirical distribution of a K-element i.i.d. sample drawn from the original distribution. Take, as $\|\cdot\|$, the norm $\|\cdot\|_p$ with $p \in [1,2]$, and show that for every $x \in \mathbf{\Delta}_n$ and every $\epsilon \in (0,1)$ one has

$$\begin{array}{l}\forall (x \in \mathbf{\Delta}_n): \\ \mathbf{E}\left\{\|x - \widehat{x}_I(\omega^K)\|_p\right\} \leq n^{\frac{1}{p}-\frac{1}{2}}K^{-\frac{1}{2}} \\ \mathrm{Prob}\left\{\|x - \widehat{x}_I(\omega^K)\|_p \leq 2\sqrt{3\ln(2n/\epsilon)}n^{\frac{1}{p}-\frac{1}{2}}K^{-\frac{1}{2}}\right\} \geq 1-\epsilon\end{array}.$$

<u>Solution:</u> In the situation in question it is immediately seen that the Υ_ℓ are just reals v_ℓ, $\phi_{\mathcal{R}}(\lambda[\Upsilon]) = \|v\|_q$, $q = \frac{p}{2-p}$, and problem (4.104) *with H set to I* boils down to

$$\min_{\Theta,v}\left\{\|v\|_q + K^{-1}\max_{x\in\mathbf{\Delta}_n}\mathrm{Tr}(\mathrm{Diag}\{x\}\Theta) : v \geq 0, \left[\begin{array}{c|c}\Theta & \frac{1}{2}I \\ \hline \frac{1}{2}I & \mathrm{Diag}\{v\}\end{array}\right]\right\}, \; q = \frac{p}{2-p}.$$

As far as the bounds (4.105) and (4.106) are concerned, we clearly lose nothing when restricting ourselves to $\Theta = \frac{1}{2}I$ and $v = \frac{1}{2}[1;...;1]$, so that the bounds become

$$\begin{array}{l}\forall (x \in \mathbf{\Delta}_n): \\ \mathbf{E}\left\{\|x - \widehat{x}_I(\omega^K)\|_p\right\} \leq n^{\frac{1}{p}-\frac{1}{2}}K^{-\frac{1}{2}}, \\ \mathrm{Prob}\left\{\|x - \widehat{x}_I(\omega^K)\|_p \leq 2\sqrt{3\ln(2n/\epsilon)}n^{\frac{1}{p}-\frac{1}{2}}K^{-\frac{1}{2}}\right\} \geq 1-\epsilon,\end{array}$$

as claimed. □

Exercise 4.20. [follow-up to Exercise 4.17] Consider the situation as follows. A retailer sells n items by offering customers via internet bundles of $m < n$ items, so that an offer is an m-element subset B of the set $S = \{1, ..., n\}$ of the items. A customer has private preferences represented by a subset P of S—the customer's *preference set*. We assume that if an offer B intersects with the preference set P of a customer, the latter buys an item drawn at random from the uniform distribution on $B \cap P$, and if $B \cap P = \emptyset$, the customer declines the offer. In the pilot stage we are interested in, the seller learns the market by making offers to K customers. Specifically, the seller draws the k-th customer, $k \leq K$, at random from the uniform distribution on the population of customers, and makes the selected customer an offer drawn at random from the uniform distribution on the set $\mathcal{S}_{m,n}$ of all m-item offers. What is observed in the k-th experiment is the item, if any, bought by the customer, and we want to make statistical inferences from these observations.

The outlined observation scheme can be formalized as follows. Let $\mathcal{S}$ be the set of all subsets of the n-element set, so that $\mathcal{S}$ is of cardinality $N = 2^n$. The population of customers induces a probability distribution p on $\mathcal{S}$: for $P \in \mathcal{S}$, p_P is the fraction of customers with the preference set being P; we refer to p as to the *preference distribution*. An outcome of a single experiment can be represented by a pair (ι, B), where $B \in \mathcal{S}_{m,n}$ is the offer used in the experiment, and ι is either 0 ("nothing is bought," $P \cap B = \emptyset$), or a point from $P \cap B$, the item which was bought, when $P \cap B \neq \emptyset$. Note that A_P is a probability distribution on the $(M = (m+1)\binom{n}{m})$-element set $\Omega = \{(\iota, B)\}$ of possible outcomes. As a result, our observation scheme is fully specified by $M \times N$ column-stochastic matrix A known to us with the columns A_P indexed by $P \in \mathcal{S}$. When a customer is drawn at random from the uniform distribution on the population of customers, the distribution of the outcome clearly is Ap, where p is the (unknown) preference distribution. Our inferences should be based on the K-element sample $\omega^K = (\omega_1, ..., \omega_K)$, with $\omega_1, .., \omega_K$ drawn, independently of each other, from the distribution Ap.

Now we can pose various inference problems, e.g., that of estimating p. We, however, intend to focus on a simpler problem—one of recovering Ap. In terms of our story, this makes sense: when we know Ap, we know, e.g., what the probability is for every offer to be "successful" (something indeed is bought) and/or to result in a specific profit, etc. With this knowledge at hand, the seller can pass from a "blind" offering policy (drawing an offer at random from the uniform distribution on the set $\mathcal{S}_{m,n}$) to something more rewarding.

Now comes the exercise:

1. Use the results of Exercise 4.17 to build a "presumably good" linear estimate

$$\widehat{x}_H(\omega^K) = H^T \left[\frac{1}{K} \sum_{k=1}^{K} \omega_k \right]$$

of Ap (as always, we encode observations ω, which are elements of the M-element set Ω, by standard basic orths in $\mathbf{R}^M$). As the norm $\|\cdot\|$ quantifying the recovery error, use $\|\cdot\|_1$ and/or $\|\cdot\|_2$. In order to avoid computational difficulties, use small m and n (e.g., $m = 3$ and $n = 5$). Compare your results with those for the "straightforward" estimate $\frac{1}{K}\sum_{k=1}^{K} \omega_k$ (the empirical distribution of $\omega \sim Ap$).
2. Assuming that the "presumably good" linear estimate outperforms the straightforward one, how could this phenomenon be explained? Note that we have no nontrivial a priori information on p!

Solution: We have no a priori information on p—all we know is that $p \in \mathbf{\Delta}_N$. Consequently, we are in the Simple case of Exercise 4.17, and the optimization problem (P) (see Exercise 4.17) responsible for building a presumably good linear estimate, $\|\cdot\|$ being $\|\cdot\|_r$, with $r \in \{1, 2\}$, after immediate simplification becomes

— in the case of $r = 1$:

$$\mathrm{Opt}_1 = \min_{H,\lambda\in\mathbf{R}^M,\Theta}\left\{\max_{P\in\mathcal{S}}\|A_P - H^TA_P\|_1 + \sum_{k=1}^M \lambda_k + \tfrac{1}{K}\max_{P\in\mathcal{S}}\mathrm{Tr}(\mathrm{Diag}\{A_P\}\Theta): \lambda\geq 0, \left[\begin{array}{c|c}\Theta & \frac{1}{2}H\\ \hline \frac{1}{2}H^T & \mathrm{Diag}\{\lambda\}\end{array}\right]\succeq 0\right\}, \quad (P_1)$$

— in the case of $r = 2$:

$$\mathrm{Opt}_2 = \min_{H,\lambda\in\mathbf{R},\Theta}\left\{\max_{P\in\mathcal{S}}\|A_P - H^TA_P\|_2 + \lambda + \tfrac{1}{K}\max_{P\in\mathcal{S}}\mathrm{Tr}(\mathrm{Diag}\{A_P\}\Theta): \lambda\geq 0, \left[\begin{array}{c|c}\Theta & \frac{1}{2}H\\ \hline \frac{1}{2}H^T & \lambda I_M\end{array}\right]\succeq 0\right\}. \quad (P_2)$$

Note that Opt is an upper bound on the risk of the linear estimate yielded by the H-component of the optimal solution.

When $m = 3$, $n = 5$ and $K = 1000$, the computation yields

$$\mathrm{Opt}_1 \approx 0.1414, \ \mathrm{Opt}_2 \approx 0.0230.$$

The results of 100 simulation runs, $K = 1000$ observations each, with a common for all 100 runs randomly selected p, are as follows:

	min	mean	median	max
$\|\widehat{x}(\cdot) - Ap\|_1$	0.0605/0.0794	0.1008/0.1232	0.0994/0.1227	0.1643/0.1867
$\|\widehat{x}(\cdot) - Ap\|_2$	0.0118/0.0190	0.0201/0.0298	0.0195/0.0300	0.0317/0.0442
$\|\widehat{x}(\cdot) - Ap\|_\infty$	0.0042/0.0073	0.0081/0.0144	0.0077/0.0137	0.0175/0.0253
$\|\widehat{x}(\cdot) - Ap\|_1$	0.0555/0.0748	0.1084/0.1101	0.1080/0.1108	0.1620/0.1600
$\|\widehat{x}(\cdot) - Ap\|_2$	0.0119/0.0210	0.0218/0.0309	0.0221/0.0309	0.0314/0.0412
$\|\widehat{x}(\cdot) - Ap\|_\infty$	0.0049/0.0100	0.0088/0.0162	0.0086/0.0156	0.0137/0.0282

The first number in a cell corresponds to the linear estimate yielded by optimal solution to (P_1) (top half of the table) or (P_2) (bottom half of the table). The second number in a cell corresponds to estimating Ap by empirical distribution of the observations.

We see that the near-optimal linear estimate indeed outperforms the straightforward one. The reason is simple: while we have no nontrivial information on p, we do have information on Ap—we know that this distribution belongs to the convex hull of vectors A_P, $P \in \mathcal{S}$, and utilize this information when building the linear estimate.

Exercise 4.21. [Poisson Imaging] *The Poisson Imaging Problem* is to recover an unknown signal observed via the Poisson observation scheme. More specifically, assume that our observation is a realization of random vector $\omega \in \mathbf{R}^m_+$ with Poisson entries $\omega_i = \mathrm{Poisson}([Ax]_i)$ independent of each other. Here A is a given entrywise nonnegative $m \times n$ matrix, and x is an unknown signal known to belong to a given compact convex subset $\mathcal{X}$ of $\mathbf{R}^n_+$. Our goal is to recover in a given norm $\|\cdot\|$ the linear image Bx of x, where B is a given $\nu \times n$ matrix.

We assume in the sequel that $\mathcal{X}$ is a subset cut off the n-dimensional probabilistic simplex $\mathbf{\Delta}_n$ by a collection of linear equality and inequality constraints. The assumption $\mathcal{X} \subset \mathbf{\Delta}_n$ is not too restrictive. Indeed, assume that we know in advance a linear inequality $\sum_i \alpha_i x_i \leq 1$ with positive coefficients which is valid on $\mathcal{X}$.[17] Introducing slack variable s given by $\sum_i \alpha_i x_i + s =$

[17]For example, in PET—see Section 2.4.3.2—where x is the density of a radioactive tracer injected into the patient taking the PET procedure, we know in advance the total amount $\sum_i v_i x_i$ of the tracer, v_i being the volumes of voxels.

1 and passing from signal x to the new signal $[\alpha_1 x_1; ...; \alpha_n x_n; s]$, after a straightforward modification of matrices A and B, we arrive at the situation where $\mathcal{X}$ is a subset of the probabilistic simplex.

Our goal in the sequel is to build a presumably good linear estimate $\widehat{x}_H(\omega) = H^T\omega$ of Bx. As in Exercise 4.17, we start with upper-bounding the risk of a linear estimate. When representing

$$\omega = Ax + \xi_x,$$

we arrive at zero mean observation noise ξ_x with $[\xi_x]_i = \omega_i - [Ax]_i$ independent of each other entries and covariance matrix $\mathrm{Diag}\{Ax\}$. We now can upper-bound the risk of a linear estimate $\widehat{x}_H(\cdot)$ in the same way as in Exercise 4.17. Specifically, denoting by $\Pi_{\mathcal{X}}$ the set of all diagonal matrices $\mathrm{Diag}\{Ax\}$, $x \in \mathcal{X}$, and by $P_{i,x}$ the Poisson distribution with parameter $[Ax]_i$, we have

$$\begin{array}{rcl}\mathrm{Risk}_{\|\cdot\|}[\widehat{x}_H|\mathcal{X}] &=& \sup_{x\in\mathcal{X}} \mathbf{E}_{\omega\sim P_{1,x}\times\ldots\times P_{m,x}}\left\{\|Bx - H^T\omega\|\right\}\\ &=& \sup_{x\in\mathcal{X}} \mathbf{E}_{\xi_x}\left\{\|[Bx - H^TA]x - H^T\xi_x\|\right\}\\ &\leq& \underbrace{\sup_{x\in\mathcal{X}} \|[B-H^TA]x\|}_{\Phi(H)} + \underbrace{\sup_{\xi:\mathrm{Cov}[\xi]\in\Pi_{\mathcal{X}}} \mathbf{E}_\xi\left\{\|H^T\xi\|\right\}}_{\Psi^{\mathcal{X}}(H)}.\end{array}$$

In order to build a presumably good linear estimate, it suffices to build efficiently computable upper bounds $\overline{\Phi}(H)$ on $\Phi(H)$ and $\overline{\Psi}^{\mathcal{X}}(H)$ on $\Psi^{\mathcal{X}}(H)$ convex in H, and then take as H an optimal solution to the convex optimization problem

$$\mathrm{Opt} = \min_H \left[\overline{\Phi}(H) + \overline{\Psi}^{\mathcal{X}}(H)\right].$$

As in Exercise 4.17, assume from now on that $\|\cdot\|$ is an absolute norm, and the unit ball $\mathcal{B}_*$ of the conjugate norm is a spectratope,

$$\mathcal{B}_* := \{u : \|u\|_* \leq 1\} = \{u : \exists r \in \mathcal{R}, y : u = My, S_\ell^2[y] \preceq r_\ell I_{f_\ell}, \ell \leq L\}.$$

Observe that

- In order to build $\overline{\Phi}$, we can use exactly the same techniques as those developed in Exercise 4.17. Indeed, as far as building $\overline{\Phi}$ is concerned, the only difference with the situation of Exercise 4.17 is that in the latter, A was a column-stochastic matrix, while now A is just an entrywise nonnegative matrix. Note, however, that when upper-bounding Φ in Exercise 4.17, we never used the fact that A is column-stochastic.
- In order to upper-bound $\Psi^{\mathcal{X}}$, we can use the bound (4.40) of Exercise 4.17.

The bottom line is that in order to build a presumably good linear estimate, we need to solve the convex optimization problem

$$\mathrm{Opt} = \min_{H,\Upsilon,\Theta} \left\{\overline{\Phi}(H) + \phi_{\mathcal{R}}(\lambda[\Upsilon]) + \Gamma_{\mathcal{X}}(\Theta) : \Upsilon = \{\Upsilon_\ell \succeq 0, \ell \leq L\} \atop \left[\begin{array}{c|c}\Theta & \frac{1}{2}HM\\ \hline \frac{1}{2}M^TH^T & \sum_\ell S_\ell^*[\Upsilon_\ell]\end{array}\right] \succeq 0\right\} \qquad (P)$$

where

$$\Gamma_{\mathcal{X}}(\Theta) = \max_{x\in\mathcal{X}} \mathrm{Tr}(\mathrm{Diag}\{Ax\}\Theta)$$

(cf. problem (4.99)) with $\overline{\Phi}$ yielded by any construction from Exercise 4.17, e.g., the least conservative Combined upper bound on Φ.

What in our present situation differs significantly from the situation of Exercise 4.17, are the bounds on probabilities of large deviations (for Discrete o.s., established in Exercise 4.19), and the goal of what follows is to establish these bounds for Poisson Imaging.

Here is what you are supposed to do:

1. Let $\omega \in \mathbf{R}^m$ be a random vector with independent entries $\omega_i \sim$ Poisson(μ_i), and let $\mu = [\mu_1; ...; \mu_m]$. Prove that whenever $h \in \mathbf{R}^m$, $\gamma > 0$, and $\delta \geq 0$, one has

$$\ln\left(\mathrm{Prob}\{h^T\omega > h^T\mu + \delta\}\right) \leq \sum_i [\exp\{\gamma h_i\} - 1]\mu_i - \gamma h^T\mu - \gamma\delta. \tag{4.107}$$

2. Taking for granted (or see, e.g., [178]) that $e^x - x - 1 \leq \frac{x^2}{2(1-x/3)}$ when $|x| < 3$, prove that in the situation of item 1 one has for $t > 0$:

$$0 \leq \gamma < \frac{3}{\|h\|_\infty} \Rightarrow \ln\left(\mathrm{Prob}\{h^T\omega > h^T\mu + t\}\right) \leq \frac{\gamma^2 \sum_i h_i^2 \mu_i}{2(1 - \gamma\|h\|_\infty/3)} - \gamma t. \tag{4.108}$$

Derive from the latter fact that

$$\mathrm{Prob}\left\{h^T\omega > h^T\mu + \delta\right\} \leq \exp\left\{-\frac{\delta^2}{2[\sum_i h_i^2\mu_i + \|h\|_\infty \delta/3]}\right\}, \tag{4.109}$$

and conclude that

$$\mathrm{Prob}\left\{|h^T\omega - h^T\mu| > \delta\right\} \leq 2\exp\left\{-\frac{\delta^2}{2[\sum_i h_i^2\mu_i + \|h\|_\infty \delta/3]}\right\}. \tag{4.110}$$

3. Extract from (4.110) the following

Proposition 4.35 *In the situation and under the assumptions of Exercise 4.21, let* Opt *be the optimal value, and H, Υ, Θ be a feasible solution to problem (P). Whenever $x \in \mathcal{X}$ and $\epsilon \in (0,1)$, denoting by P_x the distribution of observations stemming from x (i.e., the distribution of random vector ω with independent entries $\omega_i \sim \mathrm{Poisson}([Ax]_i)$), one has*

$$\begin{array}{rcl}\mathbf{E}_{\omega\sim P_x}\{\|Bx - \widehat{x}_H(\omega)\|\} & \leq & \overline{\Phi}(H) + 2\sqrt{\phi_{\mathcal{R}}(\lambda[\Upsilon])\mathrm{Tr}(\mathrm{Diag}(Ax\}\Theta)} \\ & \leq & \overline{\Phi}(H) + \phi_{\mathcal{R}}(\lambda[\Upsilon]) + \Gamma_{\mathcal{X}}(\Theta)\end{array} \tag{4.111}$$

and

$$\begin{array}{l}\mathrm{Prob}_{\omega\sim P_x}\left\{\|Bx - \widehat{x}_H(\omega)\| \leq \overline{\Phi}(H)\right. \\ \left. 4\sqrt{\frac{2}{9}\ln^2(2m/\epsilon)\mathrm{Tr}(\Theta) + \ln(2m/\epsilon)\mathrm{Tr}(\mathrm{Diag}\{Ax\}\Theta)}\sqrt{\phi_{\mathcal{R}}(\lambda[\Upsilon])}\right\} \geq 1 - \epsilon.\end{array} \tag{4.112}$$

Note that in the case of $[Ax]_i \geq 1$ for all $x \in \mathcal{X}$ and all i we have $\mathrm{Tr}(\Theta) \leq \mathrm{Tr}(\mathrm{Diag}\{Ax\}\Theta)$, so that in this case the P_x-probability of the event

$$\left\{\omega : \|Bx - \widehat{x}_H(\omega)\| \leq \overline{\Phi}(H) + O(1)\ln(2m/\epsilon)\sqrt{\phi_{\mathcal{R}}(\lambda[\Upsilon])\Gamma_{\mathcal{X}}(\Theta)}\right\}$$

is at least $1 - \epsilon$.

Solution: **1.** Taking into account that $\omega_i \sim \mathrm{Poisson}(\mu_i)$ are independent across i, we have

$$\begin{array}{rcl}\mathbf{E}\left\{\exp\{\gamma h^T\omega\}\right\} & = & \prod_i \mathbf{E}\left\{\gamma h_i\omega_i\right\} = \prod_i \exp\{[\exp\{\gamma h_i\} - 1]\mu_i\} \\ & = & \exp\{\sum_i [\exp\{\gamma h_i\} - 1]\mu_i\},\end{array}$$

whence by the Chebyshev inequality

$$\begin{array}{rcl}\mathrm{Prob}\{h^T\omega > h^T\mu + \delta\} & = & \mathrm{Prob}\{\gamma h^T\omega > \gamma h^T\mu + \gamma\delta\} \\ & \leq & \mathbf{E}\left\{\exp\{\gamma h^T\omega\}\right\}\exp\{-\gamma h^T\mu - \gamma\delta\} \\ & \leq & \exp\{\sum_i [\exp\{\gamma h_i\} - 1]\mu_i - \gamma h^T\mu - \gamma\delta\},\end{array}$$

and the required bound follows. □

2. Relation (4.108) is an immediate consequence of (4.107) combined with the inequality $\exp\{\gamma h_i\} \leq 1 + \gamma h_i + \frac{\gamma^2 h_i^2}{2(1-\gamma\|h\|_\infty/3)}$ which takes place when $|\gamma|\|h\|_\infty < 3$. Assuming w.l.o.g. that $\sum_i h_i^2 \mu_i > 0$ (otherwise (4.109) is evident, since then $h^T\omega \equiv h^T\mu = 0$), with $\gamma = \frac{t}{\sum_i h_i^2\mu_i + \|h\|_\infty t/3}$ (4.108) results in (4.109). The latter relation combines with a similar relation for $-h$ in the role of h to yield (4.110). □

3. Let $x \in \mathcal{X}$ and $\epsilon \in (0,1)$, and let $\mu = Ax$ and $\alpha = \ln(2m/\epsilon)$. Let also $\Theta = U\mathrm{Diag}\{\lambda_1, ..., \lambda_m\}U^T$ be the eigenvalue decomposition of Θ, and let u_p be p-th column in U and $h^p = \sqrt{\lambda_p} u_p$, $1 \leq p \leq m$, so that

$$\Theta = \sum_{p=1}^m h^p [h^p]^T, \quad \rho_p := \|h^p\|_\infty \leq \|h^p\|_2 = \sqrt{\lambda_p},$$

and

$$\vartheta := \sum_i \Theta_{ii}\mu_i = \sum_i \sum_p [h_i^p]^2 \mu_i = \sum_p \underbrace{\left[\sum_i [h_i^p]^2 \mu_i\right]}_{\sigma_p}.$$

Let P_x be the distribution of observation ω stemming from x, and let P'_x be the distribution of $\xi := \omega - \mu = \omega - Ax$ induced by the distribution P_x of ω, so that ξ is zero mean with the covariance matrix $\mathrm{Diag}\{Ax\}$. For $1 \leq p \leq m$, let

$$\delta_p = \tfrac{2}{3}\alpha\rho_p + \sqrt{2\alpha\sigma_p},$$

so that for every $p \leq m$ it holds $\delta_p \geq 0$ and

$$\frac{\delta_p^2}{2\sum_i [h_i^p]^2\mu_i + \frac{2}{3}\|h^p\|_\infty \delta_p} = \frac{\delta_p^2}{2\sigma_p + \frac{2}{3}\rho_p\delta_p} \geq \alpha.$$

Invoking (4.110) with $\delta = \delta_p$, $h = h^p$ we get

$$\begin{array}{rcl}\mathrm{Prob}_{\omega\sim P_x}\{|[h^p]^T\omega - [h^p]^T\mu| > \delta_p\} & \leq & 2\exp\left\{-\dfrac{\delta_p^2}{2[\sum_i [h_i^p]^2\mu_i + \|h^p\|_\infty\delta_p/3]}\right\} \\ & \leq & 2\mathrm{e}^{-\alpha} = \dfrac{\epsilon}{m},\end{array}$$

so that the P'_x-probability of the event

$$\Xi = \Xi_\Theta := \{\xi : |[h^p]^T\xi| \leq \delta_p, \ 1 \leq p \leq m\}$$

is at least $1 - \epsilon$. When this event takes place, we have

$$\begin{array}{rcl}\xi^T\Theta\xi & = & \xi^T \sum_p h^p[h^p]^T\xi \leq \sum_p \delta_p^2 \leq \sum_p 2[\frac{4}{9}\alpha^2\rho_p^2 + 2\alpha\sigma_p] \\ & \leq & 2[\frac{4}{9}\alpha^2 \sum_p \rho_p^2 + 2\alpha\sum_p \sigma_p] \leq 2[\frac{4}{9}\alpha^2\sum_p \|h^p\|_2^2 + 2\alpha\vartheta] \\ & = & 2[\frac{4}{9}\alpha^2\sum_p \lambda_p + 2\alpha\vartheta] = 2[\frac{4}{9}\alpha^2\mathrm{Tr}(\Theta) + 2\alpha\vartheta].\end{array} \tag{6.30}$$

It remains to note that for every realization of ω it holds

$$\begin{array}{rcl}\|Bx-\widehat{x}_H(\omega)\| & = & \|Bx-H^T[Ax+\xi]\| \leq \|[B-H^TA]x\|+\|H^T\xi\| \\ & \leq & \overline{\Phi}(H)+2\sqrt{\phi_{\mathcal{R}}(\lambda[\Upsilon])[\xi^T\Theta\xi]},\end{array} \tag{6.31}$$

where the concluding inequality is given by exactly the same computation as that used to derive (4.106); see the solution to Exercise 4.19. Taking expectations of both sides in (6.31) and using the fact that ξ is zero mean with covariance matrix $\mathrm{Diag}\{Ax\}$, we get (4.111). Furthermore, (6.31) combines with (6.30) to imply that when $\xi=\omega-Ax\in\Xi_\Theta$ (which happens with probability at least $1-\epsilon$), one has

$$\|Bx-\widehat{x}_H(\omega)\| \leq \overline{\Phi}(H)+4\sqrt{\tfrac{2}{9}\ln^2(2m/\epsilon)\mathrm{Tr}(\Theta)+\ln(2m/\epsilon)\mathrm{Tr}(\mathrm{Diag}\{Ax\}\Theta)}\sqrt{\phi_{\mathcal{R}}(\lambda[\Upsilon])},$$

and (4.112) follows. □

6.4.5 Numerical lower-bounding minimax risk

Exercise 4.22

4.22.A. Motivation. From the theoretical viewpoint, the results on near-optimality of presumably good linear estimates stated in Propositions 4.5, 4.16 seem to be quite strong and general. This being said, for a practically oriented user the "nonoptimality factors" arising in these propositions can be too large to make any practical sense. This drawback of our theoretical results is not too crucial—what matters in applications, is whether the risk of a proposed estimate is appropriate for the application in question, and not by how much it could be improved were we smart enough to build the "ideal" estimate; results of the latter type from a practical viewpoint offer no more than some "moral support." Nevertheless, the "moral support" has its value, and it makes sense to strengthen it by improving the lower risk bounds as compared to those underlying Propositions 4.5 and 4.16. In this respect, an appealing idea is to pass from lower risk bounds yielded by theoretical considerations to *computation-based* ones. The goal of this exercise is to develop some methodology yielding computation-based lower risk bounds. We start with the main ingredient of this methodology—the classical *Cramer-Rao* bound.

4.22.B. Cramer-Rao bound. Consider the situation as follows: we are given

- an observation space Ω equipped with reference measure Π, basic examples being (A) $\Omega=\mathbf{R}^m$ with Lebesgue measure Π, and (B) (finite or countable) discrete set Ω with counting measure Π;
- a convex compact set $\Theta\subset\mathbf{R}^k$ and a family $\mathcal{P}=\{p(\omega,\theta):\theta\in\Theta\}$ of probability densities, taken w.r.t. Π.

Our goal is, given an observation $\omega\sim p(\cdot,\theta)$ stemming from an unknown θ known to belong to Θ, to recover θ. We quantify the risk of a candidate estimate $\widehat{\theta}$ as

$$\mathrm{Risk}[\widehat{\theta}|\Theta]=\sup_{\theta\in\Theta}\left(\mathbf{E}_{\omega\sim p(\cdot,\theta)}\left\{\|\widehat{\theta}(\omega)-\theta\|_2^2\right\}\right)^{1/2}, \tag{4.113}$$

and define the "ideal" minimax risk as

$$\mathrm{Risk}_{\mathrm{opt}}=\inf_{\widehat{\theta}}\mathrm{Risk}[\widehat{\theta}],$$

the infimum being taken w.r.t. all estimates, or, which is the same, all *bounded* estimates (indeed, passing from a candidate estimate $\widehat{\theta}$ to the projected estimate $\widehat{\theta}_\Theta(\omega)=\mathrm{argmin}_{\theta\in\Theta}\|\widehat{\theta}(\omega)-\theta\|_2$ will only reduce the estimate risk).

The Cramer-Rao inequality [58, 205], which we intend to use,[18] is a certain relation between the covariance matrix of a bounded estimate and its bias; this relation is valid under mild regularity assumptions on the family $\mathcal{P}$, specifically, as follows:

1) $p(\omega,\theta) > 0$ for all $\omega \in \Omega, \theta \in U$, and $p(\omega,\theta)$ is differentiable in θ, the with $\nabla_\theta p(\omega,\theta)$ continuous in $\theta \in \Theta$;
2) The *Fisher Information matrix*

$$\mathcal{I}(\theta) = \int_\Omega \frac{\nabla_\theta p(\omega,\theta)[\nabla_\theta p(\omega,\theta)]^T}{p(\omega,\theta)} \Pi(d\omega)$$

   is well-defined for all $\theta \in \Theta$;
3) There exists function $M(\omega) \geq 0$ such that $\int_\Omega M(\omega)\Pi(d\omega) < \infty$ and

$$\|\nabla_\theta p(\omega,\theta)\|_2 \leq M(\omega) \;\forall \omega \in \Omega, \theta \in \Theta.$$

The derivation of the Cramer-Rao bound is as follows. Let $\widehat{\theta}(\omega)$ be a bounded estimate, and let

$$\phi(\theta) = [\phi_1(\theta); ...; \phi_k(\theta)] = \int_\Omega \widehat{\theta}(\omega) p(\omega,\theta)\Pi(d\omega)$$

be the expected value of the estimate. By item 3, $\phi(\theta)$ is differentiable on Θ, with the Jacobian $\phi'(\theta) = \left[\frac{\partial \phi_i(\theta)}{\partial \theta_j}\right]_{i,j\leq k}$ given by

$$\phi'(\theta)h = \int_\Omega \widehat{\theta}(\omega) h^T \nabla_\theta p(\omega,\theta)\Pi(d\omega),\; h \in \mathbf{R}^k.$$

Besides this, since $\int_\Omega p(\omega,\theta)\Pi(d\omega) \equiv 1$, invoking item 3 we have $\int_\Omega h^T \nabla_\theta p(\omega,\theta)\Pi(d\omega) = 0$, whence, in view of the previous identity,

$$\phi'(\theta)h = \int_\Omega [\widehat{\theta}(\omega) - \phi(\theta)] h^T \nabla_\theta p(\omega,\theta)\Pi(d\omega),\; h \in \mathbf{R}^k.$$

Therefore for all $g, h \in \mathbf{R}^k$ we have

$$\begin{array}{rcl} [g^T\phi'(\theta)h]^2 & = & \left[\int_\omega [g^T(\widehat{\theta} - \phi(\theta)][h^T\nabla_\theta p(\omega,\theta)/p(\omega,\theta)]p(\omega,\theta)\Pi(d\omega)\right]^2 \\ & \leq & \left[\int_\Omega g^T[\widehat{\theta} - \phi(\theta)][\widehat{\theta} - \phi(\theta)]^T g p(\omega,\theta)\Pi(d\omega)\right] \\ & & \quad\times \left[\int_\Omega [h^T\nabla_\theta p(\omega,\theta)/p(\omega,\theta)]^2 p(\omega,\theta)\Pi(d\omega)\right] \\ & & \text{[by the Cauchy Inequality]} \\ & = & \left[g^T \mathrm{Cov}_{\widehat{\theta}}(\theta) g\right] \left[h^T \mathcal{I}(\theta) h\right], \end{array}$$

where $\mathrm{Cov}_{\widehat{\theta}}(\theta)$ is the covariance matrix $\mathbf{E}_{\omega\sim p(\cdot,\theta)}\left\{[\widehat{\theta}(\omega) - \phi(\theta)][\widehat{\theta}(\omega) - \phi(\theta)]^T\right\}$ of $\widehat{\theta}(\omega)$ induced by $\omega \sim p(\cdot,\theta)$. We have arrived at the inequality

$$\left[g^T \mathrm{Cov}_{\widehat{\theta}}(\theta) g\right] \left[h^T \mathcal{I}(\theta) h\right] \geq [g^T\phi'(\theta)h]^2 \;\forall (g, h \in \mathbf{R}^k, \theta \in \Theta). \qquad (*)$$

For $\theta \in \Theta$ fixed, let $\mathcal{J}$ be a positive definite matrix such that $\mathcal{J} \succeq \mathcal{I}(\theta)$, whence by $(*)$ it holds

$$\left[g^T \mathrm{Cov}_{\widehat{\theta}}(\theta) g\right] \left[h^T \mathcal{J} h\right] \geq [g^T\phi'(\theta)h]^2 \;\forall (g, h \in \mathbf{R}^k). \qquad (**)$$

---

[18] As a matter of fact, the classical Cramer-Rao inequality dealing with unbiased estimates is not sufficient for our purposes "as is." What we need to build bounds for the minimax risk is a "bias enabled" version of this inequality. Such an inequality may be developed using Bayesian argument [99, 233].

For g fixed, the maximum of the right hand side quantity in $(**)$ over h satisfying $h^T \mathcal{J} h \leq 1$ is $g^T\phi'(\theta)\mathcal{J}^{-1}[\phi'(\theta]^T g$, and we arrive at the *Cramer-Rao inequality*

$$\forall(\theta\in\Theta, \mathcal{J}\succeq\mathcal{I}(\theta), \mathcal{J}\succ 0) : \mathrm{Cov}_{\widehat{\theta}}(\theta)\succeq\phi'(\theta)\mathcal{J}^{-1}[\phi'(\theta]^T \tag{4.114}$$

$$\left[\mathrm{Cov}_{\widehat{\theta}}(\theta)=\mathbf{E}_{\omega\sim p(\cdot,\theta)}\left\{[\widehat{\theta}-\phi(\theta)][\widehat{\theta}-\phi(\theta)]^T\right\},\ \phi(\theta)=\mathbf{E}_{\omega\sim p(\cdot,\theta)}\left\{\widehat{\theta}(\omega)\right\}\right]$$

which holds true for every bounded estimate $\widehat{\theta}(\cdot)$. Note also that for every $\theta\in\Theta$ and every bounded estimate x we have

$$\begin{array}{rcl}
\mathrm{Risk}^2[\widehat{\theta}] & \geq & \mathbf{E}_{\omega\sim p(\cdot,\theta)}\left\{\|\widehat{\theta}(\omega)-\theta\|_2^2\right\}=\mathbf{E}_{\omega\sim p(\cdot,\theta)}\left\{\|[\widehat{\theta}(\omega)-\phi(\theta)]+[\phi(\theta)-\theta]\|_2^2\right\}\\
& = & \mathbf{E}_{\omega\sim p(\cdot,\theta)}\left\{\|\widehat{\theta}(\omega)-\phi(\theta)\|_2^2\right\}+\|\phi(\theta)-\theta)\|_2^2\\
& & \underbrace{-2\,\mathbf{E}_{\omega\sim p(\cdot,\theta)}\left[[\widehat{\theta}(\omega)-\phi(\theta)]^T[\phi(\theta)-\theta)]\right\}}_{=0}\\
& = & \mathrm{Tr}(\mathrm{Cov}_{\widehat{\theta}}(\theta))+\|\phi(\theta)-\theta\|_2^2.
\end{array}$$

Hence, in view of (4.114), for every bounded estimate $\widehat{\theta}$ it holds

$$\begin{array}{l}
\forall(\mathcal{J}\succ 0:\mathcal{J}\succeq\mathcal{I}(\theta)\,\forall\theta\in\Theta):\\
\mathrm{Risk}^2[\widehat{\theta}]\geq\sup_{\theta\in\Theta}\left[\mathrm{Tr}(\phi'(\theta)\mathcal{J}^{-1}[\phi'(\theta)]^T)+\|\phi(\theta)-\theta\|_2^2\right]\\
\qquad\left[\phi(\theta)=\mathbf{E}_{\omega\sim p(\cdot,\theta)}\{\widehat{\theta}(\omega)\}\right].
\end{array} \tag{4.115}$$

The fact that we considered the risk of estimating "the entire" θ rather than a given vector-valued function $f(\theta):\Theta\to\mathbf{R}^\nu$ plays no special role, and in fact the Cramer-Rao inequality admits the following modification yielded by a completely similar reasoning:

Proposition 4.36 *In the situation described in item* **4.22.B** *and under assumptions 1)–3) of this item, let $f(\cdot):\Theta\to\mathbf{R}^\nu$ be a bounded Borel function, and let $\widehat{f}(\omega)$ be a bounded estimate of $f(\omega)$ via observation $\omega\sim p(\cdot,\theta)$. Then, setting for $\theta\in\Theta$*

$$\begin{array}{rcl}
\phi(\theta) & = & \mathbf{E}_{\omega\sim p(\cdot,\theta)}\left\{\widehat{f}(\theta)\right\},\\
\mathrm{Cov}_{\widehat{f}}(\theta) & = & \mathbf{E}_{\omega\sim p(\cdot,\theta)}\left\{[\widehat{f}(\omega)-\phi(\theta)][\widehat{f}(\omega)-\phi(\theta)]^T\right\},
\end{array}$$

one has

$$\forall(\theta\in\Theta,\mathcal{J}\succeq\mathcal{I}(\theta),\mathcal{J}\succ 0):\mathrm{Cov}_{\widehat{f}}(\theta)\succeq\phi'(\theta)\mathcal{J}^{-1}[\phi'(\theta)]^T.$$

As a result, for

$$\mathrm{Risk}[\widehat{f}]=\sup_{\theta\in\Theta}\left[\mathbf{E}_{\omega\sim p(\cdot,\theta)}\left\{\|\widehat{f}(\omega)-f(\theta)\|_2^2\right\}\right]^{1/2}$$

it holds

$$\begin{array}{l}
\forall(\mathcal{J}\succ 0:\mathcal{J}\succeq\mathcal{I}(\theta)\,\forall\theta\in\Theta):\\
\mathrm{Risk}^2[\widehat{f}]\geq\sup_{\theta\in\Theta}\left[\mathrm{Tr}(\phi'(\theta)\mathcal{J}^{-1}[\phi'(\theta)]^T)+\|\phi(\theta)-f(\theta)\|_2^2\right]
\end{array}$$

Now comes the first part of the exercise:

1) Derive from (4.115) the following:

Proposition 4.37 *In the situation of item 4.22.B, let*

- *$\Theta\subset\mathbf{R}^k$ be a $\|\cdot\|_2$-ball of radius $r>0$,*
- *the family $\mathcal{P}$ be such that $\mathcal{I}(\theta)\preceq\mathcal{J}$ for some $\mathcal{J}\succ 0$ and all $\theta\in\Theta$.*

Then the minimax optimal risk satisfies the bound

$$\mathrm{Risk}_{\mathrm{opt}}\geq\frac{rk}{r\sqrt{\mathrm{Tr}(\mathcal{J})}+k}. \tag{4.116}$$

In particular, when $\mathcal{J} = \alpha^{-1} I_k$, we have

$$\mathrm{Risk}_{\rm opt} \geq \frac{r\sqrt{\alpha k}}{r+\sqrt{\alpha k}}. \tag{4.117}$$

Hint. Assuming w.l.o.g. that Θ is centered at the origin, and given a bounded estimate $\widehat{\theta}$ with risk $\mathfrak{R}$, let $\phi(\theta)$ be associated with the estimate via (4.115). Select $\gamma \in (0,1)$ and consider two cases: (a): there exists $\theta \in \partial\Theta$ such that $\|\phi(\theta)-\theta\|_2 > \gamma r$, and (b): $\|\phi(\theta)-\theta\|_2 \leq \gamma r$ for all $\theta \in \partial\Theta$. In the case of (a), lower-bound $\mathfrak{R}$ by $\max_{\theta\in\Theta}\|\phi(\theta)-\theta\|_2$; see (4.115). In the case of (b), lower-bound $\mathfrak{R}^2$ by $\max_{\theta\in\Theta}\mathrm{Tr}(\phi'(\theta)\mathcal{J}^{-1}[\phi'(\theta)]^T)$—see (4.115)—and use the Gauss Divergence theorem to lower-bound the latter quantity in terms of the flux of the vector field $\phi(\cdot)$ over $\partial\Theta$.
When implementing the above program, you might find useful the following fact (prove it!):

Lemma 4.38 *Let Φ be an $n\times n$ matrix, and $\mathcal{J}$ be a positive definite $n \times n$ matrix. Then*

$$\mathrm{Tr}(\Phi\mathcal{J}^{-1}\Phi^T) \geq \frac{\mathrm{Tr}^2(\Phi)}{\mathrm{Tr}(\mathcal{J})}.$$

Solution: Let us act as suggested in the hint. We lose nothing when assuming that Θ is centered at the origin: $\Theta = \{\theta \in \mathbf{R}^k : \|\theta\|_2 \leq r\}$. Under the premise of the proposition, we can use $\mathcal{J}$ in (4.115). In the case of (a), (4.115) says that $\mathfrak{R} \geq \max_{\theta\in\partial\Theta}\|\phi(\theta)-\theta\|_2 \geq \gamma r$. In the case of (b), denoting by $\mathcal{S}$ the boundary of Θ (that is, the Euclidean sphere of radius r centered at the origin), by dS the element of $(k-1)$-dimensional surface of $\mathcal{S}$, by S the entire surface, and by V the volume of Θ, we have $V = k^{-1}rS$. Let now $n(\theta)$ be the unit outer normal to $\mathcal{S}$ at a point $\theta \in \mathcal{S}$. The flux $\int_{\mathcal{S}} n^T(\theta)\phi(\theta)dS$ is at least $(1-\gamma)rS$ due to

$$n^T(\theta)\phi(\theta) \geq n^T(\theta)\theta - \|\theta-\phi(\theta)\|_2 \geq r - \gamma r.$$

On the other hand, by Divergence Theorem this flux is equal to $\int_\Theta \left[\sum_i \frac{\partial\phi_i(\theta)}{\partial\theta_i}\right] d\theta$, and we arrive at the inequality

$$(1-\gamma)rS \leq \int_\Theta \left[\sum_i \frac{\partial\phi_i(\theta)}{\partial\theta_i}\right] d\theta \leq V \max_{\theta\in\Theta}\sum_i \frac{\partial\phi_i(\theta)}{\partial\theta_i},$$

whence

$$\max_{\theta\in\Theta}\sum_i \frac{\partial\phi_i(\theta)}{\partial\theta_i} \geq \frac{(1-\gamma)rS}{V} = k(1-\gamma).$$

Thus, there exists $\theta_* \in \Theta$ such that

$$\sum_i \frac{\partial\phi_i(\theta_*)}{\partial\theta_i} \geq k(1-\gamma). \tag{6.32}$$

Now, taking Lemma 4.38 for granted, (4.115) says that

$$\mathfrak{R}^2 \geq \mathrm{Tr}(\phi'(\theta_*)\mathcal{J}^{-1}[\phi'(\theta_*)]^T) \geq \mathrm{Tr}^2(\phi'(\theta_*))/\mathrm{Tr}(\mathcal{J}).$$

Invoking (6.32), we obtain

$$\mathfrak{R}^2 \geq (1-\gamma)^2k^2/\mathrm{Tr}(\mathcal{J}).$$

Therefore, in all cases we have

$$\mathfrak{R} \geq \min[\gamma r, (1-\gamma)k/\sqrt{\mathrm{Tr}(\mathcal{J})}],$$

which, setting $\gamma = \frac{k}{r\sqrt{\mathrm{Tr}(\mathcal{J})}+k}$, results in $\mathfrak{R} \geq \frac{rk}{r\sqrt{\mathrm{Tr}(\mathcal{J})}+k}$. Since this lower bound holds true for the risk $\mathfrak{R}$ of an arbitrary bounded estimate $\widehat{\theta}$, (4.116) follows.

It remains to prove Lemma 4.38. Let $\mathcal{J} = U\mathrm{Diag}\{\lambda\}U^T$ be eigenvalue decomposition of $\mathcal{J}$, let $\alpha = \sum_i \lambda_i = \mathrm{Tr}(\mathcal{J})$, and let $\Psi = U^T\Phi U$. We have

$$\begin{array}{rl}\mathrm{Tr}(\Phi\mathcal{J}^{-1}\Phi^T) = & \mathrm{Tr}([\Phi U]\mathrm{Diag}\{1/\lambda_j, j\leq n\}U^T\Phi^T) \\ = & \mathrm{Tr}([U^T\Phi U]\mathrm{Diag}\{1/\lambda_j, j\leq n\}[U^T\Phi^T U]) = \mathrm{Tr}(\Psi\mathrm{Diag}\{1/\lambda_j, j\leq n\}\Psi^T) \\ = & \sum_{i,j}\Psi_{ij}^2/\lambda_j \geq \sum_j \Psi_{jj}^2/\lambda_j \geq \min\limits_{\mu_j\geq 0:\sum_j \mu_j=\alpha} \sum_j \Psi_{jj}^2/\mu_j = \left[\sum_j |\Psi_{jj}|\right]^2/\alpha \\ \geq & \mathrm{Tr}^2(\Psi)/\alpha = \mathrm{Tr}^2(\Phi)/\mathrm{Tr}(\mathcal{J}). \qquad \square\end{array}$$

4.22.C. Application to signal recovery. Proposition 4.37 allows us to build computation-based lower risk bounds in the signal recovery problem considered in Section 4.2, in particular, the problem where one wants to recover the linear image Bx of an unknown signal x known to belong to a given ellitope

$$\mathcal{X} = \{x\in\mathbf{R}^n : \exists t\in\mathcal{T} : x^TS_\ell x\leq t_\ell, \ell\leq L\}$$

(with our usual restriction on S_ℓ and $\mathcal{T}$) via observation

$$\omega = Ax + \sigma\xi,\ \xi\sim\mathcal{N}(0, I_m),$$

and the risk of a candidate estimate, as in Section 4.2, is defined according to (4.113).[19] It is convenient to assume that the matrix B (which in our general setup can be an arbitrary $\nu\times n$ matrix) is a *nonsingular* $n\times n$ matrix.[20] Under this assumption, setting

$$\mathcal{Y} = B^{-1}\mathcal{X} = \{y\in\mathbf{R}^n : \exists t\in\mathcal{T} : y^T[B^{-1}]^TS_\ell B^{-1}y\leq t_\ell, \ell\leq L\}$$

and $\bar{A} = AB^{-1}$, we lose nothing when replacing the sensing matrix A with $\bar{A}$ and treating as our signal $y\in\mathcal{Y}$ rather than $\mathcal{X}$. Note that in our new situation A is replaced with $\bar{A}$, $\mathcal{X}$ with $\mathcal{Y}$, and B is the unit matrix I_n. For the sake of simplicity, we assume from now on that A (and therefore $\bar{A}$) has trivial kernel. Finally, let $\tilde{S}_\ell\succeq S_\ell$ be close to S_k positive definite matrices, e.g., $\tilde{S}_\ell = S_\ell + 10^{-100}I_n$. Setting $\bar{S}_\ell = [B^{-1}]^T\tilde{S}_\ell B^{-1}$ and

$$\bar{\mathcal{Y}} = \{y\in\mathbf{R}^n : \exists t\in\mathcal{T} : y^T\bar{S}_\ell y\leq t_\ell, \ell\leq L\},$$

we get $\bar{S}_\ell\succ 0$ and $\bar{\mathcal{Y}}\subset\mathcal{Y}$. Therefore, any lower bound on the $\|\cdot\|_2$-risk of recovery $y\in\bar{\mathcal{Y}}$ via observation $\omega = AB^{-1}y+\sigma\xi$, $\xi\sim\mathcal{N}(0, I_m)$, automatically is a lower bound on the minimax risk $\mathrm{Risk}_{\mathrm{opt}}$ corresponding to our original problem of interest.

Now assume that we can point out a k-dimensional linear subspace E in $\mathbf{R}^n$ and positive reals r, γ such that

[19]In fact, the approach to be developed can be applied to signal recovery problems involving Discrete/Poisson observation schemes and norms different from $\|\cdot\|_2$ used to measure the recovery error, signal-dependent noises, etc.

[20]This assumption is nonrestrictive. Indeed, when $B\in\mathbf{R}^{\nu\times n}$ with $\nu < n$, we can add to B $n-\nu$ zero rows, which keeps our estimation problem intact. When $\nu\geq n$, we can add to B a small perturbation to ensure $\mathrm{Ker}\, B = \{0\}$, which, for small enough perturbation, again keeps our estimation problem basically intact. It remains to note that when $\mathrm{Ker}\, B = \{0\}$ we can replace $\mathbf{R}^\nu$ with the image space of B, which again does not affect the estimation problem we are interested in.

(i) the $\|\cdot\|_2$-ball $\Theta = \{\theta \in E : \|\theta\|_2 \le r\}$ is contained in $\bar{\mathcal{Y}}$;
(ii) The restriction $\bar{A}_E$ of $\bar{A}$ onto E satisfies the relation

$$\mathrm{Tr}(\bar{A}_E^* \bar{A}_E) \le \gamma$$

($\bar{A}_E^* : \mathbf{R}^m \to E$ is the conjugate of the linear map $\bar{A}_E : E \to \mathbf{R}^m$).

Consider the auxiliary estimation problem obtained from the (reformulated) problem of interest by replacing the signal set $\bar{\mathcal{Y}}$ with Θ. Since $\Theta \subset \bar{\mathcal{Y}}$, the minimax risk in the auxiliary problem is a lower bound on the minimax risk $\mathrm{Risk}_{\rm opt}$ we are interested in. On the other hand, the auxiliary problem is nothing but the problem of recovering parameter $\theta \in \Theta$ from observation $\omega \sim \mathcal{N}(\bar{A}\theta, \sigma^2 I)$, which is nothing but a special case of the problem considered in item 4.22.B. As it is immediately seen, the Fisher Information matrix in this problem is independent of θ and is $\sigma^{-2}\bar{A}_E^*\bar{A}_E$:

$$e^T \mathcal{I}(\theta) e = \sigma^{-2} e^T \bar{A}_E^* \bar{A}_E e,\ e \in E.$$

Invoking Proposition 4.37, we arrive at the lower bound on the minimax risk in the auxiliary problem (and thus in the problem of interest as well):

$$\mathrm{Risk}_{\rm opt} \ge \frac{r\sigma k}{r\sqrt{\gamma} + \sigma k}. \tag{4.118}$$

The resulting risk bound depends on r, k, γ and is larger the smaller γ is and the larger k and r are.

Lower-bounding $\mathrm{Risk}_{\rm opt}$**.** In order to make the bounding scheme just outlined give its best, we need a mechanism which allows us to generate k-dimensional "disks" $\Theta \subset \bar{\mathcal{Y}}$ along with associated quantities r, γ. In order to design such a mechanism, it is convenient to represent k-dimensional linear subspaces of $\mathbf{R}^n$ as the image spaces of orthogonal $n \times n$ projectors P of rank k. Such a projector P gives rise to the disk Θ_P of the radius $r = r_P$ contained in $\bar{\mathcal{Y}}$, where r_P is the largest ρ such that the set $\{y \in \mathrm{Im}P : y^T P y \le \rho^2\}$ is contained in $\bar{\mathcal{Y}}$ ("condition $\mathcal{C}(r)$"), and we can equip the disk with γ satisfying (ii) if and only if

$$\mathrm{Tr}(P\bar{A}^T \bar{A} P) \le \gamma,$$

or, which is the same (recall that P is an orthogonal projector)

$$\mathrm{Tr}(\bar{A} P \bar{A}^T) \le \gamma \tag{4.119}$$

("condition $\mathcal{D}(\gamma)$"). Now, when P is a nonzero orthogonal projector, the simplest sufficient condition for the validity of $\mathcal{C}(r)$ is the existence of $t \in \mathcal{T}$ such that

$$\forall (y \in \mathbf{R}^n, \ell \le L) : y^T P \bar{S}_\ell P y \le t_\ell r^{-2} y^T P y,$$

or, which is the same,

$$\exists s : r^2 s \in \mathcal{T} \ \&\ P\bar{S}_\ell P \preceq s_\ell P,\ \ell \le L. \tag{4.120}$$

Let us rewrite (4.119) and (4.120) as a system of *linear* matrix inequalities. This is what you are supposed to do:

2.1) Prove the following simple fact:

Observation 4.39 *Let Q be a positive definite, R be a nonzero positive semidefinite matrix, and let s be a real. Then*

$$RQR \preceq sR$$

if and only if

$$sQ^{-1} \succeq R.$$

2.2) Extract from the above observation the conclusion as follows. Let $\mathbf{T}$ be the conic hull of $\mathcal{T}$:

$$\mathbf{T} = \mathrm{cl}\{[s;\tau] : \tau > 0, s/\tau \in \mathcal{T}\} = \{[s;\tau] : \tau > 0, s/\tau \in \mathcal{T}\} \cup \{0\}.$$

Consider the system of constraints

$$\begin{array}{c}[s;\tau] \in \mathbf{T} \ \&\ s_\ell \bar{S}_\ell^{-1} \succeq P,\ \ell \le L \ \&\ \mathrm{Tr}(\bar{A}P\bar{A}^T) \le \gamma,\\ P \text{ is an orthogonal projector of rank } k \ge 1\end{array} \tag{\#}$$

in variables $[s;\tau]$, k, γ, and P. Every feasible solution to this system gives rise to a k-dimensional Euclidean subspace $E \subset \mathbf{R}^n$ (the image space of P) such that the Euclidean ball Θ in E centered at the origin of radius

$$r = 1/\sqrt{\tau}$$

taken along with γ satisfy conditions (i)–(ii). Consequently, such a feasible solution yields the lower bound

$$\mathrm{Risk}_{\mathrm{opt}} \ge \psi_{\sigma,k}(\gamma,\tau) := \frac{\sigma k}{\sqrt{\gamma} + \sigma\sqrt{\tau} k}$$

on the minimax risk in the problem of interest.

Solution: 2.1: Since $Q \succ 0$ and $0 \preceq R \ne 0$, relation $RQR \preceq sR$ implies that $s > 0$. When $s > 0$, this relation by the Schur Complement Lemma implies that $\left[\begin{array}{c|c} R & R \\ \hline R & sQ^{-1}\end{array}\right] \succeq 0$, whence $\left[\begin{array}{c|c} R+\epsilon I & R \\ \hline R & sQ^{-1}\end{array}\right] \succeq 0$ whenever $\epsilon > 0$, implying, by the same Schur Complement Lemma, that $sQ^{-1} \succeq R(R+\epsilon I)^{-1}R \to R$ as $\epsilon \to +0$, that is, the first of the relations in question implies the second. Vice versa, assuming $sQ^{-1} \succeq R$, we clearly have $s > 0$. Let us fix $s' > s$, so that $s'Q^{-1} \succeq R + \epsilon I$ for all small enough $\epsilon > 0$, whence by the Schur Complement Lemma $\left[\begin{array}{c|c} R+\epsilon I & R+\epsilon I \\ \hline R+\epsilon I & s'Q^{-1}\end{array}\right] \succeq 0$ for all small enough positive ϵ, implying that $\left[\begin{array}{c|c} R & R \\ \hline R & s'Q^{-1}\end{array}\right] \succeq 0$, whence $RQR \preceq s'R$. Since this relation holds true for all $s' > s$, we get $RQR \preceq sR$. □

2.2: Let $[s;\tau]$, k, γ, P form a feasible solution to (#). Then clearly $s > 0$, so that $t := s/\tau$ is well-defined and $t \in \mathcal{T}$. Applying Observation 4.39 we conclude that

$$P\bar{S}_\ell P \preceq s_\ell P,\ \ell \le L \ \&\ \mathrm{Tr}(\bar{A}P\bar{A}^T) \le \gamma,$$

which implies, via (4.120), that the conditions $\mathcal{C}(1/\sqrt{\tau})$ and $\mathcal{D}(\gamma)$ take place. The claim now follows from the origin of these conditions. □

Ideally, to utilize item 2.2 to lower-bound $\mathrm{Risk}_{\mathrm{opt}}$, we should look through $k = 1,...,n$ and maximize for every k the lower risk bound $\psi_{\sigma,k}(\gamma,\tau)$ under constraints (#), thus arriving at the problem

$$\min_{[s;\tau],\gamma,P}\left\{\begin{array}{l}\frac{\sigma}{\psi_{\sigma,k}(\gamma,\tau)} = \sqrt{\gamma}/k + \sigma\sqrt{\tau} :\\ [s;\tau] \in \mathbf{T} \ \&\ s_\ell \bar{S}_\ell^{-1} \succeq P, \ell \le L \ \&\ \mathrm{Tr}(\bar{A}P\bar{A}^T) \le \gamma,\\ P \text{ is an orthogonal projector of rank } k.\end{array}\right\} \tag{P_k}$$

This problem seems to be computationally intractable, since the constraints of (P_k) include the nonconvex restriction on P to be a projector of rank k. A natural convex relaxation of this constraint is

$$0 \preceq P \preceq I_n,\ \mathrm{Tr}(P) = k.$$

The (minor) remaining difficulty is that the objective in (P) is nonconvex. Note, however, that to minimize $\sqrt{\gamma}/k+\sigma\sqrt{\tau}$ is basically the same as to minimize the convex function $\gamma/k^2+\sigma^2\tau$ which is a tight "proxy" of the squared objective of (P_k). We arrive at a convex "proxy" of (P_k)—the problem

$$\min_{[s;\tau],\gamma,P}\left\{\gamma/k^2+\sigma^2\tau:\begin{array}{l}[s;\tau]\in\mathbf{T}, 0\preceq P\preceq I_n, \mathrm{Tr}(P)=k\\ s_\ell\bar{S}_\ell^{-1}\succeq P, \ell\leq L, \mathrm{Tr}(\bar{A}P\bar{A}^T)\leq\gamma\end{array}\right\} \qquad (P[k])$$

$k=1,...,n$. Problem $(P[k])$ clearly is solvable, and the P-component $P^{(k)}$ of its optimal solution gives rise to a collection of orthogonal projectors $P_\kappa^{(k)}$, $\kappa=1,...,n$, obtained from $P^{(k)}$ by "rounding"—to get $P_\kappa^{(k)}$, we replace the κ leading eigenvalues of $P^{(k)}$ with ones, and the remaining eigenvalues with zeros, while keeping the eigenvectors intact. We can now for every $\kappa=1,...,n$ fix the P-variable in (P_k) as $P_\kappa^{(k)}$ and solve the resulting problem in the remaining variables $[s;\tau]$ and γ, which is easy—with P fixed, the problem clearly reduces to minimizing τ under the convex constraints

$$s_\ell\bar{S}_\ell^{-1}\succeq P, \ell\leq L,\ [s;\tau]\in\mathbf{T}$$

on $[s;\tau]$. As a result, for every $k\in\{1,...,n\}$, we get n lower bounds on $\mathrm{Risk}_{\mathrm{opt}}$, that is, a total of n^2 lower risk bounds, of which we select the best—the largest.

Now comes the next part of the exercise:

3) Implement the outlined program numerically and compare the lower bound on the minimax risk with the upper risk bounds of presumably good linear estimates yielded by Proposition 4.4.

   Recommended setup:

   - Sizes: $m=n=\nu=16$.
   - A, B: $B=I_n$, $A=\mathrm{Diag}\{a_1,...,a_n\}$ with $a_i=i^{-\alpha}$ and α running through $\{0,1,2\}$.
   - $\mathcal{X}=\{x\in\mathbf{R}^n: x^TS_\ell x\leq 1, \ell\leq L\}$ (i.e., $\mathcal{T}=[0,1]^L$) with randomly generated S_ℓ.
   - Range of L: $\{1,4,16\}$. For L in this range, you can generate S_ℓ, $\ell\leq L$, as $S_\ell=R_\ell R_\ell^T$ with $R_\ell=\mathtt{randn}(n,p)$, where $p=\rfloor n/L\lfloor$.
   - Range of σ: $\{1.0, 0.1, 0.01, 0.001, 0.0001\}$.

Solution: The results of typical numerical experiments implemented according to the above setup are presented in Table 6.3. We see that the presumably good linear estimates indeed are good—(theoretical upper bounds on) their risk are within factor 2 of the lower bounds on the minimax optimal risk.

Exercise 4.23 [follow-up to Exercise 4.22]

1) Prove the following version of Proposition 4.37:

Proposition 4.40 *In the situation of item 4.22.B and under Assumptions 1) – 3) from this item, let*

- $\|\cdot\|$ *be a norm on* $\mathbf{R}^k$ *such that*

$$\|\theta\|_2\leq\kappa\|\theta\|\ \forall\theta\in\mathbf{R}^k$$

- $\Theta\subset\mathbf{R}^k$ *be* $\|\cdot\|$*-ball of radius* $r>0$,
- *the family* $\mathcal{P}$ *be such that* $\mathcal{I}(\theta)\preceq\mathcal{J}$ *for some* $\mathcal{J}\succ 0$ *and all* $\theta\in\Theta$.

Then the minimax optimal risk

$$\mathrm{Risk}_{\mathrm{opt},\|\cdot\|}=\inf_{\widehat{\theta}(\cdot)}\left(\sup_{\theta\in\Theta}\mathbf{E}_{\omega\sim p(\cdot,\theta)}\left\{\|\theta-\widehat{\theta}(\omega)\|^2\right\}\right)^{1/2}$$

α	L	σ	Risk	R	α	L	σ	Risk	R	α	L	σ	Risk	R
0	1	1.0000	1.5156	1.5	0	4	1.0000	1.6949	1.6	0	16	1.0000	2.1007	1.7
0	1	0.1000	0.2860	1.7	0	4	0.1000	0.3451	1.7	0	16	0.1000	0.3796	1.8
0	1	0.0100	0.0397	1.3	0	4	0.0100	0.0399	1.2	0	16	0.0100	0.0400	1.2
0	1	0.0010	0.0040	1.0	0	4	0.0010	0.0040	1.0	0	16	0.0010	0.0040	1.0
0	1	0.0001	0.0004	1.0	0	4	0.0001	0.0004	1.0	0	16	0.0001	0.0004	1.0
1	1	1.0000	5.2537	1.4	1	4	1.0000	4.8681	1.3	1	16	1.0000	5.3097	2.0
1	1	0.1000	1.2918	1.5	1	4	0.1000	1.1947	1.9	1	16	0.1000	1.7878	1.9
1	1	0.0100	0.2724	1.7	1	4	0.0100	0.3118	2.0	1	16	0.0100	0.3667	1.9
1	1	0.0010	0.0384	1.4	1	4	0.0010	0.0386	1.3	1	16	0.0010	0.0387	1.3
1	1	0.0001	0.0039	1.2	1	4	0.0001	0.0039	1.2	1	16	0.0001	0.0039	1.2
2	1	1.0000	4.5613	1.3	2	4	1.0000	3.3401	1.1	2	16	1.0000	9.5095	1.5
2	1	0.1000	3.0733	1.4	2	4	0.1000	2.2972	1.5	2	16	0.1000	4.1331	1.7
2	1	0.0100	0.7586	1.7	2	4	0.0100	1.1169	1.5	2	16	0.0100	1.5697	2.1
2	1	0.0010	0.2730	1.7	2	4	0.0010	0.3841	2.0	2	16	0.0010	0.4406	2.0
2	1	0.0001	0.0486	1.4	2	4	0.0001	0.0492	1.4	2	16	0.0001	0.0493	1.3

Table 6.3: Sample numerical results for Exercise 4.22.3. Risk: upper bound $\sqrt{\mathrm{Opt}}$ on risk of the linear estimate; see Proposition 4.4; R: ratios of Risk to the lower risk bounds.

of recovering parameter $\theta \in \Theta$ from observation $\omega \sim p(\cdot, \theta)$ in the norm $\|\cdot\|$ satisfies the bound

$$\mathrm{Risk}_{\mathrm{opt},\|\cdot\|} \geq \frac{rk}{r\kappa\sqrt{\mathrm{Tr}(\mathcal{J})} + k}. \tag{4.121}$$

In particular, when $\mathcal{J} = \alpha^{-1} I_k$, we get

$$\mathrm{Risk}_{\mathrm{opt},\|\cdot\|} \geq \frac{r\sqrt{\alpha k}}{r\kappa + \sqrt{\alpha k}}. \tag{4.122}$$

Solution: Let us reuse the argument from the proof of Proposition 4.37. By a straightforward approximation argument, we can assume that $\|\cdot\|$ is continuously differentiable outside of the origin. Assuming w.l.o.g. that Θ is centered at the origin, let $\mathcal{S}$ be the boundary of Θ, dS be the element of $(k-1)$-dimensional surface area of $\mathcal{S}$, and $n(\theta)$ be the $\|\cdot\|_2$-unit outer normal to $\mathcal{S}$ at a point $\theta \in \mathcal{S}$. Observe that $n(\theta)$ is obtained from the gradient $g(\theta)$ of $\|\cdot\|$, taken at point $\theta \in \mathcal{S}$, by normalization

$$n(\theta) = \|g(\theta)\|_2^{-1} g(\theta),$$

whence

$$\theta \in \mathcal{S} \Rightarrow n^T(\theta)\theta = \|g(\theta)\|_2^{-1} g^T(\theta)\theta = \|g(\theta)\|_2^{-1} r.$$

Let $\widehat{\theta}$ be a bounded candidate estimate, $\phi(\theta) = \int_\Omega \widehat{\theta}(\omega)p(\omega,\theta)\Pi(d\omega)$, and $\psi(\theta) = \theta$. Denoting by $\mathfrak{R}$ the $\|\cdot\|$-risk of estimate $\widehat{\theta}$, observe that by Jensen's inequality for $\theta \in \Theta$ one has $\|\phi(\theta) - \theta\|^2 \leq \mathfrak{R}^2$. Now let $\gamma \in (0,1)$. It may happen ("case (a)") that $\mathfrak{R} \geq \gamma r$; otherwise ("case (b)") we should have $\|\phi(\theta) - \theta\| \leq \gamma r$ for $\theta \in \mathcal{S}$. Assume that (b) is the case. From the origin of $g(\cdot)$, denoting by $\|\cdot\|_*$ the norm conjugate to $\|\cdot\|$ we have $\|g(\theta)\|_* = 1$ whenever $\theta \neq 0$, implying that

$$g^T(\theta)\phi(\theta) \geq g^T(\theta)\theta - \|\theta - \phi(\theta)\| = r - \|\theta - \phi(\theta)\| \geq (1-\gamma)r.$$

It follows that the flux $\int_{\mathcal{S}} n^T(\theta)\phi(\theta)dS(\theta)$ satisfies

$$\begin{array}{l} \int_{\mathcal{S}} n^T(\theta)\phi(\theta)dS(\theta) = \int_{\mathcal{S}} \|g(\theta)\|_2^{-1} g^T(\theta)\phi(\theta)dS(\theta) \geq (1-\gamma)\int_{\mathcal{S}} \|g(\theta)\|_2^{-1} r dS(\theta) \\ = \quad (1-\gamma)\int_{\mathcal{S}} \|g(\theta)\|_2^{-1} g^T(\theta)\psi(\theta)dS(\theta) = (1-\gamma)\int_{\mathcal{S}} n^T(\theta)\psi(\theta)dS(\theta). \end{array}$$

The resulting inequality between the fluxes of ϕ and ψ combines with the Divergence Theorem to imply that

$$\int_\Theta \left[\sum_i \frac{\partial \phi_i(\theta)}{\partial \theta_i}\right] d\theta \geq (1-\gamma)\int_\Theta \left[\sum_i \frac{\partial \psi_i(\theta)}{\partial \theta_i}\right] d\theta = (1-\gamma)k\int_\Theta d\theta,$$

so that

$$\sum_i \frac{\partial \phi_i(\theta_*)}{\partial \theta_i} \geq (1-\gamma)k \tag{6.33}$$

for some $\theta_* \in \Theta$.

On the other hand, setting

$$\mathrm{Cov}_{\widehat{\theta}}(\theta) = \int_\Omega [\widehat{\theta}(\omega) - \phi(\theta)][\widehat{\theta}(\omega) - \phi(\theta)]^T p(\omega,\theta)\Pi(d\omega)$$

and observing that

$$\int_\Omega \|\widehat{\theta}(\omega) - \theta\|_2^2 p(\omega,\theta)\Pi(d\omega) \geq \int_\Omega \|\widehat{\theta}(\omega) - \phi(\theta)\|_2^2 p(\omega,\theta)\Pi(d\omega) = \mathrm{Tr}(\mathrm{Cov}_{\widehat{\theta}}(\theta)),$$

we have

$$\begin{array}{rcl}
\mathfrak{R}^2 & \geq & \int_\Omega \|\widehat{\theta}(\omega) - \theta_*\|^2 p(\omega,\theta_*)\Pi(d\omega) \\
& \geq & \kappa^{-2}\int_\Omega \|\widehat{\theta}(\omega) - \theta_*\|_2^2 p(\omega,\theta_*)\Pi(d\omega) \ [\text{since } \|\cdot\| \geq \kappa^{-1}\|\cdot\|_2] \\
& \geq & \kappa^{-2}\mathrm{Tr}(\mathrm{Cov}_{\widehat{\theta}}(\theta_*)) \\
& \geq & \kappa^{-2}\mathrm{Tr}(\phi'(\theta_*)\mathcal{J}^{-1}[\phi'(\theta_*)]^T) \ [\text{by (4.114)}] \\
& \geq & \kappa^{-2}(1-\gamma)^2k^2/\mathrm{Tr}(\mathcal{J}),
\end{array}$$

where the concluding inequality is yielded by Lemma 4.38 combined with (6.33).

Now, in case (a) we have $\mathfrak{R} \geq \gamma r$. We see that in all cases we have

$$\mathfrak{R} \geq \min[\gamma r, (1-\gamma)k/(\kappa\sqrt{\mathrm{Tr}(\mathcal{J})})].$$

Setting $\gamma = \frac{k}{r\kappa\sqrt{\mathrm{Tr}(\mathcal{J})}+k}$, we arrive at $\mathfrak{R} \geq \frac{rk}{r\kappa\sqrt{\mathrm{Tr}(\mathcal{J})}+k}$, and (4.121) follows. □

2) Apply Proposition 4.40 to get lower bounds on the minimax $\|\cdot\|$-risk in the following estimation problems:

2.1) Given indirect observation $\omega = A\theta + \sigma\xi$, $\xi \sim \mathcal{N}(0, I_m)$ of an unknown vector θ known to belong to $\Theta = \{\theta \in \mathbf{R}^k : \|\theta\|_p \leq r\}$ with given A, $\mathrm{Ker}\, A = \{0\}$, $p \in [2,\infty]$, $r > 0$, we want to recover θ in $\|\cdot\|_p$.

2.2) Given indirect observation $\omega = L\theta R + \sigma\xi$, where θ is an unknown $\mu \times \nu$ matrix known to belong to the Shatten norm ball $\Theta \in \mathbf{R}^{\mu\times\nu} : \|\theta\|_{\mathrm{Sh},p} \leq r$, we want to recover θ in $\|\cdot\|_{\mathrm{Sh},p}$. Here $L \in \mathbf{R}^{m\times\mu}, \mathrm{Ker}\, L = \{0\}$ and $R \in \mathbf{R}^{\nu\times n}, \mathrm{Ker}\, R^T = \{0\}$ are given matrices, $p \in [2,\infty]$, and ξ is random Gaussian $m \times n$ matrix (i.e., the entries in ξ are $\mathcal{N}(0,1)$ random variables) independent of each other.

2.3) Given a K-repeated observation $\omega^K = (\omega_1, ..., \omega_K)$ with i.i.d. components $\omega_t \sim \mathcal{N}(0,\theta)$, $1 \leq t \leq K$, with an unknown $\theta \in \mathbf{S}^n$ known to belong to the matrix box $\Theta = \{\theta : \beta_- I_n \preceq \theta \preceq \beta_+ I_n\}$ with given $0 < \beta_- < \beta_+ < \infty$, we want to recover θ in the spectral norm.

Solution: 2.1: We are in the case of $p(\omega,\theta) = \mathcal{N}(A\theta, \sigma^2 I)$ resulting in $\nabla_\theta p(\omega,\theta) = A^T\sigma^{-2}(\omega - A\theta)$ and $\mathcal{I}(\theta) = \sigma^{-2}A^TA$. Consequently, the premise of Proposition

4.40 holds true with $r=1$, $\kappa=k^{\frac{1}{2}-\frac{1}{p}}$ and $\mathcal{J}=\sigma^{-2}A^TA$, resulting in the bound

$$\mathrm{Risk}_{\mathrm{opt},\|\cdot\|_p}\geq\frac{r\sigma k^{\frac{1}{2}+\frac{1}{p}}}{r\|A\|_F+\sigma k^{\frac{1}{2}+\frac{1}{p}}}.$$

2.2: This is, basically, the same situation as in 2.1: rearranging the entries of θ to form a $(k=\mu\nu)$-dimensional vector, and the entries in ω to form an mn-dimensional vector, and denoting by A the matrix of the linear transformation $\theta\mapsto L\theta R$ (which now is a linear mapping of $\mathbf{R}^{\mu\nu}$ into $\mathbf{R}^{mn}$), we get $\mathrm{Ker}\,A=0$ and $\|A\|_F=\|L\|_F\|R\|_F$ (why?). As for the value of κ, note that $\|\cdot\|_2$ on the argument space (a.k.a. the Frobenius norm on $\mathbf{R}^{\mu\times\nu}$) is the Euclidean norm of the vector of singular values of a $\mu\times\nu$ matrix, while the Shatten p-norm is the ℓ_p norm of the same vector. Since the number of nonzero singular values in a $\mu\times\nu$ matrix is at most $\min[\mu,\nu]$, and every nonnegative vector of the latter dimension is the vector of singular values of an appropriate $\mu\times\nu$ matrix, we get $\kappa=\min^{\frac{1}{2}-\frac{1}{p}}[\mu,\nu]$. The resulting risk bound is therefore

$$\mathrm{Risk}_{\mathrm{opt},\|\cdot\|_{\mathrm{Sh},p}}\geq\frac{r\sigma\mu\nu\min^{\frac{1}{p}-\frac{1}{2}}[\mu,\nu]}{r\|L\|_F\|R\|_F+\sigma\mu\nu\min^{\frac{1}{p}-\frac{1}{2}}[\mu,\nu]}.$$

2.3: We are in the case where $\ln p(\omega^K,\theta)$ within an additive term independent of θ is $-\sum_{t=1}^K\frac{1}{2}[\omega_t^T\theta^{-1}\omega_t+\ln\mathrm{Det}\theta]$, resulting in

$$\langle\nabla_\theta\ln p(\omega^K,\theta),d\theta\rangle=\frac{1}{2}\sum_{t=1}^K\left[\omega_t^T\theta^{-1}d\theta\theta^{-1}\omega_t-\mathrm{Tr}(\theta^{-1}d\theta)\right],$$

$\langle\cdot,\cdot\rangle$ being the Frobenius inner product on $\mathbf{S}^n$. Consequently, the quadratic form associated with the Fisher Information matrix is

$$\begin{array}{rcl}\langle d\theta,\mathcal{I}(\theta)d\theta\rangle&=&\frac{K}{4}\mathbf{E}_{\omega\sim\mathcal{N}(0,\theta)}\left\{[\omega^T\theta^{-1}d\theta\theta^{-1}\omega-\mathrm{Tr}(\theta^{-1}d\theta)]^2\right\}\\&=&\frac{K}{4}\mathbf{E}_{\eta\sim\mathcal{N}(0,I_n)}\left\{[[\theta^{1/2}\eta]^T\theta^{-1}d\theta\theta^{-1}[\theta^{1/2}\eta]-\mathrm{Tr}(\underbrace{\theta^{-1/2}d\theta\theta^{-1/2}}_{h})]^2\right\}\\&=&\frac{K}{4}\mathbf{E}_{\eta\sim\mathcal{N}(0,I_n)}\left\{[\eta^Th\eta-\mathrm{Tr}(h)]^2\right\}.\end{array}$$

Denoting by λ_i the eigenvalues of h, the latter quantity in the chain is nothing but

$$\begin{array}{l}\frac{K}{4}\mathbf{E}_{\xi\sim\mathcal{N}(0,I_n)}\left\{[\sum_i\xi_i^2\lambda_i-\sum_i\lambda_i]^2\right\}\\=\frac{K}{4}\mathbf{E}_{\xi\sim\mathcal{N}(0,I_n)}\left\{\sum_i\lambda_i^2\xi_i^4+\sum_{i\neq j}\lambda_i\lambda_j\xi_i^2\xi_j^2-2[\sum_i\lambda_i][\sum_j\lambda_j\xi_j^2]+[\sum_i\lambda_i]^2\right\}\\=\frac{K}{4}\left[3\sum_i\lambda_i^2+\sum_{i\neq j}\lambda_i\lambda_j-[\sum_i\lambda_i]^2\right]=\frac{K}{2}\sum_i\lambda_i^2,\end{array}$$

that is,

$$\langle\theta,\mathcal{I}(\theta)d\theta\rangle=\frac{K}{2}\sum_i\lambda_i^2=\frac{K}{2}\mathrm{Tr}(\theta^{-1/2}d\theta\theta^{-1}d\theta\theta^{-1/2})=\frac{K}{2}\mathrm{Tr}(\theta^{-1}d\theta\theta^{-1}d\theta).$$

Now, the function $\mathrm{Tr}(\theta_1^{-1}d\theta\theta_2^{-1}d\theta)$ of $\theta_1\succ0$, $\theta_2\succ0$ is $\succeq$-nonincreasing in θ_1 and θ_2, meaning that replacing θ_1 with $\theta_1'\succeq\theta_1$, the value of the function does not increase, and similarly for θ_2. Indeed, to see that the first claim holds true, it suffices to rewrite the value of the function as $\mathrm{Tr}([\theta_2^{-1/2}d\theta]\theta_1^{-1}[\theta_2^{-1/2}d\theta]^T)$; to observe that the second claim is true, it suffices to rewrite the value as $\mathrm{Tr}([\theta_1^{-1/2}d\theta]\theta_2^{-1}[\theta_1^{-1/2}d\theta]^T)$. It follows that when $\theta \in \Theta$, we have

$$\langle \theta, \mathcal{I}(\theta)d\theta\rangle \leq \tfrac{K}{2}\beta_-^{-2}\mathrm{Tr}(d\theta^2) = \tfrac{K}{2}\beta_-^{-2}\|d\theta\|_F^2$$

where $\|\cdot\|_F$ is the Frobenius norm. Besides this, setting $\|\cdot\| = \|\cdot\|_{\mathrm{Sh},\infty}$, Θ is the $\|\cdot\|$-ball of radius $r = \frac{\beta_+-\beta_-}{2}$ centered at $\frac{\beta_-+\beta_+}{2}I_n$, and, as in the previous item, $\|\cdot\| \leq \|\cdot\|_F \leq \kappa\|\cdot\|$, $\kappa = \sqrt{n}$. We conclude that with the κ, r just defined, and with $\alpha = 2\beta_-^2K^{-1}$, $k = \dim \mathbf{S}_n = \frac{n(n+1)}{2}$, the situation is under the premise of Proposition 4.40, so that by this proposition

$$\mathrm{Risk}_{\mathrm{opt},\|\cdot\|_{\mathrm{Sh},\infty}} \geq \frac{\beta_-[\beta_+/\beta_- - 1]\sqrt{n+1}}{\sqrt{K}[\beta_+/\beta_- - 1] + 2\sqrt{n+1}}.$$

Note. When β_+/β_- is large enough, the lower risk bound can be improved. Indeed, it is clear that for β_+ fixed, the true minimax risk is a decreasing function of β_- (since Θ shrinks as β_- grows), which is not the case with our bound. The inverse of this bound is

$$\rho(\beta_-,\beta_+) = \frac{1}{\sqrt{n+1}}\left[\frac{\sqrt{K}}{\beta_-} + \frac{2\sqrt{n+1}}{\beta_+-\beta_-}\right],$$

which is not an increasing function of β_- in the entire range $0 < \beta_- \leq \beta_+$. Clearly, a better lower risk bound is $\frac{1}{\rho(\beta_*,\beta_+)}$ with β_* obtained by minimization of $\rho(\beta,\beta_+)$ in $\beta \in [\beta_-,\beta_+]$. The resulting bound is

$$\mathrm{Risk}_{\mathrm{opt},\|\cdot\|_{\mathrm{Sh},\infty}} \geq \left\{\begin{array}{ll} \frac{\beta_+\sqrt{n+1}}{\left(\sqrt[4]{K}+\sqrt{2}\sqrt[4]{n+1}\right)^2}, & \beta_- \leq \frac{\sqrt[4]{K}}{\sqrt[4]{K}+\sqrt{2}\sqrt[4]{n+1}}\beta_+,\\ \frac{\beta_-[\beta_+/\beta_- -1]\sqrt{n+1}}{\sqrt{K}[\beta_+/\beta_- -1]+2\sqrt{n+1}}, & \text{otherwise.}\end{array}\right.$$

Exercise 4.24 [More on Cramer-Rao risk bound] Let us fix $\mu \in (1,\infty)$ and a norm $\|\cdot\|$ on $\mathbf{R}^k$, and let $\|\cdot\|_*$ be the norm conjugate to $\|\cdot\|$, and $\mu_* = \frac{\mu}{\mu-1}$. Assume that we are in the situation of item 4.22.B and under assumptions 1) and 3) from this item; as for assumption 2) we now replace it with the assumption that the quantity

$$\mathcal{I}_{\|\cdot\|_*,\mu_*}(\theta) := \left[\mathbf{E}_{\omega\sim p(\cdot,\theta)}\left\{\|\nabla_\theta \ln(p(\omega,\theta))\|_*^{\mu_*}\right\}\right]^{1/\mu_*}$$

is well-defined and bounded on Θ; in the sequel, we set

$$\mathcal{I}_{\|\cdot\|_*,\mu_*} = \sup_{\theta\in\Theta}\mathcal{I}_{\|\cdot\|_*,\mu_*}(\theta).$$

1) Prove the following variant of the Cramer-Rao risk bound:

Proposition 4.41 *In the situation described in the beginning of item 4.22.D, let $\Theta \subset \mathbf{R}^k$ be a $\|\cdot\|$-ball of radius r. Then the minimax $\|\cdot\|$-risk of recovering $\theta \in \Theta$ via observation $\omega \sim p(\cdot,\theta)$ can be lower-bounded as*

$$\begin{array}{l}\mathrm{Risk}_{\mathrm{opt},\|\cdot\|}[\Theta] := \inf\limits_{\widehat{\theta}(\cdot)}\sup\limits_{\theta\in\Theta}\left[\mathbf{E}_{\omega\sim p(\cdot,\theta)}\left\{\|\widehat{\theta}(\omega)-\theta\|^\mu\right\}\right]^{1/\mu} \geq \frac{rk}{r\mathcal{I}_{\|\cdot\|_*,\mu_*}+k},\\ \mathcal{I}_{\|\cdot\|_*,\mu_*} = \max\limits_{\theta\in\Theta}\left[\mathcal{I}_{\|\cdot\|_*,\mu_*}(\theta) := \left[\mathbf{E}_{\omega\sim p(\cdot,\theta)}\left\{\|\nabla_\theta\ln(p(\omega,\theta))\|_*^{\mu_*}\right\}\right]^{1/\mu_*}\right].\end{array} \tag{4.123}$$

Solution: Given a bounded estimate $\widehat{\theta}(\cdot)$, let

$$\begin{array}{rcl}\mathfrak{R} & = & \sup_{\theta\in\Theta}\left[\int \|\widehat{\theta}(\omega)-\theta\|^{\mu}p(\omega,\theta)\Pi(d\omega)\right]^{1/\mu},\\ \phi(\theta) & = & \int \widehat{\theta}(\omega)p(\omega,\theta)\Pi(d\theta).\end{array}$$

From now on, we refer to $\mathfrak{R}$ as to the $\|\cdot\|$-*risk* of the estimate; note that when $\|\cdot\|=\|\cdot\|_2$ and $\mu=2$, this risk becomes what was called estimation risk in item 4.22.B of Exercise 4.22. By Jensen's inequality we have

$$\|\phi(\theta)-\theta\|\leq\mathfrak{R},$$

so that

$$\begin{array}{rl}\mathrm{Tr}(\phi'(\theta)) = \int[\nabla_\theta p(\omega,\theta)]^T\widehat{\theta}(\omega)\Pi(d\theta) = \int[\nabla_\theta p(\omega,\theta)]^T[\widehat{\theta}(\omega)-\theta]\Pi(d\theta) & \\ \leq \int \|\widehat{\theta}(\omega)-\theta\|\|\nabla_\theta\ln(p(\omega,\theta))\|_*p(\omega,\theta)\Pi(d\omega) & \\ \leq \left[\int \|\widehat{\theta}(\omega)-\theta\|^{\mu}p(\omega,\theta)\Pi(d\omega)\right]^{1/\mu}\left[\int\|\nabla_\theta\ln(p(\omega,\theta))\|_*^{\mu_*}p(\omega,\theta)\Pi(d\omega)\right]^{1/\mu_*} & \\ \leq \mathfrak{R}\mathcal{I}_{\|\cdot\|_*,\mu_*}(\theta). & \end{array}$$

On the other hand, from the proof of Proposition 4.40 (see (6.33) in Solution to Exercise 4.23) it follows that if $\Theta\subset\mathbf{R}^k$ is a $\|\cdot\|$-ball of radius r and for some $\gamma\in(0,1)$ one has $\mathfrak{R}\leq\gamma r$, then there exists $\theta_*\in\Theta$ such that

$$\mathrm{Tr}(\phi'(\theta_*))\geq(1-\gamma)k,$$

resulting in

$$(1-\gamma)k\leq\mathfrak{R}\mathcal{I}_{\|\cdot\|_*,\mu_*}(\theta_*).$$

Thus,

$$\mathfrak{R}\geq\min\left[\gamma r,\frac{(1-\gamma)k}{\mathcal{I}_{\|\cdot\|_*,\mu_*}(\theta_*)}\right]\geq\min\left[\gamma r,\frac{(1-\gamma)k}{\mathcal{I}_{\|\cdot\|_*,\mu_*}}\right].$$

We conclude that $\mathfrak{R}\geq\sup_{\gamma\in(0,1)}\min\left[\gamma r,\frac{(1-\gamma)k}{\mathcal{I}_{\|\cdot\|_*,\mu_*}}\right]$, that is,

$$\mathfrak{R}\geq\frac{rk}{r\mathcal{I}_{\|\cdot\|_*,\mu_*}+k}$$

as claimed. □

Example I: Gaussian case, estimating shift. Let $\mu=2$, and let $p(\omega,\theta)=\mathcal{N}(A\theta,\sigma^2I_m)$ with $A\in\mathbf{R}^{m\times k}$. Then

$$\begin{array}{rcl}\nabla_\theta\ln(p(\omega,\theta))=\sigma^{-2}A^T(\omega-A\theta)\Rightarrow & & \\ \int\|\nabla_\theta\ln(p(\omega,\theta))\|_*^2p(\omega,\theta)d\omega & = & \sigma^{-4}\int\|A^T(\omega-A\theta)\|_*^2p(\omega,\theta)d\omega\\ & = & \sigma^{-4}\frac{1}{[\sqrt{2\pi}\sigma]^m}\int\|A^T\omega\|_*^2\exp\{-\frac{\omega^T\omega}{2\sigma^2}\}d\omega\\ & = & \sigma^{-4}\frac{1}{[2\pi]^{m/2}}\int\|A^T\sigma\xi\|_*^2\exp\{-\xi^T\xi/2\}d\xi\\ & = & \sigma^{-2}\frac{1}{[2\pi]^{m/2}}\int\|A^T\xi\|_*^2\exp\{-\xi^T\xi/2\}d\xi\end{array}$$

whence

$$\mathcal{I}_{\|\cdot\|_*,2} = \sigma^{-1} \underbrace{\left[\mathbf{E}_{\xi\sim\mathcal{N}(0,I_m)}\left\{\|A^T\xi\|_*^2\right\}\right]^{1/2}}_{\gamma_{\|\cdot\|}(A)}.$$

Consequently, assuming Θ to be a $\|\cdot\|$-ball of radius r in $\mathbf{R}^k$, lower bound (4.123) becomes

$$\mathrm{Risk}_{\mathrm{opt},\|\cdot\|}[\Theta] \geq \frac{rk}{r\mathcal{I}_{\|\cdot\|_*}+k} = \frac{rk}{r\sigma^{-1}\gamma_{\|\cdot\|}(A)+k} = \frac{r\sigma k}{r\gamma_{\|\cdot\|}(A)+\sigma k}. \tag{4.124}$$

The case of direct observations. To see "how it works," consider the case $m=k$, $A=I_k$ of direct observations, and let $\Theta=\{\theta\in\mathbf{R}^k : \|\theta\|\leq r\}$. Then

- We have $\gamma_{\|\cdot\|_1}(I_k)\leq O(1)\sqrt{\ln(k)}$, whence the $\|\cdot\|_1$-risk bound is

$$\mathrm{Risk}_{\mathrm{opt},\|\cdot\|_1}[\Theta] \geq O(1)\frac{r\sigma k}{r\sqrt{\ln(k)}+\sigma k}; \qquad [\Theta=\{\theta\in\mathbf{R}^k : \|\theta-a\|_1\leq r\}]$$

- We have $\gamma_{\|\cdot\|_2}(I_k)=\sqrt{k}$, whence the $\|\cdot\|_2$-risk bound is

$$\mathrm{Risk}_{\mathrm{opt},\|\cdot\|_2}[\Theta] \geq \frac{r\sigma\sqrt{k}}{r+\sigma\sqrt{k}}; \qquad [\Theta=\{\theta\in\mathbf{R}^k : \|\theta-a\|_2\leq r\}]$$

- We have $\gamma_{\|\cdot\|_\infty}(I_k)\leq O(1)k$, whence the $\|\cdot\|_\infty$-risk bound is

$$\mathrm{Risk}_{\mathrm{opt},\|\cdot\|_\infty}[\Theta] \geq O(1)\frac{r\sigma}{r+\sigma}. \qquad [\Theta=\{\theta\in\mathbf{R}^k : \|\theta-a\|_\infty\leq r\}]$$

In fact, the above examples are essentially covered by the following

Observation 4.42 *Let $\|\cdot\|$ be a norm on $\mathbf{R}^k$, and let*

$$\Theta=\{\theta\in\mathbf{R}^k : \|\theta\|\leq r\}.$$

Consider the problem of recovering signal $\theta\in\Theta$ via observation $\omega\sim\mathcal{N}(\theta,\sigma^2 I_k)$. Let

$$\mathrm{Risk}_{\|\cdot\|}[\widehat{\theta}|\Theta] = \sup_{\theta\in\Theta}\left(\mathbf{E}_{\omega\sim\mathcal{N}(\theta,\sigma^2 I)}\left\{\|\widehat{\theta}(\omega)-\theta\|^2\right\}\right)^{1/2}$$

be the $\|\cdot\|$-risk of an estimate $\widehat{\theta}(\cdot)$, and let

$$\mathrm{Risk}_{\mathrm{opt},\|\cdot\|}[\Theta] = \inf_{\widehat{\theta}(\cdot)}\mathrm{Risk}_{\|\cdot\|}[\widehat{\theta}|\Theta]$$

be the associated minimax risk.

Assume that the norm $\|\cdot\|$ is absolute and symmetric w.r.t permutations of coordinates. Then

$$\mathrm{Risk}_{\mathrm{opt},\|\cdot\|}[\Theta] \geq \frac{r\sigma k}{2\sqrt{\ln(\mathrm{e}k)}r\alpha_*+\sigma k}, \quad \alpha_*=\|[1;...;1]\|_*. \tag{4.125}$$

Here is the concluding part of the exercise:

2) Prove the observation and compare the lower risk bound it yields with the $\|\cdot\|$-risk of the "plug-in" estimate $\widehat{\chi}(\omega)\equiv\omega$.

Solution: By (4.123) we have

$$\mathrm{Risk}_{\mathrm{opt},\|\cdot\|}[\Theta] \geq \frac{r\sigma k}{r\mathcal{I}_* + \sigma k}, \quad \mathcal{I}_* = \left(\mathbf{E}_{\xi\sim\mathcal{N}(0,I_k)}\left\{\|\xi\|_*^2\right\}\right)^{1/2}. \tag{6.34}$$

The norm $\|\cdot\|_*$ is absolute and symmetric along with $\|\cdot\|$, whence the maximum of $\|\cdot\|_*$ over the unit box $\{x : \|x\|_\infty \leq 1\}$ is $\alpha_* = \|[1;...;1]\|_*$. Hence, $\|\xi\|_* \leq \alpha_*\|\xi\|_\infty$, so that

$$\begin{array}{l}
\mathcal{I}_*^2 \leq \alpha_*^2\mathbf{E}_{\xi\sim\mathcal{N}(0,I_k)}\left\{\|\xi\|_\infty^2\right\} = \alpha_*^2\int_0^\infty t^2[-dF(t)] \\
\qquad\qquad \left[F(t) = \mathrm{Prob}_{\xi\sim\mathcal{N}(0,I_k)}\{\|\xi\|_\infty \geq t\} \leq \min[1, k\mathrm{e}^{-t^2/2}]\right] \\
= 2\alpha_*^2\int\limits_0^\infty tF(t)dt \leq 2\alpha_*^2\int\limits_0^\infty t\min[1, k\mathrm{e}^{-t^2/2}]dt = \alpha_*^2\left[2\ln(k) + 2k\int\limits_{\sqrt{2\ln(k)}}^\infty t\mathrm{e}^{-t^2/2}dt\right] \\
= \alpha_*^2[2\ln(k) + 2k\mathrm{e}^{-\ln(k)}] = 2\alpha_*^2[\ln(k)+1],
\end{array}$$

resulting in

$$\mathcal{I}_* \leq \sqrt{2\ln(\mathrm{e}k)}\alpha_*,$$

and (4.125) follows from (6.34). □

Observe that the $\|\cdot\|$-risk of the "plug-in" estimate $\widehat{x}(\omega) \equiv \omega$ is

$$\mathcal{I} := \sigma\left[\mathbf{E}_{\xi\sim\mathcal{N}(0,I_n)}\left\{\|\xi\|^2\right\}\right]^{1/2} \leq \sigma\alpha\sqrt{2\ln(\mathrm{e}k)}, \ \alpha = \|[1;...;1]\|,$$

where the inequality holds for exactly the same reasons as similar inequality for $\mathcal{I}_*$. We clearly have $\alpha\alpha_* = k$ (why?), implying that

$$\mathrm{Risk}_{\|\cdot\|}[\widehat{x}|\Theta] \leq \frac{\sqrt{2\ln(\mathrm{e}k)}\sigma k}{\alpha_*}.$$

In the case of "small σ," specifically,

$$\sigma k \leq \sqrt{2\ln(\mathrm{e}k)}r\alpha_*, \tag{6.35}$$

the lower bound on the minimax $\|\cdot\|$-risk given by the observation is at least $\frac{\sigma k}{2\sqrt{2\ln(\mathrm{e}k)}\alpha_*}$, that is, the plug-in estimate is optimal within the logarithmic factor $4\ln(\mathrm{e}k)$. Note that when (6.35) does not hold the lower risk bound from the observation is at least $\frac{r}{2}$, implying that in this case the trivial estimate $\widehat{x}(\omega) \equiv 0$ is optimal within factor 2. Moreover, when $\|\cdot\| = \|\cdot\|_p$ with $p \in (1,\infty)$, a straightforward refinement of the above computations shows that when $p \in [\pi_*, \pi]$, for some $\pi \in [2,\infty)$, the logarithmic in k factors in the lower bound on the minimax $\|\cdot\|_p$-risk and the upper bound on the $\|\cdot\|_p$-risk of the plug-in estimate can be replaced with factors depending solely on π.

6.4.6 Around $\mathcal{S}$-Lemma

$\mathcal{S}$-**Lemma** is a classical result of extreme importance in Semidefinite Optimization. Basically, the lemma states that when the ellitope $\mathcal{X}$ in Proposition 4.6 is an ellipsoid, (4.19) can be strengthen to $\mathrm{Opt} = \mathrm{Opt}_*$. In fact, $\mathcal{S}$-Lemma is even stronger:

Lemma 4.43. [$\mathcal{S}$-Lemma] *Consider two quadratic forms* $f(x) = x^TAx + 2a^Tx + \alpha$ *and*

$g(x) = x^TBx + 2b^Tx + \beta$ *such that* $g(\bar{x}) < 0$ *for some* $\bar{x}$. *Then the implication*

$$g(x) \leq 0 \Rightarrow f(x) \leq 0$$

takes place if and only if for some $\lambda \geq 0$ *it holds* $f(x) \leq \lambda g(x)$ *for all* x, *or, which is the same, if and only if the Linear Matrix Inequality*

$$\left[\begin{array}{c|c} \lambda B - A & \lambda b - a \\ \hline \lambda b^T - a^T & \lambda\beta - \alpha \end{array}\right] \succeq 0$$

in scalar variable λ *has a nonnegative solution.*

Proof of $\mathcal{S}$-Lemma can be found, e.g., in [15, Section 3.5.2].

The goal of subsequent exercises is to get "tight" tractable outer approximations of sets obtained from ellitopes by quadratic lifting. We fix an ellitope

$$X = \{x \in \mathbf{R}^n : \exists t \in \mathcal{T} : x^TS_kx \leq t_k, 1 \leq k \leq K\} \tag{4.132}$$

where, as always, S_k are positive semidefinite matrices with positive definite sum, and $\mathcal{T}$ is a computationally tractable convex compact subset in $\mathbf{R}^k_+$ such that $t \in \mathcal{T}$ implies $t' \in \mathcal{T}$ whenever $0 \leq t' \leq t$ and $\mathcal{T}$ contains a positive vector.

Exercise 4.25. Let us associate with ellitope X given by (4.132) the sets

$$\begin{array}{rcl} \mathcal{X} & = & \mathrm{Conv}\{xx^T : x \in X\}, \\ \widehat{\mathcal{X}} & = & \{Y \in \mathbf{S}^n : Y \succeq 0, \exists t \in \mathcal{T} : \mathrm{Tr}(S_kY) \leq t_k, 1 \leq k \leq K\}, \end{array}$$

so that $\mathcal{X}$, $\widehat{\mathcal{X}}$ are convex compact sets containing the origin, and $\widehat{\mathcal{X}}$ is computationally tractable along with $\mathcal{T}$. Prove that

1. When $K = 1$, we have $\mathcal{X} = \widehat{\mathcal{X}}$;
2. We always have $\mathcal{X} \subset \widehat{\mathcal{X}} \subset 3\ln(\sqrt{3}K)\mathcal{X}$.

Solution: We clearly have $\mathcal{X} \subset \widehat{\mathcal{X}}$, and both sets are convex compact sets containing the origin. As a result, in order to verify that $\widehat{\mathcal{X}} \subset \kappa\mathcal{X}$ for some $\kappa > 0$, it suffices to show that the support functions

$$\begin{array}{rcl} \phi_{\mathcal{X}}(W) & = & \max\limits_{Y \in \mathcal{X}} \mathrm{Tr}(YW) = \max\limits_{x \in X} x^TWx : \mathbf{S}^n \to \mathbf{R}, \\ \phi_{\widehat{\mathcal{X}}}(W) & = & \max\limits_{Y \in \widehat{\mathcal{X}}} \mathrm{Tr}(YW) \end{array}$$

satisfy the relation

$$\phi_{\widehat{\mathcal{X}}}(W) \leq \kappa\phi_{\mathcal{X}}(W)\ \forall W. \tag{6.36}$$

In the case of $K = 1$ we should prove that (6.36) holds true with $\kappa = 1$. Assuming w.l.o.g. that the largest element in $\mathcal{T} \subset \mathbf{R}_+$ is 1, we have

$$X = \{x : g(x) := x^TS_1x - 1 \leq 0\}.$$

Given $W \in \mathbf{S}^n$ and setting $f(x) = x^TWx - \phi_{\mathcal{X}}(W)$, we ensure the validity of the implication

$$g(x) \leq 0 \Rightarrow f(x) \leq 0,$$

and clearly $g(0) < 0$. Applying $\mathcal{S}$-Lemma, we conclude that $f(x) \leq \lambda g(x)$ for some $\lambda \geq 0$ and all x, whence $W \preceq \lambda S_1$ and $\phi_{\mathcal{X}}(W) \geq \lambda$, so that $\mathrm{Tr}(WY) \leq \lambda\mathrm{Tr}(S_1Y)$ for all $Y \succeq 0$, implying that

$$\forall (Y \in \widehat{\mathcal{X}}) : \mathrm{Tr}(WY) \leq \lambda\mathrm{Tr}(S_1Y) \leq \lambda \leq \phi_{\mathcal{X}}(W).$$

Thus, $\phi_{\widehat{\mathcal{X}}}(W) \leq \phi_{\mathcal{X}}(W)$, that is, (6.36) indeed holds true with $\kappa = 1$.

In the case of $K > 1$ we should prove that (6.36) holds true with $\kappa = 3\ln(\sqrt{3}K)$. We clearly have

$$\left\{W \preceq \sum_k \lambda_k S_k \ \&\ \lambda \geq 0\right\} \Rightarrow \phi_{\widehat{\mathcal{X}}}(W) \leq \phi_{\mathcal{T}}(\lambda),$$

whence

$$\phi_{\widehat{\mathcal{X}}}(W) \leq \min_\lambda \left\{\phi_{\mathcal{T}}(\lambda) : \lambda \geq 0, W \preceq \sum_k \lambda_k S_k\right\}.$$

Now, by Proposition 4.6 the right hand side in this inequality is at most

$$3\ln(\sqrt{3}K) \max_{x\in X} x^T W x = 3\ln(\sqrt{3}K)\phi_{\mathcal{X}}(W),$$

implying that (6.36) holds true with $\kappa = 3\ln(\sqrt{3}K)$. □

Exercise 4.26 For $x \in \mathbf{R}^n$ let $Z(x) = [x; 1][x; 1]^T$, $Z^o[x] = \left[\begin{array}{c|c} xx^T & x \\ \hline x^T & \end{array}\right]$. Let

$$C = \left[\begin{array}{c|c} & \\ \hline & 1 \end{array}\right],$$

and let us associate with ellitope X given by (4.132) the sets

$$\begin{array}{rcl}\mathcal{X}^+ & = & \mathrm{Conv}\{Z^o[x] : x \in X\},\\ \widehat{\mathcal{X}}^+ & = & \left\{Y = \left[\begin{array}{c|c} U & u \\ \hline u^T & \end{array}\right] \in \mathbf{S}^{n+1} : Y + C \succeq 0, \exists t \in \mathcal{T} : \mathrm{Tr}(S_k U) \leq t_k, 1 \leq k \leq K\right\},\end{array}$$

so that $\mathcal{X}^+$, $\widehat{\mathcal{X}}^+$ are convex compact sets containing the origin, and $\widehat{\mathcal{X}}^+$ is computationally tractable along with $\mathcal{T}$. Prove that

1. When $K = 1$, we have $\mathcal{X}^+ = \widehat{\mathcal{X}}^+$;
2. We always have $\mathcal{X}^+ \subset \widehat{\mathcal{X}}^+ \subset 3\ln(\sqrt{3}(K+1))\mathcal{X}^+$.

Solution: Let $\mathbf{S}^{n+1}_o$ be the space of all symmetric $(n+1)\times(n+1)$ matrices W with $W_{n+1,n+1} = 0$; thus, a generic matrix W from $\mathbf{S}^{n+1}_o$ is $W[V, v] := \left[\begin{array}{c|c} V & v \\ \hline v^T & \end{array}\right]$ with $V \in \mathbf{S}^n$ and $v \in \mathbf{R}^n$. We clearly have $\mathcal{X}^+ \subset \widehat{\mathcal{X}}^+ \subset \mathbf{S}^{n+1}_o$, and both sets are convex, compact and contain the origin. As a result, in order to verify that $\widehat{\mathcal{X}}^+ \subset \kappa\mathcal{X}^+$ for some $\kappa > 0$, it suffices to show that the support functions

$$\begin{array}{rcl}\phi_{\mathcal{X}^+}(V, v) & = & \max\limits_{Y\in\mathcal{X}^+} \mathrm{Tr}(W[V, v]Y) = \max\limits_{x\in X}[x^T V x + 2v^T x] : \mathbf{S}^{n+1}_o \to \mathbf{R},\\ \phi_{\widehat{\mathcal{X}}^+}(V, v) & = & \max\limits_{Y\in\widehat{\mathcal{X}}^+} \mathrm{Tr}(W[V, v]Y)\end{array}$$

satisfy the relation

$$\phi_{\widehat{\mathcal{X}}^+}(V, v) \leq \kappa\phi_{\mathcal{X}^+}(V, v)\ \forall(V \in \mathbf{S}^n, v \in \mathbf{R}^n). \tag{6.37}$$

In the case of $K = 1$ we should prove that (6.37) holds true with $\kappa = 1$. Assuming w.l.o.g. that the largest element in $\mathcal{T} \subset \mathbf{R}_+$ is 1, we have

$$X = \{x : g(x) := x^T S_1 x - 1 \leq 0\}.$$

Given V, v and setting

$$f(x) = x^T V x + 2v^T x - \phi_{\mathcal{X}}(W[V,v]) = x^T V x + 2v^T x - \max_{y\in X}[y^T V y + 2v^T y],$$

we ensure the validity of the implication

$$g(x) \leq 0 \Rightarrow f(x) \leq 0,$$

and clearly $g(0) < 0$. Applying $\mathcal{S}$-Lemma, we conclude that $f(x) \leq \lambda g(x)$ for some $\lambda \geq 0$ and all x, whence

$$G := \left[\begin{array}{c|c} \lambda S_1 - V & -v \\ \hline -v^T & \phi_{\mathcal{X}}(V,v) - \lambda \end{array}\right] \succeq 0.$$

Now let $Y = \left[\begin{array}{c|c} U & u \\ \hline u^T & \end{array}\right] \in \widehat{\mathcal{X}}^+$, so that $Y + C \succeq 0$, implying that $\mathrm{Tr}(G(Y+C)) \geq 0$, that is,

$$\lambda \mathrm{Tr}(S_1 U) - \mathrm{Tr}(VU) - 2v^T u + \phi_{\mathcal{X}}(V,v) - \lambda \geq 0.$$

Taking into account that $\mathrm{Tr}(S_1 U) \leq 1$ due to $Y \in \widehat{\mathcal{X}}^+$ and our normalization $\max_{t\in\mathcal{T}} t = 1$ of $\mathcal{T}$, we get

$$\mathrm{Tr}(VU) + 2v^T u \leq \phi_{\mathcal{X}}(V,v)$$

for all $Y = \left[\begin{array}{c|c} U & u \\ \hline u^T & \end{array}\right] \in \widehat{\mathcal{X}}^+$, and we see that (6.37) indeed holds with $\kappa = 1$.

Now let $K > 1$. Consider the ellitope

$$R = X \times [-1,1] = \{[x;s] \in \mathbf{R}^{n+1} : \exists t \in \mathcal{T} : x^T S_k x \leq t_k, 1 \leq k \leq K, s^2 \leq 1\},$$

and let us set

$$R^+ = \{Y \in \mathbf{S}^{n+1} : Y \succeq 0, \exists t \in \mathcal{T} : \mathrm{Tr}\left(\mathrm{Diag}\{S_k, 0\}Y\right) \leq t_k, 1 \leq k \leq K, \mathrm{Tr}(CY) \leq 1\}.$$

We clearly have

$$\begin{array}{rcl} \phi_{\widehat{\mathcal{X}}^+}(V,v) & = & \max\limits_{U,u,t} \left\{ \mathrm{Tr}(VU) + 2v^T u : \mathrm{Tr}(S_k U) \leq t_k, k \leq K, \right. \\ & & \qquad\qquad \left. Y = \left[\begin{array}{c|c} U & u \\ \hline u^T & 1 \end{array}\right] \succeq 0, t \in \mathcal{T} \right\} \\ & = & \max\limits_{Y} \left\{ \mathrm{Tr}\left(\left[\begin{array}{c|c} V & v \\ \hline v^T & \end{array}\right] Y \right) : Y \in R^+ \right\} \\ & \leq & 3\ln(\sqrt{3}(K+1)) \max\limits_{[x;s]\in R = X\times[-1,1]} \left[x^T V x + 2 s v^T x\right] \\ & & \qquad\qquad \text{[Proposition 4.6 as applied to ellitope } R] \\ & = & 3\ln(\sqrt{3}(K+1)) \max\limits_{x\in X} \left[x^T V x + 2v^T x\right] = 3\ln(\sqrt{3}(K+1))\phi_{\mathcal{X}}(V,v), \end{array}$$

and we conclude that (6.37) holds true with $\kappa = 3\ln(\sqrt{3}(K+1))$. □

6.4.7 Miscellaneous exercises

Exercise 4.27. Let $X \subset \mathbf{R}^n$ be a convex compact set, let $b \in \mathbf{R}^n$, and let A be an $m \times n$ matrix. Consider the problem of affine recovery $\omega \mapsto h^T \omega + c$ of the linear function $Bx = b^T x$

of $x \in X$ from indirect observation

$$\omega = Ax + \sigma\xi,\ \xi \sim \mathcal{N}(0, I_m).$$

Given tolerance $\epsilon \in (0,1)$, we are interested in minimizing the worst-case, over $x \in X$, width of $(1-\epsilon)$ confidence interval, that is, the smallest ρ such that

$$\mathrm{Prob}\{\xi : b^Tx - f^T(Ax+\sigma\xi) > \rho\} \leq \epsilon/2\ \&\ \mathrm{Prob}\{\xi : b^Tx - f^T(Ax+\sigma\xi) < \rho\} \leq \epsilon/2\ \forall x \in X.$$

Pose the problem as a convex optimization problem and consider in detail the case where X is the box $\{x \in \mathbf{R}^n : a_j|x_j| \leq 1,\ 1 \leq j \leq n\}$, where $a_j > 0$ for all j.

Solution: In order to get the desired probability bounds for given x and ρ we should have

$$\mathrm{Erfc}\left(\frac{[h^TAx + c - b^Tx] + \rho}{\sigma\|h\|_2}\right) \leq \epsilon/2\ \&\ \mathrm{Erfc}\left(\frac{\rho - [h^TAx + c - b^Tx]}{\sigma\|h\|_2}\right) \leq \epsilon/2,$$

that is,

$$\rho \pm [h^TAx + c - b^Tx] \geq \sigma \mathrm{ErfcInv}(\epsilon/2)\|h\|_2,$$

or, equivalently,

$$\rho \geq |[A^Th - b]^Tx + c| + \sigma \mathrm{ErfcInv}(\epsilon/2)\|h\|_2.$$

Thus, the design problem becomes

$$\min_{h,c}\left\{\psi(h,c) = \max_{x\in X}|[A^Th - b]^Tx + c| + \sigma\mathrm{ErfcInv}(\epsilon/2)\|h\|_2\right\}. \tag{\#}$$

Function ψ clearly is convex (as the supremum, over $x \in X$, of a family of convex functions of h, c parameterized by x), and is efficiently computable, provided X is a computationally tractable convex set; indeed, computing $\psi(h,c)$ reduces to maximizing over $x \in X$ two affine functions of x, namely, $[A^Th - b]^Tx + c$ and $-[A^Th - b]^Tx - c$.

Now, when X is symmetric w.r.t. the origin, we have $\psi(h,c) = \psi(h,-c)$, implying that (#) has an optimal solution with $c = 0$. In the case of $X = \{x : a_j|x_j| \leq 1\}$ we have $\psi(h,0) = \sum_{j=1}^n |\mathrm{Col}_j^T[A]h - b_j|/a_j$, and (#) becomes the convex optimization problem

$$\min_h\left\{\sum\nolimits_{j=1}^n a_j^{-1}|\mathrm{Col}_j^T[A]h - b_j| + \sigma\mathrm{ErfcInv}(\epsilon/2)\|h\|_2\right\}.$$

Exercise 4.28. Prove Proposition 4.21.

Solution: Let $\omega = A\bar{x} + \eta$ with $\bar{x} \in \mathcal{X}$ and $\|\eta\|_{(m)} \leq \sigma$. Problem ($F[\omega]$) clearly is solvable (a solution is $\bar{x}$), so that $\widehat{x} := \widehat{x}_*(\omega)$ is well-defined, and

$$\|A(\bar{x} - \widehat{x}_*(\omega))\|_{(m)} \leq \|A\bar{x} - \omega\|_{(m)} + \|A\widehat{x} - \omega\|_{(m)} \leq 2\sigma.$$

Hence, $\|B\bar{x} - B\widehat{x}\|_{(\nu)} \leq \Upsilon$ by definition of Υ; we have proved the first inequality in (4.66). On the other hand, let $x, y \in \mathcal{X}$ be such that $\|A(x-y)\|_{(m)} \leq 2\sigma$. Setting

$\bar{x} = \frac{1}{2}[x+y]$ and $\zeta = \frac{1}{2}A[x-y]$, we get $\bar{x} \in \mathcal{X}$, $\|\zeta\|_{(m)} \leq \sigma$ and $Ax-\zeta = Ay+\zeta = A\bar{x}$. It follows that the observation $\bar{\omega} = A\bar{x}$ is compatible with signals x and y:

$$\|Ax - \bar{\omega}\|_{(m)} = \|Ay - \bar{\omega}\|_{(m)} \leq \sigma,$$

implying that for every estimate $\widehat{x}(\cdot)$ it holds

$$\mathrm{Risk}_{\mathcal{H},\|\cdot\|}[\widehat{x}|\mathcal{X}] \geq \max[\|Bx - \widehat{x}(\bar{\omega})\|_{(\nu)}, \|By - \widehat{x}(\bar{\omega})\|_{(\nu)}] \geq \tfrac{1}{2}\|Bx - By\|_{(\nu)}.$$

Thus,

$$\mathrm{Risk}_{\mathrm{opt},\sigma} \geq \tfrac{1}{2}\|Bx - By\|_{(\nu)} \ \forall (x, y \in \mathcal{X} : \|A(x-y)\|_{(m)} \leq 2\sigma),$$

implying the second inequality in (4.66). □

Exercise 4.29. Prove Proposition 4.22.

Solution: When $\mathcal{X}$ is the spectratope defined in (4.63), the image of $\mathcal{X} \times \mathcal{X}$ under the linear mapping $(x,y) \mapsto \frac{1}{2}[x-y]$ is exactly $\mathcal{X}$, so that (4.66) reads

$$\begin{array}{rl}
\mathrm{Opt}_{\#} & = 2\max\limits_{x} \left\{\|Bx\|_{(\nu)} : x \in \mathcal{X}, Ax \in \sigma B_{(m)}\right\} \\
& = 2\max\limits_{x,v} \left\{v^T Bx : x \in \mathcal{X}, Ax \in \sigma B_{(m)}, v \in B^*_{(\nu)}\right\} \\
& = 2\max\limits_{x,w} \left\{w^T M^T Bx : x \in \mathcal{X}, Ax \in \sigma B_{(m)}, \exists r \in \mathcal{R} : S_\ell^2[w] \preceq r_\ell I_{f_\ell}, \ell \leq L\right\} \\
& = 2\max\limits_{x,w} \Big\{w^T M^T Bx : \exists (t \in \mathcal{T}, p \in \mathcal{P}, r \in \mathcal{R}) : R_k^2[x] \preceq t_k I_{d_k}, k \leq K, \\
& \qquad\qquad Q_j^2[Ax] \preceq \sigma^2 p_j I_{e_j}, j \leq J, S_\ell^2[w] \preceq r_\ell I_{f_\ell}, \ell \leq L\Big\} \\
& = \max\limits_{z=[w;x]} \Big\{z^T \widehat{B} z : \exists (t \in \mathcal{T}, p \in \sigma^2\mathcal{P}, r \in \mathcal{R}) : R_k^2[x] \preceq t_k I_{d_k}, k \leq K, \\
& \qquad\qquad M_j^2[x] \preceq p_j I_{e_j}, j \leq J, S_\ell^2[w] \preceq r_\ell I_{f_\ell}, \ell \leq L\Big\}
\end{array}$$

where

$$M_j[x] = Q_j[Ax], \widehat{B} = \left[\begin{array}{c|c} & M^T B \\ \hline B^T M & \end{array}\right].$$

By Proposition 4.8, the quantity

$$\begin{array}{rcl}
\mathrm{Opt} & = & \min\limits_{\substack{\Lambda=\{\Lambda_k, k\leq K\},\\ \Upsilon=\{\Upsilon_\ell, \ell\leq L\},\\ \Sigma=\{\Sigma_j, j\leq J\}}} \Big\{\phi_{\mathcal{T}}(\lambda[\Lambda]) + \phi_{\mathcal{R}}(\lambda[\Upsilon]) + \sigma^2\phi_{\mathcal{P}}(\lambda[\Sigma]) + \phi_{\mathcal{R}}(\lambda[\Sigma]) : \\
& & \Lambda_k \succeq 0, \Upsilon_\ell \succeq 0, \Sigma_j \succeq 0 \, \forall (k, \ell, j), \\
& & \left[\begin{array}{c|c} \sum_\ell \mathcal{S}_\ell^*[\Upsilon_\ell] & M^T B \\ \hline B^T M & \sum_k \mathcal{R}_k^*[\Lambda_k] + \sum_j \mathcal{M}_j^*[\Sigma_j] \end{array}\right] \succeq 0 \Big\}
\end{array}$$

upper-bounds $\mathrm{Opt}_{\#}$, and this bound is tight within the factor $2\max[\ln(2D), 1]$, $D = \sum_k d_k + \sum_\ell f_\ell + \sum_\ell e_j$. It remains to note that, as it is immediately seen, $\mathcal{M}_j^*[\Sigma_j] = A^T \mathcal{Q}_j^*[\Sigma_j] A$. □

6.5 SOLUTIONS FOR CHAPTER 5

6.5.1 Estimation by Stochastic Optimization

Exercise 5.1. Consider the following "multinomial" version of the logistic regression problem from Section 5.2.1:
For $k = 1, ..., K$, we observe pairs

$$(\zeta_k, \ell_k) \in \mathbf{R}^n \times \{0, 1, ..., m\} \tag{5.70}$$

drawn independently of each other from a probability distribution P_x parameterized by an unknown signal $x = [x^1; ...; x^m] \in \mathbf{R}^n \times ... \times \mathbf{R}^n$ as follows:

- The probability distribution of *regressor* ζ induced by the distribution S_x of (ζ, ℓ) is a once and forever fixed, and independent of x, distribution R on $\mathbf{R}^n$ with finite second order moments and a positive definite matrix $Z = \mathbf{E}_{\zeta \sim R}\{\zeta\zeta^T\}$ of second order moments;
- The conditional probability distribution of *label* ℓ given ζ, induced by the distribution S_x of (ζ, ℓ) is the distribution of a discrete random variable taking value $\iota \in \{0, 1, ..., m\}$ with probability

$$p_\iota = \left\{ \begin{array}{ll} \frac{\exp\{\zeta^T x^\iota\}}{1+\sum_{i=1}^m \exp\{\zeta^T x^i\}}, & 1 \leq \iota \leq m, \\ \frac{1}{1+\sum_{i=1}^m \exp\{\zeta^T x^i\}}, & \iota = 0. \end{array} \right. \qquad [x = [x^1; ...; x^m]]$$

Given a nonempty convex compact set $\mathcal{X} \in \mathbf{R}^{mn}$ known to contain the (unknown) signal x underlying observations (5.70), we want to recover x. Note that the recovery problem associated with the standard logistic regression model is the case $m = 1$ of the problem just defined.

Your task is to process the above recovery problem via the approach developed in Section 5.2 and to compare the resulting SAA estimate with the Maximum Likelihood estimate.

<u>Solution:</u> Let us associate with $\zeta \in \mathbf{R}^n$ the $m \times n$ matrix

$$\eta[\zeta] = \mathrm{Diag}\{\underbrace{\zeta, ..., \zeta}_{m}\} = \left[\begin{array}{c|c|c|c} \zeta & & & \\ \hline & \zeta & & \\ \hline & & \ddots & \\ \hline & & & \zeta \end{array}\right]$$

and, as always, let us associate with labels $\ell_k \in \{0, 1, ..., m\}$ vectors $y_k = y[\ell_k]$, where $y[0] = 0$ and $y[\iota] = e_\iota$, $1 \leq \iota \leq m$, $e_1, ..., e_m$ being the standard basic orths in $\mathbf{R}^m$. Finally, consider the vector field

$$\begin{array}{rcl} f([s_1; ...; s_m]) & = & \frac{[\exp\{s_1\}; ...; \exp\{s_m\}]}{1+\sum_{i=1}^m \exp\{s_i\}} \\ & = & \nabla_s \underbrace{\ln\left(1 + \sum_{i=1}^m \exp\{s_i\}\right)}_{\phi(s)} : \mathbf{R}^m \to \mathbf{R}^m. \end{array}$$

With these conventions, observations (5.70) stemming from a signal x can be converted, in a deterministic fashion, into pairs

$$\omega_k = (\eta_k = \eta[\zeta_k], y_k = y[\ell_k]) \in \mathbf{R}^{mn \times m} \times \mathbf{R}^m$$

which are i.i.d. samples drawn from distribution P_x parameterized by x of (η, y) as follows:

- the distribution Q of $\zeta \in \mathbf{R}^{mn\times m}$ is the distribution of $\eta[\zeta]$ induced by the distribution R of ζ; Q is independent of x, has finite second moments, and

$$\mathbf{E}_{\eta\sim Q}\{\eta\eta^T\} = \mathrm{Diag}\{\underbrace{Z,...,Z}_{m}\}$$

is positive definite along with $Z = \mathbf{E}_{\zeta\sim R}\{\zeta\zeta^T\}$;
- Conditional distribution of y given $\eta = \eta[\zeta]$, induced by distribution P_x of $\omega = (\eta, y)$ is a discrete distribution supported on the set $\{0, e_1, ..., e_m\} \subset \mathbf{R}^m$, and the expectation of this conditional distribution is $f(\eta^T x)$.
Note that f is the gradient field of the smooth convex function ϕ, and the Hessian of ϕ is positive definite everywhere. Consequently, f is a continuous vector field which is strongly monotone on bounded sets.

We find ourselves in the situation described in Section 5.2.1 and under Assumptions **A.1–3** (verification of these assumptions is immediate, with strong monotonicity of F on bounded sets readily given by the similar property of f combined with positive definiteness of $\mathbf{E}_{\eta\sim Q}\{\eta\eta^T\}$) and can therefore utilize the constructions and estimates developed in Section 5.2. In particular, the vector field $G_{(\eta_k,y_k)}(z)$—see (5.55)—$z = [z^1; ...; z^m] \in \mathbf{R}^{mn}$ now becomes

$$\begin{array}{rcl} G_{(\eta_k,y_k)}(z) &=& \eta_k f(\eta_k^T z) - \eta_k y_k = \eta[\zeta_k] f(\eta^T[\zeta_k]z) - \eta[\zeta_k]y[\ell_k] \\ &=& \frac{[\exp\{\zeta_k^T z^1\}\zeta_k;...;\exp\{\zeta_k^T z^m\}\zeta_k]}{1+\sum_{i=1}^m \exp\{\zeta_k^T z^i\}} - \eta[\zeta_k]y_k \\ &=& \nabla_z \left[\ln\left(1+\sum_{i=1}^m \exp\{\zeta_k^T z^i\}\right) - d_k^T z\right], \\ d_k &=& \left\{\begin{array}{ll} \underbrace{[0;...;0]}_{mn}, & \ell_k = 0; \\ [\underbrace{0;...;0}_{n(\ell_k-1)};\zeta_k;0;...;0], & \ell_k \neq 0. \end{array}\right. \end{array}$$

On the other hand, the Maximum Likelihood estimate in multinomial logistic regression, the observations being $\{(\zeta_k, \ell_k), k \le K\}$, is the minimizer, over $z = [z^1; ...; z^m] \in \mathbf{R}^m$, of the conditional negative log-likelihood $L(z)$ given $\{\zeta_k, k \le K\}$,

$$\begin{array}{rcl} L(z) &=& \sum_{k=1}^K L_k(z), \\ L_k(z) &=& \left\{\begin{array}{ll} -\ln\left(\frac{\exp\{\zeta_k z^{\ell_k}\}}{1+\sum_{i=1}^m \exp\{\zeta_k^T z^i\}}\right), & \ell_k > 0 \\ -\ln\left(\frac{1}{1+\sum_{i=1}^m \exp\{\zeta_k^T z^i\}}\right), & \ell_k = 0 \end{array}\right. \\ &=& \ln\left(1+\sum_{i=1}^m \exp\{\zeta_k^T z^i\}\right) - d_k^T z, \end{array}$$

whence

$$G_{(\eta_k,y_k)}(z) = \nabla_z L_k(z),$$

that is, the SAA estimate coincides with that by Maximum Likelihood.

Exercise 5.2 Let

$$H(x) : \mathbf{R}^n \to \mathbf{R}^n$$

be a strongly monotone and Lipschitz continuous on the entire space vector field:

$$\begin{array}{rl} \forall (x, x' \in \mathbf{R}^n) : & [H(x) - H(x')]^T[x - x'] \ge \varkappa \|x - x'\|_2, \\ & \|H(x) - H(x')\|_2 \le L\|x - x'\|_2 \end{array}$$

for some $\varkappa > 0$ and $L < \infty$.

1.1) Prove that for every $x \in \mathbf{R}^n$, the vector equation

$$H(z) = x$$

in variable $z \in \mathbf{R}^n$ has unique solution (which we denote $H^{-1}(x)$), and that for every $x, y \in \mathbf{R}^n$ one has

$$\|H^{-1}(x) - y\|_2 \leq \varkappa^{-1}\|x - H(y)\|_2. \tag{5.72}$$

1.2) Prove that the vector field

$$x \mapsto H^{-1}(x)$$

is strongly monotone with modulus

$$\varkappa_* = \varkappa/L^2$$

and Lipschitz continuous, with constant $1/\varkappa$ w.r.t. $\|\cdot\|_2$, on the entire $\mathbf{R}^n$.

Solution: **1:** Given $x, y \in \mathbf{R}^n$, let $H_x(\cdot) = H(\cdot) - x$, and let $\mathcal{X}$ be the $\|\cdot\|_2$-ball of radius $> \|H(y) - x\|_2/\varkappa$ centered at y. Let, next, z be a weak (and thus strong, since H_x is continuous) solution to $\mathrm{VI}(H_x, \mathcal{X})$ (as we know, such a solution exists). Since z is a strong solution to the VI and H_x is strongly monotone with modulus $\varkappa$ along with H, we have

$$H_x^T(y)[y - z] \geq H_x^T(z)[y - z] + \varkappa\|y - z\|_2^2 \geq \varkappa\|y - z\|_2^2,$$

whence $\|H_x(y)\|_2\|y - z\|_2 \geq \varkappa\|y - z\|_2^2$, that is, $\|y - z\|_2 \leq \|H_x(y)\|_2/\varkappa$. In particular, z is an interior point of $\mathcal{X}$, and since z is a strong solution to $\mathrm{VI}(H_x, \mathcal{X})$, we should have $H_x^T(z)[z' - z] \geq 0$ for all z' from a neighborhood of z, implying that $H_x(z) = 0$, that is $H(z) = x$. Thus, the equation $H(\cdot) = x$ has a solution, namely, z. This solution is unique, since for every two solutions z, z' to the equation we should have

$$0 = [H(z) - H(z')]^T[z - z'] \geq \varkappa\|z - z'\|_2^2.$$

As a byproduct of our reasoning, we have seen that $\|H^{-1}(x) - y\|_2 \leq \|H_x(y)\|_2/\varkappa = \|H(y) - x\|_2/\varkappa$, which is (5.72). □

2: Let $x, x' \in \mathbf{R}^n$, and let $y = H^{-1}(x)$, $y' = H^{-1}(x')$, so that $H(y) = x$ and $H(y') = x'$. By strong monotonicity of H we have

$$\begin{array}{rcl} \varkappa\|y - y'\|^2 & \leq & [H(y) - H(y')]^T[y - y'] = [x - x']^T[y - y'] \\ & = & [H^{-1}(x) - H^{-1}(x')]^T[x - x'], \end{array} \tag{$*$}$$

and by Lipschitz continuity of H we have $\|x - x'\|_2 = \|H(y) - H(y')\|_2 \leq L\|y - y'\|_2$. We conclude that $\|y - y'\|_2 \geq L^{-1}\|x - x'\|_2$ which combines with $(*)$ to imply that

$$[H^{-1}(x) - H^{-1}(x')]^T[x - x'] \geq L^{-2}\varkappa\|x - x'\|_2^2.$$

Thus, H^{-1} is strongly monotone with modulus $\varkappa/L^2$ on the entire $\mathbf{R}^n$, as claimed. Next, given $x, x' \in \mathbf{R}^n$, let us set $y = H^{-1}(x')$, so that $H(y) = x'$. Applying (5.72), we get

$$\|H^{-1}(x) - y\|_2 \leq \|H(y) - x\|_2/\varkappa$$

which is nothing but

$$\|H^{-1}(x) - H^{-1}(x')\|_2 \leq \|x - x'\|_2/\varkappa.$$

□

Let us interpret $-H(\cdot)$ as a field of "reaction forces" applied to a particle: when particle is in a position $y \in \mathbf{R}^n$ the reaction force applied to the particle is $-H(y)$. Next, let us interpret $x \in \mathbf{R}^n$ as an external force applied to the particle. An equilibrium y is a point in space where the reaction force $-H(y)$ compensates the external force, that is, $H(y) = x$, or, which is the same, $y = H^{-1}(x)$. Note that with this interpretation, strong monotonicity of H makes perfect sense, implying that the equilibrium in question is stable: when the particle is moved from the equilibrium $y = H^{-1}(x)$ to a position $y + \Delta$, the total force acting at the particle becomes $f = x - H(y + \Delta)$, so that

$$f^T\Delta = [x - H(y+\Delta)]^T\Delta = [H(y) - H(y+\Delta)]^T[\Delta] \leq -\varkappa\Delta^2,$$

that is, the force is oriented "against" the displacement Δ and "wants" to return the particle to the equilibrium position.

Now imagine that we can observe in noise equilibrium $H^{-1}(x)$ of the particle, the external force x being unknown, and want to recover x from our observation. For the sake of simplicity, let the observation noise be zero mean Gaussian, so that our observation is

$$\omega = H^{-1}(x) + \sigma\xi, \ \xi \sim \mathcal{N}(0, I_n).$$

2) Verify that the recovery problem we have posed is a special case of the "single observation" recovery problem from Section 5.2.6, with $\mathbf{R}^n$ in the role of $\mathcal{X}$ [21] and that the SAA estimate $\widehat{x}(\omega)$ from that section under the circumstances is just the root of the equation

$$H^{-1}(\cdot) = \omega,$$

that is,

$$\widehat{x}(\omega) = H(\omega).$$

Prove also that

$$\mathbf{E}\{\|\widehat{x}(\omega) - x\|_2^2\} \leq n\sigma^2 L^2. \tag{5.73}$$

Note that in the situation in question the ML estimate should be the minimizer of the function

$$f(z) = \|\omega - H^{-1}(z)\|_2^2,$$

and this minimizer is nothing but $\widehat{x}(\omega)$.

Solution: The required verification is straightforward: in the notation from Section 5.2.6 we should set $K = 1$, $m = n$, $\eta = I_n$, and $y \equiv \omega$. To verify (5.73), note that $\widehat{x}(\omega) = H(\omega) = H(\sigma\xi + H^{-1}(x))$ and $x = H(H^{-1}(x))$, so that by Lipschitz continuity of H it holds

$$\|\widehat{x}(\omega) - x\|_2 = \|H(\sigma\xi + H^{-1}(x)) - H(H^{-1}(x))\| \leq L\|\sigma\xi\|_2.$$

Exercise 5.3 [identification of parameters of a linear dynamic system] Consider the problem as follows:

A deterministic sequence $x = \{x_t : t \geq -d+1\}$ satisfies the linear finite-difference equation

$$\sum_{i=0}^{d} \alpha_i x_{t-i} = y_t, \ t = 1, 2, \ldots \tag{5.74}$$

[21] In Section 5.2.6, $\mathcal{X}$ was assumed to be closed, convex, *and bounded*; a straightforward inspection shows that when the vector field ϕ is strongly monotone, with some positive modulus, on the entire space, and η has trivial kernel, all constructions and results of Section 5.2.6 can be extended to the case of an arbitrary closed convex $\mathcal{X}$.

of given order d and is bounded,

$$|x_t| \leq M_x < \infty, \forall t \geq -d+1,$$

implying that the sequence $\{y_t\}$ also is bounded:

$$|y_t| \leq M_y < \infty, \forall t \geq 1.$$

The vector $\alpha = [\alpha_0; ...; \alpha_d]$ is unknown, all we know is that this vector belongs to a given closed convex set $\mathcal{X} \subset \mathbf{R}^{d+1}$. We have at our disposal observations

$$\omega_t = x_t + \sigma_x \xi_t, \ -d+1 \leq t \leq K, \tag{5.75}$$

of the terms in the sequence, with $\xi_t \sim \mathcal{N}(0,1)$ independent across t, with some given σ_x, and observations

$$\zeta_t = y_t + \sigma_y \eta_t \tag{5.76}$$

with $\eta_t \sim \mathcal{N}(0,1)$ independent across t and independent of $\{\xi_\tau\}_\tau$. Our goal is to recover from these observations the vector α.

Strategy. To get the rationale underlying the construction to follow, let us start with the case when there is no observation noise at all: $\sigma_x = \sigma_y = 0$. In this case we could act as follows: let us denote

$$x^t = [x_t; x_{t-1}; x_{t-2}; ...; x_{t-d}], \ 1 \leq t \leq K,$$

and rewrite (5.74) as

$$[x^t]^T \alpha = y_t, \ 1 \leq t \leq K.$$

When setting

$$A_K = \frac{1}{K}\sum_{t=1}^K x^t [x^t]^T, \ a_K = \frac{1}{K}\sum_{t=1}^K y_t x^t,$$

we get

$$A_K \alpha = a_K. \tag{5.77}$$

Assuming that K is large and trajectory x is "rich enough" to ensure that A_K is nonsingular, we could identify α by solving the linear system (5.77).

Now, when the observation noise is present, we could try to use the noisy observations of x^t and y_t we have at our disposal in order to build empirical approximations to A_K and a_K which are good for large K, and identify α by solving the "empirical counterpart" of (5.77). The straightforward way would be to define ω^t as an "observable version" of x^t,

$$\omega^t = [\omega_t; \omega_{t-1}; ...; \omega_{t-d}] = x^t + \sigma_x \underbrace{[\xi_t; \xi_{t-1}; ...; \xi_{t-d}]}_{\xi^t}$$

and to replace A_K and a_K with

$$\widetilde{A}_t = \frac{1}{K}\sum_{t=1}^K \omega^t [\omega^t]^T, \ \widetilde{a}_K = \sum_{t=1}^K \zeta_t \omega^t.$$

As far as empirical approximation of a_K is concerned, this approach works: we have

$$\widetilde{a}_K = a_K + \delta a_K, \ \delta a_K = \frac{1}{K}\sum_{t=1}^K \underbrace{[\sigma_x y_t \xi^t + \sigma_y \eta_t x^t + \sigma_x \sigma_y \eta_t \xi^t]}_{\delta_t}.$$

Since the sequence $\{y_t\}$ is bounded, the random error δa_K of approximation $\widetilde{a}_K$ of a_K is small for large K with overwhelming probability. Indeed, δa_K is the average of K zero mean random vectors δ_t (recall that ξ^t and η_t are independent and zero mean) with[22]

$$\mathbf{E}\{\|\delta_t\|_2^2\} \leq 3(d+1)\left[\sigma_x^2 M_y^2 + \sigma_y^2 M_x^2 + \sigma_x^2 \sigma_y^2\right]$$

[22]We use the elementary inequality $\|\sum_{t=1}^p a_t\|_2^2 \leq p \sum_{t=1}^p \|a_t\|_2^2$.

and δ_t is independent of δ_s whenever $|t-s|>d+1$, implying that

$$\mathbf{E}\{\|\delta a_K\|_2^2\} \leq \frac{3(d+1)(2d+1)\left[\sigma_x^2 M_y^2 + \sigma_y^2 M_x^2 + \sigma_x^2\sigma_y^2\right]}{K}. \tag{5.78}$$

The quality of approximating A_K with $\widetilde{A}_K$ is essentially worse: setting

$$\delta A_K = \widetilde{A}_K - A_K = \frac{1}{K}\sum_{t=1}^{K}\underbrace{[\sigma_x^2 \xi^t[\xi^t]^T + \sigma_x \xi^T[x^t]^T + \sigma_x x^t[\xi^t]^T]}_{\Delta_t}$$

we see that δA_K is the average of K random matrices Δ_t with *nonzero* mean, namely, the mean $\sigma_x^2 I_{d+1}$ and as such ΔA_K is "large" independently of how large K is. There is, however, a simple way to overcome this difficulty – *splitting observations* ω_t.[23]

Splitting observations. Let θ be a random n-dimensional vector with an unknown mean μ and known covariance matrix, namely, $\sigma^2 I_n$, and let $\chi \sim \mathcal{N}(0, I_n)$ be independent of θ. Finally, let $\kappa > 0$ be a deterministic real.

1) Prove that setting

$$\theta' = \theta + \sigma\kappa\chi,\ \theta'' = \theta - \sigma\kappa^{-1}\chi,$$

we get two random vectors with mean μ and covariance matrices $\sigma^2(1+\kappa^2)I_n$ and $\sigma^2(1+1/\kappa^2)I_n$, respectively, and these vectors are uncorrelated

$$\mathbf{E}\{[\theta'-\mu][\theta''-\mu]^T\} = 0.$$

Solution: θ', θ'' clearly are with mean μ, and their covariance matrices are exactly as stated above. Besides this, setting $\delta = \sigma^{-1}(\theta-\mu)$, so that δ has zero mean and covariance matrix I_n, and is independent of $\chi \sim \mathcal{N}(0, I_n)$, we have

$$\begin{array}{l}\mathbf{E}\{[\theta'-\mu][\theta''-\mu]^T\} = \sigma^2\mathbf{E}\{[\delta+\kappa\chi][\delta-\kappa^{-1}\chi]^T\}\\ = \sigma^2\left[\underbrace{\mathbf{E}\{\delta\delta^T\}}_{=I_n} + \kappa\underbrace{\mathbf{E}\{\chi\delta^T\}}_{=0} - \kappa^{-1}\underbrace{\mathbf{E}\{\delta\chi^T\}}_{=0} - \underbrace{\mathbf{E}\{\chi\chi^T\}}_{=I_n}\right]\\ = 0.\end{array}$$

□

In view of item 1, let us do as follows: given observations $\{\omega_t\}$ and $\{\zeta_t\}$, let us generate i.i.d. sequence $\{\chi_t \sim \mathcal{N}(0,1), t \geq -d+1\}$, so that the sequences $\{\xi_t\}$, $\{\eta_t\}$, and $\{\chi_t\}$ are i.i.d. and independent of each other, and let us set

$$u_t = \omega_t + \sigma_x\chi_t,\ v_t = \omega_t - \sigma_x\chi_t.$$

Note that given the sequence $\{\omega_t\}$ of actual observations, sequences $\{u_t\}$ and $\{v_t\}$ are observable as well, and that the sequence $\{(u_t, v_t)\}$ is i.i.d.. Moreover, for all t,

$$\mathbf{E}\{u_t\} = \mathbf{E}\{v_t\} = x_t,\ \mathbf{E}\{[u_t - x_t]^2\} = 2\sigma_x^2,\ \mathbf{E}\{[v_t - x_t]^2\} = 2\sigma_x^2,$$

and for all t and s

$$\mathbf{E}\{[u_t - x_t][v_s - x_s]\} = 0.$$

Now, let us put

$$u^t = [u_t; u_{t-1}; ...; u_{t-d}],\quad v^t = [v_t; v_{t-1}; ...; v_{t-d}],$$

[23]The model (5.74)– (5.76) is referred to as the Errors in Variables model [85] in the statistical literature or Output Error model in the literature on System Identification [173, 218]. In general, statistical inference for such models is difficult—for instance, the parameter estimation problem in such models is ill-posed. The estimate we develop in this exercise can be seen as a special application of the general Instrumental Variables methodology [7, 219, 241].

and let

$$\widehat{A}_K = \frac{1}{K}\sum_{t=1}^{K} u^t[v^t]^T.$$

2) Prove that $\widehat{A}_K$ is a good empirical approximation of A_K:

$$\mathbf{E}\{\widehat{A}_K\} = A_K, \quad \mathbf{E}\{\|\widehat{A}_K - A_K\|_F^2\} \leq \frac{12[d+1]^2[2d+3]\left[M_x^2+\sigma_x^2\right]\sigma_x^2}{K}, \tag{5.79}$$

the expectation being taken over the distribution of observation noises $\{\xi_t\}$ and auxiliary random sequence $\{\chi_t\}$.

Solutions: Setting

$$\begin{array}{c}\delta u_t = u_t - x_t,\ \delta v_t = v_t - x_t,\\ \delta u^t = [\delta u_t;\delta u_{t-1};...;\delta u_{t-d}],\ \delta v^t = [\delta v_t;\delta v_{t-1};...;\delta v_{t-d}],\\ \Delta = \widehat{A}_K - A_K,\end{array}$$

we have

$$\Delta = \frac{1}{K}\sum_{t=1}^{K}\underbrace{\left[\delta u^t[\delta v^t]^T + \delta_u^t[x^t]^T + x^t[\delta v^t]^T\right]}_{\Delta_t}.$$

Since the joint distribution of $\{\delta u_t, \delta v_t, -d+1 \leq t \leq K\}$ is Gaussian and these random variables are mutually uncorrelated, they are mutually independent; besides this, these variables are zero mean with variance $2\sigma_x^2$. As a result, the random matrices Δ_t are with zero mean. Besides this, when $|t-s| > d+1$, the matrices Δ_t and Δ_s are independent of each other.

We have

$$\begin{array}{l}\mathbf{E}\{\|\Delta_t\|_F^2\} \leq 3\left[\mathbf{E}\{\|\delta u^t[\delta v^t]^T\|_F^2\} + \mathbf{E}\{\|\delta u^t[x^t]^T\|_F^2\} + \mathbf{E}\{\|x^t[\delta v^t]^T\|_F^2\}\right]\\ = 3\left[\mathbf{E}\{\sum\limits_{t-d\leq r,s\leq t}[\delta u_r]^2[\delta v_s]^2\} + \mathbf{E}\{\sum\limits_{t-d\leq r,s\leq t}[\delta u_r]^2 x_s^2\} + \mathbf{E}\{\sum\limits_{t-d\leq r,s\leq t} x_r^2[\delta v_s]^2\}\right]\\ \leq \gamma^2 := 12[d+1]^2\left[M_x^2+\sigma_x^2\right]\sigma_x^2\end{array}$$

(recall we are in the case where $\delta u_t \sim \mathcal{N}(0, 2\sigma_x^2)$ with $\delta v_t \sim \mathcal{N}(0, 2\sigma_x^2)$, and δu_t and δv_s independent of each other for all t, s). It follows that

$$\begin{array}{rl}\mathbf{E}\{\|\Delta\|_F^2\} = & \frac{1}{K^2}\sum_{1\leq t,s\leq K}\mathbf{E}\{\mathrm{Tr}(\Delta_t\Delta_s^T)\}\\ = & \frac{1}{K^2}\sum_{1\leq t,s\leq K, \atop |t-s|\leq d+1}\mathbf{E}\{\mathrm{Tr}(\Delta_t\Delta_s^T)\}\\ & [\text{since } \Delta_t, \Delta_s \text{ are zero mean and mutually independent when } |t-s| > d+1]\\ \leq & \frac{1}{K^2}\sum_{1\leq t,s\leq K, \atop |t-s|\leq d+1}\gamma^2 \ [\text{by the Cauchy inequality}]\\ \leq & \frac{2d+3}{K}\gamma^2.\end{array}$$

□

Conclusion. We see that as $K \to \infty$, the differences of typical realizations of $\widehat{A}_K - A_K$ and $\widetilde{a}_K - a_K$ approach 0. It follows that if the sequence $\{x_t\}$ is "rich enough" to ensure that the minimal eigenvalue of A_K for large K stay bounded away from 0, the estimate

$$\widehat{\alpha}_K \in \mathop{\mathrm{Argmin}}_{\beta\in\mathbf{R}^{d+1}}\|\widehat{A}_K\beta - \widetilde{a}_K\|_2^2$$

will converge in probability to the desired vector α, and we can even say something reasonable about the rate of convergence. To account for a priori information $\alpha \in \mathcal{X}$, we can modify the estimate by setting

$$\widehat{\alpha}_K \in \mathop{\mathrm{Argmin}}_{\beta\in\mathcal{X}} \|\widehat{A}_K\beta - \widetilde{\alpha}_K\|_2^2.$$

Note that the assumption that noises affecting observations of x_t's and y_t's are zero mean *Gaussian* random variables independent of each other with known dispersions is not that important; we could survive the situation where samples $\{(\omega_t - x_t, t > -d\}$, $\{\zeta_t - y_t, t \geq 1\}$ are zero mean i.i.d., and independent of each other, *with a priori known variance of* $\omega_t - x_t$. Under this and mild additional assumptions (like finiteness of the fourth moments of $\omega_t - x_t$ and $\zeta_t - y_t$), the results obtained would be similar to those for the Gaussian case.

Now comes the concluding part of the exercise:

3) To evaluate numerically the performance of the proposed identification scheme, run experiments as follows:

   - Given an even value of d and $\rho \in (0,1]$, select $d/2$ complex numbers λ_i at random on the circle $\{z \in \mathbf{C} : |z| = \rho\}$, and build a real polynomial of degree d with roots λ_i, λ_i^* (* here stands for complex conjugation). Build a finite-difference equation (5.77) with this polynomial as the characteristic polynomial.
   - Generate i.i.d. $\mathcal{N}(0,1)$ "inputs" $\{y_t, t = 1, 2, ...\}$, select at random initial conditions $x_{-d+1}, x_{-d+2}, ..., x_0$ for the trajectory $\{x_t\}$ of states (5.77), and simulate the trajectory along with observations ω_t of x_t and ζ_t of y_t, with σ_x, σ_y being the experiment's parameters.
   - Look at the performance of the estimate $\widehat{\alpha}_K$ on the simulated data.

Solution: These are our results (obtained via 40 simulations for every collection of experiment's parameters with $\sigma_x = \sigma_y = \sigma$):

			$\|\cdot\|_2$-error of recovery		
d	σ	ρ	mean	median	max
2	1.0	1.0	0.0032	0.0024	0.012
2	1.0	0.8	0.0585	0.0544	0.146
2	0.5	1.0	0.0018	0.0014	0.015
2	0.5	0.8	0.0251	0.0186	0.088
2	0.1	1.0	0.0003	0.0002	0.001
2	0.1	0.8	0.0038	0.0030	0.015
6	1.0	1.0	2.5353	0.0393	47.59
6	1.0	0.8	2.9813	1.0660	19.29
6	0.5	1.0	1.3901	0.0170	27.24
6	0.5	0.8	1.5914	0.4164	19.73
6	0.1	1.0	0.3452	0.0015	13.68
6	0.1	0.8	0.0677	0.0349	0.426
10	1.0	1.0	26.362	0.4579	389.7
10	1.0	0.8	9.9474	2.9212	90.14
10	0.5	1.0	5.0241	0.1142	144.5
10	0.5	0.8	9.4643	2.8923	97.38
10	0.1	1.0	26.671	0.0132	241.8
10	0.1	0.8	2.9864	0.1886	46.50

$K = 1000$

			$\|\cdot\|_2$-error of recovery		
d	σ	ρ	mean	median	max
2	1.0	1.0	0.0000	0.0000	0.000
2	1.0	0.8	0.0061	0.0055	0.027
2	0.5	1.0	0.0000	0.0000	0.000
2	0.5	0.8	0.0028	0.0027	0.006
2	0.1	1.0	0.0000	0.0000	0.000
2	0.1	0.8	0.0004	0.0004	0.001
6	1.0	1.0	0.5370	0.0002	20.82
6	1.0	0.8	0.6703	0.0532	12.74
6	0.5	1.0	0.5929	0.0001	15.71
6	0.5	0.8	0.0704	0.0185	1.52
6	0.1	1.0	1.8412	0.0000	46.85
6	0.1	0.8	0.0078	0.0025	0.121
10	1.0	1.0	8.2724	0.0006	123.5
10	1.0	0.8	9.7451	2.1950	70.97
10	0.5	1.0	16.543	0.0002	179.9
10	0.5	0.8	5.3034	0.2000	105.5
10	0.1	1.0	27.462	0.0002	527.3
10	0.1	0.8	3.0461	0.0152	101.3

$K = 100,000$

Notice an important discrepancy in the mean and median of the recovery errors. On close inspection, these differences as well as large maximal recovery errors result from severe ill-conditioning (with condition number $\sim 10^{11}$ or more) of some realizations of the random matrix A_K.

Exercise 5.4. [more on generalized linear models] Consider a generalized linear model as follows: we observe i.i.d. random pairs

$$\omega_k = (y_k, \zeta_k) \in \mathbf{R} \times \mathbf{R}^{\nu\times\mu},\ k = 1, ..., K,$$

where the conditional expectation of the scalar label y_k given ζ_k is $\psi(\zeta_k^T z)$, z being an unknown signal underlying the observations. What we know is that z belongs to a given convex compact set $\mathcal{Z} \subset \mathbf{R}^n$. Our goal is to recover z.

Note that while the estimation problem we have just posed looks similar to those treated in Section 5.2, it cannot be straightforwardly handled via techniques developed in that section unless $\mu = 1$. Indeed, these techniques in the case of $\mu > 1$ require ψ to be a monotone vector field on $\mathbf{R}^\mu$, while our ψ is just a scalar function on $\mathbf{R}^\mu$. The goal of the exercise is to show that *when*

$$\psi(w) = \sum_{q\in\mathcal{Q}} c_q w^q \equiv \sum_{q\in\mathcal{Q}} c_q w_1^{q_1} \ldots w_\mu^{q_\mu} \qquad [c_q \neq 0, q \in \mathcal{Q} \subset \mathbf{Z}_+^\mu]$$

is an algebraic polynomial (which we assume from now on), one can use lifting to reduce the situation to that considered in Section 5.2.

The construction is straightforward. Let us associate with algebraic monomial with ν variables[24]

$$z^p := z_1^{p_1} z_2^{p_2} \ldots z_\nu^{p_\nu}$$

a real variable x_p. For example, monomial $z_1 z_2$ is associated with $x_{1,1,0,\ldots,0}$, $z_1^2 z_\nu^3$ is associated with $x_{2,0,\ldots,0,3}$, etc. For $q \in \mathcal{Q}$, the contribution of the monomial $c_q w^q$ into $\psi(\zeta^T z)$ is

$$c_q[\mathrm{Col}_1^T[\zeta]z]^{q_1}[\mathrm{Col}_2^T[\zeta]z]^{q_2}\ldots[\mathrm{Col}_\mu^T[\zeta]z]^{q_\mu} = \sum_{p\in\mathcal{P}_q} h_{pq}(\zeta) z_1^{p_1} z_2^{p_2} \ldots z_\nu^{p_\nu},$$

where $\mathcal{P}_q$ is a properly built set of multi-indices $p = (p_1, \ldots, p_\nu)$, and $h_{pq}(\zeta)$ are easily computable functions of ζ. Consequently,

$$\psi(\zeta^T z) = \sum_{q\in\mathcal{Q}} \sum_{p\in\mathcal{P}_q} h_{pq}(\zeta) z^p = \sum_{p\in\mathcal{P}} H_p(\zeta) z^p,$$

with properly selected finite set $\mathcal{P}$ and readily given functions $H_p(\zeta)$, $p \in \mathcal{P}$. We can always take, as $\mathcal{P}$, the set of all ν-entry multi-indices with the sum of entries not exceeding d, where d is the total degree of the polynomial ψ. This being said, the structure of ψ and/or the common structure, if any, of regressors ζ_k can enforce some of the functions $H_p(\cdot)$ to be identically zero. When this is the case, it makes sense to eliminate the corresponding "redundant" multi-index p from $\mathcal{P}$.

Next, consider the mapping $x[z]$ which maps a vector $z \in \mathbf{R}^\nu$ into a vector with entries $x_p[z] = z^p$ indexed by $p \in \mathcal{P}$, and let us associate with our estimation problem its "lifting" with observations

$$\overline{\omega}_k = (y_k, \eta_k = \{H_p(\zeta_k), p \in \mathcal{P}\}).$$

I.e., new observations are deterministic transformations of the actual observations; observe that the new observations still are i.i.d., and the conditional expectation of y_k given η_k is nothing but

$$\sum_{p\in\mathcal{P}} [\eta_k]_p x_p[z].$$

In our new situation, the "signal" underlying observations is a vector from $\mathbf{R}^N$, $N = \mathrm{Card}(\mathcal{P})$, the regressors are vectors from the same $\mathbf{R}^N$, and *regression is linear*—the conditional expectation of the label y_k given regressor η_k is the linear function $\eta_k^T x$ of our new signal. Given a convex compact localizer $\mathcal{Z}$ for the "true signal" z, we can in many ways find convex compact localizer $\mathcal{X}$ for $x = x[z]$. Thus, we find ourselves in the simplest possible case of the situation considered in Section 5.2 (one with scalar $\phi(s) \equiv s$), and can apply the estimation procedures developed in this section. Note that in the "lifted" problem the SAA estimate $\widehat{x}(\cdot)$ of the lifted signal $x = x[z]$ is nothing but the standard Least Squares:

$$\begin{array}{rcl}\widehat{x}(\overline{\omega}^K) & \in & \mathrm{Argmin}_{x\in\mathcal{X}} \left\{ \frac{1}{2} x^T \left[\sum_{k=1}^K \eta_k \eta_k^T\right] x - \left[\sum_{k=1}^K y_k \eta_k\right]^T x \right\} \\ & = & \mathrm{Argmin}_{x\in\mathcal{X}} \left\{ \sum_k (y_k - \eta_k^T x)^2 \right\}.\end{array} \qquad (5.80)$$

[24] note that factors in the monomial are ordered according to the indices of the variables.

Of course, there is no free lunch, and there are some potential complications:

- It may happen that the matrix $\mathcal{H} = \mathbf{E}_{\eta\sim Q}\{\eta\eta^T\}$ (Q is the common distribution of "artificial" regressors η_k induced by the common distribution of the actual regressors ζ_k) is *not* positive definite, which would make it impossible to recover well the signal $x[z]$ underlying our transformed observations, however large be K;
- Even when $\mathcal{H}$ is positive definite, so that $x[z]$ can be recovered well, provided K is large, we still need to recover z from $x[z]$, that is, to solve a system of polynomial equations, which can be difficult; besides, this system can have more than one solution.
- Even when the above difficulties can be somehow avoided, "lifting" $z \to x[z]$ typically increases significantly the number of parameters to be identified, which, in turn, deteriorates "finite time" accuracy bounds.

Note also that when $\mathcal{H}$ is not positive definite, this still is not the end of the world. Indeed, $\mathcal{H}$ is positive semidefinite; assuming that it has a nontrivial kernel L which we can identify, a realization η_k of our artificial regressor is orthogonal to L with probability 1, implying that replacing artificial signal x with its orthogonal projection onto $L^\perp$, we almost surely keep the value of the objective in (5.80) intact. Thus, we lose nothing when restricting the optimization domain in (5.80) to the orthogonal projection of $\mathcal{X}$ onto $L^\perp$. Since the restriction of $\mathcal{H}$ onto $L^\perp$ is positive definite, with this approach, for large enough values of K we will still get a good approximation of the projection of $x[z]$ onto $L^\perp$, With luck, this approximation, taken together with the fact that the "artificial signal" we are looking for is not an arbitrary vector from $\mathcal{X}$—it is of the form $x[z]$ for some $z \in \mathcal{Z}$—will allow to get a good approximation of z. Here is the first part of the exercise:

1) Carry out the outlined approach in the situation where
   - The common distribution Π of regressors ζ_k has density w.r.t. the Lebesgue measure on $\mathbf{R}^{\nu\times\mu}$ and possesses finite moments of all orders
   - $\psi(w)$ is a quadratic form, either (case A) homogeneous,

   $$\psi(w) = w^T S w \qquad [S \neq 0],$$

   or (case B) inhomogeneous:

   $$\psi(w) = w^T S w + s^T w \qquad [S \neq 0, s \neq 0]$$

   - The labels are linked to the regressors and to the true signal z by the relation

   $$y_k = \psi(\eta_k^T z) + \chi_k,$$

   where the $\chi_k \sim \mathcal{N}(0,1)$ are mutually independent and independent from the regressors.

Solution: *Case A:* Formally, we should set $\mathcal{P} = \{p = (p_1, ..., p_\nu) \in \mathbf{Z}^n_+ : \sum_i p_i = 2\}$ and write some rather messy expressions for $h_p(\zeta)$, $p \in \mathcal{P}$. To save notation, it is much better to implement our approach "from scratch", namely,

- to associate with $z \in \mathbf{R}^\nu$ a $\nu \times \nu$ symmetric matrix $x[z] = zz^T$;
- to write the relation between y_k, η_k, and z as

$$\begin{array}{rcl} y_k &=& [\zeta_k^T z]^T S[\zeta_k^T z] + \chi_k = \mathrm{Tr}(S[\zeta_k^T z][\zeta_k^T z]^T) + \chi_k = \mathrm{Tr}(S\zeta_k^T x[z]\zeta_k) + \chi_k \\ &=& \mathrm{Tr}(\underbrace{\zeta_k S \zeta_k^T}_{\eta_k} x[z]) + \chi_k. \end{array}$$

Thus, we arrive at a linear regression problem where the observed labels y_k and "artificial regressors" $\eta_k = \zeta_k S \zeta_k^T \in \mathbf{S}^\nu$ are linked by the relation

$$y_k = \langle \eta_k, x \rangle_F + \chi_k$$

($\langle\cdot,\cdot\rangle_F$ is the Frobenius inner product), signal x underlying the observations lives in $\mathbf{S}^\nu$, and η_k, χ_k, $k = 1, ..., K$, are i.i.d. Positive definiteness of the matrix $\mathcal{H}$ is equivalent to the fact that whenever $P \in \mathbf{S}^\nu$ and $P \neq 0$, the expectation of the random quantity $\langle \eta_k, P\rangle_F^2$ is positive; we are about to prove that this indeed is the case. Note that the latter quantity is nothing but $f(\zeta_k) := \mathrm{Tr}(\zeta_k S \zeta_k^T P)^2 \geq 0$, and the set Z of those ζ for which $f(\zeta) = 0$ is of Lebesgue measure zero (indeed, $f(\zeta)$ is a polynomial of ζ; as such, it is either identically zero, or the set of its zeros is of Lebesgue measure 0.[25]) In the first case $\mathrm{Tr}(\zeta S \zeta^T P) = 0$ for all ζ; assuming that it is the case and taking derivative in ζ along a direction d, we get $\mathrm{Tr}(dS\zeta^T P + \zeta S d^T P) = 0$ identically in d, ζ. Taking derivative in ζ again along direction e, we get $\mathrm{Tr}(dSe^T P + eSd^T P) = 0$, or, which is the same due to the symmetry of S and P, $\mathrm{Tr}(dSe^T P) = 0$ identically in d, e, whence $Se^T P = 0$ identically in $e \in \mathbf{R}^{\nu\times\mu}$. The latter clearly contradicts the fact that P and S are nonzero. Indeed, assuming that $g \in \mathbf{R}^\nu$ and $f \in \mathbf{R}^\mu$ are such that $Pg \neq 0$ and $Sf \neq 0$ and setting $e = [Pg]f^T$, we get $Se^T Pg = Sf[Pg]^T Pg = \|Pg\|_2^2 Sf \neq 0$, which is the desired contradiction. Thus, our $f(\zeta) \geq 0$ vanishes at the set of Lebesgue measure zero, and since the distribution of ζ_k has a density w.r.t. the Lebesgue measure, the expected value of $f(\zeta_k)$ is positive, as claimed.

From positive definiteness of $\mathcal{H}$ it follows that for large K we can recover well $x[z]$; since $x[z] = zz^T$, that means that for large K we can also recover z *up to multiplication of z by -1*. The resulting ambiguity in recovering z is quite natural—our observations do not distinguish between the cases where the signal is z or $-z$.

Case B. Here again to save notation it makes sense to repeat the derivation "from scratch." Setting $x[z] = (zz^T, z) \in \mathbf{S}^\nu \times \mathbf{R}^\nu$, we have

$$\begin{array}{rcl} y_k & = & [\zeta_k^T z]^T S[\zeta_k^T z] + s^T \zeta_k^T z + \chi_k = \mathrm{Tr}(\zeta_k S \zeta_k^T [zz^T]) + [\zeta_k s]^T z + \chi_k \\ & = & \langle \underbrace{(\zeta_k S \zeta_k^T, \zeta_k s)}_{\eta_k}, x[z]\rangle, \end{array}$$

with the inner product on the space $E := \mathbf{S}^\nu \times \mathbf{R}^\nu$ where the "lifted signal" $x[z]$ lives given by

$$\langle (P,p), (Q,q)\rangle = \mathrm{Tr}(PQ) + p^T q.$$

Positive definiteness of the matrix $\mathcal{H}$ is nothing but the fact that whenever $(P,p) \in E$ is nonzero, we have $\mathbf{E}\{\langle(\zeta_k S\zeta_k^T, \zeta_k s), (P,p)\rangle^2\} > 0$, and we are about to demonstrate that this indeed is the case. Similarly to the Case A, all we need to do is to lead to contradiction the assumption that the polynomial

$$f(\zeta) = \mathrm{Tr}(\zeta S \zeta^T P) + s^T \zeta^T p$$

is identically in $\zeta \in \mathbf{R}^{\nu\times\mu}$ equal to 0. Assuming that the latter is the case and differentiating $f(\zeta)$ in ζ along a direction d, we get

$$\mathrm{Tr}(dS\zeta^T P) + \mathrm{Tr}(\zeta S d^T P) + s^T d^T p = 0 \ \forall (\zeta, d) \in \mathbf{R}^{\nu\times\mu}.$$

The latter is impossible when $p \neq 0$ (set $\zeta = 0$ and note that $s \neq 0$). When $p = 0$, we have $P \neq 0$, whence $f(\zeta)$ is not identically zero by the same argument as in

[25]this simple claim can be easily verified by induction in the number of variables in the polynomial.

Case A.

The consequences are completely similar to those for Case A, with the only—and pleasant—difference that now z is a component of $x[z]$, so that good recovery of $x[z]$ automatically implies good recovery of z.

Now comes the concluding part of the exercise, where you are supposed to apply the approach we have developed to the situation as follows:

You are given a DC electric circuit comprised of resistors, that is, *connected* oriented graph with m nodes and n arcs $\gamma_j = (s_j, t_j)$, $1 \leq j \leq n$, where $1 \leq s_j, t_j \leq m$ and $s_j \neq t_j$ for all j; arcs γ_j are assigned with resistances $R_j > 0$ known to us. At instant $k = 1, 2, ..., K$, "nature" specifies "external currents" (charge flows from the "environment" into the circuit) $s_1, ..., s_m$ at the nodes; these external currents specify currents in the arcs and voltages at the nodes, and consequently, the power dissipated by the circuit.

Note that nature cannot be completely free in generating the external currents: their total should be zero. As a result, all what matters is the vector $s = [s_1; ...; s_{m-1}]$ of external currents at the first $m-1$ nodes, due to $s_m \equiv -[s_1 + ... + s_{m-1}]$. We assume that the mechanism of generating the vector of external currents at instant k—let this vector be denoted by $s^k \in \mathbf{R}^{m-1}$—is as follows. There are somewhere $m-1$ sources producing currents $z_1, ..., z_{m-1}$. At time k nature selects a one-to-one correspondence $i \mapsto \pi_k(i)$, $i = 1, ..., m-1$, between these sources and the first $m-1$ nodes of the circuit, and "forwards" current $z_{\pi_k(i)}$ to node i:

$$s_i^k = z_{\pi_k(i)}, \ 1 \leq i \leq m-1.$$

For the sake of definiteness, assume that the permutations π_k of $1, ..., m-1$, $k = 1, ..., K$, are i.i.d. drawn from the uniform distribution on the set of $(m-1)!$ permutations of $m-1$ elements.

Assume that at time instants $k = 1, ..., K$ we observe the permutations π_k and noisy measurements of the power dissipated at this instant by the circuit; given those observations, we want to recover the vector z.

Here is your task:

2) Assuming the noises in the dissipated power measurements to be independent of each other and of π_k zero mean Gaussian noises with variance σ^2, apply to the estimation problem in question the approach developed in item 1 of the exercise and run numerical experiments.

<u>Solution:</u> Let us associate with our circuit its *incidence matrix* B – the $m \times n$ matrix B, so that the rows of B are indexed by the nodes, and columns by the arcs; the j-th column has just two nonzero entries, namely, entry $+1$ in row s_j (s_j is the starting node of arc γ_j), and entry -1 in the row t_j (t_j is the ending node of the arc). Let node m be the ground node, and let A be the $(m-1) \times n$ matrix comprised by the first $\mu := m-1$ rows of B. Let now R be the diagonal matrix with diagonal entries R_j, $1 \leq j \leq n$, and v be the vector of voltages at the first μ nodes, the voltage at the ground node being zero. Let s be the vector of external currents at the first μ nodes, and let ι be the vector of currents in arcs. Finally, let P be the dissipated power. The Kirchhoff Laws say that

$$\begin{array}{rcll} R\iota & = & A^T v & \text{[Ohm's Law]} \\ A\iota & = & s & \text{[current conservation]} \\ P & = & \iota^T A^T v. & \end{array}$$

Note that A is of rank μ (recall that our circuit is connected). As a result, we get

$$P = s^T S s, \ S = (AR^{-1}A^T)^{-1}.$$

Now let us associate with a permutation π the $\mu \times \mu$ permutation matrix D_π; the only nonzero entry in the $\pi(i)$-th row of the matrix is in the cell i and is equal to 1, so that $(D_\pi^T x)_i = x_{\pi(i)}$, $i = 1, ..., m$. In other words, the vector s^k of external currents at time k is $D_{\pi_k}^T z$, so that our k-th label is

$$y_k = [D_{\pi_k}^T z]^T S [D_{\pi_k}^T z] + \chi_k = \langle \underbrace{D_{\pi_k} S D_{\pi_k}^T}_{\eta_k}, \underbrace{zz^T}_{x[z]} \rangle_F + \chi_k,$$

where $\chi_k \sim \mathcal{N}(0, \sigma^2)$ are independent of each other and of the $\pi_1, ..., \pi_K$ observation noises.

Observe that the matrix $\mathcal{H}$ in our situation is positive semidefinite, but not positive definite. Indeed, all matrices η_k inherit from S the trace and the sum of all elements. In other words, all realizations of our artificial regressors belong to the affine plane of codimension 2 in $\mathbf{S}^\mu$ associated with S, and thus are orthogonal to a nonzero vector f readily given by this affine plane. Since all realizations of η_k are orthogonal to f, f is in the kernel of $\mathcal{H}$. An "educated guess" (fully supported by numerical evidence) is that the kernel of $\mathcal{H}$ is the line $\mathbf{R}f$, and the orthogonal complement E to this line in $\mathbf{S}^\mu$, and this is what we used when building our Least Squares estimate

$$\widehat{x}(\{y_k, \pi_k\}_{k \le K} \in \operatorname*{Argmin}_{\substack{x \in E \\ \|x\|_F \le M}} \left\{ \tfrac{1}{2}\mathrm{Vec}^T[x] \left[\sum_{k=1}^K \mathrm{Vec}[\eta_k]\mathrm{Vec}^T[\eta_k] \right] \mathrm{Vec}[x] - \sum_{k=1}^K \mathrm{Vec}[\eta_k]^T \mathrm{Vec}[x] \right\}.$$

Here for $x \in \mathbf{S}^\mu$ $\mathrm{Vec}[x]$ is the μ^2-dimensional vector obtained by vertical concatenation of the columns of x, and M is an a priori upper bound on the Frobenius norm of $x[z]$ for $z \in \mathcal{Z}$. After $\widehat{x}$ is built, we can try to recover $x[z]$ by looking at the points of the line $\widehat{x} + \mathbf{R}f$ and selecting on this line a "nearly rank 1" matrix; the leading eigenvector of this matrix, multiplied by the square root of the corresponding eigenvalue, is our estimate $\widehat{z}$ of z (or, better said, "$\pm\widehat{z}$ is our estimate of $\pm z$"—in our problem, observations do not allow us to distinguish between z and $-z$).

Our sample numerical results are as follows:

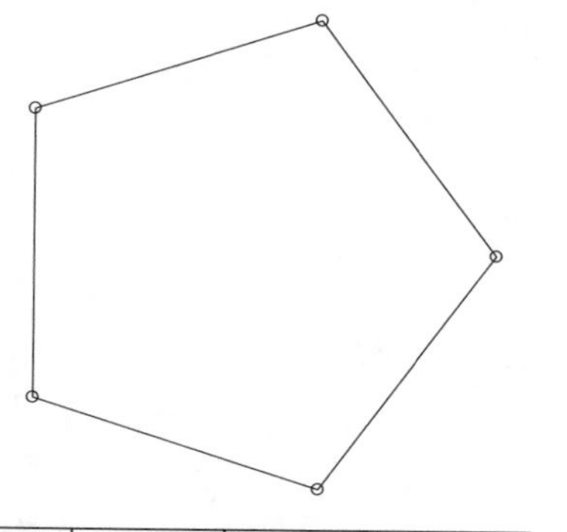

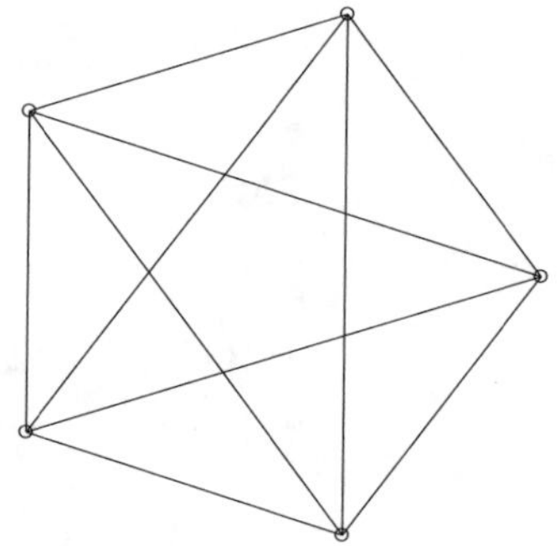

K	100	1000	10000
δ	0.0356	0.0107	0.0047

K	100	1000	10000
δ	0.0668	0.0391	0.0270

δ: $\|\cdot\|_2$-recovery error

In the reported experiments, we used $\sigma = 0.1$ and $M = 40$. The signal z was drawn at random from $\mathcal{N}(0, I_4)$, and the resistances were drawn at random from the uniform distribution on $[1, 2]$.

Exercise 5.5 [shift estimation] Consider the situation as follows: given a continuous vector field $f(u) : \mathbf{R}^m \to \mathbf{R}^m$ which is strongly monotone on bounded subsets of $\mathbf{R}^m$ and a convex compact set $\mathcal{S} \subset \mathbf{R}^m$, we observe in noise vectors $f(p - s)$, where $p \in \mathbf{R}^m$ is an observation point known to us, and $s \in \mathbf{R}^m$ is a shift unknown to us and known to belong to $\mathcal{S}$. Precisely, assume that our observations are

$$y_k = f(p_k - s) + \xi_k,\ k = 1, ..., K,$$

where $p_1, ..., p_K$ is a deterministic sequence known to us, and $\xi_1, ..., \xi_K$ are $\mathcal{N}(0, \gamma^2 I_m)$ observation noises independent of each other. Our goal is to recover from observations $y_1, ..., y_K$ the shift s.

1. Pose the problem as a single-observation version of the estimation problem from Section 5.2.
2. Assuming f to be strongly monotone, with modulus $\varkappa > 0$, on the entire space, what is the error bound for the SAA estimate?

3. Run simulations in the case of $m = 2$, $\mathcal{S} = \{u \in \mathbf{R}^2 : \|u\|_2 \leq 1\}$ and

$$f(u) = \left[\begin{array}{c} 2u_1 + \sin(u_1) + 5u_2 \\ 2u_2 - \sin(u_2) - 5u_1 \end{array}\right].$$

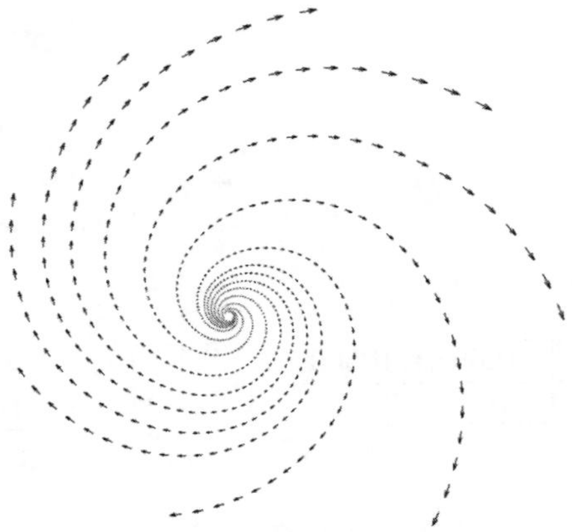

Note: Field $f(\cdot)$ is not potential; this is the monotone vector field associated with the strongly convex-concave function

$$\psi(u) = u_1^2 - \cos(u_1) - u_2^2 - \cos(u_2) + 5u_1u_2,$$

so that $f(u) = [\frac{\partial}{\partial u_1}\phi(u); -\frac{\partial}{\partial u_2}\phi(u)]$. Compare the actual recovery errors with their theoretical upper bounds.

4. Think about what can be done when our observations are

$$y_k = f(Ap_k - s) + \xi_k,\ 1 \leq k \leq K$$

with known p_k, noises $\xi_k \sim \mathcal{N}(0, \gamma^2 I_2)$ independent across k, and unknown A and s which we want to recover.

Solution: Item 1: Let us associate with $p \in \mathbf{R}^m$ the $(m+1)\times m$ matrix ("regressor")

$$\eta[p] = [p^T; -I_m],$$

and with a shift $s \in \mathbf{R}^m$ the signal $x[s] = [1; s]$. Then our observations become

$$y_k = f(\underbrace{\eta^T[p_k]}_{\eta_k^T} x[s]) + \xi_k,\ 1 \leq k \leq K,$$

and we find ourselves in the situation considered in Section 5.2.6, the signal set being $\mathcal{X} = \{[1, s] : s \in \mathcal{S}\}$.

Item 2: The SAA estimate $\widehat{x} = \widehat{x}(y)$ is the weak solution to $\mathrm{VI}(G_y, \mathcal{X})$, where

$$G_y(z) = \sum_k \left[\eta_k f(\eta_k^T z) - \eta_k y_k\right],\ \mathcal{X} = \{z = [1; s] : s \in \mathcal{S}\}.$$

Let us lower-bound the constant of strong monotonicity of $G_y(\cdot)$ on $\mathcal{X}$. Let $z = [1;r]$, $z' = [1;r']$ be two points from $\mathcal{X}$. We have

$$\begin{array}{l}[G_y(z) - G_y(z')]^T[z-z'] = \sum_k \left[f(\eta_k^T z) - f(\eta_k^T z')\right]^T \eta_k^T[z-z'] \\ \quad = \sum_k [f(p_k - r) - f(p_k - r')]^T[r' - r] \\ \quad = \sum_k [f(p_k - r) - f(p_k - r')]^T[[p_k - r] - [p_k - r']] \\ \quad \geq \sum_k \varkappa \|[p_k - r] - [p_k - r']\|_2^2 = K\varkappa\|r - r'\|_2^2 = K\varkappa\|z - z'\|_2^2,\end{array}$$

so that G_y is strongly monotone on $\mathcal{X}$ with modulus $K\varkappa$.

Now let us check the validity of Assumption **A.3$'$** from Section 5.2.6. In the notation from this assumption, we have

$$\begin{array}{rcl}\eta[y - \phi(\eta^T x[s])] & = & \sum_k \eta_k[y_k - f(\eta_k^T[1;s])] = \sum_k \eta_k[y_k - f(p_k - s)] \\ & = & \sum_k \eta_k \xi_k = \sum_k [p_k^T; -I_m]\xi_k,\end{array}$$

whence

$$\mathbf{E}\left\{\|\eta[y - \phi(\eta^T x[s])]\|_2^2\right\} = \gamma^2 \sum_k \left[\|p_k\|_2^2 + m\right].$$

Thus, Assumption **A.3$'$** is satisfied with

$$\sigma = \gamma\sqrt{\sum_k [\|p_k\|_2^2 + m]},$$

and the error bound from Proposition 5.14 reads

$$\mathbf{E}\{\|\widehat{x}(y) - [1;s]\|_2^2\} \leq \frac{\gamma^2\left[m + \frac{1}{K}\sum_k \|p_k\|_2^2\right]}{\varkappa^2}\frac{1}{K}.$$

Item 3: Let us compute the modulus of strong monotonicity of our vector field. The Jacobian of the field at a point u is $\left[\begin{array}{c|c} 2+\cos(u_1) & 5 \\ \hline -5 & 2-\cos(u_2)\end{array}\right]$, and its symmetric part is $\succeq I_2$, implying that the field is strongly monotone with modulus 1 on the entire space. Consequently, the theoretical upper bound on the recovery error for the SAA estimate as given by Proposition 5.14 becomes

$$\sqrt{\mathbf{E}\{\|\widehat{x} - s\|_2^2\}} \leq \rho := \frac{\gamma\sqrt{\left[m + \frac{1}{K}\sum_k \|p_k\|_2^2\right]}}{\sqrt{K}}.$$

The numerical experiments we are about to report were organized as follows. In a particular experiment, the "true" shift s was drawn ar random from $\mathcal{N}(0, I_2)$ and then projected onto the set $\mathcal{S} = \{s \in \mathbf{R}^2 : \|s\|_2 \leq 5\}$. The observation points $p_1, ..., p_K$ were drawn at random, independently of each other, from the uniform distribution on the boundary of $\mathcal{S}$, and the noise level γ was set to 1. In all experiments, the number of observations K was set to 100.

The median of the recovery error $\|\widehat{x} - s\|_2$ in the 10 experiments we have carried out was as small as 0.019; the largest observed ratio of the actual recovery error to its theoretical upper bound ρ was 0.089.

Item 4: Similarly to the case of an unknown shift, our measurements are of the form

$$y_k = f(\eta^T[p_k]x) + \xi_k,$$

where $\eta_k = \eta[p]$ is a properly defined 6×2 matrix affinely depending on p, and x is the six-dimensional vector obtained by writing in a single column the two columns of A and z. We can now build the SAA estimate of x similarly to what was done in the case when A was identity.

Appendix: Executive Summary on Efficient Solvability of Convex Optimization Problems

Convex Programming is a "solvable case" in Optimization: under mild computability and boundedness assumptions, a globally optimal solution to a convex optimization problem can be "approximated to any desired accuracy in reasonable time." The goal of what follows is to provide a reader with an understanding, "sufficient for all practical purposes," of what these words mean.[1]

In the sequel we are interested in computational tractability of a convex optimization problem in the form

$$\mathrm{Opt} = \min_{x\in \mathbf{B}_R^n} \left\{f_0(x) : f_i(x) \le 0,\ 1 \le i \le m\right\}, \quad \mathbf{B}_R^n = \{x \in \mathbf{R}^n : \|x\|_2 \le R\}. \tag{1}$$

We always assume that $f_i(\cdot)$, $0 \le i \le m$, are convex and continuous real-valued functions on $\mathbf{B}_R^n$, and what we are looking for is an *ϵ-accurate solution* to the problem, that is, a point $x_\epsilon \in \mathbf{B}_R^n$ such that

$$\begin{array}{rcll} f_i(x_\epsilon) & \le & \epsilon,\ i = 1, ..., m & [\epsilon\text{-feasibility}] \\ f_0(x_\epsilon) & \le & \mathrm{Opt} + \epsilon & [\epsilon\text{-optimality}] \end{array}$$

where $\epsilon > 0$ is a given tolerance. Note that when (1) is infeasible, i.e., $\mathrm{Opt} = +\infty$, every point of $\mathbf{B}_R^n$ is an ϵ-optimal (but not necessarily ϵ-feasible, and thus not necessarily ϵ-accurate) solution.

We intend to provide two versions of what "efficient solvability" means: "practical" and "scientific."

"Practical" version of efficient solvability: For most practical purposes, efficient solvability means that we can feed (1) to cvx, that is, rewrite the problem in the form

$$\mathrm{Opt} = \min_{x,u} \left\{c^T[x;u] : \mathcal{A}(x,u) \preceq 0\right\}, \tag{2}$$

where $\mathcal{A}(x,u)$ is a symmetric matrix which is affine in $[x;u]$.

"Scientific" version of efficient solvability. Let us start with the following basic fact:

(!) Assume that when solving (1), we have at our disposal R, a real V such that

$$\max_{x\in \mathbf{B}_R^n} |f_i(x)| \le V/2, \tag{3}$$

[1] For rigorous treatment of this subject in the context of continuous optimization (this is what we deal with in our book) the reader is referred to [15, Chapter 5].

and access to a *First Order oracle*—a "black box" which, given on input a query point $x \in \mathbf{B}^n_R$ and tolerance $\delta > 0$, returns δ-subgradients of f_i at x, that is, affine functions $g_{i,x}(\cdot)$, $0 \le i \le m$, such that

$$g_{i,x}(y) \le f_i(y)\ \forall y \in \mathbf{B}^n_R \ \&\ g_{i,x}(x) \ge f_i(x) - \delta,\ \ 0 \le i \le m.$$

Then for every $\epsilon \in (0, V)$, an ϵ-accurate solution to (1), or a correct claim that the problem is infeasible, can be found by an appropriate algorithm (e.g., the Ellipsoid method) at cost at most

$$N(\epsilon) = \rfloor 2n^2 \ln(2V/\epsilon) \lfloor + 1 \tag{4}$$

subsequent steps, with

- at most one call to the First Order oracle per step, the input at the t-th step being $(x_t, \epsilon/2)$, with $x_1 = 0$ and recursively computed $x_2, ..., x_{N(\epsilon)}$;
- at most $O(1)n(m+n)$ operations of precise real arithmetic per step needed to update x_t and the output of the First Order oracle (if the latter was invoked at step t) into x_{t+1}.

The remaining computational effort when executing the algorithm is just $O(1)N(\epsilon)n$ operations of precise real arithmetic needed to convert the trajectory $x_1, ..., x_{N(\epsilon)}$ and the outputs of the First Order oracle into the result of the computation—either an ϵ-accurate solution to (1), or a correct infeasibility claim.

The consequences are as follows. Consider a family $\mathcal{F}$ of convex functions of a given structure, so that every function f in the family can be identified by a finite-dimensional *data vector* Data(f). For the sake of simplicity, assume that all functions from the family are real-valued (extensions to partially defined functions are straightforward). Examples include (but, of course, do not reduce to)

1. affine functions of $n = 1, 2, ...$ variables,
2. convex quadratic functions of $n = 1, 2, ...$ variables,
3. (logarithms of) posynomials—functions of $n = 1, 2, ...$ variables of the form $\ln(\sum_{i=1}^m \exp\{\phi_i(x)\})$, each with its own m, with affine ϕ_i,
4. functions of the form $\lambda_{\max}(A_0 + \sum_{j=1}^n x_j A_j)$, where A_j, $1 \le j \le m$, are $m \times m$ symmetric matrices, $\lambda_{\max}(\cdot)$ is the maximal eigenvalue of a symmetric matrix, and m, n can be arbitrary positive integers

(in all these examples, it is self-evident what the data is). For $f \in \mathcal{F}$, denoting by $n(f)$ the dimension of the argument of f, let us call the quantity

$$\text{Size}(f) = \max[n(f), \dim \text{Data}(f)]$$

the *size* of f. Let us say that family $\mathcal{F}$

- is with *polynomial growth*, if for all $f \in \mathcal{F}$ and $R > 0$ it holds

$$\begin{aligned} V(f,R) \ &:= \ \max_{x \in \mathbf{B}^{n(f)}_R} f(x) - \min_{x \in \mathbf{B}^{n(f)}_R} f(x) \\ &\le \ \chi[\text{Size}(f) + R + \|\text{Data}(f)\|_\infty]^{(\chi \text{Size}^\chi(f))}; \end{aligned}$$

here and in what follows χ's stand for positive constants, perhaps different in different places, depending solely on $\mathcal{F}$;

- is *polynomially computable*, if there exists a code for a Real Arithmetic computer[2] with the following property: whenever $f \in \mathcal{F}$, $R > 0$, and $\delta > 0$, executing the code on input comprised of Data(f) augmented by δ, R, and a query vector $x \in \mathbf{R}^{n(f)}$ with $\|x\|_2 \leq R$, the computer after finitely many operations outputs the coefficients of affine function $g_{f,x}(\cdot)$ which is a δ-subgradient of f, taken at x, on $\mathbf{B}_R^{n(f)}$,

$$g_{f,x}(y) \leq f(y) \ \forall y \in \mathbf{B}_R^{n(f)} \ \& \ f(x) \leq g_{f,x}(x) + \delta,$$

  and the number N of arithmetic operations in this computation is upper-bounded by a polynomial in Size(f) and "the required number of accuracy digits"

$$\text{Digits}(f, R, \delta) = \log\left(\frac{\text{Size}(f) + \|\text{Data}(f)\|_\infty + R + \delta^2}{\delta}\right),$$

  that is,

$$N \leq \chi[\text{Size}(f) + \text{Digits}(f, R, \delta)]^\chi.$$

Observe that typical families of convex functions, like those in the above examples, are both with polynomial growth and polynomially computable.

In the main body of this book, the words "a convex function f is efficiently computable" mean that f belongs to a polynomially computable family (it is always clear from the context what this family is). Similarly, the words "a closed convex set X is computationally tractable" mean that the convex function $f(x) = \min_{y \in X} \|y - x\|_2$ is efficiently computable.

In our context, the role of the notions introduced is as follows. Consider problem (1) and assume that the functions f_i, $i = 0, 1, ..., m$ participating in the problem are taken from a family $\mathcal{F}$ which is polynomially computable and with polynomial growth (as is the case when (1) is a linear, or a second order conic, or a semidefinite program). In this situation a particular instance P of (1) is fully specified by its *data vector* Data(P) obtained by augmenting the "sizes" n, m, R of the instance by the concatenation of the data vectors of f_0, $f_1, ..., f_m$. Similarly to the above, let us define the size of P as

$$\text{Size}(P) = \max[n, m, \dim \text{Data}(P)],$$

so that Size(P) $\geq$ Size(f_i) for all i, $0 \leq i \leq m$. Given Data(P) and R and invoking the fact that $\mathcal{F}$ is with polynomial growth, we can easily compute V satisfying (3) and such that

$$V = V(P, R) \leq \chi[\text{Size}(P) + R + \|\text{Data}(P)\|_\infty]^{(\chi \text{Size}^\chi(P))}. \tag{5}$$

---

[2] An idealized computer capable of storing reals and carrying out operations of precise real arithmetic—the four arithmetic operations, comparison, and the computing of elementary univariate functions, like $\sin(s)$, $\sqrt{s}$, etc.

Similarly to the above, we set

$$\text{Digits}(P, R, \delta) = \log\left(\frac{\text{Size}(P) + \|\text{Data}(P)\|_\infty + R + \delta^2}{\delta}\right),$$

so that $\text{Digits}(P, R, \delta) \geq \text{Digits}(f_i, R, \delta)$, $0 \leq i \leq m$. Invoking polynomial computability of $\mathcal{F}$, we can implement the First Order oracle for problems P of the form (1) with $f_i \in \mathcal{F}$ on the Real Arithmetic Computer in such a way that executing the resulting code on input comprised of the data vector $\text{Data}(P)$ augmented by $\delta > 0$, R, and a query vector $x \in \mathbf{B}_R^n$ will produce δ-subgradients, taken at x, of f_i, $0 \leq i \leq m$; the total number $M = M(P, R, \delta)$ of real arithmetic operations in the course of computation is upper-bounded by a polynomial in $\text{Size}(P)$ and $\text{Digits}(P, R, \delta)$:

$$M(P, R, \delta) \leq \chi[\text{Size}(P) + \text{Digits}(P, R, \delta)]^\chi. \tag{6}$$

Finally, given, $\text{Data}(P)$, R, and a desired accuracy $\epsilon > 0$ and assuming w.l.o.g. that $\epsilon < V = V(P, R)$,[3] we can use the above First Order oracle (with δ set to $\epsilon/2$) in (!) in order to find an ϵ-accurate solution to problem P (or conclude correctly that the problem is infeasible). The number N of steps in this computation, in view of (4) and (5), is upper-bounded by a polynomial in $\text{Size}(P)$ and $\text{Digits}(P, R, \epsilon)$:

$$N \leq O(1)[\text{Size}(P) + \text{Digits}(P, R, \epsilon)]^\chi,$$

with computational expenses per step stemming from mimicking the First Order oracle upper-bounded by a polynomial in $\text{Size}(P)$ and $\text{Digits}(P, R, \epsilon)$ (by (6)). By (!), the overall "computational overhead" needed to process the oracle's outputs and to generate the result is bounded by another polynomial of the same type. The bottom line is that

> *When $\mathcal{F}$ is a polynomially computable family of convex functions of polynomial growth, and the objective and the constraints f_i, $i \leq m$, in* (1) *belong to $\mathcal{F}$, the overall number of arithmetic operations needed to find an ϵ-approximate solution to* (1) *(or to conclude correctly that* (1) *is infeasible) is, for every $\epsilon > 0$, upper-bounded by a polynomial, depending solely on $\mathcal{F}$, in the size* $\text{Size}(P)$ *of the instance and the number* $\text{Digits}(P, R, \epsilon)$ *of accuracy digits in the desired solution.*

For all our purposes, this is a general enough "scientific translation" of the informal claim "*an explicit convex problem with computationally tractable objective and constraints is efficiently solvable.*"

---

[3]This indeed is w.l.o.g., since, say, the origin is a V-accurate solution to P.

Bibliography

[1] M. Aizerman, E. Braverman, and L. Rozonoer. Theoretical foundations of the potential function method in pattern recognition. *Avtomatika i Telemekhanika*, 25:917–936, 1964. English translation: *Automation & Remote Control.*

[2] M. Aizerman, E. Braverman, and L. Rozonoer. *Method of potential functions in the theory of learning machines.* Nauka, Moscow, 1970.

[3] E. Anderson. *The MOSEK optimization toolbox for MATLAB Manual. Version 8.0*, 2015. http://docs.mosek.com/8.0/toolbox/.

[4] T. Anderson. The integral of a symmetric unimodal function over a symmetric convex set and some probability inequalities. *Proceedings of the American Mathematical Society*, 6(2):170–176, 1955.

[5] A. Antoniadis and I. Gijbels. Detecting abrupt changes by wavelet methods. *Journal of Nonparametric Statistics*, 14(1-2):7–29, 2002.

[6] B. Arnold and P. Stahlecker. Another view of the Kuks–Olman estimator. *Journal of Statistical Planning and Inference*, 89(1):169–174, 2000.

[7] K. Aström and P. Eykhoff. System identification—a survey. *Automatica*, 7(2):123–162, 1971.

[8] T. Augustin and R. Hable. On the impact of robust statistics on imprecise probability models: a review. *Structural Safety*, 32(6):358–365, 2010.

[9] R. Bakeman and J. Gottman. *Observing Interaction: An Introduction to Sequential Analysis.* Cambridge University Press, 1997.

[10] M. Basseville. Detecting changes in signals and systems—a survey. *Automatica*, 24(3):309–326, 1988.

[11] M. Basseville and I. Nikiforov. *Detection of Abrupt Changes: Theory and Application.* Prentice-Hall, Englewood Cliffs, N.J., 1993.

[12] T. Bednarski. Binary experiments, minimax tests and 2-alternating capacities. *The Annals of Statistics*, 10(1):226–232, 1982.

[13] D. Belomestny and A. Goldenschluger. Nonparametric density estimation from observations with multiplicative measurement errors. *arXiv 1709.00629*, 2017. `https://arxiv.org/pdf/1709.00629.pdf`.

[14] A. Ben-Tal, L. El Ghaoui, and A. Nemirovski. *Robust Optimization.* Princeton University Press, 2009.

[15] A. Ben-Tal and A. Nemirovski. *Lectures on Modern Convex Optimization: Analysis, Algorithms, and Engineering Applications.* SIAM, 2001.

[16] M. Bertero and P. Boccacci. Application of the OS-EM method to the restoration of LBT images. *Astronomy and Astrophysics Supplement Series*, 144(1):181–186, 2000.

[17] M. Bertero and P. Boccacci. Image restoration methods for the large binocular telescope (LBT). *Astronomy and Astrophysics Supplement Series*, 147(2):323–333, 2000.

[18] E. Betzig, G. Patterson, R. Sougrat, O. W. Lindwasser, S. Olenych, J. Bonifacino, M. Davidson, J. Lippincott-Schwartz, and H. Hess. Imaging intracellular fluorescent proteins at nanometer resolution. *Science*, 313(5793):1642–1645, 2006.

[19] P. Bickel and Y. Ritov. Estimating integrated squared density derivatives: sharp best order of convergence estimates. *Sankhyā: The Indian Journal of Statistics, Series A*, 50(3):381–393, 1988.

[20] P. Bickel, Y. Ritov, and A. Tsybakov. Simultaneous analysis of Lasso and Dantzig selector. *The Annals of Statistics*, 37(4):1705–1732, 2009.

[21] L. Birgé. *Approximation dans les spaces métriques et théorie de l'estimation: inégalités de Cràmer-Chernoff et théorie asymptotique des tests.* PhD thesis, Université Paris VII, 1980.

[22] L. Birgé. Vitesses maximales de décroissance des erreurs et tests optimaux associés. *Zeitschrift für Wahrscheinlichkeitstheorie und verwandte Gebiete*, 55(3):261–273, 1981.

[23] L. Birgé. Approximation dans les éspaces métriques et théorie de l'estimation. *Zeitschrift für Wahrscheinlichkeitstheorie und verwandte Gebiete*, 65(2):181–237, 1983.

[24] L. Birgé. Robust testing for independent non identically distributed variables and Markov chains. In J. Florens, M. Mouchart, J. Raoult, L. Simar, and A. Smith, editors, *Specifying Statistical Models*, volume 16 of *Lecture Notes in Statistics*, pages 134–162. Springer, 1983.

[25] L. Birgé. Sur un théorème de minimax et son application aux tests. *Probability and Mathematical Statistics*, 3(2):259–282, 1984.

[26] L. Birgé. Model selection via testing: an alternative to (penalized) maximum likelihood estimators. *Annales de l'Institut Henri Poincaré, Probabilités et Statistiques*, 42(3):273–325, 2006.

[27] L. Birgé. Robust tests for model selection. In M. Banerjee, F. Bunea, J. Huang, V. Koltchinskii, and M. Maathuis, editors, *From Probability to Statistics and Back: High-Dimensional Models and Processes—A Festschrift in Honor of Jon A. Wellner*, pages 47–64. Institute of Mathematical Statistics, 2013.

[28] L. Birgé and P. Massart. Estimation of integral functionals of a density. *The*

Annals of Statistics, 23(1):11–29, 1995.

[29] H.-D. Block. The perceptron: A model for brain functioning. I. *Reviews of Modern Physics*, 34(1):123, 1962.

[30] O. Bousquet and A. Elisseeff. Stability and generalization. *Journal of Machine Learning Research*, 2:499–526, 2002.

[31] S. Boyd, L. El Ghaoui, E. Feron, and V. Balakrishnan. *Linear Matrix Inequalities in System and Control Theory*. SIAM, 1994.

[32] E. Brodsky and B. Darkhovsky. *Nonparametric Methods in Change Point Problems*. Springer Science & Business Media, 2013.

[33] E. Brunel, F. Comte, and V. Genon-Catalot. Nonparametric density and survival function estimation in the multiplicative censoring model. *Test*, 25(3):570–590, 2016.

[34] A. Buchholz. Operator Khintchine inequality in non-commutative probability. *Mathematische Annalen*, 319(1):1–16, 2001.

[35] A. Buja. On the Huber-Strassen theorem. *Probability Theory and Related Fields*, 73(1):149–152, 1986.

[36] M. Burnashev. On the minimax detection of an imperfectly known signal in a white noise background. *Theory of Probability & Its Applications*, 24(1):107–119, 1979.

[37] M. Burnashev. Discrimination of hypotheses for Gaussian measures and a geometric characterization of the Gaussian distribution. *Mathematical Notes of the Academy of Sciences of the USSR*, 32:757–761, 1982.

[38] C. Butucea and F. Comte. Adaptive estimation of linear functionals in the convolution model and applications. *Bernoulli*, 15(1):69–98, 2009.

[39] C. Butucea and K. Meziani. Quadratic functional estimation in inverse problems. *Statistical Methodology*, 8(1):31–41, 2011.

[40] T. T. Cai and M. Low. A note on nonparametric estimation of linear functionals. *The Annals of Statistics*, 31(4):1140–1153, 2003.

[41] T. T. Cai and M. Low. Minimax estimation of linear functionals over nonconvex parameter spaces. *The Annals of Statistics*, 32(2):552–576, 2004.

[42] T. T. Cai and M. Low. On adaptive estimation of linear functionals. *The Annals of Statistics*, 33(5):2311–2343, 2005.

[43] T. T. Cai and M. Low. Optimal adaptive estimation of a quadratic functional. *The Annals of Statistics*, 34(5):2298–2325, 2006.

[44] E. Candes. Compressive sampling. In *Proceedings of the International Congress of Mathematicians*, volume 3, pages 1433–1452. Madrid, August 22-30, Spain, 2006.

[45] E. Candes. The restricted isometry property and its implications for com-

pressed sensing. *Comptes rendus de l'Académie des Sciences, Mathématique*, 346(9-10):589–592, 2008.

[46] E. Candes, J. Romberg, and T. Tao. Stable signal recovery from incomplete and inaccurate measurements. *Communications on Pure and Applied Mathematics*, 59(8):1207–1223, 2006.

[47] E. Candes and T. Tao. Decoding by linear programming. *IEEE Transactions on Information Theory*, 51(12):4203–4215, 2005.

[48] E. Candes and T. Tao. Near-optimal signal recovery from random projections: Universal encoding strategies? *IEEE Transactions on Information Theory*, 52(12):5406–5425, 2006.

[49] E. Candes and T. Tao. The Dantzig selector: statistical estimation when p is much larger than n. *The Annals of Statistics*, 35(6):2313–2351, 2007.

[50] Y. Cao, V. Guigues, A. Juditsky, A. Nemirovski, and Y. Xie. Change detection via affine and quadratic detectors. *Electronic Journal of Statistics*, 12(1):1–57, 2018.

[51] J. Chen and A. Gupta. *Parametric statistical change point analysis: with applications to genetics, medicine, and finance.* Boston: Birkhäuser, 2012.

[52] S. Chen and D. Donoho. Basis pursuit. In *Proceedings of 1994 28th Asilomar Conference on Signals, Systems and Computers*, pages 41–44. IEEE, 1994.

[53] S. Chen, D. Donoho, and M. Saunders. Atomic decomposition by basis pursuit. *SIAM Review*, 43(1):129–159, 2001.

[54] N. Chentsov. Evaluation of an unknown distribution density from observations. *Doklady Academii Nauk SSSR*, 147(1):45, 1962. English translation: *Soviet Mathematics.*

[55] H. Chernoff. A measure of asymptotic efficiency for tests of a hypothesis based on the sum of observations. *The Annals of Mathematical Statistics*, 23(4):493–507, 1952.

[56] H. Chernoff. *Sequential Analysis and Optimal Design.* SIAM, 1972.

[57] N. Christopeit and K. Helmes. Linear minimax estimation with ellipsoidal constraints. *Acta Applicandae Mathematica*, 43(1):3–15, 1996.

[58] H. Cramér. *Mathematical Methods of Statistics.* Princeton University Press, 1946.

[59] A. d'Aspremont and L. El Ghaoui. Testing the nullspace property using semidefinite programming. *Mathematical Programming Series B*, 127(1):123–144, 2011. `https://arxiv.org/pdf/0807.3520.pdf`.

[60] I. Dattner, A. Goldenshluger, and A. Juditsky. On deconvolution of distribution functions. *The Annals of Statistics*, 39(5):2477–2501, 2011.

[61] R. DeVore. Deterministic constructions of compressed sensing matrices. *Journal of Complexity*, 23(4-6):918–925, 2007.

[62] I. Devyaterikov, A. Propoi, and Y. Tsypkin. Iterative learning algorithms for pattern recognition. *Avtomatika i Telemekhanika*, 28:122–132, 1967. English translation: *Automation & Remote Control.*

[63] D. Donoho. Nonlinear wavelet methods for recovery of signals, densities, and spectra from indirect and noisy data. In I. Daubechies, editor, *Proceedings of Symposia in Applied Mathematics*, volume 47, pages 173–205. AMS, 1993.

[64] D. Donoho. Statistical estimation and optimal recovery. *The Annals of Statistics*, 22(1):238–270, 1994.

[65] D. Donoho. De-noising by soft-thresholding. *IEEE Transactions on Information Theory*, 41(3):613–627, 1995.

[66] D. Donoho. Nonlinear solution of linear inverse problems by wavelet–vaguelette decomposition. *Applied and Computational Harmonic Analysis*, 2(2):101–126, 1995.

[67] D. Donoho. Neighborly polytopes and sparse solutions of underdetermined linear equations. Technical report, Stanford University Statistics Report 2005-04, 2005. https://statistics.stanford.edu/research/neighborly-polytopes-and-sparse-solution-underdetermined-linear-equations.

[68] D. Donoho. Compressed sensing. *IEEE Transactions on Information Theory*, 52(4):1289–1306, 2006.

[69] D. Donoho, M. Elad, and V. Temlyakov. Stable recovery of sparse overcomplete representations in the presence of noise. *IEEE Transactions on Information Theory*, 52(1):6–18, 2006.

[70] D. Donoho and X. Huo. Uncertainty principles and ideal atomic decomposition. *IEEE Transactions on Information Theory*, 47(7):2845–2862, 2001.

[71] D. Donoho and I. Johnstone. Ideal spatial adaptation by wavelet shrinkage. *Biometrika*, 81(3):425–455, 1994.

[72] D. Donoho and I. Johnstone. Minimax risk over ℓ_p-balls for ℓ_p-error. *Probability Theory and Related Fields*, 99(2):277–303, 1994.

[73] D. Donoho and I. Johnstone. Adapting to unknown smoothness via wavelet shrinkage. *Journal of the American Statistical Association*, 90(432):1200–1224, 1995.

[74] D. Donoho and I. Johnstone. Minimax estimation via wavelet shrinkage. *The Annals of Statistics*, 26(3):879–921, 1998.

[75] D. Donoho, I. Johnstone, G. Kerkyacharian, and D. Picard. Wavelet shrinkage: asymptopia? *Journal of the Royal Statistical Society: Series B*, 57(2):301–337, 1995.

[76] D. Donoho and R. Liu. Geometrizing rate of convergence I. Technical report, 137a, Department of Statistics, University of California, Berkeley, 1987.

[77] D. Donoho and R. Liu. Geometrizing rates of convergence, II. *The Annals of Statistics*, 19(2):633–667, 1991.

[78] D. Donoho and R. Liu. Geometrizing rates of convergence, III. *The Annals of Statistics*, 19(2):668–701, 1991.

[79] D. Donoho, R. Liu, and B. MacGibbon. Minimax risk over hyperrectangles, and implications. *The Annals of Statistics*, 18(3):1416–1437, 1990.

[80] D. Donoho and M. Low. Renormalization exponents and optimal pointwise rates of convergence. *The Annals of Statistics*, 20(2):944–970, 1992.

[81] D. Donoho and M. Nussbaum. Minimax quadratic estimation of a quadratic functional. *Journal of Complexity*, 6(3):290–323, 1990.

[82] H. Drygas. Spectral methods in linear minimax estimation. *Acta Applicandae Mathematica*, 43(1):17–42, 1996.

[83] M. Duarte, M. Davenport, D. Takhar, J. Laska, T. Sun, K. Kelly, and R. Baraniuk. Single-pixel imaging via compressive sampling. *IEEE Signal Processing Magazine*, 25(2):83–91, 2008.

[84] L. Dumbgen and V. Spokoiny. Multiscale testing of qualitative hypotheses. *The Annals of Statistics*, 29(1):124–152, 2001.

[85] J. Durbin. Errors in variables. *Revue de l'Institut International de Statistique*, 22(1/3):23–32, 1954.

[86] S. Efromovich. *Nonparametric Curve Estimation: Methods, Theory, and Applications.* Springer Science & Business Media, 1999.

[87] S. Efromovich and M. Low. On optimal adaptive estimation of a quadratic functional. *The Annals of Statistics*, 24(3):1106–1125, 1996.

[88] S. Efromovich and M. Pinsker. Sharp-optimal and adaptive estimation for heteroscedastic nonparametric regression. *Statistica Sinica*, 6(4):925–942, 1996.

[89] T. Eltoft, T. Kim, and T.-W. Lee. On the multivariate Laplace distribution. *IEEE Signal Processing Letters*, 13(5):300–303, 2006.

[90] J. Fan. On the estimation of quadratic functionals. *The Annals of Statistics*, 19(3):1273–1294, 1991.

[91] J. Fan. On the optimal rates of convergence for nonparametric deconvolution problems. *The Annals of Statistics*, 19(3):1257–1272, 1991.

[92] G. Fellouris and G. Sokolov. Second-order asymptotic optimality in multisensor sequential change detection. *IEEE Transactions on Information Theory*, 62(6):3662–3675, 2016.

[93] J.-J. Fuchs. On sparse representations in arbitrary redundant bases. *IEEE Transactions on Information Theory*, 50(6):1341–1344, 2004.

[94] J.-J. Fuchs. Recovery of exact sparse representations in the presence of bounded noise. *IEEE Transactions on Information Theory*, 51(10):3601–3608, 2005.

[95] W. Gaffey. A consistent estimator of a component of a convolution. *The*

Annals of Mathematical Statistics, 30(1):198–205, 1959.

[96] U. Gamper, P. Boesiger, and S. Kozerke. Compressed sensing in dynamic MRI. *Magnetic Resonance in Medicine: An Official Journal of the International Society for Magnetic Resonance in Medicine*, 59(2):365–373, 2008.

[97] G. Gayraud and K. Tribouley. Wavelet methods to estimate an integrated quadratic functional: Adaptivity and asymptotic law. *Statistics & Probability Letters*, 44(2):109–122, 1999.

[98] N. Gholson and R. Moose. Maneuvering target tracking using adaptive state estimation. *IEEE Transactions on Aerospace and Electronic Systems*, 13(3):310–317, 1977.

[99] R. Gill and B. Levit. Applications of the van Trees inequality: a Bayesian Cramér-Rao bound. *Bernoulli*, 1(1-2):59–79, 1995.

[100] E. Giné, R. Latala, and J. Zinn. Exponential and moment inequalities for U-statistics. In E. Giné, D. Mason, and J. Wellner, editors, *High Dimensional Probability II*, volume 47 of *Progress in Probability*, pages 13–38. Burkhäuser, 2000.

[101] A. Goldenshluger. A universal procedure for aggregating estimators. *The Annals of Statistics*, 37(1):542–568, 2009.

[102] A. Goldenshluger, A. Juditsky, and A. Nemirovski. Hypothesis testing by convex optimization. *Electronic Journal of Statistics*, 9(2):1645–1712, 2015.

[103] A. Goldenshluger, A. Juditsky, A. Tsybakov, and A. Zeevi. Change point estimation from indirect observations. I. Minimax complexity. *Annales de l'Institut Henri Poincaré, Probabilités et Statistiques*, 44:787–818, 2008.

[104] A. Goldenshluger, A. Juditsky, A. Tsybakov, and A. Zeevi. Change point estimation from indirect observations. II. Adaptation. *Annales de l'Institut Henri Poincaré, Probabilités et Statistiques*, 44(5):819–836, 2008.

[105] A. Goldenshluger, A. Tsybakov, and A. Zeevi. Optimal change-point estimation from indirect observations. *The Annals of Statistics*, 34(1):350–372, 2006.

[106] Y. Golubev, B. Levit, and A. Tsybakov. Asymptotically efficient estimation of analytic functions in Gaussian noise. *Bernoulli*, 2(2):167–181, 1996.

[107] L. Gordon and M. Pollak. An efficient sequential nonparametric scheme for detecting a change of distribution. *The Annals of Statistics*, 22(2):763–804, 1994.

[108] M. Grant and S. Boyd. *The CVX Users' Guide. Release 2.1*, 2014. `https://web.cvxr.com/cvx/doc/CVX.pdf`.

[109] M. Grasmair, H. Li, and A. Munk. Variational multiscale nonparametric regression: Smooth functions. *Annales de l'Institut Henri Poincaré, Probabilités et Statistiques*, 54(2):1058–1097, 2018.

[110] V. Guigues, A. Juditsky, and A. Nemirovski. Hypothesis testing via Euclidean

separation. *arXiv 1705.07196*, 2017. https://arxiv.org/pdf/1705.07196.pdf.

[111] F. Gustafsson. *Adaptive Filtering and Change Detection.* John Wiley & Sons, 2000.

[112] W. Härdle, G. Kerkyacharian, D. Picard, and A. Tsybakov. *Wavelets, Approximation, and Statistical Applications.* Springer Science & Business Media, 1998.

[113] S. Hell. Toward fluorescence nanoscopy. *Nature Biotechnology*, 21(11):1347, 2003.

[114] S. Hell. Microscopy and its focal switch. *Nature Methods*, 6(1):24, 2009.

[115] S. Hell and J. Wichmann. Breaking the diffraction resolution limit by stimulated emission: stimulated-emission-depletion fluorescence microscopy. *Optics Letters*, 19(11):780–782, 1994.

[116] D. Helmbold and M. Warmuth. On weak learning. *Journal of Computer and System Sciences*, 50(3):551–573, 1995.

[117] S. Hess, T. Girirajan, and M. Mason. Ultra-high resolution imaging by fluorescence photoactivation localization microscopy. *Biophysical Journal*, 91(11):4258–4272, 2006.

[118] J.-B. Hiriart-Urruty and C. Lemarechal. *Convex Analysis and Minimization Algorithms I: Fundamentals.* Springer, 1993.

[119] C. Houdré and P. Reynaud-Bouret. Exponential inequalities, with constants, for U-statistics of order two. In E. Giné, C. Houdré, and D. Nualart, editors, *Stochastic Inequalities and Applications*, volume 56 of *Progress in Probability*, pages 55–69. Birkhäuser, 2003.

[120] L.-S. Huang and J. Fan. Nonparametric estimation of quadratic regression functionals. *Bernoulli*, 5(5):927–949, 1999.

[121] P. Huber. A robust version of the probability ratio test. *The Annals of Mathematical Statistics*, 36(6):1753–1758, 1965.

[122] P. Huber and V. Strassen. Minimax tests and the Neyman-Pearson lemma for capacities. *The Annals of Statistics*, 1(2):251–263, 1973.

[123] P. Huber and V. Strassen. Note: Correction to minimax tests and the Neyman-Pearson lemma for capacities. *The Annals of Statistics*, 2(1):223–224, 1974.

[124] I. Ibragimov and R. Khasminskii. *Statistical Estimation: Asymptotic Theory.* Springer, 1981.

[125] I. Ibragimov and R. Khasminskii. On nonparametric estimation of the value of a linear functional in Gaussian white noise. *Theory of Probability & Its Applications*, 29(1):18–32, 1985.

[126] I. Ibragimov and R. Khasminskii. Estimation of linear functionals in Gaussian

noise. *Theory of Probability & Its Applications*, 32(1):30–39, 1988.

[127] I. Ibragimov, A. Nemirovskii, and R. Khasminskii. Some problems on non-parametric estimation in Gaussian white noise. *Theory of Probability & Its Applications*, 31(3):391–406, 1987.

[128] Y. Ingster and I. Suslina. *Nonparametric Goodness-of-Fit Testing Under Gaussian Models*, volume 169 of *Lecture Notes in Statistics*. Springer, 2002.

[129] A. Juditsky, F. Kilinc-Karzan, and A. Nemirovski. Verifiable conditions of ℓ_1-recovery for sparse signals with sign restrictions. *Mathematical Programming*, 127(1):89–122, 2011.

[130] A. Juditsky, F. Kilinc-Karzan, A. Nemirovski, and B. Polyak. Accuracy guaranties for ℓ_1-recovery of block-sparse signals. *The Annals of Statistics*, 40(6):3077–3107, 2012.

[131] A. Juditsky and A. Nemirovski. Nonparametric estimation by convex programming. *The Annals of Statistics*, 37(5a):2278–2300, 2009.

[132] A. Juditsky and A. Nemirovski. Accuracy guarantees for ℓ_1-recovery. *IEEE Transactions on Information Theory*, 57(12):7818–7839, 2011.

[133] A. Juditsky and A. Nemirovski. On verifiable sufficient conditions for sparse signal recovery via ℓ_1 minimization. *Mathematical Programming*, 127(1):57–88, 2011.

[134] A. Juditsky and A. Nemirovski. On sequential hypotheses testing via convex optimization. *Automation & Remote Control*, 76(5):809–825, 2015. `https://arxiv.org/pdf/1412.1605.pdf`.

[135] A. Juditsky and A. Nemirovski. Estimating linear and quadratic forms via indirect observations. *arXiv 1612.01508*, 2016. `https://arxiv.org/pdf/1612.01508.pdf`.

[136] A. Juditsky and A. Nemirovski. Hypothesis testing via affine detectors. *Electronic Journal of Statistics*, 10:2204–2242, 2016.

[137] A. Juditsky and A. Nemirovski. Near-optimality of linear recovery from indirect observations. *Mathematical Statistics and Learning*, 1(2):101–110, 2018. `https://arxiv.org/pdf/1704.00835.pdf`.

[138] A. Juditsky and A. Nemirovski. Near-optimality of linear recovery in Gaussian observation scheme under $\|\cdot\|_2^2$-loss. *The Annals of Statistics*, 46(4):1603–1629, 2018.

[139] A. Juditsky and A. Nemirovski. On polyhedral estimation of signals via indirect observations. *arXiv:1803.06446*, 2018. `https://arxiv.org/pdf/1803.06446.pdf`.

[140] A. Juditsky and A. Nemirovski. Signal recovery by stochastic optimization. *Avtomatika i Telemekhanika*, 80(10):153–172, 2019. `https://arxiv.org/pdf/1903.07349.pdf` English translation: *Automation & Remote Control.*

[141] S. Kakade, V. Kanade, O. Shamir, and A. Kalai. Efficient learning of gener-

alized linear and single index models with isotonic regression. In J. Shawe-Taylor, R. Zemel, P. Bartlett, F. Pereira, and K. Weinberger, editors, *Advances in Neural Information Processing Systems 24*, pages 927–935. Curran Associates, Inc., 2011.

[142] A. Kalai and R. Sastry. The isotron algorithm: High-dimensional isotonic regression. In *COLT 2009 - The 22nd Conference on Learning Theory, Montreal, Quebec, Canada, June 18-21, 2009.*

[143] R. Kalman. A new approach to linear filtering and prediction problems. *Journal of Basic Engineering*, 82(1):35–45, 1960.

[144] R. Kalman and R. Bucy. New results in linear filtering and prediction theory. *Journal of Basic Engineering*, 83(1):95–108, 1961.

[145] G. Kerkyacharian and D. Picard. Minimax or maxisets? *Bernoulli*, 8:219–253, 2002.

[146] J. Kiefer and J. Wolfowitz. Stochastic estimation of the maximum of a regression function. *The Annals of Mathematical Statistics*, 23(3), 1952.

[147] J. Klemelä. Sharp adaptive estimation of quadratic functionals. *Probability Theory and Related Fields*, 134(4):539–564, 2006.

[148] J. Klemelä and A. Tsybakov. Sharp adaptive estimation of linear functionals. *The Annals of Statistics*, 29(6):1567–1600, 2001.

[149] V. Koltchinskii and K. Lounici. Concentration inequalities and moment bounds for sample covariance operators. *Bernoulli*, 23(1):110–133, 2017.

[150] V. Koltchinskii, K. Lounici, and A. Tsybakov. Nuclear-norm penalization and optimal rates for noisy low-rank matrix completion. *The Annals of Statistics*, 39(5):2302–2329, 2011.

[151] V. Korolyuk and Y. Borovskich. *Theory of U-statistics*. Springer Science & Business Media, 1994.

[152] A. Korostelev and O. Lepski. On a multi-channel change-point problem. *Mathematical Methods of Statistics*, 17(3):187–197, 2008.

[153] S. Kotz and S. Nadarajah. *Multivariate t-Distributions and Their Applications*. Cambridge University Press, 2004.

[154] C. Kraft. Some conditions for consistency and uniform consistency of statistical procedures. *University of California Publications in Statistics*, 2:493–507, 1955.

[155] J. Kuks and W. Olman. Minimax linear estimation of regression coefficients (I). *Iswestija Akademija Nauk Estonskoj SSR*, 20:480–482, 1971.

[156] J. Kuks and W. Olman. Minimax linear estimation of regression coefficients (II). *Iswestija Akademija Nauk Estonskoj SSR*, 21:66–72, 1972.

[157] V. Kuznetsov. Stable detection when signal and spectrum of normal noise are inaccurately known. *Telecommunications and Radio Engineering*, 30(3):58–

64, 1976.

[158] T. L. Lai. Sequential changepoint detection in quality control and dynamical systems. *Journal of the Royal Statistical Society. Series B*, 57(4):613–658, 1995.

[159] A. Lakhina, M. Crovella, and C. Diot. Diagnosing network-wide traffic anomalies. *ACM SIGCOMM Computer Communication Review*, 34(4):219–230, 2004.

[160] B. Laurent and P. Massart. Adaptive estimation of a quadratic functional by model selection. *The Annals of Statistics*, 28(5):1302–1338, 2000.

[161] L. Le Cam. On the assumptions used to prove asymptotic normality of maximum likelihood estimates. *The Annals of Mathematical Statistics*, 41(3):802–828, 1970.

[162] L. Le Cam. Convergence of estimates under dimensionality restrictions. *The Annals of Statistics*, 1(1):38–53, 1973.

[163] L. Le Cam. On local and global properties in the theory of asymptotic normality of experiments. *Stochastic Processes and Related Topics*, 1:13–54, 1975.

[164] L. Le Cam. *Asymptotic Methods in Statistical Decision Theory*, volume 26 of *Springer Series in Statistics*. Springer, 1986.

[165] O. Lepski. Asymptotically minimax adaptive estimation: I. Upper bounds. Optimally adaptive estimates. *Theory of Probability & Its Applications*, 36(4):645–659, 1991.

[166] O. Lepski. On a problem of adaptive estimation in Gaussian white noise. *Theory of Probability & Its Applications*, 35(3):454–466, 1991.

[167] O. Lepski. Some new ideas in nonparametric estimation. *arXiv 1603.03934*, 2016. `https://arxiv.org/pdf/1603.03934.pdf`.

[168] O. Lepski and V. Spokoiny. Optimal pointwise adaptive methods in nonparametric estimation. *The Annals of Statistics*, 25(6):2512–2546, 1997.

[169] O. Lepski and T. Willer. Oracle inequalities and adaptive estimation in the convolution structure density model. *The Annals of Statistics*, 47(1):233–287, 2019.

[170] B. Levit. Conditional estimation of linear functionals. *Problemy Peredachi Informatsii*, 11(4):39–54, 1975. English translation: *Problems of Information Transmission*.

[171] R. Liptser and A. Shiryaev. *Statistics of Random Processes I: General Theory*. Springer, 2001.

[172] R. Liptser and A. Shiryaev. *Statistics of Random Processes II: Applications*. Springer, 2001.

[173] L. Ljung. *System Identification: Theory for the User*. Prentice Hall, 1986.

[174] G. Lorden. Procedures for reacting to a change in distribution. *The Annals of Mathematical Statistics*, 42(6):1897–1908, 1971.

[175] F. Lust-Piquard. Inégalités de Khintchine dans C^p $(1 < p < \infty)$. *Comptes rendus de l'Académie des Sciences, Série I*, 303(7):289–292, 1986.

[176] M. Lustig, D. Donoho, and J. Pauly. Sparse MRI: The application of compressed sensing for rapid MR imaging. *Magnetic Resonance in Medicine: An Official Journal of the International Society for Magnetic Resonance in Medicine*, 58(6):1182–1195, 2007.

[177] L. Mackey, M. Jordan, R. Chen, B. Farrell, and J. Tropp. Matrix concentration inequalities via the method of exchangeable pairs. *The Annals of Probability*, 42(3):906–945, 2014.

[178] P. Massart. *Concentration Inequalities and Model Selection.* Springer, 2007.

[179] P. Mathé and S. Pereverzev. Direct estimation of linear functionals from indirect noisy observations. *Journal of Complexity*, 18(2):500–516, 2002.

[180] E. Mazor, A. Averbuch, Y. Bar-Shalom, and J. Dayan. Interacting multiple model methods in target tracking: a survey. *IEEE Transactions on Aerospace and Electronic Systems*, 34(1):103–123, 1998.

[181] Y. Mei. Asymptotic optimality theory for decentralized sequential hypothesis testing in sensor networks. *IEEE Transactions on Information Theory*, 54(5):2072–2089, 2008.

[182] A. Meister. *Deconvolution Problems in Nonparametric Statistics*, volume 193 of *Lecture Notes in Statistics.* Springer, 2009.

[183] G. Moustakides. Optimal stopping times for detecting changes in distributions. *The Annals of Statistics*, 15(4):1379–1387, 1986.

[184] H.-G. Müller and U. Stadtmüller. Discontinuous versus smooth regression. *The Annals of Statistics*, 27(1):299–337, 1999.

[185] A. Nemirovski. Topics in non-parametric statistics. In P. Bernard, editor, *Lectures on Probability Theory and Statistics, Ecole d'Eté de Probabilités de Saint-Flour*, volume 1738 of *Lecture Notes in Mathematics*, pages 87–285. Springer, 2000.

[186] A. Nemirovski. Interior Point Polynomial Time methods in Convex Programming. Lecture Notes, 2005. `https://www.isye.gatech.edu/~nemirovs/Lect_IPM.pdf`.

[187] A. Nemirovski. Introduction to Linear Optimization. Lecture Notes, 2015. `https://www2.isye.gatech.edu/~nemirovs/OPTI_LectureNotes2016.pdf`.

[188] A. Nemirovski, A. Juditsky, G. Lan, and A. Shapiro. Robust stochastic approximation approach to stochastic programming. *SIAM Journal on Optimization*, 19(4):1574–1609, 2009.

[189] A. Nemirovski, S. Onn, and U. Rothblum. Accuracy certificates for computa-

tional problems with convex structure. *Mathematics of Operations Research*, 35(1):52–78, 2010.

[190] A. Nemirovski, B. Polyak, and A. Tsybakov. Convergence rate of nonparametric estimates of maximum-likelihood type. *Problemy Peredachi Informatsii*, 21(4):17–33, 1985. English translation: *Problems of Information Transmission.*

[191] A. Nemirovski, C. Roos, and T. Terlaky. On maximization of quadratic form over intersection of ellipsoids with common center. *Mathematical Programming*, 86(3):463–473, 1999.

[192] A. Nemirovskii. Nonparametric estimation of smooth regression functions. *Izvestia AN SSSR, Seria Tekhnicheskaya Kibernetika*, 23(6):1–11, 1985. English translation: *Engineering Cybernetics: Soviet Journal of Computer and Systems Sciences.*

[193] Y. Nesterov and A. Nemirovskii. *Interior-Point Polynomial Algorithms in Convex Programming.* SIAM, 1994.

[194] M. Neumann. Optimal change point estimation in inverse problems. *Scandinavian Journal of Statistics*, 24(4):503–521, 1997.

[195] D. Nolan and D. Pollard. *U*-processes: Rates of convergence. *The Annals of Statistics*, 15(2):780–799, 1987.

[196] F. Österreicher. On the construction of least favourable pairs of distributions. *Zeitschrift für Wahrscheinlichkeitstheorie und verwandte Gebiete*, 43(1):49–55, 1978.

[197] J. Pilz. Minimax linear regression estimation with symmetric parameter restrictions. *Journal of Statistical Planning and Inference*, 13:297–318, 1986.

[198] M. Pinsker. Optimal filtration of square-integrable signals in Gaussian noise. *Problemy Peredachi Informatsii*, 16(2):120–133, 1980. English translation: *Problems of Information Transmission.*

[199] G. Pisier. Non-commutative vector valued l_p-spaces and completely p-summing maps. *Astérisque*, 247, 1998.

[200] M. Pollak. Optimal detection of a change in distribution. *The Annals of Statistics*, 13(1):206–227, 1985.

[201] M. Pollak. Average run lengths of an optimal method of detecting a change in distribution. *The Annals of Statistics*, 15(2):749–779, 1987.

[202] H. V. Poor and O. Hadjiliadis. *Quickest Detection.* Cambridge University Press, 2009.

[203] K. Proksch, F. Werner, and A. Munk. Multiscale scanning in inverse problems. *The Annals of Statistics*, 46(6B):3569–3602, 2018.

[204] A. Rakhlin, S. Mukherjee, and T. Poggio. Stability results in learning theory. *Analysis and Applications*, 3(4):397–417, 2005.

[205] C. R. Rao. Information and accuracy attainable in the estimation of statistical parameters. *Bulletin of Calcutta Mathematical Society*, 37:81–91, 1945.

[206] C. R. Rao. *Linear Statistical Inference and Its Applications.* John Wiley & Sons, 1973.

[207] C. R. Rao. Estimation of parameters in a linear model. *The Annals of Statistics*, 4(6):1023–1037, 1976.

[208] J. Rice. Bandwidth choice for nonparametric regression. *The Annals of Statistics*, 12(4):1215–1230, 1984.

[209] H. Rieder. Least favorable pairs for special capacities. *The Annals of Statistics*, 5(5):909–921, 1977.

[210] F. Rosenblatt. The perceptron: a probabilistic model for information storage and organization in the brain. *Psychological Review*, 65(6):386, 1958.

[211] M. Rust, M. Bates, and X. Zhuang. Sub-diffraction-limit imaging by stochastic optical reconstruction microscopy (STORM). *Nature Methods*, 3(10):793, 2006.

[212] S. Shalev-Shwartz, O. Shamir, N. Srebro, and K. Sridharan. Stochastic convex optimization. In *COLT 2009 – The 22nd Conference on Learning Theory, Montreal, Quebec, Canada*, 2009.

[213] A. Shapiro, D. Dentcheva, and A. Ruszczyński. *Lectures on Stochastic Programming: Modeling and Theory, Second Edition.* SIAM, 2014.

[214] H. Sherali and W. Adams. A hierarchy of relaxations between the continuous and convex hull representations for zero-one programming problems. *SIAM Journal on Discrete Mathematics*, 3(3):411–430, 1990.

[215] A. Shiryaev. On optimum methods in quickest detection problems. *Theory of Probability & Its Applications*, 8(1):22–46, 1963.

[216] D. Siegmund. *Sequential Analysis: Tests and Confidence Intervals.* Springer Science & Business Media, 1985.

[217] D. Siegmund and B. Yakir. *The Statistics of Gene Mapping.* Springer Science & Business Media, 2007.

[218] T. Söderström, U. Soverini, and K. Mahata. Perspectives on errors-in-variables estimation for dynamic systems. *Signal Processing*, 82(8):1139–1154, 2002.

[219] T. Söderström and P. Stoica. Comparison of some instrumental variable methods–consistency and accuracy aspects. *Automatica*, 17(1):101–115, 1981.

[220] K. Sridharan, S. Shalev-Shwartz, and N. Srebro. Fast rates for regularized objectives. In D. Koller, D. Schuurmans, B. Y., and L. Bottou, editors, *Advances in Neural Information Processing Systems 21*, pages 1545–1552. 2009.

[221] J. Stoer and C. Witzgall. *Convexity and Optimization in Finite Dimensions*

I. Springer, 1970.

[222] C. Stone. Optimal rates of convergence for nonparametric estimators. *The Annals of Statistics*, 8(6):1348–1360, 1980.

[223] C. Stone. Optimal global rates of convergence for nonparametric regression. *The Annals of Statistics*, 10(4):1040–1053, 1982.

[224] A. Tartakovsky, I. Nikiforov, and M. Basseville. *Sequential Analysis: Hypothesis Testing and Change point Detection.* CRC Press, 2014.

[225] A. Tartakovsky and V. Veeravalli. Change point detection in multichannel and distributed systems. In N. Mukhopadhyay, S. Datta, and S. Chattopadhyay, editors, *Applied Sequential Methodologies: Real-World Examples with Data Analysis*, pages 339–370. CRC Press, 2004.

[226] A. Tartakovsky and V. Veeravalli. Asymptotically optimal quickest change detection in distributed sensor systems. *Sequential Analysis*, 27(4):441–475, 2008.

[227] R. Tibshirani. Regression shrinkage and selection via the lasso. *Journal of the Royal Statistical Society. Series B*, pages 267–288, 1996.

[228] J. Tropp. An introduction to matrix concentration inequalities. *Foundations and Trends in Machine Learning*, 8(1-2):1–230, 2015.

[229] A. Tsybakov. Optimal rates of aggregation. In B. Schölkopf and M. Warmuth, editors, *Learning Theory and Kernel Machines*, volume 2777 of *Lecture Notes in Computer Science*, pages 303–313. Springer, 2003.

[230] A. Tsybakov. *Introduction to Nonparametric Estimation.* Springer Series in Statistics. Springer, New York, 2009.

[231] S. Van De Geer. The deterministic Lasso. Technical report, Seminar für Statistik, Eidgenössische Technische Hochschule (ETH) Zürich, 2007. `https://stat.ethz.ch/~geer/lasso.pdf` JSM Proceedings, 2007, paper nr. 489.

[232] S. Van De Geer and P. Bühlmann. On the conditions used to prove oracle results for the Lasso. *Electronic Journal of Statistics*, 3:1360–1392, 2009.

[233] H. Van Trees. *Detection, Estimation, and Modulation Theory, Part I: Detection, Estimation, and Linear Modulation Theory.* John Wiley & Sons, 1968.

[234] Y. Vardi, L. Shepp, and L. Kaufman. A statistical model for positron emission tomography. *Journal of the American Statistical Association*, 80(389):8–20, 1985.

[235] A. Wald. Sequential tests of statistical hypotheses. *The Annals of Mathematical Statistics*, 16(2):117–186, 1945.

[236] A. Wald. *Sequential Analysis.* John Wiley & Sons, 1947.

[237] A. Wald and J. Wolfowitz. Optimum character of the sequential probability ratio test. *The Annals of Mathematical Statistics*, 19(1):326–339, 1948.

[238] Y. Wang. Jump and sharp cusp detection by wavelets. *Biometrika*, 82(2):385–397, 1995.

[239] L. Wasserman. *All of Nonparametric Statistics.* Springer Science & Business Media, 2006.

[240] A. Willsky. *Detection of Abrupt Changes in Dynamic Systems.* Springer, 1985.

[241] K. Wong and E. Polak. Identification of linear discrete time systems using the instrumental variable method. *IEEE Transactions on Automatic Control*, 12(6):707–718, 1967.

[242] Y. Xie and D. Siegmund. Sequential multi-sensor change-point detection. *The Annals of Statistics*, 41(2):670–692, 2013.

[243] Y. Yin. Detection of the number, locations and magnitudes of jumps. *Communications in Statistics. Stochastic Models*, 4(3):445–455, 1988.

[244] C.-H. Zhang. Fourier methods for estimating mixing densities and distributions. *The Annals of Statistics*, 18(2):806–831, 1990.

Index